Fungal Pathogenesis

MYCOLOGY SERIES

Founding Editor
Paul A. Lemke

Additional Volumes in Preparation

Fungal Pathogenesis

Principles and Clinical Applications

edited by

Richard A. Calderone
Ronald L. Cihlar

Georgetown University Medical Center
Washington, D.C.

CRC Press
Taylor & Francis Group
Boca Raton London New York

CRC Press is an imprint of the
Taylor & Francis Group, an **informa** business

First published 2002 by MARCEL DEKKER, INC.

Published 2019 by CRC Press
Taylor & Francis Group
6000 Broken Sound Parkway NW, Suite 300
Boca Raton, FL 33487-2742

© 2002 by Taylor & Francis Group, LLC
CRC Press is an imprint of Taylor & Francis Group, an Informa business

First issued in paperback 2019

No claim to original U.S. Government works

ISBN-13: 978-0-367-44718-2 (pbk)
ISBN-13: 978-0-8247-0568-8 (hbk)

Visit the Taylor & Francis Web site at
http://www.taylorandfrancis.com

and the CRC Press Web site at
http://www.crcpress.com

This book is dedicated to the memory of Roy Lee Hopfer, our friend and colleague. He was taken away before his time. All of us who knew Roy miss him greatly. He preferred conversations about his family, books, music, and North Carolina basketball. We remained friends in spite of his choice in basketball.

Richard A. Calderone
Ronald L. Cihlar

Preface

Fungal infections in the immunocompromised host continue to be an important problem for the clinician. Candidiasis, for example, is the fourth leading cause of hospital-acquired (or nosocomial) infections in the United States. While not as frequent, *Aspergillus fumigatus* infections in cancer or transplant patients, nevertheless, carry a high morbidity and mortality since the infection is often not diagnosed early enough and therapy is postponed. In AIDS patients, candidiasis, *Pneumocystis carinii* pneumonia, and cryptococcosis are the most common fungal infections. In the case of candidiasis, over 80% of AIDS patients have either oral disease or both oral and vaginal infections. Esophagitis caused by *Candida* spp. is also a common manifestation in the AIDS patient. Fungal infections such as blastomycosis, histoplasmosis, coccidioidomycosis, and paracoccidioidomycosis occur in the nonimmunocompromised host, but histoplasmosis and coccidiodomycosis are commonly reported in AIDS patients as well.

In spite of their frequency, the management of these diseases is difficult because of the lack of diagnostic reagents (for candidiasis and aspergillosis) and the few antifungals available. Thus, new targets are being sought that may be exploited in the development of antifungals. In addition to antifungal therapy and vaccines, passive treatment of infected patients with specific antibody and immune reconstitution may represent new ways to treat patients. The immune approaches have been developed through extensive and exquisite studies carried out in animal models. Candidate vaccines for several of the endemic mycoses are close at hand.

The purpose of *Fungal Pathogenesis: Principles and Clinical Applications* is to acquaint the reader with research involved in virulence, immunity, diagnosis, and therapy of the most common fungal infections. The book is designed for the classroom and research and diagnostic laboratories, as well as for the clinician.

Richard A. Calderone
Ronald L. Cihlar

Introduction

Fungal Pathogenesis: Principles and Clinical Applications emphasizes the pathogenesis of human fungal infections and has been designed as a tool that can be used in teaching and research. Previous texts in medical mycology have tended to either focus on a specific topic or lack sufficient detail, especially in areas of study related to pathogenesis. Here, we hope to bridge this gap by addressing subjects related to both virulence and host responses. Emphasis has also been placed on the impact of this research on diagnosis and prevention (clinical applications). Included are chapters on the major opportunistic fungal infections as well as those that tend to have an endemic distribution. Both categories of infections are loosely defined, since at least two of the endemic mycoses are also common in the immunocompromised host such as the AIDS patient. The opportunistic infections focused on are candidiasis, pneumocystosis, aspergillosis, and cryptococcosis. Chapters on the endemic mycoses discuss histoplasmosis, blastomycosis, coccidioidomycosis and paracoccidioidomycosis.

The book is divided into four parts. Part I focuses on the virulence factors of human pathogenic fungi. This theme is especially well developed for organisms such as *Candida albicans* and other *Candida* species. Thus, host recognition, morphogenesis and signaling, phenotypic switching, and secretory enzymes that are believed to promote invasion of the organism are discussed. In addition, other chapters detail the virulence properties of *C. neoformans, Histoplasma capsulatum, Blastomyces dermatitidis, Paracoccidioides brasiliensis, Aspergillus fumigatus,* and *Pneumocystis carinii.* When possible, virulence attributes are discussed collectively—for example, dimorphism, a growth feature typical of several pathogens. Because genomics research is expected to have a major effect on our understanding of the biology of the fungal pathogens, a chapter has been included that addresses these issues.

Part II switches from features of the pathogens to host responses to fungal infections. Much has been accomplished in regard to the latter subject. Current observations not only have elucidated immune mechanisms against these fungi, but have also resulted in translational research. For example, vaccines against several of the endemic mycoses are close at hand. These possible breakthroughs have been the result of numerous basic science studies, the outcome of which has resulted in a rejuvenation of concepts on protection that may have important consequences on the design of novel therapeutic strategies. Additionally, passive transfer of antibody may be a key treatment option in the near future for cryptococcosis and candidiasis. Immune reconstitution as adjuvant therapy to the traditional use of antifungal drugs is also becoming a reality. Thus, Part II focuses on natural and acquired immunity, vaccine development, and immune reconstitution.

Part III presents descriptions of the existing antifungals, susceptibility testing, and the identification of new fungal targets that may be exploited in the development of new antifungals. A consequence of antifungal therapy is, not surprisingly, resistance. The impact of drug resistance on patient management is not completely understood; however, many mechanisms of resistance have been defined and predictions about the consequences of resistance to the patient are discussed in this section.

Part IV emphasizes issues important to the clinical mycology laboratory, such as the methods used in diagnosing fungal infections by antigen/antibody detection methodologies as well as by molecular approaches. Equally important to the diagnostic laboratory is strain variability, especially with the nonsexual fungi such as *C. albicans*. Methods for strain typing and current epidemiological observations are included in this section. Finally, although emphasis is placed on the most common of the fungal infections, the emergence of new pathogens is also discussed and the ramifications of these organisms on the clinical mycology laboratory are presented.

The purpose of this work is to present both basic science information and couple the growing knowledge base with translational research. We hope that the information will be useful in the classroom, for both graduate and undergraduate training, as well as in research and in the clinical laboratory. Most important, the book will have accomplished its objectives if new ideas for addressing old problems are developed by its readers.

Richard A. Calderone
Ronald L. Cihlar

Contents

Contents

Contributors

Darius Amjadi, M.D.   Department of Pathology and Laboratory Medicine, University of North Carolina, Chapel Hill, North Carolina

Janna L. Beckerman Ph.D.   Departments of Genetics, Cell Biology, and Development, University of Minnesota, St. Paul, Minnesota

Francesco Bistoni, M.D., Ph.D.   Microbiology Section, Department of Experimental Medicine and Biochemical Sciences, University of Perugia, Perugia, Italy

Maria Boccanera, Ph.D.   Department of Bacteriology and Medical Mycology, Istituto Superiore di Sanità, Rome, Italy

Sheldon Broedel, Jr., Ph.D.   Dorlin Pharmaceuticals, Inc., Baltimore, Maryland

Richard A. Calderone, Ph.D.   Department of Microbiology and Immunology, Georgetown University Medical Center, Washington, D.C.

Jose Antonio Calera, Ph.D.   Department of Microbiology and Genetics, University of Salamanca, Salamanca, Spain

Arturo Casadevall, M.D., Ph.D.   Department of Medicine, Albert Einstein College of Medicine, Bronx, New York

W. LaJean Chaffin, Ph.D.   Department of Microbiology and Immunology, Texas Tech University Health Sciences Center, Lubbock, Texas

Wun-Ling Chang, M.D.   Department of Medicine, Section of Infectious Diseases, Louisiana State University Health Sciences Center, Shreveport, Louisiana

Ronald L. Cihlar, Ph.D.   Department of Microbiology and Immunology, Georgetown University Medical Center, Washington, D.C.

Garry T. Cole, Ph.D.   Department of Microbiology, Medical College of Ohio, Toledo, Ohio

David C. Coleman, Ph.D., F.R.C., Patho., F.T.C.D.   School of Dental Science, University of Dublin, Trinity College, Dublin, Ireland

Chester R. Cooper, Jr., Ph.D.   Department of Biological Sciences, Youngstown State University, Youngstown, Ohio

Rebecca A. Cox, Ph.D.   Texas Center for Infectious Diseases, San Antonio, Texas

Jim E. Cutler, Ph.D.   Department of Microbiology, Montana State University, Bozeman, Montana

Flavia De Bernardis, Ph.D.   Department of Bacteriology and Medical Mycology, Istituto Superiore di Sanità, Rome, Italy

George S. Deepe, Jr., M.D.   Department of Internal Medicine/Infectious Diseases, University of Cincinnati College of Medicine, Cincinnati, Ohio

Abdelmalic El Barkani, Ph.D.   Institute of Hygiene and Microbiology, University of Würzburg, Würzburg, Germany

Evangelia Farmaki, M.D.   Third Department of Pediatrics, Aristotle University of Thessaloniki, Thessaloniki, Greece

Mahmoud A. Ghannoum, M.Sc., Ph.D.   Center for Medical Mycology, Department of Dermatology, Case Western Reserve University, Cleveland, Ohio

Bruce L. Granger, Ph.D.   Department of Microbiology, Montana State University, Bozeman, Montana

Yongmoon Han, Ph.D.   Department of Microbiology, Montana State University, Bozeman, Montana

Thomas S. Harrison, M.D.   St. George's Hospital Medical School, London, United Kingdom

Roy L. Hopfer, Ph.D.   Departments of Microbiology and Immunology and Pathology and Laboratory Medicine, University of North Carolina, Chapel Hill, North Carolina

Christina Kellogg, Ph.D.   Department of Microbiology and Immunology, Georgetown University Medical Center, Washington, D.C.

Theo N. Kirkland, M.D.   Veterans Administration, San Diego Health Care System, and Department of Pathology and Medicine, University of California, San Diego, California

Bruce S. Klein, M.D.   Department of Pediatrics, University of Wisconsin, Madison, Wisconsin

Henry Koziel, M.D. Division of Pulmonary and Critical Care Medicine, Beth Israel Deaconess Medical Center and Harvard Medical School, Boston, Massachusetts

Shankarling Krishnamurthy, Ph.D. Membrane Biology Laboratory, School of Life Sciences, Jawaharlal Nehru University, New Delhi, India

Oliver Kurzai, M.D. Institute of Hygiene and Microbiology, University of Würzburg, Würzburg, Germany

K. J. Kwon-Chung, Ph.D. Laboratory of Clinical Investigation, National Institute of Allergy and Infectious Diseases, National Institutes of Health, Bethesda, Maryland

Jean-Paul Latgé, Ph.D. Aspergillus Unit, Pasteur Institute, Paris, France

Stuart M. Levitz, M.D. Department of Medicine, Boston University School of Medicine, Boston, Massachusetts

Mark D. Lindsley, Sc.D., D(ABmm) Centers for Disease Control and Prevention, Atlanta, Georgia

Caron A. Lyman, Ph.D. Immunocompromised Host Section, Pediatric Oncology Branch, National Cancer Institute, Bethesda, Maryland

D. Mitchell Magee, Ph.D. Texas Center for Infectious Diseases, San Antonio, Texas

P. T. Magee, Ph.D. Department of Genetics, Cell Biology, and Development, University of Minnesota, St. Paul, Minnesota

Thomas G. Mitchell, Ph.D. Department of Microbiology, Duke University Medical Center, Durham, North Carolina

Christine J. Morrison, Ph.D. Centers for Disease Control and Prevention, Atlanta, Georgia

Isabelle Mouyna, Ph.D. Aspergillus Unit, Pasteur Institute, Paris, France

Fritz A. Mühlschlegel, M.D. Institute of Hygiene and Microbiology, University of Würzburg, Würzburg, Germany

Pranab K. Mukherjee, M.Sc., Ph.D. Center for Medical Mycology, Department of Dermatology, Case Western Reserve University, Cleveland, Ohio

Simon L. Newman, Ph.D. Division of Infectious Diseases, Department of Internal Medicine, University of Cincinnati College of Medicine, Cincinnati, Ohio

Gustavo Niño-Vega, Ph.D. Centro de Microbiologia y Biologia Celular, Instituto Venezolano de Investigaciones Científicas, Caracas, Venezuela

Sneh Lata Panwar, M.Sc. Membrane Biology Laboratory, School of Life Sciences, Jawaharlal Nehru University, New Delhi, India

Thomas F. Patterson, M.D., F.A.C. P.   Division of Infectious Diseases, Department of Medicine, University of Texas Health Science Center at San Antonio, San Antonio, Texas

Sofia Perea, Pharm. D., Ph.D.   Division of Infectious Diseases, Department of Medicine, University of Texas Health Science Center at San Antonio, San Antonio, Texas

Rajendra Prasad, Ph.D.   Membrane Biology Laboratory, School of Life Sciences, Jawaharlal Nehru University, New Delhi, India

Emmanuel Roilides, M.D., Ph.D.   Third Department of Pediatrics, Aristotle University of Thessaloniki, Thessaloniki, Greece

Luigina Romani, M.D., Ph.D.   Microbiology Section, Department of Experimental Medicine and Biochemical Sciences, University of Perugia, Perugia, Italy

Gioconda San-Blas, Ph.D.   Centro de Microbiologia y Biologia Celular, Instituto Venezolano de Investigaciones Científicas, Caracas, Venezuela

David R. Soll, Ph.D.   Department of Biological Sciences, University of Iowa, Iowa City, Iowa

Derek J. Sullivan, Ph.D., F.T.C.D.   Microbiology Research Unit, Department of Oral Medicine and Pathology, School of Dental Science, University of Dublin, Trinity College, Dublin, Ireland

Brian L. Wickes, Ph.D.   Department of Microbiology, University of Texas Health Science Center at San Antonio, San Antonio, Texas

Jon P. Woods, M.D., Ph.D.   Department of Medical Microbiology and Immunology, University of Wisconsin Medical School, Madison, Wisconsin

Jianping Xu, Ph.D.   Department of Biology, McMaster University, Hamilton, Ontario, Canada

Xiao-jiong Zhao, M.D., Ph.D.   Division of Clinical Pharmacology, Georgetown University Medical Center, Washington, D.C.

1
Host Recognition by Human Fungal Pathogens

W. LaJean Chaffin
Texas Tech University Health Sciences Center, Lubbock, Texas

I. INTRODUCTION

Pathogenic fungi cause a spectrum of infections from annoying to life-threatening with portals of entry that include skin contact, inhalation, and translocation across physical barriers as a result of host defects or accidental or iatrogenic abridgement of epithelial integrity. Some of these agents are also components of normal human flora or may be present in the absence of clinical symptoms. The initial obligate but not sufficient requirement for the establishment of colonization or disease is adherence of the fungus to the host. The mechanisms by which a fungus attaches and maintains contact with the host surface include nonspecific mechanisms such ionic and hydrophobic interactions as well as specific recognition between a ligand and receptor. The focus of this discussion is specific host recognition by pathogenic fungi. Viewing this interaction from the fungal perspective, the fungal component mediating the binding is considered an adhesin. Adhesins, their ligand(s), and in some cases both partners of the interaction have been identified for several fungi and are discussed in the following sections. Studies with some fungi that describe only a few general characteristics of adherence are not included.

The recognition mechanisms described to date include participation of fungal surface carbohydrates, proteins, and possibly lipids (Table 1). The host components that are partners in this recognition include the same three constituents; oligosaccharide moieties of proteins and lipids; proteins; and possibly lipids. Fungal carbohydrates implicated in adherence include mannan, β-glucan, and chitin while host carbohydrates implicated include fucose, N-acetylglucosamine, and sialic acid. Among host proteins two major groups that are recognized by fungi are serum proteins (fibrinogen, complement C3 fragments) and extracellular matrix components (ECM) such as fibronectin. In a few cases both the fungal and host components of a specific binding reaction have been identified. Even more rarely the role of the fungal component has been confirmed through genetic manipulation—i.e., deletion or mutation. With one exception for *Candida albicans* [1], these adherence reactions involve noncovalent interactions.

The most extensive examination of adherence is with *C. albicans* where there are examples of all classes of fungal-host recognition (for recent reviews see [2,3]). Despite this extensive effort, most candidal adhesins have not been purified and characterized. In addition to several extensively studied *C. albicans* adhesins, there are well-characterized examples of adhesins from *Blastomyces dermatitidis*, *C. glabrata*, *Paracccidioides brasiliensis*, and *Pneumocystis carinii*.

Table 1 Fungal Adhesins and Host Ligands

Adhesin	Host ligand	Type	Fungus
Carbohydrate	Protein	Lectinlike	C. albicans, P. carinii
Protein	Carbohydrate	Lectinlike	A. fumigatus, C. albicans, C. glabrata, C. neoformans, H. capsulatum, P. marneffei, S. schenckii
Protein	Protein	Noncovalent protein-protein	C. albicans, other Candida spp., A. fumigatus, B. dermatitidis, H. capsulatum, P. brasiliensis, P. carinii, P. marneffei, S. schenckii
Protein	Protein	Covalent protein	C. albicans
Protein?[a]	Lipid?	Protein-lipid	C. albicans
Lipid?[a]	Protein?	Protein-lipid	C. albicans

[a] Recognition involving lipids has not been extensively characterized whether as adhesin or ligand.

II. CARBOHYDRATE-PROTEIN (LECTINLIKE) BINDING SYSTEMS

The involvement of carbohydrates in the binding of fungi to the host has been demonstrated for several fungi (Table 1). The carbohydrate component of the interaction may be provided by either the fungus or the host. The host oligosaccharide involved in binding, and in some cases the fungal oligosaccharide, is the carbohydrate moiety of a glycoprotein or glycolipid. Since the participating moiety in binding is carbohydrate, adherence involving the carbohydrate component of glycoproteins or glycolipids is included in this section.

A. Carbohydrate Adhesins

Chitin, β-glucan, and specific oligosaccharides present in mannan have been implicated as adhesins (Tables 2, 3). Beta-glucan, a structural polysaccharide of fungal cell walls, binds vitronectin in a concentration-dependent and -specific manner. This has been demonstrated with β-glucan from *Saccharomyces cerevisiae* [4]. A high-molecular-weight component of *P. carinii*, consistent with β-glucan, interacts with vitronectin through the glycosaminoglycan-binding region of vitronectin [5] (Table 3). Binding to vitronectin was not inhibited by RGD-containing peptides but was reduced (90%) by treatment of *P. carinii* with periodate or heparitinase [6]. Bound vitronectin can serve as a bridge between host vitronectin receptors. Beta-glucan with bound vitronectin stimulated tumor necrosis factor alpha (TNF-α) release from macrophages [4]. Beta-glucan particles mediated TNF-α release through interaction with β-glucan receptors [7] with stimulation and suppression at low and high concentrations, respectively [4]. Release was enhanced by bound vitronectin. Beta-glucan also mediated binding of *C. albicans* to macrophages and stimulated cytokine release [8] (Table 2). Binding of vitronectin to the fungus increased the interaction with macrophages [9].

Chitin, a β-1,4-N-acetylglucosamine polymer, is less abundant in fungal cell walls than glucan polymers. Adherence of *C. albicans* to acrylic [10], human corneocytes [11], murine intestinal mucosa [12,13], and human vaginal epithelial cells [13] was inhibited by a chitin-soluble extract (Table 2). Systemic spread from the murine gut [12] and development of vaginitis [14] were also reduced by administration of the extract. *C. albicans* yeast cells treated with chitin synthesis inhibitors showed reduced adherence to buccal epithelial cells (BECs) [15,16].

Table 2 *Candida* spp. Host Recognition Interactions, Adhesins, and Ligands[a]

Organism	Type of interaction	Adhesin	Ligand
C. albicans	Lectinlike	Glucan	Vitronectin, macrophage β-glucan receptor
		Chitin	Unknown
		Mannan structures	Macrophage mannose receptor
		Acid labile and acid stabile Factor 6	Epithelial cells
		GCA1[d]	GlcNAc[d]
		Protein[d]	Fucose
		Fimbrial protein	GalNAcβ1,4 Galβ
		Unknown	Galβ1,4 Glcβ1,1-ceramide
		Unknown	Oligosaccharides of salivary proteins
	Noncovalent protein-protein	SAP3	ECM
		SAP1,2,3, others?	BECs, endothelial cells, corneocytes
		Inducible 55 kDa, ALS5/ALA1, G3PD, 60 kDa, 3 species 40–50 kDa range	Fibronectin
		Inducible 55 kDa, ALS5/ALA1, 68 and 60–62 kDa, 37 kDa, G3PDH, 21 kDa[b]	Laminin
		60 kDa, ALS5/ALA1	Collagen(s)
		50, 60, and 90 kDa; 130, 100–110, and 84 kDa integrin homologues; 30 kDa	Vitronectin
		25, 44, 65 kDa	Entacin
		Unknown	Tenascin-C
		Unknown	Serum albumin
		Unknown	Transferrin
		68 and 60–62 kDa doublet, FBP1 (58 kDa), 55 kDa	Fibrinogen
		60 kDa	C3d (complement fragment)
	—[c]	ALS1	Epithelial, endothelial cells
		INT1; 130 kDa and minor 50, 100 kDa; 42 kDa and less reactive species; 165 kDa	iC3b (complement fragment)
	Covalent protein-protein	HWP1	Buccal epithelial cell surface
C. glabrata	Lectinlike	EPA1	Asialo-lactosyl carbohydrates
	Protein-protein[c]	Unknown	Fibronectin
	[b]	Unknown	iC3b
C. krusei	Protein-protein[b]	Unknown	iC3b
	[c]	Unknown	Fibronectin
C. lusitanea	Protein-protein[b]	Unknown	iC3b
C. parapsilosis	Protein-protein[b]	Unknown	iC3b
C. stellatoidea	Protein-protein	Unknown	iC3b
		Unknown	C3d
		Unknown	Fibronectin
C. tropicalis	Protein-protein	125, 105 kDa	Fibronectin
		Unknown	iC3b

[a] Other adhesins and ligands, such as mannan and glucan, may be predicted but have not been specifically examined for some species and are not included.

[b] Inferred by reactivity with MAb OKM-1.

[c] Reactions are presumed to be protein-protein.

[d] Strain specific.

Table 3 Host Recognition and Interactions, Adhesins, and Ligands of Various Pathogenic Fungi[a]

Organism	Type of interaction	Adhesin	Ligand
A. fumigatus	Lectinlike	72 kDa	Sialic acid on laminin
		Unknown	Sialic acid fibronectin, fibrinogen
	Protein-protein[b]	Unknown	Collagen
	Protein-protein	23, 30 kDa	Fibronectin
	Protein-protein[b]	37 kDa (Asp f2)	Laminin
B. dermatitidis	Protein-protein	WI-1 (WI1)	CR3, CD14 of macrophage
C. neoformans	Lectinlike	Unknown	Galβ1,4 Glcβ1,1-ceramide
H. capsulatum	Lectinlike	Unknown	Galβ1,4 Glcβ1,1-ceramide
	Protein-protein	50 kDa	Laminin
P. brasilensis	Protein-protein	gp43	Laminin
P. carinii	Lectinlike	Glucan	Vitronectin, macrophage β-glucan receptor
		Mannan (gpA)	Macrophage mannose receptor
	Protein-protein	gpA	Fibronectin
		33 kDa	Fibronectin, laminin
P. marneffei	Lectinlike	20 kDa	Sialic acid on fibronectin, laminin
S. schenckii	Lectinlike	Unknown	Galβ1,4 Glcβ1,1-ceramide
	Protein-protein[b]	Unknown	Fibrinogen, fibronectin, laminin, type II collagen

[a] Other adhesins and ligands, such as mannan and glucan, may be predicted but have not been specifically examined for some species and are not included.
[b] Reactions are presumed to be protein-protein, although the possibility of lectinlike interaction is not excluded.

Mutant strains of the organism deficient in chitin formation were able to colonize organs of infected mice but were less virulent [17].

Yeast mannoproteins may contain both branched N-linked and short-chain O-linked oligosaccharides. Structures in both have been implicated in adherence of *C. albicans* (Table 2). A strain deficient in *MNT1*, encoding an α-1,2-mannosyl transferase that adds the second residue of O-linked mannan, showed reduced adherence to BECs and vaginal epithelial cells (VECs) and reduced virulence in murine and guinea pig models of systemic infection [18]. In the latter model, there was a decreased ability to reach and or to adhere to internal organs. O-linked mannan may also contribute to the interaction of a 58-kDa mannoprotein with fibrinogen [19]. The acid-labile portion of N-linked mannan and at least two structures in the acid-stable portion have also been implicated in adherence. A strain deficient in the acid-labile β-1,2-linked mannan showed reduced adherence to marginal zones in tissue sections of murine spleen and to kidney tissue but not liver sections [20]. A mannotetraose was identified as an adhesin [21] that mediated binding to macrophages of the marginal zone [22]. The requirements for the major contribution of the acid-stable mannan to binding were more complex as both the mannan core and oligomannosyl side chains contribute to the activity [23]. Mannan bound to the macrophage mannose receptor leads to phagocytosis [24]. Expression of this receptor was downregulated in rat macrophages following ingestion of *C. albicans* [25].

C. albicans serotype A strains are distinguished by the presence of mannose β1,2(mannose α1,2)$_3$mannose and (mannose β1,2)$_2$(mannose α1,2)$_3$mannose oligosaccharides, also known as Factor 6 determinants [26]. This structure contributed to adherence to BECs as adherence was inhibited by mannan containing this structure but not by mannan lacking the structure [27] (Table 2). A mutant serotype A strain deficient in these structures was reduced in adherence compared to the parent strain. Serotype B strains that lack the determinant adhered less well than did species such as *C. tropicalis* and *C. glabrata* that contained the determinant.

As noted above for *C. albicans*, phagocytosis of *P. carinii* was also mediated by the mannose receptor [28] (Table 3). Binding and ingestion of *P. carinii* by human and rat alveolar macrophages were reduced by 90% by inhibitors of the macrophage mannose receptor. Reduced binding and phagocytosis of the organism by macrophages from HIV-1-infected individuals correlated with downregulation of the mannose receptor [29]. A major surface protein, gpA (gp120 of MSG-1), that is mannose rich bound to alveolar macrophages in a saturable manner and binding was inhibited by addition of α-mannan [30]. gpA affinity chromatography of macrophage membrane proteins yielded a 165-kDa species that reacted with antimannose receptor antiserum. As discussed subsequently, this gpA has additional adhesin activity.

B. Adhesins Recognizing Host Carbohydrates

C. albicans expressed strain-specific lectinlike epithelial adhesins [31,32] (Table 2). Fucose inhibited binding of some strains to BECs while glucosamine or N-acetylglucosamine (GlcNAc) inhibited binding of other strains. Fucose has been shown to bind to yeast and hyphal organisms [33]. Growth on galactose increased expression of the adhesin material that was obtained from culture medium [34]. A fragment of the L-fucose-binding protein bound to the terminal trisaccharide of the blood group H antigen that terminates in fucose (Fucα1,2Galβ1,4GlcNAcβ-), supporting the suggestion that blood group antigens are the epithelial cell ligands [35]. Cells or extracellular material from the strain that recognized GlcNAc bound to glycolipids containing that sugar. Binding of L-fucose- or GlcNAc-specific strains to BECs or VECs was partially inhibited by extracellular material from the homologous strain [31]. In another study, binding to an esophageal cell line, Het-1, increased in organisms grown on galactose and the increase was associated with a 190-kDa component of cell wall extracts [36]. Binding also showed lectinlike interactions with inhibition by glucosamine or GlcNAc. Sequences of internal peptides of this 190-kDa moiety that showed homology with a glucoamylase of another yeast were used to generate probes for gene isolation [37]. *GCA1* is a glycosyl hydrolase that is transcribed to a greater extent in galactose than in sucrose- or glucose-grown cells. Reverse transcriptase (RT)-PCR detected expression of the gene in a rat model of oral infection.

Fimbriae of *C. albicans* appear to mediate binding to glycosphingolipids through the carbohydrate moiety (Table 2). The major structural unit of the long filamentous surface appendages was characterized as a glycoprotein of 66 kDa [38]. Carbohydrate, primarily D-mannose, accounted for some 80–85% of the moiety and about half of the amino acid residues were hydrophobic. Fimbriae bound directly to BECs and were able to inhibit binding of yeast cells. Fimbriae also bound to asialo-GM₁ (gangliotetraosylceramide, Galβ1,3GalNAcβ1,4-Galβ1,4Glcβ1-ceramide) and the disaccharide GalNAcβ1,4Galβ-protein conjugate [39,40]. The recognition of the carbohydrate portion rather than the lipid portion of the glycosphingolipid suggested that this was another example of a lectinlike interaction. The binding motif appeared to be shared with *Pseudomonas aeruginosa* as antibody to the bacterial binding site for asialo-GM₁ partially inhibited fimbrial and fungal cell binding to BECs [41]. Peptides derived from the bacterial sequence and antibodies to these sequences were used to demonstrate that an important part of the *C. albicans* binding motif was structurally similar to the motif formed by DEQFIPK in the bacterial pili [39].

C. albicans bound to human salivary proteins when dental acrylic, denture lining material, or hydroxylapatite beads (a model for tooth surfaces) were coated with saliva [42–45] (Table 2). Treatment of the bound salivary protein but not the fungus with neuraminidase increased binding to coated beads [42] while deglycosylation reduced binding of the fungus to saliva-coated glass slides [46]. Stratherin and proline-rich proteins (PRPs) promoted binding of yeast cells to coated beads [42]. Binding of radiolabeled yeast cells to electrophoretically separated salivary components in an overlay assay showed binding to 17-, 20-, 24-, and 27-kDa basic PRPs [47]. These proteins contain O-linked sialyated Galβ1,3GalNAc oligosaccharides [48]. Another study using an overlay assay found reactivity with a single component that was identified as a low-molecular-weight mucin, MG2 in human saliva, and rat submandibular gland (RSMG) mucin in rat saliva [49]. Further analysis of the RSMG fraction associated binding with a proteoglycan [50]. Recognition was associated with heparan sulfate side chains rather than blood group A oligosaccharides contained in the active component. Heparan sulfate has been reported in other studies to inhibit binding of *C. albicans* to ECM, but rather by binding to the ECM component and not the fungus [51].

Glycosphingolipids are recognized by several pathogenic fungi (Tables 2, 3). *Cryptococcus neoformans, C. albicans, Saccharomyces cerevisiae,* and yeast-phase *Histoplasma capsulatum* and *Sporothrix schenckii* bound to lactosylceramide (Galβ1,4Glcβ1,1ceramide) as determined by on overlay assay on a glycosphingolipid chromatogram [52]. The terminal unsubstituted galactose was essential for binding as removal of galactose abolished binding and organisms did not adhere to glycosphingolipids that contained internal lactosyl residues. Active fungal metabolism appeared to be a requirement as glucose was required in the binding medium. Binding of *C. neoformans* to cultured human glioma cells that are rich in lactosylceramide was inhibited by liposomes incorporating the glycosphingolipid. In another study (noted above), that also demonstrated binding to glycosyl moieties of sphingolipids, the binding of *C. albicans* to lactosyl ceramide was not observed [35].

C. glabrata expressed an adhesin that recognized asialo-lactosyl-containing carbohydrates [53] (Table 2). In contrast to *C. albicans,* this organism is haploid which allowed the use of forward genetics in which strains with introduced random mutations were screened for a phenotype of interest. Briefly, strains were constructed with a sequence tag and subsequently random mutations were introduced into each of 96 tagged strains. A mutant of each tagged strain was combined with mutants of other tagged strains, and the mutant pool was screened for adherence to epithelial cells. The tags of both the input and adherent cells were amplified and used to detect tags present in the input pool but absent in the adherent population. The sequence surrounding the mutation was then recovered for nonadherent strains. Fourteen strains were found defective in a gene designated epithelial adhesin 1 (*EPA1*). Deletion of *EPA1* and restoration of *EPA1* to a deletion strain ablated or restored adherence, respectively. Expression of the gene in *S. cerevisiae* conferred adherence on that normally nonadherent species. The sequence of the protein suggested a domain structure similar to *C. albicans* adhesins *HWP1* and *ALS5/ALA1,* described elsewhere. The C-terminal sequence was consistent with addition of a glycosylphosphatidylinositol (GPI) moiety. The C-terminal two-thirds of the predicted protein was also enriched in serine and threonine, and these may serve as O-glycosylation sites. Three direct repeats of a 40 amino acid sequence were noted within this region. There was some homology to the *S. cerevisiae* Ca^{2+}-dependent lectin flocculin, *FLO1,* at the amino terminus. Binding of *C. glabrata* was also Ca^{2+} dependent. Binding was inhibited by galactose or lactose but not by sialyl-lactose or other sugars. Periodate treatment of host cells but not the fungus reduced binding. Further support for the lectinlike binding of the adhesin and recognition of terminal lactose-containing carbohydrates was derived from adherence to a cell line for which variants deficient in glycosylation are available. No difference was observed between the parental and deficient strains in models of vaginal or gastrointestinal tract infection.

The interaction of *Penicillium marneffei* conidia with laminin and fibronectin was inhibited by N-acetylneuraminic acid [54,55] (Table 3). These ECM components bound to conidia but not hyphae, as shown by indirect immunofluorescence. The conidia also bound to immobilized ligand and laminin and was able to inhibit binding to fibronectin and vice versa. Binding to either ligand was abolished by addition of sialic acid. Ligand binding was age dependent as conidia from 4- and 8-day-old cultures bound more extensively than conidia from 1- and 2-day-old cultures. Ligand affinity blotting of a conidial wall extract and a conidial homogenate showed that both ligands reacted with a 20-kDa species in the conidial homogenate. The apparent discrepancy in location may reflect a low recovery of material in the cell wall extract. Since binding to both ligands appeared to be mediated through the same lectinlike reaction, the finding of a common receptor was not unexpected.

Both lectinlike recognition of sialic acid containing glycosyl moieties of fibrinogen, laminin, and fibronectin and nonlectinlike recognition of ECM proteins have been reported for *Aspergillus fumigatus* conidia (Table 3). These differences may reflect multiple adhesins and remain an unresolved issue. Binding of fibronectin and laminin was demonstrated by fluorescence-activated flow cytometry [56–58]. In one study glucose, galactose, mannose, GlcNAc, N-acetylgalactosamine, lactose, and maltose were ineffective inhibitors of fibronectin binding [56]. However, mucin was an effective inhibitor (95%) while sialylactose and asialomucin reduced binding < 30%. Binding of soluble laminin was reduced by N-acetylneuraminic acid and to a lesser extent by sialyllactose and mucin but not by asialomucin. These authors proposed a sialic acid-dependent recognition of laminin and fibronectin. A subsequent study by some of the same authors used ligand blotting to identify a laminin-binding adhesin in an extract of conidial wall proteins prepared by boiling conidia in buffer containing SDS and dithiothreitol [59]. A single 72-kDa moiety was reactive. This moiety also reacted with concanavalin A, suggesting the adhesin was a glycoprotein. Fibrinogen binding to conidia has been demonstrated by indirect immunofluorescence and electron microscopy and binding to immobilized fibrinogen [60,61]. Binding was found at the outer surface and was restricted to pathogenic species compared to strictly saprophytic or phytopathogenic species [60]. Binding was localized to fibrinogen D fragment with some 1200 saturable, specific sites per conidia [61,62]. Binding decreased during germination [61]. Binding to immobilized fibrinogen was inhibited by unlabeled fibrinogen, antifibrinogen antibody, and laminin [62]. Binding was not inhibited by RGDS or GRGDS adhesion motifs in fibrinogen. Extensive proteolysis of fibrinogen did not substantially reduce adherence, and multiple fragments supported adherence, the latter suggesting a common motif shared by several peptides [56]. Binding of soluble fibrinogen determined by FACS was reduced 80% by N-acetylneuraminic acid and 23% by sialylactose and almost complete abolished by mucin. This inhibition pattern was consistent with sialic acid recognition.

III. LIPID RECOGNITION

Recognition of lipids between fungi and the human host has not been extensively documented. The most extensively described system is adhesin recognition of host sphingolipid, although as discussed above this appeared to be a lectinlike interaction with recognition of the carbohydrate rather than the lipid portion of the structure. Evidence for the role of other lipids in candidal adherence is scant. The specificity of the lipid interaction and whether the lipid functions as an adhesin or ligand is unclear. Adherence of *C. albicans* to BECs was reduced (28–42%) by addition of total candidal lipid extract [63]. *C. tropicalis* adhered less well and the homologous lipid extract was also a somewhat less effective inhibitor. Homologous extracts from the least adherent *C. pseudotropicalis* were ineffective. On the other hand, retreatment of either yeast cells or BECS with the extracts was effective in reducing adherence. Inhibition was affected by various phospholipids, sterols, and steryl esters but not by triacylclycerols or free fatty acids. A similar study suggested that lipids on the porcine corneum reduced binding of *C. albicans* to the keratinzed epithelial surface [64]. Removal of lipid increased binding to the surface. Addition of fatty acids, sterol, and ceramide but not squalene, wax esters, cholesterol esters or triglycerides inhibited binding. Whether this inhibition represents specific recognition adhesin-ligand interaction between the fungus and the host or a nonspecific effect is unclear. Resolution of these questions awaits identification of fungal adhesins.

IV. PROTEIN-PROTEIN RECOGNITION

The majority of identified adherence interactions involve protein-protein binding. Most of the host protein ligands that mediate fungal adherence are either serum proteins or components of ECM. Serum proteins include serum albumin, fibrinogen, and fragments of complement factor C3. The ECM components include fibronectin, laminin, collagen, entactin, tenascin, and vitronectin. There has been extensive examination of the adhesins for these ligands, particularly in *C. albicans,* as will be evident from the discussion of fungal adhesins in the next two sections. However, there are several other protein-protein interactions involving additional host protein ligands, including a unique covalent binding. These are discussed in the final section on protein adhesins.

A. Adhesins Recognizing Serum Proteins

1. Serum Albumin and Transferrin

C. albicans hyphae but not yeast cells bound both serum albumin and transferrin as demonstrated by indirect immunofluorescence [65] (Table 2). Binding was variable depending on the growth medium and between germ tubes. *C. albicans* bound to immobilized bovine serum albumin (BSA) [66]. Binding was inhibited by alanine, proline, and leucine but not arginine and lysine. Pretreatment of yeast cells with amphotericin B or dithiothreitol reduced binding to BSA while only high concentrations (severalfold above MIC values) of fluconazole, ketoconazole, or flucytosine altered binding [67]. In contrast, BSA has frequently been used in studies with other potential ligands as a negative or low binding control. Thus, it would appear that binding of *C. albicans* to other ligands may be more extensive than that in other studies where the conditions were not optimized to detect binding to BSA.

2. Fibrinogen

Binding of fibrinogen to germ tubes of *C. albicans* but not to yeast cells grown on Sabouraud medium was observed by indirect immunofluorescence [65,68,69] (Table 2). Binding was localized to the 85-kDa C-terminal fibrinogen D fragment [70]. Binding was time dependent, saturable, and reversible with about 6000 binding sites per germ tube. Binding was not inhibited by several sugars tested. Ligand affinity blotting identified a 68-kDa moiety in a DTT-iodoacetamide extract. Species of 68-kDa and a 60- and 62-kDa doublet were demonstrated to bind to plastic and laminin as well as fibrinogen, suggesting a multifunctional receptor [71]. Using a different strain grown in Lee medium, another study reported binding to germ tubes and some yeast cells [19]. Ligand affinity blotting identified a 58-kDa mannoprotein in β-mercaptoethanol extracts of yeast cells and germ tubes. This species did not bind laminin, fibronectin, or complement C3 fragments. mp58 contained both N- and O-linked oligosaccharides that represented 20% and 3% respectively of the molecular mass. Removal of the O-linked oligosaccharides reduced binding suggesting a partial contribution of lectinlike interactions to binding as noted earlier. mp58 possessed two other interesting features in that it contained some sequences or domains of the type IV collagen molecule [72] as well as ubiquitinlike epitopes [73]. The surface distribution of mp58 and fibrinogen binding activity were found to be heterogeneous with patches of greater reactivity as determined by fluorescent confocal microscopy, immunoelectron microscopy, and scanning electron microscopy [19,68,69,74]. mp58 was expressed to a variable extent by clinical isolates [75] and was expressed in vivo as determined by immunocytochemical localization [76]. Screening of expression libraries with anti–cell wall antiserum [77] or anti-mp58 antibody [78] yielded cDNA clones encoding this adhesin. *FBP1* encodes a 292 amino acid protein with similarity to a family of immunodominant antigens of *Aspergillus* spp. [78]. *FBP1* expression was detected in both yeast cells and germ tubes grown on minimal medium but not in yeast cells grown on rich medium [77]. This nutritional regulation may in part explain the differences between observations of fibrinogen binding by yeast cells. Fibrinogen binding along with binding of fibronectin, laminin, and type IV collagen, to yeast cells was increased after growth in the presence of hemoglobin [79]. Increased binding of fibrinogen was associated with a 55-kDa protein that also recognized fibronectin and laminin. Yeast cells of *S. schenckii* bound fibrinogen [80]; however, the binding interaction has not been further characterized.

3. Complement Fragment C3d

C. albicans and *C. stellatoidea* adhere to complement C3 fragments as demonstrated by the ability of hyphae to rosette antibody-sensitized erythrocytes coated with C3d and iC3b [81] (Table 2). Affinity purification yielded a 60-kDa species and a 66- to 68-kDa doublet, although C3d binding appeared associated with the 60-kDa moiety that was expressed on hyphal surfaces [82–85]. A 50-kDa moiety reacting with a monoclonal antibody (MAb) to the 60-kDa species and that also had binding activity was found in the plasma membrane of yeast cells [86]. During germination reactivity was associated with the hyphal wall. Both the 50 and 60 kDa proteins inhibited rosette formation. Polyclonal antibody to the 60 kDa species reacted with additional moieties in studies by other authors [86–88]. Several studies suggested that the 50- and 60-kDa proteins shared a common peptide and reacted with concanavalin A but differed in other aspects of glycosylation [82–85]. This mp60 candidal receptor was expressed in vivo in models of murine systemic infection [86] and rat vaginal infection [89]. Although some differences in reactivity were noted reactivity was generally found in the inner cell wall, as well as at the surface, particularly in germ tubes and hyphae.

4. Complement Fragment iC3b

Since the initial report of binding of iC3b to *C. albicans* [81], the interactions and identification of the adhesin have been studied in several laboratories with some conflicting observations (Table 2). Various laboratories hypothesized that the functional similarity between the candidal receptor and the human integrin complement receptor extended to structural similarity. Human receptors CR3 and CR4 are members of the β_2 (95-kDa peptide) subset of the integrin receptor family that contain α subunits α_M (165 kDa) and α_X (150 kDa), respectively. At least eight MAbs to α_M exhibited moderate to high reactivity with the fungus and one MAb to α_X exhibited high reactivity [90–94]. Expression of iC3b binding capacity was affected by environment as germ tubes grown at 30°C showed greater expression that those grown at 38.5°C [92]. Yeast cells grown in glucose expressed a higher capacity than those grown in glutamate [94], as did cells grown in 50 mM glucose compared to 5 mM glucose [93]. *C. stellatoidea*, but not *C. tropicalis*, *C. parapsilosis*, *C. krusei*, or *C. lusitaniae*, also rosetted erythrocytes [81,91]. MAb anti-α_X (OKM-1) showed a decreasing reactivity from *C. albicans* (68%) and *C. tropicalis* (32%) to *C. parapsilosis*, *C. glabrata*, *C. lusitaniae*, *C. krusei, and S. cerevisiae* (1%–18%) that generally paralleled adherence to HeLa cells [95]. Incubation with OKM-1, iC3b, and RGD peptides from iC3b reduced adherence of *C. albicans* but not *C. tropicalis* to HeLa cells while the opposite effect was obtained with fibrinogen and fibronectin-RGD peptides [90], suggesting different mechanisms for adherence.

Immunoprecipitation of a *C. albicans* hyphal extract with OKM-1 yielded a major 130-kDa and minor 50- and 100-kDa proteins [92]. Subsequently, 70-, 66-, 55-, and 42-kDa peptides were isolated by affinity chromatography, and the three smaller peptides reacted with OKM-1 [96]. The 42-kDa species was reactive with C3, C3b, and iC3b while the other two were less reactive and appeared to differ in glycosylation.

Western blot analysis of cytosolic and membrane but not cell wall fractions with OKM-1 detected a 165-kDa species [94]. The term "integrin analog" has been applied to the *C. albicans* iC3b adhesin due to the similar functional, antigenic, and size relationships with the mammalian integrin receptor. Oligonucleotide probes based on the assumed relationship with α subunits were used to screen a genomic library and isolate a sequence, *INT1*, that contained a 1164 amino acid open reading frame (ORF) [97]. Analysis showed a putative membrane spanning region and limited homology/identity with integrins. When *INT1* was expressed in *S. cerevisiae*, cells formed structures that resembled germ tubes. *INT1* expression conferred the ability to adhere to HeLa cells on normally nonadherent *S. cerevisiae*, and adherence was reduced by anti-Int1p antibody [98]. A *C. albicans* strain deleted for *INT1* showed about a 40% reduction in adherence to HeLa cells and a wild-type strain was similarly reduced in adherence when treated with anti-Int1p antibody. Adherence was partially restored by return of one *INT1* copy. The null mutant was defective in hyphal formation on some media but not others, suggesting multiple pathways of morphogenesis. The null strain also showed reduced intestinal colonization in antibiotic-treated mice [99] and was less virulent in a murine systemic infection model [98,100]. Cells persisted in the kidneys of mice challenged with the null mutant strain and a reconstructed heterozygote strain compared to the parent strain. All strains produced hyphae in vivo [100].

B. Adhesins Recognizing ECM Proteins

In *C. albicans*, secreted aspartyl proteinases (SAPs) encoded by a family of related genes have also been implicated in adherence. Addition of inhibitors of SAP reduced adherence to endothelial cells and the formation of cavitations by yeast cells on corneocyte surfaces [101,102]. A strain deficient in *SAP3* but not *SAP1*, *SAP2*, or *SAP4-6* was reduced in adherence to ECM components found in Matrigel (a basement membrane extract from Engelbreth-Holm-Swarm cells) when cultured on glucose [103]. Adherence to BECs was reduced for Δ*sap1* when cultured in glucose and for all three single null strains when cultured in galactose. The Δ*sap4-6* strain showed enhanced adherence when grown on either glucose or galactose. These observations suggest that SAPs play a dual role in proteoloysis and adherence in host interaction.

1. Fibronectin

Adherence of *C. albicans* to fibronectin was the first of the ECM binding activities described for pathogenic fungi [104]. *C. albicans* and other *Candida* spp., *A. fumigatus*, and *P. carinii* have protein adhesins that mediate binding to fibronectin, a 440-kDa dimeric protein found in plasma, as well as ECM (Tables 2, 3). Candidal interactions with soluble, immobilized, and cell-associated fibronectin were recently reviewed [3]. The number of binding sites on *C. albicans* has been variously estimated from 8,000 sites to 5,000 high-affinity sites and 30,000 low-affinity sites, and binding to these sites can be cation dependent or independent [105–108]. Fibronectin has several binding domains characterized as fibrin, collagen, DNA, cell (containing an RGD sequence), and heparin. There is some difference among various studies as to which of these domains was recognized by candidal adhesins [51,106–110]. The cell-binding domain and its RGD sequence have been implicated in some studies while other reports demonstrated adherence to the cell-binding domain but questioned the role of the RGD sequence. Similarly, some studies found the collagen binding domain promoted little or limited binding while others found the cell-binding domain the most active domain in promoting adherence. Heparin was reported to inhibit binding to fibronectin but not to the cell-binding domain fragment. Inhibition may be mediated through binding of heparin to fibronectin. Environmental factors such as growth medium, temperature, and growth stage appeared, at least in part, to regulate expression of fibronectin binding activity [106–108,111,112]. Particularly of note, the addition of hemoglobin to growth medium increased fibronectin binding some 20- to 80-fold [112].

Several candidal proteins have been reported to have fibronectin-binding activity. Antiserum to human fibronectin integrin receptors reacted with *C. albicans* cells and inhibited binding to fibronectin [110]. Western blot analysis with the antibodies showed reactivity with three species in the 40- to 50-kDa range as well as larger polydisperse material while a ligand affinity blot identified additional species [113]. Fibronectin affinity chromatography yielded a 60-kDa species as a fibronectin adhesin candidate [105,114]. The increased binding of cells grown in the presence of hemoglobin was associated with a 55-kDa surface protein that appeared to be a multifunctional receptor, as laminin and fibrinogen also appeared to bind to this species [79,112,115]. Binding was independent of the fibronectin RGD sequence. Another cell surface protein identified as the 33-kDa glyceraldehyde-3-phophate dehydrogenase (G3PDH) was also found to be a fibronectin- and laminin-binding protein [116]. Binding to immobilized ligand was inhibited by addition of anti-G3PDH antibody or soluble G3PDH. This protein has also been implicated in fibronectin binding in bacteria [64]. The most recent candidate for a fibronectin-binding protein was identified through isolation of a *C. albicans* DNA fragment that conferred adherence to fibronectin, type IV collagen, and laminin on normally nonadherent *Saccharomyces cerevisiae* [117]. The sequence was identified as encoding a member of the ALS gene family *ALS5/ALA1*. Unlike other adhesin candidates, this protein is likely to be attached to glucan as the sequences in this family suggest that the protein contains a glycophosphatidyl-inositol structure [118]. Another member of the *ALS* family, *ALS1*, was identified as a *C. albicans* sequence that conferred increased adherence to epithelial and endothelial cells on normally nonadherent *S. cerevisiae* [119]. The ligand(s) recognized by this potential adhesin was not determined.

C. tropicalis appeared to express fewer fibronectin binding sites than *C. albicans* [120]. A 125-kDa species was detected by Western blot and a 105-kDa species by immunoprecipitation with antihuman integrin antibodies. *C. stellatoidea, C. glabrata,* and *C. krusei* bound to immobilized fibronectin and the cell-binding domains and reacted with antiintegrin antibody, although the reactivity of *C. krusei* was marginal [121]. Growth in the presence of hemoglobin also increased fibronectin binding of these three species [122].

Conidia but not germinated cells of *A. fumigatus* bound to soluble and immobilized fibronectin [57,58,123] (Table 3). Confocal fluorescence microscopy suggested that binding was localized to conidial protrusions [57]. Binding was dose dependent and saturable, and in one study, the presence of two populations that varied in the extent of binding was observed with fluorescence flow cytometry [56,58]. Laminin and the RGD peptide inhibited binding [57,123] as did mucin [56]. Ligand affinity blotting identified 23-kDa and 30-kDa peptides as potential fibronectin-binding proteins [58].

Binding of soluble fibronectin to *P. carinii* was demonstrated to be saturable with 640,000 binding sites per cell [124] (Table 3). Binding to both soluble and immobilized fibronectin was inhibited by addition of the RGDS peptide and enhanced by the presence of calcium ion [124–126]. Binding of both rat- and mouse-derived organisms to mammalian cell lines was inhibited by addition of antifibronectin antibody, soluble fibronectin, or GRDS peptide [124,127]. Antifibronectin antibody bound to the surface of the host cells [127]. The notion that fibronectin bound to host cell surfaces and mediated binding between the fungus and the host was further supported by the finding that antibodies to fibronectin-binding integrin subunits α_5 and α_v inhibited binding [128]. After incubation with *P. carinii* or the major surface glycoprotein, gpA, with one cell line, the reactivity with anti-α_5 antibody increased, suggesting that the receptor surface expression became upregulated. Expression of this α_5 subunit on the surface was decreased by exposure of the cell line to γ-interferon, as was the adherence of *P. carinii* [129]. After culture with MRC-5 cells, *P. carinii* filopodia containing cytoskeleton penetrated the target cell apparently anchoring the fungus without membrane fusion [127,130]. Addition of a microfilament inhibitor reduced binding [127,131]. The fibronectin-binding protein has been identified as gpA. Addition of gpA, which is a highly glycosylated protein containing mannose and N-acetylglucosamine residues in the oligosaccharides, inhibited binding of the fungus to host cell lines [132]. Anti-gpA antibody inhibited binding of the protein to soluble and immobilized fibronectin and binding of the fungus to cell lines [125,126,132]. A ligand affinity blot of fungal proteins and fibronectin affinity chromatography identified the 120-kDa gpA as the reactive moiety [126,132]. Another 33-kDa candidate with similarity to the mammalian nonintegrin laminin receptor has been reported [133]. The protein produced by overexpression of the cloned sequence in bacteria showed reactivity with fibronectin and laminin.

Conidia and yeast cells of *S. schenckii* bound to immobilized fibronectin [80] (Table 3). Conidia bound more abundantly than did yeast cells. The organism also bound to other ECM components. The identity and specificity of the adhesins(s) have not been reported.

2. Laminin

Laminin was recognized as a ligand by *C. albicans*, *P. brasiliensis*, and *A. fumigatus*, although this last species may recognize carbohydrate rather than protein motifs as noted previously. Several studies have examined laminin recognition by *C. albicans* and, like fibronectin binding, some differences are found between the studies (for more extensive review see [3]) (Table 2). *C. albicans* bound both soluble and immobilized laminin, and binding to immobilized laminin was inhibited by fibronectin, RGD-containing peptides, and antilaminin antibody [107,134,135]. Binding was greater to germ tubes than to yeast cells [135]. About 8000 laminin-binding sites per organism were observed and these were confined to the outer fibrillar layer. Binding was inhibited by fibrinogen but not laminin. Ligand affinity blotting with an extract of surface proteins detected reactivity with a 68-kDa protein and a 60- and 62-kDa doublet. These proteins were similar in size to ones that also recognized fibrinogen and C3d, as discussed earlier. More recently, along with binding to germ tubes, a weak binding to yeast cells was noted [136]. Binding was inhibited by fibrinogen but not fibronectin, an RGD-containing peptide, or a laminin peptide YIGSR. In this case, ligand affinity blotting with a cell extract detected reactivity with a 21-kDa protein. Yet, another study reported ligand blot reactivity with 37- and 67-kDa moieties in a cell surface extract from yeast cells [137]. Antibodies to the similar-size human high-affinity laminin receptor recognized the larger species in yeast cell surface extracts and the 37-kDa species in extracts from both yeast cells and germ tubes. However, only the yeast species bound laminin. Neither yeast species bound fibrinogen, fibronectin, or collagen type IV. Only ~ 10% of yeast cells expressed a heterogeneously distributed binding reactivity on their surface. This 37-kDa protein contained collagen domains and was ubiquitinated as was the 58-kDa fibrinogen-binding protein described earlier [72,73]. More recently, a cDNA encoding a 37-kDa protein with sequence identity to both the 37-kDa high-affinity laminin-binding protein and the Yst ribosomal proteins of *S. cerevisiae* was isolated [138]. Antibody to this protein reacted with plasma membranes and cytosol protein but not cell surfaces. In another study, as many as 10 proteins were identified by ligand blotting with an extract obtained by glucanase digestion [113]. Three other laminin-binding candidates were discussed previously as multifunctional adhesins that also bound fibronectin. *C. albicans* organisms grown in the presence of hemoglobin had enhanced binding to laminin, fibronectin, and fibrinogen, and this binding was associated with increased surface expression of a 55-kDa moiety [79]. G3PDH, a 33-kDa protein, was also a surface-located protein that mediated binding to laminin [116]. The *C. albicans ALS5/ALA1* sequence cloned in *S. cerevisiae* also conferred increased laminin binding on the latter yeast [117].

As discussed previously, *P. carinii* expressed a protein that was encoded by a gene with sequence homology to the human colon carcinoma laminin receptor [133] (Table 3). The product of the gene expressed in bacteria reacted with laminin as well as fibronectin as measured by ligand affinity blotting.

Among ECM components, laminin is the only component that has been reported to interact with *H. capsulatum* [139] (Table 3). Yeast cells interacted with both soluble and immobilized laminin. Thirty thousand saturable binding sites were observed for soluble laminin. A laminin A chain peptide (IKVAV) but not sugars of the oligosaccharides side chains inhibited binding (70%). A 50-kDa species that reacted with antibody to the human high-affinity laminin receptor was identified by ligand affinity blotting. Antibody to the 50-kDa protein bound to the cell surface and reduced laminin binding.

Laminin bound to the yeast cell surface of *P. brasiliensis* [140] (Table 3). Laminin-coated yeast cells showed increased adherence to a mammalian cell line, suggesting that laminin may bridge between fungal and host laminin receptors, as described for *P. carinii* fibronectin binding. Laminin bound in a specific and saturable manner to a purified major antigenic moiety, gp43. An antibody to a bacterial laminin binding protein reacted with the gp43 and in the presence of laminin inhibited binding of *P. brasiliensis* to a cell monolayer [141].

As noted earlier, it is unclear whether binding of *A. fumigatus* conidia to laminin is mediated by recognition of host oligosaccharides attached to the protein, by recognition of a laminin protein motif, or by both mechanisms (Table 3). Binding of laminin to conidia was observed over the entire surface with a concentration at the area of protrusions [57,59,142]. Binding of soluble laminin was saturable and lost as conidia germinated. Fibronectin and fibrinogen but not glycosaminoglycans reduced binding [56,57]. The RGD peptide and other synthetic peptides derived from laminin and fibrinogen did not reduce binding [56,57]. One study determined that N-acetylneuraminic acid reduced binding [56], and another study from the same laboratory identified a 72-kDa candidate for a lectinlike recognition of sialic acid [59]. Two other studies have implicated a 37-kDa protein in laminin binding, although whether the adhesin recognized host carbohydrate or protein motifs was not determined. A single reactive 37-kDa species was identified by ligand affinity blotting in a cell-free extract but was not observed in extracts of conidial walls [57]. The absence from cell wall extracts was attributed to poor or no solubilization of the species from the cell wall. More recently a major allergen, Asp f2, which may be found in culture filtrates, was cloned and expressed in bacteria [143]. The sequence showed homology with the *C. albicans* fibrinogen-binding protein. Both the native and recombinant protein bound laminin. Although the identity of the two 37-kDa proteins has not been investigated, it is reasonable to speculate that they are the same.

S. schenckii conidia and yeast cells bound to laminin, as well fibronectin and collagens [80] (Table 3). The binding of yeast cells and conidia was similar.

3. Collagen

C. albicans and *A. fumigatus* have been reported to bind to one or more collagens. *C. albicans* bound to immobilized type IV collagen and less well to native and denatured (gelatin) type I collagen [134,144] (Table 2). Fibronectin or the peptide GRGESP were more effective inhibitors of binding to both collagens than were two RGD-containing peptides [107]. Heparin also inhibited binding of yeast cells to collagens, apparently by binding to the collagen and blocking fungal recognition [51]. The 60-kDa peptide obtained by fibronectin affinity chromatography was also obtained by collagen affinity chromatography [105]. Addition of hemoglobin to the growth medium increased binding of *C. albicans* more to type IV collagen than to gelatin [79]. No increase was noted for native type I collagen. Adherence to immobilized type IV collagen was also observed. Increased binding to fibrinogen, fibronectin, and laminin was associated with increased expression of a promiscuous 55-kDa receptor. However, if this adhesin also recognized type IV collagen is unknown. A third collagen-binding adhesin candidate is represented by the predicted 150-kDa protein encoded by the candidal *ALS5/ALA1* that conferred increased adherence to fibronectin, laminin, and type IV collagen on nonadherent *S. cerevisiae* [117].

Conidia of *A. fumigatus* showed similar adherence to immobilized type I and type IV collagen [57,123] (Table 3). Laminin substantially reduced binding to the collagens; fibronectin was a less effective inhibitor [57]. Two studies reported that RGD did not inhibit binding to type IV collagen but differed in the observed ability of the peptide to inhibit binding to type I collagen [57,123]. The failure of RGD to inhibit binding may suggest that the interaction is a lectinlike binding as observed for laminin and fibronectin. However, that remains to be determined.

Interaction with both type I and type II collagens has been reported for *S. schenckii* conidia and yeast cells [80] (Table 3). Conidia showed greater binding to type II collagen than did yeast cells. The adhesin(s) that mediate binding of this organism to ECM proteins are unknown.

4. Vitronectin

Both lectinlike and protein-protein binding has been reported for fungal interactions with vitronectin. Vitronectin, present in serum vascular walls and dermis, is another multidomain protein that is recognized by mammalian integrins through a domain with an RGD sequence and has a glycosaminoglycan-binding region. As discussed earlier, fungal β-glucan mediated binding to vitronectin. Vitronectin binding capacity of *C. albicans* increased during late exponential growth and was optimal at acidic pH [106] (Table 2). The effect of growth temperature is unclear, as the reports differed on this issue [106,111]. Both low- and high-affinity binding sites were present, although the dissociation constant of the high-affinity sites was similar to that of low-affinity sites described for some other ligands [9]. Binding of vitronectin was reduced by treatment of the fungus with proteases or heat [106]. Binding was substantially reduced by addition of fibronectin but minimally with addition of fibrinogen, type I collagen, and GRGDS peptide; gelatin, type IV collagen, and other RGD-containing peptides had no effect. Heparin was found to be the most effective inhibitor [9,106]. In another study, binding to immobilized vitronectin was effectively inhibited by RGD-containing peptides, GRGDSP, and GRGDS [145]. The effect of proteins, peptides, and heparin is consistent with recognition of vitronectin by both protein and carbohydrate receptors. Ligand affinity blotting detected a 30-kDa species in an extract of cells obtained with detergent and reducing agent [9]. However, binding was inhibited by heparin. Antiserum to the human integrin vitronectin receptor showed major reactivity with 50-, 60-, and 90-kDa species in a Western blot of an octylglucoside extract of isolated cell walls [105]. More recently, antibodies to integrin subunits and receptors were observed to bind to yeast cell surfaces [145]. Immunoprecipitation of reactive components from a cell lysate with anti-α_v or anti-β_3 or anti-β_5 yielded 130-kDa, 110- and 100-kDa doublet, and 84-kDa proteins that corresponded to α_v, β_3, and β_5, respectively. Monoclonal antibodies to the $\alpha_v\beta_3$ and $\alpha_v\beta_5$

receptors inhibited binding to immobilized vitronectin. The addition of vitronectin inhibited yeast binding to a cell line expressing surface vitronectin.

5. Entactin

The glycoprotein entactin contains multiple domains that interact with type IV collagen and laminin to form a tight complex. Recognition of this ECM protein has been examined only for *C. albicans* [146] (Table 2). Binding of this component was observed to some yeast cells and along most hyphal extensions of germ tubes. There appeared bo no effect of growth temperature on binding of entactin [111]. Cell wall extracts from both yeast and germ tubes bound to immobilized entactin [146]. Binding was inhibited ~ 50% by addition of RGD peptide and by preincubation of the extract with laminin or fibronectin. Ligand affinity blotting detected three reactive proteins of 25, 44, and 65 kDa. The relationship among these proteins and the apparent promiscuity with laminin and fibronectin binding proteins is unknown.

6. Tenascin

There is a family of four tenascin proteins. These proteins contain multiple repeated structural motifs such as fibronectin type III and epidermal growth factor–like repeats, and a globular fibrinogen like domain. Tenascin-C is found in many developing tissues and is overexpressed in tumor cells. Heterogeneously distributed binding of tenascin-C was observed along hyphal extensions of germ tubes of *C. albicans* while yeast cells were unreactive [147] (Table 2). However, cell wall extracts from both morphologies contained material that bound to immobilized tenascin-C. Binding was inhibited by fibronectin by not by fibrinogen or RGD peptide. This suggested that one or more of the fibronectin-binding proteins may also bind tenascin through the fibronectin type III repeats.

7. Other Host Proteins

Other host proteins that are not found in the serum and ECM protein classes are also adherence targets for some fungi. *C. albicans* Hwp1p is unique adhesin in that it becomes covalently attached to the ligand on the BEC surface [1] (Table 2). A cDNA clone for *HWP1* was isolated by immunoscreening [148]. The gene was developmentally regulated, as mRNA expression was observed in germ tubes but not yeast cells and the gene product was present on hyphal surfaces. The predicted protein was a 61,122-Da acidic protein with a 10 amino acid repeat motif rich in proline and glutamine [148,149]. The C-terminal end contained abundant hydroxy-amino acids as potential sites of O-glycosylation and terminated in sequence consistent with addition of a GPI moiety. The GPI signal sequence is consistent with covalent attachment of the protein to glucan in *S. cerevisiae* [150]. Hwp1p contained sequences that were similar to substrates recognized by transglutaminases [149]. These enzymes crosslink small proline-rich proteins forming a bond between glutamine and lysine residues. Adherence of germ tubes to BECs resulted in a stable interaction that was resistant to disruption by heat or SDS but that could be inhibited by transglutaminase inhibitors [151]. The Hwp1p amino-terminus was a substrate for transglutaminase [1]. A deletion strain *hwp1/hwp1* was reduced five fold in the ability to form the stable adherence to BECs. These observations suggested that germ tubes were covalently attached to BECs through a covalent linkage between Hwp1p and unknown BEC surface protein(s) formed by the host transglutaminase.

WI-1 is a surface protein of *B. dermatitidis* that mediates binding to the CD11b/CD18 (complement receptor 3; CR3) and CD14 proteins of macrophage surfaces in the absence of serum [152] (Table 3). The relative amount of surface WI-1, a 120-kDa immunodominant antigen expressed by yeast cells, determined the extent of binding to macrophages. Binding was temperature and Mg^{2+} dependent. Blocking binding with various anti-CR3 antibodies suggested that the LPS (lipopolysaccharide) site of CR3 was the major binding site for virulent strains while less virulent strains, even though expressing WI-1 also showed binding to CD14. Analysis of the sequence encoding WI-1 suggested three structural domains [153]: (1) The amino terminus had a short hydrophobic, potentially membrane spanning sequence. (2) The central domain had 30 repeats of 24-amino acid sequence organized in two noncontiguous regions that were predicted to be exposed at the yeast cell surface. This repeat sequence was homologous to the *Yersiniae* adhesin, invasin. (3) The C-terminus contained an epidermal growth factor–like domain that might bind to extracellular matrix. A recombinant protein was able to bind to macrophage receptors as was purified native protein. A strain deficient in WI-1 expression was constructed by gene disruption [154]. This strain was reduced in the ability to bind to and be ingested by macrophages, adhered less well to lung tissue, and was avirulent in a murine infection model. Return of functional *WI1* gene to the defective strain also restored binding and virulence. This supported the role of adherence in pathogenesis of infection.

V. SUMMARY

Fungi recognize the host through multiple mechanisms (Table 1). The predominant adherence mechanisms involve lectinlike interactions with the fungus the source of either the carbohydrate or the protein and protein-protein interactions. Fungi recognize both soluble and immobilized ligands. Fungi whose adherence mechanisms have been more extensively studied employ several recognition mechanisms (Tables 2, 3). A schematic representation of several of these mechanisms in shown in Figure 1. Adhesins and their ligands usually have been studied one pair at a time. However, there is reason to believe that at various times multiple adherence interactions may be employed simultaneously. Indeed, as noted in some cases above, inhibition of a particular adhesin reduced but did not eliminate adherence. In those studies where mutant strains deficient in an adhesin have been constructed, reduced virulence has been observed in various models of infection. These findings support the basic premise that adherence of fungi to host ligands and cells is necessary to initiate colonization and infection of man by fungi.

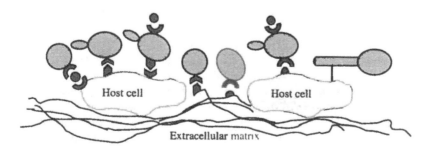

Figure 1 Schematic representation of various fungal adhesin interactions with host ligands. Protein adhesins (⌒) and may bind soluble or immobilized host ligand proteins (●, ▲). In some cases a ligand may be bound by both fungal and host adhesins (◖●). The second major class of interaction is a lectinlike interaction between a protein (❭) and carbohydrate (◢). Host cells and ECM are labeled in the figure.

ACKNOWLEDGMENTS

This work was supported in part by Public Health Service grant R01 AI23416 from the National Institute of Allergy and Infectious Diseases and Texas Higher Education Coordinating Board Advanced Research Program grant TDH012 to W.L.C.

REFERENCES

1. JF Staab, SD Bradway, PL Fidel, P Sundstrom. Adhesive and mammalian transglutaminase substrate properties of *Candida albicans* Hwp1. Science 283:1535–1538, 1999.
2. P Sundstrom. Adhesins in *Candida albicans*. Curr Opin Microbiol 2:353–357, 1999.
3. WL Chaffin, JL Lopez-Ribot, M Casanova, D Gozalbo, JP Martinez. Cell wall and secreted proteins of *Candida albicans*: identification, function, and expression. Microbiol Mol Biol Rev 62:130–180, 1998.
4. EJ Olson, JE Standing, N Griego-Harper, OA Hoffman, AH Limper. Fungal beta glucan interacts with vitronectin and stimulates tumor necrosis factor alpha release from macrophages. Infect Immun 64:3548–3554, 1996.
5. AH Limper, JE Standing, OA Hoffman, M Castro, LW Neese. Vitronectin binds to *Pneumocystis carinii* and mediates organism attachment to cultured lung epithelial cells. Infect Immun 61:4302–4309, 1993.
6. P Wisniowski, WJ Martin 2d. Interaction of vitronectin with *Pneumocystis carinii*: evidence for binding via the heparin binding domain [see comments]. J Lab Clin Med 125:38–45, 1995.
7. OA Hoffman, EJ Olson, AH Limper. Fungal beta-glucans modulate macrophage release of tumor necrosis factor-alpha in response to bacterial lipopolysaccharide. Immunol Lett 37:19–25, 1993.
8. M Castro, JA Bjoraker, MS Rohrbach, AH Limper. *Candida albicans* induces the release of inflammatory mediators from human peripheral blood monocytes. Inflammation 20:107–122, 1996.
9. AH Limper, JE Standing. Vitronectin interacts with *Candida albicans* and augments organism attachment to the NR8383 macrophage cell line. Immunol Lett 42:139–144, 1994.
10. E Segal, I Kremer, D Dayan. Inhibition of adherence of *Candida albicans* to acrylic by a chitin derivative. Eur J Epidemiol 8:350–355, 1992.
11. M Kahana, E Segal, M Schewach Millet, Y Gov. In vitro adherence of *Candida albicans* to human corneocytes. Inhibition by chitin soluble extract. Acta Derm Venereol 68:98–101, 1988.
12. MA Ghannoum, K Abu-Elteen, A Ibrahim, R Stretton. Protection against *Candida albicans* gastrointestinal colonization and dissemination by saccharides in experimental animals. Microbios 67:95–105, 1991.
13. N Lehrer, E Segal, L Barr-Nea. In vitro and in vivo adherence of *Candida albicans* to mucosal surfaces. Ann Microbiol (Paris) 134B:293–306, 1983.
14. E Segal, L Gottfried, N Lehrer. Candidal vaginitis in hormone treated mice: prevention by a chitin extract. Mycopathologia 102:157–163, 1988.
15. S Gottlieb, Z Altboum, DC Savage, E Segal. Adhesion of *Candida albicans* to epithelial cells effect of polyoxin D. Mycopathologia 115:197–205, 1991.
16. E Segal, S Gottlieb, Z Altboum, Y Gov, I Berdicevsky. Adhesion of *Candida albicans* to epithelial cells effect of nikkomycin. Mycoses 40:33–39, 1997.
17. CE Bulawa, DW Miller, LK Henry, JM Becker. Attenuated virulence of chitin-deficient mutants of *Candida albicans*. Proc Natl Acad Sci USA 92:10570–10574, 1995.
18. ET Buurman, C Westwater, B Hube, AJ Brown, FC Odds, NA Gow. Molecular analysis of CaMnt1p, a mannosyl transferase important for adhesion and virulence of *Candida albicans*. Proc Natl Acad Sci USA 95:7670–7675, 1998.
19. M Casanova, JL Lopez-Ribot, C Monteagudo, A Llombart-Bosch, R Sentandreu, JP Martinez. Identification of a 58-kilodalton cell surface fibrinogen-binding mannoprotein from *Candida albicans*. Infect Immun 60:4221–4229, 1992.

20. WL Chaffin, B Collins, JN Marx, GT Cole, KJ Morrow Jr. Characterization of mutant strains of *Candida albicans* deficient in expression of a surface determinant. Infect Immun 61:3449–3458, 1993.

21. RK Li, JE Cutler. Chemical definition of an epitope/adhesin molecule on *Candida albicans*. J Biol Chem 268:18293–18299, 1993.

22. T Kanbe, MA Jutila, JE Cutler. Evidence that *Candida albicans* binds via a unique adhesion system on phagocytic cells in the marginal zone of the mouse spleen. Infect Immun 60:1972–1978, 1992.

23. T Kanbe, JE Cutler. Minimum chemical requirements for adhesin activity of the acid-stable part of *Candida albicans* cell wall phosphomannoprotein complex. Infect Immun 66:5812–5818, 1998.

24. A Aderem, DM Underhill. Mechanisms of phagocytosis in macrophages. Annu Rev Immunol 17: 593–623, 1999.

25. VL Shepherd, KB Lane, R Abdolrasulnia. Ingestion of *Candida albicans* down-regulates mannose receptor expression on rat macrophages. Arch Biochem Biophys 344:350–356, 1997.

26. H Kobayashi, N Shibata, S Suzuki. Evidence for oligomannosyl residues containing both beta- 1,2 and alpha- 1,2 linkages as a serotype A specific epitope(s) in mannans of *Candida albicans*. Infect Immun 60:2106–2109, 1992.

27. Y Miyakawa, T Kuribayashi, K Kagaya, M Suzuki, T Nakase, Y Fukazawa. Role of specific determinants in mannan of *Candida albicans* serotype A in adherence to human buccal epithelial cells. Infect Immun 60:2493–2499, 1992.

28. RA Ezekowitz, DJ Williams, H Koziel, MY Armstrong, A Warner, FF Richards, RM Rose. Uptake of *Pneumocystis carinii* mediated by the macrophage mannose receptor. Nature 351:155–158, 1991.

29. H Koziel, Q Eichbaum, BA Kruskal, P Pinkston, RA Rogers, MY Armstrong, FF Richards, RM Rose, RA Ezekowitz. Reduced binding and phagocytosis of *Pneumocystis carinii* by alveolar macrophages from persons infected with HIV-1 correlates with mannose receptor downregulation. J Clin Invest 102:1332–1344, 1998.

30. DM O'Riordan, JE Standing, AH Limper. *Pneumocystis carinii* glycoprotein A binds macrophage mannose receptors. Infect Immun 63:779–784, 1995.

31. IA Critchley, LJ Douglas. Role of glycosides as epithelial cell receptors for *Candida albicans*. J Gen Microbiol 133:637–643, 1987.

32. IA Critchley, LJ Douglas. Isolation and partial characterization of an adhesin from *Candida albicans*. J Gen Microbiol 133:629–636, 1987.

33. G Vardar-Unlu, C McSharry, LJ Douglas. Fucose specific adhesins on germ tubes of *Candida albicans*. FEMS Immunol Med Microbiol 20:55–67, 1998.

34. J McCourtie, LJ Douglas. Extracellular polymer of *Candida albicans*: isolation, analysis and role in adhesion. J Gen Microbiol 131:495–503, 1985.

35. BJ Cameron, LJ Douglas. Blood group glycolipids as epithelial cell receptors for *Candida albicans*. Infect Immun 64:891–896, 1996.

36. E Enache, T Eskandari, L Borja, E Wadsworth, B Hoxter, R Calderone. *Candida albicans* adherence to a human oesophageal cell line. Microbiology 142:2741–2746, 1996.

37. J Sturtevant, F Dixon, E Wadsworth, JP Latge, XJ Zhao, R Calderone. Identification and cloning of GCA1, a gene that encodes a cell surface glucoamylase from *Candida albicans*. Med Mycol 37: 357–366, 1999.

38. L Yu, KK Lee, K Ens, PC Doig, MR Carpenter, W Staddon, RS Hodges, W Paranchych, RT Irvin. Partial characterization of a *Candida albicans* fimbrial adhesin. Infect Immun 62:2834–2842, 1994.

39. L Yu, KK Lee, W Paranchych, RS Hodges, RT Irvin. Use of synthetic peptides to confirm that the *Pseudomonas aeruginosa* PAK pilus adhesin and the *Candida albicans* fimbrial adhesin possess a homologous receptor-binding domain. Mol Microbiol 19:1107–1116, 1996.

40. L Yu, KK Lee, HB Sheth, P Lane-Bell, G Srivastava, O Hindsgaul, W Paranchych, RS Hodges, RT Irvin. Fimbria mediated adherence of *Candida albicans* to glycosphingolipid receptors on human buccal epithelial cells. Infect Immun 62:2843–2848, 1994.

41. KK Lee, L Yu, DL Macdonald, W Paranchych, RS Hodges, RT Irvin. Anti-adhesin antibodies that recognize a receptor-binding motif (adhesintope) inhibit pilus/fimbrial-mediated adherence of

Pseudomonas aeruginosa and *Candida albicans* to asialo-GM1 receptors and human buccal epithelial cell surface receptors. Can J Microbiol 42:479–486, 1996.

42. RD Cannon, AK Nand, HF Jenkinson. Adherence of *Candida albicans* to human salivary components adsorbed to hydroxylapatite. Microbiology 141:213–219, 1995.

43. M Edgerton, FA Scannapieco, MS Reddy, MJ Levine. Human submandibular sublingual saliva promotes adhesion of *Candida albicans* to polymethylmethacrylate. Infect Immun 61:2644–2652, 1993.

44. H Nikawa, S Hayashi, Y Nikawa, T Hamada, LP Samaranayake. Interactions between denture lining material, protein pellicles and *Candida albicans*. Arch Oral Biol 38:631–634, 1993.

45. A Vasilas, L Molina, M Hoffman, CG Haidaris. The influence of morphological variation on *Candida albicans* adhesion to denture acrylic in vitro. Arch Oral Biol 37:613–622, 1992.

46. H Nikawa, S Sadamori, T Hamada, K Okuda. Factors involved in the adherence of *Candida albicans* and *Candida tropicalis* to protein adsorbed surfaces. An in vitro study using immobilized protein. Mycopathologia 118:139–145, 1992.

47. JM O'Sullivan, RD Cannon, PA Sullivan, HF Jenkinson. Identification of salivary basic proline-rich proteins as receptors for *Candida albicans* adhesion. Microbiology 143:341–348, 1997.

48. GH Carpenter, GB Proctor. O linked glycosylation occurs on basic parotid salivary proline rich proteins. Oral Microbiol Immunol 14:309–315, 1999.

49. MP Hoffman, CG Haidaris. Analysis of *Candida albicans* adhesion to salivary mucin. Infect Immun 61:1940–1949, 1993.

50. MP Hoffman, CG Haidaris. Identification and characterization of a *Candida albicans* binding proteoglycan secreted from rat submandibular salivary glands. Infect Immun 62:828–836, 1994.

51. SA Klotz, RL Smith. Glycosaminoglycans inhibit *Candida albicans* adherence to extracellular matrix proteins. FEMS Microbiol Lett 78:205–208, 1992.

52. V Jimenez-Lucho, V Ginsburg, HC Krivan. *Cryptococcus neoformans*, *Candida albicans*, and other fungi bind specifically to the glycosphingolipid lactosylceramide (Gal beta 1-4Glc beta 1 1Cer), a possible adhesion receptor for yeasts. Infect Immun 58:2085–2090, 1990.

53. BP Cormack, N Ghori, S Falkow. An adhesin of the yeast pathogen *Candida glabrata* mediating adherence to human epithelial cells [see comments]. Science 285:578–582, 1999.

54. AJ Hamilton, L Jeavons, S Youngchim, N Vanittanakom, RJ Hay. Sialic acid-dependent recognition of laminin by *Penicillium marneffei* conidia. Infect Immun 66:6024–6026, 1998.

55. AJ Hamilton, L Jeavons, S Youngchim, N Vanittanakom. Recognition of Fibronectin by *Penicillium marneffei* Conidia via a Sialic Acid-Dependent Process and Its Relationship to the Interaction Between Conidia and Laminin. Infect Immun 67:5200–5205, 1999.

56. JP Bouchara, M Sanchez, A Chevailler, A Marot-Leblond, JC Lissitzky, G Tronchin, D Chabasse. Sialic acid-dependent recognition of laminin and fibrinogen by *Aspergillus fumigatus* conidia. Infect Immun 65:2717–2724, 1997.

57. ML Gil, MC Penalver, JL Lopez-Ribot, JE O'Connor, JP Martinez. Binding of extracellular matrix proteins to *Aspergillus fumigatus* conidia. Infect Immun 64:5239–5247, 1996.

58. MC Penalver, JE O'Connor, JP Martinez, ML Gil. Binding of human fibronectin to *Aspergillus fumigatus* conidia. Infect Immun 64:1146–1153, 1996.

59. G Tronchin, K Esnault, G Renier, R Filmon, D Chabasse, JP Bouchara. Expression and identification of a laminin-binding protein in *Aspergillus fumigatus* conidia. Infect Immun 65:9–15, 1997.

60. JP Bouchara, A Bouali, G Tronchin, R Robert, D Chabasse, JM Senet. Binding of fibrinogen to the pathogenic *Aspergillus* species. J Med Vet Mycol 26:327–334, 1988.

61. V Annaix, JP Bouchara, G Larcher, D Chabasse, G Tronchin. Specific binding of human fibrinogen fragment D to *Aspergillus fumigatus* conidia. Infect Immun 60:1747–1755, 1992.

62. P Coulot, JP Bouchara, G Renier, V Annaix, C Planchenault, G Tronchin, D Chabasse. Specific interaction of *Aspergillus fumigatus* with fibrinogen and its role in cell adhesion. Infect Immun 62:2169–2177, 1994.

63. MA Ghannoum, GR Burns, KA Elteen, SS Radwan. Experimental evidence for the role of lipids in adherence of *Candida* spp. to human buccal epithelial cells. Infect Immun 54:189–193, 1986.

64. S Law, PG Fotos, PW Wertz. Skin surface lipids inhibit adherence of *Candida albicans* to stratum corneum. Dermatology 195:220–223, 1997.

65. S Page, FC Odds. Binding of plasma proteins to *Candida* species in vitro. J Gen Microbiol 134: 2693–2702, 1988.

66. SP Hawser, K Islam. Binding of *Candida albicans* to immobilized amino acids and bovine serum albumin. Infect Immun 66:140–144, 1998.

67. K Islam, SP Hawser. Effect of antifungal agents on the binding of *Candida albicans* to immobilized amino acids and bovine serum albumin. J Antimicrob Chemother 43:583–587, 1999.

68. A Bouali, R Robert, G Tronchin, JM Senet. Characterization of binding of human fibrinogen to the surface of germ tubes and mycelium of *Candida albicans*. J Gen Microbiol 133:545–551, 1987.

69. G Tronchin, R Robert, A Bouali, JM Senet. Immunocytochemical localization of in vitro binding of human fibrinogen to *Candida albicans* germ tube and mycelium. Ann Inst Pasteur Microbiol 138:177–187, 1987.

70. V Annaix, JP Bouchara, G Tronchin, JM Senet, R Robert. Structures involved in the binding of human fibrinogen to *Candida albicans* germ tubes. FEMS Microbiol Immunol 2:147–153, 1990.

71. G Tronchin, JP Bouchara, R Robert. Dynamic changes of the cell wall surface of *Candida albicans* associated with germination and adherence. Eur J Cell Biol 50:285–290, 1989.

72. P Sepulveda, A Murgui, JL Lopez-Ribot, M Casanova, J Timoneda, JP Martinez. Evidence for the presence of collagenous domains in *Candida albicans* cell surface proteins. Infect Immun 63: 2173–2179, 1995.

73. P Sepulveda, JL Lopez-Ribot, D Gozalbo, A Cervera, JP Martinez, WL Chaffin. Ubiquitin like epitopes associated with *Candida albicans* cell surface receptors. Infect Immun 64:4406–4408, 1996.

74. JP Martinez, JL Lopez-Ribot, WL Chaffin. Heterogeneous surface distribution of the fibrinogen binding protein on *Candida albicans*. Infect Immun 62:709–712, 1994.

75. P Sepulveda, JL Lopez-Ribot, M Casanova, D Navarro, E Canton, JP Martinez. *Candida albicans* fibrinogen binding mannoprotein: expression in clinical strains and immunogeneicity in patients with systemic candidiasis. Int Microbiol 1:209–216, 1998.

76. JL Lopez-Ribot, C Monteagudo, P Sepulveda, M Casanova, JP Martinez, WL Chaffin. Expression of the fibrinogen binding mannoprotein and the laminin receptor of *Candida albicans* in vitro and in infected tissues. FEMS Microbiol Lett 142:117–122, 1996.

77. HM Alloush, JL Lopez-Ribot, WL Chaffin. Dynamic expression of cell wall proteins of *Candida albicans* revealed by probes from cDNA clones. J Med Vet Mycol 34:91–97, 1996.

78. JL Lopez-Ribot, P Sepulveda, AM Cervera, P Roig, D Gozalbo, JP Martinez. Cloning of a cDNA fragment encoding part of the protein moiety of the 58 kDa fibrinogen binding mannoprotein of *Candida albicans*. FEMS Microbiol Lett 157:273–278, 1997.

79. S Yan, RG Rodrigues, D Cahn-Hidalgo, TJ Walsh, DD Roberts. Hemoglobin induces binding of several extracellular matrix proteins to *Candida albicans*. Identification of a common receptor for fibronectin, fibrinogen, and laminin. J Biol Chem 273:5638–5644, 1998.

80. OC Lima, CC Figueiredo, BA Pereira, MG Coelho, V Morandi, LM Lopes-Bezerra. Adhesion of the human pathogen *Sporothrix schenekii* to several extracellular matrix proteins. Braz J Med Biol Res 32:651–657, 1999.

81. F Heidenreich, MP Dierich. *Candida albicans* and *Candida stellatoidea*, in contrast to other *Candida* species, bind iC3b and C3d but not C3b. Infect Immun 50:598–600, 1985.

82. RA Calderone, L Linehan, E Wadsworth, AL Sandberg. Identification of C3d receptors on *Candida albicans*. Infect Immun 56:252–258, 1988.

83. L Linehan, E Wadsworth, R Calderone. *Candida albicans* C3d receptor, isolated by using a monoclonal antibody. Infect Immun 56:1981–1986, 1988.

84. A Saxena, R Calderone. Purification and characterization of the extracellular C3d binding protein of *Candida albicans*. Infect Immun 58:309–314, 1990.

85. E Wadsworth, SC Prasad, R Calderone. Analysis of mannoproteins from blastoconidia and hyphae of *Candida albicans* with a common epitope recognized by anti-complement receptor type 2 antibodies. Infect Immun 61:4675–4681, 1993.

86. T Kanbe, RK Li, E Wadsworth, RA Calderone, JE Cutler. Evidence for expression of the C3d receptor of *Candida albicans* in vitro and in vivo obtained by immunofluorescence and immunoelectron microscopy. Infect Immun 59:1832–1838, 1991.

87. S Franzke, RA Calderone, K Schaller. Isolation of avirulent clones of *Candida albicans* with reduced ability to recognize the CR2 ligand C3d. Infect Immun 61:2662–2669, 1993.

88. JL Lopez-Ribot, JP Martinez, WL Chaffin. Comparative study of the C3d receptor and 58-kilodalton fibrinogen- binding mannoproteins of *Candida albicans*. Infect Immun 63:2126–2132, 1995.

89. A Stringaro, P Crateri, D Adriani, G Arancia, A Cassone, RA Calderone, F De Bernardis. Expression of the complement binding protein (MP60) of *Candida albicans* in experimental vaginitis. Mycopathologia 144:147–152, 1998.

90. CM Bendel, MK Hostetter. Distinct mechanisms of epithelial adhesion for *Candida albicans* and *Candida tropicalis*. Identification of the participating ligands and development of inhibitory peptides. J Clin Invest 92:1840–1849, 1993.

91. JE Edwards Jr, TA Gaither, JJ O'Shea, D Rotrosen, TJ Lawley, SA Wright, MM Frank, I Green. Expression of specific binding sites on *Candida* with functional and antigenic characteristics of human complement receptors. J Immunol 137:3577–3583, 1986.

92. A Eigentler, TF Schulz, C Larcher, EM Breitwieser, BL Myones, AL Petzer, MP Dierich. C3b-binding protein on *Candida albicans*: temperature dependent expression and relationship to human complement receptor type 3. Infect Immun 57:616–622, 1989.

93. BJ Gilmore, EM Retsinas, JS Lorenz, MK Hostetter. An iC3b receptor on *Candida albicans*: structure, function, and correlates for pathogenicity. J Infect Dis 157:38–46, 1988.

94. MK Hostetter, JS Lorenz, L Preus, KE Kendrick. The iC3b receptor on *Candida albicans*: subcellular localization and modulation of receptor expression by glucose. J Infect Dis 161:761–768, 1990.

95. CM Bendel, J St Sauver, S Carlson, MK Hostetter. Epithelial adhesion in yeast species: correlation with surface expression of the integrin analog. J Infect Dis 171:1660–1663, 1995.

96. S Alaei, C Larcher, C Ebenbichler, WM Prodinger, J Janatova, MP Dierich. Isolation and biochemical characterization of the iC3b receptor of *Candida albicans*. Infect Immun 61:1395–1399, 1993.

97. C Gale, D Finkel, N Tao, M Meinke, M McClellan, J Olson, K Kendrick, M Hostetter. Cloning and expression of a gene encoding an integrin-like protein in *Candida albicans*. Proc Natl Acad Sci USA 93:357–361, 1996.

98. CA Gale, CM Bendel, M McClellan, M Hauser, JM Becker, J Berman, MK Hostetter. Linkage of adhesion, filamentous growth, and virulence in *Candida albicans* to a single gene, *INT1*. Science 279:1355–1358, 1998.

99. KM Kinneberg, CM Bendel, RP Jechorek, EA Cebelinski, CA Gale, JG Berman, SL Erlandsen, MK Hostetter, CL Wells. Effect of *INT1* gene on *Candida albicans* murine intestinal colonization [In Process Citation] J Surg Res 87:245–251, 1999.

100. CM Bendel, KM Kinneberg, RP Jechorek, CA Gale, SL Erlandsen, MK Hostetter, CL Wells. Systemic infection following intravenous inoculation of mice with *Candida albicans* int1 mutant strains. Mol Genet Metab 67:343–351, 1999.

101. TL Ray, CD Payne. Scanning electron microscopy of epidermal adherence and cavitation in murine candidiasis: a role for *Candida* acid proteinase. Infect Immun 56:1942–1949, 1988.

102. CL Frey, JM Barone, G Dreyer, Y Koltin, SRJ Petteway, DJ Drutz. Synthetic protease inhibitors inhibit *Candida albicans* extracellular acid protease activity and adherence to endothelial cells. Abstracts of the 90th General Meeting of the American Society for Microbiology F-102, 1990.

103. HJ Watts, FS Cheah, B Hube, D Sangland, NA Gow. Altered adherence in strains of *Candida albicans* harbouring null mutations in secreted aspartic proteinase genes. FEMS Microbiol Lett 159: 129–135, 1998.

104. KG Skerl, RA Calderone, E Segal, T Sreevalsan, WM Scheld. In vitro binding of *Candida albicans* yeast cells to human fibronectin. Can J Microbiol 30:221–227, 1984.

105. SA Klotz, MJ Rutten, RL Smith, SR Babcock, MD Cunningham. Adherence of *Candida albicans* to immobilized extracellular matrix proteins is mediated by calcium dependent surface glycoproteins. Microb Pathog 14:133–147, 1993.

106. E Jakab, M Paulsson, F Ascencio, A Ljungh. Expression of vitronectin and fibronectin binding by *Candida albicans* yeast cells. Apmis 101:187–193, 1993.

107. SA Klotz, RL Smith. A fibronectin receptor on *Candida albicans* mediates adherence of the fungus to extracellular matrix. J Infect Dis 163:604–610, 1991.

108. E Negre, T Vogel, A Levanon, R Guy, TJ Walsh, DD Roberts. The collagen binding domain of fibronectin contains a high affinity binding site for *Candida albicans*. J Biol Chem 269: 22039–22045, 1994.

109. C Penn, SA Klotz. Binding of plasma fibronectin to *Candida albicans* occurs through the cell binding domain. Microb Pathog 17:387–393, 1994.

110. G Santoni, A Gismondi, JH Liu, A Punturieri, A Santoni, L Frati, M Piccoli, JY Djeu. *Candida albicans* expresses a fibronectin receptor antigenically related to alpha 5 beta 1 integrin. Microbiology 140:2971–2979, 1994.

111. TM Silva, PM Glee, KC Hazen. Influence of cell surface hydrophobicity on attachment of *Candida albicans* to extracellular matrix proteins. J Med Vet Mycol 33:117–122, 1995.

112. S Yan, E Negre, JA Cashel, N Guo, CA Lyman, TJ Walsh, DD Roberts. Specific induction of fibronectin binding activity by hemoglobin in *Candida albicans* grown in defined media. Infect Immun 64:2930–2935, 1996.

113. PM Glee, J Masuoka, WT Ozier, KC Hazen. Presence of multiple laminin and fibronectin-binding proteins in cell wall extract of *Candida albicans*: influence of dialysis. J Med Vet Mycol 34:57–61, 1996.

114. SA Klotz, RC Hein, RL Smith, JB Rouse. The fibronectin adhesin of *Candida albicans*. Infect Immun 62:4679–4681, 1994.

115. S Yan, RG Rodrigues, DD Roberts. Hemoglobin induced binding of *Candida albicans* to the cell binding domain of fibronectin is independent of the Arg-Gly Asp sequence. Infect Immun 66: 1904–1909, 1998.

116. D Gozalbo, I Gil-Navarro, I Azorin, J Renau-Piqueras, JP Martinez, ML Gil. The cell wall associated glyceraldehyde-3-phosphate dehydrogenase of *Candida albicans* is also a fibronectin and laminin binding protein. Infect Immun 66:2052–2059, 1998.

117. NK Gaur, SA Klotz. Expression, cloning, and characterization of a *Candida albicans* gene, *ALA1*, that confers adherence properties upon *Saccharomyces cerevisiae* for extracellular matrix proteins. Infect Immun 65:5289–5294, 1997.

118. GJ Smits, JC Kapteyn, H van den Ende, FM Klis. Cell wall dynamics in yeast. Curr Opin Microbiol 2:348–352, 1999.

119. Y Fu, G Rieg, WA Fonzi, PH Belanger, JE Edwards Jr, SG Filler. Expression of the *Candida albicans* gene *ALS1* in *Saccharomyces cerevisiae* induces adherence to endothelial and epithelial cells. Infect Immun 66:1783–1786, 1998.

120. GP DeMuri, MK Hostetter. Evidence for a beta 1 integrin fibronectin receptor in *Candida tropicalis*. J Infect Dis 174:127–132, 1996.

121. G Santoni, P Birarelli, LJ Hong, A Gamero, JY Djeu, M Piccoli. An alpha 5 beta 1 like integrin receptor mediates the binding of less pathogenic *Candida* species to fibronectin. J Med Microbiol 43:360–367, 1995.

122. RG Rodrigues, S Yan, TJ Walsh, DD Roberts. Hemoglobin differentially induces binding of *Candida, Trichosporon,* and *Saccharomyces* species to fibronectin. J Infect Dis 178:497–502, 1998.

123. IM Bromley, K Donaldson. Binding of *Aspergillus fumigatus* spores to lung epithelial cells and basement membrane proteins: relevance to the asthmatic lung. Thorax 51:1203–1209, 1996.

124. ST Pottratz, WJ Martin 2d. Role of fibronectin in *Pneumocystis carinii* attachment to cultured lung cells. J Clin Invest 85:351–356, 1990.

125. ST Pottratz, JR Paulsrud, JS Smith, WJ Martin 2d. Evidence for *Pneumocystis carinii* binding to a cell-free substrate: role of the adhesive protein fibronectin. J Lab Clin Med 123:273–281, 1994.

126. P Wisniowski, R Pasula, WJ Martin 2d. Isolation of *Pneumocystis carinii* gp120 by fibronectin affinity: evidence for manganese dependence. Am J Respir Cell Mol Biol 11:262–269, 1994.

127. EM Aliouat, E Dei-Cas, A Ouaissi, F Palluault, B Soulez, D Camus. In vitro attachment of *Pneumocystis carinii* from mouse and rat origin. Biol Cell 77:209–217, 1993.

128. ST Pottratz, AL Weir, PE Wisniowski. *Pneumocystis carinii* attachment increases expression of fibronectin binding integrins on cultured lung cells. Infect Immun 62:5464–5469, 1994.

129. ST Pottratz, AL Weir. Gamma inteferon decreases *Pneumocystis carini* attachment to lung cells by decreaseing expression of lung cell suface integrins. Eur J Clin Invest 27:17–22, 1997.

130. CA Itatani, GJ Marshall. Ultrastructural morphology and staining characteristics of *Pnemocystis carinii* in situ and from bronchoalveolar lavage. J Parasitol 74:700–712, 1988.

131. AH Limper, WJ Martin 2d. *Pneumocystis carinii*: Inhibition of lung cell growth mediate by parasite attachment. J Clin Invest 85:391–397, 1990.

132. ST Pottratz, J Paulsrud, JS Smith, WJ Martin 2d. *Pneumocystis carinii* attachment to cultured lung cells by pneumocystis gp 120, a fibronectin binding protein. J Clin Invest 88:403–407, 1991.

133. S Narasimhan, MY Armstrong, K Rhee, JC Edman, FF Richards, E Spicer. Gene for an extracellular matrix receptor protein from *Pneumocystis carinii*. Proc Natl Acad Sci USA 91:7440–7444, 1994.

134. SA Klotz. Adherence of *Candida albicans* to components of the subendothelial extracellular matrix. FEMS Microbiol Lett 56:249–254, 1990.

135. JP Bouchara, G Tronchin, V Annaix, R Robert, JM Senet. Laminin receptors on *Candida albicans* germ tubes. Infect Immun 58:48–54, 1990.

136. N Sakata, K Yamazaki, T Kogure. Identification of a 21 kDa laminin binding component of *Candida albicans*. Zentralbl Bakteriol 289:217–225, 1999.

137. JL Lopez-Ribot, M Casanova, C Monteagudo, P Sepulveda, JP Martinez. Evidence for the presence of a high-affinity laminin receptor like molecule on the surface of *Candida albicans* yeast cells. Infect Immun 62:742–746, 1994.

138. M Montero, A Marcilla, R Sentandreu, E Valentin. A *Candida albicans* 37 kDa polypeptide with homology to the laminin receptor is a component of the translational machinery. Microbiology 144:839–847, 1998.

139. JP McMahon, J Wheat, ME Sobel, R Pasula, JF Downing, WJ Martin 2d. Murine laminin binds to *Histoplasma capsulatum*. A possible mechanism of dissemination. J Clin Invest 96:1010–1017, 1995.

140. AP Vicentini, JL Gesztesi, MF Franco, W de Souza, JZ de Moraes, LR Travassos, JD Lopes. Binding of *Paracoccidioides brasiliensis* to laminin through surface glycoprotein gp43 leads to enhancement of fungal pathogenesis. Infect Immun 62:1465–1469, 1994.

141. AP Vicentini, JZ Moraes, JL Gesztesi, MF Franco, W de Souza, JD Lopes. Laminin binding epitope on gp43 from *Paracoccidioides brasiliensis* is recognized by a monoclonal antibody raised against Staphylococcus aureus laminin receptor. J Med Vet Mycol 35:37–43, 1997.

142. G Tronchin, JP Bouchara, G Larcher, JC Lissitzky, D Chabasse. Interaction between *Aspergillus fumigatus* and basement membrane laminin: binding and substrate degradation. Biol Cell 77:201–208, 1993.

143. B Banerjee, PA Greenberger, JN Fink, VP Kurup. Immunological characterization of Asp f 2, a major allergen from *Aspergillus fumigatus* associated with allergic bronchopulmonary aspergillosis. Infect Immun 66:5175–5182, 1998.

144. SA Klotz, RL Smith. Gelatin fragments block adherence of *Candida albicans* to extracellular matrix proteins. Microbiology 141:2681–2684, 1995.

145. E Spreghini, A Gismondi, M Piccoli, G Santoni. Evidence for alphavbeta3 and alphavbeta5 integrin like vitronectin (VN) receptors in *Candida albicans* and their involvement in yeast cell adhesion to VN. J Infect Dis 180:156–166, 1999.

146. JL Lopez-Ribot, WL Chaffin. Binding of the extracellular matrix component entactin to *Candida albicans*. Infect Immun 62:4564–4571, 1994.

147. JL Lopez-Ribot, J Bikandi, R San Millan, WL Chaffin. Interactions between *Candida albicans* and the human extracellular matrix component tenascin-C. Mol Cell Biol Res Commun 2:58–63, 1999.

148. JF Staab, CA Ferrer, P Sundstrom. Developmental expression of a tandemly repeated, proline- and glutamine-rich amino acid motif on hyphal surfaces on *Candida albicans*. J Biol Chem 271:6298–6305, 1996.

149. JF Staab, P Sundstrom. Genetic organization and sequence analysis of the hypha-specific cell wall protein gene *HWP1* of *Candida albicans*. Yeast 14:681–686, 1998.
150. K Hamada, H Terashima, M Arisawa, K Kitada. Amino acid sequence requirement for efficient incorporation of glycosylphosphatidylinositol associated proteins into the cell wall of *Saccharomyces cerevisiae*. J Biol Chem 273:26946–26953, 1998.
151. SD Bradway, MJ Levine. Do proline-rich proteins modulate a transglutaminase catalyzed mechanism of candidal adhesion? Crit Rev Oral Biol Med 4:293–299, 1993.
152. SL Newman, S Chaturvedi, BS Klein. The WI-1 antigen of *Blastomyces dermatitidis* yeasts mediates binding to human macrophage CD11b/CD18 (CR3) and CD14 [comment]. J Immunol 154:753–761, 1995.
153. LH Hogan, S Josvai, BS Klein. Genomic cloning, characterization, and functional analysis of the major surface adhesin WI-1 on *Blastomyces dermatitidis* yeasts. J Biol Chem 270:30725–30732, 1995.
154. TT Brandhorst, M Wuthrich, T Warner, B Klein. Targeted gene disruption reveals an adhesin indispensable for pathogenicity of *Blastomyces dermatitidis*. J Exp Med 189:1207–1216, 1999.

2
Genetic Basis of Pathogenicity in
Cryptococcus neoformans

Brian L. Wickes
University of Texas Health Science Center at San Antonio, San Antonio, Texas

K. J. Kwon-Chung
National Institute of Allergy and Infectious Diseases, National Institutes of Health, Bethesda, Maryland

I. INTRODUCTION

Cryptococcus neoformans is a yeastlike fungal pathogen of humans that can cause serious, life-threatening infections. It is also the only basidiomycete among the most frequently encountered human mycotic agents. Although capable of infecting healthy people, *C. neoformans* is an opportunistic fungus, which causes disease primarily in immunocompromised patients. Infections in healthy individuals are not nearly as frequent as infections in immunocompromised patients, although they occur with regular frequency. The most common types of immunosuppression which predispose patients to infection with *C. neoformans* include steroid treatment, diabetes, lymphoma, leukemia, and AIDS, with AIDS being the greatest risk factor for infection. At its peak, it was estimated that 5–10% of AIDS patients were infected with *C. neoformans* [1] although this percentage is trending downward due to more effective antiretroviral therapy and better antifungal strategies. Unfortunately, the decrease in new infections does not end the seriousness of cryptococcosis for AIDS patients. For those AIDS patients who were infected, cryptococcosis is viewed as incurable and these patients are typically treated with a lifelong antifungal regimen [2]. Further compounding this problem is recent evidence that current AIDS therapies cannot eradicate the virus from the body [3]. Patients are also failing antiretroviral therapies at an increasing rate [4]. These data suggest that management of cryptococcosis is highly dependent on the management of AIDS and since AIDS is still incurable, cryptococcosis in these patients remains a potential life-threatening complication. These problems do not begin to take into account cryptococcosis in underdeveloped countries, which have access to neither anti-HIV nor antifungal drugs. In these places, cryptococcosis remains as devastating as it was during the early stages of the AIDS epidemic in the United States.

II. ROLE OF BIOLOGY IN *C. NEOFORMANS* PATHOGENESIS

C. neoformans can infect a number of different body sites including the skin, bones, eyes, and lungs [5]. The most frequent site of infection, however, is the brain in which the disease typically is manifested as meningoencephalitis. This predilection for the brain, referred to as neurotropism, is one of a number of interesting and unanswered questions regarding *C. neoformans* pathogenesis.

Infection by *C. neoformans* almost always begins after inhalation where the primary focus of infection initiates in the lung. Although it is believed that dissemination to the brain occurs from the lungs [6], evidence from animal studies shows that it may be possible for the fungus to gain access to the brain by transiting directly through the sinus cavity [7]. In fact, *C. neoformans* has been cultured from the nasal passages of a number of animals [8–10]. In addition to the unknown mechanism mediating dissemination to the brain, the reasons for the neurotropic nature of *C. neoformans* are poorly understood. It is unclear whether the organism travels to the brain by design (a true "tropism"), or by default resulting from clearance from all other tissues and/or a reduced immunological response in the brain.

A second interesting characteristic concerning infection is the nature of the reservoir for infectious particles. Pigeon, and other avian guano, are well known sources for *C. neoformans*. In pigeon droppings, the fungus can reach massive numbers, exceeding 5×10^7 colony-forming units/g material [11]. However, in spite of the potentially massive amount of organisms present in pigeon droppings, there is little evidence that infection results from direct exposure to pigeon droppings. There have been no outbreaks reported to be associated with exposure to contaminated sites, and pigeon handlers do not show an increased incidence of cryptococcosis. These results argue for the possibility that there may be another source of infectious particles. Alternatively, host immune systems may be so effective that even massive exposure does not result in clinical disease. The inability to link infection with point-source exposure to pigeon droppings, theoretically the best scenario for infection, would appear to suggest that other factors must be in effect for infection to occur since dose does not appear to be the primary cause of infection. If dose is not a major factor in infection, what factors are? Some possible explanations include the genetics of the host which may influence susceptibility, the genetics of the fungus which could be manifested as a hypervirulent strain, a more infectious reservoir, a more infectious form, or some other factor which has yet to be investigated. Compounding these possibilities is a recent study by Garcia-Hermoso which demonstrates the need to consider even more possibilities [12]. In their study patients were capable of harboring dormant organisms for as long as 13 years before infection is activated. Their data raise additional questions such as where the organism resides in the body, how it survives for so long, and the nature of the initial infectious form of the fungus. In fact, the true nature of the infectious particle of *C. neoformans* is a third enigma concerning *Cryptococcus* and cryptococcosis.

It is generally believed that *C. neoformans* must be inhaled before infection can occur [13]. Since maximum penetration into the alveoli is required for the organism to establish infection, there must be a minimum diameter above which viable particles cannot penetrate to the depth in the lung required for colonization. Typically, particles must be <2.0 μM in size to avoid being swept out of the respiratory system [14]. Yeast cells are generally >2.0 μM in size [6], and if the capsule is included, the size can be 10–20 μM in diameter, which would be far too large to penetrate to the depth required for infection. While it is possible that capsular size could decrease under certain conditions such as dryness and starvation, these conditions are deleterious to yeast cells, at least under laboratory conditions. Basidiospores on the other hand are generally <2.0 μM in size, are resistant to drying, and require little or no nutrients once formed. As will be discussed in later sections, however, it has proven difficult to establish where or when in nature basidiospores could be formed. Therefore, it has proven difficult to reconcile the form of the fungus which initiates infection since logic suggests that the spore is the infectious particle while the data suggest that it is the yeast form.

Table 1   *Cryptococcus neoformans* as a Model Fungal Pathogen

Advantages	Comments
haploid	standard mutagenesis protocols apply
sexual cycle exists	genetic analysis is possible
excellent animal models	mimic human disease
high frequency transformation	electroporation, biolistic methods
genome project under way	facilitates molecular analysis
safe to handle	BSL2 organism, no lab fatalities
rapid growth rate	colonies visible in 24 h
multiple molecular tools	vectors, markers, reporters, mutants
Disadvantages	Comments
stable diploids not formed	essential genes difficult to study
low homologous integration	knockouts are laborious
transformation systems	expensive equipment

As researchers begin to address these questions, it is becoming increasingly clear that unraveling the nature of cryptococcosis is going to require a multidisciplinary approach. Investigating specific questions in the absence of the proper biologic context will have a minimum effect on understanding the organism and will result in a missed opportunity to apply what is learned to other human fungal pathogens. Because of a number of useful characteristics (Table 1), *C. neoformans* is evolving as one of the best models for studying human mycotic agents. Therefore, consideration of the biology of *C. neoformans* in conjunction with each research discipline will likely yield the greatest advances in our knowledge of this fungus.

III.   DEFINING VIRULENCE IN *CRYPTOCOCCUS NEOFORMANS*

As more fungal pathogens become amenable to traditional molecular and genetic analysis, "virulence" may be dissected into composite units defined as "factors." Medical mycology, being a relatively new field, typically follows the lead of those researchers studying bacterial pathogens. This strategy has proven convenient since an increasing number of bacterial virulence factors are being identified down to the DNA level. Studies of these virulence factors can now include experiments to ascertain what additional genes a given factor can or must interact with in order for an organism to be virulent. A major consideration in studying virulence in fungi and virulence in bacteria is that many model bacterial species inhabit, as their primary ecological niche, either the human body or some other animal host. These types of niches would be expected to be precise and constant in selecting for genes which are involved in virulence since the immune system would likely be a greater threat to survival than competition from other microbes. This specificity can often be identified in the laboratory because the host, which is usually the primary niche, can provide some of the clues as to what is important for pathogenesis. Furthermore, bacterial pathogens which have a human host as their primary niche can be transmitted to other hosts, resulting in the continual selection and refinement of genes required for virulence. The affected genes, such as a toxin, often can be separable from genes required for normal cellular homeostasis as they would typically be required for invasion or combating the host response, but may have little effect on growth characteristics in vitro. Most medically important fungal pathogens, on the other hand, have very few opportunities for host-driven natural selection because with few exceptions, all are saprophytes. The exposure of these organisms to evolutionary forces, which would select for specific virulence factors, is minimal since these fungi are rarely transmitted from person to person and more often than not, infection requires a diminished host response to begin with. Therefore, infections may be dependent on narrow windows of opportunity and may only occur when a particular stage of the fungal life cycle overlaps with a favorable but often transient host state. Not surprisingly, when one considers how often we are exposed to potential fungal pathogens, as much as once every few minutes for ubiquitous fungi like *Aspergillus,* it is clear that there is a very narrow time frame for infection. This narrow time frame offers little opportunity for the host immune defenses to exert evolutionary pressure above the strain level. As a result, few opportunities exist for fungal species to evolve specific genes responsible for counteracting host immune systems. Therefore, disentangling fungal virulence genes, which clearly play a role in pathogenesis, from their normal cellular function may not be possible in most cases.

When fungal infections do occur, it is rare in medical mycology to identify a singular fungal phenotype which drives pathogenesis and can be cleanly identified by Falkow's molecular version of Koch's postulates [15]. Ironically, the best and perhaps only examples are found in *C. neoformans* in the form of melanin and capsule-associated genes. Few fungal genes have been identified which so clearly effect virulence without being necessary for normal cellular growth and metabolism in vitro. More often, there are many factors, which interact to cause a particular mycosis. A good illustration of this type of multilevel interaction is dimorphism, which is exhibited by many of the major systemic fungal pathogens. This ability to exist in either (or both) a hyphal or yeast morphology during infection certainly contributes to virulence, but the phenotype itself may regulate or be associated with a multitude of other phenotypes which enhance the pathogenic process. It therefore becomes difficult or impossible to identify a "factor" required for virulence since each independent gene may be a smaller component of the overall virulent phenotype. Compounding the problem is the fact that fungal virulence genes in reality, almost certainly have some other function required for survival when the organism is in its natural environment (i.e., decaying vegetation). When expressed in the host, the apparent role in virulence is likely secondary and coincidental. Defining virulence factors in human fungal pathogens, then, becomes an exercise in trying to determine which phenotype in nature, and the associated genes, could be expressed in vivo and have a detrimental effect on the host. Unfortunately, this problem serves to blur the line between essential genes, which would not be defined as contributing to virulence yet are required for fungal viability, and nonessential genes which can be eliminated from the organism without a detrimental effect on cell viability, at least as can be best defined on laboratory media. In bacteria, it is often easy to recover known virulence genes without a detrimental effect on growth rate in vitro. In fungi, little is known regarding the effects that gene deletion has on in vitro growth rates or metabolism. Typically, only growth rates in rich medium and biochemical profiles are tested. Medical mycologists may soon have to decide if essential genes such as those encoding actin for example, should be classified as true virulence factors.

IV. ROLE OF MORPHOLOGY IN PATHOGENICITY

A. Mating and Mating Type

Mating in *C. neoformans* was first reported in 1975 when the sexual or perfect state of *C. neoformans* was discovered by crossing multiple strains in a variety of different combinations on selected media [16]. Shortly after the description of the perfect state in *C. neoformans* var. *neoformans* (*Filobasidiella neoformans*), *C. neoformans* var. *gattii* was also confirmed to possess a perfect state which was described as *Filobasidiella bacillospora* [17]. The perfect state of *C.*

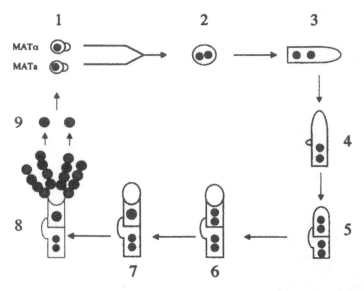

Figure 1 *C. neoformans* life cycle. (1) Budding haploid yeast cells growing vegetatively contact each other and fuse. (2) Fused cells very rapidly form dikaryotic hypha. (3) Dikaryotic hyphae with two independent nuclei, one from each mating type. (4) Hypha forms a hallmark clamp connection. (5) Clamp connection facilitates accurate segregation of nuclei into newly formed hyphal compartments. (6) Basidium forms at hyphal tip. (7) Nuclei fuse (karyogamy) in the basidium. (8) Nucleus undergoes meiosis and sporogenesis resulting in the formation of four chains of basidiospores. (9) Basidiospores are liberated and germinate to regenerate haploid yeast phase.

neoformans occurs when two cells of the opposite mating type contact each other and fuse under appropriate conditions (Fig. 1). The conjugated cell then produces a hypha which possesses typical basidiomycetous clamp connections. The hyphae are dikaryotic with each hyphal compartment containing a nucleus from each parent and are visible as a white fringe around a mixed patch of compatible cells. The nuclei within the hyphae divide in synchrony during hyphal growth until karyogamy occurs (fusion of the two nuclei) in the basidium formed at the hyphal tip. Basidiospores are produced post-meiotically and appear on the basidial surface as four long chains (Fig. 2). These spores can then be liberated from the chain, and if they land on an appropriate substrate, will initiate the vegetative phase by quickly becoming encapsulated, and then budding.

After the identification of the sexual state, a study was undertaken to determine what role, if any, mating type plays in virulence [18]. Three hundred thirty-eight isolates of *C. neoformans* were tested for mating type using four tester strains. Of these isolates, 105 were natural isolates and 233 were clinical isolates. The data shown in Table 2 revealed that there was a severe bias in both environmental isolates as well as clinical isolates for the α-mating type over the **a**-mating type. The ratio of *MAT*α : *MAT*a for environmental isolates was ~40:1 and for clinical isolates, ~30:1. A genetic explanation for this bias was not apparent since laboratory strains yield the expected 1α:1**a** ratio of offspring when crossed. The severe bias of *MAT*α to *MAT*a in clinical isolates at first glance appeared to suggest that *MAT*α cells are more infectious than *MAT*a cells. However, the *MAT*α to *MAT*a bias is conserved in environmental isolates as well. Therefore, it is possible that the predominance of the α-mating type in clinical isolates is due to a difference in exposure frequencies rather than an innate difference in virulence. To test this

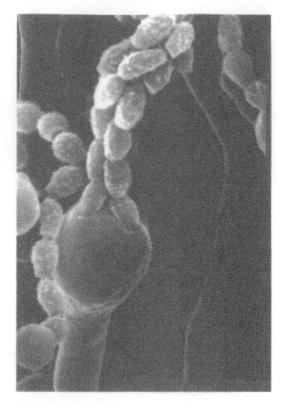

Figure 2   Basidium and basidiospores. (Photo courtesy of R. Samson.)

possibility, Kwon-Chung and Hill tested a number of *MATα* and *MATa* isolates in the mouse model [19]. Based on the data reported in their paper, the authors could not conclude that one mating type was more virulent than the other; however, under certain conditions, α-cells killed more mice than a-cells. The most important factor which led to increased virulence of α-cells was inoculum size (Table 3). Continued research on *C. neoformans* at the molecular and genetic level has shown that this fungus is extremely heterogeneous. Therefore, it was quite possible that the variation observed in the mouse study could have been due to differences in strain

Table 2   Mating Type Bias in *Cryptococcus neoformans*

| Source | Mating type | | | | Totals | Ratio α:a |
	α	a	αa[a]	NM[b]		
Environmental	87	2	0	16	105	43.5:1
Clinical	199	7	2	25	233	28.4:1
Totals	**286**	**9**	**2**	**41**	**338**	**31.8:1**

[a] Self fertile.
[b] Nonmater with any tester strain.
Source: Ref. 18.

Table 3 Comparison of *MATα* vs. *MATa* Cells for Virulence in Mice[a]

| | Inoculum (cells) | | | |
	1×10^4	1×10^5	1×10^6	1×10^7
Parental isolates				
NIH12 (α)	5[a]	8	9	10
NIH433 (a)	0	2	9	10
F$_1$ isolates (12 × 433)				
B-3501 (α)	4	4	10	10
B-3552 (α)	0	4	7	10
B-3502 (a)	0	3	4	10
B-3556 (a)	0	0	0	1
Total α deaths	9	16	35	40
Total a deaths	0	3	4	11

[a] All experiments used 10 mice per isolate per inoculum size.
Source: Ref. 19.

background, which operated independently of mating type. In order to eliminate this possibility, a congenic pair of strains was constructed which differed only at the mating type locus [20]. These two strains, JEC21 (α) and JEC20 (a), were constructed using a series of backcrosses such that the final pair represented the F$_{10}$ generation and were derived from sibling spores of different mating type. The two isolates were tested for biochemical characteristics and growth rate in order to insure that they were identical. The strains were then crossed and the progeny were scored for mating type. Five *MATα* and five *MATa* isolates for a total of 10 strains were recovered and tested for virulence in the mouse model. The data were analyzed statistically and showed that *MATα* cells killed more mice at a faster rate than *MATa* cells [20]. *MATa* cells were, however, virulent although the virulence was clearly less than *MATα* cells. These results clearly demonstrated that α-cells are indeed more virulent than a-cells. The molecular basis of this difference in virulence is being investigated in a number of laboratories. While mating type is known to play a role in virulence in other fungi, the ability of one mating type to be more virulent than the other is unexpected.

B. Monokaryotic Fruiting

Monokaryotic fruiting is defined as the ability of haploid or vegetative cells to be able to produce fruiting-body-like structures with spores [21]. This phenomenon is quite common in the higher basidiomycetes and is postulated to function as a survival mechanism for producing spores in the absence of the opposite mating type [21]. In *C. neoformans*, hyphae production by haploid cells has been known for many years. In most cases these strains were described as producing pseudohyphae or being self-fertile [22–24]. True monokaryotic fruiting in *C. neoformans* (Fig. 3) was characterized in detail in 1996 and was found to be associated with mating type [25]. In this study, it was found that *C. neoformans* can, in the traditional definition of monokaryotic fruiting, produce hyphae and basidiospores while growing as haploid yeasts. The major inducing factor for this phenotype was found to be nitrogen starvation under dry conditions. The most interesting aspect of monokaryotic fruiting in *C. neoformans* was that it was found to be linked to *MATα* cells. Cells which are *MATa* in mating type are not capable of monokaryotic fruiting

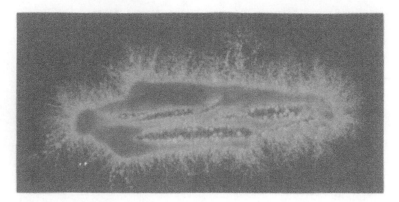

Figure 3 Monokaryotic fruiting. Haploid *MATα* vegetative yeast cells when placed on a starvation medium produce hyphae which contain only a single nucleus (monokaryotic). After prolonged growth, hyphae are visible as a fringe around the initial yeast inoculum.

and this linkage is conserved across all four serotypes. The linkage of monokaryotic fruiting, mating type, and virulence is curious and highly unusual in fungal pathogens.

It is unclear why monokaryotic fruiting is α-specific. However, it may be a contributing factor in the bias of *MATα* over *MATa* cells in environmental isolates as well as clinical isolates. The reason why this phenotype may be important in the mating type bias concerns the infectious particle of cryptococcosis. The restriction imposed by the size requirement for an infectious propagule, combined with the apparent rarity of the *MATa* mating type in the environment (implying that sexually produced spores are unlikely or do not survive), made an obvious explanation for the infectious form of *C. neoformans* difficult. It was therefore hypothesized that yeast cells were the infectious agents since in the absence of *MATa* cells, spores would not be produced [26,27]. To fit the yeast cell into the infectious particle theory, it was hypothesized that yeast cells become desiccated in the environment and shrink [27]. In this form, the cells can be easily aerosolized and are also small enough to penetrate to the alveoli. However, desiccated yeast cells, at least in the laboratory, display very poor viability [25,28]. More importantly, since *C. neoformans* can be present in pigeon droppings in massive quantities, it seems likely that there would be numerous reports of pigeon-associated outbreaks of cryptococcosis, which has not been the case. If, however, one considers monokaryotic fruiting as the source of the infectious particles, the possibility that this is the actual cause of the mating type bias seems more realistic. First, for many fungi, the spore is the infectious particle. The size, position on aerial hyphae, and resistance to environmental extremes make the spore the logical choice as the infectious agent of cryptococcosis since each of these characteristics occurs during monokaryotic fruiting. Second, monokaryotic fruiting is an extremely rare phenomenon. It is quite difficult to induce under laboratory conditions and, more importantly, does not result in an abundant production of spores. Therefore, the rare production of truly infectious particles in a large background of relatively noninfectious particles could account for the observed epidemiological data in which outbreaks due to exposure to large reservoirs have not been reported. Third, monokaryotic fruiting is linked to the α-mating type. While this mating type is more virulent than *MATa* cells, *MATa* cells are still virulent yet rarely are found in clinical isolates. Furthermore, we have not observed any difference in spores produced by monokaryotic fruiting or sexual reproduction, nor in the yeast cells which germinate from each spore type. The few *MATa* clinical isolates that are observed could be the result of basidiospore inhalation, however, in this case the source

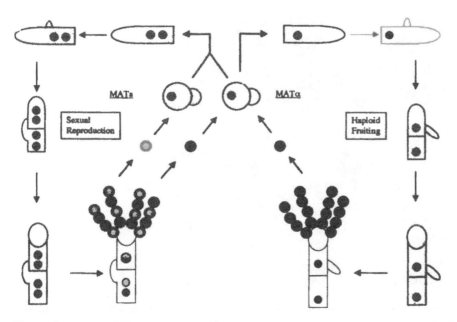

Figure 4   Amended life cycle showing both sexual reproduction and monokaryotic fruiting.

could be a traditional *MATα:MATa* mating. Fourth, monokaryotic fruiting results in a massive amount of blastospores produced off the sides of each hypha which could contribute to the disproportional representation of *MATα* cells in a local population. This bias may skew the frequency of α:a cells more so than would be expected based on differences in virulence. Fifth, infectious basidiospores have been demonstrated under controlled laboratory conditions. Zimmer has shown that basidiospores are capable of causing infection in mice [29] and recently, Sukroongreung has shown that preparations of basidiospores are infectious for mice when inhaled [30]. Whether the infectious particles are the result of sexual reproduction, monokaryotic fruiting, or even yeast cells, it is clear that this issue is by no means decided and will require additional studies before a clear source of infection is identified. The existence of monokaryotic fruiting, its mating type specificity, and its potential source of infectious particles adds another layer of complexity to the life cycle of *C. neoformans* (Fig. 4).

C. Switching

Switching was first described in *Candida albicans* in 1985 [31] and was initially detected and defined through changes in colonial morphology which was reversible at high frequency. The high frequency of the event ruled out mutation and the reversibility argued against a transposition-mediated phenomenon. The precise mechanism(s) which regulate and control switching, however, still remain elusive. Recently, *C. neoformans* has been shown to be capable of undergoing switching [32] (Fig. 5). The *C. neoformans* system bears a number of similarities to switching in *C. albicans*. It occurs at high frequency (1×10^{-3} to 1×10^{-4}) and involves changes in colony morphology which carry additional phenotypic consequences. In the *C. neoformans* system, the observed morphologies are smooth, wrinkled, and serrated and have been observed in serotypes A and D [32]. Wrinkled colonies clump more than smooth or serrated colonies in

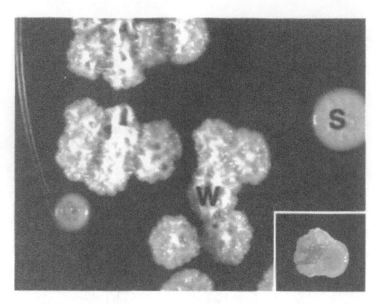

Figure 5 Switching in *C. neoformans*. A culture of wrinkled colonies (W) in which a revertant wild-type smooth colony appeared (S). Inset shows a sectored colony in which a smooth sector is produced from a wrinkled colony. (From Ref. 32.)

culture while serrated and wrinkled colonies were found to be more resistant to temperature extremes of hot and cold than the wild-type smooth colonies. Serrated colonies also appeared to be less virulent than either wrinkled or smooth colonies in the mouse model. A closer examination of switching in one of the serotype D strains revealed that it is possible that switching has a role in virulence since variations in glucuronoxylomanan structure were observed in addition to variations in actual capsule size [33]. An additional morphology described as pseudophyphal has also been identified in switching isolates [33]. While it is not exactly clear what effect a pseudophyphal morphology could have on virulence, this morphology has been observed in clinical isolates in the past [22,24]. The observation of switching in *C. neoformans* is highly significant since *C. neoformans* is more easily manipulated in the laboratory than *Candida albicans*. It will be interesting to see if there are any similarities at the molecular and or genetic level to switching in *C. albicans*. Given the large phylogenetic distance between *C. neoformans* and *C. albicans*, conserved switching-associated genes may have broad applications for many human fungal pathogens.

V. GENETIC VARIATION

The increasing number of molecular tools with which *C. neoformans* can be studied has demonstrated that this fungus is extremely heterogeneous. Initial groupings divided the fungus into two varieties which, in turn, could be subdivided into four serotypes [34]. The two varieties are *C. neoformans* var. *neoformans*, which comprises serotypes A and D, and *C. neoformans* var. *gattii*, which comprises serotypes B and C. The *gattii* variety is found in tropical and subtropical regions while the *neoformans* variety is found worldwide. Confirmed ecological niches of var. *gattii* are selected species of eucalyptus trees [35]. The primary niche of the *neoformans* variety has not been confirmed although it is occasionally isolated from wood sources. The *neoformans* variety, however, can be consistently found in pigeon and other avian guano. Since *C. neoformans* is not transmitted from person to person, infections result from exposure to environmental sources. The precise reservoir, route, and mechanism (see morphology section) of infection are uncertain; therefore, the biology of this fungus must be taken into consideration when studying the epidemiology, and implicit in any epidemiological analysis is the need to consider potential genetic variation.

Numerous methods for documenting variation have been employed over the years including DNA fingerprinting [36], RAPD analysis [37,38], isozyme analysis [39], and CHEF electrophoresis [38,40]. At present, there has not been a correlation of a specific strain with virulence. However, there are an increasing number of reports that show that clinical isolates can be, by whatever molecular test employed, similar to environmental isolates, suggesting that the isolates from the two sources could have been derived from a common ancestor [41,42]. While these reports do not prove that a person was infected from a given environmental source, they allow for this possibility.

The most useful comparisons have been performed on serial isolates from patients. These studies have been employed to document the evolution of drug-resistant isolates or for determining the cause of relapses during the course of long-term antifungal therapy [43,44]. Evidence for original strain maintenance arises from testing multiple serial isolates over a specific period of time. Techniques such as pulsed field electrophoresis or RAPD (random amplification of polymorphic DNA) analysis have shown that original banding patterns can remain relatively stable during the course of infection. These techniques are also supported by hybridization probes based on repetitive DNA. The general conclusion reached by these studies is that relapse during antifungal treatment is most often due to the original infecting strain [36,42,43,45].

Central to most mechanisms of genetic variation is the ability to undergo recombination which occurs, with the exception of mitotic recombination, during meiosis. In the laboratory, recombination in *C. neoformans* is easily demonstrated by analyzing the phenotypes of basidiospores recovered after a standard cross. Surprisingly, these results have been difficult to document in nature. The severe mating type bias observed in environmental isolates may argue that mating and meiosis occur rarely in nature. Since pigeons are not the primary niche, it is possible that the droppings are a secondary niche which, for some reason, favors survival of one mating type over the other. While nonpigeon environmental isolates fail to yield large numbers of *MATa* cells, this result could be due to not sampling from a defined, consistently populated, environmental source. Extended searches for the primary niche of *C. neoformans* may uncover a population which is in equilibrium and freely mating. In this case, recombination would be expected to contribute significantly to genetic variation. The possibility that this type of dynamic population exists has recently been suggested for the *gattii* variety of *C. neoformans* [46]. Mating type surveys of isolates recovered from single trees have revealed a bias similar to what is observed for the *neoformans* variety. However, samples containing *MATa* and *MATα* cells in ratios which approach 1:1 also occur. These isolates have yet to be analyzed in detail to confirm that they are mating; however, the data suggest that at least for the *gattii* variety, genetic recombination in a natural environment can occur. While the data would seem to make the likelihood of naturally recombining variety *neoformans* populations more plausible, there are few data supporting this possibility.

There have been a number of studies in which multiple isolates of *C. neoformans* var. *neoformans* have been tested using molecular methods and the data suggest that these isolates are clonal and not sexually reproducing [41,42,47]. These results allow for the hypothesis that *C. neoformans* may reproduce primarily by asexual means. Contributing to this hypothesis is the inability to find the *MATa* mating type from serotype A isolates, suggesting that mating opportunities may not exist, or be needed, for this serotype. Additionally, the majority of studies of environmental isolates seem to suggest that sexual reproduction in *C. neoformans* is rare; however, it is premature to draw firm conclusions without the identification of the primary niche of *C. neoformans* var. *neoformans*. It has been shown that the genetic machinery involved in the completion of mating is intact in some serotype A, *MATα* strains such as H99 (see Sec. VI.C). While some mating type-specific genes likely serve multiple functions and are therefore retained, other genes such as the pheromones should be disposable and lost from an asexual organism. Although functionality has not been tested, the *MATα* pheromone sequence has been found to be highly conserved in many serotype A, *MATα* strains [48]. While a serotype A *MATa* strain is unavailable at this writing, it is too soon to rule out the possibility that this mating type does not exist. Therefore, though sexual reproduction may not play a major role in generating variation in *C. neoformans* var. *neoformans*, it is premature to conclude that serotype A has long since evolved into an asexual organism.

VI. VIRULENCE FACTORS

As discussed previously, virulence factors can be divided into two groups—those factors which are dispensable to the organism, and those which are necessary for survival. These terms are, of course, somewhat arbitrary but are loosely based on *S. cerevisiae* precedent. Dispensable or nonessential genes can be defined as having no visible phenotypic effect on the growth or survival of the organism when cultured on standard laboratory media. Essential genes are genes that when deleted result in an inviable cell, an example of which would be actin. Their inclusion as genes which are virulence factors is debatable since, logically, a cell must be viable to be virulent. Their potential role in virulence, however, cannot be discounted since by definition, essential genes can be potential antifungal targets due to their necessity to the cell. On the other hand, *C. neoformans* is one of the few fungal pathogens which possesses clearly proven, nonessential virulence factors, two of which are capsule and melanin production.

A. Capsule

The polysaccharide capsule produced by *C. neoformans* is the hallmark of cryptococcosis and probably the major virulence factor of this fungus. It is easily visualized in an India ink stain, which is diagnostic, and is expressed in tissue (Fig. 6). Chemically, the capsule of *C. neoformans* composed of glucuronic acid, mannan, xylose, and *O*-acetyl [49], referred to collectively as GXM. In addition to the polysaccharide are two carbohydrate antigens, galactoxylomannan and mannoprotein [50]. These two antigens comprise approximately 12% of capsular mass while GXM comprises the remaining 88% [51]. The chemical composition of the capsule can be variable with the extent of *O*-acetylation and xylosyl substitution being responsible for serotype.

There is evidence that capsular composition in an individual strain may be variable [52,53]. Cherniak et al. [54] investigated relapse isolates from five patients for capsular variation. Three out of the five patients showed differences in polysaccharide structure in spite of the fact that these isolates were confirmed to be derived from the same initial strain. It is unknown what the mechanism leading to these changes is; however, it is quite possible that selection for different antigenic groups by the immune system could provide the evolutionary force which drives capsular variation. Recognition of any cellular component by the host theoretically could make any cell expressing this component less fit while cells expressing a different structure that is not as aggressively recognized would be more fit and therefore predominate in the population. What remains to be seen is if capsular variation in the absence of direct selection against capsule,

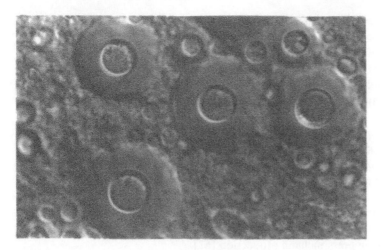

Figure 6 *C. neoformans* in tissue showing yeast cells surrounded by a large capsule.

such as in environmental isolates, occurs at a similar frequency. If not, the capsular phenotype may be variable if put under selective pressure by the immune system. Such flexibility could account for the chronic nature of cryptococcosis in AIDS patients if indeed the fungus possesses the capability to evade the immune system by varying one of its main antigens.

Progress toward understanding the genetics and molecular biology of capsule production has come as better molecular techniques have been developed for studying *C. neoformans*. The ability to perform standard genetic analysis in *C. neoformans* has contributed greatly to understanding how capsule affects virulence. Acapsular mutants can be readily made by UV or chemical mutagenesis and can be divided into different complementation groups by performing genetic crosses. Complementation of each mutant by transformation with a genomic library has made it possible to identify four capsule-associated genes: *CAP59, CAP64, CAP60,* and *CAP10* [55–58]. Each gene has been shown to be required for virulence since disruption of the gene eliminates virulence while complementation of the mutant phenotype restores virulence. Database searches with each of the translated proteins, however, have not yielded much information about the potential function of the protein in capsule biosynthesis since there is no significant homology to any other known protein. Additionally, outside of virulence and capsule size, disruption of the CAP genes has no other identifiable physiologic consequences. Information obtained from studying capsule biosynthesis will likely provide novel insights into cell wall architecture in general, and capsule production in particular. It is not known how many genes comprise the capsule pathway in *C. neoformans* or what genetic factors control variation in capsule composition. Regardless of the number of capsule-associated genes contained within this pathway, the ability to dissect this pathway using the available genetic and molecular techniques is a prime example of the utility of *C. neoformans* as a model fungal pathogen.

B. Melanin

Melanin is the second well-characterized, clearly demonstrated virulence factor. Although capsule is the more important component for virulence, melanin is probably the more studied of the two. This interest is due, in part, to the existence of many melanogenic human fungal pathogens. Additionally, melanin production is a common virulence factor in fungal plant pathogens. Finally, melanin is found throughout the animal kingdom, and the chemistry of melanin, while poorly understood, probably contributes to the versatility of the compound.

Early genetic studies of melanin production in *C. neoformans* showed that mutants incapable of producing melanin in vitro were less virulent in mice while revertants of these same mutants recovered their virulence [59,60]. As the molecular tools with which to study *C. neoformans* improved, a gene involved in melanin production was eventually identified. Williamson purified melanin from *C. neoformans* and used the purified protein sequence to clone *CNLAC1*, which was shown to be involved in the melanin biosynthetic pathway [61]. *CNLAC1* encodes a laccase which requires copper for optimum activity and is capable of oxidizing dopa or dopamine to dopaminequinone [62]. Disruption of *CNLAC1* was found to result in a melanin negative phenotype on appropriate media and a significant reduction in virulence for mice, confirming, by genetic and molecular methods, the importance of melanin in virulence [63]. Biochemical studies of melanin production in *C. neoformans* have shown that production is generally induced by starvation, lower temperatures (25–30°C), and stationary growth phase while dextrose, nitrogen, and high temperatures suppress melanin production [61].

Since the early identification of melanin production in *C. neoformans*, numerous laboratories have studied the role of this compound in pathogenesis and as a result, a number of general functions of melanin in *C. neoformans* biology have been proposed (Table 4). In *C. neoformans*, substrates for melanin production include polyphenolic compounds and catecholamines such as dopa and dopamine. These latter two compounds are found in the brain, supporting the hypothesis that the neurotropism of *C. neoformans* could be due to the brain providing the best environment for melanin production. In fungal plant pathogens, melanin plays a structural role by providing rigidity to invasive structures. In *C. neoformans*, melanin appears to play a different role which centers on protection from oxidative killing. While the exact protective mechanism is still under investigation, the probability that it occurs has recently been demonstrated by Casadevall who has developed a way to purify melanin from *C. neoformans* [64]. Cells are spheroplasted, treated with 4 M guanidinium isothiocyanate, and then boiled in 6 M HCl for 30 min. All cellular material except melanin is completely oxidized leaving a melanin "ghost" in the shape of a yeast cell. The resistance of melanin to this highly oxidative treatment is a clear indication of its antioxidative and protective role in vivo. More importantly, this technique has been used to show that *C. neoformans* produces melanin in vivo, and the residual saclike shape of the "ghost" suggests a close association with, or deposition within, the cell wall [65]. However, the produc-

Table 4 Function of Melanin in *C. neoformans*

Effect	Reference
Physical properties	
Protection from UV light	113
Protection from temperature extremes	114
Structural support	64
Immunological properties	
Antiphagocytic	115
Antioxidative	116
Anti-inflammatory	117
Biochemical properties	
Antifungal resistance	118
Electron buffering	119

tion of melanin in vivo is still somewhat unsettled as Liu et al. have not been able to find melanin in infected tissues. Instead, they argue for an alternative role for laccase in virulence such as providing iron oxidase activity, which may be responsible for the protective effect of the melanin pathway during pathogenesis [66].

C. Signal Transduction Homologs

Homologs of highly conserved signal transduction pathway genes identified in other fungi, in particular *S. cerevisiae*, are increasingly being recovered from *C. neoformans*. The *S. cerevisiae* pheromone response and pseudohyphal response pathways (reviewed in [67]) appear to be highly conserved in *C. neoformans* with some degree of functionality conserved as well. Two things have proven unique about the *C. neoformans* homologs: (1) some of the homologs appear to be mating type-specific, and (2) some of the homologs are required for virulence.

The best-studied and best-conserved signal transduction pathway in fungi is the mating or pheromone response pathway. This pathway has been dissected in detail from *S. cerevisiae* and generally serves as the starting point for the analysis of homologous or parallel pathways in other fungi. *C. neoformans* homologs of the *S. cerevisiae* mating cascade are shown in Figure 7. Three of the homologs display organization differences from their *S. cerevisiae* counterparts in that they are mating-type specific. These include homologs of *STE12* (a transcriptional activator), *STE11* (a MAP kinase kinase kinase), and *STE20* (a serine threonine kinase). These three genes have two homologs in *C. neoformans*, one for the *MATα* mating type and one for the *MATa* mating type.

STE12α was isolated during a screen for genes involved in monokaryotic fruiting [68]. This gene has been disrupted, and the disrupted strain (*ste12αΔ*) tested for virulence in mice.

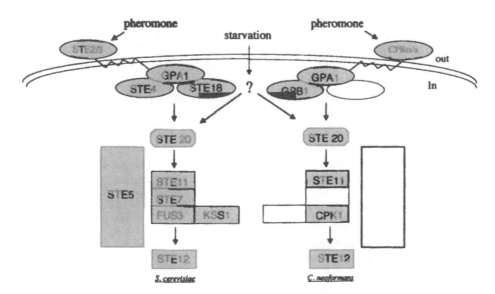

Figure 7 *C. neoformans* homologs of the *S. cerevisiae* pheromone response pathway. In *S. cerevisiae*, *STE20, STE11, STE7,* and *STE12* are also required for the production of pseudohyphae. Unfilled figures represent uncloned genes. In *C. neoformans*, *STE20, STE11,* and *STE12* have both *MATα*- and *MATa*-specific homologs.

The results of this study showed that indeed, *STE12α* is required for virulence since only 10% of the mice infected with this strain were killed after 80 days, in contrast to the wild-type strain (*STE12α*) which killed 100% of the mice during the same time period [69]. In this same study a potential explanation for these results was produced. *STE12α* was shown to be able to affect expression of two known virulence associated phenotypes, capsule and melanin. In the *ste12αΔ* mutant, expression of both capsule and melanin (*CNLAC1*) genes was reduced. Histopathological analysis of tissue from infected mice showed that the immune response was stronger in the brains of mice infected with the *ste12αΔ* strain than in mice infected with the wild-type strain. Disruptants of *STE12α* also showed other phenotypic effects; *ste12α* mutants were deficient in monokaryotic fruiting and showed a reduced mating efficiency. These phenotypes were observed in serotype A strains; however, no effect on virulence was observed for *ste12α* mutants [70]. These results are quite interesting and may delineate a cellular difference between serotype A and serotype D cells. In fact, a recent report by Cruz et al. [71] has identified another physiological difference between serotype A and serotype D strains by showing that only serotype A calcineurin disruptions were cation stress sensitive when compared to calcineurin disruptants in serotype D. Although these differences will need to be confirmed in other serotype A strains, it seems possible that serotype A and serotype D strains have a number of differences which may someday provide clues as to why serotype A strains predominate in clinical isolates.

The second mating-type-specific gene to be identified is *STE11α* [72], the first of three sequentially acting MAP kinases in the *S. cerevisiae* pheromone response pathway. *STE11α* appears to be both structurally and functionally conserved. Similar to other MAP kinase kinase kinase genes, the predicted Ste11αp has a highly conserved catalytic domain in the carboxy terminus typical of the serine threonine family of protein kinases. *ste11α* disruptants only have a modest effect on virulence in contrast to *ste12α* mutants; however, these mutants are sterile and incapable of undergoing monokaryotic fruiting. The sterile and fruiting negative phenotypes are consistent with the *S. cerevisiae* sterile and pseudohyphal negative response of *ste11* mutants.

The *C. neoformans* homolog of the *S. cerevisiae STE20* has only recently been isolated (J. Heitman, personal communication). Although phenotypic effects of a disruptant are not known, it too is mating-type specific. If its function is conserved, mutants should be sterile and unable to form hyphae making *STE20* yet another gene showing similar phenotypes to its *S. cerevisiae* counterpart. The α-specificity of these three genes leaves two possibilities for *MATa* strains when it comes to mating: (1) they do not have a conserved MAP kinase cascade, or (2) they have a *MATa*-specific cascade which is organized similar to *MATα* cells. The latter scenario appears to be the case as *MATa* homologs of *MATα* genes have been identified for *STE12* (Yun Chang, personal communication), *STE20* (Joe Heitman, personal communication), and *STE11* (unpublished data). In addition to these genes, both of the *MATα* and *MATa* pheromones have been isolated and characterized. The *MATα* pheromone was isolated in 1992 [73] and was the first mating-type-specific gene isolated from *C. neoformans*. When expressed in *MATa* cells, the *MATα* pheromone is capable of inducing a hyphal phenotype [74]. While it is unlikely that the pheromone is required for virulence, expression can be detected in rabbit CSF [75], although expression is likely due to starvation since nitrogen and carbohydrate starvation induce pheromone expression in vitro. The *MATa* pheromone has recently been isolated and shows characteristics similar to the *MATα* pheromone including the ability to induce hyphae when expressed in the opposite cell type (unpublished data).

Additional mating pathway genes which are common to both mating types have been isolated and characterized. *GPA1* was isolated by Tolkacheva [76] and shown to be highly similar to the *S. cerevisiae* α-subunit of the pheromone receptor-coupled heterotrimeric G-protein. Disruption of this gene revealed a number of phenotypic defects [77]. Mutants of this gene are sterile as is seen in other fungi, and are less virulent in the rabbit model. Causes for the reduction in virulence appear to parallel the results observed for *STE12α*. *GPA1* has been shown to affect expression of capsule and melanin since *gpa1* mutants are reduced for these two phenotypes. The effect of *gpa1* can be reversed by exogenous cAMP showing that *GPA1* has a conserved function. A second component of the receptor-associated heterotrimeric G-protein, *GPB1* (*S. cerevisiae STE4* homolog), has recently been isolated and has been shown to affect monokaryotic fruiting and mating, but not virulence [78]. These data suggest that *C. neoformans* possesses a conserved pathway which functions in both mating and the production of hyphae by haploid cells. These two pathways appear to be functionally analogous to the *S. cerevisiae* pathways, but may have a novel organization due to the existence of mating-type-specific homologs.

Calcineurin is another gene which functions in signal transduction although it is not required for mating. It is a serine/threonine phosphatase that is regulated by calmodulin [79] and is required for adaptation to pheromone in *S. cerevisiae* [80]. Calcineurin activity can be inhibited in cells treated with cyclosporin A, FK506, and rapamycin because these three compounds bind to cyclophilin A (cyclosporin) and FKBP12 (FK506, rapamycin) to form complexes which interact with calcineurin to block its function [81]. In *C. neoformans*, the gene encoding the calcineurin A catalytic subunit (*CNA1*) has been cloned and disrupted [82]. The disrupted strain was tested for growth under a variety of conditions potentially encountered during in vivo growth. It was found that under conditions of increased CO_2, elevated temperature (37°C), or alkaline pH, *cna1* mutants were unable to grow. Using an animal model of cryptococcosis, the *cna1* strain was found to be avirulent. *CNA1* falls into an interesting category because it is dispensable to the cell, but only under certain conditions. In particular, at elevated temperature, *CNA1* is required for viability and may function in multiple pathways [82].

D. Phospholipase

C. neoformans has recently been found to produce phospholipase activity [83]. Three phospholipase activities have been identified, phospholipase B, lysophospholipase, and lysophospholipase transacylase [84]. To determine if *C. neoformans* phospholipase plays a role in virulence, a collection of isolates was examined for the ability to produce phospholipase on a medium containing egg yolk. Ninety-eight percent of the isolates tested produced phospholipase activity. These isolates were then subdivided into high, intermediate, and low phospholipase producers. Representative isolates were selected and then tested for virulence in the mouse model by doing colony counts from brain and lung tissue. The data showed that virulence, as determined by colony counts, could be correlated with phospholipase activity, with the most virulent isolates producing the most phospholipase activity. Very little is known about the genes controlling phospholipase activity in *C. neoformans* since no studies have been published. However, preliminary data suggest that disruptions of this gene reduce virulence in mice (Gary Cox, personal communication).

E. Mannitol

In 1990, Wong et al. showed that *C. neoformans*, as a species, produces large amounts of D-mannitol in vitro [85]. They then tested whether mannitol was produced in vivo during active infection. Using the rabbit model of virulence, it was observed that, indeed, *C. neoformans* produces mannitol when present in high concentration in rabbit CSF. These early observations lead the authors to hypothesize two roles for mannitol in pathogenesis. The first is that mannitol increases the osmolality of the surrounding fluid which has a net effect of causing or contributing to cerebral edema, a major factor in the neurological damage caused by cryptococcal meningitis.

This hypothesis, however, has not been completely supported by other researchers. Megson et al. [86] found mannitol in patient CSF, but could not find a correlation among concentration, cryptococcal antigen titer, and CSF pressure. The second hypothetical role of mannitol in pathogenesis proposed by Wong et al. is a protective effect in which mannitol scavenges free radicals produced by the immune system for oxidative killing. This hypothesis has been supported in part by the observation that the addition of exogenous mannitol to a mixture of polymorphonuclear neutrophils and *C. neoformans* yeast cells prevented killing of *C. neoformans* [87]. The protective effect of mannitol could also be observed in a cell free system using $FeSO_4$, H_2O_2, or iodide as oxidative killing agents. It is perhaps noteworthy that the hypothesis for the role of mannitol in virulence overlaps the roles of the two other major virulence factors, melanin and capsule production. As described previously, melanin is hypothesized to protect against oxidative killing, and capsular polysaccharide has also been hypothesized to contribute to brain edema since it can be shed in large amounts into the surrounding CSF.

To date, the genetic basis for the role of mannitol in virulence is unclear. Only a single mannitol-negative mutant has been used to study virulence and this mutant was not created using molecular techniques. The mutation was created by UV mutagenesis [88]. In addition to the mannitol negative phenotype, the mutant displayed increased susceptible to heat and high salt, and showed a reduced growth rate at 37°C. It is unknown whether these secondary phenotypes are linked to the mannitol-negative phenotype or are a product of a second mutation. The mannitol-negative mutant was created in H99, a *MATα* serotype A strain which cannot be crossed, thereby making it difficult to prove that the mannitol-negative phenotype and the associated effects on virulence were due to a single gene phenomenon. Given the hypothesized role for mannitol in virulence, it would be interesting to see how combinations of double mutants containing the mannitol and *cnlac1* mutation, or mannitol mutation and any of the capsule mutations would affect virulence. These strains can be constructed if the mannitol mutation can be created in serotype D strains.

VII. ROLE OF THE HOST IMMUNE SYSTEM IN VIRULENCE

A. Race

Demographic features of patients infected with *C. neoformans* had not been fully characterized prior to the AIDS epidemic. This deficiency was because surveillance of cryptococcal infections was lacking and there had been no reports of cryptococcosis outbreaks that could be investigated to clarify epidemiological factors of the disease. Before the advent of the AIDS epidemic, reports showed that cryptococcosis occurred more commonly among Caucasians [89,90]. These reports, however, were the result of literature surveys rather than studies of racial comparisons for the prevalence of cryptococcosis [91]. In Australia, a high incidence of cryptococcosis was noted among Aborigines in the northern territory [92]. It is unclear, however, whether the high incidence is related to race or associated with social factors prevalent among Aborigines.

Ever since cryptococcosis was designated as an AIDS-defining opportunistic infection starting from the early 1980s, considerable progress has been made in our understanding of cryptococcal epidemiology. Population-based surveillance from four sites during 1992–1994 showed that the incidence of cryptococcosis in 1993 was significantly higher among African-Americans than among Caucasians [91,93]. This finding is similar to studies reported before 1992 [94,95]. A case control study from two of four surveillance sites, however, found no association between race and cryptococcosis. Although this study also showed the rate of cryptococcosis in 1993 was slightly higher for African-Americans among non-HIV-infected persons, the case study indicated that such differences were not statistically significant [93]. The higher incidence of cryptococcosis among African-Americans, therefore, does not appear to be based on a genetic association and could be due to other unidentified conditions or exposures. Therefore, the potential association of cryptococcosis and race remains largely unproven.

B. Gender

Prior to 1980, there had been several studies of a series of cases in which cryptococcosis was shown to be two to three times more common in men than in women [89,96,97]. This difference was viewed to be the result of different degrees of exposure to the ecological niche of the fungus [6]. Since the advent of the AIDS epidemic, higher incidences of cryptococcosis among males became even more pronounced which reflected gender differences among the AIDS cases. A recent epidemiological study among persons with AIDS from Atlanta and San Francisco, however, showed no significant difference in the rates of cryptococcosis between the two genders. The rates of cryptococcosis among non-HIV-infected persons from the same geographic area showed slightly higher rates among men, but the difference was not statistically significant [93]. A genetic predisposition to cryptococcosis in humans with respect to gender, therefore, has not been established.

C. Host Immune Status

Although it is known that cryptococcosis occurs in immunocompetent hosts, the disease is primarily an infection of immunocompromised individuals, especially those with defects in T-cell-mediated host defense mechanisms. As a consequence, AIDS is the greatest risk factor for cryptococcosis. A recent report by Hajjeh et al. [93] showed that 86% of cryptococcosis patients surveyed between 1992 to 1994 were HIV+ and a majority of these patients had CD4+ T-lymphocyte counts <100 mm^3. Cryptococcosis also occurs in patients with CD4+ T-lymphopenia or other cellular immune defects without HIV infection [98–100]. Before the advent of AIDS, patients with hematologic malignancies, especially Hodgkin's and non-Hodgkin's lymphomas and leukemia, belonged to the highest risk group for cryptococcosis [101]. Thirty percent to 50% of cryptococcal patients were found to have hematologic malignancies [90,101]. High-dose corticosteroid therapy [6,102] and sarcoidosis [103] are also known as predisposing factors for cryptococcosis. Organ transplantation is another frequent risk factor for cryptococcosis due mainly to the use of high-dose corticosteroids and other forms of immunosuppressive therapy [104–106]. Diabetes mellitus [96] and smoking [93,107] have been cited as predisposing factors for cryptococcosis; however, the association of diabetes with cryptococcosis is not so clear [108] compared to the more frequent risk factors listed above while the association of smoking with cryptococcosis has been reported only in AIDS patients [93,107]. These studies would appear to indicate that specific T-cell immunological defects, although caused by a variety of factors, are the major factors responsible for predisposing patients to infection with *C. neoformans*.

D. Animal Models of Virulence

Mice, rats, and rabbits have been extensively used to study cryptococcal infection. Rats and mice are most commonly used for cryptococcosis initiated by the respiratory as well as the intravenous route [109] while rabbits have been used as a meningitis model through the intrathecal route [110]. The mouse model offers an advantage over rat or rabbit models due to lower cost, ease of infecting and handling, and availability of many genetically characterized mouse strains. Mice with defined genetic differences are valuable tools for studying the role of genetic factors in resistance to cryptococcosis [111]. The rabbit model of chronic meningitis, however, offers an advantage of monitoring growth or clearance rates of *C. neoformans* from the CSF over finite periods of time. In the mouse model, either outbred or inbred strains can be infected at between 8 and 12 weeks of age (body weight >19 g), with an awareness that susceptibility to *C. neoformans* differs among inbred strains [109] (Huffnagle, personal communication). Pulmonary routes of infection are more commonly used when host-parasite interactions are being studied since the respiratory route of infection more closely mimics the route of cryptococcosis in humans. Intravenous models (via tail vein) have been more commonly used when the survival rates of mice are compared to assess the potency of virulence between congenic strains of the fungus. In addition, inoculation with an identical number of cells among mice can be more reliably achieved. For the comparison of survival rates of mice infected with strains of *C. neoformans* with known defective genes, Balb/c mice, which are moderately resistant to cryptococcal infection, have been widely used for intravenous routes of infection [20,58,63,70]. As far as the pulmonary route of infection is concerned, CBA, C.B-17, and Balb/c x DBA/2 F1 progeny can be classified as very resistant to cryptococcosis [112]. Balb/c, CB6F1, C3H (HeN and HeJ), and 129B6F2 mice, on the other hand, can be categorized as resistant while C57BL/6 can be categorized as susceptible. As in human cases, immunocompromised mice such as severe combined immunodeficient C.B-17 scid/scid (SCID) mice congenic with C.B-17 athymic nude mice in a complement deficient Balb/c background are extremely susceptible to cryptococcosis [109].

Different strains of *C. neoformans* display differences in animal susceptibility. Although a limited number of strains have been studied, H99 (serotype A strain) appears to be consistently more virulent in all mouse strains than the isogenic set of serotype D strains which are widely used for genetic studies [20]. It should be noted, however, that virulence among serotype A strains as a group varies just as greatly as serotype D strains. This variation presents difficulties in trying to select a model laboratory strain to be used for virulence studies, making it likely that investigators will continue to use strain-host combinations which are best suited for their particular research interests.

VIII.   FUTURE RESEARCH

C. neoformans has clearly arrived as the model human fungal pathogen. In spite of the decline in infections, due mainly to the decline in AIDS, virtually all of the major manipulations required to study a model organism are possible with *C. neoformans*. Furthermore, the available animal models are excellent for replicating disease conditions from inhalation-induced infections, to determining which components of the immune system are the most important during the course of infection. Additionally, in some tissues, such as the brain, *C. neoformans* reaches massive levels, theoretically making it possible to harvest in vivo expressed fungal mRNA relatively easily. This RNA could then be used to identify genes which are important for fungal growth in tissue. The funding of the *C. neoformans* genome project will greatly expand what we know about the organism and draw even more researchers into the field and the basidiomycete-phylogeny of *C. neoformans* will make genome information of broad interest to all microbiologists who work on eukaryotic organisms. It is clear, therefore, that *C. neoformans* occupies an important position in medical mycology both in the disease that it causes and in its utility as a model organism.

REFERENCES

1.   SL Chuck, MA Sande. Infections with *Cryptococcus neoformans* in the acquired immunodeficiency syndrome. N Engl J Med 321:794–799, 1989.

2. RW Pinner, RA Hajjeh, WG Powderly. Prospects for preventing cryptococcosis in persons infected with human immunodeficiency virus. Clin Infect Dis 21 (suppl 1) S:103–S107, 1995.
3. J Cohen. The daunting challege of keeping HIV suppressed. Science 277:32–33, 1997.
4. J Cohen. AIDS therapy. Failure isn't what it used to be . . . but neither is success. Science 279: 1133–1134, 1998.
5. TG Mitchell, JR Perfect. Cryptococosis in the era of AIDS-100 years after the discovery of *Cryptococcus neoformans*. Clin Microbiol Rev 8:515–548, 1995.
6. KJ Kwon-Chung, JE Bennett. Cryptococcosis. In: Medical Mycology. Philadelphia: Lea and Febiger, 1992, pp. 397–439.
7. NG Gomes, M Boni, CC Primo. The invasive behaviour of *Cryptococcus neoformans:* a possibility of direct access to the central nervous system? Mycopathologia 140:1–11, 1997.
8. JH Connolly, MB Krockenberger, R Malik, PJ Canfield, DI Wigney, DB Muir. Asymptomatic carriage of *Cryptococcus neoformans* in the nasal cavity of the koala (*Phaseolarctos cinereus*) Med Mycol 37:331–338, 1999.
9. R Malik, P Martin, DI Wigney, DB Church, W Bradley, CR Bellenger, WA Lamb, VR Barrs, S Foster, S Hemsley, PJ Canfield, DN Love. Nasopharyngeal cryptococcosis. Aust Vet J 75:483–488, 1997.
10. R Malik, DI Wigney, DB Muir, DN Love. Asymptomatic carriage of *Cryptococcus neoformans* in the nasal cavity of dogs and cats. J Med Vet Mycol 35:27–31, 1997.
11. CE Emmons. Natural occurrence of oportunistic fungi. Lab Invest 11:1026–1032, 1962.
12. D Garcia-Hermoso, G Janbon, F Dromer. Epidemiological evidence for dormant *Cryptococcus neoformans* infection. J Clin Microbiol 37:3204–3209, 1999.
13. RK Haugen, RD Baker. The pulmonary lesions in cryptococesis with special reference to subpleural nodules. Am J Clin Pathol 24:1381–1390, 1954.
14. TF Hatch. Distribution and deposition of inhaled particles in the respiratory tract. Bacteriol Rev 25:237–240, 1961.
15. S Falkow. Molecular Koch's postulates applied to microbial pathogenicity. Rev Infect Dis 10 (suppl 2) S274–S276, 1988.
16. KJ Kwon-Chung. A new genus, *Filobasidiella*, the perfect of *Cryptococcus neoformans*. Mycologia 67:1197–1200, 1975.
17. KJ Kwon-Chung. A new species of *Filobasidiella*, the sexual state of *Cryptococcus neoformans* B and C serotypes. Mycologia 68:942–946, 1976.
18. KJ Kwon-Chung, JE Bennett. Distribution of α and *a* mating types of *Cryptococcus neoformans* among natural and clinical isolates. Am J Epidemiol 108:337–340, 1978.
19. KJ Kwon-Chung, WB Hill. Sexuality and pathogenicity of *Filobasidiella neoformans* (*Cryptococcus neoformans*) In R Vanbreuseghem, C De Vroy, ed.: Sexuality and Pathogenicity of Fungi. Paris: Masson, 1981, pp. 243–250.
20. KJ Kwon-Chung, JC Edman, BL Wickes. Genetic association of mating types and virulence in *Cryptococcus neoformans*. Infect Immun 60:602–605, 1992.
21. K Esser, F Meinhardt. A common genetic control of dikaryotic and monokaryotic fruiting in the basidiomycete *Agrocybe aegerita*. Mol Gen Genet 155:113–115, 1977.
22. HJ Shadomy, JP Utz. Preliminary studies on a hypha-forming mutant of *Cryptococcus neoformans* Mycologia 58:383–390, 1966.
23. HJ Shadomy, HI Lurie. Histopathological observations in experimental. Cryptococcosis caused by a hypha producing strain of *Cryptococcus neoformans* (Coward strain) in mice. Sab 9:6–9, 1971.
24. JD Williamson, JF Silverman, CT Mallak, JD Christie. Atypical cytomorphologic appearance of *Cryptococcus neoformans:* a report of five cases. Acta Cytol 40:363–370, 1996.
25. BL Wickes, ME Mayorga, U Edman, JC Edman. Dimorphism and haploid fruiting *in Cryptococcus neoformans:* association with the alpha-mating type. Proc Natl Acad Sci USA 93:7327–7331, 1996.
26. F Farhi, GS Bulmer, JR Tacker. *Cryptococcus neoformans* IV. The not so encapsulated yeast. Infect Immun 1:526–531, 1970.
27. JB Neilson, RA Fromtling, GS Bulmer. *Cryptococcus neoformans:* size range of infectious particles from aerosolized soil. Infect Immun 17:634–638, 1977.

28. GS Bulmer. Twenty-five years with *Cryptococcus neoformans*. Mycopathologia 109:111–122, 1990.
29. B Zimmer, HO Hempel, NL Goodman. Pathogenicity of the basidiospores of *Filobasidiella neoformans*. Mycopathologia 85:149–153, 1984.
30. S Sukroongreung, K Kitiniyom, C Nilakul, S Tantimavanich. Pathogenicity of basidiospores of *Filobasidiella neoformans* var. *neoformans*. Med Mycol 36:419–424, 1998.
31. B Slutsky, J Buffo, DR Soll. High frequency switching of colony morphology in *Candida albicans*. Science 230:666–669, 1985.
32. DL Goldman, BC Fries, SP Franzot, L Montella, A Casadevall. Phenotypic switching in the human pathogenic fungus *Cryptococcus neoformans* is associated with changes in virulence and pulmonary inflammatory response in rodents. Proc Natl Acad Sci USA 95:14967–14972, 1998.
33. BC Fries, DL Goldman, R Cherniak, R Ju, A Casadevall. Phenotypic switching in *Cryptococcus neoformans* results in changes in cellular morphology and glucuronoxylomannan structure. Infect Immun 67:6076–6083, 1999.
34. KJ Kwon-Chung, JW Fell. *Filobasidiella*. In: NJW Kreger–van Rij, ed. The Yeasts: A Taxonomic Study. Elsevier: Amsterdam, 1984, pp. 472–482.
35. D Ellis, TJ Pfeiffer. Natural habitat of *Cryptococcus neoformans* var. *gatti*. J Clin Microbiol 28: 1642–1644, 1990.
36. A Varma, D Swinne, F Staib, JE Bennett, KJ Kwon-Chung. Diversity of DNA fingerprints in *Cryptococcus neoformans*. J Clin Microbiol 33:1807–1814, 1995.
37. P Ruma, SC Chen, TC Sorrell, AG Brownlee. Characterization of *Cryptococcus neoformans* by random DNA amplification. Lett Appl Microbiol 23:312–316, 1996.
38. T Boekhout, A van Belkum. Variability of karyotypes and RAPD types in genetically related strains of *Cryptococcus neoformans*. Curr Genet 32:203–208, 1997.
39. RE Safrin, LA Lancaster, CE Davis, AI Braude. Differentiation of *Cryptococcus neoformans* serotypes by isoenzyme electrophoresis. Am J Clin Pathol 86:204–208, 1986.
40. BL Wickes, TDE Moore, KJ Kwon-Chung. Comparison of the electrophoretic karyotypes and chromosomal location of ten genes in the two varieties of *Cryptococcus neoformans*. Microbiologia 140:543–550, 1994.
41. F Chen, BP Currie, LC Chen, SG Spitzer, ED Spitzer, A Casadevall. Genetic relatedness of *Cryptococcus neoformans* clinical isolates grouped with the repetitive DNA probe CNRE-1. J Clin Microbiol 33:2818–2822, 1995.
42. ME Brandt, LC Hutwagner, LA Klug, WS Baughman, D Rimland, EA Graviss, RJ Hamill, C Thomas, PG Pappas, AL Reingold, RW Pinner. Molecular subtype distribution of *Cryptococcus neoformans* in four areas of the United States. Cryptococcal Disease Active Surveillance Group. J Clin Microbiol 34:912–917, 1996.
43. ME Brandt, MA Pfaller, RA Hajjeh, EA Graviss, J Rees, ED Spitzer, RW Pinner, LW Mayer. Molecular subtypes and antifungal susceptibilities of serial *Cryptococcus neoformans* isolates in human immunodeficiency virus associated cryptococcosis. J Infect Dis 174:812–820, 1996.
44. ED Spitzer, SG Spitzer, LF Freundlich, A Casadevall. Persistence of initial infection in recurrent *Cryptococcus neoformans* meningitis. Lancet 341:595–596, 1993.
45. BP Currie, A Casadevall. Estimation of the prevalence of cryptococcal infection among AIDS patients infected with the human immunodeficiency virus in New York city. Clin Infect Dis 19: 1029–1033, 1994.
46. CL Halliday, T Bui, M Krockenberger, R Malik, DH Ellis, DA Carter. Presence of alpha and a mating types in environmental and clinical collections of *Cryptococcus neoformans* var. *gattii* strains from Australia. J Clin Microbiol 37:2920–3926, 1999.
47. T Boekhout, A van Belkum, AC Leenders, HA Verbrugh, P Mukamurangwa, D Swinne, WA Scheffers. Molecular typing of *Cryptococcus neoformans*: taxonomic and epidemiological aspects. Int J Syst Bacteriol 47:432–442, 1997.
48. KJ Kwon-Chung, M Lazera, Y Chang, BS Kang. Is *Cryptococcus neoformans* evolving into an asexual organism? The 4th International Conference on *Cryptococcus* and Cryptococcosis, London, 1999.

49. R Cherniak, E Reiss, ME Slodki, RD Plattner, SO Blumer. Structure and antigenic activity of the capsular polysaccharide from *Cryptococcus neoformans* serotype A. Mol Immunol 17:1025–1032, 1980.
50. SH Turner, R Cherniak, E Reiss. Fractionation and characterization of galatoxylomannan from *Cryptococcus neoformans*. Carbohydr Res 125:343–349, 1984.
51. R Cherniak, JB Sundstrom. Polysaccharide antigens of the capsule of *Cryptococcus neoformans*. Infect Immun 62:1507–1512, 1994.
52. C Spiropulu, RA Eppard, E Otteson, TR Kozel. Antigenic variation within serotypes of *Cryptococcus neoformans*. Infect Immun 57:3240–3242, 1989.
53. R Cherniak, LC Morris, BC Anderson, SA Meyer. Facilitated isolation, purification, and analysis of glucuronoxylomannan of *Cryptococcus neoformans*. Infect Immun 59:59–64, 1991.
54. R Cherniak, LC Morris, T Belay, ED Spitzer, A Casadevall. Variation in the structure of glucuronoxylomannan in isolates from patients with recurrent cryptococcal meningitis. Infect Immun 63: 1899–1905, 1995.
55. YC Chang, KJ Kwon-Chung. Complementation of a capsule-deficient mutation of *Cryptococcus neoformans* restores virulence. Mol Cell Biol 14:4912–4919, 1994.
56. YC Chang, LA Penoyer, KJ Kwon-Chung. The second capsule gene of *Cryptococcus neoformans*, *CAP64*, is essential for virulence. Infect Immun 64:1977–1983, 1996.
57. YC Chang, KJ Kwon-Chung. Isolation of the third capsule associated gene, *CAP60*, required for virulence in *Cryptococcus neoformans*. Infect Immun 66:2230–2236, 1998.
58. YC Chang, KJ Kwon-Chung. Isolation, characterization, and localization of a capsule-associated gene, *CAP10*, of *Cryptococcus neoformans*. J Bacteriol 181:5636–5643, 1999.
59. KJ Kwon-Chung, I Polachek, T Popkin. Melanin-lacking mutants of *Cryptococcus neoformans* and their virulence for mice. J Bacteriol 150:1414–1421, 1982.
60. KJ Kwon-Chung, JC Rhodes. Encapsulation and melanin formation as indicators of virulence in *Cryptococcus neoformans*. Infect Immun 51:218–223, 1986.
61. PR Williamson. Biochemical and molecular characterization of the diphenol oxidase of *Cryptococcus neoformans:* identification as a laccase. J Bacteriol 176:656–664, 1994.
62. PR Williamson, K Wakamatsu, S Ito. Melanin biosynthesis in *Cryptococcus neoformans*. J Bacteriol 180:1570–1572, 1998.
63. SD Salas, JE Bennett, KJ Kwon-Chung, JR Perfect, PR Williamson. Effect of the laccase gene, *CNLAC1*, on virulence of *Cryptococcus neoformans*. J Exp Med 184:377–386, 1996.
64. Y Wang, P Aisen, A Casadevall. Melanin, melanin "ghosts," and melanin composition in *Cryptococcus neoformans*. Infect Immun 64:2420–2424, 1996.
65. JD Nosanchuk, P Valadon, M Feldmesser, A Casadevall. Melanization of *Cryptococcus neoformans* in murine infection. Mol Cell Biol 19:745–750, 1999.
66. L Liu, K Wakamatsu, S Ito, PR Williamson. Catecholamine oxidative products, but not melanin, are produced by *Cryptococcus neoformans* during neuropathogenesis in mice. Infect Immun 67: 108–112, 1999.
67. I Herskowitz. MAP kinase pathways in yeast. For mating and more. Cell 80:187–197, 1995.
68. BL Wickes, U Edman, JC Edman. The *Cryptococcus neoformans STE12α* gene: a putative *Saccharomyces cerevisiae STE12* homologue that is mating type specific. Mol Microbiol 26:951–960, 1997.
69. YC Chang, BL Wickes, GF Miller, L Penoyer, KJ Kwon-Chung. A *MATα* specific gene of *C. neoformans*, *STE12α*, regulates virulence and is essential for monokaryotic fruiting. J Exp Med 2000. In press. 191:871–882.
70. C Yue, LM Cavallo, JA Alspaugh, P Wang, GM Cox, JR Perfect, J Heitman. The *STE12alpha* Homolog Is Required for Haploid Filamentation But Largely Dispensable for Mating and Virulence in *Cryptococcus neoformans*. Genetics 153:1601–1615, 1999.
71. MC Cruz, RA Sia, M Olson, GM Cox, J Heitman. Comparison of the roles of calcineurin in physiology and virulence in serotype D and serotype A strains of *Cryptococcus neoformans*. Infect Immun 68:982–985, 2000.
72. BL Wickes, U Edman, GL Woodlee, CL McClelland, DL Clarke, JC Edman. The *Cryptococcus neoformans STE11α* gene is alpha-specific and required for mating. Mol Microb (in press).

73. TDE Moore, JC Edman. The α-mating type locus of *Cryptococcus neoformans* contains a peptide pheromone gene. Mol Cell Biol 13:1962–1970, 1993.
74. BL Wickes, JC Edman. The *Cryptococcus neoformans GAL7* gene and its use as an inducible promoter. Mol Microbiol 16:1099–1109, 1995.
75. M Del Poeta, DL Toffaletti, TH Rude, SD Sparks, J Heitman, JR Perfect. *Cryptococcus neoformans* differential gene expression detected in vitro and in vivo with green fluorescent protein. Infect Immun 67:1812–1820, 1999.
76. T Tolkacheva, P McNamara, E Piekarz, W Courchesne. Cloning of a *Cryptococcus neoformans* gene, *GPA1*, encoding a G-protein α-subunit homolog. Infect Immun 62:2849–2856, 1994.
77. JA Alspaugh, JR Perfect, J Heitman. *Cryptococcus neoformans* mating and virulence are regulated by the G protein alpha subunit GPA1 and cAMP. Genes Dev 11:3206–3217, 1997.
78. P Wang, JR Perfect, J Heitman. The G-Protein beta Subunit *GPB1* Is Required for Mating and Haploid Fruiting in *Cryptococcus neoformans*. Mol Cell Biol 20:352–362, 2000.
79. MS Cyert, R Kunisawa, D Kaim, J Thorner. Yeast has homologs (*CNA1* and *CNA2* gene products) of mamalian calcineurin, a calmodulin regulated phosphoprotein phosphatase. Proc Natl Acad Sci USA 88:7376–7380, 1991.
80. MS Cyert, J Thorner. Regulatory subunit (CNB1 gene product) of yeast Ca2+/calmodulin-dependent phosphoprotein phosphatases is required for adaptation to pheromone. Mol Cell Biol 12:3460–3469, 1992.
81. J Liu, JD Farmer, WS Lane, J Friedman, I Weissman, SL Schrieber. Calcineurin is a common target of cyclophilin-cyclosporin A and FKBP-FK506 complexes. Cell 66:807–815, 1991.
82. A Odom, S Muir, E Lim, DL Toffaletti, J Perfect, J Heitman. Calcineurin is required for virulence of *Cryptococcus neoformans*. EMBO J 16:2576–2589, 1997.
83. SC Chen, M Muller, JZ Zhou, LC Wright, TC Sorrell. Phospholipase activity in *Cryptococcus neoformans:* a new virulence factor? J Infect Dis 175:414–420, 1997.
84. SC Chen, LC Wright, RT Santangelo, M Muller, VR Moran, PW Kuchel, TC Sorrell. Identification of extracellular phospholipase B, lysophospholipase, and acyltransferase produced by *Cryptococcus neoformans*. Infect Immun 65:405–411, 1997.
85. B Wong, JR Perfect, S Beggs, KA Wright. Production of the hexitol D-mannitol *by Cryptococcus neoformans* in vitro and in rabbits with experimental meningitis. Infect Immun 58:1664–1670, 1990.
86. GM Megson, DA Stevens, JR Hamilton, DW Denning. D-mannitol in cerebrospinal fluid of patients with AIDS and cryptococcal meningitis. J Clin Microbiol 34:218–221, 1996.
87. V Chaturvedi, B Wong, SL Newman. Oxidative killing of *Cryptococcus neoformans* by human neutrophils. Evidence that fungal mannitol protects by scavenging reactive oxygen intermediates. J Immunol 156:3836–3840, 1996.
88. V Chaturvedi, T Flynn, WG Niehaus, B Wong. Stress tolerance and pathogenic potential of a mannitol mutant *of Cryptococcus neoformans*. Microbiology 142:937–943, 1996.
89. GD Campbell. Primary pulmonary cryptococcosis. Am Rev Respir Dis 94:236–243, 1966.
90. ML Littman, JE Walter. Cryptococcosis: Current Status. Am J Med 45:922–932, 1968.
91. RA Hajjeh, ME Brandt, RW Pinner. Emergence of cryptococcal disease: Epidemiologic perspectives 100 years after its discovery. Epidemiol Rev 17:303–320, 1995.
92. D Lo. Cryptococosis in the northern territory. Med J Aust 27:825–826, 1976.
93. RA Hajjeh. Cryptococcosis. Population-based multistate active surveillance and risk factors in human immunodeficiency virus infected persons. J Infect Dis 179:449–454, 1999.
94. KG Castro, RM Selik, HW Jaff. Frequency of opportunistic diseases in AIDS patients by race, ethnicity, and HIV transmission categories Unites States. Abstracts, 28th Interscience Conference on Antimicrobial Agents and Chemotherapy (Los Angeles), American Society for Microbiology, 1988.
95. RW Pinner, L Hutwagner, SF Collin. Use of hospital discharge data to describe recent changes in the epidemiology of cryptococcaal disease. Abstracts, 31st Interscience Conference on Antimicrobial Agents and Chemotherapy (Chicago), American Society for Microbiology, 1991.
96. JL Lewis, S Rabinovitch. The wide spectrum of cryptococcal infections. Am J Med 53:315–322, 1972.

97. VE Edwards, JM Sutherland, JH Tyrer. Cryptococosis of the central nervous system: epidemiological, clinical and therapeutic factors. J Neurol Neurosurg Psychiatry 33:415–425, 1970.

98. A McNulty, JM Kaldor, AM McDonald, K Baumgart, DA Cooper. Acquired immunodeficiency without evidence of HIV infection; National retrospective survey. Br Med J 308:825–826, 1994.

99. RA Duncan, CF von Reyn, GM Alliegro, Z Toosi, AM Sugar, SM Levitz. Idiopathic CD4+ T lymphocytopenia four patients with opportunistic infections and no evidence of HIV infection. N Engl J Med 328:393–398, 1993.

100. DK Smith, JJ Neal, SD Holmberg. Unexplained opportunistic infections and CD4+ T-lymphocytopenia without HIV infection. An investigation of cases in the Unites States. N Engl J Med 328: 373–379, 1993.

101. VP Collins, A Gellhorn, JR Trimble. The coincidence of cryptococcosis and disease of the reticuloendothelial and lymphatic systems. Cancer 4:883–889, 1951.

102. RD Diamond, JE Bennett. Prognostic factors in cryptococal meningitis: a study in 111 cases. Ann Intern Med 80:176–181, 1974.

103. RD Diamond, JE Bennett. Disseminated cryptococcosis in man: decreased lymphocyte transformation in response to *Cryptococcus neoformans*. J Infect Dis 127:694–697, 1973.

104. HA Gallis, RA Berman, TR Cate, JD Hamilton, JC Gunnels, DL Stickel. Fungal Infection following renal transplantation. Arch Intern Med 135:1163–1171, 1975.

105. N Jabbour, J Reyes, S Kusne, M Martin, J Fung. Cryptococcal meningitis after liver transplantation. Transplantation 61:156–167, 1996.

106. GT John, M Mathew, E Snehalatha, V Anandi, A Date, CK Jacob, JCM Shastry. Cryptococcosis in renal allograft recipients. Transplantation 61:156–167, 1994.

107. PE Olson, C Earhart, RJ Rossetti, JA Newton, MR Wallace. Smoking and risk of cryptococcosis in patients with AIDS. JAMA 277:629, 1997.

108. RD Diamond. *Cryptococcus neoformans*. In: GL Mandell, JE Bennett, R Dolin, eds. Principles and Practice of Infectious Diseases. New York: Churchill Livingstone, 1995, pp. 2331–2340.

109. MF Lipscomb, CR Lyons, AA Izzo, J Lovchik, JA Wilder. Experimental pulmonary cryptococcal infection in mice. In: O Zak, M Sande, eds. Handbook of Animal Models of Infection. Orlando, FL: Academic Press, 1999, pp. 681–686.

110. JR Perfect, SDR Lang, DT Durack. Chronic cryptococcal meningitis: a new experimental model in rabbits. Am J Pathol 101:177–194, 1980.

111. JD Rhodes, LS Wicker, WJ Urba. Genetic control of susceptibility to *Cryptococcus neoformans* in mice. Infect Immun 29:494–499, 1980.

112. GB Huffnagle, MB Boyd, NE Street, MF Lipscomb. IL-5 is required for eosinophil recruitment, crystal deposition, and mononuclear cell recruitment during a pulmonary *Cryptococcus neoformans* infection in genetically "susceptible" mice (C57BL/6). J Immunol 160:2393–2400, 1998.

113. Y Wang, A Casadevall. Decreased susceptibility of melanized *Cryptococcus neoformans* to UV light. Appl Environ Microbiol 60:3864–3866, 1994.

114. AL Rosas, A Casadevall. Melanization affects susceptibility of *Cryptococcus neoformans* to heat and cold FEMS Microbiol Lett 153:265–272, 1997.

115. Y Wang, P Aisen, A Casadevall. *Cryptococcus neoformans* melanin and virulence: mechanism of action. Infect Immun 63:3131–3136, 1995.

116. ES Jacobson, SB Tinnell. Antioxidant function of fungal melanin. J Bacteriol 175:7102–7104, 1993.

117. GB Huffnagle, GH Chen, JL Curtis, RA McDonald, RM Strieter, GB Toews. Down-regulation of the afferent phase of T cell-mediated pulmonary inflammation and immunity by a high melanin-producing strain of *Cryptococcus neoformans*. J Immunol 155:3507–3516, 1995.

118. Y Wang, A Casadevall. Growth of *Cryptococcus neoformans* in presence of L-dopa decreases its susceptibility to amphotericin B. Antimicrob Agents Chemother 38:2648–2650, 1994.

119. ES Jacobson, JD Hong. Redox buffering by melanin and Fe(II) in *Cryptococcus neoformans*. J Bacteriol 179:5340–5346, 1997.

3
Secretory Proteins in Fungal Virulence

Pranab K. Mukherjee and Mahmoud A. Ghannoum
Case Western Reserve University, Cleveland, Ohio

I. INTRODUCTION

Recent years have seen a dramatic rise in the incidence of fungal infections attributable to the increasingly frequent use of anti-bacterial antibiotics, cytotoxic chemotherapeutic agents, indwelling catheters, and other predisposing factors [1,2]. Fungal-mediated host cell damage is an important component of fungal virulence and is assumed to help in invasion of host tissues by disrupting host cell membranes, resulting in membrane dysfunction and eventual cell death [3,4]. Phospholipids and proteins represent the major chemical constituents of the host cell envelope. Therefore, enzymes such as phospholipases and proteinases capable of hydrolyzing these proteins are likely to be involved in the host cell–membrane disruption processes. Many fungal pathogens secrete such extracellular proteins which target the protein or lipid component of the host membranes. Thus, secretory proteins have figured prominently among the list of factors responsible for fungal virulence. These fungal secretory proteins have been categorized into two main groups: phospholipases, which hydrolyze phospholipids [5], and proteinases, which hydrolyze peptide bonds [6,7]. This chapter focuses on the contribution of these secretory proteins to fungal virulence.

II. PHOSPHOLIPASES

The overall incidence of *Candida* bloodstream infections has increased significantly in the last two decades [2,6]. This increase ranges from 75% to as high as 487% in small to large hospitals [6,8]. A number of safe and efficacious antifungal agents have been introduced in the recent years, which has led to improving our ability to treat fungal infections including candidiasis. However, the mortality rate due to candidiasis is still unacceptably high (35–62%) [9–11], necessitating the development of novel therapeutic approaches. Candidal virulence factors attracted interest as a possible means for developing novel therapeutic interventions against candidiasis [12–16]. Such virulence factors include adherence [17–20], extracellular proteinases [7], germination [21], phospholipases [4], and phenotypic switching [22,23].

A. Phospholipases of *Candida* Species

Phospholipases are enzymes that hydrolyze one or more ester linkages of glycerophospholipids. Depending on the specific ester bond cleaved, these enzymes have been classified into phospho-

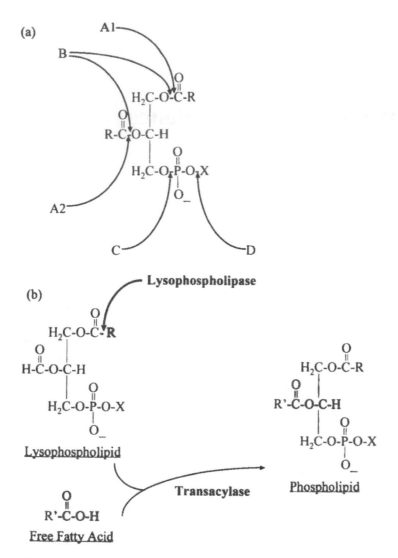

Figure 1   Sites of action of various phospholipases. (a) A1 and A2—PLA$_1$ and PLA$_2$; B—PLB; C—PLC; D—PLD. (b) Lyso-PL (lysophospholipid) and Lyso-PL transacylase.

lipases A, B, C, and D (Fig. 1a). Phospholipase B (PLB) can release the *sn*-1 and *sn*-2 fatty acids, and has both hydrolase (fatty acid release) and lysophospholipase-transacylase (LPTA) activities. The hydrolase activity cleaves fatty acids from phospholipids (PLB activity) and lysophospholipids (lysophospholipase [Lyso-PL] activity), while the transacylase activity produces phospholipid by transferring a free fatty acid to a lysophospholipid (Fig. 1b). Accordingly, PLB was also referred to as lysophospholipase or lysophospholipase-transacylase enzymes. By cleaving phospholipids, phospholipases destabilize the membrane, leading to cell lysis. Evidence implicating phospholipases in host cell penetration, injury, and lysis by micro-organisms has been reported for *Rickettsia rickettsii* [24], *Toxoplasma gondii* [25,26], *Entamoeba histolytica* [27], and *C. albicans* [5]. Among fungi, phospholipases have been identified as virulence factors for *Candida albicans, Cryptococcus neoformans, Aspergillus fumigatus*—the three most common fungal pathogens to infect the human system [5,28,29].

1. Phospholipases of *Candida albicans*

Costa et al. [30] and Werner [31] were the first to show secretion of extracellular phospholipases from *C. albicans* by growing the fungus on solid media containing egg yolk or lecithin and analyzing the lipid breakdown products. Subsequently, phospholipase activity was found in many pathogenic *C. albicans* strains using media containing blood serum and sheep red cells [30]. Pugh and Cawson [32] developed a cytochemical assay for this enzyme using lecithin as its natural substrate and used this method in a chick chorioallantoic membrane model to evaluate ultra structural details of candidal invasion and to determine the site of phospholipase production [33,34]. Invasion was initiated by placing stationary phase blastospores of *C. albicans* on the membrane, which stimulated cellular changes in the blastospores. Many of the blastospores developed hyphae with phospholipase activity concentrated at the growing tip. The activity was highest where the hyphae were in direct contact with the membrane [33].

The implication of phospholipase in tissue invasion by *C. albicans* led to the need for simple and reliable assay methods to detect candidal phospholipases. Odds and Abbott [35] described a biochemical assay to measure intracellular phospholipase activity in *C. albicans* and identified lysophosphatidylcholine (lyso-PC) as an intracellular degradation product of phosphatidylcholine (PC) [35]. Although accurate, this method had the disadvantage of being time consuming. Another method, using Sabouraud's dextrose agar (SDA) plates supplemented with phospholipid-rich egg yolk proved to be a faster method than the biochemical assay [36]. Phospholipase-positive candidal isolates grown on egg yolk–SDA plates produce a distinct, well-defined, dense white zone of precipitation around the colony. This precipitation zone is most likely due to the calcium–fatty acid complexes formed with the released fatty acids. Phospholipase activity (expressed as P_z value) was defined as the ratio of colony diameter to diameter of the zone of precipitation around phospholipase positive colonies. This easy plate method became the traditional screen for phospholipase activity for *Candida* species [37–39] and other fungi such as *C. neoformans* [28]. Samaranayake et al. [40] used this plate method to screen phospholipase-producing abilities of 41 *Candida* isolates and showed that 79% of the *C. albicans* strains tested produced extracellular phospholipases. In contrast, the *C. tropicalis, C. glabrata,* and *C. parapsilosis* isolates studied did not show any phospholipase activity [40]. In another study, it was shown that *C. albicans* isolated from blood generally produced much higher levels than isolates from wounds or urine, thus correlating levels of phospholipase activity to the site of infection [36]. Lane and Garcia [38] showed that the "star" and "ring" *C. albicans* switching variants produce similar amounts of phospholipase as the wild type, while the "stipple" variant produced between 27% to 34% more phospholipase.

These and other studies were limited by two major disadvantages of the egg yolk–based method—lack of specificity, as the egg yolk plates contain substrates for both phospholipases (phospholipids) and lipases (triglycerides), and low sensitivity, as the assay is not suitable for screening poor phospholipase-producing fungal isolates. Although sensitivity of the egg yolk–based method could be increased to some extent using culture filtrates, the issue of specificity was still a matter of concern. Therefore, this method could be used for initial screens only.

To circumvent these problems, a modification of the plate method was used in which wells were made in SDA plates supplemented with substrates specific for phospholipases A, B, or C, and the specific phospholipase activity determined by the diameter of the zone of precipitation formed around the well [41]. This method, called the radial diffusion assay, was more specific than the egg-yolk based method since it could differentiate between PLA, PLB, and

PLC activities. A more sensitive and specific colorimetric acyl-CoA-oxidase-based assay was later developed in which a concentrated culture filtrate was incubated with a PLB-specific substrate (PC or Lyso PC) and the free fatty acid released into the reaction mixture determined spectrophotometrically [42].

Although a previous study had shown the absence of extracellular phospholipase activity in *Candida* species other than *C. albicans* [40] (see above), it was recently shown that non-*albicans Candida* species also produce extracellular phospholipases as determined by both egg yolk–based and colorimetric assays [43]. However, relative to *C. albicans*, the non-*albicans* species secrete significantly lower amounts of phospholipase (e.g., *C. krusei* has ~10 times less phospholipase activity than *C. albicans* (Mukherjee and Ghannoum, unpublished data). The discrepancy observed by different workers in the phospholipase activity for the non-*albicans* species may be attributed to strain-to-strain variation or different sensitivities of the assays employed to detect phospholipase activities.

The number and specific types of phospholipase enzymes secreted by *C. albicans* have been a point of debate in the mycology community. Costa et al. [30] reported the secretion of both PLA and PLC by this clinically important yeast. Their results were based on the isolation of palmitic acid and phosphorylcholine from the proximity of candidal colonies cultured on SDA supplemented with serum and sheep erythrocytes. Since this medium is not chemically defined, the sources of the hydrolysis products are uncertain. In a different study, crude fractionation of the proteins in culture filtrates of *C. albicans* indicated the secretion of three types of phospholipases: Lyso-PL, LPTA, and PLB [44]. Further analyses led to the purification of two forms of candidal LPTA (types I and II) having different molecular mass (81 kDa for LPTA-I and 41 kDa for LPTA-II), amino acid composition, and enzymatic properties. Also, antibody raised against purified LPTA-II reacted strongly with LPTA-II, but not with LPTA-I. However, the substrate specificities of the enzymes were not determined [45]. To determine the type and substrate specificity of phospholipase(s) secreted by *C. albicans*, a high phospholipase-producing strain was grown to late log phase, and the supernatant was concentrated and assayed for phospholipase activity using the radial diffusion assay and a colorimetric assay. Only PLB activity was observed using the diffusion-based assay (Mukherjee and Ghannoum, unpublished data). In the colorimetric assay, both PLB and Lyso-PL activities were detected in the supernatant.

a. Cloning and Disruption of C. albicans Phospholipase B (caPLB1). In collaboration with Yoshinori Nozawa (Gifu University, Gifu, Japan), we purified to homogeneity the protein responsible for candidal extracellular phospholipase activity. This enzyme is a glycoprotein with a molecular weight of 84 kDa, with specific activities of 117 μmol/min/mg protein for fatty acid release (hydrolase), and 459 μmol/min/mg protein for phosphatidylcholine formation (LPTA activity) [46]. Studies with specific substrates showed that the purified enzyme has both hydrolase (PLB and Lyso-PL) and LPTA activities [46]. Phospholipases having two hydrolase activities have also been reported for *S. cerevisiae* [47–50] and *Penicillium notatum* [51].

The amino acid sequence of the purified *C. albicans* PLB protein was used to clone the gene encoding this protein. A PCR-based approach, based on degenerate oligonucleotide primers designed according to the amino acid sequences of two peptide fragments obtained from PLB, was followed to clone the ca*PLB1* gene [52]. Sequence analysis of a 6.7-kb *Eco*RI-*Cla*I genomic clone revealed a single open reading frame of 1818 bp that predicts a preprotein of 605 residues. The genomic DNA sequence encodes 17 amino acid residues that are absent from the NH_2-terminus of the mature protein. This stretch of residues represents a possible signal sequence. The predicted protein contains seven Asn-X-Ser/Thr motifs (residues 199, 261, 399, 451, 465, 492, and 573) that could potentially be N-glycosylated. One possible tyrosine phosphorylation site, Lys-Ser-Asn-Ile-Asp-Val-Ser-Ala-Tyr (residues 369–377), was also identified. Hydropathy analysis of the predicted protein sequence revealed the presence of a single stretch of hydrophobic amino acids present at the amino terminus (residues 1–18). This segment of amino acids most likely functions as a signal peptide which targets the protein to the endoplasmic reticulum for subsequent processing and ultimately secretion. Comparison of the putative candidal phospholipase with those of other proteins revealed significant homology to known fungal PLBs from *S. cerevisiae* (45%), *P. notatum* (42%), *Torulaspora delbrueckii* (48%), and *Schizosaccharomyces pombe* (38%). This gene, designated ca*PLB1*, was mapped to chromosome 6 [52]. Hoover et al. [53] cloned a homolog independently in a separate study, using degenerate oligonucleotides derived from conserved regions of *PLB1* genes from *S. cerevisiae* and other fungi. Sequence analysis of the PLB1p sequences from *S. cerevisiae*, *S. pombe*, *P. notatum*, and *T. delbrueckii* revealed the presence of a hydrophobic COOH-terminus which contained putative glycosylphosphatidylinositol (GPI) anchoring regions. In contrast, neither a hydrophobic COOH-terminus nor a GPI anchoring site was identified in the candidal PLB1p. GPI-anchored proteins may be tethered to the plasma membrane or crosslinked to the cell wall glucan, and may thus serve to regulate the release of this enzyme to the surroundings [54]. The absence of GPI anchoring in candidal PLB1p possibly leads it to be secreted directly, a characteristic which may enhance the virulence of *C. albicans*. Disruption mutants of ca*PLB1* were constructed by targeted gene disruption using the ura-blaster technique [52]. PLB1p was detected in culture filtrates from the wild-type and the heterozygous (ca*PLB1*/ca*plb1*) strain but not in the culture filtrate from the null-mutant strain (ca*plb1*/ca*plb1*). Furthermore, assay of supernatants collected from the parent and the PLB-deficient mutants for PLB and Lyso-PL activities, using specific substrates, revealed that both activities were reduced in the PLB-deficient mutant by ~99% and 80%, respectively, relative to the wild type [52].

    b.   Cloning of C. albicans *Phospholipase B* (CaPLB2).   The ca*PLB1* disrupted mutants still exhibited ~1% phospholipase and 10% lysophospholipase activities [52]. It was predicted that a second PLB gene could be the source of this activity. This hypothesis was strengthened by the identification of more than one PLB protein in *S. cerevisiae* [50,55], and *S. pombe* genome sequences (D89183 and D89204). To clone *caPLB2*, Sugiyama et al. [56] used a PCR-based approach similar to the one used to clone ca*PLB1*. A number of similarities are observed between ca*PLB1* and ca*PLB2* in size, availability of N-glycosylation sites, the presence of a single stretch of hydrophobic amino acids at the amino terminus, and the absence of a GPI anchoring site. The nucleotide sequence of ca*PLB2* contained a single open reading frame encoding a putative 608 amino acid protein with an estimated molecular mass of ~67 kDa. The predicted amino acid sequence contains six potential N-glycosylation sites (Asn-X-Ser/Thr motifs) at residues 259, 365, 450, 464, 491, and 572. The deduced amino acid sequence of *caPLB2* was homologous to that of *caPLB1* (65% identity). CaPLB2 was also found to be similar to PLBs from *S. cerevisiae*, *T. delbrueckii*, and *P. notatum* (42%, 46%, and 42% identity, respectively). Hydropathy analysis of the predicted protein revealed the presence of a cluster of hydrophobic amino acids at the N-terminus. Similar to *S. cerevisiae* and *T. delbrueckii* PLBs [50,57], ca*PLB2* possesses a potential signal sequence at the N-terminal region of *caPLB2*, where two polar amino acids (Gln-Ser) are followed by a cluster of six hydrophobic amino acids (Ile-Leu-Leu-Phe-Val-Val). Such sequence may guide proteins to the secretory pathway. Like ca*PLB1*, ca*PLB2* lacks GPI attachment site (a cluster of hydrophobic amino acids at the carboxy terminal) found in the PLBs from the non-pathogenic fungi such as *S. cerevisiae*, *P. notatum*, and *T. delbrueckii* PLBs. Since fungal PLBs and mammalian phospholipase A_2 have lysophospholipase activities, it is likely that these enzymes also share conserved amino acid regions. Three amino acid residues essential for the catalytic function have been identified in mammalian PLA$_2$ ([200]Arg, [228]Ser, and [549]Asp) [58]. Interestingly, three regions surrounding these amino acids are also conserved in PLBs from fungi. The deduced amino acid sequence of *caPLB2* contains the motifs

SGGGX^{97}RA(M/L), GL^{133}SG(G/S) and ^{381}D(S/G)G(E/L)XXXN, which may have a catalytic function [56].

To determine the phylogenetic relationship among various fungal PLBs, Sugiyama et al. [56] constructed a phylogenetic tree of PLBs, and showed that *caPLB1* and *caPLB2* are closely related to each other and ca*PLBs* are more closely related to *PLB* from *S. pombe* and *P. notatum* than *PLBs* from *S. cerevisiae* and *T. delbrueckii.*

c. Cloning of C. albicans *Phospholipase* C (CaPLC). Recently, Bennett et al. [59] cloned the *C. albicans* phospholipase C (*CAPLC1*) gene using a PCR-based approach, and sequenced it. The nucleotide sequence of the *CAPLC1* gene (2997 bp) encoded a polypeptide of 1099 amino acids with a predicted molecular mass of 124.6 kDa. The deduced amino acid sequence of this polypeptide (CAPLC1) exhibited an overall amino acid homology of >27% with PLCs previously characterized from *S. cerevisiae* and *S. pombe* and suggested that the CAPLC1 protein is a delta-form of phosphoinositide-specific PLC (PI-PLC). Although PLC was activity was detected in cell-free extracts of both yeast and hyphal forms of *C. albicans*, the protein has not been purified, and a *CAPLC1*-disrupted mutant has not been constructed. Thus, the role of *CAPLC1* in fungal virulence could not be conclusively proven. Sequences homologous to *CAPLC1* were detected in all *C. albicans* and *C. dubliniensis*, and in some *C. tropicalis*, *C. glabrata*, and *C. parapsilosis* isolates examined. In contrast, *CAPLC1* was not detected in the isolates of *C. krusei, C. kefyr, C. guillermondii,* or *C. lusitaniae* examined [59].

d. Cloning of C. albicans *Phospholipase D* (CaPLD). Phospholipase D (PLD) catalyzes the hydrolysis of phosphatidylcholine to produce phosphatidic acid (PA) and choline (see Fig. 1a). Both mammalian and fungal (*S. cerevisiae*) genes encoding PLD have been cloned and characterized [60–62]. Mammalian PLD has emerged as one of the key enzymes in intracellular signaling [63], while PLD from *S. cerevisiae* (encoded by *SPO14*) has been shown to be essential for meiosis [61,64]. The absence of meiosis in *C. albicans*, which unlike *S. cerevisiae* exists as a diploid, indicates that PLD in this clinically important yeast may play other roles than meiosis. McLain and Dolan [65] have recently shown that PLD may be an important regulator of the dimorphic transition. Since yeast-hyphal transformation is thought to be an important virulence determinant in *C. albicans* [4,12], cloning and disruption of candidal *PLD* may contribute to our understanding of the biological role of this phospholipase.

For cloning candidal *PLD*, Kanoh et al. [66] designed two oligonucleotide primers based on the conserved amino acid sequences in human PLD1a and SPO14. These primers were employed in a PCR-based approach to clone the full-length gene. The cloned *PLD* sequence had a potential open reading frame encoding a protein of 1710 amino acids with a calculated molecular mass of 196.4 kDa. The deduced amino acid sequence contained four conserved regions (I, II, III, and IV) defined by the primary structures of plant, yeast, and human PLDs. Furthermore, the HKD motif (HxKxxxxD) in regions I and IV and a serine residue in the GSRS motif in region IV, which are critical for PLD biochemical activity, were also completely conserved [66].

The number of amino acids and the calculated molecular mass of *CaPLD* closely resembled *SPO14*-coded protein (1683 amino acids, 195 kDa). Comparison of the primary structure with those of other PLDs showed the highest homology to *SPO14*. The overall homology between amino acid sequences of *CaPLD* and *SPO14* was 42%. At the four conserved regions, the homology between *CaPLD* and *SPO14* ranged from 65% to 71%, while that between *CaPLD* and rat PLDs was 42–55%. In addition to the four conserved regions found in both fungal and mammalian PLDs, candidal PLD had seven regions (A–G) of 10-91 amino acids in length that were highly homologous to *SPO14* protein. Of the seven regions only two, F and G, were conserved in mammalian PLDs. Therefore, it was speculated that these five regions (A–E) represent functional domains specific for fungi [66]. Phylogenetic analysis of PLDs from various species showed three major clusters: the first cluster composed of mammalian PLD1 and plant, nematode, and *Streptomyces* PLDs; the second cluster composed of mammalian PLD2; and the third cluster composed of fungal PLDs, including *CaPLD* and *SPO14* [66]. Because fungal PLDs are in a separate grouping from mammalian and other PLDs, it is tempting to speculate that this enzyme may perform fungal-specific functions. Unfortunately, elucidation of the biological function of candidal PLD awaits the disruption of the gene encoding it.

2. Phospholipases of *Candida glabrata*

C. glabrata is recognized increasingly as an important nosocomial pathogen [67–70]. Although it ranks third among *Candida* species in most reported cases of candidemia [68–69], one study reported *C. glabrata* to be the most common non-*albicans* species isolated, surpassing *C. tropicalis* [71]. In this study, *C. albicans* accounted for 58% of the isolates, *C. glabrata* for 32%, *C. tropicalis* for 6%, and *C. parapsilosis* for 4% [71]. *C. glabrata* has gained clinical significance not only because of an increase in its frequency but also because of the associated high number of complications and high mortality rates [1,69,72–75]. In one study, a mortality rate of 83% exceeded the observed rates from any other *Candida* species [67]. Therefore, identification of factors that contribute to the virulence of this organism may provide new attractive therapeutic targets. Recently, Clancy et al. [43] reported the detection of extracellular phospholipase activity in *C. glabrata* and provided data implicating it as a virulence factor in patients with candidemia caused by this yeast. In their study, phospholipase activity (as determined by opacity around colonies growing on egg yolk–containing media) was investigated in 51 non-*albicans Candida* species (22 *C. glabrata*, 12 *C. parapsilosis*, 10 *C. tropicalis*, 5 *C. lusitaniae*, and 2 *C. krusei* isolates) recovered from patients in a multicenter study of candidemia. Phospholipase activity was detected in 53% of isolates.

An assay employing agar-containing specific phospholipid substrates was used to identify the type of phospholipase activity secreted by *C. glabrata* [41]. As in the case of *C. albicans*, PLB and Lyso-PL accounted for 100% of phospholipase activity. No PLA or PLC activities were detected in supernatants collected from various *C. glabrata* isolates tested. To determine whether a correlation exists between phospholipase activity and clinical virulence, the authors classified the *C. glabrata* isolates into two groups: persisting isolates (PI) were strains that remained in the blood despite therapy with antifungal agents effective in vitro and despite removal of all IV catheters; nonpersistent (non-PI) strains were isolates that were cleared. Among non-*albicans Candida* species tested, the association between phospholipase activity and persistent candidemia was strongest for *C. glabrata* ($P = .004$). Moreover, the median PLB and Lyso-PL activities were higher for PI isolates (9.5 and 24 pmol/mg, respectively) than non-PI isolates (0 and 5.8 pmol/mg) ($P < .0001$, $P = .02$).

To establish the role of PLB in the pathogenesis of *C. glabrata*, an isogenic strain pair of this yeast was created which differed only in PLB activity [43]. Conserved sequences of *ScPLB1*, the gene responsible for PLB activity in *S. cerevisiae*, were used to design guessmer primers for PCR reactions with DNA from a clinical *C. glabrata* isolate. A 1030-bp PCR product which encoded part of a putative protein with 78% homology to *ScPLB1* protein was generated. To create an isogenic mutant with disruption of the PLB gene, Clancy et al. [43] first constructed a *ura3⁻* auxotroph of the parent *C. glabrata* strain. *S. cerevisiae* URA3, which is >80% identical to *C. glabrata* URA3, was amplified by PCR and *BamHI* restriction sites added to either end, and the resulting product inserted in the unique *BamHI* site of the 1030-bp PCR fragment. The disrupted product was liberated from the plasmid by *EcoRI* and was then used to transform *C. glabrata ura3⁻* auxotrophs. The resulting mutants were selected by growth on ura⁻ media. Targeted disruption was confirmed by Southern blot analysis. Colorimetric assay of phospholipase activity revealed >90% elimination of both PLB and Lyso-PL activities in the mutant isolate ($URA3^+plb^-$) compared to the parent *C. glabrata* ($URA3^+PLB^+$), confirming that the cloned gene was responsible for the vast majority of PLB and Lyso-PL activities [43]. Comparative in vivo pathogenicity studies are currently being performed to determine the role of the cloned gene in *C. glabrata* virulence.

B. Phospholipases of *Cryptococcus* Species

C. neoformans is the cause of the most common life-threatening fungal infection in patients with AIDS. Depending on the study, estimates of the frequency of cryptococcosis among AIDS patients range from 5% to 10% [76–78]. Although the occurrence has decreased in the last couple of years in the developed world, this incidence is still high in developing countries [79,80]. Given the high rate of relapse after initial antifungal therapy, the current management of *C. neoformans* infections includes lifelong suppressive therapy with antifungals.

1. Cloning of the *C. neoformans* Phospholipase B Gene

A PCR-based approach similar to the one used to clone other fungal phospholipases was used by John Perfect's group (Duke University Medical Center) in collaboration with our group to clone cryptococcal PLB [81]. Multiple degenerate PCR primers that hybridized to conserved regions from known fungal PLB genes were designed. From one of the primer combinations, a 1.2-kb amplicon was obtained and cloned into SK plasmid. This fragment was sequenced to confirm its identity and used to screen an EMBL3 *C. neoformans* genomic library in order to isolate the entire gene. Their data showed that the cryptococcal PLB gene exists in a single copy of ~2.4 kb in size and the putative protein is composed of 617 amino acids. Comparison of the cloned *C. neoformans* to other fungal PLB genes revealed amino acid homology of 37%, 36%, and 36% with *S. cerevisiae*, *P. notatum*, and *C. albicans*, respectively. The importance of PLB production for virulence properties in *C. neoformans* is still unknown and awaits the construction of an isogenic strain pair with specific deletion of PLB. Construction of this pair has already been accomplished, and animal testing is under way.

C. Phospholipases of *Aspergillus* Species

A. fumigatus is the most pathogenic member of its genus, responsible for 90% of infections caused by the aspergilli. Only two studies attempted to determine whether *Aspergillus* strains secrete phospholipase. Birch et al. [29] investigated the ability of *A. fumigatus* to produce extracellular phospholipases. Fast atom bombardment (which compares lipid-containing media, before and at intervals during *A. fumigatus* growth, for the presence of anions corresponding to major classes of phospholipids and fatty acids) enabled detection of several anions corresponding to phospholipid breakdown products. Based on specific degradation products Birch et al. [29] suggested that *A. fumigatus* secretes multiple extracellular phospholipases, including PLA, PLB, PLC, and PLD. Further characterization of PLC revealed that this activity was initially observed after 30 hr of growth and accumulated in broth cultures up to 50 hr. Maximal PLC activity coincided with fungal cultures entering the stationary phase, with greater activity occurring at 37°C than at lower temperatures [29].

Koul et al. [82] studied eight strains of *A. fumigatus* and *A. flavus* (*A. fumigatus*, n = 5; *A. flavus*, n = 3) for their ability to produce phospholipase. Following growth, for 3 days at 30°C or 37°C in Sabouraud dextrose broth supplemented with 4% glucose, supernatants were concentrated and assayed for phospholipase activity using a specific substrate radial diffusion assay capable of differentiating among PLA, PLB, and PLC. Phospholipase activity was detected in all strains regardless of temperature. For both species, PLB was the predominant enzyme. Two of three *A. flavus* strains also secreted PLA and PLC. For *A. fumigatus* grown at 30°C vs. 37°C, quantitative PLB activity ranged from 76 to 98 U/100 mL (mean ± SD = 84 ± 9) vs. 67–90 U/100 mL (mean ± SD = 79 ± 11). Similarly, for *A. flavus*, activities ranged between 50 and 81 U/100 mL (62 ± 17) vs. 33–68 U/100 mL (45 ± 20). Unlike Birch et al. [29], enzymatic activity indicative of PLD was not detected by Koul et al. [82]. Furthermore, only PLB activity was detected in *A. fumigatus* by the latter workers. This discrepancy could be due to strain-to-strain variation. Alternatively, discrepancy in the detected activities could be due to the fact that Birch et al. [29] deduced the type of phospholipase secreted by *A. fumigatus* based on phospholipids degradation products and not assays incorporating specific substrates as utilized by Koul et al. [82].

Although the role of phospholipase in the virulence of *Aspergillus* must await cloning and disruption of the gene/genes encoding these enzymes, Koul et al. [82] showed that reagent-grade PLB produced extensive cytolysis of cultured pneumocytes, offering the possibility that this enzyme enhances the virulence of *Aspergillus* by causing damage to host cells.

D. Role of Phospholipases in Fungal Virulence

The first evidence correlating phospholipase activity and candidal pathogenicity came from work published by Barrett-Bee et al. [17]. In this study, the *C. albicans* isolates that adhered most strongly to buccal epithelial cells and were most pathogenic in mice had the highest phospholipase activities. Less pathogenic isolates of *C. albicans*, *C. parapsilosis*, and *S. cerevisiae* were less adherent to epithelial cells, were less lethal to mice, and had lower phospholipase activities [17]. In another study, phospholipase activities of *C. albicans* blood isolates obtained from patients with disseminated candidiasis were compared with that of commensal isolates obtained from oral cavities of healthy volunteers. Significantly higher levels of phospholipase production were found in the blood isolates compared to the commensals ($P = .0081$). In addition, the blood isolates had significantly higher rates of germination ($P = .03$), and their germ tubes were longer than those formed by the commensal strains ($P = .016$). These findings suggest that blood isolates, unlike commensals, may have enhanced expression of several virulence factors, including both germination and phospholipase production, which enables them to invade host tissues [5].

In the same study, the pathogenicity of clinical isolates with varying phospholipase activities was compared in a murine model of disseminated candidiasis [5]. Nine blood isolates were prospectively examined for expression of virulence factors, including phospholipase and proteinase production, adherence, germination, growth rate, and ability to damage endothelial cells [5]. Additionally, the mortality of mice infected with each of these isolates was determined and the predictive value of each virulence factor for mortality was determined by Cox proportional hazards analysis. Of the virulence factors studied, only extracellular phospholipase activity was predictive of mortality [5]. In an infant mouse model of candidiasis, the high-phospholipase-producing strain CA30 was able to cross the bowel wall and disseminated hematogenously, while the low-phospholipase-producing strain CA87 failed to cross the bowel wall of the infant mouse after oral-intragastric challenge and did not cause disseminated infection [83]. Since the two strains were distinguished by their ability to invade the bowel wall with subsequent hematogenous dissemination and tissue invasion, these data further suggest that phospholipase is involved in the invasion process of *C. albicans*.

These studies provided correlative evidence implicating phospholipases in the pathogenesis of *C. albicans*. Since the candidal isolates used were not genetically related, the possibility that

Table 1 Phospholipases of Pathogenic Fungi

Organism	Enzyme	Protein	Substrate specificity/ other features	Gene	ORF	Refs.
Candida albicans	PLB1	605 aa, 84-kDa molecular mass, 7 N-glycosylation sites, no GPI-anchoring site	PC, Lyso-PC	caPLB1	1818 bp	52,53
Candida albicans	PLB2	608 aa, 67-kDa molecular mass, 6 N-glycosylation sites, lacks GPI-anchoring sites		caPLB2		56
Candida albicans	CAPLC	1099 aa, 124.6-kDa molecular mass	X and Y catalytic domains	CAPLC	2997 bp	59
Candida albicans	PLD	1710 aa, 196.4-kDa molecular mass, 42% homology to SPO14		caPLD		66
Candida glabrata	PLB	Not purified	PC, Lyso-PC	PLB	1030 bp (PCR product)	43
Cryptococcus neoformans	PLB	617 aa, molecular mass of 70–90 kDa by SDS/PAGE and 160–180 kDa by gel filtration; acidic glycoprotein-containing N-linked carbohydrate moieties	All phospholipids except phosphatidic acid, but dipalmitoyl phosphatidylcholine and dioleoyl phosphatidylcholine are the preferred substrates	PLB	2400 bp approx.	28,81, 92,93
Aspergillus fumigatus	PLB	Not purified	PC	Not cloned		29,82
Aspergillus fumigatus	PLC	Not purified		Not cloned		29,82
Aspergillus fumigatus	PLD	Not purified		Not cloned		29,82
Aspergillus flavus	PLB	Not purified	PC, Lyso-PC	Not cloned		29,82

differences observed in the virulence of these strains could be attributed to factors other than phospholipases could not be ruled out. This caveat was overcome by cloning the caPLB1 gene and constructing its PLB-disruptant isogenic strain. This isogenic strain pair facilitated detailed analyses of the role of PLB in candidal virulence (see Sec. II.A.1 and below).

1. Animal Models of Candidiasis

A reliable and standard method to establish the role of PLB (or other proteins) in candidal virulence is by using animal models of candidiasis as part of in vivo studies. Two different models of candidiasis, representing different clinical settings, were used to determine the role of caPLB1 in virulence of C. albicans: (1) a hematogenously disseminated murine model in which BALB/c mice were challenged with the parent or the PLB-deficient C. albicans strains cells and both survival as well as tissue fungal burden were used to assess the pathogenicity of the infecting strains [52,84]; and (2) an oral-intragastric infant mouse model in which inbred infant mice (crl:CFW (SW) BR), were inoculated intragastrically with blastospores of either the wild-type or PLB1-deficient strains. The latter model simulates candidal migration across the gastrointestinal tract, one of the major routes for contracting disseminated candidiasis [83].

    a.   Dissemination to Kidneys and Liver.   Using the hematogenously disseminated model, it was shown that all mice infected with the parental strain succumbed to candidal infection within 9 days, while 60% of mice challenged with the PLB-deficient strain were alive at day 15 [52]. Gross inspection of the kidneys showed that numerous visible candidal foci covered the renal cortex of mice infected with parental strain. In contrast, no candidal foci were detectable on renal surfaces of mice infected with the PLB-deficient strain. Tissue fungal burden experiments carried out to assess the severity of infection caused by wild-type and mutant strain showed that PLB1-deficient candidal strain was cleared significantly faster than the wild-type strain from the kidneys and brain [52]. This study also showed that deletion of ca*PLB1* did not render *C. albicans* strains completely avirulent, thus underscoring the notion that candidal pathogenicity is multifactorial and is regulated by more than one determinant [12].

    b.   Transmigration of C. albicans *Across Gastric Mucosa.*   The oral-intragastric infant mouse model was used to monitor transmigration of candidal cells acrossthe gastrointestinal tract using light and transmission electron microscopy. These studies revealed the presence of yeast and hyphal elements in the gastric mucosa 14 days postchallenge with both the parental and PLB-deficient strains. However, mucosal invasion in mice challenged with the PLB-deficient strain was confined to the stomach lumen and inner layers of the gastric mucosa, and a minimal neutrophil-dependent inflammatory response was elicited. Furthermore, several areas of the stomach revealed no demonstrable infection. In contrast, the parental strain invaded the submucosal tissue of the stomach, elicited a neutrophil-dependent inflammatory response, and produced systemic candidiasis to a far greater extent. That the parental strain was more efficient at crossing the GI tract and invading internal organs than its PLB-deficient counterpart suggests a role for PLB in candidal transmigration across the gastrointestinal tract and subsequent dissemination to target organs. Additionally, the parental strain was able to traverse the vasculature since numerous hyphal elements were observed within blood vessel lumens. Among the mice challenged with the parental strain, 90% showed liver colonization while 70% showed kidney colonization; only 45% and 27% of mice infected with the PLB-deficient mutant exhibited liver and kidney involvement, respectively [85].

    c.   Secretion of Phospholipase B1 During Host Tissue Invasion.   One of the criteria to prove that a particular gene or its product plays an important role in the disease process is to show that it is expressed during the infectious process [86]. Both the hematogenously disseminated intravenous [52] and the oral-intragastric mouse [85] models for candidiasis were used to determine whether PLB is secreted in vivo.

    In the hematogenously disseminated model, mice were challenged with either the parental or ca*PLB1*-deleted mutant. Kidneys were harvested and sections were processed for immunoelectron microscopy by incubating them with either PLB antiserum or goat serum, which served as a negative control [52]. Immunogold complexes were observed under the electron microscope for tissue sections infected with parental strain thus showing the secretion of PLB during infectious process. Similar complexes were not seen for PLB-deficient mutant-infected mice tissues [52]. In other experiments, *C. albicans* cells were recovered from the kidneys of mice infected with either the parental or PLB-deficient strains. Cells were prepared for immunogold electron microscopy. Sections examination showed that immunogold labeling was observed only with cells recovered from mice infected with the wild-type strain and not in association with cells recovered from animals infected with the PLB-deficient mutant (unpublished data). These data demonstrate that *C. albicans* secretes PLB during the invasion of target organs.

Using indirect immunofluorescence microscopy, Seshan et al. [85] investigated the expression of PLB during the course of candidal transmigration across the gastrointestinal tract of infant mice. Following challenge with either the parent or the PLB-deficient strains, mice were sacrificed and their gastric mucosal tissues removed and prepared for indirect immunofluorescence microscopy. Their data showed that PLB is secreted by the vast majority of candidal cells. Both hyphal and yeast forms of *C. albicans* secreted the enzyme. Although the periphery of most cells exhibited low-level fluorescence suggestive of PLB secretion, intense fluorescence was observed at the growing tips of mature and developing hyphae, suggesting that PLB secretion is concentrated at the invading hyphal tip [85].

The exact mechanism of action of phospholipases in fungal virulence is unknown. However, intuitive reasoning as well as experimental evidence indicates that these enzymes are involved in the early stages of infection—adherence, penetration, and damage.

2. *C. albicans* Phospholipase B and Adherence of *C. albicans* Host Cells

Barrett-Bee et al. [17] found a correlation between the ability of yeasts, including *C. albicans*, to adhere to buccal epithelial cells and phospholipase activity. In contrast, another study compared two *C. albicans* isolates that differed in their ability to produce phospholipase and showed that the low phospholipase producer adhered more avidly to the gastrointestinal tract of infant mice relative to the high phospholipase producer [83]. These contradictory findings could be due to the fact that the above studies utilized genetically unrelated strains. Alternatively, phospholipases may facilitate adherence in some organisms and not in others, or in certain specific settings. To determine whether *caPLB1* influences candidal adherence, the abilities of PLB-deficient mutant and parental strain were compared using a standard in vitro assay [52]. Our data showed that the percent adherence for the PLB-deficient mutant and parent strain to epithelial cells (HT-29) was 36 ± 3.9% and 35 ± 2.2%, respectively. Similarly, no significant difference in the adherence ability of the isogenic strain pairs to human umbilical vein endothelial cells was noted [52]. This suggests that PLB does not appear to directly affect the ability of *C. albicans* to adhere to host cells.

3. *C. albicans* Phospholipase B and Host Cell Injury and Penetration

Since phospholipase targets membrane phospholipids and has been shown to digest these components leading to cell lyse [87], direct host cell damage and lysis has been proposed as a main mechanism contributing to microbial virulence. Such host cell injury would be expected to facilitate the penetration of the infecting agent. In this context, Klotz et al. [88], using an in vitro model depicting the earliest events of metastatic *Candida* infection, showed that *Candida* strains first adhere to, and then penetrate, the endothelium. During transmigration, endothelial cell continuity was disrupted by the yeasts. As destruction of the endothelium progressed, the fungus penetrated deeper into the substance of the vascular tissue. These authors attributed the dissolution of a portion of the endothelial cells to phospholipase activity [88]. Similar suggestive evidence for enzymatic activity by *C. albicans* in the penetration of skin has been described [89].

More recently, evidence implicating phospholipase in host cell penetration and injury has been derived from in vitro and in vivo studies. The in vitro studies compared the ability of the parent and PLB-deficient strains to penetrate epithelial and endothelial cell monolayers [90]. Also, the ability of supernatants from cultures of these strains to cause damage to epithelial cells was compared [90]. Leidich et al. [52], using scanning electron microscopy, compared the ability of the parent and *caPLB1*-deficient mutant to penetrate both human umbilical vein endothelial (HUVEC) and HT-29 epithelial cells. Their data showed that both the parent and *caPLB1*-deficient mutant formed germ tubes which penetrated HUVEC and HT-29 cells. However, the capacity of the *caPLB1*-deficient strain to penetrate HUVEC and HT-29 host cells was significantly reduced relative to the parent. The percentage of penetrating parental hyphae was 66.7 ± 1.7 for the endothelial cell line and 57.8 ± 2.5 for the epithelial cell line. In contrast, the percentages of penetrating PLB-deficient hyphae were 37.3 ± 1.3 and 29.0 ± 2.9 for the HUVEC and HT-29 cell lines, respectively ($P < .0002$ for HUVEC; $P < .0003$ for HT-29) [52].

The ability of supernatants from cultures of the parental or PLB-deficient strains to cause damage to epithelial cells was compared using a Thiazolium Blue (MTT)-based assay [90]. Supernatant from the parental culture was about twofold more efficient in causing cell damage, compared to that from the *caPLB1*-deficient culture (supernatant from parental strain caused 62.33 ± 1.15% damage, while supernatant from *caPLB1*-deficient strain caused 34.67 ± 0.58% damage ($P < .0001$). In the same experiments, damage caused by supernatants obtained from *C. albicans* and the nonpathogenic yeast *S. cerevisiae* was compared. In contrast to the significant damage caused by *C. albicans*, only minor injury was caused by *S. cerevisiae* (3.95 ± 0.48, mean ± S.D). The above data suggest that PLB may enhance candidal virulence by directly damaging host cell membranes.

Seshan et al. [85] provided in vivo data demonstrating that candidal PLB facilitates candidal penetration of host tissues. Using an oral intragastric infant mouse model of candidiasis, these authors showed that the phospholipase-producing parental strain penetrated deep into the gastric mucosal and submucosal tissues following intragastric challenge. In contrast, the *caPLB1*-deficient mutant was not as invasive and was generally sequestered to the stomach lumen. The invasiveness of the parental strain is likely to have increased its access to the gastric vasculature, thus allowing the organism to hematogenously disseminate more efficiently than the PLB-deficient mutant. Consistent with this notion, more hyphal elements were observed in blood vessel lumina following challenge with the parental strain. These differences in penetration and dissemination were reflected by the number of candidal colony-forming units (CFU) recovered from the liver and kidneys of mice infected with either the parental or *caPLB1*-deficient strains. Organs (kidneys and liver) from mice infected with the parental strain had significantly higher candidal CFU compared to organs harvested from mice infected with the PLB-deficient mutant [85].

Therefore, a decrease in the capability of PLB1-deficient strains to cross the gastrointestinal tract and cause systemic infection, combined with their decreased capability to penetrate host cell monolayers, and the ability of an enzymatic preparation to lyse host cells indicate that PLB may play a role in candidal virulence by causing direct damage to host cell membranes. Such injury would allow fungal hyphal elements to more effectively traverse the vascular endothelium, ultimately increasing the rapidity of dissemination to and invasion of target organs.

4. *C. neoformans* Phospholipase and Virulence

The ability of *C. neoformans* to secrete phospholipase was reported by Vidotto et al. [91]. Their data showed that 22 out of 23 cryptococcal isolates tested produced phospholipase. As in the case of *Candida*, a wide variation was observed in the ability of various strains to secrete phospholipase (Pz values ranged between 0.271 and 0.49). Although these authors found a correlation between phospholipase production and the size of the capsule in the strains isolated from AIDS patients, the number of isolates analyzed for these factors was small. Thus, more isolates should be examined to confirm this notion. In another study published in the same year, Sorell and coworkers [28] examined 50 *C. neoformans* isolates for extracellular phospholipase activity, using egg yolk–based agar assay. Forty-nine of these isolates produced a pericolonial precipitate indicative of phospholipase activity. No difference in phospholipase production was observed between environmental and clinical isolates of *C. neoformans* var. *gattii*. Quantitation of cryptococci in the lungs and brains of BALB/c mice inoculated intravenously with four strains expressing high, intermediate, or low phospholipase activity revealed a correlation between phospholipase activity and virulence [28]. Based on this finding, the authors proposed that phospholipases secreted by *C. neoformans* may be associated with cryptococcal virulence.

In a second study, Chen et al. [92], using ^{1}H and ^{31}P nuclear magnetic resonance (NMR) spectroscopy combined with thin-layer chromatography (TLC) analyses, extended their work to define the nature of the phospholipase activity produced by *C. neoformans*. NMR spectroscopy revealed that the sole phospholipid degradation product of the reaction between the substrate phosphatidylcholine and cryptococcal culture supernatants was glycerophosphocholine, indicating the presence of PLB. No products indicative of PLA, PLC, PLD, or other lipase activity were detected [92]. TLC analysis confirmed that PLB and Lyso-PL activities were detected in *C. neoformans* supernatants. Additionally, LPTA activity was identified in supernatants by the formation of radioactive phosphatidylcholine from lysophosphatidylcholine. The Lyso-PL activity was 10- to 20-fold greater than PLB activity in these supernatants, with mean (± standard deviation) specific activities of 34.9 ± 7.9 and 3.18 ± 0.2 μmol of substrate hydrolyzed per min per mg of protein, respectively. Enzyme activities were stable at acid pH 3.8, with pH optima of 3.5 to 4.5. Activities were unchanged in the presence of exogenous serine protease inhibitors, divalent cations, and EDTA. Thus, *C. neoformans* secretes phospholipase with similar activity (PLB, Lyso-PL, and LPTA) to that observed in *C. albicans*. The *C. neoformans* phospholipase activities were recently purified to homogeneity [93].

E.   Role of Fungal Extracellular Phospholipases in Other Virulence Facilitating Functions

As discussed above, the evidence accumulated so far suggests that fungal phospholipases enhance virulence by damaging host cell membranes. However, based on findings derived from studies of phospholipases from bacteria, it is likely that these hydrolytic enzymes may have other functions that facilitate virulence, in addition to direct tissue damage [94]. Availability of purified fungal phospholipases, as well as isogenic strains from different pathogenic fungi, will provide clues to the functions/mechanisms employed by phospholipase during fungal pathogen interaction. While *C. albicans* strains differing in phospholipase production became available only recently and were tested in vivo for their virulence [52], isogenic sets from other pathogenic fungi are still being constructed, and in some cases (e.g., *A. fumigatus*) the genes encoding phospholipases await cloning.

A number of potential investigational areas relevant to phospholipase mechanism(s) could be proposed, including signal transduction, stimulation of host cells to release cytokines, and the host inflammatory response. Many lipids and lipid-derived products generated by phospholipases acting on phospholipids present in host cell membranes are implicated as mediators and second messengers in signal transduction [64,95]. The enzyme's substrate specificity and lipid degradation products indicate that candidal PLB could also be involved in signal transduction pathway potentiated at the lysophospholipid level. Stimulation of host cells to produce cytokines in response to soluble factors of microbial origin is well documented [96,97]. Stimulation of cytokine production in response to lytic enzymes of microbial origin has also been demonstrated in vivo [98]. Since phospholipase production is closely associated with candidal host cell injury [52], it is conceivable that this enzyme may directly or indirectly stimulate host cells to produce specific cytokines.

Microbial phospholipases have been demonstrated to be potent inflammatory agents inducing the accumulation of inflammatory cells and plasma proteins, and the release of various inflammatory mediators in vivo [99]. Additionally, microbial phospholipases are able to mobilize arachidonic acid and subsequent prostaglandin synthesis [100]. Availability of purified fungal phospholipases and isogenic strain pairs that differ only in phospholipase production makes it feasible to investigate these exciting areas of investigation.

III. SECRETORY ASPARTIC PROTEINASES

Proteolytic activities attributable to secretory aspartic proteinases (SAPs) have been found in at least three *Candida* spp. (*C. albicans, C. tropicalis,* and *C. parapsilosis*), as well as in certain *Aspergillus* species. This area has been extensively reviewed in the past [3,7,101], and therefore only a brief summary of the literature is presented here.

A. Proteinases of *Candida* Species

Staib [102] was the first to show a proteolytic activity associated with *C. albicans*. Subsequent studies have led to detailed characterization of its biochemical and biophysical properties and the elucidation of SAP crystal structure. This enzyme has been given a variety of names including *Candida* aspartyl proteinase (CAP) (103) and keratinolytic proteinase (Kpase) (104,105). The 99% nucleotide similarity with a pepsinogen gene resulted in the first cloned *Candida* proteinase gene to be called *PEP1* (106). The second cloned gene was referred to as *PRA2;* subsequently *PEP1* was renamed *PRA1* because of its similarity to the *S. cerevisiae PRA* genes. However, the absence of similarity between *Candida* proteinase and pepsinogen or Pra proteins led to the general acceptance of secretory aspartic proteinase (SAP) as the name for *Candida* proteinase [107,108]. In vitro activity of SAPs is assayed with bovine serum albumin (BSA) as the substrate in the medium, although other substrates have also been identified [103].

The following observations provided clues as to the possible role of SAPs in candidal pathogenicity:

1. The proteolytic activity at acidic pH was the highest in *C. albicans*, followed by *C. tropicalis* and *C. parapsilosis*. The order of proteolytic activity correlated with the order of virulence among these species: *C. albicans* is the most virulent, followed by *C. tropicalis* and *C. parapsilosis* [109]. Since the proteolytic activity of these species correlates with their pathogenicities, it was suggested that SAPs might be important virulence determinants of *Candida* species.

2. Studies performed in a mouse model of experimental candidiasis showed that more lethal *C. albicans* strains, which were also avid colonizers, had higher proteolytic activity than less pathogenic strains [110].

3. Clinical studies showed that *C. albicans* isolated from patients with vaginitis had 1.7-fold higher SAP production than isolates from healthy, nonvaginitis patients [111].

4. Nonproteolytic mutant strains were found to be less pathogenic in mice [112,113]. Also, a spontaneous revertant which regained half of the proteolytic activity was almost as virulent as the wild type [112]. Similarly, a proteinase-deficient strain was found to be 1000-fold less virulent as compared to the proteolytic wild-type strain [114].

5. Treatment of infected mice with pepstatin, an inhibitor of SAPs, caused a distinct protective effect against candidal infection [105,115,116].

These findings stimulated interest in secretory proteinases and led to the purification of one SAP, cloning of genes encoding various SAPs, and construction of disruption mutants.

1. Proteinases of *Candida albicans*

Attempts at purifying the *Candida* proteinase were performed early on and led to the partial purification of a 40-kDa protein with a pH optimum of 3.2 and broad substrate specificity [117]. The extracellular acid proteinase of *C. albicans* was purified from culture filtrates and shown to be a mannoprotein with carboxyl proteinase activity [118]. This protein was detected in kidney lesions of animals with experimental candidiasis using indirect immunoflourescence [118]. An extracellular proteinase of *C. albicans* was purified and characterized using pepstatin A–affinity chromatography [115]. The purified proteinase was a monomeric protein with a molecular weight of 45 kDa and isoelectric point of pH 4.4, and underwent rapid denaturation at pH >8.4 [115]. The protein had broad substrate specificity for albumin, casein, collagen, hemoglobin, and keratin [103]. Further studies based on the interactions of the enzyme with antimannan antibodies and sensitive carbohydrate detection tests ruled out glysosylation of the enzyme [119,120]. Ruchel et al. [120] also showed that while the secretory proteinase isolated from *C. albicans* and *C. tropicalis* lack glycosylation and have molecular weights in the range of 40–50 kDa, secretory proteinase from *C. parapsilosis* is a mannoprotein of 33 kDa.

    a. Cloning of SAP *Genes.*   Different approaches were followed to clone the family of differentially regulated *SAP* genes (*SAP1–SAP9*) from *C. albicans*. A conserved oligonucleotide sequence in the N-terminal of purified aspartyl proteinases was identified from the amino acid sequence and was used as probe to identify a *SAP*-specific PCR-fragment [121]. This fragment was used to screen a *C. albicans* genomic library and to isolate the *SAP1* (*PRA10*) gene. Analysis of the 1173-bp open reading frame (ORF) indicated that the *SAP1* gene is highly expressed and has homology to other aspartic proteinases. Similar strategies were used to clone the *SAP2* gene (*PRA11*) [122]. The nucleotide sequence of *SAP2* was 77% identical to that of *SAP1*, while the deduced amino acid sequences of *SAP2* and *SAP1* were 73% identical. Analyses of the mRNA levels of the two genes showed that *SAP2* expression was much higher than *SAP1* in the presence of albumin [122]. The *SAP1* gene was later mapped to chromosome 6 while *SAP2* was mapped to chromosome R [107]. Cloning of increasing number of *SAP* genes led researchers to identify the presence of a gene family for which the term "*SAP* family" was coined [7,107].

    Screening of a genomic library under low-stringency hybridization conditions with a PCR fragment from SAP1 led to the identification of SAP3, a third secreted proteinase gene. *SAP3* was 77% homologous to both *SAP1* and *SAP2*. The predicted amino acid sequence of SAP3 had 398 amino acids and a high pI of pH 5.7 [108]. Restriction fragment length polymorphisms (RFLP) using the same *C. albicans* strain WO-1, *SAP1* alleles in laboratory and clinical strains were characterized leading to the identification of *SAP4* (1254 bp) positioned upstream, in tandem to *SAP1* [123]. Screening a genomic library with a *SAP1* probe using low-stringency hybridization conditions led to the identification of three new genes: *SAP5* (1254 bp), *SAP6* (1254 bp), and *SAP7* (1764 bp) [124]. The *SAP5* and *SAP6* genes were found to be closely related and similar to the other *SAP* genes. SAP7p is the most divergent protein among the SAP protein family with only 20% similarity with SAP1p and a much longer putative prosequence. The *SAP1–7* genes have been grouped into two clusters based on nucleotide and predicted amino acid sequence similarities: *SAP1–3* and *SAP4–7* [124]. The possibility of the existence of two additional *SAP* genes not isolated after screening of the *C. albicans* gene library was also indicated in these studies. Other *Candida* species such as *C. tropicalis, C. parapsilosis*, and *C. guillermondii* were also found to possess *SAP* gene families [124].

    The screening of a *C. albicans* genomic library for the presence of other *SAP* genes led to the isolation of two new genes, *SAP8* and *SAP9*. The N-terminal amino acid sequence deduced from *SAP8* was found to be identical to the N-terminal amino acid sequence of the 41-kDa SAP1p. *SAP8* mRNA was expressed at 25°C in the presence of BSA. *SAP9* encodes an aspartic

Table 2 Secretory Aspartic Proteinases of *C. albicans*

Gene	ORF	Enzyme features	Regulation	Phylogenetic cluster	Refs.
SAP1	1173 bp	341 aa, 2 KR (lys-arg) sites, no glycosylation site	Expressed in "opaque" form, at 25°C	SAP1-3	7,124,162
SAP2	1194 bp	342 aa, 2 KR sites, 2 glycosylation sites	Most abundant, expressed in both white and opaque forms, and also in yeast at 25°C but not at 37°C in hyphal forms	SAP1-3	7,122,124
SAP3	1194 bp	340 aa, 1 KR site, 1 glycosylation site	Expressed in "opaque" form, at 25°C.	SAP1-3	7,108,124
SAP4	1254 bp	342 aa, 4 KR sites, 1 glycosylation site	Expressed in hyphae at 37°C	SAP4-6	7,123,124
SAP5	1254 bp	342 aa, 4 KR sites, no glycosylation site	Expressed in hyphae at 37°C	SAP4-6	7,124
SAP6	1254 bp	342 aa, 4 KR sites, 1 glycosylation site	Expressed in hyphae at 37°C	SAP4-6	7,124
SAP7	1764 bp	377 aa, 1 KK (lys-lys) site, 4 glycosylation sites	Silent or very weak expression	Most divergent	7,124
SAP8	1209 bp	330 aa, 2 KR sites, no glycosylation sites	Expressed in opaque and yeast forms, at 25°C	?	7,124,125
SAP9		Encodes aspartic proteinase with a Kex2p-like cleavage site and contains a putative glycophosphatidylinositol-anchor signal at the C-terminus		?	7,124,125

proteinase with a Kex2p (killer expression endopeptidse)-like cleavage site and contains a putative glycophosphatidylinositol anchor signal at the C-terminus. Although the *SAP9* gene product has not yet been isolated from cultures of *C. albicans*, transcripts of *SAP9* were observed preferentially in later growth phases when *SAP8* expression had decreased [125]. Studies so far indicate that [1] all *SAP* genes code for a 50–60 aa propeptide, [2] SAPs have a 14–21 amino acids N-terminal secretion signal linked to a propeptide containing one to four putative KR (Lys-Arg) or KK (Lys-Lys) processing sites for a Kex2-analogous regulatory proteinase (Table 2).

    b. Protein-Level Regulation of SAP Genes. Most studies conducted so far have concentrated on the regulation of *SAP* protein expression and the in vitro induction of SAP activity. SAP activity was induced in the presence of a variety of proteins, including BSA, hemoglobin, keratin, casein, or collagen, and at acidic pH [103,114,115,126,127]. In addition, peptides of more than 8 amino acids could also induce SAP activity [127,128], while ammonium ions and amino acids inhibited protease activity [114,127]. Pulse chase experiments suggested that synthesis and proteinase secretion were coupled and that the protein was not accumulated intracellularly [127,129]. Also, late-exponential-phase grown cells showed the highest levels of intracellular protease [129]. Thus, both cell cycle and environmental conditions influence SAP activity. Among *SAP1-3*, all three proteinases are expressed in the same strain, but their pattern of expression varied with the opaque-white switch phenotype of the cell. SAP1 and SAP3 were expressed in opaque cells while SAP2 was expressed in white cells of *C. albicans* strain WO-1 [108]. SAP2p was found to be the most abundant proteinase among *Candida* species; thus, all characteristics described for SAP proteins refer to the SAP2p [7].

   c. Transcriptional Regulation of SAP Genes.   Morrow et al. [106] used a differential hybridization screen to isolate an opaque-specific cDNA, *OP1A*, which was found to have >99% base homology at the nucleotide level with an acid protease (*PEP1*) gene of *C. albicans*. This cDNA is identical to *SAP1* and thus suggests that *SAP1* is most likely regulated by the switching phenotype, a finding later confirmed by Northern blot analysis [106]. mRNA studies of seven members of the *SAP* gene family (*SAP1-7*) showed that *SAP1* and *SAP3* expression varied with phenotypic switching between the white and opaque forms of *C. albicans*. The *SAP1* mRNA was expressed even in the absence of peptides. *SAP1* mRNA was downregulated in stationary phase and in presence of exogenous amino acids. *SAP2* mRNA, which was the dominant transcript in the yeast form, was found to be autoinduced by peptide products of SAP2 activity and to be repressed by amino acids. The expression of the closely related *SAP4–SAP6* genes was observed only at neutral pH during serum-induced yeast to hyphal transition. No *SAP7* mRNA was detected under any of the conditions or in any of the strains tested [130]. In a subsequent study using both an *in vitro* model of oral candidiasis and clinical samples from oral candidiasis patients, *SAP1-8* genes were found to be expressed in the order *SAP1* and *SAP3* > *SAP6* > *SAP2* and *SAP8*. Thus, although SAP2 is the abundant SAP protein in candidiasis, *SAP2* RNA is observed only late in infection. These studies proved that SAP expression is under temporal regulation [131]. Although the *SAP9* gene product has not yet been isolated from cultures of *C. albicans*, transcripts of *SAP9* were observed preferentially in later growth phases, when *SAP8* expression had decreased [125].

B. Proteinases of *Aspergillus* Species

Invasive aspergillosis due to *A. fumigatus* is thought to involve hydrolysis of structural proteins in the lung by secreted proteinases [132]. An aspartic proteinase (PEP) was purified from the culture supernatant of a clinical isolate of *A. fumigatus*. Nine of the 11 N-terminal region amino acids were identical with the corresponding sequence of the aspartic proteinase aspergillopepsin A from *A. niger* var. *awamori* (previously called *A. awamori*). The presence of the PEP antigen was demonstrated in sera of two patients with aspergillosis using dot-blot assay and in a lung infection by immunofluorescence. PEP had an estimated molecular mass of 38 kDa and a pI of pH 4.2. PEP is therefore likely to be closely related to an acid proteinase of *A. fumigatus* [133]. Lee and Kolattukudy [132] reported that *A. fumigatus* also secretes an aspartic proteinase (aspergillopepsin F) that can catalyze hydrolysis of the major structural proteins of the basement membrane. The pH optimum for the enzymatic activity was 5.0 with elastin–Congo Red as the substrate, and the activity was not significantly inhibited by pepstatin A, diazoacetyl norleucine methylester, and 1,2-epoxy-3-(p-nitrophenoxy) propane. The gene encoding this aspartic proteinase were cloned and sequenced. The open reading frame, interrupted by three introns, putatively encodes a protein of 393 amino acids composed of a 21-amino-acid signal peptide and a 49-amino-acid propeptide preceding the 323-amino-acid mature protein. The amino acid sequence of the *A. fumigatus* aspartic proteinase has 70, 66, and 67% homology to those from *A. oryzae*,

A. *awamori*, and A. *saitoi*, respectively [132]. Immunogold electron microscopy showed that the aspartic proteinase was secreted by A. *fumigatus* during lung invasion in a neutropenic mouse. Secretion was directed toward the germ tubes of penetrating hyphae [132]. Immunofluorescence studies were performed with sporulating colonies of A. *fumigatus, A. flavus,* and A. *niger* using specific polyclonal antibodies against the aspergillopepsin PEP (EC 3.4.23.18). The proteinase antigen was found mainly in developing conidiophores of aspergilli, in submerged mycelia, and on the tips of growing aerial mycelia, but not in mature aerial hyphae and spores, while sporulating conidiophores of A. *fumigatus* and A. *flavus* revealed only weak immunofluorescence [134]. In another study, sequences of the *PEP* gene of A. *fumigatus* were used as PCR primers to identify this species and differentiate it from other, coidentified, clinically important *Aspergillus* spp. [135]. Their distinct pattern of expression suggested a role for aspartic proteinase antigens in the infection process. However, studies with mutants of A. *fumigatus* deficient in PEP showed that the wild-type strain and the PEP-deficient mutants invaded tissues to a similar extent and produced comparable mortality in guinea pigs. Thus, secreted proteinases probably do not contribute decisively to tissue invasion in the pathogenesis of systemic aspergillosis [136].

C. Role of Secretory Aspartic Proteinases in Fungal Virulence

Virulent C. *albicans* strains invariably produce high amounts of SAP proteins, while SAP-deficient strains have lower virulence [110,111,113,118,137,138]. This observation led to the belief that SAPs are involved in candidal pathogenesis and are important virulence factors. Although exhaustive studies have been performed on candidal SAPs, the mechanism by which they affect virulence has not yet been determined. SAPs can affect virulence by a number of possible mechanisms: (1) digest host surface proteins, thus damaging the host tissue and aiding in adhesion; (2) overcome the host immune system by degrading host proteins; (3) replenish cellular nitrogen sources with peptide breakdown products; (4) damage endothelial cells; or (5) activate proteolytic cascades in the host [7,139]. SAP proteins have been associated with adherence [110,140,141], phenotypic switching [22,106], and hypha formation [130,142,143]. In a separate study, C. *albicans* isolates from the oral cavities of subjects at different stages of human immunodeficiency virus (HIV) infection were compared for their SAP activities [144]. Fungal isolates from symptomatic patients secreted up to eightfold more proteinase than the isolates from uninfected or HIV-infected but asymptomatic patients. Also, high-proteinase isolates were more pathogenic for mice than the low-proteinase isolates [144]. Using a rat vaginitis model of candidiasis, it was shown recently that SAP2 is an important virulence factor for *Candida* vaginitis [145]. However, as expected, SAPs cannot be regarded as the sole critical virulence factor in candidiasis, as overexpression of SAP2p (the most abundant SAP) in C. *albicans* and its expression in S. *cerevisiae* did not augment virulence of these strains in a mouse model of candidiasis [146]. Antibodies in sera from patients with systemic candidiasis were shown to react with purified acid proteinase [118]. Indirect immunofluorescence was used to show the presence of proteinase antigens in tissue sections isolated from clinical cases of mucosal and deep-seated candidiasis [147]. Recently, a monoclonal antibody (MAb; MAb CAP1) reactive with candidal aspartic proteinase was produced. The MAb showed strong sensitivity and reactivity to candidal aspartic proteinase but not to the aspartic proteinases of C. *parapsilosis,* C. *tropicalis,* and A. *fumigatus* or to human cathepsin D or porcine pepsin. This antibody could serve as an important tool in characterizing candidal aspartic proteinase and may be a valuable probe for the detection of the proteinase antigen in the sera of patients with invasive candidiasis [148].

Recently, Schaller et al. [149] showed the involvement of SAPs during host tissue invasion in an in vitro model of cutaneous candidiasis. In this study, a progressive increase in *SAP* expression in the order *SAP1* and *SAP2* > *SAP8* > *SAP6* > *SAP3* was observed [149]. Using specific polyclonal antibodies directed against the gene products of *SAP1-3* and *SAP4-6*, predominant expression of *Sap1-3* was shown [149]. Pepstatin A had a protective effect during infection of the epidermis, and *SAP*-deficient mutants exhibited an attenuated virulence phenotype, indicating possible correlation between the *SAP* expression and tissue damage in the skin [149].

IV. FUNGAL SECRETORY PROTEINS AS THERAPEUTIC AND DIAGNOSTIC TOOLS

The fact that PLB and SAPs are important virulence factors in fungal pathogenicity generates the possibility that these proteins can be used as therapeutic and/or diagnostic targets.

A. Phospholipases

PLB was shown to be a virulence factor in *C. albicans* using animal models of hematogenously disseminated [52] and gastrointestinal candidiasis [85]. This hydrolytic enzyme was detected in other pathogenic fungi including *C. neoformans*, *A. fumigatus*, and *A. flavus*. The finding that different clinically important fungi secrete phospholipases suggests that those enzymes may represent a common theme utilized by pathogenic fungi as a universal virulence factor [87]. Phospholipase secretion across various fungal genera increases their potential of being an effective therapeutic target, possibly leading to a number of drug development strategies and discovery of novel antifungals. These approaches may include development of vaccines and identification of phospholipase inhibitors.

Development of vaccines based on microbial phospholipases has been investigated for different *Clostridium* species [150–154]. In a different approach, the phospholipase D gene (*PLD*) was deleted from *Corynebacterium pseudotuberculosis* to create a PLD-negative (Toxminus) strain. Vaccination of sheep with live Toxminus *C. pseudotuberculosis* elicited strong humoral and cell-mediated immune responses and protected the animals from wild-type challenge [155].

Preliminary attempts to identify inhibitors to candidal phospholipases have been undertaken by Hanel et al. [156], who screened for synthetic phospholipase inhibitory substances using in vitro and in vivo models. Their efforts resulted in the identification of lead structures capable of inhibiting phospholipase activity in vitro [156]. Additionally, prolongation of animal survival was observed when these compounds were used in combination with fluconazole [156]. These studies provide some support to the concept of targeting fungal phospholipases for drug discovery, and warrant a systemic approach for drug discovery based on these lytic enzymes.

Another potential use of fungal extracellular phospholipases, particularly candidal phospholipase B, is as a diagnostic tool. Since it has been demonstrated that candidal phospholipase B is released during the progression of candidal infection in murine models of candidiasis [52], it is conceivable that this particular protein could be exploited as a diagnostic marker. Phospholipase B possesses a number of advantages which make it attractive for development as a diagnostic tool: (1) it is a naturally secreted protein; (2) it is a purified, well-defined antigen; and (3) blood isolates generally secrete much higher levels of enzyme than isolates from the oral cavity of healthy individuals [5] or isolates from wounds or urine [36]. Secretion of phospholipase B by a number of pathogenic fungi, including *Candida*, *Cryptococcus*, and *Aspergillus*, may pose a challenge for developing a highly specific and sensitive diagnostic test. However, this challenge could be circumvented as more information on genes encoding phospholipase B from specific pathogens become available and purification of these proteins is achieved.

B. Secretory Aspartic Proteinases

The availability of SAP inhibitor pepstatin A prompted studies to investigate its therapeutic efficacy. A murine model of disseminated candidiasis involving intranasal challenge with *C. albicans* was developed and used to explore the role of SAP in candidiasis and to assess their therapeutic utility [157]. Neutropenic mice pretreated with pepstatin A afforded strong dose-dependent protection against a subsequent lethal intranasal dose of a SAP-producing strain of *C. albicans*. This effect was comparable to the dose-dependent protection obtained with amphotericin B, which resulted in 100% survival. The reduction in mortality afforded by pepstatin A correlated with its dose-dependent blockade of *C. albicans* CFUs in the lungs, liver, and kidneys [157].

The effects of therapeutically relevant concentrations of the HIV proteinase inhibitors (PIs) saquinavir and indinavir on the in vitro proteinase activity of *C. albicans* were investigated with isolates from HIV-infected and uninfected patients with oral candidiasis. After exposure to the HIV proteinase inhibitors, proteinase activity was significantly reduced in a dose-dependent manner. These inhibitory effects, which were similar to that of pepstatin A, and the reduced virulence phenotype in experimental candidiasis after application of saquinavir indicate the usefulness of these HIV proteinase inhibitors as potential anticandidal agents [158]. Cassone et al. [159] studied the effect of indinavir and another PI—ritonavir—on the enzymatic activity of a SAP of *C. albicans* and on growth and experimental pathogenicity of this fungus. Both indinavir and ritonavir strongly ($\geq 90\%$) and dose-dependently ($0.1-10$ μM) inhibited SAP activity and production. They also significantly reduced *Candida* growth in a nitrogen-limited, SAP expression-dependent growth medium and exerted a therapeutic effect in an experimental model of vaginal candidiasis, with an efficacy comparable to that of fluconazole. Thus, besides the expected immunorestoration, patients receiving PI therapy may benefit from a direct anticandidal activity of these drugs [159]. Extensive studies in this area are going on.

Attempts have also been made to develop a reliable *SAP*-based method for the diagnosis of candidiasis. Flahaut et al. described a rapid detection method based on DNA amplification of common regions from *C. albicans*–secreted SAP genes [160]. The sensitivity of this method was 1 cell/mL from serially diluted *Candida* cultures and 1–4 cell/mL from seeded blood specimens [160]. In another study, the presence of viable cells of *C. albicans* in broth or in a reconstructed living skin equivalent was determined by the detection of amplicons of partial mRNA sequences of the genes encoding fungal actin (*ACT1*) and secreted aspartyl proteinase 2 (*SAP2*) [161]. The mRNA of both genes were amplified by reverse transcription-3' rapid amplification of cDNA ends-nested polymerase chain reaction. Single bands of *ACT1* (315 bp) and *SAP2* (162 bp) mRNA were amplified from total RNA extracts of *C. albicans*; only the former was amplified from Sabouraud broth-grown organisms. Primer pairs targeted for *ACT1* and *SAP2* were *Candida* genus-specific and *C. albicans*–specific, respectively. The sensitivity limits of the assay were 100 fg of total RNA or 10 cells of *C. albicans*, by ethidium bromide staining. Reverse transcription-3' rapid amplification of cDNA ends-nested polymerase chain reaction of the mRNA encoding specific proteins of an organism has potential application in determining the viability of the organism in tissue, thus monitoring the efficacy of an antimicrobial therapy, and in detecting mRNA expressed in very little amounts in tissue.

V. CONCLUSION

There have been many advances in the treatment and management of fungal infections, but still the rate of mortality is noticeably high. Fungal secretory proteins are among the chief virulence factors in these infections. The increasing volume of information about these proteins will enable us to develop novel and more effective strategies to treat fungal infections. These studies will also aid the development of early-diagnosis methods to detect candidiasis, thus greatly enhancing the chances of survival for a patient. Increasing advances in modern science will certainly enable mycologists to develop successful and effective therapeutic and diagnostic strategies to combat and conquer fungal infections.

ACKNOWLEDGMENTS

The work reviewed in this article which originated from the authors' laboratory was supported, in part, by grant AI35097 from the National Institutes of Health.

REFERENCES

1. CM Beck-Sagué, WR Jarvis. Secular trends in the epidemiology of nosocomial fungal infections in the United States, 1980–1990. National Nosocomial Infections Surveillance System. J Infect Dis 167:1247–1251, 1993.
2. MB Edmond, SE Wallace, DK McClish, MA Pfaller, RN Jones, RP Wenzel. Nosocomial bloodstream infections in United States hospitals: a three-year analysis. Clin Infect Dis 29:239–244, 1999.
3. B Hube, R Ruchel, M Monod, D Sanglard, FC Odds. Functional aspects of secreted *Candida* proteinases. Adv Exp Med Biol 436:339–344, 1998.
4. MA Ghannoum, KH Abu-Elteen. Pathogenicity determinants of *Candida*. Mycoses 33:265–282, 1990.
5. AS Ibrahim, F Mirbod, SG Filler, Y Banno, GT Cole, Y Kitajima, JE Edwards Jr, Y Nozawa, MA Ghannoum. Evidence implicating phospholipase as a virulence factor of *Candida albicans*. Infect Immun 63:1993–1998, 1995.
6. MA Pfaller, R Wenzel. Impact of the changing epidemiology of fungal infections in the 1990s. Eur J Clin Microbiol Infect Dis 11:287–291, 1992.
7. B Hube. *Candida albicans* secreted aspartyl proteinases. Curr Top Med Mycol 7:55–69, 1996.
8. TA Geers, SM Gordon. Clinical significance of *Candida* species isolated from cerebrospinal fluid following neurosurgery. Clin Infect Dis 28:1139–1147, 1999.
9. M Tumbarello, E Tacconelli, D de Gaetano, G Morace, G Fadda, R Cauda. Candidemia in HIV-infected subjects. Eur J Clin Microbiol Infect Dis 18:478–483, 1999.
10. C Viscoli, C Girmenia, A Marinus, L Collette, P Martino, B Vandercam, C Doyen, B Lebeau, D Spence, V Krcmery, B De Pauw, F Meunier. Candidemia in cancer patients: a prospective, multicenter surveillance study by the Invasive Fungal Infection Group (IFIG) of the European Organization for Research and Treatment of Cancer (EORTC) Clin Infect Dis 28:1071–1079, 1999.
11. SB Wey, M Mori, MA Pfaller, RF Woolson, RP Wenzel. Hospital acquired candidemia. The attributable mortality and excess length of stay. Arch Intern Med 148:2642–2645, 1988.
12. JE Cutler. Putative virulence factors of *Candida albicans*. Annu Rev Microbiol 45:187–218, 1991.
13. NH Georgopapadakou, TJ Walsh. Antifungal agents: chemotherapeutic targets and immunologic strategies. Antimicrob Agents Chemother 40:279–291, 1996.
14. JR Perfect. Fungal virulence genes as targets for antifungal chemotherapy. Antimicrob Agents Chemother 40:1577–1583, 1996.
15. R Hofbauer, D Moser, V Hammerschmidt, S Kapiotis, M Frass. Ketamine significantly reduces the migration of leukocytes through endothelial cell monolayers. Crit Care Med 26:1545–1549, 1998.
16. V Vidotto, G Picerno, S Caramello, G Paniate. Importance of some factors on the dimorphism of *Candida albicans*. Mycopathologia 104:129–135, 1988.

17. K Barrett-Bee, Y Hayes, R Wilson, J Ryley. A comparison of phospholipase activity, cellular adherence and pathogenicity of yeast. J Gen Microbiol 131:1217–1221, 1985.
18. MA Ghannoum, SS Radwan. *Candida* Adherence to Epithelial Cells. Boca Raton: CRC Press, 1990.
19. DS McKinsey, LJ Wheat, GA Cloud, M Pierce, JR Black, DM Bamberger, M Goldman, CJ Thomas, HM Gutsch, B Moskovitz, WE Dismukes, CA Kauffman. Itraconazole prophylaxis for fungal infections in patients with advanced human immunodeficiency virus infection: randomized, placebo-controlled, double-blind study. National Institute of Allergy and Infectious Diseases Mycoses Study Group. Clin Infect Dis 28:1049–1056, 1999.
20. CM Bendel, KM Kinneberg, RP Jechorek, CA Gale, SL Erlandsen, MK Hostetter, CL Wells. Systemic infection following intravenous inoculation of mice with *Candida albicans int1* mutant strains. Mol Genet Metab 67:343–351, 1999.
21. RL Taylor, MD Willcox, TJ Williams, J Verran. Modulation of bacterial adhesion to hydrogel contact lenses by albumin. Optom Vis Sci 75:23–29, 1998.
22. B Morrow, H Ramsey, DR Soll. Regulation of phase-specific genes in the more general switching system of *Candida albicans* strain 3153A. J Med Vet Mycol 32:287–294, 1994.
23. SB Wey, M Mori, MA Pfaller, RF Woolson, RP Wenzel. Risk factors for hospital-acquired candidemia. A matched case-control study. Arch Intern Med 149:2349–2353, 1989.
24. DH Walker, WT Firth, JG Ballard, BC Hegarty. Role of phospholipase-associated penetration mechanism in cell injury by *Rickettsia rickettsii*. Infect Immun 40:840–842, 1983.
25. LD Saffer, KS Long, JD Schwartzman. The role of phospholipase in host cell penetration by *Toxoplasma gondii*. Am J Trop Med Hyg 40:145–149, 1989.
26. LD Saffer, JD Schwartzman. A soluble phospholipase of *Toxoplasma gondii* associated with host cell penetration. J Protozool 38:454–460, 1991.
27. JI Ravdin, BY Croft, RL Guerrant. Cytopathogenic mechanisms of *Entamoeba histolytica*. J Exp Med 152:377–390, 1980.
28. SC Chen, M Muller, JZ Zhou, LC Wright, TC Sorrell. Phospholipase activity in *Cryptococcus neoformans:* a new virulence factor? J Infect Dis 175:414–420, 1997.
29. M Birch, G Robson, D Law, DW Denning. Evidence of multiple extracellular phospholipase activities of *Aspergillus fumigatus*. Infect Immun 64:751–755, 1996.
30. A Costa, C Costa, A Misefari, A Amato. On the enzymatic activity of certain fungi. VII. Phosphatidase activity on media containing sheep's blood of pathogenic strains of *Candida albicans*. Atti Soc Sci Fis Mat Nat XIV:93–101, 1968.
31. H Werner. Studies on the lipase activity in yeasts and yeast-like fungi. Zentralbl Bakteriol 200:113–124, 1966.
32. D Pugh, RA Cawson. The cytochemical localization of phospholipase a and lysophospholipase in *Candida albicans*. Sabouraudia 13(Pt 1):110–115, 1975.
33. D Pugh, RA Cawson. The cytochemical localization of phospholipase in *Candida albicans* infecting the chick chorio-allantoic membrane. Sabouraudia 15:29–35, 1977.
34. D Pugh, RA Cawson. The cytochemical localization of acid hydrolases in four common fungi. Cell Mol Biol 22:125–132, 1977.
35. FC Odds, AB Abbott. A simple system for the presumptive identification of *Candida albicans* and differentiation of strains within the species. Sabouraudia 18:301–317, 1980.
36. MF Price, ID Wilkinson, LO Gentry. Plate method for detection of phospholipase activity in *Candida albicans*. Sabouraudia 20:7–14, 1982.
37. H Hanel, I Menzel, H Holzmann. High phospholipase A-activity of *Candida albicans* isolated from the intestines of psoriatic patients. Mycoses 31:451–453, 1988.
38. T Lane, JR Garcia. Phospholipase production in morphological variants of *Candida albicans*. Mycoses 34:217–220, 1991.
39. MI Williamson, LP Samaranayake, TW MacFarlane. Phospholipase activity as a criterion for biotyping *Candida albicans*. J Med Vet Mycol 24:415–417, 1986.
40. LP Samaranayake, JM Raeside, TW MacFarlane. Factors affecting the phospholipase activity of *Candida species* in vitro. Sabouraudia 22:201–207, 1984.
41. E Habermann, KL Hardt. A sensitive and specific plate test for the quantitation of phospholipase. Anal Biochem 50:163–173, 1972.

42. SD Leidich, AS Ibrahim, Y Fu, W Fonzi, X Zhou, F Mirbod, S Nakashima, Y Nozawa, MA Ghannoum. Molecular, genetic and biochemical analysis of *Candida albicans PLB1* null mutant. Paper presented at Proceedings of 97th General Meeting, Washington, DC, 1997.

43. CJ Clancy, A Lewin, MA Ghannoum, MH Nguyen. Cloning of the *Candida glabrata* phospholipase B gene and construction of its disruptant mutant. Paper presented at Proceedings of 36th Annual Meeting, Denver, 1998.

44. Y Banno, T Yamada, Y Nozawa. Secreted phospholipases of the dimorphic fungus, *Candida albicans;* separation of three enzymes and some biological properties. Sabouraudia 23:47–54, 1985.

45. M Takahashi, Y Banno, Y Nozawa. Secreted *Candida albicans* phospholipases: purification and characterization of two forms of lysophospholipase-transacylase. J Med Vet Mycol 29:193–204, 1991.

46. F Mirbod, Y Banno, MA Ghannoum, AS Ibrahim, S Nakashima, Y Kitajima, GT Cole, Y Nozawa. Purification and characterization of lysophospholipase-transacylase (h-LPTA) from a highly virulent strain of *Candida albicans*. Biochim Biophys Acta 1257:181–188, 1995.

47. W Witt, HJ Bruller, G Falker, GF Fuhrmann. Purification and properties of a phospholipid acyl hydrolase from plasma membranes of *Saccharomyces cerevisiae*. Biochim Biophys Acta 711: 403–410, 1982.

48. W Witt, A Mertsching, E König. Secretion of phospholipase B from *Saccharomyces cerevisiae*. Biochim Biophys Acta 795:117–124, 1984.

49. W Witt, ME Schweingruber, A Mertsching. Phospholipase B from the plasma membrane of Saccharomyces cerevisiae. Separation of two forms with different carbohydrate content. Biochim Biophys Acta 795:108–116, 1984.

50. KS Lee, JL Patton, M Fido, LK Hines, SD Kohlwein, F Paltauf, SA Henry, DE Levin. The *Saccharomyces cerevisiae PLB1* gene encodes a protein required for lysophospholipase and phospholipase B activity. J Biol Chem 269:19725–19730, 1994.

51. K Saito, J Sugatani, T Okumura. Phospholipase B from *Penicillium notatum*. Methods Enzymol 197:446–456, 1991.

52. SD Leidich, AS Ibrahim, Y Fu, A Koul, C Jessup, J Vitullo, W Fonzi, F Mirbod, S Nakashima, Y Nozawa, MA Ghannoum. Cloning and disruption of ca*PLB1*, a phospholipase B gene involved in the pathogenicity of *Candida albicans*. J Biol Chem 273:26078–26086, 1998.

53. CI Hoover, MJ Jantapour, G Newport, N Agabian, SJ Fisher. Cloning and regulated expression of the *Candida albicans* phospholipase B (PLB1) gene. FEMS Microbiol Lett 167:163–169, 1998.

54. CF Lu, RC Montijn, JL Brown, F Klis, J Kurjan, H Bussey, PN Lipke. Glycosyl phosphatidylinositol-dependent cross-linking of alpha-agglutinin and beta 1,6-glucan in the *Saccharomyces cerevisiae* cell wall. J Cell Biol 128:333–340, 1995.

55. O Merkel, M Fido, JA Mayr, H Pruger, F Raab, G Zandonella, SD Kohlwein, F Paltauf. Characterization and function in vivo of two novel phospholipases B/lysophospholipases from *Saccharomyces cerevisiae*. J Biol Chem 274:28121–28127, 1999.

56. Y Sugiyama, S Nakashima, F Mirbod, H Kanoh, Y Kitajima, MA Ghannoum, Y Nozawa. Molecular cloning of a second phospholipase B gene, ca*PLB2* from Candida albicans. Med Mycol 37:61–67, 1999.

57. Y Watanabe, K Imai, H Oishi, Y Tamai. Disruption of phospholipase-B gene, *PLB1*, increases the survival of baker's yeast *Torulaspora delbrueckii*. FEMS Microbiol Lett 145:415–420, 1996.

58. RT Pickard, XG Chiou, BA Strifler, MR DeFelippis, PA Hyslop, AL Tebbe, YK Yee, LJ Reynolds, EA Dennis, RM Kramer, JD Sharp. Identification of essential residues for the catalytic function of 85 kDa cytosolic phospholipase A2. Probing the role of histidine, aspartic acid, cysteine, and arginine. J Biol Chem 271:19225–19231, 1996.

59. DE Bennett, CE McCreary, DC Coleman. Genetic characterization of a phospholipase C gene from *Candida albicans:* presence of homologous sequences in *Candida* species other than *Candida albicans*. Microbiology 144:55–72, 1998.

60. S Nakashima, Y Matsuda, Y Akao, S Yoshimura, H Sakai, K Hayakawa, M Andoh, Y Nozawa. Molecular cloning and chromosome mapping of rat phospholipase D genes, *Pld1a*, *Pld1b* and *Pld2*. Cytogenet Cell Genet 79:109–113, 1997.

61. M Waksman, X Tang, Y Eli, JE Gerst, M Liscovitch. Identification of a novel Ca2+-dependent, phosphatidylethanolamine-hydrolyzing phospholipase D in yeast bearing a disruption in *PLD1*. J Biol Chem 272:36–39, 1997.

62. M Waksman, Y Eli, M Liscovitch, JE Gerst. Identification and characterization of a gene encoding phospholipase D activity in yeast. J Biol Chem 271:2361–2364, 1996.

63. JH Exton. New developments in phospholipase D. J Biol Chem 272:15579–15582, 1997.

64. K Rose, SA Rudge, MA Frohman, AJ Morris, J Engebrecht. Phospholipase D signaling is essential for meiosis. Proc Natl Acad Sci USA 92:12151–12155, 1995.

65. N McLain, JW Dolan. Phospholipase D activity is required for dimorphie transition in *Candida albicans*. Microbiology 143:3521–3526, 1997.

66. H Kanoh, S Nakashima, Y Zhao, Y Sugiyama, Y Kitajima, Y Nozawa. Molecular cloning of a gene encoding phospholipase D from the pathogenic and dimorphic fungus, *Candida albicans*. Biochim Biophys Acta 1398:359–364, 1998.

67. SV Komshian, AK Uwaydah, JD Sobel, LR Crane. Fungemia caused by *Candida* species and *Torulopsis glabrata* in the hospitalized patient: frequency, characteristics, and evaluation of factors influencing outcome. Rev Infect Dis 11:379–390, 1989.

68. D Arzeni, M Del Poeta, O Simonetti, AM Offidani, L Lamura, M Balducci, N Cester, A Giacometti, G Scalise. Prevalence and antifungal susceptibility of vaginal yeasts in outpatients attending a gynecological center in Ancona, Italy. Eur J Epidemiol 13:447–450, 1997.

69. YC Chen, SC Chang, CC Sun, LS Yang, WC Hsieh, KT Luh. Secular trends in the epidemiology of nosocomial fungal infections at a teaching hospital in Taiwan, 1981 to 1993. Infect Control Hosp Epidemiol 18:369–375, 1997.

70. AL Colombo, M Nucci, R Salomao, ML Branchini, R Richtmann, A Derossi, SB Wey. High rate of non albicans candidemia in Brazilian tertiary care hospitals. Diagn Microbiol Infect Dis 34:281–286, 1999.

71. YK Kwok, YK Tay, CL Goh, A Kamarudin, MT Koh, CS Seow. Epidemiology and in vitro activity of antimycotics against candidal vaginal/skin/nail infections in Singapore. Int J Dermatol 37:145–149, 1998.

72. JA Vazquez, LM Dembry, V Sanchez, MA Vazquez, JD Sobel, C Dmuchowski, MJ Zervos. Nosocomial *Candida glabrata* colonization: an epidemiologic study. J Clin Microbiol 36:421–426, 1998.

73. S Mukherjee. *Torulopsis glabrata* esophagitis. Am J Gastroenterol 95:1106–1107, 2000.

74. J Schreiber, C Struben, W Rosahl, M Amthor. Hypersensitivity alveolitis induced by endogenous *Candida* species. Eur J Med Res 5:126, 2000.

75. JD Sobel. Vulvovaginitis due to *Candida glabrata*. An emerging problem. Mycoses 41(suppl 2):18–22, 1998.

76. SL Chuck, MA Sande. Infections with *Cryptococcus neoformans* in the acquired immunodeficiency syndrome. N Engl J Med 321:794–799, 1989.

77. RH Eng, E Bishburg, SM Smith, R Kapila. *Cryptococcal infections* in patients with acquired immune deficiency syndrome. Am J Med 81:19–23, 1986.

78. RM Selik, ET Starcher, JW Curran. Opportunistic diseases reported in AIDS patients: frequencies, associations, and trends. AIDS 1:175–182, 1987.

79. R Kappe, S Levitz, TS Harrison, M Ruhnke, NM Ampel, G Just-Nubling. Recent advances in cryptococcosis, candidiasis and coccidioidomycosis complicating HIV infection. Med Mycol 36(suppl 1):207–215, 1998.

80. J Wendisch, R Blaschke-Hellmessen, F Kaulen, R Schwarze, M Kabus. Lethal meningeal encephalitis from *Cryptococcus neoformans* var. neoformans in a girl without serious immunodeficiency. Mycoses 39(suppl 1):97–101, 1996.

81. M Gottfredsson, GM Cox, MA Ghannoum, JR Perfect. Molecular cloning of *Cryptococcus neoformans* phospholipase B gene, a putative virulence factor. Paper presented at Proceedings of ASM General Meeting, Atlanta, 1998.

82. A Koul, C Jessup, DJ Deluca, CJ Elnicky, MJ Nunez, RG Washburn, MA Ghannoum. Phospholipase production abilities of six *Aspergillus* species. Paper presented at Proceedings of ASM General Meeting, Atlanta, 1998.

83. GT Cole, KT Lynn, KR Seshan. An animal model for oropharyngeal, esophageal and gastric candidosis. Mycoses 33:7–19, 1990.
84. MA Ghannoum, B Spellberg, SM Saporito-Irwin, WA Fonzi. Reduced virulence of *Candida albicans PHR1* mutants. Infect Immun 63:4528–4530, 1995.
85. PK Mukherjee, KR Seshan, SD Leidich, J Chandra, GT Cole, MA Ghannoum: Testing of the ca*PLB1* revertant *Candida albicans* strain using hematogenously disseminated and oral-intragastric infant mice models of candidiasis confirm that phospholipase B is an important candidal virulence factor. *Microbiology* submitted:2001.
86. S Falkow. Molecular Koch's postulates applied to microbial pathogenicity. Rev Infect Dis 10(suppl 2):S274–S276, 1988.
87. MA Ghannoum. Extracellular phospholipases as universal virulence factor in pathogenic fungi. Nippon Ishinkin Gakkai Zasshi 39:55–59, 1998.
88. SA Klotz, DJ Drutz, JL Harrison, M Huppert. Adherence and penetration of vascular endothelium by *Candida* yeasts. Infect Immun 42:374–384, 1983.
89. C Scherwitz. Ultrastructure of human cutaneous candidosis. J Invest Dermatol 78:200–205, 1982.
90. PK Mukherjee, SD Leidich, J Vitullo, MA Ghannoum. Phospholipase B (PLB) enhances candidal penetration and damage of epithelial cells. Paper presented at Proceedings of 38th Interscience Conference on Antimicrobial Agents and Chemotherapy, San Diego, 1998.
91. V Vidotto, A Sinicco, D Di Fraia, S Cardaropoli, S Aoki, S Ito-Kuwa. Phospholipase activity in *Cryptococcus neoformans.* Mycopathologia 136:119–123, 1996.
92. SC Chen, LC Wright, RT Santangelo, M Muller, VR Moran, PW Kuchel, TC Sorrell. Identification of extracellular phospholipase B, lysophospholipase, and acyltransferase produced by *Cryptococcus neoformans.* Infect Immun 65:405–411, 1997.
93. SC Chen, LC Wright, JC Golding, TC Sorrell. Purification and characterization of secretory phospholipase B, lysophospholipase and lysophospholipase/transacylase from a virulent strain of the pathogenic fungus *Cryptococcus neoformans.* Biochem J 347:431–439, 2000.
94. JG Songer. Bacterial phospholipases and their role in virulence. Trends Microbiol 5:156–161, 1997.
95. CN Serhan, JZ Haeggstrom, CC Leslie. Lipid mediator networks in cell signaling: update and impact of cytokines. FASEB J 10:1147–1158, 1996.
96. AE Bryant, DL Stevens. Phospholipase C and perfringolysin O from *Clostridium perfringens* upregulate endothelial cell-leukocyte adherence molecule 1 and intercellular leukocyte adherence molecule 1 expression and induce interleukin-8 synthesis in cultured human umbilical vein endothelial cells. Infect Immun 64:358–362, 1996.
97. L Eckmann, SL Reed, JR Smith, MF Kagnoff. *Entamoeba histolytica* trophozoites induce an inflammatory cytokine response by cultured human cells through the paracrine action of cytolytically released interleukin-1α. J Clin Invest 96:1269–1279, 1995.
98. AK May, RG Sawyer, T Gleason, A Whitworth, TL Pruett. In vivo cytokine response to *Escherichia coli* α-hemolysin determined with genetically engineered hemolytic and nonhemolytic *E. coli* variants. Infect Immun 64:2167–2171, 1996.
99. DJ Meyers, RS Berk. Characterization of phospholipase C from *Pseudomonas aeruginosa* as a potent inflammatory agent. Infect Immun 58:659–666, 1990.
100. TS Walker, JS Brown, CS Hoover, DA Morgan. Endothelial prostaglandin secretion: effects of typhus rickettsiae. J Infect Dis 162:1136–1144, 1990.
101. B Hube, D Sanglard, M Schaller, A Ibrahim, FC Odds, NA Gow. What functions do six different genes for secretory proteinases have in *Candida albicans?* Mycoses 41(suppl 1):47–50, 1998.
102. F Staib. Serum proteins as nitrogen source for yeast-like fungi. Sabouraudia 4:187–193, 1965.
103. TL Ray, CD Payne. Comparative production and rapid purification of *Candida* acid proteinase from protein supplemented cultures. Infect Immun 58:508–514, 1990.
104. M Negi, R Tsuboi, T Matsui, H Ogawa. Isolation and characterization of proteinase from *Candida albicans:* substrate specificity. J Invest Dermatol 83:32–36, 1984.
105. R Tsuobi, Y Kurita, M Negi, H Ogawa. A specific inhibitor of keratinolytic proteinase from *Candida albicans* could inhibit the cell growth of *C. albicans.* J Invest Dermatol 85:438–440, 1985.
106. B Morrow, T Srikantha, DR Soll. Transcription of the gene for a pepsinogen, *PEP1,* is regulated by white opaque switching in *Candida albicans.* Mol Cell Biol 12:2997–3005, 1992.

107. BB Magee, B Hube, RJ Wright, PJ Sullivan, PT Magee. The genes encoding the secreted aspartyl proteinases of *Candida albicans* constitute a family with at least three members. Infect Immun 61: 3240–3243, 1993.

108. TC White, SH Miyasaki, N Agabian. Three distinct secreted aspartyl proteinases in *Candida albicans*. J Bacteriol 175:6126–6133, 1993.

109. F Macdonald. Secretion of inducible proteinase by pathogenic *Candida* species. Sabouraudia 22: 79–82, 1984.

110. M Ghannoum, EK Abu. Correlative relationship between proteinase production, adherence and pathogenicity of various strains of *Candida albicans*. J Med Vet Mycol 24:407–413, 1986.

111. A Cassone, F De Bernardis, F Mondello, T Ceddia, L Agatensi. Evidence for a correlation between proteinase secretion and vulvovaginal candidosis. J Infect Dis 156:777–783, 1987.

112. KJ Kwon-Chung, D Lehman, C Good, PT Magee. Genetic evidence for role of extracellular proteinase in virulence of *Candida albicans*. Infect Immun 49:571–575, 1985.

113. F Macdonald, FC Odds. Virulence for mice of a proteinase secreting strain of *Candida albicans* and a proteinase deficient mutant. J Gen Microbiol 129:431–438, 1983.

114. IK Ross, F De Bernardis, GW Emerson, A Cassone, PA Sullivan. The secreted aspartate proteinase of *Candida albicans:* physiology of secretion and virulence of a proteinase-deficient mutant. J Gen Microbiol 136:687–694, 1990.

115. R Ruchel. Properties of a purified proteinase from the yeast *Candida albicans*. Biochim Biophys Acta 659:99–113, 1981.

116. R Rüchel, B Ritter, M Schaffrinski. Modulation of experimental systemic murine candidosis by intravenous pepstatin. Zentralbl Bakteriol 273:391–403, 1990.

117. H Remold, H Fasold, F Staib. Purification and characterization of a proteolytic enzyme from *Candida albicans*. Biochim Biophys Acta 167:399–406, 1968.

118. F Macdonald, FC Odds. Inducible proteinase of *Candida albicans* in diagnostic serology and in the pathogenesis of systemic candidosis. J Med Microbiol 13:423–435, 1980.

119. CJ Morrison, SF Hurst, SL Bragg, RJ Kuykendall, H Diaz, DW McLaughlin, E Reiss. Purification and characterization of the extracellular aspartyl proteinase of *Candida albicans:* removal of extraneous proteins and cell wall mannoprotein and evidence for lack of glycosylation. J Gen Microbiol 139:1177–1186, 1993.

120. R Ruchel, B Boning, M Borg. Characterization of a secretory proteinase of *Candida parapsilosis* and evidence for the absence of the enzyme during infection in vitro. Infect Immun 53:411–419, 1986.

121. B Hube, CJ Turver, FC Odds, H Eiffert, GJ Boulnois, H Kochel, R Ruchel. Identification, cloning and characterization of the gene for the secretory aspartate protease of *Candida albicans*. Mycoses 34(suppl 1):59–61, 1991.

122. RJ Wright, A Carne, AD Hieber, IL Lamont, GW Emerson, PA Sullivan. A second gene for a secreted aspartate proteinase in *Candida albicans*. J Bacteriol 174:7848–7853, 1992.

123. SH Miyasaki, TC White, N Agabian. A fourth secreted aspartyl proteinase gene (*SAP4*) and a *CARE2* repetitive element are located upstream of the *SAP1* gene in *Candida albicans*. J Bacteriol 176:1702–1710, 1994.

124. M Monod, G Togni, B Hube, D Sanglard. Multiplicity of genes encoding secreted aspartic proteinases in *Candida* species. Mol Microbiol 13:357–368, 1994.

125. M Monod, B Hube, D Hess, D Sanglard. Differential regulation of *SAP8* and *SAP9*, which encode two new members of the secreted aspartic proteinase family in *Candida albicans*. Microbiology 144:2731–2737, 1998.

126. JO Capobianco, CG Lerner, RC Goldman. Application of a fluorogenic substrate in the assay of proteolytic activity and in the discovery of a potent inhibitor of *Candida albicans* aspartic proteinase. Anal Biochem 204:96–102, 1992.

127. A Banerjee, K Ganesan, A Datta. Induction of secretory acid proteinase in *Candida albicans*. J Gen Microbiol 137:2455–2461, 1991.

128. CG Lerner, RC Goldman. Stimuli that induce production of *Candida albicans* extracellular aspartyl proteinase. J Gen Microbiol 139:1643–1651, 1993.

129. M Homma, T Kanbe, H Chibana, K Tanaka. Detection of intracellular forms of secretory aspartic proteinase in *Candida albicans*. J Gen Microbiol 138:627–633, 1992.

130. B Hube, M Monod, DA Schofield, AJ Brown, NA Gow. Expression of seven members of the gene family encoding secretory aspartyl proteinases in *Candida albicans*. Mol Microbiol 14:87–99, 1994.

131. M Schaller, W Schafer, HC Korting, B Hube. Differential expression of secreted aspartyl proteinases in a model of human oral candidosis and in patient samples from the oral cavity. Mol Microbiol 29:605–615, 1998.

132. JD Lee, PE Kolattukudy. Molecular cloning of the cDNA and gene for an elastinolytic aspartic proteinase from *Aspergillus fumigatus* and evidence of its secretion by the fungus during invasion of the host lung. Infect Immun 63:3796–3803, 1995.

133. U Reichard, H Eiffert, R Ruchel. Purification and characterization of an extracellular aspartic proteinase from *Aspergillus fumigatus*. J Med Vet Mycol 32:427–436, 1994.

134. U Reichard, M Monod, R Ruchel. Expression pattern of aspartic proteinase antigens in aspergilli. Mycoses 39:99–101, 1996.

135. ME Kambouris, U Reichard, NJ Legakis, A Velegraki. Sequences from the aspergillopepsin PEP gene of *Aspergillus fumigatus:* evidence on their use in selective PCR identification of Aspergillus species in infected clinical samples. FEMS Immunol Med Microbiol 25:255–264, 1999.

136. U Reichard, M Monod, F Odds, R Ruchel. Virulence of an aspergillopepsin-deficient mutant of *Aspergillus fumigatus* and evidence for another aspartic proteinase linked to the fungal cell wall. J Med Vet Mycol 35:189–196, 1997.

137. FC Odds. *Candida albicans* proteinase as a virulence factor in the pathogenesis of *Candida* infections. Zentralbl Bakteriol Mikrobiol Hyg [A] 260:539–542, 1985.

138. N Lehrer, E Segal, RL Cihlar, RA Calderone. Pathogenesis of vaginal candidiasis: studies with a mutant which has reduced ability to adhere in vitro. J Med Vet Mycol 24:127–131, 1986.

139. AS Ibrahim, SG Filler, D Sanglard, JEJ Edwards, B Hube. Secreted aspartyl proteinases and interactions of *Candida albicans* with human endothelial cells. Infect Immun 66:3003–3005, 1998.

140. MA Ghannoum. *Candida albicans* antifungal-resistant strains: studies on adherence and other pathogenicity related characteristics. Mycoses 35:131–139, 1992.

141. TL Ray, CD Payne. Scanning electron microscopy of epidermal adherence and cavitation in murine candidiasis: a role for *Candida* acid proteinase. Infect Immun 56:1942–1949, 1988.

142. M Borg, R Ruchel. Expression of extracellular acid proteinase by proteolytic *Candida* spp. during experimental infection of oral mucosa. Infect Immun 56:626–631, 1988.

143. TC White, N Agabian. *Candida albicans* secreted aspartyl proteinases: isoenzyme pattern is determined by cell type, and levels are determined by environmental factors. J Bacteriol 177:5215–5221, 1995.

144. F De Bernardis, P Chiani, M Ciccozzi, G Pellegrini, T Ceddia, G D'Offizzi, I Quinti, PA Sullivan, A Cassone. Elevated aspartic proteinase secretion and experimental pathogenicity of *Candida albicans* isolates from oral cavities of subjects infected with human immunodeficiency virus. Infect Immun 64:466–471, 1996.

145. F De Bernardis, S Arancia, L Morelli, B Hube, D Sanglard, W Schafer, A Cassone. Evidence that members of the secretory aspartyl proteinase gene family, in particular *SAP2*, are virulence factors for *Candida* vaginitis. J Infect Dis 179:201–208, 1999.

146. N Dubois, AR Colina, F Aumont, P Belhumeur, L de Repentigny. Overexpression of *Candida albicans* secretory aspartyl proteinase 2 and its expression in *Saccharomyces cerevisiae* do not augment virulence in mice. Microbiology 144:2299–2310, 1998.

147. R Ruchel, F Zimmermann, B Boning-Stutzer, U Helmchen. Candidiasis visualised by proteinase-directed immunofluorescence. Virchows Arch A Pathol Anat Histopathol 419:199–202, 1991.

148. BK Na, GT Chung, CY Song. Production, characterization, and epitope mapping of a monoclonal antibody against aspartic proteinase of *Candida albicans*. Clin Diagn Lab Immunol 6:429–433, 1999.

149. M Schaller, C Schackert, HC Korting, E Januschke, B Hube. Invasion of *Candida albicans* correlates with expression of secreted aspartic proteinases during experimental infection of human epidermis. J Invest Dermatol 114:712–717, 2000.

150. ED Williamson, RW Titball. A genetically engineered vaccine against the α-toxin of *Clostridium perfringens* protects mice against experimental gas gangrene. Vaccine 11:1253–1258, 1993.

151. RW Titball, AM Fearn, ED Williamson. Biochemical and immunological properties of the C-terminal domain of the α-toxin of *Clostridium perfringens*. FEMS Microbiol Lett 110:45–50, 1993.

152. D Pavliakova, JS Moncrief, DM Lyerly, G Schiffman, DA Bryla, JB Robbins, R Schneerson. *Clostridium difficile* recombinant toxin A repeating units as a carrier protein for conjugate vaccines: studies of pneumococcal type 14, *Escherichia coli* K1, and *Shigella flexneri* type 2a polysaccharides in mice. Infect Immun 68:2161–2166, 2000.

153. F Roth, K Jansen, S Petzke. Detection of neutralizing antibodies against α-toxin of different *Clostridium septicum* strains in cell culture. FEMS Immunol Med Microbiol 24:353–359, 1999.

154. FA Uzal, JP Wong, WR Kelly, J Priest. Antibody response in goats vaccinated with liposome adjuvanted *Clostridium perfringens* type D epsilon toxoid. Vet Res Commun 23:143–150, 1999.

155. AL Hodgson, J Krywult, LA Corner, JS Rothel, AJ Radford. Rational attenuation of *Corynebacterium pseudotuberculosis:* potential cheesy gland vaccine and live delivery vehicle. Infect Immun 60:2900–2905, 1992.

156. H Hanel, R Kirsch, HL Schmidts, H Kottmann. New systematically active antimycotics from the beta blocker category. Mycoses 38:251–264, 1995.

157. K Fallon, K Bausch, J Noonan, E Huguenel, P Tamburini. Role of aspartic proteases in disseminated *Candida albicans* infection in mice. Infect Immun 65:551–556, 1997.

158. HC Korting, M Schaller, G Eder, G Hamm, U Bohmer, B Hube. Effects of the human immunodeficiency virus (HIV) proteinase inhibitors saquinavir and indinavir on in vitro activities of secreted aspartyl proteinases of *Candida albicans* isolates from HIV infected patients. Antimicrob Agents Chemother 43:2038–2042, 1999.

159. A Cassone, F De Bernardis, A Torosantucci, E Tacconelli, M Tumbarello, R Cauda. In vitro and in vivo anticandidal activity of human immunodeficiency virus protease inhibitors. J Infect Dis 180: 448–453, 1999.

160. M Flahaut, D Sanglard, M Monod, J Bille, M Rossier. Rapid detection of *Candida albicans* in clinical samples by DNA amplification of common regions from *C. albicans* secreted aspartic proteinase genes. J Clin Microbiol 36:395–401, 1998.

161. CN Okeke, R Tsuboi, M Kawai, M Yamazaki, S Reangchainam, H Ogawa. Reverse transcription-3′ rapid amplification of cDNA ends-nested PCR of *ACT1* and *SAP2* mRNA as a means of detecting viable *Candida albicans* in an in vitro cutaneous candidiasis model. J Invest Dermatol 114:95–100, 2000.

162. B Hube, CJ Turver, FC Odds, H Eiffert, GJ Boulnois, H Kochel, R Ruchel. Sequence of the *Candida albicans* gene encoding the secretory aspartate proteinase. J Med Vet Mycol 29:129–132, 1991.

4

Histoplasma capsulatum: Diary of an Intracellular Survivor

Simon L. Newman
University of Cincinnati College of Medicine, Cincinnati, Ohio

I. INTRODUCTION

Histoplasma capsulatum (Hc) is a dimorphic fungal pathogen of worldwide importance that causes a broad spectrum of disease activity. In the United States, Hc is endemic to the Ohio and Mississippi River valleys, and it has been estimated that 500,000 new infections occur in the United States each year [1]. Although the course of infection is mild in immunocompetent individuals, Hc may produce progressive disseminated infections in individuals immunocompromised by hematologic malignancies [2–4], cytotoxic therapy [5–7], or in individuals infected with human immunodeficiency virus (HIV) [8–10].

Hc lives in the saprophytic stage as a mycelial form consisting of hyphae which bear both macro- and microconidia [11,12]. Infection with Hc occurs via inhalation of microconidia or small mycelial fragments (5–8 μm) into the terminal bronchioles and alveoli of the lung. Inhaled conidia then convert into the yeast form that is responsible for the pathogenesis of histoplasmosis. Hc yeasts are phagocytized by alveolar macrophages (AM), within which they multiply [13–16]. Presumably, the dividing yeasts destroy the AM, and subsequently are ingested by other resident AM and by inflammatory phagocytes recruited to the loci of infection. Repetition of this cycle results in spread of infection to hilar lymph nodes and transient dispersal of yeasts to other organs during the acute phase of primary histoplasmosis (1–2 weeks). Thereafter, the maturation of specific cell-mediated immunity (CMI) against Hc activates MΦ to stop yeast proliferation with gradual resolution of the disease process [17,18].

The interaction of MΦ with Hc is a key event in the pathogenesis of histoplasmosis. MΦ first provide an environment for fungal replication and dissemination, and then subsequently act as the final effector cells to remove the organism from the host. Thus, the main focus of this chapter will be to review specifically what is known about the strategies of Hc yeasts to survive within human and animal MΦ, and how activated MΦ counter that strategy. Possible roles for polymorphonuclear neutrophils and dendritic cells in host defense against Hc yeasts also will be discussed at the end of the chapter.

II. RECOGNITION AND PHAGOCYTOSIS OF *H. CAPSULATUM*

During the early phase of infection, MΦ recognize, bind, and ingest Hc, thereby providing them access to a permissive intracellular environment for replication. This initial interaction between

Hc and MΦ is crucial in the pathogenesis of histoplasmosis. Binding of Hc yeasts to human cultured monocyte-derived MΦ is temperature-dependent being optimum at 37°C, and requires the presence of the divalent cations Ca and Mg [19].

MΦ recognition of unopsonized Hc yeasts and microconidia is mediated via the CD18 family of adhesion promoting glycoproteins (LFA-1 [CD11a], CR3 [CD11b], p150,95 [CD11c]) [19,20]. Each receptor molecule of the CD18 family contains a unique α-chain subunit noncovalently linked to a common β-chain subunit [21], and experiments using α- and β-chain–specific monoclonal antibodies (Mabs) suggest that the yeasts bind independently to each of the three receptors. In addition, the receptors must be mobile within the MΦ membrane for efficient attachment of yeasts, and this mobility requires intact microfilaments of the cellular cytoskeleton. Thus, preincubation of MΦ with cytochalasin D completely inhibits the binding of Hc yeasts to MΦ. Further studies using anti-CD11/CD18 Mabs demonstrate that CD18 receptors also mediate the attachment of unopsonized Hc yeasts to human AM [20] and to human neutrophils (PMN) [22]. Other MΦ receptors, such as Fc receptors, mannose-fucose receptors, or β-glucan receptors, do not appear to be involved in recognition of Hc yeasts, nor does there appear to be a role for MΦ-derived complement components [19]. Although similar studies have not been performed with murine MΦ, these cells also possess CD18 receptors [21], and, most likely, they are used in recognition of Hc.

Once bound, phagocytosis of unopsonized Hc yeasts by human cultured MΦ and freshly adherent AM is rapid, and both cell populations ingest an equivalent number of yeasts. After 1–2 hr of incubation, 85–95% of MΦ contain ingested yeasts, indicating that yeasts are not phagocytosed by a subpopulation of MΦ. In comparison to MΦ, monocytes ingest significantly fewer yeasts. However, in all three cell populations, 75–80% of attached yeasts are ingested at all time points [20]. Thus, the rate of phagocytosis of yeasts by monocyte/MΦ is determined by the rate of attachment of the yeasts to the phagocyte membrane.

Human cultured MΦ and AM also phagocytize unopsonized microconidia to a similar extent, and MΦ phagocytize microconidia almost as efficiently as yeasts [20]. Peritoneal MΦ (PM) and AM from outbred swiss albino mice also phagocytose Hc microconidia and small hyphal fragments at rates comparable to yeasts [23]. In both human and murine MΦ the microconidia convert into yeasts very rapidly (2–6 hr) after phagocytosis, whereas conversion of conidia into yeasts in tissue culture medium takes 2–3 days (S.L. Newman, unpublished observations) [23]. Thus, the phagosome of the MΦ provides microconidia with a favorable environment in which to transform into the pathogenic yeast phase.

III. STIMULATION OF THE RESPIRATORY BURST AND PHAGOLYSOSOMAL (PL) FUSION

An immediate consequence of the phagocytosis of microorganisms by MΦ is the stimulation of the cell's microbicidal armamentarium, that consists of the respiratory burst and the fusion of lysosomes with the phagocytic vacuole. Thus, a micro-organism which is confined in the phagocytic vacuole is directly exposed to toxic oxygen metabolites and various lysosomal hydrolases and cationic peptides which destroy the organism. Intracellular parasites have evolved several ways of evading these MΦ defense mechanisms including inhibition of PL fusion [24–27], escaping from the phagocytic vacuole into the cytoplasm [28], and initiating phagocytosis without stimulating the respiratory burst [29–31].

Although Hc yeasts can be killed in vitro by a combination of H_2O_2, Fe^{2+}, and iodide [32], or H_2O_2, horseradish peroxidase, and iodide [33,34], toxic oxygen radicals do not appear to mediate antifungal activity against Hc yeasts in phagocytic cells of either human or murine origin. First, Hc yeasts activate the respiratory burst of human monocyte/MΦ and neutrophils

Table 1   Macrophage Interaction with Hc Yeasts

Response	Source of macrophages			
	Human	MPM[a]	P388D1[b]	RAW[b]
Recognition via CD18	+	?	?	?
Respiratory burst	+	—	—	?
Phagolysosomal fusion	—	+	+	—
Intracellular replication	+	+	+	+
Intracellular pH 6.0–6.5	+	?	+	+
Activated by IFNγ	—	+	—[c]	+
Activated by IL-3, GM-CSF, and M-CSF	+	?	?	?
Activated by adherence to collagen matrix	+	?	?	?

[a] Mouse peritoneal macrophages.
[b] Mouse macrophage cell lines.
[c] Presumably because these cells do not make NO.

upon phagocytosis [19,22,32], but are not killed [32,35,36]. Second, the intracellular growth of Hc yeasts proceeds at a similar rate both in freshly isolated monocytes and in cultured MΦ [36] that have lost their myeloperoxidase (MPO) and exhibit a decreased ability to produce toxic oxygen metabolites [37–39]. Third, phagocytosis of serum-opsonized Hc yeasts by human neu-trophils stimulates a strong respiratory burst, and superoxide anion (O_2^-) becomes trapped within the phagocytic vacuole [22]. Thus, yeasts should be exposed directly to toxic oxygen metabolites. However, all of the fungistatic activity of human neutrophils is mediated by the contents of the azurophil granules [40].

In contrast to human MΦ, phagocytosis of unopsonized Hc yeasts by resident [41,42] or IFNγ activated [43] mouse PM does *not* stimulate the respiratory burst (Table 1). Phagocytosis of yeasts opsonized in fresh or heat-inactivated normal mouse serum or Hc immune serum does stimulate the respiratory burst of mouse PM [42], but the intracellular growth of yeasts still is not impaired [44].

The occurrence or nonoccurrence of PL-fusion upon phagocytosis of Hc yeasts by MΦ, in contrast to stimulation of the respiratory burst, does not sort out according to species (Table 1). Thus, phagocytosis of unopsonized Hc yeasts by the mouse P388D1 MΦ cell line [45] and resident PM [46] leads to normal PL fusion. In contrast, there is minimal PL fusion after ingestion of Hc yeasts by human MΦ [47] or the RAW 264.7 mouse MΦ cell line [48]. As Hc yeasts multiply readily within all of these MΦ populations, the yeasts apparently are able to evade the normally destructive effects of MΦ lysosomal hydrolases. This observation is discussed in more detail below.

IV.   STRATEGY FOR INTRACELLULAR SURVIVAL IN MACROPHAGES

In vitro studies on the intracellular fate of Hc within MΦ demonstrate that yeasts multiply readily within rabbit, mouse, and human AM [23,35,36,49], mouse, guinea pig, and human PM [35,44,50–52], and human monocytes and cultured MΦ [35,36]. The intracellular generation time of yeasts in mouse and guinea pig PM is 10–12 hr [44,52], whereas Hc yeasts multiply within human monocyte/MΦ with intracellular generation times of 14–20 hr [36]. Furthermore, although serum clearly is not required for phagocytosis of Hc yeasts by MΦ, opsonization of the yeasts with normal or Hc immune serum also does not effect the intracellular growth of Hc yeasts in either mouse or guinea pig PM [44] or in human MΦ [36].

Interestingly, monocyte-derived MΦ from individuals with HIV infection are profoundly deficient in their capacity to recognize and bind Hc yeasts, but ingestion of bound yeasts is normal, and decreased binding is associated with low CD4+ T-cell counts. Moreover, MΦ from 38% of HIV+ persons demonstrate increased permissiveness for intracellular growth compared to MΦ from HIV-negative controls [53]. Defects in the recognition and intracellular growth of Hc yeasts also is observed in normal MΦ either infected with HIV in vitro, or cultured in HIV+ serum. These defects in MΦ function are caused, in part, by the HIV envelope glycoprotein gp120. Thus, culture of normal MΦ in the presence of recombinant gp120 inhibits phagocytosis of Hc yeasts, but does not cause more rapid intracellular replication of Hc [54]. Furthermore, adsorption of gp120 from the serum of HIV+ patients removes the capacity of the serum to cause a MΦ defect in phagocytosis, but does not affect the capacity of the serum to cause accelerated intracellular replication [54]. This latter defect most likely is caused by another, as yet unknown, serum component(s) that is released into serum by HIV-infected cells.

The strategies used by Hc yeasts to adapt to the intracellular environment of human and murine MΦ have begun to be elucidated over the past several years. As discussed above, Hc yeasts survive and multiply within the phagocytic vacuoles of MΦ despite the stimulation of the respiratory burst and/or phagolysosomal fusion. In human [55] and mouse MΦ [50] that have ingested killed Hc yeasts, the organisms are digested completely, demonstrating that the yeast's thick cell wall is not sufficient by itself to enable this organism to survive within MΦ phagosomes. Further, the addition of the protein synthesis inhibitor cycloheximide to human MΦ infected with viable Hc yeasts leads to killing and digestion of the fungus [55]. As cycloheximide inhibits protein synthesis in both yeasts [56] and mammalian cells [57], the data demonstrate that synthesis of new proteins by Hc is required for intracellular survival and growth, and, conversely, that MΦ do *not* need to synthesize new proteins to kill and digest the yeasts. Thus, Hc yeasts presumably are killed and degraded by the fusion of preformed lysosomal hydrolases with phagocytic vacuoles.

Despite the fact that Hc yeasts must synthesize proteins to survive within MΦ, the yeasts do not secrete a product(s) that mediates a global inhibition of MΦ fungicidal mechanisms. Thus, in human MΦ that phagocytosed both viable yeasts and FITC-labeled heat-killed yeasts, viable yeasts survived and multiplied intracellularly over a period of 24 hr, whereas all of the heat-killed yeasts were digested [55]. Further, in P388D1 mouse MΦ that contained both Hc yeasts and fluorescent-labeled zymosan particles, phagosomes containing zymosan acidified normally, whereas phagosomes containing Hc maintained a relatively neutral pH of 6.5 [58]. Thus, the survival strategy of Hc yeasts focuses on modifying its phagosomal environment.

In mouse P388D1 MΦ [58] and human MΦ (S.L. Newman, unpublished observations), Hc yeasts maintain an intra-phagosomal pH of 6.5. Thus, the yeasts may avoid killing by lysosomal acid hydrolases. In addition, the yeasts would still be able to acquire iron from transferrin, because at pH 6.5 transferrin is half-saturated [59,60], leaving some free iron available for yeast survival and growth. Evidence supporting this idea comes from experiments designed to upset this strategy. Thus, chloroquine, which prevents release of iron from transferrin by raising endocytic and lysosomal pH [61], induces human MΦ to kill Hc yeasts [62]. The effect of chloroquine is reversed by iron nitrilo-acetate (FeNTA), an iron compound that is soluble at neutral to alkaline pH [63], but not by holotransferrin, which releases iron only in an acidic environment.

Most remarkably, chloroquine given intraperitoneally for 6 days to Hc-infected C57BL/6 mice significantly reduces the growth of Hc in spleens and livers in a dose-dependent manner.

Furthermore, treatment with chloroquine for 10 days following a lethal inoculum of Hc protected six of nine mice, whereas all control mice were dead by day 11 [62]. Thus, iron is of critical importance to the survival and multiplication of Hc yeasts in vitro in human MΦ, and in vivo in a murine model of histoplasmosis.

The molecules that Hc yeasts use to acquire iron within MΦ are unknown. It is known that Hc makes hydroxamic acid siderophores [64], but under iron starvation conditions in tissue culture media, these siderophores are not detected until after four days of culture, at which time the yeasts are in early stationary phase [65]. As Hc must respond to the hostile environment of the MΦ phagosome immediately, it is unlikely that these siderophores are used by Hc to acquire iron from transferrin. Further, in Hc-infected murine PM [66] and human MΦ (62), the iron chelator deferoxamine suppresses the intracellular growth of yeasts, and this effect is reversed by iron saturated transferrin (holotransferrin). Thus, whatever molecules are released by Hc yeasts to scavenge iron, free iron must be readily available for the yeasts to survive and multiply.

Most recently, Hc yeasts were found to produce an enzymatic ferric reductase and a nonproteinaceous ferric reductant, both of which are regulated by iron availability [67]. Whether these particular ferric reductases play a role in the intracellular survival of Hc yeasts remains to be determined. Indeed, in the environment of the MΦ phagosome, it is not yet even known whether ferric iron must be reduced to ferrous iron to be taken up by Hc yeasts.

The vacuolar ATPase (V-ATPase) is the enzyme that MΦ use to generate and maintain an acidic pH in phagosomal and endosomal pathways [68–70]. To maintain a pH of 6.5, Hc might alkalinize the phagosome to counter acidification by the V-ATPase. This mechanism might operate in murine PM or in P388D1 MΦ in which PL fusion occurs normally. Alternatively, Hc might inhibit the accumulation of functional V-ATPase into the phagosomal membrane either by inactivating the V-ATPase present in the phagosomal membrane (by direct interactions or by enhancing degradation of the enzyme), or by inhibiting fusion with vesicles containing the V-ATPase. In this case, the yeasts would be required to slightly acidify their environment to maintain a pH of 6.5.

Indeed, this latter mechanism appears to be operative in RAW 264.7 MΦ infected with Hc yeasts. Thus, bafilomycin, an inhibitor of the V-ATPase, does not inhibit the intracellular replication of Hc yeasts in RAW MΦ, nor does it affect the pH of Hc-containing phagosomes that is around 6.0. Furthermore, Hc-containing phagosomes have decreased levels of the phagosomal V-ATPase compared to phagosomes containing *Saccacharomyces cerevisae* (Sc) [48]. We hypothesize that this mechanism also will be operative in human MΦ, as PL fusion does not occur in these cells [47], nor in RAW MΦ infected with Hc yeasts [48]. Thus, Hc yeasts within MΦ phagosomes avoid destruction by preventing the insertion of the V-ATPase into the phagosomal membrane, and maintaining an intraphagosomal pH of 6.0–6.5 by an as yet unknown mechanism. This pH allows the fungus to obtain host iron, and, in those MΦ in which PL fusion does take place, the lysosomal hydrolases presumably are neutralized.

Hc yeasts also require an intact pyrimidine synthetic pathway for intracellular growth in MΦ and for virulence in mice. Inactivation of the *URA5* gene that encodes orotidine-5′-monophosphate pyrophosphorylase results in disruption of the pyrimidine biosynthetic pathway and uracil auxotrophy. These *URA5* "knockout" yeasts are avirulent in vitro both in murine RAW MΦ and in the U937 human monocyte line, as well as in vivo in mice. In fulfillment of Koch's postulates on a molecular level, virulence is restored either by supplying a functional *URA5* gene, or by adding exogenous uracil to Hc-infected MΦ [71].

In addition to the necessity to acquire iron, the acquisition of intracellular calcium also appears to be a prerequisite for the intracellular survival of Hc yeasts in MΦ. Hc yeasts, but not mycelia, release large quantities of a calcium-binding protein (CBP1) into liquid culture during exponential growth, and this CBP also allows yeasts to replicate in the presence of EGTA

[72]. In addition, CBP1 is expressed when Hc yeasts replicate within P388D1 MΦ and hamster tracheal epithelial cells. Further, CPB1 apparently is secreted in vivo in a murine model of histoplasmosis, as splenocytes from Hc-infected mice, but not uninfected mice demonstrate a CBP1-specific proliferative response [73]. Most importantly, a *CBP1* knockout strain is defective in broth culture growth when calcium is limited, is unable to colonize the lungs of mice, and fails to kill P388D1 MΦ in vitro. Restoration of the knockout to the wild-type phenotype completely restores virulence again fulfilling Koch's molecular postulates [73a]. Thus, CBP1 may serve as a virulence factor for Hc yeasts.

V. MACROPHAGE ACTIVATION AND COUNTERSTRATEGIES

The importance of cell-mediated immunity (CMI) in host defense against Hc was first demonstrated by Salvin [74]. In his early experiments, resistance to a lethal inoculum of Hc yeasts was induced in mice and guinea pigs by prior injection of dead yeasts, or by sublethal infection with viable yeasts. Resistance to a subsequent lethal inoculum of yeasts did not correlate with the titer of humoral factors (complement-fixing antibodies), and resistance could not be transferred by homologous immune serum [74]. These initial studies have been confirmed and extended by others [75–83], and it now is clear that murine host defense against Hc requires a Th-1 type of response with the production of IL-12, IFNγ, and TNFα [84–89]. The secretion of IL-12 is required for the production of IFNγ [86,88], which presumably activates MΦ to destroy the yeasts. Although required in both naive and Hc-immune animals [84–87,89], the role of TNFα still is a mystery, but it does not appear to act directly on MΦ [89,90]. The production of granulocyte-MΦ colony-stimulating factor (GM-CSF) also is necessary for survival in naive, but not Hc-immune mice, and appears to act via suppression of the Th2 cytokines IL-4 and IL-10 [91].

In vitro IFNγ activates resident mouse PM to inhibit the intracellular growth of yeasts [80], but splenic MΦ are activated by IFNγ only in the presence of LPS [90]. The mechanism(s) by which activated MΦ inhibit the intracellular growth of Hc yeasts appears to be twofold. Thus, activation of resident PM by IFNγ [66] or of splenic MΦ by IFNγ and LPS [90] can be reversed by the addition of holotransferrin or ferrous sulfate, indicating that IFNγ activation leads to restriction of intracellular iron. Further, only MΦ that produce nitric oxide (NO) have antihistoplasma activity. Thus, incubation of the RAW 264.7 mouse MΦ cell line with IFNγ leads to inhibition of the intracellular growth of Hc yeasts and the production of NO, whereas IFNγ-treated P388D1 mouse MΦ are not activated to an antihistoplasma state and do not secrete NO [92]. In addition, activation of murine splenic MΦ with IFNγ and LPS is completely blocked by N-monomethyl-L-arginine, a competitive inhibitor of nitric oxide synthase [93]. These data suggest that the mechanism of action of IFNγ is through the production of NO and the nitrosylation of iron transferrin. In vivo studies in mice confirm that the production of NO is required for successful clearance of Hc yeasts in the primary immune response. Remarkably, however, NO does *not* appear to be essential for clearance of fungi in the secondary immune response to Hc [94].

A major puzzle is the fact that although in vitro activation of PM by IFNγ or splenic MΦ by IFNγ and LPS leads to virtually complete inhibition of the intracellular replication of Hc yeasts, the yeasts are not killed. However, in in vivo models of murine histoplasmosis, 2 weeks after a sublethal inoculum the fungus has been cleared completely, suggesting that there is sterilizing immunity. Thus, other as yet unknown cytokines must be playing a role in the in vivo activation of MΦ.

In contrast to studies in murine systems, the cytokines required to activate human MΦ fungistatic or fungicidal activity against Hc remain obscure. Unlike mouse peritoneal MΦ, human MΦ are not activated to an antihistoplasma state by IFNγ or TNFα [35,95]. Human MΦ are activated by IFNγ to kill and/or inhibit the intracellular growth of *Trypanosoma cruzi* [96], *Leishmania donovani* [37], *Toxoplasma gondii* [97], and *Chlamydia psittaci* [98]. Further, the mechanism(s) of killing/growth inhibition is mediated, at least in part, through the production of toxic oxygen metabolites. The fact that Hc yeasts appear to be impervious to respiratory burst products (see discussion above), may explain, in part, why IFNγ-activated human MΦ are incapable of restricting the intracellular growth of Hc yeasts. However, this simple explanation does not explain the fact that incubation of human MΦ with IFNγ also downregulates transferrin receptors [99], which should result in decreased uptake of iron by MΦ. The logical explanation for this apparent contradiction is that after 5–10 days of culture in serum-containing medium, human MΦ have more than sufficient iron stores to support the intracellular replication of Hc yeasts, and, therefore, downregulation of transferrin receptors by IFNγ becomes irrelevant.

Although IFNγ is without effect, culture of monocytes in the presence of IL-3, GM-CSF, or M-CSF does activate MΦ fungistatic (but not fungicidal) activity against Hc yeasts. Optimal activation of MΦ by CSFs requires 5 days of coculture, and the cultures must be initiated with freshly isolated peripheral blood monocytes. Addition of IFNγ or TNFα to CSF activated MΦ during the last 24 hr prior to infection with Hc yeasts, does not further enhance MΦ fungistatic activity [95].

IL-3, GM-CSF, and M-CSF also activate human monocyte/MΦ antimicrobial activity against other intracellular pathogens including *T. cruzi* [100,101], *L. donovani* [102], *L. amazonensis* [101,103], and *Mycobacterium avium* complex [104]. In contrast to Hc, however, IL-3 acts synergistically with GM-CSF and M-CSF in stimulating MΦ antileishmania activity [101]. Furthermore, the ability of IL-3 to stimulate MΦ antileishmania activity is inhibited by antibodies to TNFα, and TNFα alone can activate MΦ to inhibit the intracellular growth of *L. amazonensis* [101]. In addition, GM-CSF activation of human MΦ to inhibit the intracellular growth of *M. avium* complex is augmented by TNFα, but suppressed by IFNγ [104].

The mechanism(s) by which CSFs stimulate human MΦ fungistatic activity against Hc yeasts is unknown. However, as discussed previously, it is unlikely that fungistasis is mediated through an oxygen-dependent mechanism, in contrast to MΦ antimicrobial activity against *L. amazonensis*, *L. donovani*, *T. cruzi*, and *C. albicans* [101,102,105].

Nitric oxide also probably does not play a role in the ability of activated human MΦ to inhibit the intracellular growth of Hc yeasts. Regardless of which CSF is used to stimulate human monocyte/MΦ anti-fungal activity, nitrite (quantified using the Greiss reagent) [106] is not detected in the culture medium. Moreover, the NO synthase inhibitor L-NMMA does not reverse the fungistatic activity of CSF-stimulated human MΦ (S.L. Newman, unpublished observations). These results agree with the reports of others that nitric oxide does not appear to play a role in the antimicrobial activity of human MΦ against *T. gondii*, *C. psittaci*, *L. donovani* [107], or *C. neoformans* [108].

Despite our lack of knowledge of the cytokines that activate human MΦ antihistoplasma activity, human monocytes and MΦ are activated to inhibit the replication of Hc yeasts upon adherence to type 1 collagen matrices [47]. Other extracellular matrix proteins, including fibronectin, laminin, and vitronectin, or nongelled collagen do not induce MΦ activation. MΦ antihistoplasma activity develops immediately upon adherence to the collagen matrices (1 hr), and fungistatic activity is maintained for up to 4 days postinfection. Most important, collagen-adherent MΦ are fungicidal, a property not even exhibited by IFNγ-activated murine PM. Interestingly, culture of collagen-adherent MΦ with IFNγ or TNFα, or IL-3, GM-CSF, or M-CSF does not augment MΦ fungicidal activity against Hc yeasts [47].

The mechanism(s) of MΦ-mediated fungistasis is not through the production of toxic oxygen radicals, NO, or the restriction of intracellular iron. However, compared to Hc-infected plastic-adherent MΦ in which PL-fusion is minimal, Hc-infected collagen-adherent MΦ demonstrate enhanced PL fusion at both 1 and 24 hr postinfection [47]. Thus, the yeasts apparently are killed by preformed lysosomal hydrolases, directly counteracting the yeasts' strategy of inhibiting PL fusion. Although not yet demonstrated, the data also suggest that adherence of MΦ to collagen matrices also compromises the ability of Hc to regulate intraphagosomal pH.

VI. DO NEUTROPHILS PLAY A ROLE?

The role of polymorphonuclear neutrophils (PMN) in the CMI response to Hc is unclear [13]. However, even the earliest studies in a murine model of histoplasmosis described PMN as being the predominant inflammatory cell type in the lungs during the first 36 hr after intranasal inoculation with Hc macroconidia. In these studies the PMN were not observed to phagocytose the macroconidia, and the few yeasts that were present at 36 hr were found only within MΦ [14]. Baughman et al. [15] observed an intense PMN response in the lung at 1 week of infection after intranasal inoculation of C57BL/6 mice with yeasts of Hc strain G217B. By the second week, PMN were largely replaced by mononuclear cells characteristic of a granulomatous inflammatory response. These in vivo observations suggest that PMN may play a role in host defense against Hc at early times postinfection.

In in vitro studies, at a high phagocyte-to-fungus ratio, human PMN apparently kill the mycelial phase, but not the yeast phase of Hc, despite the fact that opsonized yeasts induce the PMN respiratory burst [32]. Although initial reports on human [34] and murine [109] PMN demonstrated modest fungistatic activity against Hc, later studies indicate that human PMN possess very potent fungistatic activity against Hc yeasts [40]. Neutrophil-mediated fungistasis is evident as early as 2 hr, is maximal at 24 hr, and persists at high levels for up to 5 days. PMN-mediated fungistasis requires the presence of either fresh or heat-inactivated human serum, but ingestion of the yeasts per se is not required. Cytokines and agents that stimulate the respiratory burst and enhance PMN-mediated cytotoxicity (IFNγ, IL-8, IL-3, GM-CSF, G-CSF, and PMA) do not induce PMN fungicidal activity or enhance PMN-mediated fungistasis [40].

The mechanism(s) of PMN fungistatic activity is not via the production of toxic oxygen metabolites nor reactive nitrogen intermediates. Indeed, PMN from patients with chronic granulomatous disease (CGD), who lack the capacity to produce toxic oxygen metabolites due to a NADPH oxidase deficiency [110], inhibit the growth of Hc yeasts as well as PMN from normal individuals. Remarkably, all of the fungistatic activity of human PMN is localized in the azurophil granules, with no activity detected in the specific granules, membrane fraction, or the cytosol [40].

Human neutrophil azurophil granules contain two families of antimicrobial proteins each with four members: defensins and serprocidins; and two antimicrobial proteins with unique primary structures, lysozyme and bactericidal permeability increasing protein (BPI) [111–113]. These azurophil granule proteins have been found to have a broad spectrum of antimicrobial activity against gram-negative and gram-positive bacteria, protozoans, enveloped viruses, and fungi, including *Cryptococcus neoformans, Candida albicans, Aspergillus fumigatus,* and *Rhizopus oryzae.*

Human neutrophil defensins are basic peptides that are 29 to 30 amino acids in length and contain three intramolecular disulfide bonds. HNP-1, HNP-2, and HNP-3 all mediate a concentration-dependent fungistasis against Hc yeasts [113a]. The peptides have a cyclic structure with spatial segregation of charged and hydrophobic residues. This amphiphilic structure may equip defensins for insertion into the phospholipid membrane of their target organism [113–115]. However, the mechanism by which defensins might inhibit the replication of Hc yeasts is unclear, particularly because of its thick cell wall.

The second major family of azurophil granule antimicrobial proteins, the serprocidins consists of azurocidin, cathepsin G, elastase, and proteinase 3. Serprocidins are cationic glycoproteins of 25–29 kDa that exhibit considerable homology with serine proteases. Serprocidins also demonstrate a broad spectrum of antimicrobial activity against gram-positive and gram-negative bacteria and fungi [111]. Of the four members, only cathepsin G inhibits the growth of Hc yeasts, and cathepsin G activity is additive in its inhibitory activity when mixed with defensins HNP-1, 2, and 3 [113a]. Cathepsin G is a neutral protease of MW 29–31 kDa [116] and has been reported to kill *Neisseria gonorrhea* [117] and *Listeria monocytogenes* [118] by a nonenzymatic mechanism, and to kill microbes that cause dental caries by both enzymatic and nonenzymatic mechanisms [119]. Cathepsin G also can synergize with azurocidin in killing the oral microbe *Capnocytophaga sputigena* [120].

Most surprising is the fact that BPI also is fungistatic for Hc yeasts. BPI has a molecular weight of 50–60 kDa and contains a highly cationic, lysine-rich amino-terminal half and a very hydrophobic, much less charged, carboxy-terminal half [121]. BPI has a strong affinity for lipopolysaccharide (LPS), and its cytotoxic activity is therefore directed almost exclusively to gram-negative bacteria. Indeed, the susceptibility of gram-negative bacteria appears to be determined primarily by the structure of the envelope LPS, specifically the length of the polysaccharide chains [122]. However, despite the fact that Hc yeasts do not contain LPS, BPI inhibits the growth of yeasts in a concentration-dependent manner, and its activity is additive when mixed with cathepsin G or any of the three defensins [113a]. It is unclear what BPI might recognize on the surface of the yeasts to mediate these inhibitory effects.

In a murine model of histoplasmosis, depletion of neutrophils with the monoclonal antibody RB6-8C5 [123] impairs the resistance of both naive and immune mice to intranasal inoculation with Hc yeasts (S.L. Newman, manuscript submitted). Zhou et al. [94] also report that depletion of PMN is fatal in the primary immune response to Hc infection, but that PMN are not essential in the host response to a secondary reinfection. Although the two studies utilized different strains of Hc, the most likely explanation for the contrasting results is that Zhou et al. [94] inoculated the mice intravenously, while our studies utilized mice inoculated via the intranasal route. Similarly, IFNγ is *not* necessary for an effective secondary immune response when Hc yeasts are injected intravenously [94], but IFNγ is required for clearance in a pulmonary model of reinfection histoplasmosis [124]. Likewise, TNFα is *not* required for survival of mice in a systemic model of secondary infection [94], whereas TNFα is required for survival in the pulmonary model [89].

We have further found that in vitro, murine PMN exhibit a more potent fungistatic activity against Hc yeasts than do human neutrophils. In addition, unlike human PMN, mouse neutrophils mediate equal fungistasis against both opsonized and unopsonized yeasts (S.L. Newman, manuscript submitted). These results are particularly intriguing considering that mouse neutrophils do not contain defensins [125].

As most individuals possess CMI to Hc in areas where the fungus is endemic, the specific role of PMN in the immune response to Hc is unclear. Despite the fact the initial infection with Hc may cause only mild symptoms, most individuals still develop CMI. There are, however, many individuals living in areas where Hc is endemic who are skin test negative when tested with histoplasmin, but who presumably have been exposed to Hc sometime during their lifetime. Thus, it is reasonable to hypothesize that in some cases the initial inflammatory response, of which PMN are a major part, may be sufficient to clear the organism from the host, or that the initial inoculum was too small to induce long-lasting immunity.

Alternatively, neutrophils may slow the course of the infection, and, under certain circumstances, they may prevent dissemination of the yeasts from the lung. It is also possible that neutrophils may damage the yeasts in such a manner as to render them vulnerable to inflammatory macrophages. Thus, inflammatory macrophages may phagocytose yeast-containing neutrophils and subsequently kill the partially damaged yeasts. Lastly, PMN produce and secrete proinflammatory type cytokines including IL-12, GM-CSF, and TNFα [126]. Any or all of these postulated mechanisms of defense involving neutrophils during the early phase of the inflammatory response to Hc may explain, at least in part, the fact that a substantial percentage of pulmonary infections by this organism are subclinical and self-limiting.

VII. INTERACTION WITH DENDRITIC CELLS

With regard to human immunology, a major conundrum is why IFNγ doesn't activate human MΦ to an antihistoplasma state. In addition, with the exception of the CSFs, the cytokines required to activate MΦ fungistatic/fungicidal activity against Hc yeasts remain unknown. Indeed, in my own laboratory, we have not even been able to generate MΦ-activating supernates by incubating heat-killed yeasts with macrophages and lymphocytes.

Most recently, we considered the possibility that dendritic cells (DC), rather than MΦ, may be the primary antigen-presenting cells (APC) in inducing CMI to Hc. Compared to MΦ, DC are infinitely superior at processing and presenting protein antigens to T-cells [127]. However, early characterization of DC suggested that they were nonphagocytic, making them unlikely to play a role in host defense unless they could somehow pick up and process secreted microbial antigens. More recently, it has been demonstrated that so-called immature DC can phagocytose some microorganisms [128–132]. Upon phagocytosis, the immature DC are induced to become mature DC that are now fully capable of antigen presentation [133].

Human DC contain CD18 adhesion molecules on their surface, and it is known that human MΦ utilize CD18 to recognize and phagocytose Hc. Therefore, we hypothesized that human DC use these same receptors to phagocytose Hc, and, once ingested, the DC would kill and degrade the yeast, process fungal antigens, and stimulate the induction of CMI. Indeed, human DC derived from monocytes cultured in the presence of GM-CSF and IL-4 ingest heat-killed FITC-labeled Hc yeasts in a time-dependent manner, and phagocytosis of viable yeasts was confirmed by electron microscopy [133a].

Most remarkably, degraded yeasts are observed as early as 1 hr postinfection, and by 24 hr the majority of internalized yeasts are in various stages of disintegration. The mechanism of this inhibition appears to be through massive PL-fusion [133a]. Thus, although PL fusion does not occur in human MΦ that have ingested Hc yeasts, it does occur in human DC. Therefore, the strategies that Hc uses to survive in human MΦ are not operable in DC, and DC may be considered to play a role in innate immunity as well as CMI.

DC incubated with either viable or heat-killed yeasts stimulate the proliferation of autologous lymphocytes, and DC are more efficient at stimulating lymphocyte proliferation upon ingestion of *viable* yeasts than after ingestion of killed yeasts [133a]. These data correlate with the finding in a murine model of histoplasmosis that the magnitude of protection induced by viable Hc yeasts is considerably greater than the protection induced by killed yeasts [134].

In a murine model of histoplasmosis, Th1 type cytokines are observed in the lung as early as 3 days after an intranasal inoculation of Hc yeasts [135]. As the yeasts replicate extremely well in MΦ, we hypothesize that upon exposure to Hc, a small number of lung DC take up the fungus, kill it, and process fungal antigens for presentation to T-cells in nearby lymph nodes.

Stimulated T-cells then produce IFNγ and other cytokines to activate MΦ and resolve the infection.

This hypothesis is supported by the fact that ingestion of Hc yeasts by human monocytes causes downregulation of IL-12 secretion stimulated by heat-killed *Staphylococcus aureus* and IFNγ. This anergy is induced via signal transduction through CR3 (CD11b/CD18) as antibodies to CR3 also downregulate IL-12 production [136]. As IL-12 production is critical for the induction of IFNγ, this strategy may prevent MΦ from becoming activated and thereby prolong the survival of yeasts in the host. Whether downregulation of IL-12 occurs in murine MΦ-containing Hc yeasts is unknown. However, if downregulation of IL-12 did occur in murine lung MΦ, it is possible that IL-12 might be produced and secreted by lung DC that have ingested and killed Hc yeasts. Thus, lung DC might be responsible for the early production of IL-12 in mice infected intranasally with Hc yeasts. Indeed, DC that have ingested *L. major* promastigotes produce IL-12, whereas IL-12 is not detected in murine MΦ that have phagocytosed promastigotes [137].

VIII. CONCLUSIONS

Over the past several years considerable progress has been made toward understanding the interaction of Hc with human and murine MΦ. Clearly, Hc yeasts have cleverly adapted themselves to survive and multiply within the normally hostile confines of MΦ phagosomes. Although some of the mechanism(s) by which Hc yeasts accomplish this legerdemain are known, the molecules produced by Hc to survive and multiply have yet to be identified. In particular it will be important to determine how Hc acquires intracellular iron, as this is critical to the survival of the fungus.

Other important questions concerning MΦ-Hc interaction that remain unanswered include: What are the cytokines that activate human and murine MΦ to *kill* Hc yeasts? What is the mechanism(s) by which activated MΦ kill Hc yeasts? These and other questions should provide interested scientists with significant intellectual stimulation in their attempts to understand this fascinating fungus. Furthermore, insight gained from continued research on MΦ-Hc interactions may provide us with the rationale to design new drugs for the treatment of disseminated histoplasmosis in immunocompromised patients.

ACKNOWLEDGMENT

The author's research on *H. capsulatum* was supported by NIH grants AI-37639 and HL-55948.

REFERENCES

1. L Ajello. Distribution of *Histoplasma capsulatum* in the United States. In: L Ajello, EW Chick, MF Furcolow, eds. Histoplasmosis. Springfield, IL: C.C. Thomas, 1971, pp 103–122.
2. P Reddy, DF Gorelick, CA Brasher, H Larsh. Progressive disseminated histoplasmosis as seen in adults. Am J Med 48:629–636, 1970.
3. JW Smith, JP Utz. Progressive disseminated histoplasmosis. A prospective study of 26 patients. Ann Intern Med 76:557–565, 1972.
4. SF Davies, M Khan, GA Sarosi. Disseminated histoplasmosis in immunologically suppressed patients. Occurrence in a nonendemic area. Am J Med 64:94–100, 1978.
5. CA Kauffman, KS Israel, JW Smith, AC White, J Schwarz, GF Brooks. Histoplasmosis in immunosuppressed patients. Am J Med 64:923–932, 1978.

6. LJ Wheat, TG Slama, JA Norton, RB Kohler, HE Eitzen, ML French, B Sathapatayavongs. Risk factors for disseminated or fatal histoplasmosis. Analysis of a large urban outbreak. Ann Intern Med 96:159–163, 1982.

7. LJ Wheat, EJ Smith, B Sathapatayavongs, B Batteiger, RS Filo, SB Leapman, MV French. Histoplasmosis in renal allograft recipients. Two large urban outbreaks. Arch Intern Med 143:703–707, 1983.

8. PC Johnson, RJ Hamill, GA Sarosi. Clinical review: progressive disseminated histoplasmosis in the AIDS patient. Semin Respir Infect 4:139–146, 1989.

9. MA Neubauer, DC Bodensteiner. Disseminated histoplasmosis in patients with AIDS. South Med J 85:1166–1170, 1992.

10. J Wheat. Histoplasmosis and coccidioidomycosis in individuals with AIDS. A clinical review. Infect Dis Clin North Am 8:467–482, 1994.

11. WA DeMonbruen. The cultivation and characteristics of Darling's *Histoplasma capsulatum.* Am J Trop Med 14:93–125, 1934.

12. A Howell. Studies on *Histoplasma capsulatum* and similar form species. 1. Morphology and development. Mycologia 31:191–216, 1939.

13. GS Deepe, WE Bullock. Histoplasmosis: A granulomatous inflammatory response. In: JI Gallin, IM Goldstein, R Snyderman, eds. Inflammation: Basic Principles and Clinical Correlates. New York: Raven Press, 1992, pp 943–958.

14. JJ Procknow, MI Page, CG Loosli. Early pathogenesis of experimental histoplasmosis. Arch Pathol 69:413–426, 1960.

15. RP Baughman, CK Kim, A Vinegar, DE Hendricks, DJ Schmidt, WE Bullock. The pathogenesis of experimental pulmonary histoplasmosis. Correlative studies of histopathology, bronchoalveolar lavage, and respiratory function. Am Rev Respir Dis 134:771–776, 1986.

16. CL Berry. The production of disseminated histoplasmosis in the mouse: the effects of changes in reticuloendothelial function. J Pathol 97:441–457, 1969.

17. GS Deepe Jr. The immune response to Histoplasma capsulatum: unearthing its secrets. J Lab Clin Med 123:201–205, 1994.

18. GS Deepe Jr, RA Seder. Molecular and cellular determinants of immunity to Histoplasma capsulatum. Res Immunol 149:397–406; discussion 509–510, 1998.

19. WE Bullock, SD Wright. Role of the adherence promoting receptors, CR3, LFA-1, and p150,95, in binding of Histoplasma capsulatum by human macrophages. J Exp Med 165:195–210, 1987.

20. SL Newman, C Bucher, J Rhodes, WE Bullock. Phagocytosis of *Histoplasma capsulatum* yeasts and microconidia by human cultured macrophages and alveolar macrophages. Cellular cytoskeleton requirement for attachment and ingestion. J Clin Invest 85:223–230, 1990.

21. F Sanchez-Madrid, JA Nagy, E Robbins, P Simon, TA Springer. A human leukocyte differentiation antigen family with distinct alpha-subunits and a common beta-subunit: the lymphocyte function-associated antigen (LFA-1), the C3bi complement receptor (OKM1/Mac-1), and the p150,95 molecule. J Exp Med 158:1785–1803, 1983.

22. RA Schnur, SL Newman. The respiratory burst response to Histoplasma capsulatum by human neutrophils. Evidence for intracellular trapping of superoxide anion. J Immunol 144:4765–4772, 1990.

23. CL Kimberlin, AR Hariri, HO Hempel, NL Goodman. Interactions between Histoplasma capsulatum and macrophages from normal and treated mice: comparison of the mycelial and yeast phases in alveolar and peritoneal macrophages. Infect Immun 34:6–10, 1981.

24. JA Armstrong, PD Hart. Response of cultured macrophages to *Mycobacterium tuberculosis,* with observations on fusion of lysosomes with phagosomes. J Exp Med 134:713–740, 1971.

25. TC Jones, JG Hirsch. The interaction between Toxoplasma gondii and mammalian cells. II. The absence of lysosomal fusion with phagocytic vacuoles containing living parasites. J Exp Med 136:1173–1194, 1972.

26. MB Goren, P D'Arcy Hart, MR Young, JA Armstrong. Prevention of phagosome-lysosome fusion in cultured macrophages by sulfatides of Mycobacterium tuberculosis. Proc Natl Acad Sci USA 73:2510–2514, 1976.

27. MA Horwitz. The Legionnaires' disease bacterium (Legionella pneumophila) inhibits phagosome lysosome fusion in human monocytes. J Exp Med 158:2108–2126, 1983.

28. LG Tilney, DA Portnoy. Actin filaments and the growth, movement, and spread of the intracellular bacterial parasite, Listeria monocytogenes. J Cell Biol 109:1597–1608, 1989.

29. SE Anderson Jr, JS Remington. Effect of normal and activated human macrophages on Toxoplasma gondii. J Exp Med 139:1154–1174, 1974.

30. CB Wilson, V Tsai, JS Remington. Failure to trigger the oxidative metabolic burst by normal macrophages: possible mechanism for survival of intracellular pathogens. J Exp Med 151:328–346, 1980.

31. TJ Holzer, KE Nelson, RG Crispen, BR Andersen. Mycobacterium leprae fails to stimulate phagocytic cell superoxide anion generation. Infect Immun 51:514–520, 1986 [published erratum appears in Infect Immun 1986;52(1):348].

32. A Schaffner, CE Davis, T Schaffner, M Markert, H Douglas, AI Braude. In vitro susceptibility of fungi to killing by neutrophil granulocytes discriminates between primary pathogenicity and opportunism. J Clin Invest 78:511–524, 1986.

33. DH Howard. Comparative sensitivity of Histoplasma capsulatum conidiospores and blastospores to oxidative antifungal systems. Infect Immun 32:381–387, 1981.

34. E Brummer, N Kurita, S Yosihida, K Nishimura, M Miyaji. Fungistatic activity of human neutrophils against Histoplasma capsulatum: correlation with phagocytosis. J Infect Dis 164:158–162, 1991.

35. J Fleischmann, B Wu-Hsieh, DH Howard. The intracellular fate of Histoplasma capsulatum in human macrophages is unaffected by recombinant human interferon gamma. J Infect Dis 161:143–145, 1990.

36. SL Newman, L Gootee, C Bucher, WE Bullock. Inhibition of intracellular growth of Histoplasma capsulatum yeast cells by cytokine-activated human monocytes and macrophages. Infect Immun 59:737–741, 1991.

37. HW Murray, DM Cartelli. Killing of intracellular Leishmania donovani by human mononuclear phagocytes. Evidence for oxygen-dependent and independent leishmanicidal activity. J Clin Invest 72:32–44, 1983.

38. MJ Pabst, HB Hedegaard, RB Johnston Jr. Cultured human monocytes require exposure to bacterial products to maintain an optimal oxygen radical response. J Immunol 128:123–128, 1982.

39. M Sasada, A Kubo, T Nishimura, T Kakita, T Moriguchi, K Yamamoto, H Uchino. Candidacidal activity of monocyte-derived human macrophages: relationship between Candida killing and oxygen radical generation by human macrophages. J Leukoc Biol 41:289–294, 1987.

40. SL Newman, L Gootee, JE Gabay. Human neutrophil-mediated fungistasis against Histoplasma capsulatum. Localization of fungistatic activity to the azurophil granules. J Clin Invest 92:624–631, 1993.

41. LG Eissenberg, WE Goldman. Histoplasma capsulatum fails to trigger release of superoxide from macrophages. Infect Immun 55:29–34, 1987.

42. JE Wolf, V Kerchberger, GS Kobayashi, JR Little. Modulation of the macrophage oxidative burst by Histoplasma capsulatum. J Immunol 138:582–586, 1987.

43. JE Wolf, AL Abegg, SJ Travis, GS Kobayashi, JR Little. Effects of Histoplasma capsulatum on murine macrophage functions: inhibition of macrophage priming, oxidative burst, and antifungal activities. Infect Immun 57:513–519, 1989.

44. DH Howard. Intracellular growth of *Histoplasma capsulatum.* J Bacteriol 89:518–523, 1965.

45. LG Eissenberg, PH Schlesinger, WE Goldman. Phagosome lysosome fusion in P388D1 macrophages infected with Histoplasma capsulatum. J Leukoc Biol 43:483–491, 1988.

46. ML Taylor, ME Espinosa-Schoelly, R Iturbe, B Rico, J Casasola, F Goodsaid. Evaluation of phagolysosome fusion in acridine orange stained macrophages infected with Histoplasma capsulatum. Clin Exp Immunol 75:466–470, 1989.

47. SL Newman, L Gootee, C Kidd, GM Ciraolo, R Morris. Activation of human macrophage fungistatic activity against Histoplasma capsulatum upon adherence to type I collagen matrices. J Immunol 158:1779–1786, 1997.

48. JE Strasser, SL Newman, GM Ciraolo, RE Morris, ML Howell, GE Dean. Regulation of the macrophage vacuolar ATPase and phagosome-lysosome fusion by Histoplasma capsulatum. J Immunol 162:6148–6154, 1999.

49. SB De Sanchez, LM Carbonell. Immunological studies on Histoplasma capsulatum. Infect Immun 11:387–394, 1975.

50. DH Howard. Observation on tissue cultures of mouse peritoneal exudates innoculated with *Histoplasma capsulatum.* J Bacteriol 78:69–78, 1959.

51. DH Howard. Effect of Mycostatin and Fungizone on the growth of *Histoplasma capsulatum* in tissue culture. J Bacteriol 79:442–449, 1960.

52. DH Howard. Intracellular behavior of *Histoplasma capsulatum.* J Bacteriol 87:33–38, 1964.

53. S Chaturvedi, P Frame, SL Newman. Macrophages from human immunodeficiency virus-positive persons are defective in host defense against Histoplasma capsulatum. J Infect Dis 171:320–327, 1995.

54. S Chaturvedi, SL Newman. Modulation of the effector function of human macrophages for Histoplasma capsulatum by HIV-1. Role of the envelope glycoprotein gp120. J Clin Invest 100: 1465–1474, 1997.

55. SL Newman, L Gootee, R Morris, WE Bullock. Digestion of Histoplasma capsulatum yeasts by human macrophages. J Immunol 149:574–580, 1992 [published erratum appears in J Immunol 1992; 149(9):3127].

56. MR Siegel, HD Sisler. Site of action of cycloheximide in cells of Saccharomyces pastorianus. II. The nature of inhibition of protein synthesis in a cell free system. Biochim Biophys Acta 87:83–89, 1964.

57. HL Ennis, M Lubin. Cycloheximide: aspects of inhibition of protein synthesis in mammalian cells. Science 146:1474–1476, 1964.

58. LG Eissenberg, WE Goldman, PH Schlesinger. Histoplasma capsulatum modulates the acidification of phagolysosomes. J Exp Med 177:1605–1611, 1993.

59. JV Princiotto, EJ Zapolski. Difference between the two iron-binding sites of transferrin. Nature 255:87–88, 1975.

60. AN Lestas. The effect of pH upon human transferrin: selective labelling of the two iron binding sites. Br J Haematol 32:341–350, 1976.

61. DJ Krogstad, PH Schlesinger. Acid-vesicle function, intracellular pathogens, and the action of chloroquine against Plasmodium falciparum. N Engl J Med 317:542–549, 1987.

62. SL Newman, L Gootee, G Brunner, GS Deepe Jr. Chloroquine induces human macrophage killing of Histoplasma capsulatum by limiting the availability of intracellular iron and is therapeutic in a murine model of histoplasmosis. J Clin Invest 93:1422–1429, 1994.

63. GW Bates, J Wernicke. The kinetics and mechanism of iron (III) exchange between chelates and transferrin. J Biol Chem 246:3679–3686, 1971.

64. WR Burt. Identification of coprogen B and its breakdown products from Histoplasma capsulatum. Infect Immun 35:990–996, 1982.

65. DH Howard, R Rafie, A Tiwari, KF Faull. The hydroxamate siderophores of *Histoplasma capsulatum.* Infect Immun 68:2338–2343, 2000.

66. TE Lane, BA Wu-Hsieh, DH Howard. Iron limitation and the gamma interferon mediated antihistoplasma state of murine macrophages. Infect Immun 59:2274–2278, 1991.

67. MM Timmerman, JP Woods. Ferric reduction is a potential iron acquisition mechanism for Histoplasma capsulatum. Infect Immun 67:6403–6408, 1999.

68. A Pitt, LS Mayorga, PD Stahl, AL Schwartz. Alterations in the protein composition of maturing phagosomes. J Clin Invest 90:1978–1983, 1992.

69. GL Lukacs, OD Rotstein, S Grinstein. Phagosomal acidification is mediated by a vacuolar type H(+)-ATPase in murine macrophages. J Biol Chem 265:21099–21107, 1990.

70. GL Lukacs, OD Rotstein, S Grinstein. Determinants of the phagosomal pH in macrophages. In situ assessment of vacuolar H(+)-ATPase activity, counterion conductance, and H+ "leak". J Biol Chem 266:24540–24548, 1991.

71. DM Retallack, EL Heinecke, R Gibbons, GS Deepe Jr, JP Woods. The URA5 gene is necessary for histoplasma capsulatum growth during infection of mouse and human cells [In Process Citation]. Infect Immun 67:624–629, 1999.

72. JW Batanghari, WE Goldman. Calcium dependence and binding in cultures of Histoplasma capsulatum. Infect Immun 65:5257–5261, 1997.

73. JW Batanghari, GS Deepe Jr, E Di Cera, WE Goldman. Histoplasma acquisition of calcium and expression of CBP1 during intracellular parasitism. Mol Microbiol 27:531–539, 1998.

73a. TS Sebghati, JT Engle, WE Goldman. Intracellular parasitism by Histoplasma capsulatum: Fungal virulence and calcium dependence. Science 290:1368–1372, 2000.

74. SB Salvin. Acquired resistance in experimental histoplasmosis. Trans NY Acad Sci 18:462–468, 1955/56.

75. DH Howard, V Otto, RK Gupta. Lymphocyte-mediated cellular immunity in histoplasmosis. Infect Immun 4:605–610, 1971.

76. DH Howard. Further studies on the inhibition of Histoplasma capsulatum within macrophages from immunized animals. Infect Immun 8:577–581, 1973.

77. DM Williams, JR Graybill, DJ Drutz. Adoptive transfer of immunity to Histoplasma capsulatum in athymic nude mice. Sabouraudia 19:39–48, 1981.

78. B Wu-Hsieh, DH Howard. Inhibition of growth of Histoplasma capsulatum by lymphokine-stimulated macrophages. J Immunol 132:2593–2597, 1984.

79. B Wu-Hsieh, A Zlotnik, DH Howard. T-cell hybridoma produced lymphokine that activates macrophages to suppress intracellular growth of Histoplasma capsulatum. Infect Immun 43:380–385, 1984.

80. BA Wu-Hsieh, DH Howard. Inhibition of the intracellular growth of Histoplasma capsulatum by recombinant murine gamma interferon. Infect Immun 55:1014–1016, 1987.

81. GS Deepe Jr. Protective immunity in murine histoplasmosis: functional comparison of adoptively transferred T-cell clones and splenic T cells. Infect Immun 56:2350–2355, 1988.

82. AM Gomez, WE Bullock, CL Taylor, GS Deepe Jr. Role of L3T4+ T cells in host defense against Histoplasma capsulatum. Infect Immun 56:1685–1691, 1988.

83. R Allendoerfer, DM Magee, GS Deepe Jr, JR Graybill. Transfer of protective immunity in murine histoplasmosis by a CD4+ T-cell clone. Infect Immun 61:714–718, 1993.

84. JG Smith, DM Magee, DM Williams, JR Graybill. Tumor necrosis factor alpha plays a role in host defense against Histoplasma capsulatum. J Infect Dis 162:1349–1353, 1990.

85. BA Wu-Hsieh, GS Lee, M Franco, FM Hofman. Early activation of splenic macrophages by tumor necrosis factor alpha is important in determining the outcome of experimental histoplasmosis in mice. Infect Immun 60:4230–4238, 1992 [published erratum appears in Infect Immun 1992;60(12): 5324].

86. P Zhou, MC Sieve, J Bennett, KJ Kwon-Chung, RP Tewari, RT Gazzinelli, A Sher, RA Seder. IL-12 prevents mortality in mice infected with Histoplasma capsulatum through induction of IFN-gamma. J Immunol 155:785–795, 1995.

87. P Zhou, MC Sieve, RP Tewari, RA Seder. Interleukin-12 modulates the protective immune response in SCID mice infected with Histoplasma capsulatum. Infect Immun 65:936–942, 1997.

88. R Allendoerfer, GP Biovin, GS Deepe Jr. Modulation of immune responses in murine pulmonary histoplasmosis. J Infect Dis 175:905–914, 1997.

89. R Allendoerfer, GS Deepe Jr. Blockade of endogenous TNF-alpha exacerbates primary and secondary pulmonary histoplasmosis by differential mechanisms. J Immunol 160:6072–6082, 1998.

90. TE Lane, BA Wu-Hsieh, DH Howard. Gamma interferon cooperates with lipopolysaccharide to activate mouse splenic macrophages to an antihistoplasma state. Infect Immun 61:1468–1473, 1993.

91. GS Deepe Jr, R Gibbons, E Woodward. Neutralization of endogenous granulocyte macrophage colony stimulating factor subverts the protective immune response to Histoplasma capsulatum. J Immunol 163:4985–4993, 1999.

92. TE Lane, GC Otero, BA Wu-Hsieh, DH Howard. Expression of inducible nitric oxide synthase by stimulated macrophages correlates with their antihistoplasma activity. Infect Immun 62:1478–1479, 1994.

93. TE Lane, BA Wu-Hsieh, DH Howard. Antihistoplasma effect of activated mouse splenic macrophages involves production of reactive nitrogen intermediates. Infect Immun 62:1940–1945, 1994.

94. P Zhou, G Miller, RA Seder. Factors involved in regulating primary and secondary immunity to infection with Histoplasma capsulatum: TNF alpha plays a critical role in maintaining secondary immunity in the absence of IFN-gamma. J Immunol 160:1359–1368, 1998.

95. SL Newman, L Gootee. Colony stimulating factors activate human macrophages to inhibit intracellular growth of Histoplasma capsulatum yeasts. Infect Immun 60:4593–4597, 1992.

96. N Nogueira, S Chaplan, M Reesink, J Tydings, ZA Cohn. Trypanosoma cruzi: induction of microbicidal activity in human mononuclear phagocytes. J Immunol 128:2142–2146, 1982.

97. HW Murray, BY Rubin, SM Carriero, AM Harris, EA Jaffee. Human mononuclear phagocyte antiprotozoal mechanisms: oxygen dependent vs oxygen-independent activity against intracellular Toxoplasma gondii. J Immunol 134:1982–1988, 1985.

98. CD Rothermel, BY Rubin, EA Jaffe, HW Murray. Oxygen independent inhibition of intracellular Chlamydia psittaci growth by human monocytes and interferon-gamma activated macrophages. J Immunol 137:689–692, 1986.

99. TF Byrd, MA Horwitz. Interferon gamma-activated human monocytes downregulate transferrin receptors and inhibit the intracellular multiplication of Legionella pneumophila by limiting the availability of iron. J Clin Invest 83:1457–1465, 1989.

100. SG Reed, CF Nathan, DL Pihl, P Rodricks, K Shanebeck, PJ Conlon, KH Grabstein. Recombinant granulocyte/macrophage colony-stimulating factor activates macrophages to inhibit Trypanosoma cruzi and release hydrogen peroxide. Comparison with interferon-gamma. J Exp Med 166: 1734–1746, 1987.

101. JL Ho, SG Reed, J Sobel, S Arruda, SH He, EA Wick, KH Grabstein. Interleukin-3 induces antimicrobial activity against Leishmania amazonensis and Trypanosoma cruzi and tumoricidal activity in human peripheral blood-derived macrophages. Infect Immun 60:1984–1993, 1992.

102. WY Weiser, A Van Niel, SC Clark, JR David, HG Remold. Recombinant human granulocyte/macrophage colony-stimulating factor activates intracellular killing of Leishmania donovani by human monocyte-derived macrophages. J Exp Med 166:1436–1446, 1987.

103. JL Ho, SG Reed, EA Wick, M Giordano. Granulocyte-macrophage and macrophage colony-stimulating factors activate intramacrophage killing of Leishmania mexicana amazonensis. J Infect Dis 162: 224–230, 1990.

104. LE Bermudez, LS Young. Recombinant granulocyte-macrophage colony stimulating factor activates human macrophages to inhibit growth or kill Mycobacterium avium complex. J Leukoc Biol 48: 67–73, 1990.

105. PD Smith, CL Lamerson, SM Banks, SS Saini, LM Wahl, RA Calderone, SM Wahl. Granulocyte-macrophage colony-stimulating factor augments human monocyte fungicidal activity for Candida albicans. J Infect Dis 161:999–1005, 1990.

106. LC Green, DA Wagner, J Glogowski, PL Skipper, JS Wishnok, SR Tannenbaum. Analysis of nitrate, nitrite, and [15N]nitrate in biological fluids. Anal Biochem 126:131–138, 1982.

107. ML Cameron, DL Granger, JB Weinberg, WJ Kozumbo, HS Koren. Human alveolar and peritoneal macrophages mediate fungistasis independently of L-arginine oxidation to nitrite or nitrate. Am Rev Respir Dis 142:1313–1319, 1990.

108. HW Murray, RF Teitelbaum. L-arginine-dependent reactive nitrogen intermediates and the antimicrobial effect of activated human mononuclear phagocytes. J Infect Dis 165:513–517, 1992.

109. N Kurita, E Brummer, S Yoshida, K Nishimura, M Miyaji. Antifungal activity of murine polymorphonuclear neutrophils against Histoplasma capsulatum. J Med Vet Mycol 29:133–143, 1991.

110. RA Clark, HL Malech, JI Gallin, H Nunoi, BD Volpp, DW Pearson, WM Nauseef, JT Curnutte. Genetic variants of chronic granulomatous disease: prevalence of deficiencies of two cytosolic components of the NADPH oxidase system. N Engl J Med 321:647–652, 1989.

111. JE Gabay, RP Almeida. Antibiotic peptides and serine protease homologs in human polymorphonuclear leukocytes: defensins and azurocidin. Curr Opin Immunol 5:97–102, 1993.

112. T Ganz, ME Selsted, RI Lehrer. Defensins. Eur J Haematol 44:1–8, 1990.

113. SH White, WC Wimley, ME Selsted. Structure, function, and membrane integration of defensins. Curr Opin Struct Biol 5:521–527, 1995.

113a. SL Newman, L Gootee, JE Gabay, ME Selsted. Identification of constituents of human neutrophil azurophil granules that mediate fungistasis against Histoplasma capsulatum. Infect Immun 68: 5668–5672, 2000.

114. ME Selsted, SS Harwig, T Ganz, JW Schilling, RI Lehrer. Primary structures of three human neutrophil defensins. J Clin Invest 76:1436–1439, 1985.

115. WC Wimley, ME Selsted, SH White. Interactions between human defensins and lipid bilayers: evidence for formation of multimeric pores. Protein Sci 3:1362–1373, 1994.

116. LW Heck, KS Rostand, FA Hunter, A Bhown. Isolation, characterization, and amino-terminal amino acid sequence analysis of human neutrophil cathepsin G from normal donors. Anal Biochem 158: 217–227, 1986.

117. WM Shafer, VC Onunka, LE Martin. Antigonococcal activity of human neutrophil cathepsin G. Infect Immun 54:184–188, 1986.

118. CE Alford, E Amaral, PA Campbell. Listericidal activity of human neutrophil cathepsin G. J Gen Microbiol 136:997–100, 1990.

119. KT Miyasaki, AL Bodeau. In vitro killing of Actinobacillus actinomycetemcomitans and Capnocytophaga spp. by human neutrophil cathepsin G and elastase. Infect Immun 59:3015–3020, 1991.

120. KT Miyasaki, AL Bodeau. Human neutrophil azurocidin synergizes with leukocyte elastase and cathepsin G in the killing of Capnocytophaga sputigena. Infect Immun 60:4973–4975, 1992.

121. PW Gray, G Flaggs, SR Leong, RJ Gumina, J Weiss, CE Ooi, P Elsbach. Cloning of the cDNA of a human neutrophil bactericidal protein. Structural and functional correlations. J Biol Chem 264: 9505–9509, 1989.

122. P Elsbach, J Weiss. Bactericidal/permeability increasing protein and host defense against gram-negative bacteria and endotoxin. Curr Opin Immunol 5:103–107, 1993.

123. RI Tepper, RL Coffman, P Leder. An eosinophil-dependent mechanism for the antitumor effect of interleukin-4. Science 257:548–551, 1992.

124. R Allendoerfer, GS Deepe Jr. Intrapulmonary response to Histoplasma capsulatum in gamma interferon knockout mice. Infect Immun 65:2564–2569, 1997.

125. PB Eisenhauer, RI Lehrer. Mouse neutrophils lack defensins. Infect Immun 60:3446–3447, 1992.

126. AR Lloyd, JJ Oppenheim. Poly's lament: the neglected role of the polymorphonuclear neutrophil in the afferent limb of the immune response. Immunol Today 13:169–172, 1992.

127. J Banchereau, RM Steinman. Dendritic cells and the control of immunity. Nature 392:245–252, 1998.

128. CA Guzman, M Rohde, KN Timmis. Mechanisms involved in uptake of Bordetella bronchiseptica by mouse dendritic cells. Infect Immun 62:5538–5544, 1994.

129. L Filgueira, FO Nestle, M Rittig, HI Joller, P Groscurth. Human dendritic cells phagocytose and process Borrelia burgdorferi. J Immunol 157:2998–3005, 1996.

130. RA Henderson, SC Watkins, JL Flynn. Activation of human dendritic cells following infection with Mycobacterium tuberculosis. J Immunol 159:635–643, 1997.

131. M Svensson, B Stockinger, MJ Wick. Bone marrow-derived dendritic cells can process bacteria for MHC-I and MHC-II presentation to T cells. J Immunol 158:4229–4236, 1997.

132. DM Ojcius, Y Bravo de Alba, JM Kanellopoulos, RA Hawkins, KA Kelly, RG Rank, A Dautry-Varsat. Internalization of Chlamydia by dendritic cells and stimulation of Chlamydia-specific T cells. J Immunol 160:1297–1303, 1998.

133. T Kitajima, G Caceres-Dittmar, FJ Tapia, J Jester, PR Bergstresser, A Takashima. T cell mediated terminal maturation of dendritic cells: loss of adhesive and phagocytotic capacities. J Immunol 157: 2340–2347, 1996.

133a. LA Gildea, RE Morris, SL Newman. Histoplasma capsulatum yeasts are phagocytosed via very late antigen-5, killed, and processed for antigen presentation by human dendritic cells. J Immunol 166: 1049–1056, 2001.

134. S Saslow, J Schaeffer. Survival of *Histoplasma capsulatum* in experimental histoplasmosis in mice. Proc Soc Exp Biol Med 91:412–414, 1956.

135. JA Cain, GS Deepe Jr. Evolution of the primary immune response to Histoplasma capsulatum in murine lung. Infect Immun 66:1473–1481, 1998.

136. T Marth, BL Kelsall. Regulation of interleukin-12 by complement receptor 3 signaling. J Exp Med 185:1987–1995, 1997.

137. P Konecny, AJ Stagg, H Jebbari, N English, RN Davidson, SC Knight. Murine dendritic cells internalize Leishmania major promastigotes, produce IL-12 p40 and stimulate primary T cell proliferation in vitro. Eur J Immunol 29:1803–1811, 1999.

5
Dimorphism in Human Pathogens

Jon P. Woods
University of Wisconsin Medical School, Madison, Wisconsin

I. INTRODUCTION

Fungal morphologies (cellular structures) are extremely diverse, and morphogenesis (the transition between different structural morphotypes) is a complex process. "Dimorphic" is a somewhat arbitrary designation for fungi that display two predominant morphotypes. This classification is simplified, because frequently there may be more than one type of cell structure displayed by a morphotype, e.g., the mold morphotype may include mycelia (also termed "hyphae") and conidia (spores), and there may be intermediate morphologic structures, such as germ tubes and pseudohyphae. Although morphogenesis is an example of developmental changes, it is distinct from the life cycles displayed by other organisms such as some protozoan parasites. In general, there is no evidence that cycling between different morphotypes is obligatory for fungi. Instead, dimorphism is a mechanism some fungi have evolved to survive and proliferate in different environments. It follows that these fungi are capable of sensing environmental signals and of transducing these signals to activate regulatory cascades leading to morphogenesis, and moreover that different morphotypes are adaptively suited to different environments. The rapidly burgeoning field of fungal signal transduction, which clearly interfaces with morphogenesis and plays a role in controlling relevant gene expression, is covered in Chapter 6.

The relevance of dimorphism for fungal pathogenesis and infectious diseases is demonstrated by several lines of evidence. First, many fungi pathogenic for humans display dimorphism (some examples are shown in Table 1), generally with nonpathogenic and pathogenic morphotypes, although the delineation may be complex and incompletely characterized both in terms of underlying biological mechanisms and clinical disease. Second, a number of fungi pathogenic for plants (such as the corn smut agent *Ustilago maydis*) are also dimorphic, displaying nonpathogenic and pathogenic morphotypes, which emphasizes the adaptive utility of morphogenesis as one means of parasitizing the host, whether a human, other animal, or plant. Third, the ability of a dimorphic fungus to undergo the morphotypic transition is known or thought to be essential for pathogenesis and the establishment of successful infection by fungi in which it has been examined (discussed more below). Finally, there are fungal genera composed mainly of monomorphic species (e.g., *Penicillium*) but with an occasional dimorphic member (*Penicillium marneffei*). For this case, *P. marneffei* is the only *Penicillium* species that causes systemic disease in humans as well as the only dimorphic species in this genus. Thus, by a variety of criteria, fungal dimorphism and morphogenesis have been shown to be clearly essential or strongly correlated with the ability to cause human disease. This chapter is intended to provide an eclectic

Table 1 Selected Dimorphic Fungi Pathogenic for Humans

Fungus	Nonpathogenic morphotype	Nonpathogenic environmental niche	Route of infection	Pathogenic morphotype	Disease
Blastomyces dermatitidis	mold	soil	inhalation	budding yeast	pulmonary or disseminated blastomycosis
Candida albicans	yeast	gut and vaginal mucosa, skin	surface invasion	hyphae	mucocutaneous or disseminated candidiasis
Coccidioides immitis	mold	soil	inhalation	endosporulating spherule	pulmonary or disseminated coccidioidomycosis
Histoplasma capsulatum	mold	soil	inhalation	budding yeast	pulmonary or disseminated histoplasmosis
Malassezia furfur	yeast	sebaceous skin	surface invasion	hyphae	seborrheic dermatitis, dandruff, dissemination
Paracoccidioides brasiliensis	mold	soil	inhalation	budding yeast	pulmonary or disseminated paracoccidioidomycosis
Penicillium marneffei	mold	soil	inhalation	fission yeast	pulmonary or disseminated penicillosis
Sporothrix schenckii	mold	soil	direct inoculation, inhalation	budding yeast	lymphatic, mucocutaneous, or disseminated sporotrichosis

overview of fungal dimorphism, survey the diversity of pathogenic fungi that display dimorphism, and examine a subset of dimorphic fungal pathogens in more detail. Since the primary literature and also review articles on these topics would fill volumes, the material included in this chapter is necessarily limited in scope.

II. WHAT IS DIMORPHISM? TERMINOLOGY AND STRUCTURES

Figure 1 provides a schematic for the morphologies commonly observed with dimorphic fungi. The morphotypes considered here are vegetative, asexually reproducing forms. Most but not all dimorphic fungi can exist as yeast or mold. A yeast is generally a uninucleate, spherical or oval cell that may divide by budding or by fission in different fungi. (*Blastomyces dermatitidis* provides an exception, with multinucleate yeasts.) In contrast, the mold morphotype frequently displays mycelial (hyphal) and conidial (spore) elements. Mycelia or hyphae are filamentous structures that grow by apical extension and/or by lateral branching. The unidirectional growth of a hyphal tip may be considered the only example of fungal "motility" and can exert considerable physical pressure on a cellular scale. Walls between cells in a filament may be absent or incomplete, essentially leading to a multinucleate intertwined mass termed a mycelium. Condia are asexual spores that probably constitute the major infectious elements for many dimorphic fungi found in the environment due to their capacity for aerosolization and airborne dissemination, leading to inhalation or implantation on exposed host surfaces.

The terminology may seem more confusing with the addition of intermediate forms: germ tubes and pseudohyphae. Germ tubes are progressively elongating filamentous extensions from a yeast cell that represent a transition to the hyphal morphotype. At the beginning of germ tube

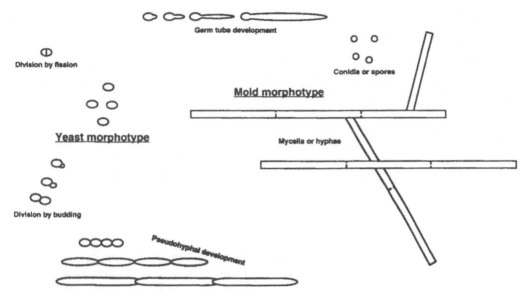

Figure 1 Dimorphic fungal morphologies.

formation, the extension may resemble the bud arising from yeast replication, but the germ tube grows until its length is many times the diameter of the parent yeast. Pseudohyphae are essentially unidirectional chains of elongated yeast cells that remain attached at the mother-daughter junction, rather than the separation of spherical mother and daughter yeasts. A pseudohyphal filament may be virtually indistinguishable from a hypha except for the presence of small indendations at the cellular junctions, in contrast to parallel hyphal walls. In clinical specimens, germ tubes and pseudohyphae may be observed most frequently for *Candida albicans*, which consequently has sometimes been termed a "polymorphic" fungus [1]. Although these structures emanate from yeast cells (germ tubes) or represent essentially a variant of yeast cells (pseudohyphae), they may appear very similar to true hyphae, and moreover, at least in the case of *Candida albicans*, these elements can be grouped more appropriately with hyphae in terms of pathogenesis and clinical diagnosis, as will be discussed more extensively below.

In addition to switching between different morphotypes, the generation of the different elements within the mold morphotype may itself be subject to environmental regulation. Generation of conidia from mycelia is termed conidiation or sporulation. Generation of mycelia from conidia is termed germination. Unlike yeasts and mycelia, conidia are generally thought not to replicate or reproduce cells of the same type. Instead, they may germinate to give rise to the other mold element, mycelia, or they may give rise to the alternate yeast morphotype, generally due to a change in environmental conditions. Additionally, mycelia can directly give rise to yeast cells when appropriate environmental changes occur. Considerable effort has been expended in characterizing different morphotypes of a variety of dimorphic fungi and in identifying environmental stimuli associated with morphotypic transitions. However, the specific biological mechanisms involved are incompletely understood, even in the most extensively studied organisms. For many dimorphic fungi, morphogenesis may be considered to occur in growing and dividing cells, with cells of one morphotype giving rise to cells of the other morphotype, rather than a quiescent single cell of one morphotype converting to the other morphotype.

Fungi that have been demonstrated to mate ("perfect" fungi) may display other morphologies during sexual reproduction under particular environmental conditions. Such conditions may include nutrient starvation and/or the presence of cells of different mating types in heterothallic fungi, which frequently is a reflection of intercellular signaling via pheromones and pheromone receptors. (Homothallic fungi display mating by cells of a single mating type.) For dimorphic perfect fungi, mating is generally practiced by only one of the predominant morphotypes. Although a linkage of particular mating types with pathogenic potential for several fungi has been demonstrated by epidemiologic studies or virulence testing in infection models, fungal mating during infection has generally not been demonstrated, and morphogenic changes associated with fungal sex have not been shown to contribute to fungal pathogenesis. Therefore, mating-associated morphologies will not be discussed further in this chapter.

III. ENVIRONMENTAL AND HOST-ADAPTED PATHOGENIC MORPHOLOGIES

A crucial concept is that human infection and disease cannot universally be associated with a particular morphotype, for fungi in general or for dimorphic fungi in particular. *Aspergillus fumigatus* and other pathogenic aspergilli are "monomorphic" molds that exist in the environment as mycelia and conidia. The conidia may be inhaled or may implant on exposed surfaces and germinate to give rise to invasive mycelia that can cause severe tissue destruction; aspergillosis is an opportunistic infection that is a significant clinical problem for particular immunocompromised individuals [2]. On the other hand, *Cryptococcus neoformans* is a monomorphic yeast (although sexual filaments involved in mating can arise under particular environmental conditions) which can be found in the environment and in the host, and is a prominent cause of fungal meningitis [3]. Likewise, while each dimorphic fungus generally shows a nonpathogenic environmental or commensal morphotype and a pathogenic host-adapted morphotype, the particular morphotype associated with human disease varies among different dimorphic fungi. For instance, dimorphic fungi acquired (demonstrably or presumptively) from the nonhuman environment display a nonpathogenic but infectious mold morphotype while living saprobically in the environment (usually the soil), and generally a pathogenic yeast morphotype while parasitizing the host. These fungi include *Histoplasma capsulatum* [4], *Blastomyces dermatitidis* [5], *Paracoccidioides brasiliensis* [6], *Sporothrix schenkii* [7], and *Penicillium marneffeii* [8]; all of these fungi replicate by budding in the yeast morphotype except for *P. marneffei*, which divides by fission as a yeast. *Coccidioides immitis* [9], the etiologic agent of "valley fever," also may be classified in this group since it displays a mold morphotype in the environment and also a host-adapted morphotype, but the latter is not yeast but rather an endosporulating spherule form. In contrast, dimorphic pathogenic fungi that inhabit human skin or mucosa as part of the normal flora, which include *Candida albicans* [10] and *Malassezia furfur* [11], display a yeast morphotype when existing as nonpathogenic commensal organisms, while disease due to these fungi is associated with the presence of an invasive hyphal morphotype or related elements. It is perhaps to be expected that different morphotypes may be observed in the same clinical specimen for these organisms, given their potential for either commensalism or pathogenesis in a human host, and also due to the complexity of environmental regulation of dimorphism, which will be discussed more later.

IV. WHY DO SOME FUNGI PRACTICE DIMORPHISM?

As mentioned above, fungal morphogenesis is an environmentally regulated process. The microenvironment of an infected host experiencing mycotic disease is simply one particular environment a fungus may encounter, albeit an important one to consider in this chapter. While the perspective is somewhat anthropomorphic, it's useful to consider how morphogenesis may be adaptive for the fungus in infecting and causing disease in the human host. For this discussion, the dimorphic fungi will be divided into two groups.

A. *Histoplasma capsulatum*

Environmental fungi principally found in the soil, as exemplified by *Histoplasma capsulatum*, generally display thermally regulated dimorphism, although other conditions can influence morphogenesis (Fig. 2, panel A). The nonpathogenic but infectious morphotype is mold, usually including mycelia and conidia. Mold is adapted to this low-temperature environment and can successfully compete and coexist with the plethora of other prokaryotic and eukaryotic soil microbes, including a variety of nonpathogenic and plant-pathogenic filamentous fungi. Conidia that are generated are readily aerosolized and may be inhaled by the mammalian host, penetrating to small airways. *H. capsulatum* displays both microconidia and tuberculate macroconidia, with the former considered the major infectious element due to their small size. Once in the lungs at 37°C, the organism converts to the yeast morphotype (or endosporulating spherule in the case of *Coccidioides immitis*), which displays a variety of features enabling successful infection leading to pulmonary or disseminated, systemic disease. *H. capsulatum* is a facultative intracellular yeast that can survive and replicate within macrophages, in spite of the professionally antimicrobial activities of this host cell. Morphogenesis is essential for virulence, in that *H. capsulatum* that is locked in the mycelial morphotype by a chemical treatment (discussed more below) is avirulent in an animal infection model [12,13]. Temperature variation is a sufficient stimulus for morphogenesis, since *H. capsulatum* can be grown in vitro in the same medium as either mycelia or yeast, by incubating at temperatures of 25°C or 37°C, respectively. However, other environmental conditions are clearly essential or can influence morphogenesis, only some of which have been characterized. For example, *H. capsulatum* has repressed sulfite reductase activity and is a functional cysteine auxotroph at 37°C, so omitting cysteine from the medium will not allow yeast growth at this temperature (this compound may play multiple roles and be involved in multiple processes in *H. capsulatum* morphogenesis) [14,15]. Additionally, for *H. capsulatum* and some other fungi in this group, as yet undefined host factors are clearly important, since there is in vivo and in vitro evidence that the morphotypic transition from mold to yeast at 37°C occurs more rapidly in the host or in the presence of host cells than it does in medium alone [16]. Therefore, a temperature of 37°C and perhaps some other host-derived environmental signal(s) are sensed by the fungus, which responds by a morphotypic transition to a form capable of exploiting the host.

B. *Candida albicans*

Candida albicans presents contrasts to the systemic dimorphic fungal pathogens in its nonpathogenic and pathogenic "lifestyles" and in the environmental signals that stimulate morphogenesis (Fig. 2, panel B). The "reservoir" for *C. albicans* is not the soil or external environment but is instead surfaces of the human body, where it lives commensally as part of the normal flora on moist skin or mucosal surfaces of the GI tract and vagina. Thus, *C. albicans* is "host adapted" even when not causing disease. In its nonpathogenic niche, it exists as a budding yeast which can be cultured or microscopically visualized from the oropharynx, rectum, or vagina of a substantial proportion of individuals. When variation occurs in the delicate host-microbe interplay, *C. albicans* undergoes a morphogenic transition to hyphae, with pseudohyphae and germ tubes also displayed. These forms are considered invasive, with the potential for directly penetrat-

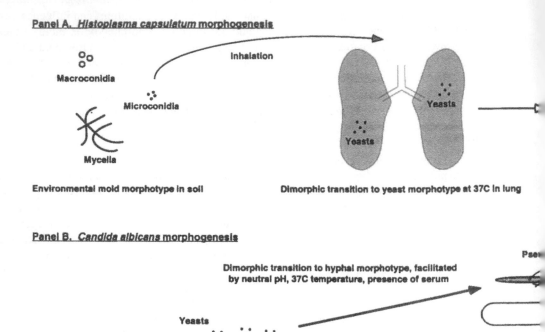

Panel A. *Histoplasma capsulatum* morphogenesis

Macroconidia

Inhalation

Microconidia

Mycelia

Yeasts

Yeasts

Environmental mold morphotype in soil

Dimorphic transition to yeast morphotype at 37C in lung

Panel B. *Candida albicans* morphogenesis

Dimorphic transition to hyphal morphotype, facilitated by neutral pH, 37C temperature, presence of serum

Pse

Yeasts

Mammalian mucosal cells

Submucosa

Blood vessel

Commensal yeast morphotype on host mucosal surface

Figure 2 Two examples of fungal dimorphism and pathogenesis. (A) *Histoplasma capsulatum* morphogenesis.

ing host mucosal cells and surfaces to reach deeper tissues and even the bloodstream, leading to dissemination in rare instances. Such outcomes are typically associated with compromise of host defense mechanisms, on a local level (e.g., skin disruption, change in the vaginal flora following antibacterial antibiotic treatment or over the menstrual cycle) or global level (e.g., HIV infection/AIDS, therapeutic immunosuppression in transplant patients). In its invasive, hyphal morphotype, *C. albicans* can be a very successful pathogen and cause quite severe and debilitating mucosal disease, but most frequently remains confined at this site. As with *H. capsulatum*, morphogenesis is critical for *C. albicans* pathogenesis, since a number of mutants that are unable to undertake the dimorphic transition (discussed more below) are generally substantially reduced in virulence in animal infection models. Although temperature also influences the morphotype of *C. albicans*, other environmental signals clearly contribute, including pH and the presence or absence of mammalian serum (a complex and undefined mixture of constituents).

Regulation of morphogenesis in this organism is incompletely understood, but in vitro conditions that mimic the mucosal (e.g., vaginal) environment (temperature <37°C, acid pH, absence of serum) promote yeast growth, while conditions that mimic the bloodstream (37°C temperature, neutral pH, presence of serum) promote invasive hyphal growth. Several different media are used for *C. albicans* in in vitro studies, and in some cases certain environmental cues can overcome the effects of others. It should also be noted that the situation is more complex and that this neat paradigm of commensal yeast and invasive hyphae for *C. albicans* is incomplete or not entirely adequate for all aspects of pathogenesis of this fungus. In cases of disseminated infection or invasion far beyond the initial mucosal barriers, such as in necrotic foci of infection in deep organs, yeast cells may be found in addition to or instead of hyphae. Individual yeasts may be more mobile (not motile) and capable of dissemination in blood and tissue fluids, and therefore this morphotype may be more adapted to this stage of disease (similar to the case for *H. capsulatum*). However, the mechanisms, regulation, and environmental signals for this process of morphogenesis from locally invasive hyphae to disseminating yeasts are not understood.

Thus, for both *H. capsulatum* and *C. albicans*, in vitro conditions that mimic the nonpathogenic environment of the organism promote existence in the nonpathogenic morphotype, and infection conditions promote the pathogenic morphotype. Some environmental cues contributing to morphogenesis, e.g., temperature, are shared between these fungi, while some may be distinct or have a different level of contribution. It is important to note that the particular morphotype associated with pathogenesis and invasive disease is different for these two fungi (yeast for *H. capsulatum*, hyphae for *C. albicans*). However, the locations of the nonpathogenic morphotypes and the mechanisms of infection and conversion to the pathogenic morphotypes also vary. In each case, the particular pathogenic morphotype is well suited for causing disease associated with the fungus (respiratory and systemic for *H. capsulatum*, mucosal invasion for *C. albicans*), so morphogenesis is an adaptive mechanism essential for the pathogenic potential of each fungus.

It is an intriguing evolutionary question as to how dimorphic fungi have evolved to be so ideally suited for parasitizing the human host. As mentioned earlier, existence as a pathogen does not appear to be obligatory or part of an invariant life cycle for fungi. In fact, it is likely that the majority of the planetary biomass of soil fungi such as *Histoplasma capsulatum* remains in the mold morphotype, does not interact with mammals, and does not undergo a morphogenic transition to yeast. Similarly, most *Candida albicans* organisms exist as commensal inhabitants of the human GI tract, moist skin, or mucosa, only occasionally causing invasive disease. This chapter seeks to describe various aspects of fungal dimorphism, but not actually account or provide an evolutionary rationale for the existence of the host-adapted morphotypes associated with disease.

V. IMPORTANCE OF MORPHOGENESIS FOR VIRULENCE

A. *Histoplasma capsulatum*

At present, no targeted *H. capsulatum* mutants defective in the expression of specific genes have been constructed that lock the fungus in a particular morphotype. Therefore, evidence for the necessity of the yeast morphotype for disease is somewhat indirect. Hallmark studies used the sulfhydryl blocking agent *p*-chloromercuriphenylsulfonate (PCMS) to lock several strains of the fungus in the mycelial morphotype, in a heritable manner [12,13]. The mechanism by which this interesting phenomenon occurs is entirely unknown. Treated mycelia remained mophotype-locked even after removal of the PCMS and a switch to 37°C (the usual signal for morphogenic transition to yeast). Yeasts that were treated with PCMS followed by removal of the chemical were able to convert to mycelia after lowering the temperature, but then subsequently did not convert back to yeast when the temperature was raised to 37°C. The irreversible and heritable nature of this phenomenon is consistent with some sort of genetic change being induced by PCMS treatment, but the genes involved or any other specific mechanisms have not been identified. At least some of the fungal strains locked in a mycelial morphotype by treatment with PCMS could still grow at 37°C, providing an exception to the usual association of a 37°C temperature and the yeast morphotype for *H. capsulatum.* These observations provided the opportunity for directly testing the role of the yeast morphotype in virulence. Morphotype-locked *H. capsulatum* mycelia were unable to cause productive infection in a mouse model, although they were able to stimulate host immune responses against the fungus. This work, although incompletely understood at a mechanistic level, provides the best evidence for the essential nature of the yeast morphotype in pathogenesis of *H. capsulatum,* or any systemic dimorphic fungus.

B. *Candida albicans*

In constrast to the situation for *H. capsulatum,* a plethora of different specific genes have been identified that influence *C. albicans* morphogenesis, and consequently a number of targeted mutants have been constructed that are defective in morphogenesis and locked in either the yeast or hyphal morphotype. Many of these genes are clearly or potentially involved in signal transduction, which is the subject of another chapter (see Chapter 6) and will not be discussed in detail here. An important point is that in virtually all cases, when Koch's molecular postulates [17] have been fulfilled for the role of a gene in morphogenesis, they have also been fulfilled for a role of the gene in virulence. For example, the *CPP1, CPH1, EFG1,* and *SSK1* genes are important for the transition of *C. albicans* from yeast to hyphae, and *cpp1, cph1, efg1,* and *ssk1* null mutants are locked in the yeast morphotype under at least some hyphae-inducing conditions, and show diminished virulence in mouse infection models, with a *cph1 efg1* double null mutant being avirulent [18–22]. Similarly, other genes have been identified that are important for the transition of *C. albicans* from hyphae to yeast, including *RBF1* and *TUP1,* and *rbf1* and *tup1* null mutants are locked in the hyphal morphotype [23,24].

Assessment of a role in virulence is influenced by the particular choice of animal infection model and the pathogenic stage(s) that is being tested. But accumulating evidence clearly indicates that when the ability of *C. albicans* to undergo morphotypic transitions is disturbed, the fungus is compromised in its ability to cause disease. The relevant genes have been identified by a variety of methods, but more candidates (particularly those involved in signal transduction) are being identified by ongoing (and nearly complete) sequencing of this organism's genome [25], and it is likely that support for the essential role of morphogenesis in virulence will continue to be generated.

VI. REGULATION OF MORPHOGENESIS

Since dimorphic fungi typically display a particular pathogenic morphotype associated with disease and there is evidence that morphogenesis is essential for virulence, a great deal of attention has been devoted to studying this process, with the rationale and motivation that blocking the morphotypic switch may prevent disease. Identification of key genes and encoded proteins may provide useful new targets for preventive (vaccine) or therapeutic (antibiotic) treatment. This idea is intellectually sound and well supported based upon available evidence. It might be noted, however, that exposure to dimorphic fungi is exceedingly common, from inhalation of soil fungi or carriage of commensal fungi on the body, and most of these interactions do not result in disease or result in disease that is contained by host responses in immunosufficient individuals. When infection has progressed to the stage of being noticed by the person and brought to the attention of a physician, generally the dimorphic transition is a past event and ongoing replication of the pathogenic morphotype is occurring. Regardless, much effort has been focused on identifying regulatory mechanisms for morphogenesis.

A. *Histoplasma capsulatum*

As implied above by the current lack of defined targeted mutants defective in morphogenesis, genes directly regulating morphotypic transitions in *H. capsulatum* have not yet been identified. Additionally, genome sequencing has not proceeded as far for this fungus as for *C. albicans*, and so identification of candidate homologs for regulatory or signal transduction genes in other organisms is not a readily available approach. It is likely that both some similar and some disparate mechanisms will be applicable for *H. capsulatum* as compared to *C. albicans*, a prediction based on the similar crucial nature of dimorphism for the biology and pathogenesis of these fungi, and on the distinctions in their environmental niches and disease progression, as described above. Although signal transductions pathways have not been extensively elucidated in *H. capsulatum*, it is known, for example, that levels of the signaling molecule cAMP differ between the mycelial and yeast morphotypes, being about fivefold higher in mycelia [26]. Moreover, exogenous cAMP and agents that raise the intracellular concentration of cAMP induce the yeast-mycelia transition, even when the temperature is maintained at 37°C [27].

These findings would be consistent with a different activity of pathways that generate or degrade cAMP, and also different activation of cAMP-dependent pathways, in the two morphotypes. However, *Paracoccidioides brasiliensis*, another systemic dimorphic fungus with much genetic and phenotypic similarity to *H. capsulatum*, shows the opposite trend in relative cAMP concentrations in the two morphotypes, with higher levels in yeast [28]. Better elucidation of regulatory mechanisms and cause-and-effect phenomena awaits identification of relevant important genes in these fungi.

B. *Candida albicans*

In constrast to the situation for *H. capsulatum*, numerous genes involved in the regulation of morphogenesis have been identified in *C. albicans*, generally genes involved in signal transduction pathways, a topic mentioned earlier and considered in another chapter (see Chapter 6). Investigators of this organism have been able to exploit the genomic sequence to search for candidate gene homologs, and also have artfully made comparisons to a framework based on the closely related model yeast *Saccharomyces cerevisiae*, which has been very extensively studied, finding both similarities and differences. Much progress has been made in identifying and characterizing components of both distinct and overlapping signal transduction pathways leading from sensation of particular environmental stimuli to different downstream effects including morphotypic transitions.

This work is addressed in another chapter (see Chapter 6), but here it will be noted that continued elegant and important studies typically, and unsurprisingly, raise further questions and emphasize the biological complexity of *C. albicans*. For instance, as mentioned above, the *CPH1* and *EFG1* genes contribute to the ability of *C. albicans* to exist and grow in the hyphal morphotype. A *cph1 efg1* double null mutant was originally shown to be extremely defective in hyphal growth when examined in vitro, and to be avirulent in mouse infection models [22]. However, Kumamota et al. [29] used a gnotobiotic piglet infection model and showed some level of virulence by this mutant in this model, and moreover observed hyphae in infected animal tissue. Upon further in vitro examination, this mutant was demonstrated to form hyphae in vitro when embedded in agar or growing on a rough agar surface, thus implicating physical or microaerophilic environmental signals that had not been extensively considered. This work demonstrates the complexity of *C. albicans* morphogenesis and the diversity of environmental signals contributing to morphotypic transitions. It also emphasizes the importance of the particular infection model being used and our incomplete understanding of relevant in vivo signals.

VII. MORPHOTYPE-SPECIFIC GENE EXPRESSION

A related but distinct area of study has been identification of fungal genes specifically expressed in yeast or mycelial/hyphal cells. By virtue of being expressed only in the pathogenic morphotype, i.e., yeast in the case of *H. capsulatum* or hyphae in the case of *C. albicans*, such genes and the proteins they encode might also provide new vaccine or antibiotics targets. They may or may not provide much information on the regulation of morphogenesis and the existence of crucial signal transduction pathways or a hypothetical essential "master switch" controlling the dimorphic transition. The recently burgeoning development of genomic approaches and microarray technology have revealed, particularly in the case of *Saccharomyces cerevisiae*, that a substantial fraction of the genes of an organism can be differentially expressed under different environmental conditions, and each such gene may or may not influence the ability of the organism to grow in the different settings [30]. Thus, determining the significance and role in virulence of any one pathogenic morphotype-specific gene requires individual focused examination. Still, there is a reasonable rationale for pursuing morphotype-specific genes that continue to be expressed while the fungi are practicing their pathogenic lifestyle. As mentioned earlier, the dimorphic transition occurs early in disease progression and has generally been accomplished by the time a patient seeks treatment, calling into question the utility of therapeutically blocking morphogenesis in an infected human. In contrast, genes that are specifically expressed by the pathogenic morphotype of a fungus, and thereby presumably adaptively advantageous in the host environment, are still available candidates for targeting by antifungal therapies. Additionally, examination of individual morphotype-specific genes, especially the identification of their regulatory signals and proteins, may provide more global clues to morphogenesis and control pathways that reach beyond the individual genes.

A. *Histoplasma capsulatum* Yeast-Specific Gene Expression

Keath and colleagues [31–34] used a differential hybridization approach to identify genes expressed by *H. capsulatum* yeasts and not by mycelia, and isolated a number of *yps* (yeast phase–specific) genes, the most extensively studied of which is *yps-3*. This gene encodes a small protein found to be released into culture medium as well as localized to the cell wall, although its surface exposure has not been demonstrated. Its function has not been determined. Recently, we have shown that expression of this gene is further upregulated, albeit modestly, during yeast infection of macrophages as compared to in vitro broth growth [35]. This finding reveals an apparently distinct level of control, since both in vitro and in vivo (infecting) *H. capsulatum* cells were of the yeast morphotype. The Keath laboratory has examined transcriptional regulation of *yps-3*, and used Southwestern blotting to identify morphotype-specific nuclear proteins that bind the upstream, presumably promoter region of the *yps-3* locus. In particular, a mycelia-specific protein designated p30M has been suggested as a candidate repressor of *yps-3* gene expression in this morphotype [34]. This approach should yield information specifically about *yps-3* regulation and also has the potential for identifying more far-reaching control pathways influencing gene expression.

Interestingly, we identified *yps-3* transcripts displaying alternate polyadenylation during our macrophage infection experiments, indicating the possibility of a different type of regulation via 3' untranslated region processing and differential transcript stability [35]. Additionally, we have noted shared features of the predicted Yps-3p amino acid sequence with epidermal growth factor (EGF) motifs from a variety of organisms, and specifically conservation of features with the carboxyl-terminal domain of the WI-1 antigen, a demonstrated virulence determinant and adhesin of the related systemic dimorphic fungus *Blastomyces dermatitidis* [35]. This finding would be consistent with possible roles of the protein in attachment to host cells, tissues, or extracellular matrix, or in signaling between host or fungal cells. Elucidation of the regulatory mechanisms and function of *yps-3* and evaluation of any role in virulence await further work.

Goldman and colleagues [36–38] identified another yeast-specific *H. capsulatum* gene by an indirect method arising from other studies. They identified and purified a small calcium-binding protein (CBP) and cloned the encoding gene (*CPB1*). Subsequently, they made the interesting observation that expression of this gene is restricted to yeast and not displayed by mycelia, and have used promoter truncations along with a *lacZ* reporter gene to identify upstream regions implicated in differential transcriptional regulation. Moreover, they found that *CBP1* plays an important role in the enhanced ability of yeast to grow in calcium-limited media relative to mycelia, raising the possibility that the encoded protein plays a role in calcium acquisition reminiscent of the role of siderophores in iron acquisition.

Significantly, preliminary work indicates that gene disruption to construct a *cbp1* null mutant reduces virulence in both cell culture and mouse models of infection [39]. Thus, this fortuitous identification of another *H. capsulatum* yeast-specific gene is leading to important findings relevant to regulation of gene expression and to pathogenesis.

A number of other *H. capsulatum* genes have been shown to be differentially expressed in yeast and mycelia, or during the morphogenic transition between the two forms, after original identification based on homology to genes from other organisms. This group includes genes encoding actin (transient decrease in expression during mycelia-yeast transition) [40], α-tubulin (expressed more highly in mycelia) [41], β-tubulin (expressed more highly in mycelia) [41], calmodulin (expressed more highly in yeast) [42], and *cdc2* (expressed more highly in yeast) [43]. As mentioned previously, many genes may be differentially expressed by an organism under different conditions, and the significance and implications of such a phenomenon are not certain. It should also be stated that estimation of gene expression is always relative, and identification of a "constitutive" control for normalization may be problematic. The regulatory mechanisms and roles in morphogenesis and in virulence for these *H. capsulatum* genes have not been characterized. Of course, the dimorphic transition clearly represents a huge morphological, metabolic, and biochemical change for the cell, so the involvement and differential expression of genes encoding structural components, signaling molecules, and cell cycle regulators are not surprising.

B. *Candida albicans* Hypha-Specific Gene Expression

Birse et al. [44] used a differential hybridization approach to identify genes expressed by *C. albicans* hyphae and not by yeasts, and isolated *ECE1*, which is not only expressed in a hypha-specific manner, but its expression correlates with the extent of cell elongation. However, disruption of this gene resulted in no detectable phenotypic effects. Several *C. albicans* genes have been identified as expressed only by the hyphal morphotype that encode cell wall proteins, including *HWP1*, *ALS3*, and *HYR1*. *HWP1* (for *h*yphal *w*all *p*rotein) encodes a surface-exposed protein that serves as a substrate for mammalian transglutaminases, thus enabling covalent attachment between the fungus and host cells. An *hwp1* null mutant was diminished in virulence in mouse infection models, and also was defective for hyphal development both in vitro and in vivo [45–48]. Thus, this gene and encoded protein are important for morphogenesis, host cell attachment and invasion, and virulence. *ALS3* encodes a member of a family of surface adhesion molecules [49], while the function of *HYR1* is unknown but apparently nonessential, since disruption and misexpression of this gene resulted in no detectable phenotypic changes [50]. These examples emphasize that a gene's identification as hypha specific may correlate with an important role in pathogenesis and perhaps in the regulation of morphogenesis (*HWP1*) or, conversely, may not have any function, role, or significance demonstrated so far (*HYR1*). A *C. albicans* gene of known function, one of the chitin synthase genes, *CHS2*, is expressed in a hypha-specific manner, but its disruption did not result in a defect in virulence in mice and the gene was not essential for morphogenesis, although hyphal development was delayed and characterized by a reduced chitin content [51]. Although the functions of such hypha-specific genes in *C. albicans* are not all known, their expression has been studied in mutants defective in various signal transduction genes such as *CPH1*, *EFG1*, and *TUP1*, in order to elucidate regulatory pathways and possible "crosstalk" between signaling cascades [52].

C. *Blastomyces dermatitidis* Yeast-Specific Gene Expression

Blastomyces dermatitidis is a systemic pathogenic fungus that is closely related to *H. capsulatum* genetically. It also displays biological and pathogenic similarities including thermal regulation of morphogenesis from an environmental mold to a pathogenic yeast, although there are also important distinctions from *H. capsulatum*. Smith and coworkers used a differential hybridization approach to identify genes expressed by *B. dermatitidis* yeasts and not by mycelia, and isolated *bys1*, the regulatory mechanisms and function of which have not yet been determined [53]. The yeast-specific expression of another gene of this fungus was identified fortuitously, similar to the case of *H. capsulatum CBP1*. Klein and coworkers [54–56] originally identified the WI-1 antigen as a large, immunodominant cell wall protein that is surface exposed and released into culture medium and that elicits both humoral and cell-mediated immune responses in the mammalian host. They described a three-domain structure of the protein, including an amino-terminal hydrophobic membrane-spanning region, a large central domain consisting of tandem repeats of a peptide sequence with homology to *Yersinia* invasin, and a carboxyl-terminal EGF-like region, the similarity of which to *H. capsulatum* Yps-3p was mentioned above.

WI-1 has been shown to mediate attachment to mammalian host cells, and moreover Koch's molecular postulates have been fulfilled for demonstrating an essential role of this protein in virulence in a mouse infection model. Recent preliminary work [57] indicates that the gene encoding the WI-1 antigen is expressed specifically by the yeast morphotype and that this regulation occurs at a transcriptional level. Moreover, conservation of some upstream sequence elements with the presumptive promoter region of the *H. capsulatum yps-3* gene were noted. It will be very interesting if studies of these two fungi can reveal common mechanisms operating for regulation of morphogenesis or at least morphotype-specific gene expression in systemic dimorphic fungi.

VIII. SUMMARY

Dimorphic fungi are successful human pathogens associated with significant morbidity and mortality. Their morphogenesis and capacity to exist in different morphotypes are critically important for their virulence. Generally each dimorphic pathogenic fungus displays an environmental or commensal morphotype and a host-adapted pathogenic morphotype. The environmental signals that trigger the transition from the former to the latter are associated with disease establishment and progression, and the switch is adaptive for successful pathogenic habitation of the fungus' new niche in the host. Regulation of morphogenesis is incompletely understood but is the subject of intense investigation, as is identification of genes expressed in a morphotype-specific fashion. Given the essential nature of morphogenesis and dimorphism for disease, such differentially expressed genes as well as the relevant regulatory mechanisms may yield useful new targets for preventive (vaccine) or therapeutic (antibiotic) treatment of fungal infections.

REFERENCES

1. FC Odds. Pathogenic fungi in the 21st century. Trends Microbiol 8:200–201, 2000.
2. J-P Latgé. *Aspergillus fumigatus* and aspergillosis. Clin Microbiol Rev 12:310–350, 1999.
3. KL Buchanan, JW Murphy. What makes *Cryptococcus neoformans* a pathogen? Emerg Infect Dis 4:71–83, 1998.
4. DM Retallack, JP Woods. Molecular epidemiology, pathogenesis, and genetics of the dimorphic fungus *Histoplasma capsulatum*. Microbes Infect 1:817–825, 1999.
5. BS Klein. Molecular basis of pathogenicity in *Blastomyces dermatitidis*: the importance of adhesion. Curr Opinion Microbiol 3:339–343, 2000.
6. E Brummer, E Castaneda, A Restrepo. Paracoccidioidomycosis: an update. Clin Microbiol Rev 6: 89–117, 1993.
7. CA Kauffman. Old and new therapies for sporotrichosis. Clin Infect Dis 21:981–985, 1995.
8. TA Duong. Infection due to *Penicillium marneffei*, an emerging pathogen: review of 155 reported cases. Clin Infect Dis 23:125–130, 1996.
9. TN Kirkland, J Fierer. Coccidioidomycosis: a reemerging infectious disease. Emerg Infect Dis 2: 192–199, 1996.
10. JF Ernst. Transcription factors in *Candida albicans*—environmental control of morphogenesis. Microbiology 146:1763–1774, 2000.
11. MJ Marcon, DA Powell. Human infections due to *Malassezia* species. Clin Microbiol Rev 5:101–119, 1992.
12. G Medoff, M Sacco, B Maresca, D Schlessinger, A Painter, GS Kobayashi, L Carratu. Irreversible block of the mycelial-to-yeast phase transition of *Histoplasma capsulatum*. Science 231:476–479, 1986.
13. G Medoff, GS Kobayashi, A Painter, S Travis. Morphogenesis and pathogenicity of *Histoplasma capsulatum*. Infect Immun 55:1355–1358, 1987.
14. DH Howard, N Dabrowa, V Otto, J Rhodes. Cysteine transport and sulfite reductase activity in a germination-defective mutant of *Histoplasma capsulatum*. J Bacteriol 141:417–421, 1980.
15. B Maresca, GS Kobayashi. Dimorphism in *Histoplasma capsulatum*: a model for the study of cell differentiation in pathogenic fungi. Microbiol Rev 53:186–209, 1989.
16. LG Eissenberg, WE Goldman. *Histoplasma* variation and adaptive strategies for parasitism: new perspectives on histoplasmosis. Clin Microbiol Rev 4:411–421, 1991.

17. S Falkow. Molecular Koch's postulates applied to microbial pathogenicity. Rev Infect Dis 10(suppl 2):S274–S276, 1988.

18. C Csank, C Makris, S Meloche, K Schroppel, M Rollinghoff, D Dignard, DY Thomas, M Whiteway. Derepressed hyphal growth and reduced virulence in a VH1 family-related protein phosphatase mutant of the human pathogen *Candida albicans*. Mol Biol Cell 8:2539–2551, 1997.

19. H Liu, J Köhler, GR Fink. Suppression of hyphal formation in *Candida albicans* by mutation of a *STE12* homolog. Science 266:1723–1726, 1994.

20. VR Stoldt, A Sonneborn, CE Leuker, J Ernst. Efg1p, an essential regulator of morphogenesis of the human pathogen *Candida albicans*, is a member of a conserved class of bHLH proteins regulating morphogenetic processes in fungi. EMBO J 16:1982–1991, 1997.

21. JA Calera, X-J Zhao, R Calderone. Defective hyphal development and avirulence caused by a deletion of the *SSK1* response regulator gene in *Candida albicans*. Infect Immun 68:518–525, 2000.

22. H-J Lo, JR Köhler, B DiDomenico, D Loebenberg, A Cacciapuoti, GR Fink. Nonfilamentous *C. albicans* mutants are avirulent. Cell 90:939–949, 1997.

23. N Ishii, M Yamamoto, F Yoshihara, M Arisawa, Y Aoki. Biochemical and genetic characterization of Rbf1p, a putative transcription factor of *Candida albicans*. Microbiology 143:429–435, 1997.

24. BR Braun, AD Johnson. Control of filament formation in *Candida albicans* by the transcriptional repressor *TUP1*. Science 277:105–109, 1997.

25. http://www-sequence.stanford.edu/group/candida/

26. J Medoff, E Jacobson, G Medoff. Regulation of dimorphism in *Histoplasma capsulatum* by cyclic AMP. J Bacteriol 145:1452–1455, 1981.

27. M Sacco, B Maresca, BV Kumar, GS Kobayashi, G Medoff. Temperature and cyclic nucleotide induced phase transitions of *Histoplasma capsulatum*. J Bacteriol 146:117–120, 1981.

28. MI Borges-Walmsley, AR Walmsley. cAMP signalling in pathogenic fungi: control of dimorphic switching and pathogenicity. Trends Microbiol 8:133–141, 2000.

29. PJ Riggle, KA Andrutis, X Chen, S. Tzipori, CA Kumamoto. Invasive lesions containing filamentous forms produced by a *Candida albicans* mutant that is defective in filamentous growth in culture. Infect Immun 67:3649–3652, 1999.

30. TR Hughes, MJ Marton, AR Jones, CJ Roberts, R Stoughton, CD Armour, HA Bennett, E Coffey, H Dai, YD He, MJ Kidd, AM King, MR Meyer, D Slade, PY Lum, SB Stepaniants, DD Shoemaker, D Gachotte, K Chakraburtty, J Simon, M Bard, SH Friend. Functional discovery via a compendium of expression profiles. Cell 102:109–126, 2000.

31. EJ Keath, AA Painter, GS Kobayashi, G Medoff. Variable expression of a yeast-phase-specific gene in *Histoplasma capsulatum* strains differing in thermotolerance and virulence. Infect Immun 57: 1384–1390, 1989.

32. EJ Keath, FE Abidi. Molecular cloning and sequence analysis of *yps-3*, a yeast phase-specific gene in the dimorphic fungal pathogen *Histoplasma capsulatum*. Microbiology 140:759–767, 1994.

33. CH Weaver, KCF Sheehan, EJ Keath. Localization of a yeast phase-specific gene product to the cell wall in *Histoplasma capsulatum*. Infect Immun 64:3048–3054, 1996.

34. FE Abidi, H Roh, EJ Keath. Identification and characterization of a phase-specific, nuclear DNA binding protein from the dimorphic pathogenic fungus *Histoplasma capsulatum*. Infect Immun 66: 3867–3873, 1998.

35. EL Heinecke, JN Fenske, JP Woods. Upregulation during macrophage infection and novel 3′ alternate transcript processing of the yeast phase-specific *YPS3* gene of *Histoplasma capsulatum*. Abstracts of 99th General Meeting of American Society for Microbiology, Chicago, 1999, poster F-75, p 310.

36. JW Batanghari, WE Goldman. Calcium dependence and binding in cultures of *Histoplasma capsulatum*. Infect Immun 65:5257–5261, 1997.

37. JW Batanghari, GS Deepe Jr, ED Cera, WE Goldman. *Histoplasma* acquisition of calcium and expression of *CBP1* during intracellular parasitism. Mol Microbiol 27:531–539, 1998.

38. JB Patel, JW Batanghari, WE Goldman. Probing the yeast phase-specific expression of the *CBP1* gene of *Histoplasma capsulatum*. J Bacteriol 180:1786–1792, 1998.

39. S Kügler, TS Sebghati, LG Eissenberg, WE Goldman. Phenotypic variation and intracellular parasitism by *Histoplasma capsulatum*. Proc Natl Acad Sci USA 97:8794–8798, 2000.

40. J el-Rady, G Shearer Jr. Cloning and analysis of an actin-encoding cDNA from the dimorphic pathogenic fungus *Histoplasma capsulatum*. J Med Vet Mycol 35:159–166, 1997.

41. GS Harris, EJ Keath, J Medoff. Expression of α and β-tubulin genes during dimorphic phase transitions of *Histoplasma capsulatum*. Mol Cell Biol 9:2042–2049, 1989.

42. J el-Rady, G Shearer Jr. Isolation and characterization of a calmodulin-encoding cDNA from the pathogenic fungus *Histoplasma capsulatum*. J Med Vet Mycol 34:163–169, 1996.

43. G Di Lallo, S Gargano, B Maresca. The *Histoplasma capsulatum cdc2* gene is transcriptionally regulated during the morphologic transition. Gene 140:51–57, 1994.

44. CE Birse, MY Irwin, WA Fonzi, PS Sypherd. Cloning and characterization of *ECE1*, a gene expressed in association with cell elongation of the dimorphic pathogen *Candida albicans*. Infect Immun 61: 3648–3655, 1993.

45. JF Staab, P Sundstrom. Genetic organization and sequence analysis of the hypha-specific cell wall protein gene *HWP1* of *Candida albicans*. Yeast 14:681–686, 1998.

46. JF Staab, SD Bradway, PL Fidel, P Sundstrom. Adhesive and mammalian transglutaminase substrate properties of *Candida albicans* Hwp1. Science 283:1535–1538, 1999.

47. LL Sharkey, MD McNemar, SM Saporito-Irwin, PS Sypherd, WA Fonzi. *HWP1* functions in the morphological development of *Candida albicans* downstream of *EFG1*, *TUP1*, and *RBF1*. J Bacteriol 181:5273–5279, 1999.

48. N Tsuchimori, LL Sharkey, WA Fonzi, SW French, JE Edwards Jr, SG Filler. Reduced virulence of *HWP1*-deficient mutants of *Candida albicans* and their interactions with host cells. Infect Immun 68:1997–2002, 2000.

49. LL Hoyer, TL Payne, M Bell, AM Myers, S Scherer. *Candida albicans ALS3* and insights into the nature of the *ALS* gene family. Curr Genet 33:451–459, 1998.

50. DA Bailey, PJ Feldmann, M Bovey, NA Gow, AJ Brown. The *Candida albicans HYR1* gene, which is activated in response to hyphal development, belongs to a gene family encoding yeast cell wall proteins. J Bacteriol 178:5353–5360, 1996.

51. NA Gow, PW Robbins, JW Lester, AJ Brown, WA Fonzi, T Chapman, OS Kinsman. A hyphal-specific chitin synthase gene (*CHS2*) is not essential for growth, dimorphism, or virulence of *Candida albicans*. Proc Natl Acad Sci USA 91:6216–6220, 1994.

52. BR Braun, AD Johnson. *TUP1*, *CPH1*, and *EFG1* make independent contributions to filamentation in *Candida albicans*. Genetics 155:57–67, 2000.

53. EF Burg 3d, LH Smith Jr. Cloning and characterization of *bys1*, a temperature-dependent cDNA specific to the yeast phase of the pathogenic dimorphic fungus *Blastomyces dermatitidis*. Infect Immun 62:2521–2528, 1994.

54. LH Hogan, S Josvai, BS. Klein. Genomic cloning, characterization, and functional analysis of the major surface adhesin WI-1 on *Blastomyces dermatitidis* yeasts. J Biol Chem 270:30725–30732, 1995.

55. SL Newman, S Chaturvedi, BS Klein. The WI-1 antigen on *Blastomyces dermatitidis* yeasts mediates binding to human macrophage CD18 and CD14 receptors. J Immunol 154:753–761, 1995.

56. TT Brandhorst, M Wüthrich, T Warner, B Klein. Targeted gene disruption reveals an adhesin indispensable for pathogenicity of *Blastomyces dermatitidis*. J Exp Med 189:1207–1216, 1999.

57. PJ Rooney, BS Klein. Yeast-phase specific expression and regulation of WI-1, an essential virulence determinant of *Blastomyces dermatitidis*. Abstracts of 99th General Meeting of American Society for Microbiology, Chicago, 1999, abstract F-98, p 315.

6

Signaling and the Biology of Human Fungal Pathogens

Jose Antonio Calera
University of Salamanca, Salamanca, Spain

Richard A. Calderone
Georgetown University Medical Center, Washington, D.C.

I. INTRODUCTION

While virulence factors of several pathogenic fungi have been proposed, little is known about the regulation of their expression or the environmental factors (signals) in the host, be it animal or human, that trigger their expression. In *Candida albicans*, for example, extracellular enzymes (secretory aspartyl proteinases [SAPS] and phospholipase B), morphogenesis (yeast to hyphae transition), host recognition by cell surface molecules (adhesins), and phenotypic switching have been considered among the virulence attributes of this organism [1]. However, of all of these phenotypic properties, only signal pathways that regulate morphogenesis have been partially defined [2]. Recent literature indicates that deletions in genes of the morphogenesis signal transduction pathways in *Candida albicans* result in altered morphogenesis patterns and a reduction in the virulence of this organism [2–13]. At least two morphogenesis signal pathways have been proposed, but it is likely that other pathways exist. Like *C. albicans*, *Cryptococcus neoformans* has multiple virulence determinants, including capsule and melanin synthesis [14], and there also is an association of mating type and virulence [15,16]. However, until recently, little was known about the regulation of these virulence attributes, although an association of a nutrient-deprivation induced signal pathway and virulence has been suggested [17,18]. These and other examples of signal transduction as a means by which pathogenic fungi regulate their virulence and growth will be the focus of this chapter.

Before beginning the discussion of signaling in pathogenic fungi, a few general comments should be emphasized.

1. There are many similarities among the fungal and mammalian cell signal pathways. The mitogen-activated protein kinase (MAPK) cascade is common to both cell systems. In fact, there is a strong belief that in some ways, fungi, with their much simpler systems, can be used as models for the study of mammalian signal pathways [19].

2. There is cross-talk among different pathways, although this subject has not been well addressed in studies to date.

3. Sequence homology of signal proteins among different organisms does not imply a similar function.

Host

Signals   »→   Signal perception   »→   Transduction   »→   Gene transcription   »→   Adaptive response

Figure 1   A signal transduction pathway composed of a membrane-bound protein which perceives signal, the transduction of that signal by phosphotransfer through a MAP kinase pathway, and activation of gene transcription leading to an adaptive response. Host signals which trigger this pathway might include pH, temperature, or host proteins (as, for example, proteins of phagocytic cells). Receptors of signal such as G-proteins or histidine kinases become phosphorylated and initiate the phosphorylation of MAP kinases and, in the case of the *C. albicans* morphogenesis pathway, *CPH1* or *EFG1* transcription factors.

4.   Membrane receptors used in fungal signal pathways are not of the tyrosine kinase family [20]. Instead, it would appear that fungi use either G-protein coupled or histidine sensor kinases, which modulate the transfer of signals to the cell nucleus through a phosphorylation cascade.

There are no functional equivalents of the latter group of membrane receptors in mammalian cells, suggesting that they might offer unique targets for the development of anti-fungals. The components of a prototypical signal pathway involving either a histidine kinase membrane receptor or a G-protein receptor and their relationship to the MAPK and target gene activation are depicted in Figure 1. For an extensive discussion of signaling pathways in yeast, the reader is directed to reviews by Banuett [20], Gustin et al. [21], and Bolker [22].

II.   SIGNALING AND MORPHOGENESIS IN *C. ALBICANS*

C. albicans is a common pathogen of the immunocompromised patient, causing infections of mucosal surfaces as well as invasive disease. The type of candidiasis that occurs appears to depend upon the underlying host defect. For example, mucosal infections are common in the AIDS patient (vaginitis, oral, esophageal) but systemic disease is rare. On the other hand, invasive disease more often occurs in the neutropenic patient. There does not appear to be any association between strains of the organism and specific types of infection. As a pathogen of diverse body sites, the organism should have adapted mechanisms which it can utilize to survive under a variety of stress conditions in the host—i.e., pH, nutrient deprivation, microbial competition, or protective proteins of the host (blood, salivary, or those of phagocytic cells).

While most often a pathogen in the immunocompromised host, there is good evidence that *C. albicans* requires a number of factors which promote its colonization and invasion [1]. One of these factors is morphogenesis, or the conversion of unicellular yeast forms (isotropic growth) to a filamentous (polar) growth pattern, the latter referred to as hyphae or pseudohyphae. Operationally, a model that might explain the importance of each growth form in pathogenesis might include the following scenario. Yeast cells are believed to play a major role in the colonization of host surfaces and establishing a commensal relationship with the human host. From its mucosal location, such as the gastrointestinal tract, and as host defense mechanisms are subjugated by one of several factors, the organism converts to a filamentous form, invades the gut mucosa, and enters the lymphatics and blood. Intuitively, it would seem that the conversion to filamentous growth might better promote the invasive capabilities of the organism. While the progression of disease and spread of the organism at this point can only be guessed, more than likely, during invasion by filamentous growth, either hyphal fragments or perhaps newly formed yeasts are carried by the blood to target organs, where extensive hyphal/pseudohyphal (polar) growth ensues. Growth of the organism in tissues is followed by (or occurs simultaneously with) a cycle (or several cycles) of isotropic growth (budding) and subsequent dissemination to other sites. Thus, the pattern of pathogenesis (at least as it relates to morphogenesis of *C. albicans*) probably involves many rounds of yeast-hyphal/pseudohyphal-yeast interconversions resulting in dissemination and invasion of the host. Of relevance to the growth habits of *C. albicans* in the host, the rather low sensitivity of blood cultures [23] in the systemically invaded host might be explained by the discontinuous release of the smaller fungal elements (yeasts, hyphal fragments) into the bloodstream. Alternatively, the low sensitivity of blood cultures might be explained by the continuous release of a few organisms from infected tissues into the blood, but in numbers that escape detection.

It is very likely that not only are host signals different for triggering induction of yeast or filamentous growth but also host signals will vary according to the specific site of invasion. In any case, host signals which induce these interconversions are unknown. On the other hand, much is known about in vitro signals (see below). However, there is very little information about mechanisms that might enable the organism to sense its environment and regulate its genetic machinery to adapt to stress conditions in the host. It is certain that the morphologic transition from yeast to hyphae appears to be one of the most important factors that enables *C. albicans* to survive and grow in host tissues. Signal transduction pathways involved in morphogenesis of *C. albicans* have recently been partially identified [2–13]. However, most of the signal studies that have been completed have not defined the proteins involved in the environmental sensing of signals and how these proteins regulate morphogenesis.

A. Signals That Trigger Morphogenesis In Vitro

For *C. albicans*, the conversion of yeast cells to filamentous growth can be induced by a number of conditions in vitro [24]. Certainly, environmental factors (pH and temperature) are critical and, in general, nutrient-poor media and specific nutrients which are poor substrates for the organism (N-acetyl glucosamine and serum) also induce germination (polar growth). On the other hand, nutrient-rich tissue culture media (RPMI, M-199) also induce germination. Temperature affects germination, which occurs to a greater extent above 30°C than at lower temperatures, while the percent of germinating yeasts increases with increasing pH to an optimum at or around neutrality. Thus, the input signals for germination are many, a situation which might imply that a number of diverse signal pathways are utilized for transcription of genes required for morphogenesis. Current data indicate that there are at least two independent signal pathways which regulate morphogenesis (yeast-hyphal interconversion), although the components of both pathways are not completely known [2]. Of great importance, there is mounting evidence that each of these signal pathways is important in disease development, at least in a murine model of hematogenously disseminated candidiasis.

B. Methods for Studying the Morphogenesis Signal Transduction Pathways in *C. albicans*

The wealth of information on signaling and morphogenesis of *C. albicans* has been achieved through the development of efficient gene cloning procedures. For example, many genes of *C. albicans* (including putative signal genes) can either complement or suppress the corresponding mutations of genes which function in the pseudohyphal pathways of *Saccharomyces cerevisiae* [25,26]. Therefore, using this approach, it has been fairly straightforward to identify homologs of the *C. albicans* MAP kinase filamentation-invasion pathway as well as downstream genes which encode putative transcription factors. Thus, gene function can be ascertained based upon complementation of *S. cerevisiae* mutants with *C. albicans* libraries or specific isolated gene homologs. Further, PCR-based technology using primers designed according to sequences of known genes has resulted in the use of cloned PCR products as probes which can then be used in screening libraries of *C. albicans* for complete genes. However, a critical caveat to the studies just described is the need to establish function of a specific gene in *C. albicans*, especially in situations where the *C. albicans* homolog does not complement a specific *S. cerevisiae* mutant (sometimes for unknown reasons or when an *S. cerevisiae* homolog does not exist in *C. albicans*). In this regard, a number of homologs of *C. albicans* have turned out to be functionally different from the *S. cerevisiae* counterpart.

Thus, the analysis of gene function depended upon the validation of a method which would generate single and double-copy deletion mutants. This important experimental method is achieved through the "urablaster" gene disruption approach [27]. In this procedure, sequential disruption of a specific gene is accomplished by transforming the *C. albicans* Ura3⁻ CAI4 strain with a disruption cassette composed of the *C. albicans* URA3 flanked by *hisG* sequences of *Salmonella* and sequences of the specific *C. albicans* gene one is seeking to disrupt. As an example, in Figure 2 (deletion) is shown the disruption cassette for a gene encoding a two-component response regulator gene (*SSK1*) of *C. albicans* [28]. Ura3⁺ transformants are isolated and homologous recombination in one of the two *SSK1* alleles is verified by Southern hybridization (Fig. 2, *CSSK11*, lower). Isolation of single gene-deleted strains which are Ura3⁻ is accomplished by selection on 5-FOA plates (Fig. 2, ura⁻ segregants); such strains (*CSSK12*) are again transformed with the same deletion cassette, Ura3⁺ transformants are isolated, and deletion of the second allele is again verified by Southern hybridization (Fig. 2, *CSSK21*, lower). Of importance is the construction of a reconstituted (revertant) strain to ensure that the phenotype of the double deletant (homozygous null) is due to the specific deletion and not as a consequence of the manipulations which occur during transformation. Such a revertant (*CSSK23*) is isolated by transformation of the null strain (Ura3⁻ *CSSK22*) with a copy of the wild-type allele in juxtaposition with the URA3 (Fig. 2, reintegration).

Using this procedure, a number of *C. albicans* genes (including genes encoding signal proteins) have been functionally characterized. Further, the additional utilization of expression vectors has enabled investigators to align genes in signal pathways. Epistasis experiments rely on the rescue of the wild-type phenotype by the expression of downstream genes in specific null strains. Using the latter approach as well as phenotypic clustering of individual mutants, as stated above, at least two morphogenesis signal transduction pathways have been identified which activate hyphal growth in *C. albicans* [2,3,28].

Equally important to the establishment of molecular methods to assess gene function is the need for a standardized animal model which can be used to evaluate mutations for their effect on virulence. Of the many animal models that are available, the intravenous-inoculated, hematogenously disseminated, murine model of candidiasis remains the "gold standard" [28,29]. The set of parental, heterozygote, null, and revertant (reconstituted) strain constructs is usually compared by single-dose inoculation (10^6 yeast cells), and mortality and tissue load of organism are determined at specific time points. Some investigators use up to 10^7 yeasts per animal but such an amount of cells is too high to discern clearly between the mortality caused by the growth of cells or by the excess number of cells itself. Comparisons of growth, morphogenesis, and virulence are then possible with the otherwise isogenic set of parental, single-copy allelic strains (heterozygote and revertant), and null strains.

In conclusion, these methods have contributed to our understanding of gene function in this organism and have suggested an approach to the identification of virulence factors. The two pathways that have been the most completely characterized (the Cph1p and Efg1p pathways) and which, on the basis of epistasis studies, are thought to be parallel and independently regulated, will be discussed initially. Also, the role of sensor histidine kinases and the Ssk1p response regulator, which likely regulate morphogenesis in a Cph1p-Efg1p-independent pathway, as well

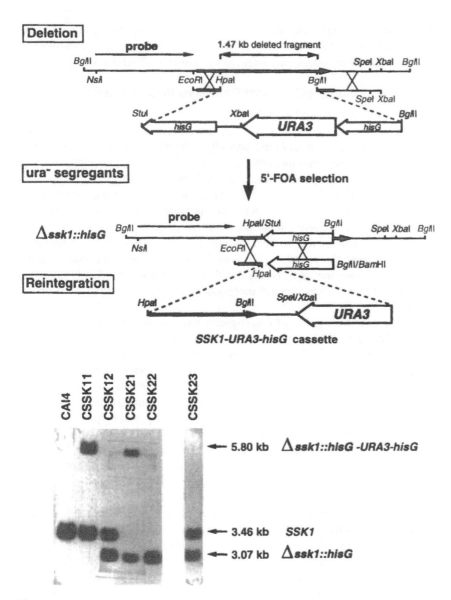

Figure 2 Construction of a Δssk1 of *C. albicans* using the "urablaster" technique [27,28]. The disruption cassette, consisting of the hisG-*URA3-hisG* and flanking sequences of the *SSK1*, is shown in the upper (Deletion) section of the figure. Ura3+ transformants are selected and ura3⁻ segregants obtained by growth on 5'-FOA. A second round of disruption is used to generate the *ssk1* null strain. A strain reconstituted with one copy of *SSK1* is constructed using the cassette depicted in the figure (Reintegration). Southern hybridizations are used to confirm the correct integration of cassettes following each transformation.

as the cell integrity pathway of *C. albicans* are discussed. However, as is the case with many of the signal pathways characterized in *S. cerevisiae,* crosstalking among pathways should be an important regulatory mechanism, but is only vaguely described in *C. albicans* [2].

C. Cph1p and Efg1p Morphogenesis Signal Pathways in *C. albicans*

In *S. cerevisiae,* pseudohyphal growth and agar invasion by pseudohyphae are induced under low nitrogen availability. The regulation of this growth appears to be associated with the mating MAPK pathway since strains mutated in *ste12* are partially defective in pseudohyphal growth [10]. However, this partial defect suggests that other genes may be involved in regulation, and therefore, *PHD1* was investigated since previous data indicated that its overexpression suppresses the filamentation and agar invasion defect of *ste12/ste12* mutants. Since it was thought that *PHD1* might be responsible for the residual regulatory activity because *ste12* mutants still form filaments, a *ste12/ste12 phd1/phd1* double mutant was constructed [11]. Such mutants did not form filaments and invade the agar medium. Transformation of the double mutant with either *STE12* or *PHD1* partially restored the pseudohyphal growth and invasion of the agar medium. Taken together, these data indicate that in *S. cerevisiae* there are two independently regulated signal pathways (Ste12p and Phd1) which control morphogenesis when the organism is grown under starvation conditions.

Likewise, studies by this same group of investigators have shown that mutations in the *C. albicans* homolog of *STE12* (*CPH1*) are only partially defective in their ability to switch from yeast to filamentous forms [10,11]. Thus, *cph1/cph1* strains have reduced hyphal formation on hyphal-inducing agar but with time form pseudohyphae or hyphae and germ tubes in serum. To examine the role of *EFG1,* the *C. albicans* homolog of *PHD1,* null strains (*efg1/efg1*) were constructed and examined phenotypically for growth under filamentation-inducing conditions [11]. On serum agar, the *efg1/efg1* null formed higher proportions of pseudohyphae (long filaments) which were morphologically different from those of parental cells which make hyphae predominantly. In serum, the *efg1/efg1* strain failed to form germ tubes and hyphae. However, because the *efg1/efg1* did form filaments under some of the conditions described above, as in *S. cerevisiae,* double mutants (*cph1/cph1 efg1/efg1*) were constructed and investigated for in vitro phenotype, interactions with macrophages, and virulence [11].

Such double mutants failed to form germ tubes under any conditions that were used. The fate of parent, the *CPH1/cph1* mutant and the double mutant were evaluated following their ingestion by a murine macrophage cell line [11]. While all strains were equally phagocytized, parental cells and the *CPH1/cph1* mutant each germinated within and eventually killed the macrophages. On the other hand, the double mutant remained as a yeast and did not kill the macrophages even though the macrophages contained as many as 80 cells of the *cph1/cph1* strain. Finally, each strain was evaluated for virulence and, while single deletant strains of *CPH1* and *EFG1* retained some degree of virulence, the double mutant was nearly avirulent (10^6 yeasts, no deaths, vs. 10^7 yeasts where 50% of the animals survived). Thus, the *C. albicans* double mutant behaves in vitro similarly to the *S. cerevisiae* double mutant, implying that there are also two independently regulated signal pathways in *C. albicans* and these pathways are at least partially required for virulence and morphogenesis.

The role of Efg1p in the morphogenesis of *C. albicans* has also been reported by Stoldt et al. [12]. These investigators have shown that when yeast cells are incubated in media which induce hyphae, transcription of *EFG1* declined but remained unchanged in cells incubated in YPD medium. In the YPD medium, cells remained as yeasts. These data indicate that Efg1p may not be associated with the conversion of yeast cells to hyphae. While double deletion mutants in *EFG1* were not obtained, heterozygotes were constructed in which one copy of *EFG1* was deleted and the promoter of the second allele was replaced with the promoter for the phosphoenolpyruvate carboxylase gene (*PKC1*), the latter of which is a glucose-repressible promoter. When cells were grown in the presence of glucose, moderate levels of *EFG1* transcript were noted which increased in amount when cells were grown in media which allowed derepression of the *PKC1* promoter. In both inducing and noninducing media, cells appeared elongated and resembled pseudohyphae, and the addition of 5% horse serum did not induce the formation of true hyphae. These observations indicate that reduced *EFG1* expression induced pseudohyphal formation and prohibited the formation of true hyphae. *EFG1* was also overexpressed in yeast cells; the morphology of these cells again indicated that *EFG1* induces pseudohyphal formation. However, when 5% horse serum was added to cells with a pseudohyphal phenotype, true hyphae were formed, so that *EFG1* does not inhibit the formation of true hyphae under the appropriate conditions.

Both *CPH1* and *EFG1* encode putative transcription factors, which are downstream of two independent signal pathways [2,3,10,12]. However, while the activation of *EFG1* has not been described, the transduction of signal and subsequent activation of *CPH1* has been shown to follow the MAP kinase pathway, the latter of which transfers signals from the cell surface to the nucleus for a variety of responses in eukaryotic cells. This pathway has been described in detail by Csank et al. [3,6] as well as by Brown and Gow [2]. Also, the genes encoding proteins of both pathways, as well as the in vitro phenotypes and virulence of strains deleted in each signal pathway gene, are indicated in Table 1. A brief description of both the Efg1p and Cph1p pathways is described below.

C. albicans homologs of the *S. cerevisiae* MAP kinase pathway include, in order of sequential phosphorylation, *CST20* (MAPKKKK), *HST7* (MAPKK), and *CEK1* (MAPK) [2–4]. Ste11p (MAPKKK) has been assigned as an intermediate kinase which is presumably phosphorylated by Cst20p and in turn phosphorylates Hst7p [2]. However, while identified by DNA sequencing, a Ste11p function in morphogenesis has not been determined as yet. That these genes lie in the

Table 1 Phenotypes of Null Strains with Deletions in Genes (Δ) Encoding Signal Transduction Proteins and Transcription Factors

| Strain | Morphogenesis[a] | | | | | | | |
| | Agar media | | | | Liquid medium serum | Virulence | Pathway[b] Cph1 or Efg 1 | Reference |
	Lee	Spider[c]	SLAD	Serum				
wt	H	I	I	H	H	V	Both	10,11
Δ*ras 2*	nd	nd	nd	Y	Y	nd	Efg1	33
Δ*ras1*	nd	nd	nd	nd	Y	nd	?	37
Δ*cst 20*	Y	Y	I	H	H	av	Cph1	3
Δ*hst7*	Y	Y	I	H	H	nd	Cph1	3
Δ*cek1*	Y	Y	I	H	H	att	Cph1	3
Δ*cpp1*[d]	H	ni	nd	nd	H	nd	Cph1	6
Δ*cph1*	Y	ni	nd	ni	H	V	Cph1	10
Δ*efg1*	nd	nd	nd	PH	Y	att	Efg1	11
Δ*efg1/cph1*	Y	Y	Y	Y	Y	av	(—)	11
Δ*mkc1*	H	ni	nd	nd	H	att	Neither	80,81

[a] H, hyphae; Y, yeasts; PH, pseudohyphae; V, virulent; att, attenuated; I, invasive; ni, noninvasive.
[b] Tentative assignment to either the Cph1p or Efg1p morphogenesis pathways.
[c] Spider without glucose.
[d] *cpp1* null when grown at 23°C under hyphal-inducing conditions.

same pathway has been postulated by two types of studies, as described previously in this chapter. First, null mutations in each of these genes result in similar phenotypes, i.e., defects in conversion to hyphae (usually on agar media). Second, epistasis studies have demonstrated that overexpression of *CPH1* when placed under control of the *ADH* promoter rescues the wild type phenotype in *cst20/cst20*, *hst7/hst7*, and *cek1/cek1* null strains, while overexpression of *HST7*, again driven by the *ADH* promoter, only rescues the *cst20* and the *hst7* null strains. This observation aligns *HST7* as upstream of *CST20* and *CPH1*. The targets of Cph1p activation are unknown in *C. albicans* but are presumed to be similar to a *STE12*-like pathway target. Regulatory controls are predicted to exist such as a protein phosphatase (Cpp1), which is thought to influence the phosphorylation of Cek1p [6]. Interestingly, null strains in *cpp1* form filaments in noninducing conditions only [6].

The Efg1p pathway is less well defined, but indications are that it is distinct from the Cph1 pathway [2]. The components of the Efg1 pathway have been suggested based upon studies with the CaCct8p chaperonin subunit protein of *C. albicans* [32]. Suppressors of morphogenesis were identified by transforming a *S. cerevisiae* strain, which expressed the *C. albicans EFG1*, with a *C. albicans* genomic library [32]. Transformants that did not produce pseudohyphae were identified, the plasmid was recovered, and its genomic insert was sequenced. One such transformant contained an insert whose sequence corresponded to the *CaCCT8*, which was previously shown to encode a subunit of the cytostolic chaperonin complex. While the transformed *S. cerevisiae* grew as yeasts at a rate identical to the recipient strain, pseudohyphal formation was completely blocked [32]. The specificity of the suppression exhibited by *CaCCT8* was examined. In addition to *EFG1*, *CaCCT8* also suppressed the $RAS2^{Val119}$ but not *CPH1*, *TEC1* (a transcription factor which acts together with *STE12*) or *CDC42*-induced pseudohyphal formation. The key observation that $RAS2^{Val119}$ or *EFG1*-induced pseudohyphal formation was equally suppressed most likely points to Ras2p as a trigger of filamentation through the Efg1p pathway [32]. While the above-mentioned observations indicate that the effect of *CaCCT8* is specifically on the Efg1p morphogenesis pathway, suppression of other Ras2p activities unrelated to morphogenesis by CaCct8p was noted. Thus, heat sensitivity, glycogen-trehalose accumulation, and lack of sporulation in *S. cerevisiae*, phenotypes that are related to a Ras2p function, were suppressed by coexpression of *CaCCT8* and a deleted version of *CaCCT8* (*cacct8-Δ1*).

Recent studies indicate that Ras signaling is required for the induction of germination of *C. albicans* [33,34]. Deletion mutants (*ras1-2/ras1-3*) of *C. albicans* were viable but grew at slower rates than parental cells with a doubling time threefold higher than parental cells at 23°C. Strains lacking *RAS1* failed to germinate in 10% serum at 37°C after 24 hr and formed round colonies without extending hyphae on serum agar medium, while the parental and heterozygote colonies were similar in appearance, forming myceliated colonies. Dominant-negative (RAS^{V16}) and dominant-active (RAS^{V13}) alleles were generated and plasmids constructed in which each of these alleles were expressed from the inducible maltose promoter, and used to transform parental cells (*RAS1/RAS1*) [33]. Cells transformed with a dominant-active allele of *RAS1* produced abundant hyphae in a sucrose (maltose promoter is induced) medium (pH 5.0) within 3 days at 30°C, while cells transformed with the dominant-negative allele failed to form hyphae after 2 weeks of incubation. Both parental (on sucrose medium) and transformed strains (on the glucose-repressible medium where the maltose promoter is repressed) failed to form hyphae. In *S. cerevisiae*, Ras2p also triggers filamentation via a cAMP-signal pathway. Since cAMP is a known inducer of hyphae formation in *C. albicans*, it is tempting to speculate that the Efg1p pathway precedes via a Ras2p and cAMP transduction, as suggested by Brown and Gow [2]. Verification of these components in a filamentation pathway of *C. albicans* awaits further study.

Finally, it is suggested that filamentation via Efg1p proceeds through a Gα-heterotrimeric protein(s), a conclusion which has been established for *S. cerevisiae*, but which is only speculative for *C. albicans*. Thus far, the only Gα-encoding gene isolated and mutated (*CAG1*) does not appear to encode a protein which is essential for morphogenesis [35]. However, a *GPA2* homolog of the Gα-heterotrimeric protein of *S. cerevisiae* has been recently described from *C. albicans* and may be functionally related to the *S. cerevisiae* Gpa2p which is essential for pseudohyphal formation [2].

The model presented by Brown and Gow [2] also includes proteins that have been shown to be required for morphogenesis. For instance, Int1p, Tup1p and Rbf1p (the latter two are transcriptional repressors of filamentation) have been described but linkages to any specific site in the Efg1p or Cph1p pathways are uncertain [30,31]. Also, while Ras2p is a plasma membrane protein and may provide an initial signal perception function, as stated above, the identity of an upstream, transmembrane protein(s) with a signal receptor function awaits further study.

D. Connections of the Morphogenesis Pathways with Effector Molecules

Transition of a yeast cell to a hyphal cell requires that cells change from a isotropic growth pattern to a polar type of growth. It can be fairly certain that the signal pathways described above are involved in this regulation. Equally important, there are a number of structural, hyphal-specific proteins that have been described, including Ece1p (function unknown [36]), Hwp1p (an adhesin [37–39]), Hyr1p (nonessential cell wall protein [40]), the ALS family of proteins (adhesins [41–45]), and Int1p [46]. However, the regulation of these proteins and their association with either of the two signal pathways described above has not been described.

E. Sensor Histidine Kinases and Response Regulators of
C. albicans

The two-component histidine kinases of bacteria and lower eukaryotes have been shown to be critical to either the expression of virulence genes in the host, growth, or the adaptation of organisms to environmental stresses [47–78]. The term ''two-component'' refers to the bacterial signal system which includes both a sensor protein (histidine kinase) and a response regulator protein, the latter of which is used by the organism as a transcriptional activator of genes involved in pathogenesis or growth [47,53]. Since homologs of bacterial histidine kinases and response regulators were discovered in *S. cerevisiae,* a search for their counterparts has been performed in other yeasts, including the pathogen *C. albicans.*

The *S. cerevisiae* two-component branch of the HOG (hyperosmotic glycerol) pathway of *S. cerevisiae* has been studied extensively [49,53,55,56,63,65,66]. In this system, phosphorylations of histidine and aspartate residues occur sequentially on the proteins Sln1p, Ypd1p, and Ssk1p [56]. Sln1p is a transmembrane, hybrid histidine kinase that is anchored to the plasma membrane of *S. cerevisiae*. It possesses both a key histidine residue (H-box), which is initially autophosphorylated, and a key aspartate residue within its response regulator domain to which the autophosphorylated His transfers its phosphate residue. Subsequent phosphorylations occur on a histidine residue of Ypd1p and an aspartate residue of Ssk1p [56]. Under normal growth conditions only, Sln1p becomes autophosphorylated and initiates a phosphorylation that ends with the phosphorylation of the Ssk1p response regulator. Phosphorylated Ssk1p fails to bind to a MAPKKK kinase protein (Ssk2p/Ssk22p), which prevents the activation of target genes. However, when cells are grown under high-osmolarity conditions, the unphosphorylated Ssk1p binds to Ssk2p, which then induces the autophosphorylation of Ssk2p and, in turn, activates the Ssk2p/Ssk22p-Pbs2p-Hog1p MAP kinase pathway resulting in the induction of target genes involved in glycerol production. The intracellular accumulation of glycerol allows cells to adapt to the high osmotic conditions in which it is confronted. It should be pointed out that the homolog of the Ssk1p protein in *Schizosaccharomyces pombe* (Mcs4) has a broader function. Mcs4 allows cells to grow under several restrictive conditions, such as oxidative, temperature, and protein inhibitor stress [68–71].

Three sensory histidine kinase genes (*CHK1, CaSLN1,* and *NIK1/COS1*) and a single response regulator gene (*SSK1*) have been identified in *C. albicans* [29,72–77]. The proteins encoded by the sensor kinase genes are, in fact, hybrid histidine kinases, which means that they possess both an H-box (containing a key histidine residue) and a regulator domain with a key aspartyl residue, which putatively undergo sequential phosphorylation. CaSln1p is a 150-kDa protein with a 37% overall identity to the *S. cerevisiae* Sln1p [72], while Nik1p/Cos1p is a 119-kDa protein with an overall 70% similarity to the *Neurospora crassa* Cos1p [72–74]. The function of both genes in *C. albicans* has been studied by constructing gene-deleted strains using the "urablaster" procedure described earlier in the chapter. A Δ*sln1* strain of *C. albicans* has reduced growth on high-osmolarity media [72]. *SLN1* was able to complement the *sln1* strain of *S. cerevisiae*, indicating that it may provide *C. albicans* cells with an adaptation function during osmotic stress. The second of the three hybrid histidine kinases of *C. albicans* (*NIK1/COS1*), as stated above, is a homolog of the *Neurospora crassa* Nik1, which in that organism is essential for the development of hyphae. Strains of *C. albicans* deleted in *nik1/cos1* do not form hyphae on agar media which are known to induce hyphal formation, such as serum and Spider agar, as readily as parental cells [74]. Thus, Nik1p/Cos1p may be part of a morphogenesis signal pathway.

CHK1 is, likewise, a hybrid histidine kinase and has been the focus of study in our laboratory [29,75,77]. The carboxy-terminus of Chk1p resembles other hybrid histidine kinases in having the typical histidine and aspartyl residues that are involved in phosphotransfer. On the other hand, the N-terminal half of Chk1p most closely resembles two putative sensor kinases from *S. pombe* (Genbank accession nos. Z98978 and AL031543). Because the function of the two histidine kinases from *S. pombe* has not been determined, similarities of Chk1p were sought with other proteins using standard searches of databases. We found that the N-terminus of Chk1p shows significant homology with the putative PknB Ser/Thr kinase from *Mycobacterium tuberculosis,* but the homology is limited to the VIa-XI domains of PknB and to a lesser extent with the calcium/calmodulin-dependent kinase type I from different species [77]. Thus, Chk1p contains a partial Ser/Thr kinase, a characteristic which is different from all other histidine kinases described for *C. albicans.* Domains VIa–XI of Ser/Thr kinases are thought to be involved in peptide binding and initiating phosphotransfer. Additionally, a P-loop, which could potentially be involved in ATP or GTP binding, was also identified in Chk1p. Thus, the features described for Chk1p are unique among the *C. albicans* histidine kinases but are similar to those of *S. pombe* [77].

Functional studies of *CHK1* of *C. albicans* were done using the "urablaster" technique to construct deletion mutants in *CHK1* [29,77]. Strains deleted of *CHK1* are flocculent in liquid media which induce hyphal formation, indicating that a change in the cell surface of the deleted strain has occurred such that cells now flocculate [77]. Thus, it would seem as if Chk1p is part of a signal system which regulates the expression of a cell surface determinant of hyphal forms of *C. albicans.* In comparison, yeast growth in the Δ*chk1* is similar to wild-type cells. The mutation, therefore, is specific for the hyphal form of the organism such that Chk1p is probably an important component of a signal pathway which is required for normal morphogenetic events. Colony morphology of Δ*chk1* is also different from colonies formed by parental cells. On agar media, which normally induce hyphal growth, wild-type colonies have a fuzzy appearance, while Δ*chk1* forms smooth colonies, composed of hyphae unable to extend beyond the periphery of the colony itself. We have also observed that *CHK1* is essential for the virulence of *C. albicans* in a murine hematogenously disseminated candidiasis model [29]. However, interestingly, Δ*chk1*

Table 2   In Vitro Phenotype of *C. albicans chk1* and *ssk1* Deletion Mutants Compared to CAF-2 (Parent)

Media	CAF-2	Δcssk1	Δchk1
m-199 broth, pH 7.5, 37°C	hyphae	hyphae[b]	hyphae[b]
m-199 broth, pH 3.5, 30°C	yeast	yeast	yeast
Colonies on Spider agar	fuzzy, myceliated	smooth	smooth
Colonies on m-199 agar	fuzzy, myceliated	smooth	smooth
Colonies on SLAD agar	invasive[a]	hyperinvasive	invasive

[a] Invasion or penetration of the agar media by filamentous growth.
[b] The hyphae flocculate extensively.
Source: Refs. 34, 35.

mutants are as virulent as parental strains in the rat vaginal candidiasis model [29]. As yet, we have not determined the signal which induces the phosphorylation of Chk1p, but it would appear that there are tissue-specific signals since *CHK1* would seem to be essential for invasion systemically but not in the vaginal canal.

The essential role of *CHK1* in a murine systemic candidiasis model has also recently been reported by others [76]. These authors also showed that *SLN1* and *COS1/NIK1* were each required for virulence in the same animal model, but the *CHK1* deletion caused a much more drastic reduction in virulence than the other two histidine kinases.

In addition to *CHK1*, we have characterized the first response regulator gene of *C. albicans* (*CaSSK1*), which is homologous to the *Saccharomyces cerevisiae SSK1* as well as the *Schizosaccharomyces pombe mcs4+* gene [28,78]. As with *CHK1*, gene-deleted strains in *CaSSK1* have been constructed [28]. The Δ*ssk1* mutants have altered morphogenesis patterns similar to that described above for the Δ*chk1* mutants. For example, the Δ*ssk1* colonies on agar-inducing media are smooth instead of fuzzy and hyphae formed in liquid culture (M-199) also flocculate extensively [28]. Unlike Δ*chk1*, cells within colonies on agar-inducing media grow as yeasts. Interestingly, on SLAD agar, a medium which is used to evaluate the ability of a strain to invade agar, the Δ*ssk1* strain forms extensive invasive hyphae, much more so than parental cells. Nevertheless, for both mutants, yeast growth appears to be similar in appearance to parental cells. As with the Δ*chk1*, Δ*ssk1* is avirulent in a murine model of hematogenously disseminated candidiasis [28], and both strains are rapidly cleared from both in the liver and kidney. A comparison of the in vitro phenotypes of the *CHK1* and *SSK1* nulls of *C. albicans* is depicted in Table 2.

F.   Cell Integrity Pathway of *C. albicans*

As with the mating/pseudohyphal signal transduction pathway described above, the cell integrity signal pathway (*PKC1*) of *S. cerevisiae* has also been partially identified in *C. albicans* [79–81]. The functions associated with this pathway are diverse and include the regulation of genes associated with polarized cell growth and cell wall construction. For *C. albicans*, a homolog of the *S. cerevisiae PKC1*-mediated *SLT2* (*MPK1*) has been isolated [79]. This gene, designated *MKC1* (mitogen kinase of *Candida*), was cloned by complementation of the lytic phenotype of a *S. cerevisiae slt2* mutant. Deletion mutants have been evaluated to establish the role of *MKC1* in *C. albicans* [80]. Phenotypically, null strains appeared elongated and remained attached compared to parental cells. A more drastic alteration of cell phenotype was observed when cells were grown at 42°C. By scanning electron microscopy, the cell surface of such cells had a ruffled appearance with changes occurring within 1 hr of incubation at the high temperature, in comparison to the smooth cell surface of parental cells. Of a variety of inhibitors tested for their effect on the null strain, only caffeine and inhibitors of glucan and chitin synthesis appeared to be more inhibitory to the null strain than parental cells. The augmented inhibitory activity depended upon the inhibitor tested, varying from twofold (cilofungin, echinocandin) to 20-fold (nikkomycin Z).

The overall composition of the cell wall in the null strain was only slightly altered compared to parental cells. However, O-glycosylation (as measured by antibody reactivity) was consistently

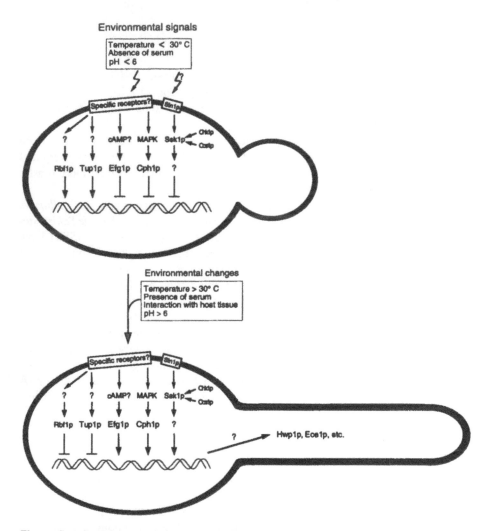

Figure 3   The yeast to hyphal transition in *C. albicans* is shown as a series of signal transduction pathways (cAMP, MAPK, Ssk1p) which are induced by environmental signals. Specific receptors for the cAMP and MAPK are unknown while Sln1p is tentatively assigned as the receptor for the histidine kinase-response regulator pathway(s). Transcription factors (Rbf1p, Tup1p, Efg1p, Cph1p) are indicated. Rbf1p or Tup1p induce yeast growth (upper). Deletion of either *TUP1* or *RBF1* leads to constitutive production of hyphae. In contrast, Efg1p and Cph1p, and perhaps other transcription factors regulated by Ssk1p, induce hyphal growth and their deletion leads to constitutive yeast growth.

higher in the null strain, indicating that the cell wall of the mutant was altered such that O-mannosylated residues were more readily exposed. Further, the null strain did not appear to be more susceptible to oxidative stress or nutrient deprivation but appeared to be defective in acquired thermotolerance. In regard to the effect of the deletion on morphogenesis, null strains were unable to invade Spider medium indicating that *MKC1* is required for morphogenesis under certain nutritional conditions. Finally, *MKC1* was able to complement the caffeine or thermosensitive phenotype of a *S. cerevisiae slt2* mutants and interact with the MAP kinase of *S. cerevisiae*, thus establishing that *MKC1* functions in that pathway. The importance of *MKC1* in virulence was assessed in a murine model of candidiasis [81]. Null strains were found to be less pathogenic as shown by longer survival times, percent mortality and reduced fungal load in tissues.

In Figure 3, we have summarized the MAP kinase signaling pathways for nutrient deprivation-induced filamentation in *C. albicans*. These pathways are remarkably similar to that of *S. cerevisiae*, at least for most of the genes described thus far. Not unexpectedly, some of the homologs differ in gene function. Examples of the latter include *TUP1*, a suppressor of hyphal formation in *C. albicans*, which is required for morphogenesis in *S. cerevisiae* [30]. Likewise, the *SSK1* homologs of these two organisms functionally are unrelated [80]. These observations indicate that *S. cerevisiae* will continue to be a good model to use in understanding processes in *C. albicans*, with the caveat that one can expect functional differences among gene homologs. Finally, it would appear that the SSK1 pathway of *C. albicans* may be independent of the Cph1/Efg1p signal pathways since epistasis transformation experiments with *CPH1* under the ADH promoter failed to rescue the wild-type phenotype in Δ*ssk1* null strains [3].

III. SIGNALING: GROWTH, MATING, AND VIRULENCE IN *CRYPTOCOCCUS NEOFORMANS*

Cryptococcus neoformans is a basidiomycetous yeast which is a common cause of meningoencephalitis in the AIDS patient. While primarily an opportunistic organism, like *C. albicans*, *C. neoformans* has several defined factors which contribute to its virulence, including capsule formation, melanization, growth at 37°C, and mating-type behavior/filamentation [14–18]. Unlike *C. albicans*, *C. neoformans* has a sexual system with defined mating types which permit genetic approaches to the study of gene function [82]. Transformation and gene disruption via homologous recombination have been used as approaches to define the virulence factors described above. A brief description of the putative virulence factors is given below.

C. neoformans produces an extracellular, glucuronoxylomannan polysaccharide capsule which provides an apparent escape mechanism for the organism from uptake by human (and animal) phagocytic cells and may also prevent cells from death by desiccation. Several genes which are essential for capsule formation have been identified, including *CAP59*, *CAP64*, and *CAP60* [83–85], and acapsular, avirulent phenotypes occur as a result of each gene deletion. While specific functions in capsule biosynthesis for each of these genes has not been defined, sequence information indicates that a putative transmembrane domain located at the N-terminus and a Gly324 residue in the center of the Cap59p are essential for functional activity. *CAP59* and *CAP64* are unlinked to each other but are linked to convergently transcribed genes [84,85]. Both genes were isolated by complementation of acapsular mutants, and their requirement for capsule formation has been determined through gene deletions. More recently, *CAP60* has been isolated and its function determined using an approach similar to that used with *CAP59* and *CAP64* [83]. Cap60p has a putative transmembrane domain and is similar in its sequence to

CAP59 at the center of its gene, although *CAP59* and *CAP60* could not be substituted for each other. Cap60p appears to be located on the nuclear membrane of cells [83].

Mutants of *C. neoformans* in melanin formation are attenuated in virulence in a murine model of cryptococcosis [14]. More recently, using molecular approaches, strains lacking *CNLAC1*, which encodes a phenoloxidase (laccase) protein and a rate-limiting enzyme in melanin formation, are also attenuated in virulence, while reintroduction of the *CNLAC1* restores virulence to a Mel⁻ mutant of *C. neoformans* [86]. It is postulated that the production of melanin protects cells from UV light (in nature) while perhaps acting as a oxidant scavenger to protect cells against the toxic effects of host phagocytes.

Mating type of *C. neoformans* has been suggested as a factor which contributes to virulence [15]. Such an observation is based upon the fact that most clinical and environmental isolates are of the α-mating type rather than the a-mating type. Additionally, the observation has been made that in congenic strains (which differ only in their mating type locus), the MATα strain of *C. neoformans* serotype D was more virulent than the MATa strain [15]. Interestingly, the MATα strain makes more melanin than MATa and has a 45-kb region which includes genes for pheromone production as well as a homolog of the *STE12* gene of *S. cerevisiae*, which, as stated above, is essential for starvation-induced filamentation in *S. cerevisiae* and *C. albicans* (*CPH1*).

Curiously, only MATα cells form filaments when starved for nitrogen and produce haploid MATα spores, a phenomenon referred to as haploid fruiting. The change from a yeast to a filament growth probably does not occur in the host, but its production in the environment may be responsible for the higher occurrence of MATα strains in disease since the spores which are formed are smaller than yeast cells (1–4 μm) and perhaps their small size makes them more accessible to penetration to the alveoli following their inhalation. It is very clear that *C. neoformans* utilizes several factors to adapt to the host environment, not the least of which is its ability to grow at 37°C. For each of these virulence factors, environmental signals which induce their expression have been suggested [18]. The signal which seems to be common to melanin and capsule synthesis, haploid filamentation, and mating is nutrient deprivation. Thus, capsule synthesis is regulated by iron deprivation [18] but also requires high CO_2/HCO_3^- ratios for optimum production, while melanin synthesis is induced by carbohydrate deprivation and haploid/filamentation and mating are induced by nitrogen deprivation [18].

A. Signaling Pathways in the Expression of Virulence Factors in *C. neoformans*

1. The Calcineurin (CNA1) Signal

C. neoformans must be able to grow at 37°C to survive in the host, a hypothesis which has been proven by an observation that a ts mutant in an unidentified gene which was unable to grow at 37°C was avirulent in murine cryptococcosis model [15]. Beyond this initial observation, the question to be asked is whether this organism uses a signal mechanism to sense temperature. In fact, it appears that the signal protein calcineurin is used to regulate growth of *C. neoformans* at 37°C [87]. Calcineurin is a highly conserved serine-threonine Ca^{2+}-calmodulin-activated protein phosphatase which is found in a variety of animal cells, including mammals [88]. The current model of calcineurin function in mammalian cells indicates that the protein forms a complex with and then dephosphorylates a phosphoprotein, the transcription factor NFAT1, upon activation of T-cells [88]. The hypophosphorylated NFAT1 is then translocated to the nucleus and forms a complex with other proteins (*fos* and *jun*) which then bind to target DNA. These interactions require at least two signal cascades (Ca^{2+} and PKC pathways). The association of calcineurin signaling activity with temperature regulation of growth in *C. neoformans* was suggested by the observation that inhibitors of calcineurin activity in mammalian cells (cyclosporin A and FK506) vary in their effect upon growth of *C. neoformans* and that the effect is temperature dependent [87]. For example, the MIC for growth inhibition by cyclosporin A at 22°C was >100 μg/mL but was only 1–5 μg/mL at 37°C with *C. neoformans* strain H99; hence, an association of temperature regulation with calcineurin [87]. To gain insight into its function in *C. neoformans*, the calcineurin gene (*CNA1*) was isolated and disrupted. The resulting *cna1* mutant was viable at 24°C but not at 39°C and, not unexpectedly, was avirulent in a rabbit cryptococcosis model as measured by the number of viable organisms present in the CSF of infected animals from 0 to 12 days postinfection [87]. However, the strain reconstituted with wild-type *CNA1* was not as virulent as the parental strain, indicating that complete restoration of the wild-type phenotype did not occur after reintroduction of the gene. In addition to an inability to grow at elevated temperatures, the *cna1* mutant was growth inhibited in 5% CO_2 and at an alkaline pH of 7.3 compared to wild-type cells, suggesting that the growth of the mutant is compromised under conditions that the organism is likely to encounter in the host. A model similar to that presented above for mammalian cells has been proposed, which suggests that increases in the intracellular levels of calcium and calmodulin occur in response to an unknown signal(s) and result in an activation of calcinuerin and the subsequent dephosphorylation of an unknown protein(s). Thus, *CNA1* is required for survival under the growth conditions described above (elevated temperature, CO_2, and alkaline pH). The substrate protein for calcineurin is unknown but could conceivably include ion pump proteins or transcription factor(s) [87].

2. The GPA1/cAMP-and GPB1-Regulated Pathways in *C. neoformans*

In contrast to the paucity of information on the role of heterotrimeric, G-proteins in the morphogenesis pathway of *C. albicans*, the genes encoding two G-protein subunits (GPA1[Gα subunit] and GPB1 [Gβ subunit]) have been identified and their functions determined in *C. neoformans* [89,90]. In brief, the GPA1 (Gα) subunit along with the participation of cAMP are critical to mating as well as the expression of melanin and capsule expression, i.e., virulence [89] (Fig. 4). It is very likely that the signal which triggers the GPA1 response is nutrient deprivation (starvation). On the other hand, the GPB1 (Gβ) subunit is essential for mating and filamentation (haploid fruiting) but not virulence [90]. The extracellular signal which triggers the GPB1 pathway is probably the mating pheromone signal (MFα1). The cell/virulence functions of both GPA1 and GPB1 are shown in Figure 4. *GPB1* was isolated by PCR using primers designed to conserved sequences (two conserved peptides) of other Gβ genes [90]. The PCR product was sequenced to confirm identity and was then used to screen a gDNA library to obtain the entire *GPB1* locus. The *C. neoformans* GPB1 protein shared identity with human (68%), *S. pombe* (40%), and *S. cerevisiae* (38%) G-protein β subunits. To determine function, a *gpb1* mutant was constructed by gene deletion via homologous recombination with an *ADE2* disruption cassette. The *gpb1* mutant was sterile while a strain containing the reintroduced *GPB1* regained its mating behavior, indicating the essential role that *GPB1* plays in mating.

As stated above, *GPA1* also is required for mating, but differences exist between *GPA1* and *GPB1* in regard to their requirements for cAMP in mating and the mating phenotype of each gene-deleted strain [90]. First, the *gpb1* mutant has an absolute mating defect whereas the *gpa1* mutant eventually will mate after prolonged incubation. Second, the addition of 2 mM cAMP suppresses the mating defect of *gpa1* but not *gpb1* mutants. There does not appear to be any interactions between these two proteins when the two-hybrid system was used for analyses. While essential for mating, *GPB1* is not required for virulence since the *gpb1* mutant made

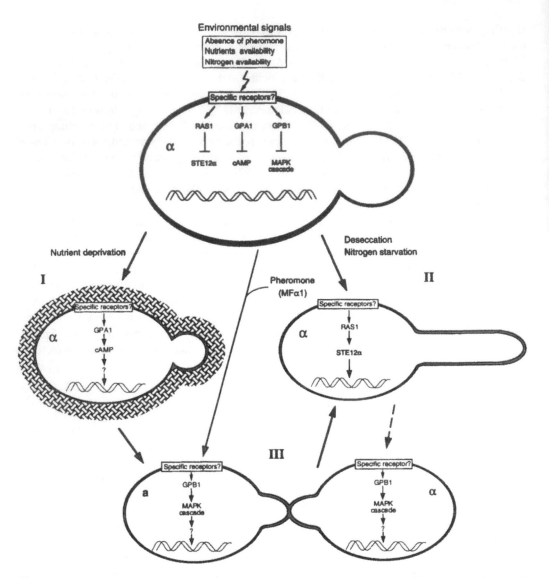

Figure 4   A *C. neoformans* yeast cell (upper) is depicted as responding to environmental signals via specific receptors (Ras1p, Gpb1p, or Gpa1p). Nutrients that induce melanin and/or capsule formation do so through the Gpa1 signal pathway (I), while under conditions of nitrogen starvation and desication, haploid filaments are formed via the Ras1 pathway (II). MATα cells of *C. neoformans* produce pheromone (MFα1), which induces mating when cells of the opposite compatibility type (MATa) are present (III). The mating process is thought to occur through the Gpb1 signal pathway. Capsule is indicated by the interlacing structure depicted in (I).

wild-type levels of melanin and capsule and was as virulent as parent cells in the rabbit model of meningoencephalitis. Epistasis studies were done to align GPB1 with a MAP kinase pathway. Two constructs were made in which expression of *CPK1* or *STE12* was placed under the control of the *C. neoformans GAL7* promoter (*GAL7-CPK1* and *GAL7-STE12*). *CPK1* and *STE12* were chosen for study since both are MAP kinase cascade proteins which have been identified in *C. neoformans*. Transformation of the *gpb1* mutant with *GAL7-CPK1* but not *GAL7-STE12* suppressed the mating defect, indicating that *CPK1* functions downstream of *GPB1*. On the other hand, the wild-type phenotype was not restored in the *gpa1* mutant by transformation with *GAL7-CPK1*, indicating that each subunit (GPA1 and GPB1) is very likely a component of parallel but independent pathways. The relationships between signal pathways and expression of virulence factors in *C. neoformans* are shown in Figure 4.

B. STE12 Functions in *C. neoformans* and *C. albicans*

A key component in signaling events which is common to both *C. albicans* and *C. neoformans* is Ste12p. For *C. albicans*, as described above, Cph1p (Ste12p) is part of the starvation-induced filamentation signal pathway and is associated with a virulence function; mutants lacking both *CPH1* and *EFG1* are avirulent, while single mutations in either gene attenuate virulence [10–12]. Recent observations on the function of the Ste12p in *C. neoformans* have been reported [91,92]. Overexpression of the *C. neoformans STE12α* stimulated production of the monokaryotic fruiting, mating factor α, and melanin [91]. Interestingly, *STE12* was identified only in α-mating types of *C. neoformans*. It is proposed that the Ste12p pathway is essential for these functions. As another approach to understanding the role of Ste12p in *C. neoformans* cell functions, deletion mutants were constructed and evaluated [92]. *STE12* was isolated from serotype A cells using the serotype D *STE12* as a probe of a gDNA library prepared from serotype A cells. Sequence analysis of hybridizing clones revealed the typical conserved domains of this family of proteins, including an N-terminal homeobox DNA binding domain and a C-terminal domain similar to other zinc finger proteins [92]. Homologous recombination was used to delete the wild-type *STE12* using the *ADE2* as a selectable marker, and reintroduction of the *STE12* into null strains allowed for evaluations of function in a gene-restored strain. The Δ*ste12* and wild-type strains were mated with a *MATα* and compared for filament (a result of cell fusion between mating types) and basidiospore formation typical of the sexual reproductive phase of the organism. The effect of the mutation on mating appeared to be quantitative, in that fewer filaments and basidia were formed at early times compared to wild-type cells, but within a few days the overall amount of mating was similar to wild-type cells. The mating response in *ste12* was dependent upon the medium used in mating such that the results described above occurred on V8 agar but not on SLAD or filament agar. On SLAD or filament agar, the *ste12* mutant displayed a more prominent mating defect, although with prolonged incubation time the mutant produced extensive mating structures.

The *ste12* mutant was also evaluated for its ability to undergo haploid fruiting [91]. For these experiments strain *MATα* serotype A H99 was used since this strain is unable to form haploid fruiting structures on several starvation media. Interestingly, the gene encoding the small G-protein Ras1 (*RAS1-Q67L*) stimulated haploid fruiting when expressed in strain H99 and, similarly, restored this process in the reconstituted strain but not the *ste12* mutant (personal communication). Thus, *STE12* functions in haploid fruiting and is apparently downstream of *RAS2*. Likewise, the transformation of a *gpb1* mutant with the *GAL7-STE12* fusion gene in cells expressing the Ras1-Q67L protein resulted in haploid fruiting in the *gpb1* mutant but did not restore the mating phenotype. Thus, Ste12p, while a component of the Ras2 signal pathway, does interact with the *GPB1* mating/haploid fruiting pathway but only for the induction of haploid fruiting (Fig. 4). Finally, the *ste12* mutant was evaluated for the production of melanin and capsule formation. While the levels of melanin were equivalent among wild-type, mutant, and reintroduced *STE12* strains, the level of capsule was estimated to be reduced by 50% (cell size estimated by photomicroscopy) in the mutant strain. In spite of this reduction in capsule production, differences in virulence among the strains tested were not observed in both rabbit and murine models of the infection [92].

IV. PROSPECTUS

Host environmental signals may be quite critical to the expression of genes that are essential for invasion by human pathogenic fungi. However, little is known about the cell receptors that perceive these signals and initiate an adaptive response for the organism. With *C. albicans*, for example, much is known of at least two MAP kinase signal pathways which regulate morphogenesis, but receptor and effector proteins (i.e., the beginning and the end of these pathways) have not been linked to any specific signal pathway. It is fairly certain that host signals may vary from tissue to tissue, since as stated above, *CHK1* of *C. albicans* is essential for intravascular disease but is not needed in the vaginal canal. There are other examples of environmentally regulated (tissue-specific) genes that encode virulence factors, as, for example, the pH-regulated genes *PHR1* and *PHR2* of *C. albicans* [93–95]. The relevance of these two genes to the virulence of this organism is tissue specific—i.e., systemic disease (*PHR1*) and vaginal infection (*PHR2*).

Much of the data discussed in this chapter have focused upon the fungal proteins that putatively detect environmental signals as well as the phosphotransfer proteins that are key to the regulation of an adaptive response to an environmental insult. These studies have resulted in the identification of signal proteins which appear to be functionally unique to fungi, bacteria, and higher plants, as for example the histidine kinase and response regulator proteins of *C. albicans*. This observation suggests that sufficient specificity may exist such that antifungal drugs may be developed which inhibit the activity of these proteins while being relatively nontoxic to mammalian cells [96–100]. In fact, there are reports on the identification of two-component histidine kinase inhibitors [101,102]. Along with their specificity, the histidine kinases and response regulator proteins are required for disease development, at least in murine models of systemic candidiasis. The importance of continued research on signal transduction pathways in the human pathogenic fungi is apparent from a recent observation which indicates that *cph1/efg1* mutants can still form hyphae and are virulent in a piglet model of candidiasis [103]. These data mean that additional signal pathways exist which thus far have not been identified using in vitro growth conditions. The picture becomes more complex, but hopefully, we are closer to finding the "magic bullet" which can be used to treat infections such as candidiasis and cryptococcosis.

REFERENCES

1. WL Chaffin, JL Lopez-Ribot, M Casanova, D Gozalbo, D, JP Martinez. Cell wall and secreted proteins of *Candida albicans*: identification, function, and expression. Microbiol Mol Biol Rev 62: 130–180, 1998.
2. AJP Brown, NAR Gow. Regulatory networks controlling *Candida albicans* morphogenesis. Trends Microbiol 7:333–338, 1999.
3. C Csank, K Schroppel, E Leberer, D Harcus, O Mohamed, S Meloche, DY Thomas, M Whiteway. Roles of the *Candida albicans* mitogen-activated protein kinase homolog, Cek1p, in hyphal development and systemic candidiasis. Infect Immun 66:2713–2721, 1998.

4.   E Leberer, D Harcus, ID Broadbent, KL Clark, D Dignard, K Ziegelbauer, A Schmidt, NA Gow, AJ Brown, DY Thomas. Signal transduction through homologs of the Ste20p and Ste7p protein kinases can trigger hyphal formation in the pathogenic fungus *Candida albicans*. Proc Natl Acad Sci USA 93:13217–13222, 1996.

5.   K Malathi, K Ganesan, A Datta. Identification of a putative transcription factor in *Candida albicans* that can complement the mating defect of *Saccharomyces cerevisiae*. J Biol Chem 269:22945–22951, 1994.

6.   C Csank, C Makris, S Meloche, K Schroppel, M Rollinghoff, D Dignard, DY Thomas, M Whiteway. Derepressed hyphal growth and reduced virulence in a VH1 family-related protein phosphatase mutant of the human pathogen *Candida albicans*. Mol Biol Cell 8:2539–2551, 1997.

7.   CJ Gimeno, PO Ljungdahl, CA Styles, GR Fink. Unipolar cell divisions in the yeast *Saccharomyces cerevisiae* lead to filamentous growth: regulation by starvation and RAS. Cell 68:1077–1090, 1992.

8.   JR Kohler, GR Fink. *Candida albicans* strains heterozygous and homozygous for mutations in mitogen-activated protein kinase signaling components have defects in hyphal development. Proc Natl Acad Sci USA 93:13223–13228, 1996.

9.   E Leberer, K Ziegelbauer, A Schmidt, D Harcus, D Dignard, J Ash, L Johnson, DY Thomas. Virulence and hyphal formation of *Candida albicans* require the Ste20p-like protein kinase CaCla4p. Curr Biol 7:539–546, 1997.

10. H Liu, J Kohler, GR Fink. Suppression of hyphal formation in *Candida albicans* by mutation of a *STE12* homolog. Science 266:1723–1726, 1994.

11. HJ Lo, JR Kohler, B DiDomenico, D Loebenberg, A Cacciapuoti, GR Fink. Nonfilamentous *Candida. albicans* mutants are avirulent. Cell 90:939–949, 1997.

12. VR Stoldt, A Sonneborn, CE Leuker, JF Ernst. Efg1p, an essential regulator of morphogenesis of the human pathogen *Candida albicans*, is a member of a conserved class of bHLH proteins regulating morphogenetic processes in fungi. EMBO J 16:1982–1991, 1997.

13. L Yaar, M Mevarech, Y Koltin. A *Candida albicans* RAS-related gene (*CaRSR1*) is involved in budding, cell morphogenesis and hypha development. Microbiology 143:3033–3044, 1997.

14. KJ Kwon-Chung, I Polacheck, TJ Popkin. Melanin-lacking mutants of *Cryptococcus* neoformans and their virulence for mice. J Bacteriol 150:1414–1421, 1982.

15. KJ Kwon-Chung, JC Edman, BL Wickes. Genetic association of mating types and virulence in *Cryptococcus neoformans*. Infect Immun 60:602–605, 1992.

16. B Wickes, ME Mayorga, U Edman, JC Edman. Dimorphism and haploid fruiting in *Cryptococcus neoformans*: association with the α-mating type. Proc Natl Acad Sci USA 93:7327–7331, 1996.

17. JA Alspaugh, JR Perfect, J Heitman. *Cryptococcus neoformans* mating and virulence are regulated by the G-protein α subunit GPA1 and cAMP. Genes Dev 11:3206–3217, 1997.

18. JA Alspaugh, JR Perfect, J Heitman. Signal transduction pathways regulating differentiation and pathogenicity of *Cryptococcus neoformans*. Fungal Genet Biol 25:1–14, 1998.

19. ME Cardenas, A Sanfridson, NS Carter, J Heitman. Signal transduction cascades as targets for therapeutic intervention by natural products. Trends Biotechnol 16:427–433, 1998.

20. F Banuett. Signaling in the yeasts: an informational cascade with links to the filamentous fungi. Microbiol Mol Biol Rev 62:249–274, 1998.

21. M Gustin, J Albertyn, M Alexander, K Davenport. MAP kinase pathways in the yeast *Saccharomyces cerevisiae*. Microbiol Mol Bio Rev 62:1264–1300, 1998.

22. M Bolker. Sex and crime: heterotrimeric G proteins in fungal mating and pathogenesis. Fungal Genet Biol 25:143–156, 1998.

23. PE Verweij, D Poulain, T Kobayashi, TF Patterson, DW Denning, J Ponton. Current trends in the detection of antigenemia, metabolites and cell wall markers for the diagnosis and therapeutic monitoring of fungal infections. Med Mycol 36S:146–155, 1998.

24. F Odds. *Candida* and Candidosis. 2nd ed. London: Bailliere Tindall, 1988, pp 42–59.

25. M Whiteway, D Dignard, DY Thomas. Dominant negative selection of heterologous genes: isolation of *Candida albicans* genes that interfere with *Saccharomyces cerevisiae* mating factor-induced cell cycle arrest. Proc Natl Acad Sci USA 89:9410–9414, 1992.

26. J Perez-Martin, JA Uria, AD Johnson. Phenotypic switching in *Candida albicans* is controlled by a *SIR2* gene. EMBO J 18:2580–2592, 1999.

27. WA Fonzi, MY Irwin. Isogenic strain construction and gene mapping in *Candida albicans*. Genetics 134:717–728, 1993.

28. JA Calera, R Calderone. Defective hyphal development and avirulence caused by a deletion of the *CSSK1* response regulator gene in *Candida albicans*. Infect Immun 68:518–525, 2000.

29. JA Calera, X-J Zhao, F De Bernardis, M Sheridan, R Calderone. Avirulence of *Candida albicans* CaHK1 mutants in a murine model of hematogenously disseminated candidiasis. Infect Immun 67: 4280–4284, 1999.

30. BR Braun, AD Johnson. Control of filament formation in *Candida albicans* by the transcriptional repressor TUP1. Science 277:105–109, 1997.

31. N Ishii, M Yamamoto, F Yoshihara, M Arisawa, Y Aoki. Biochemical and genetic characterization of Rbf1p, a putative transcription factor of *Candida albicans*. Microbiology 143:429–435, 1998.

32. F Rademacher, V Kehren, VR Stoldt, JF Ernst. A *Candida albicans* chaperonin subunit (CaCct8p) as a suppressor of morphogenesis and Ras phenotypes in *C. albicans* and *Saccharomyces cerevisiae*. Microbiology 144:2951–2960, 1998.

33. Q Feng, E Summers, B Guo, G Fink. Ras signaling is required for serum-induced hyphal differentiation in *C. albicans*. J Bacteriol 181:6339–6346, 1999.

34. L Yaar, M Mevarech, Y Koltin. A *Candida albicans RAS*-related gene (*CaRSR1*) is involved in budding, cell morphogenesis and hypha development. Microbiology 143:3033–3044, 1997.

35. C Sadhu, D Hoekstra, MJ McEachern, SI Reed, JB Hicks. A G-protein α subunit from asexual *Candida albicans* functions in the mating signal transduction pathway of *Saccharomyces cerevisiae* and is regulated by the a1-α2 repressor. Mol Cell Biol 12:1977–1985, 1992.

36. CE Birse, M Irwin, W Fonzi, P Sypherd. Cloning and characterization of *ECE1*, a gene expressed in association with cell elongation of the dimorphic pathogen, *Candida albicans*. Infect Immun 61: 3648–3655, 1993.

37. JF Staab, P Sundstrom. Genetic organization and sequence analysis of the hypha-specific cell wall protein *HWP1* of *Candida albicans*. Yeast 14:681–686, 1998.

38. JF Staab, SD Bradway, PL Fidel, P Sundstrom. Adhesive and mammalian transglutaminase substrate properties of *Candida albicans* Hwp1. Science 283:1535–1538, 1999.

39. JF Staab, CA Ferrer, P Sundstrom. Developmental expression of a tandemly repeat, proline- and glutamine-rich amino acid motif on hyphal surfaces of *Candida albicans*. J Biol Chem 271: 6298–6305, 1996.

40. D Bailey, P Feldman, M Bovey, N Gow, A Brown. The *Candida albicans HYR1* gene, which is activated on response to hyphal development, belongs to a gene family encoding yeast cell wall proteins. J Bacteriol 178:5353–5360, 1996.

41. LL Hoyer, S Scherer, AR Shatzman, GP Livi. *Candida albicans ALS1*: domains related to a *Saccharomyces cerevisiae* sexual agglutinin separated by a repeating motif. Mol Microbiol 15:39–54, 1995.

42. LL Hoyer, TL Payne, M Bell, AM Myers, S Scherer. *Candida albicans ALS3* and insights into the nature of the *ALS* gene family. Curr Genet 33:451–459, 1998.

43. Y Fu, SG Filler, BJ Spellberg, W Fonzi, AS Ibrahim, T Kanbe, MA Ghannoum, JE Edwards. Cloning and characterization of CAD1/AAF1, a gene from *Candida albicans* that induces adherence to endothelial cells after expression in *Saccharomyces cerevisiae*. Infect Immun 66:2078–2084, 1998.

44. Y Fu, G Rieg, WA Fonzi, PH Belanger, JE Edwards, SG Filler. Expression of the *Candida albicans* gene *ALS1* in *Saccharomyces cerevisiae* induces adherence to endothelial and epithelial cells. Infect Immun 66:1783–1786, 1998.

45. NK Gaur, SA Klotz. Expression, cloning, and characterization of a *Candida albicans* gene, ALA1, that confers adherence properties upon *Saccharomyces cerevisiae* for extracellular matrix proteins. Infect Immun 65:5289–5294, 1997.

46. CA Gale, D Finkel, N Tao, M Meinke, M McClellan, J Olson, K Kendrick, M Hostetter. Cloning and expression of a gene encoding an integrin-like protein in *Candida albicans*. Proc Natl Acad Sci USA 93:357–361, 1996.

47. JL Appleby, JS Parkinson, RB Bourret. Signal transduction via the multi-step phosphorelay: not necessarily a road less traveled. Cell 86:845–848, 1996.

48. B Arico, JF Miller, C Roy, S Stibitz, D Monack, S Falkow, R Gross, R Rappuoli. Sequences required
     for expression of *Bordetella pertussis* virulence factors share homology with prokaryotic signal
     transduction proteins. Proc Natl Acad Sci USA 86:6671–6675, 1989.
49. LJ Brown, H Bussey, RC Steward. Yeast Skn7p functions in a eukaryotic two-component regulatory
     pathway. EMBO J 13:5186–5194, 1994.
50. C Chang, SF Kwok, AB Bleecker, EM Meyerowitz. Arabidopsis ethylene-response gene ETR1:
     similarity of product to two-component regulators. Science 262:539–544, 1993.
51. JA Hoch, TJ Silhavy, eds. Two-Component Signal Transduction. Washington: ASM Press,
52. EM Hrabak, DK Willis. The lemA gene required for pathogenicity of *Pseudomonas syringae* pv.
     syringae on bean is a member of a family of two-component regulators. J Bacteriol 174:3011–3020,
     1992.
53. T Maeda, SM Wurgler-Murphy, H Saito. A two-component system that regulates an osmosensing
     MAP kinase cascade in yeast. Nature 369:242–245, 1994.
54. S Nagasawa, S Tokishita, H Aiba, T Mizuno. A novel sensor-regulator protein that belongs to the
     homologous family of signal transduction proteins involved in adaptive responses in *Escherichia
     coli*. Mol Microbiol 6:799–807, 1992.
55. IM Ota, A Varshavsky. A yeast protein similar to bacterial two-component regulators. Science 262:
     566–569, 1993.
56. F Posas, SM Wurgler-Murphy, T Maeda, EA Witten, TC Thai, H Saito. Yeast HOG1 MAP kinase
     cascade is regulated by a multi step phosphorelay mechanism in the SLN1-YPD1-SSK1 "two-
     component" osmosensor. Cell 86:865–875, 1996.
57. SC Schuster, AA Noegel, F Oehme, G Gerisch, MI Simon. The hybrid histidine kinase DokA is
     part of the osmotic response system of *Dictyostelium*. EMBO J 15:3880–3889, 1996.
58. MA Uhl, JF Miller. Integration of multiple domains in a two-component sensor protein: the *Borde-
     tella pertussis* BvgAS phosphorelay. EMBO J 15:1028–1036, 1996.
59. N Wang, G Shaulsky, R Escalante, WF Loomis. A two-component histidine kinase gene that func-
     tions in *Dictyostelium* development. EMBO J 15:3890–3898, 1996.
60. WT Chang, PA Thomason, JD Gross, PC Neweil. Evidence that the RdeA protein is a component
     of a multistep phosphorelay modulating rate of development in *Dictyostelium*. EMBO J 17:
     2809–2816, 1998.
61. LA Alex, KA Borkovich, MI Simon. Hyphal development in *Neurospora crassa:* involvement of
     a two-component histidine kinase. Proc Natl Acad Sci USA 93:3416–3421, 1996.
62. MJ Zinda, CK Singleton. The hybrid histidine kinase dhkB regulates spore germination in *Dictyostel-
     ium discoideum*. Dev Biol 196:171–183, 1998.
63. T Ketela, JL Brown, RC Stewart, H Bussey. Yeast Skn7p activity is modulated by the Sln1p-Ypd1p
     osmosensor and contributes to regulation to the HOG pathway. Mol Gen Genet 259:372–378, 1998.
64. WF Loomis, G Shaulsky, N Wang. Histidine kinases in signal transduction pathways of eukaryotes.
     J Cell Sci 110:1141–1145, 1997.
65. F Posas, H Saito. Activation of the yeast SSK2 MAP kinase kinase kinase by the SSK1 two-
     component response regulator. EMBO J 17:1385–1394, 1998.
66. C Schüller, JL Brewster, MR Alexander, MC Gustin, H Ruis. The HOG pathway controls osmotic
     regulation of transcription via the stress response element (STRE) of the *Saccharomyces cerevisiae*
     CTT1 gene. EMBO J 13:4382–4389, 1994.
67. BA Morgan, GR Banks, WM Toone, D Raitt, S Kuge, LH Johnston. The Skn7 response regulator
     controls gene expression in the oxidative stress response of the budding yeast *Saccharomyces cerevis-
     iae*. EMBO J 16:1035–1044, 1997.
68. G Cottarel. Mcs4, a two-component system response regulator homologue, regulates the *Schizosac-
     charomyces pombe* cell cycle control. Genetics 147:1043–1051, 1997.
69. JC Shieh, MG Wilkinson, V Buck, BA Morgan, K Makino, JB Millar. The Mcs4 response regulator
     coordinately controls the stress-activated Wak1-Wis1-Sty1 MAP kinase pathway and fission yeast
     cell cycle. Genes Dev 11:1008–1022, 1997.
70. S Stettler, E Warbrick, S Prochnik, S Mackie, P Fantes. The Wis1 signal transduction pathways is
     required for expression of cAMP-repressed gene in fission yeast. J Cell Sci 109:1927–1935, 1996.

71. K Shiozaki, M Shiozaki, P Russell. Mcs4 mitotic catastrophe suppressor regulates the fission yeast cell cycle through the Wik1-Wis1-Spc1 kinase cascade. Mol Biol Cell 8:409–419, 1997.

72. S Nagahashi, T Mio, N Ono, T Yamada-Okabe, M Arisawa, H Bussey, H Yamada-Okabe. Isolation of *CaSLN1* and *CaNIK1*, the genes for osmosensing histidine kinase homologues, from the pathogenic fungus *Candida albicans*. Microbiology 144:425–432, 1998.

73. T Srikantha, L Tsai, K Daniels, L Enger, K Highley, DR Soll. The two-component hybrid kinase regulator CaNIK1 of *Candida albicans*. Microbiology 144:2715–2729, 1998.

74. LA Alex, C Korch, CP Selitrennikoff, MI Simon. *COS1*, a two-component histidine kinase that is involved in hyphal development in the opportunistic pathogen *Candida albicans*. Proc Natl Acad Sci USA 95:7069–7073, 1998.

75. JA Calera, GH Choi, R Calderone. Identification of a putative histidine kinase two-component phosphorelay gene (*CaHK1*) in *Candida albicans*. Yeast 14:665–674, 1998.

76. T Yamada-Okabe, T Mio, N. Ono, Y Kashima, M Matsui, M Arisawa, H Yamada-Okabe. Roles of three histidine kinases in hyphal development and virulence of the pathogenic fungus, *Candida albicans*. J Bacteriol 181:7243–7247, 1999.

77. JA Calera, R Calderone. Flocculation of hyphae is associated with a deletion in the putative CaHK1 two-component histidine kinase gene from *Candida albicans*. Microbiology 145:1431–1442, 1999.

78. JA Calera, R Calderone. Identification of a putative response regulator two-component phosphorelay gene (*CaSSK1*) from *Candida albicans*. Yeast 15:1243–1254, 1999.

79. F Navarro-Garcia, M Sanchez, J Pla, C Nombela. Functional characterization of the *MKC1* gene of *Candida albicans*, which encodes a mitogen-activated protein kinase homolog related to cell integrity. Mol Cell Biol 15:2197–2206, 1995.

80. F Navarro-Garcia, R Alonso-Monge, H Rico, J Pla, R Santandreu, C Nombela. A role for the MAP kinase in cell wall construction and morphological transitions in *Candida albicans*. Microbiology 144:411–424, 1998.

81. R Diez-Orejas, G Molero, F Navarro-Garcia, J Pla, C Nombela, M Sanchez-Perez. Reduced virulence of *Candida albicans MKC1* mutants: a role for mitogen-activated protein kinase in pathogenesis. Infect Immun 65:833–837, 1997.

82. KJ Kwon-Chung. Morphogenesis of *Filobasidiella neoformans*, the sexual state of *Cryptococcus neoformans*. Mycologia 68:942–946, 1976.

83. YC Chang, KJ Kwon-Chung. Isolation of the third capsule-associated gene, *CAP60*, required for virulence in *Cryptococcus neoformans*. Infect Immun 66:2230–2236, 1998.

84. YC Chang, LA Penoyer, KJ Kwon-Chung. The second capsule gene of *Cryptococcus neoformans*, *CAP64*, is essential for virulence. Infect Immun 64:1977–1983, 1996.

85. YC Chang, BL Wickes, KJ Kwon-Chung. Further analysis of the *CAP59* locus of *Cryptococcus neoformans:* structure defined by forced expression and the description of a new ribosomal protein gene. Gene 167:179–183, 1995.

86. SD Salas, JE Bennett, KJ Kwon-Chung, JR Perfect, PR Williamson. Effect of the laccase gene, *CNLAC1*, on virulence of *Cryptococcus neoformans*. J Exp med 184:377–386, 1996.

87. A Odom, S Muir, E Lim, DL Toffaletti, J Perfect, J Heitman. Calcinuerin is required for virulence of *Cryptococcus neoformans*. EMBO J 16:2576–2589, 1997.

88. CS Hemenway, J Heitman. Calcineurin: structure, function, and inhibition. Cell Biochem Biophys 30:115–151, 1999.

89. T Tolkacheva, P McNamara, E Piekarz, W Courchesne. Cloning of a *Cryptococcus neoformans* gene, *GPA1*, encoding a G-protein α subunit homolog. Infect Immun 62:2849–2856, 1994.

90. P Wang, JR Perfect, J Heitman. The G-protein β subunit GPB1 is required for mating and haploid fruiting in *Cryptococcus neoformans* Mol Cell Biol, 20:352–362, 2000.

91. BL Wickes, U Edman, JC Edman. The *Cryptococcus neoformans STE12α* gene: a putative *Saccharomyces cerevisiae STE12* homologue that is mating type specific. Mol Microbiol 26:951–960, 1997.

92. C Yue, LM Cavallo, JA Alspaugh, P Wang, G Cox, JR Perfect, J Heitman. The *STE12α* homolog is required for haploid filamentation but largely dispensable for mating and virulence in *Cryptococcus neoformans*. Genetics 153:1601–1615, 1999.

93. FA Muhlschlegel, WA Fonzi. PHR2 of *Candida albicans* encodes a functional homolog of the pH-regulated gene *PHR1* with an inverted pattern of pH-dependent expression. Mol Cell Biol 17: 5960–5967, 1997.
94. SM Saporito-Irwin, CE Birse, PS Sypherd, WA Fonzi. *PHR1*, a pH-regulated gene of *Candida albicans,* is required for morphogenesis. Mol Cell Biol 15:601–613, 1995.
95. F De Bernardis, FA Muhlschlegel, A Cassone, WA Fonzi. The pH of the host niche controls gene expression in and virulence of *Candida albicans.* Infect. Immun. 66:3317–3325, 1998.
96. JR Davie, RM Wynn, M Meng, YS Huang, G Aalund, DT Chuang, KS Lau. Expression and characterization of branched-chain alpha-ketoacid dehydrogenase kinase from the rat. Is it a histidine-protein kinase? J Biol Chem 270:19861–19867, 1995.
97. R Diez-Orejas, G Molero, F Navarro-Garcia, J Pla, C Nombela, M Sanchez-Perez. Reduced virulence of *Candida albicans* MKC1 mutants: a role for mitogen-activated protein kinase in pathogenesis. Infect Immun 65:833–837, 1997.
98. RA Harris, JW Hawes, KM Popov, Y Zhao, Y Shimomura, J Sato, J Jaskiewicz, TD Hurley. Studies on the regulation of the mitochondrial alpha-ketoacid dehydrogenase complexes and their kinases. Adv Enzyme Regul 37:271–293, 1997.
99. KM Popov, NY Kedishvili, Y Zhao, Y Shimomura, DW Crabb, RA Harris. Primary structure of pyruvate dehydrogenase kinase establishes a new family of eukaryotic protein kinases. J Biol Chem 268:26602–26606, 1993.
100. KM Popov, Y Zhao, Y Shimomura, MJ Kuntz, RA Harris. Branched-chain alpha-ketoacid dehydrogenase kinase. Molecular cloning, expression, and sequence similarity with histidine protein kinases. J Biol Chem 267:13127–13130, 1992.
101. JF Barrett, RM Goldschmidt, LE Lawrence, B Foleno, R Chen, JP Demers, S Johnson, R Kanojia, J Fernandez, J Bernstein, L Licata, A Donetz, S Huang, DJ Hlasta, MJ Macielag, K Ohemeng, R Frechette, MB Frosco, DH Klaubert, JM Whiteley, L Wang, JA Hoch. Antibacterial agents that inhibit two-component signal transduction systems. Proc Natl Acad Sci USA 95:5317–5322, 1998.
102. JF Barrett, JA Hoch. Two-component signal transduction as a target for microbial anti-infective therapy. Antimicrob Agents Chemother 42:1529–1536, 1998.
103. PJ Riggle, KA Andrutis, X Chen, SR Tzipori, CA Kumamoto. Invasive lesions containing filamentous forms produced by a *Candida albicans* mutant that is defective in filamentous growth in culture. Infect Immun 67:3649–3652, 1999.

7
Adaptation of Fungi to Alterations in Ambient pH

Oliver Kurzai, Abdelmalic El Barkani, and Fritz A. Mühlschlegel
University of Würzburg, Würzburg, Germany

I. ADAPTATION AND PATHOGENICITY

Adaptation refers to capabilities or structures of an organism that improve its chances for survival in a specific environment. Many bacteria, especially archaebacteria, have attracted attention because of their ability to withstand extreme conditions in their ecological niches, be it high temperatures in boiling geysers, enormous salt concentrations, or the sea bed. In the case of human pathogens, adaptation means adjustment to the various sorts of defenses that have been developed to avoid colonization and infection. These span from very unspecific to highly sophisticated and precisely regulated mechanisms. An extraordinary number of strategies to escape the human body's defenses have been developed by the great variety of organisms that can cause infection. Infectious micro-organisms owe their success to their ability to adapt to and therefore escape the hostile environment of the human body. Despite this, mechanisms responsible for adaptation to host niches are poorly understood in human pathogenic fungi.

II. *METARHIZIUM ANISOPLIAE* AND THE PROBLEM OF DIGESTING AN INSECT

Degradation of several insects integument is an important part in the life of the entomopathogenic fungus *Metarhizium anisopliae*. This hyphomycete is responsible for the green muscardine disease of >200 insect species and is of great importance as an agent used for biological control of different pests and mosquito larvae [1,2]. To invade its hosts, *M. anisopliae* is equipped with an armament of different secretory proteolytic and chitinolytic enzymes working together in digesting the host's cuticle (Fig. 1) [3]. Expression of these enzymes has been shown to be carefully regulated during the course of an infection to ensure precise dissection of the hosts outer layers. During these studies, ambient pH turned out to be a central environmental stimulus responsible for up- and downregulation within the enzyme battalion. Thus, *M. anisopliae* can react to a changing environmental condition by differentially expressing several extracellular enzymes, each under its optimal working condition [3].

The control of virulence determinants by environmental signals is a feature well known for bacterial pathogens. The evolutionary advantage gained by coordinate expression of such

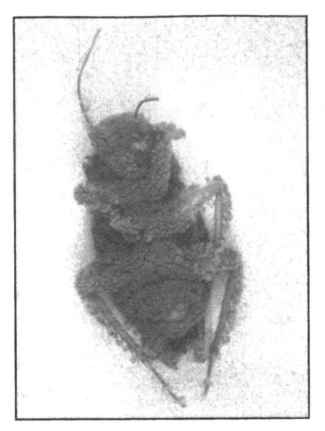

Figure 1 A cockroach (*Periplaneta americana*) is shown which is infected by the entomopathogenic fungus *Metarhizium anisopliae*. The host's cuticule is digested by different pH-regulated proteolytic and chitinolytic enzymes of *M. anisopliae*. (Courtesy of R. J. St. Leger.)

genes has led to the development of highly sophisticated signal transduction cascades in the prokaryotic kingdom [4]. Besides temperature, iron concentration, availability of carbon or nitrogen sources, and numerous other environmental signals, pH has long been considered a central determinant for controlled expression of virulence-associated genes in different bacteria [5–7]. It seems logical that adaptation of potentially pathogenic fungi, which are in most cases opportunistic pathogens and display low levels of pathogenic potential, should be equally important to their survival as pathogens. Only recently significant efforts have been undertaken in elucidating the regulatory cascades involved in fungal adaptation. For many of these studies of fungal pathogens, pH has turned out to be an important mediator of fungal virulence and adaptation to the host niche.

III. FROM INSECT TO MAN: pH IN THE HUMAN HOST

pH in the human body is one of the most variable and at the same time one of the most tightly controlled conditions. In different body compartments, pH values range from 1 to alkaline. The reason for this extraordinary variability is the fact that regulation of the pH serves more than one purpose. Blood pH is strictly regulated within its physiological boundaries of 7.35–7.45. Several buffer systems contribute to this homeostasis, which is maintained via a metabolic and a respiratory pathway in kidneys and lung. In contrast to blood, pH conditions in the stomach vary between neutral and highly acidic. If acid production by stomach epithelial cells is blocked either by iatrogenic intervention or as a result of an autoimmune process, the normally sterile environment of the stomach becomes a portal for infection and colonization.

Only a few pathogens have acquired mechanisms to resist the normal gastric acid. For example, *Mycobacterium tuberculosis* can survive in this hostile environment because of the unusual properties of its cell wall, whereas *Helicobacter pylori* profits from its motility and excessive urease production [8]. Thus, maintenance of an acidic environment serves as a natural protection against colonization and infection. This principle is also employed on the human skin and in the vagina. In both cases the normal microflora contributes to the development and maintenance of an acidic environment which in turn protects against the growth of nonphysiological and potentially pathogenic micro-organisms. pH of the skin is maintained at ~5.5, especially through production of fatty acids by the anaerobic microflora [8].

Among other mechanisms, natural protection against vaginal infection is provided by the metabolic products of lactobacilli. These bacteria constitute part of the normal vaginal flora during reproductive life. In degrading glycogen, provided by the vaginal epithelium, to lactic acid, the normal vaginal pH is regulated to 4.5 by the lactobacilli. All external conditions that result in alterations of the normal vaginal flora and consequently in an elevation of the pH predispose to colonization and infection [8]. Regulation of pH in the human body serves different purposes, one of which is the protection against infection. Thus, micro-organisms of the human host will in most cases be confronted with variable or even hostile pH conditions—a challenge for adaptation.

IV. FUNGI AND pH

Fungi in general are more resistant to alterations in the ambient pH than bacteria, and most fungi can stand considerably acid environments. This is especially true for saprophytic fungi. *Aspergillus nidulans*, for example, has been shown to be capable of sustained growth in a pH range between 2.5 and 9. This ability leads to a need for an efficient system which ensures intracellular pH-homeostasis and increases the evolutionary sense of pH-regulated expression of proteins.

In the 1980s, studies on pH-regulated gene expression in a number of fungi were reported, including *Penicillium charlessii* and *Neurospora crassa* [9,10]. *Aspergillus nidulans*, however, was the first fungal species (and actually the first organism) for which regulatory genes involved in pH-dependent expression of effector genes were described [11]. The knowledge about the *A. nidulans* cascade that links extracellular pH to differential gene expression has led to important progress in the understanding of how gene regulation works in other fungi.

V. A CASCADE REGULATING pH-DEPENDENT GENE EXPRESSION IN ASPERGILLUS

A. Effects of Ambient pH on *Aspergillus nidulans*

Observing the effects of the growth medium pH on gene expression in *A. nidulans*, several genes and phenotypic features were shown to be regulated in a pH-dependent manner. Utilization of γ-amino-*n*-butyrate (GABA) is optimal at pH 5 and not observed at pH 8 due to differential regulation of GABA-permease [11]. Kanamycin and neomycin exhibit increased toxicity at pH 8 compared to pH 5 [11]. In contrast to this, molybdate is less toxic at pH 8, and its toxic effect increases with acidification up to pH 6.5 [11].

Expression of the phosphate repressible alkaline phosphatase is increased with a rise in external pH [11]. An inverse pattern is found for phosphate-repressible acid phosphatase [11,12]. An alkaline protease coding gene (*prtA*) has been shown to be induced under alkaline conditions [12]. Acid phosphodiesterases are repressed under alkaline growth conditions [12]. Two genes encoding xylanases (*xlnA* and *xlnB*) display inverted patterns of pH-dependent regulation. In the presence of D-xylose, *xlnA* is expressed under alkaline, whereas *xlnB* is expressed under acidic conditions [13].

Finally, *ipnA* is predominantly expressed at alkaline pH [14]. This gene encodes isopenicillin-N-synthetase, an enzyme catalyzing the second step in penicillin biosynthesis. Therefore, numerous pathways in *A. nidulans* which mediate diverse functions, including penicillin biosynthesis and metabolism, are influenced by the ambient pH.

B. From Extracellular pH to Regulation of Gene Expression

Based on the knowledge of differential gene expression in *A. nidulans* in response to the ambient pH, Caddick and colleagues were able to analyze mutations altering the regulation of extracellular enzymes by pH [11]. Two groups of mutations were isolated, the first mimicking alkaline growth conditions and the second mimicking growth in an acidic environment. Only one gene was affected by the alkali-mimicking mutations whereas the acidity-mimicking mutations were dispersed over several independent genes.

C. PacC as the Central Regulator

Whereas different acidity-mimicking mutations affected several independent genes, only one gene was responsible for the alkali-mimicking mutations. This was named *pacC*. Notably, mutations in *pacC* were able to suppress all acidity-mimicking mutations for all aspects of their phenotypes, suggesting a central role of *pacC* within the cascade [11]. Later, it was shown that mutations in *pacC* actually give rise to two opposite phenotypes. Mutations removing only the carboxy-terminal residues from the transcript product mimic growth at alkaline pH (*pacC^c* mutations) as described by Caddick et al. In contrast to this, mutations affecting more N-terminal residues result in the opposite phenotype, mimicking growth at acidic pH (*pacC^{+/−}* mutations) (Fig. 2). *pacC^c* mutations are always dominant to *pacC^{+/−}* mutations if introduced into the same strains [12].

Based on these observations, it was suggested that PacC is the heart of the *A. nidulans* signal transduction cascade responsible for pH-dependent expression of effector genes. *pacC* encodes a protein of 678 amino acid residues containing three zinc-finger domains and is interrupted by two introns of 85 and 53 nucleotides, respectively. The zinc-finger domains are located near the N-terminus of the protein and belong to the Cys_2His_2 class. *pacC* itself is an alkaline-induced gene and mutations in *pacC* either induce or repress its own expression depending on the locus of truncation [12]. Footprinting analyses with PacC and the promoter of the alkaline-expressed gene *ipnA* revealed the core consensus sequence GCCARG responsible for binding of PacC to DNA [12,15]. Three of these sites are necessary and sufficient for pH-dependent regulation of the *ipnA*-promoter [16]. Two forms of PacC are detectable in *A. nidulans* cell extracts. One matches the expected full-length transcription product and predominates at acidic growth conditions or in acidity-mimicking mutants. The other is, though still containing all zinc-finger domains, truncated at the C-terminus and consists of ~40% of the full-length protein

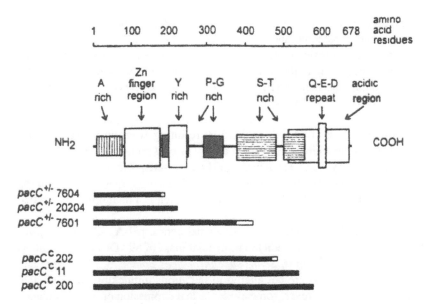

Figure 2 A scheme of PacC protein showing selected features. The portions of the protein remaining in certain mutants are indicated by solid bars, with open bar extensions denoting approximate lengths of abnormal sequence due to frameshift mutations. (From Ref. 12.)

[17]. The full-length form is inactive as a transcriptional activator or repressor, whereas the truncated form represses the expression of acidity-induced genes and enhances the expression of alkaline-induced genes. The processing of the inactive form to the active truncated form occurs at alkaline pH, which apparently induces a conformation in full-length PacC accessible to carboxy-terminal proteolysis [18].

D. Acidity-Mimicking Mutations and the *pal* Genes

The acidity-mimicking mutations have been assigned to genes whose products are essential for assuring processing of PacC. Though all genes affected by acidity-mimicking mutations known so far, have been cloned and sequenced, their interaction and functions within the cascade are yet unclear.

Two introns interrupt the open reading frame of *palB* that encodes a protein of 842 amino acids [19]. The transcription product contains a site highly homologous to the catalytic sites of members of the calpain family of calcium-activated cysteine proteases. In contrast to the calpains, PalB lacks the typical calcium-binding domains. Thus it has been suggested that PalB, despite having a similar substrate specificity as the calpains, is not calcium regulated. PalB is required for growth at alkaline but not acidic pH. Its mRNA levels, however, are not dependent on the external pH. PalB is not responsible for processing of PacC [19].

The *palH* open reading frame is interrupted by a single intron and encodes a putative 760 amino acid protein [20,21]. Interestingly, this protein is predicted to contain seven transmembrane domains and thus might well reside in the cytoplasmic membrane. *palH* disruption mutants totally prevent PacC processing, even at neutral pH, thus resembling mutations in *palA*, *-B*, *-C*, and *-F* [21]. The putative seven transmembrane domains could suggest an interaction with a heterotrimeric GTP-binding protein as in the case of the G-protein-coupled superfamily of receptors which all have seven transmembrane domains [22]. However, the involvement of a G-protein in pH-dependent gene expression in *A. nidulans* has not been shown as yet. Mutational evidence suggests that the function of PalH is mainly exerted within the N-terminal moiety containing the seven transmembrane helices [21]. PalH homologs have been described in *S. cerevisiae* and in *Yarrowia lipolytica* [21,23]. Although *palI* deletion mutants are phenotypically very similar and resemble the typical features of the other mutants, considerable growth can be detected at pH 8. The complete lack of growth at this pH, resulting from mutations in *palA*, *-B*, *-C*, *-F*, and *-H*, is epistatic to the maintained growth ability of *palI* mutations [24]. The level of *palI* transcription is not affected by PacC. The putative 601 amino acid protein with a molecular weight of 65 kDa contains four hydrophobic regions that could resemble membrane-spanning domains and displays high homology to Rim9 of *S. cerevisiae*, which is a membrane-associated sensor (24–26).

Only little is known about the functions of the other pal genes, pal-A, -C, and -F. *palA* encodes a protein of 798 amino acids with a predicted molecular mass of 89 kDa [27]. Homologs of *palA* are found in *Caenorhabditis elegans*, *Schizosaccharomyces pombe*, *S. cerevisiae*, and *C. albicans* (27–29). *palA* mRNA levels are not altered in mutations mimicking either an acid or an alkaline environment. There is, however, some evidence that expression of *palA* is reduced at pH 4. No alterations in *palA* expression were observed between pH 6.5 and pH 8, and *palA* expression is unlikely to be controlled by PacC [27].

The *palC* open reading frame is interrupted by a single intron and encodes a putative 507 residue protein with a predicted molecular mass of 55 kDa [21]. Of the six pal genes involved in *A. nidulans* pH signaling, *palC* is the only one lacking a homolog in *S. cerevisiae* based on gene bank comparisons. The only known homolog of *palC* so far has been detected in expressed sequence tags in the *Neurospora crassa* sequencing project. It has been suggested based on mutational analyses that at least one of the 142 C-terminal residues is essential for PalC function [21].

The sequence of *palF* gives no immediate clue of its function. Its uninterrupted open reading frame encodes a 775 amino acid protein with a predicted molecular mass of 84 kDa. The putative PalF protein shares homologies with two open reading frames in the *S. cerevisiae* genome (YGLO45w and YGLO46w) [30].

VI. IMPACT OF THE *ASPERGILLUS NIDULANS* CASCADE WITH RESPECT TO OTHER FUNGAL SPECIES

When the full-length sequence of *pacC* was revealed, only one homologous gene could be detected in the databases, apart from the zinc-finger domains. This gene (*RIM1*, regulator of *IME2* of *S. cerevisiae*) had been described as a positive acting regulator of meiosis [12,25,26,31]. Due to the existence of another gene in *S. cerevisiae*, that had previously been also named *RIM1* (replication in mitochondria), the homolog of *pacC* might more appropriately be referred to as *RIM101* [32]. *RIM101* was found during the analysis of *S. cerevisiae* mutants defective in coordinate *IME2* expression. Expression of this early sporulation-specific gene is normally regulated by *IME1*, a central regulator of meiosis and spore formation in *S. cerevisiae*. Thus *RIM101* deletion mutants are impaired in sporulation and meiotic gene expression [31]. Furthermore, these mutants display reduced growth at low temperatures (17°C) and an altered colony morphology [31]. Finally, *RIM101* has also been described as a positive regulator of invasive growth of haploid *S. cerevisiae* cells [26].

As described for PacC, Rim101 is activated by carboxyterminal cleavage. Proteolysis of Rim101 is regulated by the external pH and induced under alkaline conditions [26]. Interestingly, processing of the *Aspergillus* PacC can occur in *S. cerevisiae*, indicating conserved regulatory pathways between these two species [18]. In contrast to *pacC*, *RIM101* expression is not regulated by the ambient pH. It has recently been shown that *RIM101* is, in addition to its numerous other functions, involved in the *S. cerevisiae* response to ambient pH. This observation was accompanied by the detection of a *palB* homolog in *S. cerevisiae*, designated *CPL1* (for *c*alpain-like *p*rotease) [33]. Deletion of *CPL1* results in an impaired growth at alkaline pH without observable effects under acidic growth conditions. Noticeably, this phenotype could be suppressed by constitutive expression of *RIM101*. Furthermore, deletion of *CPL1* affected sporulation and promoted degradation of *RIM101* [33].

Aspergillus niger is a close relative of *A. nidulans*, and several of its extracellular enzymes have been shown to be regulated in response to the ambient pH. The *A. niger pacC* homologue is capable of complementing a *pacC* deletion in *A. nidulans*, and both proteins are predicted to be very similar with the greatest differences located at the C-terminus [34] (Table 1)

Based on the knowledge that penicillin biosynthesis in *A. nidulans* is strongly pH dependent and isopenicillin-N-synthase (*ipnA*) is the prototype of an alkaline expressed gene, the conservation of this regulational pattern in other penicillin-producing fungi was examined. *Penicillium chrysogenum* is utilized for industrial penicillin production due to its excessive biosynthesis of the antibiotic. As in *A. nidulans*, the genes encoding the enzymes catalyzing the first two steps in penicillin biosynthesis are regulated by a bidirectional promoter, and the *ipnA* homolog *pcbC* is strongly induced under alkaline conditions. The *pacC* homolog in *P. chrysogenum* responsible for this regulation pattern is capable of complementing an *A. nidulans pacC* deletion mutant and has been shown to recognize the same core consensus sequence (GCCARG) [35]. However, modes of regulation of penicillin biosynthesis in these two species are not identical. In *P. chrysogenum* much more than in *A. nidulans*, the availability of carbon sources influences penicillin production. Interestingly, the region between the two genes involved in penicillin biosynthesis in *P. chrysogenum* contains even more PacC-binding sites than in *A. nidulans*. Seven of these sites have been deduced from its sequence, at least six of which have been shown to efficiently bind to PacC [35].

The yeast *Yarrowia lipolytica* secretes an acidic (*AXP*) and an alkaline (*XPR2*) protease depending on the pH of the growth medium [36]. *YlRIM101*, the homolog of *S. cerevisiae RIM101* and *A. nidulans pacC* in *Y. lipolytica*, is essential for mating and sporulation as well

Table 1 Features of *A. nidulans* PacC and Its Homologs in Other Fungi

Gene homologs	Species	Zinc-fingers	ORF	Amino acids	Expression	Regulation	C-terminal negative acting domain
PacC	*A. nidulans*	3	2034 bp	678	neutral, alkaline pH	*palA, B, C, F, H, I*	~60%
Rim101	*S. cerevisiae*	3	1884 bp	628	alkaline pH	*RIM8, 9, 13*	~11%
YlRim101	*Y. lipolytica*	3	1755 bp	585	neutral pH	*PAL1, 2, 3, 4*	~40%
PacC	*A. niger*	3	2031 bp	677	alkaline pH	not known	~21%
PacC	*P. chrysogenum*	3	1923 bp	641	alkaline pH	not known	not known
Rim101	*C. albicans*	3	1986 bp	661	alkaline pH	*PRR1/RIM8, RIM20*	~12%

as for induction of alkaline protease [37]. The promoter of *XPR2* contains the *pacC* site consensus sequence within a decameric repeat involved in pH- and carbon/nitrogen source-dependent induction [38]. Although the 5′ noncoding region of *AXP* also contains YIRim101 consensus sequences, the effects of this regulator on the expression of *AXP* are yet unclear [37].

VII. *CANDIDA ALBICANS*

Candida albicans is the most frequently isolated fungal pathogen in the clinical mycology laboratory [39]. The source of this pleomorphic yeast is in most cases endogenous, and infection occurs after the fungus has spread from sites of colonization. Accordingly, *C. albicans* is very well adapted to the human host, as it is capable of colonization as well as a cause of disseminated and often lethal infections. The variable patterns of this adaptation are best symbolized by the high morphogenetic flexibility of *C. albicans.* Switching between yeast and hyphal forms, as well as the formation of elongated projections termed germ tubes, has long been considered one of several factors contributing to the virulence of *C. albicans* [40–42]. A variety of environmental conditions have been implicated in *C. albicans* morphogenesis, each characterized as favoring either the yeast or the hyphal growth form [41]. Chemical as well as physical signals have been shown to influence morphogenesis. Recent results have suggested that most of these extracellular signals converge on two central regulators of gene transcription in *C. albicans* (Fig. 3, and see Chapter 6). These two transcription factors, encoded by *CPH1* and *EFG1*, are responsible for filamentation under a broad range of conditions [43–45]. On the contrary, mutants which have been constituted by deleting both *CPH1* and *EFG1* are unable to form filaments under most

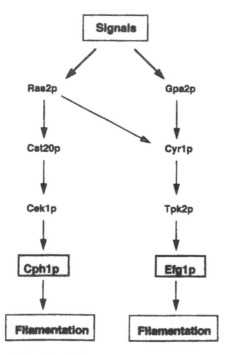

Figure 3   Main signal transduction pathways controlling dimorphism of *C. albicans.* (Modified from Ref. 45.)

environmental and nutritional conditions. Strikingly, these mutants have been shown to be avirulent in an animal model. Thus filamentation of *C. albicans* is likely to be directly linked to
virulence [42].

A.   pH-Regulated Dimorphism in *Candida albicans*

pH and temperature are among the external signals that trigger morphogenesis in *C. albicans*.
The effects caused by alterations of these important host-niche determinants, environmentally
regulated genes have been extensively studied since 1970 [40]. The landmarks of pH-regulated
dimorphism were thoroughly characterized by Buffo and colleagues in 1984. Accordingly, a
shift of temperature above 35.5°C and of pH >6.0 are necessary for induction of pH-regulated
dimorphism (Fig. 4) [46]. Two factors account for the interest in this morphogenic feature. pH
and temperature conditions inducing filamentation are found in the human body and alterations
in both conditions are likely to occur to a pathogen during infection.

B.   pH-Regulated Genes and Their Functions in *Candida albicans*

pH-regulated genes of *C. albicans* that have been described so far belong to two classes of
proteins. The first of these are secreted enzymes, especially *SAP2* encoding a secreted aspartyl
proteinase [47]. Most of the pH-regulated genes described in *C. albicans*, however, encode for
cell wall-associated proteins. One of the obvious functions of the fungal cell wall is maintaining
cell shape. This claims maximal rigidity and resistance to environmental stimuli as well as highly
dynamic and interactive responses to environmental stimuli. Therefore, it seems reasonable that
genes responsible for cell-wall organization are controlled by external conditions [47,48].

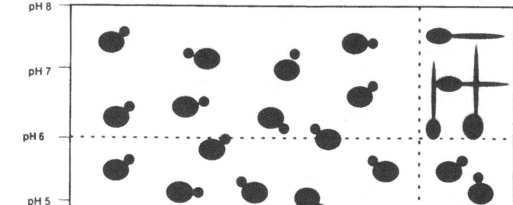

Figure 4 Temperature and pH regulate dimorphism in *C. albicans*. A temperature >35.5°C together
with a pH ≥6.0 induces filamentation.

1. Secreted Aspartyl Proteinases (Saps)

pH-dependent expression of secreted enzymes is a very common feature and has been described for *M. anisopliae* and *A. nidulans* above. For the *SAP2* gene, which encodes a secreted aspartyl proteinase in *C. albicans*, expression, monitored by the steady-state mRNA level, was observed only at pH conditions ~4.0 but not <pH 2.5 or >pH 5.5. This has been shown to correlate with the detection of Sap2p in Western blots. However, in another study, these pH conditions were not sufficient for expression, such that the induction and repression of *SAP2* is also dependent on the growth phase and an external inducer [49,50].

A current hypothesis is that induction of *SAP2* by self-generated cleavage products is optimal in the pH range 3.0–5.0 as this is the pH optimum of Sap2p. Thus, the amount of inducing peptides that is generated is likely to be at a maximum. Notably, induction of *SAP2* by these peptides is poor at pH 2.5 or 6.0 even if they are artificially added to the growth medium. It has been speculated that this may be explained by a pH-dependent mechanism of monitoring these peptides on the cell surface or translocating the peptides into the cell. Accordingly, pH is directly involved in regulation of *SAP2* expression but is not the dominating signal in a number of stimulating and repressing conditions [49]. In media with protein as the sole source of nitrogen, *SAP2* is generally found to be the most abundantly expressed gene among a family of at least 10 secreted aspartyl proteinases of *C. albicans* [49].

These extracellular proteinases contribute significantly to virulence of this yeast during mucosal and disseminated infections. Deletion mutants of *SAP2* were almost completely avirulent in a vaginal infection model [51]. Furthermore Sap2 has been shown to contribute to virulence in an animal model of systemic infection possibly by mediating endothelial cell damage [52]. In a recent report Sap2 has been shown to contribute significantly to tissue damage in an in vitro model of human oral candidosis [53].

2. The pH-regulated antigen *PRA1*

PRA1 (for *p*H-regulated *a*ntigen1) was identified by testing cDNA clones of *C. albicans* against mycelium-specific antiserum. Its expression is maximal at pH 7 and does not occur <pH 6 [54]. In addition, expression of *PRA1* is media dependent and can be delayed or totally absent depending on the composition of the medium. Homologs of *PRA1* have been described in *Aspergillus fumigatus*, *A. nidulans*, and *S. cerevisiae*; their functions, however, are unknown. Furthermore, *PRA1* displays some homology with the deuterolysin family of zinc metalloproteinases, which might suggest a proteolytic function. Deletion of *PRA1* results in a loss of germ tube formation and an aberrant chitin distribution at elevated temperatures [54].

3. *PHR1* and *PHR2*

PHR1 and *PHR2* are two functionally homologous genes that are characterized by their inverse pattern of pH dependent expression [55]. *PHR1* is only expressed at neutral to alkaline pH and is not detectable under acidic conditions <pH 5.0, whereas *PHR2* is induced by acidic ambient pH and not expressed under alkaline conditions [56,57] (Fig. 5a). Both encode for putative glycosylphosphatidylinositol-anchored proteins. Deletion of either gene results in defects of virulence and morphogenesis under the pH conditions promoting the expression of the respective gene [56,57] (Fig. 5b). The impaired cellular integrity of the deletion mutants is paralleled by an alteration in the cell wall structure [58,59]. The glucan fraction is reduced mainly as a result of a lack in β-1,6-glucan. β-1,3-Glucan is released into the medium, and the glucan fraction displays an elevated alkali solubility. It is hypothesized that the cells compensate for these changes by increasing the chitin content of the cell wall. Further, in mutant cells, 40% in contrast to only 4% of the β-1,6-glucosylated mannoproteins become connected to chitin [58,59].

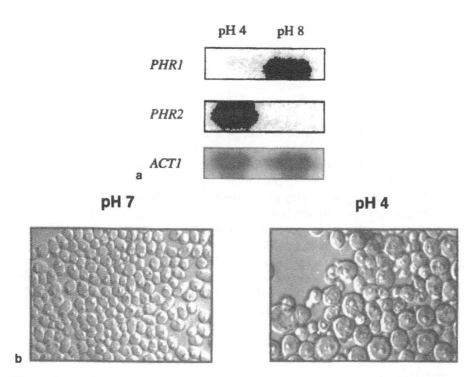

Figure 5 (a) Northern analysis of the *C. albicans* pH-regulated genes *PHR1* and *PHR2*. (b) The *phr2* deletion mutant (CFM-2) displays defects in growth and morphology only at acid pH. *phr1* mutants show a similar phenotype with an opposite pH-dependent pattern.

Taking advantage of the pH-regulated expression of *PHR1* and *PHR2*, it has been shown that the corresponding enzymes are likely to be involved in the proper crosslinking of the cell wall glucan [58]. Phr1p and Phr2p very likely contribute to the correct organization of the cell wall glucan fraction by crosslinking β-1,3- and β-1,6-glucans [58]. Together with chitin, the cell wall glucan forms the structural backbone of the fungal cell wall, ensuring cellular integrity and mediating morphogenic plasticity during dimorphic switching [47,48]. This may well explain the reduced virulence of PHR deletion mutants in host niches with the restrictive pH. Thus, pH-dependent regulation of *PHR1* and *PHR2* is not only an effect observed in vitro. The *PHR1* and the *PHR2* deletion mutants have been shown to exhibit severe pH-conditional defects in virulence. A Δ*phr1* mutant is unable to cause systemic infection in a mouse model, whereas it is not impaired in a model of vaginal candidiasis. In contrast, the Δ*phr2* mutant does not cause sustained vaginal candidiasis but is not impaired in the systemic infection [60]. These in vivo conditions reflect the restrictive in vitro pH. The pH of blood is ~7.4 whereas the pH in the vagina is 4.0.

PHR homologous genes have been described in a number of fungal species including *Candida maltosa, Candida glabrata, S. cerevisiae,* and *A. fumigatus* [61–64]. However, all of the characterized homologs differ from *PHR1* and *PHR2* in that they do not possess the pH-dependent expression profile.

C. The pH Transduction Cascade in *Candida albicans*

Though *PHR1* and *PHR2* have been described and studied as a unique model of pH-regulated switching between functional homologous genes, the molecular mechanisms responsible for their tight regulation are unknown. The first steps toward the description of a signal transduction cascade responsible for pH-dependent gene expression were taken by the discovery of sequences in the *C. albicans* genome that were homologous to parts of the *Aspergillus* pH-regulated cascade [28,29,65,66].

1. pH-Response Regulator 1 (*PRR1*)

The pH response regulator 1 (*PRR1*) was identified by Porta and colleagues based on homology screening of the *C. albicans* genome with *A. nidulans palF* [65]. The 621 amino acid protein Prr1p is 24% identical to PalF. Expression of *PRR1* is regulated by the ambient pH in that it is maximal at pH 4 and reduced about threefold under alkaline conditions. Deletion of both *PRR1* alleles results in an acidity-mimicking phenotype as shown for Δ*palF* in *A. nidulans*. In contrast to the mutations described in *palF,* however, which result in a growth deficit at alkaline pH, *PRR1* disruption mutants were capable of sustained growth over the whole pH range (4.0–7.0). In the homozygous null mutant, *PHR1* is repressed independent of the ambient pH whereas expression of *PHR2* occurs at high levels under acidic as well as under alkaline growth conditions. Both effects are lost upon reintroduction of a single wild-type allele of *PRR1* into the mutants and cannot be observed in a heterozygous deletion mutant. Deletion of *PRR1,* however, not only has effects on pH-regulated gene expression but also affects morphogenesis of the mutants. These effects are most clearly observed on agar-solidified media. Deletion of *PRR1* not only prevents filamentation on medium 199 at pH 7.5 and, thus, pH-regulated dimorphism, but also filamentation in a number of other environmental conditions including the response to serum [65].

2. pH Response Regulator 2 (*PRR2 = RIM101*)

Parallel to the detection of *PRR1,* Ramon and colleagues reported the existence of a *pacC* homolog in the *C. albicans* genome which was designated *RIM101* or *PRR2* (for *p*H *r*esponse regulator 2), hereafter referred to as *RIM101* [66]. Like *pacC* and its other known homologs, *RIM101* contains three zinc-finger domains of the Cys_2His_2 class with all the conserved critical residues as well as several putative nuclear localization signals. Expression of *RIM101* is itself pH dependent and maximal under alkaline growth conditions. Deletion of *RIM101* results in a loss of alkaline-induced *RIM101* expression and constant low mRNA levels [66].

Deletion of *RIM101* results in a phenotype resembling that of the *PRR1* deletion mutants with respect to alterations in pH-regulated gene expression. In Δ*prr2* mutants, we find constitutive, pH-independent expression of *PHR2* and no detectable expression of *PHR1* regardless of the environmental pH conditions. Interestingly, deletion of *RIM101* also prevents repression of *PRR1* at alkaline pH. Thus *PRR1* expression is controlled by *RIM101* in a putative feedback mechanism [66].

As seen with *PRR1,* deletion of *RIM101* also results in filamentation defects under a variety of conditions. On medium 199 (pH 7.5) no filamentation of the homozygous deletion mutants could be observed, indicating severe defects in pH-regulated dimorphism as described for the *PRR1* disruption. Filamentation and invasive growth is also affected under a number of other conditions in these mutants including the response to serum [29,66].

3. Unlocking pH-Regulated Dimorphism

The description of *PRR1* and *RIM101* suggested a high degree of conservation between the cascade in *A. nidulans* and a putative regulatory pathway in *C. albicans.* This was further supported by the finding of *ENX3,* a *C. albicans* homolog of the *A. nidulans* gene *palA.* Characterization of *ENX3* deletion mutants has shown that they are defective in filamentation on Spider medium [28]. The common mechanisms of fungal adaptation to changes in the environmental pH presented so far are accompanied by an understanding of function of this regulative cascade that is very specific for each fungal species. That is, in the example of the human pathogen *C. albicans*, we can see an entirely new aspect of this cascade, which is that pH is directly linked to morphogenic plasticity and thus to an important determinant of *C. albicans* virulence.

Based on the knowledge of the *Aspergillus* pH response cascade, it seemed likely that *RIM101* would play an important part in the regulation of pH-dependent gene expression. This could be confirmed by phenotypic analysis of *RIM101* deletion mutants of *C. albicans*. Deletion mutants for *RIM101* and its upstream inducer *PRR1*, however, suggested that both genes also were directly linked to dimorphism.

The key experiment for the elucidation of the central position of *RIM101* within this regulatory network was the isolation of phenotypically reverted *PHR2* deletion mutants [67]. These revertants had regained the ability to grow at acidic pH in spite of the fact that both alleles of *PHR2*, which is normally required for growth in an acidic environment, were deleted (Fig. 6a). It could be shown that expression of *PHR1*, normally restricted to alkaline and neutral conditions, occurs over the whole pH spectrum in these revertants (Fig. 6b). This could well explain their viability at acidic pH as *PHR1* and *PHR2* are functionally homologous. Further-

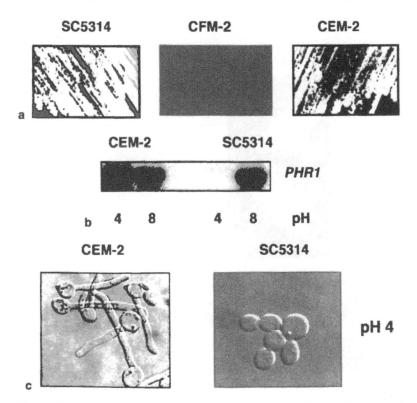

Figure 6 (a) Growth on YNB agar at pH 4: SC5314 (*C. albicans* wild-type strain), CFM-2 (*phr2* deletion mutant), and CEM-2 (CFM-2 revertant strain). (b) Northern analysis of CEM-2 and SC5314. CEM-2 shows aberrant expression of *PHR1* at pH 4. (c) Morphology of CEM-2 and SC5314 cells, grown in liquid media at pH 4 and 37°C. In contrast to the wild-type strain, CEM-2 displays filamentation at acidic pH.

more, the aberrant cellular morphology normally observed for *PHR2* deletion mutants exposed to acidic pH was absent in the revertants again very likely due to *PHR1* misexpression [67].

Notably, these revertants displayed an additional phenotype at pH 4.0 that could not be explained by aberrant expression of *PHR1*. Filamentous growth was equally observed in a number of different media in liquid culture as well as on agar-solidified plates (Fig. 6c). From a broader perspective, the revertants displayed typical alkaline features of *C. albicans* (pH-induced filamentation, *PHR1* expression) at an acidic environmental pH. Taken together, these results suggested that a mutation occurring within the regulative cascade responsible for differential expression of *PHR1* is responsible for the revertant's phenotype [67].

4. Deletion of a Single *RIM101* Allele in the Revertants Causes Two Phenotypes

Introduction of a knockout cassette into *RIM101* should theoretically affect both alleles with a similar probability, and in normal wild-type strains this should result in heterozygous genotypes that cannot be phenotypically differentiated from each other. However, when a *RIM101* knockout cassette was introduced into revertant strains this resulted in two grossly different phenotypes. One type of transformant (CEM-5) did not differ from the original revertant (that is the parental) phenotype (Fig. 7a).

In contrast, the other type of transformant (CEM-6) was unable to grow under acidic conditions and no longer revealed a disturbed pH-regulated dimorphism. The latter resembled the phenotype of the original *PHR2* deletion mutant, and aberrant expression of *PHR1* could

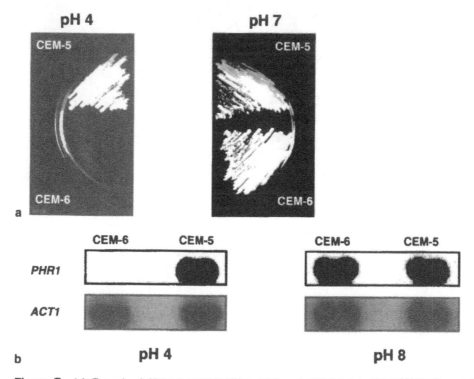

Figure 7 (a) Growth of CEM-5 (*RIM101ᵐ/rim101*) and CEM-6 (*rim101ᵐ/RIM101*) on YNB agar at pH 4 and pH 7. (b) Northern analysis of CEM-5 and CEM-6. CEM-5 (*RIM101ᵐ/rim101*) shows in contrast to CEM-6 (*rim101ᵐ/RIM101*) a pH-independent expression of *PHR1*.

no longer be detected (Fig. 7b). These results suggest that the revertant phenotype could be due to a dominant event in a single *RIM101* allele, which was not affected by the knockout cassette in the first group of transformants but was deleted in the second group [67].

5. Point Mutations Result in Dominant Active Gain of Function *RIM101* Derivatives

Sequencing of the remaining undisrupted *RIM101* allele in the transformants with sustained growth at pH 4.0 revealed point mutations resulting in the introduction of premature stop-codons,

mutated *RIM101*

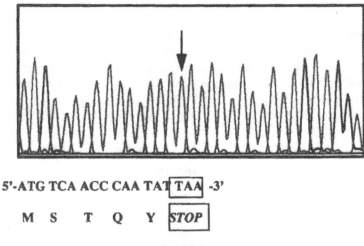

5'-ATG TCA ACC CAA TAT TAA -3'

M S T Q Y STOP

wild type *RIM101*

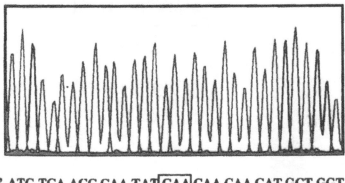

5'-ATG TCA ACC CAA TAT CAA CAA CAA CAT GCT GGT-3'

M S T Q Y Q Q Q H A G

Figure 8 DNA sequence chromatogramm showing a point mutation leading to an early stop codon in one *RIM101* allele at position 1426 of the open reading frame in the revertant strain CEM-2.

resulting in the expression of putatively truncated Rim101p derivatives (Rim101p^m). In contrast to this, the remaining *RIM101* allele in the transformants which failed to grow at pH 4.0 were unaltered and 100% identical to the wild-type sequence [67] (Fig. 8). Davis and colleagues were also able to show that truncated forms of Rim101 can constitutively suppress the phenotype of a homozygous *RIM101* deletion mutant [29].

Integration of *RIM101^m* but not the wild-type allele into a *PHR2* deletion mutant resulted in a phenotype identical to that described for the revertants [67]. Interestingly, these strains were able to filament at 37°C and pH 4, and the introduction of multiple copies of *RIM101* substituted for a 37°C temperature signal, allowing filamentation to occur at pH 4 and temperatures <30°C. Thus, *RIM101^m* can override temperature limits in pH dependent dimorphism as well as the boundaries of pH. This indicates that pH might not be the only environmental signal exerting its effects via the *RIM101* cascade [67].

6. *RIM101* in the Context of Other Transcriptional Regulators

We have described above that filamentation in *C. albicans* is mainly regulated by the two transcriptional activators, Cph1 and Efg1. Whereas Cph1 acts in a mitogen-activated protein kinase-dependent cascade, Efg1 is situated downstream of adenylcyclase in a cAMP-dependent pathway [45]. To determine whether *RIM101^m*-induced filamentation was dependent, on one of these factors, *RIM101^m*, was expressed in mutants lacking these central regulators [67]. In these experiments *RIM101^m* induced filamentation at pH 4 and 37°C in a Δ*cph1* but not in a Δ*efg1* mutant or a Δ*cph1*Δ*efg1* mutant (Fig. 9a,b). Thus, *RIM101^m* stimulates filamentation in an *EFG1*-dependent manner. In contrast to this, the regulation of *PHR1* and *PHR2* by *RIM101^m*

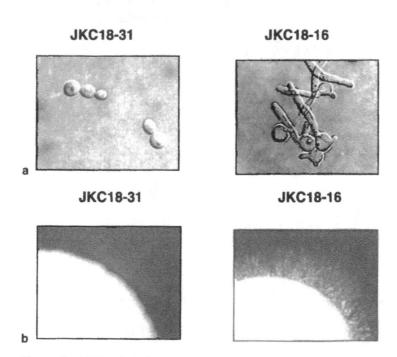

Figure 9 (a) Morphology of *cph1* mutants transformed with *RIM101* (JKC18-31) and *RIM101^m* (JKC18-16) after growth in liquid medium with pH 4. Only the *RIM101^m* induces filamentation. (b) Colony morphology of *cph1* mutants transformed with *RIM101* (JKC18-31) and *RIM101^m* (JKC18-16) after growth on solidified medium at pH 4.

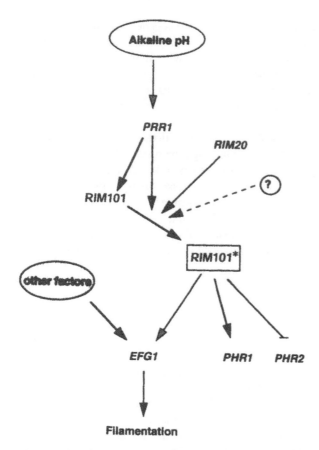

Figure 10   Model showing a putative pH-regulatory pathway in *C. albicans*. Alkaline pH induces activation of *RIM101* via a signal cascade including *PRR1* and *RIM20*. The active form of *RIM101* (*RIM101**) regulates the induction/repression of pH-responsive genes and controls pH-regulated dimorphism in an *EFG1*-dependent manner.

was not dependent on Efg1 [67]. In all of the *RIM101^m*-containing strains, regardless of whether or not *CPH1* or *EFG1* or both were deleted, *RIM101^m* induced expression of *PHR1* and prevented expression of *PHR2*, independent of the ambient pH [66]. Therefore, we can differentiate at least two entirely different functions of *RIM101*. Like most of the other *pacC* homologous transcription factors, *RIM101* controls pH-dependent gene expression, probably in a direct manner. Remarkably, it also is involved in morphogenesis and possibly acting upstream of Efg1 [67] (Fig. 10).

VIII.   JUST A LITTLE BIT DIFFERENT: *CANDIDA DUBLINIENSIS*

Candida dubliniensis is the *Candida* species most closely related to *C. albicans* [68–70]. *C. dubliniensis* has been shown to be a common cause of oropharyngeal candidiasis in HIV-infected patients [68–71]. Isolation of this species from other sites of infection is rare, and it has been suggested that this is due to a reduced level of virulence in comparison to *C. albicans*. However, *C. dubliniensis* is the only species in the genus *Candida* that closely resembles the pH-dependent dimorphism described above for *C. albicans* [72]. PHR-homologous genes have been described in the *C. dubliniensis* genome, and *CdPHR1* has been shown to be functionally equivalent to the *C. albicans* homolog [72,73]. Remarkably, the extraordinary pattern of expressional regulation found with the *C. albicans* homologs is conserved in *C. dubliniensis* [73]. The existence of two pairs of very closely related genes from two species offered a unique ability to learn more about regulation and interspecies variation. By heterologous expression of *CdPHR1* in an ectopic location in *C. albicans* it could be shown that a 1-kb 5′ noncoding sequence is actually sufficient to maintain pH-regulated expression [72]. This suggests that, similar to the situation described for *Aspergillus,* signal sequences in the promoter might be responsible for the interaction with transcriptional regulators, supposedly a *C. dubliniensis RIM101* homolog. These sequences may differ from the *A. nidulans* PacC consensus sequence. Interestingly, *CdPHR1* is strongly expressed under control of its native promoter in *S. cerevisiae* [72]. However, expression of the gene is no longer regulated by the ambient pH. The same results were found for *C. albicans PHR2.* This observation might indicate a divergent evolution of *S. cerevisiae,* a species profiting from a functional mating pathway, and *C. albicans/dubliniensis,* which require perfect adaptation to the human host.

IX. CONCLUDING REMARKS

Adaptation is an important feature in fungal pathogenesis. Fungi—mostly opportunistic pathogens—have to try and find their host niches and adapt to the conditions within. We have shown how pH is an important determinant of different host niches, and, consequently, many fungal pathogens have developed means of adapting to pH conditions they encounter during infection. In the example of *C. albicans,* a very common parasite of humans, pH is linked not only to differential expression of secreted factors but also to cell wall organization and morphogenesis.

pH is not the only important environmental condition defining a host niche. Numerous other environmental conditions challenge adaptation. However, the pH response system of ascomycete fungi has been extensively studied compared to all the other host niche signals. Thus, the knowledge about the pH response cascade in fungi is likely to serve as a model for a better understanding of adaptation.

ACKNOWLEDGMENTS

The authors wish to thank Matthias Frosch for his support. William Fonzi, Bernhard Hube, Amalia Porta, and Ana Ramon contributed to the work in many stimulating discussions. We thank G.R. Fink for strains JKC18 and HLC67, and D.P. Bockmühl and J.F. Ernst for strain CDB1. Experiments in the lab of F.M. would not have been possible without the expert technical assistance of Steffi Müksch. O.K. wants to thank Monika Brand for deliberately improving the manuscript. F.M. gratefully acknowledges support by the Deutsche Forschungsgemeinschaft through grant MU 1212/2-1.

REFERENCES

1.   P Ferron. Fungal Control. In: GA Kerkut, LI Gilbert, eds. Comprehensive Insect Physiology, Biochemistry and Pharmacology. Vol 12. New York: Pergamon Press, 1985, pp 314–346.

2. RJ St Leger, L Joshi, MJ Bidochka, DW Roberts. Construction of an improved mycoinsecticide overexpressing a toxic protease. Proc Natl Acad Sci USA 93:6349–6354, 1996.

3. RJ St Leger, L Joshi, D Roberts. Ambient pH is a major determinant in the expression of cuticle-degrading enzymes and hydrophobin by *Metarhizium anisopliae*. Appl Environ Microbiol 64: 709–713, 1998.

4. JJ Mekalanos. Environmental signals controlling expression of virulence determinants in bacteria. J Bacteriol 174:1–7, 1992.

5. K Skorupski, RK Taylor. Control of the *ToxR* virulence regulon in *Vibrio cholerae* by environmental stimuli. Mol Microbiol 25:1003–1009, 1997.

6. J Behari, P Youngman. Regulation of *hly* expression in *Listeria monocytogenes* by carbon sources and pH occurs through separate mechanisms mediated by PrfA. Infect Immun 66:3635–3642, 1998.

7. JA Carroll, CF Garon, TG Schwan. Effects of environmental pH on membrane proteins in *Borrelia burgdorferi*. Infect Immun 67:3183–3187, 1999.

8. C Mims, J Playfair, I Roitt, D Wakelin, R Williams. Medical Microbiology. 2nd ed. London: Mosby, 1998, pp 91–109.

9. E Nahas, HF Terenzi, A Rossi. Effect of carbon source and pH on the production and secretion of acid phosphatase (EC 3.1.3.2) and alkaline phosphatase (EC 3.1.3.1) in *Neurospora crassa*. J Gen Microbiol 128:2017–2021, 1982.

10. JE Gander, S Janovec. Regulation of metabolism in *Penicillium charlessii* by organic acids: role of L-tartaric acid. Curr Top Cell Regul 24:99–109, 1984.

11. MX Caddick, AG Brownlee, HN Arst Jr. Regulation of gene expression by pH of the growth medium in *Aspergillus nidulans*. Mol Gen Genet 203:346–353, 1986.

12. J Tilburn, S Sarkar, DA Widdick, EA Espeso, M Orejas, J Mungroo, MA Peñalva, HN Arst Jr. The Aspergillus PacC zinc finger transcription factor mediates regulation of both acid- and alkaline-expressed genes by ambient pH. EMBO J 14:779–790, 1995.

13. AP MacCabe, M Orejas, JA Pérez-González, D Ramón. Opposite patterns of expression of two *Aspergillus nidulans* xylanase genes with respect to ambient pH. J Bacteriol 180:1331–1333, 1998.

14. EA Espeso, J Tilburn, HN Arst Jr, MA Peñalva. pH-regulation is a major determinant in expression of a fungal penicillin biosynthetic gene. EMBO J 12:3947–3956, 1993.

15. EA Espeso, J Tilburn, L Sánchez-Pulido, CV Brown, A Valencia, HN Arst Jr, MA Peñalva. Specific DNA recognition by the *Aspergillus nidulans* three zinc finger transcription factor PacC. J Mol Biol 274:466–480, 1997.

16. EA Espeso, MA Peñalva. Three binding sites for the *Aspergillus nidulans* pacC zinc-finger transcription factor are necessary and sufficient for regulation by ambient pH of the isopenicillin N synthase gene promoter. J Biol Chem 271:28825–28830, 1996.

17. M Orejas, EA Espeso, J Tilburn, S Sarkar, HN Arst Jr, MA Peñalva. Activation of the *Aspergillus* PacC transcription factor in response to alkaline ambient pH requires proteolysis of the carboxy-terminal moiety. Genes Dev 9:1622–1632, 1995.

18. JM Mingot, J Tilburn, E Diez, E Bignell, M Orejas, DA Widdick, S Sarkar, CV Brown, MX Caddick, AE Espeso, HN Arst Jr, MA Peñalva. Specificity determinants of proteolytic processing of *Aspergillus* PacC transcription factor are remote from the processing site and processing occurs in yeast if pH signalling is bypassed. Mol Cell Biol 19:1390–1400, 1999.

19. SH Denison, M Orejas, HN Arst Jr. Signaling of ambient pH in *Aspergillus* involves a cysteine protease. J Biol Chem 270:28519–28522, 1995.

20. HN Arst Jr, E Bignell, J Tilburn. Two new genes involved in signaling ambient pH in *Aspergillus nidulans*. Mol Gen Genet 245:787–790, 1994.

21. S Negrete-Urtasun, W Reiter, E Diez, SH Denison, J Tilburn, EA Espeso, MA Peñalva, HN Arst Jr. Ambient pH signal transduction in *Aspergillus*: completion of gene characterization. Mol Microbiol 33:994–1003, 1999.

22. L Stryer. Biochemistry. 4th ed. New York: W.H. Freeman, 1995, pp 325–360.

23. B Treton, M Lambert, A Lépingle, S Blanchin-Roland, C Gaillardin. Two distinct *Saccharomyces cerevisiae* genes encode homologs of *Yarrowia lipolytica* Pal2p and Pal3p involved in the ambient pH signal transduction pathway. Curr Genet. 35:410, 1999.

24. SH Denison, S Negrete-Urtasun, JM Mingot, J Tilburn, WA Mayer, A Goel, EA Espeso, MA Peñalva, HN Arst Jr. Putative membrane components of signal transduction pathways for ambient pH regulation in *Aspergillus* and meiosis in *Saccharomyces* are homologous. Mol Microbiol 30:259–264, 1998.
25. SSY Su, AP Mitchell. Identification of functionally related genes that stimulate early meiotic gene expression in yeast. Genetics 133:67–77, 1993.
26. W Li, AP Mitchell. Proteolytic activation of Rim1p, a positive regulator of yeast sporulation and invasive growth. Genetics 145:63–73, 1997.
27. S Negrete-Urtasun, SH Denison, HN Arst Jr. Characterization of the pH signal transduction pathway gene *palA* of *Aspergillus nidulans* and identification of possible homologs. J Bacteriol 179: 1832–1835, 1997.
28. RB Wilson, D Davis, AP Mitchell. Rapid hypothesis testing with *Candida albicans* through gene disruption with short homology regions. J Bacteriol 181:1868–1874, 1999.
29. D Davis, RB Wilson, AP Mitchell. *RIM101*-dependent and -independent pathways govern pH responses in *Candida albicans*. Mol Cell Biol. 20:971–978, 2000.
30. W Maccheroni Jr, GS May, NM Martinez-Rossi, A Rossi. The sequence of *palF*, an environmental pH resonse gene in *Aspergillus nidulans*. Gene 194:163–167, 1997.
31. SSY Su, AP Mitchell. Molecular characterization of the yeast meiotic regulator gene *RIM1*. Nucleic Acids Res 21:3789–3797, 1993.
32. E Van Dyck, F Foury, B Stillman, SJ Brill. A single-stranded DNA binding protein required for mitochondrial DNA replication in *S. cerevisiae* is homologous to *E. coli* SSB. EMBO J 11:3421–3430, 1992.
33. E Futai, T Maeda, H Sorimachi, K Kitamoto, S Ishiura, K Suzuki. The protease activity of a calpain-like cysteine protease in *Saccharomyces cerevisiae* is required for alkaline adaptation and sporulation. Mol Gen Genet 260:559–568, 1999.
34. AP MacCabe, JPTW van den Hombergh, J Tilburn, HN Arst Jr, J Visser. Identification, cloning and analysis of the *Aspergillus niger* gene *pacC*, a wide domain regulatory gene responsive to ambient pH. Mol Gen Genet 250:367–374, 1996.
35. T Suárez, MA Peñalva. Characterization of a *Penicillium chrysogenum* gene encoding a *PacC* transcription factor and its binding sites in the divergent *pcbAB-pcbC* promoter of the penicillin biosynthetic cluster. Mol Microbiol 20:529–540, 1996.
36. DG Ahearn, SP Meyers, RA Nichols. Extracellular proteinases of yeasts and yeast like fungi. Appl Microbiol 16:1370–1374, 1968.
37. M Lambert, S Blanchin-Roland, F Le Louedec, A Lépingle, C Gaillardin. Genetic analysis of regulatory mutants affecting synthesis of extracellular proteinases in the yeast *Yarrowia lipolytica*: identification of a *RIM101/pacC* homolog. Mol Cell Biol 17:3966–3976, 1997.
38. C Madzak, S Blanchin-Roland, RR Cordero Otero, C Gaillardin. Functional analysis of upstream regulating regions from the *Yarrowia lipolytica XPR2* promoter. Microbiology 145:75–87, 1999.
39. SK Fridkin, WR Jarvis. Epidemiology of nosocomial fungal infections. Clin Microbiol Rev 9: 499–511, 1996.
40. P Auger, J Joly. Factors influencing germ tube production in *Candida albicans*. Mycopathologia 61: 183–186, 1977.
41. FC Odds. *Candida* and Candidosis. London: Bailliere Tindall, 1988.
42. H Lo, JR Kohler, B Di Domenico, D Loebenberg, A Cacciapuoti, GR Fink. Nonfilamentous *C. albicans* mutants are avirulent. Cell 90:939–949, 1997.
43. H Liu, J Kohler, G Fink. Suppression of hyphal formation in *Candida albicans* by mutation of a *STE12* homolog. Science 266:1723–1726, 1994.
44. VR Stoldt, A Sonneborn, CE Leuker, JF Ernst. Efg1p, an essential regulator of morphogenesis of the human pathogen *Candida albicans*, is a member of a conserved class of bHLH proteins regulating morphogenetic processes in fungi. EMBO J 16:1982–1991, 1997.
45. AJP Brown, NAR Gow. Regulatory Networks controlling *Candida albicans* morphogenesis. Trends Microbiol 7:333–338, 1999.
46. J Buffo, MA Herman, DR Soll. A characterization of pH-regulated dimorphism in *Candida albicans*. Mycopathologia 85:21–30, 1984.

47. WL Chaffin, JL Lopéz-Ribot, M Casanova, D Gozalbo, JP Martínez. Cell wall and secreted proteins of *Candida albicans:* identification, function and expression. Microbiol Mol Biol Rev 62:130–180, 1998.

48. GJ Smits, JC Kapteyn, H van den Ende, FM Klis. Cell wall dynamics in yeast. Curr Opinion Microbiol 2:348–352, 1999.

49. B Hube, M Monod, DA Schofield, AJP Brown, NAR Gow. Expression of seven members of the gene family encoding secretory aspartyl proteinases in *Candida albicans.* Mol Microbiol 14:87–99, 1994.

50. TC White, N Agabian. *Candida albicans* secreted aspartyl proteinases: isoenzyme pattern is determined by cell type and levels are determined by environmental factors. J Bacteriol 177:5215–5221, 1995.

51. F De Bernardis, S Arancia, L Morelli, B Hube, D Sanglard, W Schäfer, A Cassone. Evidence that members of secretory aspartyl proteinases gene family, in particular *SAP2*, are virulence factors for *Candida vaginitis.* J Infect Dis 179:201–208, 1999.

52. B Hube, D Sanglard, FC Odds, D Hess, M Monod, W Schäfer, AJ Brown, NAR Gow. Gene disruption of each of the secreted aspartyl proteinase genes *SAP1*, *SAP2* and *SAP3* in *Candida albicans* attenuates virulence. Infect Immun 65:3529–3538, 1997.

53. M Schaller, HC Korting, W Schäfer, J Bastert, WC Chen, B Hube. Secreted aspartic proteinase (Sap) activity contributes to tissue damage in a model of human oral candidosis. Mol Microbiol 34:169–180, 1999.

54. M Sentandreu, MV Elorza, R Sentandreu, WA Fonzi. Cloning and characterization of *PRA1*, a gene encoding a novel pH-regulated antigen of *Candida albicans.* J Bacteriol 180:282–289, 1998.

55. F Mühlschlegel, W Fonzi, L Hoyer, T Payne, FM Poulet, J Clevenger, JP Latgé, J Calera, A Beauvais, S Paris, M Monod, J Sturtevant, M Ghannoum, Y Nozawa, R Calderone. Molecular mechanisms of virulence in fungus-host interactions for *Aspergillus fumigatus* and *Candida albicans.* Med Mycol 36(S):238–248, 1998.

56. SM Saporito-Irwin, CE Birse, PS Sypherd, WA Fonzi. *PHR1* a pH-regulated gene of *Candida albicans* is required for morphogenesis. Mol Cell Biol 15:601–613, 1995.

57. FA Mühlschlegel, WA Fonzi. *PHR2* of *Candida albicans* encodes a functional homolog of the pH-regulated gene *PHR1* with an inverted pattern of pH-dependent expression. Mol Cell Biol 17: 5960–5967, 1997.

58. WA Fonzi. *PHR1* and *PHR2* of *Candida albicans* encode putative glycosidases required for proper cross-linking of β-1.3- and β-1,6-glucans. J Bacteriol 181:7070–7079, 1999.

59. L Popolo, M Vai. Defects in assembly of the extracellular matrix are responsible for altered morphogenesis of a *Candida albicans phr1* mutant. J Bacteriol 180:163–166, 1998.

60. F De Bernardis, FA Mühlschlegel, A Cassone, WA Fonzi. The pH of the host niche controls gene expression in and virulence of *Candida albicans.* Infect Immun 66:3317–3325, 1998.

61. T Nakazawa, H Horiuchi, A Ohata, M Takagui. Isolation and characterization of *EPD1*, an essential gene for pseudohyphal growth of a dimorphic yeast, *Candida maltosa.* J Bacteriol 180:2079–2086, 1998.

62. M Weig, F Mühlschlegel. Unpublished data.

63. M Vai, E Gatti, E Lacana, L Popolo, L Alberghina. Isolation and deduced amino acid sequence of the gene encoding gp115, a yeast glycophospholipid-anchored protein containing a serine rich region. J Biol Chem 266:12242–12248, 1991.

64. I Mounya, T Fontaine, M Vai, M Monod, M Diaquin, L Popolo, WA Fonzi, B Henrissat, RP Hartland, JP Latge. The glucanosyltransferase of *Aspergillus fumigatus* responsible for the elongation of cell wall bete (1–3) glucan. GenBank AF072700, 1999.

65. A Porta, AM Ramon, WA Fonzi. *PRR1*, a homolog of *Aspergillus nidulans palF*, controls pH-dependent gene expression and filamentation in *Candida albicans.* J Bacteriol 181:7516–7523, 1999.

66. AM Ramon, A Porta, WA Fonzi. Effect of environmental pH on morphological development of *Candida albicans* is mediated via the PacC-related transcription factor encoded by *PRR2.* J Bacteriol 181:7524–7530, 1999.

67. A El Barkani, O Kurzai, WA Fonzi, A Ramon, A Porta, M Frosch, FA Mühlschlegel. Dominant active alleles of *RIM101* (*PRR2*) bypass the pH restriction on filamentation of *Candida albicans*. Mol Cell Biol 20:4635–4647, 2000.
68. DJ Sullivan, TJ Westerneng, KA Haynes, DE Bennett, DC Coleman. *Candida dubliniensis* sp. nov.: phenotypic and molecular characterization of a novel species associated with oral candidosis in HIV-infected individuals. Microbiology 141:1507–1521, 1995.
69. SR Schorling, HC Korting, M Frosch, FA Mühlschlegel. The role of *Candida dubliniensis* in oral candidiasis in human immunodeficiency virus-infected individuals. Crit Rev Microbiol. 26:59–68, 2000.
70. O Kurzai, WJ Heinz, F Mühlschlegel. ASM-conference on *Candida* and candidiasis. Mycoses 42: 427–430, 1999.
71. O Kurzai, HC Korting, D Harmsen, W Bautsch, M Molitor, M Frosch, FA Mühlschlegel. Molecular and phenotypic identification of the yeast pathogen *Candida dubliniensis*. J Mol Med 78:521–529, 2000.
72. WJ Heinz, O Kurzai, AA Brakhage, WA Fonzi, HC Korting, M Frosch, FA Mühlschlegel. Molecular responses to environmental changes are conserved between *Candida dubliniensis* and *Candida albicans*. Int J Med Microbiol 290:231–238, 2000.
73. O Kurzai, WJ Heinz, DJ Sullivan, DC Coleman, M Frosch, FA Mühlschlegel. Rapid PCR test for discriminating between *Candida albicans* and *Candida dubliniensis* isolates using primers derived from the pH-regulated *PHR1* and *PHR2* genes of *C. albicans*. J Clin Microbiol 37:1587–1590, 1999.

8

Molecular Biology of Switching in *Candida*

David R. Soll
University of Iowa, Iowa City, Iowa

I. INTRODUCTION

Candida albicans is an obligate diploid, apparently unable to undergo mating or meiosis [1,2]. Therefore, the variability afforded a sexual organism through meiotic recombination is not part of its arsenal for phenotypic variability which is apparently so essential for the success of pathogenic as well as nonpathogenic organisms. Another way an organism can obtain phenotypic variability is through developmental programs. *C. albicans* and related species are capable of undergoing the bud-hypha transition, which provides them with at least two alternative growth forms [3]. The capacity to form a hypha appears to be basic to some forms of infection, and mutants that cannot form hyphae exhibit diminished virulence in a systemic model for pathogenesis [4–6]. However, two alternative growth forms do not seem to represent a great enough repertoire of phenotypic diversity to account for the extraordinary success *C. albicans* has had as both a pathogen and commensal [7,8]. It should therefore have been no surprise to discover that most strains of *C. albicans* and related species undergo spontaneous high-frequency switching between a number of general phenotypes that can be distinguished by colony morphology [9–15], and in some cases cell morphology [11,14].

II. GENERAL CHARACTERISTICS OF SWITCHING

Switching in *C. albicans* strain 3153A was discovered by plating cells on defined nutrient agar limiting for zinc [9]. Cells of this strain expressing the basic white phenotype were demonstrated to switch spontaneously between seven colony morphologies (Fig. 1) at frequencies ranging between 10^{-3} and 10^{-4}, and cells expressing variant phenotypes were demonstrated to switch at frequencies between 10^{-2} and 10^{-3} [9,12]. It was subsequently discovered that although all tested strains of *C. albicans* were capable of switching, the variant colony morphologies in the switching repertoires of different strains could vary quite dramatically. For instance, *C. albicans* strain WO-1 switched primarily between a white hemispherical colony morphology and a gray flat colony morphology, which were originally designated "white" and "opaque," respectively. This phase transition has not been demonstrated in *C. albicans* strain 3153A. Although variant colony phenotypes can differ between strains, several general characteristics emerged that were common to switching in a variety of tested strains [12]. First, switching occurs spontaneously

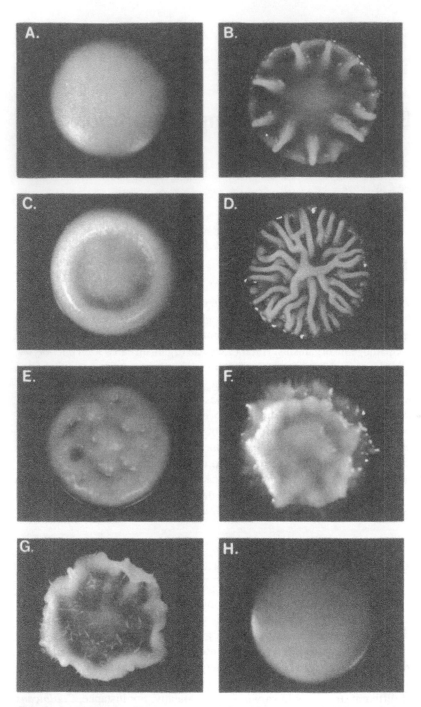

Figure 1 Original switching system of *C. albicans* strain 3153A revealed on zinc-limiting agar. (A) Original smooth; (B) star; (C) ring; (D) irregular wrinkle; (E) stippled; (F) hat; (G) fuzzy; (H) revertant smooth. (From Ref. 9.)

[9–14]. Second, switching is reversible at high frequency [9–13]. Third, a low dose of UV stimulates an increase in the frequency of switching of cells originally in a low-frequency mode of switching [16]. The UV-stimulated increase is heritable; i.e., progeny of cells treated with UV continue to switch at high frequency through many generations after UV treatment [16]. Other insults or extreme environmental perturbations, such as an increase or a decrease in temperature, affect switching frequencies [11,17–19]. Fourth, infecting strains on average switch at higher frequencies than commensals [15,20–22]. Fifth, variant phenotypes on average switch at higher frequencies than the basic smooth white phenotype in most strains [9]. Finally, switching involves the differential expression of combinations of phase-specific genes [12,19,23–34].

III. SWITCHING REGULATES SEVERAL VIRULENCE TRAITS

When switching in *C. albicans* strain 3153A was first discovered, it was suggested that it played a role in pathogenesis [9]. This suggestion was reinforced when it was demonstrated that switching affects (1) susceptibility to antifungals [21,35,37], (2) adhesion and cohesion [38,39], (3) release of aspartyl proteinase activity [24,27,28,40], (4) constraints on the bud-hypha transition [41], (5) antigenicity [42], (6) sensitivity to white blood cells and oxidants [17], and (7) virulence in animal models of pathogenesis [40,43] (Table 1). The capacity to switch between a limited number of general phenotypes, each with a different combination of virulence traits, would appear to represent a strategy for variability in which switching represents a higher-order regulatory process [8,44]. Each population would contain cells in alternate switch phenotypes positioned to enrich in response to a rapid alteration in the environment.

IV. "WHITE-OPAQUE TRANSITION": AN EXPERIMENTAL MODEL FOR SWITCHING

To study the molecular basis of switching, the "white-opaque transition" in *C. albicans* strain WO-1 was selected for a number of reasons. First, this reversible transition involves two major

Table 1 Differences Between White and Opaque Phase Cells in Phenotypic Characteristics and Gene Expression

Phenotypic characteristics	Reference	Gene expression	Reference
1. Cellular morphology	11,14	1. Expression of *PEP1 (SAP1)*	24,27,28
2. Accessibility of dyes	14,	2. Expression of *OP4*	25
3. Antigenicity	42	3. Expression of *WH11*	19
4. Adhesion and cohesion	38,39	4. Expression of *CDR3*	26
5. Constraints on the bud-hypha transition	41	5. Expression of *CaNIK1*	29
		6. Expression of *EFG1*	30,31
6. Sensitivity to white blood cells and oxidants	17	7. Expression of *PGM1*	(unpublished observation)
7. Susceptibility to anifungals	36,98	8. Expression of *PFK2*	(unpublished observation)
8. Sugar assimilation pattern	97		
9. Virulence in a mouse tail injection model for systemic infections	43	9. Expression of *SAP3*	27,28
10. Virulence in a mouse skin colonization model for cutaneous infections	40		

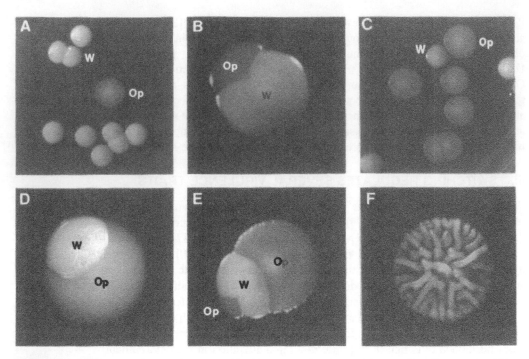

Figure 2 "White-opaque transition" in *C. albicans* strain WO-1. (A) A switch from white (W) to opaque (Op); (B) opaque sector in a white colony; (C) switch from opaque (Op) to white (W); (D) white sector in an opaque colony; (E) multiple sectoring; (F) the minor wrinkled phenotype. (From Ref. 11.)

phases (Fig. 2). The colonies of the two major phases can be discriminated on all tested media. By adding phloxine B to the agar medium, differences between the two phases are accentuated [14]. On this agar, white phase colonies and sectors are white or very light pink, while opaque phase colonies and sectors stain red. Although several minor phenotypes other than white and opaque were identified (e.g., wrinkled in Fig. 2F) in large plating experiments [11], the frequencies at which they appeared were low and, therefore, their infrequent appearance did not interfere in quantitative analyses of transitions between white and opaque. Second, this reversible transition has a dramatic effect on cellular phenotype. In the white phase, cells are round and bud like diploid *Saccharomyces cerevisiae*, while in the opaque phase, cells are twice as big, elongate, or bean-shaped and bud in an aberrant fashion [11,14]. When examined by scanning electron microscopy, the surfaces of white phase cells were demonstrated to be smooth like most budding cells of *C. albicans* and *S. cerevisiae*, while opaque phase cells exhibited pimples, sometimes with a single small bleb emanating from the pimple center [14]. When examined by transmission electron microscopy, the wall of white phase cells was uniform, while that of opaque phase cells again exhibited pimples containing central pores [14]. In addition, opaque phase cells contain a large vacuole which in turn contains vesicles [14].

The use of the white-opaque transition in strain WO-1 as a model for studying switching has been enhanced by the recent development of auxotrophic strains and transformation constructs for efficient two-step gene knockouts of any nonessential gene of interest [31]. In addition, it has been demonstrated that phase-specific and inducible genes integrated at the ADE2 locus are regulated at that ectopic location in the same manner as at their normal genomic location [45]. For these and additional reasons that will emerge later in the review, the white-opaque transition represents the model of choice for studying the molecular basis of switching in *C. albicans*.

V. SWITCHING TURNS PHASE-SPECIFIC GENES ON AND OFF

The pleiotrophic effects initially revealed in the biological characterization of the phase transition in strain WO-1 (Table 1) suggested that switching involved the regulation of a variety of phase-specific genes. In 1992, the first phase-specific gene was cloned by a differential hybridization screen of an opaque phase-specific cDNA library with white and opaque phase cDNA [24]. The opaque phase-specific gene proved to be a pepsinogin cloned a year earlier by Hube et al. [46]. This phase-specific gene was named *PEP1* [24]. It was subsequently renamed *SAP1*, since it proved to be a member of a family of secreted aspartyl proteinases [27,28]. In 1993, the second phase-specific gene, *OP4*, was cloned by a similar differential hybridization screen [25]. The deduced *OP4* protein contained 402 amino acids, with an estimated molecular mass of 41.3 kDa. Neither the DNA nor the deduced amino acid sequence exhibited significant homology to either DNA or protein sequences in respective databases. The deduced protein did, however, possess a number of noteworthy characteristics [25]. First, 38% of protein mass is composed of only three amino acids—alanine, leucine, and serine. Second, the amino terminal 26 amino acids represent a hydrophobic region. Third, the carboxy-terminal 100 amino acids has a pH of 10.73. Fourth, the deduced sequence contains two serine clusters, each bordered by an alanine-rich cluster, in regions with strong potentials for α-helices. Fifth, the protein contains seven potential glycosylation sites.

A similar approach was used in an attempt to clone a white phase-specific gene, but it proved unsuccessful. However, when white phase cDNA libraries were subtracted with opaque phase cDNA, the differential hybridization screen worked, and the white phase-specific gene *WH11* was cloned [19]. *WH11* was expressed in the white budding phase, but it was not expressed in either the opaque phase or in the white hyphal phase. Therefore, *WH11* expression is specific to the white budding phenotype. The deduced protein of *WH11* contains 65 amino acids with a molecular mass of 7.8 kDa [19]. The deduced sequence is similar to that of the glucose/lipid-regulated protein G1p1p of *S. cerevisiae* [50], a low-molecular-weight heat shock protein [48,49]. Antiserum raised against an rWh11 protein was used to demonstrate that in white phase cells, Wh11p is distributed throughout the cytoplasm, but is excluded from the plasma membrane, nucleus, vacuoles, and vesicles [50]. Recently, a *WH11* homolog, referred to as *AnWH11*, was cloned from *Aspergillus nidulans* [51].

Subsequent to the cloning and characterization of *PEP1(SAP1)*, *OP4*, and *WH11*, a number of additional genes were demonstrated to be phase specific in *C. albicans*. Balan et al. [26] demonstrated that expression of the ABC transporter gene CDR3 is regulated by the white-opaque transition. This gene is a member of the family of ABC transporters that function as drug resistance genes in *C. albicans*. Srikantha et al. [29] have demonstrated that the two-component regulator *CaNIK1* is also differentially expressed in the opaque phase. Two component regulators are found in bacteria and lower eukaryotes, and play a role as environmental or intracellular sensors [52,53]. Finally, it was recently reported that the *trans*-acting factor gene *EFG1* was differentially expressed in the white phase [30]. Subsequently, it was demonstrated that the *EFG1* gene was transcribed in the white and opaque phases, but from different transcription start sights of the same gene, resulting in phase-specific mRNAs of different molecular mass [31]. The smaller opaque phase transcript (2.2 kb) was 20 times less abundant than the larger white phase transcript (3.2 kb) [31].

In a recent screen using an opaque phase-specific *cis*-acting activation sequence functionally identified in an opaque phase gene promoter, the white phase-specific gene *PGM1* was cloned (S. Lockhart and D.R. Soll, unpublished observations). *PGM1* encodes phosphoglycerate mutase. This screen has also identified a number of additional white and opaque phase-specific genes, now under analysis. Because there has been no systematic search for all opaque phase-specific and white phase-specific genes, it seems likely that the final number of phase-specific genes will be significantly higher than that reviewed here (Table 1). However, even with the limited number so far identified, it is noteworthy that a significant number of them by their nature could play roles in pathogenesis. It seems very unlikely that by chance the first cloned gene was *PEP1 (SAP1)* [24], a major aspartyl proteinase that may be involved in tissue penetration, and that subsequent cloned phase-specific genes included an ABC transporter [26] that may be involved in drug resistance, and a two-component histidine kinase regulator [29] that may be involved in sensing intracellular or environmental changes. It therefore seems likely that when the full repertoire of phase-specific genes is elucidated, we will find that it includes many additional genes that are directly or indirectly involved in pathogenesis. Although the phenotypic characterizations of variant phenotypes of the switching systems of *Candida glabrata* [54] and *Cryptococcus neoformans* [58,59] are still in their infancy, the results so far acquired suggest that the same will hold true for these systems as well, supporting the suggestion that high-frequency phenotypic switching represents a higher-order virulence trait in the pathogenic fungi [8,44].

VI. THE SWITCH EVENT, MASS CONVERSION, AND GENE EXPRESSION

In the transition back and forth between the white and the opaque phases, phase-specific genes are turned on and off. Because an increase in temperature induces mass conversion from the opaque to white phase [11,18,19,25], one can pinpoint in time in a semisynchronously switching population the flip between opaque and white phase gene expression, and correlate it with the point of phenotypic commitment to the white phase [19] (Fig. 3). By shifting opaque phase cells from 25°C to >37°C (e.g., 42°C), one initiates the process of mass conversion, and then by returning cells to 25°C at 1-hr intervals, one can pinpoint the time at which cells commit to generating white phase cells at the higher temperature. Prior to phenotypic commitment, cells returned to 25°C continue to generate opaque phase daughter cells, and after phenotypic commit-

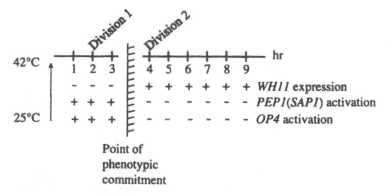

Figure 3 Temperature-induced semisynchronous mass conversion from the opaque to white phase.

ment, cells returned to 25°C form white phase daughter cells. The point of phenotypic commitment was demonstrated to occur between 3 and 4 hr for the majority of cells, which coincided with the second semisynchromous round of cell division [19] (Fig. 3). Opaque phase cells shifted from 25°C to 42°C did not express *WH11* prior to the point of phenotypic commitment [19]. Cells shifted after the point of phenotypic commitment expressed *WH11* (Fig. 3). Activation of *WH11*, therefore, occurs at the time of phenotypic commitment. Opaque phase cells shifted from 25°C to 42°C immediately turned off *OP4* and *PEP1 (SAP1)* transcription [25]. Within 30 min at 42°C, the levels of both transcripts were negligible. However, if these cells are returned to 25°C prior to the point of commitment, the levels of both transcripts returned to the original opaque phase levels within 30 min. Expression could therefore be rapidly reactivated prior to the commitment event. However, if cells were returned to 25°C after the point of commitment, transcription of neither gene is turned back on [25]. Therefore, neither *PEP1 (SAP1)* or *OP4* could be reactivated after the point of phenotypic commitment (Fig. 3). These results together demonstrate that at the point of phenotypic commitment, white phase genes are developmentally activated and opaque phase genes are developmentally deactivated. That this occurs coincidentally with the second cell doubling in a diploid cell is reminiscent of imprinting [57], a phenomenon in which a recessive change in one allele cannot be expressed until the second cell doubling, and then in only one of the two daughter cells. In the semisynchronous kinetics of phenotypic commitment in *C. albicans* [19], there is a hint that the latter characteristic also holds true (D.R. Soll, in preparation).

VII.   REGULATION OF PHASE-SPECIFIC GENE EXPRESSION

In the white-opaque transition, a clear "yin-yang" process of gene regulation is observed. When white phase-specific genes are turned on in the transition from the opaque to white phase, opaque phase-specific genes are turned off, and when opaque phase-specific genes are turned on in the transition from the white to opaque phase, white phase-specific genes are turned off. The simplest hypothesis for yin-yang regulation is that all white phase-specific genes are activated by a single white phase-specific *trans*-acting factor, and all opaque phase-specific genes are activated by a single opaque phase-specific *trans*-acting factor. In its simple form, this hypothesis predicts that all white phase cells would share the same *cis*-acting activation sequence in their promoter and all opaque phase cells would share the same *cis*-acting activation sequence in their promoter. The reality may be a bit more complex than this.

   Although the upstream 5′ regions preceding the open reading frames (ORFs) of the first phase-specific genes had been sequenced and putative regulatory regions identified through homology with known *cis*-acting sequences in the promoters of genes of other organisms [19,24,25], the identification of true regulatory sequences requires a functional analysis of deletion derivatives of the promoter fused to a sensitive reporter gene. For *C. albicans*, this strategy was hampered by the absence of an effective reporter gene, due in large part to the alteration in codon usage [58,59]. In *C. albicans*, serine residues rather than leucine residues are encoded by CUG. Therefore, heterologous genes encoding proteins with leucines at functional sites will not be synthesized in a functional form in *C. albicans* [59]. The first functional analysis of a phase-specific gene promoter was performed on *WH11* [60]. In this analysis, phase-specific regulation was assessed by Northern blot hybridization, and integration of the deletion derivative constructs was targeted to the *ADE2* locus. This locus had been proven to be relatively neutral in that a variety of promoters integrated at this sight exhibited normal regulation and strength [45]. A deletion analysis of the *WH11* promoter revealed two major activation sequences that function synergistically [60].

To increase the sensitivity of this analysis, a reporter system was developed that employed the luciferase of the sea pansy *Renilla reniformis* [45]. The *Renilla* luciferase contains no leucines and, therefore, is expressed in a functional form in *C. albicans*. Using this reporter, the two major activation sequences were confirmed, the distal activation sequence DAS and the proximal activation sequence PAS [61]. In addition, a minor activation domain was identified proximal to PAS [61]. The major *cis*-acting activation sequences in the *WH11* promoter have no homology with identified promoter sequences in other organisms, including *S. cerevisiae* and were therefore deemed unique [60,61]. Gel retardation experiments were performed with white phase and opaque phase cell extracts and the distal transcription activation sequences DAS and PAS [60]. One white phase-specific complex was identified for DAS and two for PAS. There was no suggestion in the deletion derivatives of the *WH11* promoter that there existed negative regulation of *WH11* in the opaque phase [60,61]. These results suggest that switching is accompanied by activation or synthesis of white phase-specific *trans*-acting factors that activate white phase-specific gene transcription.

In order to obtain a fuller picture of phase-specific regulation, the promoter of the opaque phase-specific gene *OP4* was functionally characterized using the *Renilla* luciferase as a reporter system [62]. A strong 17-bp transcription activation sequence was identified that contained a MADS box consensus binding site that was most closely related to the Mcm1 binding site of *Saccharomyces cerevisiae* [63,64]. Mcm1 in conjunction with other proteins functions as a universal regulator in *S. cerevisiae*, both activating and repressing a variety of genes involved in mating and the cell cycle [65]. Point mutations in the Mcm1 binding site of an Mcm1/α2-activated promoter of *S. cerevisiae* that led to loss of function were engineered in the activation site of the *OP4* promoter [62]. In each case, there was a similar loss of function.

Using the MADS box consensus sequence binding site, a number of additional phase-regulated genes have been cloned from *C. albicans* including the phosphofructokinase gene *PFK2* (S. Lockhart and D.R. Soll, in preparation), suggesting that a phase-specific *trans*-acting factor coordinately regulates a number of opaque phase-specific genes through this binding site. In addition, the secreted aspartyl proteinase 3 gene (*SAP3*) [27,28] and the *Candida* drug resistance 3 gene (CDR3) [26], both expressed exclusively in the opaque phase, contain a complete MADS box consensus sequence binding site in their promoters. From these results, one might conclude that all opaque phase-specific genes are coordinately regulated through the same *cis*-acting MADS box consensus sequence binding site, but that is not the case. The functionally identified *cis*-acting sequences for the opaque phase-specific gene of *PEP1(SAP1)* do not contain a recognizable MADS box consensus sequence binding site (S. Lockhart, H. Fang, and D.R.

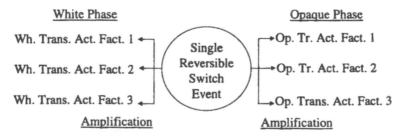

Figure 4 A model for gene regulation during the white-opaque transition in which the single switch event in each direction is amplified through the formation of multiple phase-specific *trans*-acting factors. A switch also involves the deactivation of *trans*-acting factors of the previous phenotype.

Soll, in preparation), suggesting that the regulation of opaque phase-specific genes involves more than one opaque phase-specific *trans*-acting factor.

The results so far in hand suggest that the switch event in the white-to-opaque direction results in the synthesis or activation of two or more opaque phase-specific *trans*-acting factors that activate opaque phase-specific genes, and that the switch in the opaque-to-white direction may also result in the synthesis or activation of one or more white phase-specific *trans*-acting factors that activate white phase-specific genes (Fig. 4). The cessation of synthesis, rapid turn-over, and/or deactivation of *trans*-acting factors of the original phenotype must accompany each switch. These results suggest that the switch in each direction may be a singular event, but is amplified in complexity to regulate multiple genes in a single phase through different *cis*-acting sequences (Fig. 4).

VIII. GENES INVOLVED DOWNSTREAM IN THE EXPRESSION OF A SWITCH PHENOTYPE CAN BE MISINTERPRETED AS INVOLVED IN THE SWITCH EVENT

It should be evident in the model in Figure 4 that both mutations affecting the switch event and mutations affecting downstream gene regulation can block expression of the white or opaque phase phenotype. Therefore, genes involved in downstream gene regulation can be misinterpreted as affecting the switch event. For instance, if a mutant cell undergoes the switch event but can't express the terminal switch phenotype, the mutated gene can mistakenly be interpreted as playing a direct role in the switch event. A case in point is the *EFG1* gene, which encodes a putative transcription factor homologous to *PHD1* in *S. cerevisiae* [66] and *STUA* in *Aspergillus nidulans* [67], both of which are involved in cellular morphogenesis. Initially, it was demonstrated that overexpression of *EFG1* in strain WO-1 stimulated the opaque-to-white transition, and under-expression in strain CAI8 caused a cell morphology similar to that of opaque phase cells in strain WO-1, but without pimples [30]. These results led to the logical suggestion that *EFG1* was involved in the switch event in the white-opaque transition. The phenotype of *EFG1* null mutants in strain WO-1 superficially supported this view [31]. Mutants formed what appeared to be opaque colonies at 25°C with a majority of pimpled, elongate cells. However, when mutant opaque phase cells were subjected to the temperature shift that initiates mass conversion, they flipped from a pattern of opaque phase gene expression to a pattern of white phase gene expres-sion after the point of commitment, and on close observation were found to form a cellular morphology that straddled the white and opaque phase phenotypes [31].

Like normal opaque phase cells, mutant cells formed daughter cells after the point of commitment that were smooth (pimpleless), but unlike normal daughter cells, mutant daughter cells were elongate, or bean-shaped, like opaque phase cells. Together, these results revealed that *EFG1* null mutant cells switch from the opaque phase to the white phase after a shift from 25°C to 42°C, but they cannot express the complete white budding phase phenotype. Therefore, *EFG1* functions downstream of the switch event, and is necessary for the round phenotypic characteristic of white phase cells, but not for the regulation of phase-specific gene expression or the removal of opaque phase pimples. These results were consistent with the observation that reduced expression of *EFG1* in strain CAI8 resulted in a pimpleless elongate phenotype [30].

The above set of observations demonstrate that without a method for distinguishing be-tween the effects on the switch event and effects on the terminal phenotype, mutations affecting the latter can be misinterpreted as mutations affecting the former. The capacity to monitor the commitment event both through daughter cell phenotype and gene regulation provides a method for discriminating between the involvement of a gene in the switch event or in the downstream expression of phenotype.

IX. REGULATION OF THE SWITCH EVENT

In bacterial phase transitions elucidated at the molecular level, the basic switch event usually occurs at a single locus and, therefore, involves a single event [68]. A priori, one would expect the white-opaque transition to also represent a single event. The question is, what is the event? In bacterial phase transitions, a change occurs in DNA sequence in the form of either an inversion or recombination that alters gene expression [68]. Although the bacterial mechanism of reversible DNA reorganization remains a viable model for switching in the pathogenic fungi, an equally viable model involves alterations in chromatin state that are heritable, but metastable. With these two models in mind, let us first consider mutations that affect the switching process for hints of mechanism.

The first gene that was demonstrated to affect the frequency of switching in *C. albicans* was the white phase-specific gene *WH11* [43]. *WH11* was ligated downstream of the opaque phase-specific *OP4* promoter, in the plasmid pCWOP16. This plasmid was linearized at the *ADE2* gene in order to target integration at the *ADE2* locus. Integration resulted in misexpression of the white phase-specific gene *WH11* in the opaque phase phenotype. A result of misexpression was a dramatic increase in the frequency of white phase cells in opaque phase colonies [43]. In other words, misexpression of *WH11* resulted in an apparent instability of the opaque phase, leading to a dramatic increase in the frequency of the opaque-to-white transition. The deduced *WH11* protein is homologous to a low-molecular-weight heat shock protein of *S. cerevisiae* [48]. Some low-molecular-weight heat shock proteins function as chaperon proteins [49]. Wh11p is localized in the cytoplasm and is excluded from the nucleus [50]. Because Wh11p is excluded from the nucleus, it is highly unlikely that it is directly involved in the switch event per se. However, it may indirectly affect switching frequency by regulating the availability of a factor directly involved in the switch event.

The second gene demonstrated to affect the frequency of switching is *SIR2*. Perez-Martin et al. [69] cloned a gene from *C. albicans* that was a homolog of the *S. cerevisiae SIR2* gene. They deleted *CaSir2* in strain CAI4, which undergoes switching between a number of phenotypes comparable to those in strain 3153A [29], but which does not undergo the white-opaque transition. The *SIR2* deletion mutant exhibited a dramatic increase in the frequency of switching, as high as 10^{-1} [69]. In contrast to *WH11*, *Sir2p* is a nuclear protein with demonstrated involvement in gene silencing and suppression of recombination [70,71]. Therefore, one could hypothesize a direct role for *Sir2p* in switching through the modulation of chromatin structure.

Finally, deletion of the two-component hybrid kinase regulator *CaNIK1* in strain CAI4 resulted in a small reduction in the frequency of variant phenotype formations in both untreated and UV-treated cell cultures [29]. Again, it is not obvious how a two-component hybrid kinase would function in switching except through a regulatory phosphorylation cascade. It therefore seems highly unlikely that the three disparate gene products—Wh11p, Sir2p, and CaNik1p—all function directly in the actual switch event, or affect switching indirectly in a similar fashion.

It has also been demonstrated that temperature can cause mass conversion in the opaque-to-white direction [11,18,19], that ultraviolet irradiation can affect the frequency of switching in both directions [16], and that gas exchange can affect the frequency of switching at the edge of a colony [12]. It has also been demonstrated in temperature-induced mass conversion that the commitment event coincides with the second cell doubling and a yin-yang change in the expression of phase-specific genes. These observations as well as the general characteristics of switching must contain clues to the puzzle. One might consider whether the misexpression of *WH11*, the deletion of *CaNIK1* and the effects of temperature, irradiation, and gas exchange would more likely affect the frequency of a specific reorganization of DNA or the frequency of a specific change in chromatin state. There is little doubt that the deletion of *SIR2* would more likely affect the latter.

Let us next consider the precision of a switch. In the white-opaque transition, one can follow switching in a lineage. After 15 reversible switches back and forth between white and opaque, cells from the first and last white phenotypes are phenotypically indistinguishable, and cells from the first and last opaque phenotypes are phenotypically indistinguishable (D.R. Soll, unpublished observations). They have the same phase-specific cellular morphologies, switch to the alternate phenotypes at the same frequencies and express the same genes. This result suggests that switching represents a single, reversible, and precise event. We then consider whether the two major models for switching are compatible with this characteristic.

First, let us consider mechanisms involving DNA reorganization. DNA inversion is used to regulate fimbriae variation in *E. coli* [72], flagellar phase variation in *S. typhimurium* [73], pilia variation in *Moraxella bovis* [74], and phage host variation [75]. In all of these cases, the sequential rearrangement of DNA is precise and reversible. Although the model can be applied to the white-opaque transition, alone it cannot accommodate the more complex switching systems, like those in strain 3153A [9] (Fig. 1) and CAI8 [29], which involve multiple variant phenotypes. The substitution of DNA at the expressed mating type locus in *S. cerevisiae* is also precise and reversible [71]. Two silent cassettes, "a" and "∝," are precisely substituted at an expression locus so that cells are either "a" or "∝." This model can accommodate a simple phase transition like the white-opaque transition in strain WO-1, as well as a complex switching system like that in strain 3153A or CAI8. In the latter case, one can simply increase the number of silent cassettes. The elucidation of the DNA rearrangement models in both bacteria and yeast were initially facilitated by mutational analyses of haploid organisms. Unfortunately, because *C. albicans* is diploid, a mutational analysis of switching has not been feasible. However, the recent discovery of high-frequency phenotypic switching in the haploid yeast *C. glabrata* [54] (see Sec. XI) provides us for the first time with an experimental system for a mutational approach to the switching mechanism.

The second model involves heritable, reversible, and precise changes in chromatin structure leading to the activation or deactivation of an associated locus. Pilus and Rine [76] demonstrated that a mutant of the silent information regulator 1 gene (sir1-1) was leaky, resulting in a genetically homogeneous but phenotypically heterogeneous population for the expression of HMLα, the silent locus for mating type α in wild-type cells. They demonstrated that this locus switched back and forth at a frequency of 4×10^{-3}. Subsequently, Gottshling and coworkers [77] demonstrated that when the *ADE2, URA3, HIS3,* or *TRP1* gene was positioned close to a telomeric region of a chromosome, each gene switched between an expressed and repressed state at similar frequencies. The positional effects depended on functional *SIR2, SIR3,* and *SIR4* genes [77]. It is easy to see how switching could be regulated by this type of positional effect. A master regulatory gene would be positioned next to a telomeric region. The position would lead to spontaneous flips between an expressed and unexpressed state. The two states would correlate with switch phenotypes. Several observations on switching are compatible with a model involving metastable changes in chromatin. First, Ramsey et al. [78] demonstrated that when cells of *C. albicans* strain 3153A jumped from a low frequency of phenotypic switching in the original smooth phase to a high frequency of phenotypic switching in a variant phase, they also underwent frequent reorganization of ribosomal cistrons. This observation paralleled that of Gottlieb and Esposito [70], who found that a mutation in *SIR2* led to an increase in recombination in ribosomal DNA in *S. cerevisiae*. Second, the observation by Johnson and colleagues [69] that deletion of

SIR2 results in an increase in the switching frequency of *C. albicans* strain CAI8 again implicates silencing, and presumably heterochromatinization in switching. Third, the observation that phenotypic commitment during mass conversion coincides with the second cell doubling suggests that a metastable recessive change in a single allele may become homozygous at the second cell doubling, a mechanism suggested for "gene imprinting," which shares these cell division-associated characteristics and has been explained in terms of heritable changes in chromatin state [57]. Unfortunately, both DNA reorganization and heritable changes in chromatin state can be precise, and it is conceivable that all of the environmental perturbations and mutations that affect switching frequency can directly or indirectly affect either type of mechanism. Therefore, we may actually not yet have any strong clues that distinguish between the two models.

X. INVESTIGATING THE ROLES OF PHASE-SPECIFIC GENES IN PATHOGENESIS

The white-opaque transition provides a unique system for assessing the role of phase-specific genes in virulence. In most pathogenic systems, the expression of a phase-specific gene either facilitates pathogenesis in one pathogenic situation (e.g., the formation of type 1 fimbriae in *E. coli* [72]), or leads to repeated replacement of the same antigen, in response to a sequence of host immune responses (e.g., antigen variability in *Borrelia hermsii* [78], *Neisseria genorrhoeae* [86], and *Trypanosoma brucei* [81]). Both the switching process and performance in animal models suggest that switching plays a far more complex role in *C. albicans* and related species than it does in the aforementioned pathogens. Because switching in *C. albicans* regulates diverse genes in a combinatorial fashion in switching systems with multiple switch phenotype [82], each switch phenotype may be more highly adapted for a different anatomical niche, host physiology, immune response, drug therapy, or disease state [8].

The comparative virulence of white and opaque phase cells in two alternative animal models supports this suggestion. In a mouse tail injection model that assesses virulence through death and/or organ colonization caused by an induced systemic infection, white phase cells are far more virulent than opaque phase cells [40,43] (Fig. 5). Injection of 10^6 white phase cells

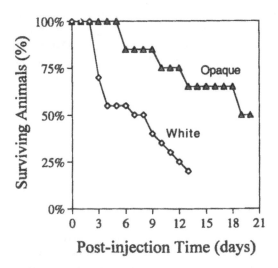

Figure 5 Differences in the virulence of white and opaque phase cells of *C. albicans* in infection. (From Refs. 40, 43.)

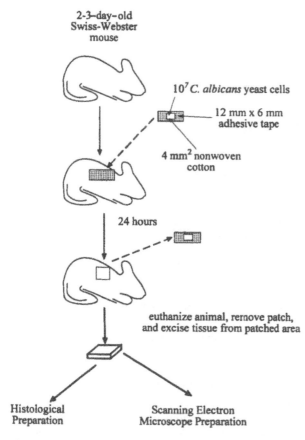

2-3–day–old
Swiss-Webster
mouse

$10^7 C.$ *albicans* yeast cells

12 mm x 6 mm
adhesive tape

4 mm^2 nonwoven
cotton

24 hours

euthanize animal, remove patch,
and excise tissue from patched area

Histological
Preparation

Scanning Electron
Microscope Preparation

Figure 6 Protocol for the only mouse skin model for cutaneous infection.

killed 50% of mice after ~6 days, while injection of 10^6 opaque phase cells killed 50% of mice after 18 days [40]. An analysis of colonization in the kidney showed 2×10^6 versus 8×10^4 cfu per gram of tissue after 12 days for mice injected with white phase and opaque phase cells, respectively [43]. In addition, the cells ultimately and belatedly colonizing the kidneys of mice injected with opaque phase cells were predominately white, suggesting that the delay in kidney colonization and death in the latter animals was due to the time necessary for the opaque phase cell population to switch to the white phase, and that opaque phase cells per se may not be virulent in a systemic model when expressing that phenotype.

In contrast, opaque phase cells are more virulent than white phase cells in a mouse model of cutaneous infection [40] (Fig. 6). In this model, a cotton patch is saturated with cells, then fixed to the skin on the back of a newborn mouse for 24 hr. The patch is then removed and the cells adhering to and/or imbedded in the skin are counted in electron micrographic sections. While 17% of sections through skin treated with white phase cells contained >25 adhering yeast cells, 72% of sections through skin treated with opaque phase cells contained >25 adhering yeast cells [40]. The dramatic difference in colonization between white and opaque phase cells is illustrated in the representative sections in Figure 7A and B, respectively. In addition, while white phase cells adhered to the skin surface without affecting skin cell morphology, opaque phase cells sunk into the skin cells, causing cavities [40]. These results demonstrate that in

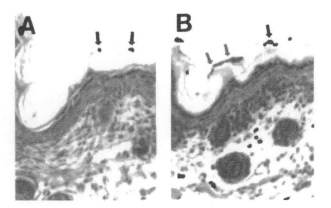

Figure 7 Difference in colonization of white (A) and opaque (B) phase cells of *C. albicans* in the baby mouse skin model for cutaneous infection. Arrows point to yeast cells.

contrast to systemic infections, opaque phase cells are far more virulent than white phase cells in a mouse model of cutaneous infection [40].

Several strategies can be used to test the role of select genes in pathogenesis. These include overexpression, deletion, and misexpression. Because a knockout strategy was not available for strain WO-1 until recently [31], the first strategy that was employed was misexpression. In this strategy, the open reading frame of a phase-specific gene is ligated downstream of a promoter of the opposite phase in a *C. albicans* transforming plasmid containing the *C. albicans* ADE2 gene. This plasmid is linearized at the *ADE2* gene, which targets transformation to the *ADE2* locus of the yeast [83]. The adenine auxotroph *C. albicans* Red 3/6 was used. The plasmid provides the *ADE2* gene for transformant selection. Previous analyses of promoter function using constructs in which a variety of regulated promoters were fused upstream of the *Renilla* luciferase reporter gene demonstrated that all tested promoters were regulated in a normal fashion and that relative promoter strengths were consistent with expression levels at normal genomic locations [45].

The first misexpression experiment was performed on the white phase-specific gene *WH11* [43]. The *WH11* ORF was placed downstream of the promoter region of the opaque phase gene *OP4* in the transforming vector pCWOP16, generating a mutant *WH11* (*mWH11*) transcript 20 nucleotides longer than the transcript of the endogenous gene, thus allowing discrimination of the two transcripts by Northern analysis. Transformants expressed both the *mWH11* transcript and the *mWH11* protein in the opaque phase [43]. The *mWH11* protein exhibited the same cytoplasmic distribution in opaque phase cells previously demonstrated for the *WH11* protein in white phase cells [50]. Misexpression of *WH11* had no effect on the unique opaque phase phenotype [43]. Opaque phase cells misexpressing *mWH11* were elongate and asymmetric and formed pimples in the cell wall. These results clearly demonstrate that *WH11*, which has been implicated in generating a round cell phenotype [19,50], is not sufficient to suppress the opaque phase phenotype.

Cells that misexpress *WH11* in the opaque phase switch between the white and opaque phase phenotypes, and form white and opaque phase cells with apparently normal morphology. However, misexpression of *WH11* as previously noted in this review exerts a dramatic effect on the stability of the opaque phase phenotype and, hence, on the frequency of the opaque-to-white transition. The effect is demonstrated in the frequencies of white phase cells in clonally derived opaque phase cell populations and opaque phase cells in clonally derived white phase cell populations. In control cells, the frequencies were measured as 6.0×10^{-4} and 1.3×10^{-3}, respectively [43]. In misexpression mutants, the frequencies were 2.0×10^{-1} and 1.7×10^{-3}. Hence, the frequency of opaque phase cells in a white phase population, which reflects the frequency of the switch from the white to the opaque phase, was the same in mutant and control cells. However, the frequency of white phase cells in an opaque phase population, which reflects the frequency of the switch from opaque to white, was 333 times higher in the misexpression mutant. The impact of *WH11* misexpression on switching in turn affected pathogenesis in the mouse tail injection model for systemic infections. Opaque phase cells of the misexpression mutant exhibited increased virulence paralleling that of white phase cells [43]. An analysis of cells infecting the kidney of test animals demonstrated that cells of the misexpression mutant rapidly switched back to the white phenotype which in turn led to an acceleration of kidney colonization.

The second misexpression experiment took full advantage of the alternative models of virulence. The *PEP1 (SAP1)* open reading frame was placed downstream of the *WH11* promoter in the transforming vector pCPW7 [40]. Since the *WH11* promoter is weak compared to the endogenous *PEP1 (SAP1)* promoter [45], a transformant containing approximately 10 tandem repeats of the plasmid was used to assess the effect of *PEP1 (SAP1)* misexpression on phenotype and virulence [40]. The level of *PEP1 (SAP1)* transcript and extracellular protease activity was comparable in opaque phase cells of either the control or mutant and white phase cells of the mutant. Misexpression of *PEP1 (SAP1)* had no effect on either the white phase phenotype or switching frequencies. Misexpression also had no effect on the already high level of virulence of white phase cells in the mouse tail injection model of systemic infection [40]. However, misexpression had a dramatic impact on the virulence of white phase cells in the mouse model of cutaneous infection. White phase cells of the control strains exhibited a low level of colonization and did not form cavities in the surface skin cells. White phase cells of the misexpression mutant, however, colonized skin at levels even higher than opaque phase cells and caused cavitation in the superficial skin layers, like opaque phase cells [40]. The addition of pepstatin, an inhibitor of pepsinogens like Sap1, blocked cavitation but not increased colonization [40]. Since pepsinogen functions extracellularly, this result suggests that Sap1 plays two virulence roles in the skin model, one that is cellular (not sensitive to pepstatin) and involved in increased adhesion, and one that is extracellular (sensitive to pepstatin) and involved in cavitation [40].

XI. SWITCHING IN *CANDIDA GLABRATA:* A POTENTIALLY MORE AMENABLE SYSTEM FOR ELUCIDATING THE SWITCHING MECHANISM

Switching has been demonstrated in *C. albicans* [9–11], *C. tropicalis* [13,84], *C. dubliniensis* (S. Joly, C. Pujol, and D.R. Soll, submitted), *C. parapsilosis* (L. Enger, S. Joly, and D.R. Soll, in preparation), *C. lusitanea* (S. Lockhart, A. Hill, and D.R. Soll, in preparation), *C. glabrata* [54], and *Cryptococcus neoformans* [55,56]. The majority of what we know concerning switching in general and the regulation of phase-specific genes in particular, has been obtained in *C. albicans*. However, *C. albicans* may not be the best system for elucidating the molecular basis of the switch event. *C. albicans* is diploid and because of that a mutational analysis of switching has been stymied. Because of this, there has been no direct method for identifying genes in *C. albicans* that are directly involved in switching.

C. glabrata, on the other hand, is haploid and for that reason may prove to be a superior system for elucidating the molecular mechanisms that regulate phenotypic switching. *C. glabrata*

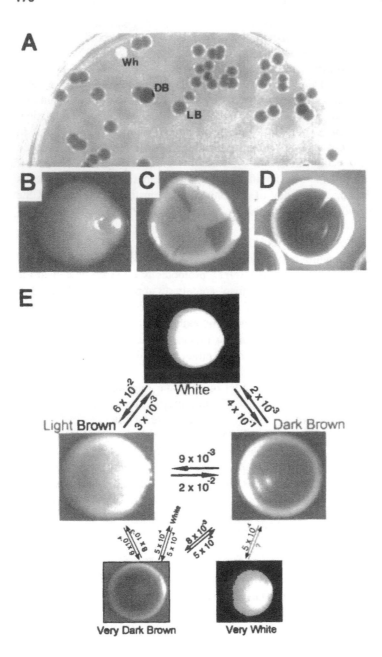

Figure 8 Switching in *C. glabrata*. Cells were incubated on agar containing 1 mM CuSO₄. (A) Cells from a light brown (LB) culture plated at low density on CuSO₄ agar. Note switches to dark brown (DB) and white (Wh). (B, C, D) A white, light brown, and dark brown colony, respectively, with sectors. (E) Frequencies of alternative phenotypes in clonally derived populations of the various phenotypes in the switching repertoire.

Color ─────────────────── vWh < Wh < LB < DB < vDB

Frequency of Switching .... Wh > LB > DB

Transcription of *MT-II* .... Wh < LB < DB

Transcription of *HLP* .... Wh < LB < DB

Figure 9 Phenotypes of *C. glabrata* switching show gradation for a number of parameters.

has recently been demonstrated to switch among three major phenotypes—white, light brown, and dark brown [54] (Fig. 8). Two minor phenotypes have also been identified, very white (vWh) and very dark brown (vDB) (Fig. 8). Coloration of the phenotypes is obtained by growing cells on agar containing 1 mM $CuSO_4$. Switching occurs at high frequency, is reversible, and is spontaneous [54]. It is assumed that the differences in color result from differences in the capacity of the three pehnotypes to reduce $CuSO_4$ to Cu_2S. Since it has already been shown that *C. glabrata* grown on $CuSO_4$ amplifies the metallothionein gene *MT-II* in order to achieve resistance to copper [85], the three major phenotypes (Wh, LB, DB) were tested for amplification by Southern blot analysis. They were found to possess the same level of *MT-II* amplification (nine copies each) [54], thus excluding amplification as the basis of switching [54]. Northern analysis demonstrated that transcription of a second metallothionein gene *MT-I* [86], the transcription factor gene *AMT1* involved in the regulation of metallothionein genes [87], the ABC transporter gene *PDHI* [88], the gene *TRPI* involved in tryptophan metabolism [89], the gene *HIS3* involved in histidine metabolism [89], and the gene *EPA1* that encodes an adhesin [90], are not regulated by switching.

In contrast, transcription levels of *MT-II* and a newly discovered hemolysin gene *HLP* are highly regulated by switching [54]. While the levels of the unregulated genes approximate a 1:1:1 ratio in white, light brown, and dark brown cells, respectively, the levels of *MT-II* are 1:27:81, respectively, and the levels of *HLP* 1:20:35, respectively [54]. Therefore, just as in the case of *C. albicans*, switching in *C. glabrata* involves the selective regulation of phase-specific genes. However, the switching mechanisms appear to differ in the two species. While switching in *C. albicans* is usually between two or more quite different phenotypes, switching in *C. glabrata* is graded. The graded phenotype is apparent in coloration, general switching frequency, expression of *MT-II*, and expression of *HLP* [54] (Fig. 9). Because *C. glabrata* is haploid, mutations affecting switching are immediately expressed, and because the three switch phenotypes are expressed at the colony level, mutants can be easily screened. The switching mechanism can therefore now be analyzed in *C. glabrata* using a random insertional mutagenesis strategy such as Restriction Enzyme Mediated Integration (REMI) [91,92].

XII. CONCLUSIONS

Organisms that are successful in a variety of environments must have machinery that facilitates rapid adaptation. There are two basic sources of such variability. First, there is evolutionary change that involves mutational changes in the genetic material. These changes lead to allelic differences that are combinatorially shuffled during sexual mating and through recombination. Second, there is developmental change, which is programmed into the life cycle of an organism.

Differentiation can lead to alternative phenotypes that are more highly suited for different environmental challenges. Switching in bacterial pathogens provides cells with a rapid method of adaptation, and involves DNA reorganization. However, these reorganizations are specific to gene loci that encode specific cell surface antigens. In effect, the regulation of switching genes in pathogenic bacteria represents a form of differentiation that provides each infecting population with alternative variant phenotypes. Switching in *C. albicans* and related species also can be categorized as a developmental solution to variability. It too provides a number of alternative phenotypes that may be more highly adapted to handle rapid environmental challenges.

However, switching in *C. albicans* and related species differs from bacterial switching systems in the level of complexity inherent in the expression of alternative phenotypes. As we delve deeper into the regulation of switch phenotypes in *Candida* spp., we discover that the switch event is amplified through regulatory circuitry to turn on and off a variety of genes with disparate functions, leading to the genesis of variant phenotypes that differ in far more than a single surface antigen. Before the identification of phase-regulated genes, it was demonstrated that switch phenotypes could differ in a variety of putative pathogenic traits including adhesion [38,39], antigenicity [42], sensitivity to white blood cells and white blood cell metabolites [17], constraints on the bud-hypha transition [41], sensitivity to antifungals [35–37], and sugar metabolism [93]. When phase-specific genes were identified, a surprising majority proved to encode proteins that either represented putative virulence factors, such as secreted aspartyl proteinases [24,27,28], an ABC transporter [26], and a hemolysin [54], or regulatory proteins such as the two-component hybrid kinase regulator [29] or a *trans*-acting factor [30,31]. Switching in *C. albicans* and related species therefore represents a mechanism for combinatorially regulating a variety of genes implicated in pathogenesis. It represents a complex and highly evolved system which cannot be trivialized. We therefore can reasonably conclude that a system this sophisticated that has evolved in a number of loosely related pathogens must be involved in pathogenesis.

It should be clear in this review that there are two major aspects of switching that are being investigated at the molecular level—the actual switch event, and the circuitry that regulates phase-specific gene expression. Although the two are obviously connected, the two can be confused, leading to misinterpretations of the role of genes in the switching process. It is clear from the characteristics of the white-opaque transition that this switching system provides an excellent experimental system for elucidating both molecular aspects of switching, and provides a unique set of virulence models for assessing the roles of phase-specific genes in particular and switching in general in virulence. The recent development of a knockout system for generating null mutants in strain WO-1 should facilitate all of the above objectives. Finally, the discovery of switching in *C. glabrata* now provides us with a system amenable to a traditional mutational analysis of the switching process.

ACKNOWLEDGMENTS

The author is indebted to Dr. T. Srikantha, Dr. S. Lockhart, and Mr. S. Lachke for discussions of unpublished data and the mechanisms of switching. The work in the author's laboratory is funded by Public Health Service grant AI39735 from NIH.

REFERENCES

1. S Scherer, PT Magee. Genetics of *Candida albicans*. Microbiol Rev 54:226–241, 1990.
2. WL Whelan. The genetics of medically important fungi. Crit Rev Microbiol 21:99–170, 1987.

3.   NA Gow. Germ tube growth in medical mycology. Curr Top Med Mycol 8:43–55, 1997.
4.   JD Sobel, G Muller, HR Buckley. Critical role of germ tube formation in the pathogenesis of candidal vaginitis. Infect Immun 44:576–580, 1984.
5.   MG Shepherd. Pathogenicity of morphological and auxotrophic mutants of *Candida albicans* in experimental infections. Infect Immun 50:541–544, 1985.
6.   HJ Lo, JR Kohler, B DiDomenico, D Loebenberg, A Cacciapuoti, GR Fink. Nonfilamentous *C. albicans* mutants are avirulent. Cell 90:939–949, 1997.
7.   FC Odds. *Candida* and Candidiasis. London: Baillière Tindall, 1988.
8.   DR Soll. Switching and its possible role in *Candida* pathogenesis. In: New Fungal Strategies. Edinburgh: Churchill Livingstone, 1992, pp 156–172.
9.   B Slutsky, J Buffo, DR Soll. High frequency "switching" of colony morphology in *Candida albicans*. Science 230:666–669, 1985.
10. C Gil, R Pomes, C Nombela. A complementation analysis by parasexual recombination of *Candida albicans* morphological mutants. J Gen Microbiol 134:1587–1595, 1988.
11. B Slutsky, M Staebell, J Anderson, L Risen, M Pfaller, DR Soll. "White-opaque transition": a second high-frequency switching system in *Candida albicans*. J Bacteriol 169:189–197, 1987.
12. DR Soll. High frequency switching in *Candida albicans*. Clin Microbiol Rev 5:183–203, 1992.
13. DR Soll, M Staebell, CJ Langtimm, M Pfaller, J Hicks, TVG Rao. Multiple *Candida* strains in the course of a single systemic infection. J Clin Microbiol 26:1448–1459, 1988.
14. JM Anderson, DR Soll. The unique phenotype of opaque cells in the "white-opaque transition" in *Candida albicans*. J Bacteriol 169:5579–5588, 1987.
15. J Hellstein, H Vawter-Hugart, P Fotos, J Schmid, DR Soll. Genetic similarity and phenotypic diversity of commensal and pathogenic strains of *Candida albicans* isolated from the oral cavity. J Clin Microbiol 31:3190–3199, 1993.
16. B Morrow, J Anderson, E Wilson, DR Soll. Bidirectional stimulation of the white-opaque transition of *Candida albicans* by ultraviolet irradiation. J Gen Microbiol 135:1201–1208, 1989.
17. MP Kolotila, RD Diamond. Effects of neutrophils and in vitro oxidants on survival and phenotypic switching of *Candida albicans* WO-1. Infect Immun 58:1174–1179, 1990.
18. EHA Rikkerink, BB Magee, PT Magee. Opaque-white phenotypic transition: a programmed morphological transition in *Candida albicans*. J Bacteriol 170:895–899, 1988.
19. T Srikantha, DR Soll. A white-specific gene in the white-opaque switching system of *Candida albicans*. Gene 131:53–60, 1993.
20. S Jones, G White, PR Hunter. Increased phenotypic switching in strains of *Candida albicans* associated with invasive infections. J Clin Microbiol 32:2869–2870, 1994.
21. KG Vargas, SA Messer, MA Pfaller, S.R. Lockhart, JT Stapleton, DR Soll. Elevated switching and drug resistance of Candida from HIV-positive individuals prior to thrush. J Clin Microbiol 38:3595–3607, 2000.
22. DR Soll, CJ Langtimm, J McDowell, J Hicks, R Galask. High frequency switching in *Candida* strains isolated from vaginitis patients. J Clin Microbiol 25:1611–1622, 1987.
23. DR Soll, T Srikanthan, B Morrow, A Chandrasekhar, K Schoppel and S. Lockhart. Gene regulation in the white-opaque transition of *Candida albicans*. Can J Bot 73(suppl. 1):S1049–S1057, 1995.
24. B Morrow, T Srikantha, DR Soll. Transcription of the gene for a pepsinogen, PEP1, is regulated by white-opaque switching in *Candida albicans*. Mol Cell Biol 12:2997–3005, 1992.
25. B Morrow, T Srikantha, J Anderson, DR Soll. Coordinate regulation of two opaque-specific genes during white-opaque switching in *Candida albicans*. Infect Immun 61:1823–1828, 1993.
26. I Balan, A Alareo, M Raymond. The *Candida albicans* CDR3 gene codes for an opaque-phase ABC transporter. J Bacteriol 179:7210–7218, 1997.
27. B Hube, M Monod, D Schofield, A Brown, N Gow. Expression of seven members of the gene family encoding aspartyl proteinases in *Candida albicans*. Mol Microbiol 14:87–99, 1994.
28. T White, S Miyasaki, N Agabian. Three distinct secreted aspartyl proteinases in *Candida albicans*. J Bacteriol 175:6126–6133, 1993.
29. T Srikantha, DR Soll. The two component histidine kinase regulator Ca*NIK1* of *Candida albicans*. Microbiology 144:2715–2729, 1998.

30. A Sonneborn, B Tebarth, JF Ernst. Control of white-opaque phenotypic switching in *Candida albicans* by the Efg1p morphogenetic regulator. Infect Immun 67:4655–4660, 1999.

31. T Srikantha, L Tsai, K Daniels, DR Soll. *EFG1* null mutants of *Candida albicans* can switch, but can not express the complete phenotype of white-phase budding cells. J Bacteriol 182, 1580–1591, 2000.

32. T Srikantha, DR Soll. The two component histidine kinase regulator Ca*NIK1* of *Candida albicans*. Microbiology 144:2715–2729, 1998.

33. DR Soll. The emerging molecular biology of switching in *Candida albicans*. ASM News 62:415–420, 1997.

34. DR Soll. Gene regulation during high frequency switching in *Candida albicans*. Microbiology 143: 279–288, 1997.

35. B Slutsky. A characterization of two high frequency switching systems in the dimorphic yeast *Candida albicans*. Ph.D. thesis, University of Iowa, Iowa City, 1986.

36. DR Soll, J Anderson, M Bergen. The developmental biology of the white-opaque transition in *Candida albicans*. In: Prasad, ed. *Candida albicans:* Cellular and Molecular Biology. New York: Springer Verlag, 1991, pp 20–45.

37. DR Soll, R Galask, S Isley, TVG Rao, D Stone, J Hicks, J Schmid, K Mac, C Hanna. "Switching" of *Candida albicans* during successive episodes of recurrent vaginitis. J Clin Microbiol 27:681–690, 1989.

38. K Vargas, PW Wertz, D Drake, B Morrow, DR Soll. Differences in adhesion of *Candida albicans* 3153A cells exhibiting switch phenotypes of buccal epithelium and stratum corneum. Infect Immun 62:1328–1335, 1994.

39. MJ Kennedy, AL Rogers, LA Hanselman, DR Soll, RJ Yancey. Variation in adhesion and cell surface hydrophobicity in *Candida albicans* white and opaque phenotypes. Mycopathologia 102:149–156, 1988.

40. C Kvaal, T Srikantha, K Daniels, J McCoy, DR Soll. Misexpression of the opaque phase-specific gene *PEP1 (SAP1)* in the white phase of *Candida albicans* enhances growth in serum and virulence in a cutaneous model. Infect Immun 67:6652–6662, 1999.

41. J Anderson, L Cundiff, B Schnars, M Gao, I Mackenzie, DR Soll. Hypha formation in the white-opaque transition of *Candida albicans*. Infect Immun 57:458–467, 1989.

42. J Anderson, R Mihalik, DR Soll. Ultrastructure and antigenicity of the unique cell wall "pimple" of the *Candida* opaque phenotype. J Bacteriol 172:224–235, 1990.

43. CA Kvaal, T Srikantha, DR Soll. Misexpression of the white phase-specific gene WH11 in the opaque phase of *Candida albicans* affects switching and virulence. Infect Immun 65:4468–4475, 1997.

44. FC Odds. Switch of phenotype as an escape mechanism of the intruder. Mycoses 40(suppl 2):9–12, 1997.

45. T Srikantha, A Klapach, WW Lorenz, LK Tsai, LA Laughlin, JA Gorman, DR Soll. The sea pansy *Renilla reniformis* luciferase serves as a sensitive bioluminescent reporter for differential gene expression in *Candida albicans*. J Bacteriol 178:121–129, 1996.

46. B Hube, CJ Turner, FC Odds, H Eiffert, GJ Boulnois, H Kochel, R Ruichel. Sequence of the *Candida albicans* gene encoding the secretory aspartate proteinase. J Med Vet Mycol 29:129–132, 1991.

47. RL Stone, V Matarese, BB Magee, PT Magee, DA Bernlohr. Cloning, sequencing and chromosomal assignment of a gene from *Saccharomyces cerevisiae* which is negatively regulated by glucose and positively by lipids. Gene 96:171–176, 1990.

48. UM Prackett, PA Meacock. *HSP12*, a new small heat shock gene of *Saccharomyces cerevisiae:* analysis of structure, regulation and function. Mol Gen Genet 223:97–106, 1990.

49. KB Merck, PJ Groenen, CE Voorter, WA de Haard-Hockman, J Horwitz, H Bloemendal, WW deJong. Structural and functional similarities of bovine alpha-crystallin and mouse small heat-shock protein. A family of chaperones. J Biol Chem 268:1046–1052, 1993.

50. K Schröppel, T Srikantha, M DeCock, D Wessels, S Lockhart, DR Soll. Cytoplasmic localization of the white phase-specific *WH11* gene product of *Candida albicans*. Microbiology 142:2245–2254, 1996.

51. JR Dutton, S Johns, BL Miller. StuAp is a sequence-specific transcription factor that regulates developmental complexity in *Aspergillus nidulans*. EMBO J 16:5710–5721, 1997.

52. LA Alex, MI Simon. Protein histidine kinases and signal transduction in prokaryotes an eukaryotes. Trends Genet 10:133–138, 1994.

53. WF Loomis, G Shaulsky, N Wang. Histidine kinases in signal transduction pathways of eukaryotes. J Cell Sci 110:1141–1145, 1997.

54. S Lachke, T Srikantha, C Kvaal, L Tsai, DR Soll. Phenotypic switching in *Candida glabrata* involves phase-specific regulation of the metallothionein gene *MT-11* and the newly discovered hemolysin gene *HLP*. Infect Immun 68:884–895, 2000.

55. D Goldman, B Fries, S Franzot, L Montella, A Casadevall. Phenotypic switching in the human pathogenic fungus *Cryptococcus neoformans* is associated with changes in virulence and pulmonary inflammatory response in rodents. Proc Natl Acad Sci USA 95:14967–14972, 1998.

56. BC Fries, DC Goldman, R Cherniak, R Ju, A Casadevall. Phenotypic switching in *Cryptococcus neoformans* results in changes in cellular morphology and glucuronoxylomannan structure. Infect Immun 67:6076–6083, 1999.

57. LP Villarreal. Relationship of eukaryotic DNA replication to committed gene expression: general theory for gene control. Microbiol Rev 55:512–542, 1991.

58. MA Santos, G Keith, MF Tuite. Non-standard translational events in *Candida albicans* mediated by an unusual seryl-tRNA with a 5'-CAG-3' (leucine) anticodon. EMBO J 12:607–616, 1993.

59. DR Soll, T Srikantha. Reporters for the analysis of gene regulation in fungi pathogenic to man. Curr Opin Microbiol 1:400–405.

60. T Srikantha, A Chandrasekhar, DR Soll. Functional analysis of the promoter of the phase-specific *WH11* gene of *Candida albicans*. Mol Cell Biol 15:1797–1805, 1995.

61. T Srikantha, L Tsai, DR Soll. The WH11 gene of *Candida albicans* is regulated in two distinct development programs through the same transcription activation domains. J Bacteriol 179: 3837–3844, 1997.

62. SR Lockhart, M Nguyen, T Srikantha, DR Soll. A MADS box protein consensus binding site is necessary and sufficient for activation of the opaque-phase specific gene *OP4* of *Candida albicans*. J Bacteriol 180:6607–6616, 1998.

63. T Acton, H Zhong, A Vershon. DNA-binding specificity of Mcm1: operator mutations that alter DNA-bending and transcriptional activities by a MADS box protein. Mol Cell Biol 17:1881–1889, 1997.

64. C Christ, B Tye. Functional domains of the yeast transcription/replication factor MCM1. Genes Dev 5:751–763, 1991.

65. R Treisman, G Ammerer. The SRF and MCM1 transcription factors. Curr Opin Genet Dev 2:221–226, 1992.

66. CJ Gimeno, GR Fink. Induction of pseudohyphal growth by overexpression of PHD1, a *Saccharomyces cerevisiae* gene related to transcriptional regulators of fungal development. Mol Cell Biol 14: 2100–2112, 1994.

67. KY Miller, J Wu, BL Miller. StuA is required for cell pattern formation in *Aspergillus*. Genes Devel 6:1770–1782, 1992.

68. AC Glasgow, KT Hughes, MI Simon. Bacterial DNA inversion systems. In: DE Berg, MM Howe, eds. Mobile DNA. Washington: American Society for Microbiology, 1989, pp 637–660.

69. J Perez-Martin, JA Uria, AD Johnson. Phenotypic switching in *Candida albicans* is controlled by a *SIR2* gene. EMBO J 18:2580–2592, 1999.

70. S Gottleib, S Esposito. A new role for a yeast transcriptional silencer gene, SIR 2, in regulation of recombination in ribosomal DNA. Cell 56:771–776, 1989.

71. AJS Klar, JN Strathern, JA Abraham. Involvement of double-stranded chromosomal breaks for mating type switching in *Saccharomyces cerevisiae*. Cold Spring Harbor Symp Quant Biol 49:77–82, 1984.

72. JM Abraham, CS Freitag, JR Clements, BI Eisenstein. An invertible element of DNA controls phase variation of type I fimbriae of *Escherichia coli*. Proc Natl Acad Sci USA 82:5724–5727, 1985.

73. J Zieg, MI Simon. Analysis of the nucleotide sequence of an invertible controlling element. Proc Natl Acad Sci USA 77:4196–4200, 1980.

74. CF Marrs, WW Reuhl, GK Schoolnik, S Falkow. Pilin gene phase variation of *Moraxella bovis* is caused by an inversion of the pilin gene. J Bacteriol 170:3032–3039, 1988.

75. E Daniell, W Boram, J Abelson. Genetic mapping of the inversion loop in bacteriophage Mu DNA. Proc Natl Acad Sci USA 70:2153–2156, 1973.

76. L Pilus, J Rine. Epigenetic inheritance of transcriptional states of *S. cerevisiae.* Cell 59:637–647, 1989.

77. DE Gottschling, OM Aparacio, BL Billington, VA Zakian. Position effect at *S. cerevisiae* telomeres: reversible repression of Pol H transcription. Cell 63:751–762, 1990.

78. H Ramsey, B Morrow, DR Soll. An increase in switching frequency correlates with an increase in recombination of the ribosomal cistrons of *Candida albicans* strain 3153A. Microbiology 140: 1525–1531, 1994.

79. A Barbour. Antigenic variation in relapsing fever *Borrelia* species: genetic aspects. In: DE Berg, MM Howe, eds. Mobile DNA. Washington: American Society for Microbiology, 1989, pp 783–790.

80. J Swanson, JM Koomey. Mechanisms for variation of pili and outer membrane protein H in *Neisseria gonorrhoea.* In: DE Berg, MM Howe, eds. Mobile DNA. Washington: American Society for Microbiology, 1989, pp 743–762.

81. JE Donelson. DNA rearrangements and antigenic variation in African trypanosomes. In: DE Berg, MM Howe, eds. Mobile DNA. Washington: American Society for Microbiology 1989, pp 763–772.

82. B Morrow, H Ramey, DR Soll. Regulation of phase-specific genes in the more general switching system of *Candida albicans* strain 3153A. J Med Vet Mycol 32:287–394, 1994.

83. T Srikantha, B Morrow, K Schröppel, DR Soll. The frequency of integrative transformation correlates with the transcriptional state of phase-specific genes of *Candida albicans.* Mol Gen Genet 246: 342–352, 1995.

84. S Joly, C Pujol, DR Soll. Development and verification of two species fingerprinting probes for *Candida tropicalis* amenable to computer analysis. J Clin Microbiol 34:3063–3071, 1996.

85. RK Mehra, JR Garey, R Winge. Selective and tandem amplification of a member of the metallothionein gene family in *Candida glabrata.* J Biol Chem 265:6369–6375, 1990.

86. RK Mehra, JR Garey, TR Butt, WR Gray, DR Winge. *Candida glabrata* metallothioneins cloning and sequence of the genes and characterization of proteins. J Biol Chem 264:19747–19753, 1989.

87. P Zhou, DJ Thiele. Isolation of a metal-activated transcription factor gene from *Candida glabrata* by complementation in *Saccharomyces cerevisiae.* Proc Natl Acad Sci USA 88:6112–6116, 1991.

88. H Miyazaki, Y Miyazaki, A Geber, T Parkinson, C Hitchcock, DJ Falconer, DJ Ward, K Marsden, JE Bennett. Fluconazole resistance associated with drug efflux and increased transcription of a drug transporter gene, PDH1, in *Candida glabrata.* Antimicrob Agents Chemother 42:1695–1701, 1998.

89. K Kitada, E Yamaguchi, M Arisawa. Cloning of the *Candida glabrata TRP1* and *HIS3* genes, and construction of their disruption strains by sequential integrative transformation. Gene 165:203–206, 1995.

90. BP Cormack, N Ghori, S Falkow. An adhesion of the yeast pathogen *Candida glabrata* mediating adherence to human epithelial cells. Science 285:578–582, 1999.

91. FJ Maier, W Schäfer. Mutagenesis via insertional- or restriction enzyme-mediated-integration (REMI) as a tool to tag pathogenicity related genes in plant pathogenic fungi. Biol Chem 380:855–864, 1999.

92. A Kuspa, WF Loomis. Tagging developmental genes in *Dictyostelium* by restriction enzyme-mediated integration of plasmid DNA. Proc Natl Acad Sci USA 89:8803–8807, 1992.

93. DR Soll. Dimorphism and high frequency switching in *Candida albicans.* In: DR Kirsch, R Kelly, MB Kurtz, eds. The Genetics of *Candida albicans.* Boca Raton, FL: CRC press, 1990, pp 147–176.

9
Pathogenic Properties of *Blastomyces dermatitidis*

Bruce S. Klein
University of Wisconsin, Madison, Wisconsin

Wun-Ling Chang
Louisiana State University Health Sciences Center, Shreveport, Louisiana

I. INTRODUCTION

Blastomycosis is a systemic fungal infection of humans and animals caused by the dimorphic fungus *Blastomyces dermatitidis*. The disease and organism were first described by Gilchrist in 1894 in Baltimore. The infection, initially known as "Gilchrist's disease," was thought to originate in the skin until pathologic descriptions by Schwarz and Baum [1] revealed the primary pulmonary nature of the illness. In the early 1900s, the illness became known as "Chicago's disease" because of the number of cases noted from that area. Later, it became known as "North American blastomycosis" because of its wide distribution in the United States and Canada. Reports of sporadic cases worldwide have lead to use of the more appropriate term, "blastomycosis."

The disease has a wide spectrum of manifestations that include asymptomatic self-limiting pulmonary illness, focal pulmonary or cutaneous illness, and disseminated disease. There are many unanswered questions regarding the pathogenesis of blastomycosis, and virulence factors of *B. dermatitidis*. Recent advances in molecular biological techniques have helped in answering some of these questions. This chapter provides an overview of what is currently known about the fungal agent *B. dermatitidis*, its virulence attributes, and the pathogenesis of the resulting disease blastomycosis.

II. AGENT

B. dermatitidis is a dimorphic fungus able to grow in two forms depending on environmental conditions. It assumes a yeast form at 35–37°C, and a mold form at 25–30°C. The mold is the infective form, whereas the yeast is the tissue or so-called parasitic form. Therefore, without exception, *B. dermatitidis* appears in the yeast form in histopathological specimens of tissue or in body fluid specimens. On the other hand, *B. dermatitidis* is most commonly isolated as a mold from clinical specimens that are cultured at 25–30°C. Conversion of the mold to the yeast form is required to confirm the presence of a dimorphic fungus. Understanding the mechanism of dimorphism has been the thrust of research into the ultrastructure and biochemistry of *B. dermatitidis.*

A. Dimorphism

Under light microscopy, the yeasts are spherical, thick-walled organisms between 6 and 15 μm in diameter. They are easily distinguishable from other pathogenic yeast by their size, shape, and broad-based budding (Fig. 1). The mycelia are typically thin, septate hyphae, which produce conidiophores that branch at right angles from the main hyphal segment (Fig. 1). The conidiophore produces single conidia 2–10 μm in diameter that are usually oval or pyriform, but occasionally "dumbbell" in configuration. The conidia appear to have smooth walls under light microscopy, but under electron microscopy they have a rough or echinulate surface. Conidia from different *B. dermatitidis* isolates are not distinguishable. In addition, because the conidia and mycelia of *B. dermatitidis* do not have a diagnostic appearance, definitive identification requires conversion to the yeast form by culturing at 37°C or exoantigen testing of mycelia if conversion is unsuccessful.

Dimorphism may hold clues to the organism's virulence. Ultrastructural and biochemical studies have been directed at comparing the yeast and mold forms to determine the mechanism of dimorphism and possibly virulence. In these studies, media and incubation conditions have been shown to influence the appearance of the cell wall and cytoplasm, the development of intracytoplasmic membranes, and even glycogen and lipid content [2]. Ultrastructural studies have not identified unique features of *B. dermatitidis* yeast and hyphae cytoplasm [2,3]. Both yeast and hyphae are multinucleated. The yeast cell has about two to five nuclei per cell. Nuclear division seems to occur independently from cell division [4]. The hyphal cells also have different numbers of nuclei per cell. Further studies into the cellular anatomy of the yeast form indicate that the outer membrane of the nuclei in each cell are interconnected. The nuclear membrane is also contiguous with the plasma membrane and membranes of other intracellular organelles [5]. The role of this extensive membrane system is not known, although it may facilitate change from the yeast to the mycelial form.

 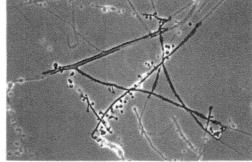

Figure 1   Dimorphism of *Blastomyces dermatitidis.* (Left) Large yeast form cells from a KOH preparation of a patient sputum, illustrating the distinctive features of a thick, refractile cell wall and unipolar budding of daughter cells from parent cells at a broad base. (Right) Mold form growing at room temperature, illustrating conidiophores that produces single conidia 2–10 μm in diameter that are usually oval or pyriform. (From http://fungusweb.utmb.edu/mycology/fungi/koh-blastomyces-dermatitidis2.jpg and http://fungusweb.utmb.edu/mycology/fungi/blader1.jpg.)

During conversion from yeast to mold form, multiple intracytoplasmic infolding and layering up of plasma membranes can be seen on electron microscopy as early as 6 hr after switching the temperature of yeast cultures to 24–26°C. After 12–18 hr, the cells appear to have an intracellular septum with woronin bodies on both sides. The cell wall of the intermediate forms is thinner than that of yeast cells. By 18–24 hr, hyphal forms are seen. These have increased amounts of glycogen granules that tend to associate with the intracytoplasmic membrane systems formed in the early yeast to mycelial transition [3]. Overall knowledge of the ultrastructure and its functional role is incomplete.

What is known about the physiology of *B. dermatitidis* is derived from classical biochemical techniques. Such studies have provided some information on nutritional requirements and chemical composition of yeast and mycelia. Studies on nutritional requirement have resulted in conflicting results probably due to different experimental methods. In one study, biotin was found to be an absolute requirement for growth of both mycelia and yeast of six isolates [6]. Other studies indicated no requirement for biotin or other vitamins [7]. Given the observation that both yeast and mycelia can grow and undergo phase transition in tap water, either nutritional requirements for maintenance of this fungus are low or mechanisms for storing necessary nutrients are available. Similarly, absolute requirement for an amino acid has not been demonstrated. Yet, not unexpectedly, organic nitrogen sources can stimulate growth of *B. dermatitidis* [8]. The question of whether specific nutritional substances can enhance or suppress growth and how this may effect fungal virulence needs further investigation.

Comparative studies on the chemical composition of yeast and mycelia have been the source of our present understanding of the biochemistry of *B. dermatitidis*. As with previous ultrastructural studies, media and incubation conditions affect the content of individual chemical moieties. Different growth media specifically affect the lipid and fatty acid content of both yeast and mycelia, while not affecting protein and carbohydrate content [9]. Lipids make up ~9% of the dry weight of yeast cells [10]. About 75–80% of extractable lipids from yeast and mycelia are oleic and linoleic acids. Oleic acid predominates in yeasts, while a higher percentage of linoleic acid is recovered from mycelia [11]. Studies to localize lipids found that ~2% of the yeast cell wall was lipid, while negligible lipid was found in the mycelial cell wall. The lipid content of yeasts is interesting as several studies have noted differences in lipid content between virulent and avirulent isolates of *B. dermatitidis*. Lipid content directly correlated with virulence of the isolate in mice in one study [12]. However, in another study, a spontaneous avirulent mutant had a three- to fourfold increase in total lipid content, primarily composed of an increase in fatty acids, when compared to the virulent strain [13].

Other differences in cell wall composition between yeast and mycelia have been studied in an effort to understand dimorphism and virulence. The cell walls of both yeast and mycelia contain chitin and glucan. The glucan in yeast cells is 95% α-glucan, while that of mycelia is 60% α- and 40% β-glucan [14]. This shift in predominant type of glucan is similar to *Paracoccidiodes braziliensis* and is thought to be secondary to activation of α-glucan synthase with increased temperature [15]. The chitin isolated from *B. dermatitidis* mycelia is similar to chitin obtained from *H. capsulatum* and *P. brasiliensis* [16]. Chitin is a stable cell wall component during dimorphic change. Its role in dimorphism is suggested by electron microscopic studies showing different chitin fibril arrangements in yeast and mycelial cell walls. In yeast, the chitin fibrils are noted to be tightly interwoven in a random orientation, while in mycelia they are arranged in more longitudinal manner in many areas of the cells wall [17]. Isolation of the enzyme chitin synthetase from the yeast and mycelial phases indicates similar biochemical characteristics, but the yeast phase enzyme is in latent form, requiring activation by protease, whereas the mycelial phase enzyme is in active form [18].

The dimorphism of *B. dermatitidis* is primarily temperature dependent. However, studies into the mechanism of dimorphic transition indicate a more complex regulation other than by temperature alone. Yeasts grown at 37°C can revert to mycelia in the presence of c-AMP phosphodiesterase inhibitors, theophylline or 3-isobutyl-1-methylxanthine. Using electron microscopy to study yeasts at 37°C, c-AMP phosphodiesterase is localized to nuclear and mitochondrial membranes and to endoplasmic reticulum and vacuolar areas. In contrast, similar studies of the yeast 24 hr after lowering the incubation temperature to 25°C, show decreased enzyme activity of c-AMP phosphodiesterase in vacuolar areas. This suggests that increased levels of intracellular c-AMP may be required for dimorphic transition to the yeast phase [19]. Other physiologic changes that occur with mycelial to yeast transition upon change of temperature are related to mitochondrial respiration. Three stages in the transition are described: (1) immediately after an increase in temperature, partial or complete uncoupling of oxidative phosphorylation, leading to decreases in ATP levels, respiration rate and electron transport components; (2) a dormant period of 4–6 days with absent to low rates of respiration; and (3) a recovery phase marked by increase in respiration, increased ATP and electron transport components and the appearance of yeast cell morphology. During stage 2, the presence of cysteine is required for the operation of the sulfhydryl-induced shunt pathway. These stages of phase transition are shared by *H. capsulatum* and *P. brasiliensis* [20]. Nutritionally dependent dimorphism is also indicated by reported conversion of mycelial to yeast form at 26°C on partially defined media.

B. Serotypes

B. dermatitidis strains from different geographic areas have been studied by exoantigen analysis. Two serotypes of *B. dermatitidis* were identified. Serotype 1 included all the 88 North American strains, one of 12 African strains, and one of each strain from India and Israel. These strains displayed A-antigen and K-antigen. Serotype 2 included the rest of the African strains. They were deficient in A-antigen but had K-antigen [21]. African serotype 2 strains also were noted to have different colonial morphology and microscopic features. When the African isolates were mated with positive and negative tester strains of *A. dermatitidis*, they were found to be sexually incompatible [21,22]. The geographic distribution of positive and negative isolates of *A. dermatitidis* in North America is evenly distributed and these isolates mate with each other, but not with African isolates.

C. Ecological Niche

Despite knowing the general area of endemicity of blastomycosis, the ecological niche of *B. dermatitidis* is incompletely understood [23]. This is largely due to the inherent difficulty of culturing the organism from the environment. Many attempts at isolating the fungus from its natural habitat have resulted in only seven successful studies. The first reported isolation was from one of 600 samples taken from the floor of a tobacco curing barn in Kentucky [24]. Two additional isolations were made from an abandoned kitchen, chicken houses, and other animal pens or stalls in Georgia [25,26]. None of these isolations were at sites of human or animal infections. *B. dermatitidis* was successfully isolated from a bag of pigeon manure that served as the source of infection of a horticulturist in Minnesota [27]. The organism was also isolated from the earthen floor of a shed of an oil field worker who developed blastomycosis in Ontario, Canada [28].

    B. dermatitidis was first isolated from the soil at a site of human infection following the Eagle River, Wisconsin, outbreak in 1984. Samples yielding the fungus were taken from an abandoned beaver lodge and from the beaver dam at a pond visited by children who acquired blastomycosis [29]. Subsequently, *B. dermatitidis* was isolated from soil along the Tomorrow River in Wisconsin, where another outbreak occurred in May 1985 [30]. These and other studies suggest that the fungus grows in microfoci of moist, rich soil with acid pH, containing decaying and rotting vegetation or animal manure [29]. Such foci appear to be located at the banks of waterways in endemic zones. Studies in the laboratory have shown that moisture releases conidia from their conidiophore. Thus, rain or mist may promote infectivity of *B. dermatitidis* via aerosolization of conidia from a microfocus. Difficulty in isolating *B. dermatitidis* from its natural habitat may relate to its lysis in soil by *Streptomyces* sp. and *Bacillus* sp. [31].

III. HOST-FUNGUS DYNAMICS

A. Host Genetic Factors

Certain host factors appear to determine susceptibility to infection. For example, clinically apparent blastomycosis has been diagnosed more frequently in males than females. Male preponderance of disease could, in part, reflect hormone-related factors. This is consistent with laboratory studies in hamsters that demonstrated females were more resistant to the lethal effects of *B. dermatitidis* infection (52% survival vs. 7%), and that ovariectomy increased the resistance of females (80% survival) and castration increased the resistance of males (40% survival) [32]. Additionally, the improved survival that ovariectomy conferred on females was nullified by treatment with testosterone (47% survival).

Racial distribution of blastomycosis has inconsistently suggested a higher occurrence among blacks. Studies purporting this association were of an early series of cases, possibly reflecting the demographic features of populations seen in charity hospitals in rural and agricultural southern states or the predominant occupation of manual laborer among blacks in those states. This has not been confirmed in other large case series, particularly those in the midwestern United States.

Inbred strains of mice show marked differences in susceptibility to experimental *B. dermatitidis* infection after pulmonary challenge with yeast-form organisms. Of eight strains with various *H-2* backgrounds evaluated in one study [33], the C3H/HeJ strain had the highest mortality (88%) and the DBA/1J strain had the lowest (37%). The resistance of the C3H/HeN strain, which differs from the C3H/HeJ in that it exhibits sensitivity to lipopolysaccharide and lacks a defect in macrophage cytotoxicity, suggests that the susceptibility of C3H/HeJ mice is not related to the C3H background or *H-2* locus.

B. Pathogenesis

An understanding of the evolution of the inflammatory response to *B. dermatitidis* in humans is incomplete because the preponderance of histopathological information has been obtained from autopsy series or descriptions of isolated rather than serial biopsy specimens. Recent work in inbred C57BL/6 mice has helped characterize the sequence of events that take place in the inflammatory response after an intravenous injection of 10^6 *B. dermatitidis* yeasts [34]. The tissue response progressed from acute neutrophilic invasion during the first 7 days of infection to pyogranuloma formation by the fifth week of infection. The lungs, brains, superficial fascia, livers, and spleens of the mice were involved. By the fifth week, the greatest burden of infection and inflammation was found in the lungs and brains, other visceral organs were less involved, and only small granulomas containing few yeasts were seen in the liver parenchyma. The lymph nodes and spleens were relatively spared.

A histological picture including suppuration and granuloma formation has also been reported in the tissues of congenitally athymic nude mice inoculated intravenously with *B. dermatitidis* yeasts [35]. By day 12 of infection, a pyogranulomatous reaction was evident in the brain, with lesions consisting of three zones—central necrosis, an intermediate region of intense suppuration, and an accumulation of mononuclear cells around the periphery of the lesion. Although much larger numbers of yeast cells were seen in the lesions of nude mice than in their immunologically intact littermates, this study supports the hypothesis that granuloma formation is not entirely under the control of mature T-cells (which nude mice lack because they have no thymus).

At least two products of *B. dermatitidis* have been associated with the influx of inflammatory cells found in lesions of blastomycosis. The first is a serum-independent chemotactic factor, detectable in the culture filtrate of growing yeasts, which stimulates the migration of both human neutrophils and monocytes [36]. The factor resists heat treatment at 100°C for 60 min and alkaline conditions, but is inactivated below pH 7. The molecular size of the chemotactic factor is unknown; however, it is retained after dialysis using a size cutoff of 3500 daltons. The second product of the organism is cell wall material found in an alkali-soluble, water-soluble extract of virulent *B. dermatitidis* [37]. Intraperitoneal inoculation of this extract elicits formation of granulomas that are histologically indistinguishable from the lesions seen following inoculation of experimental animals with live yeasts. An identically prepared extract from the cell wall of an avirulent strain of *B. dermatitidis* elicited suppuration but no granuloma formation, leading investigators to postulate that since the most striking chemical difference between extracts of the two strains was the amount of covalently bound phospholipid [38], lipid in the cell wall might be responsible for the granulomatous reaction.

C. Virulence

Work on virulence factors has been limited, but selected cell wall constituents, and particularly the WI-1 adhesin/antigen, have been implicated.

1. Cell Wall Fractions

The study of virulence factors in *B. dermatitidis* began with the early work of DiSalvo and Denton [12] examining total extractable lipid and phospholipid fractions of whole yeast cells. They correlated increased lipid content directly with increased virulence in four strains. No correlation of virulence was seen with phospholipid content. Later, Cox and Best [38] examined a different fraction, trypsin-treated cell walls, from two strains. They associated decreased virulence (strain GA-1) with increased α-1,3-glucan content, and they associated increased virulence (strain KL-1) with increased chitin content and with glucan-containing covalently attached phospholipid. Subsequent studies continue to implicate high phospholipid content of cell walls as a characteristic of highly virulent strains [13].

Evidence for the role of phospholipid in virulence came from intraperitoneal inoculation of mice with cell wall preparations. Cell walls from virulent (KL-1) and avirulent (GA-1) strains elicit different histological reactions, which are hypothesized to be due to different phospholipid contents of the alkali-soluble fraction of trypsin-treated cell walls [37]. Trypsin-treated cell walls produce tissue necrosis and death in this assay, but only cell walls from the virulent isolate elicit a characteristic granulomatous response. Fractionation of the cell wall into alkali-soluble and -insoluble fractions separates tissue necrosis and mortality from granulomatous responses. Injection of alkali-soluble fractions (10–30 mg) produces a granulomatous response but is not lethal and does not produce necrosis. Only alkali-soluble fractions from the virulent strains produce a granulomatous response at the 10-mg dose. A 30-mg dose is required to see any polymorphonu-clear cell (PMN) infiltration when alkali-soluble material from the avirulent strain is injected. The greater ability to produce a granulomatous response correlates with the increased phospholipid component of the alkali-soluble fraction in the virulent strain. Injection of 25 mg of alkali-insoluble cell walls from either strain intraperitoneally is lethal and produces tissue necrosis with only occasional PMN infiltration. This finding suggests that alkali-insoluble fractions contain an endotoxinlike activity.

One difficulty in interpreting previous studies of virulence has been the diverse origin of many standard laboratory strains. Work with a set of related strains was described by Stevens and coworkers which confirms and extends observations made with older isolates. They de-scribed two strains spontaneously derived from a highly virulent strain, ATCC 26199, with altered lethality in murine models of infection [13]. Attenuated ATCC 60915 arose after in vitro culture, is 10,000-fold less virulent by intranasal challenge, and is cleared from the lung very early after infection. Avirulent ATCC 60916 is nonlethal in the same model and is also cleared from the lung within 4 days of infection [39]. The wild-type and mutant yeasts differ in their cell wall composition, especially in the expression of α-1,3-glucan and WI-1.

2. α-1,3-Glucan

Previous studies of *Paracoccidioides brasiliensis* and *Histoplasma capsulatum* have shown an association between loss of the α-1,3-glucan carbohydrate polymer and attenuated virulence of those fungi [40–42]. These observations prompted study of the genetically related strains of *B. dermatitidis* for expression of the polymer. The avirulent mutant (ATCC 60916) has lost all detectable α-1,3-glucan, and the attenuated strain (ATCC 60915) has greatly reduced amounts of α-1,3-glucan on the cell surface compared with wild-type ATCC 26199 [43] (Fig. 2). Although the precise role of the polymer in virulence is unknown for *B. dermatitidis* and other dimorphic fungi, it has been hypothesized that α-1,3-glucan may mask other cell wall components such as the WI-1 adhesin/antigen on the surface of the yeast, at least in *B. dermatitidis* (see below).

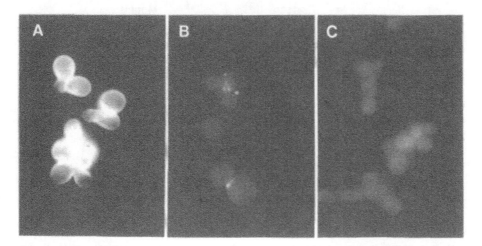

Figure 2 Surface expression of α-1,3-glucan on three genetically related strains of *B. dermatitidis* yeasts. Binding of mAb MOPC 104e was used to assess the amount of α-1,3-glucan on wild-type strain 26199 (Panel A), and mutant strains 60915 (Panel B), and 60916 (Panel C). (From Ref. 43.)

3. WI-1 Adhesin/Antigen

WI-1 is a 120-kDa cell wall adhesin that has been isolated from the surface of all North American *B. dermatitidis* strains examined thus far [44]. The molecule contains 34 copies of a 25 amino acid tandem repeat, which are highly homologous to invasin, an adhesin of *Yersinia* spp. [45,46]. The tandem repeats mediate attachment to human monocyte-derived macrophages, mainly through binding complement type 3 (CR3) receptors [46,47]. WI-1 is also a key antigenic target of humoral [44,45,48] and cellular [49] responses during human infection; the tandem repeat is the chief site of antibody recognition.

Examination of the related strains ATCC 26199, ATCC 60915, and ATCC 60916 showed that WI-1 expression is significantly altered on the surface of mutant yeasts, along with other changes in protein expression [50]. These mutant strains have increased amounts of expressed or exposed WI-1 on the cell wall (Fig. 3), and they shed less WI-1 into culture supernatants during in vitro growth [50]. The mutants bind more avidly to human macrophages in a manner corresponding to the expression of WI-1 (Fig. 4). Anti-WI-1 Fab fragments block the attachment of yeast to the cells (Table 1), and coating of latex beads with WI-1 (Fig. 5) promotes their attachment to macrophages (Table 2). These data establish the role of WI-1 in mediating adherence to macrophages. Other work has shown that the mutants also stimulate a respiratory burst that is 20-fold greater than that seen with the wild-type yeasts upon contact with murine neutrophils [39]. It is speculated that the copious shedding of WI-1 by wild-type yeasts may influence the ability of the virulent strain to escape from recognition and inhibition of replication by

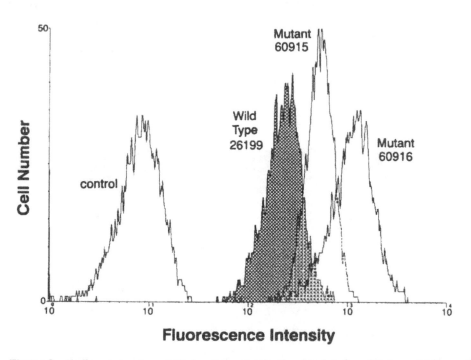

Figure 3 Surface expression of WI-1 on three genetically related strains of *B. dermatitidis* yeasts. Binding of mAb AD3-BD6 was used to quantify the amount of WI-1 on wild-type and mutant strains. Profile designated control represents staining of strain 26199 with an irrelevant isotype control antibody Leu 5b. Mutant strains 60915 and 60916 stained with this antibody yielded a nearly identical profile. (From Ref. 50.)

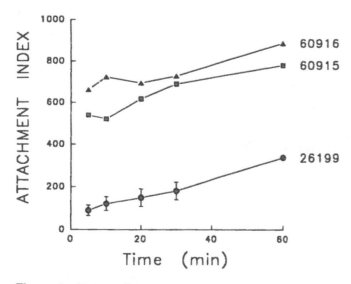

Figure 4 Binding of three genetically related strains of *B. dermatitidis* yeast to human macrophages. Attachment index (number of yeasts bound per 100 macrophages) was determined serially over 60 min during incubation at 37°C. Data are the mean ± SEM of four experiments with wild-type strain 26199, and mean of two experiments each with mutants 60915 and 60916. (From Ref. 50.)

phagocytes, as has been described during 72-hr cocultures of these yeasts with murine macrophages (see below) [51].

An intriguing feature of the attachment of *B. dermatitidis* yeasts to macrophages is the difference in receptors that bind wild-type and mutant yeasts (Table 3). Mutant yeasts (and purified WI-1 on latex microspheres) bind CR3 (CD11b/CD18) and CD14 receptors on the macrophages, whereas wild-type yeasts bind CR3 alone. CD14 is a receptor for binding lipopolysaccharide alone or complexed to lipopolysaccharide-binding protein [52]. Binding CD14 on phagocytes markedly increases the affinity of CR3 receptors for its ligand [53]. Thus, the binding of CD14 by WI-1 on mutant yeasts and the cooperative interaction of CD14 and CR3 on the macrophage might help explain the greatly enhanced binding of mutants to the macrophages.

Table 1 Inhibition of the Binding of *B. dermatitidis* Yeasts to Human Macrophages by Fab Anti-WI-1

Strain	Control percent of		Anti-WI-1 percent of		Percent of inhibition of AI
	Binding	AI	Binding	AI	
ATCC 26199	77 ± 6[a]	297 ± 37	31 ± 4	91 ± 12	69[b]
ATCC 60915	90 ± 2	795 ± 38	37 ± 7	172 ± 34	78[b]
ATCC 60916	91 ± 4	777 ± 49	48 ± 1	187 ± 19	76[b]

Yeasts were incubated with Fab anti-WI-1 for 30 min at 37°C. The yeasts then were incubated with Mø adhered in Terasaki tissue culture plates for 60 min (26199) or 10 min (60915 and 60916) at 37°C. The % binding and attachment index (AI) was quantified via phase-contrast microscopy.
[a] Mean ± SEM, n = 2, 26199; n = 4, 60915; n = 2, 60916.
[b] *P* < .01 compared to the control (Student's t-test).
Source: Ref. 47.

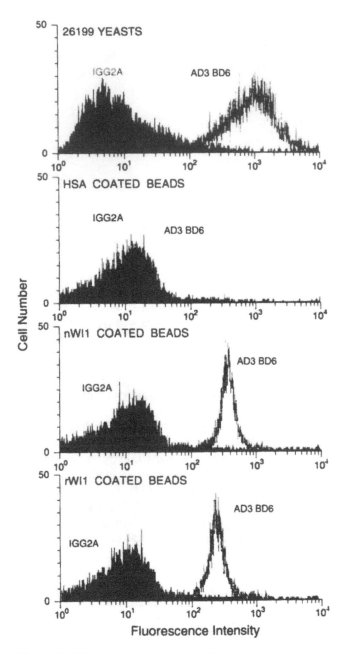

Figure 5 Flow cytometry analysis of WI-1 on *B. dermatitidis* yeasts and WI-1 coated microspheres. Binding of WI-1 specific monoclonal antibody AD3-BD6 was used to quantify the amount of bound native WI-1 (nWI-1) or recombinant WI-1 (rWI-1) adhered to latex microspheres. Staining of *B. dermatitidis* strain 26199 (positive control) and human serum albumin (HSA)-coated beads (negative control) with monoclonal AD3-BD6 is shown for comparison. Staining of yeast and beads with the IgG2a isotype control mAb Leu 5b is shown. Coated beads were analyzed for binding to macrophages, as depicted in Table 2. (From Ref. 46.)

Table 2 Binding of WI-1 Coated Beads to Human Macrophages

Latex beads[b]	Binding to macrophages[a]	
	Percentage of binding	Attachment index
HSA coated	25 ± 4[c]	75 ± 5
nWI-1 coated	73 ± 3	597 ± 81
rWI-1 coated	80 ± 4	629 ± 76

[a] Percentages of binding and attachment index were determined as in Ref. 47.
[b] HSA = human serum albumin; nWI-1 = native WI-1; rWI-1 = recombinant WI-1.
[c] Results are mean ± SEM ($n = 5$).
Source: Ref. 46.

There are two possible explanations for why WI-1 on mutants but not wild-type yeasts mediates attachment of CD14. First, α-1,3-glucan may mask the CD14-binding site of WI-1 on wild-type yeasts whereas a WI-1 binding site for CD14 may be exposed on mutant yeasts devoid of the α-1,3-glucan polymer. An alternative explanation is that the high density and possible clustering of WI-1 on the surface of the mutants increase its affinity for CD14 receptors just as these properties enhance the affinity of other ligands for macrophage receptors [54]. In contrast, the lower density of WI-1 on wild-type yeasts may have insufficient affinity to bind CD14. The binding of wild-type and mutant yeasts to different receptors on human macrophages, especially the binding of mutant yeasts to CD14 receptors on phagocytes, could influence the inflammatory response to the fungus and may be important for understanding the pathogenesis of blastomycosis.

Table 3 Inhibition of the Binding of *B. dermatitidis* Yeast Strains 26199 and 60915 to Human Macrophages by Monoclonal Antibodies

Antibody	Specificity	Strain 26199		Strain 60915	
		AI	Percent of inhibition	AI	Percent of inhibition
Buffer	—	330 ± 86[a]	—	629 ± 48[b]	—
Ts1/22	CD11a (LFA-1)	316 ± 80	4	617 ± 16	2
MN-41	CD11b (CR3-iC3b)	250 ± 75	24	474 ± 47	25
904	CD11b (CR3-LPS)	114 ± 50	65	56 ± 9	91
Leu-M5	CD11c (p150,95)	196 ± 50	41	265 ± 55	58
IB4	CD18	45 ± 16	86	40 ± 7	94
3G8	CD16 (FcRIII)	249 ± 65	24	464 ± 69	26
3C10	CD14	328 ± 85	0	241 ± 21	62

Mø adhered in Terasaki tissue culture plates were incubated with 5 μL of mAb (50 μg/mL) for 30 min at 4°C, and then were incubated with 1×10^5 yeasts of strain 26199 for 1 hr at 37°C and with strain 60915 for 10 min at 37°C. The attachment index (AI) was quantified via phase-contrast microscopy.
[a] Mean ± SEM, $n = 4$.
[b] Mean ± SEM, $n = 3$.
Source: Ref. 47.

It is not clear from the foregoing studies how WI-1 serves as a virulence factor on wild-type yeast, but is greatly increased in surface expression and promotes enhanced phagocyte recognition in the hypovirulent mutants. It has been hypothesized that WI-1 must be regulated in its expression on the cell surface versus its secretion into the environment in order to achieve its optimal function as a virulence factor. Beachey [55] described good versus bad adherence for the role of *Escherichia coli* and *Proteus mirabillis* pili in the pathogenesis of these gram-negative infections. On mucosal surfaces, the bacteria must posses surface adhesins to adhere. After invading deeper tissues, they must shed their adhesin or produce masking capsules to avoid phagocyte recognition [55]. Thus, pathogens must regulate adhesion expression to survive and produce disease in the host.

4. African Blastomycosis

WI-1 is immunologically similar to the previously described A antigen in that both antigens display the 25 amino acid repeat as an immunodominant B-cell epitope [48,56]. African strains of *B. dermatitidis* have many antigens in common with North American strains, but a distinguishing feature is that most African strains lack the A antigen present on North American strains [21]. African strains also tend to cause a distinctive pattern of blastomycosis, which is thought to be less severe and in which chronic cutaneous lesions dominant the clinical picture [57]. This pattern contrasts with the primarily pulmonary and systemic symptoms typically seen with North American strains. Analysis of the most widely prevalent serotype of African *B. dermatitidis* (serotype 2) demonstrated that it lacks WI-1 protein expression (Fig. 6) because the WI-1 coding sequence is not present in the genome (Fig. 7) [58]. Thus, the loss of expression of WI-1

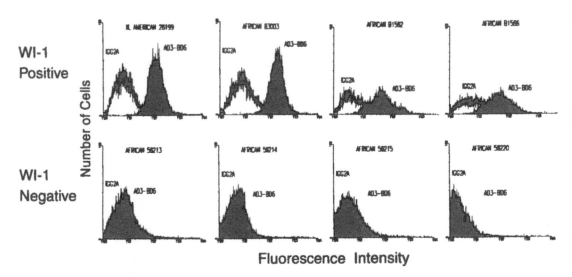

Figure 6 Flow cytometry analysis of WI-1 on African strains of *B. dermatitidis*. Binding of mAb AD3-BD6 was used to quantify WI-1 on the surface of yeasts. Staining with an IgG2A isotype control mAb also is shown. African strains B3003, B1562, and B1566 are serotype 1 isolates. African strains 56213, 56214, 56215, 56216, and 56220 are serotype 2. The upper panel shows WI-1 positive North American and African serotype 1 isolates. The lower panel shows WI-1 negative serotype 2 African isolates. (From Ref. 58.)

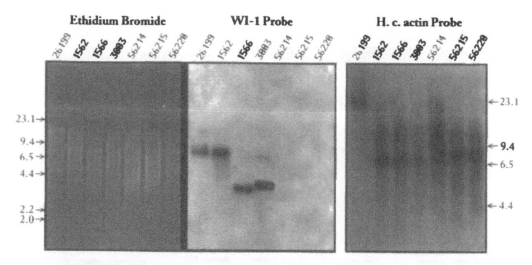

Figure 7 Southern analysis of WI-1 DNA sequences in African strains of *B. dermatitidis*. The identity of strains tested is described in the legend of Figure 6. For Southern analyses, 10 μg of restricted, genomic DNA was loaded per gel lane. Ethidium bromide was used to visually inspect gels and confirm that a comparable amount of DNA was present for each strain (left panel). Genomic DNA from the gel in the left panel was transferred and probed as described with full-length genomic WI-1 from ATCC 26199 (middle panel). Genomic DNA from a separate gel was probed with actin from *Histoplasma capsulatum* (right panel). The hybridizing band for 26199 actin migrates at a position with chromosomal DNA because that DNA sample was incompletely digested. Molecular size markers (kb) are depicted in the left- and right-hand margins of the panels. (From Ref. 58.)

could be linked with the lower-virulence phenotype and distinct disease pattern of African blastomycosis.

5. Genetic Manipulation of *B. dermatitidis*

A transformation system has recently been developed for *B. dermatitidis* [59]. Yeast cells are electrotransformed with linear plasmid DNA expressing dominant selectable markers such as *Escherichia coli* hygromycin phospotransferase. This system yields low transformation frequencies in the range of 10–100 transformants/μg transforming DNA. Transforming DNA typically integrates ectopically in multiple linked or unlinked copies. Despite the high frequency of illegitimate recombination in this and related dimorphic fungi, the transformation system for *B. dermatitidis* provides the opportunity to genetically alter expression of determinants in the fungus and test their role in pathogenicity and virulence. Such efforts rest on the feasibility of accomplishing homologous recombination and gene disruption, an achievement that has been difficult to accomplish in the systemic dimorphic fungi.

6. Homologous Gene Targeting of WI-1 in *B. dermatitidis*

In a recent study [60], the WI-1 locus in *B. dermatitidis* was mutated by allelic replacement, and expression reconstituted by means of gene transfer. This was accomplished in two independent strains of the fungus, ATCC 26199 and ATCC 60915. Gene targeting occurred at a frequency of 1–2%, and appeared to be facillitated by the use of a promoter trapping strategy [61] (Fig. 8A, B). Thus, a short stretch of the WI-1 upstream sequence, placed in front of the selectable

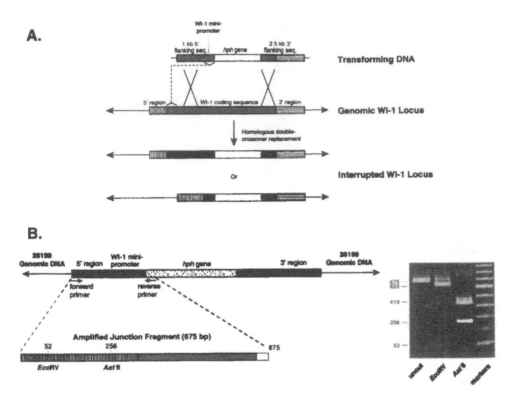

Figure 8 (A) Strategy for targeting and replacing the WI-1 locus in *B. dermatitidis*. Anticipated homologous crossover leading to a gene replacement event at the WI-1 locus. Linearized pQWhph can direct the hph gene to the WI-1 locus through homologous crossover in two ways. X's depict crossing over at 1 kB of WI-1 5′ flanking sequence (all WI-1 coding, designated in black) and at 2.5 kB of WI-1 3′ flanking sequence (881 bp of WI-1 coding in black and 1395 bp of noncoding in horizontal lines). Alternatively, the broken line depicts crossing over at a short region of the 375 bp WI-1 minipromoter in front of the hph gene. Either crossover event would replace most of the WI-1 coding region with the hph cassette. (B) PCR analysis for targeted gene replacement of WI-1. Candidates for a WI-1 null phenotype were analyzed for targeted recombination by PCR amplification of a novel junction fragment spanning the WI-1 5′ region (not present on the targeting vector) and the hph gene. The forward primer is present only in the endogenous locus, whereas the reverse primer is present in the hph gene on transforming DNA. The resulting 675-bp fragment is amplified if WI-1 sequences have been interrupted by hph in the manner depicted by a dashed line in (A), yielding an interrupted locus shown at the bottom of (A) and in (B). The product's authenticity is confirmed by the presence of EcoRV and Aat II restriction sites in the WI-1 upstream segment of the product, as shown in an agarose gel, and in accord with published WI-1 genomic sequence [51]. (From Ref. 60.)

marker *E. coli hph,* was used to facillitate a crossover at the 5′ flanking region and, in effect, capture the full endogenous WI-1 promoter to enhance transcription of the selectable marker and survival of transformants in the presence of hygromycin B. WI-1 expression was reconstituted by co-transformation, using a second dominant selectable marker encoding resistance to chlorimuron ethyl.

7. Defective Binding and Entry of WI-1 Knockout Yeast

Because WI-1 has been shown to bind yeasts to macrophages [46,47,50], WI-1 knockout yeasts were investigated for binding. Knockout yeasts (strain 55) bound and entered murine macrophages poorly (Fig. 9A). The number of knockout yeasts bound to or inside of macrophages in vitro was far lower than that for wild-type yeasts at all time points between 15 min and 6 hr of incubation. At 6 hr, the association index for the knockout yeast was one-sixth the value for the wild-type yeast, and the ingestion index was one-eighth as high. The defect in binding and entering macrophages was restored fully in the knockout when 10% serum was added as a source of complement (Fig. 9B) or after WI-1 was reexpressed in strain 4/55 (Fig. 9C). The findings imply that WI-1 mediates binding and entry of yeast into macrophages in the opsonin-poor environment of the lung early in the course of infection, before tissues are inflamed and serum exudes into the alveolus.

To determine whether knockout yeasts bound poorly within the noninflamed lung itself, the number of yeasts that adhered to thin sections of lung in an ex vivo binding assay was quantified. The knockout bound poorly, whereas both the wild-type isolate and the WI-1 reconstituted isolate 4/55 bound avidly (Fig. 9D), emphasizing the key role of WI-1 in promoting interactions directly with constituents of the lung alveolus.

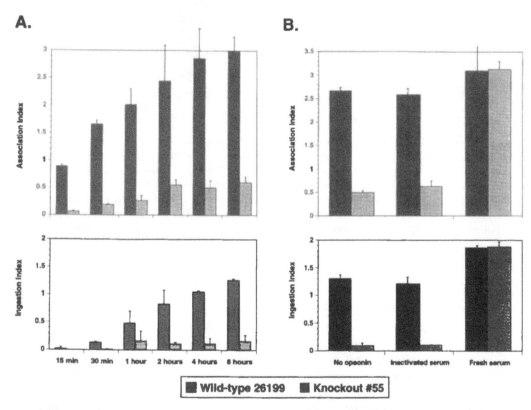

Figure 9 WI-1 mediated binding of *B. dermatitidis* to macrophages and lung tissue. (A) Binding and entry of macrophages by *B. dermatitidis* wild-type and knockout yeast. The upper portion of the panel shows binding of yeasts to macrophage J774.16 cells. The lower portion shows entry of yeasts into the macrophages. Binding and entry of unopsonized yeasts were quantified after incubation for varying intervals as shown. The association index is defined as the number of attached and ingested yeasts per macrophage. The ingestion index is defined as the number of yeasts internalized per macrophage. (B) Complement reverses defects in binding and entry in the WI-1 knockout strain. Assays were done as in (A), except that they included 10% fresh mouse serum (as a source of complement) or heat inactivated mouse serum (as a control).

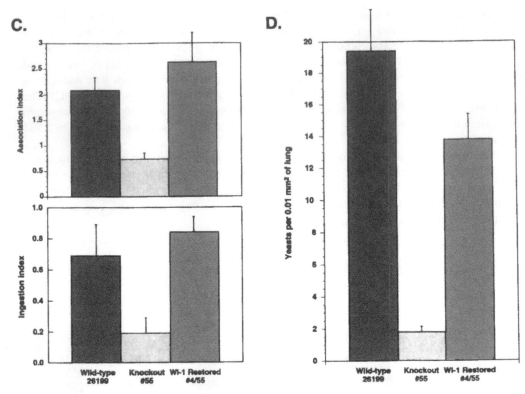

Figure 9 *(Continued)* (C) Defective binding and entry of macrophages by WI-1 knockout yeasts is reversed following reexpression of WI-1. Unopsonized yeasts were incubated with macrophages for 6 hr; results are the mean ± SEM of three separate experiments. (D) Binding of yeasts to lung tissue ex vivo. Lungs of normal mice were removed and cryopreserved and 6-μm sections were applied to glass slides. The number of unopsonized yeasts attached to lung tissue sections was quantified. The extent of binding is defined as the number of yeasts attached per 0.01 mm² of lung tissue, which was determined by inspecting a high-power field at 600× magnification and counting 30 fields per strain in each experiment. (From Ref. 60.)

8. Indispensable Role of WI-1 in Virulence

It was hypothesized that WI-1 knockout yeast unable to bind lung tissues and enter macrophages during infection would be less virulent. This hypothesis was tested in a murine model of lethal pulmonary blastomycosis [62]. All mice infected with the wild-type strain ATCC 26199 at a dose of 10^2 yeasts died from an overwhelming pulmonary and disseminated infection several weeks after inoculation, whereas all mice infected with the same dose of isogenic, WI-1 knockout strain 55 lived and appeared healthy during observation over 72 days (Fig. 10A). Mice infected with either 10^3 or 10^4 yeasts of knockout strain 55, which are respectively 10 and 100 times the lethal dose of wild-type yeast, also survived and appeared healthy during the 72-day observation period. On postmortem examination, the lungs of mice infected with the knockout were macroscopically normal, but contained a small number of well-formed granulomas with sequestered organisms (Fig. 10B). The lungs of mice infected with the wild-type strain were filled with organisms and widespread inflammation. To confirm that the loss of WI-1 is directly responsible for reduced virulence of strain 55 in vivo, it had to be shown that restored expression

A.

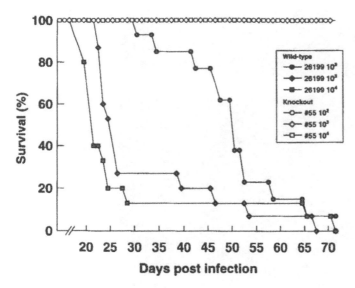

C.

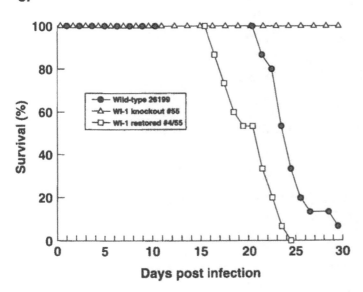

Figure 10 Targeted gene replacement of WI-1 reduces the pathogenicity of *B. dermatitidis*. (A) Survival following infection with wild-type strain 26199 and WI-1 knockout strain 55. Male BALB/c mice (*n* = 15 mice/group) were infected intranasally with yeast cells of each strain, in varied doses. Survival was monitored for 72 days after infection. The two groups differ significantly (*P* ≪ .001) at each infectious dose tested. The experiment shown is representative of three independent experiments. The phenotype of knockout yeasts was stable; no revertants were identified among yeasts grown from mice infected with strain 55. (C) Survival after infection with wild-type strain 26199, WI-1 knockout strain 55, and WI-1 reconstituted strain 4/55. Survival experiments were done as in (A), using a dose of 10^4 yeasts to establish infection. The wild-type and WI-1 reconstituted strains were significantly different from the WI-1 knockout strain (*P* ≪ .001 for each comparison). (From Ref. 60.)

B.

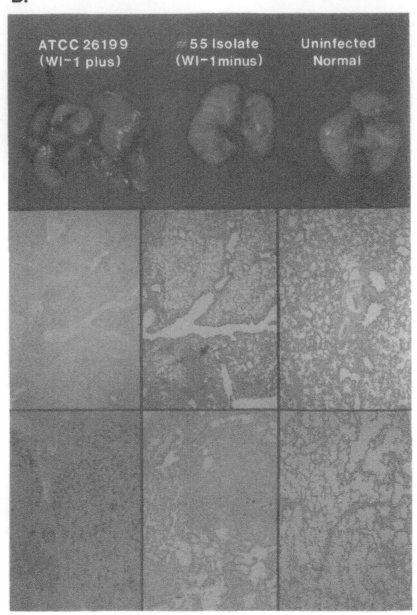

Figure 10   *(Continued)* (B) Gross and microscopic pathology of mice infected with ATCC 26199 wild-type yeasts or WI-1 knockout yeasts. Mice were analyzed 3 weeks postinfection. Lungs were stained with hematoxylin and eosin (H&E) to assess inflammation, and with Gomori Methenamine Silver (GMS) to visualize yeasts. The arrow denotes an isolated granuloma surrounded by normal lung tissue in a mouse infected with the WI-1 knockout strain 55.

of WI-1 conferred pathogenicity on the knockout strain. The WI-1 reconstituted strain 4/55 indeed killed 100% of infected mice, just as did the wild-type strain 26199 (Fig. 10C).

IV. CONCLUSIONS

To summarize, these studies demonstrated successful homologous gene targeting and allelic replacement at the WI-1 locus in *B. dermatitidis,* and unambiguously established the pivotal role of WI-1 in both adherence and virulence.

Formal proof of the importance of a virulence determinant requires fulfillment of a "molecular" Koch's postulate: loss of virulence upon gene disruption, and restoration of virulence upon gene reconstitution [63]. Prior studies of *B. dermatitidis* and other systemic dimorphic fungi have implicated candidate virulence factors through correlative and other means [64], but none of the observations have been substantiated by genetic tests, leaving their validity open to debate. Past studies that have implicated WI-1 in adherence, pathogenesis, and virulence are one such example, using exclusively nongenetic methods. Genetic intractability of *B. dermatitidis* has, until very recently, prevented definitive studies. Work described above illustrates a genetically proven virulence factor in *B. dermatitidis,* and also emphasizes the power of reverse genetics for studies of pathogenesis in these complex microorganisms.

REFERENCES

1. J Schwarz, GL Baum. Blastomycosis. Am J Clin Pathol 21:999–1029, 1951.
2. MH Zahree, TE Wilson, ER Scheer. The effect of growth media upon the ultrastructure of *Blastomyces dermatitidis.* Can J Microbiol 28:211–218, 1982.
3. RG Garrison, JW Lane, MF Field. Ultrastructural changes during the yeastlike to mycelial phase conversion of *Blastomyces dermatitidis* and *Histoplasma capsulatum.* J Bacteriol 101:628–635, 1970.
4. ED DeLamater. The nuclear cytology of *Blastomyces dermatitidis.* Mycologia 40:430–445, 1948.
5. GA Edwards, MR Edwards. The intracellular membranes of *Blastomyces dermatitidis.* Am J Bot 47: 622–632, 1960.
6. WJ Halliday, E McCoy. Biotin as a growth requirement for *Blastomyces dermatitidis.* J Bacteriol 70:464–468, 1955.
7. GL Giraldi, NC Laffer. Nutritional studies on the yeast phase of *Blastomyces dermatitidis* and *Paracoccidioides brasiliensis.* J Bacteriol 83:219–227, 1962.
8. S Levine S, ZJ Ordal. Factors influencing the morphology of *Blastomyces dermatitidis.* J Bacteriol 52:687–694, 1946.
9. A Massoudnia, ER Scheer. The influence of medium on the chemical composition of *Blastomyces dermatitidis.* Curr Microbiol 7:25–28, 1982.
10. RL Beck, CR Hauser. Chemical studies of certain pathogenic fungi. I. The lipids of *Blastomyces dermatitidis.* J Am Chem Soc 60:2599–2603, 1938.
11. JE Domer, JG Hamilton. The readily extracted lipids of *Histoplasma capsulatum* and *Blastomyces dermatitidis.* Biochim Biophys Acta 231:465–478, 1971.
12. AF DiSalvo, JF Denton. Lipid content of four strains of *Blastomyces dermatitidis* of different mouse virulence. J Bacteriol 85:927–931, 1963.
13. C Brass, CM Volkman, DE Philpott, HP Klein, CJ Halde, DA Stevens. Spontaneous mutant of *Blastomyces dermatitidis* attenuated in virulence for mice. Curr Microbiol 7:25–28, 1982.
14. F Kanetsuna, LM Carbonell. Cell wall composition of the yeast-like and mycelial forms of *Blastomyces dermatitidis.* J Bacteriol 106:946–948, 1971.

15. F Kanetsuna, LM Carbonell, I Azuma, et al. Biochemical studies on the thermal dimorphism of *Paracoccidiodes brasiliensis.* J Bacteriol 110:208–218, 1972.

16. NAR Gow, GW Gouday. Ultrastructure of chitin in hyphae of *Candida albicans* and other dimorphic and mycelial fungi. Protoplasma 115:52–58, 1983.

17. F Kanetsuna. Ultrastructural studies on the dimorphism of *Paracoccidiodes brasiliensis, Blastomyces dermatitidis* and *Histoplasma capsulatum.* Sabouraudia 19:275–286, 1981.

18. G Shearer, LW Larsh. Chitin synthetase from yeast and mycelial phases of *Blastomyces dermatitidis.* Mycopathologia 90:91–96, 1985.

19. S Paris, RG Garrison. Cyclic adenosine 3′,5′ monsphosphate as a factor in phase morphogenesis of *Blastomyces dermatitidis.* Mykosen 27(7):340–345, 1984.

20. G Medoff, A Painter, GS Kobayashi. Mycelial to yeast phase transitions of dimorphic fungi *Blastomyces dermatitidis* and *Paracoccidiodes brasiliensis.* J Bacteriol 169:4055–4060, 1987.

21. L Kaufman, PG Standard, RJ Weeks, AA Padhye. Detection of two *Blastomyces dermatitidis* serotypes by exoantigen analysis. J Clin Microbiol 18:110–114, 1983.

22. KJ Kwon-Chung. Genetic analysis on the incompatibility system of *Ajellomyces dermatitidis.* Sabouraudia 9:231–238, 1971.

23. BS Klein, JM Vergeront, JP Davis. Epidemiologic aspects of blastomycosis, the enigmatic systemic mycosis. Semin Respir Infect 1:29–39, 1986.

24. JF Denton, ES McDonough, L Ajello, et al. Isolation of *Blastomyces dermatitidis* from soil. Science 133:1126–1127, 1961.

25. JF Denton, AF DiSalvo. Isolation of *Blastomyces dermatitidis* from natural sites in Augusta, Georgia. Am J Trop Med Hyg 13:716–722, 1964.

26. JF Denton, AF DiSalvo. Additional isolation of *Blastomyces dermatitidis* from natural sites. Am J Trop Med Hyg 28:697–700, 1979.

27. GA Sarosi, DS Serstock. Isolation of *Blastomyces dermatitidis* from pigeon manure. Am Rev Respir Dis 114:1179–1183, 1976.

28. A Bakerspigel, J Kane, D Schans. Isolation of *Blastomyces dermatitidis* from an earthen floor in southwestern Ontario, Canada. J Clin Microbiol 24:890–891, 1986.

29. BS Klein, JM Vergeront, RJ Weeks, et al. Isolation of *Blastomyces dermatitidis* in soil associated with a large outbreak of blastomycosis in Wisconsin. N Engl J Med 314:529–534, 1986.

30. BS Klein, JM Vergeront, AF DiSalvo, L Kaufman, JP Davis. Two outbreaks of blastomycosis along rivers in Wisconsin. Am Rev Respir Dis 136:1333–1338, 1987.

31. ES McDonough, JJ Dubats, TR Wisniewski. Soil *Streptomycetes* and bacteria related to lysis of *Blastomyces dermatitidis.* Sabouraudia 11:244–250, 1973.

32. ME Landay, J Mitten, J Millar. Disseminated blastomycosis in hamsters II. Effect of sex on susceptibility. Mycopathol Mycol Appl 42:73–80, 1970.

33. PA Morozumi, JW Halpern, DA Stevens. Susceptibility differences of inbred strains of mice to blastomycosis. Infect Immun 32:160–168, 1981.

34. GS Deepe Jr, CL Taylor, TE Taylor. Evolution of inflammatory response and cellular immune responses in murine model of disseminated blastomycosis. Infect Immun 50:183–189, 1985.

35. M Miyaji, K Nishimura. Granuloma formation and killing functions of granuloma in congenitally athymic nude mice infected with *Blastomyces dermatitidis* and *Paracoccidiodes brasiliensis.* Mycopathologia 82:129–141, 1983.

36. LM Thurmond, TG Mitchell. *Blastomyces dermatitidis* chemotactic factor: kinetics of production and biological characterization evaluated by a modified neutrophil chemotaxis assay. Infect Immun 46:87–93, 1984.

37. RA Cox, LR Mills, GK Best, et al. Histologic reaction to cell walls of an avirulent and a virulent strain of *Blastomyces dermatitidis.* J Infect Dis 129:179–186, 1974.

38. RA Cox, GK Best. Cell wall composition of two strains of *Blastomyces dermatitidis* exhibiting differences in virulence for mice. Infect Immun 5:449–453, 1972.

39. CJ Morrison, DA Stevens. Mechanisms of fungal pathogenicity correlation of virulence in vivo, susceptibility to killing by polymorphonuclear neutrophils in vitro, and neutrophil superoxide anion induction among *Blastomyces dermatitidis* isolates. Infect Immun 59:2744–2749, 1991.

40. KR Klimpel, WE Goldman. Cell walls from avirulent variants of *Histoplasma capsulatum* lack α-1,3-glucan. Infect Immun 56:2997–3000, 1988.

41. J Hallak, F San-Blas, and B San-Blas. Isolation and wall analysis of dimorphic mutants of *Paracoccidiodes brasiliensis*. Sabouraudia 20:51–62, 1982.

42. G San-Blas, F San-Blas. *Paracoccidiodes brasiliensis:* cell wall structure and virulence. Mycopathologia 62:77–86, 1977.

43. LH Hogan, BS Klein. Altered expression of α-(1,3)-glucan on genetically related strains of *Blastomyces dermatitidis* that differ in virulence. Infect Immun 62:3543–3546, 1994.

44. BS Klein, JM Jones. Isolation, purification and radiolabeling of a novel 120-KD surface protein on *Blastomyces dermatitidis* yeasts to detect antibody in infected patients. J Clin Invest 85:152–161, 1990.

45. BS Klein, LH Hogan, JM Jones. Immunologic recognition of a 25-amino acid repeat arrayed in tandem on a major antigen of *Blastomyces dermatitidis*. J Clin Invest 92:330–337, 1993.

46. LH Hogan, S Josvai, BS Klein. Genomic cloning, characterization, and functional analysis of the major surface adhesin WI-1 on *Blastomyces dermatitidis* yeasts. J Biol Chem 270:30725–30732, 1995.

47. SL Newman, S Chaturvedi, BS Klein. WI-1 on *Blastomyces dermatitidis* yeast mediates binding to human macrophage CR3(CD11b/CD18) and CD14. J Immunol 154:753–761, 1995.

48. BS Klein, JM Jones. Purification and characterization of the major antigen WI-1 from *Blastomyces dermatitidis*, yeasts, and immunological comparison with A antigen. Infect Immun 62:3890–3900, 1994.

49. BS Klein, PM Sondel, JM Jones. WI-1, a novel 120-KD surface protein on *Blastomyces dermatitidis* yeast cells, is a target antigen of cell mediated immunity in human blastomycosis. Infect Immun 60:4291–4300, 1992.

50. BS Klein, S Chaturvedi, LH Hogan, JM Jones, SL Newman. Altered expression of the surface protein WI-1 in genetically-related strains of *Blastomyces dermatitidis* that differ in virulence regulates recognition of yeasts in human macrophages. Infect Immun 62:3536–3542, 1994.

51. E Brummer, PA Morozumi, DE Philpott, et al. Virulence of fungi: correlation of virulence of *Blastomyces dermatitidis* in vivo with escape from macrophage inhibition of replication in vitro. Infect Immun 32:864–871, 1981.

52. SD Wright, RA Ramos, PS Tobias, RJ Ulevitch, JC Mathison. CD14, a receptor for complexes of lipopolysaccharide (LPS) and LPS binding protein. Science 249:1431, 1990.

53. SD Wright, RA Ramos, A Hermanowski-Vosatka, P Rockwell, and PA Detmers. Activation of the adhesive capacity of CR3 on neutrophils by endotoxin: dependence on lipopolysaccharide binding protein and CD14. J Exp Med 173:1281, 1991.

54. A Hermanowski-Vosatka, PA Detmers, O Gotze, SC Silverstein, SD Wright. Clustering of ligand on the surface of a particle enhances adhesion to receptor-bearing cells. J Biol Chem 263:17822, 1988.

55. EH Beachey. Bacterial adherence: adhesin-receptor interactions mediating the attachment of bacteria to mucosal surfaces. J Infect Dis 143:325–345, 1981.

56. SF Hurst, L Kaufman. Western immunoblot analysis and serologic characterization of *Blastomyces dermatitidis* yeast from extracellular antigen. J Clin Microb 30:3043–3049, 1992.

57. Lancet. Blastomycosis—one disease or two? Lancet i:25–26, 1989. (Editorial.)

58. BS Klein, BA Aizenstein, LH Hogan. African strains of *Blastomyces dermatitidis* that lack the major surface adhesin WI-1. Infect Immun 65:1505–1509, 1997.

59. LH Hogan, BS Klein. Transforming DNA integrates at multiple sites in the dimorphic fungal pathogen *Blastomyces dermatitidis*. Gene 186:219–226, 1997.

60. T Brandhorst, M Wuethrich, T Warner, BS Klein. Targeted gene disruption reveals an adhesin indispensable for pathogenicity of *Blastomyces dermatitidis*. J Exp Med 189:1207–1216, 1999.

61. H te Riele, ER Maandag, A Clarke, M Hooper, A Berns. Consecutive inactivation of both alleles of the pim-1 proto-oncogene by homologous recombination in embryonic stem cells. Nature 348:649–651, 1990.

62. M Wüthrich, WL Chang, BS Klein. Immunogenicity and protective efficacy of the WI-1 adhesin of
     Blastomyces dermatitidis. Infect Immun 66:5443–5449, 1998.
63. S Falkow. Molecular Koch's postulates applied to microbial pathogenicity. Rev Infect Dis 10:
     S274–S276, 1988.
64. LH Hogan, BS Klein, SM Levitz. Virulence factors of medically important fungi. Clin Microbiol
     Rev 9:469–488, 1996.

10

Paracoccidioides brasiliensis: Virulence and Host Response

Gioconda San-Blas and Gustavo Niño-Vega
Instituto Venezolano de Investigaciones Científicas, Caracas, Venezuela

I. INTRODUCTION

Many zoopathogenic fungi grow in more than one morphology, with a saprobic phase usually consisting of hyphae, and a parasitic phase, frequently expressed as yeast. The phenotypic duality of form has been termed dimorphism [1]. Dimorphism is a useful model to examine the basis of morphogenesis and cellular differentiation in eukaryotes, leading to a better understanding of the pathogenic behavior of a given species, and the virulence of strains within it; therefore providing clues for the control of an important group of mycoses which afflict man and animals. Temperature, or nutritional factors, or both are usually the agents which trigger a change in fungal morphology [1]. Few fungi, among them *Paracoccidioides brasiliensis*, have been studied systematically to search for the molecular and biochemical events leading to dimorphism and eventual expression of pathogenicity.

First described by Lutz in 1908 [2], *P. brasiliensis* is the causative agent of paracoccidioido-mycosis, a human systemic mycosis geographically confined to Latin America, where it consti-tutes one of the most prevalent deep mycoses [3]. The frequency of mucocutaneous lesions suggests the respiratory route by inhalation of airborne propagules as the portal of entry of the fungus [4]. After penetrating the host, the fungus must convert to its Y form, a fundamental step for the successful establishment of the infection [5]. In nature, the fungus has been isolated from liver, spleen, and lungs of armadillos (*Dasypus novemcinctus*) [6]. The geographical distri-bution of both the vertebrate and the fungus superimpose very closely, from southern Mexico to Argentina, suggesting that the armadillo may well be the main natural reservoir for *P. bra-siliensis* [6].

Most infected people develop only asymptomatic or subclinical paracoccidioidomycosis, with progression to disease depending on host factors, fungal virulence, and environmental conditions. Skin tests indicate an important variation in the frequency of reactive tests (2–60%), the higher proportion of positive reactions occurring in individuals living in rural communities, where they perform agricultural pursuits. Endemic areas are usually found at altitudes between 1300 and 1700 m, temperatures between 18 and 24°C, high annual rain falls (2000–3500 mm), and humid or very humid premountainous forests [4]. High positive reactivity (60–75%) in the adult population of endemic areas [4] points to the relevance of this mycosis in Latin American soil. The disease presents a variety of signs and symptoms which have been the basis for the classification of its clinical forms into acute or subacute form (juvenile type) and chronic form (adult type), with presentations as systemic mycosis with a rapid course, or as localized mycosis with a chronic evolution [3]. As discussed later, the diversity of clinical forms may have to do not only with the immunological conditions of the host, but also with the variability of fungal strains.

Although the contact with *P. brasiliensis* is essentially the same for the two sexes, paracoccidioidomycosis is 13–87 times more common in men than in women, the reason for which being the possibility that the hormonal milieu of the host might directly influence *P. brasiliensis*, affecting its pathogenicity [7]. The possible biochemical reasons for this will be discussed later in this review.

II. THE ORGANISM

A. Morphology

The fungus is a pleomorphic organism that depends on the temperature of incubation as the only requirement for the expression of its morphology [5]. In vitro at 37°C and in infected tissues, *P. brasiliensis* grows in the form of spherical or oval cells that vary in size from a few nanometers in diameter in young, recently separated buds, to 10–30 μm or more in a mature yeast (Y) cell [8]. Yeast cells are multinucleate [9] and multiply by polar or multipolar budding. Buds are connected to the mother cell by narrow necks, giving the whole structure the appearance of a ship's wheel (Fig. 1A), this being the most important taxonomic and diagnostic characteristic of *P. brasiliensis*. The mycelial (M) form grows slowly at room temperature. Microscopic

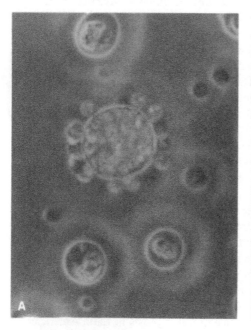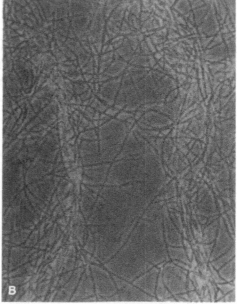

Figure 1 *Paracoccidioides brasiliensis*. (A) Y phase; (B), M phase. Bar: 5 μm.

observations show septate, slender, and freely branching hyphae, 1–3 μm in width [9], with the appearance of interwoven threads (Fig. 1B). *P. brasiliensis* is classified as imperfect fungus (deuteromycete) due to the lack of sexual structures which would allow a more precise classification. However, as hyphae of *P. brasiliensis* are known to have simple septal pores and two-layered cell walls [8], this fungus might be ascomycetous. The Y phase, multiplying by simultaneous multiple budding, has a mode of reproduction which is well known in zygomycetes but not in asco- or basidiomycetes. But the occasional rod-shaped invaginations visualized in the freeze-etching preparations of the Y plasma membrane of *P. brasiliensis* may suggest an asco- or basidiomycetous origin [10]. More reliable evidence pointing to an ascomycetous classification of *P. brasiliensis* comes from molecular studies of large subunit ribosomal RNA sequence comparisons of dimorphic fungi [11] and comparison of the deduced amino acid sequences of fungal ornithine decarboxylase (ODC) genes [12]. In both cases, the generated dendrograms place *P. brasiliensis* in close neighborhood with *Coccidioides immitis*, an ascomycetous belonging to the order Onygenales, suggesting a similar classification for the former.

B. Cell Wall as Virulence Factor

The cell wall has been considered the main structure affected by those morphogenetic changes intimately linked to pathogenicity in dimorphic fungi. Perhaps one of the best-defined examples of cell wall chemistry, morphogenesis, and virulence is precisely *P. brasiliensis*, in which its cell wall vary in composition according to the morphological phase, particularly with regard to the carbohydrate constituents, which account for >80% of its dry weight [13,14]. Of them, chitin is common to both phases, although three times more abundant in the Y phase. Glucose polymers, instead, are arranged mainly as α-1,3-glucan in the cell wall of the pathogenic Y phase (95%) plus a small amount (5%) of β-1,3-glucan, whereas the latter is the only glucose polymer found in the M cell wall. α-1,3-Glucan plays an important role in host-parasite relationships in *P. brasiliensis* [13,15,16] as it is found in large proportions (∼ 40% of the Y cell wall) in virulent strains and in proportionally lower amounts (down to 3%) as the strain virulence decreases. Hence, it has been proposed [5,15,16] that the peripheral α-glucan of *P. brasiliensis* may play a role as a protective layer against the host defense mechanisms, this being a kind of biochemical control of pathogenicity, relying on the inability of phagocytic cells to digest α-1,3-glucan. Later on, this mechanism of virulence was also proposed for *Histoplasma capsulatum* [17] and *Blastomyces dermatitidis* [18].

C. Cell Wall Biosynthesis as Prospective Target for Specific Antifungal Drugs

Besides a role as regulator of virulence and in the inflammatory process observed in paracoccidioidomycosis (see Sec. III.C), *P. brasiliensis* cell wall has an additional interest as a likely candidate for the site of action of highly specific antifungal drugs that would block the correct synthesis of wall polysaccharides, namely, glucan and chitin, without interfering with any metabolic process in the host. In pursue of this goal, studies on biochemical details of glucan synthesis, and more recently, on the search and sequencing of genes codifying for glucan and chitin synthetases, are being done, as a first step in the search for such antibiotics.

The synthesis in vitro of *P. brasiliensis* β-glucan requires UDP-glucose as the preferred nucleotide precursor [19] and the particulate enzyme located in the plasma membrane [20]. In line with some ascomycetes [21], the reaction is inhibited by GTP and other nucleotides [22], in sharp contrast to the role assigned to these compounds as general stimulators of fungal cell wall synthesis [23]. So far, only one gene (*FKSPb1*), homologous to the β-glucan synthetase genes *FKS1* and *FKS2* from *Saccharomyces cerevisiae*, and *FKSa*, from *Aspergillus nidulans*, have been cloned and sequenced from *P. brasiliensis* [24]. To date, no gene for α-glucan synthetase has been reported.

In all pathogenic fungi, chitin represents a major structural component of the cell wall [25,26]. In *P. brasiliensis*, this polysaccharide represents 43% of the dry weight of the wall of the pathogenic Y form and 13% of the M cell wall [5]. The absence of this polysaccharide from human cells makes its biosynthesis a logical target for the development of antifungal antibiotics, and this is why chitin synthetases are being extensively studied. Based on differences in regions of high sequence conservation, chitin synthetases have been organized into five classes [27,28] whose functional implications are not yet clear. In *P. brasiliensis*, five genes (*PbrCHS1* in class I, *PbrCHS2* in class II, *PbrCHS3* in class IV, and *PbrCHS4* and *PbrCHS5* in class V) have been reported [29,30]. The complexity of chitin synthesis has only recently become apparent, as some fungi have large multigene families encoding chitin synthetase isoenzymes [31]. This raises the possibility that more chitin synthetase genes might be present in *P. brasiliensis*.

Expression of *P. brasiliensis* chitin synthetase genes was explored through the temperature-induced dimorphic transitions from Y to M and reverse, by means of Northern analysis [30]. Transcripts of *PbrCHS3* were not detectable perhaps due to the presence of a putative intron within the sequence of the probe, which may have reduced the sensitivity of detection for its transcript. All other transcripts were detected in both Y and M cultures where they exhibited a similar pattern of expression during the M to Y transition, with a decrease in the level of expression during the first 8 hr, and an increase in expression after 24 hr to a lower level than that in the M form. The exception to this pattern was *PbrCHS5* whose level of expression in the Y cells returned to that of M used as the inoculum, suggesting that during this transition, the remodeling of the cell wall does not require activity of these gene product. The preferential expression of these chitin synthetase genes in the M form of the fungus is surprising, since the Y form of the organism has a higher chitin content than the M form [5].

Although higher levels of expression of *PbrCHS1*, *PbrCHS2*, *PbrCHS4*, and *PbrCHS5* correlated with RNA isolations from M cultures, it is possible that transcription of the genes responds to changes in temperature or other environmental changes such as the production of metabolites by the fungus, and also to posttranscriptional modifications [32,33]. The temporal pattern of transcriptional changes correlated with the developmental timing in the generation of Y or M forms. Still, the absence of expression of some *PbrCHS* genes at 4 and 8 hr coincides with the stage 1 heat shock responses; hence, short-term changes in gene expression may be related to heat shock rather than morphogenesis per se. Also, heat shock elements within the promoters of *PbrCHS* genes may play a direct role in linking their expression to changes in temperature associated with colonization of the human body [30].

The development of systems for reverse genetics in dimorphic fungi such as *P. brasiliensis* will further clarify the role of the complex multi-*CHS* gene families—by targeted mutagenesis—in the regulation of cell wall assembly and cellular morphogenesis, as is currently being explored for filamentous fungi [28,34,35]. This approach could lead to the design of specific antifungal antibiotics.

D. Antibiotics in Use

Azoles are another group of compounds whose use as antifungal drugs has been promoted by their effectiveness against many pathogens [36]. Their interference with the correct synthesis of ergosterol in the fungal membrane leads to structural deformations and eventual cell death. *P. brasiliensis* is extremely sensitive in vitro to azoles [37]. Minimal inhibitory concentration (MIC) values for the Y phase of *P. brasiliensis* ranged from 10^{-10} to 10^{-8} M, in the following sequence: saperconazole > itraconazole > ketoconazole.

Currently, azoles, mainly ketoconazole and itraconazole, are extensively used in the treatment of paracoccidioidomycosis. The former is effective in severe cases and relapses, although side gastrointestinal or endocrinal effects may appear. Itraconazole is 10 times more active against *P. brasiliensis* than ketoconazole and does not interfere with endocrine metabolism. A daily dosage of 100–200 mg for 6 months reduced the proportion of relapses to 3–5%. It has become the drug of choice for the treatment of paracoccidioidomycosis [38].

Amphotericin B desoxycholate is also indicated for the treatment of severe cases of disseminated disease, although follow-up therapy with sulphonamide is indicated to prevent relapses [38]. Lipid-based preparations of amphotericin B usually have a therapeutic-toxic advantage, but this does not seems to be the case in paracoccidioidomycosis, as preliminary reports on the use of amphotericin B colloidal dispersions suggest a failure in curing the infection [39]. Supplementation of therapy with yeast β-glucan as an inespecific immunostimulant of the reticuloendotelial system has also been used with encouraging results (Fig. 2) [40].

Ajoene [(E,Z)-4,5,9-trithiadodeca-1,6,11-triene-9-oxide] is another compound currently being explored as a potential antifungal drug. Clinical studies indicate its effectiveness in the treatment of tinea pedis, cruris, and corporis [41], comparable to that of terbinafine. Ajoene is a synthetic compound derived from allicin [42], a natural product of garlic (*Allium sativum*). Growth in vitro of *P. brasiliensis* is inhibited by ajoene, more effectively in the Y phase (90%) than in the M phase (60%) at concentrations of 50 μM ajoene [43]. M-to-Y transition, but not

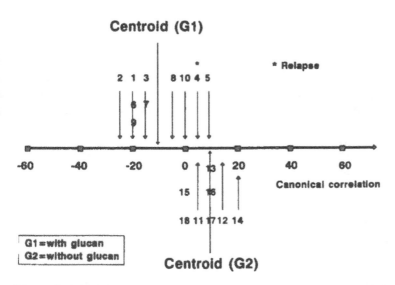

Figure 2 Canonical correlation showing the difference in the values of five variables (tumor necrosis factor, erythrocyte sedimentation rate, double immunodiffusion, phytohemagglutination skin test, and paracoccidioidin skin test) before and after treatment with glucan. 1–10, G1 group (juvenile, acute, or subacute, form of paracoccidioidomycosis); 11–18, G2 group (disseminated adult form of the disease). The centroid is the point that represents the average values for the canonical variables of the patients in each group. (From Ref. 40.)

the reverse, is blocked indefinitely in the presence of ajoene [44]. The proof that ajoene interfered with a correct synthesis of membranous compounds, therefore leading to the disturbance of its architecture and functionality, came from lipid studies. Ajoene induced alterations in phospholipid and fatty acid proportions so that phosphatidyl choline (PC) was reduced to ~ 18% in either phase, and phosphatidyl ethanolamine (PE) increased to 38% (Y phase) or 44% (M phase). This suggests inhibition of PC synthesis [45], and determines a novel mechanism of antifungal action.

E. Hormones in Dimorphism and Pathogenicity

In a variety of pathogenic and nonpathogenic fungi, mammalian hormones can be regulators of fungal growth and morphogenic transition, probably through interactions in which the molecular similarity between the mammalian hormone and a putative fungal ligand may be playing a role [7]. *P. brasiliensis* was one of such cases in which sex steroid binding proteins and a specific binding of 17-β-[^{3}H]estradiol (E2) to cytosolic proteins led to inhibition of M or conidia transition to the Y phase [7]. This result may help to explain why paracoccidioidomycosis is more prevalent among the male population, inasmuch as circulating estradiol in females could block M-to-Y transition of the infecting *P. brasiliensis*, the initial step in the development of the disease. However, contradictory results come from experimental animal models. While female mice seem more resistant to paracoccidioidomycosis than male animals [46], either the reverse [47] or no difference according to sex [48,49] has also been reported. Further studies [50] suggested that female mice at estrus and metestrus I and II stages were more susceptible than male mice but as susceptible at proestrus or diestrus stages. These results, as a whole, put a note of caution on extrapolations from experiments on animal models into the natural host.

F. Strain Variability

A problem which arises once and again when recapitulating studies on *P. brasiliensis* or for that matter, on any fungal pathogen, is the diversity of behaviors of strains. This is observed in the changing degree of virulence [5,51], discrepancies in nutritional requirements [52,53], and cell wall composition and its possible consequence in morphology and virulence [5,15,16], among other factors. In searching for answers to this problem, several groups have started genetic studies on an extensive number of *P. brasiliensis* strains.

Marked geographical distinction among strains of *P. brasiliensis* has been revealed by RAPD [54] and RFLP analyses [54a]. With few exceptions, genetic patterns generated by RAPD analyses of 33 *P. brasiliensis* isolates led to their grouping into five well-defined clusters, according to their geographical origin (Fig. 3) but not to virulence of strains or pathology of disease. These results were later confirmed by RFLP analyses [54a]. Other researchers [55] succeeded in clustering 15 isolates into two well-differentiated groups, according to virulence, also by means of RAPD analysis. Group I (slightly virulent), subdivided into three subgroups (A, B, and C) to comprise 12 isolates, while group II (highly virulent) encompassed three isolates. In this work, no correlation with geographical origin was taken into account.

Comparison of partial gp43-kDa gene nucleotide sequence (539 bp) among 10 clinical isolates and nine armadilloes isolates produced seven different sequences, suggesting that the fungus may be a heterogeneous species [56], results that are matched by those of Morais et al. [57] who compared gp43 sequences of six strains and concluded that the promoter and the 3′ regions were generally conserved among different strains, while differences were found in the region between nucleotides 550 and 1150.

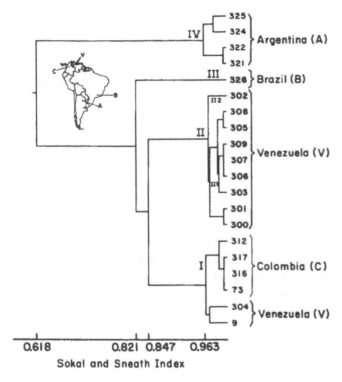

Figure 3 Dendrogram of genetic relationships among 33 *P. brasiliensis* strains. (From Ref. 54.)

III. THE HOST

Paracoccidioidomycosis produces major pathological tissue alterations characteristic of a chronic granulomatous disease. The host immune response is associated with hypergammaglobulinemia, elevated levels of immune complexes, activation of the complement system, and impaired cellular immunity. In the disseminated form, granulomas are not formed and T-cell hyporeactivity is a hallmark, while in the localized form, patients mount a granulomatous response and possess functional T lymphocytes. To establish the true prevalence of paracoccidioidomycosis (the disease as well as the infection), clinical diagnosis and serological tests help greatly in establishing an unequivocal diagnosis. The former requires the demonstration of characteristic multiple budding yeast cells in biological fluids or tissue sections; the latter requires the existence of specific antigens for reliable results. Immunological methods rely on the identification of the host's humoral responses, which are usually impaired or absent in patients with severe juvenile forms of the disease and in immunocompromised patients. Determining disease activity or assesing treatment responses by measuring antibody levels is difficult, since antibody titers may remain elevated or persist at stationary levels, even in the presence of clinical improvement.

A. Serodiagnosis

According to their origin, *P. brasiliensis* antigens have been clasified as (1) exocellular (metabollic) antigens, (2) intracellular (somatic or cytoplasmic) antigens, and (3) cell wall antigens [58].

By far, the most important and extensively studied are the exocellular antigens, followed at the distance by cytoplasmic antigens.

The only reliable serological tests of high specificity that can currently be used alone for the diagnosis of paracoccidioidomycosis is the immunodiffusion test [59] with the use of a specific exocellular glycoprotein antigen called gp43, obtained from concentrated dialyzed supernatant fluids of yeast cultures [60]. Gp43 is an immunodominant antigen for antibody-dependent and immune cellular responses in patients with paracoccidioidomycosis. Its partial chemical structure is shown in Figure 4 [61,62]. This structure was deduced from experiments in which treatment with some lectins suggested the presence of O-linked oligosaccharides, whose nonreducing end units are β-galactopyranosyl residues, while gp43 treatment with endo-β-N-acetyl-glucosaminidase H gave rise to a nonglycosylated protein of MW 38 kDa [60]. The mature protein has a single N-linked oligosaccharide chain [61] which in itself is poorly immunogenic, though it is responsible for the cross-reactivity of the gp43 with sera from patients with histoplasmosis, when the ELISA test is used [63]. This oligosaccharide chain is conformed by a neutral high-mannose core with a N-acetyl-glucosamine dimer (Man$_7$GlcNAc$_2$) to which a (1→6)-linked α-D-mannopyranose (Man$p$) chain of variable length, substituted at the 2-O positions by single α-D-Man$p$ residue, is attached. A terminal unit of β-D-galactofuranose (Gal$f$) is (1→6)-linked to one of the (1→2)-linked mannosyl residues, either in the A or in the C arm of the oligossacharide (Fig. 4). The heterogeneity of the oligosaccharide is determined by the

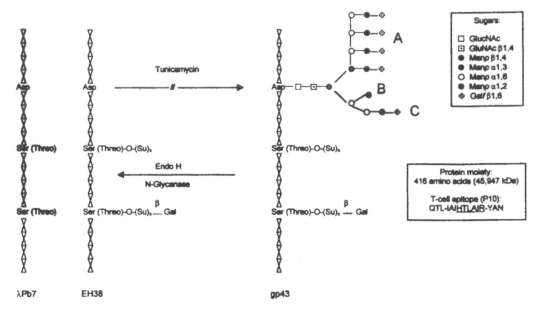

Figure 4 Structure of *P. brasiliensis*–specific antigen gp43. Asparagine (Asp), serine (Ser), and threonine (Threo) are amino acids belonging to the peptide chain of the molecule. Sugar symbols as described in the figure; *p* and *f:* sugars in the pyranose and furanose forms, respectively. Abbreviations: GluNAc, N-acetyl glucosamine; Man, mannose; Gal, galactose; α and β, linkage configurations; Su, undetermined sugar; *x,* <5 sugar residues; *n,* 25–30 sugar residues. The possible sites of insertion of one β-galactofuranosyl unit per oligosaccharide molecule are indicated (dashed lines). Characteristics of the protein chain and T-cell epitope (P10) in it which induces a Th-1 response protective against *P. brasiliensis* infection are also signaled (underlined amino acids: essential domain of the epitope). (From Refs. 61–64, 122.)

different sizes of the A arm and the sites of insertion of the β-D-Gal$f$ unit. The average size of the most frequent subtype is Hex$_{13.6}$GlcNAc$_2$ [61]. The deduced amino acid sequence showed similarities of 56–58% with exo-1,3-β-D-glucanases from *Saccharomyces cerevisiae* and *Candida albicans*. However, the gp43 is devoid of hydrolase activity and does not crossreact immunologically with the fungal glucanases [64]. In order to identify the peptide epitopes involved in the immunological reactivities of the gp43 and to obtain highly specific recombinant molecules for diagnosis of the infection, Cisalpino et al. [64] sequenced genomic and cDNA clones representing the entire coding region of the antigen. The gp43 open reading frame was found in a 1329-bp fragment with 2 exons interrupted by an intron of 78 nucleotides, and a leader sequence of 105 bp. The gene encodes a polypeptide of 416 amino acids (Mr 45,947) with a leader peptide of 35 residues [64]. In it, a peptide (P10) was found that behaves as a T-cell epitope (see Sec. III.C). Aiming to design more sensitive methods of diagnosis, PCR amplification of gp43, using primers localizing in low homology regions of the sequence, was carried out. With it, the presence of *P. brasiliensis* DNA in 11 suspected patients, with a sensitivity level of 10 cells/mL sputum was achieved [65].

Gp43 is stored inside large dense core vesicles, which flow into the plasma membrane and extrude from cell membrane into the cell wall, whereby the antigen is exported to the extracellular space at scattered sites interspersed along the cell surface [66]. Besides its function as a major antigen, gp43 performs duties as specific binder to the extracellular matrix protein laminin, therefore increasing adhesion of the fungus to epithelial cells and enhancement of fungal virulence [67,68].

Although gp43 is currently the most reliable antigen for paracoccidioidomycosis, a note of caution must be given as a few false negatives are reported [69]. Crossreactivities with histoplasmosis [63] and lobomycosis [70] sera have been documented. Also, strain variability in this antigen has been reported [71], so that the expression and isoforms of gp43 varied in 13 different strains of *P. brasiliensis* when tested against a panel of 50 sera from patients with paracoccidioidomycosis. Eight of the 13 strains produced high amounts of gp43, two produced small amounts, and three were unable to secrete the glycoprotein [71]. At least four gp43 isoforms were found, with pI's ranging from 5.8 to 8.5. By using the monoclonal antibody (Mab) capture immunoassay, and the anti-gp43 Mab 17c as the capturing antibody, recognition of pI 8.5 gp43 isoform was significantly lower for both acute (56%) and chronic patients (71%), compared with gp43 isoforms from the standard strain B-339. But use of anti-gp43 Mab 8a, which recognized a different epitope, led to a similar reaction in all patients' sera, suggesting that not all the antigenic epitopes expressed by gp43 are equally present in all *P. brasiliensis* strains [72]. The possibility that such differences in performance may be due to variability in the polysaccharide conformation of the glycoprotein should be considered. Mabs 17c and 24a were able to partially inhibit the laminin-dependent fungal adhesion to epithelial cells [73]. Mab 24a significantly reduced *P. brasiliensis* infection in vivo, while Mab 8a increased both laminin-dependent fungal adhesion and pathogenesis [68]. Additional results suggest that some strains are able to produce gp43 in both morphological phases of the fungus, while others refrain from doing so in its mycelial phase [74].

The dot immunoblotting assay [75] was tested with sera from patients with paracoccidioidomycosis, histoplasmosis, Jorge Lobo disease, aspergillosis, candidiasis, and cryptococcosis, and from healthy subjects. Native gp43 gave positive results with all sera from paracoccidioidomycosis patients and 31% of sera from those with Jorge Lobo disease. Patients undergoing antimycotic treatment showed a decreased antibodies titers in their sera. No false-positive reactions were obtained. Capture enzyme immunoassay with Mabs 17C, 21A, 21F, and 32B raised against gp43 led to the detection of specific human anti-gp43 immunoglobulin G in patients with paracoccidioidomycosis. Specific antibodies in the sera from patients were detected at dilutions as high as 1:819,200, greatly improving ELISA assays [76].

Salina et al. [77] evaluated immunoblotting and competition enzyme immunoassay (EIA-c) for the detection of circulating antigen in urine samples. Bands of 70 and 43 kDa were detected more often in urine samples from patients before treatment. The immunoblot method detected gp43 and gp70 separately or concurrently in 11 of 12 patients, whereas the competition enzyme immunoassay detected antigenuria in only nine. The EIA-c test was highly specific (100%) and sensitive (75%), with a limit of antigen detection of 2.3 ng of protein per mL urine. Gp43 remained present in the urine samples collected during the treatment period, though at low levels during clinical recovery and a significant increase in relapses.

The diagnosis of paracoccidioidomycosis by indirect serological methods which rely on antibody detection is highly valuable. However, antibody levels may be absent in immunocompromised patients or may remain present months after successful therapy. Alternative tests aimed primarily at the identification of circulating antigens are needed, especially when the disease is severe and the results of therapy need to be assessed. With a modification of the standard hybridoma production method, Mab P1B was produced [78] that was directed against a 87-kDa determinant. When used to develop an inhibition ELISA, it was capable of detecting as little as 5.8 ng of circulating antigen per mL serum. Sera from 46 patients with paracoccidioidomycosis or other mycoses, and sera from healthy individuals were evaluated with this technique; overall sensitivity was 80.4% (37 of 46 patients positive), and specificity was 81.4%, with important crossreactivity from sera of aspergillosis (10 of 10) and histoplasmosis (4 of 10) patients. The technique detected circulating antigen in 100% of patients with the acute form of the disease, and in 83.3% and 60% of patients with the chronic multifocal and unifocal forms, respectively [79]. Monoclonal antibodies, used as a cocktail (Mab BJ1, BJ18) against several and relevant circulating antigens, including the immunodominant gp43 antigen, would constitute an improved method to achieve higher sensitivity and specificity [80].

A 58-kDa antigenic determinant was also detected by ELISA. This is an exocellular glycoprotein, with a pI of 5.2, and no homology with known proteins, as revealed by N-terminal amino acid sequence analysis [81]. Eighty-one percent of immune human sera from paracoccidioidomycosis patients reacted by immunoblotting to the partially purified preparation. In addition, a 22 to 25-kDa antigen [82] of possible use as immunochemical marker, and a 27-kDa specific antigen [83,84] have been reported, the latter being also cloned and recombined. This antigen is recognized by 91% of the patients when immunoblotting is used and shows no crossrectivity with sera derived from patients with various mycoses. This observation is of interest because recombinant proteins provide a highly reproducible source of defined antigen, which could lead to improvements in the serodiagnosis of paracoccidioidomycosis [85].

Cloning and characterizing genes that codify for antigens in *P. brasiliensis* has been an approach recently incorporated to the search for improved methods of diagnosis. Besides the cloning of gp43, mentioned above [64], Soares et al. [86] obtained a cDNA clone, PbY11, highly and preferentially expressed in yeast cells, that code for a 40-kDa protein with low homology to other sequences in the Gene Bank. A gene corresponding to N-acetyl-β-D-glucosaminidase was also detected whose protein is highly expressed in two forms of MWs 114 and 203 kDa that were reactive with paracoccidioidomycosis sera [87].

Of the cytoplasmic antigens under study, a glycosylphosphoryl lipid extracted from *P. brasiliensis* showed a strong reaction with sera from paracoccidioidomycosis patients while lacking crossreactivity with histoplasmosis sera [88]. The compound is a glycosylinositol phosphorylceramide with the structure Man$p\alpha$1 $\rightarrow$ 3[Gal$f\beta$1 $\rightarrow$ 6]Man$p\alpha$1 $\rightarrow$ 2Ins1-P-1Cer. Initial characterization established that the galactofuranose (Gal$f$) residue was responsible for its serological activity [89]. However, the specifity and sensitivity of this glycosylphosphoryl lipid as a potential antigen of choice in diagnosis remains to be tested.

B. Humoral Immunity

Patients with paracoccidioidomycosis have a hyperactive humoral immunity, with increased levels of serum IgG, IgE, and IgA [90]. This increment in antibody production is partly related to the virulence of the fungal strain, inasmuch as intermediate and slightly virulent isolates induce weak IgG antibody production, whereas the most virulent isolates induce stronger specific humoral responses [91]. Hypergammaglobulinemia is also induced by the alkali-insoluble fraction and β-glucan present in *P. brasiliensis* cell wall [92].

Susceptible mice produced significant anti–*P. brasiliensis* IgM antibodies during the early stages of the infection and increased titers of IgG were detected later in the evolution of the disease; resistant mice had low titers of specific IgG during the whole course of the infection and a delayed peak of specific IgM at week 16 [93]. Therefore, as in the profile of immune response in human disease, susceptible mice responded with an increase in IgG antibody levels to the dissemination of the disease and its clinical severity.

Mendes-Giannini et al. [94] showed that sera from 39 patients contained IgG-, IgA-, and IgM-specific antibodies to the 43-kDa exoantigen. By the fourth month of chemotherapy, there was a decay of IgG, IgA, and IgM antibody titers to this antigen, correlating with clinical improvement. Therefore, it was suggested that changes in antibody titers may be useful indicators of the extent of active disease and valuable for determining the prognosis of the infection. Further tests on IgG, IgM, and IgA antibodies to gp43 were carried out [95] in samples from controls and 23 patients with paracoccidioidomycosis, before and after chemotherapy started. IgG and IgM ELISA antibodies were more often found in patients with acute than chronic disease. Four to six months after treatment, their levels were decreased in acute cases. Therefore, IgG-ELISA anti-gp43 may represent a good marker to monitor clinical response to therapy. The technique has also been useful to determine differential antibody isotype expression in juvenile and adult forms of paracoccidioidomycosis. Patients with the juvenile form had higher IgG4 but lower IgG2 levels than patients with the adult form. IgG4, regulated by interleukin-4, was found in all juvenile form patients but only in 12% of the adult form patients. IgG2, regulated by IFN-γ, was found in 41% of the adult form patients, and only in 12% of the juvenile patients. These results suggested that the switch to the IgG subclasses in paracoccidioidomycosis is regulated by the patient's T-helper subset (Th1 or Th2) dominant cytokine profile [96].

The significant differences of IgG asymmetric antibodies between control and paracoccidioidomycosis serum may be the cause of the lack of correlation observed frequently between immunodiffusion tests and ELISA tests [97]. Immunoblot tests in sera from patients with the acute and chronic forms of paracoccidioidomycosis [98] produced a reactivity profile of IgG, but not of IgM or IgA, similar for both types of patients, gp43 being recognized by all the sera. The effect of idiotypic modulation in the immune response to gp43 in mice and humans was studied by Souza et al. [73] with the use of three distinct anti-gp43 monoclonal antibodies (17c, 8a, and 24a). It was possible to confirm the expected presence of anti-idiotypic antibodies (anti-Id) (AB2) in the sera of mice, and also an increasing amount of anti-anti-Id (AB3) antibodies, after the third course of immunization. Moreover, human paracoccidioidomycosis patients' sera with high titers of anti-gp43 antibodies were able to generate anti-Id antibodies. These data suggested that the immune response to *P. brasiliensis* could be spontaneously modulated by the idiotypic network.

C. Cellular Immunity

Cellular immune response represents the main defense mechanism against *P. brasiliensis* infection. Recent work on the defensive role of polymorphonuclear neutrophils (PMNs) in *P. brasiliensis* infection suggests that PMN from human peripheral blood could be fungistatic, depending on the strain under study [99,100]. The authors also reported the enhancing effect IFN-γ has on the antifungal activity of human PMN against *P. brasiliensis*. This work is linked to a previous study by Goihman et al. [101] who found differences in the lytic performance of PMN against several *P. brasiliensis* strains. Some of them were strains with remarkable modifications in their cell walls, namely, the lack of the main neutral polysaccharide (α-1,3-glucan) in the yeast cell wall. These strains were avirulent [15,16] and were readily destroyed by phagocytes, whereas other strains whose cell wall was correctly structured survived the PMN attack and were highly virulent. This was interpreted [15,16] as the mechanism of defense the fungus exhibited against lysis by PMN, which were unable to digest the wide layer of α-1,3-glucan present only in the surviving yeast cells.

Macrophages obtained from mice infected with *P. brasiliensis* yeast cells and treated in vitro with IFN-γ showed a modification of the pattern of class II major histocompatibility complex (Ia) antigen expression [102]. This effect was related to the production of nitric oxide (NO) which is the major microbicidal agent of macrophages, capable of controlling many infections caused by intracellular parasites. Thus, the immunosuppression that occurs during the course of infection may be explained in part by increased NO production that promotes suppression of Ia antigen, consequently inhibiting antigen presentation by macrophages with a deficient lymphoproliferative response. NO production is also associated to immunosuppression of T lymphocytes [103].

Distinct subpopulations of helper T cells stimulate different immune response pathways. Immunity to *P. brasiliensis* could be regulated by Th1 or Th2 subsets, which would ultimately determine the outcome of the infection. Lymphokines produced by Th1 lymphocytes are correlated to resistance against certain diseases, whereas lymphokines produced by Th2 cells can be associated with susceptibility. These two activation pathways are usually antagonistic, resulting either in the activation of a cellular immune response, which in paracoccidioidomycosis would confer protection, or in high production of antibodies that would be implicated in pathology [104]. In the development of the disease, the activation of Th2 subsets leads to B cell stimulation and inefficient macrophage activation in experimental murine paracoccidioidomycosis [105]. Instead, resistance is associated to a Th1 predominant response. *P. brasiliensis* stimulated T-cell expansion, interleukin-2 production, and differentiation into cytotoxic T cells when *P. brasiliensis* lysates were confronted with cord blood T cells. Involved in the response were *P. brasiliensis* proteins of 10^5 to 10^6 kDa. T-cells may regulate B-cell activity in paracoccidioidomycosis through the γ/δ subset of the T-cell receptor [106].

Susceptibility in mice to *P. brasiliensis* pulmonary infection is associated with acquired immune mechanisms that allow fungal dissemination to other organs and disease progression [107]. This type of immune response fails to efficiently activate pulmonary macrophages, induces low levels of DTH reactivity, and stimulates the production of IgG1-, IgG2a-, and IgG2b-specific antibodies. On the contrary, resistance is linked to DTH responses, pulmonary macrophage activation, and antibody response in which IgG2a and IgG3 titers were always significantly higher than those observed in susceptible mice. Since all these activities are regulated mainly by IFN-γ, it may follow that resistance to pulmonary paracoccidioidomycosis is under the control of the Th1 subset of T-lymphocytes. Resistance has also been associated with immunological activities and protective roles governed by IFN-γ [108], whose in vivo depletion increases the severity of the desease developed by resistant and susceptible mice [109]. This was associated with DTH anergy, exacerbated pulmonary infection, and early fungal dissemination to liver and spleen. Treatment with anti-IL-4 Mab induced a significant decrease in the dissemination of yeasts to liver and spleen and reversal of DTH anergy, effect also seen in IL-12-treated animals. Hence, immunoprotection in pulmonary paracoccidioidomycosis was associated with production of cytokines with enhanced macrophage activation and cellular immunity, whereas uncontrolled secretion of type 2 cytokines or macrophage-inhibiting cytokines characterizes the progressive disease. Host defense is impaired if depletion of CD8(+)T cells is induced, suggesting a role of these cells in clearance of the fungus from tissues, and demonstrating their prominent protective activity in the immune responses mounted by susceptible animals [110].

Cell wall components of *P. brasiliensis* have a fundamental role in the inflammatory process observed in paracoccidioidomycosis [111]. When a cell wall fraction (F1), composed mainly of β-glucan and chitin, was inoculated into mice, it provoked a granulomatous reaction, toxic and macrophage-stimulating activity, high levels of secretion of immunological mediators such as TNF-α and fragments of complement system, and also have a nonspecific modulatory effect on the production of antibodies against unrelated antigens. F1 also induced an accumulation of inflammatory (CD11b/C18)+ macrophages that related to immunological disturbances [112], and a strong proliferative response in CD4+ lymphocytes [113]. There is a clear-cut lymphocyte hyporesponsiveness to the cell wall antigen in patients before treatment (more profound in the acute form of the disease than in the chronic form), compared with healthy *P. brasiliensis*–sensitized individuals, and a clear restoration of the cell wall antigen responses after clinical cure [113]. All these effects are suggestive of a cell wall role as immunoregulator of the disorders observed in paracoccidioidomycosis. These characteristics may lie at the bottom of the successful use of β-1,3-glucan as an adjuvant in antifungal treatment against the disease (Fig. 2) [40]. F1 fractions of different *P. brasiliensis* strains activated the human complement system in different manners; namely, F1 from a low-virulence strain was more efficient than the fractions from more virulent strains in inhibiting both the classical and the alternative complement pathways [114]. The authors suggest alterations in the structure and/or composition of the different F1 fractions as responsible for the variations observed in the activities tested, mainly due to greater concentrations of β-glucan in F1 from the less virulent strain. These results agree with previous reports suggesting important modifications in the contents of α- and β-glucan in the cell walls of *P. brasiliensis* strains with different degrees of virulence [15,16].

In patients, the great majority of cells at the site of skin tests with paracoccidioidin were T-cells (48%) with a T-helper phenotype (CD45 RO-positive cells), 25% were macrophages (CD68-positive cells), and very few were B lymphocytes (CD20-positive cells), consistent with a DHT pattern [115]. Choosing to stimulate human cord blood mononuclear cells as a source for naive T cells, with lysates of *P. brasiliensis* yeast cells, Munk and Kaufmann [116] found that the fungal cell stimulated T-cell expansion, interleukin-2 production and differentiation into cytotoxic T cells, mainly through a high and a low molecular weight proteic fractions (100,000 and <1,000, respectively). They were able to do so, independent from previous exposure to cross-reactive antigens.

The main antigenic component of *P. brasiliensis*, gp43, presents T cell epitopes [117]. It stimulates B cells [118], is involved in *P. brasiliensis* pathogenicity [119], and is responsible for induction of delayed-type hypersensitivity in animals and humans [120]. Almeida et al. [121] compared the T cell populations stimulated in resistant and susceptible mice when gp43 is presented by macrophages or B cells. In resistant mice, purified gp43 seems to have been preferentially presented by macrophages and stimulated Th1 lymphokine production. In susceptible animals, gp43 was presented by B cells, which led to stronger activation of Th2 subsets. T cells from resistant mice responded as those from susceptible animals when stimulated by gp43 presented by macrophages or B cells from susceptible mice or viceversa, indicating that there are no significant differences in the T cell repertoires from resistant and susceptible mice. Apart from eliciting high antibody titers, gp43 is also immunodominant in DTH reactions in infected animals and humans. Its cellular immune response involves CD4+ Th-1 lymphocytes, secreting IFN-γ and interleukin 2 (IL-2) but not IL-4 and IL-10. The T-cell epitope of this antigen was mapped to a 15-amino acid peptide (P10) whose sequence QTL-IAIHTLAIR-YAN, contained the underlined hexamer sequence as the essential domain of the epitope [122]. Immunization of mice with either gp43 or P10 led to vigorous protection against intratracheal challenge by virulent *P. brasiliensis,* with a >200-fold decrease in lung CFU and halting of dissemination to the spleen and liver. This protection was attributed mainly to an IFN-γ-mediated cellular immune response. Unlike gp43, which induced an antibody response compatible with both Th-1 and Th-2 activation in infected BALB/c mice, P10 did not induce a humoral response [122].

Genetic immunization against paracoccidioidomycosis using a plasmid DNA containing the gp43 gene is an approach which is being tested [123]. The antibody response against gp43 increased after immunization of male BALB/c mice with the vector VR-gp43. The specificity of the response was demonstrated by immunoblotting, with IgG1 and IgG2a being the predominant isotype elicited, suggesting a type 2 cellular immune response to DNA immunization. Another gene product currently under study for immunoprotection is a genomic clone for HSP60, highly homologous to *H. capsulatum* and *C. immitis,* reported to contain two introns and codify for a 592 amino acid protein, with a molecular weight of 62 kDa and pI 5.44 [124]. A second approach in this line is that followed by Camargo and coworkers [125], who have addressed the question of the possible immunomodulatory properties of nonmethylated CpG sequences from fungal DNA, like that found in bacterial DNA. Using B10.A mouse strain, susceptible to *P. brasiliensis,* the authors observed that mice treated with DNA + fungus showed a significant decrease of anti-gp43 antibodies, in comparison with the control animals, the decrease probably due to a Th1 shift in the immune response to the main antigen, gp43.

Impairment of the cell-mediated immune response, a common finding among patients with the disseminated form of paracoccidioidomycosis may be associated with inhibiting effects of plasma [126]. In this study, the levels of *P. brasiliensis* antigens detected in the plasma of patients with paracoccidioidomycosis correlated with the suppression index detected by the low mitogenic response of peripheral blood mononuclear cells to phytohemaglutinin. This inhibitory effect on lymphoproliferation was observed in the plasma of 58% of patients. Although no plasma factor was identified, previous work points to polypeptides of molecular masses of 34 kDa [127] and 43 and 62 kDa [128].

IV. CONCLUDING REMARKS

In the last decade, research on paracoccidioidomycosis and its causal agent has seen an exponential growth in all fields. Particularly, the introduction of molecular methods to study epidemiology and immunology has provided new insights that look promising to further developments in the prevention, diagnosis, and treatment of the disease. From this point of view, special efforts should be given to research on immunization, a field in which some results are already appearing in the literature, either by using peptides fragmented from gp43, the main *P. brasiliensis* antigen, or plasmid DNA containing this antigen's gene, as well as other genes, or gene products. Experiments on this line will be complemented by those aimed at the understanding of the complex cellular and humoral immunity processes, to which much effort has been devoted in the last decade both in experimental animal models and in the human host.

Gp43, the best-studied antigen, both structurally and functionally, should be promoted in all clinical mycological laboratories of the region as the antigen of choice, substituting the old preparations of low specificity still in use in remote Latin American laboratories. In this respect, the design of a ready-to-use diagnostic kit, highly specific and sensitive, whose detection could be achievable with equipment of low sophistication, is highly desirable. Once comparable results are obtained, a better knowledge of the epidemiology of paracoccidioidomycosis should be achieved. This would also lead to early diagnosis and thus better chances for a successful treatment.

Deciphering the biology of *P. brasiliensis* and its morphological transition, a process pivotal for the understanding of pathogenicity and establishment of the disease, is another field in which research should be encouraged. The last decade has witnessed a remarkable development in this area, as molecular methods have provided the tools for exploring fields that were unattainable with the traditional biochemical methods used in the past. The synergy achieved by the joint use of molecular and biochemical approaches is providing knowledge on genes involved in the differential process of transition; on mechanisms of knowledge on genes involved in the differential process of transition; on mechanisms of cell wall construction and architecture and the most probable development of highly selective antifungal drugs; and on molecular taxonomy to identify the correct classification of *P. brasiliensis,* or the possible crypticity hidden in an apparent single species. This latter point is of paramount importance to understand the wide spectrum of pathological manifestations of paracoccidioidomycosis and the different behaviors of strains.

Surely, the years to come will see intense research focused in all these aspects that concern the most relevant systemic mycosis in Latin America.

ACKNOWLEDGMENTS

To Consejo Nacional de Investigaciones Científicas y Tecnológicas (CONICIT) for grant No. G-97000615.

REFERENCES

1. PJ Szaniszlo, CW Jacob, PA Geis. Dimorphism: morphological and biochemical aspects. In: DH Howard, ed. Fungi Pathogenic for Human and Animals, Part A, Biology. New York: Marcel Dekker, 1983, pp 323–436.
2. A Lutz. Uma micose pseudo-coccídica localizada na boca e observada no Brasil: contribução ao conhecimento das hiphoblastomicoses americanas. Brasil Med 22:121–124, 1908.
3. MF Franco. Host-parasite relationships in paracoccidioidomycosis. J Med Vet Mycol 25:5–18, 1987.
4. A Restrepo. Ecology of *Paracoccidioides brasiliensis.* In: M Franco, CS Lacaz, A Restrepo-Moreno, G Del Negro, eds. Paracoccidioidomycosis. Boca Raton, FL: CRC Press, 1994, pp 121–130.
5. G San-Blas, F San-Blas. Biochemistry of *Paracoccidioides brasiliensis* dimorphism. In: M Franco, CS Lacaz, A Restrepo-Moreno, G Del Negro, eds. Paracoccidioidomycosis. Boca Raton, FL: CRC Press, 1994, pp 49–66.
6. E Bagagli, A Sano, KI Coelho, S Alquati, M Miyaji, ZP Camargo, GM Gomes, M Franco, MR Montenegro. Isolation of *Paracoccidioides brasiliensis* from armadillos (*Dasypus novemcinctus*) captured in an endemic area of paracoccidioidomycosis. Am J Trop Med Hyg 58:505–512, 1998.
7. DA Stevens. The interface of mycology and endocrinology. J Med Vet Mycol 27:133–140, 1989.
8. LM Carbonell. Ultrastructure of dimorphic transformation in *Paracoccidioides brasiliensis.* J Bacteriol 100:1076–1082, 1969.
9. F Queiroz-Telles. *Paracoccidioides brasiliensis* ultrastructural findings. In: M Franco, CS Lacaz, A Restrepo-Moreno, G Del Negro, eds. Paracoccidioidomycosis. Boca Raton, FL: CRC Press, 1994, pp 27–47.

10. K Takeo, A Sano, K Nishimura, M Miyaji, M Franco, F Kanetsuna. Cytoplasmic and plasma membrane ultrastructure of *Paracoccidioides brasiliensis* yeast phase cells as revealed by freeze-etching. Mycol Res 94:1118–1122, 1990.

11. MC Leclerc, H Phillipe, E Gueho. Phylogeny of dermatophytes and dimorphic fungi based on large subunit ribosomal RNA sequence comparisons. J Med Vet Mycol 32:331–341, 1994.

12. JC Torres-Guzmán, B Xoconostle-Cazares, L Guevara-Olvera, L Ortiz, G San-Blas, A Domínguez, J Ruiz-Herrera. Comparison of fungal ornithine decarboxylases. Curr Microbiol 33:390–392, 1996.

13. G San-Blas, F San-Blas. Molecular aspects of fungal dimorphism. CRC Crit Rev Microbiol 11:101–127, 1984.

14. F Kanetsuna, LM Carbonell, I Azuma, Y Yamamura. Biochemical studies on thermal dimorphism of *Paracoccidioides brasiliensis*. J Bacteriol 110:208–218, 1972.

15. G San-Blas. The cell wall of fungal human pathogens: its possible role in host-parasite relationship. A review. Mycopathologia 79:159–184, 1982.

16. G San-Blas. *Paracoccidioides brasiliensis:* cell wall glucans, pathogenicity, and dimorphism. In: M McGinnis, ed. Current Topics in Medical Mycology. Vol. 1. New York: Springer-Verlag, 1986, pp 235–257.

17. KR Klimpel, WE Goldman. Cell walls from avirulent variants of *Histoplasma capsulatum* lack α-(1,3)-glucan. Infect Immun 56:2997–3000, 1988.

18. LH Hogan, BS Klein. Altered expression of surface α-(1,3)-glucan in genetically related strains of *Blastomyces dermatitidis* that differ in virulence. Infect Immun 62:3536–3542, 1994.

19. G San-Blas. Biosynthesis of glucans by subcellular fractions of *Paracoccidioides brasiliensis*. Exp Mycol 3:249–258, 1979.

20. F Sorais-Landáez, G San-Blas. Localization of β-glucan synthetase in membranes of *Paracoccidioides brasiliensis*. J Med Vet Mycol 31:421–426, 1993.

21. DR Quigley, CP Selitrennikoff. β-(1,3)-glucan synthetase activity of *Neurospora crassa:* stabilization and partial characterization. Exp Mycol 8:202–214, 1984.

22. G San-Blas, F San-Blas. Effect of nucleotides on glucan synthesis in *Paracoccidioides brasiliensis*. Sabouraudia 24:241–243, 1986.

23. PJ Szaniszlo, MS Kang, E Cabib. Stimulation of β-1,3-glucan synthetase of various fungi by nucleoside triphosphates. A generalized regulatory mechanism for cell wall biosynthesis. J Bacteriol 161:1188–1194, 1985.

24. M Pereira, MSS Felipe, MM Brigido, CMA Soares, MO Azevedo. Molecular cloning and characterization of a glucan synthase gene from the human pathogenic fungus *Paracoccidioides brasiliensis*. Yeast 16:451–462, 2000.

25. CA Munro, NAR Gow. Chitin biosynthesis as a target for antifungals. In: GK Dixon, LG Copping, DW Hollomon, eds. Antifungal Agents. Oxford: Bios Scientific Publishers, 1995, pp 161–171.

26. CE Bulawa. Genetics and molecular biology of chitin synthesis in fungi. Annu Rev Microbiol 47:505–534, 1993.

27. A Bowen, J Chen-Wu, M Momarny, R Young, P Szaniszlo, PW Robbins. Classification of fungal chitin synthetases. Proc Natl Acad Sci USA 89:519–523, 1992.

28. CA Specht, Y Liu, PW Robbins, CE Bulawa, N Jartchouk, KR Winter, PJ Riggle, JC Rhodes, CL Dodge, DW Culp, PT Borgia. The chsD and chsE genes of *Aspergillus nidulans* and their role in chitin synthesis. Fungal Genet Biol 20:153–167, 1996.

29. GA Niño-Vega, ET Buurman, GW Gooday, G San-Blas, NAR Gow. Molecular cloning and sequencing of a chitin synthetase gene (CHS2) of *Paracoccidioides brasiliensis*. Yeast 14:181–187, 1998.

30. GA Niño-Vega, CA Munro, G San-Blas, GW Gooday, NAR Gow. Differential expression of chitin synthase genes during temperature-induced dimorphic transition in *Paracoccidioides brasiliensis*. Med Mycol 38:31–39, 2000.

31. A Miyazaki, T Ootaki. Multiple genes for chitin synthetase in the zygomycete fungus *Phycomyces blakesleeanus*. J Gen Appl Microbiol 43:333–340, 1997.

32. W-J Choi, B Santos, A Duran, E Cabib. Are yeast chitin synthetases regulated at the transcriptional or the posttranslational level? Mol Cell Biol 14:7685–7694, 1994.

33. CA Munro, DA Schofield, GW Gooday, NAR Gow. Regulation of chitin synthesis during dimorphic growth of *Candida albicans*. Microbiology 144:391–401, 1998.

34. M Fujiwara, H Horiuchi, A Ohta, M Takagi. A novel fungal gene encoding chitin synthetase with a myosin motor-like domain. Biochem Biophys Res Commun 236:75–78, 1997.

35. E Mellado, A Aufauvre-Brown, NAR Gow, DW Holden. The *Aspergillus fumigatus* chsC and chsG genes encode class III chitin synthetases with different functions. Mol Microbiol 20:667–679, 1996.

36. JR Graybill, SG Revankar, TF Patterson. Antifungal agents and antifungal susceptibility testing. In: L Ajello, RJ Hay, eds. Medical Mycology. Vol. 4. Topley and Wilson's Microbiology and Microbial Infections. London: Arnold, 1998, pp 163–175.

37. G San-Blas, AM Calcagno, F San-Blas. A preliminary study of in vitro antibiotic activity of saperconazole and other azoles on *Paracoccidioides brasiliensis*. J Med Vet Mycol 31:169–174, 1993.

38. B Wanke, AT Londero. *Paracoccidioides brasiliensis*. In: L Ajello, RJ Hay, eds. Medical Mycology. Vol. 4. Topley and Wilson's Microbiology and Microbial Infections. London: Arnold, 1998, pp 395–407.

39. R Dietze, VG Fowler, TS Steiner, PM Pecanha, GR Corey. Failure of amphotericin B colloidal dispersion in the treatment of paracoccidioidomycosis. Am J Trop Med Hyg 60:837–839, 1999.

40. DA Meira, PCM Pereira, J Marcondes-Machado, RP Mendes, B Barraviera, J Pellegrino, MT Rezkallah-Iwasso, MTS Peraçoli, LM Castilho, I Tomazini, CL Silva, N Tiraboschi, PR Curi. The use of glucan as immunostimulant in the treatment of paracoccidioidomycosis. Am J Trop Med Hyg 55: 496–503, 1996.

41. E Ledezma, JC López, P Marin, H Romero, G Ferrara, L De Sousa, A Jorquera, R Apitz Castro. Ajoene in the topical short-term treatment of tinea cruris and tinea corporis in humans. Randomized comparative study with terbinafine. Arzneimittelforschung 49:544–547, 1999.

42. E Block, S Ahmad, JL Catafalmo, MK Jain, R Apitz-Castro. Antithrombic organosulfur compounds from garlic: structural, mechanistic, and synthetic studies. J Am Chem Soc 108:7045–7055, 1986.

43. G San-Blas, F San-Blas, F Gil, L Mariño, R Apitz-Castro. Inhibition of growth of the dimorphic fungus *Paracoccidioides brasiliensis* by ajoene. Antimicrob Agents Chemother 33:1641–1644, 1989.

44. G San-Blas, L Mariño, F San-Blas, R Apitz-Castro. Effect of ajoene on dimorphism of *Paracoccidioides brasiliensis*. J Med Vet Mycol 31:133–141, 1993.

45. G San-Blas, JA Urbina, E Marchán, LM Contreras, F Sorais, F San-Blas. Inhibition of *Paracoccidioides brasiliensis* by ajoene is associated with blockade of phosphatidylcholine biosynthesis. Microbiology 143:1583–1586, 1997.

46. VLG Calich, LM Singer-Vermes, AM Siqueira, E Burger. Susceptibility and resistance of inbred mice to *Paracoccidioides brasiliensis* strains. Br J Exp Pathol 66:585–594, 1985.

47. JG McEwen, V Bedoya, M Patiño, ME Salazar, A Restrepo. Experimental murine paracoccidioidomycosis induced by the inhalation of conidia. J Med Vet Mycol 25:165–175, 1987.

48. J Defaveri, MF Rezkallah-Iwaso, MF Franco. Experimental pulmonary paracoccidioidomycosis in mice: morphology and correlation of lesions with humoral and cellular immune response. Mycopathologia 77:3–11, 1982.

49. MA Robledo, JR Graybill, J Ahrens, A Restrepo, DJ Drutz, M Robledo. Host defense against experimental paracoccidioidomycosis. Am Rev Respir Dis 125:563–567, 1982.

50. A Sano, M Miyaji, N Nishimura. Studies on the relationship between the estrous cycle of BALB/c mice and their resistance to *Paracoccidioides brasiliensis* infection. Mycopathologia 119:141–145, 1992.

51. SS Kashino, VL Calich, E Burger, LM Singer-Vermes. Alterations in the pathogenicity of one *Paracoccidioides brasiliensis* isolate do not correlate with its in vitro growth. Mycopathologia 111: 173–180, 1990.

52. S Paris, S Durán, F Mariat. Nutritional studies on *Paracoccidioides brasiliensis:* the role of organic sulfur in dimorphism. J Med Vet Mycol 23:85–92, 1985.

53. F San-Blas, S Centeno. Isolation and preliminary characterization of auxotrophic and morphological mutants of the yeastlike form of *Paracoccidioides brasiliensis*. J Bacteriol 129:138–144, 1977.

54. AM Calcagno, G Niño-Vega, F San-Blas, G San-Blas. Geographic discrimination of *Paracoccidioides brasiliensis* strains by randomly amplified polymorphic DNA analysis. J Clin Microbiol 36: 1733–1736, 1998.

54a. GA Niño-Vega, AM Calcagno, G San-Blas, F San-Blas, GW Gooday, NAR Gow. RFLP analysis reveals marked geographical isolation between strains of *Paracoccidioides brasiliensis*. Med Micol 38:437–441, 2000.

55. EEWI Molinari-Madlum, MSS Felipe, CMA Soares. Virulence of *Paracoccidioides brasiliensis* isolates can be correlated to groups defined by random amplified polymorphic DNA analysis. Med Mycol 37:269–276, 1999.

56. A Sano, J Defaveri, R Tanaka, K Yokoyama, N Kurita, M Franco, KIR Coelho, E Bagagli, MR Montenegro, M Miyaji, K Nishimura. Pathogenicities and gp43 kDa gene of three *Paracoccidioides brasiliensis* isolates originated from a nine-banded armadillo (*Dasypus novemcinctus*). Mycopathologia 144:61–65, 1999.

57. FV Morais, TF Barros, MK Fukada, PS Cisalpino, R Puccia. Polymorphism in the gene coding for the immunodominant antigen gp43 from the pathogenic fungus *Paracoccidioides brasiliensis*. J Clin Microbiol 38:3960–3966, 2000.

58. G San-Blas, F San-Blas. Antigenic structure of *Paracoccidioides brasiliensis*. In: Kurstak E, ed. Immunology of Fungal Diseases. New York: Marcel Dekker, 1989. pp 171–192.

59. ZP Camargo, C Unterkircher, LR Travassos. Identification of antigenic polypeptides of *Paracoccidioides brasiliensis* by immunoblotting. J Med Vet Mycol 27:407–412, 1989.

60. R Puccia, DT Takaoka, LR Travassos. Purification of the 43 kDa glycoprotein from exocellular components excreted by *Paracoccidioides brasiliensis* in liquid culture (TOM medium). J Med Vet Mycol 29:57–60, 1991.

61. IC Almeida, DCA Neville, A Mehlert, A Treumann, MAJ Ferguson, JO Previato, LR Travassos. Structure of the N-linked oligosaccharide of the main diagnostic antigen of the pathogenic fungus *Paracoccidioides brasiliensis*. Glycobiology 6:507–515, 1996.

62. G San-Blas, F San-Blas. Immune and chemical responses to paracoccidioidomycosis. In: Jacobs PH, Nall L, eds. Fungal Disease, Biology, Immunology, and Diagnosis. New York: Marcel Dekker, 1997, pp 219–235.

63. R Puccia, LR Travassos. 43-Kilodalton glycoprotein from *Paracoccidioides brasiliensis*: immunological reactions with sera from patients with paracoccidioidomycosis, histoplasmosis, and Jorge Lobo disease. J Clin Microbiol 29:1610–1615, 1991.

64. PS Cisalpino, R Puccia, LM Yamauchi, MIN Cano, JF Silveira, LR Travassos. Cloning, characterization, and epitope expression of the major diagnostic antigen of *Paracoccidioides brasiliensis*. J Biol Chem 271:4553–4560, 1996.

65. GM Gomes, PS Cisalpino, CP Taborda, ZP Camargo. PCR for diagnosis of paracoccidioido mycosis. J Clin Microbiol 38:3478–3480, 2000.

66. AH Straus, E Freymüller, LR Travassos, HK Takahashi. Immunochemical and subcellular localization of the 43 kDa glycoprotein antigen of *Paracoccidioides brasiliensis* with monoclonal antibodies. J Med Vet Mycol 34:181–186, 1996.

67. AP Vicentini, JZ Moraes, JL Gesztesi, MF Franco, W Souza. Laminin-binding epitope on gp43 from *Paracoccidioides brasiliensis* is recognized by a monoclonal antibody raised against *Staphylococcus aureus* laminin receptor. J Med Vet Mycol 35:37–43, 1996.

68. JL Gesztesi, R Puccia, LR Travassos, AP Vicentini, JZ Moraes, MF Franco, JD Lopes. Monoclonal antibodies against glycoprotein gp43 from *Paracoccidioides brasiliensis* modulate laminin-mediated fungal adhesion to epithelial cells and pathogenesis. Hybridoma 15:415–22, 1996.

69. GMB Del Negro, G Benard, CM Assis, MSM Vidal, NM Garcia, C Otani, MA Shikanai-Yasuda, CS Lacaz. Lack of reactivity of paracoccidioidomycosis sera in the double immunodiffusion test with the gp43 antigen: report of two cases. J Med Vet Mycol 33:113–116, 1995.

70. ZP Camargo, RG Baruzzi, SM Maeda, MC Floriano. Antigenic relationship between *Loboa loboi* and *Paracoccidioides brasiliensis* as shown by serological methods. J Med Vet Mycol 36:413–417, 1998.

71. MCR Moura-Campos, JL Gesztesi, AP Vincentini, JD Lopes, ZP Camargo. Expression and isoforms of gp43 in different strains of *Paracoccidioides brasiliensis*. J Med Vet Mycol 33:223–227, 1995.

72. MC Souza, JL Gesztesi, AR Souza, JZ Moraes, JD Lopes, ZP Camargo. Differences in reactivity of paracoccidioidomycosis sera with gp43 isoforms. J Med Vet Mycol 35:13–18, 1997.

73. AR Souza, JL Gesztesi, JZ Moraes, CRB Cruz, J Sato, M Mariano, JD Lopes. Evidence of idiotypic modulation in the immune response to gp43, the major antigenic component of *Paracoccidioides brasiliensis* in both mice and humans. Clin Exp Immunol 114:40–48, 1998.

74. R Mattar-Filho, MO Azevedo, M Pereira, RSA Jesuino, SM Salem-Izacc, WA Brito, JL Gesztesi, RBA Soares, MSS Felipe, CMA Soares. Expression of glycoprotein gp43 in stage-specific forms and during dimorphic differentiation of *Paracoccidioides brasiliensis*. J Med Vet Mycol 35: 341–345, 1997.

75. CP Taborda, ZP Camargo. Diagnosis of paracoccidioidomycosis by dot immunobinding assay for antibody detection using the purified and specific antigen gp43. J Clin Microbiol 32:554–556, 1994.

76. ZP Camargo, JL Gesztesi, ECO Saraiva, CP Taborda, AP Vicentini, JD Lopes. Monoclonal antibody capture enzyme immunoassay for detection of *Paracoccidioides brasiliensis* antibodies in paracoccidioidomycosis. J Clin Microbiol 32:2377–2381, 1994.

77. MA Salina, MA Shikanai-Yashuda, RP Mendes, B Barraviera, MJS Mendes-Giannini. Detection of circulating *Paracoccidioides brasiliensis* antigen in urine of paracoccidioidomycosis patients before and during treatment. J Clin Microbiol 36:1723–1728, 1998.

78. BL Gómez, JI Figueroa, AJ Hamilton, B Ortiz, MA Robledo, RJ Hay, A Restrepo. Use of monoclonal antibodies in diagnosis of paracoccidioidomycosis: new strategies for detection of circulating antigens. J Clin Microbiol 35:3278–3283, 1997.

79. BL Gómez, JI Figueroa, AJ Hamilton, S Diez, M Rojas, AM Tobón, RJ Hay, A Restrepo. Antigenemia in patients with paracoccidioidomycosis: detection of the 87-kilodalton determinant during and after antifungal therapy. J Clin Microbiol 36:3309–3316, 1998.

80. BL Gómez, AJ Hamilton, JI Figueroa, S Diez, M Rojas, AM Tobón, RJ Hay, A Restrepo. Antigenemia in paracoccidioidomycosis: its use in diagnosis and therapy follow up. Abstracts of the VII International Meeting on Paracoccidioidomycosis, Campos do Jordão, São Paulo, Brazil, 1999, p 41.

81. JI Figueroa, AJ Hamilton, MH Allen, R Hay. Isolation and partial characterization of a *Paracoccidioides brasiliensis* 58 kDa extracellular glycoprotein which is recognized by human immune sera. Trans R Soc Trop Med Hyg 89:566–572, 1995.

82. JI Figueroa, A Hamilton, M Allen, R Hay. Immuno-histochemical detection of a novel 22- to 25-kilodalton glycoprotein of *Paracoccidioides brasiliensis* in biopsy material and partial characterization by using species-specific monoclonal antibodies. J Clin Microbiol 32:1566–1574, 1994.

83. JG McEwen, BL Ortiz, AM García, AM Florez, S Botero, A Restrepo. Molecular cloning, nucleotide sequencing and characterization of a 27 kDa antigenic protein from *Paracoccidioides brasiliensis*. Fungal Genet Biol 20:125–131, 1996.

84. BL Ortiz, AM García, A Restrepo, JG McEwen. Immunological characterization of a recombinant 27-kilodalton antigenic protein from *Paracoccidioides brasiliensis*. Clin Diag Lab Immunol 3: 239–241, 1996.

85. AJ Hamilton. Serodiagnosis of histoplasmosis, paracoccidioidomycosis and penicilliosis marneffei; current status and future trends. Med Mycol 36:351–364, 1998.

86. CMA Soares, SM Salem-Izacc, RSA Jesuino, MSS Felipe, FJ Gomez, GS Deepe. Isolation and characterization of a cDNA coding for an antigenic 40 kDa protein from the fungus *Paracoccidioides brasiliensis*. Abstracts of the VII International Meeting on Paracoccidioidomycosis, Campos do Jordão, São Paulo, Brazil, 1999, p 78.

87. RBA Soares, MSS Felipe, CJ Ulhoa, CMA Soares. Purificação e caracterização de N-acetil-β-D-glicosaminidase do fungo patogênico humano *Paracoccidioides brasiliensis*. Abstracts of the VII International Meeting on Paracoccidioidomycosis, Campos do Jordão, São Paulo, Brazil, 1999, p 197.

88. MS Toledo, E Suzuki, AH Straus, HK Takahashi. Glycolipids from *Paracoccidioides brasiliensis*. Isolation of a galactofuranose-containing glycolipid reactive with sera of patients with paracoccidioidomycosis. J Med Vet Mycol 33:247–251, 1995.

89. SB Levery, MS Toledo, AH Straus, HK Takahashi. Structure elucidation of sphingolipids from the mycopathogen *Paracoccidioides brasiliensis*: an immunodominant β-galactofuranose residue is carried by a novel glycosylinositol phosphorylceramide antigen. Biochemistry 37:8764–8775, 1998.

90. M Arango, L Yarzábal. T-cell dysfunction and hyperimmunoglobulinemia E in paracoccidioido-
    mycosis. Mycopathologia 79:115–123, 1982.
91. LM Singer-Vermes, E Burger, MF Franco, M Moscardi-Bacchi, MJS Mendes-Giannini, VLG Calich.
    Evaluation of the pathogenicity and immunogenicity of seven *Paracoccidioides brasiliensis* isolates
    in susceptible inbred mice. J Med Vet Mycol 27:71–82, 1989.
92. SL Oliveira, MF Silva, AM Soares, CL Silva. Cell wall fractions from *Paracoccidioides brasiliensis*
    induce hypergammaglobulinemia. Mycopathologia 121:1–5, 1993.
93. CAC Vaz, LM Singer-Vermes, VLG Calich. Comparative studies on the antibody repertoire pro-
    duced by susceptible and resistant mice to virulent and nonvirulent *Paracoccidioides brasiliensis*
    isolates. Am J Trop Med Hyg 59:971–977, 1998.
94. MJ Mendes-Giannini, JP Bueno, MA Shikanai-Yashuda, AMS Stolf, A Masuda, VA Neto, AW
    Ferreira. Antibody response to the 43 kDa glycoprotein of *Paracoccidioides brasiliensis* as a marker
    for the evaluation of patients under treatment. Am J Trop Med Hyg 43:200–206, 1990.
95. JP Bueno, MJS Mendes-Giannini, GMB Del Negro, CM Assis, CK Takiguti, MA Shikanai-Yashuda.
    IgG, IgM and IgA antibody response for the diagnosis and follow-up of paracoccidioidomycosis:
    conmparison of counterimmunoelectrophoresis and complement fixation. J Med Vet Mycol 35:
    213–217, 1997.
96. H Baida, PJ Biselli, M Juvenale, GM Del Negro, MJ Mendes-Giannini, AJ Duarte, G Benard.
    Differential antibody isotype expression to the major *Paracoccidioides brasiliensis* antigen in juve-
    nile and adult form of paracoccidioidomycosis. Microbes Infect 1:273–278, 1999.
97. EN Itano, RA Margni, ZP Camargo, MS Kaminami, NJ Carlos, AS Machado. Problemas dos
    anticorpos assimétricos no diagnóstico da paracoccidioidomicose. Abstracts of the VII International
    Meeting on Paracoccidioidomycosis, Campos do Jordão, São Paulo, Brazil, 1999, p 88.
98. MH Botta, ZP Camargo. Immunological response to cell-free antigens of *Paracoccidioides bra-
    siliensis:* relationship with clinical forms of paracoccidioidomycosis. J Clin Microbiol 31:671–676,
    1993.
99. N Kurita, SK Biswas, M Oarada, A Sano, K Nishimura, M Miyaji. Fungistatic and fungicidal
    activities of murine polymorphonuclear leucocytes against yeast cells of *Paracoccidioides bra-
    siliensis*. Med Mycol 37:19–24, 1999.
100. N Kurita, M Oarada, E Ito, M Miyaji. Antifungal activity of human polymorphonuclear leucocytes
    against yeast cells of *Paracoccidioides brasiliensis*. Med Mycol 37:261–267, 1999.
101. M Goihman-Yahr, J Pereira, G Istúriz, N Viloria, M Carrasquero, N Saavedra, MH Gómez, A
    Román, B San-Martin, MCB Albornoz, B Fernández, E Avila-Millán. Relationship between diges-
    tive and killing abilities of neutrophils against *Paracoccidioides brasiliensis*. Mycoses 35:269–274,
    1992.
102. AL Bocca, MF Silva, CL Silva, FQ Cunha, F Figueiredo. Macrophage expression of class II major
    histocompatibility complex gene products in *Paracoccidioides brasiliensis*–infected mice. Am J
    Trop Med Hyg 61:280–287, 1999.
103. CL Silva, MF Silva. Participation of CD11B$^+$ and CD23$^+$ macrophages in the inflammatory reaction
    induced by *P. brasiliensis*. Abstracts of the VII International Meeting on Paracoccidioidomycosis,
    Campos do Jordão, São Paulo, Brazil, 1999, p 53.
104. SR Almeida, JZ Moraes, ZP Camargo, JL Gesztesi, M Mariano, JD Lopes. Pattern of immune
    response to gp43 from *Paracoccidioides brasiliensis* in susceptible and resistant mice is influenced
    by antigen-presenting cells. Cell Immunol 190:68–76, 1998.
105. VLG Calich, CAC Vaz, E Burger. Immunity to *Paracoccidioides brasiliensis* infection. Res Immunol
    149:407–417, 1998.
106. ME Munk, RA Fazioli, VLG Calich, SHE Kaufmann. *Paracoccidioides brasiliensis*–stimulated
    human γ/δ T cells support antibody production by B cells. Infect Immun 63:1608–1610, 1995.
107. LE Cano, LM Singer-Vermes, CAC Vaz, M Russo, VLG Calich. Pulmonary paracoccidioidomycosis
    in resistant and susceptible mice: relationship among progression of infection, bronchoalveolar cell
    activation, cellular immune response, and specific isotype patterns. Infect Immun 63:1777–1783,
    1995.

108. LE Cano, SS Kashino, C Arruda, D André, CF Xidieh, LM Singer-Vermes, CAC Vaz, E Burger, VLG Calich. Protective role of gamma interferon in experimental pulmonary paracoccidioido-mycosis. Infect Immun 66:800–806, 1998.

109. VLG Calich, C Arruda. Natural and acquired immunity in the pulmonary model of paracoccidioido-mycosis. Abstracts of the VII International Meeting on Paracoccidioidomycosis, Campos do Jordão, São Paulo, Brazil, 1999, p 51.

110. LE Cano, LM Singer-Vermes, TA Costa, JO Mengel, CF Xidieh, C Arruda, DC Andre, CA Vaz, E Burger, VL Calich. Depletion of CD8(+)T cells in vivo impairs host defense of mice resistant and susceptible to pulmonary paracoccidioidomycosis. Infect Immun 68:352–359, 2000.

111. MF Silva, CL Silva. The role of somatic structures of the fungus *Paracoccidioides brasiliensis* upon B cell activation in experimental paracoccidioidomycosis. Clin Exp Immunol 101:321–327, 1995.

112. MF Silva, AL Bocca, R Ferracini, F Figueiredo, CL Silva. Cellular requirements for immunomodula-tory effects caused by cell wall components of *Paracoccidioides brasiliensis* on antibody production. Clin Exp Immunol 109:261–271, 1997.

113. G Benard, MA Hong, GMB Del Negro, L Batista, MA Shikanai-Yashuda, AJS Duarte. Antigen-specific immunosuppression in paracoccidioidomycosis. Am J Trop Med Hyg 54:7–12, 1996.

114. LSP Crotts, YM Lucisano-Valim, CL Silva, JE Barbosa. Interactions of F1 fractions from different strains of *Paracoccidioides brasiliensis* with human complement and with human neutrophils. Myco-pathologia 140:19–27, 1997.

115. M Marques, M Moscardi-Bacchi, S Marques, M Franco. Immunohistochemical characterization of mononuclear cells in delayed hypersensitivity reactions to *Paracoccidioides brasiliensis* (paracoc-cidioidin test). Mycopathologia 124:7–11, 1993.

116. ME Munk, SHE Kaufmann. Human cord blood T-cell receptor αβ cell responses to protein antigens of *Paracoccidioides brasiliensis* yeast forms. Immunology 84:98–104, 1995.

117. LR Travassos, R Puccia, P Cisalpino, C Taborda, EG Rodrigues, M Rodrigues, JF Silveira, IC Almeida. Biochemistry and molecular biology of the main diagnostic antigen of *Paracoccidioides brasiliensis*. Arch Med Res 26:297–304, 1995.

118. ZP Camargo, C Unterkircher, SP Campoy, LR Travassos. Production of *Paracoccidioides bra-siliensis* exoantigens for immunodiffusion tests. J Clin Microbiol 26:2147–2151, 1988.

119. AP Vicentini, JL Gesztesi, MF Franco, W Souza, JZ Moraes, LR Travassos, JD Lopes. Binding of *Paracoccidioides brasiliensis* to laminin through surface glycoprotein gp43 leads to enhancement of fungal pathogenesis. Infect Immun 62:1465–1469, 1994.

120. EG Rodrigues, LR Travassos. Nature of the reactive epitopes in *Paracoccidioides brasiliensis* poly-saccharide antigen. J Med Vet Mycol 32:77–81, 1994.

121. SR Almeida, CS Unterkircher, ZP Camargo. Involvement of the major glycoprotein (gp43) of *Paracoccidioides brasiliensis* in attachment to macrophages. Med Mycol 36:405–411, 1998.

122. CP Taborda, MA Juliano, R Puccia, M Franco, LR Travassos. Mapping of the T-cell epitope in the major 43-kilodalton glycoprotein of *Paracoccidioides brasiliensis* which induces a Th-1 response protective against fungal infection in BALB/c mice. Infect Immun 66:786–793, 1998.

123. AR Pinto, R Puccia, SN Diniz, MF Franco, LR Travassos. DNA-based vaccination against murine paracoccidioidomycosis using the gp43 gene from *Paracoccidioides brasiliensis*. Vaccine 18: 3050–3058, 2000.

124. SM Salem-Izacc, GS Deepe, RSA Jesuino, MSS Felipe, FJ Gomez, CMA Soares. Cloning and characterization of HSP60 from *Paracoccidioides brasiliensis*. Abstracts of the VII International Meeting on Paracoccidioidomycosis, Campos do Jordão, São Paulo, Brazil, 1999, p 76.

125. MC Souza, M Corrêa, JD Lopes, ZP Camargo. Study of the immune response at immunization with DNA from *Paracoccidioides brasiliensis* in susceptible mice. Abstracts of the VII International Meeting on Paracoccidioidomycosis, Campos do Jordão, São Paulo, Brazil, 1999, p 161.

126. MF Sugizaki, MTS Peraçoli, MJ Mendes-Giannini, AMVC Soares, CS Jurokawa, RP Mendes, SA Marques, DV Freire-Maia. Correlation between antigenemia of *Paracoccidioides brasiliensis* and inhibiting effects of plasma in patients with paracoccidioidomycosis. Med Mycol 37:277–284, 1999.

127. D Chequer-Bou-Habib, MF Ferreira-da-Cruz, B Galvão-Castro. Immunosuppressive effect of para-coccidioidomycosis sera on the proliferative response of normal mononuclear cells. Mycopathologia 119:65–71, 1992.
128. CS Unterkircher, SC Yazaki, MT Shimizu, AO Jorge, ZP Camargo. Specific components found in circulating immune complexes (CIC) in paracoccidioidomycosis. J Med Vet Mycol 34:273–277, 1996.

11

Pneumocystis carinii

Henry Koziel
Beth Israel Deaconess Medical Center and Harvard Medical School, Boston, Massachusetts

I. INTRODUCTION

Pneumocystis carinii is a unique opportunistic pathogen which primarily causes pneumonia in persons with congenital or acquired immunodeficiency. Historically considered a protozoan, recent studies suggest that *P. carinii* may represent a type of fungus, although classification remains controversial. While identified as an organism in the early 1900s, *P. carinii* was first associated with human disease in the 1940s. By the 1960s, this pathogen was increasingly recognized as the cause of pneumonia in patients receiving chemotherapeutic and other immuno-suppressive agents as these drugs were more widely prescribed. *P. carinii* pneumonia heralded the onset of the AIDS epidemic in 1981, and has since assumed a prominent role as an important agent of opportunistic pulmonary disease, particularly in HIV-infected persons. However, al-though recognized as a cause of opportunistic pneumonia for >50 years, our understanding of *P. carinii* remains somewhat rudimentary.

Virulence refers to the capacity of a microbe to produce disease in a host, and is greatly dependent on both microbial factors and host factors. In the case of *P. carinii*, several important observations exemplify possible virulence factors. Exposure to this organism occurs early in life, and may be frequent and repeated throughout life, but clinically apparent disease such as pneumonia remains rare in the absence of congenital or acquired immunodeficiency. The nonmo-tile *P. carinii* is likely transmitted via an airborne route, and once inhaled, disease is generally limited to the lungs of the susceptible host. Pneumonia is established as a consequence of closely apposed parasitic interaction of *P. carinii* with the alveolar epithelial cells, an apparent requirement for disease development. The paucity of immune cells observed in the airways of animals with *P. carinii* pneumonia suggests that this extracellular organism is perhaps capable of evading host immune cell recognition. These important observations may provide vital clues into the pathogenesis of *P. carinii* pneumonia, and suggest possible microbial virulence factors which may include mechanisms of overcoming host defenses, procuring nutrients or essential components from the host cells, and evading immune recognition to allow survival and spread of the organisms from host to susceptible host. Numerous investigators have recently attempted to define and characterize the factors necessary for establishing *P. carinii* pneumonia.

Several recent excellent publications are available for more detailed and comprehensive review of *P. carinii* and *P. carinii* pneumonitis [1–3]. This chapter will provide a summary of major aspects of *P. carinii*, highlighting important (and often controversial) aspects of this opportunistic organism and the pathogenesis of disease, with emphasis on *P. carinii* as a fungus.

The focus of this chapter will be to provide an overview of both recognized or proposed microbe-related factors which may contribute to virulence, and host-related factors which may contribute to the pathogenesis of *P. carinii* pneumonia, highlighting some of the important recent investigations which have greatly advanced our understanding of this fascinating pathogen.

II. BIOLOGY OF *P. CARINII*

For investigators of *P. carinii*, one major obstacle challenging research of this opportunistic pathogen is the lack of a reliable system for *P. carinii* propagation in continuous ex vivo culture. Although some investigators have propagated *P. carinii* in short-term culture [4,5], successfully generating organisms which maintain infectivity [6], the development of a culturing system which allows propagation of large quantities of enriched organisms remains elusive. This one fact greatly limits our ability to study this opportunist, where much of what we know about this organism relies on cumbersome animal models, limited in vitro investigations, and observations from human cases. However, despite this significant limitation, much new and exciting information has been recently gained through innovative investigations employing genetically altered animal models and exploiting powerful techniques in cell and molecular biology.

A. Structure

P. carinii is a unicellular eukaryote. Based primarily on light microscopic appearance and tinctural analysis, investigators have traditionally described two different forms of *P. carinii*: (1) the trophozoite, or trophic form; and (2) the cyst form. These descriptive terminology are based on the historical notion that *P. carinii* represents a protozoon. However, as the taxonomic classification is challenged, alternative terminology more appropriate for yeastlike fungal life stages may be proposed and adopted.

The trophic form represents the smaller life form, ranging in size from 1 to 5 μm. The trophic forms are pleomorphic and amoeboid in shape, and generally represent the majority of organism forms identified in the lungs of animals with *P. carinii* pneumonia. They frequently exist in large clusteres [7]. Investigators have further described small trophic forms (1–2 μm) which may represent haploid progeny perhaps from asexual binary fission or spore release by cyst forms, and large trophic forms (3–5 μm) which may represent diploid cells. By electron microscopy, the trophic form cell membrane is bound by a thin (20–50 nm), electron-dense cell wall which is likely rich in polysaccharide antigens that represent sites for lectin binding. The cell membrane appears as a characteristic trilaminar unit membrane containing prominent intramembranous particles [2] and cholesterol.

The cytoplasm of the trophic form is characterized by few recognizable organelles. These organelles include a single mitochondrion (with lammellar cristae), rough endoplasmic reticulum, free ribosomes, a Golgi-like apparatus, vacuoles, lipid droplets, glycogen granules, lysosomes, and dense "round bodies" of uncertain significance [8–13]. The trophic form contains a single nucleus surrounded by a double-layer nuclear envelope, with an electron-dense nucleolus [14]. The apparent paucity of organelles may represent a true characteristic of *P. carinii*, or may represent an artifact related to sample preparation and fixation techniques [2,11].

The cyst form is the largest and most readily recognized form of the organism, although representing a small percentage of organisms in the lungs of animals with *P. carinii* pneumonia. The smooth spherical cyst form measures 5–8 μm in diameter and is surrounded by a 70- to 160-nm trilaminar cell wall: a 20- to 50-nm outer unit membrane covered by an electron-dense layer, a thin electron-lucent layer, and an inner trilaminar unit membrane characteristic of a plasmalemma [2].

The cytoplasm of the cyst form contains mitochondria, rough endoplasmic reticulum membrane, free ribosomes, glycogen granule, and characteristic intracystic bodies (or spores). The intracystic bodies are 1- to 2-μm spherical structures surrounded by a double unit membrane. A cyst form may contain up to eight intracystic bodies, and the bodies contain a single nucleus, a single mitochondrion, abundant endoplasmic reticulum, ribosomes, microtubules, vacuoles, and glycogen granules. The intracystic bodies may be attached to each other and the cyst wall by a stalklike structure. These intracystic bodies are also referred to as "daughter forms" and may represent precursors to tropic forms upon release by the cyst form.

Other forms of *P. carinii* have been described, including the precyst, or intermediate form, early precyst, later precyst, thin-walled cysts, empty or "ghost" cysts, and crescent or "helmet"-shaped forms [15]. However, the relationship of these various forms to one another and to virulence and disease pathogenesis remains uncertain and controversial.

B. Life Cycle

In the absence of a reliable long-term in vitro culture system for *P. carinii*, the life cycle of the organism remains poorly understood. The available information on the life cycle is primarily based on static histochemical and ultrastructural analysis of different forms (and presumably different life stages) of the organism in the lungs of the rodent model or the lungs of humans infected with *P. carinii*, or observations in short-term culture of organisms.

A number of different life cycles for *P. carinii* have been proposed most of which include both an asexual and sexual phase of development (Fig. 1) [16]. Current concepts suggest that the trophic forms develop into cyst forms, undergoing meiotic division and subsequent mitosis as the organisms differentiate from precyst to mature cyst, with the development of intracystic spores or "daughter cells." The "mature" cyst forms subsequently rupture to release the eight spores, each of which is thought to develop into trophic forms (with the cycle then repeating) [13]. Furthermore, trophic forms may be capable of replication by an asexual process such as binary fission. However, the proposed life cycles have yet to be firmly established, and the exact relationship of the trophic and cyst forms to the other described forms of *P. carinii* remain uncertain. Also, whether the proposed life cycles occur in the environment or only in the host has not been determined, and which form(s) exist in the environmental reservoir has not been established. Furthermore, which form represents the "latent" stage of the organism, and what factor(s) trigger or mediate stage development and what mechanism or trigger promotes sporozoite (or spore) release from the cyst forms remains to be determined [17].

C. Taxonomy

Historically, *P. carinii* was classified as a protozoan although increasingly newer evidence suggests it to be a fungus, perhaps related to the Ascomycetes. Table 1 summarizes some of the significant findings favoring classification of *P. carinii* as a protozoan or a fungus. *P. carinii* was first identified as a type of trypanosome by Chagas [18] but was later reclassified as a different organism by Delanoe and Delanoe [19,20]. *P. carinii* was traditionally considered a protozoan on the basis of the appearance of structures common to protozoa, the amoeboid appearance of the trophic form, structural similarities of the mitochondria, analogous cell wall structures, and susceptibility to antiprotozoan agents such as pentamidine and trimethoprim-sulfamethoxazole [8,10,21,22].

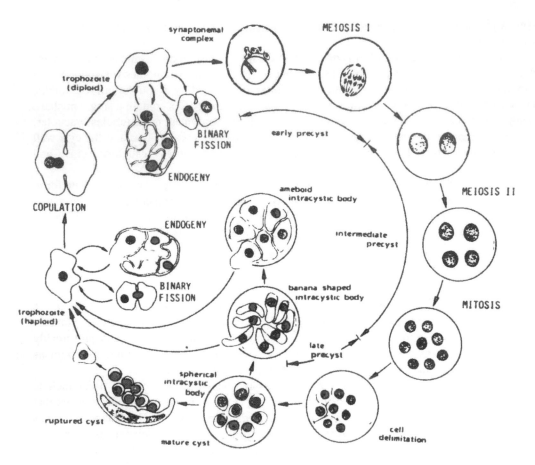

Figure 1   A proposed life cycle of *P. carinii*.

More recent evidence supports classification of *P. carinii* as a fungus. Most compelling is the molecular biological sequence analysis of *P. carinii* genes, including the 18s ribosomal RNA [23,24] and mitochondrial DNA [25] sequences which place the organism taxonomically among the fungi rather than the protozoa. The ultrastructure of the cyst wall of *P. carinii* resembles that of fungi [26], and *P. carinii* and fungal cell walls share a common epitope identified by a monoclonal antibody [27]. *P. carinii* also lacks some of the characteristic protozoan organelles, such as rhoptries, subpellicular tubules, and conoids [21]. *P. carinii* shows DNA sequence homology with the red yeast fungi [28]. The gene for elongation factor 3 (EF-3), which is found exclusively in fungi, has been found in *P. carinii* [29]. In addition, thymidylate synthase (TS) and dihydrofolate reductase (DHFR) are two distinct enzymes in *P. carinii* [30,31], whereas in protozoa those activities are contained within a single bifunctional protein [32].

Although *P. carinii* may be more closely related to fungi, the precise location remains to be determined [23,24,33]. Furthermore, some of the features which historically place *P. carinii* as a protozoan, such as its microscopic appearance, susceptibility to antiprotozoan drugs, lack of ergosterol, and lack of growth in fungal media, features which are unusual for fungi, remain to be fully explained. These observations suggest the possibility that *P. carinii* represents a

Table 1 Controversies in the Taxonomy of *P. carinii*: Fungus, Protozoan, or Unique Hybrid?

Support for classification as a protozoan	Support for classification as a fungus
Microscopic appearance, morphologic features (ex. trophozoite, sporozoite, cyst)	Staining characteristics similar to fungi Ultrastructural similarities to fungi
Susceptibility to antiprotozoan agents[a]	Poorly developed mitochondria containing lamellar cristae[b]
Lack of susceptibility to anti-fungal agents[c]	rRNA sequence homology to fungi
Lack of growth on fungal media	Mitochondrial DNA sequence homology to fungi
Absence of ergosterol in cell wall	Translocation elongation factor-3 gene[d]
	Cell wall contains β-1,3-glucan
	Thymidylate synthase (TS) and dihydrofolate reductase (DHFR) are distinct proteins[e]
	TS and DHFR homologous to *S. cerevisiae*
	Cross-reactivity of *P. carinii* monoclonal antibodies with fungi
	Sporogenous state of *P. carinii* similar to ascospore formation of yeast

[a] Trimethoprim-sulfamethoxazole, sulfonamides, pentamidine isethionate, alpha-difluoromethylornithine.
[b] Protozoan mitochondria have tubular cristae.
[c] Targeted against sterol synthesis pathways.
[d] Found exclusively in fungi.
[e] TS and DHFR are a single protein in protozoa.

unique hybrid, possessing features of both fungi and protozoa, or perhaps belongs in a unique taxonomic classification.

D. Biochemistry

The current understanding of the biochemistry of *P. carinii* is incomplete, again in part due to the inability to reliably propagate sufficient numbers of the organism in culture free of host cell contamination [34]. Evidence suggests that the organisms consume oxygen, although nutritional requirements and metabolic pathways are poorly understood. The outermost bilayer membrane of the cyst cell wall may function as a selective permeability barrier, a regulator of macromolecule uptake and transport, and/or function as an anchor for surface antigens as described for other organisms in which two bilayer membranes exist [35], although this remains to be established. The periplasmic space between these membranes may also be a metabolically important region although this has not been directly validated.

Studies demonstrate that the cell surface of *P. carinii* cyst forms contain abundant carbohydrates [36–38], where carbohydrates rich in glycosyl/manosyl, N-acetyl-D-glucosamine (GlcNAc), and galactose/N-acetyl-galactosamine (Gal/GalNAc) residues may constitute 8% of the outer cyst wall of *P. carinii* [39]. Glucose is the most abundant sugar and likely exists in the form of glucan [26,40]. The carbohydrate composition of human- and rat-derived *P. carinii* are similar, although differences in the relative-proportions of individual sugars are apparent [41].

Lipids may constitute 50% of cyst wall preparations, and include predominantly palmitate (16:0) and other fatty acids such as oleate [42]. Although *P. carinii* and host lung preparations both contain steryl esters, free fatty acids, monoglycerides, triglycerides, free sterols, and steryl esters, *P. carinii* contains more esterified sterols. *P. carinii* also contains unique phytosterols [43]. Although cholesterol appears to be present, ergosterol has not been detected [38,43]. The lack of ergosterol explains the resistance of *P. carinii* to sterol synthesis inhibitors, such as amphotericin B and ketoconazole.

E. Genetic Organization

A comprehensive project to sequence the *P. carinii* genome is currently in progress [44]. Isolation and identification of specific *P. carinii* genes will provide some insight into function and nutritional/metabolic requirements for this organism, and importantly identify targets for new novel therapeutic interventions. Among the identified genes, *P. carinii* has one copy of thymidylate synthase (TS), one copy of β-tublulin (BTU), and one copy of internal transcribed spacer (ITS) region of nuclear rRNA genes [45]. Other identified genes include cdc2 cyclin-dependent kinase (an essential gene involved in cell cycle regulation), cdc13 (cofactor for cdc2) [46], HSP70 homologue BiP (grp78), endoplasmic reticulum resident chaperonins [47], and prohibitin, which may regulate *P. carinii* growth [17]. Little is known, however, about gene regulation in *P. carinii*.

Comparison of *P. carinii* isolates from different mammalian hosts has revealed a broad degree of genetic diversity and host specificity. *P. carinii* infects many mammalian hosts and had initially been considered a single species. Recently, it has been demonstrated that *P. carinii* strains affecting different hosts may be distinct subspecies, such as *P. carinii* sp. f. *carinii* (in rats) and *P. carinii* sp. f. *hominis* (in humans) [48]. Furthermore, there appear to be many different genotypes affecting humans [49–51]. The significance of genotypic diversity with respect to virulence, infectivity, and drug susceptibility has not been established.

III. *P. CARINII* VIRULENCE FACTORS

Several observations may provide insight into *P. carinii* virulence. Pulmonary infection with *P. carinii* is present throughout the world and in a variety of mammal species in addition to humans [52,53]. The organism may be ubiquitous within the environment, although an environmental reservoir has not been identified. Data from experimental animals [54] and outbreaks of *P. carinii* described among immunocompromised patients [55] support transmission of infection through inhalation, although the infectious form of *P. carinii* has not been established. Furthermore, *P. carinii* may exhibit tissue tropism, as disease caused by *P. carinii*, with rare exception, is generally limited to the lungs. *P. carinii* interaction with host cells, an important component of microbial virulence, appears to be a critical step in the pathogenesis of *P. carinii* pneumonia. Furthermore, *P. carinii* may elaborate molecules with adverse consequences to the host and may incorporate host cell molecules to fulfill metabolic requirements and to evade or impair host immune recognition.

Recent investigations have provided important insights into *P. carinii* virulence. Certain traditional virulence factors such as mechanisms to facilitate *P. carinii* binding to host epithelial cells are the focus of active investigation, whereas factors for *P. carinii* which regulate growth in the host, or elaborated factors which contribute to pathogenesis remain incompletely characterized. Specific host factors which contribute to *P. carinii* pneumonia pathogenesis are discussed in greater detail in the following section of this chapter. This section will focus on our current

Table 2 Summary of Recognized or Proposed *P. carinii* Virulence Factors in the Susceptible Host

P. carinii virulence factor (?)	Pathophysiologic consequence
Supported by data from clinical, in vitro, or animal model investigations	
Genetic variability of surface antigenic determinants	Evade immune recognition by host cells upon repeated exposures
Small size of trophic form	Aerosol inhaled by host
Host fibronectin, vitronectin	Facilitate binding to host epithelial cells
Absorption of SP-A	Evade host phagocytic cell recognition
Absorption of sMR	Evade host phagocytic cell recognition
Enhance HIV-1 replication	Contribute to HIV-1 pathogenesis and pulmonary immune cell dysfunction
Suggested by data from clinical, in vitro, or animal model investigations	
Latent or dormant life form	Reactivation with immunosuppression
P. carinii extracellular matrix receptor	Facilitate binding to host cells
Release of chymase	Direct injury to lung epithelial cells
Release of reactive oxygen species	Direct toxicity to lung cells
Altered host surfactant components	Disturbance of lung physiology
Altered host cholesterol component	Alter macrophage phagocytosis
DHPG gene polymorphism	Anti-Pneumocystis drug resistance
Downregulation of lung transcription factor (ex. GATA-2)	Impair lung cell immune function
Proposed, but not established	
Tubular extensions (trophic forms)	Anchor organism to host cells or facilitate transportation of host-derived nutrients or structural molecules to *P. carinii*

understanding of *P. carinii* virulence factors. Table 2 provides a summary of some of the recognized and possible factors contributing to *P. carinii* virulence discussed in this chapter.

A. Epidemiological Features

P. carinii pneumonia has been described in children and adults of all ages in a global distribution [53], with evidence for some geographic variation in prevalence. Increased rates in countries previously reporting lower prevalence [56,57] may relate to actual increases, changes in reporting, or the use of more sensitive diagnostic tools, although alterations in an environmental reservoir for the organisms cannot be excluded. The natural habitat of *P. carinii* is not currently known.

Countries or regions with colder ambient climate may have a higher incidence of *P. carinii* pneumonia, although the influence of climate appears limited [58,59]. There is no apparent seasonal variation for *P. carinii* pneumonia [60].

P. carinii pneumonia has been described in both wild and domestic animals [52]. Evidence suggests a host specificity for *P. carinii*, as animals challenged with *P. carinii* organisms from other species develop limited or no evidence of *P. carinii* pneumonia. There is no evidence for animal to human transmission.

Different strains of *P. carinii* may exist in different geographical locations, as determined by the demonstration of different PCR products and variations in the nucleotide sequences of certain genetic loci, such as the DNA internal transcribed spacers (ITS) and rRNA genes [61].

Other studies have failed to demonstrate differences in strains from geographically removed regions [62,63]. The possible influence of strain differences on virulence is unknown.

Serologic evidence suggests that most healthy individuals are exposed to *P. carinii*, and exposure likely occurs early in life [64–66]. Serologic studies from various countries, including Africa, suggest that 60–83% of children and adults demonstrate antibodies to *P. carinii* [67,68]. The consequences of this apparent early exposure to *P. carinii* remain to be determined. Whether these antibodies represent effective and compete clearance of the organisms, a latent or persistent asymptomatic state, or are protective against subsequent exposure to *P. carinii* have not been determined.

B. Transmission of *P. carinii*

The available data support an aerosol route of transmission for *P. carinii* [69]. Except in rare cases, disease attributable to *P. carinii* in the susceptible host is generally limited to the lungs. Investigators have detected *P. carinii* DNA in ambient air samples from rural England [70,71], and recent studies describe geographic clustering of *P. carinii* cases [71a,71b] although an environmental reservoir of *P. carinii* has not been identified.

P. carinii pneumonia is reported worldwide in humans and in a variety of mammalian hosts, including rodents, ferrets, rabbits, horses, dogs, and cats [52]. Animal-to-animal transmission has been demonstrated within the same species [69,72–74], but transmission from one animal species to another has not been described [75]. Animal-to-human transmission of *P. carinii* has not been documented, which may in part reflect differences in surface antigens and *P. carinii* DNA recovered in lung tissues from animals and humans [76–78]. Reported epidemiological cluster outbreaks of *P. carinii* pneumonia among immunocompromised patients suggest that human *P. carinii* pneumonia may be transmitted by inhalation [55,79] from infected humans or from an environmental source. Animals cohoused with mice with active *P. carinii* pneumonia develop detectable *P. carinii* DNA in lung and sputum specimens, both of which clear following resolution of *P. carinii* pneumonia in the infected animals. In human studies, investigators have detected *P. carinii* DNA in ambient air samples from hospital wards [80], although serocoversion in hospital personnel caring for patients with *P. carinii* pneumonia have not been demonstrated [81].

C. Latency, Reactivation, and Acquisition of New *P. carinii* Infection

The controversy in the development of *P. carinii* pneumonia generally does not relate to the initial acquisition of the organism, which is most likely by the aerosolized route. Initial acquisition generally does not result in overt clinical disease in the immunocompetent host, and the development of serum anti–*P. carinii* antibodies supports exposure and activation of the host adaptive immune response. The controversy relates to whether subsequent pneumonia represents reactivation of a dormant phase of the organism in the lungs of susceptible hosts (analogous to *Mycobacterium tuberculosis* disease pathogenesis), or whether each episode of *P. carinii* pneumonia represents the acquisition of new organisms in the setting of congenital or acquired immunosuppression.

The high prevalence of serum antibodies to *P. carinii* in the general population, and the experimental induction of *P. carinii* pneumonia by administration of immunosuppressive agents to animals without direct exogenous inoculation [82–85] have in part supported the concept that *P. carinii* pneumonia represents the reactivation of latent organisms resident in lung tissue. However, available data do not support respiratory tract colonization in the immunocompetent hosts, as characteristic cyst or trophic forms and intracellular forms of *P. carinii* have not been demonstrated in the lungs of immunocompetent hosts, and efforts to detect *P. carinii* DNA in the lungs or respiratory tract specimens of immunocompetent hosts have generally been unsuccessful [86–89]. Taken together, these observations suggest that chronic asymptomatic carriage of *P. carinii* in the lungs of immunocompetent hosts is unlikely. However, the possibility that a transient carrier state exists, cannot be excluded.

Accumulating evidence, however, suggests that clinically apparent episodes of *P. carinii* pneumonia may represent newly acquired infection. Sequencing genetic markers such as internal transcribed spacer (ITS) regions (the regions located between 18S and 5.8S and between 5.8S and 26S of *P. carinii* rRNA) demonstrate that ITS sequences vary among different *P. carinii* isolates. Thus, the detection of two or more types of ITS nucleotide sequences in clinical respiratory tract specimens suggests that different *P. carinii* organisms were acquired by the susceptible hosts, perhaps from different sources or different exposures [49,51,90–93], although genetic mutations in a single species cannot be excluded; whether *P. carinii* contains a single copy of ITS remains to be conclusively established.

Accumulating data suggest that animals effectively and completely clear *P. carinii* following an episode of pneumonia. Susceptible SCID mice clear *P. carinii* pneumonia following reconstitution with spleen cells from immunocompetent mice, without evidence for *P. carinii* organisms or *P. carinii* DNA at 3 weeks following reconstitution [94]. Furthermore, the subsequent induction of immunosuppression of these same recovered SCID mice by either CD4 + T-lymphocytes depletion or a combination of CD4 + T-lymphocyte depletion and corticosteroids (both recognized methods for inducing susceptibility to *P. carinii* pneumonia) failed to induce *P. carinii* pneumonia in these animals [94], suggesting that no viable or latent form of the organisms remained in these animals. These important observations indicate that the immune response to *P. carinii* can completely clear the organism from the infected host, and supports the hypothesis that *P. carinii* pneumonia that develops in immunocompromised patients likely represents acquisition of new organisms infection resulting from exposure to an exogenous source of *P. carinii* and not necessarily from reactivation of latent infection.

D. Interaction with Lung Epithelial Cells

Binding of a potential pathogen to host epithelial cells represents an important and common component of microbial virulence for many pathogens. In the case of *P. carinii*, current concepts support the notion that interaction with alveolar epithelial cells appears to be a critical step in establishing disease in the susceptible host [95,96]. Experimental evidence demonstrates that trophic forms of the *P. carinii* tightly bind to the surface of alveolar type 1 pneumocytes, interdigitating membranes without membrane fusion with the host cells, and without apparent invasion of the host cells by the organisms [97]. *P. carinii* attachment may be mediated in part by fibronectin [98], vitronectin [99], laminin [100], and mannose receptors [101]. Extracellular matrix protein-mediated binding to host epithelial cells may be in part facilitated by *P. carinii* ECM receptors [102]. The mechanism preventing *P. carinii* interaction with alveolar epithelial cells in the immunocompetent host and the precise role of immunosuppression which is permissive for the interaction of *P. carinii* with alveolar epithelial cells remains to be established.

The precise nature of the closely apposed binding of *P. carinii* to host alveolar epithelial cells remains to be elucidated. The interaction may serve to allow transfer, exchange, or acquisition of host cell nutrients, regulatory or structural molecules [103] for use by *P. carinii*. The finding of specialized structures on the surface of mature trophic forms raises the possibility that these tubular expansions or filipodia may serve a transport function for the organism (Fig. 2), although these peculiar structures may represent an artifact of preparation and fixation.

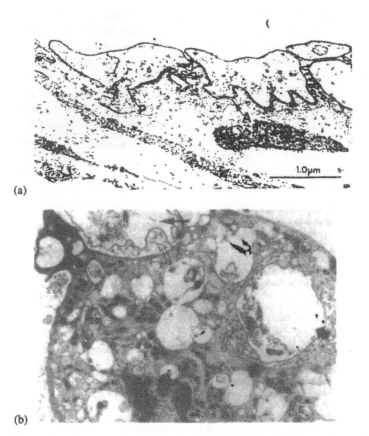

Figure 2 Electron micrographs demonstrating *P. carinii* interaction with (a) alveolar epithelial cells and (b) alveolar macrophages. (a) Close apposition of *P. carinii* trophic forms with alveolar epithelial cell surfaces in the immunosuppressed rodent model. Note interdigitation of *P. carinii* membranes with epithelial cell membranes (as identified at points P and A), without fusion of the respective membranes. (b) One internalized and partially degraded *P. carinii* trophic form (short solid arrow) and one trophic form bound to the macrophage surface and undergoing phagocytosis (long solid arrow). The characteristic external tubular structures of *P. carinii* (observed in cross section) are apparent immediately adjacent to the organism cell membrane (curved hollow arrows). (From Ref. 97.)

Attachment of *P. carinii* to alveolar epithelial cells in the susceptible host is associated with organism proliferation [83], although the factors influencing proliferation are not known [17]. The host consequences of *P. carinii* attachment to alveolar epithelial cells includes impaired cell cycle progression of host cells [104]. *P. carinii* may penetrate or affect the lung epithelial barrier, perhaps impairing alveolar epithelial cell line proliferation [104], although recent data suggest that *P. carinii* does not disrupt primary alveolar type-II epithelial cell metabolic, structural or barrier function [105].

E. Role of *P. carinii* Surface Components

The outer surface of the organism contains a 95- to 115-kDa glycoprotein structure, referred to as the major surface glycoprotein (MSG), *P. carinii* gp-120, or *P. carinii* glycoprotein-A (gp-A). This glycoprotein represents an important antigenic determinant, as enzymatic treatment of *P. carinii* cysts with zymolyase, a β-1-3-glucanase, disrupts the cyst wall and liberates the surface antigens [106]. In addition, this glycoprotein exhibits considerable genetic variability, a characteristic which may represent an important virulence factor, as host antibodies derived from prior exposure to *P. carinii* may only be partially protective against subsequent infection with *P. carinii* containing a different antigenic variant [107]. In addition, surface glycoprotein antigenic variability may contribute to the observed species specificity for *P. carinii* and serves to evade host recognition [108], although the incorporation of host cell molecules or other factors may account for the species specificity.

The major surface glycoprotein (MSG) mediates *P. carinii* binding to host cells. Fibronectin and vitronectin-mediated *P. carinii* binding to lung epithelial cells [99,109] as well as binding to the alveolar macrophage mannose receptor occur via recognition of the *P. carinii* MSG [110]. Furthermore, *P. carinii* major glycoprotein can stimulate cytokine release [111] and stimulate respiratory burst response in macrophages [112]. As the MSG of *P. carinii* is recognized by pulmonary immune cells such as alveolar macrophages, assimilation of host molecules on the *P. carinii* surface, thus concealing important antigenic determinants and pattern recognition molecules, may represent a mechanism for eluding host defenses [113].

The cell wall of *P. carinii* is also rich in β-glucans [26,114], major structural components of many related fungal cell walls [115,116]. Purified β-glucans from several fungal species phylogenetically related to *P. carinii* stimulate the release of TNF-α and IL-1β from cultured mononuclear cells [117–119]. Isolated *P. carinii* cell wall fractions rich in glucan carbohydrate potently induce TNF-α and macrophage inflammatory protein-2 (MIP-2) from alveolar macrophages [120], presumably through interaction with cognate β-glucan receptors. The β-glucan-mediated release of proinflammatory cytokines may in part account for the increase in TNF-α and neutrophillic infiltration in the *P. carinii* infected rodent lung.

F.   *P. carinii* Genetic Factors

Multiple strains of *P. carinii* may exist, as determined by differences in variations in the nucleotide sequences of certain genetic foci, as described above. Studies suggest that individuals with *P. carinii* pneumonia may be infected with one or more types of the organism [61], although the influence of strain variation on virulence is unknown. Furthermore, the influence of different *P. carinii* strains on primary or recurrent infection and lung tropism remains unknown.

G.   Elaboration of *P. carinii* Factors

Preliminary studies suggest that *P. carinii* may secrete proteolytic enzymes, such as chymase, which may have direct adverse effects on lung epithelial cells [121]. However, *P. carinii* is not known to secrete products which degrade antibodies or impair phagocytic cell function (such as inhibitors of adherence to phagocytic cells, suppression of phagocytosis, impairment of oxidative burst, or prevention of phogosome-lysosomal fusion), nor has resistance to lysosomal enzymes and other host factors been described.

H.   *P. carinii* Motility

The morphologic descriptions of *P. carinii* trophic form as amoeboid or pleomorphic infer motility. However, cytoskeletal elements are either poorly developed or poorly preserved in *P. carinii*, and neither microtubules nor microfilaments have been demonstrated in nondividing trophic forms. Based on available data, *P. carinii* does not appear to be a motile organism.

I. Subversion of Host Detection by Using Host Molecules

P. carinii can acquire host molecules such as lipids [103]. Other organisms are capable of acquiring host molecules which may allow the organism to appear as "self" and thus evade immune recognition by the host cells. However, whether *P. carinii* utilizes such a mechanism of evasion has not been established.

SP-A may impair detection of *P. carinii* by alveolar macrophages [113]. More recent data suggest that the soluble form of the macrophage mannose receptor, released in response to *P. carinii*, binds to the organism and impairs macrophage uptake [122]. Thus, this pathogen-driven subversion of an important innate recognition mechanism in the lungs could help swing the host-pathogen balance in favor of the pathogen. By coating itself in host-derived glycoprotein molecules, *P. carinii* potentially delays recognition and destruction by the host immune system, enabling the organism to survive and proliferate.

J. Drug Resistance

Drug resistance may be emerging as a significant problem in *P. carinii* pneumonia, based on dihodropteroate synthase (DHPS) polymorphisms [123,124] and mutations in the ubiquinone (coenzyme Q) binding sites of *P. carinii* [125]. Recent reports suggest that pneumonia caused by *P. carinii* with mutations in the DHPS gene are associated with higher clinical mortality, presumably attributable to resistance to the anti-*Pneumocystis* agents [126]. The potential problem of drug resistance may become more apparent with the routine administration of anti–*P. carinii* prophylaxis to immunocompromised patients considered at high clinical risk. Recent reports suggest that for HIV-infected persons, treatment with HAART is accompanied by a rise in CD4+ T-lymphocyte counts to >200 cells/mm,^{3} permitting the discontinuation of anti–*P. carinii* prophylaxis and eliminating the development of resistance. However, for non-HIV patients, guidelines for the prescription of prophylaxis are less standardized [127].

K. Organism Proliferation in the Alveolar Airspace

When allowed to proliferate in the alveolar airspace of a susceptible host, *P. carinii* forms large clusters of cells [7] which may create a physical challenge for their elimination by phagocytic cells such as macrophages. Furthermore, whether these *P. carinii* trophic forms lack a specific surface antigen which does not allow recognition by immune cells, or whether coated with a molecule which inhibits he ability to mount a host immune response (analogous to the bacterium *P. aeruginosa*) has not been established.

L. Host Lung Cell Transcription Factors

P. carinii pneumonia in the rat results in downregulation of GATA-2 mRNA [128], an important transcription factor in hematopoietic cell development [129], specifically in alveolar macrophages, ciliated bronchoepithelial cells, and interstitial monocytes [128]. The specific effect of this inhibition on pulmonary immune cells function as it relates to *P. carinii* remains to be established.

M. Release of Reactive Oxygen Species by *P. carinii*

P. carinii is capable of inducing release of detectable (although low) levels of reactive oxygen species in vitro [112,130]. Whether this capacity results in host cell oxidant damage remains to be established. This finding also presents an interesting paradox, as *P. carinii* has limited antioxidant capacity and appears quite sensitive to oxidants [131].

N. HIV-1 Replication

Analysis of BAL specimens from HIV-infected persons suggests that *P. carinii* is associated with enhanced HIV-1 replication in the lungs of these individuals [132], and this replication is compartmentalized relative to blood cells. The enhanced HIV-1 replication may not be unique to *P. carinii*, as *M. tuberculosis* pulmonary disease is associated with enhanced HIV-1 replication in the lungs [133]. Enhanced HIV-1 replication may contribute to local pulmonary immune cell dysfunction observed in HIV-infected persons [134]. *P. carinii* associated enhanced HIV-1 replication may contribute to HIV-1 disease progression and shorten the survival time of patients [135].

IV. HOST RESPONSE TO *P. CARINII*

Many recent investigations have provided important insights into identifying the host response to *P. carinii*. In this regard, genetic knockout animal models have been especially informative (Table 3).

Histologically, *P. carinii* pneumonia in the susceptible host appears as a diffuse or patchy alveolar filling process with a foamy, eosinophilic exudate composed of extensive clusters of predominantly trophic forms and few cyst forms in a matrix of cellular debris and fibrin [136]. Trophic forms may be intimately adherent to host alveolar epithelial cells, but *P. carinii* appears to be an extracellular pathogen [21]. This alveolar exudate is accompanied by type II epithelial cell proliferation, but most notably with limited host inflammatory response [137]. The lack of an inflammatory host response in the susceptible host is incompletely understood. Determining

Table 3 Animal Models for the Study of *P. carinii* Pneumonia

	Reference
Increased susceptibility to *P. carinii* pneumonia	
Corticosteroid treated rodent model	82
Athymic (nude) mice	74
SCID mice	141, 256
CD4 + T-lymphocyte-depleted mice	138
CD4 + T-lymphocyte-depleted mice	257
Transgenic B-cell deficient mice	149
RAG-1 −/− knockout mice	158
MHC class II molecule knockout Aβ −/− mice	158
TCRβ −/− knockout mice	158
TNF-α RI/II & IFN-γ knockout mice	175
uPA-depleted mice	258
Reduced pulmonary clearance of *P. carinii*	
Alveolar macrophage depletion in rat	153
CD40L blocking in SCID mice	259
Increased lung inflammation in *P. carinii* pneumonia	
INF-γ knockout mice	196

the cellular and molecular components participating in the host response to *P. carinii* remains an area of active investigation.

A. Lymphocytes

Most of the immunocompromised conditions associated with *P. carinii* susceptibility share a common, predominant T-lymphocyte dysfunction. The importance of T-lymphocytes, and in particular the CD4+ T-lymphocyte, in host defense against *P. carinii* has been clearly established both clinically and in animal models. CD4+ T-lymphocyte-depleted mice are susceptible to *P. carinii* pneumonia following direct inoculation of exogenous organisms into the lungs [138]. Impaired recruitment of lymphocytes into the lungs [139], occurs in contrast to immunocompetent mice which effectively clear the organisms [138]. Severe combined immunodeficiency (SCID) mice, genetically lacking T- and B-lymphocytes, are naturally susceptible to *P. carinii* pneumonia when raised outside of barrier isolation, but infected animals effectively clear the organisms following reconstitution with CD4+ T-lymphocytes from immunologically intact mice [140,141]. Adoptive transfer of purified CD4+ T-lymphocytes, pretreated with *P. carinii* major surface glycoprotein, into rodents with active *P. carinii* pneumonia reduced the *P. carinii* burden in the lungs [142]. Although clearly established as an important cell in the host response, the precise mechanism of the CD4+ T-lymphocyte-mediated response to *P. carinii* remains to be fully elucidated.

Other T-lymphocytes and other immune cells may be able to effectively defend the host against this organism. CD8+ T-lymphocytes may also contribute to an effective host response to *P. carinii*. Progressive *P. carinii* infection in CD4-depleted mice is associated with marked accumulation of CD8+ T-lymphocytes in the lungs [139,143], and loss of both CD4+ and CD8+ T-lymphocyte cells results in more intense pneumonia than just loss of CD4+ T-cells [144]. CD8+ T-cells provide partial protection against *P. carinii*, perhaps in part through elaboration of IFN-γ [144], as recent studies have demonstrated that IFN-γ and CD8+ T-lymphocytes restore host defenses against *P. carinii* even in the setting of CD4+ T-lymphocyte depletion [145]. In addition, preliminary data suggest that CD8+ T-lymphocytes may have direct cytotoxic effects on *P. carinii*. Aerosolized heat-killed *E. coli* (acting as an immune response modifier) facilitates resolution of *P. carinii* pneumonia in the absence of CD4+ T-lymphocytes [146], which may in part be mediated by TNF-α and some unidentified Thy-1+ CD4-CD8 α/β or γ/δ T-cells [146].

B-lymphocytes also contribute in the host response to *P. carinii*. Most adults demonstrate detectable serum antibodies to *P. carinii* [66], and increased anti–*P. carinii* antibody titers have been demonstrated in some patients with active *P. carinii* pneumonia [147,148]. In animal models, transgenic mice deficient in B-lymphocytes are susceptible to *P. carinii* pneumonia [149], and immunization-mediated increases in circulating anti–*P. carinii* antibodies may protect against subsequent *P. carinii* challenge, even in the setting of CD4+ T-lymphocyte depletion [150]. Furthermore, reconstitution of SCID mouse model of *P. carinii* pneumonia, using spleen cells from immunocompetent mice but depleted of B-lymphocytes, are less able to restore host defense [151].

B. NK Cells

P. carinii clearance from the lungs of CD4+ T-lymphocyte-depleted mice treated with IFN-γ gene transfer into the lungs is associated with the influx of CD8+ T-lymphocytes and NK cells into the lungs [145]. The specific role of NK cells, however, has not been established.

C. Macrophages

Alveolar macrophages are the primary phagocytic cells within the alveolar space, and several observations support an important role for alveolar macrophages in the successful host response to *P. carinii* [152]. Rats selectively depleted of alveolar macrophages demonstrate impaired clearance of *P. carinii* pneumonia [153]. Resolution of pneumonia in various animal models of *P. carinii* pneumonia [138,140,154] is associated with an influx of alveolar macrophages, accompanied by increased *P. carinii* phagocytosis by alveolar macrophages [155] and release of macrophage-derived cytokines [146]. In vitro exposure of *P. carinii* to macrophages results in phagocytosis and rapid digestion of the organisms [156,157]. Macrophage activation (as measured by cytokine and NO release) is not sufficient for mediating clearance of *P. carinii* in the setting of immunodeficiency [158], suggesting that macrophages require the presence of functional CD4+ T-lymphocytes.

    P. carinii binding by alveolar macrophages is mediated by extracellular matrix proteins (fibronectin, vitronectin) and receptors [111,159], immunoglobulin (Fc) [156] receptors, and mannose receptors [160], whereas *P. carinii* phagocytosis is mediated by immunoglobulin [156] and mannose receptors [134,160]. Complement receptors may enhance Fc-mediated phagocytosis. Binding and phagocytosis may be modified by alveolar lining materials such as surfactant protein-A (SP-A) [113,161].

    P. carinii induces a macrophage oxidative burst response [112,130,160], whereas nonoxidative mechanisms such as nitric oxide production may not contribute significantly to the host response [162]. *P. carinii* induces the release of arachidonic acid and its metabolites, prostaglandin E2 and leukotriene B4, from alveolar macrophages [118]. These molecules may participate in the effective host response to *P. carinii*. However, whether this response is altered in immunocompromised persons susceptible to *P. carinii* pneumonia has not been fully investigated.

D. Neutrophils/Polymorphonuclear Leukocytes

The role of neutrophils in the defense against *P. carinii* remains unclear. Neutrophils may be present and increased in BAL fluid from AIDS patients with *P. carinii* pneumonia [163,164], although this is not a consistent finding [165]. BAL neutrophilia may be more common in non-HIV *P. carinii* pneumonia [166]. The presence of neutrophils may correlate with a poor prognosis [164,167], perhaps attributable to the generation and release of toxic radicals and proteolytic enzymes [168], although this has not been established with certainty. Migration of neutrophils may be a relatively late event in *P. carinii* pneumonia among AIDS patients, and the poor prognosis may be related to advanced *P. carinii* infection. Previous bacterial pneumonia in immunosuppressed animal models (accompanied by neutrophilia) appears protective against subsequent *P. carinii* pneumonia [169]. In vitro, interaction of *P. carinii* with neutrophils from healthy persons results in *P. carinii* phagocytosis and degradation, and an oxidative burst response facilitated by antibody [170,171] but not complement [171]. GM-CSF further enhances the oxidative response [170]. Whether similar events occur in vivo has not been determined.

E. Cytokines

Experimental and clinical evidence indicates an important role for local cytokine production in the host response and perhaps host susceptibility to *P. carinii* pneumonia. Both TH1 and TH2 type responses appear to play important roles [172,173]. The emerging data support an important role for several cytokines, although studies have not definitively identified the cellular source for the release of *P. carinii*–mediated cytokines in the lungs. Furthermore, studies have generally acknowledged the complexity and redundancy of the host cytokine network. However, the mechanisms by which cytokines influence *P. carinii* clearance, or perhaps influence the *P. carinii* life cycle, have not been defined. Some of the cytokines examined in the host response to *P. carinii* include the following:

1. TNF-α

The release of TNF-α by alveolar macrophages in response to *P. carinii* appears to be critical for elimination of organisms from the host. In animal models, *P. carinii* clearance is significantly impaired in the presence of anti-TNF neutralizing antibodies [174], or in animals with genetically deleted TNF receptors [175]. Stimulation of endogenous TNF-α release accelerated *P. carinii* clearance despite continued immunosuppression [146], an effect which was reversed with anti-TNF antibody. Furthermore, immunocompetent mice fail to effectively clear *P. carinii* after treatment with recombinantly expressed TNF-α inhibitor [176].

The cellular source for TNF-α is likely alveolar macrophages [111,177,177a]. The mechanisms of *P. carinii* induced TNF-α release from alveolar macrophages may be mediated by the *P. carinii* cyst wall binding to macrophage β-glucan receptors [177]. This enhanced release of TNF-α is further augmented by opsonization of *P. carinii* with immune serum, or with the adhesion molecules fibronectin and vitronectin [111], molecules which are elevated in the lungs of patients with *P. carinii* pneumonia [111].

Exaggerated in vivo release of TNF-α has been observed in bronchoalveolar lavage fluid or alveolar macrophages recovered from animals [178] and human subjects [179–181] with *P. carinii* pneumonia. Furthermore, alveolar macrophages from persons with *P. carinii* pneumonia may be primed to release increased amounts of TNF-α in response to LPS stimulation [179,180], although this is not a consistent observation [181]. TNF-α concentrations in lung lavage fluid inversely correlate with *P. carinii* cyst concentrations [179]. In asymptomatic HIV-infected persons, TNF-α release by alveolar macrophages may be elevated [181], although more recent data suggest that TNF-α release is not enhanced [132,182]. In vitro HIV-1 infection of monocyte-derived macrophages blunted the TNF-α release in response to *P. carinii* compared to uninfected cells [183].

The specific mechanism(s) of TNF-α mediated host defense against *P. carinii* have not been established, but may involve macrophage activation and enhanced recruitment of inflammatory cells to the lungs [184] or perhaps direct toxicity to *P. carinii* [185], a unique function for a cytokine [186]. Defining the role of TNF-α remains an area of active investigation.

2. Interleukin-1 (IL-1)

Endogenous IL-1 is an important proinflammatory cytokine in the host defense against *P. carinii*. Clearance of *P. carinii* organisms following immune reconstitution of *P. carinii*–infected SCID mice is associated with increase in IL-1 in lung tissue [187], and clearance is abrogated with IL-1 receptor-blocking antibodies. IL-1 mRNA is increased in lung tissue from *P. carinii*–infected SCID mice, and the source appears to be alveolar macrophages in contact with *P. carinii* organisms [172]. IL-1 release is diminished in alveolar macrophages infected in vitro with HIV [188], raising the possibility that IL-1 release may contribute to *P. carinii* susceptibility in the HIV-infected host.

3. Interleukin-6 (IL-6)

The role of IL-6 in the host response to *P. carinii* remains unclear. IL-6 is detected in bronchoalveolar lavage fluid of corticosteroid-treated rats with *P. carinii* pneumonia [189], and treatment of mice with a TNF inhibitor gene decreased mRNA for IL-6 in lung tissue and resulted in more severe infection in mice depleted of CD4 + T-lymphocytes [176]. In contrast, IL-6 mRNA levels are not elevated in the lungs of SCID mice during progressive infection [172,190]. However, reconstitution and subsequent clearance of *P. carinii* organisms are associated with IL-6 detection in serum and the lung [190], and IL-6 mRNA in the lung [172]. However, neutralization of IL-6 does not influence outcome, but is associated with increased lung inflammation [190].

IL-6 was not present in bronchoalveolar lavages or plasma from immunosuppressed patients with *P. carinii* pneumonia [191], and IL-6 production by bronchoalveolar lavage cells or peripheral blood cells in response to LPS is decreased compared with cells obtained from uninfected subjects. Whether this defect was caused by *P. carinii* infection, corticosteroids, or immunosuppressive drugs has not been determined. In vitro *P. carinii*–mediated IL-6 release is greatly enhanced in human monocyte-derived macrophages infected in vitro with HIV-1 compared to uninfected cells [183].

4. Interferon-γ (IFN-γ)

By one of several mechanisms, IFN-γ may have a protective effect in the immunocompetent host. *P. carinii* antigens elicit IFN-γ production by T-cells [192,193], which in turn may prime other effector cell functions. IFN-γ may activate the phagocytic cell host defenses against *P. carinii*, as IFN-γ increases alveolar macrophage TNF release following exposure to *P. carinii* [194], enhances alveolar macrophage killing of *P. carinii* [186], and enhances neutrophil oxidative burst response following exposure to *P. carinii* [195].

IFN-γ mRNA does not accumulate in the lungs of susceptible SCID mice during progressive *P. carinii* pneumonia although IFN-γ mRNA is present in the lungs during recovery following reconstitution [172]. Neutralization of IFN-γ in *P. carinii*–infected SCID does not affect clearance of the organisms following reconstitution with immunocompetent splenocytes [174]. IFN-γ knockout mice are not susceptible to *P. carinii* pneumonia [196]. Furthermore, *P. carinii*–infected SCID mice effectively clear the infection following reconstitution with splenocytes from either intact or from IFN-γ knockout mice [196].

Exogenous administration of IFN-γ can modulate *P. carinii* pneumonia, reducing the intensity of *P. carinii* pneumonia when delivered via aerosol to CD4 + T-lymphocyte-depleted mice [197], or delivered systemically to corticosteroid-treated rats infected with *P. carinii* [198]. Furthermore, IFN-γ inhibits attachment of *P. carinii* to alveolar epithelial cells [199], which may interfere with an important step in disease pathogenesis. Thus, IFN-γ may modulate the inflammatory response in *P. carinii* pneumonia and may provide a protective role in *P. carinii* pneumonia. Whether this protective effect is absent in the immunocompromised host remains to be established.

5. Granulocyte/Macrophage Colony-Stimulating Factor (GM-CSF)

GM-CSF may enhance the host response to *P. carinii*. GM-CSF added to cultures of neutrophils and *P. carinii* enhances release of reactive oxygen species [170]. Systemic administration of GM-CSF to mice depleted of CD4 + T-lymphocytes reduced the intensity of *P. carinii* infection [200]. Recent experiments demonstrate that GM-CSF plays a critical role in the inflammatory response to *P. carinii*; in a GM-CSF knockout mouse model, depletion of CD4 T-lymphocytes increased lung inflammation and intensity of *P. carinii* infection compared to wild-type animals [201]. Furthermore, alveolar macrophages from GM-CSF knockout animals have impaired phagocytosis of *P. carinii* [201].

F. Urokinase-Type Plasminogen Activator

Urokinase-type plasminogen activator (uPA), a protease intimately involved in inflammatory-cell migration and activation, participates in the degradation of matrix proteins by leukocytes during recruitment to inflammatory sites. uPA is expressed by macrophages and activated lymphocytes [202]. uPA receptor (CD87) is present on monocytes, neutrophils, and activated lymphocytes [203]; clusters at the leading edge of migrating monocytes during chemotactic response; and is important to the clearance of *C. neoformans* [204]. uPA knockout mice develop uniformly heavy *P. carinii* pneumonia when challenged intratracheally. uPA activity is markedly decreased in alveolar macrophages and in BAL fluid obtained from HIV-infected persons, and plasminogen activator inhibitor activity is increased in BAL fluid obtained from HIV-infected individuals during *P. carinii* pneumonia [205].

G. Extracellular Matrix Proteins

Interaction of *P. carinii* organisms with host-defense cells occurs within the alveolar lining material (ALM). Fibronectin and vitronectin are present in the ALM and can facilitate binding of *P. carinii* to lung epithelial cells. *P. carinii* binding to lung epithelial cells may be toxic to the epithelial cells and thus contribute to pathogenesis of disease [104]. Fibronectin may also facilitate *P. carinii* binding to alveolar macrophages but does not trigger phagocytosis or respiratory burst [159]. Fibronectin and laminin-mediated attachment of *P. carinii* to host cells may be facilitated by an ECM receptor on the organism [102].

H. Surfactant Proteins

P. carinii induces selective alterations in component expression and biophysical activity of lung surfactant [206], resulting in decreased protein and mRNA levels of hydrophobic SP-B and SP-C, increased hydrophilic SP-D, and no change in SP-A. The specific consequences of these changes with regard to *P. carinii* pathogenesis remain to be established. SP-A is elevated in the BAL specimens from HIV-infected persons with active *P. carinii* pneumonia [207]. SP-A levels are elevated in rodents with active *P. carinii* pneumonia, and also in the lungs of these animals prior to development of *P. carinii* pneumonia [208] (although the animals are receiving corticosteroids). SP-A effect on *P. carinii* binding and phagocytosis by alveolar macrophages is controversial. In vitro investigations demonstrate that SP-D enhances *P. carinii* binding to alveolar macrophages [209], although the influence on phagocytosis is not known.

I. Inflammatory Host Response

Inflammation plays a critical role in the host response to *P. carinii* during the course of effective treatment, and may in part contribute to morbidity and mortality associated with *P. carinii* pneumonia [166,167,172,210]. In the animal model of *P. carinii* pneumonia, the inflammatory response is preserved although the profile of the cells recruited into the lungs is different from that of immunocompetent animals [139], and this inflammatory response is not sufficient to effect clearance or resolution of pneumonia. The lung inflammatory response associated with immune reconstitution results in significant decrease in lung compliance and oxygenation [211] and is associated with the presence of CD8+ T-lymphocytes and neutrophils in the BAL fluid. The lower respiratory tract inflammatory response may be in part mediated by the *P. carinii* cell wall β-glucan [120] through interaction with macrophages. These observations provide direct evidence that the host's response to *P. carinii* directly impairs pulmonary function and contributes to the pathogenesis of *P. carinii* pneumonia.

J. Host Genetic Factors

In the murine models for *P. carinii* pneumonia, there are strain differences in the severity of *P. carinii* pneumonia, suggesting important but unrecognized host factors [212]. Whether such host genetic factors contribute to *P. carinii* pneumonia in humans has not been determined.

V. MANAGEMENT OF *P. CARINII* INFECTIONS

P. carinii is an opportunistic pathogen, and human disease is generally limited to immunocompromised hosts. Immunosuppressive states permissive for *P. carinii* pneumonia are characterized by profound impairment of lymphocyte function; these include chronic corticosteroid therapy, cancer chemotherapy, chronic lymphocytic leukemia, malnutrition, and HIV-1 [213]. *P. carinii* pneumonia is not common in patients with immunosuppression associated with neutropenia or defective humoral immunity, and there are rare reports of *P. carinii* pneumonia in subjects without apparent recognized risk factors [214,215].

HIV-1 has had the most dramatic impact on the incidence and prevalence of *P. carinii* pneumonia on a global scale, as *P. carinii* pneumonia occurs in >75% of HIV-infected persons in the absence of antiretroviral therapy and prophylaxis [216], and accounted for 46% of the AIDS index diagnoses in 1992 [217,218]. The development of *P. carinii* pneumonia inversely correlates with the levels of peripheral blood CD4+ T-lymphocytes [219]. Recent advances in the diagnosis and management of HIV-1 infection have been associated with a marked reduction in the incidence of AIDS-related *P. carinii* pneumonia [220] and a decline in morbidity and mortality for AIDS-associated *P. carinii* pneumonia [220,221]. However, for immunocompromised individuals without HIV infection [221], morbidity and mortality remain quite high [221]. Furthermore, mortality related to respiratory failure for both HIV and non-HIV cases exceeds 50% [221].

A. Clinical Features of *P. carinii* Pneumonia

Patients diagnosed with *P. carinii* pneumonia often present with fever, nonproductive cough, and exertional dyspnea, often associated with fatigue, malaise, and weight loss [222,223]. The duration of symptoms prior to diagnosis is significantly shorter for non-HIV patients, with a mean duration of 5 days [222], but may be associated with a rapid development of acute respiratory distress [223]. In contrast, for HIV-infected persons, the mean duration of symptoms is 25 days [222], and the development is often more indolent [222]. Physical examination may demonstrate fine inspiratory crackles, and laboratory testing may demonstrate elevated serum LDH [224] and hypoxemia at rest or with exercise [225]. Pulmonary function testing may reveal a reduced diffusing capacity [226]. Chest radiographs may demonstrate characteristic bilateral perihilar interstitial infiltrates, although atypical features are increasingly common [227–233]. Normal radiographic findings may occur in up to 19% of cases [234].

B. Diagnosis

Confirmation of clinically suspected cases of *P. carinii* pneumonia requires the demonstration of the characteristic cyst and/or trophic forms in respiratory tract specimens. Examination of spontaneous expectorated sputum rarely reveals *P. carinii* organisms, although a diagnosis can often be made by examination of stained smears of sputum induced through ultrasonic nebulization of saline [235]. A diagnosis may also be made through examination of specimens obtained by bronchoalveolar lavage, transbronchial biopsy, or open lung biopsy [236]. Specimens may be stained with Gomori's Methenamine Silver stain to identify the cyst form [7,237,238]; Giemsa and modified Giemsa stains (e.g., Diff-Quik) to identify trophozoites, intracystic bodies, and all intermediate forms [7]; or direct and indirect fluorescent monoclonal antibody stains [239].

    P. carinii organisms may persist by staining characteristics (although they may not represent viable organisms) in subsequent bronchoscopic specimens during the course of treatment for *P. carinii* and up to 50 days [240]. Sequential bronchoscopic examination to monitor the course of treatment is not recommended. Determination of *P. carinii* organism burdens in the airways remains a research tool [166]. Serum levels of LDH, generally elevated in most cases of *P. carinii* pneumonia, generally decline with clinical improvement [241].

    For known HIV-infected persons with peripheral blood CD4 + T-lymphocytes <200 cells/ mm^3 and oral candidiasis, who present with >2 weeks of fever and nonproductive cough, a presumptive diagnosis of *P. carinii* may warrant a trial of anti-*Pneumocystis* therapy in the absence of diagnostic specimens [242], although empirical treatment for *P. carinii* pneumonia is controversial. Patients who fail to respond to empirical therapy should undergo further testing to establish a diagnosis. Empirical therapy for non-HIV patients has not been investigated. The use of PCR technology to diagnose *P. carinii* pneumonia is highly sensitive and has contributed to our understanding of the epidemiology of *P. carinii*. Detection of *P. carinii* DNA [243] may represent active infection, resolving infection, a transient carrier state, or perhaps subclinical infection, and thus has limited clinical utility.

C. Treatment

As a carrier state has not been well defined, the demonstration of *P. carinii* in a respiratory specimen from persons with newly acquired symptoms is generally regarded as pathogenic and warrants treatment. A number of antimicrobials are available for the treatment of *P. carinii* pneumonia, including trimethoprim-sulfamethoxazole (TMP-SMX), pentamidine isethionate, atovaqone, primaquine and clindamycin, and DFMO [244]. TMP-SMX remains the agent of choice for mild, moderate, and severe cases of *P. carinii* pneumonia. Although pentamidine demonstrates equal efficacy to TMP-SMX [245], a more severe side-effects profile relegates this agent to second-line therapy. The selection of alternative regimens is based in part on the clinical severity of pneumonia, as some agents are only available in oral form. Most available antimicrobials target *P. carinii* DHFR or DHPS enzyme. High rates of adverse drug reactions or intolerance [246] and treatment failure rates with second-line therapies remain a considerable problem [247].

D. Adjunctive Therapies

This type of therapy can be initiated at 3–5 days following the appropriate treatment (described above) for *P. carinii* pneumonia, Many patients diagnosed with *P. carinii* pneumonia experience progressive respiratory distress and hypoxemia within 5 days of initiation of appropriate antimicrobial medications. This observation is attributed to excessive lung inflammation related to the killing of *P. carinii* organisms and is consequence of an effective host immune response [211]. For HIV-infected persons with resting hypoxemia at the time of admission, the administration of adjunctive corticosteroids within 72 hr of establishing the diagnosis of *P. carinii* pneumonia [248] may prevent clinical deterioration, presumably by modifying the acute host inflammatory response, possibly by influencing TNF-α [249,250], and IL-1 release by alveolar macrophages [249]. The use of adjunctive corticosteroids for non-HIV PCP has not yet been established, although preliminary observations suggest that recovery may be accelerated with high-dose corticosteroids [223].

E. Chemoprophylaxis

Current guidelines recommend the initiation of chemoprophylaxis against *P. carinii* pneumonia for all HIV-infected persons with peripheral blood CD4+ T-lymphocyte counts <200 cells/ mm^3 [250a]. Trimethoprim-sulfamethoxazole remains the agent of choice, although dapsone, aerosolized pentamidine, and atovaqone represent alternatives if intolerance develops. For HIV-infected persons, recent evidence suggests that primary prophylaxis for *P. carinii* pneumonia may be discontinued once peripheral blood CD4+ T-lymphocytes are consistently >200 cells/ mm^3 [251]. For immunocompromised patients without HIV infection, the prescription of anti–*P. carinii* prophylaxis is less standardized [127].

F. Complications

Recognized complications include respiratory failure, with an associated high mortality rate for both AIDS-related and non-HIV *P. carinii* pneumonia [221]. Recent studies support the continued use of an aggressive approach to management of *P. carinii* in the HIV-infected population, including ICU admission, intubation, and mechanical ventilatory support [252]. For HIV-associated *P. carinii* pneumonia, other associated complications include pneumothorax [253], persistence of *P. carinii* pneumonia (i.e., nonresponder), or recurrence of *P. carinii* pneumonia [240], whereas these complications generally are not associated with non-HIV *P. carinii* pneumonia. *P. carinii* pneumonia enhances HIV-1 replication in the lungs [132] and may contribute to the accelerated progression of HIV disease [135]. Overall mortality rates ~10% for AIDS-related *P. carinii* pneumonia [254] and 35–43% for non-HIV-related *P. carinii* pneumonia [221,223,255].

VI. CONCLUSIONS

Much has been learned with regard to the pathogenesis of this opportunistic pathogen which causes serious disease in the immunocompromised patient. As with many microbes, infection with *P. carinii* may be common, but disease is unusual in the immunocompetent host. Recent studies have improved our understanding of *P. carinii* virulence factors, although many virulence factors remain speculative.

P. carinii is related to fungi but may represent a unique organism. Although the environmental reservoir for *P. carinii* has not been identified, *P. carinii* is likely acquired by the host via the respiratory route, although the infectious form and the inoculum size required to induce disease have not been established. For the healthy, immunocompetent host, although exposure is likely frequent, the combined human and animal data suggest that the organism is cleared effectively, without evidence for persistence of a latent phase of infection. However, in the setting of immunodeficiency, infection with *P. carinii* (perhaps through exposure to infected individuals) leads to disease, and recurrent episodes of *P. carinii* in the immunocompromised host likely represent newly or recently acquired organisms rather than reactivation of latent infection. Both the innate and acquired immune systems are important in the host response to effective clearance of this organism, but the generation of *P. carinii*–specific antibodies developed via acquired immunity have limited efficacy as newly acquired organisms may have different antigenic determinants on the surface and thus may not be recognized by preformed antibodies. The host inflammatory response to *P. carinii* may be responsible for the severe clinical manifestations of the disease, especially following the initiation of appropriate antimicrobials, whereas corticosteroids may favorably modify this inflammatory response.

To better understand *P. carinii* pathogenesis, many challenges remain, including culturing the organism and determining genetic factors, virulence factors, and metabolic growth requirements. Learning more about the organism and identifying specific or novel genes will allow more specific antimicrobial targeting. Identifying defects in host defense mechanisms may allow immunomodulation as adjunctive therapy for *P. carinii* pneumonia. Questions which remain to be established include: What/where is the environmental reservoir for *P. carinii*? What is the infectious form of the organism? What is the incubation period for disease? What are the nutritional/metabolic requirements? What determines virulence? Is *P. carinii* latent/reactivation or acquisition of new organism? What affects organism viability? Determining the answers to these and other questions will allow a more thorough understanding of the pathogenesis of pneumonia, and allow the development of novel therapeutic agents.

REFERENCES

1. WT Hughes. *Pneumocystis carinii* Pneumonitis. Boca Raton, FL: CRC Press, 1987.
2. AG Smulian, PD Walzer. The biology of *Pneumocystis carinii*. Crit Rev Microbiol 18:191–216, 1992.
3. PD Walzer, ed. *Pneumocystis carinii* Pneumonia. New York: Marcel Dekker, 1994.
4. M Armstrong, F Richards. Propogation and purification of rat *Pneumocystis carinii* in short-term cell culture. J Protozool 36:24S–27S, 1989.
5. S Merali, U Frevert, JH Williams, K Chin, R Bryan, AB Clarkson Jr. Continuous axenic cultivation of *Pneumocystis carinii*. Proc Natl Acad Sci USA 96:2402–2407, 1999.
6. JM Beck, RL Newbury, BE Palmer. *Pneumocystis carinii* pneumonia in scid mice induced by viable organisms propogated in vitro. Infect Immun 64:4643–4647, 1996.
7. J Ruffolo, M Cushion, P Walzer. Microscopic techniques for studying *Pneumoncystis carinii* in fresh specimens. J Clin Microbiol 23:17–21, 1986.
8. WG Campbell. Ultrastructure of *Pneumocystis carinii*. Arch Pathol 93:312–324, 1972.
9. CA Itatani, J Marshall. Ultrastructure morphology and staining characteristics of *Pneumocystis carinii* in situ and form bronchoalveolar lavage. J Parasitol 74:700–712, 1988.
10. JJ Ruffolo, MT Cushion, PD Walzer. Ultrastructural observations on life cycle stages of *Pneumocystis carinii*. J Protozool 36:535–545, 1989.
11. CWM Bedrossian. Ultrastructure of *Pneumocystis carinii*: a review of internal and surface characteristics. Semin Diagn Pathol 6:212–237, 1989.
12. F Palluault, E Dei-Cas, C Slomianny, B Soulez, D Camus. Golgi complex and lysosomes in rabbit derived *Pneumocystis carinii*. Biol Cell 70:73–82, 1990.
13. M Cushion, J Stringer, P Walzer. Cellular and molecular biology of *Pneumocystis carinii*. Int Rev Cytol 131:59–107, 1991.
14. P Hasleton, A Curry, E Rankin. *Pneumocystis carinii* pneumonia: a light microscopical and ultrastructural study. J Clin Pathol 34:1138–1146, 1981.
15. ME Vossen, PJ Beckers, JH Meuwissen, AM Stadhouders. Developmental biology of *Pneumocystis carinii*, an alternative view on the life cycle of the parasite. A Parasitenkd 55:101–118, 1978.
16. M Cushion, J Ruffolo, P Walzer. Analysis of the developmental stages of *Pneumocystis carinii* in vitro. Lab Invest 58:324–331, 1988.
17. S Narasimhan, MYK Armstrong, JK McClung, FF Richards, EK Spicer. Prohibitin, a putative negative control element present in *Pneumosytis carinii*. Infect Immun 65:5125–5130, 1997.
18. C Chagas. Nova pripanozomiaza humana. Mem Inst Oswaldo Cruz 1:159–218, 1909.

19. P Delanoe, M Delanoe. Sur les rapports des kystos de *carinii* le *Trypanosoma lewis*. CR Acad Sci 155:658–660, 1912.

20. P Delanoe, M Delanoe. De la rarete de *Pneumocystis carinii* chez cobayes de la region de Paris: absence de kysts chez d'autres animaux lapin, grenouille, zanguilles. Bull Soc Pathol Exot 7: 271–274, 1914.

21. J Vavra, K Kucera. *Pneumocystis carinii*: its ultrastructure and ultrastructural affinities. J Protozool 17:463–483, 1970.

22. EK Ham, SD Greenberg, RC Reynolds, DB Singer. Ultrastructure of *Pneumocystis carinii*. Exp Mol Pathol 14:362–372, 1971.

23. S Stringer, J Stringer, M Blase, P Walzer, M Cushion. *Pneumocystis carinii*: sequence from ribosomal RNA implies a close relationship with fungi. Exp Parasitol 68:450–461, 1989.

24. JC Edman, JA Kovacs, H Mansur, DV Santi, HJ Elwood, ML Sogin. Ribosomal RNA sequence show *Pneumocystis carinii* to be a member of the fungi. Nature 334:519–522, 1988.

25. F Pixley, A Wakefield, S Banerji, J Hopkin. Mitochonidrial gene sequences show fungal homology for *Pnemocystis carinii*. Mol Microbiol 5:1347–1351, 1991.

26. Y Matsumoto, S Matsuda, T Tegoshi. Yeast glucan in the cyst wall of *Pneumocystis carinii*. J Protozool 36(suppl):21S–22S, 1989.

27. B Lundgren, JA Kovacs, NN Nelson, F Stock, A Martinez, VJ Gill. *Pneumocystis carinii* and specific fungi have a common epitope, identified by a monoclonal antibody. J Clin Microbiol 30: 391–395, 1992.

28. AE Wakefield, SE Peters, S Banerji, PD Bridge, GS Hall, DL Hawksworth, LA Guiver, AG Allen, JM Hopkin. *Pneumocystis carinii* shows DNA homology with the ustomycetous red yeast fungi. Mol Microbiol 6:1903–1911, 1992.

29. MF Ypma-Wong, WA Fonzi, PS Sypherd. Fungus-specific translation elongation factor 3 gene present in *Pneumocystis carinii*. Infect Immun 60:4140–4145, 1992.

30. J Edman, U Edman, M Cao, B Lundgren, J Kovacs, D Santi. Isolation and expression of the *Pneumocystis carinii* dihydrofolate reductase gene. Proc Natl Acad Sci USA 86:8625–8629, 1989.

31. U Edman, J Edman, B Lundgren, D Santi. Isolation and expression of *Pneumocystis carinii* thymidylate synthase gene. Proc Natl Acad Sci USA 86:6503–6507, 1989.

32. I Toth, G Lazar, H Goodman. Purification and immunochemical characterization of a dihydrofolate reductase-thymidylate synthase enzyme complex from wild-carrot cells. EMBO J 6:1853–1858, 1987.

33. J-I Watanabe, H Hori, K Tanabe, Y Nakamura. 5s ribosomal RNA sequence of *Pneumocystis carinii* and its phylogenetic association with *Rhizopoda/Myxomycota/Zygomycota* group. Mol Biochem Parasitol 32:163–167, 1989.

34. E Sloand, B Laughon, M Armstrong, M Barlett, W Blumenfeld, M Cushion, A Kalica, J Kovacs, W Martin, E Pitt, E Pesanti, F Richards, R Rose, P Walzer. The challenge of *Pneumocystis carinii* culture. J Euk Microbiol 40:188–195, 1993.

35. RE Calhoon, S Aaronson. Evidence for affinities among the major taxa from the analysis of amino acid frequencies in glycoproteins from cell surface structures. Ann NY Acad Sci 361:472–480, 1981.

36. M Cushion, J DeStefano, P Walzer. *Pneumocystis carinii*: surface reactive carbohydrates detected by lectin probes. Exp Parasitol 67:137–147, 1988.

37. H Yoshikawa, T Tegoshi, Y Yoshida. Detection of surface carbohydrates on *Pneumocystis carinii* by fluorescein-conjugated lectins. Parasitol Res 74:43–49, 1987.

38. H Yoshikawa, H Morioka, Y Yoshida. Ultrastructural detection of carbohydrates in the pellicle of *Pneumocystis carinii*. Parasitol Res 74:537–543, 1988.

39. JA DeStefano, MT Cushion, V Puvanwsarajah, PD Walzer. Analysis of *Pneumocystis carinii* cyst wall. II. Sugar composition. J Protozool 37:436–441, 1990.

40. DM Schmatz, MA Romanchek, LA Pittarelli, RE Schwartz, RA Fromtling, KH Nollstadt, FL Vanmiddlesworth, KE Wilson, MJ Turner. Treatment of *Pneumocystis carinii* pneumonia with 1,3-B-glucan synthesis inhibitors. Proc Natl Acad Sci USA 87:5950–5954, 1990.

41. B Lundgren, GY Lipschik, JA Kovacs. Purification and characterization of a major human *Pneumocystis carinii* surface antigen. J Clin Invest 87:163–170, 1991.

42. ES Kaneshiro, MT Cushion, PD Walzer, K Jayasimhulu. Analysis of *Pneumocystis* fatty acids. J Protozool 36:69S–72S, 1989.

43. ST Furlong, JA Samia, RM Rose, JA Fishman. Phytosterols are present in *Pneumocystis carinii*. Antimicrob Agents Chemother 38:2534–2540, 1994.

44. Q Mei, RE Turner, V Sorial, D Klivington, CW Angus, JA Kovacs. Characterization of major glycoprotein genes of human *Pneumocystis carinii* and high-level expression of a conserved region. Infect Immun 66:4268–4273, 1998.

45. X Tang, MS Bartlett, JW Smith, JJ Lu, CH Lee. Determination of copy number of rRNA genes in *Pneumocystis carinii* f. sp. *hominis*. J Clin Microbiol 36:2491–1494, 1998.

46. TJ Kottom, CF Thomas, KK Mubarak, EB Leof, AH Limper. *Pneumocystis carinii* uses a functional Cdc13 B-type cyclin complex during its life cycle. Am J Respir Cell Mol Biol 22:722–731, 2000.

47. TT Stedman, GA Buck. Identification, characterization, and expression of the BiP endoplasmic reticulum resident chaperonins in *Pneumocystis carinii*. Infect Immun 64:4463–4471, 1996.

48. JR Stringer. *Pneumocystis carinii*: what is it, exactly? Clin Microbiol Rev 9:489–498, 1996.

49. S Latouche, E Ortona, E Mazars, P Margutti, E Tamburrini, A Siracusano, K Guyot, M Nigou, P Roux. Biodiversity of *Pneumocystis carinii hominis*: typing with different DNA regions. J Clin Microbiol 35:383–387, 1997.

50. AG Tsolaki, RF Miller, AP Underwood, S Banerji, AE Wakefield. Genetic diversity at the internal transcribed spacer regions of the rRNA operon among isolates of *Pneumocystis carinii* from AIDS patients with recurrent pneumonia. J infect Dis 174:141–156, 1996.

51. CH Lee, J Helweg-Larsen, B Lundgren, JD Lundgren, X Tang, S Jin, B Li, MS Bartlett, JJ Lu, M Olsson, SB Luvas, P Roux, A Cargnel, C Atzori, O Matos, JW Smith. Update on *Pneumocystis carinii* f. sp. *hominis* typing based on nucleotide sequence variations in internal transcribed spacer regions of rRNA genes. J Clin Microbiol 36:734–741, 1998.

52. W Hughes. Natural occurrences in animals. In: *Pneumocystis carinii* Pneumonitis. Boca Raton: CRC Press, 1987, pp 57–70.

53. WT Hughes. Geographic distribution. In: Hughes WT, ed. *Pneumocystis carinii* Pneumonitis. Boca Raton, FL: CRC Press, 1987, pp 33–56.

54. B Soulez, F Palluault, J Cesbron, E Dei-Cas, A Carpon, D Camus. Introduction of *Pneumocystis carinii* in a colony of SCID mice. J Protozool 38:123S–125S, 1991.

55. M Chusid, K Heyrman. An outbreak of *Pneumocystis carinii* at a pediatric hospital. Pediatrics 62: 1031–1035, 1978.

56. AS Malin, LDZ Gwanzura, S Klein, et al. *Pneumocystis carinii* in Zimbabwe. Lancet 2:1258–1262, 1995.

57. JD Lundgren, C Pedersen, N Clumeck, JM Gatell, AM Johnson, B Ledergerber, S Vella, A Phillips, JO Nielsen. Survival differences in European patients with AIDS, 1979–89. Br Med J 308(6936): 1068–1073, 1994.

58. DR Hoover, NMH Graham, H Bacellar, et al. Epidemologic patterns of upper respiratory illnesses and *Pneumocystis carinii* pneumonia in homosexual men. Am Rev Respir Dis 143:756–759, 1991.

59. M-C Delmas, V Schwoebel, et al. Recent trends in *Pneumocystis carinii* pneumonia as AIDS-defining disease in nine European countries. J Acquir Immune Defic Syndr 9:74–80, 1995.

60. PD Walzer, DP Perl, DJ Krogstead, PG Rawson, MG Schultz. *Pneumocystis carinii* pneumonia in the United States: epidemiologic, diagnostic, and clinical features. Ann Intern Med 80:83–93, 1974.

61. C-H Lee, J-J Lu, X Tang, et al. Prevalence of various *Pneumocystis carinii* sp. f. *hominis* types in different geographical locations. J Euk Microbiol 43:37S, 1996.

62. S Latouche, E Ortona, E Mazars, et al. Biodiversity of French and Italian human *Pneumoncystis carinii*. J Euk Microbiol 43:54S, 1996.

63. AE Wakefield, CC Fritscher, AS Malin, et al. Genetic diversity in human-derived *Pneumocystis carinii* isolated from four geographic locations: analysis of mitochondrial ribosomal RNA gene sequences. J Clin Microbiol 32:2959–2961, 1994.

64. L Pifer, W Hughes, S Stagno, D Woods. *Pneumocystis carinii* infection: evidence for high prevalence in normal and immunosuppressed children. Pediatrics 62:35–41, 1978.

65. J Meuwissen, I Tauber, A Leewenberg, P Beckers, J Shiehen. Parasitologic and serologic observations of infection with *Pneumocystis* in humans. J Infect Dis 136:43–49, 1977.

66. S Peglow, G Smulian, M Linke, et al. Serologic responses to specific *Pneumocystis carinii* antigens in health and disease. J Infect Dis 161:296–306, 1990.
67. AG Sumulian, DW Sullivan, MJ Linke, et al. Geographic variation in the humoral response to *Pneumocystis carinii*. J Infect Dis 167:1243–1247, 1990.
68. AE Wakefield, TJ Stewart, ER Moxon, et al. Infection with *Pneumocystis carinii* is prevalent in healthy Gambian children. Trans R Soc Trop Med Hyg 84:800–802, 1990.
69. W Hughes. Natural mode of acquisition for de novo infection with *Pneumocystis carinii*. J Infect Dis 145:842–848, 1982.
70. AE Wakefield. Detection of DNA sequences identical to *Pneumocystis carinii* in samples of ambient air. J Euk Microbiol 41:116S, 1994.
71. AE Wakefield. DNA sequences identical to *Pneumocystis carinii* f. sp. *carinii* and *Pneumocystis carinii* f. sp. *hominis* in samples of air spore. J Clin Microbiol 34:1754–1759, 1996.
71a. MN Dohn, ML White, EM Vigdorth, CR Buncher, VS Hertzberg, RP Baughman, AG Smulian, PD Walzer. Geographic clustering of *Pneumocystis carinii* pneumonia in patients with HIV infection. Am J Respir Crit Care Med 162:1617–1621, 2000.
71b. AM Morris, M Swanson, H Ha, L Huang. Geographic distribution of human immunodeficiency virus–associated *Pneumocystis carinii* pneumonia in San Francisco. Am J Respir Crit Care Med 162:1622–1626, 2000.
72. WT Hughes, DL Bartley, BM Smith. A natural source of infection due to *Pneumocystis carinii*. J Infect Dis 147:595, 1983.
73. J Hendley, T Weller. Activation and transmission in rats of infection with *Pneumocystis*. Proc Soc Exp Biol Med 137:1401–1404, 1971.
74. PD Walzer, V Schnelle, D Armstrong, PP Rosen. Nude mouse: a new experimental model for *Pneumocystis carinii* infection. Science 197:177–179, 1977.
75. F Gigliotti, AG Harmsen, CG Haidaris, PJ Haidaris. *Pneumocystis carinii* is not universally transmissible between mammalian species. Infect Immun 61:2886–2890, 1993.
76. J Kovacs, J Halpern, B Lundgren, J Swan, J Parillo, H Masur. Monoclonal antibodies to *Pneumocystis carinii*: identification of specific antigens and characterization of antigenic differences between rat and human isolates. J Infect Dis 159:60–70, 1989.
77. N Bauer, J Paulsrud, M Bartlett, J Smith, CI Wilde. *Pneumocystis carinii* organisms obtained from rats, ferrets, and mice are antigenically different. Infect Immun 61:1315–1319, 1993.
78. F Gigliotti, P Haidaris, C Haidaris, T Wright, K van der Meid. Further evidendce of host species-specifica variation in antigens of *Pneumocystis carinii* using the polymerase chain reaction. J Infect Dis 168:191–194, 1993.
79. C Singer, D Armstrong, P Rosen, D Schottenfield. *Pneumocystis carinii* pneumonia: a cluster of eleven cases. Ann Intern Med 82:772–777, 1975.
80. Ms Bartlett, J-J Lu, C-H Lee, PJ Durant, SF Queener, JW Smith. Types of *Pneumocystis carinii* detected in air samples. J Eukaryot Microbiol 43:44S, 1996.
81. B Lundgren, I Jungstrom, LP Rothman. Antibody responses to *Pneumocystis carinii* in hospital personnel with and without exposure to patients with *P. carinii* pneumonitis. J Euk Microbiol 43:10S, 1996.
82. J Frenkel, J Good, J Schultz. Latent *Pneumocystis* infections of rats, relapse, and chemotherapy. Lab Invest 15:1559–1577, 1966.
83. PD Walzer, RD Powell, K Yoneda, ME Rutledge, JE Milder. Growth characteristics and pathogenesis of experimental *Pneumocystis carinii* pneumonia. Infect Immun 27:928–937, 1980.
84. DC Stokes, F Gigliotti, JE Rehg, RL Snellgrove, WT Hughes. Experimental *Pneumocystis carinii* pneumonia in the ferret. Br J Exp Pathol 68:267–276, 1987.
85. M Armstrong, A Smith, F Richards. *Pneumocystis carinii* pneumonia in the rat model. J Protozool 38:136S–138S, 1991.
86. O Settnes, J Genner. *Pneumocystis carinii* in human lungs at autopsy. Scand J Infect Dis 18:489–496, 1986.
87. PR Millard, AR Heryet. Observations favouring *Pneumocystis carinii* pneumonia as a primary infection: a monoclonal antibody study on paraffin sections. J Pathol 154:365–370, 1988.

88. AE Wakefield, FJ Pixley, S Banerji, K Sinclair, RF Miller, ER Maxon, JM Hopkin. Detection of *Pneumocystis carinii* with DNA amplification. Lancet 336:451–453, 1990.

89. S Peters, A Wakefield, K Sinclair, P Millard, J Hopkins. A search for *Pneumocystis carinii* in post mortem lungs by DNA amplification. J Pathol 166:195–198, 1992.

90. SP Keely, JR Stringer. Sequences of *Pneumocystis carinii* f. sp. *hominis* strains associated with recurrent pneumonia vary at multiple loci. J Clin Microbiol 35:2745–2747, 1997.

91. S Latouche, J-L Poirot, C Bernard, P Roux. Study of internal transcribed spacer and mitochondrial large subunit genes of *Pneumocystis carinii* hominis isolated by bronchoalveolar lavage from human immunodeficiency virus infected patients during one or several episodes of pneumonia. J Clin Microbiol 35:1687–1690, 1997.

92. JJ Lu, MS Bartlett, MM Shaw, SF Queener, JW Smith, M Oritz-Rivera, MJ Leibowitz, CH Lee. Typing of *Pneumocystis carinii* strains that infect humans based on nucleotide sequence vatiations of internal transcribed spacers of rRNA genes. J Clin Microbiol 32:2904–2912, 1994.

93. JJ Lu, MS Bartlett, JW Smith, CH Lee. Typing of *Pneumocystis carinii* strains with type-specific oligonucleotide probes derived from nucleotide sequences of internal transcribed spacers of rRNA genes. J Clin Microbiol 33:2973–2977, 1995.

94. W Chen, F Gigliotti, AG Harmsen. Latency is not an inevitable outcome of infection with *Pneumocystis carinii*. Infect Immun 61:5406–5409, 1993.

95. PN Lanken, M Minda, GG Pietra, AP Fishman. Alveolar response to experimental *Pneumocystis carinii* pneumonia in the rat. Am J Pathol 99:561–588, 1980.

96. K Yoneda, P Walzer. Attachment of *Pneumocystis carinii* to type I alveolar cells studied by freeze-fracture electron microscopy. Infect Immun 40:812–815, 1983.

97. EG Long, JS Smith, JL Meier. Attachment of *Pneumocystis carinii* to rat pneumocytes. Lab Invest 54:609–615, 1986.

98. ST Pottratz, J Paulsrud, JS Smith, WJ Martin. *Pneumocystis carinii* attachment to cultured lung cells by *Pneumocystis* gp 120, a fibronectin binding protein. J Clin Invest 88:403–407, 1991.

99. AH Limper, JE Standing, OA Hoffman, M Castro, LW Neese. Vitronectin binds to *Pneumocystis carinii* and mediates organism attachment to cultured lung epithelial cells. Infect Immun 61:4302–4309, 1993.

100. JA Fishman, JA Samia, J Fuglestad, RM Rose. The effects of extracellular matrix (ECM) proteins on the attachment of *Pneumocystis carinii* to lung cell lines in vitro. J Protozool 38:34S–37S, 1991.

101. AH Limper, ST Pottratz, WJ Martin. Modulation of *Pneumocystis carinii* adherence to cultured lung cells by a mannose-dependent mechanism. J Lab Clin Med 118:492–499, 1991.

102. S Narasimhan, MY Armstrong, K Rhee, JC Edman, FF Richards, E Spiccr. Gene for an extracellular matrix receptor protein from *Pneumocystis carinii*. Proc Natl Acad Sci USA 91(16):7440–7444, 1994.

103. ST Furlong, H Koziel, MS Bartlett, GL McLaughlin, MM Shaw, RM Jack. Lipid transfer from human epithelial cells to *Pneumocystis carinii* in vitro. J Infect Dis 175:661–668, 1997.

104. AH Limper, WJ Martin. *Pneumocystis carinii*: Inhibition of lung cell growth mediated by parasite attachment. J Clin Invest 85:391–396, 1990.

105. JM Beck, AM Preston, JG Wagner, SE Wilcoxen, P Hossler, SR Meshnick, RI Paine. Interaction of rat *Pneumocystis carinii* and rat alveolar epithelial cells in vitro. Am J Physiol 275:L118–L125, 1998.

106. MJ Linke, MT Cushion, PD Walzer. Properties of the major antigens of rat and human *Pneumocystis carinii*. Infect Immun 57:1547–1555, 1989.

107. F Gigliotti, JA Wiley, AG Harmsen. Immunization with *Pneumocystis carinii* gpA is immunogenic but not protective in a mouse model of *P. carinii* pneumonia. Infect Immun 66:3179–3182, 1998.

108. CW Angus, A Tu, P Vogel, M Qin, JA Kovacs. Expression of variants of the major surface glycoprotein of *Pneumocystis carinii*. J Exp Med 183:1229–1234, 1996.

109. ST Pottratz, WJ Martin. Role of fibronectin in *Pneumocystis carinii* attachment to cultured lung cells. J Clin Invest 85:351–356, 1990.

110. DM O'Riordan, JE Standing, AH Limper. *Pneumocystis carinii* glycoprotein A binds macrophage mannose receptors. Infect Immun 63:779–784, 1995.

111. LW Neese, JE Standing, EJ Olson, M Castro, AH Limper. Vitronectin, fibronectin, and gp120 antibody enhance macrophage release of TNF-alpha in response to *Pneumocystis carinii*. J Immunol 152:4549–4556, 1994.

112. H Koziel, X Li, MYK Armstrong, FF Richards, RM Rose. Alveolar macrophages from human immunodeficiency virus–infected persons demonstrate impaired oxidative burst response to *Pneumocystis carinii* in vitro. Am J Respir Cell Mol Biol 23:452–459, 2000.

113. H Koziel, DS Phelps, JA Fishman, MYK Armstrong, FF Richards, RM Rose. Surfactant protein-A reduces binding and phagocytosis of *Pneumocystis carinii* by human alveolar macrophages in vitro. Am J Respir Cell Mol Biol 18:834–843, 1998.

114. DJ Williams, JA Radding, A Dell, K-H Khoo, ME Rogers, FF Richards, MYK Armstrong. Glucan synthesis in *Pneumocystis carinii*. J Protozool 38:427–437, 1991.

115. EM Shematek, JA Braatz, E Cabib. Biosynthesis of the yeast cell wall. I. Preparation and properties of beta-(1 leads to 3) glucan synthetase. J Biol Chem 255:888, 1980.

116. JH Duffus, C Levi, DJ Manners. Yeast cell-wall glucans. Adv Microb Physiol 23:151, 1982.

117. G Abel, JK Czop. Stimulation of human monocyte beta-glucan receptors by glucan particles induces production of TNF-alpha and IL-1b. Int J Immunopharmacol 14:1363–1373, 1992.

118. M Castro, TI Morgenthaler, OA Hoffman, JE Standing, MS Rohrbach, AH Limper. *Pneumocystis carinii* induces the release of arachidonic acids and its metabolites from alveolar macrophages. Am J Respir Cell Mol Biol 9:73–81, 1993.

119. EJ Olson, JE Standing, N Griego-Harper, OA Hoffman, AH Limper. Fungal beta-glucan interacts with vitronectin and stimulates tumor necrosis factor alpha release from macrophages. Infect Immun 64:3548–3554, 1996.

120. R Vassallo, JE Standing, AH Limper. Isolated *Pneumocystis carinii* cell wall glucan provokes lower respiratory tract inflammatory responses. J Immunol 164:3755–3763, 2000.

121. WM Breite, AM Bailey, WJ Martin. *Pneumocystis carinii* chymase is capable of altering lung epithelial cell permeability. Am Rev Respir Dis 147:A33 [abstract], 1993.

122. IP Fraser, K Takahashi, H Koziel, B Fardin, A Harmsen, RAB Ezekowitz. *Pneumocystis carinii* enhances the production of soluble mannose receptor by macrophages. Microb Infect 2:1305–1310, 2000.

123. BR Lane, JC Ast, PA Hossler, DP Mindell, MS Bartlett, JW Smith, SR Meshnick. Dihydropteroate synthase polymorphisms in *Pneumocystis carinii*. J Infect Dis 175:482–485, 1997.

124. L Ma, L Borio, H Masur, JA Kovacs. *Pneumocystis carinii* dihydropteroate synthase but not dihydrofolate reductase gene mutations correlate with prior trimethoprim-sulfamethoxazole or dapsone use. J Infect Dis 180:1969–1978, 1999.

125. DJ Walker, AE Wakefield, MN Dohn, RF Miller, RP Baughman, PA Hossler, MS Bartlett, JW Smith, P Kazanjian, SR Meshnick. Sequence polymorphisms in the *Pneumocystis carinii* cytochrome b gene and their association with atovaquone prophylaxis failure. J Infect Dis 178:1767–1775, 1998.

126. J Helweg-Larsen, TL Benfield, J Eugen-Olsen, JD Lundgren, B Lundgren. Effects of mutations in *Pneumocystis carinii* dihydropteroate synthase gene on outcome of AIDS-associated *P. carinii* pneumonia. Lancet 354:1347–1351, 1999.

127. NG Mansharmani, D Balachandran, I Vernovsky, R Garland, H Koziel. Peripheral blood CD4 + T-lymphocyte counts during *P. carinii* pneumonia in immunocompromised patients without HIV infection. Chest 118:712–720, 2000.

128. X Tang, ME Lasbury, DD Davidson, MS Bartlett, JW Smith, C-H Lee. Down-regulation of GATA-2 transcription during *Pneumocystis carinii* infection. Infect Immun 68:4720–4724, 2000.

129. FY Tsai, G Keller, FC Kuo, M Weiss, J Chen, M Rosenblatt, FW Alt, SH Orkin. An early haematopoietic defect in mice lacking the transcription factor GATA-2. Nature 371:221–226, 1994.

130. HA Hidalgo, RJ Helmke, VF German, JA Mangos. *Pneumocystis carinii* induces an oxidative burst in alveolar macrophages. Infect Immun 60:1–7, 1992.

131. E Pesanti. *Pneumocystis carinii*: oxygen uptake, antioxidant enzymes, and susceptibility to oxygen-mediated damage. Infect Immun 44:7–11, 1984.

132. H Koziel, S Kim, C Reardon, X Li, R Garland, P Pinkston, H Kornfeld. Enhanced in vivo HIV-1 replication in the lungs of HIV-infected persons with *Pneumocystis carinii* pneumonia. Am J Respir Crit Care Med 160:2048–2055, 1999.

133. K Nakata, WN Rom, Y Honda, R Condos, S Kanegasaki, Y Cao, M Weiden. *Mycobacterium tuberculosis* enhances human immunodeficiency virus-1 replication in the lung. Am J Respir Crit Care Med 155:996–1003, 1997.

134. H Koziel, Q Eichbaum, BA Kruskal, P Pinkston, RA Rogers, MYK Armstrong, FF Richards, RM Rose, RAB Ezekowitz. Reduced binding and phagocytosis of *Pneumocystis carinii* by alveolar macrophages from persons infected with HIV-1 correlates with mannose receptor downregulation. J Clin Invest 102, 1332–1344, 1998.

135. DH Osmond, DP Chin, J Glassroth, PA Kvale, JM Wallace, MJ Rosen, LB Reichman, WK Poole, PC Hopewell. Impact of bacterial pneumonia and *Pneumocystis carinii* pneumonia on human immunodeficiency virus disease progression. Pulmonary Complications of HIV Study Group. Clin Infect Dis 29:536–543, 1999.

136. S Huang, K Marshall. *Pneumocystis carinii* infection: a cytologic, histologic and electron microscopic study of the organism. Am Rev Respir Dis 102:623–629, 1970.

137. J Watts, F Chandler. Evolving concepts of infection by *Pneumocystis carinii*. In: P Rosen, R Fechner, ed. Pathology Annual. Norwalk, CT: Appleton & Lange, 1991, pp 93–138.

138. JE Shellito, VV Suzara, W Blumenfeld, JM Beck, HJ Steger, TH Ermak. A new model of *Pneumocystis carinii* infection in mice selectively depleted of helper T lymphocytes. J Clin Invest 85: 1686–1693, 1990.

139. JM Beck, ML Warnock, JL Curtis, MJ Sniezek, SM Arraj-Peffer, HB Kaltreider, JE Shellito. Inflammatory responses to *Pneumocystis carinii* in mice selectively depleted of helper T lymphocytes. Am J Respir Cell Mol Biol 5:186–197, 1991.

140. A Harmsen, M Stankiewicz. Requirement for CD4+ cells in resistance to *Pneumocystis carinii* pneumonia in mice. J Exp Med 172:937–945, 1990.

141. JB Roths, JD Marshall, RD Allen, GA Carlson, CL Sidman. Spontaneous *Pneumocystis carinii* pneumonia in immunodeficient mutant scid mice. Am J Pathol 136:1173–1186, 1990.

142. S Theus, R Andrews, P Steele, P Walzer. Adoptive transfer of lymphocytes sensitized to the major surface glycoprotein of *Pneumocystis carinii* confers protection in the rat. J Clin Invest 95: 2587–2593, 1995.

143. T Ishimine, K Kawakami, A Nakamoto, A Saito. Analysis of cellular response and gamma interferon synthesis in bronchoalveolar lavage fluid and lung homogenate of mice infected with *Pneumocystis carinii*. Microbiol Immunol 39:49–58, 1995.

144. JM Beck, R Newbury, B Palmer, M Warnock, P Byrd, H Kaltreider. Role of CD8+ lymphocytes in host defense against *Pneumocystis carinii* in mice. J Lab Clin Med. 128:477–487, 1996.

145. JK Kolls, S Habetz, MK Shean, C Vazquez, JA Brown, D Lei, P Schwarzenberger, P Ye, S Nelson, WR Summer, JE Shellito. IFN-gamma and CD8+ T-cells restore host defense against *Pneumocystis carinii* in mice depleted of CD4+ T-cells. J Immunol 162:2890–2894, 1999.

146. AG Harmsen, W Chen. Resolution of *Pneumocystis carinii* pneumonia in CD4+ lymphocyte-depleted mice given aerosols of heat-treated *Escherichia coli*. J Exp Med 176:881–886, 1992.

147. B Lundgren, J Lundgren, L Mathiesen, J Nielsen, T Nielsen, J Kovacs. Antibody responses to a major *Pneumocystis carinii* antigen in human immunodeficiency virus–infected patients with and without *Pneumocystis carinii* pneumonia. J Infect Dis 165:1151–1155, 1992.

148. B Lundgren, M Lebech, K Lind, J Nielsen, J Lundgren. Antibody response to a major human *Pneumocystis carinii* surface antigen in patients without evidence of immunosuppression and in patients with suspected atypical pneumonia. Eur J Clin Microbiol Infect Dis 12:105–109, 1993.

149. H Marcotte, D Levesque, K Delaney, A Bourgeault, R de la Durantaye, S Brochu, MC Lavoie. *Pneumocystis carinii* infection in transgenic B cell deficient mice. J Infect Dis 173:1034–1037, 1996.

150. AG Harmsen, W Chen, F Gigliotti. Active immunity to *Pneumocystis carinii* reinfection in T cell depleted mice. Infect Immun 63:2391–2395, 1995.

151. AG Harmsen, M Stankiewicz. T cells are not sufficient for resistance to *Pneumocystis carinii* pneumonia in mice. J Protozool 38:44–45S, 1991.

152. R Vassallo, CF Thomas Jr, Z Vuk-Paklovic, AH Limper. Alveolar macrophage interactions with *Pneumocystis carinii*. J Lab Clin Med 133:535–540, 1999.

153. AH Limper, JS Hoyte, JE Standing. The role of alveolar macrophages in *Pneumocystis carinii* degradation and clearance from the lung. J Clin Invest 99:2110–2117, 1997.

154. E Barton, W Campbell. *Pneumocystis carinii* in lungs of rats treated with cortisone acetate. Am J Pathol 54:209–236, 1969.

155. K Yoneda, PD Walzer. The interaction of *Pneumocystis carinii* with host cells: an ultrastructural study. Infect Immun 29:692–703, 1980.

156. H Masur, TC Jones. The interaction in vitro of *Pneumocystis carinii* with macrophages and L-cells. J Exp Med 147:157–170, 1978.

157. LA von Behren, EL Pesanti. Uptake and degradation of *Pneumocystis carinii* by macrophages in vitro. Am Rev Respir Dis 118:1051–1059, 1978.

158. R Hanano, K Reifenberg, SH Kaufmann. Activated pulmonary macrophages are insufficient for resistance against *Pneumocystis carinii*. Infect Immun 66:305–314, 1998.

159. ST Pottratz, WJ Martin. Mechanism of *Pneumocystis carinii* attachment to cultured rat alveolar macrophages. J Clin Invest 86:1678–1683, 1990.

160. RAB Ezekowitz, DJ Williams, H Koziel, MYK Armstong, A Warner, FF Richards, RM Rose. Uptake of *Pneumocystis carinii* mediated by the macrophage mannose receptor. Nature 351:155–158, 1991.

161. MD Williams, JR Wright, KL March, WJ Martin II. Human surfactant protein A enhances attachment of *Pneumocystis carinii* to rat alveolar macrophages. Am J Respir Cell Mol Biol 14:232–238, 1997.

162. JE Shellito, JK Kolls, R Olariu, JM Beck. Nitric oxide and host defense against *Pneumocystis carinii* infection in a mouse model. J Infect Dis 173:432–439, 1996.

163. KR Young, JA Rankin, GP Naegel, ES Paul, HY Reynolds. Bronchoalveolar lavage cells and proteins in patients with the acquired immunodeficiency syndrome. An immunologic analysis. Ann Intern Med 103:522–533, 1985.

164. G Mason, C Hasimota, P Dickman, L Foutty, C Cobb. Prognostic implications of bronchoalvelar lavage neutrophilia in patients with *Pneumocystis carinii* pneumonia and AIDS. Am Rev Respir Dis 139:1336–1342, 1989.

165. B Jensen, I Lisse, J Gerstoft, S Borgeskov, P Skinhoj. Cellular profiles in bronchoalveolar lavage fluid of HIV-infected patients with pulmonary symptoms: relation to diagnosis and prognosis. AIDS 5:527–533, 1991.

166. AH Limper, KP Offord, TF Smith, WJ Martin II. Pneumocystis carinii pneumonia: differences in lung parasite number and inflammation in patients with and without AIDS. Am Rev Respir Dis 140:1204–1209, 1989.

167. RL Smith, WM El-Sadir, ML Lewis. Correlation of bronchoalveolar lavage cell populations with clinical severity of *Pneumocystis carinii* pneumonia. Chest 92:60–64, 1988.

168. SJ Weiss. Tissue distruction by neutrophils. N Engl J Med 320:365–376, 1989.

169. EL Pesanti. Effects of bacterial pneumonitis on development of pneumocystosis in rats. Am Rev Respir Dis 125:723–726, 1982.

170. MB Taylor, M Phillips, CSF Easmon. Opsonophagocytosis of *Pneumocystis carinii*. J Med Microbiol 36:223–228, 1992.

171. AL Laursen, N Obel, J Rungby, PL Andersen. Phagocytosis and stimulation of the respiratory burst in neutrophils by *Pneumocystis carinii*. J Infect Dis 168:1466–1471, 1993.

172. TW Wright, CJ Johnston, AG Harmsen, JN Finkelstein. Analysis of cytokine mRNA profiles in the lungs of *Pneumocystis carinii*–infected mice. Infect Immun 17:491–500, 1997.

173. BA Garvy, JA Wiley, F Gigliotti, AG Harmsen. Protection against *Pneumocystis carinii* pneumonia by antibodies generated from either T helper 1 or T helper 2 responses. Infect Immun 65:5052–5056, 1997.

174. W Chen, EA Havell, AG Harmsen. Importance of endogenous tumor necrosis factor alpha and gamma interferon in host resistance against *Pneumocystis carinii* infection. Infect Immun 60:1279–1284, 1992.

175. DG Rudmann, AM Preston, MW Moore, JM Beck. Susceptibility to *Pneumocystis carinii* in mice is dependent on simultaneous deletion of IFN-gamma and type 1 and 2 TNF receptor genes. J Immunol 161:360–366, 1998.

176. J Kolls, D Lei, C Vazquez, et al. Exacerbation of murine *Pneumocystis carinii* infection by adenoviral-mediated gene transfer of TNF inhibitor. Am J Respir Cell Mol Biol 16:112–118, 1997.

177.   OA Hoffman, JE Standing, AH Limper. *Pneumocystis carinii* stimulates tumor necrosis factor-alpha release from alveolar macrophages through a beta-glucan-medicated mechanism. J Immunol 150: 3932–3940, 1993.

177a.   E Corsini, C Dykstra, W Craig, R Tidwell, G Rosenthal. *Pneumocystis carinii* induction of tumor necrosis factor-α by alveolar macrophages: modulation by pentamidine isethionate. Immunol Lett 34:303–308, 1992.

178.   J Kolls, J Beck, S Nelson, W Summer, J Shellito. Alveolar macrophage release of tumor necrosis factor during murine *Pneumocystis carinii* pneumonia. Am J Respir Cell Mol Biol 8:370–376, 1993.

179.   VL Krishnan, A Meager, DM Mitchell, AJ Pinching. Alveolar macrophages in AIDS patients: increased spontaneous tumour necrosis factor-alpha production in *Pneumocystis carinii* pneumonia. Clin Exp Immunol 80:156–160, 1990.

180.   A Millar, R Miller, N Foley, A Meager, S Semple, G Rook. Production of tumor necrosis factor-alpha by blood and lung mononuclear phagocytes from patients with human immunodeficiency virus-related lung disease. Am J Respir Cell Mol Biol 5:144–148, 1991.

181.   D Israel-Biet, J Cadranel, K Beldjord, JM Andrieu, A Jeffrey, P Even. Tumor necrosis factor production in HIV-seropositive subjects. Relationship with lung opportunistic infections and HIV expression in alveolar macrophages. J Immunol 147:490–494, 1991.

182.   R Buhl, HA Jaffe, KJ Holroyd, Z Borok, JH Roum, A Mastrangeli, FB Wells, M Kirby, C Saltini, RG Crystal. Activation of alveolar macrophages in asymptomatic HIV-infected individuals. J Immunol 150:1019–1028, 1993.

183.   O Kandil, JA Fishman, H Koziel, P Pinkston, RM Rose, HG Remold. Human immunodeficiency virus type 1 infection of human macrophages modulates the cytokine response to *Pneumocystis carinii*. Infect Immun 62:644–650, 1994.

184.   K Tracey, A Cerami. Tumor necrosis factor: a pleiotropic cytokine and therapeutic target. Annu Rev Med 45:491–503, 1994.

185.   E Pesanti, T Tomicic, S Donta. Binding of ^{125}I-labelled tumor necrosis factor to *Pneumocystis carinii* and an insoluble cell wall fraction. J Protozool 38:28S–29S, 1991.

186.   EL Pesanti. Interaction of cytokines and alveolar cells with *Pneumocystis carinii* in vitro. J Infect Dis 163:611–616, 1991.

187.   W Chen, EA Havell, LL Moldawer, KW McIntyre, RA Chizzonite, AG Harmsen. Interleukin 1: an important mediator of host resistance against *Pneumocystis carinii*. J Exp Med 176:713–718, 1992.

188.   S Roy, L Fitz-Gibbon, L Poulin, M Wainberg. Infection of human monocyte/macrophages by HIV-1: effect on secretion of IL-1 activity. Immunology 64:233–239, 1988.

189.   R Perenbroom, P Beckers, J van der Meer, et al. Pro-inflammatory cytokines in lung and blood during steroid-induced *Pneumocystis carinii* pneumonia in rats. J Leukoc Biol 60:710–715, 1996.

190.   W Chen, E Havell, F Gigliotti, A Harmsen. Interleukin-6 production in a murine model of *Pneumocystis carinii* pneumonia: relation to resistance and inflammatory response. Infect Immun 61:97–102, 1993.

191.   R Perenbroom, A van Schijndel, P Beckers, et al. Cytokine profiles in bronchoalveolar lavage fluid and blood in HIV-seronegative patients with *Pneumocystis carinii* pneumonia. Eur J Clin Invest 26:159–166, 1996.

192.   SA Theus, DW Sullivan, PD Walzer, AG Smulian. Cellular responses to a 55-kilodalton recombinant *Pneumocystis carinii* antigen. Infect Immun 62(8):3479–3484, 1994.

193.   S Theus, R Andrews, M Linke, P Walzer. Characterization of rat CD4 T cell clones specific for the major surface glycoprotein of *Pneumocystis carinii*. J Euk Microbiol 44:96–100, 1997.

194.   E Corsini, C Dykstra, W Craig, R Tidwell, G Rosenthal. *Pneumocystis carinii* induction of tumor necrosis factor-alpha by alveolar macrophages: modulation by pentamidine isethionate. Immunol Lett 34:303–308, 1992.

195.   M Taylor, C Easmon. The neutrophil chemiluminescence response to *Pneumocystis carinii* is stimulated by GM-CSF and gamma interferon. FEMS Microbiol Immunol 4:41–44, 1991.

196.   G Garvy, A Ezekowitz, A Harmsen. Role of gamma interferon in the host immune and inflammatory responses to *Pneumocystis carinii* infection. Infect Immun 65:373–379, 1997.

197. JM Beck, HD Liggitt, EN Burnette, HJ Fuchs, JE Shellito, RJ Debs. Reduction in intensity of *Pneumocystis carinii* pneumonia in mice by aerosol administration of gamma interferon. Infect Immun 59:3859–3862, 1991.

198. HL Shear, G Valladares, MA Narachi. Enhanced treatment of *Pneumocystis carinii* pneumonia in rats with interferon-gamma and reduced doses of trimethoprim/sulfamethoxazole. J Acquir Immune Defic Syndr 3:943–948, 1990.

199. S Pottratz, A Weir. Gamma interferon inhibits *Pneumocystis carinii* attachment to lung cells by decreasing expression of lung cell-surface integrins. Eur J Clin Invest 27:17–22, 1997.

200. J Mandujano, N D'Souza, S Nelson, W Summer, R Beckerman, J Shellito. Granulocyte-macrophage colony stimulating factor and *Pneumocystis carinii* pneumonia in mice. Am J Respir Crit Care Med 151(4):1233–1238, 1995.

201. RI Paine, AM Preston, S Wilcoxen, H Jin, BB Siu, SB Morris, JA Reed, G Ross, JA Whitsett, JM Beck. Granulocyte-macrophage colony stimulating factor in the innate response to *Pneumocystis carinii* pneumonia in mice. J Immunol 164:2602–2609, 2000.

202. E Bianchi, E Ferrero, F Fazioli, F Mangeili, J Wang, J Bender, F Blasi, R Pardi. Integrin-dependent induction of functional urokinase receptors in primary T-lymphocytes. J Clin Invest 98:1133–1141, 1996.

203. A Nykjaer, B Moller, RF Todd, T Christensen, PA Andreasen, J Gliemann, CM Petersen. Urokinase receptor: an activation antigen in human T-lymphocytes. J Immunol 152:505–516, 1994.

204. MR Gyetko, G-H Chen, RA McDonald, R Goodman, GB Huffnagle, CC Wilkinson, JA Fuller, GB Toews. Urokinase is required for the pulmonary inflammatory response to *Cryptococcus neoformans*: a murine transgenic model. J Clin Invest 97:1818–1826, 1996.

205. E Angelici, C Contini, R Romani, O Epifano, P Serra, R Canipari. Production of plasminogen activator and plasminogen activator inhibitors by alveolar macrophages in control subjects and AIDS patients. AIDS 10:283–290, 1996.

206. EN Atochina, MF Beers, ST Scanlon, AM Preston, JM Beck. *P. carinii* induces selective alterations in component expression and biophysical activity of lung surfactant. Am J Physiol Lung Cell Mol Physiol 278:L599–L609, 2000.

207. DS Phelps, RM Rose. Increased recovery of surfactant protein A in AIDS-related pneumonia. Am Rev Respir Dis 143:1072–1075, 1991.

208. DS Phelps, JA Fishman, RM Rose. Surfactant protein A levels in glucocorticoid-immunosupressed rats infected with *Pneumocystis carinii*. Am Rev Respir Dis 145:A246, 1992.

209. D O'Riordan, J Standing, K Kwon, D Chang, E Crouch, A Limper. Surfactant protein D interacts with *Pneumocystis carinii* and mediates organism adherence to alveolar macrophages. J Clin Invest 95:2699–2710, 1995.

210. TW Wright, CJ Johnston, AG Harmsen, JN Finkelstein. Chemokine gene expression during *Pneumocystis carinii*–driven pulmonary inflammation. Infect Immun 67:3452, 1999.

211. TW Wright, F Gigliotti, JN Finkelstein, JT McBride, CL An, AG Harmsen. Immune-mediated inflammation directly impairs pulmonary function, contributing to the pathogenesis of *Pneumocystis carinii* pneumonia. J Clin Invest 104:1307–1317, 1999.

212. MYK Armstrong, MT Cushion. Animal models. In: PD Walzer, ed. *Pneumocystis carinii* Pneumonia. 2d ed. New York: Marcel Dekker, 1993, pp 182–187.

213. H Masur, C Lane, JA Kovacs, CJ Allegra, JC Edman. *Pneumocystis* pneumonia: from bench to clinic. Ann Intern Med 111:813–826, 1989.

214. J Jacobs, D Libby, R Winters, D Gelmont, E Fried, B Hartman, J Laurence. A cluster of *Pneumocystis carinii* pneumonia in adults without predisposing illnesses. N Engl J Med 324:246–250, 1991.

215. P Palanage, P Serra, F Di Sabato, C Contini, M Giacovazzo. *Pneumocystis carinii* pneumonia in a patient with chronic obstructive pulmonary disease but no evident immunoincompetence. Clin Infect Dis 19:543–544, 1994.

216. A Fauci. The human immunodeficiency virus: infectivity and mechanisms of pathogensis. Science 239:617–622, 1988.

217. J Murray, J Mills. Pulmonary infectious complications of human immunodeficiency virus infection. Part I. Am Rev Respir Dis 141:1356–1372, 1990.

218. S Safrin. *Pneumocystis carinii* pneumonia in patients with the acquired immunodeficiency syndrome. Semin Respir Infect 8:96–103, 1993.

219. J Phair, A Munoz, R Detels, R Kaslow, C Rinaldo, A Saah. The risk of *Pneumocystis carinii* pneumonia among men infected with human immunodeficiency virus type 1. N Engl J Med 322:161–165, 1990.

220. FJ Palella Jr, KM Delaney, AC Moorman, MO Loveless, J Fuhrer, GA Satten, DJ Aschman, SD Holmberg. Declining morbidity and mortality among patients with advanced human immunodeficiency virus infection. HIV Outpatient Study Investigators. N Engl J Med 338:853–860, 1998.

221. N Mansharamani, R Garland, D Delaney, H Koziel. Management and outcome patterns for adult *Pneumocystis carinii* pneumonia 1985–95: comparison of HIV-associated cases to other immunocompromised states. Chest 118:704–711, 2000.

222. J Kovacs, J Hiemenz, A Macher, D Stover, H Murray, J Shelhamer, H Lane, C Urmacher, C Honig, D Longo, M Parker, C Natanson, J Parrillo, A Fauci, P Pizzo, H Masur. *Pneumocystis carinii* pneumonia: a comparison between patients with the acquired immunodeficiency syndrome and other immunodeficiencies. Ann Intern Med 100:663–671, 1984.

223. JG Pareja, R Garland, H Koziel. Use of adjunctive corticosteroids in severe adult non-HIV *Pneumocystis carinii* pneumonia. Chest 113:1215–1224, 1998.

224. R Smith, C Ripps, M Lewis. Elevated lactate dehydrogenase values in patients with *Pneumocystis carinii* pneumonia. Chest 93:987–992, 1988.

225. D Stover, R Greeno, A Gagliardi. The use of a simple exercise test for the diagnosis of *Pneumocystis carinii* pneumonia in patients with AIDS. Am Rev Respir Dis 139:1343–1346, 1989.

226. P Hopewell. *Pneumocystis carinii* pneumonia: diagnosis. J Infect Dis 157:1115–1119, 1988.

227. F Askin, A Katzenstein. Pneumocystic infection masquerading as diffuse alveolar damage: a potential source of diagnostic error. Chest 79:420–422, 1981.

228. W Blumenfeld, N Basgoz, WJ Owen, D Schmidt. Granulomatous pulmonary lesions in patients with the acquired immunodeficiency syndrome (AIDS) and *Pneumocystis carinii* infection. Ann Intern Med 109:505–507, 1988.

229. J Barrio, M Suarez, J Rodriguez. *Pneumocystis carinii* pneumonia presenting as cavitating and noncavitating solitary pulmonary nodules in patients with the acquired immunodeficiency syndrome. Am Rev Respir Dis 135:1094–1097, 1986.

230. MJ Saldana, JM Mones. Cavitation and other atypical manifestations of *Pneumocystis carinii* pneumonia. Semin Diagn Pathol 6:273–286, 1989.

231. M Lee, R Schinella. Pulmonary calcification caused by *Pneumocystis carinii* pneumonia. A clinicopathological study of 13 cases in acquired immune deficiency syndrome patients. Am J Surg Pathol 15:376–380, 1991.

232. A Gal, M Koss, S Strigle, P Angritt. *Pneumocystis carinii* infection in the acquired immunodeficiency syndrome. Semin Diagn Pathol 6:2387–2395, 1989.

233. H Edelstein, R McCaber. Atypical presentations of *Pneumocystis carinii* pneumonia in patients receiving inhaled pentamodine prophylaxis. Chest 98:1366–1369, 1990.

234. M Katz, R Braon, D Grady. Risk stratification of ambulatory patients suspected of *Pneumocystis carinii* pneumonia. Arch Intern Med 156:177–188, 1991.

235. T Bigby, D Margolskee, J Curtis, et al. The usefulness of induced sputum in the diagnosis of *Pneumoncystis carinii* pneumonia in patients with the acquired immunodeficiency syndrome. Am Rev Respir Dis 133:515–518, 1986.

236. J Fraser, C Lilly, E Israel, P Hulme, P Hanaff. Diagnostic yield of bronchoalveolar lavage and bronchoscopic lung biopsy for detection of *Pneumocystis carinii*. Mayo Clin Proc 71:1025–1029, 1996.

237. PD Walzer, CK Kim, MT Cushion. *Pneumocystis carinii*. In: PD Walzer, RM Genta, eds. Parasitic Infections in the Immunocompromised Host. New York: Marcel Dekker, 1989, pp 83–178.

238. VS Baselski, MK Robison, LW Pifer, DR Woods. Rapid detection of *Pneumocystis carinii* in bronchoalveolar lavage samples by using Cellufluor staining. J Clin Microbiol 28:393–394, 1990.

239. J Kovacs, V Ng, H Masur, G Leong, K Hadley, G Evans, H Lane, F Ognibene, J Shelhamer, J Parillo, V Gill. Diagnosis of *Pneumocystis carinii* pneumonia: improved detection in sputum with use of monoclonal antibodies. N Engl J Med 318:589–593, 1988.

240. JH Shelhamer, FP Ognibene, AM Macher, C Tuazon, R Steiss, D Longo, JA Kovacs, MM Parker, C Natanson, NC Lane, AS Fauci, JE Parrillo, H Masur. Persistence of *Pneumocystis carinii* in lung tissue of acquired immunodeficiency syndrome patients treated for *Pneumocystis carinii* pneumonia. Am Rev Respir Dis 130:1161–1165, 1984.

241. MK Zaman, DA White. Serum lactate dehydrogenase levels and *Pneumocystis carinii* pneumonia. Am Rev Respir Dis 137:796–800, 1988.

242. Anonymous. USPHS/IDSA guidelines for the prevention of opportunistic infection in persons infected with human immunodeficiency virus: disease specific recommendations. Clin Infect Dis 25(suppl):S313–S335, 1997.

243. J Torres, M Goldman, LJ Wheat, X Tang, MS Bartlett, JW Smith, SD Allen, CH Lee. Diagnosis of *Pneumocystis carinii* pneumonia in human immunodeficiency virus–infected patients with polymerase chain reaction: a blinded comparison to standard methods. Clin Infect Dis 30:141–145, 2000.

244. PA Tietjen, DE Stover. *Pneumocystis carinii* pneumonia. Semin Respir Crit Care Med 16:173–186, 1995.

245. NC Klein, FP Duncanson, TH Lenox, C Forszpanial, CB Sherer, H Quentzel, M Nunez, M Suarez, O Kawwaff, A Pitta-Alvarez, et al. Trimethoprim-sulfamethoxazole versus pentamidine for *Pneumocystis carinii* pneumonia in AIDS patients: results of a large prospective randomized treatment trial. AIDS 6:301–305, 1992.

246. FM Gordin, GL Simon, CB Wofsy, J Mills. Adverse reactions to trimethoprim-sulfamethoxazole in patients with the acquired immunodeficiency syndrome. Ann Intern Med 100:495–499, 1984.

247. D Smith, S Davies, J Smithson, I Harding, BG Gazzard. Enflornithine versus cotrimoxazole in the treatment of *Pneumocystis carinii* pneumonia in AIDS patients. AIDS 6:1489–1493, 1992.

248. Anonymous. National Institutes of Health–University of California expert panel for cortocosteroids as adjunctive therapy for pneumocystis pneumonia. Consensus statement on the use of corticosteorid as adjunctive therapy for pneumocystis pneumonia in the acquired immunodeficiency syndrome. N Engl J Med 323:1500–1504, 1990.

249. ZB Huang, E Eden. Effect of corticosteroids on IL1-beta and TNF-alpha release by alveolar macrophages from patients with AIDS and *Pneumocystis carinii* pneumonia. Chest 104:751–755, 1993.

250. A Waage, A Bakke. Glucocorticoids suppress the production of TNF by LPS-stimulated human monocytes. Immunology 53:299–302, 1988.

250a. Centers for Disease Control and Prevention. 1997 USPHS/IDSA guildlines for the prevention of opportunistic infections in persons infected with human immunodeficiency virus. MMWR 46(R-12):1–46, 1997.

251. H Furrer, M Egger, M Opravil, E Bernasconi, B Hirschel, M Battegay, A Telenti, PL Vernazza, M Rickenbach, M Flepp, R Malinverni. Discontinuation of primary prophylaxis against *Pneumocystis carinii* pneumonia in HIV-1-infected adults treated with combination antiretroviral therapy. N Engl J Med 340:1301–1306, 1999.

252. MJ Rosen, K Clayton, RF Schneider, W Fulkerson, RA Vijaya, et al. Intensive care of patients with HIV infection. Utilization, critical illnesses, and outcomes. Am J Respir Crit Care Med 155:67–71, 1997.

253. KA Sepkowitz, EE Telzak, JWM Gold, EM Bernard, S Blum, M Carrow, M Dickmeyer, D Armstrong. Pneumothorax in AIDS. Ann Intern Med 114:455–459, 1991.

254. RD Horner, CL Bennett, D Rodriquez, RA Weinstein, HA Kessler, GM Dickinson, JL Johnson, SE Cohn, WL George, SC Gilman, F Shapiro. Relationship between procedures and health insurance for critically ill patients with *Pneumocystis carinii* pneumonia. Am J Respir Crit Care Med 152:1435–1442, 1995.

255. SH Yale, AH Limper. *Pneumocystis carinii* pneumonia in patients without acquired immunodeficiency syndrome: associated illnesses and prior corticosteroid therapy. Mayo Clin Proc 71:5–13, 1996.

256. W Chen, J Mills, A Harmsen. Development and resolution of *Pneumocystis carinii* pneumonia in severe combined immunodeficient mice: a morphological study of host inflammatory responses. Int J Exp Pathol 73:709–720, 1992.

257. DC McFadden, MA Powles, JG Smith, AM Flattery, K Bartizal, DM Schmatz. Use of anti-CD4+
      hybridoma cells to induce *Pneumocystis carinii* in mice. Infect Immun 62(11):4887–4892, 1994.
258. JM Beck, AM Preston, MR Gyetko. Urokinase-type plasminogen activator in inflammatory cell
      recruitment and host defense against *Pneumocystis carinii* in mice. Infect Immun 67:879–884,
      1999.
259. JA Wiley, AG Harmsen. CD40 ligand is required for resolution of *Pneumocystis carinii* pneumonia
      in mice. J Immunol 155:3525–3529, 1995.

12

Genomics in Fungal Pathogenesis

Janna L. Beckerman and P. T. Magee
University of Minnesota, St. Paul, Minnesota

I. INTRODUCTION

By the end of the year 2000, completed genome sequences of over 75 micro-organisms will be available. Although the majority of sequenced genomes will be prokaryotic, fungal genomes will be the next most represented group of eukaryotic organisms. To date, the only completed fungal genome is of *Saccharomyces cerevisiae*; however, fungal genome projects are under way for the zoopathogens *Candida albicans*, *Pneumocystis carinii*, *Aspergillus fumigatus*, and *Cryptococcus neoformans*. Projects are in various stages of completion for model organisms *Neurospora crassa* and *Aspergillus nidulans*, as well as for the model plant pathogen *Magnaporthe grisea*. Upon completion of the genome projects of these seven organisms, an enormous amount of information will be available to the fungal biologist. The sheer volume of information that genome projects present has redefined biological research as an information science called "genomics." Although the word genomics was coined in 1986 [1], the field of genomics has already been divided into "structural genomics" and "functional genomics." Whereas structural genomics concerns itself with the elucidation of the physical map of the genome, functional genomics involves experimental approaches to examine global gene expression ("the transcriptome") through large scale experimental methodologies and is dependent on structural genomics. Both approaches require that the biologists organize (and reorganize) data via statistical and computational analyses in an attempt to discover patterns and relationships. These relationships may exist among species, within the genome of single species (e.g., gene families), or within a single DNA sequence.

II. GOALS OF THE FUNGAL GENOME PROJECTS

The development of genetic and molecular analysis of fungi has greatly advanced our understanding of the biology of this very large kingdom. However, many of the fungi which are of the most intrinsic or practical interest are refractory to genetic analysis (e.g., the obligate parasites such as the human pathogen *P. carinni*, and the plant pathogenic rusts and powdery mildews), and even those which are experimentally accessible contain large genomes, so that a gene-by-gene analysis of their biology becomes unfeasible, even with a very large research community. For example, 25 years' work by the very extensive and active community of geneticists and biochemists working on *Saccharomyces cerevisiae* failed to discover about half the putative genes revealed by the genomic sequence of this organism [2]. Even with an international consortium devoted to the identification of every open reading frames (ORF) in the yeast genome, this approach failed to identify 15 unique RNA transcripts, several of which are likely to contain small ORFs [3].

With the advent of whole genome analysis, both general approaches to discover the functions of unknown ORFs and specific targeting of genes chosen for their potential interest become possible, and much can be learned in a relatively short time. Among the information that can be derived from the annotated sequence of a fungal genome are candidate genes for antifungal drugs, identification of gene families which may be refractory to normal genetic or molecular analysis, and clues to undiscovered aspects of the organism's life style. For this reason, pharmaceutical companies are focusing a great deal of attention on the emerging information about the genomes of fungi.

III. STRUCTURAL GENOMICS

Structural genomics concerns itself with a physical map of the genome of an organism. Although the ultimate genomic map is the completed DNA sequence, that sequence is often not completely informative, since it does not define individual chromosomes, repeated DNA, and unclonable segments. For this reason, sequence determination must be accompanied by the construction of a physical as well as a genetic map. To make a physical map, the genomic DNA is subdivided into a series of clones that are then ordered along the chromosome in such a way that all clones are contiguous (a contig map). Preparation of this map is simpler in fungi than higher eukaryotes because fungi have smaller chromosomes than most eukaryotes. These smaller chromosomes readily lend themselves to CHEF (contour clamped homogenous electric field) electrophoresis and allow the separation and resolution of specific chromosomes for karyotypic analysis. DNA probes of either known genes or random fragments of DNA are then hybridized to membranes that contain the DNA of both the collection of clones and the karyotype. This identifies a subset of overlapping clones from the library that contain the DNA sequence of the probe and reveals the position of the probe within the karyotype. These overlapping clones are referred to as a contig (Fig. 1). Subsequent probes produce new contigs that extend previous contigs or may exist as isolated minicontigs. Contigs may also be extended by a process called "chromosome walking." Walking can be accomplished by subcloning the ends of a given contig and using the subcloned fragment to reprobe the library and identify new clones to extend the contig. Unless directionality has been previously established, only one subclone will extend the contig whereas the other will reidentify the previously identified clones. In the end, the goal is to connect all the contigs along each chromosome and provide a set of overlapping clones spanning the entire chromosome.

For smaller genomes, shotgun sequencing is the most efficient approach. In this method, random clones covering several genomes' worth of DNA are sequenced. The complete sequence is assembled using specialized computer programs. The deep coverage that results from multiple overlapping clones permits the assignment of the final sequence and overlaps to formulate sequence contigs. This approach is extremely efficient for fungi, most of which have small genomes. The alternative (and "classical") method is to use the physical map to generate contigs of each chromosome, then sequence the individual clones in each contig. Since the order of the clones is known, the sequence is easily aligned along the chromosome. This approach diminishes the computing power required to assemble the final sequence, since a single chromosome is necessarily less complex than the genome. This method has been essential for determining the

Making the Fosmid Map

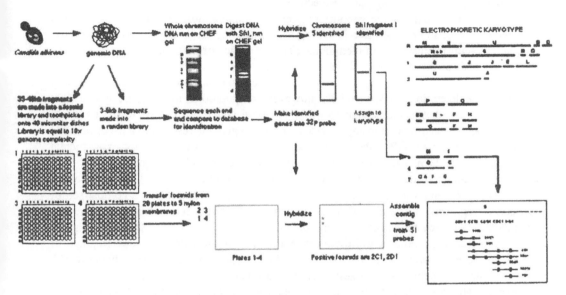

Figure 1 Overview of the physical mapping portion of the *Candida albicans* Genome Project. An example is shown of the assembly of a contig on chromosome 5. In the later stages of the map construction, probes are not sequenced. As more probes are tested, larger and larger contigs are constructed. The object is to develop complete contigs (overlapping sets of clones) for each chromosome. The ultimate goal, of course, is one contig per chromosome. This will allow the analysis of the location of repeated DNA and of the frequent chromosomal translocations. The finished map will facilitate the sequencing project by allowing the assignment of alleles with significant heterozygosity and locating any areas where sequence is unavailable.

sequence of large genomes, but the increasing capacity of computers has meant that even the human genome sequencing is now being carried out by the shotgun approach [4].

The physical map is very useful for the study of synteny between genomes. Information provided by synteny studies can provide inferences about the phylogenic relationships and the type of rearrangements that may have occurred since divergence from a common ancestor. Over evolutionary time, the gene order on chromosomes can be reorganized by deletion, translocation, inversion, and gene duplication. This reorganization can be so extreme that little or no conservation of gene order can be detected, as is the case with *Haemophilus influenzae* compared to *Escherichia coli*. Conversely, almost total conservation may be easily recognized, as is seen between *Mycoplasma genitalium* and *M. pneumonia* (http://www.zmbh.uni-heidelberg.de/M__pneumoniae/genome/G__Comparison.html). Conservation of gene order has been reported among ascomycetous fungi; not surprisingly, gene order conservation at the loci studied decreases with evolutionary distance [5]. In the study by Keogh et al. [5], it was found that 59% of adjacent gene pairs in *Kluveromyces lactics* or *K. marxianus* are also adjacent in *S. cerevisiae*. Furthermore, 16% of *Kluyveromyces* gene order can be inferred from ancestral gene order in *Saccharomyces* prior to the ancient whole genome duplication event [6]. This is not surprising when one considers the close evolutionary relationship of these two genera, a relationship that has been supported by several independent phylogenetic analyses. Conversely, the phylogenetically distinct *C. albicans*, and even more distant *Schizosaccharomyces pombe*, genomes contain 13% linkage and no linkage supporting data, respectively [5].

One of the first steps of the genome project is the annotation of genes. This involves locating the ORFs usually defined as a DNA sequence encoding an uninterrupted string of 100 or more amino acids beginning with methionine [7]. The choice of 100 amino acids is arbitrary and certainly leads to the omission of some functional genes; these may be identified as the functional genomic analysis progresses. Ultimately, the cellular role of the protein encoded by each ORF will be determined experimentally; however, computational methods can give insight into the function by identifying sequences in the database which are similar to the unknown gene. The annotation accompanying any such sequences can give the researcher a list of possible functions for the unknown ORF, thereby providing direction for the necessary experimental research to determine the protein's role. The most commonly used tool for sequence comparison is the database search tool BLAST (http://www.ncbi.nlm.nih.gov/BLAST/) [7]. BLAST utilizes an algorithm that produces a High-scoring Segment Pair (HSP). The HSP is a comparison of two sequence fragments of arbitrary but equal length that can produce an alignment score that meets or exceeds a predetermined cutoff score for similarity. Analysis of BLAST results requires thoughtfulness and care on the part of the investigator. Utilizing too stringent a scoring cut-off will result in the elimination of genes that are, in fact, similar to the sequence being tested, whereas a too-lax alignment may suggest a relationship that doesn't exist [7].

There are additional limitations to BLAST searches, the most important of which is misleading or erroneous sequence annotation in the database, which allows one to go "wrong with confidence" as recent work by Pallen [8] illustrates. Pallen [8] points out that misinterpretation of BLAST results in one paper led the authors to identify an unknown gene from *Campolobacter jejuni* as producing a protein homologous to one involved in type III secretion systems. The homology found was relatively small, and the complete genome sequence of *C. jejuni* lacks any of the other genes required for this system. Pallen [8] concludes, "Claims that sequences are sufficiently similar to be considered homologous . . . should only be made after unprejudiced searches of sequence databases using sequence analysis methods that provide estimates of statistical significance of sequence similarities." Because pairwise comparisons permit the alignment of any given sequence with any other given sequence depending upon gaps and definitions of sequence similarities [7], a distribution of HSP scores (listed in decreasing values of significance) can result from a given search. One should avoid imputing highly significant values in favor of expected (or desired) matches.

Misassignment of sequence is but one problem associated with BLAST results. Many sequence similarity results are simply uninformative. For example, BLAST search results may show a significant level of sequence identity to an orthologous gene annotated as a suppressor of another function but of unknown mechanism. The rapidly increasing number of ORFs of unknown function arising from genomic sequencing projects means that identification of even very similar orthologs may provide little or no information. A major problem due to the volume of data that genome projects have produced is the assignment of function to ORFs based solely upon sequence similarity searches without empirical evidence.

Additional Web-based database search engines have been created to provide more information from sequence data. One of the first and most widely used programs, PIR (Protein Information Resources; http://pir.georgetown.edu) provides researchers the computation necessary to examine protein domains and structure, molecular evolution, and functional genomics. Another, SMART (Simple Modular Architecture Research Tool) is designed to identify and annotate functional domains [9]. By examining the yeast genome with SMART (http://www.bork.embl-heidelberg.de/Modules/sinput.shtml), Schultz et al. [9] were able to identify 350 more signaling domains than were previously identified. A graphical display is provided showing the positions of these domains within the query sequence. Domains are linked to Medline and the Molecular Modeling Database via Entrez. This provides easy access to the data relating to homology/structure/function. Such programs, like BLAST searches, are not reliable indicators of gene function when experimental data do not exist to allow these domains to be identified. An example appropriate for a discussion of fungal genome projects is the analysis by Schultz et al. [9] of the *INT1* gene from *Candida albicans*. This gene was isolated as an integrin ortholog by using a cDNA probe to the transmembrane domain of human αM-integrin. The authors then compared the sequence of *INT1* to the characteristic motifs of several integrin α-subunits using BESTFIT [10]. In doing so, they identified a putative integrin I-domain with 18% sequence identify to the human αM [10]. Within this domain they identified an area with 25% sequence identity to the fibrinogen binding protein of *Staphylococcus aureus* [10]. These values are on the cusp of significant probability [7]. Additionally, a hydrophobic sequence and a putative membrane spanning domain were identified [10]. Schultz et al. [9] argued that the absence of a detectable transmembrane domain and the identification of a PH-domain suggests that *INT1* is not an integrin-like protein but is more homologous to *S. cerevisiae BUD4* [9]. However, experimental evidence refutes this contention (C. Gale, personal communication).

First, the role of *INT1* in adhesion, filamentous growth, and virulence has been demonstrated experimentally [11], and continued work suggests it neither functions as a *BUD4* homolog in bud site selection nor rescues *bud4* mutants in *S. cerevisiae* (C. Gale, personal communication). Interestingly, the *C. albicans* genome project has generated annotated sequence equivalent to five genomes worth of coverage and a *BUD4* homolog in *Candida* has not been identified. A number of possibilities remain. Could *INT1* have a specific developmental role in the yeast to hyphal transition? One intriguing hypothesis is that as the production of true hyphae evolved, the *INT1* gene evolved to fulfill additional functions in *Candida* beyond that of *BUD4*, and that the new properties prevent it from functioning in a homologous fashion in *Saccharomyces*. In spite of this conjecture, recent BLASTX searches of *INT1* suggests that a similar gene exists in *Schizosaccharomyces pombe*. If *INT1* is necessary for additional functions in *C. albicans*, what is an *INT1*-like gene doing in *S. pombe*? In summary, the inference of function by sequence homology alone is a hazardous undertaking.

IV. FUNCTIONAL GENOMICS

The publication of the completed genome of many bacteria and the yeast *Saccharomyces cerevisiae* provides a framework for the development of functional genomics. To date, all published genome sequences have included the identification, classification, and explanation of genes by cellular role, if known. The status of the number of known and unknown genes is compared and finally, unique features of a given organism are highlighted. Upon completion of the preliminary analysis of the structural genome, the functional genomics of many organisms has faltered [1]. A major exception to this statement has been the analysis of the *S. cerevisiae*.

The first published examination of total gene expression (transcriptome) in yeast was done by serial analysis of gene expression (SAGE) [12]. This process is based on three concepts: (1) A short sequence tag contains sufficient information to uniquely identify a transcript provided that the tag is obtained from a unique position; (2) this sequence can be used to identify a given transcript; and (3) quantitation of the number of times a given tag appears provides the level of expression of the corresponding transcript. Through the use of polylinkers, tags of all the transcripts in the cell can be concatenated to create a chain of DNA fragments, which is then sequenced. The number of times a given tag appears in the sequence enables the researcher to quantitate the number of mRNA copies of a given gene per cell. In the original study [12], cDNA libraries of yeast cells in log-phase, S-phase-arrested, and C2/M were created. From these libraries, the authors examined 60,633 transcripts that were produced by 4665 genes. Transcription levels ranged from 0.3 to >200 transcripts per cell [12]. Only 1981 of the ORFs found in this study had previously known functions at the time of publication, compared to the 2684 ORFs that were uncharacterized. Although some genes had high levels of expression, clusters of specific expression levels were not readily apparent. Telomeric position effect was identified, with an average of ~3.2 tags/gene seen within 10 kb of the telomere compared to 12.4 tags/gene outside of the telomeric region [12]. Although some limitations to the SAGE method exist (small or extremely low copy number transcripts may escape detection; transcripts with long 3' untranslated regions may be misassigned), these data set the stage for future forays into functional genomics [12].

SAGE was instrumental in the development of functional genomics. However, it is the creation of microarrays (and their reproducibility) that has redefined the field. Microarray technology is, quite simply, the miniaturization of filter-based assays (macroarrays) like those used to probe libraries. Key differences between macroarrays and microarrays include reproducibility, parallelism, and automation [13]. To date, most microarrays are used to assay gene expression under a variety of conditions. Unlike previous methods such as differential display, which determined a threshold level of gene expression but did not give quantitative data, microarray technology shows the relative level of expression for all genes being expressed (the transcriptome). For this reason, microarray technology is producing enormous amounts of data to be analyzed. The sheer volume of data requires both statistical and computational analysis. To date, microarrays of the *S. cerevisiae* genome have been used to study strain variation, gene expression under differing conditions, such as heat and cold shock, and glucose derepression [14] and variation in ploidy [15]. Microarray technology has been used to examine genome-wide patterns of gene expression in strains undergoing adaptive evolution [16]. Ultimately, one would anticipate that changes in gene expression would be determined for multigenic phenotypes that have been previously difficult to characterize, and that this would provide an understanding of a myriad of biological processes. Examples of such complicated multigenic traits include the phenotypic switching of *C. albicans* [17,18], regulation of conidiation of *A. nidulans* and *A. fumigatus* [19], and appressorial development and surface recognition in *M. grisea* [20–22].

A third approach to functional genomics is the disruption of every gene in the genome, followed by examination of the phenotype of each mutant. This requires the construction of a library of strains, each of which has one gene disrupted. For *Saccharomyces cerevisiae*, which has ~6000 genes, an international consortium has undertaken this effort, with separate genetic regions assigned to different laboratories. Yeast biologists have constructed a library of 6000 signature-tagged deletion mutant strains (one for each ORF identified) [23]. These signature-tagged deletion mutant strains allow yeast biologists to examine the role of the protein encoded by each ORF under specific conditions. This approach provides a system to identify genes essential under a particular conditions (those mutant strains that did not survive) and permits characterization of "orphan" ORFs—i.e., ORFs without an assigned function.

V. THE *SACCHAROMYCES CEREVISIAE* GENOME DELETION PROJECT

A Web page exists for this project (http://www-sequence.stanford.edu/group/yeast__ deletion__project/deletions3.html). Such an international project requires agreement on the method of gene disruption so that the various mutants can be compared directly. For the *Saccharomyces cerevisiae* Genome Deletion Project, the approach was the use of polymerase chain reaction (PCR) technology with the dominant selectable marker being gentamycin resistance

[23]. Primers with ~50 bp of homology to the genomic sequence on each side of the gene to be deleted and with 25 bp of homology to the selectable marker were used. This approach works because *Saccharomyces cerevisiae* has a very high rate of homologous recombination. Obviously, the phenotypic analysis of these deletion mutants is a daunting undertaking, especially since many of the unknown ORFs are not essential for viability under a variety of laboratory conditions. Additionally, less drastic phenotypes may manifest themselves only under specific growth conditions. It had been assumed that expression of a gene under a specific growth condition is indicative of the gene's importance in that condition. However, in one comprehensive *S. cerevisiae* study, deletion of a gene which was highly expressed under particular growth conditions did not affect the phenotype shown by the cells under those conditions [23]. This paradox has been seen in other fungi in less comprehensive studies [19].

One important goal of all fungal genome projects has been the identification of potential antifungal drug targets. Essential genes, preferably those without human homologs, are considered to be the best potential antifungal drug targets. Hence, in any genome project, the identification of the vital genes is very important. In the *S. cerevisiae* gene disruption project, 6925 strains, each with a precise deletion in one of 2026 ORFs [23], have been described. Winzeler et al. [23] found that 17% (or 356) of the deleted ORFs were essential for viability under the conditions tested. Although the paper claims that the list of essential genes is available, the cited list of the 356 ORFs is not available for the general scientific community.

Functional genomic approaches exist that do not require the complete sequence of the genome. Two such approaches include Signature-Tagged Mutagenesis (STM) and In Vivo Expression Technology (IVET). Both of these approaches identify genes through a random process; both can effectively scan the genome and both are labor intensive. Signature-tagged mutagenesis uses a mix of mutants constructed by random mutagenesis to infect a host. Each mutant has a distinguishable molecular marker, usually a unique sequence of DNA placed between two primer sites that are the same in all strains. The multiple PCR products prepared from DNA isolated from the mix of mutants used for the infection are compared with products from the mix isolated from the moribund host. Any strains present in the input pool but absent in the mix isolated at the end of the infection will be identifiable by the absence of their PCR products in the second pool. Mutants that are lost during the infection process are inferred to have lost the function of a gene that is essential for infection. This procedure has been used to identify a gene that is required for adherence in *Candida glabrata* [24]. Curiously, the identified gene was not found to have a role in initial colonization or persistance in multiple in vivo infection models [24].

In vivo expression technology uses a promoterless copy of a gene that is essential for growth in the host. A library of fragments of genomic DNA is inserted 5' to the gene and the library is transformed into a strain lacking the gene. The mix of transformants is used to infect a host, and the cells that are able to establish an infection are assumed to have constructs with a promoter for a gene that is normally expressed during infection. The promoter is then rescued and its normally adjacent gene is identified. This approach has been successfully applied to *Histoplasma capsulatum* [25].

VI. FUNGAL GENOME PROJECTS PRESENTLY UNDER WAY

In addition to the very well advanced *Saccharomyces cerevisiae* genome project, several other fungal genome projects are under way. It should be noted that the genomes of several fungi (*C. albicans*, *A. nidulans*, and *M. grisea*) have probably already been determined several times on a proprietary basis by chemical companies. Genome projects for these same fungi and *Pneumocystis carinii* and *Cryptococcus neoformans* are also being carried out in the public domain.

A. Candida albicans

The first pathogenic fungus to be subjected to genomic analysis was *Candida albicans*. *C. albicans* is the single most important human fungal pathogen, and it is diploid as usually isolated. It has eight pairs of homologous chromosomes, and the size of the haploid equivalent of its genome is 16 Mb, about one-third larger than that of *S. cerevisiae*. With the development of parasexual genetics in *Candida* in 1981, and its similarity to *Saccharomyces*, investigators were encouraged to attempt similar kinds of experiments. Electrophoretic karyotypes of various strains were produced in the late 1980s [26–28], but the assignments of genes to chromosomes did not begin until around 1990. The question of chromosome number was resolved in 1991 [29], and in 1993 the first macrorestriction map of the *Candida* genome was published [30]. This map was based on the fragments produced by the 8-b-p-specific enzyme *Sfi*I, which cuts 34 times in the diploid *C. albicans* genome. The genomes of two well-studied strains, 1006 [31], and WO-1 [32], were analyzed in Chu et al. [30]. This map showed several properties of the genome which have proved to be fundamental. The genomes of both strains showed polymorphisms with respect to *Sfi*I restriction sites, so that in several cases sites present on one homolog were absent on the other. This suggested that there is a reasonable level of polymorphism at the level of DNA sequence. The two homolog of chromosome 7 of 1006 [33] are separable by pulsefield gel electrophoresis, although the *Sfi*I restriction patterns are identical. Compared to 1006 and most other *Candida* strains, WO-1 has an aberrant karyotype, with 10 electrophoretic bands instead of the usual eight. Chu et al. showed that the extra bands in WO-1 were reciprocal translocation products of the chromosomes found in most strains [30]. A similar finding had been made by Thrash-Bingham and Gorman, using entirely different isolates [34]. Chu et al. also showed that the three translocations in WO-1 have all occurred in the areas possessing the *Sfi*I restriction sites [30]. The macrorestriction map used the chromosomal nomenclature agreed upon in Palm Springs, at the first ASM meeting on *Candida* and candidiasis. This resulted in a standard nomenclature in which the ribosomal DNA chromosome was designated R, since it changes size and has no reproducible position on a pulse-field electrophoresis gel, and the others being numbered 1 to 7 in decreasing order of size.

Extensive work in the laboratory of Tanaka in Japan showed that a middle repeat sequence, since named the Major Repeat Sequence (MRS), occurred several times in the *Candida* genome and contained multiple sites for the *Sfi*I enzyme [33,35–37]. It is now clear that at least 22 of the *Sfi*I sites are located in MRS sequences. The MRS sequence has three parts, called HOK1, RPS, and Rb2. HOK1 and RB2 appear once in each MRS cluster; RPS can be repeated up to 10 times. Chromosome length polymorphisms between homologs and between strains are often due to differences in the number of RPS repeats in the MRS on the two homologs [37a]. Repeats have been shown to undergo recombination and change in size [38].

The observation in WO-1 that the translocations occur at *Sfi*I sites seems to be true for a variety of strains. In fact, although the karyotypes of isolates can vary significantly from the karyotype of the standard strains (such as 1006 or SC5314 [39], the size and pattern of *Sfi*I fragments are remarkably constant in *C. albicans*. This is due in large part to the fact that RPS has several *Sfi*I sites, so that the products of *Sfi*I digestion have lost all but the flanking regions of the MRS, and these are common to all MRSs. The RPS repeats, which are the basis of chromosome length polymorphism, are digested away. Chu et al. [30] labeled the *Sfi*I fragments in increasing order of size from AA (at 50 kb the smallest) to U (at 2200 kb the largest). The size range and even the individual sizes of these fragments are so consistent that it is almost impossible to distinguish one strain from another based on the *Sf*I digestion pattern despite the large number and broad size range of the fragments. Hence, the striking divergence of electrophoretic karyotypes found in clinical isolates [40] seems due to translocations which have occurred at the MRS. The effect of these translocations on gene expression is not known at this time.

To understand fully the complex biology and pathogenicity of *Candida albicans* as well as to identify potential drug targets, development of a physical map began in 1995 at the University of Minnesota (Fig. 1). In 1998 the complete map of chromosome 7 appeared [33]. The map is based upon a library of genomic DNA from the strain 1161 (derived from strain 1006 by selection on 2-deoxygalactose to yield a gal⁻ strain) contained in the fosmid vector. This vector, developed to clone human DNA, contains up to 40 kb of insert and is maintained in *E. coli* at one copy per cell, so that it allows one to clone repeated DNA like the MRS. The library consists of 3840 clones, about 10 genome equivalents. The partial map has been useful in demonstrating the existence of gene families, like that for the secreted aspartyl proteinase (SAP) genes [41], for locating the MTL locus [42], and for identifying the chromosomal location of a variety of genes important in pathogenesis. The map is expected to be complete by the summer of 2001.

The map of chromosome 7 was constructed as described above, but to increase the precision of the map, Chibana et al. used random breakage mapping to determine the exact size of the fragments produced by *Sf*I digestion [33]. The final map revealed several important features of *C. albicans* chromosomes as well as specific aspects of chromosome 7. First, the map was complete, so that there is no region on chromosome 7 that is unclonable in *Escherichia coli*, using the fosmid vector. Second, there is a subtelomeric repeat which has homology to the reverse transcriptase of LINE-type retrotransposons [33]. This repeat is found adjacent to many, if not all, of the telomeres in *C. albicans*. Chromosome 7 has two MRS sequences; these are arranged in an inverted repeat orientation [33]. The separable homologs in strain 1006 result from a difference in the number of repeats of RPS in one of the MRS regions. It seems likely that this mechanism is responsible for much of the karyotypic variation seen in pulse-field gel electrophoresis of *Candida albicans* chromosomes (Chibana et al., submitted).

In addition to the physical map, the complete sequence of the *Candida* genome is being determined. This effort, supported by the Burroughs Wellcome Fund and the National Institute for Dental and Craniofacial Research, is being carried out at the Stanford Genome Center. The sequence of strain SC5314, the parent of CAI4 [39], the strain most commonly used for molecular genetic experiments, is being determined by shotgun sequencing of random clones followed by assembly of the sequence into contigs. An assembly based on sequence amounting to 5.5 genome equivalents is available (http://www-sequence.stanford.edu/group/candida/) and an assembly based on 10× coverage is expected by the summer of 2000. At that time, the effort will turn to finishing the sequence (closing the gaps between the contigs) and resequencing areas of ambiguous sequence. At about the same time, efforts will be turned to anchoring the sequence in the map, so that the two will be colinear and one will be able to move from map to sequence and reverse with ease. Although the two strains used in the map and sequencing projects are not related, their karyotypes are very similar, and so far no divergences in gene order or chromosome organization have appeared. Hence, the two databases will be complementary.

The availability of the complete sequence of the *Saccharomyces cerevisiae* genome has permitted a number of interesting inferences to be drawn from comparisons of the sequence data from the two species (S. Scherer, personal communication). One is the significantly greater diversity of catabolic pathways found in *Candida* (Table 1). This most likely reflects the diverse niches in the body that *Candida* can occupy compared to the relatively uniform life style of *Saccharomyces*. Other differences which may relate to pathogenicity are gene families such as SAP [41] and ALS [43–45], the latter encoding proteins which affect adherence to mammalian cell surfaces. The identification of genes, essential in *Saccharomyces*, which are shared by the

Table 1   Putative Catabolic Genes Located in *Candida albicans* by Sequence Homology

Gene name	Putative function	Homolog found in
HQD1, HQD2	Hydroxyquinol-1,2 dioxygenase	*Ralstonia pickettii*
TCX1, TCX2	Phenol hydroxylase	*Trichosporon cutaneum*
PSX1	Dioxygenase	*Pseudomonas* sp.
ALK1-ALK8	Alkane inducible cytochrome P450	*Candida maltosa*
AMI1, AMI3	Acetamidase	*Emericella nidulans*
AMI2	Formamidase	*Methylophilus methylotrophus*
ASM1	Acid sphingomyelinase	*Caenorhabditis elegans*
ECH1	Enoyl-CoA hydratase	*Caenorhabditis elegans*
POX18	Oleate-inducible peroxisomal protein	*Candida tropicalis*

These are genes whose closest homolog (listed) is in an organism other than *Saccharomyces cerevisiae*. Further information about these genes can be found on the *Candida albicans* Information Page (http://alces.med.umn.edu/candida.html).

two fungi will provide a list of potential drug targets. Ultimately, the full effect of the *Candida* genome project will be felt as previously unknown virulence factors are identified and as new drug targets are discovered and exploited.

The partial coverage presently available has already yielded two very interesting observations. The first is the identification of a sequence homologous to the *MAT* locus in *Saccharomyces cerevisiae* [42]. These idiomorphs are located in a heterozygous region of about 10 kb on chromosome 5. They were found because the sequencing project identified a sequence with high homology to *MAT*a1, and cloning of the gene and its adjacent regions yielded a sequence on the other homolog related to *MAT*α. The second observation is the identification of a number of sequences that seem to be complete or partial retrotransposons. These were identified in a genomics-like approach by Goodwin and Poulter [46]. By screening the *C. albicans* sequencing database at Stanford, Goodwin and Poulter were able to identify at least 350 retrotransposon insertions [46]. They found that the retrotransposon population differs significantly from the *Saccharomyces* population in both diversity and the fact that they retain functional examples. They raise the possibility that the nonstandard codon usage of *C. albicans* may be responsible for generating the retrotransposon diversity in this organism. An equally plausible hypothesis is that the nonstandard codon usage prevented new retrotransposon insertions by mistranslating CUG, thereby stabilizing a genome that was threatened with being overrun by transposons.

B.   *Aspergillus nidulans*

Aspergillus nidulans has been considered a model system for the study of both fungal genetics and fungal development [19]. Prior to the development of molecular biology, the *A. nidulans* researcher was able to use classical genetics and tetrad analysis, parasexual genetics, mitotic crossing over, and the mapping of translocation break points in isogenic strain backgrounds for genetic analysis [47]. Combining these data in the production of a fine-resolution linkage map provided an independently verified ordered framework of markers. These markers provide anchorage for a physical map.

The first step in the construction of the physical map was resolution of the chromosomes using pulse-field gel electrophoresis. Isolated chromosomes were used as probes of two genomic libraries consisting of >5,000 clones [48]. Clones were divided into subsets based upon the frequency of hybridization to each chromosome; one of these subsets consisted of chromosome-

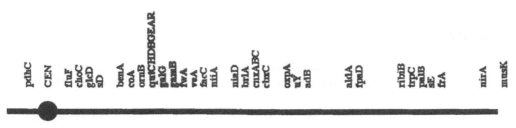

Linkage Map of Chromosome VIII

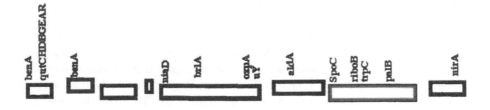

Physical Map of Chromosome VIII

Figure 2 Comparison between the linkage and physical maps of *A. nidulans*. Chromosome 8 of *A. nidulans* is compared between the linkage data [47] and the physical mapping data [48]. Gaps in the physical map of chromosome 8 could be cloned by complementation of known mutants that have linkage assignments in the general vicinity of the gaps, or by chromosome walking.

specific clones, a second of clones that were positive with multiple chromosomes, and a third of clones containing sequences common to all chromosomes [48]. Chromosome-specific probes were then used to probe the library, and identified clones were not used to reprobe the library [48]. This method is referred to as sampling without replacement. Clones are given a binary tag to identify the probes which hybridize to them [48]. The collection was then ordered and contigs created. Contigs were then further compressed to make minimum contigs.

The resulting physical map encompasses 94% of the library and contains >49 contigs that represent both chromosomes and chromosome fragments [48]. Because a high-resolution linkage map exists, cosmids needed to close the map can be easily identified by traditional genetics. Identified cosmids were then used to begin chromosome walking to collapse contigs into discrete chromosomes (Fig. 2). Many of the classical genetic markers have not yet been cloned, and not all molecular probes have been mapped. However, a repository of *A. nidulans* strains exists at the Fungal Genetics Stock Center (http//:www.fgsc.net).

C. *Pneumocystis carinii*

Pneumocystis carinii was first described in 1909; at that time it was thought to be a form of lung-specific trypanosome. It was not until 1989, with the increasing incidence of *Pneumocystis carinii* pneumonia (PCP) due to the AIDS epidemic that molecular techniques identified *P. carinii* as a member of the fungal kingdom [49,50]. That this organism was misclassified for 75 years underscores its intractability as an experimental organism. The single most important problem confronting the *P. carinii* researcher is the inability to culture the organisms in vitro;

however, a recent report suggests that an axenic culture system may be possible [51]. At this point, however, genomic analysis holds the greatest promise as a technique to understand its biology.

Today, *Pneumocystis carinii* is recognized as having many special forms or formae speciales (f. sp.). The most highly studied of these is *Pneumocystis carinii* f. sp. *carinii*, which is found in laboratory rats, and is the organism whose genome sequence is being determined [52–54]. The genome size is 8 Mb and is divided into 13–15 linear chromosomes; the organism as isolated is haploid. Chromosome sizes range from 300 to 700 kb and karyotypes are variable between and within formae speciales. Interestingly, many of the genes characterized to date have small introns; as many as nine introns may occur in a single gene [54]. There are three gene families of repeat DNA, two of which encode proteins and the third of which is telomeric. This telomeric repeat is typical of eukaryotic chromosomes, but is nontypical in that it is found at some nontelomeric loci, including upstream of the α-tubulin gene [55]. One of the more surprising discoveries of the *Pneumocystis carinii* f. sp. *carinii* genome project is that the ribosomal DNA cluster (18S, 26S, and 5.8S rRNA) is not repeated more than twice [55].

The *Pneumocystis carinii* f. sp. *carinii* Genome Project (PCGP) is under way. The small sizes of the chromosomes will permit karyotype resolution by CHEF and ease in the preparation of the physical map. The PCGP is expected to advance research on this obligate parasite in a variety of ways. Because the lack of a consistent and reliable method of *P. carinii* propagation is the most serious obstacle in the development of diagnostic methods, drug testing, and basic biological studies, one obvious impact the PCGP would be the identification of the metabolic pathway deficiencies that presently prevent in vitro culture.

D. *Magnaporthe grisea*

The rice blast fungus, *M. grisea*, has become a model organism for the study of plant pathology at all levels, from population [56] to molecular genetics [57,58]. The fungus is haploid and usually reproduces asexually, and population studies suggest that reproduction is clonal [56]. In 1974, the perfect state of this fungus was identified, and fertile strains that can be crossed in the laboratory have been developed [59]. Unlike the zoopathogens mentioned here, the life cycle of *M. grisea* includes the development of a specialized infection structure, the appressorium, required for infection [21]. Development of chemical or genetic methods that prevent appressorium formation has become a target of *Magnaporthe* research. Infection by *M. grisea* requires melanin, and inhibition of melanin biosynthesis results in loss of pathogenicity [58]. Although numerous genes involved in the infection process have been identified, no uniformly successful method of control has been found, and this pathogen is still a major problem in developing countries. The *M. grisea* genome project will accelerate progress in understanding the mechanisms of plant pathogenesis exemplified by this fungus. Additionally, information from this project may provide an opportunity to advance the understanding of fungal zoopathogens by identification of those processes that may have been conserved among fungal pathogens irrespective of host.

M. grisea, like *A. nidulans* and *C. albicans*, contains many repetitive DNA elements. Romao and Hamer [60] were able to construct a map consisting of eight linkage groups through the mapping of repetitive elements in the *M. grisea* genome. A genetic map was constructed by Skinner et al. [61] that was based on segregation of 61 progeny. Additionally, they identified nine classes of repetitive DNA elements. This work provided a foundation for the construction of the physical mapping part of the *M. grisea* genome project.

Construction of the *Magnaporthe grisea* physical map is under way. The map of chromosome 7, the smallest of *M. grisea*'s chromosomes (4.2 Mb), has been completed, using a large insert bacterial artificial chromosome (BAC) library [62]. Because of the presence of repetitive elements, a single-copy sublibrary possessing $12\times$ genome coverage was created [62]. To construct this map, chromosomes were separated by CHEF electrophoresis and chromosome 7 DNA was recovered. This DNA was used to probe the single-copy sublibrary to identify those inserts containing chromosome 7 DNA [62]. The identified clones were then used as probes against both the BAC library and the CHEF-separated chromosomes for additional confirmation that the identified BAC clones did in fact contain chromosome 7 DNA [62]. One hundred forty-seven BAC clones specific to chromosome 7 were identified [62]. To locate additional BAC clones, a strategy of hybridization without replacement was used [48] to bring the number of identified chromosome 7–containing BAC clones to 585 [62]. The final 40 BAC clones were identified using 20 RFLP markers found on chromosome 7 [62]. Data were analyzed using a random cost algorithm to the developed binary matrix as per Prade et al. [48]. This produced nine contigs. However, with only a single clone linking some contigs, and multiple clone assignments to different contigs, contig assembly by DNA fingerprint was employed. All 625 BAC clones were subjected to DNA fingerprinting, and data were analyzed by software using different thresholds or cutoffs to determine if fragments were contiguous. Results still showed anomalies and deficiencies. Ultimately, the data from both the binary matrix and BAC fingerprinting were combined, and chromosome walking was undertaken to seal gaps between adjacent contigs. This resulted in closure of four of the eight gaps. The remaining four gaps could not be closed due to the presence of repetitive elements. As found with *A. nidulans* and *C. albicans*, repetitive elements tended to cluster [48,62].

At this point, the entire BAC library is being fingerprinted and assembled into contigs. Contigs will be anchored to chromosomes by hybridization of available genetic markers and ESTs produced by the parallel EST project. Finally, end sequences of BAC clones combined with the ESTs are expected to reveal a large number of the ~10,000 genes in the *M. grisea* genome and provide anchors between the sequence data and the physical map [62].

VII. COMPARATIVE GENOMICS

One of the primary justifications in the undertaking of fungal genome projects is to develop a biorationale approach for the development of antifungal drugs. Ideally, drug targets would be essential for fungal viability but not have a mammalian homolog that the drug could possibly antagonize. Genes that are essential for the production of the fungal cell wall (β1–3 glucanase, chitinase, etc. and cell membrane (ergosterol biosynthesis) are obvious targets. Other possibilities include genes involved in polyunsaturated fatty acid production and the unique biosynthetic pathways fungi use for amino acid catabolism, as is the case for lysine. A similar approach with a different tack would be the examination and comparison of the genomes of the pathogenic fungi *C. albicans*, *A. fumigatus*, and *M. grisea* to determine if specific genes exist that facilitate a pathogenic lifestyle. Comparison between *A. fumigatus* and the closely related *A. nidulans* can provide even more interesting avenues of research in understanding both the pathogenic and asexual lifestyle of *A. fumigatus* compared to *A. nidulans*.

In addition to the development of novel antimycotics, fungi such as *A. nidulans* are important industrially for the production of citric acid and industrial enzymes such as amylases, proteases, and pectinases. Members of the Aspergilli are also utilized in soy fermentation (*A. oryzae*), and for the production of novel pharmacological compounds such as griseofulvin and cephalosporin. *M. grisea*, with its plethora of adhesive glycomolecules which adhere even to Teflon is another fungus that a genomic approach may elucidate the mechanism behind the production of these compounds.

Table 2 Distribution of Repetitive Elements in Fungi

Fungus	Retrotransposon	Line-like	Sine-like	Other
Saccharomyces cerevisiae	331 LTR-like fragments Ty1(217),2(34),3(41),	None	None	None
Candida albicans	>350 LTR-like fragments	Subtelomere repeat CARE/REL-2	Unknown	MRS (RPS, HOK, RB)
Aspergillus nidulans	Unknown	Unknown	Unknown	F2PO8 (Fot1/ Pogo-like)
Aspergillus fumigatus	Afut1	Unknown	Unknown	Unknown
Magnaporthe grisea	Fosbury, Grasshopper, Magay	MGR583	MGR-SINE, MGSR1	Unknown
Pneumocystis carinii	4 families of 11–23 nucleotide repeats	Unknown	Unknown	9 nucleotide repeat 16-kb fragment MSG and PRT

In examining fungal genomes, some similarities readily present themselves. One of these is the presence of repetitive DNA, particularly the ubiquitous nature of LTR-retrotransposon families. Table 2 lists the types of retrotransposable elements present in the genomes of fungi that we have examined here. To date, LTRs appear to be present in all the genomes except for *A. nidulans*. Many groups [46,63,64] have proposed that the presence of multiple repeats may act as scaffolds for translocation events. The physical map of *Candida* will enable researchers to critically evaluate this hypothesis through the mapping and cloning of translocation breakpoints.

Table 3 Current Web Site Addresses for Fungal Genome Projects

Organism	Web site	Salient features
Candida albicans	http://alces.med.umn.edu/ Candida.html	Physical maps, sequence data, review of *C. albicans* biology, techniques, programs for molecular analyses
Aspergillus nidulans	http://fungus.genetics.uga.edu:5080/Physical_Maps.html	Physical maps, software for genome analysis
Aspergillus nidulans	http://www.genome.ou.edu/ asper_blast.html	Mixed cDNA library of vegetative and asexual development
Aspergillus fumigatus	http://www.aspergillus.man.ac.uk/ index.htm	Links to genome sequencing, reviews, protocols, bibliographic database
Pneumocystis carinii	http://biology.uky.edu/Pc/	EST data, karyotype, review of *P. carinii* information
Cryptococcus neoformans	http://www.genome.ou.edu/ cneo.html	EST database of *C. neoformans*
Magnapothe grisea	http://ascus.cit.cornell.edu/ blastdb/	Genetic and physical maps, marker data, bibliography

Current work is underway (Chibana et al., in preparation) to map the MRS in *C. albicans* to provide a foundation for such an undertaking. The repository of *A. nidulans* translocation mutants also permits similar studies in a filamentous fungus [47].

In conclusion, the fungal genome projects currently being undertaken provide a wealth of data that was unimaginable just a decade ago (Table 3). The sheer volume of data that genome projects produce has made the use of the computer a biological necessity in order to organize the data in a systematic fashion. Using this computer-generated organization, bioinformaticists will be able to discern the patterns and relationships within a given species and between species, and try to elucidate biological phenomena and processes. Although the tools to examine the rapidly increasing volume of data are not perfect, many new insights are already appearing because of this global approach to biology. The fungi are a diverse group, and comparisons within this kingdom will be extremely fruitful in expanding our knowledge both of the properties its members share with other organisms and of their particularities. Among the latter will surely be new and interesting mechanisms of pathogenesis as well as presently undiscovered drug targets.

ACKNOWLEDGMENTS

The work from this laboratory described here was supported by grants AI16567, AI35109, and AI46351 awarded to P.T.M.

REFERENCES

1. P Hieter, M. Boguski. Functional genomics: it's all how you read it. Science 278:601–602, 1986.
2. A Goffeau, BG Barrell, H Bussey, RW Davis, B Dujon, H Feldmann, F Galibert, JD Hoheisel, C Jacq, M Johnston. Life with 6000 genes [see comments]. Science 274:546, 563–567, 1996.
3. W Olivas, D Muhlrad, R Parker. Analysis of the yeast genome: identification of new non-coding and small ORF-containing RNAs. Nucleic Acids Res 25:4619–4625, 1997.
4. JC Venter, MD Adams, GG Sutton, AR Kerlavage, HO Smith, M Hunkapiller. Shotgun sequencing of the human genome. Science 280:1540–1542, 1998.
5. RS Keogh, C Seoighe, KH Wolfe. Evolution of gene order and chromosome number in *Saccharomyces*, *Kluyveromyces* and related fungi. Yeast 14:443, 1998.
6. K Wolfe, D Shields. Molecular evidence for an ancient duplication of the entire yeast genome. Nature 387:708–713, 1997.
7. SF Altshul, W Gish, W Miller, EW Myers, DJ Lipman. Basic local alignment search tool. J Mol Biol 215:403–410, 1990.
8. M Pallen, B Wren, J Parkhill. Going wrong with confidence. Mol Microbiol 34:195, 1999.
9. J Schultz, F Milpetz, P Bork, C Ponting. SMART, a simple modular architecture research tool: identification of signaling domains. Proc Natl Acad Sci USA 95:5857–5864, 1998.
10. C Gale, D Finkel, N Tao, M Meinke, M McClellan, J Olson, K Kendrick, M Hostetter. Cloning and expression of a gene encoding an integrin-like protein in *Candida albicans*. Proc Natl Acad Sci USA 93:357–361, 1996.
11. CA Gale, CM Bendel, M McClellan, M Hauser, JM Becker, J Berman, MK Hostetter. Linkage of adhesion, filamentous growth, and virulence in *Candida albicans* to a single gene, INT1. Science 279:1355–1358, 1998.
12. V Velculescu, L Zhang, W Zhou, J Vogelstein, M Basrai, DJ Bassett, P Hieter, B Vogelstein, K Kinzler. Characterization of the yeast transcriptome. Cell 88:243, 1997.
13. M Schena, RA Heller, K Konrad, E Lachenmeier, RW Davis. Microarrays: biotechnology's discovery platform for functional genomics., Trends Biotechnol 301:6, 1998.

14. DA Lashkari, JL DeRisi, JH McCusker, AF Namath, C Gentile, SY Hwang, PO Brown, RW Davis. Yeast microarrays for genome wide parallel genetic and gene expression analysis. Proc Natl Acad Sci USA 94:13057–13062, 1997.

15. T Galitski, AJ Saldanha, CA Styles, ES Lander, GR Fink. Ploidy regulation of gene expression. Science 285:251–254, 1999.

16. TL Ferea. Systematic changes in gene expression patterns following adaptive evolution in yeast. Proc Natl Acad Sci USA 96:9721, 1999.

17. DR Soll, B Morrow, T Srikantha. High-frequency phenotypic switching in *Candida albicans*. Trends Genet 9:61–65, 1993.

18. DR Soll, B Morrow, T Srikantha, K Vargas, P Wertz. Developmental and molecular biology of switching in *Candida albicans*. Oral Surg Oral Med Oral Pathol 78:194–201, 1994.

19. TH Adams. Asexual sporulation in higher fungi. In: The Growing Fungus *Gow and Gadd*, 1994.

20. JL Beckerman, DJ Ebbole. MPG1, a gene encoding a fungal hydrophobin of *Magnaporthe grisea*, is involved in surface recognition. MPMI 9:450–456, 1996.

21. RA Dean. Signal pathways and appressorium morphogenesis. Annu Rev Phytopathol 35:211–234, 1997.

22. RJ Howard. Cell biology of pathogenesis. In: RS Zeigler, SA Leong, PS Teng, eds. Rice Blast Disease. Wallingford, CT: CAB International, 1994, pp 3–22.

23. E Winzeler, DD Shoemaker, H Liang, et al. Functional characterization of the *S. cerevisiae* genome by gene deletion and parallel analysis. Science 285:901–906, 1999.

24. B Cormack, N Ghori, S Falkow. An adhesin of the yeast pathogen *Candida glabrata* mediating adherence to human epithelial cells. Science 285:578–582, 1999.

25. D Retallack, G Deepe, J Woods. Applying in vitro expression technology to the fungal pathogen *Histoplasma capsulatum*. Microb Pathog 28:169–182, 2000.

26. BB Magee, PT Magee. Electrophoretic karyotypes and chromosome numbers in *Candida* species. J Gen Microbiol 133:425–430, 1987.

27. RG Snell, IF Hermans, RJ Wilkins, BE Corner. Chromosomal variations in *Candida albicans*. Nucleic Acids Res 15:362, 1987.

28. RG Snell, RJ Wilkins. Separation of chromosomal DNA molecules from *C. albicans* by pulsed field gel electrophoresis. Nucleic Acids Res 14:4401–4406, 1986.

29. B Wickes, J Staudinger, BB Magee, KJ Kwon-Chung, PT Magee, S Scherer. Physical and genetic mapping of *Candida albicans*: several genes previously assigned to chromosome 1 map to chromosome R, the rDNA-containing linkage group. Infect Immun 59:2480–2484, 1991.

30. WS Chu, BB Magee, PT Magee. Construction of an SfiI macrorestriction map of the *Candida albicans* genome. 175:6637–6651, 1993.

31. AK Goshorn, S Scherer. Genetic analysis of prototrophic natural variants of *Candida albicans*. Genetics 123:667–673, 1989.

32. B Slutsky, M Staebell, J Anderson, L Risen, M Pfaller, DR Soll. "White-opaque transition": a second high-frequency switching system in *Candida albicans*. J Bacteriol 169:189–197, 1987.

33. H Chibana, BB Magee, S Grindle, Y Ran, S Scherer, PT Magee. A physical map of Chromosome 7 of *Candida albicans*. Genetics 149:1739–1752, 1998.

34. C Thrash-Bingham, JA Gorman. DNA translocations contribute to chromosome length polymorphisms in *Candida albicans*. Curr Genet 22:93–100, 1992.

35. S Iwaguchi, M Homma, H Chibana, K Tanaka. Isolation and characterization of a repeated sequence (RPS1) of *Candida albicans*. J Gen Microbiol 138:1893–1900, 1992.

36. A Chindamporn, Y Nakagawa, M Homma, H Chibana, M Doi, K Tanaka. Analysis of the chromosomal localization of the repetitive sequences (RPSs) in *Candida albicans*. Microbiology 141:469–476, 1995.

37. A Chindamporn, Y Nakagawa, I Mizuguchi, H Chibana, M Doi, K Tanaka. Repetitive sequences (RPSs) in the chromosomes of *Candida albicans* are sandwiched between two novel stretches, HOK and RB2, common to each chromosome, Microbiology 144:849–857, 1998.

37a. H Chibana, J Beckerman, PT Magee. Fine-resolution physical mapping of genomic diversity in *Candida albicans*. Genome Research 10:1865–1877, 2000.

38. C Pujol, S Joly, B Nolan, T Srikantha, DR Soll. Microevolutionary changes in *Candida albicans* identified by the complex Ca3 fingerprinting probe involve insertions and deletions of the full-length repetitive sequence RPS at specific genomic sites [In Process Citation]. Microbiology 145:2635–2646, 1999.

39. WA Fonzi, MY Irwin. Isogenic strain construction and gene mapping in *Candida albicans*. Genetics 134:717–728, 1993.

40. PT Magee, L Bowdin, J Staudinger. Comparison of molecular typing methods for *Candida albicans*. J Clin Microbiol 30:2674–2679, 1992.

41. BB Magee, B Hube, RJ Wright, PJ Sullivan, PT Magee. The genes encoding the secreted aspartyl proteinases of *Candida albicans* constitute a family with at least three members. Infect Immun 61:3240–3243, 1993.

42. CM Hull, AD Johnson. Identification of a mating type-like locus in the asexual pathogenic yeast *Candida albicans*. Science 285:1271–1275, 1999.

43. LL Hoyer, TL Payne, M Bell, AM Myers, S Scherer. *Candida albicans* ALS3 and insights into the nature of the ALS gene family. Curr Genet 33:451–459, 1998.

44. LL Hoyer, TL Payne, JE Hecht. Identification of candida albicans ALS2 and ALS4 and localization of als proteins to the fungal cell surface [In Process Citation]. J Bacteriol 180:5334–5343, 1998.

45. LL Hoyer, S Scherer, AR Shatzman, GP Livi. *Candida albicans* ALS1: domains related to a *Saccharomyces cerevisiae* sexual agglutinin separated by a repeating motif. Mol Microbiol 15:39–54, 1995.

46. TJ Goodwin, PT Poulter. Multiple LTR-retrotransposon families in the asexual yeast *Candida albicans*. Genome Res 10:174–191, 2000.

47. AJ Clutterbuck. The validity of the *Aspergillus nidulans* linkage map. Fungal Genet Biol 21:267–277, 1997.

48. RA Prade, J Griffith, K Kochut, J Arnold, WE Timberlake. In vitro reconstruction of the *Aspergillus* (= *Emericella*) *nidulans* genome, Proc Natl Acad Sci USA 94:14564–14569, 1997.

49. JC Edman, JA Kovacs, H Masur, DV Santi, HJ Elwood, ML Sogin. Ribosomal RNA sequence shows *Pneumocystis carinii* to be a member of the fungi. Nature 334:519–522, 1988.

50. SL Stringer, K Hudson, MA Blase, PD Walzer, MT Cushion, JR Stringer. Sequence from ribosomal RNA of *Pnuemocystis carinii* compared to those of four fungi suggests an ascomycetous affinity. J Protozool 36:14S–16S, 1989.

51. S Merali, U Frevert, JH Williams, K Chin, R Bryan, AB Clarkson Jr. Continuous axenic cultivation of *Pneumocystis carinii*. Proc Natl Acad Sci USA 96:2402–2407, 1999.

52. J Arnold, MT Cushion. Constructing a physical map of the *Pneumocystis* genome. J Eukaryot Microbiol 44:8S, 1997.

53. MT Cushion, J Arnold. Proposal for a *Pneumocystis* genome project, J Eukaryot Microbiol 44:7S, 1997.

54. JR Stringer, MT Cushion. The genome of *Pneumocystis carinii*. FEMS Immunol Med Microbiol 22:15–26, 1998.

55. J Watanabe, K Nakata, H Nashimoto. Cloning and characterization of a repetitive sequence from *Pneumocystis carinii*. Parasitol Res 78:23–27, 1992.

56. J Hamer, L Farrall, MJ Orbach, B Valent, FG Chumley. Host species-specific conservation of a family of repeated DNA sequences in the genome of a fungal plant pathogen. Proc Natl Acad Sci USA 86:9981–9985, 1989.

57. B Valent. Rice blast as a model system for plant pathology. Phytopathology 80:33–36, 1990.

58. B Valent, FG Chumley. Molecular genetic analysis of the rice blast fungus, *Magnaporthe grisea*. Annu Rev Phytopathol 29:443–467, 1991.

59. TT Hebert. The perfect stage of *Pyricularia grisea*. Phytopathology 61:83–87, 1971.

60. J Romao. Genetic organization of a repeated DNA sequence family in the rice blast fungus. Proc Natl Acad Sci USA 89:5316–5320, 1992.

61. DZ Skinner, AD Budde, ML Farman, JR Smith, H Leung, SA Leong. Genome organization of *Magnaporthe grisea*: genetic map, electrophoretic karyotype, and occurrence of repeated DNAs. Theor Appl Genet 87:545–557, 1993.

62. H Zhu, M Sasinowski, RA Dean. Physical map and organization of chromosome 7 in the rice blast fungus, *Magnaporthe grisea*. Genome Res 9:739–750, 1999.

63. F Kempken, U Kuck. Transposons in filamentous fungi—facts and perspectives. Bioessays 20: 652–659, 1998.

64. JM Kim, S Vanguri, JD Boeke, A Gabriel, D Voytas. Transposable elements and genome organization: a comprehensive survey of retrotransposons revealed by the complete *Saccharomyces cerevisiae* genome sequence. Genet Res 8:464–478, 1998.

13

Cell-Mediated Immunity and Endemic Mycoses

D. Mitchell Magee and Rebecca A. Cox
Texas Center for Infectious Diseases, San Antonio, Texas

I. INTRODUCTION

The first century of study of the endemic mycotic diseases is coming to a close. While nomenclature of the organisms varied during the early years, the systemic infections that we now know are caused by the four primary endemic fungi; *Coccidioides immitis*, *Blastomyces dermatitidis*, *Histoplasma capsulatum*, and *Paracoccidioides brasiliensis*, which were first reported in 1891, 1894, 1901, and 1908, respectively [1–4]. The diseases caused by these pathogenic fungi are primarily acquired by inhalation of the infectious particle from the mycelial phase of the organisms growing in the soil of indigenous areas. Once inhaled, these organisms undergo a morphological conversion into the parasitic phase, which, for *B. dermatitidis*, *H. capsulatum*, and *P. brasiliensis*, is typically budding yeast. The *C. immitis* parasitic cycle is more complex and has multiple forms ranging from the 3- to 5-μm endospore to the 60- to 80-μm spherule [5]. This dimorphic life cycle is interwoven in that, the parasitic phase cells can convert back into the mycelial form when culture conditions are changed or the organism is seeded to the soil. Increased temperature and/or CO_2 concentration appear to be the primary factors in the saprobic to parasitic phase conversion [6,7]. During this primary conversion, the host is exposed to antigens from both saprobic and parasitic phase cells. Thereafter, with rare exception, the host is primarily responding to the parasitic phase cells and derived antigens. This review will attempt to coalesce current observations on induction of protective cell-mediated immunity against the endemic fungi, with special emphasis on *C. immitis* and *H. capsulatum*, and with contributions from other fungal model systems, notably *Cryptococcus neoformans* and *Candida albicans*.

Disease presentation during infection by the endemic fungi can be quite varied, though in most cases infections caused by the endemic fungi localize in the lungs and go unnoticed or result in mild flulike symptoms. Persons with localized lesions often do not require any antifungal therapy, but those with progressive disease need appropriate therapy and often extensive follow-up [8–10]. Notwithstanding, these fungi are pathogens and fully immunocompetent people can develop severe, life-threatening, disseminated disease. When dissemination occurs, it is thought to occur early in the disease process; sites most often associated with systemic disease include skin, bone, and meninges. In these cases, the optimal treatment is still Amphotericin B that can be followed by oral azoles for suppressive therapy [11]. The duration of antifungal therapy in these severe cases can range from many months or years, to lifelong. Unfortunately, these therapeutic approaches are not fungicidal and the host's immune system is needed for ultimate control of the infection.

Multiple host factors, including race, age, and sex, can influence the progression of the disease. The influence of race and sex indicates that genetic factors are an important component in the ultimate determination of disease severity. For example, in coccidioidomycosis there has been a longstanding correlation that Asians, especially those from the Philippines, and African-Americans are at increased risk of dissemination. This was most recently observed in an analysis of the coccidioidomycosis outbreak in the early 1990s that confirmed the dissemination rate among these racial groups was 3–10 times that of Caucasians [12,13]. It had been previously shown that pregnant women, who acquired coccidioidomycosis, particularly in the third trimester, were at significantly increased risk of severe disease [14,15]. It was thought that the increased disease severity could be a result of hormonal changes that might augment fungal growth or immunosuppression during the later stages of pregnancy [16]. In a study of 32 pregnancies in Kern County, California, during a 1993 epidemic, the dissemination rate was considerably less than that previously reported and, although the numbers in this outbreak were small, the number of disseminated cases were threefold that of nonpregnant females [17]. The influence of sex is most pronounced in the diseases caused by P. brasiliensis and B. dermatitidis and H. capsulatum with each showing marked increased dissemination rates in males compared to females [18–20]. Exposure differences between males and females, as a result of occupation, have been argued as one of the possible reasons to explain the differences in dissemination between the sexes. However, in defined outbreaks when exposure is equal among the sexes, there have been increased rates of dissemination of blastomycosis in males [21]. Interestingly in P. brasiliensis infections, and opposite to observations for C. immitis [16], female hormones may play a role in preventing the conversion to the yeast phase, which may ameliorate the infection in women [22].

The importance of T-cell-mediated immunity is well documented in both experimentally infected animals and humans in these infections. A preponderance of the data show that control of infections correlates with the presence of delayed-type hypersensitivity (DTH) to fungal antigens and low fungal-specific antibody titers. Conversely, in the progressive forms of the disease, DTH reactivity diminishes or is completely ablated, and antibody titers markedly increase. This evidence is underscored by the increased incidence of these endemic fungal diseases in those infected with the human immunodeficiency virus (HIV). For those with HIV infection who live in the respective endemic area, C. immitis and H. capsulatum are major problems [23–26]. The devastating effects of loss of T-cell number and/or function renders patients highly susceptible to developing severe disseminated disease. In one prospective study, Ampel et al. showed that a $CD4^+$ T-cell count of <250 cells/μL was the most important risk factor for developing severe coccidioidomycosis [27]. Many of these patients also lacked T-cell reactivity as measured by skin test reactivity or in vitro proliferative responses to Coccidioides and non-Coccidioides antigens. Although B. dermatitidis and P. brasiliensis infection in HIV individuals has not been observed at the same level as for C. immitis or H. capsulatum, there are reports of small numbers of cases in this immunodeficient group [28,29].

II. PARADIGM OF INDUCTION OF PROTECTIVE T-CELL-MEDIATED IMMUNITY

T-cell-mediated immunity in the systemic mycoses results from a complex interaction among fungus, innate host defenses, host genetics, and cytokines which ultimately influences T-cell differentiation. With respect to the host response against the endemic fungi, two phenotypic outcomes, as determined by cytokine secretion patterns, can result. High-level production of interferon-γ (IFN-γ) and interleukin-12 (IL-12), with low IL-4 and IL-10, indicates a T-helper 1 (Th1) phenotype and is associated with protective immunity; conversely, low-level production of IFN-γ and IL-12 with high IL-4 and IL-10 indicates a T-helper 2 (Th2) phenotype response and is associated with suppressive or nonprotective immunity. These phenotypes were initially defined by in vitro responses of CD4$^+$ T-cell clones and were thought to be mutually exclusive [30]. However, in vivo results in infectious models reveal that both pathways of T-helper cell differentiation are often engaged and that the outcome is a preferential expression of one T-helper type over another. Moreover, the dichotomy in cytokine secretion profile has also been demonstrated to be no longer limited to CD4$^+$ Th1 and Th2 cells. Studies have established the existence of CD8$^+$ T-cell subsets, designated Tc1 and Tc2, with the capacity to differentially secrete IFN-γ and IL-4, respectively [31]. Dominant Th1 immunity, with its concomitant IFN-γ production, results in activation of macrophages to increased antifungal activity. Macrophage activation was first characterized as protective in models of facultative intracellular parasite infection. Dominant Th2 immunity, along with its influence and enhancement of antibody production, is protective against some helminths and pyogenous bacteria. That in vivo responses show induction of both Th arms is seemingly counterintuitive. In a model in which antifungal host defenses are augmented by IFN-γ, what would be the benefit of cytokine production that might diminish the effectiveness of IFN-γ? However, evidence gathered in cytokine knockout mice infected with *C. albicans* revealed that both IL-4 and IL-10, which are Th2-associated cytokines, are required for induction of Th1 protective immunity [32,33]. Thus, the induction, expression, and maintenance of protective immunity are clearly under complex multiple levels of regulatory control.

For induction of cognate T-cell-mediated immunity, one of the important initial interactions is the uptake of fungal antigens by antigen presenting cells (APCs) and subsequent presentation to the naïve T-cells in association with MHC molecules. Fungal antigens presented in context with class I major histocompatibility antigens (MHC) molecules stimulate CD8$^+$ T-cells while those associated with class II MHC molecules stimulate CD4$^+$ T-cells. This indicates that APCs exert remarkable control over the initiation of T-cell-mediated immunity. There are three major potential APCs: macrophages, B-cells, and dendritic cells. The influence of the cell type on the subsequent T-helper type differentiation indicates that macrophages may be more efficient APCs for Th1 responses whereas B-cells are more efficient APCs for Th2 responses [34]. Dendritic cells are the most potent antigen-presenting cells on a cell-by-cell basis and stimulate both CD4$^+$ and CD8$^+$, possibly as a result of the increased expression of MHC class I and class II molecules on their cell surface [35]. Further, these cells are capable of stimulating both Th1 and Th2 CD4$^+$ T-cells and thus can induce both arms of T-helper immunity.

Once the important antigen is processed and displayed on the APC surface with the attendant MHC molecules, other accessory molecules are needed to provide critical signals to fully engage T-cell responsiveness. The B7/CD28 complex is a key set of ligands between APC and T-cell, respectively. The B7 molecules on the APC are actually a family of molecules and the most extensively studied are the molecules designated B7-1 and B7-2 [36]. In one study of B7 induction during the interaction of human monocytes and fungi, it was observed that acapsular *C. neoformans* and *C. albicans* induced marked levels of B7-1 expression with only modest levels of B7-2 induction [37]. Of note in this study, an encapsulated strain of *C. neoformans* was significantly less stimulatory for B7-1 expression than the acapsular strain; however, addition of anticapsular glucuronoxylomannan antibodies significantly increased B7-1 induction by encapsulated *C. neoformans*. The differential induction of one B7-1 over B7-2 may be one component in the ability to stimulate a protective T-cell response since B7-2 induction may promote Th2 immunity [34].

The CD40/CD40 Ligand (CD40L) interaction is a second important set of signals for T-cell activation. The results obtained in a model of human monocyte/lymphocyte interactions with *C. neoformans* revealed that the acapsular strains of *C. neoformans* and *C. albicans* induced monocytes to make significantly higher levels of IL-12 than encapsulated *C. neoformans* [38]. Further, addition of stimulated T-cells markedly enhanced IL-12 production, and this enhancement was significantly diminished if the anti-CD40L was included in the culture conditions to block receptor interaction. While CD40/CD40L interactions may be beneficial, they are not essential for fully developed T-cell immunity. Experiments in CD40L knockout mice showed that the CD40L$^{-/-}$ mice were not less susceptible to lethal challenge *H. capsulatum* than CD40L$^{+/+}$ mice [39]. Thus, the APC provides co-stimulatory molecules that can enhance the early induction of protective immunity. Yet, there are factors from infecting organisms, possibly virulence or pathogenicity related, that may also influence the developing immune response.

After the initial stimulation of T cells, immune responses are enhanced by the recruitment of additional lymphocytes and monocytes to the foci of infection. Colony-stimulating factors are released that induce proliferation and differentiation of bone marrow progenitor cells to increase the number of granulocytes and macrophages in the peripheral blood [40]. Chemokines are secreted and are the primary chemotactic substances that initiate the cascade of events for lymphocytes and phagocytic cells to diapedese into sites of inflammation. The chemokines are a large family of low-molecular weight proteins that are grouped by the spacing of a pair of amino terminal cysteines into either CC chemokines or CXC chemokines [41]. For example, in the murine model of *C. neoformans* infection, the CC chemokine, macrophage inflammatory protein 1α (MIP-1α), is a key component in protective immunity. Treatment of mice with neutralizing anti-MIP-1α antiserum resulted in less neutrophil and macrophage recruitment to lungs and significantly increased fungal burden in the lungs after challenge as compared to control antibody-treated mice [42]. The importance of MIP-1α was supported by subsequent studies in which knockout mice that were deficient in CCR5, one of the receptors for MIP-1α, were challenged with *C. neoformans*. These mice were significantly more susceptible to disease than mice with functional CCR5 receptors [43]. This indicates that the ability to recruit specific receptor-positive cells may be critical for the induction and maintenance of protection. These receptors may be useful as indicators of response type since it has been recently shown that CCR5 receptor is a marker for Th1 cells and the CCR3 receptor is a marker for Th2 cells [44]. The extensive nature of this chemokine family and their corresponding receptors could provide a mechanism whereby precisely defined lymphocyte subsets could be recruited to the site of infection.

During initial interactions with APC, T-cells undergo differentiation from an uncommitted precursor capable of secreting many cytokines into the defined T-helper subtype [30]. Along with the parameters described above, the local cytokine milieu may dramatically influence the T-helper phenotype that emerges. If the initial cellular interactions occur in an environment rich in IFN-γ, then there is a preferential induction of Th1 cells; while if IL-4 is present, there is a preferential induction of Th2 cells [30,34]. Several innate host defense mechanisms are potential regulators for early elaboration of IFN-γ. From the initial interaction of infecting fungus and macrophages or dendritic cells, IL-12 and IL-18 are elaborated. Both IL-12, a heterodimeric cytokine consisting of 35-kDa and 40-kDa subunits, and IL-18, which is structurally related to IL-1β, strongly promote Th1 differentiation [45,46].

The most likely process by which both IL-12 and IL-18 promote Th1 responses is by inducing natural killer (NK) cells to elaborate IFN-γ. Potential innate sources of IL-4 include mast cells, basophil, and NK-T cells [30]. The latter cell population is unique in that it expresses surface antigens of both the NK and T-cells, and is capable of secreting both IL-4 and IFN-γ [47]. IL-10 is another potent downregulator of Th1 responses [48]. In studies of coccidioidomycosis, high levels of IL-10 message were correlated with susceptibility in an analysis of susceptible and resistant mice [49]. High levels of IL-10 could function to inhibit activation and antigen presentation function of macrophages by downregulating MHC and co-stimulatory molecule expression of APC [34]. However, as noted earlier in the model of *C. albicans*, IL-10 was necessary for end-stage Th1 immunity in IL-12 knockout mice [33]. Mecacci et al. hypothesizes that IL-10 control of anti-*Candida* Th1 immunity may be dose dependent with high IL-10 associated with disease exacerbation and low IL-10 associated with amelioration. It was delineated that the low-dose IL-10 treatment resulted in a slight increase of B7-1 and marked inhibition of B7-2 expression by macrophages, which would promote Th1 immunity. This observation confirms that differential expression of co-stimulatory signals on APC may also influence end stage T-helper cell differentiation.

III. T-CELLS IN THE ENDEMIC MYCOSES

Immune responses are often divided and categorized by function or phenotype. For example, there are primary divisions between innate and acquired, humoral and cellular, CD4$^+$ and CD8$^+$ T-cells, Th1 and Th2 cytokines, and primary and secondary immune responses. These classifications are an attempt to reduce and explain protective immunity, but such simplifications at times have caused confusion. Protective immunity is the culmination of many convergent factors containing potential contributions from the multiple components of the immune armamentarium. Nevertheless, in the endeavor to define protective immunity against the endemic fungi, it is clear that T-cells, with contributions from both CD4$^+$ and CD8$^+$ cells, are the center of the protective response.

Multiple model systems have been employed to define the cellular nature of protection, which include studies via passive transfer of cells and sera, in vivo depletion of T cells and subsets, and the use of genetically immunodeficient animals. For example, endemic mycoses in congenitally athymic mice are much more severe than in euthymic littermates, which defines the fundamental requirement for T-cells in protection [50–54]. In host defenses in *C. immitis* infection, it was shown via passive transfer experiments that immune lymphocytes, and not sera, were the most important factor in immune mice [50,55]. It was subsequently shown that passive transfer of protection was dependent upon both CD4$^+$ and CD8$^+$ T-cells [56].

Against *H. capsulatum*, in vivo depletion of T-cell subsets has revealed that CD4$^+$ T cells are dominant in protection in primary immunity [57]. This observation is supported by studies revealing that transfer of a *H. capsulatum*–specific CD4$^+$ T cell clone protected irradiated mice against challenge [58]. However, CD8$^+$ T-cells were also shown to be necessary for optimal protection as determined by in vivo depletion and the use of MHC class I deficient knockout mice [59]. In studies of blastomycosis, passive transfer of immune T-cells but not sera has demonstrated the importance of cellular immunity against this fungus [52]. Against *P. brasiliensis* infection, in vivo depletion of CD8$^+$ T-cell subsets during primary infection has been analyzed in both genetically susceptible and resistant mice with markedly different results [60].

In genetically resistant mice, CD8$^+$ depletion had no effect on fungal colony forming units in the lungs, but did lead to increased dissemination of *P. brasiliensis* to the liver and spleen late in infection. In susceptible mice, however, CD8$^+$ depletion had a deleterious effect and resulted in increased fungal lung burden as well as dissemination both early and late during infection. The dependence of protection upon both T-cell subsets indicates that multiple complex mechanisms of protective immunity are required to fully control these fungal pathogens.

IV. MECHANISMS OF CELLULAR REGULATION OF PROTECTION IN THE ENDEMIC MYCOSES

A. Production and Regulation of Cytokine Elaboration

It is hypothesized that the primary mechanism for the action of CD4$^+$ T-cells in protective immunity against the endemic fungi is through the elaboration of cytokines, with IFN-γ being the most important cytokine associated with protection. As mentioned earlier, during the discussion of the regulation of induction of Th1 mediated immunity, there is a cross-regulating cascade that leads to the high level of IFN-γ that characterizes protective primary immunity. The emphasis on IFN-γ has resulted from experiments in which this cytokine was used as an immunotherapeutic agent or alternatively depleted via specific antibody or by gene disruption. In our model of *C. immitis* infection, the role of IFN-γ has been examined in genetically susceptible and resistant mouse strains [61]. It was initially noted that resistant mice produce IFN-γ earlier in the course of disease and at higher levels late in disease than do genetically susceptible mice. It was further noted that disease control was significantly reduced when IFN-γ production was abrogated in vivo by treating resistant mice with specific antibody; conversely, disease control was enhanced by immunotherapy with recombinant IFN-γ in susceptible mice. The importance of IFN-γ has also been shown in models of *H. capsulatum* [62] and *P. brasiliensis* [63] infection. IL-12 was another key cytokine defined as protective for the most of the endemic fungi [64–66]. This was most elegantly defined in the *H. capsulatum* model in which it was shown that the primary action of IL-12 was actually mediated through IFN-γ production [64]. Notwithstanding the importance of IFN-γ, tumor necrosis factor alpha (TNF-α) (67,68) and granulocyte-macrophage colony-stimulating factor (GM-CSF) [69] were other cytokines important in the expression of protection against *H. capsulatum*.

The mechanisms of controlling protective immunity against the endemic fungi by CD8$^+$ T-cells are not clearly defined. As mentioned earlier, CD8$^+$ function differs in the *P. brasiliensis* model depending upon host genetics [60]. In resistant mice, CD8$^+$ T-cells are critical for controlling dissemination of infection but not at the lung level, whereas in susceptible mice, CD8$^+$ T-cells are important for controlling fungal burden locally, in the lungs, and at sites of dissemination. Interestingly, CD8$^+$ depletion led to marked increases in DTH reactivity in the susceptible mice. As shown in the *H. capsulatum* model using in vivo depletion of T cells subsets, CD4$^+$ depletion lead to marked diminution of the IFN-γ response, whereas CD8$^+$ depletion resulted in marked increases in IFN-γ production [70]. It is somewhat surprising that, since CD8$^+$ cells were a component of optimal control of fungal burden in the tissues, there was diminished control of fungal burden even in the presence of high levels of IFN-γ. However, this is concordant with the increase in DTH reactivity during *P. brasiliensis* infection in CD8$^+$-depleted animals and supports the hypothesis that CD4$^+$ Th1 cells are critical for DTH/IFN-γ production [71]. In the aggregate, it appears that the CD8$^+$ T-cell functions either directly or by control of CD4$^+$ function for protection. In addition to control of CD4$^+$ activity or direct cytokine production, cytolytic activity is another potential CD8$^+$ function. While there is no direct evidence for a cytotoxic effect of CD8$^+$ T cells against the endemic fungi, this area clearly needs investigation.

B. Macrophage Activation

Macrophage activation occurs in an IFN-γ and/or TNF-α rich environment that results in increasing antimicrobial properties of macrophages against the invading fungus. It was established, as reviewed above, that an effective Th1 immune response provides IFN-γ for macrophage activation. TNF-α is produced innately during the initial macrophage fungus interaction in *H. capsulatum* [72] and *C. immitis* [73–75] and is thus available early during infection for increasing macrophage function. The murine model of macrophage-fungus interaction has been the best studied and shows that cytokine-activated macrophages kill *C. immitis* [76], *P. brasiliensis*, and *B. dermatitidis* [77] and inhibit the growth of *H. capsulatum* [78]. Antifungal activity in murine macrophages was most prominently related to the production of nitric oxide [79,80], with the limitation of available iron as an important mechanism against *H. capsulatum* [81]. Antifungal mechanisms of human monocytes or monocyte-derived macrophages have not been as clearly defined. For example, stimulation of monocytes with either TNF-α or IFN-γ imparted significant killing of *C. immitis* arthroconidia [82,83]. Yet, in human monocyte/macrophage interactions with *H. capsulatum*, IFN-γ and TNF-α failed to induce fungal inhibition. Alveolar, peritoneal, or monocyte-derived human macrophages could not be activated with IFN-γ or TNF-α to inhibit *H. capsulatum*, but colony-stimulating factors did induce fungistasis [84,85].

Interestingly, when monocytes were cultured on type I collagen matrices, as compared to plastic, they acquired fungistatic properties and addition of cytokines did not augment fungal inhibition [86]. The mechanism(s) of human macrophage-mediated fungistasis are also not well delineated. Inhibition of *H. capsulatum* is not associated with generation of reactive oxygen intermediates [87]. The capacity of human macrophages to generate reactive nitrogen intermediates has been a point of controversy, but the inability to detect nitric oxide in human monocyte/ macrophage cultures may be a result of the sensitivity of the assay systems [88,89]. The action of reactive nitrogen intermediates is not limited to fungal inhibition and may account for a significant downregulation of immunity. For example, as previously noted, high levels of IFN-γ produced were in CD8⁺-depleted mice, yet there was a diminished capacity to control *H. capsulatum* [70]. This high level of IFN-γ may result in high levels of reactive nitrogen intermediate production which can then inhibit lymphocyte proliferation or induce lymphocyte apoptosis [88,90].

C.   Antibody Augmentation of Cellular Host Defenses

The role of antibody in protection against the endemic fungi has not been firmly established. In fact, early studies using polyclonal antisera to passively transfer protection have not met with success. While there are pathways in which antibody responses develop independent of T-cells, there are important mechanisms whereby T-cells influence antibody responses. The action of cytokines IFN-γ and IL-4 markedly influence the isotype of antibody produced, with IFN-γ enhancing IgG2a isotypes and IL-4 enhancing IgG1 and IgE [91]. The potential contribution of antibody against pathogens in which cell-mediated immunity is the primary protective mechanism has been elegantly shown in other fungal models.

Monoclonal antimannotriose IgM and IgG3 antibodies have been derived and shown to be protective against systemic and mucosal candidiasis [92,93]. Monoclonal antibodies were developed against *C. neoformans* glucuronoxylomannan and have been shown to fall into categories that result in enhanced disease, have no effect (nonprotective), or are protective [94]. Several important observations from this seminal series of experiments are that the isotype of the monoclonal antibody is one of the key factors in determining protection, with IgG2a antibodies appearing to confer the highest level of protection against systemic challenge [95]. This last report revealed that antibodies could function as opsonins since they functioned to increase phagocytosis of *C. neoformans*. In other studies it was shown that monoclonal anti–*C. neoformans* could result in enhanced in vitro monokine secretion and T-cell proliferation [38,96]. These results indicate that there are antibodies that could aid in promoting a Th1 immune response which, if engaged, would result in IFN-γ production that would then stimulate isotype switching toward the protective IgG2a isotype. Thus, a feedback cycle could be engaged an reinforces the hypothesis that cellular and humoral immunity could both be components of Th1 protective immunity against the endemic fungi.

V. IMMUNOMODULATION OF HOST DEFENSES

Because of all the regulatory interactions, the choreography of the cascade of immunity is amenable to modulation. Establishing protective immune responses could be accomplished by active immunization using a fungal vaccine, targeted to persons living in, or traveling to, an endemic area. As an alternative approach, therapeutic use of cytokines, used as either recombinant proteins or by gene therapy, might shift T-helper potential to augment failing host immune responses or, when necessary, downregulate hyperresponsive immunity. The end result would be a method by which those with severe disseminated disease, those with known genetic susceptibility, or those traveling to an endemic area would undergo a successful course of treatment or develop protection from severe disease.

A plethora of antigens have been isolated and evaluated for their ability to induce T-cell responses against the endemic fungi and are discussed in greater detail in other chapters. Many of these antigens are potential vaccine candidates [97,98] and the following only highlights those that are prominent. From *C. immitis*, the proteins under investigation include: antigen 2 [99], which has the same coding sequence as the gene for the proline-rich antigen [100], and, hence, is collectively designated Ag2/PRA; recombinant Urease [101]; recombinant spherule outer wall glycoprotein (SOWgp) [102]; T-cell-reactive protein (TCRP) [103]; and a recombinant 60-kDa heat-shock protein (HSP) [104]. Two HSPs, HSP60 [105] and HSP70 [106], and the calcium-binding protein (CBP-1) [107] have been cloned and the recombinant products shown to be reactive with *H. capsulatum*–immune T-cells. In studies of *B. dermatitidis*, the adhesin WI-1 is a virulence factor and T-cell immunodominant antigen [108]. For *P. brasiliensis*, the 43-kDa exocellular glycoprotein has been shown to be immunodominant [109]. Our reports of vaccination have revealed that recombinant Ag2/PRA is only modestly protective in mice challenged by an intraperitoneal route. In marked contrast, immunization of mice with Ag2/PRA cDNA induced a significantly higher level of protection as compared to the recombinant Ag2/PRA, as measured by the ability of mice to survive intraperitoneal challenge and to have decreased fungal dissemination after pulmonary challenge with *C. immitis* [110]. We subsequently utilized a combination of Ag2/PRA with a gene construct for IL-12 p35 and p40 chains as an adjuvant [111]. Combination of Ag2/PRA with IL-12 significantly increased the ability of mice to control dissemination; however, the control at the lung level remained unchanged. While these antigens delineated above show promise at further defining T-cell reactivity in regard to protective immunity, the internal processing of fungal antigens by APCs has not been extensively studied. Moreover, the limited genetic information obtained from the endemic mycoses warrants further studies to identify antigens that have not yet been discovered but which may have important immunostimulatory capacity. Genome projects for these organisms would provide much needed information that would further vaccine development and generate molecular tools for understanding the basic biology of these organisms and their complicated life cycles.

Therapeutic interventions to increase the immune response have been undertaken with the endemic fungi. We have reported that recombinant cytokine therapy with IFN-γ and IL-12 augments murine host defenses against *C. immitis* [61,65], but the effectiveness of immunotherapy was only observed when treatment was started early in the course of disease. Once disease was established, recombinant cytokine therapy was ineffective in altering host defenses in this acute model of infection. It may take the development of chronic disease models to fully explore the capacity of immunotherapy to modulate host responses against these systemic mycoses. For prospects of treating human disease, it is clear that immunotherapy will be used as an adjunct to conventional antifungal chemotherapy. Studies of cryptococcosis revealed that either IFN-γ or IL-12 augments suboptimal antifungal chemotherapy of an acute model of infection [112]. In addition to IFN-γ and IL-12, the colony-stimulating factors are being evaluated which will augment both the number and function of phagocytic cells [113].

The use of recombinant cytokines has been somewhat problematic in early clinical human trials with many potential side effects noted. One therapeutic modification would be to utilize gene therapy to ameliorate these potential side effects. We tested this hypothesis by creating a macrophage cell line that was transfected to produce IL-12 and used as an immunotherapy to treat coccidioidomycosis [114]. The cell line provided significant protection against a pulmonary challenge of *C. immitis* and modulated immune responses to increase IFN-γ production. Taking this strategy further in an animal model of leishmaniasis, studies report on immunotherapeutic intervention with bone marrow–derived dendritic cells that are transiently transfected to express IL-12 and then pulsed with *Leishmania* antigens [115]. This therapy was useful both as a vaccine to induce immunity and as a therapeutic intervention in established disease. As we learn more and gain an understanding of the control of the immune response, it will be possible to develop therapies that could reverse a failing immune response.

VI. SUMMARY

The majority of people who acquire an infection by one of the endemic fungi may not ever know that they have had the encounter. Thus, the fully functioning immune system can protect the host and result in acquired immunity that provides long-lived protection. There are many potential contributors to the development of this protective immunity that include innate and acquired mechanisms with optimal control mediated through both CD4$^+$ and CD8$^+$ T-cells. In contrast, those who are unable to control the disease face potentially toxic, prolonged, and expensive chemotherapy. Gaining a better understanding of the critical innate and acquired immune parameters associated with protection will eventually lead to immunotherapeutic strategies to augment chemotherapy and lead to more effective treatments for those with life-threatening fungal infections.

ACKNOWLEDGMENTS

This work was supported by grants AI32134 from the National Institutes of Health and the California HealthCare Foundation. The California HealthCare Foundation, based in Oakland, California, is a nonprofit philanthropic organization whose mission is to expand access to affordable, quality health care for underserved individuals and communities, and to promote fundamental improvements in the health status of the people of California. Supported by the Department of Health Services of the State of California and by California State University, Bakersfield.

The authors would like to thank Dr. Chengyong Jiang, Dr. F. Douglas Ivey, and Melanie Woitaske for critical comments.

REFERENCES

1.	A Posadas. Ensayo anatomopatologico sobre una neoplasia considerada como micosis fungoidea. An Circ Med Argent 15:8, 1892.

2. T Gilchrist. Protozoan dermatitidis. J Curan Genitourin Dis 12:496, 1894.

3. S Darling. A protozoan general infection producing pseudotubercles in the lungs and focal necrosis in the liver, spleen, and lymph nodes. JAMA 46:1283–1285, 1906.

4. A Lutz. Uma mycose pseudococcidica localisada na boca e observada no Brasil. Contrubicao ao conhecimento das hyphoblastomycoses americanas. Brasil-Med 22:121–124, 1908.

5. SH Sun, SS Sekhon, M Huppert. Electron microscopic studies of saprobic and parasitic forms of *Coccidioides immitis*. Sabouraudia 17:265–273, 1979.

6. ZB Sheiban. Morphological conversion of 13 strains of three dimorphic fungi in tap and stream water. Bull Soc Pathol Exot Filiales 68:46–50, 1975.

7. SA Klotz, DJ Drutz, M Huppert, SH Sun, PL DeMarsh. The critical role of CO_2 in the morphogenesis of *Coccidioides immitis* in cell-free subcutaneous chambers. J Infect Dis 150:127–134, 1984.

8. J Wheat, G Sarosi, D McKinsey, R Hamill, R Bradsher, P Johnson, J Loyd, C Kauffman. Practice guidelines for the management of patients with histoplasmosis. Clin Infect Dis 30:688–695, 2000.

9. SW Chapman, RW Bradsher Jr, GD Campbell Jr, PG Pappas, CA Kauffman. Practice guidelines for the management of patients with blastomycosis. Clin Infect Dis 30:679–683, 2000.

10. JN Galgiani, NM Ampel, A Catanzaro, RH Johnson, DA Stevens, PL Williams. Practice guidelines for the treatment of coccidioidomycosis. Clin Infect Dis 30:658–661, 2000.

11. O Lortholary, DW Denning, B Dupont. Endemic mycoses: a treatment update. J Antimicrob Chemother 43:321–331, 1999.

12. R Johnson, J Caldwell, G Welch, H Einstein. The great coccidioidomycosis epidemic: clinical features. In: H Einstein, A Catanzaro, eds. Coccidioidomycosis: Proceedings of the 5th international conference. Washington: National Foundation for Infectious Diseases, 1996, pp 77–87.

13. E Asura, J Caldwell, R Johnson, H Einstein, G Welch, R Talbot, H Affentranger. Coccidioidomycosis epidemic of 1991: epidemiological features. In: H Einstein, A Catanzaro, eds. Coccidioidomycosis: Proceedings of the 5th international conference. Bethesda, MD: National Foundation for Infectious Diseases, 1996, pp 98–107.

14. JN Galgiani. Coccidioidomycosis. West J Med 159:153–171, 1993.

15. D Pappagianis. Coccidioidomycosis. Semin Dermatol 12:301–309, 1993.

16. DJ Drutz, M Huppert, SH Sun, WL McGuire. Human sex hormones stimulate the growth and maturation of *Coccidioides immitis*. Infect Immun 32:897–907, 1981.

17. J Caldwell, E Arsura, W Kilgore, A Garcia, V Reddy, R Johnson. Coccidioidomycosis in pregnancy during an epidemic in California. Obstet Gynecol 95:236–239, 2000.

18. KJ Kwon-Chung, JE Bennett. Histoplasmosis. In: C Cann, ed. Medical Mycology. Malvern, PA: Lea and Febiger, 1992, pp 464–513.

19. KJ Kwon-Chung, JE Bennett. Blastomycosis. In: C Cann, ed. Medical Mycology. Malvern, PA: Lea and Febiger, 1992.

20. KJ Kwon-Chung, JE Bennett. Paracoccidioidomycosis. In: C Cann, ed. Medical Mycology. Malvern, PA: Lea and Febiger, 1992, pp 594–619.

21. G Sarosi, SF Davies. Self-limited blastomycosis: a report of 39 cases. Semin Respir Infect 1:40–44, 1986.

22. ME Salazar, A Restrepo, DA Stevens. Inhibition by estrogens of conidium-to-yeast conversion in the fungus *Paracoccidioides brasiliensis*. Infect Immun 56:711–713, 1988.

23. DG Fish, NM Ampel, JN Galgiani, CL Dois, PC Kelly, CH Johnson, D Pappagianis, JE Edwards, RB Wasserman, RJ Clark. Coccidioidomycosis during human immunodeficiency virus infection. A review of 77 patients. Medicine 69:384–391, 1990.

24. JN Galgiani, NM Ampel. Coccidioidomycosis in human immunodeficiency virus-infected patients. J Infect Dis 162:1165–1169, 1990.

25. GA Sarosi, PC Johnson. Disseminated histoplasmosis in patients infected with human immunodeficiency virus. Clin Infect Dis 14(suppl 1):S60–S67, 1992.

26. SH Salzman, RL Smith, CP Aranda. Histoplasmosis in patients at risk for the acquired immunodeficiency syndrome in a nonendemic setting. Chest 93:916–921, 1988.

27. NM Ampel. Delayed-type hypersensitivity, in vitro T-cell responsiveness and risk of active coccidioidomycosis among HIV-infected patients living in the coccidioidal endemic area. Med Mycol 37:245–250, 1999.

28. PG Pappas, JC Pottage, WG Powderly, VJ Fraser, CW Stratton, S McKenzie, ML Tapper, H Chmel, FC Bonebrake, R Blum. Blastomycosis in patients with the acquired immunodeficiency syndrome. Ann Intern Med 116:847–853, 1992.

29. LZ Goldani, AM Sugar. Paracoccidioidomycosis and AIDS: an overview. Clin Infect Dis 21: 1275–1281, 1995.

30. TR Mosmann, S Sad. The expanding universe of T-cell subsets: Th1, Th2 and more. Immunol Today 17:138–146, 1996.

31. L Li, S Sad, D Kagi, TR Mosmann. CD8 Tc1 and Tc2 cells secrete distinct cytokine patterns in vitro and in vivo but induce similar inflammatory reactions. J Immunol 158:4152–4161, 1997.

32. A Mencacci, G Del Sero, E Cenci, CF d'Ostiani, A Bacci, C Montagnoli, M Kopf, L Romani. Endogenous interleukin 4 is required for development of protective CD4$^+$ T helper type 1 cell responses to *Candida albicans*. J Exp Med 187:307–317, 1998.

33. A Mencacci, E Cenci, G Del Sero, C Fe d'Ostiani, P Mosci, G Trinchieri, L Adorini, L Romani. IL-10 is required for development of protective Th1 responses in IL-12-deficient mice upon *Candida albicans* infection. J Immunol 161:6228–6237, 1998.

34. PA Morel, TB Oriss. Crossregulation between Th1 and Th2 cells. Crit Rev Immunol 18:275–303, 1998.

35. J Banchereau, RM Steinman. Dendritic cells and the control of immunity. Nature 392:245–252, 1998.

36. AK Abbas, AH Sharpe. T-cell stimulation: an abundance of B7s. Nat Med 5:1345–1346, 1999.

37. A Vecchiarelli, C Monari, C Retini, D Pietrella, B Palazzetti, L Pitzurra, A Casadevall. *Cryptococcus neoformans* differently regulates B7-1 (CD80) and B7-2 (CD86) expression on human monocytes. Eur J Immunol 28:114–121, 1998.

38. C Retini, A Casadevall, D Pietrella, C Monari, B Palazzetti, A Vecchiarelli. Specific activated T cells regulate IL-12 production by human monocytes stimulated with *Cryptococcus neoformans*. J Immunol 162:1618–1623, 1999.

39. P Zhou, RA Seder. CD40 ligand is not essential for induction of type 1 cytokine responses or protective immunity after primary or secondary infection with *Histoplasma capsulatum*. J Exp Med 187:1315–1324, 1998.

40. SH Gregory, DM Magee, EJ Wing. The role of colony-stimulating factors in host defenses. Proc Soc Exp Biol Med 197:349–360, 1991.

41. TN Wells, AE Proudfoot, CA Power. Chemokine receptors and their role in leukocyte activation. Immunol Lett 65:35–40, 1999.

42. GB Huffnagle, RM Strieter, LK McNeil, RA McDonald, MD Burdick, SL Kunkel, GB Toews. Macrophage inflammatory protein-1alpha (MIP-1alpha) is required for the efferent phase of pulmonary cell-mediated immunity to a *Cryptococcus neoformans* infection. J Immunol 159:318–327, 1997.

43. GB Huffnagle, LK McNeil, RA McDonald, JW Murphy, GB Toews, N Maeda, WA Kuziel. Cutting edge: role of C-C chemokine receptor 5 in organ-specific and innate immunity to *Cryptococcus neoformans*. J Immunol 163:4642–4646, 1999.

44. P Loetscher, M Uguccioni, L Bordoli, M Baggiolini, B Moser, C Chizzolini, JM Dayer. CCR5 is characteristic of Th1 lymphocytes. Nature 391:344–345, 1998.

45. G Trinchieri, P Scott. Interleukin-12: basic principles and clinical applications. Curr Top Microbiol Immunol 238:57–78, 1999.

46. CA Dinarello. IL-18: A TH1-inducing, proinflammatory cytokine and new member of the IL-1 family. J Allergy Clin Immunol 103:11–24, 1999.

47. KJ Hammond, SB Pelikan, NY Crowe, E Randle-Barrett, T Nakayama, M Taniguchi, MJ Smyth, IR van Driel, R Scollay, AG Baxter, DI Godfrey. NKT cells are phenotypically and functionally diverse. Eur J Immunol 29:3768–3781, 1999.

48. KW Moore, A O'Garra, R de Waal Malefyt, P Vieira, TR Mosmann. Interleukin-10. Annu Rev Immunol 11:165–190, 1993.

49. J Fierer, L Walls, L Eckmann, T Yamamoto, TN Kirkland. Importance of interleukin-10 in genetic susceptibility of mice to *Coccidioides immitis*. Infect Immun 66:4397–4402, 1998.

50. L Beaman, D Pappagianis, E Benjamini. Significance of T cells in resistance to experimental murine coccidioidomycosis. Infect Immun 17:580–585, 1977.

51. KV Clemons, CR Leathers, KW Lee. Systemic *Coccidioides immitis* infection in nude and beige mice. Infect Immun 47:814–821, 1985.

52. E Brummer, PA Morozumi, PT Vo, DA Stevens. Protection against pulmonary blastomycosis: adoptive transfer with T lymphocytes, but not serum, from resistant mice. Cell Immunol 73:349–359, 1982.

53. DM Williams, JR Graybill, DJ Drutz. Adoptive transfer of immunity to *Histoplasma capsulatum* in athymic nude mice. Sabouraudia 19:39–48, 1981.

54. E Burger, CC Vaz, A Sano, VL Calich, LM Singer-Vermes, CF Xidieh, SS Kashino, K Nishimura, M Miyaji. *Paracoccidioides brasiliensis* infection in nude mice: studies with isolates differing in virulence and definition of their T cell–dependent and T cell–independent components. Am J Trop Med Hyg 55:391–398, 1996.

55. LV Beaman, D Pappagianis, E Benjamini. Mechanisms of resistance to infection with *Coccidioides immitis* in mice. Infect Immun 23:681–685, 1979.

56. RA Cox, DM Magee. Protective immunity in coccidioidomycosis. Res Immunol 149:417–428, 1998.

57. AM Gomez, WE Bullock, CL Taylor, GS Deepe Jr. Role of L3T4+ T cells in host defense against *Histoplasma capsulatum*. Infect Immun 56:1685–1691, 1988.

58. R Allendoerfer, DM Magee, GS Deepe Jr, JR Graybill. Transfer of protective immunity in murine histoplasmosis by a CD4+ T-cell clone. Infect Immun 61:714–718, 1993.

59. GS Deepe Jr. Role of CD8+ T cells in host resistance to systemic infection with *Histoplasma capsulatum* in mice. J Immunol 152:3491–3500, 1994.

60. LE Cano, LM Singer-Vermes, TA Costa, JO Mengel, CF Xidieh, C Arruda, DC Andre, CA Vaz, E Burger, VL Calich. Depletion of CD8+ T cells in vivo impairs host defense of mice resistant and susceptible to pulmonary paracoccidioidomycosis. Infect Immun 68:352–359, 2000.

61. DM Magee, RA Cox. Roles of gamma interferon and interleukin-4 in genetically determined resistance to *Coccidioides immitis*. Infect Immun 63:3514–3519, 1995.

62. R Allendoerfer, GS Deepe Jr. Intrapulmonary response to *Histoplasma capsulatum* in gamma interferon knockout mice. Infect Immun 65:2564–2569, 1997.

63. LE Cano, SS Kashino, C Arruda, D Andre, CF Xidieh, LM Singer-Vermes, CA Vaz, E Burger, VL Calich. Protective role of gamma interferon in experimental pulmonary paracoccidioidomycosis. Infect Immun 66:800–806, 1998.

64. P Zhou, MC Sieve, J Bennett, KJ Kwon-Chung, RP Tewari, RT Gazzinelli, A Sher, RA Seder. IL-12 prevents mortality in mice infected with *Histoplasma capsulatum* through induction of IFN-gamma. J Immunol 155:785–795, 1995.

65. DM Magee, RA Cox. Interleukin-12 regulation of host defenses against *Coccidioides immitis*. Infect Immun 64:3609–3613, 1996.

66. VL Calich, SS Kashino. Cytokines produced by susceptible and resistant mice in the course of *Paracoccidioides brasiliensis* infection. Braz J Med Biol Res 31:615–623, 1998.

67. R Allendoerfer, GS Deepe Jr. Blockade of endogenous TNF-alpha exacerbates primary and secondary pulmonary histoplasmosis by differential mechanisms. J Immunol 160:6072–6082, 1998.

68. P Zhou, G Miller, RA Seder. Factors involved in regulating primary and secondary immunity to infection with *Histoplasma capsulatum*: TNF-alpha plays a critical role in maintaining secondary immunity in the absence of IFN-gamma. J Immunol 160:1359–1368, 1998.

69. GS Deepe Jr, R Gibbons, E Woodward. Neutralization of endogenous granulocyte-macrophage colony-stimulating factor subverts the protective immune response to *Histoplasma capsulatum*. J Immunol 163:4985–4993, 1999.

70. R Allendorfer, GD Brunner, GS Deepe Jr. Complex requirements for nascent and memory immunity in pulmonary histoplasmosis. J Immunol 162:7389–7396, 1999.

71. TA Fong, TR Mosmann. The role of IFN-gamma in delayed-type hypersensitivity mediated by Th1 clones. J Immunol 143:2887–2893, 1989.

72. JG Smith, DM Magee, DM Williams, JR Graybill. Tumor necrosis factor–alpha plays a role in host defense against *Histoplasma capsulatum*. J Infect Dis 162:1349–1353, 1990.

73. DC Slagle, RA Cox, U Kuruganti. Induction of tumor necrosis factor alpha by spherules of *Coccidioides immitis*. Infect Immun 57:1916–1921, 1989.

74. NM Ampel. In vitro production of tumor necrosis factor-alpha by adherent human peripheral blood mononuclear cells incubated with killed coccidioidal arthroconidia and spherules. Cell Immunol. 153:248–255, 1994.

75. RA Cox, DM Magee. Production of tumor necrosis factor alpha, interleukin-1 alpha, and interleukin-6 during murine coccidioidomycosis. Infect Immun 63:4178–4180, 1995.

76. L Beaman. Fungicidal activation of murine macrophages by recombinant gamma interferon. Infect Immun 55:2951–2955, 1987.

77. E Brummer, LH Hanson, A Restrepo, DA Stevens. In vivo and in vitro activation of pulmonary macrophages by IFN-gamma for enhanced killing of *Paracoccidioides brasiliensis* or *Blastomyces dermatitidis*. J Immunol 140:2786–2789, 1988.

78. BA Wu-Hsieh, DH Howard. Inhibition of the intracellular growth of *Histoplasma capsulatum* by recombinant murine gamma interferon. Infect Immun 55:1014–1016, 1987.

79. E Brummer, DA Stevens. Antifungal mechanisms of activated murine bronchoalveolar or peritoneal macrophages for *Histoplasma capsulatum*. Clin Exp Immunol 102:65–70, 1995.

80. A Gonzalez, W de Gregori, D Velez, A Restrepo, LE Cano. Nitric oxide participation in the fungicidal mechanism of gamma interferon–activated murine macrophages against *Paracoccidioides brasiliensis* conidia. Infect Immun 68:2546–2552, 2000.

81. TE Lane, BA Wu-Hsieh, DH Howard. Iron limitation and the gamma interferon-mediated antihistoplasma state of murine macrophages. Infect Immun 59:2274–2278, 1991.

82. L Beaman. Effects of recombinant gamma interferon and tumor necrosis factor on in vitro interactions of human mononuclear phagocytes with *Coccidioides immitis*. Infect Immun 59:4227–4229, 1991.

83. NM Ampel, GC Bejarano, JN Galgiani. Killing of *Coccidioides immitis* by human peripheral blood mononuclear cells. Infect Immun 60:4200–4204, 1992.

84. J Fleischmann, B Wu-Hsieh, DH Howard. The intracellular fate of *Histoplasma capsulatum* in human macrophages is unaffected by recombinant human interferon-gamma. J Infect Dis 161:143–145, 1990.

85. SL Newman, L Gootee. Colony-stimulating factors activate human macrophages to inhibit intracellular growth of *Histoplasma capsulatum* yeasts. Infect Immun 60:4593–4597, 1992.

86. SL Newman, L Gootee, C Kidd, GM Ciraolo, R Morris. Activation of human macrophage fungistatic activity against *Histoplasma capsulatum* upon adherence to type 1 collagen matrices. J Immunol 158:1779–1786, 1997.

87. SL Newman. Macrophages in host defense against *Histoplasma capsulatum*. Trends Microbiol 7:67–71, 1999.

88. C Bogdan, M Rollinghoff, A Diefenbach. Reactive oxygen and reactive nitrogen intermediates in innate and specific immunity. Curr Opin Immunol 12:64–76, 2000.

89. C Jagannath, JK Actor, RL Hunter Jr. Induction of nitric oxide in human monocytes and monocyte cell lines by *Mycobacterium tuberculosis*. Nitric Oxide 2:174–186, 1998.

90. BA Wu-Hsieh, W Chen, HJ Lee. Nitric oxide synthase expression in macrophages of *Histoplasma capsulatum*–infected mice is associated with splenocyte apoptosis and unresponsiveness. Infect Immun 66:5520–5526, 1998.

91. RL Coffman, TR Mosmann. Isotype regulation by helper T cells and lymphokines. Monogr Allergy 24:96–103, 1988.

92. Y Han, JE Cutler. Antibody response that protects against disseminated candidiasis. Infect Immun 63:2714–2719, 1995.

93. Y Han, MH Riesselman, JE Cutler. Protection against candidiasis by an immunoglobulin G3 (IgG3) monoclonal antibody specific for the same mannotriose as an IgM protective antibody. Infect Immun 68:1649–1654, 2000.

94. A Casadevall. Antibody immunity and invasive fungal infections. Infect Immun 63:4211–4218, 1995.

95. S Mukherjee, SC Lee, A Casadevall. Antibodies to *Cryptococcus neoformans* glucuronoxylomannan enhance antifungal activity of murine macrophages. Infect Immun 63:573–579, 1995.

96. A Vecchiarelli, C Retini, C Monari, A Casadevall. Specific antibody to *Cryptococcus neoformans* alters human leukocyte cytokine synthesis and promotes T-cell proliferation. Infect Immun 66: 1244–1247, 1998.

97. GS Deepe Jr. Prospects for the development of fungal vaccines. Clin Microbiol Rev 10:585–596, 1997.

98. DM Dixon, A Casadevall, B Klein, L Mendoza, L Travassos, GS Deepe Jr. Development of vaccines and their use in the prevention of fungal infections. Med Mycol 36:57–67, 1998.

99. Y Zhu, C Yang, DM Magee, RA Cox. Molecular cloning and characterization of *Coccidioides immitis* antigen 2 cDNA. Infect Immun 64:2695–2699, 1996.

100. KO Dugger, KM Villareal, A Ngyuen, CR Zimmermann, JH Law, JN Galgiani. Cloning and sequence analysis of the cDNA for a protein from *Coccidioides immitis* with immunogenic potential. Biochem Biophys Res Commun 218:485–489, 1996.

101. JJ Yu, SL Smithson, PW Thomas, TN Kirkland, GT Cole. Isolation and characterization of the urease gene (URE) from the pathogenic fungus *Coccidioides immitis*. Gene 198:387–391, 1997.

102. CY Hung, NM Ampel, L Christian, KR Seshan, GT Cole. A major cell surface antigen of *Coccidioides immitis* which elicits both humoral and cellular immune responses. Infect Immun 68:584–593, 2000.

103. EE Wyckoff, EJ Pishko, TN Kirkland, GT Cole. Cloning and expression of a gene encoding a T-cell reactive protein from *Coccidioides immitis*: homology to 4-hydroxyphenylpyruvate dioxygenase and the mammalian F antigen. Gene 161:107–111, 1995.

104. PW Thomas, EE Wyckoff, EJ Pishko, JJ Yu, TN Kirkland, GT Cole. The hsp60 gene of the human pathogenic fungus *Coccidioides immitis* encodes a T-cell reactive protein. Gene 199:83–91, 1997.

105. FJ Gomez, R Allendoerfer, GS Deepe Jr. Vaccination with recombinant heat shock protein 60 from *Histoplasma capsulatum* protects mice against pulmonary histoplasmosis. Infect Immun 63: 2587–2595, 1995.

106. R Allendoerfer, B Maresca, GS Deepe Jr. Cellular immune responses to recombinant heat shock protein 70 from *Histoplasma capsulatum*. Infect Immun 64:4123–4128, 1996.

107. JW Batanghari, GS Deepe Jr, E Di Cera, WE Goldman. Histoplasma acquisition of calcium and expression of CBP1 during intracellular parasitism. Mol Microbiol 27:531–539, 1998.

108. BS Klein, PM Sondel, JM Jones. WI-1, a novel 120-kilodalton surface protein on *Blastomyces dermatitidis* yeast cells, is a target antigen of cell-mediated immunity in human blastomycosis. Infect Immun 60:4291–4300, 1992.

109. CP Taborda, MA Juliano, R Puccia, M Franco, LR Travassos. Mapping of the T-cell epitope in the major 43-kilodalton glycoprotein of *Paracoccidioides brasiliensis* which induces a Th-1 response protective against fungal infection in BALB/c mice. Infect Immun 66:786–793, 1998.

110. C Jiang, DM Magee, TN Quitugua, RA Cox. Genetic vaccination against *Coccidioides immitis*: comparison of vaccine efficacy of recombinant antigen 2 and antigen 2 cDNA. Infect Immun 67: 630–635, 1999.

111. C Jiang, DM Magee, RA Cox. Coadministration of interleukin 12 expression vector with antigen 2 cDNA enhances induction of protective immunity against *Coccidioides immitis*. Infect Immun 67:5848–5853, 1999.

112. DA Stevens. Combination immunotherapy and antifungal chemotherapy. Clin Infect Dis 26: 1266–1269, 1998.

113. E Roilides, MC Dignani, EJ Anaissie, JH Rex. The role of immunoreconstitution in the management of refractory opportunistic fungal infections. Med Mycol 36:12–25, 1998.

114. C Jiang, DM Magee, RA Cox. Construction of a single-chain interleukin-12-expressing retroviral vector and its application in cytokine gene therapy against experimental coccidioidomycosis. Infect Immun 67:2996–3001, 1999.

115. SS Ahuja, RL Reddick, N Sato, E Montalbo, V Kostecki, W Zhao, MJ Dolan, PC Melby, SK Ahuja. Dendritic cell (DC)-based anti-infective strategies: DCs engineered to secrete Il-12 are a potent vaccine in a murine model of an intracellular infection. J Immunol 163:3890–3897, 1999.

14

Humoral Immunity and *Cryptococcus neoformans*

Arturo Casadevall
Albert Einstein College of Medicine, Bronx, New York

I. INTRODUCTION

The mechanisms of host defense against *Cryptococcus neoformans* have been extensively studied since the 1950s. There is consensus in the field that cellular immunity is the primary host defense mechanism against *C. neoformans* infection (reviewed in [1]). In contrast, it has been difficult to unequivocally establish a role for humoral immunity against *C. neoformans*. This is perplexing given that this pathogen has a polysaccharide capsule and antibody responses are usually required for protection against encapsulated pathogens. The importance of humoral immunity to a pathogen is generally established by one or more of the following criteria:

1. Demonstrating that passive transfer of immune sera to a naive host protects against infectious challenge.
2. Demonstrating an inverse correlation between the titer of specific antibody titer and susceptibility to infection.
3. Associating susceptibility to infection with particular deficits in humoral immunity.

Application of these criteria to assess the role of humoral immunity against *C. neoformans* produced many inconclusive or negative results (see below). As a result there has been considerable uncertainty as to the role of humoral immunity in protection against *C. neoformans* infection (for prior reviews see [2–5]). In fact, it was not until the late 1980s when monoclonal antibodies (mAbs) to the capsular polysaccharide were used in protection experiments that antibody administration was shown to consistently modify the course of experimental *C. neoformans* infection to the benefit of the host (reviewed in [2]). In recent years, studies from independent laboratories have provided convincing evidence that some antibodies are useful in host defense. As a result, a new consensus has arisen that humoral immunity can contribute to host defense. The emerging view is that humoral immunity collaborates with cellular immunity and that the host benefits from having both types of immune responses (reviewed in [6,7]). This chapter will review the available published information on humoral immunity against *C. neoformans* and argue that there is now overwhelming evidence that antibody can make an important contribution to host defense.

II. SPECIAL CHARACTERISTICS OF *C. NEOFORMANS* AS A PATHOGEN

C. neoformans has several unique characteristics which, in combination, present formidable problems to the immune system.

A. Cell Wall

Like all fungi, *C. neoformans* has a cell wall which provides structural strength to the fungal cell. The cell wall protects the cell from complement-mediated lysis by the membrane attack complex [8]. As a result, *C. neoformans* is impervious to cell damage by activation of complement cascade proteins.

B. Polysaccharide Capsule

C. neoformans is unique among the medically important fungi in that it has a large polysaccharide capsule exterior to the cell wall that is composed primarily of glucuronoxylomannan (GXM) (reviewed in [9]). The polysaccharide capsule is a major virulence determinant since acapsular mutants are not virulent [10]. Numerous studies have shown that the polysaccharide capsule is antiphagocytic in vitro, such that phagocytic cells are unable to ingest *C. neoformans* cells in the absence of antibody or complement opsonins. The capsule appears to prevent phagocytosis in vitro by presenting a surface that is not recognized by phagocytic cell receptors [11]. In contrast, nonencapsulated strains readily attach to and are ingested by phagocytic cells. These observations would suggest that the capsule promotes virulence by inhibiting phagocytosis. However, ultrastructural studies of *C. neoformans* pulmonary infection have shown that encapsulated cells are rapidly phagocytosed in the lung, possibly as a result of opsonization by complement or surfactant derived opsonins [12]. Thus, it is unlikely that capsular polysaccharide functions in virulence by simply preventing phagocytosis. Instead, it is likely that the polysaccharide capsule and shed polysaccharide (see below) interfere with the ability of the host to mount an effective immune response to cryptococcal infection. The capsule blocks binding of antibodies to the cell wall and prevents such antibodies from opsonizing yeast cells directly or activating the classical complement activation. The *C. neoformans* capsule imparts a high negative charge to the yeast cells [13,14] such that the zeta potential of encapsulated cells is approximately 10 times higher than that of acapsular mutants [14]. The contribution of this high negative charge to virulence is uncertain. Cell charge alone does not appear to be responsible for the antiphagocytic properties of the *C. neoformans* capsule [13].

The capsular polysaccharide is a type 2 T-cell-independent antigen that is weakly immunogenic. Injection of small amounts of capsular polysaccharide into mice elicits low-titer IgM responses whereas injection of larger doses produce a state of antibody unresponsiveness characterized by the absence of a significant antibody response [15–17]. The net result is that neither *C. neoformans* infection nor immunization with GXM elicits high titer responses. This lack of antibody responsiveness may be clinically significant because the presence of antibodies to the polysaccharide capsule has been correlated with improved prognosis in patients [18].

C. Exopolysaccharides

C. neoformans cells release capsular polysaccharides into tissue during infection [19]. Although a causal relationship between virulence and polysaccharide release has not been proven there are many studies showing that exopolysaccharides are potent immunomodulators that interfere

Table 1 Biological Effects Reported for the Capsular Polysaccharide of *C. neoformans*

Effects	References
Binding to CD18 on human neutrophils	175
Blocks-access of opsonic antibody to yeast surface	47,46
Brain cell edema	176–179
Confers high negative charge to cell	13,14
Dose related enhancement or suppression of mixed-lymphocyte reaction in mice	58
Dose-related immunological nn-responsiveness	58,56
Enhancement of S. cerevisiae phagocytosis by macrophage	58
Hastens mortality in mice	20
Induction of L-selectin shedding from human neutrophils	180
Induction of TNF receptor loss from human neutrophils	180
Induction of T-suppressor Cells	181
Inhibition of phagocytosis	11,182–184
Inhibition of leucocyte migration	185,186
Suppression of lymphocyte proliferation	187

with the development of effective immune responses (Table 1). Intravenous injection of polysaccharide to mice shortens their survival when challenged with lethal *C. neoformans* infection [20]. Mechanisms by which exopolysaccharides may interfere with humoral immunity include: (1) binding antibody and preventing it from mediating phagocytosis of yeast cells; (2) induce the phenomenon of antibody unresponsiveness whereby little or no antibody responses are made to infection; and (3) interfering with the development of cellular immunity which may be essential for antibody-mediated protection [21].

D. Intracellular Pathogenesis

C. neoformans is a facultative intracellular pathogen that can replicate inside phagocytic cells in vitro [22,23]. Phagocytosis of *C. neoformans* by macrophages results in acidification of the phagolysosomal vacuole [24]. The mechanisms by which *C. neoformans* escapes intracellular killing are not understood. Latent infection in rats is associated with persistence of *C. neoformans* in macrophages [25]. With regard to humoral immunity, the ability of this fungus to survive in macrophages is important because intracellular residence places it outside the reach of humoral immune mechanisms.

E. Proteolytic Enzymes

C. neoformans produces proteolytic enzymes capable of digesting immunoglobulins [26]. Whether such enzymes function in vivo to degrade immunoglobulins or complement is unknown. Evidence for proteolytic damage to tissue has been reported in mice infected with *C. neoformans* [27].

F. Melanin Production

C. neoformans cells have a laccase that can catalyze the production of melanin-like pigments [28], and this enzymatic activity is important for virulence [29]. Deposition of melanin-like pigments in *C. neoformans* cells reduces antibody-mediated phagocytosis [30] and protects the cells against fungicidal effects of oxidants and defensins [31,32]. *C. neoformans* cells melanize during murine infection [33,34], a process which probably enhances survival in tissue. Melanization may interfere with humoral immune mechanisms by protecting *C. neoformans* against killing by host effector cells.

III. THE ANTIGENS OF *C. NEOFORMANS*

C. neoformans has many antigens that can elicit antibody responses including proteins, polysaccharides, and pigments. The antigenicity of each of these antigens and the efficacy of the various antibody responses will be considered separately.

A. Protein Antigens

C. neoformans infection elicits strong antibody responses to protein antigens in both humans and rodents [35–37]. Kakeya et al. reported that a 77-kDa heat shock protein was the major target of the humoral immune response during murine infection [38]. Comparison of humoral responses to protein antigens in humans, mice, and rats revealed that all three species recognized similar immunodominant protein antigens, but there were also significant interspecies differences in the patterns of antigens recognized [39]. Interestingly, there was also considerable individual variation in the antibody responses among animals within inbred strains [39]. Although the cause for the variation among mice with the same genetic background is not understood, studies in rats have also shown major differences in the antibody response to *C. neoformans* proteins from variants of *C. neoformans* derived from a single strain by phenotypic switching [40].

Relatively little work has been done to study the efficacy of antibody responses to *C. neoformans* protein antigens in host defense. Antibody responses to enzymes involved in tissue invasion such as phospholipases [41,42] and proteinases [26] could enhance host resistance by neutralizing these enzymatic activities. The first evidence that antibodies to protein antigens may be useful for host defense was provided by Merkel and Scoffield [43], who reported that an IgG2a mAb to secreted protein antigens of 75-kDa and 100-kDa molecular mass could opsonize *C. neoformans* for phagocytosis by macrophages. The study suggests that antibody responses to protein antigens may be useful in host defense.

B. Polysaccharide Antigens

The polysaccharides of *C. neoformans* can be divided into cell wall and capsular polysaccharide antigens. Antibodies to cell wall glucans are ubiquitous in human sera [44] presumably due to prior infection with *C. neoformans* or other fungi that have crossreacting antigens. These antibodies are primarily of the IgG2 class and recognize glucan β-$(1 \rightarrow 3)$- or β-$(1 \rightarrow 6)$ linkages [44]. IgG to cell wall polysaccharides are opsonic for murine macrophages [45,46] but not human monocytes [44]. However, these antibodies are not opsonic for encapsulated strains presumably because the polysaccharide capsule prevents interactions between surface IgG and Fc receptors in phagocytic cells [47,45]. Ultrastructural studies of antibody localization by antibody derived from normal human serum have shown binding deep inside the capsule in locations where they cannot interact with phagocytic cells [48]. Hence, the function of antibodies to cell wall antigen

Table 2 Effects Reported for Specific Antibody to *C. neoformans*

Effect	References
Activation of classical complement pathway and suppression of alternative complement pathway	188,189
Antibody-dependent cell mediated cytotoxicity	143,144
Antibody-dependent natural killer cell-mediated antifungal activity	152,151
Capsular (quellung) reaction	54,190,191
Change in cell charge upon binding yeast cells	14
Clearance of serum polysaccharide antigen	115,163,164
Enhanced antifungal activity by murine macrophagelike cell line J774	138,155
Enhancement of B7-1 costimulatory molecule and MHC class II expression	168
Enhancement of IL-8 release by human neutrophils	161
Enhancement of IL-1β, IL-2, and TNF-α production by human peripheral mononuclear cells	156
Enhancement of IL-12 release from human monocytes	192
Enhancement of fluconazole efficacy in mice	138
Enhancement of amphotericin B efficacy in mice	133,193,194
Enhancement of flucytosine efficacy in mice	135
Enhancement of T-cell proliferation	156,157
Enhances killing by human neutrophils and monocytes	145
Increase in nitrite production my murine macrophages	195,196
Increase in oxygen-related oxidant production by murine macrophages	197
Increased cellularity in rabbit meningitis	95
Induction of IL-6	162
Prolongation of survival in mice	63,109,111, 112,114,115, 117,121,198
Promotes phagocytosis by eosinophils	199
Promotes phagocytosis by microglia	154,200
Promotes phagocytosis by neutrophils	48
Promotes phagocytosis by macrophages and monocytes	12,45,48, 61,138,155, 162, 201
Reversal of T-cell suppression mediated by antibody to B7-2 in suspensions of lymphocytes and *C. neoformans*	166

in protection is uncertain. In contrast, antibodies to the capsule have been shown by several independent research groups to be opsonic and active in infection (Table 2).

The major components of the *C. neoformans* capsule are glucuronoxylomannan (GXM), galactoxylomannan (GalXM), and mannoprotein (MP). Most studies of antibody responses to polysaccharide antigen have focused on GXM. Structural differences in GXM produce antigenic differences that lead to the serologic grouping of *C. neoformans* into five serotype classifications known as A, B, C, D, and AD [49–51]. GXM is the major polysaccharide antigen and antibodies to GXM have been shown to be biologically active (Table 1). The immunogenicity of capsular polysaccharide has been extensively studied since the 1930s. In interpreting the literature it is important to note that some studies have used total capsular polysaccharide fractions whereas others have employed purified GXM, which constitutes 80–90% of the total capsular polysaccharide. Regardless of the polysaccharide preparation used, the theme of all studies is that capsular polysaccharides are weak immunogens that seldom elicit strong antibody responses.

Studies in the 1940s and 1950s revealed that the *C. neoformans* capsular polysaccharides are weakly antigenic [49,52–55]. In the 1970s several investigators carried out rigorous studies of polysaccharide immunogenicity using newer serological methods. Murphy and Cozad studied the B-cell response to cryptococccal capsular polysaccharide in mice by simultaneously measuring B-cells using the hemolytic plaque technique and the serum titer of agglutinating antibody [17]. Immunization with capsular polysaccharide resulted in a transient antibody response that was followed by the induction of tolerance [17]. Immunological unresponsiveness was attributed to tolerance resulting from terminal differentiation of antibody producing cells without proliferation [17]. Kozel and Cazin showed that the optimally immunogenic dose of purified polysaccharide in mice was 10 μg/mouse with higher and lower amounts eliciting little or no antibody response [16]. Immunization with polysaccharide in Freund's adjuvant or complexed to protein significantly enhanced the immunogenicity of the polysaccharide antigen [16]. Mice given polysaccharide in Freund's adjuvant made antibody responses over a wide range of immunizing doses [16]. A later study by the same group showed that a state of immunological unresponsiveness was induced in mice given 100–800 μg, such that subsequent challenge with polysaccharide antigen in Freund's adjuvant resulted in reduced antibody production [56]. The state of immunological responsiveness remained over the length of the study (12 weeks) and was attributed to persistence of polysaccharide in tissue [56]. For the GXM component of capsular polysaccharide, tolerance in mice occurs after infusion of both high and low doses, and these effects are T-cell-dependent and T-cell-independent, respectively [57].

Analysis of the antibody response elicited by injection of purified capsular polysaccharide into mice by a B-cell assay revealed that the magnitude of the antibody response was highly dependent on the immunizing dose [11,58]. Maximal responses were noted at 1 μg/mouse with larger doses resulting in significantly smaller B-cell responses [58]. Depletion of T-cells with antithymocyte serum resulted in a sixfold increase in B-cell response to GXM which was interpreted as indicative of suppressor T-cell regulation of the antibody response [58]. The optimal dose for eliciting an antibody response is a function of the route for immunization. In C3H mice the dose that elicited the highest titer after SC and IV immunization was 1 and 10 μg/mouse, respectively [59]. Given that polysaccharides from *C. neoformans* strains are structurally diverse, it is likely that the optimal amount of polysaccharide required to elicit a maximal antibody response differs depending on the strain.

The magnitude of the antibody response to *C. neoformans* capsular polysaccharide in mice is under genetic control [59]. Analysis of antibody responses to capsular polysaccharide in 23 inbred strains of mice revealed strain-related differences in serum antibody titer after IV immunization [59]. Among the mouse strains studied BALB/c and CeH/HeJ were the best antibody responders with titers >1:1000. The antibody response to cryptococcal polysaccharide appeared to be controlled by four autosomal genes that included the Igh locus and the E_α subregion of the H-2 locus [59]. CBA/N mice produced no antibody after inoculation with polysaccharide indicating that this antigen is a type 2 T-cell-independent antigen [57,59].

Conjugation of polysaccharide to protein carriers markedly enhances the immunogenicity of carbohydrate epitopes. Goren and Middlebrook synthesized a variety of conjugates and showed that they were highly immunogenic in mice [60]. Similarly Kozel and Follete used a conjugate of capsular polysaccharide to methylated bovine serum albumin to elicit high-titer antisera that was opsonic for *C. neoformans* [61]. Devi et al. conjugated GXM to toxoids and showed that these were highly immunogenic in mice and elicited protective antibody responses [62,63]. Molecular studies of antibodies elicited in infection and by conjugate immunization revealed similar variable gene usage [64].

Immunization of laboratory animals with intact cells seldom elicits strong antibody responses to *C. neoformans* polysaccharide (reviewed in [53]). A comparison of the relative immunogenicity of formalin killed cells from various strains revealed that strains with large capsules were significantly less likely to elicit antibody responses in rabbits than strains with small capsules [53].

Several studies have documented that antibodies reactive with capsular polysaccharide are common in the sera of normal individuals. The percentage of individuals with serum antibodies to capsular polysaccharide varies with the study presumably as a result of differences in the polysaccharide preparation, the methodology, and the human population studied. Henderson et al. reported that 89% of normal individuals have antibody to cryptococcal polysaccharide by radioimmunoassay using a total polysaccharide preparation as antigen [65]. Dromer et al. reported that the prevalence of IgM and IgG to cryptococcal polysaccharide in human sera was 60% and 27%, respectively, by ELISA using a total polysaccharide preparation [66]. Houpt et al. measured the antibody levels to the GXM fraction by ELISA and found that the prevalence of IgM and IgG in normal sera was 98% and 28%, respectively. DeShaw and Pirofski reported that antibodies to *C. neoformans* GXM were ubiquitous in patients with and without HIV infection, with sera from HIV + individuals having the higher level of IgA [67]. The antigenic stimulation responsible for the high prevalence of antibodies reactive with *C. neoformans* polysaccharide in normal sera has not been established but may result from environmental exposure to *C. neoformans*, exposure to other microbes with crossreactive antigens, or prior subclinical cryptococcal infections [67,68]. Serological crossreactions of *C. neoformans* polysaccharide have been reported with antigens from *Streptococcus pneumoniae* [69], DF-2 bacteria [70], and *Trichosporon* sp. [71]. Analysis of complement activation by *C. neoformans* cells in the presence of sera with high or low levels of naturally occurring antibody to GXM revealed no differences, suggesting that the level of antibody was insufficient to affect this process [68]. Low levels of antibodies reactive with *C. neoformans* capsular polysaccharide appear to be common in normal human sera, but their contribution to host defense is unknown.

The majority of patients with cryptococcosis (pre-AIDS) lack serum antibody at the time of diagnosis [72]. For example, only 29 of 75 (39%) patients with active cryptococcal infections had low serum antibody titer by indirect immunofluorescence [72]. The antigenicity of *C. neoformans* capsular polysaccharide is of clinical interest because patients with serum antibody are more likely to respond to therapy [72]. Individuals who survive cryptococcal meningitis have long-lasting immunological unresponsiveness to *C. neoformans* polysaccharide antigen [65,73]. Whether this deficit in antibody production was acquired or preceded infection is not known [65,73]. The ability of normal individuals to respond to pneumococcal polysaccharide has been shown to be under genetic control [74], and the same may apply to cryptococcal polysaccharide antigen. Cryptococcal infection elicits antibody responses in the CSF, but relatively little work has been done to study the human intrathecal antibody response to *C. neoformans* antigens during cryptococcal meningitis. Two case reports have described intrathecal antibody production [75,76], but the functional efficacy of those antibodies are unknown.

Galactoxylomannan (GalXM) constitutes a minor component of the cryptococcal polysaccharide which is antigenically different than GXM. GalXM is immunogenic and IgM to GalXM was found in 22% of patients with *C. neoformans* infection [77]. Antibodies reactive with *C. neoformans* GalXM are also reactive with polysaccharides from other fungi, including *C. albicans*. The functional efficacy of antibodies to GalXM defense against *C. neoformans* infection is unknown.

In summary, the cryptococcal capsular polysaccharide is a weak antigen regardless of whether immunization uses purified antigen or whole cells. The weak immunogenicity of the capsular polysaccharide is probably responsible for the relatively low levels of antibody titers observed in human cryptococcosis and experimental infection. Since antibody responses to the polysaccharide antigen are biologically active, lack of immunogenicity appears to be an important element of the mechanism by which this fungal component contributes to virulence.

C. Pigment Antigens

C. neoformans has a laccase that can catalyze the synthesis of melaninlike pigments from a variety of phenolic substrates [28]. Fungal melanin synthesized from L-dopa can elicit strong antibody responses in mice [78], and high titers of antibodies to melanin are elicited in experimental cryptococcal infection [33]. The functional efficacy of antibodies to melanin in defense against *C. neoformans* is unknown.

IV. MOLECULAR GENETICS OF ANTIBODY RESPONSES TO *C. NEOFORMANS* GXM

The antibody response to *C. neoformans* GXM has been characterized at the molecular level by studying the variable (V) region gene sequences used in assembling GXM-binding mAbs. In addition, anti-idiotypic reagents reactive with well-characterized mAbs have been used to investigate the types of antibodies found in polyclonal sera. Both murine and human antibody responses have been studied.

Sequence analysis of murine antibodies that bind GXM indicates that this antibody response is highly restricted in variable gene usage [64,74–82]. Variable gene restriction means that a relatively few gene elements are used to assemble antibodies to GXM. The molecular and cellular mechanisms responsible for variable gene restriction are not well understood, but the result is a largely homogenous antibody response that is often oligoclonal (e.g., derived from only a few B-cells). Heavy-chain variable region (V_H) usage in antibodies to GXM involves primarily gene elements from the 3' variable gene families [80]. Similarly, only a few light-chain variable region gene elements (V_L) have been described in antibodies to GXM [79,80]. Antibody responses to GXM have been analyzed in seven genetically different mouse strains, and the results suggest that restriction in V region usage and idiotype is a frequent phenomenon [79]. Antibodies to GXM are often somatically mutated and there is evidence that somatic mutation can influence the fine specificity of antigen binding, resulting in differences in protective efficacy [83]. Variable region analysis has allowed the classification of mAbs to GXM into molecular groups based on V_H and V_L gene element usage [80]. Antibodies assigned to a molecular group display similar binding characteristics with regard to their GXM specificity [84,80].

Murine antibody responses to *C. neoformans* infection and to immunization with a GXM-TT conjugate vaccine have been studied by generating mAbs and then characterizing the individual mAbs with regard to variable gene usage and serological properties [64,85]. Both responses elicit antibodies that use similar V_H and V_L gene elements, but there are differences in the antibody isotype distribution and affinity [64]. The majority of mAbs made in infected mice are IgM whereas the majority of mAbs generated from mice immunized with GXM-TT are IgG [79,81,85]. Antibodies elicited by the conjugate vaccine tend to have more somatic mutations and higher affinity for polysaccharide antigen consistent with the T-dependent nature of that antibody response.

Information on the molecular genetics of the human antibody response to GXM has been obtained by the Pirofski laboratory [86,87]. Although fewer details are known about the human response, the available evidence indicates that this antibody response is also restricted with regard to variable gene usage. The variable region of two human mAbs generated from peripheral lymphocytes of individuals vaccinated with the GXM-TT conjugate vaccine (see below) has been reported [86]. Sequence analysis of the two mAbs revealed that both used an identical V_λ 1a-J_λ genetic element with different, yet related, V_H3 gene elements [86]. Idiotipic analysis of human immune sera using rabbit antisera to one mAb revealed high prevalence of the idiotype consistent with a restricted response [86]. Furthermore, there was considerable homology between the human mAb sequences and those of the class II murine mAb 2H1, suggesting conservation of structural motifs during evolution.

In summary, the antibody response to *C. neoformans* GXM is among the best-characterized antibody responses at the molecular level. The available evidence indicates that the antibody response is highly restricted in both humans and mice, suggesting considerable uniformity of antibody variable region structure. The mechanism(s) responsible for variable region restriction are not well understood but comparative analysis of murine mAbs suggests that it may arise from the need for a specific binding site structural requirement for binding GXM [79].

V. ANTIBODY-GXM INTERACTIONS

The majority of mAbs generated to *C. neoformans* capsular polysaccharide bind to the GXM fraction. Although many mAbs have been the subject of extensive functional and serological studies, the exact epitope recognized by these antibodies has not been determined. A problem in this area of study is that no oligopolysaccharides containing the entire GXM repeat structure have been synthesized for crystal structure studies. Hence, our knowledge of molecular-GXM interactions is derived primarily from solution studies with native and chemically modified GXM. De-*O*-acetylation of GXM results in loss of reactivity for many mAbs, indicating that this group is either a component of the epitope or necessary for epitope conformation [84,85,88]. The most detailed studies of antibody binding to GXM in solution have been carried out using mAb 436 [82], a class II antibody that binds an immunodominant epitope found in all serotypes [80]. MAb 436 binds to GXM from all four serotypes with an apparent dissociation constant (K_D) of $\sim -1.7 \times 10^6$ M [82]. When GXM is de-*O*-acetylated the K_D increased 31-fold, indicating that *O*-acetyl groups make an important contribution to the affinity of the interaction [82]. Carboxyl groups in GXM appear to be essential for antibody binding to GXM [82]. Analysis of the binding of mAb 439 to *C. neoformans* GXM revealed enhanced fluorescence from tryptophan residues consistent with the loss of water from the antigen-antibody combining site [82].

An innovative approach to the problem of fine specificity mapping in mAbs to GXM was to use peptide libraries expressed in phage to identify peptides that serve as GXM mimetics [89]. Peptides were identified that could discriminate between protective and nonprotective mAbs on the basis of their ability to bind to the antigen-binding site [89]. The protective mAb 2H1 was crystallized with and without a peptide mimetic in the binding site and the crystal structure revealed that most peptide-antibody interactions occurred with heavy-chain residues [90].

VI. EFFICACY OF HUMORAL IMMUNITY AGAINST *C. NEOFORMANS*

A. Susceptibility and Antibody Responses

Cryptococcosis has been extensively studied in humans and in several laboratory animals including mice, rabbits, and rats. Comparison of the relative susceptibility of the various species to *C. neoformans* with their ability to mount antibody responses to the capsular polysaccharide provides circumstantial evidence for the importance of humoral immunity in host defense. The paucity of human infection in normal individuals implies that humans are highly resistant to *C. neoformans* infection. Serum antibody to *C. neoformans* polysaccharide and protein antigen is prevalent in humans with and without a history of infection [37,66,67,91,92]. Rabbits are extremely resistant to infection even when fungi are inoculated directly into the cerebrospinal fluid [93]. The resistance of rabbits is not due to a higher basal temperature because steroid administration renders this species susceptible to infection [93]. Rabbits mount strong antibody responses to infection and have historically served as the source of immune serum for antibody studies because this species reliably produces antibody when inoculated with *C. neoformans* [49,54,94]. Hobbs et al. demonstrated a temporal association between that clearance of cryptococcal infection in rabbits and the appearance of opsonic antibody in the CSF [95]. Rats can be infected with *C. neoformans* without immunosuppression, but this species can control the infection such that rats seldom die from cryptococcosis [25,96]. Rats also mount antibody responses to infection, and clearance of *C. neoformans* pulmonary infection is temporally associated with granuloma formation and the appearance of serum opsonins [96]. Persistence of latent infection in rats has been associated with downregulation of both humoral and cellular immune responses [25,96]. Hence, humans, rabbits, and rats consistently produce antibody responses during infection and each are relatively resistant species to *C. neoformans* infection.

The argument is often made that humoral immunity is not critical for defense against *C. neoformans* because individuals with B-cell defects are not at increased risk for infection. However, this argument may not be correct because the literature includes several well-documented cases of cryptococcosis in patients with hypogammaglobulinemia [97–99] and hyper-IgM syndrome [100–102]. Although some of these conditions can be mixed humoral and cellular disorders, they are characterized by major defects in antibody function. The relative paucity of cases in the literature may reflect the difficulty in making an association between two relatively rare disorders: antibody deficiency and cryptococcosis. In this regard it is noteworthy that only a minority of patients with severe defects in cell mediated immunity develop cryptococcosis.

In contrast, most mouse strains are highly susceptible to infection. Many studies have documented that mice seldom make high-titer antibody responses to polysaccharide antigens [81,85,103]. One study revealed that infected mice become sick and die when their serum antibody levels declined during the course of infection [104]. Whether there is a causal relationship between murine susceptibility to infection and lack of antibody responsiveness during infection is not known. Given that some antibodies can be protective in mice (see below), it is possible that the vulnerability of mice to *C. neoformans* infection is associated with weak humoral responses to *C. neoformans* polysaccharide antigens.

Evidence that some antibody responses may be deleterious against *C. neoformans* comes from comparison of pulmonary infection in BALB/c and C.B-17 mice [105]. These two mouse strains are congenic except for the Ig heavy-chain-containing locus in chromosome 12. C.B-17 mice are more resistant than BALB/c mice to *C. neoformans* infection despite the fact that infected BALB/c mice have significantly higher serum antibody titers [105]. Of particular interest was the observation that serum IgG3 levels were higher in the more susceptible BALB/c mice, since this isotype has been reported to be either nonprotective or disease enhancing in mice [106,107]. Although the differences in susceptibility between these two strains may or may not be due to differences in quantitative or qualitative differences in the antibody response, this study illustrates that higher serum antibody levels do not necessarily translate into increased resistance to infection in mice. To date no systematic study of susceptibility to infection versus serum anybody levels for multiple strains has been done.

Shahar et al. studied the sequence of events which follow intraperitoneal infection in rabbits and mice [108]. The cellular reaction to *C. neoformans* infection involves migration of both polymorphonuclear and mononuclear cells to the peritoneum and their interaction with yeast cells to form ring structures which surround and destroy cryptococci. Injection of rabbit immune sera into the mouse peritoneum enhances the intensity of ring formation. In the rabbit, this cellular reaction is more intense than in the mouse and does not need the injection of exogenous antibody because the rabbits make their own antibody responses. In rabbits, the omentum contained large numbers of plasma cells whereas in mice no plasma cells were observed. This study suggests that one explanation for the marked difference in susceptibility between mice and rabbits to intraperitoneal infection is that rabbits mount a strong humoral response which enhances the efficacy of the cellular response.

In summary, humans, rats, and rabbits make antibody responses to *C. neoformans* infection and are considered resistant (or relatively resistant) species. In contrast, mice seldom make antibody responses to *C. neoformans* GXM and are considered a highly susceptible species favored for laboratory studies. Overall, these studies suggest that hosts which respond to infection by mounting strong antibody responses are less susceptible to cryptococcal infection.

B. Correlations of Antibody Prevalence and Susceptibility

Serological studies in patients with cryptococcosis in the pre-AIDS era provide suggestive evidence for the usefulness of humoral immunity. In 1968, Bindschadler and Bennett reported a statistically significant correlation between the presence of serum antibody and the likelihood of cure with amphotericin B in patients with cryptococcosis [72]. Diamond and Bennett subsequently identified the presence of serum antibodies to *C. neoformans* during infection as a favorable prognostic marker [18]. However, it was unclear whether the favorable prognosis in these patients reflected a beneficial effect of antibody or the ability to detect antibody in patients with a less severe infection and lower serum antigen.

Serological studies of patients with AIDS with and without cryptococcosis have demonstrated both quantitative and qualitative differences in the prevalence of serum antibody [66,67,87,91]. Measurement of serum IgG to the capsular polysaccharide by ELISA revealed that French patients with AIDS complicated by cryptococcosis had significantly lower levels that those without AIDS or AIDS without cryptococcosis [91]. In contrast, there was no difference in the level of IgM antibody to the capsular polysaccharide among the three groups [91]. This result was interpreted as indicative of a defect in humoral immunity in patients with AIDS and cryptococcosis which made them vulnerable to infection [91]. Another study, involving American patients, confirmed that patients with HIV infection at risk for *C. neoformans* infection had lower levels of serum IgG to GXM [67]. Although these studies are consistent with a role for antibody in defense it was not clear whether the quantitative defect in IgG levels in patients with AIDS and cryptococcosis preceded the infection or represented an acquired humoral defect during infection such as antibody unresponsiveness following polysaccharide antigen stimulation. There does not appear to be a correlation between antibody levels to GXM in patients with AIDS complicated by cryptococcosis and their clinical course [66]. However, qualitative differences in the antibodies response to GXM in the sera of patients with HIV infection may also contribute to the increase susceptibility for cryptococcal infection in this population [92]. In this regard, HIV-infected patients at risk for infection have significantly lower levels of antibodies expressing a certain V_H3 determinant reported in a protective human antibody to *C. neoformans* GXM [92].

In summary, several studies have attempted to establish the relationship between serum antibody level to polysaccharide antigens and susceptibility to infection. In humans the presence of serum antibody to *C. neoformans* was associated with improved outcome. Normal individuals at very low risk of infection have naturally occurring antibody to GXM. Serum antibody responses to GXM in individuals who are HIV + and at risk for infection are quantitatively and qualitatively different from those in HIV-negative individuals. Taken together, these findings suggest that humoral immunity contributes to protection against infection in humans.

C. Passive Studies with Polyclonal Sera

Most studies that have investigated the protective efficacy of polyclonal immune sera in protection against *C. neoformans* infection have reported a beneficial effect, provided the antibody is administered before or during infection (Table 3). Furthermore, the majority of studies that have shown a benefit to passive immunization have used rabbit sera. In contrast, passive immunization with sera from mice that have survived *C. neoformans* infection was not protective [109,110]. The inability of sera from survivor mice to protect against infection can be explained by the absence of antibodies reactive with the capsule [110]. In 1981, Graybill et al. reported that rabbit antibody to *C. neoformans* protected both normal and thymus-deficient mice when administered at the same site as infection [109]. One study evaluated the efficacy of passive immunization with rabbit immune sera in steroid-treated rabbits given intracisternal infection [95]. Administration of rabbit antibody to naive rabbits achieved serum agglutinating titer of 1:32 but did not prevent dissemination of infection from the CNS or reduce CFU in the brain [95]. Interestingly, antibody treated rabbits had higher leukocyte counts in the CSF but the increased cellularity did not translate into reduced fungal burden [95]. Polyclonal preparations are complex mixtures of specific and nonspecific immunoglobulins which can differ in the amount, isotype, and specificity of antibody. The variable efficacy observed during passive studies with polyclonal sera may reflect quantitative and qualitative differences between the sera employed.

Table 3 Summary of Passive Antibody Studies with Polyclonal Sera Preparations

Study	Antibody preparation	Animal model	Time of serum administration	Protection
Gadebusch et al. [198, 202]	Rabbit immune	Mouse IP	After; multiple doses	Yes
Louria and Kaminsky [203]	Rabbit and mouse	White Swiss IV	After	No
Lim and Murphy [110]	Mouse immune	CBA/J IV	After	No
Graybill et al. [109, 142]	Rabitt immune	Balb/c IP	Before	Yes
	Rabbit immune	Balb/c Nude (nu/nu)		Yes
	Mouse survivor	Balb/c IP		No
Graybill et al. [142]	Rabbit immune	C3H/HeJ	Before	Yes
		C3H/HeJ C depleted		No
Perfect et al. [95]	Rabbit immune	Rabbit intracisternal	Before	No
Devi [63]	Mouse immune after GXM-TT vaccination	Swiss albino	Before	Yes

D. Passive Studies with Monoclonal Antibodies

In 1987, Dromer et al. demonstrated that passive administration of an mAb to GXM known as E1 prolonged survival in lethally infected mice [111]. This was a seminal paper in *C. neoformans* field because it stimulated interest in humoral immunity and ushered in a new era of experimentation using mAbs. Administration of mAb E1 1 day prior to lethal infection resulted in approximately fivefold increase in average survival relative to DBA/2 mice given a control mAb [111]. Mice given mAb E1 had lower organ CFUs and no cryptococcal antigen in their sera. The efficacy of mAb E1 was dose dependent and required the administration of at least 10 μg/mouse to prolong survival [111].

Subsequent studies revealed that the antibody had no inherent antifungal activity in vitro with or without complement [112]. Administration of E1 resulted in protection against *C. neoformans* infection in DBA/2 but not Balb/c mice [112]. DBA/2 mice are partially deficient in complement component C5a, a defect that results in rapid and fatal pulmonary congestion after intravenous inoculation with *C. neoformans*. Administration of mAb E1 to DBA/2 mice prior to infection converted the pathogenesis of intravenous infection from a fulminant pneumonia to a more chronic meningoencephalitis [112]. Although the mechanisms involved in this phenomenon are not fully understood, the presence of antibody may alter the pathogenesis of infection by reducing the size of the inoculum through more effective phagocytosis. In this regard, antibody-treated mice had lower CFUs in the spleen and lung at all time intervals studied in the first 48 hr of infection [112]. Interestingly, no reduction was observed in brain CFUs, suggesting that yeast cells reaching the CNS are not susceptible to the effects of serum antibody. In contrast, Balb/c mice given IV infection die of meningoencephalitis regardless of whether antibody was given before infection. According to this model Balb/c mice were not protected because they died of chronic meningoencephalitis irrespective of antibody administration.

In 1990, the Kozel group used a family of isotype switched IgG1, IgG2a, and IgG2b mAbs in passive protection studies against *C. neoformans* in Swiss Webster mice [113]. This group confirmed that mAbs to GXM were active against *C. neoformans* by reducing lung and spleen CFUs but did not observe prolonged survival in antibody-treated mice. No reduction in CFUs was observed in brain tissue. Opsonization studies using these mAbs revealed that their relative opsonic power was IgG2a > IgG1 > IgG2b. One of these antibodies was subsequently shown to prolong survival in a murine model of intraperitoneal infection in A/J mice [114].

Passive protection experiments with other mAbs confirmed that antibody could be protective. In 1992, Mukherjee et al. reported passive protection experiments in an intraperitoneal model of infection using both A/J and Balb/c mice which showed that antibody administration could significantly prolong survival and reduce serum antigen level [115]. The efficacy of mAbs was dependent on the isotype with protection efficacy, being IgG1, IgA > IgM > IgG3 [115]. Subsequent studies with these mAbs revealed that they were protective in mouse models of intravenous [116], intracerebral [117], and intratracheal [12,118] infection when the antibody was administered before infection. Isotype switching of a nonprotective (or minimally protective) IgG3 mAb to IgG1, IgG2a, and IgG2b significantly enhanced its protective efficacy, proving that isotype was a critical determinant of antibody efficacy against *C. neoformans* in murine systemic infection [107,119]. Comparison of two IgM mAbs derived from a single pre-B-cell in vivo which differed in specificity as a result of somatic mutations revealed that one was protective but the other was not [83,120], providing the first conclusive evidence that fine specificity was also a critical determinant of antibody efficacy.

A human mAb to *C. neoformans* GXM has been shown to prolong survival in mice given intraperitoneal infection [121]. However, protection was abolished if the mice were depleted of complement by injection with cobra venom toxin. This mAb uses a heavy-chain variable region composed of V_H3 which has been shown to be deficient in the antibodies found in the sera from patients with HIV infection at risk for cryptococcosis [92]. A mouse-human IgG1 chimeric antibody constructed using the variable region of a protective murine mAb was shown to prolong survival in mice given lethal infection [122]. These results suggest that human IgM and IgG1 constant regions are effective against *C. neoformans* in murine models of infection.

In summary, four independent research groups have shown that passive administration of mAbs to the *C. neoformans* GXM can prolong survival and/or reduce organ CFUs. The results obtained with mAbs are more consistent than those reported with polyclonal reagents. This may reflect the fact that the activity of polyclonal sera is the aggregate of its components whereas mAbs represent one specificity and one isotype. Hence, mAbs have inherently higher specific activity than polyclonal sera and represent homogeneous and invariant preparations for the study of antibody efficacy against infectious agents. Studies with mAbs have shown that a complex structure-function relationship exists for mAbs such that protective efficacy is dependent on antibody amount, specificity, and isotype.

E. Protection by Conjugate Vaccines

By the 1960s, rabbit antibodies to *C. neoformans* had been shown to prolong survival in mice given lethal infection, but it was not clear whether the active antibodies were specific for the polysaccharide capsule (Table 3). The question of whether antibodies to the polysaccharide were active against *C. neoformans* polysaccharide was difficult to answer because polysaccharide alone did not elicit high-titer responses and the rabbit immune sera used in passive protection experiments presumably contained antibodies to non-polysaccharide antigens. To address this question, Goren and Middlebrook synthesized several polysaccharide-protein conjugate vaccines which elicited high levels of antibodies to polysaccharide when used to immunize mice [60,123]. However, experimental infection of immunized mice with high serum levels of antibody to polysaccharide did not result in protection against lethal infection. In fact, there was a suggestion for enhanced infection in many immunized mice. This experience led Goren to conclude that humoral immunity had little or no role in the control of *C. neoformans* infection [123]. The efficacy of conjugate vaccines in protection against *C. neoformans* was not revisited again until the 1990s when Devi et al. reported the synthesis of several conjugate vaccines made by linking sonicated serotype GXM derivatized with adipic acid to tetanus toxoid (TT) or *Pseudomonas aeruginosa* exoportein A (rEPA) [62]. GXM-TT and GXM-rEPA conjugate vaccines prepared through hydroxyl activation were more immunogenic than the compounds prepared through carboxyl activation. This vaccine elicited high antibody titers with specificity to a GXM antigenic determinant that included *O*-acetyl residues on the mannan backbone [62,85]. Mice immunized with GXM-TT were protected against lethal infectious challenge [63], and mAbs generated from immunized mice were protective in mice [115]. Furthermore, sera from mice immunized with GXM-TT were protective when used in passive protection experiments (Table 3).

How does one account for the very different protective efficacies exhibited by these two vaccine preparations that were each highly immunogenic? Although all potential explanations must be considered speculation, there are several differences between the two vaccine preparations that could have contributed to the divergent results. The Goren and Middlebrook vaccine was made from autoclaved whole capsular polysaccharide and probably included other polysaccharide components in addition to GXM. In contrast, the Devi et al. vaccine was made from purified sonicated GXM. The chemical conjugation strategies used to link protein and polysaccharide in the two vaccines were different, and this may have elicited antibodies with subtle differences in specificity. The Goren and Middlebrook vaccines [60,123] produced extremely high levels of agglutinating antibody, raising the possibility of lack of efficacy through a prozone effect. Serological studies of mice immunized with GXM-TT revealed high levels of IgM and IgG1 [85], which have been shown to be effective isotypes against *C. neoformans* in mice [115], but no comparable information is available for the earlier vaccine. In retrospect, both studies provide important insights into the complexity of humoral immunity against this fungus. The Goren and Middlebrook vaccine demonstrated that not all antibody responses to the capsule were protective, whereas the GXM-TT vaccine established the precedent that it was possible to protect against infection by antibodies to certain polysaccharide fractions.

The Pirofski laboratory has carried out detailed studies on the isotype composition, specificity, and opsonic efficacy of human sera from individuals vaccinated with the experimental GXM-TT vaccine [86,87,121,124,125]. Vaccination of human volunteers with GXM-TT elicits high titer of IgM and IgG antibodies to GXM [125]. Human antibodies elicited by GXM-TT are opsonic for *C. neoformans* in assays using both human and murine phagocytic cells [125]. Most of the opsonic activity was associated with antibodies of the IgG2 isotype [125]. Two human IgM mAbs with specificity for GXM have been generated from peripheral lymphocytes of individuals vaccinated with GXM-TT [86]. One of these mAbs (mAb 2E9) was used to isolate a peptide mimotope that was used to probe the fine specificity of the human antibody response in GXM-TT vaccines [87].

F. Susceptibility in Mice with Deficiencies in Humoral Immunity

Studies in mice with humoral immune defects have produced tantalizing hints for the importance of humoral immunity in host defense. CBA/N *xid* mice are more susceptible to *C. neoformans* infection than CBA/Ca mice [126]. Since CBA/N *xid* mice have a humoral deficiency characterized by lack of Lyb 5+ B-cells, this result is consistent with an important role for antibody-mediated immunity.

Administration of cyclophosphamide in doses which impair humoral immune responses resulted in impaired resistance to *C. neoformans* IV infection in mice [127]. These findings were interpreted as consistent with an important role for humoral immunity in host resistance against *C. neoformans* infection [127]. However, conclusions from cyclophosphamide depletion experiments are complicated by the fact that this agent could have uncontrolled effects in other aspects of immune function. Two studies have analyzed the contribution of B-cells to host defense in mice. One study compared susceptibility of mice depleted of B-cells to normal mice and found no differences, leading to the conclusion that antibodies were not important in protection [128]. Another study used mice made B-cell deficient by targeted gene disruption of IgM and concluded that B-cells had a role for containment of infection in the brain under conditions of impaired cell-mediated immunity [129]. The interpretation of results in B-cell-deficient mice with regard to the role of humoral immunity is complicated by several factors including (1) mice seldom mount antibody significant responses to infection [81], which means that the comparison may involve no specific antibody in the control group versus no antibody in the B-cell-deficient group (e.g., comparing two negatives); and (2) B-cells probably influence the cellular response to infection through their capacity for antigen presentation. One study has compared *C. neoformans* infection in mice deficient in FcRγ chain (knockout, FcRγ −/−) and heterozygous congenic mice (FcRγ +/−) revealed no difference in survival or organ CFUs between mice with and mice without functional Fc receptors [130]. Although this result would appear to argue against a role for IgG in host defense, this conclusion cannot be made because murine infection seldom elicits significant IgG antibody [81]. Administration of neither IgG1 or IgG3 were protective in FcRγ −/− mice when given before infection [130]. However, mice given IgG3 had significantly reduced survival and higher tissue CFU relative to control mice which received no antibody, indicating that this isotype was disease enhancing in this model even in the absence of functional Fc receptors.

Studies of the role of humoral immunity using mice with defects in B-cell function have produced conflicting results with regard to a useful role for antibody in protection. Three studies have produced results suggesting that antibody responses are important, but two studies have found no evidence that B-cells have a role in protection. A central problem with studies using mice to study this question is that this species seldom mounts significant antibody responses to infection and is inherently highly susceptible. As a result, it is difficult to make definitive conclusions from negative results.

G. Antibody Enhancement of Antifungal Therapy

Several studies have shown that specific antibody can enhance the efficacy of antifungal therapy against *C. neoformans* in experimental animal infection. There is also limited experience in humans given antibody in addition to amphotericin therapy. In 1964, Gordon and Lapa demonstrated that addition of rabbit immune sera to amphotericin B enhanced the efficacy of the antifungal drug in murine cryptococcosis. In the 1960s, three patients were treated with rabbit immune sera in addition to amphotericin B [131]. Administration of rabbit serum to patients with cryptococcosis led to a reduction in serum antigen level without significant toxicity [131]. These reports were proceeded by the use of normal human immunoglobulin as adjunctive therapy to amphotericin B therapy in the late 1950s [132]. The rationale for the use of normal human immunoglobulin was the observation that most individuals have antibody whereas patients with cryptococcosis often lack such antibody. Although the human trials were too small to draw conclusions about the benefit of antibody therapy with either human or rabbit immunoglobulin, it appears that antibody infusion was well tolerated [131,132].

In recent years the difficulties involved in treating cryptococcosis have renewed interest in adjunctive antibody therapy using monoclonal antibodies to the capsular polysaccharide. Dromer et al. [133] demonstrated that addition of an IgG1 mAb to amphotericin B enhanced the therapeutic effect of the antifungal drug in DBA/2 mice given lethal intravenous infection. Although the mechanism of synergy is not fully understood, these investigators observed that more antibody was bound by amphotericin B–treated yeast cells than control cells [133]. Recently, Nosanchuk et al. reported that amphotericin altered the capsule charge of *C. neoformans* and amphotericin B–treated cells were more easily phagocytosed when opsonized by a mAb than control cells [134]. Specific antibody has also been shown to enhance the efficacy of fluconazole and 5-flucytosine in murine models of *C. neoformans* infection [135,136].

VII. VARIABLES IN ANTIBODY PROTECTION STUDIES

Although simple in concept, passive protection experiments to evaluate the protective efficacy of an antibody preparation are complex in practice because many variables can affect the results. Hence, one must be cautious in drawing definitive conclusions from negative data.

A. Animal Model

The selection of an animal model can have a profound effect in the outcome of antibody protection studies. The route of infection, the route of antibody administration, the inoculum, and the timing of antibody administration can each affect the outcome of the experiments (reviewed in [2]). Antibody-mediated protection can vary depending on the mouse strain used. In general, antibody is most effective when given before infection against small inocula. The selection of mouse strains to test the protective efficacy of antibody reagents can influence whether protection is observed. For example, administration of the mAb E1 prolonged survival in DBA/2 mice but not Balb/c mice [112]. When evaluating the potential efficacy of an antibody reagent it is probably wise to carry out small pilot studies using several conditions to determine the contribution of individual variables and optimize experimental design.

B. Antibody Amount

To observe antibody-mediated protection it is necessary to provide a threshold antibody concentration in serum. That amount may vary depending on the strain and the animal model used. In an IV model of *C. neoformans* infection using a serotype A strain, 10 μg of an IgG1 mAb was protective but 1 μg was not [111,112]. Hence, negative results with passive administration of polyclonal sera may reflect insufficient amounts of specific antibody. Conversely, the experience with antibody studies against *Streptococcus pneumoniae* indicates that too much antibody may also abolish protective efficacy in a prozonelike phenomenon [137]. The relationship between the antibody dose required for protection against *C. neoformans* and infective inoculum has not been systematically explored.

C. Antibody Isotype

The isotype of anticryptococcal antibodies appears to be an important variable in determining whether antibody-mediated protection is observed. In the IP and IV models of infection, mAbs of the IgG3 class have consistently been either not protective or disease enhancing in various mouse strains [83,106,130]. Among the murine IgG isotypes, IgG2a and IgG1 are among the most protective in systemic models of infection [113,119,138]. Antibody isotype can also affect experimental design since there are major differences in serum half-life ($t_{1/2}$). For example, the $t_{1/2}$ of murine IgM and IgA antibodies is only hours while that of IgG is up to 8 days [139].

D. Antibody Specificity

To mediate protection against *C. neoformans*, an antibody must have specificity for certain epitopes. The comparison of two IgM mAbs that bound GXM revealed that they differed in protective efficacy [83]. The protective and nonprotective IgMs produced annular and punctate indirect immunofluorescence patterns (Fig. 1) on the capsule of *C. neoformans*, respectively

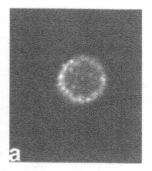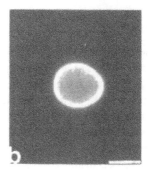

Figure 1   Annular (a) and punctate (b) patterns indirect immunofluorescence after antibody binding to a *C. neoformans* cell. Annular patterns have been associated with protective antibodies whereas punctate binding have not been associated with antibody-mediated protection. Bar represents 10 μm. (From Refs. 83, 120.)

[83,120]. The protective IgM is opsonic even in the absence of serum complement whereas the nonprotective IgM promotes attachment but not ingestion [120,140]. For these IgMs, the type of indirect immunofluorescence pattern varied with the serotype of the strain such that mAb 13F1 gave punctate and annular patterns with serotype D and A strains, respectively [140]. Figure 1 shows representative annular and punctate indirect immunofluorescence patterns after antibody binding to a *C. neoformans* cell. The protective efficacy of mAb 13F1 correlated with the type of pattern observed such that it was protective against serotype A strains but not against serotype D strains [120]. These results indicate that reactivity with the polysaccharide capsule is not a sufficient condition for antibody protective efficacy against *C. neoformans*.

E.   *C. neoformans* Strain

The ability of antibody to protect against cryptococcal infection varies with the *C. neoformans* strain [141]. For some strains antibody administration before infection prolonged survival in mice. However, for other strains antibody administration did not prolong survival despite reducing organ CFUs and binding to the capsule [141]. This phenomenon is not understood but the results suggest that yet uncharacterized strain factors affect antibody efficacy.

F.   Immunological Status of the Host

The efficacy of antibodies to *C. neoformans* is dependent on the immunological status of the host. Some antibodies are not effective in the absence of an intact complement system [121,142]. Antibodies that are protective in normal mice are not protective in mice lacking CD4 T cells [21], IFN-γ [21], or Fc receptors [130]. Interestingly, an IgG3 mAb that is not protective in normal mice can mediate protection in mice lacking CD8 T-cells [21]. The mechanism for this phenomenon is unexplained.

VIII.   MECHANISMS OF ANTIBODY-MEDIATED PROTECTION

Numerous studies have established that antibody alone has no fungicidal or fungistatic effect against *C. neoformans* in the absence of host immune cells [143–145]. Hence, the mechanisms of antibody-mediated protection are likely to result from the enhancement of cellular defense mechanisms.

A.   Phagocytosis and Ringlike Structures

Phagocytosis of *C. neoformans* in vitro by neutrophils and macrophages is very inefficient in the absence of complement-derived or antibody opsonins. Numerous studies have established that antibody to the capsular polysaccharide is a potent opsonin for *C. neoformans* (Table 2). Phagocytosis may be an important mechanism for host defense because ingestion of yeast cells enhances the antifungal efficacy of macrophages.

The relative efficacy of complement-derived and antibody opsonins differs depending on the size of the *C. neoformans* polysaccharide capsule. Comparison of complement- and antibody-mediated phagocytosis of *C. neoformans* cells for two strains that differ in capsule size revealed that complement efficiently opsonized the weakly encapsulated strain but that antibody was required for opsonization of the strain with large capsules [146]. Griffin studied the relative efficacy of phagocytosis of *C. neoformans* by Fc or C3b receptors and noted that significantly more yeast cells were ingested by antibody-mediated phagocytosis [147]. However, the two mechanisms appeared to have different and cooperative functions [147]. Fc receptor phagocytosis was highly efficient but could be blocked with antigen-antibody complexes whereas C3b required lymphokine activation [147].

The opsonic efficacy of murine antibodies to the capsular polysaccharide is dependent on the immunoglobulin isotype. Schlagetter and Kozel studied a family of isotype switched mAbs and found that the relative opsonic efficacy for murine macrophages was IgG2a > IgG1 > IgG2b [148]. Mukherjee et al. studied a different family of isotype switched mAbs and reported that the relative opsonic efficacy was IgG1 and IgG3 > IgG2a > IgG2b using J774 murine macrophagelike cells [138]. The differences in the relative opsonic order of the various IgG isotypes in the two studies may reflect the use of different mAbs or experimental conditions [138]. Antibodies of the IgM and IgA classes are also opsonic for murine J774 cells [138]. For some murine IgM mAbs opsonization does not appear to require complement [120,138].

Many *C. neoformans* cells are too large to be phagocytosed by leukocytes. Inoculation of *C. neoformans* into the peritoneal cavity of rabbits and guinea pigs results in the formation of ringlike structures around large encapsulated yeast cells consisting of neutrophils and/or monocytes [149]. This process was hastened by the presence of specific antibody in the peritoneal cavity, and electron microscopy revealed enhanced contacts between yeast cells and phagocytic cells [150]. In this model antibody promoted interactions between yeast cells and phagocytic cells, leading to the formation of histiocytic rings, which may be precursors of granulomatous inflammation.

B. Antibody-Mediated Enhancement of Antifungal Activity by Immune Cells

Addition of capsule-binding antibody to suspensions of *C. neoformans* and monocytes or neutrophils cells can promote fungal killing by these phagocytic cells. This effect may or may not be a consequence of increased phagocytosis. In 1974, Diamond showed that addition of rabbit specific antibody to cultures of *C. neoformans* and human mononuclear cells reduced the original cryptococcal inoculum by >90% [143]. Addition of cytochalasin B to block phagocytosis had only a minor effect on mononuclear cell antifungal efficacy, indicating that the fungicidal effect was not the result of ingestion by phagocytic cells [143]. A later study, by Diamond and Allison, demonstrated that several leukocyte populations were capable of effecting antibody-dependent cell-mediated killing of *C. neoformans* including granulocytes, monocytes, and lymphocytes [143,144].

Although the lymphocyte population responsible for the antibody-mediated antifungal activity was not evaluated, two subsequent studies from different laboratories showed that NK cells can inhibit *C. neoformans* growth in vitro [151,152]. Both human peripheral blood neutrophils and monocytes kill *C. neoformans* in vitro when incubated in the presence of rabbit immune sera using a chromium release assay [145]. Neutrophils were significantly more effective in killing *C. neoformans* than monocytes [145]. The presence of antibody to GXM restores the antifungal efficacy of PMNs from patients with AIDS against *C. neoformans* [153]. Similarly, human microglia mediate transient fungistasis against *C. neoformans* in the presence of specific antibody to the capsule [154]. Macrophagelike J774 cells efficiently reduce the size of the inoculum when incubated with *C. neoformans* in the presence of antibody [138,155].

C. Lymphocyte Proliferation

Two research groups have reported that addition of antibody to *C. neoformans* can stimulate lymphocyte proliferation when added to suspensions of yeast cells and peripheral blood mononuclear cells. Vecchiarelli et al. showed that addition of a mAb to GXM to a suspension of human mononuclear cells and *C. neoformans* resulted in a significant increase in lymphocyte proliferation [156]. Syme et al. used a different mAb to demonstrate that this effect was the result of increased phagocytosis and consequent antigen presentation to lymphocytes [157]. These observations suggest a mechanism by which antibody can enhance cellular responses essential for the control of this infection.

D. Cytokine Expression

Fc receptor activation triggers signal transduction events which results in production of certain cytokines including IL-2 [158], IL-1β [159], and IL-8 [160]. For *C. neoformans*, coincubation of yeast cells and peripheral blood mononuclear cells in the presence and absence of specific antibody to the capsular polysaccharide elicits differences in cytokine production [156]. Addition of mAb to GXM to suspensions of human PBMCs and *C. neoformans* increased the concentration of IL-1β, IL-2, and TNF-α and reduced the concentration of IL-10 in cell supernatants [156]. Similarly, addition of IgG1 mAb to suspensions of PMNs and *C. neoformans* resulted in increased concentrations of IL-8 in supernatant, probably as a result of activation of classical complement pathway with release of C3a and C5a [161]. Incubation of rat alveolar macrophages with *C. neoformans* and rabbit polyclonal antibody resulted in phagocytosis of yeast cells and induced IL-6 mRNA [162]. Interestingly, no IL-6 mRNA was induced by phagocytosis of complement-opsonized *C. neoformans*, indicating differences in the cytokine response to antibody and complement-derived opsonins [162]. Changes in pro-inflammatory cytokine expression mediated by Fc receptor activation and/or enhanced phagocytosis can translate into more effective cellular responses against *C. neoformans*.

E. Clearance of Polysaccharide Antigen

C. neoformans infections are characterized by the accumulation of polysaccharide antigen in tissue. Polysaccharide antigen has been described to have many deleterious effects on host immune function (Table 1). Hence, clearance of polysaccharide antigen can be expected to be beneficial to the host. Antibody administration to mice and rats with serum levels of polysaccharide antigen results in rapid clearance by formation of antigen-antibody complexes which are deposited in reticuloendothelial cells [163–165]. All isotypes are effective in promoting clearance of serum polysaccharide antigen [163].

   The contribution of this effect to the mechanism of antibody-mediated protection is unknown. Protective and nonprotective mAbs appear be equally effective in promoting clearance of serum antigen deposition of antigen-antibody complexes in the liver. However, it is possible that antibody-mediated clearance of polysaccharide antigen from tissue is an essential aspect of antibody-mediated protection and that the lack of efficacy of nonprotective antibodies reflect other physiologic effects despite antigen clearance. Mice given mAb have been shown to have lower brain weights than control mice despite similar brain fungal burden [116]. This phenomenon may reflect lower brain edema as a result of antibody-mediated clearance of tissue antigen [116]. Under certain conditions formation of antigen-antibody complexes can be deleterious to the host and lead to acute lethal toxicity in mice (see below).

F. Induction of Molecules Important for Immune Response

The Vecchiarelli group has pioneered the study of effects of GXM-binding antibodies on immune recognition molecules [166–168]. Addition of GXM-binding mAb to suspensions of lymphocytes and *C. neoformans* can result in enhanced expression of B7-1 co-stimulatory molecule and MHC class II expression [168]. Furthermore, GXM-binding antibody reverses the T-cell suppression that is induced when lymphocytes and *C. neoformans* are incubated with an antibody to B7-2 costimulatory molecule, possibly by enhancing B7-1 expression [166]. Antibody-mediated phagocytosis has been shown to enhance the expression of CD4 molecules on the surface of monocytes [167]. These observations suggest a mechanism by which the presence of antibody can enhance or facilitate a cellular immune response to infection. In this regard, antibody administration to mice with intratracheal infection has been shown to promote granulomatous inflammation [12,118].

IX. ANTIBODY-MEDIATED LETHAL TOXICITY IN *C. NEOFORMANS* INFECTIONS

Infusion of antibody to certain strains of mice with *C. neoformans* infection can result in acute lethal toxicity (ALT) characterized by scratching, lethargy, and respiratory distress [169]. ALT was most pronounced in Swiss Webster mice given antibody at days 10–18 postinfection [169]. Although this phenomenon does not appear to be immediately relevant to the function of humoral immunity in infection, it is a reminder of the powerful effects of antigen-antibody complexes on host physiology. Furthermore, this effect has raised concern about similar effects in humans, given ongoing efforts to develop passive antibody therapy for human cryptococcosis [169]. ALT appears to be the result of release of platelet-activating factor (PAF) from phagocytic cells as a result of Fc receptor activation by antibody-GXM complexes [169,170]. Death results as a consequence of rapid hemoconcentration from a PAF-induced capillary leak syndrome [170]. Among murine antibody isotypes, ALT was observed with IgG1, IgG2a, and IgG2b but not with IgG3, IgM, or IgA [169,170]. For IgG1 and IgG3 the difference in toxicity has been attributed to differential expression of tissue PAF in response to IgG1-GXM and IgG3-GXM complexes [171]. The likelihood of ALT reaction in mice is a function of antibody dose, time of infection, and serum GXM level [169,170]. The phenomenon of ALT has been described only in mice, and mouse strains differ greatly in their susceptibility to this reaction. In fact, there is even intrastrain variation depending on the mouse supplier: Swiss Webster mice from Charles River Laboratories (Wilmington, MA) are susceptible whereas Swiss Webster mice from Simonsen Laboratories [Gilroy, CA] are resistant. The cause of this variability is not understood but may reflect inherent differences in susceptibility to PAF. The relevance of this observation to the possible use of antibody therapy in humans is unclear, since administration of antibody to humans [131] and rats [164] has not been accompanied by toxicity.

X. INTERDEPENDENCE OF HUMORAL AND CELLULAR IMMUNITY

There is an emerging view in the field that successful defense against *C. neoformans* requires both effective cellular and humoral immune responses (for reviews that espouse this view see [5–7,172,173]. Furthermore, there is accumulating evidence that the efficacy of humoral and cellular responses are interdependent. The efficacy of passive antibody is dependent on the presence of CD4 T cells [21]. On the other hand, antibody administration enhances granulomatous inflammation in murine pulmonary infection [12,118]. Granuloma formation is a hallmark of cell-mediated immunity, and granulomatous inflammation is associated with control of *C. neoformans* infection [1].

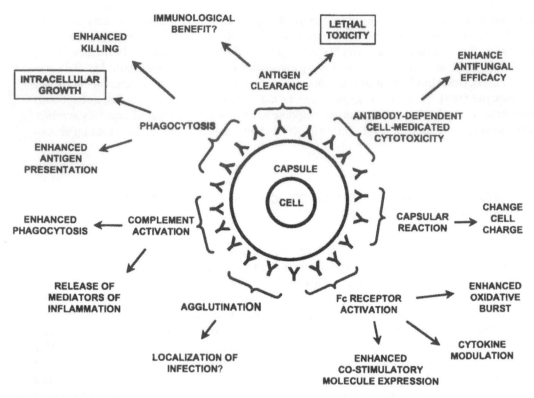

Figure 2   Schematic representation of the many consequences of antibody binding to the capsule of *C. neoformans*. The center of the figure represents a yeast cell surrounded by antibody molecules, and the brackets represent effects of antibody binding. For references to these effects see Table 2. Effects with boxed border are potentially deleterious to the host. Effects followed by a question mark have not been demonstrated but represent potential outcomes based on what is known about antibody function and this system.

There are many mechanisms by which humoral and cellular immunity can collaborate to control *C. neoformans* infection. T-cells can activate phagocytic cells to enhance their antifungal efficacy against ingested yeast cells. T-cells are also essential in the formation of giant cells which can engulf and ingest large *C. neoformans* cells [174]. On the other hand, antibody-mediated phagocytosis can result in enhanced expression of costimulatory molecules and release of pro-inflammatory cytokines, and promote lymphoproliferation as a result of enhanced antigen presentation (Table 2). Furthermore, antibody may indirectly enhance cellular responses by promoting the clearance of capsular polysaccharide, which has been associated with many detrimental effects to immune function (Table 2). Figure 2 shows a schematic representation of the many potential consequences of antibody binding to the capsule of *C. neoformans*.

XI. SUMMARY

There is an overwhelming body of published literature that humoral immunity can contribute to host defense against *C. neoformans*. Although there is widespread agreement that cellular immunity is the primary host defense against *C. neoformans*, it appears that the availability of certain types of antibodies during infection can tilt the balance decisively in favor of the host. Much of the what we know about host defense against *C. neoformans* infection is based on murine studies. Mice are highly susceptible to cryptococcal infection and seldom mount strong antibody responses to capsular polysaccharide. Whether the remarkable susceptibility of the mouse to *C. neoformans* is linked to its inability to mount a strong antibody response has not been established. However, the congruence of these two facts, combined with knowledge that some antibody responses are protective, provides a tantalizing hint that this animal system is an inadequate model for more resistant hosts, such as humans.

The application of hybridoma technology to generate mAbs and their use to study the potential of humoral immunity has revolutionized this field since the first description of a protective mAb by Dromer et al. in 1987 [111]. MAbs have shown that the relationship among structure, function, and protective efficacy for antibodies to *C. neoformans* is extremely complex. At this time, the challenge is to understand how antibodies function against *C. neoformans*.

REFERENCES

1. A Casadevall, JR Perfect. *Cryptococcus neoformans* 1998. American Society for Microbiology Press, Washington, D.C.
2. A Casadevall. Antibody immunity and invasive fungal infections. Infect Immun 63:4211–4218, 1995.
3. A Casadevall. Antibody immunity and *Cryptococcus neoformans*. Can J Bot 73(suppl 1): S1180–S1185, 1995.
4. L Pirofski, A Casadevall. Antibody immunity to *Cryptococcus neoformans*: paradigm for antibody immunity to the fungi? Zbl Bakt 284:475–495, 1996.
5. A Vecchiarelli, A Casadevall. Antibody-mediated effects against *Cryptococcus neoformans*: evidence for interdependency and collaboration between humoral and cellular immunity. Res Immunol 149:321–333, 1998.
6. A Casadevall, A Cassone, F Bistoni, JE Cutler, W Magliani, JW Murphy, L Polonelli, L Romani. Antibody and/or cell-mediated immunity, protective mechanisms in fungal disease: an ongoing dilemma or an unnecessary dispute? Med Mycol 36:95–105, 1998.
7. A Cassone, S Conti, F De Bernardis, L Polonelli. Antibodies, killer toxins and antifungal immunoprotection: a lesson from nature? Immunol Today 18:164–169, 1997.
8. SM Levitz. Overview of host defenses in fungal infections. Clin Infect Dis 14(suppl 1):S37–S42, 1992.
9. R Cherniak, JB Sundstrom. Polysaccharide antigens of the capsule of *Cryptococcus neoformans*. Infect Immun 62:1507–1512, 1994.
10. YC Chang, KJ Kwon-Chung. Complementation of a capsule-deficient mutation of *Cryptococcus neoformans* restores its virulence. Mol Cell Biol 14:4912–4919, 1994.
11. TR Kozel, E Gotschlich. The capsule of *Cryptococcus neoformans* passively inhibits phagocytosis of the yeast by macrophages. J Immunol 129:1675–1680, 1982.
12. M Feldmesser, A Casadevall. Effect of serum IgG1 against murine pulmonary infection with *Cryptococcus neoformans*. J Immunol 158:790–799, 1997.
13. TR Kozel, E Reiss, R Cherniak. Concomitant but not casual association between surface charge and inhibition of phagocytosis by cryptococcal polysaccharide. Infect Immun 29:295–300, 1980.
14. JD Nosanchuk, A Casadevall. Cellular charge of *Cryptococcus neoformans*: contributions from the capsular polysaccharide, melanin, and monoclonal antibody binding. Infect Immun 65:1836–1841, 1997.
15. TR Kozel, J Cazin Jr. Immune response to *Cryptococcus neoformans* soluble polysaccharide. Infect Immun 5:35–41, 1972.

16. TR Kozel, J Cazin Jr. Induction of humoral antibody response by soluble polysaccharide of *Cryptococcus neoformans*. Mycopathol Mycol Appl 54:21–30, 1974.

17. JW Murphy, GC Cozad. Immunological unresponsiveness induced by cryptococcal polysaccharide assayed by the hemolytic plaque technique. Infect Immun 5:896–901, 1972.

18. RD Diamond, JE Bennett. Prognostic factors in cryptococcal meningitis. Ann Intern Med 80: 176–181, 1974.

19. SC Lee, A Casadevall, DW Dickson. Immunohistochemical localization of capsular polysaccharide antigen in the central nervous system cells in cryptococcal meningoencephalitis. Am J Pathol 148: 1267–1274, 1996.

20. JE Bennett, HF Hasenclever. *Cryptococcus neoformans* polysaccharide: studies of serologic properties and role in infection. J Immunol 94:916–920, 1965.

21. R Yuan, A Casadevall, J Oh, MD Scharff. T cells cooperate with passive antibody to modify *Cryptococcus neoformans* infection in mice. Proc Natl Acad Sci USA 94:2483–2488, 1997.

22. RD Diamond, JE Bennett. Growth of *Cryptococcus neoformans* within human macrophages in vitro. Infect Immun 7:231–236, 1973.

23. SC Lee, Y Kress, M-L Zhao, DW Dickson, A Casadevall. *Cryptococcus neoformans* survive and replicate in spacious phagosomes in human microglia. Lab Invest 73:871–879, 1995.

24. SM Levitz, SH Nong, KF Seetoo, TS Harrison, RA Speizer, ER Simons. *Cryptococcus neoformans* resides in an acidic phagolysosome of human macrophages. Infect Immun 67:885–890, 1999.

25. DL Goldman, SC Lee, AJ Mednic, L Montell, A Casadevall. Persistent *Cryptococcus neoformans* infection in the rat is associated with intracellular parasitism, decreased inducible nitric oxide synthase expression and altered antibody responsiveness. Infect Immun 68:832–838, 2000.

26. L-C Chen, E Blank, A Casadevall. Extracellular proteinase activity of *Cryptococcus neoformans*. Clin Diagn Lab Immunol 3:570–574, 1996.

27. CA Salkowski, E Balish. Cutaneous cryptococcosis in athymic and beige-athymic mice. Infect Immun 59:1785–1789, 1991.

28. PR Williamson. Biochemical and molecular characterization of the diphenol oxidase of *Cryptococcus neoformans*: identification as a laccase. J Bacteriol 176:656–664, 1994.

29. SD Salas, JE Bennett, KJ Kwon-Chung, JR Perfect, PR Williamson. Effect of the laccase gene, *CNLAC1*, on virulence of *Cryptococcus neoformans*. J Exp Med 184:377–386, 1996.

30. Y Wang, P Aisen, A Casadevall. *Cryptococcus neoformans* melanin and virulence: mechanism of action. Infect Immun 63:3131–3136, 1995.

31. Y Wang, A Casadevall. Susceptibility of melanized and nonmelanized *Cryptococcus neoformans* to nitrogen- and oxygen-derived oxidants. Infect Immun 64:3004–3007, 1994.

32. TL Doering, JD Nosanchuk, WK Roberts, A Casadevall. Melanin as a potential cryptococcal defense against microbiocidal proteins. Med Mycol 37:175–181, 1999.

33. JD Nosanchuk, P Valadon, M Feldmesser, A Casadevall. Melanization of *Cryptococcus neoformans* in murine infection. Mol Cell Biol 19:745–750, 1999.

34. AL Rosas, JD Nosanchuk, M Feldmesser, GM Cox, HC McDade, A Casadevall. Synthesis of polymerized melanin by *Cryptococcus neoformans* in infected rodents. Infect Immun 68:2845–2853, 2000.

35. L-C Chen, L Pirofski, A Casadevall. Extracellular proteins of *Cryptococcus neoformans* and host antibody response. Infect Immun 65:2599–2605, 1997.

36. AJ Hamilton, J Goodley. Purification of the 115-kilodalton exoantigen of *Cryptococcus neoformans* and its recognition by immune sera. J Clin Microbiol 31:335–339, 1993.

37. AJ Hamilton, JI Figueroa, L Jeavons, RA Seaton. Recognition of cytoplasmic yeast antigens of *Cryptococcus neoformans* and *Cryptococcus neoformans* var. *gattii* by immune human sera. FEMS Immunol Med Microbiol 17:111–119, 1997.

38. H Kakeya, H Udono, N Ikuno, Y Yamamoto, K Mitsutake, T Miyazaki, K Tomono, H Koga, T Tashiro, E Nakayama, S Kohno. A 77-kilodalton protein of *Cryptococcus neoformans* a member of the heat shock protein 70 family, is a major antigen detected in the sera of mice with pulmonary cryptococcosis. Infect Immun 65:1653–1658, 1997.

39. L-C Chen, DL Goldman, TL Doering, L Pirofski, A Casadevall. Antibody response to *Cryptococcus neoformans* proteins in rodents and humans. Infect Immun 67:2218–2224, 1999.

40. DL Goldman, BC Fries, SP Franzot, L Montella, A Casadevall. Phenotypic switching in the human pathogenic fungus *Cryptococcus neoformans* is associated with changes in virulence and pulmonary inflammatory response in rodents. Proc Natl Acad Sci USA 95:14967–14972, 1998.

41. SCA Chen, LC Wright, RT Santangelo, M Muller, VR Moran, PW Kuchel, TC Sorrell. Identification of extracellular phospholipase B, lysophospholipase, and acyltransferase produced by *Cryptococcus neoformans*. Infect Immun 65:405–411, 1997.

42. SCA Chen, M Muller, JZ Zhou, L Wright, TC Sorrell. Phospholipase activity in *Cryptococcus neoformans*: a new virulence factor? J Infect Dis 175:414–420, 1997.

43. GJ Merkel, BA Scofield. An opsonizing monoclonal antibody that recognizes a noncapsular epitope expressed on *Cryptococcus neoformans*. Infect Immun 67:4994–5000, 1999.

44. RG Keller, GS Pfrommer, TR Kozel. Occurrences, specificities, and functions of ubiquitous antibodies in human serum that are reactive with the *Cryptococcus neoformans* cell wall. Infect Immun 62:215–220, 1994.

45. TR Kozel, TG McGaw. Opsonization of *Cryptococcus neoformans* by human immunoglobulin G: role of immunoglobulin G in phagocytosis by macrophages. Infect Immun 25:255–261, 1979.

46. T McGaw, TR Kozel. Opsonization of *Cryptococcus neoformans* by human immunoglobulin G: masking of immunoglobulin G by cryptococcal polysaccharide. Infect Immun 25:262–267, 1979.

47. TR Kozel. Non-encapsulated variant of *Cryptococcus neoformans*. II. Surface receptors for cryptococcal polysaccharide and their role in inhibition of phagogocytosis by polysaccharide. Infect Immun 16:99–106, 1977.

48. TR Kozel, B Highison, CJ Stratton. Localization on encapsulated *Cryptococcus neoformans* of serum components opsonic for phagocytosis by macrophages and neutrophils. Infect Immun 43:574–579, 1984.

49. RW Benham. Cryptococci—their identification by morphology and by serology. J Infect Dis 57:255–274, 1935.

50. R Ikeda, T Shinoda, Y Fukuzawa, L Kaufman. Antigenic characterization of *Cryptococcus neoformans* serotypes and its application to serotyping of clinical isolates. J Clin Microbiol 36:22–29, 1982.

51. DE Wilson, JE Bennett, JW Bailey. Serologic grouping of *Cryptococcus neoformans*. Proc Soc Exp Biol Med 127:820–823, 1968.

52. JM Einbinder, RW Benham, CT Nelson. Chemical analysis of the capsular substance of *Cryptococcus neoformans*. J Invest Dermatol 22:279–283, 1954.

53. JM Neill, I Abrahams, CE Kapros. A comparison of the immunogenicity of weakly encapsulated and of strongly encapsulated strains of *Cryptococcus neoformans (Torula histolytica)*. J Bacteriol 59:263–275, 1950.

54. JM Neill, CG Castillo, RH Smith, CE Kapros. Capsular reactions and soluble antigens of *Torula histolytica* and *Sporotrichum schenckii*. J Exp Med 89:93–106, 1949.

55. CL Hoff. Immunity studies of *Cryptococcus hominis (Torula histolytica)* in mice. J Lab Clin Med 27:751–754, 1942.

56. TR Kozel, WF Gulley, JJ Cazin. Immune response to *Cryptococcus neoformans* soluble polysaccharide: immunological unresponsiveness. Infect Immun 18:701–707, 1977.

57. JB Sundstrom, R Cherniak. A glucuronoxylomannan of *Cryptococcus neoformans* serotype A is a type 2 T-cell-independent antigen. Infect Immun 60:4080–4087, 1992.

58. JF Breen, IC Lee, FR Vogel, H Friedman. Cryptococcal capsular polysaccharide-induced modulation of murine immune responses. Infect Immun 36:47–51, 1982.

59. F Dromer, P Yeni, J Charreire. Genetic control of the humoral response to cryptococcal capsular polysaccharide in mice. Immunogenetics 28:417–424, 1988.

60. MB Goren, GM Middlebrook. Protein conjugates of polysaccharide from *Cryptococcus neoformans*. J Immunol 98:901–913, 1967.

61. TR Kozel, JL Follete. Opsonization of encapsulated *Cryptococcus neoformans* by specific anticapsular antibody. Infect Immun 31:978–984, 1981.

62. SJN Devi, R Schneerson, W Egan, TJ Ulrich, D Bryla, JB Robbins, JE Bennett. *Cryptococcus neoformans* serotype A glucuronoxylomannan-protein conjugate vaccines: synthesis, characterization, and immunogenicity. Infect Immun 59:3700–3707, 1991.

63. SJN Devi. Preclinical efficacy of a glucuronoxylomannan-tetanus toxoid conjugate vaccine of *Cryptococcus neoformans* in a murine model. Vaccine 14:841–842, 1996.

64. J Mukherjee, A Casadevall, MD Scharff. Molecular characterization of the antibody responses to *Cryptococcus neoformans* infection and glucuronoxylomannan-tetanus toxoid conjugate immunization. J Exp Med 177:1105–1106, 1993.

65. DK Henderson, JE Bennett, MA Huber. Long-lasting, specific immunologic unresponsiveness associated with cryptococcal meningitis. J Clin Invest 69:1185–1190, 1982.

66. F Dromer, DW Denning, DA Stevens, A Nobel, JR Hamilton. Anti–*Cryptococcus neoformans* antibodies during cryptococcosis in patients with the acquired immunodeficiency syndrome. Serodiagn Immunother Infect Disease 7:181–188, 1995.

67. M DeShaw, L-A Pirofski. Antibodies to the *Cryptococcus neoformans* capsular glucuronoxylomannan are ubiquitous in serum from HIV+ and HIV− individuals. Clin Exp Immunol 99:425–432, 1995.

68. DC Houpt, GST Pfrommer, BJ Young, TA Larson, TR Kozel. Occurrences, immunoglobulin classes, and biological activities of antibodies in normal human serum that are reactive with *Cryptococcus neoformans* glucuronoxylomannan. Infect Immun 62:3857–2864, 1994.

69. PA Rebers, SA Barker, M Heidelberger, Z Dische, EE Evans. Precipitation of the specific polysaccharide of *Cryptococococcus neoformans* A by types II and XIV antipneumococcal antisera. J Am Chem Soc 80:1135–1137, 1958.

70. MA Westerink, D Amsterdam, RJ Petell, MN Stram, MA Apicella. Septicemia due to DF-2. Cause of a false-positive cryptococcal latex agglutination result. Am J Med 83:155–158, 1987.

71. SJN Devi, PG Reddy, CA Lyman, TJ Walsh, CE Frasch, AC Bush. Immunohistochemical properties of a polysaccharide antigen of *Trichosporon beigelii* that cross-reacts with the capsular polysaccharide of *Cryptococcus neoformans*. Immunol Infect Dis 6:87–92, 1996.

72. DD Bindschadler, JE Bennett. Serology of human cryptococcosis. Ann Intern Med 69:45–52, 1968.

73. DK Henderson, VL Kan, JE Bennett. Tolerance to cryptococcal polysaccharide in cured cryptococcosis patients: failure of antibody secretion in vitro. Clin Exp Immunol 65:639–646, 1986.

74. DM Musher, JE Groover, DA Watson, JP Pandey, MC Rodriguez-Barradas, RE Baughn, MS Pollack, EA Graviss, M de Andrade, CI Amos. Genetic regulation of the capacity to make immunoglobulin G to pneumococcal capsular polysaccharides. J Invest Med 45:57–68, 1997.

75. L La Mantia, A Salmaggi, L Tajoli, D Cerrato, E Lamperti, A Nespolo, G Bussone. Cryptococcal meningoencephalitis: intrathecal immunological response. J Neurol 233:362–366, 1986.

76. KG Porter, DG Sinnamon, RR Gillies. *Cryptococcus neoformans*–specific oligoclonal immunoglobulins in the cerebrospinal fluid in cryptococcal meningitis. Lancet 1:1262–1262, 1977.

77. E Reiss, R Cherniak, R Eby, L Kaufman. Enzyme Immunoassay detection of IgM to galactoxylomannan of *Cryptococcus neoformans*. Diagn Immunol 2:109–115, 1984.

78. JD Nosanchuk, AL Rosas, A Casadevall. The antibody response to fungal melanin in mice. J Immunol 160:6026–6031, 1998.

79. G Nussbaum, S Anandasabapathy, J Mukherjee, M Fan, A Casadevall, MD Scharff. Molecular and idiotypic analysis of the antibody response to *Cryptococcus neoformans* glucuronoxylomannan-protein conjugate vaccine in autoimmune and non-autoimmune mice. Infect Immunity 67:4469–4467, 1999.

80. A Casadevall, M DeShaw, M Fan, F Dromer, TR Kozel, L Pirofski. Molecular and idiotypic analysis of antibodies to *Cryptococcus neoformans* glucuronoxylomannan. Infect Immun 62:3864–3872, 1994.

81. A Casadevall, MD Scharff. The mouse antibody response to infection with *Cryptococcus neoformans*: V_H and V_L usage in polysaccharide binding antibodies. J Exp Med 174:151–160, 1991.

82. EW Otteson, WH Welch, TR Kozel. Protein-polysaccharide interactions. A monoclonal antibody specific for the capsular polysaccharide of *Cryptococcus neoformans*. J Biol Chem 269:1858–1864, 1994.

83. J Mukherjee, G Nussbaum, MD Scharff, A Casadevall. Protective and non-protective monoclonal antibodies to *Cryptococcus neoformans* originating from one B-cell. J Exp Med 181:405–409, 1995.

84. T Belay, R Cherniak, TR Kozel, A Casadevall. Reactivity patterns and epitope specificities of anti–*Cryptococcus neoformans* monoclonal antibodies by enzyme-linked immunosorbent assay and dot enzyme assay. Infect Immun 65:718–728, 1997.

85. A Casadevall, J Mukherjee, SJN Devi, R Schneerson, JB Robbins, MD Scharff. Antibodies elicited by a *Cryptococcus neoformans* glucuronoxylomannan-tetanus toxoid conjugate vaccine have the same specificity as those elicited in infection. J Infect Dis 65:1086–1093, 1992.

86. L Pirofski, R Lui, M DeShaw, AB Kressel, Z Zhong. Analysis of human monoclonal antibodies elicited by vaccination with a *Cryptococcus neoformans* glucuronoxylomannan capsular polysaccharide vaccine. Infect Immun 63:3005–3014, 1995.

87. H Zhang, Z Zhong, L Pirofski. Peptide epitopes recognized by a human anti-cryptococcal glucuronoxylomannan antibody. Infect Immun 65:1158–1164, 1997.

88. C Spiropulu, RA Eppard, E Otteson, TR Kozel. Antigenic variation within serotypes of *Cryptococcus neoformans* detected by monoclonal antibodies specific for the capsular polysaccharide. Infect Immun 57:3240–3242, 1989.

89. P Valadon, G Nussbaum, LF Boyd, DH Margulies, MD Scharff. Peptide libraries define the fine specificity of anti-polysaccharide antibodies to *Cryptococcus neoformans*. J Mol Biol 261:11–22, 1996.

90. ACM Young, P Valadon, A Casadevall, MD Scharff, JC Sacchettini. The three-dimensional structures of a polysaccharide binding antibody to *Cryptococcus neoformans* and its complex with a peptide from a phage display library: implication for the identification of peptide mimotopes. J Mol Biol 274:622–634, 1997.

91. F Dromer, P Aucouturier, J-P Clauvel, G Saimot, P Yeni. *Cryptococcus neoformans* antibody levels in patients with AIDS. Scand J Infect Dis 20:283–285, 1988.

92. R Fleuridor, RH Lyles, L Pirofski. Quantitative and qualitative differences in the serum antibody profiles of human immunodeficiency virus-infected persons with and without *Cryptococcus neoformans* meningitis. J Infect Dis 180:1526–1535, 1999.

93. JR Perfect, SDR Lang, DT Durack. Chronic cryptococcal meningitis. Am J Pathol 101:177–193, 1980.

94. ED Evans, JF Kessel. The antigenic composition of *Cryptococcus neoformans*. J Immunol 67:109–114, 1951.

95. JR Perfect, SDR Lang, DT Durack. Influence of agglutinating antibody in experimental cryptococcal meningitis. B J Exp Pathol 62:595–599, 1981.

96. D Goldman, SC Lee, A Casadevall. Pathogenesis of pulmonary *Cryptococcus neoformans* infection in the rat. Infect Immun 62:4755–4761, 1994.

97. S Gupta, M Ellis, T Cesario, M Ruhling, B Vayuvegula. Disseminated cryptococcal infection in a patient with hypogammaglobulinemia and normal T-cell function. Am J Med 82:129–131, 1987.

98. GA Sarosi, JD Parker, IL Doto, FE Tosh. Amphotericin B in cryptococcal meningitis. Ann Intern Med 71:1079–1087, 1992.

99. JA Wahab, MJ Hanifah, KE Choo. Bruton's agammaglobulinaemia in a child presenting with cryptococcal empyema thoracis and periauricular pyogenic abscess. Singapore Med J 36:686–689, 1995.

100. M Iseki, M Anzo, N Yamashita, N Matsuo. Hyper-IgM immunodeficiency with disseminated cryptococcosis. Acta Paediatr 83:780–782, 1994.

101. CU Kyong, G Virella, HH Fudenberg, CP Darby. X-linked immunodeficiency with increased IgM: clinical, ethnic, and immunologic heterogeneity. Pediatr Res 12:1024–1026, 1978.

102. M-C Tabone, G Leverger, J Landman, C Aznar, L Boccon-Gibod, G Lasfargues. Disseminated lymphonodular cryptococcosis in a child with X-linked hyper-IgM immunodeficiency. Pediatr Infect Dis J 13:77–79, 1994.

103. DB Louria. Specific and non-specific immunity in experimental cryptococcosis in mice. J Exp Med 111:643–665, 1960.

104. EN Scott, HG Muchmore, FG Felton. Enzyme-linked immunosorbent assays in murine cryptococcosis. Sabouraudia 19:257–265, 1981.

105. JA Lovchik, JA Wilder, GB Huffnagle, R Riblet, CR Lyons, MF Lipscomb. Ig heavy chain complex-linked genes influence the immune response in a murine cryptococcal infection. J Immunol 163: 3907–3913, 1999.

106. G Nussbaum, R Yuan, A Casadevall, MD Scharff. Immunoglobulin G3 blocking antibodies to *Cryptococcus neoformans*. J Exp Med 183:1905–1909, 1996.

107. R Yuan, A Casadevall, G Spira, MD Scharff. Isotype switching from IgG3 to IgG1 converts a non-protective murine antibody to *C. neoformans* into a protective antibody. J Immunol 154:1810–1816, 1995.

108. A Shahar, Y Kletter, M Aronson. Granuloma formation in cryptococcosis. Isr J Med Sci 5: 1164–1172, 1969.

109. JR Graybill, M Hague, DJ Drutz. Passive immunization in murine cryptococcosis. Sabouraudia 19: 237–244, 1981.

110. TS Lim, JW Murphy. Transfer of immunity to cryptococcosis by T-enriched splenic lymphocytes from *Cryptococcus neoformans*–sensitized mice. Infect Immun 30:5–11, 1980.

111. F Dromer, J Charreire, A Contrepois, C Carbon, P Yeni. Protection of mice against experimental cryptococcosis by anti–*Cryptococcus neoformans* monoclonal antibody. Infect Immun 55:749–752, 1987.

112. F Dromer, C Perrone, J Barge, JL Vilde, P Yeni. Role of IgG and complement component C5 in the initial course of experimental cryptococcosis. Clin Exp Immunol 78:412–417, 1989.

113. JE Sanford, DM Lupan, AM Schlagetter, TR Kozel. Passive immunization against *Cryptococcus neoformans* with an isotype-switch family of monoclonal antibodies reactive with cryptococcal polysaccharide. Infect Immun 58:1919–1923, 1990.

114. J Mukherjee, TR Kozel, A Casadevall. Monoclonal antibodies reveal additional epitopes of serotype D *Cryptococcus neoformans* capsular glucuronoxylomannan that elicit protective antibodies. J Immunol 161:3557–3568, 1998.

115. J Mukherjee, MD Scharff, A Casadevall. Protective murine monoclonal antibodies to *Cryptococcus neoformans*. Infect Immun 60:4534–4541, 1992.

116. S Mukherjee, S Lee, J Mukherjee, MD Scharff, A Casadevall. Monoclonal antibodies to *Cryptococcus neoformans* capsular polysaccharide modify the course of intravenous infection in mice. Infect Immun 62:1079–1088, 1994.

117. J Mukherjee, L Pirofski, MD Scharff, A Casadevall. Antibody mediated protection in mice with lethal intracerebral *Cryptococcus neoformans* infection. Proc Nat Acad Sci USA 90:3636–3640, 1993.

118. M Feldmesser, Y Kress, A Casadevall. Effect of antibody to capsular polysaccharide on eosinophilic pneumonia in murine infection with *Cryptococcus neoformans*. J Infect Dis 177:1639–1646, 1998.

119. R Yuan, G Spira, J Oh, M Paizi, A Casadevall, MD Scharff. Isotype switching increases antibody protective efficacy to *Cryptococcus neoformans* infection in mice. Infect Immun 66:1057–1062, 1998.

120. G Nussbaum, W Cleare, A Casadevall, MD Scharff, P Valadon. Epitope location in the *Cryptococcus neoformans* capsule is a determinant of antibody efficacy. J Exp Med 185:685–697, 1997.

121. R Fleuridor, Z Zhong, L Pirofski. A human IgM monoclonal antibody prolongs survival of mice with lethal cryptococcosis. J Infect Dis 178:1213–1216, 1998.

122. SL Zebedee, RK Koduri, J Mukherjee, S Mukherjee, S Lee, DF Sauer, MD Scharff, A Casadevall. Mouse-human immunoglobulin G1 chimeric antibodies with activity against *Cryptococcus neoformans*. Antimicrob Agents Chemother 38:1507–1514, 1994.

123. MB Goren. Experimental murine cryptococcosis: effect of hyperimmunization to capsular polysaccharide. J Immunol 98:914–922, 1967.

124. Z Zhong, L Pirofski. Antifungal activity of a human anti-glucuronoxylomannan antibody. Clin Diagn Lab Immunol 5:58–64, 1998.

125. Z Zhong, L Pirofski. Opsonization of *Cryptococcus neoformans* by human anticryptococcal glucuronoxylomannan antibodies. Infect Immun 64:3446–3450, 1996.

126. G Marquis, S Montplaisir, M Pelletier, S Mousseau, P Auger. Genetic resistance to murine cryptococcosis: increased susceptibility in the CBA/N XID mutant strain of mice. Infect Immun 47:282–287, 1985.

127. SS Duke, RA Fromtling. Effects of diethylstilbestrol and cyclophosphamide on the pathogenesis of experimental *Cryptococcus neoformans* infections. J Med Vet Mycol 22:125–135, 1984.

128. DP Monga, R Kumar, LN Mahapatra, AN Malaviya. Experimental cryptococcosis in normal and B-cell deficient mice. Infect Immun 26:1–3, 1979.

129. KM Aguirre, LL Johnson. A role for B cells in resistance to *Cryptococcus neoformans* in mice. Infect Immun 65:525–530, 1997.

130. R Yuan, R Clynes, J Oh, JV Ravetch, MD Scharff. Antibody-mediated modulation of *Cryptococcus neoformans* infection is dependent on distinct Fc receptor function and IgG subclass. J Exp Med 187:1–8, 1998.

131. MA Gordon, A Casadevall. Serum therapy of cryptococcal meningitis. Clin Infect Dis 21: 1477–1479, 1995.

132. ML Littman. Cryptococcosis (Torulosis). Am J Med 27:976–998, 1959.

133. F Dromer, J Charreire. Improved amphotericin B activity by a monoclonal anti–*Cryptococcus neoformans* antibody: study during murine cryptococcosis and mechanisms of action. J Infect Dis 163:1114–1120, 1991.

134. JD Nosanchuk, W Cleare, SP Franzot, A Casadevall. Amphotericin B and fluconazole affect cellular charge, macrophage phagocytosis, and cellular morphology of *Cryptococcus neoformans* at subinhibitory concentrations. Antimicrob Agents Chemother 43:233–239, 1999.

135. M Feldmesser, J Mukherjee, A Casadevall. Combination of 5-flucytosine and capsule binding monoclonal antibody in therapy of murine *Cryptococcus neoformans* infections and in vitro. J Antimicrob Chemother 37:617–622, 1996.

136. J Mukherjee, M Feldmesser, MD Scharff, A Casadevall. Monoclonal antibodies to *Cryptococcus neoformans* glucuronoxylomannan enhance fluconazole activity. Antimicrob Agents Chemother 39: 1398–1405, 1995.

137. K Goodner, FL Horsfall. The protective action of type I antipneumococcus serum in mice. I. Quantitative aspects of the mouse protection test. J Exp Med 62:359–374, 1935.

138. S Mukherjee, SC Lee, A Casadevall. Antibodies to *Cryptococcus neoformans* glucuronoxylomannan enhance antifungal activity of murine macrophages. Infect Immun 63:573–579, 1995.

139. RR Pollock, DL French, JP Metlay, B Birshtein, MD Scharff. Intravascular metabolism of normal and mutant mouse immunoglobulin molecules. Eur J Immunol 20:2021–2027, 1993.

140. W Cleare, A Casadevall. The different binding patterns of two IgM monoclonal antibodies to *Cryptococcus neoformans* serotype A and D strains correlates with serotype classification and differences in functional assays. Clin Diagn Lab Immunol 5:125–129, 1998.

141. J Mukherjee, MD Scharff, A Casadevall. Variable efficacy of passive antibody administration against diverse *Cryptococcus neoformans* strains. Infect Immun 63:3353–3359, 1995.

142. JR Graybill, J Ahrens. Immunization and complement interaction in host defense against murine cryptococcosis. J Reticuloendothel Soc 30:347–357, 1981.

143. RD Diamond. Antibody-dependent killing of *Cryptococcus neoformans* by human peripheral blood mononuclear cells. Nature 247:870–874, 1974.

144. RD Diamond, AC Allison. Nature of the effector cells responsible for antibody-dependent cell-mediated killing of *Cryptococcus neoformans*. Infect Immun 14:716–720, 1976.

145. GPG Miller, S Kohl. Antibody-dependent leukocyte killing of *Cryptococcus neoformans*. J Immunol 131:1455–1459, 1983.

146. R Ikeda, T Shinoda, K Kagaya, Y Fukazawa. Role of serum factors in the phagocytosis of weakly or heavily encapsulated *Cryptococcus neoformans* strains by guinea pig blood leukocytes. Microbiol Immunol 28:51–61, 1984.

147. FM Griffin. Roles of macrophage Fc and C3b receptors in phagocytosis of immunologically coated *Cryptococcus neoformans*. Proc Natl Acad Sci USA 78:3853–3857, 1980.

148. AM Schlageter, TR Kozel. Opsonization of *Cryptococcus neoformans* by a family of isotype-switch variant antibodies specific for the capsular polysaccharide. Infect Immun 58:1914–1918, 1990.

149. M Kalina, Y Kletter, A Sharar, M Aronson. Acid phosphatase release from intact phagocytic cells surrounding a large-sized parasite. Proc Soc Exp Biol Med 136:407–410, 1971.

150. JM Papadimitriou, TA Robertson, Y Kletter, M Aronson, MN-I Walters. An ultrastructural examination of the interaction between macrophages and *Cryptococcus neoformans*. J Pathol 124:103–109, 1978.

151. MF Miller, TG Mitchell, WJ Storkus, JR Dawson. Human natural killer cells do not inhibit growth of *Cryptococcus neoformans* in the absence of antibody. Infect Immun 58:639–645, 1990.

152. N Nabavi, JW Murphy. Antibody-dependent natural killer cell-mediated growth inhibition of *Cryptococcus neoformans*. Infect Immun 51:556–562, 1986.

153. C Monari, A Casadevall, C Retini, F Baldelli, F Bistoni, A Vecchiarelli. Antibody to capsular polysaccharide enhances the function of neutrophils from patients with AIDS against *Cryptococcus neoformans*. AIDS 13:653–660, 1999.

154. SC Lee, Y Kress, DW Dickson, A Casadevall. Human microglia mediate anti–*Cryptococcus neoformans* activity in the presence of specific antibody. J Neuroimmunol 16:152–161, 1995.

155. S Mukherjee, M Feldmesser, A Casadevall. J774 murine macrophage-like cell interactions with *Cryptococcus neoformans* in the presence and absence of opsonins. J Infect Dis 173:1222–1231, 1996.

156. A Vecchiarelli, C Retini, C Monari, A Casadevall. Specific antibody to *Cryptococcus neoformans* alters human leukocyte cytokine synthesis and promotes T cell proliferation. Infect Immun 66:1244–1247, 1998.

157. RM Syme, TF Bruno, TR Kozel, CH Mody. The capsule of *Cryptococcus neoformans* reduces T-lymphocyte proliferation by reducing phagocytosis, which can be restored with anticapsular antibody. Infect Immun 67:4620–4627, 1999.

158. I Anegon, MA Cuturi, G Trinchieri, B Perussia. Interaction of Fc receptor (CD16) ligands induces transcription of interleukin 2 receptor (CD25) and lymphokine genes and expression of their products in human natural killer cells. J Exp Med 167:452–472, 1988.

159. CB Marsh, MP Lowe, BH Rovin, JM Parker, Z Liao, DL Knoell, MD Wewers. Lymphocytes produce IL-1β in response to Fcγ Receptor cross-linking: effects on parenchymal IL-8 release. J Immunol 160:3942–3948, 1998.

160. CB Marsh, JE Gadek, GC Kindt, SA Moore, MD Wewers. Monocyte Fc-gamma receptor cross-linking induces IL-8 production. J Immunol 155:3161–3167, 1995.

161. A Vecchiarelli, C Retini, A Casadevall, C Monari, D Pietrella, TR Kozel. Involvement of C3a and C5a in interleukin-8 secretion by human polymorphonuclear cells in response to capsular polysaccharide material of *Cryptococcus neoformans*. Infect Immun 66:4324–4330, 1998.

162. R-K Li, TG Mitchell. Induction of interleukin-6 mRNA in rat alveolar macrophages by in vitro exposure to both *Cryptococcus neoformans* and anti–*C. neoformans* antiserum. J Med Vet Mycol 35:327–334, 1997.

163. N Lendvai, A Casadevall, Z Liang, DL Goldman, J Mukherjee, L Zuckier. Effect of immune mechanisms on the pharmacokinetics and organ distribution of cryptococcal polysaccharide. J Infect Dis 177:1647–1659, 1998.

164. DL Goldman, A Casadevall, LS Zuckier. Pharmacokinetics and biodistribution of a monoclonal antibody to *Cryptococcus neoformans* capsular polysaccharide antigen in a rat model of cryptococcal meningitis: implications for passive immunotherapy. J Med Vet Mycol 35:271–278, 1997.

165. DL Goldman, SC Lee, A Casadevall. Tissue localization of *Cryptococcus neoformans* glucuronoxylomannan in the presence and absence of specific antibody. Infect Immun 63:3448–3453, 1995.

166. C Monari, TR Kozel, A Casadevall, D Pietrella, B Palazzetti, A Vecchiarelli. B7 costimulatory ligand regulates development of the T-cell response to *Cryptococcus neoformans*. Immunology 98:27–35, 1999.

167. D Pietrella, C Monari, C Retini, B Palazzetti, TR Kozel, A Vecchiarelli. *Cryptococcus neoformans* and *Candida albicans* regulate CD4 expression on human monocytes. J Infect Dis 178:1464–1471, 1998.

168. A Vecchiarelli, C Monari, C Retini, D Pietrella, B Palazzetti, L Pitzurra, A Casadevall. *Cryptococcus neoformans* differently regulates B7-1 (CD80) and B7-2 (CD86) expression on human monocytes. Eur J Immunol 28:114–121, 1998.

169. AC Savoy, DM Lupan, PB Mananlo, JS Roberts, AM Schlageter, LC Weinhold, TR Kozel. Acute lethal toxicity following passive immunization for treatment of murine cryptococcosis. Infect Immun 65:1800–1807, 1997.

170. N Lendvai, A Casadevall. Antibody mediated toxicity in *Cryptococcus neoformans* infection: mechanism and relationship to antibody isotype. J Infect Dis 180:791–801, 1999.

171. N Lendvai, X Qu, W Hsueh, A Casadevall. Mechanism for the isotype dependence of antibody-mediated toxicity in *Cryptococcus neoformans* infected mice. J Immunol 164:4367–4374, 2000.

172. GPG Miller. The immunology of cryptococcal disease. Semin Respir Infect 1:45–52, 1986.

173. RA Fromtling, HJ Shadomy. Immunity in cryptococcosis: an overview. Mycopathologia 77: 183–190, 1982.

174. JO Hill. CD4$^+$ T cells cause multinucleated giant cells to form around *Cryptococcus neoformans* and confine the yeast within the primary site of infection in the respiratory tract. J Exp Med 175: 1685–1695, 1992.

175. ZM Dong, JW Murphy. Cryptococcal polysaccharide bind to CD18 on human neutrophils. Infect Immun 65:557–563, 1997.

176. A Hirano, HM Zimmerman, S Levine. Fine structure of cerebral fluid accumulation. V. Transfer of fluid from extracellular compartments in acute phase of cryptococcal polysaccharide lesions. Arch Neurol 11:632–641, 1964.

177. A Hirano, HM Zimmerman, S Levine. Fine structure of cerebral fluid accumulation. VI. Intracellular accumulation of fluid and cryptococcal polysaccharide in oligodendria. Arch Neurol 12:189–196, 1965.

178. A Hirano, HM Zimmerman, S Levine. The fine structure of cerebral fluid accumulation. III. Extracellular spread of cryptococcal polysaccharides in the acute stage. Am J Pathol 46:1–11, 1964.

179. A Hirano, HM Zimmerman, S Levine. The fine structure of cerebral fluid accumulation. VII. Reactions of astrocytes to cryptococcal polysaccharide implantation. J Neuropathol Exp Neurol 24: 386–396, 1965.

180. ZM Dong, JW Murphy. Cryptococcal polysaccharides induce L-selectin shedding and tumor necrosis receptor loss from the surface of human neutrophils. J Clin Invest 97:689–698, 1996.

181. R Blackstock. Cryptococcal capsular polysaccharide utilizes an antigen-presenting cell to induce a T-suppressor cell to secrete TsF. J Med Vet Mycol 34:19–30, 1996.

182. TR Kozel, GST Pfrommer, AS Guerlain, BA Highison, GJ Highison. Role of the capsule in phagocytosis of *Cryptococcus neoformans*. Rev Infect Dis 10:S436–S439, 1988.

183. FJ Swenson, TR Kozel. Phagocytosis of *Cryptococcus neoformans* by normal and thioglycolate-activated macrophages. Infect Immun 21:714–720, 1978.

184. GS Bulmer, JR Tacker. Phagocytosis of *Cryptococcus neoformans* by alveolar macrophages. Infect Immun 11:73–79, 1975.

185. ZM Dong, J Murphy. Intravascular cryptococcal culture filtrate (CneF) and its major component, glucuronoxylomannan, are potent inhibitors of leukocyte accumulation. Infect Immun 63:770–778, 1995.

186. E Drouhet, G Segretain. Inhibition de la migration leucocytaire in vitro par un polyoside capsulaire de *Torulopsis (Cryptococcus) neoformans*. Ann Inst Pasteur 81:674, 1951.

187. CH Mody, RM Syme. Effect of polysaccharide capsule and methods of preparation on human lymphocyte proliferation in response to *Cryptococcus neoformans*. Infect Immun 61:464–469, 1993.

188. TR Kozel, RS MacGill, KK Wall. Bivalency is required for anticapsular monoclonal antibodies to optimally suppress activation of the alternative complement pathway by the *Cryptococcus neoformans* capsule. Infect Immun 66:1547–1553, 1998.

189. TR Kozel, BCH deJong, MM Grinsell, RS MacGill, KK Wall. Characterization of anti-capsular monoclonal antibodies that regulate activation of the complement system by *Cryptococcus neoformans*. Infect Immun 66:1538–1546, 1998.

190. EE Evans. Capsular reactions of *Cryptococcus neoformans*. Ann NY Acad Sci 89:184–192, 1960.

191. J Mukherjee, W Cleare, A Casadevall. Monoclonal antibody mediated capsular reactions (quellung) in *Cryptococcus neoformans*. J Immunol Methods 184:139–143, 1995.

192. C Retini, A Casadevall, D Pietrella, C Monari, B Palazetti, A Vecchiarelli. Specific activated T cells regulate interleukin-12 production by human monocytes stimulated with *Cryptococcus neoformans*. J Immunol 162:1618–1623, 1999.

193. MA Gordon, E Lapa. Serum protein enhancement of antibiotic therapy in cryptococcosis. J Infect Dis 114:373–378, 1964.

194. J Mukherjee, L Zuckier, MD Scharff, A Casadevall. Therapeutic efficacy of monoclonal antibodies to *Cryptococcus neoformans* glucuronoxylomannan alone and in combination with amphotericin B. Antimicrob Agents Chemother 38:580–587, 1994.

195. N Mozaffarian, JW Berman, A Casadevall. Enhancement of nitric oxide synthesis by macrophages represents an additional mechanism of action for amphotericin B. Antimicrob Agents Chemother 41:1825–1829, 1997.

196. N Mozaffarian, JW Berman, A Casadevall. Immune complexes increase nitric oxide production by interferon-gamma-stimulated murine macrophage-like J774.16 cells. J Leukoc Biol 57:657–662, 1995.

197. A Vazquez-Torres, J Jones-Carson, E Balish. Peroxynitrite contributes to the candidacidal activity of nitric oxide-producing macrophages. Infect Immun 64:3127–3133, 1996.

198. HH Gadebusch. Passive immunization against *Cryptococcus neoformans*. Proc Soc Exp Biol Med 98:611–614, 1958.

199. M Feldmesser, A Casadevall, Y Kress, G Spira, A Orlofski. Eosinophil–*Cryptococcus neoformans* interactions in vivo and in vitro. Infect Immun 65:1899–1907, 1997.

200. R Mazzola, R Barluzzi, A Brozzetti, JR Boelaert, T Luna, S Saleppico, F Bistoni, E Blasi. Enhanced resistance to *Cryptococcus neoformans* infection induced by chloroquine in a murine model of meningoencephalitis. Antimicrob Agents Chemother 41:802–807, 1997.

201. SM Levitz, TP Farrell, RT Maziarz. Killing of *Cryptococcus neoformans* by human peripheral blood mononuclear cells stimulated in culture. J Infect Dis 163:1108–1113, 1991.

202. HH Gadebusch. Specific degradation of *Cryptococcus neoformans* 3723 capsular polysaccharide by a microbial enzyme. III. Antibody stimulation by partially decapsulated cells. J Infect Dis 107:406–409, 1960.

203. DB Louria, T Kaminski. Passively-acquired immunity in experimental cryptococcosis. Sabouraudia 4:80–84, 1965.

15

Vaccines, Antibodies, and Passive Immunity in Candidiasis

Jim E. Cutler, Bruce L. Granger, and Yongmoon Han
Montana State University, Bozeman, Montana

I. INTRODUCTION

The title of this chapter reflects two important concepts that have resulted from research on *Candida* and candidiasis, namely that a vaccine against the disease is a possibility and that antibodies can be protective. Neither tells us about protective mechanisms, but the second dispels the dogma that antibodies against *Candida* are of no benefit to the host. These concepts arise even though our basic understanding of *Candida*-host interactions is far from complete. Our lack of knowledge does not, of course, preclude clinical application of a vaccine formulation, as attested by many examples, the ultimate being the use of the vaccine against smallpox near the end of the 18th century. Indeed, clinical application inspires further basic science, which in turn leads to new insights into host-parasite interactions and vaccine improvements. In this chapter we will examine this scenario as it pertains to antibodies that protect against various forms of candidiasis.

Although vaccine development is useful in our quest for understanding *Candida*-host interactions, the rationale for promoting clinical application of such vaccines may be questioned. Rationale for efforts toward a vaccine against any infectious disease should include at least the following criteria.

1. The incidence and prevalence of the disease is substantial on either a worldwide basis or in defined populations. The latter would include people who are at high risk because of occupation or military circumstances.

2. Suitable therapeutic alternatives to immunization are unavailable, are inadequate, or do not prevent suffering.

3. Recovery from disease results in resistance to recurrence of the disease.

The first two criteria can be readily addressed with respect to candidiasis. Regarding the first one, there is no question as to the public health importance of candidiasis, which is the basis of the chapters on candidiasis in this book. The second criterion is also met, as the effectiveness of antifungal drugs is becoming limited, due to the emergence of strains resistant to existing drugs and the slow development of new antifungals. In addition, patients usually receive antifungal drugs after the onset of candidiasis, after experiencing some degree of suffering.

The third criterion is an interesting one. Whether or not recovery from candidiasis results in immunity to disease recurrence is difficult to assess, because people generally have a high degree of "natural" resistance to the disease throughout most of their lives. This resistance is complex and is likely due to a combination of innate defenses and acquired specific immunity. This complexity is due in part to the fact that the primary cause of candidiasis, *Candida albicans* [1–7], is often a normal commensal inhabitant of human mucosal surfaces. Other *Candida* species of clinical importance also may be transiently associated with human surfaces [8,9]. Whether the mere presence of the fungi on mucosal surfaces is sufficient for immunologic sensitization is unknown, as perhaps virtually everyone experiences at least subclinical transient tissue invasion by *C. albicans*. The complexity of *Candida*-host interactions is discussed in more detail below. These considerations lead to at least two important questions about the usefulness of vaccines against candidiasis.

If most people are naturally sensitized to *Candida*, what is the benefit of vaccination? The etiologic agents of candidiasis are of enormous antigenic complexity, such that immunologic sensitization through natural immunization may not necessarily result in protective immunity. As we will discuss later in this chapter, antibodies against certain *Candida* antigens but not others protect against experimental candidiasis. Other chapters in this book will allude to this same concept when considering protective cell-mediated immunity against candidiasis.

Who should receive immunopreventive measures in the form of preformed antibodies or active immunization against development of candidiasis? Short-term protection via passive transfer of protective antibodies should be considered for those patients who will undergo medical procedures that could place them at high risk of developing hematogenously disseminated candidiasis. People who will be subjected to abdominal surgery [10–13], transplantations [14–18], and other major surgical interventions [9,19–23] should be considered. Provided that a safe, effective, and long-lasting immunity can be achieved through active immunization, all women should be vaccinated. Over 70% of women in the United States will experience at least one bout of *Candida* vaginitis; about one-half of these will have more than one episode, and many will suffer chronic disease [24,25]. The availability of effective antifungal drugs does not offset the benefits of disease prevention through immunization. Because the number and incidence of risk factors leading to development of candidiasis increases with age, immunization should also be considered for everyone over the age of 65. These risk factors include, but are not limited to, use of dentures [26,27], diabetes [28], surgical intervention procedures, placement of indwelling catheters [29–31], and long-term stays in intensive care units [30,32,33].

In this chapter, we will emphasize vaccine developments as they relate to induction of specific antibodies. We will not attempt to give a comprehensive compilation or cataloging of studies over the past 40 years or so. Rather, we will take the liberty of considering selected works on *Candida* antigens or antibodies that are important from a historical perspective or are strong candidates for Phase I clinical trials. Our emphasis on protective antibodies should not be taken to mean that cell-mediated immune responses are unimportant in host defense against various forms of candidiasis, however; other chapters of this book will consider this important arm of the immune system in considerable detail.

II. THE CASE AGAINST ANTIBODIES

A role for antibodies in host defense against candidiasis will continue to be met with skepticism until supportive direct clinical evidence is obtained. This negative perception is deep rooted and may have had its beginnings during the studies on coccidioidomycosis by Charles E. Smith in the 1930s through the 1950s. Among his many contributions made on this disease (as a result of data collected from almost 40,000 patients) was the astute observation that changes in the complement fixation titer [34–36], later determined as the IgG response against *Coccidioides immitis* chitinase [37], could be used as a prognostic indicator in cases of primary pulmonary disease. A rising complement fixation antibody titer is an ominous sign indicative of impending extrapulmonary dissemination, whereas a waning titer suggests either a favorable host response to the primary pulmonary infection or beneficial antifungal therapy against hematogenously disseminated disease [36]. Others have concluded from these observations that antibodies do not protect the host against coccidioidomycosis, even though precise antibody specificities, isotypes, and titers against specific epitopes could not have been considered at that time. An analysis of Smith's data, interestingly enough, suggests that a precipitin response early in the pathogenesis of disease, later determined as an IgM response against fungal antigens rich in mannose and an unusual sugar, 3-*O*-methyl mannose [36,38], may correlate with a favorable outcome for some patients. Nonetheless, the negative correlation between complement fixation antibody titers and effective host response against *Coccidioides immitis* seems to have primed the medical community toward the belief that antibodies against fungi are not helpful in host defense against mycotic diseases in general. This belief has recently been challenged [39].

Reasoning based on extrapolation of clinical and experimental data on candidiasis has similarly led to the belief that antibodies do not protect against various forms of this disease. The presence of *Candida*-reactive antibodies in the blood of people is very common [40]. Immunologic sensitization undoubtedly occurs because of the presence of viable *C. albicans* as a member of the mucosal normal flora [9] and through clinically apparent and unapparent infections due to this fungus. Thus, the presence of antibodies does not preclude development of candidiasis either as disseminated life-threatening disease or as superficial lesions at mucosal and dermal sites [41–43]. In a now classic paper, Kirkpatrick et al. reported the results of clinical studies on 12 patients with chronic mucocutaneous candidiasis [41]. In this 1971 hallmark contribution, the authors observed normal or elevated serum agglutinin and precipitin titers, and normal parotid fluid agglutinin levels in all 12 patients. Eight of the 12 patients were, however, severely depleted in cell mediated immune functions, which naturally led to the conclusion that this arm of the immune system is responsible for host defense against mucocutaneous candidiasis. The antibody findings in these patients seemed incompatible with the idea that immunoglobulins offer protection against mucocutaneous disease. But, as in the case of serology studies on coccidioidomycosis, antibody titers against specific epitopes were not determined.

In animal experimental work, mice immunized against whole dead *C. albicans* yeast cells plus a novel mucosal adjuvant had heightened resistance to hematogenously disseminated disease, but the evidence supported cell-mediated immunity as the primary host defense [44]. In one study on susceptibility of various kinds of immunodeficient mice to hematogenously disseminated candidiasis, the importance of candidal-specific antibodies was largely dismissed and, instead, T-cell-mediated immunity was concluded as the important acquired specific host defense [45]. These conclusions are in line with those made by Bistoni and co-workers, who studied the reasons why mice became resistant to candidiasis if they were infected with an avirulent strain (CA-2 or PCA-2) of *C. albicans* prior to challenge with a virulent strain [46–49]. Many of these studies linked resistance to Th1-type responses, which is a theme supported by several investigators [49–53]; others have obtained support for Th1 and Th2 responses in host defense against experimental candidiasis [54]. In another study, evidence was obtained that antibodies specific for the fungus may actually enhance disease severity [55]. In an in vitro experiment, antibodies specific for *C. albicans* were reported to block phagocytosis of the fungus [56,57]. In a more analytical investigation, Costantino et al. [58] compared the antibody responses in a variety of inbred strains of mice with respect to predominant isotypes, titers, and antigenic specificities. The antigenic specificities recognized by polyclonal responses to a systemic infection with *C. albicans* were ascertained by Western blot analysis of denatured protein extracts from yeast cells of the fungus. These were interesting studies because the strains of mice chosen for investigation differed with regard to antibody titers, predominant yeast cell antigen recognized, and relative susceptibility or resistance to disseminated candidiasis. The authors concluded that antibody responses did not correlate with either susceptibility or resistance to disseminated candidiasis.

III. A HYPOTHESIS: ANTIBODIES MAY BE PROTECTIVE AGAINST CANDIDIASIS

Given the above evidences against protective antibodies, how do we reconcile the thrust of this chapter which defends the existence of protective antibodies? We reason that yeast cells have an enormous number of epitopes associated with their cell surface, and antibody to only a finite number of these epitopes will show protection against candidiasis. We hypothesize that people who are exposed to *C. albicans* will not necessarily respond by making "protective" antibodies in sufficient concentration to be of any benefit against development of candidiasis. If this hypothesis is correct, merely finding antibodies in the sera of patients or experimental animals with candidiasis is not reason to reject a positive role for antibodies in host defense against various forms of candidiasis. The more important questions to consider are the specificities and titers of antibodies that are present in the serum of each patient. The ideal antibody-inducing vaccine formulation against candidiasis will consist in part of either a cocktail of a select number of purified antigens or an extract of the fungus that is highly enriched for so-called protective epitopes. The remainder of the formulation will consist of appropriate carriers or adjuvants to ensure responses of appropriate titer and isotype. These considerations are delineated below as criteria for antibody protection against candidiasis.

IV. COMPLEXITY OF *CANDIDA*-HOST INTERACTIONS

An understanding of our hypothesis requires a brief consideration of the complexity of *Candida*-host interactions. Under normal circumstances, *C. albicans* is usually a noninvasive commensal that has a set of properties which allow it to survive on host mucosal surfaces occupied by large numbers of other competing micro-organisms [59]. The properties that allow this fungus to survive in this hostile environment become its so-called virulence traits when host conditions arise that allow tissue invasion. The breakdown of from one to several of a multitude of possible host defense mechanisms may allow for development of candidiasis. Events that occur naturally, such as onset of diabetes or AIDS, or as a result of medical procedures, such as various surgical interventions or use of central venous lines, are broadly classed as risk factors leading to a compromised immune state and a heightened probability of development of one or more forms of candidiasis [17,23,30,60–68]. The immunologic defense breach may involve innate barriers, such as destruction of mucosal surfaces or decreased neutrophil activities, or acquired specific immunity most notably exemplified by abnormal T-cell function, as alluded to above. These examples are oversimplifications of the problem, as they do not address a multitude of factors such as those associated with endocrine dysfunction [69,70], hypercalcemia [71], thyroiditis [70], or malignancies [72,73], or the fact that people without apparent immunodeficiency also may develop candidiasis [70,74,75]. Clearly, the interactions of *C. albicans* or other *Candida* spp. with the host are enormously complex.

It should be no surprise that an individual defective in one defense system may not necessarily have greater susceptibility to candidiasis, because other systems are sufficient to compensate for the defect and suppress these fungi of low virulence. The relative importance of a given defense mechanism also will be dependent on the specific site within the body to which the fungus has gained access. Neutropenia is clearly a high risk factor for development of hematogenously disseminated candidiasis [68,76–80], but the nonneutropenic patient is also susceptible [81–87]. This may indicate that susceptibility to disseminated candidiasis depends on both access of the fungus to the circulatory system and the inoculum dose. In experimental animals, severe reduction in circulating neutrophils renders the animal more susceptible to fatal disseminated candidiasis, but in these experiments the animals usually receive the challenge dose of viable yeast cells directly into the bloodstream [88–94], thus bypassing host defense systems on mucosal surfaces. The clinical parallel to this situation may be the neutropenic patient with an indwelling central venous line on which biofilms of *C. albicans* develop and serve as a continuous inoculum source of fungi into the circulatory system [68,78].

An additional level of complexity in this consideration is the virulence state of the fungus. Growing as a biofilm on plastic, *C. albicans* will likely be in various stages of development with respect to formation of hyphae, pseudohyphae, and yeast forms. During such morphogenetic changes, the fungal cell surface varies dramatically. The degree of cell surface hydrophobicity (CSH) changes as an inverse function of cell wall protein mannosylation, with high CSH associated with low mannosylation and low CSH (i.e., hydrophilicity) associated with high levels of protein mannosylation [95,96]. An interesting feature is that high-CSH yeast forms of a given isolate of this species are more virulent than are their hydrophilic counterparts [97–99]. The definitive reason for the increased virulence of hydrophobic yeast forms is speculative and may relate to increased adherence properties of these fungal cells for tissues [98,100–104]. A surface change from hydrophilicity to hydrophobicity precedes germination [105]; thus, yeast forms with high CSH also germinate into hyphal forms more readily than hydrophilic yeast forms. This latter observation is consistent with recent work on gene knockout/complementation experiments which suggest [106] that the hyphal form is more virulent than are yeast forms [107–115].

The preponderance of investigations on host defense mechanisms against candidiasis has focused on acquired resistance, mainly on cell-mediated immunity. Based on the complexity of *Candida*-host interactions, however, we propose that under the appropriate clinical or experimental conditions, virtually any host defense mechanism plays an important role. Importantly, the mere detection of polyclonal antibodies in the circulation of people who are suffering from candidiasis does not preclude a possible beneficial role of antibodies against the disease. Instead, the following minimum criteria need to be considered when assessing a functional role for antibodies.

1. *Specificity*. A polyclonal antibody response against an enormously complex antigenic array, such as whole cells of *C. albicans*, may not necessarily contain antibodies specific for the "protective" epitopes. For example, protective antibodies against cryptococcosis are specific for glucuronoxylomannan [116–118], whereas for protection against candidiasis, anti-β-1,2-mannotriose [119], and anti-HSP 90 responses are effective [120].

2. *Ig isotype*. The effector function of particular immunoglobulin isotypes may be critical for the protective activity of antibodies. For example, a polyclonal antibody response against *Cryptococcus neoformans* would not be expected to protect the host if the primary isotype consists of IgG3, whereas a strong IgG1 response against the glucuronoxylomannan should be protective for the host [121–123]. IgE responses against *C. albicans* [124–128] would not be expected to be of benefit to the host, but strong complement-fixing antibodies, such as IgM and IgG3 are protective [119,129–131].

3. *Titer*. A patient with serum antibodies against *C. albicans* may contain antibodies with the correct specificity and Ig isotype, but the titer of this antibody may be insufficient for protective benefits.

4.   *Effector function components.* If the patient has antibodies of the correct specificity, isotype, and titer, this individual may not benefit from these antibodies if effector function components are deficient. For example, antibodies that protect by promoting rapid deposition of complement factor C3 onto the fungal cell surface will be of little benefit to a patient who is deficient in C3 or phagocytic cells.

These criteria are not meant for comparing the relative merits of antibodies with cell-mediated immunity, or with any other host defense mechanism. A pertinent conceptual reminder is that candidiasis is a disease resulting from tissue invasion by fungal species of low virulence in normal people. Because there is not a single frank virulence factor, there is no reason to suspect that a single host defense mechanism should stand out as the most important explanation for susceptibility or resistance to disease. Indeed, there are a multitude of innate and acquired host defenses that collectively form an effective barrier to development of clinically apparent candidiasis. The breakdown or dysfunction of one or more of these barriers can render the host susceptible to candidiasis. Host defense mechanisms against candidiasis are not mutually exclusive, but are cumulative and integrated. A deficiency in one mechanism may well be compensated for by embellishment of another, whether CMI, antibodies, phagocytic cell functions, or combinations of these and others. We have embraced this view, the essentials of which were proclaimed by Casadevall et al. during a symposium on antibody and CMI defense against candidiasis and cryptococcosis [132].

V. EARLY EVIDENCE THAT ANTIBODIES CAN PROTECT

Prior to the 1960s, a few anecdotal observations suggested that antibodies might be of benefit to the host in overcoming disseminated candidiasis. In one study, a description of patients with apparent pulmonary candidiasis was instructive in that, contrary to the widely accepted belief that candidiasis patients have circulating antibodies against the fungus [9], some of these patients were without detectable *Candida*-specific agglutinins [133]. Furthermore, the recovery of one patient in this study was correlated with administration of rabbit serum containing a high titer of antibodies specific for the fungus. As alluded to later in this chapter, a group of investigators in Sweden has been administering polyclonal bovine anti-*Candida* antibodies per os to bone marrow transplant recipients in an attempt to reduce *C. albicans* colonization in the oral cavity [134].

A. The Mourad-Friedman Experiments

Mourad and Friedman generated the first strong experimental data for protective antibodies, reported in two short papers in the 1960s [135,136]. These investigators immunized mice with sonically ruptured *C. albicans* yeast cells. A priming dose and boosters were given subcutaneously at days 1, 5, 12, 20, 30, and 40. Sera from animals with a high antibody titer (slide agglutinin titers >1280 against whole yeast cells) were pooled and each of 33 mice received an IP inoculation of 0.5 mL 1 day after IV infection with a lethal dose of viable yeast cells. The animals received an additional 0.2 mL of the antiserum IP every 3 days for 15 days, for a total dose of 2 mL antiserum. An equal number of control animals received similar dosing of either normal mouse serum or saline. Cumulative deaths were recorded daily for the 140-day observation period. Deaths began occurring in the normal serum control mice by day 8 after infection, but in the test group, deaths were not observed until day 18. Whereas all of the control animals died by day 50 after challenge, 33% of the test animals survived the 140-day period and all of these were culture negative for *C. albicans* in their organs. These results compared favorably with actively immunized mice, in which 50% of the animals survived the challenge [136]. These studies were striking because work by previous investigators [137] failed to show a positive effect of antibodies in host defense against disseminated candidiasis.

There are two aspects of the Mourad-Friedman work that deserve further comment. The hyperimmunization schedule and nature of the antigen (viz., yeast cell sonicates) used in these investigations induced an enormous agglutinin response in mice. Over the past three decades, we have immunized mice with an array of *Candida* whole-cell and subcellular antigenic preparations, but have never observed such profound agglutinin titers (our unpublished data). Considering the antigenic complexity of whole yeast cells, the titers obtained by these researchers may have been key to their success as opposed to others who have done similar experiments but obtained negative results. That is, the fortuitous responses of mice may well have resulted in production of antibodies that fulfill the criteria we proposed above for protective responses. A second interesting point about their work is that the mice received antisera one day after challenge, which suggests that antibodies may have therapeutic potential in patients who develop hematogenously disseminated candidiasis.

B. Pearsall's Experiments Corroborate Mourad-Friedman Observations

In an initial investigation, Pearsall and Lagunoff studied experimental candidiasis by use of "a mouse-thigh lesion model for experimental candidiasis" [138]. Mice received a large inoculum of *C. albicans* yeast cells into the thigh muscle, which allowed the investigators to readily follow the development of the resulting lesion over a period of 4–6 weeks and define related immunologic events. Considering the already accepted doctrine that CMI was the prevalent mechanism of acquired specific immunity in host defense against candidiasis, it was of interest that there was little evidence of a granulomatous response indicative of CMI, yet the lesions resolved by 6 weeks after infection.

In follow-up work, Pearsall et al. used the same model to gain insights into acquired resistance [139]. They first obtained evidence for induction of acquired immunity in their model, as mice that recovered from a primary infection developed smaller lesions when challenged again in the thigh muscle with a large inoculum of yeast cells. To ascertain which limb of the immune system was responsible for the acquired resistance, naive mice received varying amounts of immune lymphocytes or immune serum, they were infected, and the growth of the thigh muscle lesion was measured. Four inbred strains of mice were tested—C57Bl/Ks, C57Bl/6J, AKR, and C3H. The AKR mice showed a slight indication that transfer of immune lymphocytes enhanced resistance to the development of the thigh lesion, but this effect did not correlate with induction of a delayed-type hypersensitivity response. Transfer of lymphocytes to the other mouse strains had no effect.

In contrast to the lack of transfer of protection by lymphocytes, naive mice that received immune serum grew significantly smaller thigh lesions following the infection [139]. Although we may infer that antibodies explained the enhanced resistance, the investigators did not show that immune serum absorbed with yeast cells removed the protective factor. In addition, the conclusions reached by the authors that CMI was not responsible for enhanced resistance must be viewed with caution. Whereas mice that recovered from a primary thigh lesion were used as the lymphocyte donors, immune serum for passive transfer was obtained from a different set of animals that were immunized as described in the Mourad-Friedman experiments. The possibility certainly remains that an insufficient number of immune lymphocytes were adoptively transferred and that cell-mediated immunity also may contribute to defense in the mouse thigh model of candidiasis. Regardless, these experiments tend to support a possible role for antibodies in host defense against candidiasis, which corroborates the studies of Mourad-Friedman that took place a full 10 years before.

C. Other Studies

In 1978, Domer's group determined that *C. albicans* cutaneous infection provoked mice to produce antibodies specific for the fungus, and such animals were less susceptible to disseminated candidiasis than controls [140]. Further experiments supported a specific protective effect. If B-cells were depleted by anti-μ therapy, the mice were unable to make antibody in response to the cutaneous infection, their T-cell activities appeared unaffected, but these animals were more susceptible to disseminated disease than controls [141]. Other investigators essentially confirmed these experiments [142].

VI. RECENT DIRECT EVIDENCE FOR PROTECTIVE ANTIBODIES

A. Anti-Heat Shock Protein 90

In a series of highly interesting and provocative reports, Matthews and coworkers have provided evidence that the development of antibodies against *C. albicans* heat shock protein 90 (hsp90) is protective against both human and laboratory animal candidiasis. Reports on this protein began appearing in the mid-1980s; thus, these investigations marked the start of the recent era of papers on protective antibodies against candidiasis. Evidence that favors a beneficial effect of anti-hsp90 merits special emphasis in this chapter, because a recent presentation by this group of scientists indicates that a genetically engineered human antibody against hsp90 is entering into pilot scale production with the prospect of use in Phase IIa clinical trials in patients with disseminated candidiasis [143]. Although we will refer to this protein throughout as "hsp90" in accordance with the Matthews group, the nomenclature is a bit confusing as homology of the *C. albicans* protein to the cognate heat shock protein of *S. cerevisiae* and work by others [144–146] indicates that the protein in question might best be referred to as hsp82.

The first observations leading to the conclusion that anti-hsp90 responses are protective were based on surveys of antibody and candidal antigens recoverable from sera of patients who had disseminated candidiasis [147,148]. Forty-five candidiasis patients were studied in the first analysis [148], and 92 patients in the second [147]. The correlation was imperfect, but patients who recovered also tended to produce a relatively high and protracted titer of antibody against a 47-kDa fungal antigen, whereas patients who succumbed produced either no antibody, antibody of low titer, or transient antibody to this antigen. The investigators also concluded that the 47-kDa antigen was a breakdown product of a larger fungal antigen as evidenced by consistent reactivity of patient antisera with the 47-kDa and with larger antigens [147]. Based on their finding that 92% of the patients who had antibody against *Candida* antigens also had antibody specific for the 47-kDa antigen, they concluded that this antigen is immunodominant [147–149], and subsequent analysis showed it is not the same as other similar-size immunodominant antigens reported by others [150,151]. The initial inference by the authors that an antibody response to the 47-kDa antigen was protective and that the antigen might be a vaccine candidate was quite speculative, because recovery from disease may have been due to immune responses unrelated to the anti-47-kDa response. Fortunately, the observations were pursued and these investigators produced data that more convincingly supported a protective role for antibodies to the 47-kDa antigen.

Over the ensuing 2 years, further basic characterization of the 47-kDa antigen was done, which included immunoelectron microscopic analysis of the antigen's cellular location in fungal cells [149] and cloning of part of the gene that encodes the antigen [152]. For the cellular localization of antigen, the investigators eluted specific polyclonal antibody from the antigen band on Western blots for subsequent use in immunocolloidal gold electron microscopic analysis. The gold labeling revealed the presence of the 47-kDa antigen in the cytosol primarily near the plasmalemma and in discrete areas on the outer aspects of the cell wall, which should be a good location for host immunologic recognition. The antigen was also found across the cell wall in a manner consistent with hypothetical *C. albicans* cell wall channels first reported by Poulain and coworkers [153]. Subsequent cloning and sequencing of a portion of the DNA coding region for the 47-kDa antigen was consistent with the earlier hypothesis that the antigen is a breakdown product of a larger protein species; that is, there was a high degree of similarity (83.5% identity) of the deduced 395 amino acid sequence of the *C. albicans* antigen with hsp90 of *S. cerevisiae* [146,152].

The protective ability of anti-hsp90 antibodies in host defense against disseminated candidiasis was substantially strengthened by further work that led to the identification of an hsp90 epitope that is purportedly responsible for induction of the most significant protective antibodies. The conserved epitope, composed of the amino acid sequence LKVIRK, was originally identified through the use of patient sera who recovered from candidiasis and epitope mapping technology [154]. LKVIRK is located at positions 317–322 from the carboxy terminus of 47-kDa and hsp90 is found in mammalian hsp90, including mouse and man [154]. In these same studies, the investigators obtained an important monoclonal antibody, CA-Str7-1, specific for LKVIRK, by use of hybridoma technology in which splenocytes were obtained from mice hyperimmunized with the peptide *LKVIRK*NIVKKMIE conjugated to keyhole limpet hemocyanin. Mice had increased resistance to a lethal challenge with *C. albicans* if before challenge they received either patient's serum containing anti-hsp90, an Ig fraction of patient's serum or MAb CA-Str7-1.

The protective monoclonal antibody, CA-Str7-1, has since been lost (R. Matthews, personal communication), but protection was also reported by these investigators when they pretreated mice with phage displaying a human antibody fragment specific for LKVIRK [155]. Appropriate control phage were used as negative controls, but the quality of the test and control phages, with regard to contaminants such as bacterial endotoxin, was not evaluated. The phage experiments were done using two types of assessments. One was an unconventional "acute lethality model" in which the mice were challenged with extremely high IV doses (6×10^7 and 1×10^8 yeast cells) of *C. albicans* and the other was a more conventional "chronic model" in which the animals were challenged IV with 10^6 yeast cells. The rapid lethality observed in the first assessment may have been due mostly to toxicity, whereas morbidity and mortality associated with the second assessment correlated with fungal growth in the kidneys. Importantly, even though the animals in both types of assessment received a very small IV dose (10^8 phage particles) of the phage-displayed antibody 2 hours after challenge, significant extended survival times were obtained in the acute model, and reduced kidney CFUs were recorded from animals in the chronic model as compared to mice that received appropriate negative control phage preparations. These results are especially interesting for two reasons: (1) a very small amount of anti-LKVIRK is apparently required for a protective effect; and (2) the antibody can be given after the initiation of disease, which implies a possible therapeutic application.

The investigators hypothesized that the fungal hsp90 is a "toxin" that binds to and either inactivates host proteins or prevents normal protein degradation [120,155]. Furthermore, they reasoned that during pathogenesis of candidiasis the hsp90 becomes released following death of fungal cells in a manner similar to release of endotoxin during sepsis due to gram-negative bacteria [120]. The toxin release hypothesis is consistent with their finding cited above that protective antibody can be given after initiation of infection.

The toxin hypothesis may also be consistent with work by these investigators on *Saccharomyces cerevisiae*. Overexpression of its hsp82, which is >78% identical to the *C. albicans* hsp90, led to increased virulence of the *S. cerevisiae* [146]. The investigators used strains PMY1-3A and PMY1-82 that were isogenic except that PMY1-82 contained multiple episomal copies of the homologous heat-inducible HSP82 gene (the episomal construct of which was described previously [144]). The two strains were grown for 48 hr at 30°C and heat shocked at 39°C for 75 min just prior to virulence testing in mice. Unfortunately, virulence comparisons of the two strains grown at 30°C were not done, which precludes evaluation of possible non-HSP82 differences between the strains. Extremely high IV doses of yeast (at least 10^8 cells) were required to produce a lethal infection in mice, but in all three experiments significant differences were recorded between the two strains with respect to resultant tissue burden of *S. cerevisiae* (viz., colony-forming units) and mortality [146]. The authors concluded that these results provided direct evidence that hsp90 is a virulence factor.

In summary, the overall data support the contention by the Matthews group that antibodies specific for the LKVIRK epitope of hsp90 are of benefit to experimental animals with disseminated candidiasis, and indirect evidence supports a positive effect of such antibodies in patients, as well. Scaleup production of a recombinant human anti-LKVIRK produced in *Escherichia coli* is in process for projected use in Phase IIa clinical trials. The investigators believe that *C. albicans* hsp90 acts as a toxin; thus, anti-hsp90 would be expected to have therapeutic advantage, as indicated by the experimental animal studies. Because human hsp90 also contains the LKVIRK epitope, it will be of interest to observe both short and long-term benefits and possible untoward effects of administering what may be considered an autoantibody to patients.

B. Antimannan Responses

1. Work from the Cutler Lab

Several years ago, Cutler and coworkers initiated studies on adherence characteristics of *C. albicans* in an attempt to better understand pathogenesis of hematogenously disseminated candidiasis. Mannans were found to be entirely responsible for the attachment of yeast cells to mouse splenic marginal zone macrophages [156–158]. Furthermore, antibodies specific for certain mannan epitopes enhanced resistance of mice against both hematogenously disseminated candidiasis and vaginal infection [119,130,131,159–161].

a. Adhesins. An elegant adherence method developed for investigating leukocyte homing [162,163] was adapted for studies on *C. albicans* adherence to mouse tissues [164,165]. By use of this ex vivo assay, hydrophilic yeast cells were found to adhere specifically to mouse splenic marginal zones and to subcapsular and trabecular sinus areas of peripheral lymph nodes [164–167]. More specifically, the yeast cells adhered to specific macrophage populations in both of these tissue locations [156,157,166–169]. Adherence of yeast cells to these macrophage populations is not limited to the ex vivo assay. Administration of yeast cells in vivo by routes to allow an interaction with splenic tissue or peripheral lymph nodes results in a yeast cell-macrophage association at these tissue sites in a pattern identical to that observed in the ex vivo assay [164,170] (Cutler, Riesselman, Tripp, unpublished results).

The nature of the host cell (i.e., macrophage) receptor for yeast cell attachment has not been defined. Evidence, however, was obtained that the receptor is not sialoadhesin [167], but may be a unique kind of mannose receptor [169] or galectin-3 [171].

The use of the ex vivo assay enabled studies that led to a definition of the chemical nature of the *Candida* adhesin responsible for adherence to splenic and lymph node macrophages. Two approaches were used to identify the adhesins. First, yeast cell extracts and fractions thereof were used to treat splenic or lymph node tissue sections prior to addition of yeast cells, and inhibition of yeast cell adherence was noted. The extracts were especially rich in phosphomannans, termed the phosphomannan complex (PMC), of which a structural model has been proposed by Suzuki and coworkers [172]. According to the model, the PMC is N-linked to cell wall protein, which accounts for nearly 10% of the PMC by weight [156,168]. The PMC is composed of an acid-stable part, which consists of a backbone mannan whose mannose units are linked by α-1,6-glycosidic linkages, and oligomannosyl side chains primarily of α-linked mannose units. When the PMC is heated with 10 mM HCl, oligomannosyl residues attached to the acid-stable part of the PMC through phosphodiester bonds are released. The mannose units of these various-size oligomannosyl chains are linked exclusively by β-1,2-glycosidic bonds [172]. Because these phosphodiester-linked oligomannosyl chains are released by treatment with dilute acid, they are termed the acid labile part of the PMC. Second, latex beads approximately the same diameter as yeast cells were coated with putative yeast cell adhesins and used in the ex vivo assay in place of yeast cells. Both approaches led to the same conclusions, that the PMC is responsible for adherence of yeast cells to the macrophages. Adhesin activity was lost when the extracts were treated with agents that destroyed the glycan part of the extract, whereas protein-destroying treatments had no effect [156]. Further analysis showed the presence of a strong adhesin site associated with certain moieties within the acid-stable part of the PMC [157,168]; a β-1,2-mannotetraose, which is one of the acid-labile components, is another adhesin site [158].

   b. Vaccine Development. The results of the adhesin studies led to investigations on the effect of antibodies specific for the PMC in pathogenesis of experimental candidiasis. Initial immunization attempts showed that PMC extracts were poorly immunogenic in mice. Subsequently, encapsulation of the PMC into conventional-type liposomes resolved the immunogenicity problem as mice responded to IV doses of the preparation by making high serum titers of polyclonal antibody against mannan epitopes [130,159]. Mice that received a priming dose of mannan liposomes and followed by four weekly boosters were more resistant to an IV challenge of a lethal dose of live *C. albicans* yeast cells than animals that received liposomes alone or buffer diluent prior to the challenge.

   Several lines of evidence support the existence of protective antibodies in the sera of mannan-liposome-vaccinated mice. Whole sera and Ig fractions of the sera obtained from immune animals transferred protection to naive mice. Immune sera preabsorbed with *C. albicans* yeast cells, however, did not transfer protection [130]. Additional evidence was obtained in experiments where attempts to improve the vaccine formulation were made.

   The unstable nature of the liposomes in the vaccine preparation led to investigations into the possibility that a protein carrier, covalently coupled to the mannan extract, would obviate the need for the liposomes. Indeed, mannan conjugated to bovine serum albumin (BSA), when given with an adjuvant, was consistently immunogenic in mice. Antisera obtained from these animals had the same ability to transfer protection to naive animals as described above for sera obtained from mannan-liposome-immunized mice [161]. In addition to eliminating the need for liposomes, the mannan-BSA conjugate vaccine formulation is superior because the vaccine can be administered intraperitoneally (IP) instead of IV and enhanced resistance to candidiasis can be induced by a priming dose followed by a single booster.

   c. Protective Monoclonal Antibodies (MAbs). Hybridoma technology was used to isolate an IgM monoclonal antibody from splenocytes that originated from either mice immunized against the mannan-liposome preparation (MAb B6.1) or from mice immunized against whole yeast cells (MAb B6). Both antibodies are IgM, both agglutinate whole yeast cells and latex beads coated with mannan extract, and both are specific for *Candida* mannan [119,130]. Only MAb B6.1, however, enhances resistance of mice to hematogenously disseminated candidiasis and this antibody is markedly more effective in reducing vaginal colonization by *C. albicans* than MAb B6 [159]. To determine more precisely the specificity of the two monoclonal antibodies, the PMC was isolated from *C. albicans* cell walls as before [156] and hydrolyzed by boiling in 10 mM HCl to liberate the acid-labile phosphodiester-linked oligomannosyl residues from the acid-stable part of the PMC. The acid-labile residues of varying mannose chain lengths were separated by size exclusion column chromatography and tested for their ability to inhibit agglutination of mannan-latex beads and reactivity with *C. albicans* yeast cells by MAbs B6.1 and B6. These approaches allowed for the definition of the epitope recognized by MAb B6.1 as an acid-labile β-1,2-mannotriose and the B6 epitope as an acid-stable moiety in the PMC [119]. Further analysis indicated that the B6.1 epitope is expressed in vivo and in vitro and is distributed uniformly and continuously along the cell surface of hydrophilic yeast cells, whereas the B6 epitope tends to have a patchy distribution [119].

   d.   Speculations on Mechanisms by Which Antibodies Protect Against Candidiasis.   The protective antibody MAb B6.1 is an IgM, but this isotype is not a requisite for protection against disease. Fractionation of the polyclonal antiserum from mice vaccinated against the mannan-liposomal preparation revealed protective activities associated with both the IgM and IgG fractions [130]. Furthermore, an IgG3 monoclonal antibody (MAb C3.1), which is specific for the same epitope as MAb B6.1, shows a similar protective activity as MAb B6.1 [131].

   The mechanism of protection by antibodies appears to be dependent on complement activation. In an in vitro assay, mouse neutrophils showed the highest ability to ingest and kill *C. albicans* yeast cells when in the presence of MAb B6.1 plus complement, as compared to either MAb B6.1 in the absence of complement or MAb B6 with or without complement [129].

   The Cutler lab is testing the hypothesis that MAbs B6.1 and C3.1 induce very rapid deposition of complement onto the surface of yeast cells, which favors an interaction of the opsonized yeast cells with host phagocytic cells. In addition to an in vitro requirement of complement for enhanced neutrophil candidacidal activity by MAb B6.1, compelling evidence has been obtained that the protective activities of MAbs B6.1 and C3.1 require the presence of complement in vivo (Han and Cutler, manuscript in preparation).

2.   Other Protective Antibodies That may be Specific for Mannan and/or Protein

In a series of studies on experimental vaginal infection in oophorectomized, estrogen-treated rats, Cassone and coworkers concluded that antibodies against a *Candida* mannan and antibodies specific for the protein portion of secreted aspartyl proteinase 2 (Sap 2) are protective [173–175]. The presumption that an antimannan response is protective was based on the results of an ELISA in which a mannoprotein (MP) extract was used as the capture antigen to evaluate antibody titers that correlated with both protective responses to active immunization and transfer of protection to naive animals. Unfortunately, the MP extract was crude. Although it was composed mostly of mannose, the preparation also contained an undefined amount of protein [175,176], thus making impossible the definitive claim that only antimannan responses were measured. Following intra-vaginal infection at days 14 and 28, the investigators detected the presence of IgM, IgG and IgA antibodies with specificity for mannan and secreted proteinase [175], and evidence was obtained that of the antibody responses, part may be occurring locally in the vaginal tissue [173]. Protection was transferred to naive rats by instilling vaginal lavage material, or an Ig fraction of the material, intravaginally before an intravaginal infection with yeast cells. The protective antibodies were apparently specific for *C. albicans* because immune lavage fluid preabsorbed with *C. albicans*, but not with *S. cerevisiae*, lost the ability to transfer protection [175].

   The mechanism by which antibodies protected the host in these studies is unclear. The widely touted virulence factor-associated trait of secreted proteinases [59,177–184] may intuitively lead to the conclusion that anti-Sap neutralizes the proteinase activity [173,174]. The cell wall–associated nature of the secreted proteinases [185] could, however, promote formation of agglutinates in the presence of antibody with resultant reduction of infectious units. In addition, the binding of antibody to cell wall surface proteinases may facilitate yeast-phagocyte interactions through antibody- and antibody-complement-mediated opsonic activity. Clearly, more work is required for a full understanding of how antibodies assist the host in resolving candidiasis. As alluded to earlier in this chapter, any mechanism that reduces the virulence or number of fungal viable units to a certain threshold number should enhance resistance. The mechanisms may, thus, vary depending on the predominant tissue location of the infection and immunologic status of the host. A wide variety of mechanisms could satisfy criteria for clearance that may include direct effects of antibodies on the ability of the fungus to germinate [186,187], neutralization of toxins [120,155,188,189], reduction of infectious units by agglutination [190], inhibition of adherence to host tissue sites, and opsonic activity (as discussed above). A few of these aspects have been studied and are considered in the following sections. Other aspects, such as toxin neutralization and inhibition of adherence, are still speculative but speculation on some candidate *C. albicans* targets for protective antibody responses is provided at the end of this chapter.

VII. ANTIBODIES THAT DIRECTLY INHIBIT *C. ALBICANS*

A. Antibodies That Inhibit Morphogenesis

Reports from two independent laboratories indicate that certain antibodies are capable of preventing a yeast to hyphal morphogenesis of *C. albicans*. As alluded to earlier, circumstantial evidence relates morphogenesis and pathogenicity of this fungus, such that morphogenetic inhibitory antibodies should help the host prevent onset and development of various forms of candidiasis. Based on observations by Grappel et al. that antibodies against dermatophytes affects morphological development and may even inhibit growth of these fungi [191], Grappel and Calderone [186] extended these observations to *C. albicans* [186]. They grew one serotype A strain and two serotype B strains of *C. albicans* in a chemically defined medium as yeast form cells at room temperature for 16 hr on a rotary shaker. Under these growth conditions the cells would presumably be in late log phase and expected to have relatively thin cell walls. The investigators washed and suspended the yeast cells in buffer and added twice the yeast cell suspension volume of either pooled rabbit anti–*C. albicans* serum (agglutinin titers of at least 320), a gamma globulin fraction of the antiserum, normal rabbit serum, or buffer. The various suspensions were incubated at 37°C for 30 min, then followed for induction of germination at 27 and 37°C or analyzed for oxygen consumption at 27°C. Whereas normal rabbit serum–treated yeast cells of the three isolates had increased oxygen consumption and robust development of germ tubes, yeast cells treated with either pooled antiserum or the gamma globulin fraction showed a dramatic decrease in respiratory function and ability to undergo morphogenesis. No yeast cell–absorbed antiserum controls were run in these studies. These observations should be pursued for determination of the specificity and isotypes of antibodies required for these effects. Neither mouse polyclonal antisera nor monoclonal antibodies obtained in work from the Cutler lab have negative effects on the ability of *C. albicans* yeast cells to germinate or replicate (Han and Cutler, unpublished observations).

In work that may corroborate some of the Grappel and Calderone findings, Casanova et al. reported that Fab fragments of monoclonal antibody 4C12 inhibited germination of *C. albicans* [187]. In previous work, MAb 4C12 was described as an IgG1 that has specificity for the protein portion of a 260-kDa cell wall mannoprotein of *C. albicans* [192]. Of high interest is that apparently only hyphal forms of the fungus express the protein. The investigators grew a strain of *C. albicans* (strain 3153A) to stationary phase in the chemically defined medium of Lee et al. [193], suspended the cells to 5×10^7/mL of the same medium, incubated the cells for 2 hr at 37°C, added the appropriate antibody or control material, and looked for the presence of germ tubes following an additional 3-hr incubation at 37°C. This method of germination is somewhat unusual because the high yeast cell concentration should be inhibitory for germination [194,195] or at least short germlings should be apparent by 1.5–2.5 hr incubation [196]. Nonetheless, the investigators found that Fabs, but not intact IgG1, of MAb 4C12 inhibited germ tube formation. Although a critical negative control, which consisted of irrelevant Fabs from human IgG, did not inhibit germination, the control material was apparently obtained from a commercial source and not prepared by the investigators. Because yeast cell germination is readily inhibited by a multitude of environmental factors and contaminants [197] (Hazen and Cutler, unpublished data), a more ideal control would have been irrelevant Fabs prepared exactly as the test Fabs. The specificity of MAb 4C12 for the surface of germ tubes was shown by indirect fluorescence antibody techniques, and this result was in agreement with the lack of detection of the 4C12 epitope in yeast cell extracts [192]. Whereas the distribution of the epitope reactive with MAb 4C12 appeared rather uniformly distributed over the hyphal cell surface, yeast inhibited from germinating showed a punctate appearance of the epitope [187], possibly representing the point of germ tube emergence. The authors hypothesized that the Fabs of MAb 4C12 may alter the conformation of the 260-kDa protein, which would adversely affect the ability of the yeast cell to germinate [187].

The above observations are provocative and warrant further investigation. The results suggest that antibodies specific for certain epitopes may have direct inhibitory effects on *C. albicans*, and thus these epitopes may be considered in the formulation of a vaccine against candidiasis. Because such antibodies may act, at least in part, by inhibiting a yeast to hyphal transition, these antibodies also should be useful in experiments designed to test the importance of morphogenesis in the candidal pathogenesis.

B. Antibodies Recognized by a Killer Factor Receptor

Among the most intriguing observations is the finding by Polonelli and coworkers that antibody raised against the combining site of antibody specific for a yeast killer factor (i.e., anti-idiotypic antibody) behaves as native killer factor and is, thus, directly toxic to the *C. albicans* cell expressing the killer factor receptor.

Seminal studies on yeast killer factor–producing strains and killer factors were done by Bevan and coworkers, who described the phenomenon in *Saccharomyces cerevisiae* [198,199]. They found that strains of the same species of yeast could be divided into killer, sensitive, and neutral on the basis of whether they secreted killer factor into the growth medium, were sensitive to the killer factor, or were not affected by the factor, respectively. The investigators also determined that production of the polypeptide by the killer yeast correlated with the presence of cytoplasmic double-stranded RNA [198]. Sensitivity of a yeast strain to killer factor is dependent on expression of a cell wall killer factor receptor; the mode of action with the K28 killer toxin of *Saccharomyces cerevisiae* may involve arrest of DNA synthesis [200]. Killer factor systems are not restricted to *S. cerevisiae*, as they have now been described for *Ustilago, Kluyveromyces, Pichia, Hansenula,* and *Candida* species [201,202]. Furthermore, the action of killer factors is not restricted to yeasts and even extends to prokaryotes [202,203].

In addition to a flood of basic science reports that Bevan's work spawned on this interesting phenomenon, Polonelli's group explored the potential epidemiologic application of killer factors in biotyping strains of *C. albicans* [204,205] and other pathogenic yeasts [206,207]. One hundred

C. albicans isolates were surveyed for susceptibility patterns against different factor-producing *Hansenula* and *Pichia* species [204,205]. All *C. albicans* isolates were sensitive to at least one killer yeast. In one study, 25 biotypes or strain types were distinguished, 52% of the isolates were sensitive to all nine killer yeasts, and 100% were sensitive to *H. canadensis* [204]. Sensitivity of the *C. albicans* isolates to the killer yeast species *Pichia (Hansenula) anomala* varied between 7% and 98%, depending on the killer strain.

During the past few years, the Polonelli group has been investigating the potential use of yeast killer factor activities in the treatment of fungal diseases. For these studies, *P. anomala* UCSC 25F (ATCC 96603) was apparently chosen because of its toxicity for a broad range of unrelated micro-organisms [208]. Ultrafiltration-concentrated *P. anomala* culture medium containing killer factor 25F that inhibits *Malassezia furfur* and *M. pachydermatis* in vitro was applied topically to experimental epidermal lesions induced in rabbits, guinea pigs, and dogs [207]. Clinical improvement and negative cultures occurred more rapidly on the area of the lesions treated with the 50X-concentrated culture filtrate than on untreated parts of the lesions. The investigators did not use as negative controls either concentrated culture medium from uninoculated growth medium or concentrated culture medium from a killer factor–negative strain of *Pichia* or *Hansenula*. Nonetheless, Polonelli and coworkers pursued the concept of killer factor therapy for experimental *Candida* vaginitis in the rat.

Monoclonal antibodies specific for the UCSF 25F killer factor polypeptide were initially obtained for use in affinity purification of the factor [208,209] with the intent of using the factor for treatment of candidiasis and, perhaps, other infectious diseases. Direct use of the toxin for in vivo application has potential limitations, however, such as the tedium of large-scale purification, possible in vivo instability of the factor, suboptimal in vivo conditions for antifungal activity, and a probable host immune response that neutralizes the factor.

Instead of employing the monoclonal antibodies for factor purification, the investigators used one of them as an immunogen for obtaining antiidiotypic antibodies with the hope that such antibodies would mimic killer factor activity. BALB/c mice were immunized with a crude preparation of the UCSF 25F killer toxin for subsequent isolation of an IgG1 monoclonal antibody, mAbKT4, that precipitated and neutralized the killer toxin in vitro [208]. mAbKT4 was then used as an immunogen in rabbits to obtain antiidiotypic antibodies [209]. The choice of using rabbits, rather than syngeneic mice, with respect to the origin of mAbKT4, is perplexing. Although anti-idiotypic antisera have been raised in animals unrelated to the source of the original antibody [210,211], this approach should induce significant responses to nonidiotypic determinants, which would introduce an additional level of complexity. The only evidence presented by the investigators that they, in fact, obtained rabbit antiidiotypic antibodies was the reduction in the formation of a precipitin band in agar (Ouchterlony double-diffusion test) between a well containing the rabbit antiserum and a well containing mAbKT4 precombined with the KT4 killer factor [209]. The latter well should contain immune complexes or a mixture of complexes and uncombined reactants. The complexes would not be expected to diffuse very well through the agarose, which might explain the reduced amount of precipitin band which developed. Regardless, evidence was obtained that the rabbit antiserum and affinity column-purified antibody directly inhibited the growth of *C. albicans* [209], which is a unique finding. In these studies the investigators apparently did not determine whether the reduction in colony-forming units on the plate was in fact due to direct toxicity of the antiserum for fungal cells or simple agglutination of the cells. It would be interesting to determine if the mode of inhibitory action of the antiserum is the same as that of the killer factor.

In follow-up experiments, the Polonelli group used the mAbKT4 as an immunogen in BALB/c mice (i.e., syngeneic animals with respect to the origin of mAbKT4) and in rats for testing induction of enhanced resistance against disseminated and vaginal candidiasis, respectively. The mice were immunized with varying doses of mAbKT4, then challenged with two different lethal doses of *C. albicans* yeast cells given IV [213b]. The average survival time of mice challenged with 1×10^6 yeast cells was extended from ~4.5 days (nonimmunized mice) to ~10 days (mAbKT4-immunized mice), and from ~2.5 days to 4 days when challenged with 5×10^6 yeast cells IV. The possible direct beneficial effects of mAbKT4 on survival of mice were not addressed. Passive transfer of sera from mAbKT4-immunized mice was not tested, but immune sera showed some ability to inhibit growth of *C. albicans* in vitro, and this inhibition was blocked by mAbKT4. In another study, oophorectomized rats were immunized by an intra-vaginal placement on days 0, 7, 14, and 21 of mAbKT4 emulsified in Freund adjuvant [214]. On day 30, the rats were infected intravaginally with *C. albicans*, some animals were reinfected on day 124, and the vaginal lavages were collected at various times for culture and evaluation of anti-idiotypic antibody. The investigators found evidence that *C. albicans* infection boosted the anti-idiotypic antibody response in mAbKT4-immunized rats. Nonimmunized rats infected several times intravaginally developed antikiller toxin antibodies in their vaginal tracts. In other work, a monoclonal anti-idiotypic antibody has been isolated and characterized [213a]; this rat IgM binds to *C. albicans* and inhibits its growth. Concomitant with the above rat infection studies, human volunteers who were either asymptomatic or were symptomatic for *Candida* vaginitis were analyzed by a competitive inhibition ELISA [213b] for titers of antikiller toxin and anti-idiotypic killer toxin antibodies in their vaginal fluids. Nine of 13 symptomatic patients had evidence of the antiidiotypic antibodies (designated as KTIdAb for antibodies that mimic the killer toxin) in their vaginal fluid. KTIdAb was not detectable in asymptomatic volunteers [214]. Finally, immunoaffinity-purified human KTIdAb given intravaginally to rats prior to intravaginal infection with *C. albicans* had ~50% fewer fungal CFU's recoverable from their vaginal contents as compared to animals given negative control material. These results imply that KTIdAb responses occur in humans during a natural infection with *C. albicans*, and such antibodies may be of benefit to the host. Because anti-idiotypic antibodies may be difficult to raise, it might be of interest to determine whether toxic antibodies can be readily induced in response to isolated killer factor receptor as an immunogen.

The observations were further extended by showing that anti-idiotypic single-chain Fv antibody constructs (ScFv) function similarly as the native intact KTIdAb [215]. To obtain these constructs by methods originally described by others [216], the IgG1 mouse mAbKT4 was used as an immunogen for BALB/c mice, specific immune splenocytes were selected by panning on plates coated with mAbKT4, cDNAs for KTIdAb ScFv's were cloned into *Escherichia coli*, and bacteria expressing the correct ScFv were selected on plates coated with mAbKT4. The recombinant ScFv KTIdAb was recovered from culture supernatant material or from periplasmic extracts of the host *E. coli* by the use of affinity chromatography [215], but no mention was given regarding probable endotoxin contamination of the final preparation. The ScFv KTIdAb directly inhibited in vitro growth of up to 90% of *C. albicans* yeast cells, and this inhibitory activity was neutralized by preincubation of the ScFv preparation with mAbKT4. Again, how-ever, there was no mention of whether the ScFv preparation directly agglutinated the yeast cells. When tested in vivo by the rat vaginitis model, the greatest reduction in recoverable *C. albicans* CFU's from vaginal material occurred when animals were treated with the recombinant anti-idiotypic preparation at the time of challenge. Appropriate negative control treatments had no effect as compared to untreated infected controls. It would, of course, be interesting to learn of the protective potential of these recombinant molecules against disseminated candidiasis.

The application of the antibody killer factor mimic approach may be confounded by the fact that the specific etiologic agent of disease must express the killer factor receptor. A generally useful vaccine for generation of antibodies that directly kill *C. albicans*, then, would require a formulation that consists of a mixture of monoclonal antibodies with sufficient diversity to affect most isolates of *C. albicans* or other agents proposed as targets for this approach.

VIII.   FUTURE PERSPECTIVES AND CONCLUSIONS

The literature contains compelling evidence supporting the hypothesis that antibodies are protective against both disseminated and cutaneous forms of candidiasis. The data also indicate that a useful vaccine formulation should consist of a limited number of *Candida* antigens, rather than the use of a whole-cell preparation. The erratic success of the latter approach is likely due to the enormous antigenic complexity of whole *C. albicans* and the relatively few antigens that may lead to a protective antibody response. In this chapter we have reviewed several targets for antibodies that may be helpful to the host in resisting candidiasis, and may thus be useful in the formulation of vaccines. Several additional antigens that have been recently described appear to merit close examination as well.

Immunosuppressive and immunoenhancing activities of *C. albicans* antigens have been reported as being associated with whole cells, cell wall fractions, proteins, and mannans from this fungus [217–226]. Of these, the antigen most defined is a p43 that apparently nonspecifically stimulates B-cells, but suppresses the ability of experimental animals to respond to sheep erythrocytes and candidal antigens [226,227]. Although it is difficult to assess, some of the activities of p43 appear similar to those of so-called suppressor B-cells induced in mice by soluble extracts of *C. albicans* as reported by Rogers and coworkers [220]. Both *C. albicans* factors suppress antibody responses, and the activities of both may be associated with production of certain cytokines [226,228]. Since mice that were either immunized against p43 or received antisera from immune animals were protected against disseminated candidiasis [217], the molecular and functional characterization of the p43 antigen is important and it must be considered as a possible vaccine candidate.

Certain *C. albicans* adhesins should also represent suitable antigenic targets for induction of protective antibodies. Inhibition of all adhesins is not necessarily desirable, as we may presume that *C. albicans* PMC adhesins, alluded to above, are beneficial to the host because they direct yeast cells from the blood to host phagocytic cells. Candidal adhesins that allow for mucosal colonization, adherence to endothelial cells, and attachment to extracellular matrix proteins, however, are likely to benefit the fungus. A *C. albicans* hyphal-specific surface protein, Hwp1, serves as a substrate for mammalian transglutaminases, which leads to a covalent association between the fungus and human buccal epithelial cells [229]. Strains with disrupted *HWP1* were less virulent, implying a role of this protein in pathogenesis of candidiasis. A *C. albicans ALS1* (agglutininlike sequence) has been cloned and expressed in *S. cerevisiae*, which resulted in marked enhancement of the transformant's ability for adhering to human endothelial cells [230]. Using a similar approach, *C. albicans ALA1* (agglutininlike adhesin) was reported to be involved in attachment to extracellular matrix proteins [231]. This adhesin, however, appears to be a member of the *ALS* gene family [232]. Finally, antibodies against certain *C. albicans* hydrophobic cell wall proteins inhibit attachment of the fungus to extracellular matrix proteins [102]. A potentially useful vaccine may be composed of an appropriately balanced concoction of all of these surface proteins.

Pursuit of an antibody-inducing vaccine against candidiasis in humans is not only rational, but with identification of potential vaccine candidates already in place, such a vaccine should be realized in the near future. The experimental and clinical evidence favors a role of antibodies, and recent application of bovine anti–*C. albicans* polyclonal antisera to reduce oral carriage of *C. albicans* in bone marrow transplant recipients [134] suggests a role for antibodies against *C.*

albicans in human medicine. Even if an antibody at a given titer may not be capable by itself of curing a given infection, it might tip the balance toward prevention or more rapid resolution of candidiasis.

ACKNOWLEDGMENTS

Research reported from the Cutler lab was supported by grants from the National Institute of Allergy and Infectious Diseases (RO1 AI24912, RO1 AI41502, and PO1 AI37194).

REFERENCES

1. YF Berrouane, LA Herwaldt, MA Pfaller. Trends in antifungal use and epidemiology of nosocomial yeast infections in a university hospital. J Clin Microbiol 37(3):531–537, 1999.
2. MA Pfaller, RN Jones, GV Doern, HS Sader, RJ Hollis, SA Messer. SENTRY Participation Group. International surveillance of bloodstream infections due to *Candida* species: frequency of occurrence and antifungal susceptibilities of isolates collected in 1997 in the United States, Canada, and South America for the SENTRY program. J Clin Microbiol 36(7):1886–1889, 1998.
3. DLR Yamamura, C Rotstein, LE Nicolle, S Ioannou. Candidemia at selected Canadian sites: results from the fungal disease registry, 1992–1994. CMA J 160(4):493–499, 1999.
4. RE Lewis, ME Klepser, MA Pfaller. Update on clinical antifungal susceptibility testing for *Candida* species. Pharmacotherapy 18(3):509–515, 1998.
5. V Kremery, P Fuchsberger, J Trupl, M Blahova, A Danizovicova, J Svec, L Drgona. Fungal pathogens in etiology of septic shock in neutropenic patients with cancer. Zbl Bakt 278:562–565, 1993.
6. K Kralovicova, S Spanik, E Oravcova, M Mrazova, E Morova, V Gulikova, E Kukuchova, P Koren, P Pichna, J Nogova, A Kunova, J Tripl, V Kremery. Fungemia in cancer patients undergoing chemotherapy versus surgery: risk factors, etiology and outcome. Scand J Infect Dis 29:301–304, 1997.
7. Y-KC Kwok, Y-K Tay, C-L Goh, A Kamarudin, M-T Koh, C-S Seow. Epidemiology and in vitro activity of antimycotics against candidal vaginal/skin/nail infections in Singapore. Int J Dermatol 37:145–149, 1998.
8. Y-C Huang, T-Y Lin, H-S Leu, J-L Wu, J-H Wu. Yeast carriage on hands of hospital personnel working in intensive care units. J Hosp Infect 39:47–51, 1998.
9. FC Odds. *Candida* and Candidosis. 2d ed. London: Bailliere Tindall, 1988.
10. P Eggimann, P Francioli, J Bille, R Schneider, M-M Wu, G Chapuis, R Chiolero, A Pannatier, J Schilling, S Geroulanos, MP Glauser, T Calandra. Fluconazole prophylaxis prevents intra-abdominal candidiasis in high-risk surgical patients. Crit Care Med 27(6):1066–1072, 1999.
11. A Rantala, J Niinikoski, O-P Lehtonen. Early *Candida* isolations in febrile patients after abdominal surgery. Scand J Infect Dis 25:479–485, 1993.
12. A Hoerauf, S Hammer, B Müller-Myhsok, H Rupprecht. Intra-abdominal *Candida* infection during acute necrotizing pancreatitis has a high prevalence and is associated with increased mortality. Crit Care Med 26(12):2010–2015, 1998.
13. A Rantala. Postoperative candidiasis. Ann Chir Gynaecol Suppl 205:6–52, 1993.
14. RA Gladdy, SE Richardson, HD Davies, RA Superina. *Candida* infection in pediatric liver transplant recpients. Liver Transpl Surg 5(1):16–24, 1999.
15. L Klingspor, G Stintzing, A Fasth, J Tollemar. Deep *Candida* infection in children receiving allogeneic bone marrow transplants: incidences, risk factors and diagnosis. Bone Marrow Transpl 17: 1043–1049, 1996.
16. P Castald, RJ Stratta, RP Wood, RS Markin, KD Patil, MS Shaefer, AN Langnas, EC Reed, S Li, TJ Pillen, BW Shaw. Clinical spectrum of fungal infections after orthotopic liver transplantation. Arch Surg 126:149–156, 1991.

17. DW Warnock. Fungal complications of transplantation: diagnosis, treatment and prevention. J Anti-microb Chemother 36:73–90, 1995.
18. LA Collins, MH Samore, MS Roberts, R Luzzati, RL Jenkins, WD Lewis, AW Karchmer. Risk factors for invasive fungal infections complicating orthotopic liver transplantation. J Infect Dis 170: 644–652, 1994.
19. S Verghese, A Mullasari, P Padmaja, P Sudha, MC Sapna, KM Cherian. Fungal endocarditis follow-ing cardiac surgery. Indian Heart J 50:418–422, 1998.
20. Y Berrouane, H Bisiau, F Le Baron, C Cattoen, P Duthilleul, E Dei Cas. Candida albicans blastoconi-dia in peripheral blood smears from non-neutropenic surgical patients. J Clin Pathol 51:337–338, 1998.
21. R Rabinovici, D Szewczyk, P Ovadia, JR Greenspan, JJ Sivalingam. Candida pericarditis: clinical profile and treatment. Ann Thorac Surg 63:1200–1204, 1997.
22. RJ Fass, DA Goff, SJ Sierawski. Candida infections in the surgical intensive care unit. J Antimicrob Chemother 38:915–916, 1996.
23. DA Dean, KW Burchard. Fungal infection in surgical patients. Am J Surg 171:374–382, 1996.
24. G Irving, D Miller, A Robinson, S Reynolds, AJ Copas. Psychological factors associated with recurrent vaginal candidiasis: a preliminary study. Sex Transm Infect 74:334–338, 1998.
25. PL Fidel, JD Sobel. Immunopathogenesis of recurrent vulvovaginal candidiasis. Clin Microbiol Rev 9:335–348, 1996.
26. SR Lockhart, S Joly, K Vargas, J Swails-Wenger, L Enger, DR Soll. Natural defenses against Candida colonization breakdown in the oral cavities of the elderly. J Dent Res 78(4):857–868, 1999.
27. BC Webb, CJ Thomas, MDP Willcox, DWS Harty, KW Knox. Candida-associated denture stomati-tis. Aetiology and management: a review. Aust Dent J 43(1):45–50, 1998.
28. JA Vazquez, JD Sobel. Fungal infections in diatetes. Infect Dis Clin North Am 9:97–116, 1995.
29. M Nucci, AL Colombo, F Silveira, R Richtmann, R Salomão, ML Branchini, N Spector. Risk factors for death in patients with candidemia. Infect Control Hosp Epidemiol 19:846–850, 1998.
30. MJ Richards, JR Edwards, DH Culver, RP Gaynes, National Nosocomial Infections Surveillance System. Nosocomial infections in coronary care units in the United States. Am J Cardiol 82:789–793, 1998.
31. JH Rex. Editorial response: catheters and candidemia. Clin Infect Dis 22:467–470, 1996.
32. MA Pfaller, SA Messer, A Houston, MS Rangel-Frausto, T Wiblin, HM Blumberg, JE Edwards, W Jarvis, MA Martin, HC Neu, L Saiman, JE Patterson, JC Dibb, CM Roldan, MG Rinaldi, RP Wenzel. National epidemiology of mycoses survey: a multicenter study of strain variation and antifungal susceptibility among isolates of Candida species. Diagn Microbiol Infect Dis 31(1): 289–296, 1998.
33. J-L Vincent, E Anaissie, H Bruining, W Demajo, M El-Ebiary, J Haber, Y Hiramatsu, G Nitenberg, D Pittet, T Rogers, P Sandven, G Sganga, M-D Schaller, J Solomkin, P-O Nystrom. Epidemiology, diagnosis and treatment of systemic Candida infection in surgical patients under intensive care. Intens Care Med 24(206):216, 1998.
34. CE Smith, MT Saito, SA Simons. Pattern of 39,500 serologic tests in coccidioidomycosis. JAMA 160:546–552, 1956.
35. CE Smith, D Pappagianis, HB Levine, M Saito. Human coccidioidomycosis. J Bacteriol Rev 25: 310–320, 1961.
36. D Pappagianis, BL Zimmer. Serology of coccidioidomycosis. Clin Microbiol Rev 3(3):247–268, 1990.
37. CR Zimmermann, SM Johnson, GW Martens, AG White, D Pappagianis. Cloning and expression of the complement fixation antigen-chitinase of Coccidioides immitis. Infect Immun 64(12): 4967–4975, 1996.
38. GT Cole. Models of cell differentiation on conidial fungi. Microbiol Rev 50:95–132, 1986.
39. A Casadevall. Antibody immunity and invasive fungal infections. Infect Immun 63(11):4211–4218, 1995.

40. JE Domer, RI Lehrer. Introduction to *Candida*: Systemic candidiasis. In: JW Murphy, H Friedman, M Bendinelli, eds. Infectious Agents and Pathogenesis: Fungal Infections and Immune Responses. New York: Plenum Press, 1993, pp 49–116.

41. CH Kirkpatrick, RR Rich, JE Bennett. Chronic mucocutaneous candidiasis: model-building in cellular immunity. Ann Intern Med 74:955–978, 1971.

42. JC García-Ruiz, M Del Carmen Arilla, P Regúlez, G Quindos, A Álvarez, J Pontón. Detection of antibodies to *Candida albicans* germ tubes for diagnosis and therapeutic monitoring of invasive candidiasis in patients with hematologic malignancies. J Clin Microbiol 35(12):3284–3287, 1997.

43. IK Aronson, K Soltani. Chronic mucocutaneous candidosis: a review. Mycopathologia 60(1):17–25, 1976.

44. L Cardenas-Freytag, E Cheng, P Mayeux, JE Domer, JD Clements. Effectiveness of a vaccine composed of heat-killed *Candida albicans* and a novel mucosal adjuvant, LT(R192G), against systemic candidiasis. Infect Immun 67(2):826–833, 1999.

45. MT Cantorna, E Balish. Acquired immunity to systemic candidiasis in immunodeficient mice. J Infect Dis 164:936–943, 1991.

46. F Bistoni, A Vecchiarelli, E Cenci, P Puccetti, P Marconi, A Cassone. Evidence for macrophage-mediated protection against lethal *Candida albicans* infection. Infect Immun 51:668–674, 1986.

47. L Romani, A Mencacci, U Grohmann, S Mocci, P Mosci, P Puccetti, F Bistoni. Neutralizing antibody to interleukin 4 induces systemic protection and T helper type 1–associated immunity in murine candidiasis. J Exp Med 176:19–25, 1992.

48. L Romani, P Puccetti, A Mencacci, E Cenci, R Spaccapelo, L Tonnetti, U Grohmann, F Bistoni. Neutralization of IL-10 up-regulates nitric oxide production and protects susceptible mice from challenge with *Candida albicans*. J Immunol 152:3515–3521, 1994.

49. P Puccetti, L Romani, F Bistoni. A T_H1-T_H2-like switch in candidiasis: new perspectives for therapy. Trends Microbiol 3(6):237–240, 1995.

50. L Romani. Immunity to *Candida albicans*: Th1, Th2 cells and beyond. Curr Opin Microbiol 2: 363–367, 1999.

51. E Cenci, A Mencacci, G Del Sero, F Bistoni, L Romani. Induction of protective Th1 responses to *Candida albicans* by antifungal therapy alone or in combination with an interleukin-4 antagonist. J Infect Dis 176:217–226, 1997.

52. E Cenci, A Mencacci, R Spaccapelo, L Tonnetti, P Mosci, K-H Enssle, P Puccetti, L Romani, F Bistoni. T helper cell type 1 (Th1)- and Th2-like responses are present in mice with gastric candidiasis but protective immunity is associated with Th1 development. J Infect Dis 171:1279–1288, 1995.

53. PL Fidel, KA Ginsburg, JL Cutright, NA Wolf, D Leaman, K Dunlap, JD Sobel. Vaginal-associated immunity in women with recurrent vulvovaginal candidiasis: evidence for vaginal Th1-type responses following intravaginal challenge with *Candida* antigen. J Infect Dis 176:728–739, 1997.

54. A Vazquez-Torres, J Jones-Carson, RD Wagner, T Warner, E Balish. Early resistance of interleukin-10 knockout mice to acute systemic candidiasis. Infect Immun 67(2):670–674, 1999.

55. C Bromuro, R La Valle, S Sandini, F Urbani, CM Ausiello, L Morelli, C Fe d'Ostiani, L Romani, A Cassone. A 70-kilodalton recombinant heat shock protein of *Candida albicans* is highly immunogenic and enhances systemic murine candidiasis. Infect Immun 66(5):2154–2162, 1998.

56. FM LaForce, DM Mills, K Iverson, R Cousins, ED Everett. Inhibition of leukocyte candidacidal activity by serum from patients with disseminated candidiasis. J Lab Clin Med 86:657–666, 1975.

57. SM Walker, SJ Urbaniak. A serum-dependent defect of neutrophil function in chronic mucocutaneous candidiasis. J Clin Pathol 33:370–372, 1980.

58. PJ Costantino, NF Gare, JR Warmington. Humoral immune responses to systemic *Candida albicans* infection in inbred mouse strains. Immunol Cell Biol 73:125–133, 1995.

59. JE Cutler. Putative virulence factors of *Candida albicans*. Ann Rev Microbiol 45:187–218, 1991.

60. JH Rex, TJ Walsh, EJ Anaissie. Fungal infections in iatrogenically compromised hosts. Adv Intern Med 43:321–371, 1998.

61. GP Bodey. The emergence of fungi as major hospital pathogens. J Hosp Infect 11(suppl A):411–426, 1988.

62. L Marodi. Local and systemic host defense mechanisms against *Candida*: immunopathology of candidal infections. Pediatr Infect Dis J 16(8):795–801, 1997.

63. WT Hughes, D Armstrong, GP Bodey, AE Brown, JE Edwards, R Feld, P Pizzo, KVI Rolston, JL Shenep, LS Young. 1997 Guidelines for the use of antimicrobial agents in neutropenic patients with unexplained fever. Clin Infect Dis 25:551–573, 1997.

64. A Voss, JLML le Noble, FM Verduyn Lunel, NA Foudraine, JFGM Meis. Candidemia in intensive care unit patients: risk factors for mortality. Infection 25(1):8–10, 1997.

65. SK Fridkin, WR Jarvis. Epidemiology of nosocomial fungal infections. Clin Microbiol Rev 9: 499–511, 1996.

66. WR Jarvis. Epidemiology of nosocomial fungal infections, with emphasis on *Candida* species. Clin Infect Dis 20:1526–1530, 1995.

67. RP Wenzel. Nosocomial candidemia: Risk factors and attributable mortality. Clin Infect Dis 20: 1531–1534, 1995.

68. SG Filler, JE Edwards. When and how to treat serious candidial infections: concepts and controversies. Curr Clin Top Infect Dis 15:1–18, 1995.

69. K Arulanantham, JM Dwyer, M Genel. Evidence for defective immunoregulation in the syndrome of familial candidiasis endocrinopathy. N Engl J Med 300(4):164–168, 1979.

70. JM Dwyer. Chronic mucocutaneous candidiasis. Annu Rev Med 32:491–497, 1981.

71. HM Kantarjian, MF Saad, EH Estey, RV Sellin, NA Samaan. Hypercalcemia in disseminated candidiasis. Am J Med 74:721–724, 1983.

72. EJ Anaissie, GP Bodey. Fungal infections in patients with cancer. Pharmacotherapy 10:164–169, 1990.

73. R Martino, R Lopez, A Sureda, S Brunet, A Domingo-Albós. Risk of reactivation of a recent invasive fungal infection in patients with hematological malignancies undergoing further intensive chemo-radiotherapy. a single-center experience and review of the literature. Haematologica 82: 297–304, 1997.

74. B Christensson, L Ryd, L Dahlberg, S Lohmander. *Candida albicans* arthritis in a nonimmunocompromised patient. Acta Orthop Scand 64:695–698, 1993.

75. A Kapur, R Vasudeva, CW Howden. *Candida* splenic abscess in the absence of obvious immunodeficiency. Am J Gastroenterol 92(3):509–512, 1997.

76. RE Warren. Chemoprophylaxis for candidosis and aspergillosis in neutropenia and transplantation: a review and recommendations. J Antimicrob Agents Chemother 32:5–21, 1993.

77. O Lortholary, B Dupont. Antifungal prophylaxis during neutropenia and immunodeficiency. Clin Microbiol Rev 10(3):477–504, 1997.

78. PA Pizzo, M Rubin, A Freifeld, TJ Walsh. The child with cancer and infection. I. Empiric therapy for fever and neutropenia, and preventive strategies. J Pediatr 119:679–693, 1991.

79. TJ Walsh, JW Lee. Prevention of invasive fungal infections in patients with neoplastic diseases. Clin Infect Dis 17(suppl 2):S468–S480, 1993.

80. E Anaissie. Opportunistic mycoses in the immunolcompromised host: experience at a cancer center and review. Clin Infect Dis 14(suppl 1):S43–S53, 1992.

81. JE Edwards, SG Filler. Current strategies for treating invasive candidiasis: emphasis on infections in nonneutropenic patients. Clin Infect Dis 14(suppl 1):S106–S113, 1992.

82. British Society for Antimicrobial Chemotherapy. Management of deep *Candida* infection in surgical and intensive care unit patients. Intens Care Med 20:522–528, 1994.

83. M El-Ebiary, A Torres, N Fàbregas, JP de la Bellacasa, J González, J Ramirez, D del Bano, C Hernandez, MT Jimmenez de Anta. Significance of the isolation of *Candida* species from respiratory samples in critically ill, non-neutropenic patients. an immediate postmortem histologic study. Am J Respir Crit Care Med 156:583–590, 1997.

84. JS Solomkin. Pathogenesis and management of *Candida* infection syndromes in nonneutropenic patients. New Horizons 1(2):202–213, 1993.

85. BJ Kullberg, MG Netea, JHAJ Curfs, M Keuter, JFGM Meis, JMW Van der Meer. Recombinant murine granulocyte colony-stimulating factor protects against acute disseminated *Candida albicans* infection in nonneutropenic mice. J Infect Dis 177:175–181, 1998.

86. J Nolla-Salas, A Sitges-Serra, C León-Gil, J Martínez-González, MA León-Regidor, P Ibáñez-Lucía, JM Torres-Rodriguez. Candidemia in non-neutropenic critically ill patients: analysis of prognostic factors and assessment of systemic antifungal therapy. Intens Care Med 23(1):23–30, 1997.

87. P Phillips, S Shafran, G Garber, C Rotstein, F Smaill, I Fong, I Salit, M Miller, K Williams, JM
     Conly, J Singer, S Ioannou. Multicenter randomized trial of flucanazole versus amphotericin B for
     treatment of candidemia in non-neutropenic patients. Eur J Clin Microbiol Infect Dis 16:337–345,
     1997.
88. Y Han, JE Cutler. Assessment of a mouse model of neutropenia and the effect of an anti-candidiasis
     monoclonal antibody in these animals. J. Infect Dis 175:1169–1175.
89. A Fulurija, RB Ashman, JM Papadimitriou. Neutrophil depletion increases susceptibility to systemic
     and vaginal candidiasis in mice, and reveals differences between brain and kidney in mechanisms
     of host resistance. Microbiology 142:3487–3496, 1995.
90. J Jensen, T Warner, E Balish. The role of phagocytic cells in resistance to disseminated candidiasis
     in granulocytopenic mice. J Infect Dis 170:900–905, 1994.
91. M Matsumoto, S Matsubara, T Matsuno, M Tamura, K Hattori, H Nomura, M Ono. Protective
     effect of human granulocyte colony-stimulating factor on microbial infection in neutropenic mice.
     Infect Immun 55:2715–2720, 1993.
92. S Steinshamn, K Bergh, A Waage. Effects of stem cell factor and granulocyte colony-stimulating
     factor on granulocyte recovery and *Candida albicans* infection in granulocytopenic mice. J Infect
     Dis 168:1444–1448, 1993.
93. JW Van't Wout, H Mattie, R van Furth. Comparison of the efficacies of amphotericin B, Fluconazole,
     and Itraconazole against a systemic *Candida albicans* infection in normal and neutropenic mice.
     Antimicrob Agents Chemother 33:147–151, 1989.
94. F Bistoni, A Vecchiarelli, E Cenci, G Sbaraglia, S Perito, A Cassone. A comparison of experimental
     pathogenicity of *Candida* species in cyclophosphamide-immunodepressed mice. Sabouraudia: J Med
     Vet Mycol 22:409–418, 1984.
95. KC Hazen, PM Glee. Cell surface hydrophobicity and medically important fungi. In: M Borgers,
     R Hay, MG Rinaldi, eds. Current Topics in Medical Mycology. Barcelona: Prous Science, 1995,
     pp 1–31.
96. J Masuoka, KC Hazen. Cell wall protein mannosylation determines *Candida albicans* cell surface
     hydrophobicity. Microbiology 143:3015–3021, 1997.
97. PP Antley, KC Hazen. Role of yeast cell growth temperature on *Candida albicans* virulence in
     mice. Infect Immun 56:2884–2890, 1988.
98. ANB Ellepola, LP Samaranayake. The effect of limited exposure to antimycotics on the relative
     cell-surface hydrophobicity and the adhesion of oral *Candida albicans* to buccal epithelial cells.
     Arch Oral Biol 43:879–887, 1998.
99. PM Glee, P Sundstrom, KC Hazen. Expression of surface hydrophobic proteins by *Candida albicans*
     in vivo. Infect Immun 63(4):1373–1379, 1995.
100. KC Hazen, DL Brawner, MH Riesselman, MA Jutila, JE Cutler. Differential adherence between
     hydrophobic and hydrophilic yeast cells of *Candida albicans*. Infect Immun 59:907–912, 1991.
101. KC Hazen. Participation of yeast cell surface hydrophobicity in adherence of *Candida albicans* to
     human epithelial cells. Infect Immun 57:1894–1900, 1989.
102. J Masuoka, G Wu, PM Glee, KC Hazen. Inhibition of *Candida albicans* attachment to extracellular
     matrix by antibodies which recognize hydrophobic cell wall proteins. FEMS Immunol Med Micro-
     biol 24:421–429, 1999.
103. TMJ Silva, PM Glee, KC Hazen. Influence of cell surface hydrophobicity on attachment of *Candida
     albicans* to extracellular matrix proteins. J Med Vet Mycol 33:117–122, 1995.
104. B Ener, LJ Douglas. Correlation between cell-surface hydrophobicity of *Candida albicans* and
     adhesion to buccal epithelial cells. FEMS Microbiol Lett 99:37–42, 1992.
105. BW Hazen, KC Hazen. Dynamic expression of cell surface hydrophobicity during initial yeast cell
     growth and before germ tube formation of *Candida albicans*. Infect Immun 56:2521–2525, 1988.
106. SD Kobayashi, JE Cutler. *Candida albicans* hyphal formation and virulence: is there a clearly
     defined role? Trends Microbiol 6(3):92–94, 1998.
107. R Alonso-Monge, F Navarro-Garcia, G Molero, R Diez-Orejas, M Gustin, J Pla, M Sanchez, C
     Nombela. Role of the mitogen-activated protein kinase Hog1p in morphogenesis and virulence of
     Candida albicans. J Bacteriol 181(10):3058–3068, 1999.

108. BE Corner, PT Magee. *Candida* pathogenesis: unravelling the threads of infection. Curr Biol 7(11): R691–R694, 1997.

109. C Csank, C Makris, S Meloche, K Schröppel, M Röllinghoff, D Dignard, DY Thomas, M Whiteway. Derepressed hyphal growth and reduced virulence in a VH1 family-related protein phosphatase mutant of the human pathogen *Candida albicans*. Mol Biol Cell 8:2539–2551, 1997.

110. C Csank, K Schroppel, E Leberer, D Harcus, O Mohamed, S Moleche, DY Thomas, M Whiteway. Roles of *Candida albicans* mitogen-activated protein kinase homolog, Cek1p, in hyphal development and systemic candidiasis. Infect Immun 66(6):2713–2721.

111. CA Gale, CM Bendel, M McClellan, M Hauser, JM Becker, J Berman, MK Hostetter. Linkage of adhesion, filamentous growth, and virulence in *Candida albicans* to a single gene, INT1. Science 279:1355–1358, 1998.

112. MK Hostetter. Linkage of adhesion, morphogenesis, and virulence in *Candida albicans*. J Lab Clin Med 132:258–263, 1998.

113. E Leberer, K Ziegelbauer, A Schmidt, D Harcus, D Dignard, J Ash, L Johnson, DY Thomas. Virulence and hyphal formation of *Candida albicans* require the Ste20p-like protein kinase CaCla4p. Curr Biol 7:539–546, 1997.

114. H-J Lo, JR Kohler, B DiDomenico, D Loebenberg, A Cacciapuoti, GR Fink. Nonfilamentous *Candida albicans* mutants are avirulent. Cell 90:939–949, 1997.

115. AP Mitchell. Dimorphism and virulence in *Candida albicans*. Curr Opin Microbiol 1:687–692, 1998.

116. F Dromer, J Charreire, A Contrepois, C Carbon, P Yeni. Protection of mice against experimental cryptococcosis by anti–*Cryptococcus neoformans* monoclonal antibody. Infect Immun 55:749–752, 1987.

117. J Mukherjee, MD Scharff, A Casadevall. Protective monoclonal antibodies to *Cryptococcus neoformans*. Infect Immun 60:4534–4541, 1992.

118. G Nussbaum, W Cleare, A Casadevall, MD Scharff, P Valadon. Epitope location in the *Cryptococcus neoformans* capsule is a determinant of antibody efficacy. J Exp Med 185:685–694, 1997.

119. Y Han, T Kanbe, R Cherniak, JE Cutler. Biochemical characterization of *Candida albicans* epitopes that can elicit protective and nonprotective antibodies. Infect Immun 65(10):4100–4107, 1997.

120. R Matthews, J Burnie. The role of hsp90 in fungal infection. Immunol Today 13:345–348, 1992.

121. R Yuan, A Casadevall, G Spira, MD Scharff. Isotype switching from IgG3 to IgG1 converts a nonprotective murine antibody to *Cryptococcus neoformans* into a protective antibody. J Immunol 154:1810–1816, 1995.

122. RR Yuan, G Spira, J Oh, M Paizi, A Casadevall, MD Scharff. Isotope switching increases efficacy of antibody protection against *Cryptococcus neoformans* infection in mice. Infect Immun 66(3): 1057–1062, 1998.

123. S Mukherjee, SC Lee, A Casadevall. Antibodies to *Cryptococcus neoformans* glucuronoxylomannan enhance antifungal activity of murine macrophages. Infect Immun 63(2):573–579, 1995.

124. T Kanbe, M Morishita, K Ito, K Tomita, K Utsunomiya, A Ishiguro. Evidence for the presence of immunoglobulin E antibodies specific to the cell wall phosphomannoproteins of *Candida albicans* in patients with allergies. Clin Diagn Lab Immunol 3:645–650, 1996.

125. M Nermes, J Savolaninen, K Kalimos, K Lammintausta, M Viander. Determination of IgE antibodies to *Candida albicans* mannan with nitrocellulose-RAST in patients with atopic diseases. Clin Exp Allergy 24:318–323, 1994.

126. A Pacheco, M Cuevas, B Carbelo, L Maiz, MJ Pavon, I Perez, S Quirce. Eosinophilic lung disease associated with *Candida albicans* allergy. Eur Respir J 12:502–504, 1998.

127. E Roig, JL Malo, S Montplaisir. Anti-*Candida albicans* IgE and IgG subclasses in sera of patients with allergic bronchopulmonary aspergillosis (ABPA). Allergy 52:394–403, 1997.

128. J Savolainen, J Kosonen, P Lintu, M Viander, J Pene, K Kalimo, EO Terho, J Bousquet. *Candida albicans* mannan- and protein-induced humoral, cellular and cytokine responses in atopic dermatitis patients. Clin Exp Allergy 29:824–831, 1999.

129. TC Caesar-TonThat, JE Cutler. A monoclonal antibody to *Candida albicans* enhances mouse neutrophil candidacidal activity. Infect Immun 65(12):5354–5357, 1997.

130. Y Han, JE Cutler. Antibody response that protects against disseminated candidiasis. Infect Immun 63(7):2714–2719, 1995.

131. Y Han, M Riesselman, JE Cutler. Protection against candidiasis by an immunoglobulin G3 (IgG3) monoclonal antibody specific for the same mannotriose as an IgM protective antibody. Infect Immun 68(3):1649–1654, 2000.

132. A Casadevall, A Cassone, F Bistoni, JE Cutler, W Magliani, JW Murphy, L Polonelli, L Romani. Antibody and/or cell-mediated immunity, protective mechanisms in fungal disease: an ongoing dilemma or an unnecessary dispute? Medical Mycology 36(suppl 1):95–105, 1998.

133. DS Martin. The practical application of some immunologic principles to the diagnosis and treatment of certain fungus infections. J Invest Dermatol 4, 471–481, 1941.

134. J Tollemar, N Gross, N Dolgiras, C Jarstrand, O Ringden, L Hammarstrom. Fungal prophylaxis by reduction of fungal colonization by oral administration of bovine anti-Candida antibodies in bone marrow transplant recipients. Bone Marrow Transpl 23:283–290, 1999.

135. S Mourad, L Friedman. Passive immunization of mice against Candida albicans. Sabouraudia 6: 103–105, 1968.

136. S Mourad, L Friedman. Active immunization of mice against Candida albicans. Proc Soc Exp Biol Med 106:570–572, 1961.

137. RC Hurd, CH Drake. Candida albicans infections in actively and passively immunized animals. Mycopathol Mycologia Appl 6:290–297, 1953.

138. NN Pearsall, D Lagunoff. Immunological responses to Candida albicans. Infect Immun 9(6): 999–1002, 1974.

139. NN Pearsall, BL Adams, R Bunni. Immunologic responses to Candida albicans. III. Effects of passive transfer of lymphoid cells or serum on murine candidiasis. J Immunol 120:1176–1180, 1978.

140. DK Giger, JE Domer, JT McQuitty. Experimental murine candidiasis: pathological and immune responses to cutaneous inoculation with Candida albicans. Infect Immun 19:499–509, 1978.

141. U Kuruganti, LA Henderson, RE Garner, R Asofsky, PJ Baker, JE Domer. Nonspecific and Candida-specific immune responses in mice suppressed by chronic administration of anti-μ. J Leukoc Biol 44:422–433, 1988.

142. PK Maiti, A Kumar, R Kumar, LN Mohapatra. Role of antibodies and effect of BCG vaccination in experimental candidiasis in mice. Mycopathologia 91:79–85, 1985.

143. RC Matthews, G Rigg, S Hodgetts, J Nelson. Human recombinant antibody to hsp90 in the treatment of disseminated candidiasis. ASM Proceedings, General Meeting, 2000.

144. L Cheng, K Hirst, PW Piper. Authentic temperature-regulation of a heat shock gene inserted into yeast on a high copy number vector. Influences of overexpression of HSP90 protein on high temperature growth and thermotolerance. Biochim Biophys Acta 1132:26–34, 1992.

145. KA Borkovich, FW Farrelly, DB Finkelstein, J Taulien, S Lindquist. HSP82 is an essential protein that is required in higher concentrations for growth of cells at higher temperatures. Mol Cell Biol 9:3919–3930, 1989.

146. S Hodgetts, R Matthews, G Morrissey, K Mitsutake, P Piper, J Burnie. Over-expression of Saccharomyces cerevisiae hsp90 enhances the virulence of this yeast in mice. FEMS Immunol Med Microbiol 16:229–234, 1996.

147. RC Matthews, JP Burnie, S Tabaqchali. Isolation of immunodominant antigens from sera of patients with systemic candidiasis and characterization of serological response to Candida albicans. J Clin Microbiol 25:230–237, 1987.

148. RC Matthews, JP Burnie, S Tabaqchali. Immunoblot analysis of the serological response in systemic candidosis. Lancet 2:1415–1418, 1984.

149. R Matthews, C Wells, JP Burnie. Characterisation and cellular localisation of the immunodominant 47-Kda antigen of Candida albicans. J Med Microbiol 27:227–232, 1988.

150. RA Greenfield, JM Jones. Purification and characterization of a major cytoplasmic antigen of Candida albicans. Infect Immun 32(2):469–477, 1981.

151. KM Franklyn, JR Warmington, AK Ott, RB Ashman. An immunodominant antigen of Candida albicans shows homology to the enzyme enolase. Immunol Cell Biol 68:173–178, 1990.

152. R Matthews, J Burnie. Cloning of a DNA sequence encoding a major fragment of the 47 kilodalton stress protein homologue of *Candida albicans*. FEMS Microbiol Lett 60:25–30, 1989.

153. D Poulain, J-C Cailliez, J-F Dubremetz. Secretion of glycoproteins through the cell wall of *Candida albicans*. Eur J Cell Biol 50:94–99, 1989.

154. RC Matthews, JP Burnie, D Howat, T Rowland, F Walton. Autoantibody to heat-shock protein 90 can mediate protection against systemic candidosis. Immunology 74:20–24, 1991.

155. R Matthews, S Hodgetts, J Burnie. Preliminary assessment of a human recombinant antibody fragment to hsp90 in murine invasive candidiasis. J Infect Dis 171:1668–1671, 1995.

156. T Kanbe, Y Han, B Redgrave, MH Riesselman, JE Cutler. Evidence that mannans of *Candida albicans* are responsible for adherence of yeast forms to spleen and lymph node tissue. Infect Immun 61(6):2578–2584, 1993.

157. T Kanbe, JE Cutler. Minimum chemical requirements for adhesin activity of the acid-stable part of *Candida albicans* cell wall phosphomannoprotein complex. Infect Immun 66(12):5812–5818, 1998.

158. R-K Li, JE Cutler. Chemical definition of an epitope/adhesin molecule on *Candida albicans*. J Biol Chem 268(24):18293–18299, 1993.

159. Y Han, RP Morrison, JE Cutler. A vaccine and monoclonal antibodies that enhance mouse resistance to *Candida albicans* vaginal infection. Infect Immun 66(12):5771–5776, 1998.

160. Y Han, JE Cutler. Assessment of a mouse model of neutropenia and the effect of an anti-candidiasis monoclonal antibody in these animals. J Infect Dis 175:1169–1175, 1997.

161. Y Han, MA Ulrich, JE Cutler. *Candida albicans* mannan extract-protein conjugates induce a protective immune response against experimental candidiasis. J Infect Dis 179:1479–1484, 1999.

162. HB Stamper, JJ Woodruff. Lymphocyte homing into lymph nodes: in vitro demonstration of the selective affinity of recirculating lymphocytes for high-endothelial venules. J Exp Med 144:828–833, 1976.

163. EC Butcher, RG Scollay, IL Weissman. Lymphocytes adherence to high endothelial venules: characterization of a modified in vitro assay, and examination of the binding of syngeneic and allogeneic lymphocyte populations. J Immunol 123:1996–2003, 1979.

164. JE Cutler, DL Brawner, KC Hazen, MA Jutila. Characteristics of *Candida albicans* adherence to mouse tissue. Infect Immun 58:1902–1908, 1990.

165. MH Riesselman, T Kanbe, JE Cutler. Improvements and important considerations of an ex vivo assay to study interactions of *Candida albicans* with splenic tissue. J Immunol Methods 1450:153–160, 1991.

166. Y Han, N van Rooijen, JE Cutler. Binding of *Candida albicans* yeast cells to mouse popliteal lymph node tissue is mediated by macrophages. Infect Immun 61:3244–3249, 1993.

167. Y Han, S Kelm, MH Riesselman, PR Crocker, JE Cutler. Mouse sialoadhesin is not responsible for *Candida albicans* yeast cell binding to splenic marginal zone macrophages. Infect Immun 62:2115–2118, 1994.

168. T Kanbe, JE Cutler. Evidence for adhesin activity in the acid-stable moiety of the phosphomannoprotein complex of *Candida albicans*. Infect Immun 62(5):1662–1668, 1994.

169. T Kanbe, MA Jutila, JE Cutler. Evidence that *Candida albicans* binds via a unique adhesion system on phagocytic cells in the marginal zone of the mouse spleen. Infect Immun 60(5):1972–1978, 1992.

170. D Tripp, JE Cutler. Evidence for complement independent in vivo adherence of *Candida albicans* yeast cells to mouse splenic marginal zone. Proc ASM Meetings, 1994.

171. C Fradin, D Poulain, T Jouault. Beta-1,2-linked oligomannosides from *Candida albicans* bind to a 32-kilodalton macrophage membrane protein homologous to the mammalian lectin galectin-3. Infect Immun 68:4391–4398, 2000.

172. N Shibata, K Ikuta, T Imai, Y Satoh, R Satoh, A Suzuki, C Kojima, H Kobayashi, K Hisamichi, S Suzuki. Existence of branched side chains in the cell wall mannan of pathogenic yeast, *Candida albicans*. J Biol Chem 270(3):1113–1122, 1995.

173. F De Bernardis, G Santoni, M Boccanera, E Spreghini, DA Adriani, L Morelli, A Cassone. Local anticandidal immune responses in a rat model of vaginal infection by and protection against *Candida albicans*. Infect Immun 68:3297–3304, 2000.

174. F De Bernardis, M Boccanera, DA Adriani, E Spreghini, G Santoni, A Cassone. Protective role of antimannan and anti-aspartyl proteinase antibodies in an experimental model of *Candida albicans* vaginitis in rats. Infect Immun 65:3399–3409, 1997.

175. A Cassone, M Boccanera, D Adriani, G Santoni, F De Bernardis. Rats clearing a vaginal infection by *Candida albicans* acquire specific, antibody-mediated resistance to vaginal reinfection. Infect Immun 63:2619–2624, 1995.

176. A Torosantucci, C Palma, M Boccanera, CM Ausiello, GC Spagnoli, A Cassone. Lymphoproliferative and cytotoxic responses of human peripheral blood mononuclear cells to mannoprotein constituents of *Candida albicans*. J Gen Microbiol 136:2155–2163, 1990.

177. F De Bernardis, S Arancia, L Morelli, B Hube, D Sanglard, W Schäfer, A Cassone. Evidence that members of the secretory aspartyl proteinase gene family, in particular *SAP2*, are virulence factors for *Candida* vaginitis. J Infect Dis 179:201–208, 1999.

178. PR Hunter, CAM Frase. Application of a numerical index of discriminatory power to a comparison of four physiochemical typing methods for *Candida albicans*. J Clin Microbiol 27:2156–2160, 1989.

179. M Ghannoum, KA Elteen. Correlative relationship between proteinase production, adherence and pathogenicity of various strains of *Candida albicans*. J Med Vet Mycol 24:407–413, 1986.

180. L Hoegl, M Ollert, HC Korting. The role of *Candida albicans* secreted aspartic proteinase in the development of candidoses. J Mol Med 74:135–142, 1996.

181. Y Kondoh, K Shimizu, K Tanaka. Proteinase production and pathogenicity of *Candida albicans*. II. Virulence for mice of *C. albicans* strains of different proteinase activity. Microbiol Immunol 31:1061–1069, 1987.

182. MW Ollert, C Wende, M Gorlich, CG McMullan-Vogel, M Borg-Von Zepelin, C-W Vogel, HC Korting. Increased expression of *Candida albicans* secretory proteinase, a putative virulence factor, in isolates from human immunodeficiency virus-positive patients. J Clin Microbiol 33(10): 2543–2549, 1995.

183. F Macdonald. Secretion of inducible proteinase by pathogenic *Candida* species. Sabouraudia: J Med Vet Mycol 22:79–82, 1984.

184. TL Ray, CD Payne. Scanning electron microscopy of epidermal adherence and cavitation in murine candidiasis: a role for *Candida* acid proteinase. Infect Immun 56:1942–1949, 1988.

185. RC Goldman, DJ Frost, JO Capobianco, S Kadam, RR Rasmussen, C Abad-Zapatero. Antifungal drug tests: *Candida* secreted aspartyl protease and fungal wall β-glucan synthesis. Infect Agents Dis 4:228–247, 1995.

186. SF Grappel, RA Calderone. Effect of antibodies on the respiration and morphology of *Candida albicans*. S Afr Med J 14:51–60, 1976.

187. M Casanova, JP Martinez, WL Chaffin. Fab fragments from a monoclonal antibody against a germ tube mannoprotein block the yeast-to-mycelium transition in *Candida albicans*. Infect Immun 58: 3810–3812, 1990.

188. MA Ghannoum. Potential role of phospholipases in virulence and fungal pathogenesis. Clin Microbiol Rev 13:122–143, 2000.

189. MA Ghannoum, KH Abu-Elteen. Pathogenicity determinants of *Candida*. Mycoses 33:265–282, 1990.

190. DB Louria, JK Smith, RG Brayton, M Buse. Anti-*Candida* factors in serum and their inhibitors. J Infect Dis 125(2):102–114, 1972.

191. SF Grappel, A Ferthiere, F Blank. Effect of antibodies on growth and structure of *Trichophyton mentagrophytes*. Sabouraudia 9:50–55, 1971.

192. M Casanova, ML Gil, L Cardenoso JP Martinez, R Sentandreu. Identification of wall-specific antigens synthesized during germ tube formation by *Candida albicans*. Infect Immun 57:262–271, 1989.

193. KL Lee, HR Buckley, CC Campbell. An amino acid liquid synthetic medium for the development of mycelial and yeast forms of *Candida albicans*. Sabouraudia 13:148–153, 1975.

194. KC Hazen, JE Cutler. Autoregulation of germ tube formation by *Candida albicans*. Infect Immun 24:661–666, 1979.

195. KC Hazen, JE Cutler. Isolation and purification of morphogenic autoregulatory substance produced by *Candida albicans*. J Biochem 94:777–783, 1983.

196. EGV Evans, FC Odds, MD Richardson, KT Holland. The effect of growth medium on filament production in *Candida albicans*. Sabouraudia 12:112–119, 1974.

197. JE Cutler, KC Hazen. Yeast/mold dimorphism. In: JW Bennett, A Ciegler, eds. Secondary Metabolism and Differentiation in Fungi. New York: Marcel Dekker, 1983, pp 267–306.

198. EA Bevan, AJ Herring, DJ Mitchell. Preliminary characterization of two species of dsRNA in yeast and their relationship to the "killer" character. Nature 142:81–86, 1973.

199. DR Woods, EA Bevan. Studies on the nature of the killer factor produced by *Saccharomyces cerevisiae*. J Gen Microbiol 51:115–126, 1968.

200. MJ Schmitt, P Klavehn, J Wang, I Schonig, DJ Tipper. Cell cycle studies on the mode of action of yeast K28 killer toxin. Microbiology 142:2655–2662, 1996.

201. L Polonelli, S Conti, M Gerloni, W Magliani, C Chezzi, G Morace. Interfaces of the yeast killer phenomenon. Crit Rev Microbiol 18(1):47–87, 1991.

202. L Polonelli, G Morace. Reevaluation of the yeast killer phenomenon. J Clin Microbiol 24:866–869, 1986.

203. F Izgu, D Altinbay. Killer toxins of certain yeast strains have potential growth inhibitory activity on gram-positive pathogenic bacteria. Microbios 89:15–22, 1997.

204. L Polonelli, C Archibusacci, M Sestito, G Morace. Killer system: a simple method for differentiating. J Clin Microbiol 17:774–780, 1983.

205. L Polonelli, M Castagnola, DV Rossetti, G Morace. Use of killer toxins for computer-aided differentiation of *Candida albicans* strains. Mycopathologia 91:175–179, 1985.

206. G Morace, C Archibusacci, M Sestito, L Polonelli. Strain differentiation of pathogenic yeasts by the killer system. Mycopathologia 84:81–85, 1984.

207. L Polonelli, R Lorenzini, F De Bernardis, G Morace. Potential therapeutic effect of yeast killer toxin. Mycopathologia 96:103–107, 1986.

208. L Polonelli, G Morace. Production and characterization of yeast killer toxin monoclonal antibodies. J Clin Microbiol 25:460–462, 1987.

209. L Polonelli, G Morace. Yeast killer toxin-like anti-idiotypic antibodies. J Clin Microbiol 26:602–604, 1988.

210. H Kohler, S Kaveri, T Kieber-Emmons, WJW Morrow, S Muller, S Raychaudhuri. Idiotypic networks and nature of molecular mimicry: an overview. Methods Enzymol 178:3–35, 1989.

211. NR Farid. Anti-idiotypic antibodies as probes of hormone receptor structure and function. Methods Enzymol 178:191–211, 1989.

212. L Polonelli, S Conti, M Gerloni, W Magliani, M Castagnola, G Morace, C Chezzi. 'Antibiobodies': antibiotic-like anti-idiotypic antibodies. J Med Vet Mycol 29:235–242, 1991.

213a. L Polonelli, N Seguy, S Conti, M Gerloni, D Bertolotti, C Cantelli, W Magliani, JC Caelliez. Monoclonal yeast killer toxin-like candidacidal anti-idiotypic antibodies. Clin Diagn Lab Immunol 4:142–146, 1997.

213b. L Polonelli, R Lorenzini, F DeBernardis, M Gerloni, S Conti, G Morace, W Magliani, C Chezzi. Idiotypic vaccination: immunoprotection mediated by anti-idiotypic antibodies with antibiotic activity. Scand J Immunol 37:105–110, 1993.

214. L Polonelli, F De Bernardis, S Conti, M Boccanera, W Magliani, M Gerloni, C Cantelli, A Cassone. Human natural yeast killer toxin-like candidacidal antibodies. J Immunol 156:1880–1885, 1996.

215. W Magliani, S Conti, F De Bernardis, M Gerioni, D Bertolotti, P Mozzoni, A Cassone, L Polonelli. Therapeutic potentials of antiidiotypic single chain antibodies with yeast killer toxin activity. Nat Biotechnol 15:155–158, 1997.

216. G Winter, AD Griffiths, RE Hawkins, HR Hoogenboom. Making antibodies by phage display technology. Annu Rev Immunol 12:433–455, 1994.

217. D Tavares, P Ferreira, M Vilanova, A Videira, M Arala-Chaves. Immunoprotection against systemic candidiasis in mice. Int Immunol 7(5):785–796, 1995.

218. SP Li, S-I Lee, JE Domer. Alterations in frequency of interleukin-2 (IL-2)-, gamma interferon-, or IL-4-secreting splenocytes induced by *Candida albicans* mannan and/or monophosphoryl lipid A. Infect Immun 66(4):1392–1399, 1998.

219. JE Domer. *Candida* cell wall mannan: a polysaccharide with diverse immunologic properties. CRC Crit Rev Microbiol 17:33–51, 1989.
220. CF Cuff, BJ Packer, TJ Rogers. A further characterization of *Candida albicans*-induced suppressor B-cell activity. Immunol 68:80–86, 1989.
221. JE Domer, LG Human, GB Andersen, JA Rudbach, GL Asherson. Abrogation of suppression of delayed hypersensitivity induced by *Candida albicans*–derived mannan by treatment with monophosphoryl lipid A. Infect Immun 61:2122–2130, 1993.
222. A Cassone. Immunogenic and immunomodulatory properties of mannoproteins from *Candida albicans*. Can J Bot 73:1192–1198, 1995.
223. E Piccolella, G Lombardi, R Morelli. Generation of suppressor cells in the response of human lymphocytes to a polysaccharide from *Candida albicans*. J Immunol 126(6):2151–2155, 1981.
224. E Piccolella, G Lombardi, R Morelli. Mitogenic response of human peripheral blood lymphocytes to a purified *C. albicans* polysaccharide fraction: lack of helper activities is responsible for the in vitro unresponsiveness to a second antigenic challenge. J Immunol 126(6):2156–2160, 1981.
225. JE Cutler, RK Lloyd. Enhanced antibody responses induced by *Candida albicans* in mice. Infect Immun 38:1102–1108, 1982.
226. D Tavares, P Ferreira, M Arala-Chaves. Increased resistance to systemic candidiasis in athymic or interleukin-10-depleted mice. J Infect Dis 182:266–273, 2000.
227. D Tavares, A Salvador, P Ferreira, M Arala-Chaves. Immunological activities of a *Candida albicans* protein which plays an important role in the survival of the microorganism in the host. Infect Immun 61(5):1881–1888, 1993.
228. N Vazquez, HR Buckley, TJ Rogers. Production of IL-6 and TNF-α by the macrophage-like cell line RAW 264.7 after treatment with a cell wall extract of *Candida albicans*. J Interferon Cytokine Res 16:465–470, 1996.
229. JF Staab, SD Bradway, PL Fidel, P Sundstrom. Adhesive and mammalian transglutaminase substrate properties of *Candida albicans* Hwp 1. Science 283:1535–1538, 1999.
230. Y Fu, G Rieg, WA Fonzi, PH Belanger, JE Edwards, SG Filler. Expression of the *Canddia albicans* gene ALS1 in *Saccharomyces cerevisiae* induces adherence to endothelial and epithelial cells. Infect Immun 66(4):1783–1786, 1998.
231. NK Gaur, SA Klotz, RL Henderson. Overexpression of the *Candida albicans ALA1* gene in *Saccharomyces cerevisiae* results in aggregation following attachment of yeast cells to extracellular mattrix proteins, adherence properties similar to those of *Candida albicans*. Infect Immun 67:6040–6047, 1999.
232. LL Hoyer, TL Payne, JE Hecht. Identification of *Candida albicans ALS2* and *ALS4* and localization of Als proteins to the fungal cell surface. J Bacteriol 180:5334–5343, 1998.

16

Vaccines for Histoplasmosis

George S. Deepe, Jr.
University of Cincinnati College of Medicine, Cincinnati, Ohio

I. CLINICAL MANIFESTATIONS

Human infection with the pathogenic fungus *Histoplasma capsulatum* poses a significant health risk to those who reside or have resided in the endemic region. This area encompasses the midwestern and southeastern United States and Central and South America. It is one of several endemic pathogenic fungi present in the Americas. Rough estimates of new infections with *H. capsulatum* within the United States range from 200,000 to 500,000 annually [1]. Infection is acquired via inhalation of microconidia or mycelial fragments into the alveoli and terminal bronchioles where they convert into the yeast phase [2]. It is this form that is responsible for clinical and pathological manifestations of the disease. The vast majority of primary cases of histoplasmosis (>95%) produce mild, influenza-like symptoms or are asymptomatic [3,4]. Infected individuals infrequently seek medical attention and if they do, rarely is histoplasmosis considered a possible explanation for their symptoms, even in the endemic region.

Although principally self-resolving, primary infection can cause severe symptoms that require hospitalization and treatment. A minority of cases produce high fever, marked pulmonary interstitial infiltrates on chest X-ray, and shortness of breath. It is presumed that these individuals inhale a large inoculum of conidia and mycelial fragments, but the definition of "large" remains obscure since the infective dose for man has never been determined. Hence, primary histoplasmosis can cause a serious illness that requires medical attention, and it must be recognized by practicing physicians.

Once the conversion has transpired, the organism replicates intracellularly until cell-mediated immunity is activated, usually within 1–3 weeks after mild to moderate infection [5]. One particular feature of *H. capsulatum* that makes it a nettlesome pathogen is its ability to maintain a dormant state. Following the concerted interaction of T-cells and professional phagocytes, the organism is restricted in its growth and becomes surrounded by granulomas that eventually calcify [6]. Within this environment, the organism most likely remains metabolically active yet does not replicate or replicates at a much slower rate than during active growth (generation time 9–19 hr) [2]. The yeasts can remain dormant for many years and never produce symptomatic disease. Reactivation of infectious foci can develop usually in individuals whose immune system, in particular the T-cell arm, becomes dysfunctional by either pharmacological agents that depress T-cell function or viral infections such as HIV [7–9].

II. MECHANISMS OF HOST RESISTANCE

In order to understand how a vaccine might protect against *H. capsulatum*, it is important to define the mechanisms of natural host resistance. In the following sections, the contribution of lymphocytes, phagocytes, and cytokines on elimination of the fungus will be examined.

A. T-Cell Regulation of Host Defenses

Cell-mediated immunity is the principal mechanism by which the host restricts growth of this fungus [10–12]. Ample evidence both in experimental models and in man indicates that T-cells are a key component of the cellular control of this infection. Depletion of either α/β T-cell receptor-bearing (TCR)$^+$ cells or CD4$^+$ cells in mice abrogates the protective immune response [13,14]. Elimination of the CD8$^+$ subpopulation of cells blunts the efficiency in which *H. capsulatum* is cleared, but experimental animals still possess the capacity to sterilize tissues [13,15]. The contribution of this T-cell subset to the human condition is not known. Thus, in experimental animals and most likely in humans, there is a hierarchy among T-cell subpopulations in protective immunity: CD4$^+$ > CD8$^+$ cells.

In secondary immunity, the contribution of T-cells to the protective immune response differs somewhat. Elimination of CD4$^+$ cells impairs clearance but does not produce a fatal infection as is observed in primary exposure whereas the lack of CD8$^+$ cells is accompanied by less efficient clearance (intravenous model) or no change at all (pulmonary model) [13,15]. Alternatively, if both CD4$^+$ and CD8$^+$ cells are depleted, the mice succumb to the infectious process. However, there is an unequivocal disparity in the fungal burden between visceral and lymphoid organs. Mice deficient in both T-cell subsets are able to restrict growth of *H. capsulatum* in the lungs over a 30-day period, but they have a pronounced progression of infection in spleens which ultimately causes their demise [13]. In the absence of T-cells, lungs are capable of controlling infection whereas spleens are not.

The chief mechanism by which T-cells contribute to clearance of *H. capsulatum* is via production of cytokines. Both lung CD4$^+$ and CD8$^+$ T-cells can synthesize interferon-γ (IFN-γ) ex vivo in mice infected with *H. capsulatum* [16–18]. However, in primary pulmonary histoplasmosis, elimination of α/β-bearing TCR$^+$ cells or CD4$^+$ cells causes a dramatic reduction in IFN-γ in lungs [13]. Although CD8$^+$ cells have been shown to generate this cytokine [18], their elimination in primary disease is associated surprisingly with an elevation in IFN-γ levels in lungs [13]. Thus, despite their synthetic capacity, these cells appear to exert a regulatory role in IFN-γ production. Production of tumor necrosis factor (TNF)-α, granulocyte-macrophage colony-stimulating factor (GM-CSF), interleukins-4, -10, and -12 are unchanged in mice whose T-cells or T-cells subsets have been depleted [13]. Similarly, elimination of these T-cells or CD4$^+$ cells from immune mice depresses only IFN-γ production [13]. Depletion of CD8$^+$ cells does not alter release of any of the above referenced cytokines [13]. Despite the relative importance of CD8$^+$ T-cells to host control, there is no evidence that they contribute to the protective immune response by expressing cytotoxic activity against yeast-laden phagocytes.

B. B-Cells, Antibody, and Histoplasmosis

Unlike candidiasis and cryptococcosis, in which humoral immunity is central in host defenses [19–21], neither B-cells nor antibody impact host defenses against *H. capsulatum* [13,22,23]. Studies in which immune sera are passively transferred to mice have failed to show significant alteration in fungal burden of recipient mice [13]. B-cell knockout mice control primary and secondary infection as well as B-cell-replete animals [13]. In secondary infection, the lungs of mice lacking B and T-cells can reduce the fungal burden [13]. These studies obviously suggest that a humoral immune response does not impact the protective immune response. One explanation for the failure of immune sera to confer protection is that sera contain no protective antibodies. Another explanation is that the number of "protective" antibody molecules is small in comparison to those that either express no protective function or exacerbate infection. Hence, it is possible that particular clones of immunoglobulin can mediate protection either prophylactically or therapeutically. Although the concept of protective antibodies to an intracellular pathogen may seem unconventional, monoclonal antibodies that protect mice infected with *Mycobacterium tuberculosis* have been created [24]. Thus, it is quite possible that distinct clones of B-cells may express protective activity. Until that is proven, a vaccine for histoplasmosis should focus on arming T-cells to protect. Antibody detection may be useful as a marker of vaccination rather than as a measure of the elaboration of a protective immune response.

C. Innate Immunity

Macrophages and neutrophils are critically important for clearance of *H. capsulatum*, although the latter cells are not typically targets of vaccines for intracellular pathogens because of their short life span in the circulation and tissues. The pivotal nature of the interaction between phagocytes and *H. capsulatum* is much more fully described in Chapter 4. Because of the tight junction between the innate and acquired immune systems [25], it is imperative that the appropriate soluble signals from the innate system be generated that direct the components of acquired immunity to generate a protective immune response. For *H. capsulatum*, as with other intracellular pathogens, those signals must drive the immune system toward a Th1 type of response [18,26]. In nascent immunity, any vaccine should direct the innate system to generate IL-12 which will in turn prompt the production of IFN-γ. Although it is possible that a vaccine could stimulate this cytokine independent of IL-12, it appears that the most efficient mechanism is through IL-12. Furthermore, a vaccine for prevention of histoplasmosis also may have to trigger production of TNF-α and/or GM-CSF by constituents of the innate immune system since both of these cytokines are necessary for host control of primary infection [26–32].

Activation of natural killer (NK) cells or γ/δ TCR⁺ is not a common means by which vaccines induce immunity since the cells do not have memory. Yet, they are a rich source of cytokines including IFN-γ and TNF-α and thus if engaged by vaccine constituents may amplify the immune response to a vaccine [33].

D. Cytokines and Histoplasmosis

1. Primary Infection

Much of what we know regarding the influence of endogenous cytokines and host regulation of histoplasmosis stems from experiments in mice. Since these molecules are the likely mediators of successful resolution of infection, knowledge of their relative importance in host defenses is requisite. Although animal models do not mimic the human condition precisely, the studies in mice are particularly germane since this species is able to sterilize tissues infected with *H. capsulatum* whereas humans do not. Thus, by exploiting the differences between mouse and man, it is possible to develop a better vaccine that will be successful in humans. In the following section, the influence of endogenous cytokines on both primary and secondary infection will be reviewed.

In primary histoplasmosis in which the host recovers spontaneously, a Th1 response is the dominant phenotype in lungs and spleens of mice [18,26,27,34]. Thus, relatively small inocula that do not produce overwhelming disease provoke production of high levels of IL-12 and IFN-γ but not IL-4 or IL-10 [18]. Activation of the IL-12/IFN-γ axis is critical for survival in primary infection. Neutralization of endogenous IL-12 using monoclonal antibodies (mAb) to this cytokine is accompanied by a dramatic increase in the number of colony-forming units (CFU) in lungs and spleens and results in death of animals given an otherwise sublethal infection [26,27]. In addition, treatment with mAb to IL-12 sharply reduces IFN-γ but not GM-CSF and TNF-α levels, two cytokines that are important in host defenses in primary infection [27]. IL-12 functions by inducing IFN-γ release in mice infected with *H. capsulatum*. Administration of recombinant IL-12 promotes survival of mice given a lethal challenge with this fungus, and the effect of IL-12 is dependent on generation of IFN-γ [26,35]. Whether mice injected with an large inoculum of this fungus have a deficiency of IFN-γ that is restored by IL-12 or such mice require more of the former cytokine to cope with the large numbers of yeasts is not known. Nevertheless, IL-12 certainly holds promise for boosting immune responses either in vaccination or in adjunctive therapies for histoplasmosis as well as other infectious diseases.

IFN-γ is vitally important. Based on numerous in vitro studies, it is the chief cytokine that mediates growth inhibition by murine macrophages [26,27,34,36]. Studies in mice given mAb to IFN-γ and mice genetically deficient in this cytokine reveal the same finding. A sublethal inoculum causes a rapid demise of animals [26,36]. In IFN-γ knockout mice, the levels of TNF-α and IL-12 are not perturbed [36]. Generation of these two cytokines appears to be independent of IFN-γ release. The findings stress that the absence of IFN-γ and its ability to activate phagocytes is devastating.

Another necessary cytokine in this form of the infection is TNF-α [28–31]. Inhibition of its in vivo activity abrogates protective immunity. Similar to neutralization of either IL-12 or IFN-γ, blockade of endogenous TNF-α results in increased CFU in organs, and mice succumb to infection. The lack of TNF-α does not impair generation of IFN-γ, GM-CSF, or IL-12 but is associated with depressed nitric oxide levels [30]. This molecule is central to clearance of primary infection. Mice given inhibitors of nitric oxide or nitric oxide knocknout mice fail to contain infection [26]. Thus, the IL-12 IFN-γ axis and TNF-α appear to operate independently in mediating protective immunity. A deficiency of IFN-γ does not impact TNF-α levels and conversely, neutralization of TNF-α does not impact either IL-12 or IFN-γ. The two cytokines do converge on a common final mediator. They are both necessary for synthesis of nitric oxide, and the lack of one sharply diminishes the production of this gas (Fig. 1).

A fourth cytokine involved in the protective immune response is endogenous GM-CSF. Administration of mAb to GM-CSF abolishes the protective immunity [32]. A high proportion

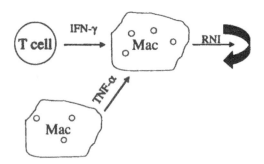

Figure 1 Pathway of IL-12, IFN-γ, and TNF-α regulation of protective immunity to *H. capsulatum*. RNI-reactive nitrogen intermediates.

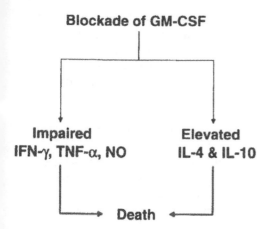

Figure 2 Pathway of GM-CSF regulation of protective immunity to *H. capsulatum*.

of mice given this mAb succumb to overwhelming infection as compared to controls. In the lungs of GM-CSF-neutralized animals, there are a concomitant reduction in IFN-γ and TNF-α and marked elevations in IL-4 and IL-10 [32]. The alteration in the Th2 cytokines contribute to the impaired host defenses since treatment with mAb to IL-4 or IL-10 restores protective immunity [32]. Lavage cells from mice given mAb to GM-CSF produce less nitric oxide than those from infected controls [32]. Thus, mice lacking in IFN-γ, TNF-α, or GM-CSF manifest a common defect-poor production of nitric oxide (Fig. 2).

2. Secondary Infection

Each of the cytokines mentioned in the preceding sections has been analyzed for their influence in secondary immunity. As is the case with other models of intracellular infections [37,38], IL-12 does not contribute to the generation of protective immunity in secondary infection. Neutralization of IL-12 fails to change the course of secondary infection [27]. The influence of IFN-γ in this phase of infection is dependent on the model employed. Studies utilizing an intravenous model have consistently demonstrated that blockade of IFN-γ with a combination of anti-IFN-γ, anti-IL-12, and anti-IFN-γ receptor does not modulate protective immunity [31]. On the other hand, in secondary pulmonary histoplasmosis, IFN-γ knockout mice cannot restrict growth [36]. Differences in mice (e.g., genetically deficient mice vs. mAb-neutralized mice) and the route of infection certainly can explain the discrepant findings. Thus, IFN-γ may be dispensable in hematogenous histoplasmosis but is necessary when the lung is the first organ to contact the fungus.

Perhaps the most crucial cytokine necessary for survival of mice from rechallenge is TNF-α. In both the systemic and the pulmonary model, neutralization of TNF-α leads to fatal histoplasmosis [30,31]. In the systemic model, the influence of TNF-α is not observed in immunocompetent mice, but only if this cytokine is neutralized in IFN-γ knockout mice [31]. The effect of mAb to TNF-α in pulmonary histoplasmosis is manifest in otherwise immunocompetent mice. Unlike primary infection, neutralization of TNF-α does not alter nitric oxide levels [30], thus supporting previous evidence that nitric oxide is not involved in secondary infection [26]. In the pulmonary model, neutralization of TNF-α is associated with increases in the levels of IL-4 and IL-10 in lungs of mice. The bioactivity of these two cytokines contributes to the downregulation of immunity. Administration of the combination of anti-IL-4 plus anti-IL-10 mAb, but not either alone, restores the protective immune response [30]. The mechanism by which the elevated levels of IL-4 and IL-10 promote exacerbation is not known.

Endogenous GM-CSF regulates host control of secondary infection. Unlike primary histoplasmosis, neutralization of this cytokine in mice reexposed to yeasts blunts clearance during weeks 2 and 3 of infection but does not produce an overwhelming infection [32]. Moreover, the effect of mAb to GM-CSF in secondary histoplasmosis is not associated with alterations in levels of IFN-γ, TNF-α, IL-4, or IL-10 [32], thus distinguishing its activity from that of primary infection.

At the cellular and molecular level, regulation of protective immunity in secondary infection contrasts with that of primary infection. The influence of T-cell subsets and cytokine control of protective immunity demonstrate clear and distinct differences between the two phases of infection. The studies of nascent and memory immunity are important for the development of prophylactic and therapeutic vaccines in humans.

E.   Cellular and Molecular Determinants of Immunity in Humans

Much of what we know about the human condition is derived from in vitro studies and clinical observations. Humans who have been to exposed to *H. capsulatum* and resolved their infection exhibit the hallmarks of a Th1 response: granulomas characterized by infiltration with lymphocytes and macrophages and delayed-type hypersensitivity responses. On the other hand, disseminated histoplasmosis has features consistent with a Th2 response, i.e., rising antibody responses, whereas delayed-type hypersensitivity responses become absent, i.e., elevated IgE responses and disorganized inflammatory responses [39,40].

The role of CD4$^+$ cells in controlling histoplasmosis in humans has been surmised from observations of HIV-infected patients who rarely develop histoplasmosis unless the CD4$^+$ cells are <200 cells/μL [41]. Much of the disease that is observed is believed to reactivation rather than reinfection, although it is difficult to discern the difference. One of the intriguing questions is why all HIV patients do not develop histoplasmosis when their CD4$^+$ cells fall below 200. In many areas of the midwestern and southeastern United States, the proportion of individuals who are infected approximates 90% [1]. Based on this value, many more patients would be expected than are diagnosed. Two explanations may explain these findings. First, the small number of CD4$^+$ may be sufficient to control histoplasmosis in most patients. Second, CD8$^+$ cells may contribute more than anticipated to host control. These two contentions are not mutually exclusive and both may be operative.

One of the central paradoxes regarding human studies is that despite the accumulated data in mice and indirectly in humans, it has been difficult to prove that IFN-γ is a potent activator of human macrophages. Two studies have demonstrated unequivocally that human macrophages exposed to this cytokine are not activated to express fungistatic or fungicidal activity [42,43]. On the other hand, one study has shown that in a short-term assay (2 hr), IFN-γ can arm macrophages to inhibit the growth of yeasts [44]. No explanation exists to explain the discrepancy between the in vitro studies and the indirect in vivo evidence. No evidence exists for the influence of IL-12. Such information may not be available until subjects appear who are deficient in IFN-γ or are missing IL-12 or IFN-γ receptors, as is the case with *Mycobacteria* infections [45–47]. Unfortunately, many of those subjects have resided in areas in which *H. capsulatum* is not endemic.

The family of cytokines that is known to activate human macrophages to express fungistatic activity in vitro are CSF, IL-3, GM-CSF, and M-CSF which stimulate macrophages to express anti-*Histoplasma* activity [42]. How this family promotes antifungal activity is not known. One possible explanation is that these cytokines enhance phagolysosomal fusion in *H. capsulatum*–infected macrophages. In humans, it is the one mechanism that has been identified as an anti-*Histoplasma* mechanism in human phagocytes [48]. The paucity of information regarding the human condition is disconcerting and more data will have to be gathered to ensure that vaccine development proceeds in a rational manner.

III. PROTECTIVE IMMUNOGENS FROM *H. CAPSULATUM*

A. Early Studies of Immunogens

The history of protective immunogens from *H. capsulatum* is limited. Until recently, there has not been the intensive effort to develop a vaccine as has been the case with coccidioidomycosis. Initially, it was reported that mice exposed to viable or heat-killed yeasts would survive a second challenge. In that model, mice were challenged intracerebrally, an unusual route to say the least [49]. That report did establish that yeasts could confer protection and that the effect of viable yeasts was significantly greater than that of killed yeasts. Subsequently, several reports indicated that yeast extracts could induce a protective immune response. Garcia and Howard [50] demonstrated that an ethylenediamine extract of yeast cells was protective in mice challenged intravenously. Unfortunately, no additional studies were performed to identify if there was a single molecule within that extract that could mediate protection. Another constituent that appeared to be promising was a ribosomal-protein complex from *H. capsulatum* [51]. The protective activity of this admixture was dependent on the integrity of both the ribosomal and the protein portion [51]. The reports regarding the ribosomal-protein complex from *H. capsulatum* transpired during a time when this topic was being studied in prokaryotes. Although the protein content was necessary for the protective activity, no further studies were reported that determined that a single protein within the complex could induce a protective response. In addition, exposure of mice to mycelia that cannot transform also mediate a protective response against yeast cells. Thus, mycelia treated with p-chloromercuryphenylsulfonic acid, a sulfhydryl blocking agent which inhibits the mycelial to yeast transition, can protect outbred mice from a lethal challenge with yeast cells [52]. These results connote that there are shared immunogens between the two phases of *H. capsulatum*.

B. Cell Wall and Cell Membrane and Heat Shock Protein 60

Several years ago, our laboratory began to revisit the concept of protective immunity. This work arose out of studies that sought to determine the nature of epitopes recognized by monoclonal populations of T-cells. We prepared a detergent extract from the cell wall and cell membrane (CW/M) of actively growing yeast cells. This complex is composed of numerous proteins as assessed by polyacrylamide gel electrophoresis. Immunobiologically, it is a target of the T-cell-dependent immune response by mice that had recovered from a challenge with yeast cells [53]. Injection of this material emulsified in adjuvant could induce a T-cell-mediated immune responses. Moreover, this extract protects mice against a lethal challenge delivered intravenously [53].

Further inquiries regarding this extract focused on the possibility that a single molecule, most likely protein containing, could mimic the activity of the extract. To pursue this postulate, CW/M was subjected to T-cell Western blot analysis using T-cell clones as targets for antigen mapping. Although several regions of activity were identified, we concentrated on an area that encompassed 53 to 64 kDa. When acrylamide gels of CW/M were reviewed, a prominent Coomassie-stained band was evident at ~62 kDa [53]. This protein, termed HIS-62, was isolated by electroelution and analyzed for its immunobiological activity. Like CW/M, this protein induced reactivity by T-cell clones and was recognized by the clones that responded to the 53- to 64-kDa region [54]. Injection of HIS-62 elicited cell-mediated immune responses and conferred protection against a lethal intravenous challenge in three genetically disparate strains of mice. In this limited analysis, there was no evidence of genetic restriction of reactivity [54]. This point is important since any vaccine should be able to evoke a response in the outbred human population. This point was pursued in in vitro studies with human peripheral blood mononuclear cells in which seven individuals from the endemic region were tested for their capacity to respond to this protein. Cells from all seven proliferated to this molecule, and in two individuals who were not from the region, it did not stimulate their cells [55]. Thus, this antigen stimulated reactivity in an outbred population. Although that finding does not signify that it would have protective activity, it does indicate that there were no obvious genetic restrictions on recognition.

The nature of HIS-62 was determined by molecular techniques. The purified protein was subjected to Edman degradation and amino acid sequencing provided the first clues that HIS-62 was a member of the heat shock protein (hsp) 60 family [56]. Cloning of the gene from genomic DNA confirmed the amino acid data. Once it had been isolated, a recombinant protein was generated in a bacterial expression system and analyzed for its ability to confer protection. Indeed, in a pulmonary model of histoplasmosis, hsp 60 when combined with adjuvant protected mice against a lethal intranasal challenge [56]. The identification of HIS-62 as hsp 60 was perplexing because it was unanticipated that such a highly conserved molecule would elicit a protective immune response. However, studies with *Legionella pneumophila* also reported that cognate hsp 60 was protective in an experimental model of Legionnaire's disease [57]. Thus, evidence began to accrue that highly conserved molecules could confer protection in disparate infectious diseases.

C. Mapping Immunogenic Determinants of Hsp 60

Following the identification of hsp 60 as a protective antigen, we sought to determine if the protective activity required the whole molecule or it was segregated into a fragment of the protein. By molecular techniques, we generated four recombinant fragments each encompassing ~250 amino acids. Each fragment was tested for its ability to stimulate murine splenocytes from immunized mice and to confer protection in two strains of mice. All fragments induced a proliferative response by spleen cells; however, only fragment 3, which spanned amino acids 172–443, significantly reduced CFU in lungs, livers, and spleens of mice and reduced mortality in mice challenged intranasally with a lethal inoculum of *H. capsulatum* [58]. Thus, the protective activity was segregated into a domain of the whole protein.

D. HIS-80

During the T-cell Western blot studies, an 80-kDa molecule, HIS-80, which was reactive with antibody to hsp 70, was found to be immunogenic. This molecule was isolated by electroelution from CW/M and tested for immunogenicity. HIS-80 was also a target of the cell-mediated immune response and did reduce CFU at week 1 of infection in mice given a sublethal intravenous inoculation [59]. However, it did not protect against a lethal challenge. The complete identity of this protein remains unknown. It is not hsp 70 but a member of the family. Based on its molecular mass it is most likely BiP [60].

E. Immunogenic but Nonprotective Antigens from *H. capsulatum*

Our laboratory has isolated the genes encoding two other antigens, the H antigen and hsp 70, and analyzed the immunogenicity of the recombinant forms. Both antigens evoke cell-mediated immune responses in mice, but neither of them induces a protective immune response in mice to a lethal or sublethal challenge [61,62]. These findings signify that not all proteins that induce cell-mediated immune responses possess the capacity for inducing protection. What makes a molecule protective and what distinguishes protective from nonprotective moieties? Another query is whether there are motifs or canonical sequences within proteins that make them protective. Those are critically germane questions raised by the findings and cannot be easily answered. Our laboratory is pursuing a comparison of hsp 70 and hsp 60 with a focus on the cytokine response in tissues and the inflammatory response they evoke.

IV. FUTURE DIRECTIONS

The present review has analyzed much of the known data regarding *H. capsulatum* and protective immunity. The findings recounted here represent only a beginning toward the goal of vaccine development for histoplasmosis. Several unknowns still exist. For example, virtually nothing has been accomplished with genetic vaccination. Exciting evidence has been presented with coccidioidomycosis and a DNA vaccine [63,64], although it did not reduce CFU as dramatically in lungs as it did in spleens. The use of cytokines or cytokine genes as adjuvants also remains an exciting but unknown prospect. IL-12 and GM-CSF have been demonstrated in model systems to exhibit adjuvantlike activity [65,66]. The potency of vaccines may be improved by integration of these two molecules.

Another unexplored arena is therapeutic vaccination. Recently, investigators have reported that a DNA vaccine containing *M. tuberculosis* hsp 60 can be used therapeutically in mice as well as prophylactically. Our own studies utilizing the protective recombinant fragment as a therapeutic vaccine have not been as salutary as those in experimental tuberculosis [67,68], but work is ongoing in this area to determine the optimal amounts of protein and the timing, which is probably critical. One disadvantage with the mouse model is that a chronic infectious model does not exist. Thus, the timing for intervention therapeutically is quite limited.

V. WHO SHOULD BE VACCINATED?

One of the central considerations in a vaccine for histoplasmosis is who should be vaccinated. It is certainly not necessary for everyone. Those who require vaccination would be immunosuppressed individuals residing in or moving to an endemic region. Histoplasmosis is an occupational hazard, especially for those who do outside construction or who farm. Others to be considered are neonates who might be exposed since the disease is often fulminant in them. Additionally, individuals who enjoy spelunking or caving frequently may be candidates for a vaccine.

REFERENCES

1. L Ajello. Distribution of *Histoplasma capsulatum* in the United States. In: L Ajello, EW Chick, MF Furculow, eds. Histoplasmosis. Springfield, IL: Charles C. Thomas, 1971, pp 103–122.
2. J Schwarz. Histoplasmosis. New York: Praeger, 1981, pp 1–102.
3. RA Goodwin, RM Des Prez. Histoplasmosis. Am Rev Respir Dis 117:929–956, 1979.
4. J Wheat. Histoplasmosis. Experience during outbreaks in Indianapolis and review of the literature. Medicine 76:339–354, 1997.

5. RA Goodwin, RM Des Prez. Histoplasmosis in normal hosts. Medicine 60:231–266, 1981.
6. J Vanek, J Schwarz. The gamut of histoplasmosis. Am J Med 50:89–104, 1971.
7. SF Davies, M Khan, GA Sarosi. Disseminated histoplasmosis in immunologically suppressed patients. Occurrence in a nonendemic area. Am J Med 64:94–100, 1978.
8. LJ Wheat, TG Slama, JA Norton, RB Kohler, HE Eitzen, MLV French, B Sathapatayavongs. Risk factors for disseminated histoplasmosis. Analysis of a large urban outbreak. Ann Intern Med 96: 159–163, 1982.
9. LJ Wheat, PA Connolly-Stringfield, RL Baker, MF Curfman, ME Eads, KS Israel, SA Norris, DH Webb, ML Zeckel. Disseminated histoplasmosis in the acquired immune deficiency syndrome: clinical findings, diagnosis and treatment, and review of the literature. 69:361–374, 1990.
10. DM Adamson, GC Cozad. Effect of antilymphocyte serum on animals experimentally infected with *Histoplasma capsulatum* or *Cryptococcus neoformans*. J Bacteriol 100:1271–1276, 1969.
11. DM Williams, JR Graybill, DJ Drutz. *Histoplasma capsulatum* infection in nude mice. Infect Immun 21:973–977, 1978.
12. RP Artz, WE Bullock. Immunoregulatory responses in experimental disseminated histoplasmosis: depression of T-cell dependent and T-effector responses by activation of splenic suppressor cells. Infect Immun 23:893–902, 1979.
13. R Allendörfer, GD Brunner, GS Deepe Jr. Complex requirements for nascent and memory immunity in pulmonary histoplasmosis. J Immunol 162:7389–7396, 1999.
14. AM Gomez, WE Bullock, CL Taylor, GS Deepe Jr. Role of L3T4+ T cells in host defense against *Histoplasma capsulatum*. Infect Immun 56:1685–91, 1988.
15. GS Deepe Jr. Role of CD8+ T cells in host resistance to systemic infection with *Histoplasma capsulatum* in mice. J Immunol 152:3491–500, 1994.
16. B Wu-Hsieh, DH Howard. Inhibition of growth of *Histoplasma capsulatum* by lymphokine stimulated macrophages. Infect Immun 132:2593–2597, 1984.
17. GS Deepe Jr, JG Smith, G Sonnenfeld, D Denman, WE Bullock. Development and characterization of *Histoplasma capsulatum*–reactive murine T-cell lines and clones. Infect Immun 54:714–722, 1986.
18. JA Cain, GS Deepe Jr. Evolution of the primary immune response to *Histoplasma capsulatum* in murine lung. Infect Immun 66:1473–1481, 1998.
19. A Casadevall. Antibody immunity and invasive fungal infections. Infect Immun 63:4211–4218, 1995.
20. J Mukherjee, MD Scharff, A Casadevall. Protective murine monoclonal antibodies to *Cryptococcus neoformans*. Infect Immun 60:4534–4541, 1992.
21. Y Han, MA Ulrich, JE Cutler. *Candida albicans* mannan extract-protein conjugates induce a protective immune response against experimental candidiasis. J Infect Dis 179:1477–1484, 1999.
22. N Khardori, S Chaudhary, P McConnachie, RP Tewari. Characterization of lymphocytes responsible for protective immunity to histoplasmosis in mice. Mykosen 26:523–532, 1983.
23. DA Rowley, M Huber. Growth of *Histoplasma capsulatum* in normal, superinfected and immunized mice. J Immunol 77:15–20, 1956.
24. R Teitelbaum, A Glatman-Freedman, B Chen, JB Robbins, E Unanue, A Casadevall, BR Bloom. A mAb recognizing a surface antigen of *Mycobacterium tuberculosis* enhances host survival. Proc Natl Acad Sci USA 95:15688–15693, 1998.
25. DT Fearon, RM Locksley. The instructive role of innate immunity on the acquired immune response. Science 272:50–53, 1996.
26. P Zhou, MC Sieve, J Bennett, KJ Kwon-Chung, RP Tewari, RT Gazzinelli, A Sher, RA Seder. IL-12 prevents mortality in mice infected with *Histoplasma capsulatum* through induction of IFN-γ. J Immunol 155:785–795, 1995.
27. R Allendoerfer, GP Boivin, GS Deepe Jr. Modulation of immune responses in murine pulmonary histoplasmosis. J Infect Dis 175:905–914, 1997.
28. JG Smith, DM Magee, DM Williams, JR Graybill. Tumor necrosis factor-a plays an important role in host defense against *Histoplasma capsulatum*. J Infect Dis 162:1349–1353. 1990.
29. BA Wu-Hsieh, GS Lee, M Franco, FM Hofman. Early activation of splenic macrophages by tumor necrosis factor alpha is important in determining the outcome of experimental histoplasmosis in mice. Infect Immun 60:4230–4238, 1992.

30. R Allendoerfer, GS Deepe Jr. Blockade of endogenous TNF-α exacerbates primary and secondary pulmonary histoplasmosis by differential mechanisms. J Immunol 160:6072–6082, 1998.

31. P Zhou, G Miller, RA Seder. Factors involved in regulating primary and secondary immunity to infection with *Histoplasma capsulatum*: TNF-α plays a critical role in maintaining secondary immunity in the absence of IFN-γ. J Immunol 160:1359–1368, 1998.

32. GS Deepe Jr, R Gibbons, E Woodward. Neutralization of endogenous GM-CSF subverts the protective immune response to *Histoplasma capsulatum*. J Immunol 163:4985–4993, 1999.

33. CA Biron. Activation and function of natural killer cell responses during viral infections. Curr Opin Immunol 9:24–34, 1997.

34. B Wu-Hsieh. Relative susceptibilities of inbred mouse strains C57BL/6 and A/J to infection with *Histoplasma capsulatum*. Infect Immun 57:3788–3792, 1989.

35. P Zhou, MC Sieve, RP Tewari, RA Seder. Interleukin-12 modulates the protective immune response to SCID mice infected with *Histoplasma capsulatum*. Infect Immun 65:936–942, 1997.

36. R Allendoerfer, GS Deepe Jr. Intrapulmonary response to *Histoplasma capsulatum* in gamma interferon knockout mice. Infect Immun 65:2564–2569, 1997.

37. CS Tripp, O Kanagawa, E Unanue. Secondary response to *Listeria* infection requires IFN-γ but is partially independent of IL-12. J Immunol 155:3427–3432, 1995.

38. RT Gazzinelli, M Wysocka, S Hayashi, EY Denkers, S Hieny, P Caspar, G Trinchieri, A Sher. Parasite-induced IL-12 stimulates early IFNγ synthesis and resistance during acute infection with *Toxoplasma gondii*. J Immunol 153:2533–2543, 1994.

39. GS Deepe Jr, WE Bullock. Histoplasmosis: a granulomatous inflammatory response. In: JI Gallin, IM Goldstein, R Snyderman, eds. Inflammation: Basic Principles and Clinical Correlates. 2nd ed. New York: Raven Press, 1992, pp 943–958.

40. RA Cox, DR Arnold. Immunoglobulin E in histoplasmosis. Infect Immun 29:290–293, 1980.

41. LJ Wheat, PA Connolly-Springfield, RL Baker, MF Curfman, ME Eads, KS Israel, SA Norris, DH Webb, ML Zeckel. Disseminated histoplasmosis in the acquired immune deficiency syndrome: clinical findings, diagnosis and treatment, and a review of the literature. Medicine 69:361–374, 1990.

42. SL Newman, L Gootee. Colony-stimulating factors activate human macrophages to inhibit intracellular growth of *Histoplasma capsulatum* yeasts. Infect Immun 60:4593–4597, 1992.

43. J Fleischman, B Wu-Hsieh, DH Howard. The intracellular fate of *Histoplasma capsulatum* in human macrophages is unaffected by recombinant human interferon-γ. J Infect Dis 161:143–145, 1990.

44. E Brummer, N Kurita, S Yoshida, K Nishimura, M Miyaji. Killing of *Histoplasma capsulatum* by γ-interferon-activated human monocyte-derived macrophages: evidence for a superoxide anion-dependent mechanism. J Med Microbiol 35:29–34, 1991.

45. F Altare, A Durandy, D Lammas, J-F Emile, S Lamhamedi, F Le Deist, P Drysdale, E Jouanguy, R Döffinger, F Bernaudin, O Jeppson, JA Gollob, E Meinl, AW Segal, A Fischer, D Kumararatne, J-L Casanova. Impairment of mycobacterial immunity in human interleukin-12 receptor deficiency. Science 280:1432–1435, 1998.

46. R de Jong, F Altare, I-A Haagen, DG Elferink, T de Boer, PJC van Breda Vriesman, PJ Kabel, JMT Draasima, JT van Dissel, FP Kroon, J-L Casanova, THM Ottenhoff. Severe mycobacterial and salmonella infections in interleukin-12 receptor-deficient patients. Science 280:1435–1438, 1998.

47. E Jouanguy, F Altare, S Lamhamedi, P Revy, J-F Emile, M Newport, M Levin, S Blanche, E Seboun, A Fischer, J-L Casanova. Interferon-γ-receptor deficiency in an infant with fatal Calmette-Guerin infection. N Engl J Med 335:1956–1961, 1996.

48. SL Newman, L Gootee, C Kidd, GM Ciraolo, R Morris. Activation of human macrophage fungistatic activity against *Histoplasma capsulatum* upon adherence to type I collagen matrices. J Immunol 158:1779–1786, 1997.

49. S Saslow, J Schaeffer. Survival of *Histoplasma capsulatum* in experimental histoplasmosis in mice. Proc Soc Exp Biol Med 91:412–414, 1956.

50. JP Garcia, DH Howard. Characterization of antigens from the yeast phase of *Histoplasma capsulatum*. Infect Immun 4:116–125, 1971.

51. RP Tewari, C Feit. Immunogenicity of ribosomal preparations from yeast cells of *Histoplasma capsulatum*. Infect Immun 10:1091–1097, 1974.

52. G Medoff, M Sacco, B Maresca, D Schlessinger, A Painter, GS Kobayashi, L Carratu. Irreversible block of the mycelial-to-yeast phase transition of *Histoplasma capsulatum*. Science 231:476–479, 1986.

53. AM Gomez, JC Rhodes, GS Deepe Jr. Antigenicity and immunogenicity of an extract from the cell wall and cell membrane of *Histoplasma capsulatum* yeast cells. Infect Immun 59:330–336, 1991.

54. FJ Gomez, AM Gomez, GS Deepe Jr. Protective efficacy of a 62-kilodalton antigen, HIS-62, from the cell wall and cell membrane of *Histoplasma capsulatum* yeast cells. Infect Immun 59:4459–4464, 1991.

55. HM Henderson, GS Deepe Jr. Recognition of *Histoplasma capsulatum* yeast-cell antigens by human lymphocytes and human T-cell clones. J Leukoc Biol 51:432–436.

56. FJ Gomez, R Allendoerfer, GS Deepe Jr. Vaccination with recombinant heat shock protein 60 from *Histoplasma capsulatum* protects mice against pulmonary histoplasmosis. Infect Immun 63: 2587–2595, 1995.

57. SJ Blander, MA Horwitz. Major cytoplasmic membrane protein of *Legionella pneumophila*, a genus common antigen and member of the hsp 60 family of heat shock proteins, induces protective immunity in a guinea pig model of Legionnaire's disease. J Clin Invest 91:717–723, 1993.

58. GS Deepe Jr, R Gibbons, GD Brunner, FJ Gomez. A protective domain of heat-shock protein 60 from *Histoplasma capsulatum*. J Infect Dis 174:828–834, 1996.

59. FJ Gomez, AM Gomez, GS Deepe Jr. An 80-kilodalton antigen from *Histoplasma capsulatum* that has homology to heat shock protein 70 induces cell-mediated immune responses and protection in mice. Infect Immun 60:2565–2571, 1992.

60. S Munro, HRB Pelham. An hsp 70–like protein in the ER: identify with the 78 kDa glucose regulated protein and immunoglobulin heavy chain binding protein. Cell 46:291–300, 1986.

61. GS Deepe Jr, GG Durose. Immunobiological activity of recombinant H antigen from *Histoplasma capsulatum*. Infect Immun 63:3151–3157, 1995.

62. R Allendoerfer, B Maresca, GS Deepe Jr. Cellular immune responses to recombinant heat shock protein 70 from *Histoplasma capsulatum*. Infect Immun 64:4123–4128, 1996.

63. C Jiang, DM Magee, TN Quitugua, RA Cox. Genetic vaccination against *Coccidioides immitis*: comparison of vaccine efficacy of recombinant antigen 2 and antigen 2 cDNA. Infect Immun 67: 630–635, 1999.

64. RO Abuodeh, LF Shubitz, E Siegel, S Snyder, T Peng, KI Orsborn, E Brummer, DA Stevens, JN Galgiani. Resistance to *Coccidioides immitis* in mice after immunization with recombinant protein or a DNA vaccine of a proline-rich antigen. Infect Immun 67:2935–2940, 1999.

65. S Gurunathan, C Prussin, DL Sacks, RA Seder. Vaccine requirements for sustained cellular immunity to an intracellular parasitic infection. Nat Med 4:1409–1415, 1998.

66. M-H Tao, R Levy. Idiotype/granulocyte-macrophage colony-stimulating factor fusion protein as a vaccine for B-cell lymphoma. Nature 362:755–758, 1993.

67. RE Tascon, MJ Colston, S Ragno, E Stavropoulos, D Gregory, DB Lowrie. Vaccination against tuberculosis by DNA injection. Nat Med 2:888–892, 1996.

68. DB Lowrie, RE Tascon, VLD Bonato, MMF Lima, LH Faccioli, E Stavropoulos, MJ Colston, RG Hewinson, K Moelling, CL Silva. Therapy of tuberculosis in mice by DNA vaccination. Nature 400: 269–271, 1999.

17

Coccidioidomycosis: Pathogenesis, Immune Response, and Vaccine Development

Theo N. Kirkland
Veterans Administration, San Diego Health Care System, and University of California, San Diego, California

Garry T. Cole
Medical College of Ohio, Toledo, Ohio

I. INTRODUCTION

Coccidioides immitis is a primary fungal pathogen, which resides in soil of the desert Southwest of the United States. Like most medically important fungi that cause systemic disease, *C. immitis* demonstrates different morphologies in its saprobic and parasitic phases, but is distinguished from other fungal pathogens by the unique morphogenetic features of its growth in host tissue (Fig. 1).

Coccidioidomycosis, the disease caused by this pathogenic fungus, is also known as Valley Fever because the organism is prevalent in the San Joaquin Valley of central California. *C. immitis* infections are contracted almost exclusively by the respiratory route [1]. The clinical spectrum of disease is broad, ranging from an asymptomatic infection to a rapidly fatal mycosis. The most common clinical presentation is self-limited pneumonia, but in some cases the fungus can cause chronic cavitary pulmonary disease or disseminate beyond the lungs to the skin, bones, meninges, and other body organs. Coccidioidomycosis can also present as erythema nodosum or as a reactive arthritic condition, which is commonly referred to as desert rheumatism [2].

II. AGENT

A. Ecology

C. immitis is found in soil of the bioclimatic region known as the Lower Sonoran Life Zone [3]. The climatic features of this region are high temperatures (mean of 26–32°C in the summer and only rare winter frosts) and annual rainfall of 5–20 in. The soil is typically alkaline and has a high salt content. *C. immitis* seems to be better equipped to survive in this harsh environment than other organisms competing for the same ecological niche. Soil that has been tilled and fertilized is much less likely to contain *C. immitis* since enrichment for agriculture leads to a more favorable environment for competing micro-organisms. The distribution of *C. immitis* in

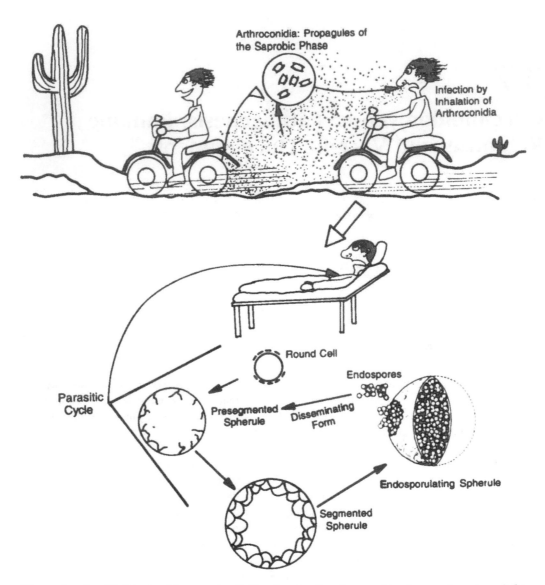

Figure 1 *Coccidioides immitis* grows in alkaline desert soil as the saprobic phase, and produces infectious, airborne spores (arthroconidia). Inhaled arthroconidia convert into round cells (spherule initials), which ultimately give rise to endosporulating spherules. Endospores differentiate into second-generation spherules and can disseminate from the lungs to other body organs.

the desert soil is sparse and unpredictable. In the past it has been thought that *C. immitis* flourished near rodent burrows, but more recent studies have disputed this notion [4]. Ancient Indian burial sites, known as middens, have been found to be a particularly rich source of the fungal pathogen [4]. Coccidioidomycosis infects wild animals (mammals) and domesticated dogs [5]. *C. immitis* is usually found at between 4 and 12 in. below the surface of the soil. Temperatures as high as 60°C and ultraviolet light probably kill the organisms within the first 4 in. of the soil surface. The mycelia can tolerate drought but thrive after rain [6], and the fungus can be more easily isolated from soil immediately after the rainy season. Nevertheless, after prolonged periods of drought *C. immitis* persists while competing organisms in the same ecologic niche die out.

B. Geographic Distribution

Coccidioidomycosis is primarily found in the desert regions of Southern California, Arizona, Nevada, New Mexico, and west Texas (Fig. 2). This large area is home to ~20% of the population of the United States. It also includes some of the most rapidly expanding cities in the nation and attracts large numbers of visitors each year. The urban perimeters are extending further into the desert, as exemplified by the Bakersfield region of California and the Phoenix-Tucson area of Arizona. However, many cases of coccidioidomycosis are also found in regions that are not hyperendemic, such as San Diego and Los Angeles. Outbreaks of coccidioidomycosis have been reported among archaeology students digging in prehistoric Indian sites in northern California [7,8]. In 1977, a major dust storm blew soil from the San Joaquin Valley to northern California, including San Francisco, Marin County, Santa Clara, and Monterey County. Immediately following the storm, many cases of coccidioidomycosis were reported in nonendemic regions of middle and northern California [9]. At the time, there was some concern that *C. immitis* might be able to seed and persist in the soil in these areas, but that has not occurred.

The map in Figure 2 shows that the range of *C. immitis* includes west Texas and a large part of the desert regions of northern Mexico. A few cases of coccidioidomycosis have also

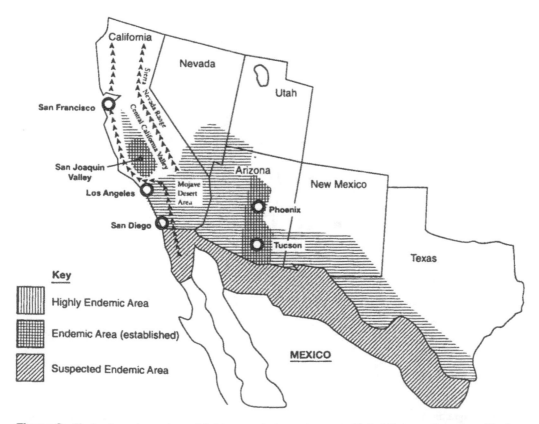

Figure 2 Endemic regions of coccidioidomycosis in southwestern United States and northern Mexico.

been reported in Central and South America [6]. The largest South American endemic region is in Argentina where the climate is dry and the soil conditions are similar to those in the desert southwest. New endemic regions have also been recently identified in Brazil (D. Pappagianis, personal communication, 2000). Despite these geographic limitations, physicians outside the endemic regions should consider coccidioidomycosis if the patient has ever traveled through the desert Southwest or lived in an endemic area. Reactivation of a prior asymptomatic *C. immitis* infection is of potential concern for immunocompromised individuals.

C. Mycology

1. Arthroconidium Formation

The soil-inhabiting mycelial phase of *C. immitis* gives rise to infectious, airborne arthroconidia by what appears to be a simple process of fragmentation of hyphal elements [10]. Upon close examination, this mode of conidium formation involves four distinct and apparently coordinated events pivotal for differentiation of the propagules (Fig. 3) [11]. These include (1) arrest of apical growth of aerial hyphae (Fig. 3A,B), followed by progressive septation of the filaments

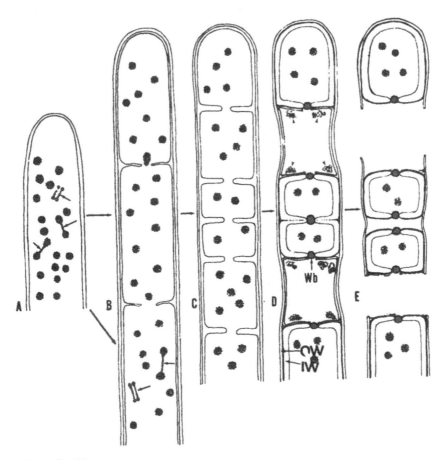

Figure 3 Diagrammatic sequence of stages of arthroconidium formation. Arrows indicate mitotic nuclei. IW, inner conidial wall layer; OW, outer conidial wall layer; Wb, Woronin body (septal plug).

(Fig. 3B,C) which typically leaves each compartment with at least two nuclei; (2) condensation of cytoplasm in certain hyphal compartments and autolysis of cytoplasm in adjacent cells of the chain, while the original outer hyphal wall remains intact (Fig. 3C,D); (3) synthesis of a new inner-wall layer that encompasses each viable and sealed compartment (i.e., septal pores are plugged), completing its conversion into a conidium (Fig. 3D); and (4) disarticulation of the chain of arthroconidia by mechanical fracture of the wall of the alternating, empty cells (Fig. 3E). Single, cylindrical arthroconidia (~3–6 × 2–4 μm) are small enough to pass down the respiratory tree and reach the alveoli of the host [12]. Most conidia, however, probably impact the mucosal lining of the upper respiratory tract. The ciliated epithelial cells are capable of sweeping the mucus and entrapped fungal cells proximally toward the pharynx, where they are removed via the digestive tract. *C. immitis* conidia neither adhere to nor germinate in the gastrointestinal tract.

Those arthroconidia that reach the terminal air sacs encounter alveolar macrophages. Evidence from experimental animal studies of coccidioidomycosis has indicated that host phagocytes are inefficient in the clearance of arthroconidia (Fig. 4A). Arthroconidia of *C. immitis* appear to be well equipped with barriers to host defenses [13]. The outer conidial wall layer (OW; Fig. 4 A,B), which is derived from the original hyphal wall (Fig. 3), is a hydrophobic sleeve that may have evolved as an adaptation for air dispersal of the soil saprobe. It may also serve as a passive barrier to destructive enzymes and oxidative products released by the host defense cells, thereby, contributing to survival of the pathogen in vivo. The hydrophobic outer wall layer has also been suggested to have antiphagocytic properties [14]. Stripping conidia of their outer wall by a cell fractionation procedure (Fig. 4C) [13] prompted an increase in phagocytosis (>25%), but not a significant increase in killing of arthroconidia by human polymorphonuclear neutrophils (PMNs) [14].

Wall-associated protein arrays have been revealed in freeze fracture preparations of *C. immitis* arthroconidia (Fig. 4D) [15], and are apparently localized at the interface of the outer and inner conidial wall layers. These crystallinelike complexes have been termed "rodlets" [16] and identified as hydrophobins, which are well documented protein components of microbial cell walls that contribute to cell surface hydrophobicity [17]. Hydrophobins have been detected in walls of several fungi [18]. Although the peptide sequences of hydrophobins are quite diverse, they share three common characteristics: all are small proteins (96–157 amino acids), all have eight cysteine residues arranged in a conserved pattern, and all have a similar profile of hydrophobic domains [17]. The rodlet layer of arthroconidia probably contributes to the hydrophobic surface barrier and prevents premature hydration of the cell wall that could otherwise lead to untimely germination of the propagule.

2. Morphogenetic Transition of Arthroconidia to Spherule Initials

Successful colonization of the host respiratory mucosa by *C. immitis* depends on conversion of arthroconidia to "round cells" or spherules. Little is known of the morphogenetic features of the initial stages of this saprobic to parasitic phase transition, and limited information is available on growth factors, which influence this process. A common requirement for such transitions among dimorphic fungal pathogens is an abrupt rise in temperature, typically from 25°C to 37°C. Thermal dimorphism [19] in *Histoplasma capsulatum* has been examined extensively [20]. Morphogenetic events associated with transition from the hyphal to yeast phase in this pathogen are part of a complex of physiological changes referred to as the heat shock response. Investigations of dimorphism in *H. capsulatum* have focused on identification of the "primary sensor" capable of monitoring temperature shifts, and on the mechanisms by which molecular signals are transferred from the sensor to the nucleus where they ultimately influence expression

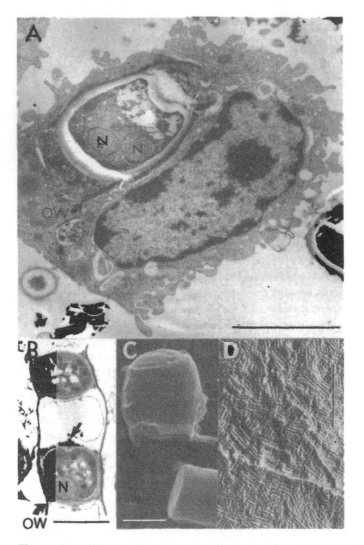

Figure 4   (A) Thin section of arthroconidium engulfed by murine alveolar macrophage. N, nucleus; OW, outer conidial wall. (B) Thin sections of arthroconidia chain separated by degenerate cell. Arrow indicates septal plug; N, nucleus; OW, outer conidial wall. (C) Arthroconidia examined by scanning electron microscopy after stripping of the outer wall by a mechanical shearing process. (D) Replica of wall surface of freeze-fractured arthroconidium showing rodlet fascicles. Bars in A–D represent 3.0 μm, 3.0 μm, 3.0 μm, and 0.5 μm, respectively.

of specific genes. These studies provide a model for examinations of thermal dimorphism in other fungal pathogens.

Early conversion of the saprobic to parasitic phase in *C. immitis* involves "rounding-up" of the cylindrical arthroconidia (Fig. 5A). The outer, hydrophobic cell wall fractures as the inner wall increases in thickness. The initial thickening of the inner conidial wall is probably due to hydration. Individual propagules or chains of arthroconidia may be dispersed into the air and inhaled by the host. Soon after contact with the respiratory mucosa, these cells typically undergo nuclear division as isotropic growth proceeds (Fig. 5B,C).

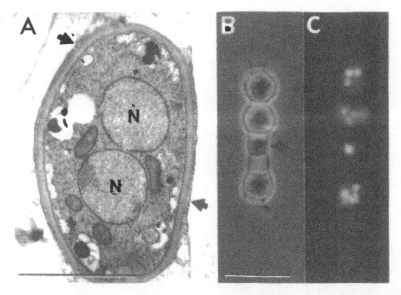

Figure 5 Thin-section (A) and light micrographs (B, C) of spherule initials. The thin-sectioned cell is at an early stage of isotropic growth, and shows fractures of its outer wall (arrows). N, nucleus. The light micrograph of a chain of spherule initials (B) and corresponding ethidium bromide-stained cells in (C) shows the multinucleate nature of these young, parasitic cells. The arrow indicates an undeveloped arthroconidium. Bars in A and B represent 2.0 μm and 10.0 μm, respectively.

3. Spherule Growth and Segmentation

The isotropic growth phase of the spherule initially results in a coenocyte that may be 60 μm or more in diameter. A large central vacuole is commonly observed in sectioned preparations of parasitic cells at this stage of development. It has been suggested that the vacuole may be a source of internal turgor pressure [21], and is most likely an important reservoir for ions and macromolecules [22]. Diametric growth of spherules, like hyphal tip elongation, most likely relies on coordinated processes of turgor-driven expansion, plasticization of the preexisting wall, and biosynthesis and intussusception of new wall polymers [23]. Isotropic growth of spherules is complete or nears completion when the process of segmentation is initiated. We have demonstrated a temporal correlation between peaks of ornithine decarboxylase (ODC) activity and induction of the isotropic growth phase and then the segmentation phase [24]. An increase in ODC activity appears to trigger these two morphogenetic events in the parasitic cycle. Conversion to the round-cell and segmentation stages of development is blocked in the presence of 1,4-diamino-2-butanone, an inhibitor of ODC activity. Reports of the influence of ODC activity on fungal cell differentiation in other dimorphic fungi are well documented [25–27].

Spherule segmentation is a process initiated by centripetal growth of the innermost wall layer of the cell envelope, as revealed by the three-dimensional reconstruction of a parasitic cell in Figure 6A. Morphological details of progressive stages of development of the intracellular segmentation apparatus in intact spherules are not clearly visible by light microscopy. Sectioned cells also do not provide an adequate appreciation of the complex pattern of endogenous wall growth during spherule segmentation. Therefore, a computer visualization method was developed to interpret the steps of wall differentiation that lead to compartmentalization of the spherule cytoplasm [21]. Computer reconstructions of serially sectioned hemispherules, stained with

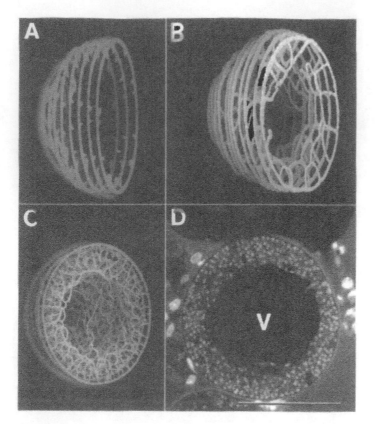

Figure 6 Stages of spherule segmentation wall differentiation. Serial sections stained with WGA/FITC reveal early, synchronous ingrowth of segmentation wall from spherule envelope (A), followed by progressive stages of the compartmentalization of cytoplasm at the perimeter of the central vacuole (B, C). Thin section of segmented spherule (D) was stained with mithramycin to show multinucleate compartments. V, vacuole. Bar in D represents 30.0 μm.

wheat germ agglutinin-FITC, are shown in Figure 6A–C. The lectin binds to N-acetylglucosamine polymers in the cell wall. The images represent three successive stages of development of the segmentation apparatus. In the initial stage (Fig. 6A), much of the central region of the spherule is occupied by a vacuole. The extent of ingrowth from various sites on the inner circumference of the spherule appears to be equal, which suggests synchrony in the initiation of segmentation wall differentiation. A distinct pattern of compartmentalization is evident in Figure 6B. Invaginated regions of newly synthesized wall fuse and give rise to isolated cytoplasmic compartments. The latter are further subdivided by cross-walls, to generate a multitude of small cytoplasmic units within the parental spherule (Fig. 6C). The central vacuole is still prominent at this stage. Figure 6D shows a stage of spherule development, which is comparable to that revealed by the computer reconstruction in Figure 6C. The sectioned spherule in Figure 6D was stained with mithramycin [28] to show the multinucleate nature of the compartmentalized cytoplasm adjacent to the prominent, central vacuole.

4. Endosporulation

The final stages of spherule differentiation are shown in Figure 7A and B. Segmentation is apparently arrested when uninucleate cytoplasmic compartments are formed [29]. The segmenta-

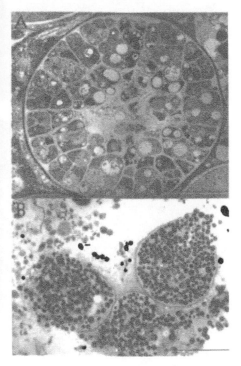

Figure 7 (A) Thin section of spherule at early stage of endospore differentiation, showing partial digestion of segmentation wall, and rupture of central vacuole. (B) Light micrograph of sectioned endosporulating spherules. Note that endospores have undergone isotropic growth. Bars in A and B represent 10.0 μm and 50.0 μm, respectively.

tion wall then begins to undergo autolysis and the central vacuole disappears as the uninucleate cells initiate differentiation into endospores. This event gives rise to 200–300 endospores in the typical spherule (Fig. 7A). The endospores at this stage are typically 2–4 μm in diameter. Prior to endospore release from the maternal spherule, only fragments of the original segmentation wall remain. Breakdown of the segmentation apparatus may be partially due to mechanical fractures of rigid wall polymers as isotropic growth of the endospores is initiated. More likely, however, wall hydrolases secreted by the endospores are responsible for autolysis of the segmentation wall [21,30]. Results of preliminary electron-microscopic studies of endosporulating spherules stained with wheat germ agglutinin-gold label has revealed that chitin largely disappears within the spherule as endospores begin to form [31]. We have shown that at least two chitinases are expressed by *C. immitis* [32]. An understanding of their function(s) awaits molecular analyses of morphogenetic events in *C. immitis*. The endospores, while still contained within the maternal spherule, begin to undergo isotropic growth, which results in rupture of the spherule wall (Fig. 7B). Under appropriate growth conditions, the endospores that are released give rise to a second generation of spherules, completing the parasitic cycle of *C. immitis*.

5. In Vitro Growth Factors

Early studies by Converse [33] and Levine and coworkers [34] established defined media for growth of the parasitic phase of *C. immitis* in axenic culture. In vitro conversion of arthroconidia to spherules provides a simple method for distinguishing *C. immitis* from related, but nonpatho-

Table 1 Composition of Medium, Inoculation,
and Incubation Conditions Used for Growth of the
Parasitic Phase of *C. immitis*

Component	Concentration (g/L)
Ammonium acetate	1.230
KH_2PO_4	0.510
K_2HPO_4	0.520
$MgSO_4 \cdot 7H_2O$	0.400
$ZnSO_4 \cdot 7H_2O$	0.400
NaCl	0.014
$CaCl_2 \cdot 2H_2O$	0.003
$NaHCO_3$	0.012
Tamol[a]	0.500
Glucose	4.000

[a] Anionic detergent (sodium salt of a condensed aryl sulfonic
acid); Rohm and Haas, Philadelphia.
Inoculation and incubation: Arthroconidia (1×10^8) grown
on glucose-yeast extract agar plates (30°C, 30 days) are harvested by the spin-bar method [10], suspended in 1.0 mL of
the above medium, and used to inoculate 100 mL of fresh
liquid growth medium contained in a 250-mL Ehrlenmeyer
flask. The flasks are plugged with a rubber stopper provided
with two membrane-sealed holes for addition of 20% CO_2/
80% air [17]. The inoculated flasks are purged daily with
20% CO_2, and maintained in gyratory incubator (100 rpm)
at 39°C.

genic arthroconidium-producing fungi (see Sec. II.C.6). We have modified the formula of the
original glucose-salts medium and the incubation conditions to optimize production of endosporulating spherules (Table 1). It would be advantageous to establish cultures with synchronously
developing cells, since this would make it easier to study temporal expression of developmentally
related genes [35]. Liquid cultures inoculated with arthroconidia obtained from glucose yeast
extract agar plates which had been incubated at 30°C for 30 days convert to first generation
spherules in a fairly synchronous manner. However, parasitic cells harvested after 96–120 hr
are no longer synchronized, primarily because of variation in the time of rupture and release of
endospores from mature spherules. Alternatively, endospores collected from endosporulating
stock cultures and isolated by differential centrifugation may be used as the inoculum for in
vitro growth studies of the parasitic phase [30,36]. Under these conditions, the cultures remain
in virtually synchronous spherule-endospore development through one generation. Certain pecul-
iarities in the composition of the defined medium and growth conditions for *C. immitis* warrant
comment. The detergent Tamol appears to assist in the suspension of the hydrophobic conidia
and may contribute to spherule maturation and endospore release [37]. CO_2 is essential for
morphogenesis of *C. immitis* [38], although its role in the parasitic cycle is not known. If the
inoculum size of viable arthroconidia is high ($\sim 1 \times 10^8$ pr 100 mL medium), sufficient metabolic CO_2 is produced in a sealed-250 mL flask so that exogenous CO_2 is unnecessary. Ammonium acetate is the only nitrogen source in the defined medium, which suggests that the organism
is capable of efficient turnover of amino substrates in order to maintain protein synthesis during
extended periods of spherule growth and endosporulation.

The pH of the defined medium is ~6.4. Spherule cultures incubated at 40°C and purged with 20% CO_2/80% air daily for 7 days show a fall in pH to ~5.2. The ability of both the saprobic and parasitic phases of *C. immitis* to produce ammonia has long been known [39] but not investigated. Ammonia and ammonium ions appear to be released from parasitic cells by secretion of the contents of cytoplasmic transport vesicles [11]. Using a fluorescent pH indicator (seminapthofluorescein; SNAFL), discrete cytoplasmic vesicles with alkaline contents have been observed. These vesicles appear to be more abundant in endospores and young spherules than in mature, segmented spherules. Bump [39] pointed out that greater amounts of ammonia are produced by *C. immitis* as the pH of the growth medium is reduced. We tested this observation by determination of relative amounts of ammonia/ammonium released into tissue culture media (RPMI 1640) by spherules and endospores isolated separately from the defined glucose-salts medium, and then transferred to RPMI and incubated for 12 hr at 37°C (5% CO_2). At pH 4.0, little growth of either inoculum occurred, but at pH 5.0 the endospores had clearly released more ammonia into the culture medium than the mature spherules. Ammonia production by both endospores and spherules decreased sharply above pH 6.0. An intriguing possibility is that this enhanced production of ammonia by endospores in an acidic environment may be related to their ability to survive in the presence of host phagocytes [14]. Both macrophages and PMNs are inefficient in killing *C. immitis* endospores. *H. capsulatum* yeast survival within the phagolysosome of host macrophages is at least partly due to their ability to modulate the pH within these organelles, which in turn may interfere with the activity of many lysosomal enzymes [40]. Whether alkalinization of the phagolysosome is also a factor associated with in vivo survival of *C. immitis* endospores is unknown.

In vitro differentiation of *C. immitis* appears to be identical to its development in vivo [41,42]. However, the higher growth rate and percentage of spherules that endosporulate in tracheal explant cultures compared to the defined glucose-salts medium suggest that host factors influence *C. immitis* morphogenesis. Very little is known about host-derived growth factors in coccidioidomycosis. Drutz and coworkers [43,44] have demonstrated that 17β-estradiol, progesterone, and testosterone can bind to *C. immitis* parasitic phase cells by a specific-hormone-binding system. The authors suggested that the presence of these sex hormones may stimulate spherule maturation and endosporulation. The nature of the putative signal transduction pathways has not been determined. A reasonable hypothesis is that fungal genes, which encode steroid receptorlike proteins, are evolutionarily related to the mammalian steroid receptor gene super-family. These receptor proteins may represent ancestral molecules that are highly conserved and associated with mating, reproduction, and development in fungi as well as in higher organisms [45,46].

6. Taxonomic Affinity of *C. immitis*

Emmons [47] reported that certain soil ascomycetous fungi that belong to the Gymnoascaceae have common habitat preferences with *C. immitis*, demonstrate similar modes of conidiogenesis, and can survive passage through animals. Sigler and Carmichael [48] reintroduced the genus *Malbranchea* for soil saprobes with arthroconidia, which form by a process identical to that of *C. immitis*. The authors used the name *Malbranchea* state of *C. immitis* for the saprobic phase of the respiratory pathogen, and described *M. dendritica* and *M. gypsea* to accommodate nonpathogenic isolates that had been considered atypical variants of *C. immitis*. Many of the true fungal pathogens of humans are placed in separate families of the order Onygenales, including agents of cutaneous infection in the Arthrodermataceae (*Trichophyton* and *Microsporum*), and human respiratory pathogens in the Onygenaceae (*Coccidioides, Histoplasma*, and *Blastomyces*).

Currah [49] has argued that these two families include natural groups of related ascomycetous fungi characterized by ascospore cell walls that are smooth (Arthrodermataceae), or punctuate-reticulate (Onygenaceae), conidia which form by lytic dehiscence mechanisms, and mycelia that have the ability to degrade keratin. Although strong morphologic evidence points to a close relationship between *C. immitis* and certain members of the Onygenaceae, confirmatory evidence was lacking. Bowman and coworkers [50] showed that the 1713 base pair (bp) nucleotide sequence of the 18S ribosomal gene of *C. immitis* differed from that of *H. capsulatum* and *B. dermatitidis* by only 35 and 33 bp substitutions, respectively. However, these two species produce solitary aleurioconidia rather than alternate arthroconidia as described in *C. immitis* and *Malbranchea* spp. Other investigators of the onygenalean fungi [51] have argued that differences in conidiogenesis should not exclude taxa from this family, and that nonpathogens should be included with related pathogens in the Onygenales.

Emmons [47] described a fungus that produced barrel-shaped arthroconidia which are morphologically identical to those of *C. immitis*. However, the fungus also produced a sexual phase, and the teleomorph was named *Uncinocarpus reesii* [48]. This nonpathogenic, filamentous fungus together with teleomorphs of other morphologically similar *Malbranchea* species is classified in the Onygenales [49]. Molecular evidence suggests that a close phylogenetic connection exists between *C. immitis* and *U. reesii* [52]. Alignments of 18S rDNA sequences of *C. immitis*, *U. reesii*, and seven additional members of the Onygenaceae were compared, including *H. capsulatum*, *B. dermatitidis*, and five arthroconidium-producing soil saprobes accommodated in the genus *Malbranchea*. A summary of the number of base pair differences between *C. immitis* and each of these taxa is presented in Table 2. The 1713-bp sequence of *U. reesii* rDNA differs from that of *C. immitis* by only five substitutions. Wagener parsimony analysis of these nine 18S rDNA sequences was performed together with comparisons of sequences from seven additional pathogenic and nonpathogenic fungi obtained from the Genbank database. A strict consensus cladogram from eight equally parsimonious trees was constructed (Fig. 8). The bracketed taxa are all accommodated in the Onygenaceae. The branch, which unites *C. immitis* with *U. reesii*, was strongly supported at 93% confidence value. Most other branches were less strongly supported. The 18S rDNA sequences of *M. gypsea* revealed the highest number of base substitutions compared to *C. immitis* (Table 2), and this species (Mg in Fig. 8) was separated in the phylogenetic tree from other members of the Onygenaceae. *Cryptococcus neoformans* (Cn), a basidiomycetous fungus, was chosen as an outgroup for construction of the tree. Selection of an alternative outgroup or combination of outgroups (e.g., *Sporothrix schenckii*, *Saccharomyces cerevisiae*,

Table 2 Number of Base-Pair Differences in Aligned 18S rDNA Sequences Between *C. immitis* and Selected Members of the Onygenaceae

Taxon	Number of base-pair differences
C. immitis	—
Malbranchea state of *Uncinocarpus reesii*	5
M. state of *Auxarthron zuffianum*	28
M. dendritica	33
M. filamentosa	30
M. albolutea	23
M. gypsea	44
B. dermatitidis	33
H. capsulatum	35

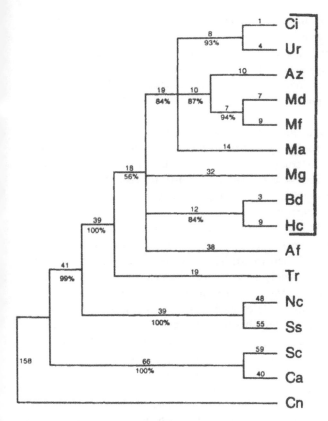

Figure 8 Phylogenetic tree showing relationships between 16 taxa based on comparison of 18S rDNA gene sequences. The tree was rooted with the basidiomycetous fungus *Cryptococcus neoformans* (Cn) as the outgroup. The bracket includes taxa accommodated within the Onygenaceae. (From Ref. 52.)

and/or *Candida albicans*) did not affect the alignment of taxa within the Onygenaceae shown in Figure 8. The maximum parsimony tree, therefore, strongly supports a close relationship between *C. immitis* and *U. reesii*, and argues that these taxa represent a monophyletic pair within the Onygenaceae. Despite this close evolutionary relationship, *U. reesii* is not an animal pathogen [68].

III. MYCOSIS

A. Epidemiology

Because *C. immitis* infections usually arise from inhalation of arthroconidia, most patients contract coccidioidomycosis as a result of exposure to dust in endemic regions either during recreational activities (Fig. 1) or occupational encounters such as construction or farm work. This requirement for exposure to airborne particles probably explains the seasonality of coccidioidomycosis, and, epidemiologists have noticed that the highest incidence of coccidioidal infections is from mid to late-summer and through the fall [53]. Alternatively, people can be infected with *C. immitis* by percutaneous inoculation. This usually happens as a laboratory accident, but is occasionally seen in traumatic injuries that are contaminated with desert soil. Person-to-person transmission of *C. immitis* is not known to occur.

The second obvious requirement for *C. immitis* infection is the availability of a susceptible host. Secondary infections with this pathogen are extraordinarily rare. The only truly protected population is the group of people who have recovered from symptomatic coccidioidomycosis. In the 1940s, 80–90% of the population in the San Joaquin Valley became infected, as determined by positive skin tests [54]. This represents an annual attack rate of ~10% per year. More recent estimates in the same region have indicated a much lower infection rate, ~3% per year. Changes in lifestyle, such as introduction of air conditioning and better methods for avoiding dust, may explain this drop in the rate of infection. A study of a random population in Tucson showed that 30% of the people in the sample population were skin test positive for coccidioidomycosis [55]. Such reports underscore the probability that there are a large number of people within the endemic region who remain susceptible to infection.

As far as we know, there is no difference in the incidence of primary infection with *C. immitis* between racial groups. However, there is very good evidence that some racial groups are more prone to disseminated infection than others [6,56]. The most convincing data are shown in Table 3. These results were derived from a study conducted at the Lemoore U.S. Naval Air Station between 1961 and 1977. As indicated, people of caucasian background have a dissemination rate of ~2.3%, compared to a dissemination rate of 23.5% among African-Americans and 21% in the case of Filipinos. Similar data have been obtained from a study of patients infected during a dust storm in December 1977, which temporarily spread *C. immitis* arthroconidia to northern California. The dissemination rate among caucasians as a result of this incident was 11% versus 54% for African-Americans and 67% for Filipinos [6,57].

Sex also seems to play a role in the risk of *C. immitis* dissemination. Males are significantly more likely to develop disseminated disease than nonpregnant white females [14]. White females are much more likely to develop erythema nodosum than white males, which is a good prognostic sign [56]. Pregnancy substantially increases the risk of dissemination, especially during the third trimester [58,59], and advanced age may slightly increase the risk of disseminated coccidioidomycosis.

Immunosuppression can be a major factor contributing to development of disseminated *C. immitis* infection. People who are immunosuppressed with cytotoxic drugs and/or steroids for

Table 3 Coccidioidomycosis at Lemoore U.S. Naval Air Station, 1961–1977

Ethnic derivation	Number of patients	% of total patients	Disseminated disease	
			Number	%
Caucasian	173	77	4	2.3
Black	17	7.6	4	23.5
Filipino	19	8.4	4	21
Mexican	6	2.7		
Asian	4	1.8		
Guamanian	2	0.9		
Caucasian/Indian	1	0.4		
Filipino/Italian	1	0.4	1	
Hawaiian	1	0.4		
Spanish	1	0.4		
Total	225	100	13	46.8

autoimmune disease or organ transplantation have a much higher risk of developing disseminated coccidioidomycosis than immunocompetent individuals [60]. Patients with AIDS also show a higher frequency of disseminated infections because of their profound suppression of cell-mediated immunity [61]. A survey of coccidioidomycosis cases among university students in an endemic region of Arizona indicated that infected individuals developed symptomatic illness that necessitated medical intervention at a rate of ~0.4% per year [62]. On the other hand, the annual frequency of symptomatic coccidioidomycosis among HIV-infected patients in the Tucson area has been reported to be as high as 27% [61].

B. Clinical Manifestations

1. Natural History

Once arthroconidia are inhaled, they transform into spherules in the terminal bronchi and begin to endosporulate. The fecundity of *C. immitis* contributes significantly to its pathogenicity as one arthroconidium can give rise to 100–300 endospores. Although pneumonia may not be clinically apparent, the organism has the potential to proliferate in the lungs within 1–2 weeks. *C. immitis* has been recovered during this early period from scalene lymph nodes [63] and urine [64] in patients who ultimately present with self-limited disease. We know from a variety of studies that the incubation period from conidial inhalation to development of clinical disease is ~1–3 weeks. The host responds to the infection in ways that determine whether or not the infection will result in subclinical disease, pneumonia, or disseminated disease. The most critical arm of the immune response to *C. immitis* is thought to be T-cell-mediated immunity [65]. Clinically, this is measured as a positive skin test to coccidioidin or spherulin (>10 mm of induration at 24 or 48 hr). The lack of a positive skin test is a poor prognostic sign. The genetic makeup, sex, state of health, and adequacy of the immune response of the patient all play roles in determining the clinical form of the disease caused by *C. immitis* infection.

2. Clinical Forms

Erythema nodosum, erythema multiforme, and a diffuse erythematous rash referred to as toxic erythroderma are three of the skin manifestations of coccidioidomycosis [66]. Erythema nodosum occurs in ~2% of infected men, but 20–25% of infected women, and is a good prognostic sign [56]. Erythema multiforme is more common in children than adults. Toxic erythroderma has been observed in only ~5–10% of the cases of confirmed coccidioidomycosis. The rash is evanescent and commonly persists for only 1–2 days.

Pneumonia is the most common manifestation of symptomatic coccidioidomycosis. The clinical features are not particularly distinctive; fatigue, cough with or without mild hemoptysis, chest pain, and dyspnea [67]. Fever is present in about half the cases, and arthralgia or myalgia and headache are present in approximately one-fifth. Pneumonia can be severe and prolonged, even in self-limited cases. Involvement of the pleura is fairly common, as revealed by pleuritic chest pain and pleural effusions [68]. The chest X-ray picture is variable with patchy or confluent infiltrates. Cavities or nodules are uncommon in primary pneumonia. One radiographic clue of coccidioidal infection is prominent hilar adenopathy, which is more often unilateral than bilateral [69]. In routine laboratory tests an indicator of infection is eosinophilia, which is noted in ~25% of patients with coccidioidomycosis [70]. In 95% of the cases, coccidioidal pneumonia resolves spontaneously. In a few patients, however, lung involvement can be progressive causing respiratory failure, or the pneumonia can develop into a chronic cavitary disease with multiple complica-

Table 4 Complications of *C. immitis* Pneumonia

Respiratory insufficiency
Chronic cavities
Pleural effusions
Bronchopleural fistulae
Residual nodules

tions (Table 4). The latter is most common in diabetics and resembles cavitary tuberculosis, except that the cavities may have thinner walls and little surrounding infiltrate. Chronic cavitary coccidioidomycosis may also be associated with hemoptysis, spilling of the contents of a cavity into a bronchus, which may result in the production of diffuse infiltrates, or bronchopleural fistulae.

IV. HOST-FUNGUS DYNAMICS

When an arthroconidium is inhaled and begins to convert to a round cell, a sequence of innate immune responses is initiated in the naive host. The micro-organism comes into contact first with epithelial cells in the respiratory tract, which may elicit response of PMNs [71] and macrophages. In the majority of individuals who are able to clear the infectious agent, a granuloma is formed which either kills the organism or limits its growth. We will first discuss these aspects of the host response in the context of genetically determined resistance to infection.

Epidemiologic evidence indicates that genetic differences exist in the risk of humans developing disseminated coccidioidomycosis. There are also differences in susceptibility to lethal infection in inbred mouse strains challenged with *C. immitis*. Experiments with mice have been conducted by intraperitoneal (IP) or intranasal (IN) routes of inoculation. In both models, certain mouse strains are resistant to infection, while others are very susceptible [72]. The numbers of organisms found in the spleen and peritoneum of susceptible and resistant mice are comparable for the first 7–10 days postinoculation. By day 14, however, there are many more spherules found in the lungs of genetically susceptible mice than in the resistant strains. Genetically susceptible animals can be protected against infection by immunization with a live, attenuated mutant of *C. immitis*, indicating that they are capable of making an appropriate immune response under optimal circumstance [72]. Resistance can be abrogated by X-irradiation and adoptively transferred with spleen cells [73]. This suggests that a specific population of cells in the spleen must express the resistance gene.

Cox and coworkers have confirmed that resistance and susceptibility to coccidioidomycosis are genetically determined in DBA/2 and BALB/c mice, respectively, after *C. immitis* infection via the IN or IP routes [74]. The authors determined that the ability of genetically susceptible mice to make a delayed-type hypersensitivity (DTH) response to *C. immitis* antigen was equivalent to that of the genetically resistant mice through day 12 postinfection. Between day 12 and day 15, however, the DTH response in the susceptible animals declined precipitously. The decline in DTH was not specific for *C. immitis* since it was also demonstrated with irrelevant, recall antigens in the *C. immitis*–infected animals.

A. Humoral Defenses

C. immitis antigens have been shown to activate complement by both the alternative and classical pathways. However, the organism is resistant to complement-mediated killing. Antibody response to *C. immitis* infection has always been thought to be of very little benefit since high antibody titers typically correlate with a poor clinical outcome [75]. Antibody and complement undoubtedly provide opsonic activity, which enhances ingestion by PMNs and macrophages. Complement activation leads to release of molecular fragments that are chemotactic for PMNs [76] and opsonize arthroconidia and endospores for phagocytosis by PMNs. We cannot exclude the possibility that a subset of antibodies to certain *C. immitis* antigens is protective, but there is no evidence to support this hypothesis.

B. Cellular Defenses

Polymorphonuclear neutrophils are only able to partially inhibit growth of the pathogen, and are unable to kill the organism [14]. PMNs are slightly more effective in inhibiting growth of arthroconidia than mature spherules [14,77]. Since mature spherules are typically 30–80 μm in diameter, a single PMN is unable to phagocytose the fungal cell. Endospores, on the other hand, are more sensitive to growth inhibition by these host cells. Most investigators of cellular immune response to *C. immitis* believe that macrophages are the pivotal effector cells in coccidioidomycosis. This concept has arisen from the general paradigm for granulomatous diseases: activated T-lymphocytes secrete cytokines, which activate macrophages, inducing the formation of a granuloma, which kills or contains the organism.

As previously indicated in this chapter, the evidence that T-cells are pivotal in host defense against coccidioidomycosis is very persuasive [78,79]. However, demonstration of killing of *C. immitis* by macrophages in vitro has been problematic. Some workers have found that arthroconidia ingested by macrophages in the presence of immune lymphocytes are killed [80]. In addition, these same workers have reported that pretreatment of macrophages with interferon-γ and/or tumor necrosis factor (TNF)-α enhances the ability of the host cells to kill arthroconidia and endospores in vitro [81]. Others have reported very modest killing of arthroconidia (20–30%) by peripheral blood mononuclear cells, and that this process could not be modulated by interferon-γ and/or TNF-α [82]. As the spherule develops and matures it becomes more resistant to macrophages, so that <10% of mature spherules are killed [82]. Taken together, these experiments suggest that there is very little killing of the organism in vitro. These inconsistencies in the literature may in part be due to the use of different methods and sources of mononuclear phagocytes, as well as different strains of *C. immitis*.

Some investigators who have focused their research efforts on immunological aspects of coccidioidomycosis have been fascinated by the idea that specific immunologic suppression elicited by *C. immitis* antigens may prevent the host from mounting an effective T-cell response. One of the first reports of such immunosuppression involved an endosporulation antigen that inhibited lymphocyte proliferation [83]. The inhibitory material in this mixture has not been identified. We have shown that the isolated outer arthroconidial wall (Fig. 4B) is capable of killing T cells in vitro and can inhibit production of superoxide anion by alveolar macrophages [13]. As pointed out earlier, the conidial wall contains hydrophobins that are present as crystalline complexes, or rodlets at the interface of the outer and inner wall layers (Fig. 4D). The effect of exposure of host cells to hydrophobins released from arthroconidia of *C. immitis* is unknown.

Suspension in distilled water of intact *C. immitis* arthroconidia that contain low concentrations of nonionic detergents or commercial surfactant (L-α-phosphatidylcholine dipalmitoyl) has been shown to result in partial breakdown of the hydrophobic wall barrier, hydration of the inner wall, and release of wall-associated proteins [84]. Hydration of the inner conidial wall of *C. immitis* appears to occur naturally in vivo as the outer wall fractures during cell germination [Fig. 5A], resulting in release of wall-associated products collectively referred to as the soluble conidial wall fraction (SCWF) [85]. Subfractions of the SCWF include several proteases [86], some of which have been shown to be capable of cleaving secretary IgA and thereby may compromise localized mucosal immune response to arthroconidia and derived spherule initials. Two of these proteases have been isolated and characterized as 19-kDa and 34-kDa components of the SCWF [86]. These and other macromolecules of the SCWF may negatively influence host defense mechanisms and contribute to survival of *C. immitis* in vivo.

The Th1/Th2 paradigm has also intrigued investigators in coccidioidomycosis research. It is suggested that T-helper cells can be divided into at least two classes: those that secrete interleukin (IL-2) and interferon-γ (Th1), and those that secrete IL-4, IL-5, IL-6, and IL-10 (Th2). The Th1 cells secrete cytokines that activate macrophages, while the Th2 cytokines stimulate B-cells to produce antibodies and do not participate in the DTH response. There is evidence that a correlation exists between resistance/susceptibility to *C. immitis* and the profile of cytokine response that is elicited in mice [87–91]. Relevant to this point is a study conducted by Magee and Cox [91] on the administration of recombinant IL-12 to mice prior to challenge with *C. immitis*. IL-12 promotes production of IFN-γ and thus influences the balance of Th2 and Th1 responses. Administration of IL-12 to susceptible mice either before or after intraperitoneal challenge with *C. immitis* significantly reduced the fungal load in these animals compared to controls. Analysis of cytokine gene expression in the lungs of animals immunized with IL-12 indicated a shift from a Th2 to a Th1 pathway of immune response. The authors also showed that neutralization of IL-12 in infected, resistant animals using anti-IL-12 antibody resulted in a significant increase in the fungal burden. Recognition of the pivotal nature of the Th1/Th2 balance in host defense against coccidioidal infection has helped to explain the elevated risk of pregnant females in their third trimester to contract disseminated coccidioidomycosis. Th2 cytokines, which are secreted at the maternal-fetal interface, may suppress cell-mediated immunity and thereby provide a privileged environment for development of a coccidioidal infection.

Susceptible strains of mice (BALB/c, C57BL/6 [B6], and CAST/Ei) infected with *C. immitis* have been shown to make more IL-10 and IL-4 mRNA than resistant strains [92]. It is known that IL-10 inhibits stimulation of the Th1 pathway, blocks cytokine release from macrophages, and plays a regulatory role in the production of IL-12 [93]. A particularly interesting observation is that *C. immitis*–infected, IL-10-deficient mice were as resistant to coccidioidomycosis as DBA/2 mice, which suggests that upregulation of IL-10 expression may contribute to susceptibility to coccidioidal infection in certain strains of inbred mice [92]. Fierer and coworkers [92] have suggested that in both human and murine coccidioidomycosis, too much IL-10 can direct the immune response along the Th2 pathway. However, little is known about patterns of cytokines produced by humans during coccidioidal infections. It is reasonable to speculate, nevertheless, that immunodominant antigens of *C. immitis* which stimulate a profound increase in IL-10 and IL-4 production may compromise the effectiveness of the immune response to clear the pathogen. Mice that are genetically resistant to *Leishmania major* preferentially mount a Th1 response to the organism; those mouse strains that are genetically susceptible to the parasite predominately mount a Th2 response [94]. The idea that similar events occur in coccidioidomycosis is reasonable since antibody responses and DTH seem to be inversely correlated with activation of cell-mediated immunity. However, the myriad factors which influence the ratio, mobilization, and cytokine expression of T-cells in genetically susceptible versus resistant mice, or in patients with limited infection versus disseminated disease, is still not totally appreciated.

V. VACCINE RESEARCH

A. Methods Used to Identify T-Cell-Reactive Antigens

For many years, researchers have attempted to develop a human vaccine against coccidioidomycosis without success. However, several *C. immitis* vaccines have proved to be effective in animal systems. A killed spherule vaccine was shown to protect mice and a variety of other animals from experimental infection with *C. immitis* [95]. Between 1980 and 1985 a double-blind human study was conducted using a vaccine consisting of formalin-killed spherules versus a placebo [96]. The study involved almost 3000 people, but only a minority of the vaccinated volunteers became skin test positive to *C. immitis*. There was no difference in the number of cases of coccidioidomycosis or the severity of the disease in the formalin-killed spherules-vaccinated group compared to the placebo group. One explanation for the ineffectiveness of this vaccine may be that relatively small numbers of killed organisms could be injected into humans without unacceptable local side effects of pain and swelling. Nevertheless, the vaccine trial made it clear that whole killed spherules do not provide immunoprotection against coccidioidomycosis in humans. More recently, the formalin-killed spherules vaccine product was fractionated to yield an immunoreactive, soluble, subcellular fraction which was isolated as a 27,000 xg pellet [97]. This multicomponent preparation, when accompanied by adjuvant, was shown to be immunoprotective in mice against lethal intranasal and intravenous challenge with *C. immitis*. With the exception of this antigenic preparation, and an alkali-soluble, water-soluble extract of the spherule wall described by Lecara and coworkers [98], none of the crude, soluble T-cell-reactive antigens of *C. immitis* which have been generated over the years have been shown to be immunoprotective in experimental animal models. It is entirely reasonable to expect, however, that a mixture of *C. immitis* antigens may be necessary to confer protective immunity rather than a single antigen.

Since the cell wall of *C. immitis* contains a large amount of material that is nonantigenic for T-lymphocytes, the whole organism is not the ideal vaccine candidate. One would like to immunize patients with selected, immunogenic antigens that have been purified, biochemically characterized, and shown to protect experimental animals from infection challenge. It has been difficult to identify these antigens, and there is no consensus at the moment as to which antigens should be selected. Several different approaches have been used to try to obtain antigenic fractions. The first was to fractionate lysates of the mycelial (coccidioidin) and spherule (spherulin) phases [99]. Other investigators have used alkali treatment to extract antigens from the arthroconidial and spherule walls [98]. Another approach has been to use mechanical disruption to obtain *C. immitis* antigens [85]. The advantages of this last method are that it avoids chemical extraction and autolysis, which may alter the native structure of antigens, and it yields reproducible preparations of intact proteins. Cole and coworkers [85] found that by removing the outer conidial wall from arthroconidia, the organism released a variety of proteins in the SCWF described above. In addition to proteolytic antigens, which could have a negative impact on the host response, this mixture of proteins has been shown to be extraordinarily effective in stimulating immune murine T-cell proliferation. In addition, the spherule phase of the organism spontaneously releases a membranous material consisting primarily of glycosylated proteins and lipids known as the spherule outer wall (SOW) [100]. This wall fraction has been shown to be a potent antigen in T-cell-mediated immune responses in both mice and humans [101].

Another approach to the problem of generating pure *C. immitis* antigens is to use techniques of molecular biology. The major advantage of the molecular approach is that once antigens are cloned and a suitable prokaryotic or eukaryotic vector expresses proteins, an endless supply of completely defined antigen is available. Therefore, one would not have to repeatedly grow *C. immitis*, extract the antigen, and purify it from a complex mixture. In addition, the molecular approach has the advantage of being able to deliver antigens as part of a viable vaccine system, such as via an attenuated bacterial or viral vector, should that be required to effectively immunize people against coccidioidomycosis. We believe that the systematic identification and evaluation of *C. immitis* T-cell-reactive antigens in experimental animals is a rational approach to the ultimate development of a vaccine.

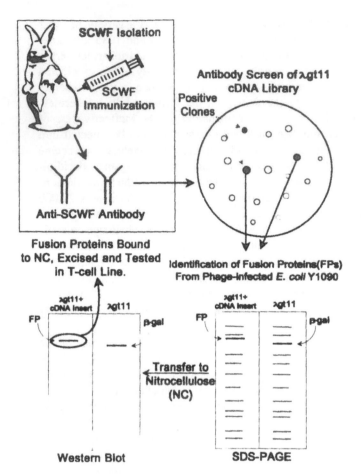

Figure 9 Diagrammatic summary of T-cell immunoblot methodology used to identify, isolate, and test immunoreactivity of *C. immitis* fusion proteins.

We initially used a modified T-cell immunoblot procedure, which was developed to identify mycobacterial antigens [102], for isolation of the first *C. immitis* gene that encodes a T-cell-reactive antigen [103]. A diagrammatic summary of this procedure is presented in Figure 9. The antigen of interest was expressed as a fusion protein with β-galactosidase. The T-cell-reactive fusion protein (TCRP) consisted of a 66-amino acid recombinant peptide encoded by the *C. immitis* cDNA. The TCRP was shown to be a potent stimulant of immune T-cells derived from mice which were immunized with an attenuated strain of *C. immitis*. Antibody raised against this fusion protein was used to immunolabel the antigen in *C. immitis* cells. The TCRP was localized primarily in the wall of arthroconidia and cytoplasm of parasitic cells. Development of molecular strategies for identification of T-cell antigens has proved to be a productive approach toward the generation of a human vaccine against *C. immitis* infection.

B. Candidate Antigens for an Experimental Vaccine

Four *C. immitis* antigens selected on the basis of their ability to stimulate immune T-cell proliferation have been cloned, expressed as recombinant proteins, and evaluated as vaccine candidates

Table 5 Features of *C. immitis* Antigens Which Have Been Identified as Candidates for a Vaccine Against Disseminated Coccidioidomycosis

Antigen	Source	Localization	MW (kDa) Native Ag	MW (kDa) Recombinant Ag	Immunoresponse	Efficacy of immunoprotection	References
T-cell-reactive protein (TCRP)	Saprobic and parasitic phases	Cytoplasmic, but also detected in filtrate of parasitic phase	50.4	48	rTCRP recognized by patient Ab; stimulates murine immune T-cell (CD4+) proliferation; stimulates IFN-γ production in vitro	Approx. 10-fold fewer organisms in lungs than control mice after IP challenge with 50 viable arthroconidia (strain C735)	104,105
Spherule outer wall glycoprotein (SOWgp)	Parasitic phase specific	Cell wall	58–82	29.1–39.5	rSOWgp elicits both B cell and PBMC response in skin test-positive volunteers; stimulates IFN-γ production in vitro	Clears organism from lungs and spleen of 60% and 20% of immunized mice, respectively after IP, challenge with 50 viable arthroconidia (strain C735)	100,101, 106,107
Urease (URE)	Saprobic and parasitic phase	Cytoplasmic, but also detected at cell surface endospores	101	63	rURE stimulates murine immune T-cell (CD4+) proliferation; stimulates IFN-γ production in vitro	Reduces CFUs of organism in lungs and spleen by 3–4 $\log_{10}$ units; 40% of immunized mice survive IP, challenge with 50 viable arthroconidia (strain C735) for > 40 days postinoculation	108
Protein-rich antigen 2 (PRA/Ag2)	Saprobic and parasitic phase	Cell wall	33	19.4	rPRA/Ag2is recognized by patient Ab and stimulates murine immune T-cell proliferation	rPRA/Ag2 expressed by GST vector showed little to no protection of mice challenged with 250 arthroconidia (Silveira strain) by IP route; rPRA/Ag2 expressed by pET vector showed protection (reduction of > 2 $\log_{10}$ units of CFUs in lungs and spleen of immunized mice challenged IP [50 arthroconida] with the R.S. strain)	120,121, 124,128

in murine models of coccidioidomycosis. A summary of the biochemical and immunogenic features of these antigens is presented in Table 5.

1. Recombinant TCRP Antigen

One of these vaccine candidates is the recombinant T-cell-reactive protein (rTCRP) described above. The original cDNA, which encoded the fusion peptide, was used to screen a *C. immitis* genomic library, and the full-length *TCRp* gene was sequenced [104]. The translated hydrophobic protein revealed 50% identity and 70% homology to a mammalian cytoplasmic enzyme, 4-hydroxyphenylpyruvate dioxygenase, involved in the degradation of phenylalanine to tyrosine. The 48-kDa recombinant TCRP was expressed by *E. coli* transformed with pET28b-*TCRp* and shown to be a potent stimulant of immune T cell proliferation in vitro [105]. BALB/c mice immunized with formalin-fixed spherules also showed a good cellular response to the purified rTCRP in the in vitro T-cell proliferation assays. This suggested that the native antigen is presented to the host upon exposure to the parasitic phase of *C. immitis*. Phenotypic analysis of the responsive cells indicated that they were primarily CD4 T cells. The rTCRP stimulated production of IFN-γ by the responsive T-cells at a level comparable to that of immune T-cells exposed to the crude SCWF antigen. On the basis of these collective data, we concluded that the TCRP is an immunogenic component of *C. immitis* spherules which elicits a Th1 response in the immunized host.

The results of our evaluation of the immunoprotective efficacy of rTCRP were disappointing [105]. The data derived from T-cell proliferation assays, therefore, are not necessarily good predictors of which parasitic cell proteins are most likely to elicit a protective response in the murine model. The immunoprotective capacity of the rTCRP was determined by use of the same immunization schedule that elicited a strong murine T-cell proliferation response. The adjuvants used in combination with the recombinant protein may have significantly influenced the outcome of this experiment. The first two immunizations were conducted subcutaneously with the rTCRP (4 μg) plus 100 μL AdjuPrime (Pierce, Rockford, IL) on days 1 and 14. The final immunization was performed 14 days later in the footpads and base of the tail with 7.5 μg of rTCRP in 100 μL complete Freund's adjuvant (CFA; Sigma). Neither AdjuPrime nor CFA is an ideal adjuvant for stimulation of the Th1 pathway of immune response. An inoculum of *C. immitis*, known to cause lethal disease in BALB/c mice, was used [72]. The inoculum was delivered by the IP route, which may not be as rigorous a test of potential vaccine candidates as intranasal challenge. However, we used the IP route of infection because we have found that it is more reproducible than intranasal challenge. Furthermore, mice do not begin to die from IP challenge with arthroconidia for at least 10–14 days postinoculation, which should permit the specific T-cell immune response to control the spread of the pathogen. In three separate experiments, the immunized mice had ~10-fold fewer organisms in their lungs than did the control animals. Immunization with formalin-killed spherules [96], on the other hand, led to sterile lungs.

2. Recombinant SOWgp Antigen

The membranous outer wall layer released from the surface of in vitro–grown spherules (SOW) [106] induces a potent proliferative response of immune murine T-cells [100,101]. The crude SOW was used to immunize BALB/c mice subcutaneously to evaluate its ability to immunoprotect against a lethal challenge of *C. immitis* by the IP route. The immunized mice showed a dramatic reduction of 5 $\log_{10}$ colony-forming units (CFUs) in lung homogenates at 2 weeks postchallenge compared to infected control mice (Fig. 10A). To identify the antigenic component(s) of SOW, the crude wall material was first subjected to Triton X-114 extraction, and a water-soluble fraction derived from this treatment was examined for protein composition and reactivity in humoral and cellular immunoassays [101]. Protein electrophoresis revealed that the aqueous fraction of three different isolates of *C. immitis* each contained one or two major glycoproteins (SOWgp), distinguished by their molecular sizes which ranged from 58 to 82 kDa. The SOWgps, however, showed identical N-terminal amino acid sequences, and each was recognized by sera from patients with *C. immitis* infection. Antibody raised against the 58-kDa glycoprotein (SOWgp58) of the Silveira isolate was used for Western blot and immunolocaliza-

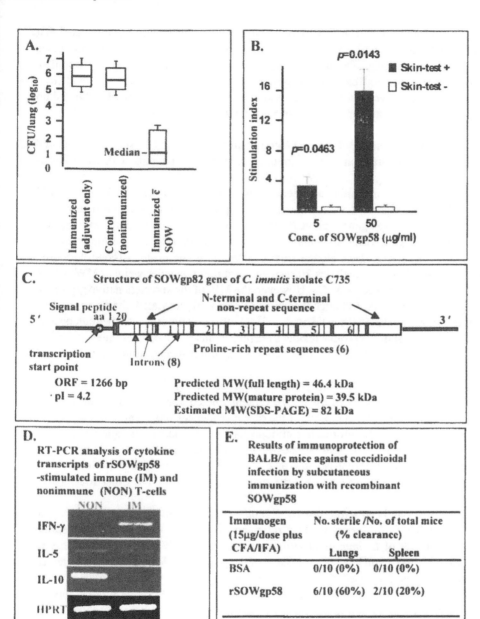

Figure 10 Evaluation of spherule outer wall (SOW) and spherule outer wall glycoprotein (SOWgp) as candidate vaccine antigens. (A) CFUs of *C. immitis* in lungs of mice immunized with SOW compared to control mice. (B) In vitro proliferation of PBMCs of skin test positive and negative human volunteers in response to native SOWgp isolated from the Silveira strain of *C. immitis*. (C) Structure of the SOWgp gene isolated from the C735 strain. (D) Expression of selected cytokines (IFN-γ, IL-5, IL-10) by splenocytes of mice immunized with recombinant SOWgp protein from Silveira strain, compared to nonimmunized mice. Expression of the hypoxanthine phosphoribosyltransferase (HPRT) gene was used to control for amount of cDNA template used in the RT-PCR. (E) Evaluation of immunoprotective efficacy of native SOWgp isolated from Silveira strain.

tion analyses. The antiserum recognized SOWgps of other *C. immitis* strains, demonstrated that expression of the SOWgp is parasitic phase specific, and localized the antigen in the membranous SOW at the surface of spherules. The purified SOWgp was tested for its ability to stimulate proliferation of human peripheral blood monocytic cells (PBMC). These were obtained from healthy volunteers with positive skin test reaction to spherulin, and volunteers who showed no skin test reaction to the same antigen. The native SOWgp was shown to stimulate proliferation of PBMC from skin test–positive but not skin test–negative donors (Fig. 10B), and the activated cells secreted IFN-γ. It was concluded that SOWgp is a major parasitic cell surface–expressed antigen that elicits both humoral and cellular immune responses in patients with coccidioidal infections.

The N-terminal and internal amino acid sequences of the purified SOWgp from the C735 strain of *C. immitis* were used to design oligonucleotide primers for PCR amplification of a genomic fragment of the *SOWgp* gene. The PCR product was then used as a hybridization probe to isolate the full-length gene from a *C. immitis* genomic library [107]. A summary of the structural features of the *SOWgp* gene cloned from *C. immitis* isolate C735 is shown in Figure 10C. The predicted signal peptide includes 20 residues, and the molecular weight (MW) of the mature protein was estimated to be 39.5 kDa. However, the MW of the secreted, native glycoprotein in SDS-PAGE gels is 82 kDa. Although the native protein is glycosylated, the sugar residues do not account for the difference in the estimated molecular mass. This is due to the high proline content of the SOWgp. Most of the proline residues are present in the six repeat sequences shown in Figure 10C. The significance of these proline-rich repeats with respect to immunoreactivity of SOWgp is still unknown.

The cDNA sequences of the *SOWgp* genes of three isolates of *C. immitis* were obtained and aligned. The differences in MW of the native proteins, noted above, is due to differences in the number of proline-rich, tandem repeats. If the repeat sequences represent immunoreactive epitopes (i.e., T-cell- rather than B-cell-reactive epitopes), then the sequence variation of SOWgp may influence host response to different isolates of the pathogen. Preliminary studies of the *SOWgp* gene structure in 60 different isolates of *C. immitis* has revealed that the number of tandem repeats ranges from three to eight, but the majority of isolates have either four or five repeats.

The *SOWgp* gene has been expressed by *E. coli* and antiserum was raised against the recombinant protein (rSOWgp). Northern hybridization and Western blot analyses of expression of the gene during growth of *C. immitis* in vitro confirmed that *SOWgp* is expressed only in the parasitic phase. The maximum level of expression of the gene-specific transcript was detected in presegmented spherules, decreased during spherule maturation, and was then elevated again in second-generation, presegmented spherules. The production of the SOWgp antigen appears to be cyclic, is highly regulated during the parasitic phase, and is most likely presented to the host immune system throughout the course of the disease. Based on these observations and the confirmed immunoreactivity of SOWgp as demonstrated by human PBMC proliferation assays, we initiated evaluations of the recombinant protein as a vaccine candidate using our murine model of coccidioidomycosis.

BALB/c mice were immunized by the subcutaneous route with 15 μg rSOWgp of the Silveira strain plus incomplete Freund's adjuvant. Two and four weeks later, the immunization protocol was repeated first with IFA and then CFA. Control mice were immunized with adjuvant alone. The two groups of animals were challenged with 50 arthroconidia of *C. immitis* (Silveira strain) by the IP route at 42 days after initiation of the immunization protocol, and then sacrificed and evaluated for clearance of the organism 2 weeks postinoculation. Two additional groups of mice were immunized with rSOWgp or adjuvant alone as above and sacrificed at 42 days without challenge with *C. immitis*. These animals were used to compare the cytokine profiles of spleen cells of nonimmune to rSOWgp-immunized mice. The results of these preliminary studies are presented in Figure 10D and E. Analyses of transcript levels of selected cytokines expressed by splenocytes indicate a significant increase in IFN-γ but not IL-5 or IL-10 expression in mice immunized with rSOWgp compared to mice immunized with adjuvant alone (Fig. 10D). These results suggest that subcutaneous immunization of mice with rSOWgp stimulated a Th1 pathway of immune response. Evaluation of the immunoprotective efficacy of rSOWgp indicated that the recombinant protein provides a modest level of protection against lethal challenge (Fig. 10E). A significant number of the SOWgp-immunized mice showed total clearance of the pathogen from their lungs (60%) and spleen (20%), while all control mice were infected with large numbers of CFUs of *C. immitis* in both organs. The recombinant SOWgp antigen appears to be another potential candidate for a coccidioidomycosis vaccine cocktail.

3. Recombinant Urease

The full-length urease (*URE*) gene of *C. immitis* has been cloned and a 1.7-kb fragment of the gene has been expressed by *E. coli* [108]. Antibody raised against the 63-kDa recombinant protein (rURE) has been used to show that the level of production of the native urease is highest during the endosporulation phase of the parasitic cycle (Fig. 11A). The peak of expression of the *URE* gene apparently correlates with the release of ammonia from endosporulating spherules, which was discussed earlier in this chapter (see Sec. II.C.6). *Helicobacter pylori*, which also produces a urease, is the causative agent of gastritis and peptic and duodenal ulcers, and is associated through its chronic infection of mucosal tissue with gastric cancer [109,110]. The urease of *H. pylori* generates a layer of ammonia at the bacterial cell surface which neutralizes its immediate environment and permits the bacteria to survive the acid pH of the stomach long enough to reach and invade the less acidic mucin layer. As expected, the ammonia produced by *H. pylori* urease activity is cytotoxic to gastric epithelial cells [111] and is therefore at least partly responsible for necrosis of host tissue at the site of colonization. A feature of the bacterial urease which influences the host response to *H. pylori* is that the cytosolic enzyme is apparently cotransported across the cell membrane with a heat-shock protein [112], at least in late logarithmic phase, and is thereby presented to host immune cells. The *H. pylori* urease has been shown to be a potent stimulus of mononuclear phagocyte activation and inflammatory cytokine production [113]. The local production of cytokines by urease-stimulated mononuclear phagocytes may play a central role in the development of *H. pylori* gastroduodenal inflammation.

The amino acid sequence of the *C. immitis* urease shows 52% identity to that of the *H. pylori* urease [108]. T-cells from BALB/c mice immunized with the 63-kDa rURE showed a strong in vitro proliferative response to the homologous antigen. The mRNA isolated from rURE-stimulated T-cells showed a significant rise in transcript levels of Th1-type cytokines, including IFN-γ and Il-2. A proliferative response was also induced in vitro by rURE in the presence of T-cells isolated from mice that had been immunized with formalin-killed, endosporulating spherules (FKES). These results confirm that the native urease is presented to the host by the parasitic cells, perhaps at the endospore cell surface. On the basis of these in vitro data, we elected to test the immunoprotective capacity of rURE in the mouse model challenged with *C. immitis* by the IP route.

Earlier immunization protocols used for evaluating rTCRP and rSOWgp as vaccine candidates employed CFA and IFA as adjuvants. As previously mentioned, these are not the most appropriate reagents for priming the Th1 pathway of host immune response. Recent studies have shown that the innate immune system is able to distinguish certain prokaryotic DNAs from vertebrate DNAs by detection of unmethylated bacterial CpG dinucleotides in particular base

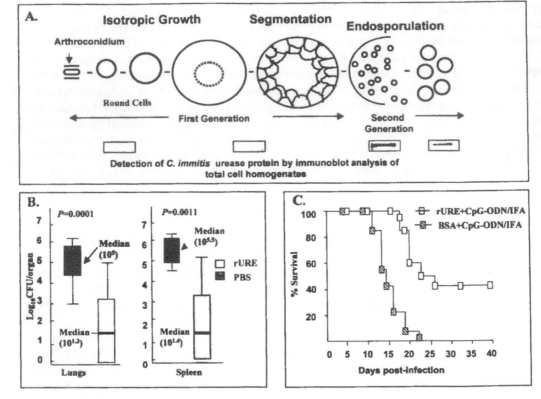

Figure 11 Evaluation of *C. immitis* urease as candidate vaccine antigen. (A) Detection of urease protein in total cell homogenates of different stages of parasitic cycle by immunoblot analysis using antiserum raised to purified recombinant protein (rURE). (B, C) Immunoprotective efficacy of rURE plus adjuvant (CpG-ODN/IFA) based on residual fungal burden in lungs and spleen (B) and survival of mice (C) after IP challenge (50 viable arthroconidia; 735 strain).

contexts (i.e., CpG motifs) [114]. CpG DNA is thought to directly activate dendritic cells and macrophages, which function as antigen-presenting cells (APCs), to produce cytokines that create a Th1-like milieu in lymphoid tissue [115]. Administration of synthesized CpG-containing oligonucleotides (CPG-ODNs) to mice has been shown to induce cytokine secretion by the activated APCs, including IL-6, TNF-α, IL-1, IL-12, and IFN-γ [116]. An important observation relevant to the use of CpG-ODN as an adjuvant is that coadministration of CpG-ODNs with soluble protein antigens in IFA promotes Th1 responses [117]. CpG-ODNs have been used effectively as adjuvants to enhance immunoprotection against several infectious diseases. For example, infection of BALB/c mice with the parasite *Leishmania major* in the presence of CpG-ODNs shifted the T-cell response toward Th1 cytokines and resulted in protection against a normally lethal infection [118].

　　　BALB/c mice were immunized either with rURE plus CpG-ODN/IFA, or CpG-ODN/IFA alone, and then challenged by the IP route with a lethal inoculum of *C. immitis* arthroconidia. The results of analysis of the residual fungal burden and survival of these two groups of mice are shown in Figure 11B and C, respectively. The CFUs in the lungs and spleen of immunized mice were reduced by >3 Log_{10} units compared to controls, and 50% of the immunized animals had cleared the organism from these two organs at the time of sacrifice. In addition, 44% of immunized animals survived for >40 days postchallenge in contrast to none of the control mice, which died between 12 and 22 days after inoculation. These data provide support for further investigations of the rURE as a vaccine reagent by itself or in combination with other vaccine candidates.

4. Recombinant Proline-Rich Antigen 2

Earlier studies showed that immunoreactive components of the spherule cell wall could be extracted by alkali treatment [119] or toluene exposure, followed by deglycosylation with hydrogen fluoride (HF) [120]. In both cases, the crude extracts or subfractions of these products were shown to be immunoprotective in mice against a lethal challenge of *C. immitis* [98,121]. Zhu and coworkers [122,123] cloned a spherule protein that was present in the alkali-soluble, water-soluble preparation of the parasitic cell wall, and referred to it as antigen 2 (Ag2). This nomenclature was based on a reference system, which relies on two-dimensional immunoelectrophoretic separation of coccidioidin and spherulin antigens [99]. The *Ag2* gene was reported to encode a protein with a predicted molecular mass of 19.4 kDa. The deduced amino acid sequence was reported to contain 10 proline repeats (TXXP), a predicted N-terminal signal sequence and C-terminal glycosyl phosphatidylinositol (GPI) anchor, eight cysteine residues which could be associated with disulfide linkages, and 24 potential sites of *O*-glycosylation. At about the same time as the *Ag2* gene was isolated, Galgiani's laboratory cloned an identical gene encoding an immunoreactive protein which was shown to stimulate human T-cells [124]. The MW of this T-cell-reactive polypeptide, originally isolated by HF deglycosylation of components of a lysate of *C. immitis* spherules, is 33 kDa [120]. Specific antibodies reactive with the 33-kDa proline-rich antigen (PRA) have been detected both in sera [125] and CSF of patients with coccidioidal infections [126]. The heavily glycosylated PRA/Ag2, therefore, presumably contains mainly *O*-linked carbohydrate, is spherule wall bound [127], and elicits both T-cell and antibody responses in patients with coccidioidomycosis. Recombinant PRA/Ag2, (rPRA/Ag2) was expressed by *E. coli* and the protein was tested for antigenic activity. Serum IgG antibodies in 37 of 42 patients recognized the purified recombinant protein with culture-proven, progressive pulmonary or extrapulmonary coccidioidal disease. None of the patients with self-limited coccidioidomycosis had detectable antibodies in serum samples collected up to 141 days after illness began. It appeared, therefore, that the PRA/Ag2 might have prognostic value in clinical studies to identify patients with disseminated *C. immitis* infections.

T cells from rPRA/Ag2-immunized BALB/c mice showed a modest in vitro proliferative response to the recombinant protein [121]. Furthermore, immunization with rPRA/Ag2 confirmed protection against IP challenge. Figure 12A shows that the numbers of organisms in the lung and spleen were reduced by 2–2.5 Log_{10} units in mice immunized with rPRA/Ag2 + IFA/CFA compared to mice immunized with adjuvant alone [121]. These data suggest that the rPRA/Ag2 does elicit an immunoprotective response in mice, and underscore the earlier caution that the magnitude of the T-cell-proliferative response to the antigen in vitro may not be a good predictor of its efficacy as a vaccine candidate.

Studies of the immunoprotective efficacy of rPRA/Ag2 in Cox's laboratory using an alternative bacterial expression vector revealed a modest but significant decrease of CFUs in tissues of immunized BALB/c mice after IP challenge with 250 arthroconidia [128]. However, the rPRA/Ag2-immunized mice did not show increased survival compared to control mice during a 30-day period postinfection. In contrast, however, the authors showed that immunization of

A.

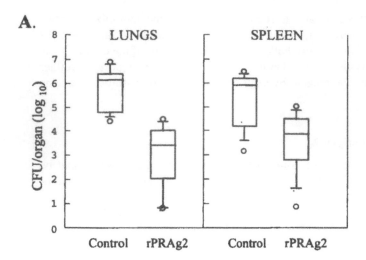

B.

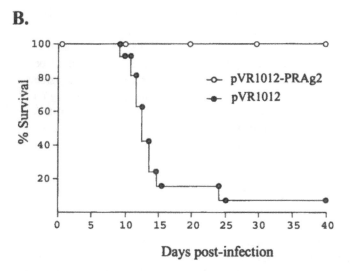

Days post-infection

Figure 12 Evaluation of PRA/Ag2 as candidate vaccine antigen. (A) Box plot of CFUs found in lungs and spleen of immune or control mice 14 days after IP challenge (50 arthroconidia; R.S. strain) (121). (B) Survival of BALB/c mice immunized by the intramuscular route with pVR1012-PRA/Ag2 mammalian plasmid vector construct and challenged IP with 2500 arthroconidia (Silveira strain). (From Ref. 128.)

mice with *PRA/Ag2* DNA ligated into the pVR1012 mammalian plasmid vector elicited 100% protection against IP challenge with 2500 arthroconidia, and against IN challenge with 50 arthroconidia (Fig. 12B). Mice immunized with the plasmid construct also mounted a strong DTH response to the parent alkali-soluble, water-soluble fraction, and splenocytes isolated from these same animals secreted IFN-γ in vitro upon exposure to the crude antigenic material. The use of DNA for immunizations against infectious diseases introduces new strategies for modulation of Th1 and Th2 pathways of immune response. The exciting results demonstrated by this first study of a DNA vaccine against *C. immitis* infection provide hope that immunoprotection of humans against coccidioidomycosis might be an achievable goal within the next decade.

VI. SUMMARY AND FUTURE RESEARCH

In spite of the fact that *C. immitis* has been recognized as a human pathogen for more than a century, relatively little is known about the biology of the fungus or details of its interaction with host innate and acquired immune defenses. There is renewed national interest in coccidioidomycosis, however, because of the rise in density of the population in endemic regions, and the increasing numbers of immunosuppressed patients who are highly susceptible to this disease [129]. *C. immitis* is a diphasic microbe, which lives in the soil as a hyphal form, and resides in the host as coenocytes. The parasitic cells each give rise to 200–300 endospores by an unusual process of segmentation of their multinucleate cytoplasm. The cell biology of this fungus is largely unexplored and offers challenging research problems related to its intriguing mechanisms of regulation of such morphogenetic events as the formation of alternate conidia and autolytic cells from fertile hyphae, synchronous initiation of segmentation wall growth in young spherules, and differentiation of endospores from uninucleate, cytoplasmic compartments of the parasitic cells. Unfortunately, the number of researchers who work with this micro-organism is small. This is largely due to the fact that *C. immitis* is a primary human pathogen that must be handled in a biological level 3 containment laboratory. The fungus cannot be easily exchanged between investigators because of its inclusion on a list of microbes that could presumably be used in acts of bioterrorism.

Diagnosis of coccidioidal infection is typically based on positive cultures and serology. Specific DNA/RNA detection methods have significantly improved the sensitivity of detection of *C. immitis*, but these tests have limited use in clinical laboratories. There is need for a specific, sensitive, and inexpensive antigen detection method which could be applied to patient urine, serum, and cerebrospinal fluid. Treatment of patients with coccidioidal infections has limitations that are typical of other systemic fungal diseases. Not all patients who need antifungal therapy respond to treatment, and many of those who do respond relapse once the drug therapy is stopped. New drug targets need to be discovered to encourage development of less toxic and more efficacious compounds to treat coccidioidomycosis.

The major goal of at least five research laboratories in the United States, which are currently focused on studies of coccidioidomycosis, is to develop a human vaccine against this respiratory disease. The feasibility of producing a prophylactic vaccine is based on the observation that individuals who have recovered from symptomatic coccidioidomycosis retain lifelong immunity against the disease. Current research on development of a coccidioidomycosis vaccine is based on a systematic, molecular approach to the identification of antigens that stimulate T-cell-mediated responses in vivo. The final experimental product to be tested in the next human trial may consist of a cocktail of recombinant antigens, perhaps including a cytokine antagonist (e.g., IL-12) [130], or may involve an immunization protocol which includes a combination of an initial DNA vaccination to prime the host T-helper 1 pathway of immune response followed by subcutaneous or intranasal delivery of a T-cell-reactive recombinant protein [131]. The beginning of the 21st century is marked by an increased understanding of the nature of immune response to fungal diseases, and this has led to renewed interest amongst medical mycologists in the feasibility of development of fungal vaccines [132]. Technological advances in the areas of new adjuvant development, viable vector delivery of immunogens to specific host tissues, and DNA immunization [133] have offered hope that the generation of a human vaccine against San Joaquin Valley fever may be achieved.

ACKNOWLEDGMENTS

The authors gratefully acknowledge the research support of the National Institutes of Health and the California Healthcare Foundation.

REFERENCES

1. A Polesky, CM Kirsch, LS Snyder, P LoBue, FT Kagawa, BJ Dykstra, JH Wehner, A Catanzaro, NM Ampel, DA Stevens. Airway coccidioidomycosis-report of cases and review. Clin Infect Dis 28:1273–1280, 1999.
2. SC Deresinski. Coccidioidomycosis of bone and joints. In: DA Stevens, ed. Coccidioidomycosis. A Test. New York: Plenum, 1980, pp 195–211.
3. KT Maddy. The geographic distribution of *Coccidioides immitis* and possible ecologic implications. Ariz Med 15:178–188, 1958.
4. GH Lacy and FE Swatek. Soil ecology of *Coccidioides immitis* at amerindian middens in California. Appl Microbiol 27:379–388, 1974.
5. TM Millman, TR O'Brien, PF Suter, and AM Wolf. Coccidioidomycosis in the dog: its radiographic diagnosis. J Am Vet Radiol Soc 20:52–65, 1979.
6. D Pappagianis. Epidemiology of coccidioidomycosis. In: DA Stevens, ed. Coccidioidomycosis, A Text. New York: Plenum, 1980, pp 63–85.
7. SB Werner, D Pappagianis. Coccidioidomycosis in northern California—an outbreak among archeology students near Red Bluff. Calif Med 119:16–20, 1973.
8. SB Werner, D Pappagianis, I Heindl, A Mickel. An epidemic of coccidioidomycosis among archeology students in northern California. N Engl J Med 286:507–512, 1972.
9. D Pappagianis, H Einstein. Tempest from tehachapi takes toll or *Coccidioides* conveyed aloft and afar. West J Med 129:527–530, 1978.
10. GT Cole, SH Sun. Arthroconidium-spherule-endospore transformation in *Coccidioides immitis*. In: PJ Szaniszlo, L Harris eds. Fungal dimorphism: with emphasis on fungi pathogenic for humans. New York: Plenum Press, 1985, pp 281–333.
11. GT Cole. *C. immitis*: resistance to host defense mechanisms. Clin Adv Treatment Fungal Infec 4: 1–12, 1993.
12. GT Cole, RA Samson. The conidia. In: Y Al-Doory, JF Domson, eds. Mould Allergy. Philadelphia: Lea Febiger, 1984 pp 66–103.
13. GT Cole, TN Kirkland, SH Sun. Conidia of *Coccidioides immitis*: their significance in disease initiation. In: GT Cole, HC Hoch, eds. The Fungal Spore and Disease Initiation in Plants and Animals. New York: Plenum Press, 1991, pp 403–443.
14. DJ Drutz, M Huppert. Coccidioidomycosis: factors affecting the host-parasite interaction. J Infect Dis 147:372–390, 1983.
15. GT Cole, LM Pope, M Huppert, SH Sun, P Starr. Ultrastructure and composition of conidial wall fractions of *Coccidioides immitis*. Exp Mycol 7:297–318, 1983.
16. RE Beever. Purification and chemical characterization of the rodlet layer of *Neurospora crassa* conidia. J Bacteriol 140:1063–1070, 1979.
17. MA Stringer, WE Timberlake. Cerato-ulmin, a toxin involved in Dutch elm disease, is a fungal hydrophobin. Plant Cell 5:145–146, 1993.
18. JGH Wessels. Wall growth, protein excretion and morphogenesis in fungi. New Phytol 123:397–413, 1993.
19. WJ Nickerson, GE Edwards. Studies on the physiological bases of morphogenesis in fungi. I. The respiratory metabolism of dimorphic pathogenic fungi. J Gen Physiol 33:41–55, 1949.
20. B Maresca, G DiLallo, CL Pardini, G Petrella, L Carratu. Addition of unsaturated fatty acids down-modulates heat shock gene expression and produces attenuated strains of the fungus *Histoplasma capsulatum*. In: B Baresca, GS Kobayashi, H Yamaguchi, eds. Molecular Biology and Its Application to Medical Mycology. Berlin: Springer-Verlag, 1993, pp 243–250.
21. GT Cole, D Kruse, KR Seshan, S Pan, PJ Szaniszlo, J Richardson, B Bian. Factors regulating morphogenesis in *Coccidioides immitis*. In: H Vanden Bossche, FC Odds, D Kerridge, eds. Dimorphic Fungi in Biology and Medicine. New York: Plenum Press, 1993, pp 191–212.
22. DJ Klionsky, PK Herman, SD Emr. The fungal vacuole: composition, function and biogenesis. Microbiol Rev 54:266–292, 1990.

23. S Bartnicki-Garcia. Fundamental aspects of hyphal morphogenesis. In: JM Ashworth, JE Smith, eds. Microbial Differentiation: Twenty-third Symposium of the Society for General Microbiology. London: Cambridge University Press, 1973, pp 245–267.

24. L Guevara-Olvera, CY Hung, JJ Yu, GT Cole. Sequence, expression and functional analysis of the *Coccidioides immitis* ODC (ornithine decarboxylase) gene. Gene 242:437–448, 2000.

25. WA Fonzi, PS Sypherd. Expression of the gene for ornithine decarboxylase of *Saccharomyces cerevisiae* in *Escherichia coli*. Mol Cell Biol 5:161–166, 1985.

26. MA Pfaller, J Riley, T Gerarden. Polyamine depletion and growth inhibition in *Candida albicans* and *Candida tropicalis* by a-difluoromethylornithine and cycloheylamine. J Med Vet Mycol 26: 119–126, 1988.

27. L Guevara-Olvera, C Calvo-Mendez, J Ruiz-Herrera. The role of polyamine metabolism in dimorphism of *Yarrowia lipolytica*. J Gen Microbiol 193:485–493, 1993.

28. CW Jacobs, RL Roberts, PJ Szaniszlo. Reversal of multicellular-form development in a conditional morphological mutant of the fungus *Wangiella dermatitidis*. J Gen Microbiol 131:1719–1728, 1985.

29. SH Sun, M Huppert. A cytological study of morphogenesis in *Coccidioides immitis*. Sabouraudia 14:185–198, 1976.

30. SM Johnson, D Pappagianis. The coccidioidal complement fixation and immunodiffusion-complement fixation antigen is a chitinase. Infect Immun 60:2588–2592, 1992.

31. GT Cole, EJ Pishko, KR Seshan. Possible roles of wall hydrolases in the morphogenesis *Coccidioides immitis*. Can J Bot 73(suppl):S1132–S1141, 1995.

32. EJ Pishko, T Kirkland, GT Cole. Isolation and characterization of two chitinase-encoding genes (cts1 and cts2) from the fungus *Coccidioides immitis*. Gene 167:173–177, 1995.

33. JL Converse. Effect of surface active agents on endosporulation of *Coccidioides immitis* in a chemically defined medium. J Bacteriol 74:106–107, 1957.

34. HB Levine. Purification of the spherule-endospore phase of *Coccidioides immitis*. Sabouraudia 1: 112–115, 1961.

35. GT Cole, SW Zhu, LL Hsu, D Kruse, KR Seshan, F Wang. Isolation and expression of a gene which encodes a wall-associated proteinase of *Coccidioides immitis*. Infect Immun 60:416–427, 1992.

36. RF Hector, D Pappagianis. Enzymatic degradation of the wall of spherules of *Coccidioides immitis*. Exp Mycol 6:136–152, 1982.

37. JL Converse, AR Besemer. Nutrition of the parasitic phase of *Coccidioides immitis* in a chemically defined liquid medium. J Bacteriol 78:231–239, 1959.

38. SA Klotz, DJ Drutz, M Huppert, SH Sun, PL Demarsh. Critical role of CO_2 in the morphogenesis of *Coccidioides immitis* in cell-free subcutaneous chambers. J Infect Dis 150:127–134, 1984.

39. WS Bump. Observation on growth of *Coccidioides immitis*. J Infect Dis 36:561–565, 1925.

40. LG Eissenberg, WE Goldman, PH Schlesinger. *Histoplasma capsulatum* modulates the acidification of phagolysosomes. J Exp Med 177:1605–1611, 1993.

41. M Miyaji, K Nishimura, L Ajello. Scanning electron microscope studies on the parasitic cycle of *Coccidioides immitis*. Mycopathologia 89:51–57, 1985.

42. SH Sun, GT Cole DJ Drutz, JL Harrison. Electron-microscopic observations of the *Coccidioides immitis* parasitic cycle in vivo. J Med Vet Mycol 24:183–192, 1986.

43. DJ Drutz, M Huppert, SH Sun, W Mcguire. Human sex hormones stimulate growth and maturation of *Coccidioides immitis*. Infect Immun 32:897–907, 1981.

44. BL Powell, DJ Drutz, M Huppert, SH Sun. Relationship of progesterone- and estradiol-binding proteins in *Coccidioides immitis* to coccidioidal dissemination in pregnancy. Infect Immun 40: 478–485, 1983.

45. RM Evans. The steroid and thyroid hormone receptor gene superfamily. Science 240:889–895, 1988.

46. PJ Mallow, X Zhao, ND Madani, D Feldman. Cloning and expression of the gene from *Candida albicans* that encodes a high affinity corticosterioid-binding protein. Proc Natl Acad Sci USA 90: 1902–1906, 1993.

47. CW Emmons. Fungi which resemble *Coccidioides immitis*. In: L Ajello, ed. Coccidioidomycosis. Tucson: University of Arizona Press, 1967, 333–337.

48. L Sigler, JW Carmichael. Taxonomy of *Malbranchea* and some other hyphomycetes with arthroconidia. Mycotaxon 4:349–488, 1976.
49. RS Currah. Taxonomy of the Onygenales: Arthrodermataceae, Gymnoascacease, Myxotrichaceae and Onygenaceae. Mycotaxon 24:1–216, 1985.
50. BH Bowman, JW Taylor, AG Brownlee, J Lee, S-D Lu, TJ White. Molecular evolution of the fungi: relationships of basidiomycetes, ascomycetes, and chytridiomycetes. Mol Biol Evol 9: 285–296, 1992.
51. BH Bowman, JW Taylor. Molecular phylogeny of pathogenic and nonpathogenic onygenales. In: DR Reynolds, JW Taylor, eds. The Fungal Holomorph: Mitotic, Meiotic and Pleomorphic Speciation in Fungal Systematics. Wallingford: CAB International, pp 169–178, 1993.
52. S Pan, L Sigler, GT Cole. Evidence for a phylogenetic connection between *Coccidioides immitis* and *Uncinocarpus reesii* (Onygenaceae). Microbiology 140:1481–1494, 1994.
53. CE Smith, RR Beard, HG Rosenberger, EG Whiting. Effect of season and dust control on coccidioidomycosis. JAMA 132:833–838, 1946.
54. D Pappagianis. Epidemiology of coccidioidomycosis. Curr Top Med Mycol 2:199–238, 1988.
55. RR Dodge, MD Lebowitz, RA Barbee, B Burrows. Estimates of *C. immitis* infection by skin test reactivity in an endemic community. Am J Public Health 75:863–865, 1985.
56. CE Smith. Epidemiology of acute coccidioidomycosis with erythema nodosum ("SanJoaquin" or "Valley Fever"). Am J Public Health 30:600–611, 1940.
57. PL Williams, DL Sable, P Mendez, LT Smyth. Symptomatic coccidioidomycosis following a severe natural dust storm-an outbreak at the Naval Air Station, Lemoore, Calif. Chest 76:566–570, 1979.
58. CM Peterson, K Schuppert, PC Kelly, D Pappagianis. Coccidioidomycosis and pregnancy. Obstet Gynecol Surv 48:149–156, 1993.
59. EE Wack, NM Ampel, JN Galgiani, DA Bronnimann. Coccidioidomycosis during pregnancy—an analysis of ten cases among 47,120 pregnancies. Chest 94:376–379, 1988.
60. IM Cohen, JN Galgiani, D Potter, DA Ogden. Coccidioimomycosis in renal replacement therapy. Arch Intern Med 142:489–494, 1982.
61. DA Bronnimann, RD Adam, JN Galgiani, MP Habib, EA Peterson, B Porter, JW Bloom. Coccidioidomycosis in the acquired immunodeficiency syndrome. Ann Intern Med 106:372–379, 1987.
62. SS Kerrick, LL Lundergan, Galgiani. Coccidioidomycosis at a university health service. Am Rev Respir Dis 131:100–102, 1985.
63. JW Coburn. Scalene lymph node involvement in primary and disseminated coccidioidomycosis. Ann Intern Med 56:911–924, 1962.
64. R DeFelice, MA Wieden, JN Galgiani. The incidence and implications of coccidioidouria. Am Rev Respir Dis 125:49–52, 1982.
65. CE Smith, EG Whiting, EE Baker, HG Rosenberger, RR Beard, MT Saito. The use of coccidioidin. Am Rev Tuberculosis 57:330–351, 1948.
66. ER Hobbs. Coccidioidomycosis. Dermatol Clin 7:227–239, 1989.
67. JN Galgiani. Coccidioidomycosis. West J Med 159:153–171, 1993.
68. SA Lonky, A Catanzaro, K Moser, H Einstein. Acute coccidioidal pleural effusion. Am Rev Respir Dis 114:681–688, 1976.
69. DW Jenkins Jr. Persistent adenopathy in coccidioidomycosis. South Med J 70:531–532, 1977.
70. RM Echols, DL Palmer, GW Long. Tissue eosinophilia in human coccidioidomycosis. Rev Infect Dis 4:656–664, 1882.
71. N Williams. T cells on the mucosol frontline. Science 280:198–200, 1998.
72. TN Kirkland, J Fierer. Inbred mouse strains differ in resistance to lethal *Coccidioides immitis* infection. Infect Immun 40:912–917, 1983.
73. TN Kirkland, J Fierer. Genetic control of resistance to *Coccidioides immitis*: a single gene that is expressed in spleen cells determines resistance. J Immunol 135:548–552, 1985.
74. RA Cox, W Kennell, L Boncyk, JW Murphy. Induction and expression of cell-mediated immune responses in inbred mice infected with *Coccidioides immitis*. Infect Immun 56:13–17, 1988.
75. CE Smith, MT Saito, SA Simons. Pattern of 39,500 serologic tests in coccidioidomycosis. JAMA 160:546–552, 1956.

76. JN Galgiani, RA Isenberg, DA Stevens. Chemotaxigenic activity of extracts from the mycelial and spherule phases of *Coccidioides immitis* for human polymorphonuclear leukocytes. Infect Immun 21:862–865, 1978.
77. JN Galgiani, CM Payne, JF Jones. Human polymorphonuclear-leukocyte inhibition of incorporation of chitin precursors into mycelia of *Coccidioides immitis*. J Infect Dis 149:404–412, 1984.
78. L Beaman, D Pappagianis, E Benjamini. Significance of T cells in resistance to experimental murine coccidioidomycosis. Infect Immun 17:580–585, 1977.
79. L Beaman, D Pappagianis, E Benjamini. Mechanisms of resistance to infection with *Coccidioides immitis* in mice. Infect Immun 23:681–685, 1979.
80. L Beaman, E Benjamini, D Pappagianis. Role of lymphocytes in macrophage-induced killing of *Coccidioides immitis* in vitro. Infect Immun 34:347–353, 1981.
81. L Beaman. Fungicidal activation of murine macrophages by recombinant gamma interferon. Infect Immun 55:2951–2955, 1987.
82. NM Ampel, GC Bejarano, JN Galgiani. Killing of *Coccidioides immitis* by human peripheral blood mononuclear cells. Infect Immun 60:4200–4204, 1992.
83. C Brass, HB Levine, DA Stevens. Stimulation and suppression of cell-mediated immunity by endosporulation antigens of *Coccidioides immitis*. Infect Immun 35:431–436, 1982.
84. GT Cole. The fungal propagule: infectious fungal propagules. In: D Schlessinger, ed. Microbiology. Washington: American Society for Microbiology, 1983, pp 245–248.
85. GT Cole, TN Kirkland, SH Sun. An immunoreactive, water-soluble conidial wall fraction of *Coccidioides immitis*. Infect Immun 55:657–667, 1987.
86. GT Cole, SW Zhu, S Pan, L Yuan, D Kruse, SH Sun. Isolation of antigens with proteolytic activity from *Coccidioides immitis*. Infect Immun 57:1524–1534, 1989.
87. RA Cox. Cell-mediated immunity. In: D Howard, ed. Fungi Pathogenic for Humans and Animals. Part B. New York: Marcel Dekker, 1983, pp 61–98.
88. RA Cox, DM Magee. Production of tumor necrosis factor alpha, interleukin-1a, and interleukin-6 during murine coccidioidomycosis. Infect Immun 63:4178–4180, 1995.
89. RA Cox, DM Magee. Protective immunity in coccidioidomycosis. Res Immunol 149:4178–4180, 1998.
90. DM Magee, RA Cox. Roles of gamma interferon and interleukin-4 in genetically determined resistance to *Coccidioides immitis*. Infect Immun 63:3514–3519, 1995.
91. DM Magee, RA Cox. Interleukin-12 regulation of host defenses against *Coccidioides immitis*. Infect Immun 64:3609–3613, 1996.
92. J Fierer, L Walls, L Eckmann, T Yamamoto, TN Kirkland. Importance of IL-10 in genetic susceptibility of mice *to Coccidioides immitis*. Infect Immun 66:4397–4402, 1998.
93. CA Janeway, P Travers, M Walport, JD Capra. Immunobiology. In: The Immune System in Health and Disease. New York: Garland Publishing, 1999, pp 263–305.
94. FP Heinzel, MD Sadick, BJ Holaday, RL Coffman, RM Locksley. Reciprocal expression of interferon gamma or interleukin 4 during the resolution or progression of murine leishmaniasis. J Exp Med 169:59–72, 1989.
95. HB Levine, JM Cobb, CE Smith. Immunogenicity of spherule-endospore vaccines of *Coccidioides immitis* for mice. J Immunol 87:218–227, 1961.
96. D Pappagianis, Valley Fever Study Group. Evaluation of the protective efficacy of the killed *Coccidioides immitis* spherule vaccine in humans. Am Rev Respir Dis 148:656–660, 1993.
97. CR Zimmermann, SM Johnson, GW Martens, AG White, BL Zimmer, D Pappagianis. Protection against lethal murine coccidioidomycosis by a soluble vaccine from spherules. Infect Immun 66:2342–2345, 1998.
98. G Lecara, RA Cox, RB Simpson. *Coccidioides immitis* vaccine: potential of an alkali-soluble, water-soluble cell wall antigen. Infect Immun 39:473–475, 1983.
99. M Huppert, NS Spratt, KR Vukovich, SH Sun, EH Rice. Antigenic analysis of coccidioidin and spherulin determined by two-dimensional immunoelectrophoresis. Infect Immun 20:541–551, 1978.
100. GT Cole, T Kirkland, M Franco, SW Zhu, L Yuan, SH Sun, VN Hearn. Immunoreactivity of a surface wall fraction produced by spherules of *Coccidioides immitis*. Infect Immun 56:2695–2701, 1988.

101. C Hung, NM Ampel, L Christian, KR Seshan, GT Cole. A major cell surface antigen of *Coccidioides immitis* which elicits both humoral and cellular immune responses. Infect Immun 68:584–593, 2000.

102. SP Lee, NG Stoker, KA Grant, ZT Handzel, R Hussain, KPWJ McAdam, HM Dockrell. Cellular immune responses of leprosy contacts to fractionated *Mycobacterium leprae* antigens. Infect Immun 57:2475–2480, 1989.

103. TN Kirkland, SW Zhu, D Kruzse, LL Hsu, K Seshan, GT Cole. *Coccidioides immitis* fractions which are antigenic for immune T lymphocytes. Infect Immun 59:3952–3961, 1991.

104. EE Wyckoff, EJ Pishko, TN Kirkland, GT Cole. Cloning and expression of a gene encoding a T-cell reactive protein from *Coccidioides immitis*: homology to 4-hydroxyphenylpyruvate dioxygenase and the mammalian F antigen. Gene 161:107–111, 1995.

105. TN Kirkland, PW Thomas, F Finley, GT Cole. Immunogenicity of a 48-kilodalton recombinant T-cell reactive protein of *Coccidioides immitis*. Infect Immun 66:424–431, 1998.

106. GT Cole, KR Seshan, M Franco, E Bukownik, SH Sun, VN Hearn. Isolation and morphology of an immunoreactive outer wall fraction produced by spherules of *Coccidioides immitis*. Infect Immun 56:2686–2694, 1988.

107. CY Hung, JJ Yu, GT Cole. Expression and immunoreactivity of the spherule outer wall antigen of *Coccidioides immitis*. *Abstracts, 995th General Meeting of American Society for Microbiology, Chicago*, F-83, 312, 1993.

108. JJ Yu, SL Smithson, PW Thomas, TN Kirkland, GT Cole. Isolation and characterization of the urease gene (URE) from the pathogenic fungus *Coccidioides immitis*. Gene 198:387–391, 1997.

109. AA Salyers, DD Whitt. Bacterial Pathogenesis. A Molecular Approach. *Washington: American Society for Microbiology*, 1994, pp 274–277.

110. TL Cover, MJ Blaser. *Helicobacter pylori*: a bacterial cause of gastritis, peptic ulcer disease, and gastric cancer. ASM News 61:21–26, 1995.

111. DT Smoot, HLT Mobley, GR Chippendale, JF Lewison, JH Resau. *Helicobacter pylori* urease activity is toxic to human gastric epithelial cells. Infect Immun 58:1992–1994, 1990.

112. P Krishnamurthy, M Parlow, JB Zitzer, NB Vakil, HLT Mobley, M Levy, SH Phadnis, BE Dunn. *Helicobacter pylori* containing only cytoplasmic urease is susceptible to acid. Infect Immun 66:5060–5066, 1998.

113. PR Harris, HLT Mobley, GI Perez-Perez, MJ Blaser, PD Smith. *Helicobacter pylori* urease is a potent stimulus of mononuclear phagocyte activation and inflammatory cytokine production. Gastroenterology 111:419–425, 1996.

114. AM Krieg. Mechanisms and applications of immune stimulatory CpG oligodeoxynucleotides. Biochim Biophys Acta 93321:1–10, 1999.

115. AM Krieg, A-K Yi, J Schorr, HL Davis. The role of CpG dinucleotides in DNA vaccines. Trends Microbiol 6:23–27, 1998.

116. GB Lipford, M Bauer, C Blank, R Reiter, H Wagner, K Heeg. CpG-containing synthetic oligonucleotides promote B and cytotoxic T-cell responses to protein antigen: a new class of vaccine adjuvants. Eur J Immunol 27:2340–2344, 1997.

117. RS Chu, OS Targoni, AM Krieg, PV Lehmann, CV Harding. CpG oligodeoxynucleotides act as adjuvants that switch on T helper 1 (Th1) immunity. J Exp Med 186:1623–1631, 1997.

118. S Zimmerman, O Egeter, S Hausmann, GB Lipford, M Rocken, H Wagner, K Heeg. Cutting edge: CpG oligodeoxynucleotides trigger protective and curative TH1 responses in lethal murine leishmaniasis. J Immunol 160:3627–3630, 1998.

119. RA Cox, LA Britt. Antigenic heterogeneity of an alkali-soluble, water-soluble cell wall extract of *Coccidioides immitis*. Infect Immun 50:365–369, 1985.

120. KO Dugger, JN Galgiani, NM Ampel, SH Sun, DM Magee, JL Harrison, JH Law. An immunoreactive apoglycoprotein purified from *Coccidioides immitis*. Infect Immun 59:2245–2251, 1991.

121. TN Kirkland, F Finley, KI Orsnorne, JN Galgiani. Evaluation of the proline-rich antigen of *Coccidioides immitis* as a vaccine candidate in mice. Infect Immun 66:3519–3522, 1998.

122. Y Zhu, Y Chunmu, M Magee, RA Cox. *Coccidioides immitis* antigen 2: analysis of gene and protein. Gene 181:121–125, 1996.

123. Y Zhu, C Yang, DM Magee, RA Cox. Molecular cloning and characterization of *Coccidioides immitis* antigen 2 cDNA. Infect Immun 64:2695–2700, 1996.

124. KO Dugger, KM Villareal, A Nguyen, CR Zimmermann, JH Law, JN Galgiani. Cloning and sequence analysis of the cDNA for a protein from *Coccidioides immitis* with immunogenic potential. Biochem Biophys Res Commun 218:485–489, 1996.

125. MA Wieden, LL Lundergan, J Blum, KL Delgado, R Coolbaugh, R Howard, T Peng, E Pugh, N Reis, J Theis. Detection of coccidioidal antibodies by 33-kDa spherule antigen, *Coccidioides* EIA, and standard serologic tests in sera from patients evaluated for coccidioidomycosis. J Infect Dis 173:1273–1277, 1996.

126. JN Galgiani, T Peng, ML Lewis, GA Cloud, D Pappagianis. Cerebrospinal fluid antibodies detected by ELISA against a 33-kDa antigen from spherules of *Coccidioides immitis* in patients with coccidioidal meningitis. J Infect Dis 173:499–502, 1996.

127. JN Galgiani, SH Sun, KO Dugger, NM Ampel, GC Grace, JL Harrison, MA Wieden. An arthroconidial-spherule antigen of *Coccidioides immitis*: differential expression during in vitro fungal development and evidence for humoral response in humans after infection or vaccination. Infect Immun 60:2627–2635, 1992.

128. C Jiang, DM Magee, TN Quitugua, RA Cox. Genetic vaccination against *Coccidioides immitis*: comparison of vaccine efficacy of recombinant antigen 2 and antigen 2 cDNA. Infect Immun 67:630–635, 1999.

129. JN Galgiani. Coccidioidomycosis: a regional disease of national importance. Rethinking approaches for control. Anals of Int Med 130:293–300, 1999.

130. MC Jiang, DM Magee, RA Cox. Coadministration of interleukin-12 expression vector with antigen 2 cDNA enhances induction of protective immunity against *Coccidioides immitis*. Infect Immun 67:5848–5853, 1999.

131. J Schneider, SC Gilbert, TJ Blanchard, T Hanke, KJ Robson, CM Hannan, M Becker, R Sinden, GL Smith, AVS Hill. Enhanced immunogenicity for CD8 + T cell induction and complete protective efficacy of malaria DNA vaccination by boosting with modified vaccinia virus Ankara. Nat Med 4:397–403, 1998.

132. GS Deepe. Prospects for the development of fungal vaccines. Clin Microbiol Rev 10:585–596, 1997.

133. MA Liu. Vaccine developments. Nat Med 4(suppl):515–519, 1998.

18

Innate Immunity Against Fungal Pathogens

Luigina Romani
University of Perugia, Perugia, Italy

I. INTRODUCTION

The kingdom of Fungi comprises a number of species that are associated with a wide spectrum of diseases in humans and animals, ranging from allergy and autoimmunity to life-threatening infections. Most fungi (such as the pathogenic *Histoplasma capsulatum, Paracoccidioides brasiliensis, Coccidioides immitis, Blastomyces dermatitidis, Cryptococcus neoformans, Aspergillus fumigatus,* and *Pneumocystis carinii*) are ubiquitous in the environment. Some, including *Candida albicans*, establish lifelong commensalism on human body surfaces. Not surprisingly, therefore, human beings are constantly exposed to fungi, primarily through inhalation or traumatic implantation of fungal elements.

Nevertheless, beside the frequent occurrence of allergic hypersensitivity, only a very limited number of fungi cause severe infections. It is often argued that fungi have evolved for a saprophytic existence and that mammalian infection is not necessary for the survival of any fungal species. However, in this context, it would make little teleological sense that fungi have low inherent virulence. Inversely, the observation that most are opportunists that show infectivity only in patients with a variety of immunologic defects, indicates that a high degree of coexistence has evolved between fungi and their mammalian hosts which deviates into overt disease only under certain circumstances.

Although not unique among infectious agents, fungi possess complex and unusual relationships with the vertebrate immune system, partly due to some prominent features. Among these is their ability to exist in different forms and to reversibly switch from one to the other in infection. Examples are the dimorphic fungi (*H. capsulatum, P. brasiliensis, C. immitis,* and *B. dermatitidis*) which transform from saprobic filamentous molds to unicellular yeasts in the host, the filamentous fungi (such as *Aspergillus* spp.) that, inhaled as unicellular conidia, may transform into a multicellular mycelium, and some species of *Candida*, capable of growing in different forms such as yeasts (blastospores), pseudohyphae, and hyphae. This implicates the existence of a multitude of recognition and effector mechanisms to oppose fungal infectivity at the different body sites. For commensals, two prominent features are also important, including the highly effective strategies of immune evasion they must have evolved to survive in the host environment and the prolonged antigenic stimulation of the host that can have profound immunoregulatory consequences. Thus, in the context of the antagonistic relationships that characterize the host-pathogen interactions, the strategies used by the host to limit fungal infectivity are necessarily disparate ([1] and references therein) and, in retaliation, fungi have developed their own elaborate tactics to evade or overcome these defenses [2–5]. This may have resulted in an expanded repertoire of crossregulatory and overlapping antifungal host responses that makes it extremely difficult to define the relative contribution of individual components of the immune system in antifungal defense.

Within the limitations imposed by these considerations, this chapter is an advanced attempt to analyze the role of natural immunity in resistance to pathogenic fungi. Innate and acquired cell-mediated immunity have been acknowledged as the primary mediators of mammalian resistance to fungi ([1] and references therein). Traditionally considered only as a first line of defense, it is now clear that innate immunity plays an instructive role in the adaptive response to pathogens, by inducing key costimulatory molecules and cytokines [6,7].

Through the involvement of different pattern recognition receptors [8], cells of the innate immune system not only discriminate between the different forms of fungi, but also contribute to discrimination between self and pathogens at the level of the adaptive T-helper (Th) immunity. Thus, the traditional dichotomy between the functions of innate and adaptive immunity in response to fungi has been recently challenged by the concept of an integrated immune response to fungi ([1] and references therein). These important and novel aspects of innate immunity to fungi will be emphasized throughout the chapter.

II. NATURAL IMMUNITY AGAINST PATHOGENIC FUNGI

In general, humans have a high level of resistance to fungi, and most infections are mild and self-limiting [9,10]. The majority of fungi are detected and destroyed within hours by innate defense mechanisms that are not antigen specific and do not require a prolonged period of induction. The innate immune mechanisms act immediately and are followed some hours later by an early induced response, which must be activated by infection but does not generate lasting protective immunity. These early phases help to keep infection under control. Thus, the seminal observation of Metchnikoff, in the 19th century, on the innate immune defense of the water flea (*Daphnia*) against fungal spores is fully acknowledged. In vertebrates, however, if the infectious organism can breach these early lines of defense an adaptive immune response will ensue, with generation of antigen-specific Th effector cells that specifically target the pathogen and memory cells that prevent subsequent infection with the same microorganism. As different Th cell subsets exist that are endowed with the ability to release a distinct panel of cytokines, capable of activating and deactivating signals to effector phagocytes, the activation of the appropriate Th subset is instrumental in the generation of a successful immune response to fungi ([1] and references therein).

III. INNATE IMMUNITY AS A FIRST LINE OF DEFENSE AGAINST FUNGI

Figure 1 illustrates cells, factors, and functions of the innate immune resistance to fungi.

A. The Natural Barriers

Adherence to host tissues is considered the pivotal first step in the pathogenesis of fungal infections [9]. Fungi possess a variety of complementary structures which adhere to cell surfaces and the extracellular matrix [3,4]. They also secrete a variety of enzymes, such as phospholipases and proteases, considered to be major virulent factors, as they cause host cell damage and lysis and impair antifungal host defenses [5]. Nevertheless, surface epithelia make up a natural barrier

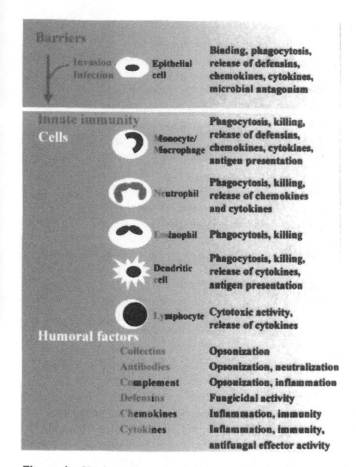

Figure 1 The innate immune resistance to fungi: cells, factors, and functions.

to fungal infections. Intact skin serves as a primary barrier to any infection caused by fungi that primarily colonize the superficial, cutaneous, and subcutaneous layers of skin. The epithelia lining the respiratory, gastrointestinal, and genitourinary tracts acts as a mechanical, microbiological, and chemical barrier to fungi. Nonpathogenic microorganisms attached to epithelia may prevent the overgrowth of fungi, thereby decreasing the likelihood of mucosal and systemic spread by several mechanisms, which include nutritional competition, blocking receptors for fungal adhesins on epithelial cells, production of antifungal compounds, increasing epithelial cell renewal rates, alteration of pH, and production of an anaerobic oxidation-reduction potential. Indeed, lactobacilli and bifidobacteria have shown efficacy in the biotherapy of candidiasis [11].

The attachment of fungi to epithelial cells appears to occur through lectin-, laminin-, and fibronectin/vitronectin-mediated binding to complementary receptors on epithelial surfaces, and does not necessarily disrupt the metabolic, structural, or barrier function of epithelia [3,4,9,10]. Inversely, a number of important mediators, including defensins, chemokines, and cytokines, are released upon epithelial cell encounter with fungi [12–14]. These findings suggest that the epithelial cells are crucially involved in orchestration of the subsequent inflammatory and immunological responses to fungi. In addition, epithelial cells themselves may be endowed with some antifungal effector activities, as shown by the ability of epithelial cells from the vagina to inhibit *C. albicans* growth ([1] and references therein) and of those from lungs to internalize *A. fumigatus* conidia [15]. However, internalization of *Aspergillus* conidia by type II pneumocytes may not prevent hyphal formation, suggesting that in pulmonary defense mechanisms against filamentous fungi, epithelial cells do not participate in the host control of conidial germination and hyphal growth.

When hematogenous dissemination to visceral organs is an important step in the pathogenesis of the infection, adherence and penetration to the endothelial lining of blood vessels to invade the deep tissues may likely occur ([1] and references therein). Moreover, as binding of fungi to platelets has been observed [15–18], it is also likely that platelets may participate in host defense within the intravascular compartment.

A variety of factors contribute to innate antifungal defense on mucosal surfaces, among which are secretory IgA (see Sec. III.C) and locally produced enzymes and mediators with antifungal and proinflammatory activities, such as antileukoprotease, intestinal peptides, defensins, and pulmonary surfactant [1,12,19–23].

Defensins are small antimicrobial peptides (containing <100 aminoacids) that act, at least in part, by disrupting the structure or function of the microbial cell membrane, by virtue of their cationic nature. Although structurally and functionally similar, defensinlike peptides also abound in invertebrates and plants, but only recently has the relevance of vertebrate defensins in host defense been appreciated [21]. In the case of fungi, defensins are known to have direct antimicrobial activity against *C. albicans* [12] and *C. neoformans* [20] and to enhance that of phagocytes against *H. capsulatum* [24].

Pulmonary surfactant is a complex mixture of lipids and proteins that lowers surface tension at the air-liquid interface of the lung. Surfactant proteins (SPs) are members of the collectin family, known to be an important constituent of the innate immunity through microbe opsonization and activation of phagocytic cells [23]. Through recognition of carbohydrate structures, SPs can bind both acapsular and encapsulated *C. neoformans* [25], *A. fumigatus* conidia [26], and *P. carinii* [27]. Binding to SPs, however, may result in disparate effects, depending on fungal species. Phagocytosis is enhanced against *A. fumigatus* [26], but prevented against *C. neoformans* [25] and *P. carinii* [27]. Therefore, the increased level of SP-A may also act as a pathogenetic determinant of the infection. This should not be taken to contradict the beneficial role SPs may have in pulmonary host defense against fungi. SPs have long been known to beneficially dampen the persistent inflammatory response evoked by inhaled antigens; indeed, SP-A downregulated the production of proinflammatory cytokines by alveolar macrophages in response to *C. albicans* [28]. Interestingly, SP-A shares sequence homology with the mannose-binding protein, a serum lectin that binds different fungi and promotes complement deposition [29]. Thus, collectins represent a common first-line antifungal defense that is operative from the mucosal to distal levels.

B. The Complement

The complement system is a group of proteins, activated in a cascade fashion to provide a rapid humoral defense against micro-organisms. There are three pathways of complement activation. The classical pathway is triggered by the antibody (mainly IgG and IgM), and is thus part of the adaptive humoral immune response; the alternative pathway initiates directly on pathogen surfaces, does not depend on antibody, and is thus an important natural defense in nonimmune host; the lectin pathway is activated by the binding of mannose-binding lectin to mannans [30]. Despite different activation mechanisms, all three lead to the three main effector functions of complement: opsonization of microbes, through deposition of C3b and its breakdown product iC3b on particles, recruitment of inflammatory cells, and direct killing of pathogens.

The importance of the complement system in host resistance to fungi has been extensively described elsewhere [31]. C5 and C3 deficiencies greatly reduce resistance to *C. neoformans, C. albicans, A. fumigatus,* and *P. brasiliensis.* The absence of complement activation by cryptococci in brain tissue may thus provide an explanation for the niche that the central nervous system provides to *C. neoformans.*

All fungi studied to date have the ability to activate the alternative pathway leading to deposition of opsonic C3 on cell surfaces. This process is relatively low. However, C3 deposition is greatly accelerated through activation of the classical pathway by antibodies specific for antigens on the cell surface. This finding, while suggesting that different pathways normally operate together, also points to the important role served by antibodies directed to the cell surface in amplifying C3b deposition and opsonization, once either the classical or lectin pathway has been activated. This may occur with anticapsular monoclonal antibodies that, depending on their epitope specificity, differentially influence C3 binding and deposition on encapsulated cryptococci and, most likely, with naturally occurring antibodies, such as the antimannan IgG in candidiasis or antibodies reactive to β-glucan in blastomycosis [31].

It is interesting to note that activation of the different pathways also depends on the morphologic and structural characteristics of fungi. Encapsulated cryptococci are powerful activators of the alternate complement system, while nonencapsulated cryptococci initiate the classical pathway. The capsular size and serotype are major variables that influence activation of the complement system. Likewise, resting conidia, swollen conidia, and hyphae of *Aspergillus* differ in the mode of initiation of the complement cascade [15].

C3b deposition on fungal surfaces signals the ultimate destruction of the pathogen by phagocytic cells. This occurs by the specific recognition of bound complement components by complement receptors (CRs) on phagocytes. There are several known types of CRs on phagocytic cells, of which the CR3 mediates the binding of fungi. The importance and consequences of the CR3 recognition in the innate response of fungi will be dealt with below. CR3 is also expressed on *C. albicans* yeasts and hyphae [3], and evidence exists for its role in virulence. Indeed, complement receptors act as fungal adhesins [3] and play a role in iron acquisition by the fungus [32]. Expression of iC3b binding activity by *C. albicans* enabled the organism to bind complement-coated erythrocytes and provided a means of obtaining heme-derived iron for growth.

As we will discuss further below, there is an emerging concept that the complement system and CRs on phagocytes not only fulfill the requirement of a first-line defense against pathogens, but also control and promote subsequent adaptive immunity. Not surprisingly, therefore, complement components are involved in the secretion of cytokines by polymorphonuclear cells in response to fungi [33].

C. The Antibodies

There are many ways in which antibodies contribute to protection against fungi: prevention of adherence, neutralization of extracellular enzymes, inhibition of yeast-to-mycelium transition, agglutination, opsonization, complement activation, and activation of a variety of effector cells bearing Fc receptors (FcRs), specific for the Fc portion of antibodies (reviewed in [34,35]). Different accessory cells bear FcR for antibodies of different isotypes, and therefore the isotype of the antibody determines the final effector mechanism recruited in response to fungi. That isotypes determine the protective efficacy of antibodies is best illustrated in a murine model of cryptococcal infection. Both polysaccharides and proteins of *C. neoformans* elicit antibody responses, but only antibodies to the capsular glucuronoxylomannan (GXM) are opsonic and mediate protection upon passive transfer. However, among anti-GXM monoclonal antibodies, protective, nonprotective, and deleterious antibodies could be distinguished [34,35]. This finding is particularly intriguing, as it suggests that antibodies may increase fungal infectivity as well. The demonstration that isotype switching converts a nonprotective IgG3 antibody to an IgG1 protective antibody [36] is relevant. In the same model, it is shown that epitope specificity is an important determinant of the protective efficacy [37] and also influences C3 activation and deposition on fungal surface. It is important to emphasize here that the specific antibody influences cytokine production and lymphoproliferation [38], and the protective efficacy of antibodies relies on an intact T-cell function [39]. These findings clearly point to the interdependency between humoral and cellular immunity in fungal infections.

The existence of protective and nonprotective monoclonal antibodies to *C. neoformans* and *C. albicans* have greatly helped in clarifying the controversial issue on the role of antibodies in fungal infections. It is conceivable that differences in the composition and proportion of protective and nonprotective antibodies produced in infection may underlie the variable results obtained with polyclonal sera in experimental infections, as well as the lack of correlation of the presence of specific antibody with protection against infection [34]. Indeed, while studies in B-cell-deficient mice have clearly shown that B-cells are important in the host response against fungi [40–42], the elevated serum levels of polyclonally activated and specific antibodies frequently observed in mice and humans with progressive fungal infections [34] suggest that the levels of circulating antibodies do not correlate with protection. However, the presence of serum antibody is associated with favorable prognosis in cryptococcal meningitis [34], and antibodies to a 47-kDa breakdown product of heat shock protein 90 are associated with recovery from *C. albicans* infections and protection against disseminated disease in patients with AIDS [43]. Sera from patients who had recovered from disseminated candidiasis confer protection upon passive transfer into infected mice. Epitope mapping with sera from infected patients identified an immunodominant linear epitope, LKVIRK, which is recognized by immune sera. Both murine monoclonal and human recombinant antibodies produced against this epitope are protective in murine models of disseminated candidiasis. Immunoelectron microscopy suggested that the 47-kD antigen is present in the cytoplasm but also in the cell wall of *C. albicans*, thus being accessible to antibody ([1] and references therein).

Additional evidence points to the existence of protective antibody. Firstly, both polyclonal sera and IgM monoclonal antibodies to a mannan adhesin fraction passively transfer protection against disseminated and vaginal candidiasis in mice ([1] and references therein) [44]. A crude extract of the phosphomannan complex encapsulated into a liposome and, given to intact or immunocompromized mice, elicited a protective response, clearly attributable to agglutinating antibodies [44]. Here again, the sharing of the same isotype, specificity, and agglutinating function is not sufficient to confer a protective value to monoclonal antibodies.

Secondly, antibodies specific for the idiotype of antibodies reactive with yeast killer toxin (anti-idiotypic antibodies) are lethal to strains of *C. albicans* expressing the receptor for toxin ([1] and references therein) [45]. Naturally occurring antibodies with killer toxin-like activity are present in the vaginal fluid and serum of patients with vaginal candidiasis. Anti-idiotypic antibodies as well as single-chain variable antibodies confer protection upon passive transfer in a rat model of vaginal candidiasis [46]. Thus, antibody-dependent protection can also be operative in mucosal defense. Here again, evidence exists for and against a protective role of locally produced antibodies. IgA deficiency is not usually associated with *C. albicans* infections, and levels of vaginal IgA and IgG to *C. albicans* are similar in women with and without vaginal candidiasis ([34] and references therein). Secretory IgA both enhance [47] and reduce [48] adherence of fungi to epithelial cells. However, vaginal vaccination with a monoclonal antibody with specificity for yeast killer toxin elicited a secretory IgA anti-idiotypic response which protects rats from challenge with *C. albicans* [49]. The secretory anti-idiotype IgAs are fungicidal for *C. albicans*, presumably as a result of molecular mimicry of yeast killer toxin. Moreover, passive protection occurs with vaginal fluid containing antibodies to mannan constituents and the aspartyl proteinase of *C. albicans* [50].

Finally, indirect evidence for the protective role of antibodies, at least in candidal and cryptococcal infections, comes from the observation that *C. albicans* and *C. neoformans* have evolved strategies to evade antibody immunity, such as the production of proteases which may degrade immunoglobulin, antigenic variation, and release of capsular polysaccharide antigen, causing antibody unresponsiveness and tolerance [34].

D. The Phagocytes

Fungal surfaces undergo profound changes during morphogenesis that modify the host phagocytic response [51]. In vertebrates, the professional phagocytes against fungi are represented by polymorphonuclear leukocytes (neutrophils and eosinophils) and mononuclear leukocytes (monocytes and macrophages). Macrophages, continuously maturing from circulating monocytes and circulating neutrophils, once in tissues, engulf opsonized and unopsonized fungi through various receptors, including the CRs, FcRs, and the mannose receptors (MRs).

The MRs belong to a family of lectins that mediate nonopsonic phagocytosis of fungi ([52] and references therein). It is a 175-kDa type I membrane glycoprotein expressed in most differentiated macrophages, sinusoid endothelium, and dendritic cells (DCs). The MR preferentially recognizes α-linked oligomannoses with branched structures. MR ligation by fungi can be linked to induction of effector functions, but the link is dependent on activation.

CR3 is a heterodimeric molecule (CD11b/CD18) which belongs to the β2-integrin family of cell adhesion molecules and is expressed on monocytes and macrophages, polymorphonuclear leukocytes, and natural killer (NK) cells. Integrins constitute the critical link between cells and the extracellular matrix with the ability to activate intracellular signaling complexes. A remarkable characteristic of CR3 is its broad capacity for recognition of diverse ligands, such as iC3b, β-glucan, and zymosan. Thus, the multiplicity of binding sites for CR3 results in both opsonic and nonopsonic binding. CR1 and CR3 mediate binding of fungi and, when appropriately activated, they also mediate phagocytosis. However, they are incapable of initiating other cellular activities that ensue upon FcR ligation, such as the generation of oxidative burst.

The FcR comprises a family of molecules that bind to the Fc portion of immunoglobulin molecules, each member of the family recognizing immunoglobulin of one or a few closely related isotypes. Like T-cell receptors (TCRs), FcRs are multisubunit proteins in which only one of the chains is required for specific recognition. The other chains are required for transport to the cell surface and for signal transduction when Fc is bound. Crosslinking of FcRs by antibody-coated pathogens activates phagocytosis with enhanced antigen presentation, increases the production of fungicidal oxidants and promotes cytokine and chemokine release by phagocytic cells, such as macrophages and polymorphonuclear neutrophils. Release of stored mediators by other cells, such as NK cells, eosinophils, and mast cells, also occurs as a result of FcR engagement.

As further discussed in Section IV.E, these receptors govern the ability of phagocytes to discriminate not only among fungi but also between the different forms of fungi and to originate discriminative responses to them. The interaction of fungi with phagocytes has many purposes and important consequences. The main purpose is the destruction of the pathogen through either a phagocytic process that provides an immediate innate cellular immune response against fungi residing intracellularly or through the secretion of microbicidal compounds against uningestible fungal elements. Fungal infectivity, however, can also be efficiently opposed by interfering with transition from less virulent to more virulent fungal forms. One important consequence of the interaction of phagocytes with fungi is the release of chemokines and cytokines, antigen uptake and processing, and the induction of costimulatory activity on phagocytic cells, all processes necessarily impacting on the induction of the adaptive immune response.

Polymorphonuclear neutrophils (PMNs) represent an important line of defense against hyphae [53]. Not surprisingly, therefore, quantitative or qualitative defects of neutrophils are major predisposing factors to disseminated candidiasis and aspergillosis ([1] and references therein). However, human PMNs are also able to exert inhibitory effects against other forms of fungi, such as swollen conidia of A. fumigatus [15,54], conidia of B. dermatitidis [55], arthroconidia and spherules of C. immitis ([1] and references therein), and yeasts of Candida [56], C. neoformans [57], Trichosporon beigelii [57], and the dimorphic fungi ([1] and references therein), [58]. In general, killing efficiency against these fungal forms is relatively low, unless PMNs have been activated previously with activating cytokines (see below).

Against C. albicans, for instance, although phagocytosis and killing of blastospores by PMNs are superior to that of germ tubes and longer hyphae [59], only a fraction of engulfed Candida yeasts are digested [60,61]. Phagocytosis of C. albicans yeasts by PMNs occurs in the absence of serum opsonins, but opsonins are required for optimal intracellular killing of the fungus [60]. Factors found to limit the intracellular killing of C. albicans by PMNs include, among others, the ability of some ingested cells to produce germ tube, a defective sealing of phagolysosomes, inhibition of PMN degranulation by fungal products, and defective opsonization ([61] and references therein). Other studies have shown that phagocytosis of unopsonized fungi by PMNs stimulates only limited oxidative burst when occurring through a mannose-inhibitable mechanism [62]. Thus, it appears that phagocytosis through a mannose-inhibitable receptor may not be coupled in PMNs with the activation of the metabolic burst. In contrast, internalization through CR3 activates the respiratory burst of, and the release of azurophilic granules by, PMNs against opsonized H. capsulatum yeasts [63]. It seems therefore that binding and internalization of fungi by PMNs may occur through the MR [62], CR3 [62–65], and FcR [64,66], although effective phagocytosis and killing of fungi by PMNs occurs via targeting FcγRI or FcαRI.

In contrast to ingestible yeasts, extracellular hyphal elements are very susceptible to attack by PMNs [67,68]. PMNs are considered an essential line of defense against germinating conidia and hyphae of Aspergillus [68]. Human neutrophils are highly efficient in clearing germinating Candida cells from endothelial cell monolayers ([56] and references therein). Studies in mice confirm the important role of PMNs in the innate control of fungal infections, as shown by increased susceptibility to fungal infections of mice with quantitative [56,69–72] or qualitative [73] defects of neutrophils, by the ability of neutrophils to exert fungistatic and fungicidal activities [58,60,62,64,74,75] and by the ability of fungal cells or their products to affect neutrophil functions [76,77].

Macrophages are a heterogeneous population that vary according to their species of origin, anatomic location, state of activation, and culture conditions. All these variables greatly influence macrophage interactions with fungal cells [78–80]. Binding and internalization of fungi by mononuclear phagocytes may occur through the MR [81–85], CR1 [64,86,87], CR4 [88], FcγRs [89], and CD14 [86] receptors.

Macrophages serve as a protected environment in which the pathogenic dimorphic yeasts multiply and disseminate from the lung to other organs [1,80,86]. They fail to restrict the intracellular growth of these fungi by several mechanisms, which include defective phagosome-lysosome fusion ([1] and references therein), regulation of phagosomal pH [90], release of calcium-binding protein [91], lack of inhibition of conidia to yeast transition [92], and suppression of the respiratory burst [89]. However, it is interesting to note that FcR-mediated phagocytosis rescues the suppression of the respiratory burst [89], which may suggest a possible mechanism through which T-cells enhance the antifungal activity of macrophages and may explain the failure of macrophages from HIV-infected subjects to efficiently oppose fungal infectivity [93–95].

As already emphasized, the engagement of each receptor may result in profoundly different downstream intracellular events that have important consequences in the expression of the anti-fungal effector functions of phagocytosis. For instance, the internalization of *P. carinii* through the macrophage MR [82] accounts for the ability of alveolar macrophages to restrict the growth of *P. carinii* to such an extent that the downregulated mannose receptor expression in alveolar macrophages of HIV-infected persons is considered to contribute to susceptibility to *P. carinii* pneumonia [96]. However, internalization via constitutively competent MR does not represent an effective way of clearing yeasts, such as *C. neoformans* and *C. albicans*. The ingestion of unencapsulated cryptococcal yeasts by murine macrophages in vitro occurs via the mannose and beta-glucan receptors [97], but in vivo opsonins are required for uptake of the organism and fungal growth inhibition [98,99]. SP-A and -D, and mannose-binding proteins are effective opsonins of unencapsulated cryptococci in the lungs [100]. However, the high levels of CO_2 in the lungs favor capsule formation by *C. neoformans* [101], which impairs phagocytosis [102] by increasing the negative cellular charge [103], and these opsonins are ineffective against encapsulated organisms [100]. Encapsulated cryptococci can be opsonized by C3b and by antic-apsular antibodies [27], and binding of encapsulated cryptococci may occur through multiple receptors [88]. Once internalized, however, the fungus may survive in an acidic phagolysosome [104] or secrete putative virulence molecules [105], which may account for its intracellular parasitism [106].

Interaction and phagocytosis of human macrophages with *Candida* yeasts is influenced by many factors, including C3 [107], soluble mannose-binding proteins [108], and vibronectin [109]. Even unopsonized *C. albicans* yeasts can be internalized through the macrophage MR [83]; however, the highest killing activity is observed against opsonized yeasts [110,111]. Actually, the enhancement of phagocytosis and killing of *C. albicans* by macrophages correlated with a decreased number of mannose receptors [111].

The candidacidal activity of macrophages largely depends on the anatomical location and the forms of the fungus [112]. Thus, fresh murine peritoneal macrophages have lower fungicidal capacity than either murine or human monocytes and alveolar macrophages. The low fungicidal activity of macrophage precursors is restored by culture in the presence of macrophage colony-stimulating factor (M-CSF) [113].

In infected tissues, *Candida* yeasts and hyphae are seen both within cells and extracellu-larly. Studies in vitro have shown that the two forms of the fungus differ in their mode of entry in and killing by macrophages. The filamentous form of *Candida* is not phagocytosed by a murine macrophage cell line [114], but could penetrate intact macrophages even when phagocytosis is blocked [115].

The two forms of the fungus also show different susceptibility to proteolytic activity of macrophages [116]. Although the two forms of the fungus utilize different mechanisms of entry, phagocytosis of yeasts and germ tubes is followed by the rapid recruitment of late endosomes and lysosomes in macrophages [115].

Little is known about mechanisms of recognition and entry of *A. fumigatus* conidia in pulmonary alveolar macrophages, known to represent the first line of defence against *Aspergillus* conidia [15,68]. Murine macrophages have been shown to recognize and attach to *A. fumigatus* conidia through lectin-like attachment sites in the absence of opsonins and also human mononu-clear cells bind conidia by a β-1,4-glucoside-inhibitable receptor [15]. Alveolar macrophages ingest inhaled conidia very rapidly, destroy them intracellularly, and prevent germination to hyphae, the invasive form of the fungus. Resting conidia are to some extent more resistant that swollen conidia to both oxygen-dependent and oxygen-independent metabolites (see below). In terminal airways, complement and antibodies cannot be readily available, and therefore alveolar macrophages are able to recognize and bind conidia even in the absence of opsonins. In addition, conidia poorly activate the complement system by the classical pathway, and, even when opsonized, they only trigger a modest oxidative burst.

Various enzymes, such as elastases and proteases produced by the fungus, might play an important role in the ability of conidia to evade phagocytosis by alveolar macrophages, to resist hydrolysis by endogenous peptides, and in the ability of germinating conidia to cross anatomical barriers [15]. As the second line of defense, circulating poly- and mono-nuclear cells will then be recruited at the site of infection [15]. Overall, impaired macrophage function, combined with qualitative or quantitative defects of polymorphonuclear neutrophils, results in invasion of tissue and vessels, causing the development of invasive pulmonary aspergillosis.

Accumulating evidence points to a unique role of DCs in infections, as they are regarded as both essential for innate recognition and initiation of Th cell differentiation and functional commitment [117]. DCs serve as antigen-presenting cells in cryptococcosis [118] and aspergillosis [119]. Recent studies indicate that DCs phagocytose *C. albicans* [120,121]. Interestingly, the yeasts and hyphae of *C. albicans* are internalized through different receptors and phagocytic mechanisms [121]. Engulfment of yeasts occurs via coiling, overlapping phagocytosis, eventually leading to phagolysosome formation, where different stages of progressive yeast degradation are seen. In contrast, internalization of hyphae appears to occur through a more conventional zipper-type phagocytosis. Internalization of either form is inhibitable, although to a different extent, by soluble mannan. Once inside the cells, hyphae appear to promote rupture of the phagosomal membrane and escape into the cytoplasm. Thus, not only are yeasts and hyphae ingested through different forms of phagocytosis, but, once inside the cells, they reside in distinct cellular compartments. After phagocytosis of yeasts or hyphae, the downstream cellular events are clearly different (see Sec. IV.D.). Ingestion of yeasts, but not hyphae, activates DCs for candidacidal activity, including nitric oxide (NO) production. Thus, dendritic cells may fulfill the requirement of a cell uniquely capable of sensing the different forms of a fungus in terms of antifungal effector functions and type of Th immune responses elicited (further discussed in Sec. IV.D.).

E. Nonphagocytic Cells

NK cells serve as an early defense against some intracellular pathogens, by their ability to kill infected cells and to produce cytokines activating phagocytic cells. A number of studies suggests that NK and T-cells participate in early innate defence against fungi, through direct antimicrobial activity and by the release of cytokines activating phagocytes. Human NK cells bind to fungi [10]. The interaction of NK cells with fungi may result in inhibition of fungal growth [122] and either stimulation [123] or inhibition [124] of release of cytokines that activate macrophages to a fungicidal state. Studies in mice, however, failed to demonstrate an essential role for NK cells in the early host response to fungi, although fungi may activate NK and large granular lymphocytes in vivo [56,125,126] and stimulation of NK cells may increase resistance to fungal infections [127,128].

Among T-cells, those bearing the γδ chain of the TCR are found in most epithelia and may contribute to immunosurveillance at the body surface. Human T-cells bearing the αβ or the γδ TCR respond to *P. brasiliensis* and *P. carinii* and are crucially involved in supporting polyclonal B-cell activation [129]. TCR γδ+ cells enhance the candidacidal activity of macrophages at the mucosal surface through release of activating cytokines [130], but they are dispensable in host resistance to *P. carinii* [131]. Direct antimicrobial activity, reduction of fungal adherence, and release of interferon (IFN)-γ have all been observed upon binding of fungi to human CD3$^+$ cells [132,133].

IV. NATURAL IMMUNITY IN ACTION

A. Effector Mechanisms

Freshly isolated mononuclear and polymorphonuclear cells express intrinsic antifungal activity. However, numerous studies have clearly demonstrated that expression of the full antifungal function requires activation by either CSFs or by T-cell-derived cytokines (see Sec. IV.B.).

As already pointed out, the antifungal functions of effector phagocytes include mechanisms of killing and growth inhibition of fungi as well as pathways to oppose fungal infectivity, including effects on dimorphism [134] and phenotypic switching [135]. The restriction of fungal growth occurs by both oxygen-dependent [55,83,84,111,136–138] and oxygen-independent mechanisms [64,136,139–141], intracellular or extracellular release of effector molecules [140], defensins [23], neutrophil cationic peptides [64,139–141], and iron sequestration [142].

The oxidative killing occurs through production of toxic reactive oxygen intermediates (ROIs), the nature of which varies depending on the nature of pathogens and type of phagocytic cells. In retaliation, fungi have evolved strategies to selectively inhibit the respiratory burst [143]. The production of ROIs is initiated by the nicotinamide adenine dinucleotide phosphate (NADPH)-oxidase complex. The activation of NADPH-oxidase can be elicited by microbial products (lipolysaccharide; LPS), by IFN-γ, interleukin (IL)-8, or by IgG binding to FcR. The primary product of the reaction catalyzed by the NADPH-oxidase is superoxide (O_2^-), which can be converted to H_2O_2 by superoxide dismutase (SOD), to hydroxyl radicals (·OH) and hydroxyl anions (OH$^-$) by the iron-catalyzed Haber-Weiss reactions or, after dismutation to H_2O_2, to hypochlorous acid (HOCl) and chloramines by either myeloperoxidase-dependent or -independent mechanisms [144]. Precisely which molecules are responsible for killing specific fungi are difficult to determine experimentally. One approach to implicate specific ROIs is to examine the effects of specific inhibitors or scavengers on oxidative killing by phagocytes, such as SOD to scavenge O_2^-, catalase to destroy H_2O_2, mannitol or DMSO to scavenge OH$^-$ and istidine to scavenge single oxygen ($^1O_2$). It is worth mentioning that some of these inhibitory factors represent important virulence determinants of fungi [2–4,137].

The hydrogen peroxide–myeloperoxidase–halide system is the main oxidative pathway utilized by human PMNs to kill fungi [137,145,146]. Production of ROIs is sufficient for hyphal killing by PMNs, even though oxidant generation may not always be sufficient to mediate hyphal killing without complementary nonoxidative mechanisms [141]. It is worth emphasizing that opsonization can greatly affect the early pertussis-sensitive events in the activation of the respiratory burst of PMNs, such as phospholipase C activation, the rise in cytosolic free Ca^{2+} ions, and actin polymerization [147–149].

Myeloperoxidase is a lysosomal hemoprotein found in azurophilic granules of neutrophils and in monocytes but not monocyte-derived macrophages, which lack myeloperoxidase [83]. It acts by itself as a mediator in the oxygen-dependent killing of fungi [107,150,151]. However, myeloperoxidase deficiency, while predisposing to candidiasis [150], does not predispose to invasive aspergillosis [152], suggesting that a defective myeloperoxidase-independent pathway may predispose to invasive aspergillosis. That defective NADPH-oxidase complex likely represents a predisposing factor to infection is demonstrated by the high susceptibility to aspergillosis of patients with chronic granulomatous disease [153,154] and by the failure of PMNs from these patients to kill *C. neoformans* [155].

Oxidant release also represents one major mechanism of fungal growth inhibition by activated human monocytes or monocyte-derived macrophages [1,83,84,111]. Human macrophages respond to *H. capsulatum* engulfment with a vigorous respiratory burst, an activity potentiated by binding to collagen [156] and inhibited by the yeast [157]. Monocyte/macrophages from nonimmune persons also effectively phagocytose *C. immitis*, but fail to restrict its intracellular growth by lack of phagosome-lysosome fusion, an activity restored by exposure to T-cell-derived cytokines ([1] and references therein). The expression of optimal antifungal activity against *P. carinii* [158] and *C. neoformans* [159] mainly occurs through the production of superoxide. However, production of oxidant scavengers such as melanin and mannitol can protect *C. neoformans* from killing by mediators such as NO, HOCl, and H_2O_2 [137,160].

Additional toxic molecules produced by phagocytes are the reactive nitrogen intermediates (RNIs). Different NO synthases convert the amino acid L-arginine and molecular oxygen to L-citrulline and NO. It is remarkable that the NO synthase of macrophages is stimulated by LPS and oppositely regulated by Th1 and Th2 cytokines (it is induced by IFN-γ and TNF-α and inhibited by IL-4, IL-10, IL-13, and TGF-β) [144]. Production of NO occurs in rat [161–163] and murine [164–167] macrophages in response to fungi; in retaliation, fungi also inhibit NO production [168,169]. The pleiotropic activity of NO, which includes a direct activity on fungal growth and on transition through different forms [170], as well as a negative immunomodulatory activity [144], precludes a general conclusion on the significance of NO production in fungal infections. It is not surprising, therefore, that a positive correlation [161–164,166] or no correlation [171–173] has been found between NO production and inhibition of fungal growth. Since there is no definitive evidence that human macrophages synthesize NO, the contribution of this nitrogen intermediate to human host defenses is still unknown.

Iron is an essential nutrient for fungi. Iron sequestration in response to infection is a demonstrated host defense mechanism, and thus, iron acquisition is an important pathogenetic determinant [142]. Iron restriction by activated macrophages represents a mechanism to control not only fungal viability, but also intracellular transition from conidia to yeasts [174] and synthesis of cryptococcal polysaccharide capsule [175]. Tight regulation of phagolysosomal pH is important to limit iron acquisition by fungi, as the element is not accessible at a pH > 6.5. Thus, raising the phagolysosomal pH increases the antifungal activity of human mononuclear phagocytes [176].

B. Chemokine and Cytokine Release

Upon contact with a pathogen, cells of the innate immune system release a battery of cytokines that have profound effects on the functional activity of the innate response and on subsequent events. The local release of these effector molecules serves to regulate cell trafficking of various types of leukocytes, thus initiating the inflammatory response, and to activate phagocytic cells to a microbicidal state.

Some of the cytokines released belong to a family of closely related proteins called chemokines, small polypeptides (~8–14 kDa) that are synthesized by PMNs, macrophages, endothelial cells, keratinocytes of the skin, fibroblasts of connective tissue, and T-lymphocytes. About 40 chemokines have now been identified in humans. They mainly act on neutrophils, monocytes, lymphocytes, and eosinophils and play a pivotal role in host defense mechanisms. However, the chemokines have a wide range of effects in many different cell types beyond the immune system, including various cells of the central nervous system and endothelial cells. Chemokines have been divided into the two major subfamilies on the basis of the arrangement of the two N-terminal cysteine residues, CXC and CC, depending on whether the first two cysteine residues have an amino acid between them (CXC) or are adjacent (CC).

The long-standing observation that fungi have the ability to attract leukocytes at the site of infection [10] is now supported by the findings that fungi stimulate release of chemokines from many cells [177–186]. Monocyte chemotactic protein-1 (MCP-1) and macrophage inflammatory protein − 1 alpha (MIP-1 alpha) and TCA3, together with other chemokines, are essential in mediating local leukocyte recruitment in the lung of mice with cryptococcosis [177,178,180,181,187], *P. carinii* pneumonia [183], and aspergillosis [184]. Local production of chemokines are also observed in mice with vaginal candidiasis [185]. In vitro studies have indicated that the macrophage chemokine response to *C. albicans* is not mediated by the MR [182]. Of interest, the chemokine response of macrophages to *C. neoformans* and *C. albicans* is not different in persons with and without HIV infection [186], a finding suggesting that the scant inflammatory response often seen in AIDS patients with cryptococcosis and candidiasis is not secondary to suboptimal beta-chemokine release. Studies in mice with genetically disrupted chemokine receptors [188–192] or chemokine gene [193] further add to the important role of chemokines in the early inflammatory response as well as to the type 1 and type 2 cytokine balance.

Cytokines are small protein or glycoprotein messenger molecules that convey information from one cell to the other. Most are secreted but some can be expressed on the cell surface or held in reservoirs in the extracellular matrix. More than 200 have now been identified including the interleukins, growth factors, chemokines, interferons, and a host of others. All cytokines bind to specific receptors expressed on the surface of the target cell, thereby triggering complex intracellular signaling cascades, which ultimately control gene expression required for the cellular response.

Interaction of host cells with fungi results in the release of many cytokines. However, what is important is to define the biological significance and the role of a given cytokine in the context of the infection. This may not be an easy task, given the pleiotropic, redundant, synergistic, and antagonistic activities of cytokines. The definition of the biological significance also depends on parameters of infection selected relative to the cytokine effect, as, for instance, levels of cytokines may [194] or may not [195] correlate with clinical severity of the infection. Moreover, a particular cytokine by itself is beneficial or deleterious, its ultimate effect depending on the context in which it is produced or operates.

The pattern of early cytokine production distinguishes susceptible and resistant mice with cryptococcosis, candidiasis, aspergillosis, histoplasmosis, paracoccidioidomycosis, and coccidioidomycosis ([1] and references therein). The CSFs augment the number of circulating phagocytes and their precursors. Moreover, CSFs have been shown to enhance activation of the fungicidal capacity of phagocytic cells in vitro ([1] and references therein) Ablation of granulocyte-macrophage CSF increases susceptibility of mice to fungal infections and is associated with depressed macrophage activation [196,197].

TNF-α, IL-1β, and IL-6 are proinflammatory cytokines readily produced upon interaction of phagocytes with fungal cells, in vitro and in vivo, in mice and humans. TNF-α is an essential cytokine in the innate control of infections caused by *C. albicans* [1,198–200], *C. neoformans* [1,201], *A. fumigatus* [184,190,202], *P. carinii* [127], and *H. capsulatum* [203]. It regulates the recruitment of inflammatory cells [190,198,200] through induction of chemokine release [190], triggers the respiratory burst [204] and NO production [203] in phagocytes, and induces costimulatory molecules [199]. TNF-α is not required for IL-12 production [199,201], as it is for IL-6 [127,199]. TNF-α production by PMNs [205] and monocytes [206] is opsonic dependent and inhibited by cryptococcal polysaccharide [207].

Although IL-1 shares many properties with TNF-α, IL-1β, similarly to IL-6, mainly acts through recruitment of PMNs. IL-1β does not seem to be as essential as TNF-α in the innate antifungal response. Administration of IL-1β has been beneficial in mice with candidiasis;

nevertheless, IL-1 deficiency does not increase the susceptibility of mice to the infection as it does IL-6 deficiency [1,208,209].

IFN-γ is a key cytokine in the innate control of fungal infections [210,211]. It is produced by T-cells and NK cells in response to both IL-12 and IL-18 released by PMNs, macrophages, and DCs upon exposure to fungi. IFN-γ, in turn, stimulates migration, adherence, phagocytosis, and oxidative killing of PMNs and activates macrophages for fungicidal activity. In addition, IFN-γ may stimulate the antifungal activity of endothelial cells and decrease the expression of lung-surface integrins, thus reducing fungal attachment [212]. Studies in mice [1,56,213] and humans [214–216] have clearly shown the importance of IFN-γ as one major determinant of resistant to fungal infections. However, studies in IFN-γ-deficient mice have revealed the dispensability of IFN-γ as effector molecule [217,218], thus unraveling the complex role of IFN-γ in the intimate intricacy between the innate and the adaptive immunity to fungi. It was found that IFN-γ promotes Th-1 reactivity by its ability to maintain IL-12 responsiveness on CD4$^+$ cells [218].

Although IL-12 enhances the antifungal activity of phagocytes independently of IFN-γ [219], the induction of IFN-γ is the main activity of IL-12 [220] and IL-18 [221] in fungal infections. Upon phagocytosis of fungi, IL-12 is produced by different cells, including PMNs [71,222], macrophages [223], and DCs [121]. The production of IL-12 very likely depends on the mode of entry of the fungus (see Sects. IV.D and IV.E). Studies in mice have shown that IL-12 is essential for the induction of protective Th1 responses to fungi [202,220,224–226]. Studies in humans, however, have failed to unequivocally confirm the association of IL-12 with the occurrence of protective responses to fungi [220]. For instance, IL-12 expression and production in response to *C. neoformans* and *C. albicans* were similar in persons with and without HIV infection [227].

One most important cytokine produced in the course of fungal infections is IL-10. It is readily produced by PMNs [71,228] and macrophages [228,229] upon phagocytosis of fungi and plays a crucial role in determining susceptibility to fungal infections ([1] and references therein). IL-10 acts by impairing the antifungal effector functions of phagocytes [230,231], including secretion of inflammatory cytokines [230] and IL-12 [232], and by inhibiting the development of protective cell-mediated immunity [233]. It also acts as one major cytokine discriminating between virulent and less virulent forms of fungi [71,233,234]. However, later on the course of the infection, high-level production of IL-10 may contribute to resolution of the infection and the inflammatory response ([1] and references therein).

Similar to IL-10, transforming growth factor (TGF)-β is produced by macrophages in mice with candidiasis, and appears to discriminate between virulent and low-virulence *C. albicans* [235].

Numerous studies indicate that, in addition to IL-10, IL-4 may act as one major discriminative factor of susceptibility and resistance in certain fungal ([1] and references therein). Ablation of IL-4 renders susceptible mice resistant to candidiasis and aspergillosis [236,237]. Although IL-4 may inhibit the fungicidal activity of phagocytes ([1] and references therein), it also activates phagocytes [224,238]. Thus, one likely mechanism of the inhibitory activity of IL-4 in infections relies on its ability to promote Th2 reactivity, thus dampening protective Th1 responses. One major unresolved issue concerns the cell source of IL-4 in infection. Only recently has decisive evidence been obtained in this regard. It has been found that DCs produce IL-4 upon phagocytosis of hyphae but not yeasts of *C. albicans* [121], thus sensing the different forms of the fungus in terms of production of directive cytokines. This issue will be further discussed below (Section IV.D.).

C. The Inflammatory Response

The recruitment of inflammatory cells at sites of the infection requires the coordinated action of proteolytic enzymes [239], leukotriens [240,241], chemokines [177,178,181,183,184,187], and cytokines [213,242]. Beside PMNs and macrophages, eosinophils may be an important component of the inflammatory reaction to fungi [242,243]. Because of their toxic granule contents, eosinophils are potentially toxic to host tissues. Thus, it is difficult to distinguish between eosinophils as antimicrobial effector cells and eosinophils as participants in tissue damage after recruitment by inflammatory signals. Indeed, their presence is indicative of a poor prognosis in human coccidioidomycosis [243]. Nevertheless, evidence indicate that eosinophils phagocytose opsonized *C. albicans* [244] and *C. neoformans* [245], thus suggesting that eosinophils may behave as effector antifungal cells. However, the paucity of information implicating eosinophils in human infections precludes a better definition of the role of eosinophils in the fungus-host interaction.

The inflammatory response to fungi may serve to limit the infection but may also represent an important determinant of pathogenicity [246]. Recovery from infection does not always correlate with resolution of inflammation. In experimental candidiasis, the course and outcome of the infection in different strains of mice correlate with fungal load but also with immunopathology [56]. Ashman has provided evidence of the existance of at least two Mendelian-type resistance genes (*Carg1* and *Carg2*) that control the host response to systemic *C. albicans* infection in mice. The genes appear to affect distinct parameters of infection, *Carg1* determining the extent of tissue distruction and *Carg2* influencing the susceptibility of the kidney. Allocation of presumptive "resistant" and "susceptible" alleles of these genes among various inbred strains gives an excellent correlation with the various measures of infection ([1] and references therein).

D. The Instructive Role

Data are now accumulating for an essential role of the innate immunity in orchestrating the subsequent adaptive immunity to pathogens [6–8]. In the vertebrate host, the innate and adaptive immune responses are now considered to be integrated as a single immune system, with the innate response preceding, and being necessary for, the adaptive immune response. Through recognition of invariant molecular structures shared by large groups of pathogens (also known as PAMPs, pathogen-associated molecular patterns) by a set of germline-encoded receptors (referred to as pattern recognition receptors, PRRs) of host cells, the innate immune system fulfills the requirement of discrimination between different types of pathogens [6–8,52,247–249]. The products of these recognition events are the production of various antimicrobial peptides in mammals and insects as well as the expression of cytokines, chemokines, and costimulatory molecules through which the innate immune system instructs and contributes to the discrimination of pathogens from self in adaptive immunity.

The instructive role of the innate immune system in the adaptive immune responses to fungi occurs at different levels. The initial handling of a fungal pathogen by cells of the innate immune system plays a major role in determining CD4$^+$ Th development. Indeed, qualitative or quantitative defects of antifungal effector and immunoregulatory functions of phagocytic cells results in the development of anticandidal Th2, rather than Th1, cell responses [71]. As CD4$^+$ Th cell differentiation in vivo is critically affected by the tissue fungal load, an important role of phagocytic cells relies on their ability to control the fungal growth through various antifungal effector mechanisms. In addition to this, the instructive role of the innate immune system in the adaptive immune response to fungi may be operative at the levels of expression of costimulatory molecules [250,251], and chemokine and cytokine production (see above).

In candidiasis, neutrophils, more than macrophages, are endowed with the ability to produce directive cytokines, such as IL-10 and IL-12. Most importantly, IL-12 appears to be released in response to a *Candida* strain that initiates Th1 development in vivo but IL-10 is released in response to a virulent strain [71]. By producing directive cytokines, such as IL-10 and IL-12, neutrophils influence antifungal Th cell development, as evidenced by the inability of neutropenic mice to mount protective anticandidal Th1 responses. However, Th1-mediated resistance is increased upon IL-12 administration in neutropenic mice or IL-10 neutralization in nonneutropenic mice. Thus, neutrophils, through the differential production of directive cytokines, may directly contribute to discriminative Th responses to virulent and nonvirulent forms of the fungus. Human neutrophils also produce bioactive IL-12 in response to a mannoprotein fraction of *C. albicans*, capable of inducing Th1 cytokine expression in peripheral blood mononuclear cells [222]. Because of the large number of neutrophils present in the blood or inflammatory tissues in infection, it is likely that neutrophil production of cytokines may influence the development and/or maintenance of the Th repertoire to *C. albicans*. Interestingly, it has recently been reported that neutrophils quickly release *Candida* antigens upon phagocytosis [252]. Thus, it is likely that the immunoregulatory role of neutrophils in infections may go beyond their cytokine production, to include signaling through antigen presentation and costimulation. Ultimately, this would be a likely expectation shared with other cells of the innate immune system.

DCs are exquisitely sensitive to the different forms of *C. albicans*. One important virulence factor of the fungus is its ability to reversibly switch from a unicellular yeast form into various filamentous forms, all of which can be found in tissues [3]. However, whether it is the yeast or the hyphal form that is responsible for pathogenicity is still an open question. One possibility is that the filamentous growth form is required to evade the cells of the immune system, whereas the yeast form may be the mode of proliferation in infected tissues. To make it likely, a cell should exist that finely discriminates between the two forms of the fungus in terms of class of immune response elicited. DCs phagocytose both yeasts or hyphae of the fungus. However, the downstream cellular events after internalization of either form of the fungus, are clearly different. In vitro, ingestion of yeasts activates dendritic cells for IL-12 and NO production and priming of Th1 cells, while ingestion of hyphae inhibits IL-12, NO, and Th1 priming. In vivo, generation of antifungal protective immunity is observed upon injection of DCs ex vivo pulsed with *C. albicans* yeasts but not hyphae. The immunization capacity of yeast-pulsed dendritic cells is lost in the absence of IL-12, whereas that of hypha-pulsed dendritic cells is gained in the absence of IL-4 [121]. These results indicate that DCs fulfill the requirement of a cell uniquely capable of sensing the virulent and nonvirulent forms of *C. albicans* in terms of type of immune responses elicited. Thus, DCs appear to meet the challenge of Th priming and education in *C. albicans* saprophytism and infections.

E.   The Discrimination: PRRs and Toll-Like Receptors

Unlike the adaptive response, the innate responses to pathogens involve recognition mechanisms that are based on relatively invariant receptors, and they do not lead to the lasting protective immunity against the inducing pathogen that is the hallmark of adaptive immunity. The Toll-like receptors (TLRs) can be considered as PRRs that subserve this recognition [8]. Toll was originally defined as a *Drosophila* gene important for ontogenesis and antimicrobial resistance. Interestingly enough, Toll and other genes are all required for the rapid transcriptional induction of the gene encoding an antifungal peptide, drosomycin, in response to infection [247]. The recognition of sequence similarity between the cytoplasmic portion of Toll and that of signaling

IL-1 receptor represented the merging point of *Drosophila* work with more conventional cytokine/innate immunity research.

TLRs are type I transmembrane proteins that are grouped into the same gene family based on their sequence similarity. Six human members have been cloned and partially characterized, but there is evidence for 14 TLRs in the human genome. TLRs cross the cytoplasmic membrane once, and their intracellular portion is extremely similar to the cytoplasmic domain of the IL-1 receptor and related molecules. TLRs activate specialized antifungal genes through the activation of NF-kB family members.

Distinct TRLs now appear to be involved in the recognition of PAMPs on different classes of pathogens: LPS of gram-negative bacteria, lipoteichoic acid and peptidoglycan of gram-positive bacteria, mannan and glucan of fungi, and phosphoglycan of parasites [8]. TLR4, identified as an essential component of the LPS receptor signaling complex, is expressed on cells of the immune system, and its overexpression causes the release of cytokines (IL-1 and IL-6) and the expression of costimulatory molecules that are essential for activation of the adaptive immunity. The evidence suggests that CD14 recruits LPS to TLR proteins, thereby facilitating optimal signal transduction. TRL2, in contrast, mediates responses to gram-positive bacteria and yeasts [253].

A striking common feature of microbial patterns is their polysaccharide chains. It is therefore not surprising that a large number of TLRs identified in vertebrates are lectin. TLRs are borne by particular types of cells such as macrophages, DC and NK cells, and probably also epithelial and endothelial cells in the lung, kidney, skin, and gastrointestinal tract. The recognition of pathogens by TLRs triggers their engulfment and the release of cytokines IL-1, IL-6, and TNF-α. Mammalial TLRs include MRs, scavenger receptors, integrins, collectins, and some clusters of differentiation antigens [52].

MRs and other β-glucan receptors are known receptors on neutrophils, macrophages, and DCs that may work as TLRs recognizing PAMPs of fungi. The early observation that phagocytes from different anatomical districts exhibit an heterogeneous secretory response to the different forms of *C. albicans* [112] may thus be revisited in terms of cell heterogeneity in the use of TLRs upon contact with fungal PAMPs.

Although the study of the involvement of TLRs and of other recognition receptors in response to fungi is a relatively novel and emerging field, several considerations can be drawn. CR3 engagement is the most efficient in uptake of opsonized yeasts. However, for effective killing and cytokine release to occur, the concomitant engagement of the FcγRI or FcαRI receptors is required [60,64,66,89]. Thus, signaling through CR3 may not lead to phagocyte activation. It has long been known that interaction with macrophage CR3 leads to suppression of the immune response to *C. albicans* [84]. Recently, it has been demonstrated that signaling via CR3 downregulates IL-12 production in response to *H. capsulatum* [254]. Thus, failure to induce IL-12 may account for the lack of development of a protective response to fungi. This means that the use of the different recognition receptors, either alone or simultaneously, may result in coupling or uncoupling phagocytosis with inflammatory and immunological responses and that this enables the innate immune system to decide whether or not to implicate the adaptive immune system.

It has recently been shown that suppression of IL-12 production also occurs in dendritic cells in response to the hyphae of *Candida* [121]. This finding suggests that the use of different PRRs by different forms of fungi may represent one important strategy of immune evasion and parasitism. If different PRRs are used by opsonized and unopsonized fungi, this implies that the activation of phagocytes to a fungicidal state in vivo may vary under divergent conditions at specific localized sites of infection.

One further important observation is that the functioning of the TLRs and PRRs is sensitive to cytokines. The mannose receptor-mediated uptake of unopsonized *Candida* yeasts may lead to phagocyte abuse if not accompanied by the coordinate activation of the cell cytotoxic machinery [84]. Increased expression of the receptor with no induced cytotoxicity is induced upon exposure of macrophages to IL-4 [238]. In contrast, IFN-γ downregulates the expression of macrophage mannose receptors and nevertheless results in effective killing, presumably via increased coupling of the receptor to cytotoxic functions [84]. Thus, cooperation between Th1 and Th2 cytokines may be required for optimal stimulation of mannose receptor-mediated phagocytosis [255].

Production of cytokines and chemokines in response to fungi also appears to involve different PRRs. Production of some cytokines, such as IL-1β, IL-6, and GM-CSF, is mediated by the MRs, while some chemokine responses may be mediated by other receptors [182]. This may help to explain why hyphae of *Candida*, although not internalized by a macrophage cell line, are capable of inducing production of inflammatory cytokines to an higher extent than phagocytosed yeasts [114].

Finally, the relative importance of each PRR in the antifungal innate reactivity is highlighted by several findings, as already discussed in section IV.A. The internalization of *P. carinii* through the macrophage MR accounts for the ability of alveolar macrophages to restrict the growth of the fungus [82], such that the downregulated mannose receptor expression in alveolar macrophages of HIV-infected persons was considered to contribute to susceptibility to *P. carinii* pneumonia [96]. However, internalization via constitutively competent macrophage MR does not represent an effective way of clearance of yeasts, such as *C. neoformans* and *C. albicans* (see Sec. IV.A.). Against these yeasts, it is interesting to note that signaling through CD14 greatly increases the fungicidal and secretory activity of effector phagocytes [256,257]. Thus, it is clear that TLRs are differently involved in recognition of PAMPs on fungi.

It would appear that the different PRRs have hierarchical organization. In terms of activation potential, at least tiers can be identified. The lower tier comprises diverse PRRs, including CR3, MRs, scavenger receptors, and CD14, receptors which tend to be abundant on the surfaces of first-response leukocytes (such as PMNs, mononuclear phagocytes, and NK cells) and are called upon to serve as early-warning systems that rapidly recognize and respond to microbial invaders in an antigen-independent manner. Nevertheless, although the ability to react rapidly to an infectious challenge is important, it is equally important that the response be measured and in proportion to the magnitude of the threat. Thus, it is not surprising that the ability of the first-tier receptors to activate, in isolation, various effector functions is limited; in this regard, CR3 is clearly representative of the entire class. Cell activation therefore requires additional signals, such as receptor clustering, the presence of cytokines, and cooperation with other types of receptors.

The second tier of recognition receptor is adaptive and antigen specific; the prototype is FcγRIII. Ligation of FcγRIII alone is usually sufficient to trigger phagocytosis, a vigorous oxidative burst, and generation of proinflammatory signals. Ultimately, recognition of antibody-opsonized particles represents a high-level threat.

V. EVOLUTION OF THE IMMUNE RESPONSE TO FUNGI

Not only has the innate immune system a decisive role in the orchestration of subsequent adaptive Th immunity to fungi, but cells of the adaptive immune system may feedback on the innate immune system. The optimal expression of the antifungal effector activity of cells of the innate immune system is attained in association with antigen-driven immune responses ([1] and references therein). This may occur through the release of cytokines with activating (IFN-γ, TNF-α) and deactivating (IL-4/IL-10) signals to effector phagocytes [211], which may include an activity on phagosome to lysosome fusion within phagocytes [258]. The finding that Th1 and Th2 cytokines may affect the expression of TLRs [255] provides us with a novel ground of interaction between the innate and adaptive immune system to fungi.

VI. CONCLUSIONS

For many years, innate immunity has been considered as a separate entity from the adaptive immune response to pathogens and has been regarded to be of secondary importance in the hierarchy of immune functions. It is now clear that the innate and adaptive immune systems cooperate in response to fungi, being the functioning of lymphocytes bearing clonally rearranged receptors dependent on signal provided by the innate recognition system (Fig. 2). The identification of PRRs as a critical component of the innate recognition system is a fundamental breakthrough in immunology, but much remains to be learned. The discoveries of PRRs, including TLRs, may have important clinical implications. Neutrophils from patients with advanced HIV infection have impaired CR function and preserved FcγR function [259]. It appears that targeting the PRRs is an attractive strategy of vaccination and therapy of fungal diseases.

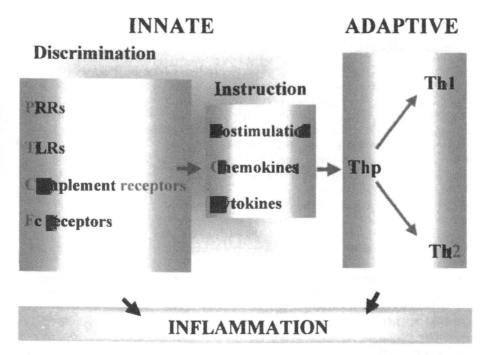

Figure 2 The integrated view of the innate and adaptive immune responses to fungi. The innate immune system discriminates between different fungi and different forms of them, and produce sets of chemokines, cytokines, and costimulatory molecules through which signals to the adaptive T-helper (Th) immune system. Together, they contribute to the inflammatory response.

ACKNOWLEDGMENTS

Many thanks are due Drs. Antonella Mencacci and Elio Cenci, and Jo-Anne Rowe for invaluable secretarial support. This work was supported by grants from the AIDS 50B.33 project.

REFERENCES

1. L Romani, SH Kaufmann. Immunity to fungi. Res Immunol 149:277–281, 1998.
2. SE Vartivarian. Virulence properties and nonimmune pathogenetic mechanisms of fungi. Clin Infect Dis 14:S30–S36, 1992.
3. R Calderone. Molecular pathogenesis of fungal infections. Trends Microbiol 2:461–463, 1994.
4. LH Hogan, BS Klein, SM Levitz. Virulence factors of medically important fungi. Clin Microbiol Rev 9:469–488, 1996.
5. MA Ghannoum. Potential role of phospholipases in virulence and fungal pathogenesis. Clin Microbiol Rev 13:122–143, 2000.
6. DT Fearon, RM Locksley. The instructive role of innate immunity in the acquired immune response. Science 272:50–53, 1996.
7. R Medzhitov, CA Janeway Jr. Innate immunity: impact on the adaptive immune response. Curr Opin Immunol 9:4–9, 1997.
8. KV Anderson. Toll signaling pathways in the innate immune response. Curr Opin Immunol 12: 13–19, 2000.
9. JW Murphy. Mechanisms of natural resistance to human pathogenic fungi. Annu Rev Microbiol 45:509–538, 1991.
10. J Murphy, H Fiedman, M Bendinelli, eds. Fungal Infections and Immune Responses. New York: Plenum Press, 1993.
11. E Balish, RD Wagner. Probiotic bacteria for prophylaxis and therapy of candidasis. Revista Iberoamericana de Micologia 15:261–264, 1998.
12. JM Schroder, J Harder. Human beta-defensin-2. Int J Biochem Cell Biol 31:645–651, 1999.
13. TL Benfield, B Lundgren, JH Shelhamer, JD Lundgren. *Pneumocystis carinii* major surface glycoprotein induces interleukin-8 and monocyte chemoattractant protein-1 release from a human alveolar epithelial cell line. Eur J Clin Invest 29:717–722, 1999.
14. P Borger, GH Koeter, JA Timmerman, E Vellenga, JF Tomee, HF Kauffman. Proteases from *Aspergillus fumigatus* induce interleukin (IL)-6 and IL-8 production in airway epithelial cell lines by transcriptional mechanisms. J Infect Dis 180:1267–1274, 1999.
15. JP Latge. *Aspergillus fumigatus* and aspergillosis. Clin Microbiol Rev 12:310–350, 1999.
16. RM Des Prez, S Steckley, RM Stroud, J Hawiger. Interaction of *Histoplasma capsulatum* with human platelets. J Infect Dis 142:32–39, 1980.
17. L Christin, DR Wysong, T Meshulam, R Hastey, ER Simons, RD Diamond. Human platelets damage *Aspergillus fumigatus* hyphae and may supplement killing by neutrophils. Infect Immun 66: 1181–1189, 1998.
18. R Robert, S Nail, A Marot-Leblond, J Cottin, M Miegeville, S Quenouillere, C Mahaza, JM Senet. Adherence of platelets to *Candida* species in vivo. Infect Immun 68:570–576, 2000.
19. JF Tomee, AT Wierenga, PS Hiemstra, HK Kauffman. Proteases from *Aspergillus fumigatus* induce release of proinflammatory cytokines and cell detachment in airway epithelial cell lines. J Infect Dis 176:300–303, 1997.
20. YQ Tang, J Yuan, CJ Miller, ME Selsted. Isolation, characterization, cDNA cloning, and antimicrobial properties of two distinct subfamilies of alpha-defensins from rhesus macaque leukocytes. Infect Immun 67:6139–6144, 1999.
21. RI Lehrer, T Ganz. Antimicrobial peptides in mammalian and insect host defence. Curr Opin Immunol 11:23–27, 1999.
22. M Cutuli, S Cristiani, JM Lipton, A Catania. Antimicrobial effects of alpha-MSH peptides. J Leukoc Biol 67:233–239, 2000.

23. P Eggleton, KB Reid. Lung surfactant proteins involved in innate immunity. Curr Opin Immunol 11:28–33, 1999.

24. MA Couto, L Liu, RI Lehrer, T Ganz. Inhibition of intracellular *Histoplasma capsulatum* replication by murine macrophages that produce human defensin. Infect Immun 62:2375–2378, 1994.

25. AM Walenkamp, AF Verheul, J Scharringa, IM Hoepelman. Pulmonary surfactant protein A binds to *Cryptococcus neoformans* without promoting phagocytosis. Eur J Clin Invest 29:83–92, 1999.

26. T Madan, P Eggleton, U Kishore, P Strong, SS Aggrawal, PU Sarma, KB Reid. Binding of pulmonary surfactant proteins A and D to *Aspergillus fumigatus* conidia enhances phagocytosis and killing by human neutrophils and alveolar macrophages. Infect Immun 65:3171–3179, 1997.

27. H Koziel, DS Phelps, JA Fishman, MY Armstrong, FF Richards, RM Rose. Surfactant protein-A reduces binding and phagocytosis of *pneumocystis carinii* by human alveolar macrophages in vitro. Am J Respir Cell Mol Biol 18:834–843, 1998.

28. S Rosseau, P Hammerl, U Maus, A Gunther, W Seeger, F Grimminger, J Lohmeyer. Surfactant protein A down-regulates proinflammatory cytokine production evoked by *Candida albicans* in human alveolar macrophages and monocytes. J Immunol 163:4495–4502, 1999.

29. O Neth, DL Jack, AW Dodds, H Holzel, NJ Klein, MW Turner. Mannose-binding lectin binds to a range of clinically relevant microorganisms and promotes complement deposition. Infect Immun 68:688–693, 2000.

30. JE Volanakis. Overview of the complement system. In: JE Volanakis, MM Frank, eds. The Human Complement System in Health and Disease. New York: Marcel Dekker, 1998, pp 9–32.

31. TR Kozel. Activation of the complement system by pathogenic fungi. Clin Microbiol Rev 9:34–46, 1996.

32. MA Moors, TL Stull, KJ Blank, HR Buckley, DM Mosser. A role for complement receptor–like molecules in iron acquisition by *Candida albicans*. J Exp Med 175:1643–1651, 1992.

33. A Vecchiarelli, C Retini, A Casadevall, C Monari, D Pietrella, TR Kozel. Involvement of C3a and C5a in interleukin-8 secretion by human polymorphonuclear cells in response to capsular material of *Cryptococcus neoformans*. Infect Immun 66:4324–4330, 1998.

34. A Casadevall. Antibody immunity and invasive fungal infections. Infect Immun 63:4211–4218, 1995.

35. A Casadevall. Antibody-mediated protection against intracellular pathogens. Trends Microbiol 6: 102–107, 1998.

36. R Yuan, A Casadevall, G Spira, MD Scharff. Isotype switching from IgG3 to IgG1 converts a nonprotective murine antibody to *Cryptococcus neoformans* into a protective antibody. J Immunol 154:1810–1816, 1995.

37. J Mukherjee, G Nussbaum, MD Scharff, A Casadevall. Protective and nonprotective monoclonal antibodies to *Cryptococcus neoformans* originating from one B cell. J Exp Med 181:405–409, 1995.

38. A Vecchiarelli, C Retini, C Monari, A Casadevall. Specific antibody to *Cryptococcus neoformans* alters human leukocyte cytokine synthesis and promotes T-cell proliferation. Infect Immun 66: 1244–1247, 1998.

39. RR Yuan, A Casadevall, J Oh, MD Scharff. T cells cooperate with passive antibody to modify *Cryptococcus neoformans* infection in mice. Proc Natl Acad Sci USA 94:2483–2488, 1997.

40. H Marcotte, D Levesque, K Delanay, A Bourgeault, R de la Durantaye, S Brochu, MC Lavoie. *Pneumocystis carinii* infection in transgenic B cell–deficient mice. J Infect Dis 173:1034–1037, 1996.

41. RD Wagner, A Vazquez-Torres, J Jones-Carson, T Warner, E Balish. B cell knockout mice are resistant to mucosal and systemic candidiasis of endogenous origin but susceptible to experimental systemic candidiasis. J Infect Dis 174:589–597, 1996.

42. KM Aguirre, LL Johnson. A role for B cells in resistance to *Cryptococcus neoformans* in mice. Infect Immun 65:525–530, 1997.

43. R Matthews, J Burnie. Antibodies against *Candida*: potential therapeutics? Trends Microbiol 4: 354–358, 1996.

44. Y Han, MA Ulrich, JE Cutler. *Candida albicans* mannan extract-protein conjugates induce a protective immune response against experimental candidiasis. J Infect Dis 179:1477–1484, 1999.

45. L Polonelli, F De Bernadis, S Conti, M Boccanera, W Magliani, M Gerloni, C Cantelli, A Cassone. Human natural yeast killer toxin-like candidacidal antibodies. J Immunol 156:1880–1885, 1996.

46. W Magliani, S Conti, F de Bernardis, M Gerloni, D Bertolotti, P Mozzoni, A Cassone, L Polonelli. Therapeutic potential of antiidiotypic single chain antibodies with yeast killer toxin activity. Nat Biotechnol 15:155–158, 1997.

47. L Polonelli, M Gerloni, S Conti, P Fisicaro, C Cantelli, P Portincasa, F Almondo, PL Barea, FL Hernando, J Ponton. Heat-shock mannoproteins as targets of secretory IgA in *Candida albicans*. J Infect Dis 169:1401–1405, 1994.

48. A Jalil, P Moja, C Lambert, M Perol, L Cotte, JM Livrozet, A Boibieux, JM Vergnon, F Lucht, R Tran, C Contini, C Genin. Decreased production of local immunoglobulin A to *Pneumocystis carinii* in bronchoalveolar lavage fluid from human immunodeficiency virus–positive patients. Infect Immun 68:1054–1060, 2000.

49. L Polonelli, F De Bernardis, S Conti, M Boccanera, M Gerloni, G Morace, W Magliani, C Chezzi, A Cassone. Idiotypic intravaginal vaccination to protect against candidal vaginitis by secretory, yeast killer toxin-like anti-idiotypic antibodies. J Immunol 152:3175–3182, 1994.

50. F De Bernardis, M Boccanera, D Adriani, E Spreghini, G Santoni, A Cassone. Protective role of antimannan and anti-aspartyl proteinase antibodies in an experimental model of *Candida albicans* vaginitis in rats. Infect Immun 65:3399–3405, 1997.

51. RD Diamond. Fungal surfaces: effects of interactions with phagocytic cells. Rev Infect Dis 10: S428–S431, 1988.

52. A McKnight, S Gordon. Forum in immunology: innate recognition systems. Microbes Infect 2: 239–336, 2000.

53. A Schaffner, CE Davis, T Schaffner, M Markert, H Douglas, AI Braude. In vitro susceptibility of fungi to killing by neutrophil granulocytes discriminates between primary pathogenicity and opportunism. J Clin Invest 78:511–524, 1986.

54. SM Levitz, RD Diamond. Mechanisms of resistance of *Aspergillus fumigatus* conidia to killing by neutrophils in vitro. J Infect Dis 152:33–42, 1985.

55. CJ Morrison, E Brummer, DA Stevens. In vivo activation of peripheral blood polymorphonuclear neutrophils by gamma interferon results in enhanced fungal killing. Infect Immun 57:2953–2958, 1989.

56. RB Ashman, JM Papadimitriou. Production and function of cytokines in natural and acquired immunity to *Candida albicans* infection. Microbiol Rev 59:646–672, 1995.

57. CA Lyman, TJ Walsh. Phagocytosis of medically important yeasts by polymorphonuclear leukocytes. Infect Immun 62:1489–1493, 1994.

58. E Brummer, JG McEwen, DA Stevens. Fungicidal activity of murine inflammatory polymorphonuclear neutrophils: comparison with murine peripheral blood PMN. Clin Exp Immunol 66:681–690, 1986.

59. C Scherwitz, R Martin. The phagocytosis of *Candida albicans* blastospores and germ tubes by polymorphonuclear leukocytes. Dermatologica 159:12–23, 1979.

60. HA Pereira, CS Hosking. The role of complement and antibody in opsonization and intracellular killing of *Candida albicans*. Clin Exp Immunol 57:307–314, 1984.

61. G Marquis, S Garzon, S Montplaisir, H Strykowski, N Benhamou. Histochemical and immunochemical study of the fate of *Candida albicans* inside human neutrophil phagolysosomes. J Leukoc Biol 50:587–599, 1991.

62. DL Danley, AE Hilger. Stimulation of oxidative metabolism in murine polymorphonuclear leukocytes by unopsonized fungal cells: evidence for a mannose-specific mechanism. J Immunol 127: 551–556, 1981.

63. RA Schnur, SL Newman. The respiratory burst response to *Histoplasma capsulatum* by human neutrophils. Evidence for intracellular trapping of superoxide anion. J Immunol 144:4765–4772, 1990.

64. SL Newman, L Gootee, JE Gabay. Human neutrophil-mediated fungistasis against *Histoplasma capsulatum*. Localization of fungistatic activity to the azurophil granules. J Clin Invest 92:624–631, 1993.

65. ZM Dong, JW Murphy. Cryptococcal polysaccharides induce L-selectin shedding and tumor necrosis factor receptor loss from the surface of human neutrophils. J Clin Invest 97:689–698, 1996.

66. AB van Spriel, IE van den Herik-Oudijk, NM van Sorge, HA Vile, JA van Strijp, JG van de Winkel. Effective phagocytosis and killing of Candida albicans via targeting FcgammaRI (CD64) or FcalphaRI (CD89) on neutrophils. J Infect Dis 179:661–669, 1999.

67. A Cockayne, FC Odds. Interactions of Candida albicans yeast cells, germ tubes and hyphae with human polymorphonuclear leucocytes in vitro. J Gen Microbiol 130:465–471, 1984.

68. A Schaffner, H Douglas, A Braude. Selective protection against conidia by mononuclear and against mycelia by polymorphonuclear phagocytes in resistance to Aspergillus. Observations on these two lines of defense in vivo and in vitro with human and mouse phagocytes. J Clin Invest 69:617–631, 1982.

69. J Jensen, T Warner, E Balish. Resistance of SCID mice to Candida albicans administered intravenously or colonizing the gut: role of polymorphonuclear leukocytes and macrophages. J Infect Dis 167:912–919, 1993.

70. A Fulurija, RB Ashman, JM Papadimitriou. Neutrophil depletion increases susceptibility to systemic and vaginal candidiasis in mice, and reveals differences between brain and kidney in mechanisms of host resistance. Microbiology 142:3487–3496, 1996.

71. L Romani, F Bistoni, P Puccetti. Initiation of T-helper cell immunity to Candida albicans by IL-12: the role of neutrophils. Chem Immunol 68:110–135, 1997.

72. R Allendoerfer, GSJ Deepe Jr. Infection with Histoplasma capsulatum: host-fungus interface. Revista Iberoamericana de Micologia 15:256–260, 1998.

73. J Jones-Carson, A Vazquez-Torres, E Balish. Defective killing of Candida albicans hyphae by neutrophils from beige mice. J Infect Dis 171:1664–1667, 1995.

74. D Drutz, C Frey. Intracellular and extracellular defenses of human phagocytes against Blastomyces dermatitidis conidia and yeasts. J Lab Clin Med 105:737–750, 1985.

75. N Kurita, SK Biswas, M Oarada, A Sano, K Nishimura, M Miyaji. Fungistatic and fungicidal activities of murine polymorphonuclear leucocytes against yeast cells of Paracoccidioides brasiliensis. Med Mycol 37:19–24, 1999.

76. ZM Dong, JW Murphy. Effects of the two varieties of Cryptococcus neoformans cells and culture filtrate antigens on neutrophil locomotion. Infect Immun 63:2632–2644, 1995.

77. C Retini, A Vecchiarelli, C Monari, C Tascini, F Bistoni, TR Kozel. Capsular polysaccharide of Cryptococcus neoformans induces proinflammatory cytokine release by human neutrophils. Infect Immun 64:2897–2903, 1996.

78. R Calderone, J Sturtevant. Macrophage interactions with Candida. Immunol Ser 60:505–515, 1994.

79. S Levitz. Macrophage-Cryptococcus interactions. Immunol Ser 60:533–543, 1994.

80. S Newman. Macrophages in host defense against Histoplasma capsulatum. Trends Microbiol 7:67–71, 1999.

81. RA Ezekowitz, K Sastry, P Bailly, A Warner. Molecular characterization of the human macrophage mannose receptor: demonstration of multiple carbohydrate recognition-like domains and phagocytosis of yeasts in Cos-1 cells. J Exp Med 172:1785–1794, 1990.

82. RA Ezekowitz, DJ Williams, H Koziel, MY Armstrong, A Warner, FF Richards, RM Rose. Uptake of Pneumocystis carinii mediated by the macrophage mannose receptor. Nature 351:155–158, 1991.

83. L Marodi, H Korchak, R Jonhston. Mechanisms of host defense against Candida species. I. Phagocytosis by monocytes and monocyte-derived macrophages. J Immunol 146:2783–2789, 1991.

84. L Marodi, S Schreiber, DC Anderson, RP MacDermott, HM Korchak, RB Johnston Jr. Enhancement of macrophage candidacidal activity by interferon-gamma. Increased phagocytosis, killing, and calcium signal mediated by a decreased number of mannose receptors. J Clin Invest 91:2596–2601, 1993.

85. I Szabo, L Guan, TJ Rogers. Modulation of macrophage phagocytic activity by cell wall components of Candida albicans. Cell Immunol 164:182–188, 1995.

86. BS Klein, SL Newman. Role of cell-surface molecules of Blastomyces dermatitidis in hostpathogen interactions. Trends Microbiol 4:246–251, 1996.

87. SM Levitz, A Tabuni, TR Kozel, RS MacGill, RR Ingalls, DT Golenbock. Binding of *Cryptococcus neoformans* to heterologously expressed human complement receptors. Infect Immun 65:931–935, 1997.

88. SM Levitz, A Tabuni. Binding of *Cryptococcus neoformans* by human cultured macrophages. Requirements for multiple complement receptors and actin. J Clin Invest 87:528–535, 1991.

89. JE Wolf, V Kerchberger, GS Kobayashi, JR Little. Modulation of the macrophage oxidative burst by *Histoplasma capsulatum*. J Immunol 138:582–586, 1987.

90. JE Strasser, SL Newman, GM Ciraolo, RE Morris, ML Howell, GE Dean. Regulation of the macrophage vacuolar ATPase and phagosome-lysosome fusion by *Histoplasma capsulatum*. J Immunol 162:6148–6154, 1999.

91. JW Batanghari, GSJ Deepe, E Di Cera, WE Goldman. *Histoplasma* aquisition of calcium and expression CBP1 during intracellular parasitism. Mol Microbiol 27:531–539, 1998.

92. LE Cano, E Brummer, DA Stevens, A Restrepo. Fate of conidia of *Paracoccidioides brasiliensis* after ingestion by resident macrophages or cytokine-treated macrophages. Infect Immun 60: 2096–2100, 1992.

93. S Chaturvedi, P Frame, SL Newman. Macrophages from human immunodeficiency virus–positive persons are defective in host defense against *Histoplasma capsulatum*. J Infect Dis 171:320–327, 1995.

94. JM Orenstein, C Fox, SM Wahl. Macrophages as a source of HIV during opportunistic infections. Science 276:1857–1861, 1997.

95. C Monari, F Baldelli, D Pietrella, C Retini, C Tascini, D Francisci, F Bistoni, A Vecchiarelli. Monocyte dysfunction in patients with acquired immunodeficiency syndrome (AIDS) versus *Cryptococcus neoformans*. J Infect 35:257–263, 1997.

96. H Koziel, Q Eichbaum, B Kruskal, P Pinkston, R Rogers, M Armstrong, F Richards, R Rose, R Ezekowitz. Reduced binding and phagocytosis of *Pneumocystis carinii* by alveolar macrophages from persons infected with HIV-1 correlates with mannose receptor downregulation. J Clin Invest 102:1332–1344, 1998.

97. CE Cross, GJ Bancroft. Ingestion of acapsular *Cryptococcus neoformans* occurs via mannose and beta-glucan receptors, resulting in cytokine production and increased phagocytosis of the encapsulated form. Infect Immun 63:2604–2611, 1995.

98. SM Levitz, A Tabuni, R Wagner, H Kornfeld, EH Smail. Binding of unopsonized *Cryptococcus neoformans* by human bronchoalveolar macrophages: inhibition by a large-molecular-size serum component. J Infect Dis 166:866–873, 1992.

99. S Lee, Y Kress, D Dickson, A Casadevall. Human microglia mediate anti–*Cryptococcus neoformans* activity in the presence of specific antibody. J Neuroimmunol 62:43–62, 1995.

100. S Schelenz, R Malhotra, RB Sim, U Holmskov, GJ Bancroft. Binding of host collectins to the pathogenic yeast *Cryptococcus neoformans*: human surfactant protein D acts as an agglutinin for acapsular yeast cells. Infect Immun 63:3360–3366, 1995.

101. DL Granger, JR Perfect, DT Durack. Virulence of *Cryptococcus neoformans*. Regulation of capsule synthesis by carbon dioxide. J Clin Invest 76:508–516, 1985.

102. A Yasuoka, S Kohno, H Yamada, M Kaku, H Koga. Influence of molecular sizes of *Cryptococcus neoformans* capsular polysaccharide on phagocytosis. Microbiol Immunol 38:851–856, 1994.

103. JD Nosanchuk, A Casadevall. Cellular charge of *Cryptococcus neoformans*: contributions from the capsular polysaccharide, melanin, and monoclonal antibody binding. Infect Immun 65:1836–1841, 1997.

104. S Levitz, S Nong, K Seetoo, T Harrison, R Speizer, E Simons. *Cryptococcus neoformans* resides in an acidic phagolysomes of human macrophages. Infect Immun 67:885–819, 1999.

105. J Nosanchuk, P Valadon, M Feldmesser, A Casadevall. Melanization of *Cryptococcus neoformans* in murine infection. Mol Cell Biol 19:745–750, 1999.

106. D Goldman, S Lee, A Mednick, L Montella, A Casadeval. Persistent *Cryptococcus neoformans* pulmonary infection in the rat is associated with intracellular parasitism decreased inducible nitric oxide synthase expression and altered antibody responsiveness to cryptococci polysaccharide. Infect Immun 68:832–838, 2000.

107. SS Lefkowitz, MP Gelderman, DL Lefkowitz, N Moguilevsky, A Bollen. Phagocytosis and intracellular killing of *Candida albicans* by macrophages exposed to myeloperoxidase. J Infect Dis 173: 1202–1207, 1996.

108. DJ Kitz, PD Stahl, JR Little. The effect of a mannose binding protein on macrophage interactions with *Candida albicans*. Cell Mol Biol 38:407–412, 1992.

109. A Limper, J Standing. Vitronectin interacts with *Candida albicans* and augments organism attachment to the NR8383 macrophage cell line. Immunol Lett 42:139–144, 1994.

110. M Sasada, RB Johnston Jr. Macrophage microbicidal activity. Correlation between phagocytosis-associated oxidative metabolism and the killing of *Candida* by macrophages. J Exp Med 152:85–98, 1980.

111. L Marodi, JR Forehand, RB Johnston Jr. Mechanisms of host defense against *Candida* species. II. Biochemical basis for the killing of *Candida* by mononuclear phagocytes. J Immunol 146: 2790–2794, 1991.

112. E Blasi, M Puliti, L Pitzurra, A Bartoli, F Bistoni. Heterogeneous secretory response of phagocytes from different anatomical districts to the dimorphic fungus *Candida albicans*. Cell Immunol 153: 239–247, 1994.

113. M Baccarini, F Bistoni, ML Lohmann-Matthes. In vitro natural cell-mediated cytotoxicity against *Candida albicans*: macrophage precursors as effector cells. J Immunol 134:2658–2665, 1985.

114. E Blasi, L Pitzurra, M Puliti, L Lanfrancone, F Bistoni. Early differential molecular response of a macrophage cell line to yeast and hyphal forms of *Candida albicans*. Infect Immun 60:832–837, 1992.

115. R Kaposzta, L Marodi, M Hollinshead, S Gordon, RP da Silva. Rapid recruitment of late endosomes and lysosomes in mouse macrophages ingesting *Candida albicans*. J Cell Sci 112:3237–3248, 1999.

116. E Blasi, L Pitzurra, A Chimienti, R Mazzolla, M Puliti, R Barluzzi, F Bistoni. Differential susceptibility of yeast and hyphal forms of *Candida albicans* to proteolytic activity of macrophages. J Infect Immun 63:1253–1257, 1995.

117. C Reis e Sousa, A Sher, P Kaye. The role of dendritic cells in the induction and regulation of immunity to microbial infection. Curr Opin Immunol 11:392–399, 1999.

118. BJ Masten, MF Lipscomb. Comparison of lung dendritic cells and B cells in stimulating naive antigen-specific T cells. J Immunol 162:1310–1317, 1999.

119. E Cenci, A Mencacci, A Bacci, F Bistoni, WP Kurup, L Romani. T-cell vaccination in mice with invasive pulmonary aspergillosis. J Immunol 2000. In press.

120. B Chen, Y Shi, JD Smith, D Choi, JD Geiger, JJ Mulè. The role of tumor necrosis factor alpha in modulating the quantity of peripheral blood–derived, cytokine-driven human dendritic cells and its role in enhancing the quality of dendritic cell function in presenting soluble antigens to CD4 + T cells in vitro. Blood 91:4652–4661, 1998.

121. C Fè d'Ostiani, G Del Sero, A Bacci, C Montagnoli, A Spreca, A Mencacci, P Ricciardi-Castagnoli, L Romani. Dendritic cells discriminate between yeast and hyphae of the fungus *Candida albicans*: implications for initiation of T helper cell immunity in vitro and in vivo. J Exp Med 191:1661–1674, 2000.

122. SM Levitz, MP Dupont, EH Smail. Direct activity of human T lymphocytes and natural killer cells against *Cryptococcus neoformans*. Infect Immun 62:194–202, 1994.

123. J Djeu. Cytokines and anti-fungal immunity. Adv Exp Med Biol 319:217–223, 1992.

124. JW Murphy, A Zhou, SC Wong. Direct interactions of human natural killer cells with *Cryptococcus neoformans* inhibit granulocyte-macrophage colony-stimulating factor and tumor necrosis factor alpha production. Infect Immun 65:4564–4571, 1997.

125. L Scaringi, P Cornacchione, E Rosati, M Boccanera, A Cassone, F Bistoni, P Marconi. Induction of LAK-like cells in the peritoneal cavity of mice by inactivated *Candida albicans*. Cell Immunol 129:271–287, 1990.

126. H Warschkau, H Yu, AF Kiderlen. Activation and suppression of natural cellular immune functions by *Pneumocystis carinii*. Immunobiology 198:343–360, 1998.

127. JK Kolls, D Lei, C Vazquez, G Odom, WR Summer, S Nelson, J Shellito. Exacerbation of murine *Pneumocystis carinii* infection by adenoviral-mediated gene transfer of a TNF inhibitor. Am J Respir Cell Mol Biol 16:112–118, 1997.

128. MH Qureshi, T Zhang, Y Koguchi, K Nakashima, H Okamura, M Kurimoto, K Kawakami. Combined effects of IL-12 and IL-18 on the clinical course and local cytokine production in murine pulmonary infection with *Cryptococcus neoformans*. Eur J Immunol 29:643–649, 1999.

129. L Romani. The T cell response against fungal infections. Curr Opin Immunol 9:484–490, 1997.

130. J Jones-Carson, A Vazquez-Torres, HC van der Heyde, T Warner, RD Wagner, E Balish. Gamma delta T cell–induced nitric oxide production enhances resistance to mucosal candidiasis. Nat Med 1:552–557, 1995.

131. R Hanano, S Kaufmann. *Pneumocystis carinii* pneumonia in mutant mice deficient in both TCRalphabeta and TCRgammadelta cells: cytokine and antibody responses. J Infect Dis 179:455–459, 1999.

132. SM Levitz, HL Mathews, JW Murphy. Direct antimicrobial activity of T cells. Immunol Today 16: 387–391, 1995.

133. SM Levitz, EA North. gamma Interferon gene expression and release in human lymphocytes directly activated by *Cryptococcus neoformans* and *Candida albicans*. Infect Immun 64:1595–1599, 1996.

134. J Galgiani, R Hayden, C Payne. Leukocyte effects on the dimorphism of *Coccidioides immitis*. J Infect Dis 146:56–63, 1982.

135. MP Kolotila, RD Diamond. Effects of neutrophils and in vitro oxidants on survival and phenotypic switching of *Candida albicans* WO-1. Infect Immun 58:1174–1179, 1990.

136. RD Diamond, RA Clark. Damage to *Aspergillus fumigatus* and *Rhizopuz oryzae* hyphae by oxidative and nonoxidative microbicidal products of human neutrophils in vitro. Infect Immun 38:487–495, 1982.

137. V Chaturvedi, B Wong, SL Newman. Oxidative killing of *Cryptococcus neoformans* by human neutrophils. Evidence that fungal mannitol protects by scavenging reactive oxygen intermediates. J Immunol 156:3836–3840, 1996.

138. L Christin, DR Wysong, T Meshulam, S Wang, RD Diamond. Mechanisms and target sites of damage in killing of *Candida albicans* hyphae by human polymorphonuclear neutrophils. J Infect Dis 176:1567–1578, 1997.

139. SM Levitz, ME Selsted, T Ganz, RI Lehrer, RD Diamond. In vitro killing of spores and hyphae of *Aspergillus fumigatus* and *Rhizopus oryzae* by rabbit neutrophil cationic peptides and bronchoalveolar macrophages. J Infect Dis 154:483–489, 1986.

140. IE Flesch, G Schwamberger, SH Kaufmann. Fungicidal activity of IFN-gamma-activated macrophages. Extracellular killing of *Cryptococcus neoformans*. J Immunol 142:3219–3224, 1989.

141. DK Stein, SE Malawista, G Van Blaricom, D Wysong, RD Diamond. Cytoplasts generate oxidants but require added neutrophil granule constituents for fungicidal activity against *Candida albicans* hyphae. J Infect Dis 172:511–520, 1995.

142. DH Howard. Acquisition, transport, and storage of iron by pathogenic fungi. Clin Microbiol Rev 12:394–404, 1999.

143. EH Smail, DA Melnick, R Ruggeri, RD Diamond. A novel natural inhibitor from *Candida albicans* hyphae causing dissociation of the neutrophil respiratory burst response to chemotactic peptides from other post-activation events. J Immunol 140:3893–3899, 1988.

144. C Bogdan, M Rollinghoff, A Diefenbach. Reactive oxygen and reactive nitrogen intermediates in innate and specific immunity. Curr Opin Immunol 12:64–76, 2000.

145. RD Diamond, RA Clark, CC Haudenschild. Damage to *Candida albicans* hyphae and pseudohyphae by the myeloperoxidase system and oxidative products of neutrophil metabolism in vitro. J Clin Invest 66:908–917, 1980.

146. SM Levitz, RD Diamond. Killing of *Aspergillus fumigatus* spores and *Candida albicans* yeast phase by the iron–hydrogen peroxide–iodide cytotoxic system: comparison with the myeloperoxidase–hydrogen peroxide–halide system. Infect Immun 43:1100–1102, 1984.

147. MP Kolotila, RD Diamond. Stimulation of neutrophil actin polymerization and degranulation by opsonized and unopsonized *Candida albicans* hyphae and zymosan. Infect Immun 56:2016–2022, 1988.

148. DR Wysong, CA Lyman, RD Diamond. Independence of neutrophil respiratory burst oxidant generation from the early cytosolic calcium response after stimulation with unopsonized *Candida albicans* hyphae. Infect Immun 57:1499–1505, 1989.

149. RD Diamond, L Noble. Patterns of guanine nucleotide exchange reflecting disparate neutrophil activation pathways by opsonized and unopsonized *Candida albicans* hyphae. J Infect Dis 162: 262–265, 1990.

150. RI Lehrer, MJ Cline. Leukocyte myeloperoxidase deficiency and disseminated candidiasis: the role of myeloperoxidase in resistance to *Candida* infection. J Clin Invest 48:1478–1488, 1969.

151. E Brummer, N Kurita, S Yoshida, K Nishimura, M Miyaji. A basis for resistance of *Blastomyces dermatitidis* killing by human neutrophils: inefficient generation of myeloperoxidase system products. J Med Vet Mycol 30:233–243, 1992.

152. RG Washburn, JI Gallin, JE Bennett. Oxidative killing of *Aspergillus fumigatus* proceeds by parallel myeloperoxidase-dependent and -independent pathways. Infect Immun 55:2088–2092, 1987.

153. AW Segal. The NADPH oxidase and chronic granulomatous disease. Mol Med Today 2:129–135, 1996.

154. DE Morgenstern, MA Gifford, LL Li, CM Doerschuk, MC Dinauer. Absence of respiratory burst in X-linked chronic granulomatous disease mice leads to abnormalities in both host defense and inflammatory response to *Aspergillus fumigatus*. J Exp Med 185:207–218, 1997.

155. RD Diamond. Antibody-dependent killing of *Cryptococcus neoformans* by human peripheral blood mononuclear cells. Nature 247:148–150, 1974.

156. S Newman, L Gootee, C Kidd, J Ciraolo, R Morris. Activation of human macrophage fungistatic activity against *Histoplasma capsulatum* upon adherence to type 1 collagen matrixes. J Immunol 158:1779–1786, 1997.

157. JN Galgiani. Differences in oxidant release by human polymorphonuclear leukocytes produced by stimulation with different phases of *Coccidioides immitis*. J Infect Dis 172:199–203, 1995.

158. A Laursen, N Obel, J Rungby, P Andersen. Phagocytosis and stimulation of the respiratory burst in neutrophils by *Pneumocystis carinii*. J Infect Dis 168:466–471, 1993.

159. RD Diamond, RK Root, JE Bennett. Factors influencing killing of *Cryptococcus neoformans* by human leukocytes in vitro. J Infect Dis 125:367–376, 1972.

160. Y Wang, A Casadevall. Susceptibility of melanized and nonmelanized *Cryptococcus neoformans* to nitrogen- and oxygen-derived oxidants. Infect Immun 62:3004–3007, 1994.

161. D Goldman, Y Cho, M Zhao, A Casadevall, SC Lee. Expression of inducible nitric oxide synthase in rat pulmonary *Cryptococcus neoformans* granulomas. Am J Pathol 148:1275–1282, 1996.

162. JF Downing, DL Kachel, R Pasula, WJ Martin 2d. Gamma interferon stimulates rat alveolar macrophages to kill *Pneumocystis carinii* by L-arginine- and tumor necrosis factor–dependent mechanisms. Infect Immun 67:1347–1352, 1999.

163. NT Gross, K Nessa, P Camner, C Jarstrand. Production of nitric oxide by rat alveolar macrophages stimulated by *Cryptococcus neoformans* or *Aspergillus fumigatus*. Med Mycol 37:151–157, 1999.

164. E Cenci, L Romani, A Mencacci, R Spaccapelo, E Schiaffella, P Puccetti, F Bistoni. Interleukin-4 and interleukin-10 inhibit nitric oxide–dependent macrophage killing of *Candida albicans*. Eur J Immunol 23:1034–1038, 1993.

165. A Vazquez-Torres, J Jones-Carson, T Warner, E Balish. Nitric oxide enhances resistance of SCID mice to mucosal candidiasis. J Infect Dis 172:192–198, 1995.

166. JA Lovchik, CR Lyons, MF Lipscomb. A role for gamma interferon–induced nitric oxide in pulmonary clearance of *Cryptococcus neoformans*. Am J Respir Cell Mol Biol 13:116–124, 1995.

167. AL Bocca, EE Hayashi, AG Pinheiro, AB Furlanetto, AP Campanelli, FQ Cunha, F Figueiredo. Treatment of *Paracoccidioides brasiliensis*–infected mice with a nitric oxide inhibitor prevents the failure of cell-mediated immune response. J Immunol 161:3056–3063, 1998.

168. K Kawakami, T Zhang, MH Qureshi, A Saito. *Cryptococcus neoformans* inhibits nitric oxide production by murine peritoneal macrophages stimulated with interferon-gamma and lipopolysaccharide. Cell Immunol 180:47–54, 1997.

169. T Chinen, MH Qureshi, Y Koguchi, K Kawakami. *Candida albicans* suppresses nitric oxide (NO) production by interferon-gamma (IFN-gamma) and lipopolysaccharide (LPS)-stimulated murine peritoneal macrophages. Clin Exp Immunol 115:491–497, 1999.

170. F Abaitua, A Rementeria, R San Millan, A Eguzkiza, JA Rodriguez, J Ponton, MJ Sevilla. In vitro survival and germination of *Candida albicans* in the presence of nitrogen compounds. Microbiology 145:1641–1647, 1999.

171. E Michaliszyn, S Senechal, P Martel, L de Repentigny. Lack of involvement of nitric oxide in killing of *Aspergillus fumigatus* conidia by pulmonary alveolar macrophages. Infect Immun 63: 2075–2078, 1995.

172. S Mukherjee, M Feldmesser, A Casadevall. J774 murine macrophage-like cell interactions with *Cryptococcus neoformans* in the presence and absence of opsonins. J Infect Dis 173:1222–1231, 1996.

173. J Lovchik, M Lipscomb, CR Lyons. Expression of lung inducible nitric oxide synthase protein does not correlate with nitric oxide production in vivo in a pulmonary immune response against *Cryptococcus neoformans*. J Immunol 158:1772–1778, 1997.

174. LE Cano, B Gomez, E Brummer, A Restrepo, DA Stevens. Inhibitory effect of deferoxamine or macrophage activation on transformation of *Paracoccidioides brasiliensis* conidia ingested by macrophages: reversal by holotransferrin. Infect Immun 62:1494–1496, 1994.

175. SE Vartivarian, EJ Anaissie, RE Cowart, HA Sprigg, MJ Tingler, ES Jacobson. Regulation of cryptococcal capsular polysaccharide by iron. J Infect Dis 167:186–190, 1993.

176. SM Levitz, TS Harrison, A Tabuni, X Liu. Chloroquine induces human mononuclear phagocytes to inhibit and kill *Cryptococcus neoformans* by a mechanism independent of iron deprivation. J Clin Invest 100:1640–1646, 1997.

177. GB Huffnagle, RM Strieter, TJ Standiford, RA McDonald, MD Burdick, SL Kunkel, GB Toews. The role of monocyte chemotactic protein-1 (MCP-1) in the recruitment of monocytes and CD4+ T cells during a pulmonary *Cryptococcus neoformans* infection. J Immunol 155:4790–4797, 1995.

178. GB Huffnagle, RM Strieter, LK McNeil, RA McDonald, MD Burdick, SL Kunkel, GB Toews. Macrophage inflammatory protein-1 alpha (MIP-1 alpha) is required for the efferent phase of pulmonary cell-mediated immunity to a *Cryptococcus neoformans* infection. J Immunol 159:318–327, 1997.

179. GB Huffnagle, LK McNeil. Dissemination of *C. neoformans* to the central nervous system: role of chemokines, Th1 immunity and leukocyte recruitment. J Neurovirol 5:76–81, 1999.

180. SM Levitz, EA North, Y Jiang, SH Nong, H Kornfeld, TS Harrison. Variables affecting production of monocyte chemotactic factor 1 from human leukocytes stimulated with *Cryptococcus neoformans*. Infect Immun 65:903–908, 1997.

181. HA Doyle, JW Murphy. MIP-1 alpha contributes to the anticryptococcal delayed-type hypersensitivity reaction and protection against *Cryptococcus neoformans*. J Leukoc Biol 61:147–155, 1997.

182. Y Yamamoto, TW Klein, H Friedman. Involvement of mannose receptor in cytokine interleukin-1 beta (IL-1beta), IL-6, and granulocyte-macrophage colony-stimulating factor responses, but not in chemokine macrophage inflammatory protein 1 beta (MIP-1 beta), MIP-2, and KC responses, caused by attachment of *Candida albicans* to macrophages. Infect Immun 65:1077–1082, 1997.

183. TW Wright, CJ Johnston, AG Harmsen, JN Finkelstein. Chemokine gene expression during *Pneumocystis carinii*–driven pulmonary inflammation. Infect Immun 67:3452–3460, 1999.

184. S Schelenz, DA Smith, GJ Bancroft. Cytokine and chemokine responses following pulmonary challenge with *Aspergillus fumigatus*: obligatory role of TNF-alpha and GM-CSF in neutrophil recruitment. Med Mycol 37:183–194, 1999.

185. M Saavedra, B Taylor, N Lukacs, PL Fidel Jr. Local production of chemokines during experimental vaginal candidiasis. Infect Immun 67:5820–5826, 1999.

186. C Huang, SM Levitz. Stimulation of macrophage inflammatory protein-1 alpha, macrophage inflammatory protein-1 beta, and RANTES by *Candida albicans* and *Cryptococcus neoformans* in peripheral blood mononuclear cells from persons with and without human immunodeficiency virus infection. J Infect Dis 181:791–794, 2000.

187. HA Doyle, JW Murphy. Role of the C-C chemokine, TCA3, in the protective anticryptococcal cell-mediated immune response. J Immunol 162:4824–4833, 1999.

188. JL Gao, TA Wynn, Y Chang, EJ Lee, HE Broxmeyer, S Cooper, HL Tiffany, H Westphal, J Kwon-Chung, PM Murphy. Impaired host defense, hematopoiesis, granulomatous inflammation and type 1–type 2 cytokine balance in mice lacking CC chemokine receptor 1. J Exp Med 185:1959–1968, 1997.

189. B Mehrad, RM Strieter, TA Moore, WC Tsai, SA Lira, TJ Standiford. CXC chemokine receptor-2 ligands are necessary components of neutrophil-mediated host defense in invasive pulmonary aspergillosis. J Immunol 163:6086–6094, 1999.

190. B Mehrad, RM Strieter, TJ Standiford. Role of TNF-alpha in pulmonary host defense in murine invasive aspergillosis. J Immunol 162:1633–1640, 1999.

191. GB Huffnagle, LK McNeil, RA McDonald, JW Murphy, GB Toews, N Maeda, WA Kuziel. Cutting edge: role of C-C chemokine receptor 5 in organ-specific and innate immunity to *Cryptococcus neoformans*. J Immunol 163:4642–4646, 1999.

192. TR Traynor, WA Kuziel, GB Toews, GB Huffnagle. CCR2 expression determines T1 versus T2 polarization during pulmonary *Cryptococcus neoformans* infection. J Immunol 164:2021–2027, 2000.

193. E Balish, RD Wagner, A Vazquez-Torres, J Jones-Carson, C Pierson, T Warner. Mucosal and systemic candidiasis in IL-8Rh-/- BALB/c mice. J Leukoc Biol 66:144–150, 1999.

194. CL Silva, MF Silva, LH Faccioli, RC Pietro, SA Cortez, NT Foss. Differential correlation between interleukin patterns in disseminated and chronic human paracoccidioidomycosis. Clin Exp Immunol 101:314–320, 1995.

195. RM Perenboom, AC van Schijndel, P Beckers, R Sauerwein, HW Van Hamersvelt, J Festen, H Gallati, JW van der Meer. Cytokine profiles in bronchoalveolar lavage fluid and blood in HIV-seronegative patients with *Pneumocystis carinii* pneumonia. Eur J Clin Invest 26:159–166, 1996.

196. GS Deepe Jr, R Gibbons, E Woodward. Neutralization of endogenous granulocyte-macrophage colony-stimulating factor subverts the protective immune response to *Histoplasma capsulatum*. J Immunol 163:4985–4993, 1999.

197. R Paine 3d, AM Preston, S Wilcoxen, H Jin, BB Siu, SB Morris, JA Reed, G Ross, JA Whitsett, JM Beck. Granulocyte-macrophage colony-stimulating factor in the innate immune response to *Pneumocystis carinii* pneumonia in mice. J Immunol 164:2602–2609, 2000.

198. S Steinshamn, MH Bemelmans, LJ van Tits, K Bergh, WA Buurman, A Waage. TNF receptors in murine *Candida albicans* infection: evidence for an important role of TNF receptor p55 in antifungal defense. J Immunol 157:2155–2159, 1996.

199. A Mencacci, E Cenci, G Del Sero, C Fe d'Ostiani, P Mosci, C Montagnoli, A Bacci, F Bistoni, VF Quesniaux, B Ryffel, L Romani. Defective co-stimulation and impaired Th1 development in tumor necrosis factor/lymphotoxin-alpha double-deficient mice infected with *Candida albicans*. Int Immunol 10:37–48, 1998.

200. MG Netea, LJ van Tits, JH Curfs, F Amiot, JF Meis, JW van der Meer, BJ Kullberg. Increased susceptibility of TNF-alpha lymphotoxin-alpha double knockout mice to systemic candidiasis through impaired recruitment of neutrophils and phagocytosis of *Candida albicans*. J Immunol 163:1498–1505, 1999.

201. N Rayhane, O Lortholary, C Fitting, J Callebert, M Huerre, F Dromer, JM Cavaillon. Enhanced sensitivity of tumor necrosis factor/lymphotoxin-alpha-deficient mice to *Cryptococcus neoformans* infection despite increased levels of nitrite/nitrate, interferon-gamma, and interleukin-12. J Infect Dis 180:1637–1647, 1999.

202. E Cenci, A Mencacci, C Fe d'Ostiani, G Del Sero, P Mosci, C Montagnoli, A Bacci, L Romani. Cytokine- and T helper–dependent lung mucosal immunity in mice with invasive pulmonary aspergillosis. J Infect Dis 178:1750–1760, 1998.

203. R Allendoerfer, GS Deepe Jr. Blockade of endogenous TNF-alpha exacerbates primary and secondary pulmonary histoplasmosis by differential mechanisms. J Immunol 160:6072–6082, 1998.

204. RD Diamond, CA Lyman, DR Wysong. Disparate effects of interferon-gamma and tumor necrosis factor–alpha on early neutrophil respiratory burst and fungicidal responses to *Candida albicans* hyphae in vitro. J Clin Invest 87:711–720, 1991.

205. F Bazzoni, MA Cassatella, C Laudanna, F Rossi. Phagocytosis of opsonized yeast induces tumor necrosis factor–alpha mRNA accumulation and protein release by human polymorphonuclear leukocytes. J Leukoc Biol 50:223–228, 1991.

206. SM Levitz, A Tabuni, H Kornfeld, CC Reardon, DT Golenbock. Production of tumor necrosis factor alpha in human leukocytes stimulated by *Cryptococcus neoformans*. Infect Immun 62:1975–1981, 1994.

207. A Vecchiarelli, C Retini, D Pietrella, C Monari, C Tascini, T Beccari, TR Kozel. Downregulation
      by cryptococcal polysaccharide of tumor necrosis factor alpha and interleukin-1 beta secretion from
      human monocytes. Infect Immun 63:2919–2923, 1995.
208. L Romani, A Mencacci, E Cenci, R Spaccapelo, C Toniatti, P Puccetti, F Bistoni, V Poli. Impaired
      neutrophil response and CD4+ T helper cell 1 development in interleukin 6–deficient mice infected
      with *Candida albicans*. J Exp Med 183:1345–1355, 1996.
209. FH van Enckevort, MG Netea, AR Hermus, CG Sweep, JF Meis, JW Van der Meer, BJ Kullberg.
      Increased susceptibility to systemic candidiasis in interleukin-6 deficient mice. Med Mycol 37:
      419–426, 1999.
210. DA Stevens, TJ Walsh, F Bistoni, E Cenci, KV Clemons, G Del Sero, C Fe d'Ostiani, BJ Kullberg,
      A Mencacci, E Roilides, L Romani. Cytokines and mycoses. Med Mycol 36:174–182, 1998.
211. JW Murphy, F Bistoni, GS Deepe, RA Blackstock, K Buchanan, RB Ashman, L Romani, A Men-
      cacci, E Cenci, C Fe d'Ostiani, G Del Sero, VL Calich, SS Kashino. Type 1 and type 2 cytokines:
      from basic science to fungal infections. Med Mycol 36:109–118, 1998.
212. ST Pottratz, AL Weir. Gamma-interferon inhibits *Pneumocystis carinii* attachment to lung cells by
      decreasing expression of lung cell-surface integrins. Eur J Clin Invest 27:17–22, 1997.
213. BA Garvy, F Gigliotti, AG Harmsen. Neutralization of interferon-gamma exacerbates *Pneumocystis*-
      driven interstitial pneumonitis after bone marrow transplantation in mice. J Clin Invest 99:
      1637–1644, 1997.
214. CM Ausiello, F Urbani, S Gessani, GC Spagnoli, MJ Gomez, A Cassone. Cytokine gene expression
      in human peripheral blood mononuclear cells stimulated by mannoprotein constituents from *Candida
      albicans*. Infect Immun 61:4105–4111, 1993.
215. DB Corry, NM Ampel, L Christian, RM Locksley, JN Galgiani. Cytokine production by peripheral
      blood mononuclear cells in human coccidioidomycosis. J Infect Dis 174:440–443, 1996.
216. ML Grazziutti, JH Rex, RE Cowart, EJ Anaissie, A Ford, CA Savary. *Aspergillus fumigatus* conidia
      induce a Th1-type cytokine response. J Infect Dis 176:1579–1583, 1997.
217. Q Qian, J Cutler. γ-Interferon is not essential in host defense against disseminated candidiasis in
      mice. Infect Immun 65:1748–1756, 1997.
218. E Cenci, A Mencacci, G Del Sero, CF d'Ostiani, P Mosci, A Bacci, C Montagnoli, M Kopf, L
      Romani. IFN-gamma is required for IL-12 responsiveness in mice with *Candida albicans* infection.
      J Immunol 161:3543–3550, 1998.
219. E Roilides, S Tsaparidou, I Kadiltsoglou, T Sein, TJ Walsh. Interleukin-12 enhances antifungal
      activity of human mononuclear phagocytes against *Aspergillus fumigatus*: implications for a gamma
      interferon-independent pathway. Infect Immun 67:3047–3050, 1999.
220. L Romani, P Puccetti, F Bistoni. Interleukin-12 in infectious diseases. Clin Microbiol Rev 10:
      611–636, 1997.
221. H Okamura, H Tsutsui, S Kashiwamura, T Yosimoto, K Nakanishi. Interleukin-18: a novel cytokine
      that augments both innate and acquired immunity. Adv Immunol 17:281–312, 1998.
222. A Cassone, P Chiani, I Quinti, A Torosantucci. Possible participation of polymorphonuclear cells
      stimulated by microbial immunomodulators in the dysregulated cytokine patterns of AIDS patients.
      J Leukoc Biol 62:60–66, 1997.
223. C Retini, A Casadevall, D Pietrella, C Monari, B Palazzetti, A Vecchiarelli. Specific activated T
      cells regulate IL-12 production by human monocytes stimulated with *Cryptococcus neoformans*. J
      Immunol 162:1618–1623, 1999.
224. A Mencacci, E Cenci, G Del Sero, C Fe d'Ostiani, P Mosci, G Trinchieri, L Adorini, L Romani.
      IL-10 is required for development of protective Th1 responses in IL-12-deficient mice upon *Candida
      albicans* infection. J Immunol 161:6228–6237, 1998.
225. K Decken, G Kohler, K Palmer-Lehmann, A Wunderlin, F Mattner, J Magram, MK Gately, G
      Alber. Interleukin-12 is essential for a protective Th1 response in mice infected with *Cryptococcus
      neoformans*. Infect Immun 66:4994–5000, 1998.
226. P Zhou, RA Seder. CD40 ligand is not essential for induction of type 1 cytokine responses or
      protective immunity after primary or secondary infection with *Histoplasma capsulatum*. J Exp Med
      187:1315–1324, 1998.

227. TS Harrison, SM Levitz. Role of IL-12 in peripheral blood mononuclear cell responses to fungi in persons with and without HIV infection. J Immunol 156:4492–4497, 1996.

228. G Del Sero, A Mencacci, E Cenci, CF d'Ostiani, C Montagnoli, A Bacci, P Mosci, M Kopf, L Romani. Antifungal type 1 responses are upregulated in IL-10-deficient mice. Microbes Infect 1: 1169–1180, 1999.

229. A Vecchiarelli, C Retini, C Monari, C Tascini, F Bistoni, TR Kozel. Purified capsular polysaccharide of *Cryptococcus neoformans* induces interleukin-10 secretion by human monocytes. Infect Immun 64:2846–2849, 1996.

230. SM Levitz, A Tabuni, SH Nong, DT Golenbock. Effects of interleukin-10 on human peripheral blood mononuclear cell responses to *Cryptococcus neoformans, Candida albicans*, and lipopolysaccharide. Infect Immun 64:945–951, 1996.

231. E Roilides, A Dimitriadou, I Kadiltsoglou, T Sein, J Karpouzas, PA Pizzo, TJ Walsh. IL-10 exerts suppressive and enhancing effects on antifungal activity of mononuclear phagocytes against *Aspergillus fumigatus*. J Immunol 158:322–329, 1997.

232. K Kawakami, MH Qureshi, Y Koguchi, K Nakajima, A Saito. Differential effect of *Cryptococcus neoformans* on the production of IL-12p40 and IL-10 by murine macrophages stimulated with lipopolysaccharide and gamma interferon. FEMS Microbiol Lett 175:87–94, 1999.

233. R Blackstock, KL Buchanan, AM Adesina, JW Murphy. Differential regulation of immune responses by highly and weakly virulent *Cryptococcus neoformans* isolates. Infect Immun 67:3601–3609, 1999.

234. J Xiong, K Kang, L Liu, Y Yoshida, K Cooper, M Ghannoum. *Candida albicans* and *Candida krusei* differentially induce human blood mononuclear cell interleukin-12 and gamma interferon production. Infect Immun 68:2464–2469, 2000.

235. R Spaccapelo, L Romani, L Tonnetti, E Cenci, A Mencacci, G Del Sero, R Tognellini, SG Reed, P Puccetti, F Bistoni. TGF-beta is important in determining the in vivo patterns of susceptibility or resistance in mice infected with *Candida albicans*. J Immunol 155:1349–1360, 1995.

236. A Mencacci, G Del Sero, E Cenci, CF d'Ostiani, A Bacci, C Montagnoli, M Kopf, L Romani. Endogenous interleukin 4 is required for development of protective CD4+ T helper type 1 cell responses to *Candida albicans*. J Exp Med 187:307–317, 1998.

237. E Cenci, A Mencacci, G Del Sero, A Bacci, C Montagnoli, CF d'Ostiani, P Mosci, M Bachmann, F Bistoni, M Kopf, L Romani. Interleukin-4 causes susceptibility to invasive pulmonary aspergillosis through suppression of protective type 1 responses. J Infect Dis 180:1957–1968, 1999.

238. M Stein, S Keshav, N Harris, S Gordon. Interleukin 4 potently enhances murine macrophage mannose receptor activity: a marker of alternative immunologic macrophage activation. J Exp Med 176: 287–292, 1992.

239. MR Gyetko, GH Chen, RA McDonald, R Goodman, GB Huffnagle, CC Wilkinson, JA Fuller, GB Toews. Urokinase is required for the pulmonary inflammatory response to *Cryptococcus neoformans*. A murine transgenic model. J Clin Invest 97:1818–1826, 1996.

240. M Castro, NV Ralston, TI Morgenthaler, MS Rohrbach, AH Limper. *Candida albicans* stimulates arachidonic acid liberation from alveolar macrophages through alpha-mannan and beta-glucan cell wall components. Infect Immun 62:3138–3145, 1994.

241. AI Medeiros, CL Silva, A Malheiro, CM Maffei, LH Faccioli. Leukotrienes are involved in leukocyte recruitment induced by live *Histoplasma capsulatum* or by the beta-glucan present in their cell wall. Br J Pharmacol 128:1529–1537, 1999.

242. GB Huffnagle, MB Boyd, NE Street, MF Lipscomb. IL-5 is required for eosinophil recruitment, crystal deposition, and mononuclear cell recruitment during a pulmonary *Cryptococcus neoformans* infection in genetically susceptible mice (C57BL/6). J Immunol 160:2393–2400, 1998.

243. WB Harley, MJ Blaser. Disseminated coccidioidomycosis associated with extreme eosinophilia. Clin Infect Dis 18:627–629, 1994.

244. T Ishikawa, AC Dalton, CE Arbesman. Phagocytosis of *Candida albicans* by eosinophilic leukocytes. J Allergy Clin Immunol 49:311–315, 1972.

245. M Feldmesser, A Casadevall, Y Kress, G Spira, A Orlofsky. Eosinophil–*Cryptococcus neoformans* interactions in vivo and in vitro. Infect Immun 65:1899–1907, 1997.

246. TW Wright, F Gigliotti, JN Finkelstein, JT McBride, CL An, AG Harmsen. Immune-mediated inflammation directly impairs pulmonary function, contributing to the pathogenesis of *Pneumocystis carinii* pneumonia. J Clin Invest 104:1307–1317, 1999.
247. EB Kopp, R Medzhitov. The Toll-receptor family and control of innate immunity. Curr Opin Immunol 11:13–18, 1999.
248. B Beutler. Endotoxin, Toll-like receptor 4, and the afferent limb of innate immunity. Curr Opin Microbiol 3:23–28, 2000.
249. A Bendelac, D Fearon. Receptors and effectors of innate immunity. Curr Opin Immunol 12:11–12, 2000.
250. A Vecchiarelli, C Monari, C Retini, D Pietrella, B Palazzetti, L Pitzurra, A Casadevall. *Cryptococcus neoformans* differently regulates B7-1 (CD80) and B7-2 (CD86) expression on human monocytes. Eur J Immunol 28:114–121, 1998.
251. L Romani. Immunity to *Candida albicans*: Th1, Th2 cells and beyond. Curr Opin Microbiol 2: 363–367, 1999.
252. C Ashley, M Morhart, R Rennie, B Ziola. Release of *Candida albicans* yeast antigens upon interaction with human neutrophils in vitro. J Med Microbiol 46:747–755, 1997.
253. DM Underhill, A Ozinsky, AM Hajjar, A Stevenson, CB Wilson, M Bassetti, A Aderem. The Toll-like receptor 2 is recruited to macrophage phagosomes and discriminates between pathogens. Nature 401:811–815, 1999.
254. T Marth, BL Kelsall. Regulation of interleukin-12 by complement receptor 3 signaling. J Exp Med 185:1987–1995, 1997.
255. D Raveh, BA Kruskal, J Farland, RA Ezekowitz. Th1 and Th2 cytokines cooperate to stimulate mannose-receptor-mediated phagocytosis. J Leukoc Biol 64:108–113, 1998.
256. C Palma, A Cassone, D Serbousek, CA Pearson, JY Djeu. Lactoferrin release and interleukin-1, interleukin-6, and tumor necrosis factor production by human polymorphonuclear cells stimulated by various lipopolysaccharides: relationship to growth inhibition of *Candida albicans*. Infect Immun 60:4604–4611, 1992.
257. W Chaka, AF Verheul, VV Vaishnav, R Cherniak, J Scharringa, J Verhoef, H Snippe, IM Hoepelman. *Cryptococcus neoformans* and cryptococcal glucuronoxylomannan, galactoxylomannan, and mannoprotein induce different levels of tumor necrosis factor alpha in human peripheral blood mononuclear cells. Infect Immun 65:272–278, 1997.
258. L Beaman, E Benjamini, D Pappagianis. Role of lymphocytes in macrophage-induced killing of *Coccidioides immitis* in vitro. Infect Immun 34:347–353, 1981.
259. C Monari, A Casadevall, D Pietrella, F Bistoni, A Vecchiarelli. Neutrophils from patients with advanced human immunodeficiency virus infection have impaired complement receptor function and preserved Fcgamma receptor function. J Infect Dis 180:1542–1549, 1999.

19

Immune Reconstitution Against Human Mycoses

Emmanuel Roilides and Evangelia Farmaki
Aristotle University of Thessaloniki, Thessaloniki, Greece

Caron A. Lyman
National Cancer Institute, Bethesda, Maryland

I. INTRODUCTION

An effective immune response is the major contributor to host defense against opportunistic mycoses. This is best illustrated by the finding that invasive mycoses (IM) primarily occur when immune components are absent or dysfunctional. Profound neutropenia, such as develops in patients with cancer and/or bone marrow transplantation (BMT) secondary to the underlying disease or to the treatment, is a major risk factor for IM. Other risk factors include corticosteroid treatment and phagocytic dysfunction resulting from immunocompromising disorders such as infection with human immunodeficiency virus (HIV), chronic granulomatous disease, and graft versus host reaction.

While infections due to *Candida* spp. and *Aspergillus* spp. are the most common, previously rarely encountered fungi have emerged as important opportunistic pathogens. As examples, *Trichosporon beigelii*, *Fusarium* spp., *Scedosporium* spp., and *Penicillium marneffei* can all cause invasive disease. During neutropenia, overall mortality due to candidiasis exceeds 60% [1] while aspergillosis and fusariosis can cause a mortality of >90% [2,3]. Unfortunately, these infections are often only minimally responsive to antifungal therapy, especially in patients with prolonged neutropenia where even the most efficient antifungal agents have a limited efficacy [4].

During the last decade, advances in understanding the host response to these infections, combined with the discovery and availability of a number of cytokines have opened new avenues for antifungal therapy. Based on in vitro and in vivo models of mycoses, the critical roles of intact innate and acquired immunities have become evident and reconstitution of immune function has been attempted. This has been experimentally achieved by either reconstitution of effector cells numerically and/or functionally with cytokines and/or white blood cell transfusions (WBCTx), or by manipulation of cytokine dysbalance. The results from these studies examining the potential role of immunoreconstitution as therapy for opportunistic mycoses are described in this chapter.

II. IN VITRO MODULATION OF PHAGOCYTIC CELL FUNCTIONS IN RESPONSE TO FUNGI

A number of cytokines have been found to influence the host response to fungi such as *Candida* and *Aspergillus* spp. by modulating cellular functions. Th1 cytokines are important in deep-seated candidiasis, when the recruitment of antigen-specific lymphocytes and locally high cytokine concentrations are required to stimulate the anticandidal activities of nonspecific effector cells, including neutrophils (PMN) and macrophages [5]. On the other hand, downregulation of Th1 responses may allow *Candida* to evade intracellular destruction by phagocytes [6,7]. In addition, hemopoietic, Th1, and Th2 cytokines have been shown to be significant regulators of host defenses against aspergillosis in vitro and in vivo [8–10]. As described in detail below, granulocyte colony-stimulating factor (G-CSF), granulocyte-macrophage colony-stimulating factor (GM-CSF), macrophage colony-stimulating factor (M-CSF), interferon-γ (IFN-γ), tumor necrosis factor-α (TNF-α), interleukin (IL)-1, IL-4, and IL-10 have all been shown to be important factors in the host response to these organisms.

A.   *Candida* spp.

G-CSF has been shown to exert many effects on PMN. In addition to increasing the number of mature circulating PMN, G-CSF has been found to enhance their oxidative burst in response to *Candida albicans* blastoconidia and pseudohyphae [11], as well as their antifungal activity against *C. albicans* pseudohyphae. Of note, the fungicidal PMN activity against blastoconidia was unaffected by G-CSF in studies where PMN were incubated with blastoconidia for <2 hr [12,13], whereas enhancement was detected in studies using incubation periods ranging from 18 to 24 hr [14,15]. The latter enhancement may be due to the antifungal effect of calprotectin which is released upon PMN death [16] and/or to an impact of G-CSF on phagolysosome formation and the enclosure of the organism within the phagolysosome [13,14]. G-CSF also has been shown to enhance PMN-induced damage of pseudohyphae of certain non-*albicans Candida* species such as *Candida tropicalis* and *Candida parapsilosis* [17]. In agreement with these results, G-CSF administered to healthy volunteers significantly enhanced PMN-mediated damage of *C. albicans* pseudohyphae during the treatment period [18]. PMN from HIV-infected patients have been found to exhibit reduced candidacidal activity which is partially corrected by G-CSF treatment [19], although others were not able to show an anticandidal defect of PMN from HIV-infected patients [20,21].

GM-CSF was one of the earliest hemopoietic cytokines reported to enhance anti-*Candida* activities of both human PMN [22–24] and mononuclear cells (MNC) [25]. Specifically, it has been shown to enhance fungicidal activity of human PMN [13,15] and MNC [25,26] against *Candida* blastoconidia.

M-CSF is known to accelerate proliferation and differentiation of MNC, to activate mature macrophages, and to recruit peripheral blood monocytes to sites of infection [27]. This cytokine has been shown to modulate such mononuclear phagocyte functions as H_2O_2 production, phagocytosis, and killing of *C. parapsilosis* blastoconidia [27]. In addition, M-CSF enhances production of IL-1, IFN-γ, and TNF-α. Murine macrophages exhibit enhanced fungicidal activity against *C. albicans* blastoconidia when treated with M-CSF [28,29]. In vitro treatment of rabbit splenic adherent cells with M-CSF increased their phagocytic activity against *C. albicans* blastoconidia [30].

During a Phase I study of M-CSF administration to cancer patients, MNC were elutriated from the blood of patients before initiation of therapy and 3 days after the first dose of M-CSF. Superoxide anion (O_2^-) production by these MNC in response to N-formyl methionyl leucyl phenylalanine (FMLP) and fungicidal activity against blastoconidia of *C. albicans* were significantly enhanced at day 3 of therapy as compared to both controls and pretherapy values [31]. In another study, single intravenous administration of M-CSF was found to enhance phagocytosis and intracellular killing of *Candida* blastoconidia by MNC as well as O_2^- production in response to FMLP [32].

IFN-γ, a potent Th1 immunomodulator, enhances phagocytic function against fungi [33] and, in particular, the anti-*Candida* activity of both PMN and MNC [34,35]. Depending on the experimental conditions such as time of incubation and assay used, IFN-γ has been shown to have an enhancing effect [34] or no effect [36,37] on fungicidal activities of human or murine PMN against *C. albicans* blastoconidia.

Despite its toxicity, TNF-α is a promising agent for adjunctive therapy because of its potency as an immunoenhancing cytokine. It can augment the production of other cytokines, such as GM-CSF, as well as enhance several PMN functions. Fungicidal activity of PMN against blastoconidia of *C. albicans* and *Candida glabrata* have been shown to be enhanced [34,38], whereas the results with pseudohyphae of *C. albicans* have been equivocal [35]. The variable effects of TNF-α on fungicidal activity of PMN against different growth forms of the same organism underscore the complicated interactions of cytokines with the antifungal mechanisms of phagocytes [39,40].

The Th2-type cytokines, such as IL-4, primarily exert a suppressive effect on immune cells. IL-4 has been shown to suppress the oxidative burst of MNC, as measured by O_2^- production in response to phorbol myristate acetate (PMA), and the killing of *C. albicans* blastoconidia [41]. In addition, IL-10, a similar potent Th2 cytokine, affects PMN function by suppressing their phagocytic activity against blastoconidia and by reducing PMN-induced damage of *C. albicans* pseudohyphae [42].

Newer Th1-type interleukins such as IL-12 and IL-15 appear to be important in upregulation of host defenses against resistant opportunistic mycoses [43–45]. In particular, IL-15 has been shown to augment anti-*Candida* activities of human PMN [46] and MNC [47] and is a promising cytokine for adjunctive antifungal therapy. The in vitro results described above are summarized in Table 1.

B. *Aspergillus fumigatus*

As with *Candida*, G-CSF has been shown to modulate several antifungal functions of PMN against *A. fumigatus*. These findings are summarized in Table 2. Specifically, it enhances PMN O_2^- production in response to serum opsonized or unopsonized hyphae of *A. fumigatus* and increases PMN-mediated damage of *A. fumigatus* hyphae [48]. Incubation of corticosteroid-treated PMN with G-CSF prevents the corticosteroid-induced suppression of antifungal activities [49]. While PMN from HIV-infected patients possess decreased ability to cause damage to *A. fumigatus* hyphae, in vitro treatment of these PMN with G-CSF has been shown to restore their antifungal ability to normal levels [50].

In a study utilizing PMN harvested from healthy adult volunteers treated with five daily doses of 300 μg G-CSF there was enhanced anti-*A. fumigatus* activity as compared to activity of PMN harvested before therapy [51]. Antifungal activity was evaluated as oxidative burst (chemiluminescence) in response to extracts of *A. fumigatus* and as PMN-mediated killing of *A. fumigatus* conidia. Both G-CSF and GM-CSF were found to enhance PMN-mediated damage of *A. fumigatus* hyphae, and voriconazole exhibited a synergistic effect with these cytokines [52].

Incubation of human MNC with GM-CSF has been shown to enhance O_2^- production in response to PMA and to increase MNC-mediated damage of *A. fumigatus* hyphae [53]. On the

Table 1 In Vitro Effects of Cytokines on the Antifungal Function of Phagocytes in Response to *Candida* spp.

Cytokines	PMN	MNC/macrophages
G-CSF		
Phagocytosis	↑	
Oxidative burst	↑	
Conidial damage	↑	
Hyphal damage	↑	
GM-CSF		
Phagocytosis	↑	↑
Conidial damage	↑	↑
Hyphal damage	↑	
M-CSF		
Phagocytosis		↑
O_2^-		↑
Conidial damage		↑
IFN-γ		
Phagocytosis	↑	↑
Conidial damage	equivocal	No
TNF-α		
Conidial damage	↑	
Hyphal damage	equivocal	
IL-4		
Oxidative burst		↓
Conidial damage		↓
IL-10		
Phagocytosis	↓	↑
Oxidative burst		↓
Hyphal damage	↓	↓

↑: increase, ↓: decrease, No: no effect.

other hand, while treatment of MNC with dexamethasone (DEX) suppresses O_2^- release and hyphal damage, GM-CSF appears to prevent these deleterious effects, suggesting a potential therapeutic role for this cytokine in patients at risk for or suffering from invasive aspergillosis (IA) due to corticosteroid treatment [54].

Similar potent enhancement of antifungal activities of phagocytes has been achieved with M-CSF. In one study, MNC-induced damage of *A. fumigatus* hyphae was significantly increased by treatment with M-CSF and this was associated with enhanced O_2^- production in response to PMA. Phagocytosis of *A. fumigatus* conidia by MNC-derived macrophages and pulmonary alveolar macrophages (PAM) was strongly enhanced by M-CSF [55]. Mononuclear cells harvested from patients who received prolonged courses of M-CSF showed enhanced O_2^- production and antifungal activity by MNC against *A. fumigatus* hyphae [31]. These results demonstrate that M-CSF augments antifungal activity of mononuclear phagocytes against *A. fumigatus* conidia and hyphae.

Incubation of PMN with IFN-γ has been shown to increase oxidative burst and PMN-mediated damage in response to both serum-opsonized and unopsonized hyphae of *A. fumigatus* [48]. The combination of IFN-γ and G-CSF exhibited an additive effect compared to either

Table 2 In Vitro Effects of Cytokines on the Antifungal Function of Phagocytes in Response to *Aspergillus* spp.

Cytokines	PMN	MNC/macrophages
G-CSF		
O_2^--production	↑	
Hyphal damage	↑	
Corticosteroid immunosuppression	restoration	
GM-CSF		
O_2^- production	↑	↑
Hyphal damage	↑	↑
Corticosteroid immunosupression		restoration
M-CSF		
Phagocytosis		↑
O_2^- production		↑
Conidial damage		↑
Hyphal damage		↑
IFN-γ		
Oxidative burst	↑	↑
Hyphal damage	↑	↑
Corticosteroid immunosupression	restoration	restoration
TNF-α		
Phagocytosis		↑
O_2^- production	↑	↑
Conidial damage		No
Hyphal damage	↑	↑
IL-4		
Phagocytosis		No
Conidial damage		No
Hyphal damage		↓
IL-10		
Phagocytosis		↑
O_2^- production		↓
Conidial damage		No
Hyphal damage		↓

↑: increase, ↓: decrease, No: no effect.

cytokine alone. Similarly, treatment of human MNC with IFN-γ resulted in enhanced O_2^- production in response to and increased damage of *A. fumigatus* hyphae, and there was an additive effect when combined with GM-CSF [53]. These findings led to the conclusion that the effects of IFN-γ on anti-*Aspergillus* activities of human PMN and MNC can be combined with the effects of either G-CSF or GM-CSF, and serve as a basis for potential experimental animal and clinical use of combinations of these cytokines.

In addition to suppressing hyphal damage, DEX prevents macrophages from inhibiting germination of *Aspergillus* conidia. While IFN-γ enhanced the antimicrobial activity of DEX-treated macrophages against bacteria, it did not enhance activity against *Aspergillus* conidia, which are handled mostly by nonoxidative antimicrobial mechanisms [56,57]. However, while hydrocortisone and DEX suppressed the antifungal activity of human PMN and MNC against *A. fumigatus* hyphae as measured by decreased O_2^- production and decreased hyphal damage, IFN-γ was able to restore these activities [49,54].

Of note, Gaviria et al. [58] compared the three cytokines, G-CSF, GM-CSF, and IFN-γ. Specifically, the antifungal activities of PMN and MNC taken from healthy donors were evaluated against *C. albicans*, *A. fumigatus*, and *F. solani*. The study showed that IFN-γ exhibited the broadest antifungal activity and enhanced hyphal damage of *A. fumigatus*.

In general, TNF-α has enhancing effects on the antifungal activities of human PMN, MNC, and PAM against *A. fumigatus*. While the effects on MNC functions are moderate, incubation of PMN with TNF-α has been shown to significantly enhance O_2^- production in response to and hyphal damage of unopsonized *A. fumigatus* hyphae. When PAM were incubated with TNF-α, there was increased phagocytosis of *A. fumigatus* conidia but no significant increase in intracellular killing of conidia (Table 2) [59].

On the other hand, treatment of cells with the Th2-type cytokines generally resulted in some suppression of antifungal activity of phagocytes. IL-4, for example, significantly suppressed MNC-induced damage of *A. fumigatus* hyphae. In contrast, it did not alter phagocytic activity or inhibition of conidial germination. These results suggest a lack of pathogenic role of IL-4 on the early phase and a suppressive role on the host response in the late phase of IA [60]. MNC pretreated with IL-10, another Th2-type cytokine, exhibited suppressed O_2^- production in response to PMA, FMLP, and unopsonized *A. fumigatus* hyphae [61]. Anti-IL-10 antibody neutralized this suppressive effect. Furthermore, treatment of MNC with similar concentrations of IL-10 decreased MNC-mediated damage of *Aspergillus* hyphae. In comparison, the phagocytosis of conidia was enhanced and the intracellular conidiocidal activity was not significantly affected by IL-10. Thus, IL-10 suppresses oxidative burst and antifungal activity of MNC against *A. fumigatus* hyphae (a late host response event), whereas it increases their phagocytic activity (an early host response event). In this study, IFN-γ, and GM-CSF but not M-CSF appeared to counteract suppressive IL-10 effects [61].

C. Other Fungi

T. beigelii, a drug-resistant fungus, is ingested and killed by PMN much less efficiently than *C. albicans* [62] and can cause fatal deep infections in patients with prolonged neutropenia [63]. While the host defenses against *Trichosporon* are incompletely understood, the cytokines GM-CSF, IFN-γ, and M-CSF all have been shown to enhance fungicidal activity of human MNC against this organism [62].

Fusarium spp. have emerged as important pathogens in immunocompromised patients [3,64]. Risk factors for fusariosis include prolonged neutropenia and treatment with corticosteroids. A recent study showed that G-CSF, GM-CSF, and IFN-γ enhance antifungal activity of human phagocytes against *F. solani* [58]. However, administration of G-CSF in volunteers did not enhance the ability of their PMN to damage *F. solani* hyphae [18].

Scedosporium apiospermum (*Pseudallescheria boydii*) and *S. prolificans* both cause invasive infections that are resistant to conventional antifungal therapy. Risk factors for infection by these rare fungi are prolonged neutropenia and corticosteroid therapy [65–68]. Incubation of PMN with IFN-γ and GM-CSF has been shown to enhance O_2^- production in response to *S. prolificans* hyphae [69].

P. marneffei is another emerging filamentous fungus that may cause fatal disease in immunocompromised patients. M-CSF enhanced O_2^- production by MNC in response to serum-opsonized and unopsonized *P. marneffei* conidia similar to the M-CSF-enhanced response to *A. fumigatus* [70].

III. IMMUNORECONSTITUTION: EXPERIMENTAL ANIMAL STUDIES

Cytokines either increase the number of phagocytes or regulate (enhance or suppress) the function of these cells. Examples in the first category are G-, GM-, and M-CSF; examples in the second category are IFN-γ, IL-4, and IL-10. While hemopoietic cytokines have been extensively used to shorten the duration of chemotherapy-induced neutropenia [71–75] and as adjunctive therapy in patients with fever and neutropenia [76,77], few data on their utility in defined mycoses are available [78]. Thus, current interpretations of the potential utility of cytokines are substantially based on animal models and on a minimal amount of uncontrolled clinical data.

A. Administration of Immunoenhancing Cytokines

Several animal models have been used to study the role of cytokines in the management of IM in immunocompromised (neutropenic or immunosuppressed) hosts. In one of the earliest studies of cytokines in animal models of mycoses, G-CSF given to cyclophosphamide-treated mice was reported to offer protection from subsequent development of aspergillosis. In that study, steroid-treated mice were protected only when simultaneous antifungal chemotherapy was administered [79]. In a more recent study, mice immunosuppressed either with hydrocortisone or with 5-fluorouracil were infected intranasally with A. fumigatus. Beginning 3 days before infection, groups of mice were given either recombinant G-CSF or the antifungal triazole posaconazole, or both. In the corticosteroid-pretreated mice, G-CSF strongly antagonized the antifungal activity of posaconazole. While the drug achieved reduced counts in lung tissue and prolonged survival, the combination therapy with G-CSF reversed this effect. In contrast, mice made neutropenic with 5-fluorouracil and then infected with A. fumigatus benefited from either G-CSF or triazole, and the effect of the combination was additive. These findings suggest that host factors contribute in different ways to the outcome of cytokine therapy in aspergillosis [80].

The effect of G-CSF was investigated in a nonneutropenic murine model of acute disseminated C. albicans infection. Mice treated with a single dose of G-CSF showed a significant reduction in mortality and in fungal growth from kidneys, spleen, and liver [81]. On the other hand, the combination of G-CSF with fluconazole had a synergistic effect on clearance of C. albicans from organs of infected nonimmunocompromised mice but not on the survival of these animals [82]. These variant results may have been due to inconsistent effects of G-CSF, or to adverse effects of G-CSF affecting overall activity.

In neutropenic mice with disseminated trichosporonosis, G-CSF (30–100 μg/kg/day) administered either before or after infection improved the survival rate from <25% up to 100% and led to organ clearance. However, GM-CSF (0.8–2 μg/kg/day) decreased the survival rate. This may be related at least in part to the finding that GM-CSF increased PMN counts less significantly than did G-CSF, and that there was a highly elevated level of TNF-α in bronchoalveolar lavage fluid which may have had an adverse effect on local phenomena. These results suggest that other host defense mechanisms, such as TNF-α overproduction in the lungs, have an important role in the prognosis of trichosporonosis [83].

Unlike the findings with trichosporonosis, administration of recombinant murine GM-CSF to cyclophosphamide-treated mice enhanced their resistance to lethal challenge with C. albicans [84]. Similarly, in mice with acute candidiasis, Cenci et al. found that M-CSF administration protected them from subsequent lethal challenge with C. albicans. In this model, increased survival and reduced recovery of fungi from the organs was observed [29].

Further, the effects of M-CSF (100–600 μg/kg/day) in augmenting pulmonary host defense against A. fumigatus have been studied in neutropenic rabbits with pulmonary aspergillosis. In these studies, rabbits were given M-CSF starting 3 days preinoculation and then throughout neutropenia. Rabbits receiving prophylactic M-CSF had significantly increased survival and decreased pulmonary injury as evidenced by computerized tomographic scanning and histopathology. Microscopic studies demonstrated greater numbers and more activated PAM in lung tissue of rabbits receiving M-CSF in comparison to controls. PAM harvested from M-CSF-treated rabbits exhibited significantly greater phagocytosis of *A. fumigatus* conidia as compared to PAM from control rabbits [85].

In vivo studies using IFN-γ to enhance immune responses against invading fungi have had somewhat disparate results. One intravenous dose of IFN-γ shortly before or simultaneously with the challenge of nonimmunosuppressed mice with *C. albicans* decreased the growth of the organism from kidneys, spleen, and liver. However, it did not reduce the growth of *C. albicans* in cyclophosphamide-pretreated mice. These findings suggest that IFN-γ enhances host resistance against acute disseminated *C. albicans* infection in mice through activated PMN [86].

In another study, administration of IFN-γ to naive mice and their subsequent challenge with *C. albicans* resulted in a higher infectious burden and increased mortality [87]. Of interest, this detrimental effect of IFN-γ was not observed if the mice had been immunized with *C. albicans* as infants. The same exacerbation of murine candidiasis was observed after treatment of mice with IL-12, a cytokine that augments secretion of IFN-γ [43]. Clearly other factors are also important for in vivo modulation of host defenses, and the administration of cytokines may be accompanied by significant toxicity.

TNF-α is a cytokine that may be beneficial but whose toxicities are very limiting. In corticosteroid-treated mice with IA, IFN-γ and TNF-α have been shown to decrease mortality (from 40–60% to 0%) and fungal burden in organs as demonstrated by culture and histology [88]. In addition, TNF-α was shown to have a protective role in a murine model of systemic infection by *C. albicans* [89]. In contrast, neutralization of TNF-α and GM-CSF by treatment of mice with monoclonal antibodies reduced the influx of PMN into the lungs and delayed fungal clearance following pulmonary challenge with *A. fumigatus* [90]. In another study, intratracheal challenge of both neutropenic and nonneutropenic mice with *A. fumigatus* conidia resulted in an increase of lung TNF-α levels, which correlated with the histologic development of a peribronchial infiltration of PMN and MNC. Neutralization of TNF-α resulted in increased mortality in both normal and cyclophosphamide-treated animals, which was associated with increased lung fungal burden. Depletion of TNF-α resulted in a reduced lung PMN influx in both normal and cyclophosphamide-treated animals, which occurred in association with decreased lung levels of the chemokines macrophage inflammatory protein-2 and macrophage inflammatory protein-1α. In cyclophosphamide-treated animals, intratracheal administration of a TNF-α agonist peptide 3 days before the administration of *Aspergillus* conidia resulted in improved survival. This study indicated that TNF-α is a critical component of innate immunity in both immunocompromised and immunocompetent hosts, and that pretreatment with a TNF-α agonist peptide in a compartmentalized fashion can significantly enhance resistance *to A. fumigatus* in neutropenic animals [91]. Unfortunately, however, due to excessive toxic adverse effects it is difficult to evaluate the antifungal effects of TNF-α clinically.

Studies have also shown a beneficial role for IL-1 against systemic candidiasis in immunocompromised mice. Pretreatment of mice with IL-1 protected them from acute disseminated candidiasis [92] and decreased the number of *C. albicans* in kidneys and spleens of the animals [93,94]. Although the mechanism of IL-1-induced protection is not well understood, IL-1 in combination with antifungal agents administered during neutropenia could improve outcome from difficult-to-treat cases of candidiasis [94]. The main findings of the experimental animal studies are summarized in Table 3.

Understanding the synergistic activity of phagocytes, cytokines, and antifungal agents is an area of active research with potential clinical usefulness. In vitro studies have shown additive effects of G-CSF or GM-CSF and azoles (fluconazole and voriconazole) when combined with PMN or MNC against either *C. albicans* or *A. fumigatus* [15,52]. Moreover, additive effects have been found between M-CSF-treated PAM and amphotericin-B lipid complex [100]. In murine models of IM, synergism has been shown between G-CSF and posaconazole in neutro-

ble 3 Use of Cytokines in Animal Models of IM

tokines	Animal model	Organism	Antifungal Rx	Outcome
CSF	Cyclophosphamide-treated mice	*Aspergillus*		Protective effect [79]
	N mice	*Aspergillus*	posaconazole	Additive effect [80]
	Immunocompetent and N mice	*Candida*	fluconazole	Additive effect, ↑ survival and renal clearance [95]
	Cyclophosphamide-treated mice	*Candida*	alone	↑ survival, ↓ the number of candida in kidneys
			+ amphotericin-B	↑↑ survival [96]
	Non-N mice	*Candida*		↓ mortality and fungal growth of organs [81]
	Cyclophosphamide-treated mice	*Candida* *Aspergillus*		Protective effect ↓ candida growth of kidney [97]
	N mice	*Trichosporon*		Improved survival and organ clearance [83]
M-CSF	Cyclophosphamide-treated mice	*Candida*		↑ resistance, ↓ candida recovery from the organs [84]
-CSF	Non-N mice	*Candida*		↑ survival ↓ *Candida* recovery [29]
	Non-N rats	*Candida*	fluconazole	Improved survival [98]
	Cyclophosphamide-treated mice	*Candida*	+ fluconazole + amphotericin B	↑ survival ↑↑ survival [99]
	N rabbits	*Aspergillus*		↑ survival ↑ PAM activation [85]
N-γ	Nonimmunosuppressed mice	*Candida*		↓ *Candida* growth [86]
	Naive mice	*Candida*		↑ mortality [87]
N-γ, TNF-α	Corticosteroid-treated mice	*Aspergillus*		↓ mortality and organ clearance [88]
NF-α	Non-N mice	*Candida*		Protective effect ↑ resistance [88]
NF-α, GM-CSF	Non-N mice	*Aspergillus*		Role in recruitment of neutrophils (↑ fungal clearance) [90]
NF-α agonist	N mice	*Aspergillus*		↑ survival [91]
-1	Cyclophosphamide-treated mice	*Candida*		↑ survival [92]
	Immunosuppressed mice	*Candida*		↓ number of *Candida* in organs [93]
	N mice	*Candida*	fluconazole	↓↓ number of *Candida* in organs [94]

: neutropenia.

penic mice with aspergillosis [80] and between M-CSF and amphotericin-B against candidiasis [99].

B. Modulation of Immunosuppressive Cytokines

Recent studies suggested that IL-4 and IL-10 regulate phagocytic cell function in vivo. For example, in a murine model of IA, resistance to *A. fumigatus* was correlated with intact functions of conidial killing and hyphal damage by lung phagocytes [10]. In these mice, resistance could be augmented either by neutralization of IL-4 or by IL-10 deficiency accomplished with gene knockout methodology. Conversely, susceptibility to IA was correlated with impaired conidial killing and hyphal damage and with increased IL-4 and IL-10 production.

Furthermore, administration of Th2 cytokines to mice with IA supports the above results. For example, administration of IL-4 and IL-10 increases susceptibility to the challenging fungus and reduces survival after infectious challenge. In contrast, administration of soluble IL-4 receptor, a blocking agent for IL-4, enhances the resistance of mice to *A. fumigatus* and improves their survival [101].

IV. IMMUNORECONSTITUTION—CLINICAL STUDIES

A. Cytokines

1. Chronic Granulomatous Disease (CGD)

a. Prevention.   Patients with CGD are at high risk for IA. Indeed, IA is common cause of pneumonia and mortality in these patients [102]. IFN-γ has been shown to enhance the oxidative burst of PMN and MNC derived from patients with CGD [103–105]. In a large prospective, randomized, placebo-controlled trial, the incidence of serious infections in the group of CGD patients who received 50 μg/m² IFN-γ three times a week was reduced from 24% of controls to 4% in 2 years [106]. In an ex vivo accompanying study of some patients from the above investigation, PMN from IFN-γ-treated patients caused more damage to *Aspergillus* hyphae than PMN from controls [107]. In a subsequent double-blind randomized study, PMN harvested from CGD patients treated with 100 μg/m² IFN-γ for 2 consecutive days exhibited higher oxidative burst and increased damage to *A. fumigatus* hyphae as compared to cells harvested from patients treated with 50 μg/m² IFN-γ or before initiation of IFN-γ administration [108]. These findings may be relevant to the decreased incidence of serious infections observed in CGD patients treated with IFN-γ prophylactically.

b. Therapy.   Randomized clinical studies are very difficult to perform with unusual moulds, making controlled in vitro and animal studies very important. Based on in vitro and animal data, some clinicians have chosen to treat these life-threatening mold infections with combination therapy. The majority of IM cases have been due to *Aspergillus* spp., many of whom were cured with combined antifungal therapy and IFN-γ (Table 4).

These patients were not receiving IFN-γ prophylactically. Use of other immunotherapies has been rare in CGD patients. One patient with invasive multifocal infection due to *A. nidulans* was reported to be successfully treated by HLA-genoidentical BMT, G-CSF-elicited PMN, G-CSF, and liposomal amphotericin-B. The infection had been unresponsive to treatment with amphotericin-B and IFN-γ. At 2 years post-BMT, the patient was well with full immune reconstitution and no sign of *Aspergillus* infection [111].

2. HIV Infection

Invasive mycoses, especially aspergillosis, have become important opportunistic infections in HIV-infected patients [118–120]. PMN and MNC-derived macrophages from HIV-infected pa-

Table 4 Case Reports of CGD Patients with Fungal Infections Who were Treated with Combined Antifungal Therapy and a Cytokine

Reference	Type of fungal infection	Fungal isolate	Antifungal therapy	Cytokine
Williamson [109]	Soft tissue infection on the right heel	*Paecilomyces varioti*	Amphotericin-B followed by itraconazole	IFN-γ
Phillips [110]	Disseminated infection	*Pseudallescheria boydii*	Amphotericin-B	IFN-γ
Ozsahin [111]	Invasive multifocal infection	*A. nidulans*	Liposomal amphotericin-B	G-CSF G-CSF-elicited PMN
Roilides [112]	Lobar pneumonia and tibia osteomyelitis	*Chrysosporium zonatum*	Liposomal amphotericin-B	IFN-γ
Cohen-Abbo [113]	Multifocal osteomyelitis	*Paecilomyces varioti*	Amphotericin-B followed by itraconazole	IFN-γ
Pasic [114]	Osteomyelitis	*A. fumigatus*	Itraconazole	IFN-γ
Touza Rey [115]	Brain abscesses	*A. fumigatus*	Various antifungal agents and surgery	IFN-γ
Bernhisel-Broadbent [116]	Pulmonary infection	*A. fumigatus*	Amphotericin-B and flucytosine	IFN-γ
Tsumura [117]	Tibia osteomyelitis	*A. fumigatus*	Amphotericin-B	IFN-γ

tients possess decreased ability to damage hyphae and to ingest conidia of *A. fumigatus*, respectively [50,121]. These defects may be related to the well-described CD4 + T-helper lymphocyte dysfunction in these patients [122]. Of interest, the PMN defect is partially correctable by G-CSF in vitro [50]. In addition, in a randomized, multicenter, controlled trial of patients with advanced HIV infection, G-CSF decreased incidence and duration of severe neutropenia, incidence of bacterial infections, number of hospital days, and duration of intravenous antibiotics [123]. No mention, however, of any mycoses was included in the study.

3. Cancer-Related Therapy

Malignancies and the related immunodeficiencies constitute the broadest field of acquired defects in host defenses and the greatest need for immune reconstitution. Thus, most of the preclinical and clinical studies have focused on this patient population. To date, although the cytokines have been extensively studied preclinically, no randomized studies have been designed to specifically examine the role of cytokines against IM in immunocompromised patients. The main reason for this is that the number of IM as infectious sequelae of immunocompromise is relatively small, and no study performed with other objectives has had the statistical power to demonstrate a significant difference in the proportion of IM. Thus, conclusive clinical data are still missing, making the issue of cytokine use in the management of IM controversial.

Table 5 Case Reports of Cancer Patients with Fungal Infections Who were Treated with Combined Antifungal Therapy and a Cytokine

Reference	No. of pts	Underlying condition	Fungal infection	Cytokines	Outcome
Grauer [124]	1	AML & AlloBMT	*T. beigelii* septicemia	G-CSF	Cured
Poynton [125]	2	AL	Hepatosplenic candidosis	IFN-γ + GM-CSF	Cured
Spielberger [126]	1	Neutropenia	Disseminated *Fusarium* infection	GM-CSF (PMN transfusion)	Cured
Gonzalez [127]	1	Neutropenia	Invasive zygomycosis	G-CSF	Cured
Hennequin [128]	1	Neutropenia	*Fusarium oxysporum* infection	G-CSF	Cured
Fukushima [129]	1	ALL	Invasive thoracopulmonary mucormycosis	G-CSF	Cured
Dornbusch [130]	5	Neutropenia	Invasive pulmonary aspergillosis	G-CSF	2 died 2R (with S) 1 CR (without S)

R: response; CR: complete response; S: surgery.

a. Case Reports.   Several case reports have suggested the value of hemopoietic cytokines in treating IM in cancer patients (Table 5).

b. Studies of Cytokine Administration.   There have been two studies reported using GM-CSF as adjuvant therapy in neutropenic patients with controversial conclusions. In an initial pilot study of eight neutropenic patients with IM there were six responses with four of them being complete responses. Among the eight patients, there were two patients with *Aspergillus* (one responded), five *Candida* spp. cases (four responded), and one *Trichosporon* (responded) [131]. The dose of GM-CSF used in this study was excessively high, ranging up to 700 $\mu g/m^2$, and was followed by increased toxicity.

A subsequent open study of GM-CSF in proven IM in neutropenic cancer patients conducted in 17 patients did not show similar favorable results. Eight of the patients suffered from candidemia, eight from pulmonary aspergillosis, and one from fusariosis. They were treated with GM-CSF 5 μg/kg and amphotericin-B 1 mg/kg. There were six deaths and the authors felt that this regimen failed to improve outcome [132].

A multicenter study examining the utility of the administration of G-CSF in nonneutropenic patients with invasive candidiasis has been conducted in Europe. Although no significant difference in survival was observed [133], the patients who had higher numbers of PMN due to G-CSF had more favorable outcome than the patients who did not increase their PMN counts.

In another randomized trial, the efficacy of G-CSF in 119 severely neutropenic patients with hematological malignancies after intensive chemotherapy and infection was investigated. Patients received either ceftazidime plus amikacin, or the same antimicrobial regimen plus G-CSF (5 μg/kg/day). Patients who received antibiotics plus G-CSF had more clinical responses

(82% vs. 60%), less superinfections, fewer days in hospital, reduced antibiotic use, and reduced mortality. Although only four fungal infections occurred, they were all encountered in the group receiving antibiotics alone. Toxicity secondary to G-CSF was absent [134]. Unfortunately, this study is a typical example of the problem of lack of power for determining a significant difference.

Similarly, in a retrospective study of patients with autologous bone marrow transplantation (ABMT) for lymphoid cancer, the incidence of infections 28 days post-ABMT was compared in patients who had or had not taken GM-CSF. Overall, fewer infections were observed (13% vs. 40% of controls; $P = .001$), a trend of fewer IM (4% vs. 14%; $P = .09$), and fewer days of amphotericin-B usage (median 2 vs. 8.5 days; $P = .03$) with GM-CSF treatment [135].

The strongest data supporting adjunctive cytokine administration in IM come from the Eastern Cooperative Oncology Group study, a prospective, randomized, placebo-controlled Phase III study of GM-CSF. In this study, 124 older patients with acute myelogenous leukemia, 55–70 years of age, were given GM-CSF or placebo from day 11 postinduction. A higher rate of complete response and longer overall survival were observed among the patients treated with GM-CSF. In addition, lower infection-related toxicity was observed. Among patients with pneumonia, mortality was as follows: 2/14 patients (14%) in the GM-CSF group, and 7/13 (54%) in the placebo group ($P = .046$). Fungal infection-related mortality in the GM-CSF group was 1/52 (2%) as compared to the placebo group, which was 9/47 (19%; $P = .006$) [136,137].

In a more recent publication from this trial, the etiological agents present at the time of death were reported [138]. Among the patients who received GM-CSF or placebo, there were 11 patients with aspergillosis, seven with candidiasis, and two with infections due to other fungi. Only one of eight patients who had been randomized to receive GM-CSF and developed fungal infection died (13%). This contrasted to nine among 12 patients on placebo (75%; $P = .02$). No apparent difference between aspergillosis and candidiasis was noted with these very small numbers of patients.

In one of the early nonrandomized studies, M-CSF was administered to 46 neutropenic cancer patients with IM. Among them, 30 patients suffered from invasive candidiasis, 15 from IA, and one from mucormycosis. Overall, there was a trend of decreased incidence of mycoses. There was, however, a significantly greater survival in patients with candidiasis and Karnofsky score >20% compared with historical controls ($P < .05$) [139]. This study was followed by a double-blind controlled study of M-CSF in patients with acute myelogenous leukemia and febrile neutropenia. Although no impact on disease-free survival was found, an outcome heavily dependent on many confounding factors in these high-risk patients, a decreased incidence and duration of febrile neutropenia and a significant decrease in the use of systemic antifungals were observed [140]. The results of the above preventive and therapeutic studies on outcome of the patients are summarized in Table 6.

The issue of cost-effectiveness of G-CSF with regard to IM was addressed in a recently published study. Neutropenic patients with presumed IM were randomized to receive either amphotericin-B alone or in combination with G-SCF (3–5 mg/kg daily). There were 62% responders to combination versus 33% to amphotericin B-alone ($P = .027$). Of note, all nonresponders received liposomal amphotericin-B. Based on drug acquisition, hospital stay, and treatment duration, the combination regimen was more cost-effective [141].

B. Cytokine-Elicited White Blood Cell Transfusions [142]

WBCTx therapy for patients with neutropenia and infections was first attempted >60 years ago [143], but it was not until the 1960s that this therapy was frequently used and studied. As

Table 6 Clinical Studies of Use of Cytokines for Prevention or Treatment of IM in Cancer Patients

Cytokine	Outcome	References
Prevention		
GM-CSF	Overall ↓ infections, a trend of ↓ IM and ↓ days of amphotericin-B usage	[135]
GM-CSF	More complete responses, ↑ survival, ↓ fungal infection-related mortality	[136,137]
G-CSF	More clinical responses, ↓ superinfections, hospitalization, antibiotic use, mortality	[134]
	Fungal infections only in the group treated with antibiotic alone	
M-CSF	↓ incidence and duration of febrile neutropenia, ↓↓ use of systemic antifungals	[140]

Cytokine	Fungal infections	Outcome	References
Treatment			
GM-CSF	*Candida* 5	6/8 responses, 4/8 complete responses	[131]
	Aspergillus 2		
	Trichosproron 1		
GM-CSF	*Candida* 8	No effect on outcome (6 deaths)	[132]
	Aspergillus 8		
	Fusarium 1		
GM-CSF	*Candida* 3	↓ mortality (13% in GM-CSF group, 75% in placebo group, $P = .02$)	[138]
	Aspergillus 4		
	Other 1		
G-CSF	*Candida*	No significant difference in survival	[133]
M-CSF	*Candida* 30	Trend of ↓ mycoses	[140]
	Aspergillus 15	Significant improvement of survival of patients with candidiasis and Karnofsky score > 20%	
	Other 1		

discussed by Schiffer [144], early studies described benefits of WBCTx using large doses of PMN from donors with chronic myeloid leukemia (CML). Although these were uncontrolled trials, dramatic clinical changes were observed in patients who were doing poorly with antibiotic treatment alone. As CML donors are not always readily available, more sophisticated apheresis techniques were developed that resulted in collection of higher numbers of PMN. With this, the focus shifted to the use of WBCTx obtained from normal donors.

Seven experimental animal studies demonstrated the efficacy of WBCTx therapy for the eradication of infection [145–151]. Based on encouraging results from these studies, a series of clinical trials were conducted in patients with neutropenia-related infections [152–158]. In these studies, most patients had refractory bacterial infections, and they were randomized to receive treatment with WBCTx plus antibiotics versus antibiotics alone. Most of them showed that WBCTx could be lifesaving among those patients with prolonged neutropenia [152,153,156,157] or could improve overall survival [154]; two showed no benefit of WBCTx therapy [155,158]. Data from three uncontrolled studies [159–161] showed a response rate >60% when WBCTx were given as adjunctive therapy to patients who were not responding to antibiotic treatment alone.

However, methodological problems, issues related to the high cost of the procedure, significant adverse effects to the recipient (fever, chills, hypotension, pulmonary infiltrates, respiratory distress, transmission of cytomegalovirus, graft-versus-host disease, alloimmunization, and haemolytic reactions), and insufficient quantities of PMN used for transfusion [162] made

WBCTx less attractive. Because of these problems and the development of newer and more effective antibiotics, WBCTx were almost abandoned in the late 1980s.

In the 1990s, a new need for WBCTx therapy became apparent due to the increased incidence of neutropenia-related mycoses and the limited efficacy of antifungal agents in this setting. However, in order to have a good response to WBCTx it was necessary to evaluate and modify those factors that could have been responsible for the marginal efficacy and severe toxicity of the WBCTx reported previously. Those factors were impaired quality of PMN obtained from older leukapheresis techniques, alloimmunization status of some WBCTx recipients, and insufficient quantities of PMN used for transfusion [162]. Indeed, all reported data suggested that adequate WBCTx therapy depends on selection of the right recipient, the best collection technique, and the best donor. By optimizing these factors, transfusion of a high number of good-quality PMN can be achieved.

Patients with evidence of alloimmunization (platelet refractoriness, antileukocyte antibodies, repeated febrile transfusion reactions, or posttransfusion pulmonary infiltrates) may not benefit from WBCTx [163]. They usually have a low posttransfusion increment [164] and more pulmonary reactions secondary to longer retention of PMN in the lungs [165]. In addition, transfused PMN seem to be unable to migrate to the sites of infection in alloimmunized infection [166].

One experimental [150] and three clinical [160,161,167] studies have shown that the higher the number of cells transfused per square meter of body surface area, the better the clinical response to WBCTx. The filtration leukapheresis technique used in the 1970s was inexpensive but yielded damaged PMN [168]. In addition, transfusion of cells collected with this procedure was associated with a high incidence of transfusion reactions and poor recovery of cells after transfusion [169]. More recently, continuous-flow centrifugation with the addition of a rouleauxing agent such as hydroxyethyl starch to effectively separate PMN from erythrocytes is the preferred leukapheresis technique and produces a higher yield of better-quality PMN [170].

The best donor would be a person who is blood related (to avoid toxicity related to incompatibility) and who can yield a high number of PMN. Even with the best leukapheresis technique, the number of PMN that can be collected is still relatively small (Table 7).

Under optimal conditions, transfusion of 10^{10} PMN should produce a posttransfusion increment of 10^3 PMN/μL in a 70-kg adult [169]. This value is close to 1.12×10^{11}, the estimated daily production of PMN for a 70-kg person [171]. Alterations in leukocyte kinetics due to fever, infection [169], and profound neutropenia [167] may further increase the transfusion requirements, and it has been suggested that doses of 10^{11} PMN may be required to give a higher proportion of successes after WBCTx [167].

Table 7 Number of PMN at the Time of Starting White Blood Cell Transfusion Therapy

	PMN $\times$ 10^{10}	References
Daily production of PMN by healthy person	11	[171]
Number of PMN needed to achieve an ANC > 1000/μL	10	[169]
Mean yield of PMN per WBCTx collection using:		
CML donors	5.8	[167]
Dexamethasone-treated healthy donors	2.5	[172]
G-CSF-treated healthy donors	3.9	[173]

Source: Ref. 142.

Given the apparent limits of collection methodology, additional work has focused on finding donors that might provide higher yields. Patients with CML were successfully used as donors by Freireich et al., and provided an average of 5.8×10^{10} PMN per transfusion [167]. However, patients with CML are not convenient as donors, and further work focused on treating healthy donors with steroids to increase their circulating WBC. Dexamethasone was the most promising alternative (mean PMN yield was 2.5×10^{10}) [172] but never produced the very high yields thought desirable for consistent success [167].

Subsequently, recombinant G-CSF has been shown to stimulate the proliferation and differentiation of PMN in vitro and to increase the absolute neutrophil count (ANC) safely in patients with neutropenia [174]. These results raised the possibility that administration of G-CSF to WBC donors would increase their ANC to levels that would lead to a higher yield of better-quality PMN (Table 7). Bensinger et al. treated eight donors with the same dosage of G-CSF and obtained comparable PMN yields (mean 4.1×10^{10}) and 24-hr posttransfusion ANC (median 570/μL), but no data about clinical efficacy of WBCTx were provided [173]. These results are consistent with in vitro studies that have demonstrated that cytokine regimens containing G-CSF can prolong the survival and function of PMN, even if the PMN are irradiated [175–177].

In a pilot study, G-CSF-elicited WBCTx were used to treat 15 adult neutropenic patients (<500/μL) with documented and refractory mycoses [178]. All 15 adult patients had hematological malignancies, and seven had been treated with BMT. The cases included 11 mould infections (seven aspergillosis, three fusariosis, one unidentified) and four yeast infections (three candidiasis, one trichosporonosis). Eight of the patients had evidence for widely disseminated infection. At the time of institution of WBCTx, all of the infections had been progressive despite appropriate antifungal treatment for a median of 10 days and the patients had been neutropenic for a median of 23 days. At the end of WBCTx therapy, 11 of 15 patients had a favorable response (nine improved and two stabilized), and in seven of these patients, the favorable response appeared substantially due to WBCTx. Those patients in whom WBCTx therapy was started early during neutropenia and shortly after diagnosis of fungal infection were more likely to respond. The favorable responses were still seen 3 weeks after the end of WBCTx therapy in eight of 11 responders. Due to special efforts to select only blood-related donors and nonalloimmunized recipients, the frequency of adverse reactions was <5%. Donors tolerated the treatment well and achieved a median PMN count of 29.4×10^3/μL that led to a mean PMN yield of 4.1×10^{10}, and a mean 1-hr and 24-hr posttransfusion PMN count of 594 and 396/μL, respectively. This pilot study demonstrated that G-CSF-enhanced WBCTx therapy can be lifesaving for patients with refractory neutropenia-related mycoses. More information about long-term safety of donors' treatment with G-CSF is needed before recommendations on use of G-CSF-elicited WBCTx are made [142].

More recently, Peters et al. reported the results of a prospective study of WBCTx in patients with malignancies or hematological disorders and serious bacterial and fungal infections during neutropenia. WBCTx were derived from blood-related donors and after G-CSF or prednisolone treatment. The WBCTx were safe and there were greater numbers of WBC after G-CSF elicitation than prednisolone elicitation. Nine patients suffered from IA and 4 from *Candida* infections. Among the nine patients with IA, five had a complete clearance of infection, and among the four patients with *Candida* infection, two recovered completely following WBCTx therapy [179].

In another report, G-CSF-primed PMN transfusions were administered to two neutropenic patients with *C. tropicalis* fungemia in combination with amphotericin-B, resulting in elimination of the infection [180]. Furthermore, Catalano et al. used combined treatment with amphotericin-B and G-CSF-stimulated WBCTx in an aplastic patient with IA undergoing BMT with complete recovery from the infection [181].

In contrast, a retrospective study of administration of WBCTx to neutropenic patients with IM did not find any benefit showing an equally low infection resolution rate of 29% in the transfused group and 23% in the nontransfused group for noncandidal mycoses and 56% versus 50% for candidal sepsis, respectively [182]. Clearly, more investigations are needed before this technique becomes a standard of care for life-threatening IM in neutropenic patients.

V. RESTORATION OF LYMPHOCYTE RESPONSE POST PERIPHERAL BLOOD STEM CELL TRANSPLANT BY ADDITION OF AUTOLOGOUS DENDRITIC CELLS

Patients who have received BMT or peripheral blood stem cell transplant (PBSCT) are susceptible to IA even after engraftment [183,184]. It is known that patients receiving chemotherapy or BMT/PBSCT have decreased numbers of dendritic cells (DC) [185]. These patients also display lymphopenia and poor proliferative response to mitogenic stimuli for approximately 4–6 months posttransplant, as well as a prolonged impairment of specific T-lymphocyte response to antigens [186]. The lack of response to antigens is multifactorial, but one factor could be the decreased number of DC. Therefore, to evaluate whether DC could improve the proliferative response of lymphocytes 1 month after transplant, Grazziutti et al. prepared DC from a cryopreserved sample of apheresis blood collected pre-PBSCT. When peripheral blood MNC from five PBSCT recipients were collected 1 month posttransplant and exposed to heat-killed *A. fumigatus* in the absence of exogenously added DC, no proliferative response was observed in vitro. However, if pre-PBSCT DC were added to these cultures, there was a significant proliferative response to heat killed *A. fumigatus* in two of five patients. Similar experiments using MNC from pre-PBSCT apheresis blood in place of DC showed no improvement in the proliferative response to heat killed *A. fumigatus* compared with the response of peripheral blood MNC alone. This inefficiency of MNC may have been due to an increased sensitivity of MNC to the deleterious effects associated with cryopreservation/thawing [187].

VI. T-CELL ADOPTIVE THERAPY—PROSPECTS OF VACCINATION

Studies have begun to assess the ability of fungal antigens to induce Th1-type reactivity as potential candidates for fungal vaccines. Similar to what has occurred upon nasal exposure to viable *A. fumigatus* conidia, treatment of immunocompetent mice with *Aspergillus* crude culture filtrate antigens resulted in the development of local and peripheral protective Th1 memory responses. These were mediated by antigen-specific CD4 + T cells producing IFN-γ and IL-2 capable of conferring protection upon adoptive transfer to naive recipients. Protective Th1 responses could not be observed in mice deficient of IFN-γ or IL-12. These results showed that *Aspergillus* antigens exist with the ability to induce Th1-type reactivity during infection [188], which suggests the existence of fungal antigens useful as a potential candidate vaccine against invasive pulmonary aspergillosis [189].

VII. CONCLUSIONS

The increasing incidence of IM and the emergence of previously rare opportunistic fungal pathogens is of alarming importance in the management of immunocompromised patients. Normalization of the host defenses along with antifungal therapy is the cornerstone of successful treatment.

In the last decade, much has been learned about phagocytic cell and cytokine involvement in immunopathogenesis of mycoses by in vitro and animal model studies. In parallel with destroying fungi using potent antifungal agents, reconstitution of immune response by either exogenous modulation of enhancing/regulatory cytokines or transfusion of cytokine-elicited allogeneic phagocytes appears to be a promising adjunct to antifungal chemotherapy for these life-threatening diseases. In particular, G-CSF-elicited WBCTx may represent a valuable new tool for use in management of these infections. Further evaluation of the safety and efficacy of these immunotherapeutic modalities is an urgent priority for research during the start of the new decade. One may have to distinguish between patients with no PMN or with dysfunctional PMN in order to choose which cytokines to use [190]. Since even large randomized studies of unselected patients are not expected to have the statistical power necessary to prove the beneficial role of cytokine modulation, only very high-risk patients need to be targeted in randomized studies of single or combined cytokines as potential beneficiaries of immunomodulation. It is only through such studies that guidelines may be established as to the appropriate use of these agents. A better understanding of the synergy between cytokines and specific antifungal agents may provide additional powerful tools for managing these serious infections in the twenty-first century.

REFERENCES

1. EJ Anaissie, JH Rex, O Uzun, S Vartivarian. Predictors of adverse outcome in cancer patients with candidemia. Am J Med 104:238–245, 1998.
2. DW Denning. Invasive aspergillosis. Clin Infect Dis 26:781–803, 1998.
3. EI Boutati, EJ Anaissie. *Fusarium*, a significant emerging pathogen in patients with hematologic malignancy: ten years' experience at a cancer center and implications for management. Blood 90: 999–1008, 1997.
4. A Bohme, M Karthaus. Systemic fungal infections in patients with hematologic malignancies: indications and limitations of the antifungal armamentarium. Chemotherapy 45:315–324, 1999.
5. P Puccetti, L Romani, F Bistoni. A Th1-Th2-like switch in candidiasis: new perspectives for therapy. Trends Microbiol 3:237–240, 1995.
6. E Cenci, L Romani, A Mencacci, R Spaccapelo, E Schiaffella, P Puccetti, F Bistoni. Interleukin-4 and interleukin-10 inhibit nitric oxide-dependent macrophage killing of *Candida albicans*. Eur J Immunol 23:1034–1038, 1993.
7. P Puccetti, A Mencacci, E Cenci, R Spaccapelo, P Mosci, K-H Enssle, L Romani, F Bistoni. Cure of murine candidiasis by recombinant soluble interleukin-4 receptor. J Infect Dis 169:1325–1331, 1994.
8. E Roilides, H Katsifa, TJ Walsh. Pulmonary host defences against *Aspergillus fumigatus*. Res Immunol 149:454–465, 1998.
9. E Cenci, S Perito, K-H Enssle, P Mosci, J-P Latge, L Romani, F Bistoni. Th1 and Th2 cytokines in mice with invasive aspergillosis. Infect Immun 65:564–570, 1997.
10. E Cenci, A Mencacci, C Fè d'Ostiani, G Del Sero, P Mosci, C Montagnoli, A Bacci, L Romani. Cytokine- and T helper-dependent lung mucosal immunity in mice with invasive pulmonary aspergillosis. J Infect Dis 178:1750–1760, 1998.
11. E Roilides, K Uhlig, D Venzon, PA Pizzo, TJ Walsh. Neutrophil oxidative burst in response to blastoconidia and pseudohyphae of *Candida albicans*: augmentation by granulocyte colony-stimulating factor and interferon-γ. J Infect Dis 166:668–673, 1992.
12. E Roilides, TJ Walsh, PA Pizzo, M Rubin. Granulocyte colony-stimulating factor enhances the phagocytic and bactericidal activity of normal and defective human neutrophils. J Infect Dis 163: 579–583, 1991.
13. LA Bober, MJ Grace, C Pugliese-Sivo, A Rojas-Triana, T Walters, LM Sullivan, SK Narula. The effect of GM-CSF and G-CSF on human neutrophil function. Immunopharmacology 29:111–119, 1995.

14. Y Yamamoto, TW Klein, H Freidman, S Kimura, H Yamaguchi. Granulocyte colony-stimulating factor potentiates anti–*Candida albicans* growth inhibitory activity of polymorphonuclear cells. FEMS Immunol Med Microbiol 7:15–22, 1993.

15. S Vora, N Purimetla, E Brummer, DA Stevens. Activity of voriconazole, a new triazole, combined with neutrophils or monocytes against *Candida albicans*: effect of granulocyte colony-stimulating factor and granulocyte-macrophage colony-stimulating factor. Antimicrob Agents Chemother 42: 907–910, 1998.

16. RB Ashman, JM Papadimitriou. Production and function of cytokines in natural and acquired immunity to *Candida albicans* infection. Microbiol Rev 59:646–672, 1995.

17. E Roilides, A Holmes, C Blake, PA Pizzo, TJ Walsh. Effects of granulocyte colony-stimulating factor and interferon-γ on antifungal activity of human polymorphonuclear neutrophils against pseudohyphae of different medically important *Candida* species. J Leukoc Biol 57:651–656, 1995.

18. JM Gaviria, JA van Burik, DC Dale, RK Root, WC Liles. Modulation of neutrophil-mediated activity against the pseudohyphal form of *Candida albicans* by granulocyte colony-stimulating factor (G-CSF) administered in vivo. J Infect Dis 179:1301–1304, 1999.

19. A Vecchiarelli, C Monari, F Baldelli, D Pietrella, C Retini, C Tascini, D Francisci, F Bistoni. Beneficial effect of recombinant human granulocyte colony-stimulating factor on fungicidal activity of polymorphonuclear leukocytes from patients with AIDS. J Infect Dis 171:1448–1454, 1995.

20. A Cassone, C Palma, JY Djeu, F Aiuti, I Quinti. Anticandidal activity and interleukin-1β and interleukin-6 production by polymorphonuclear leukocytes are preserved in subjects with AIDS. J Clin Microbiol 31:1354–1357, 1993.

21. E Roilides, S Mertins, J Eddy, TJ Walsh, PA Pizzo, M Rubin. Impairment of neutrophil chemotactic and bactericidal function in HIV-infected children and partial reversal after in vitro exposure to granulocyte-macrophage colony-stimulating factor. J Pediatr 117:531–540, 1990.

22. D Metcalf, CJ Begley, GR Johnson, NA Nicola, MA Vadas, AF Lopez, DJ Williamson, GG Wong, SC Clark, EA Wang. Biologic properties in vitro of a recombinant human granulocyte-macrophage colony-stimulating factor. Blood 67:37–45, 1986.

23. AF Lopez, DJ Williamson, JR Gamble, CJ Begley, JM Harlan, SJ Klebanoff, A Waltersdorph, GG Wong, SC Clark, MA Vadas. Recombinant human granulocyte-macrophage colony-stimulating factor stimulates in vitro mature human neutrophil and eosinophil function, surface receptor expression, and survival. J Clin Invest 78:1220–1228, 1986.

24. Y Kletter, I Bleiberg, DW Golde, I Fabian. Antibody to Mol abrogates the increase in neutrophil phagocytosis and degranulation induced by granulocyte-macrophage colony-stimulating factor. Eur J Haematol 43:389–396, 1989.

25. PD Smith, CL Lamerson, SM Banks, SS Saini, LM Wahl, RA Calderone, SM Wahl. Granulocyte-macrophage colony-stimulating factor augments human monocyte fungicidal activity for *Candida albicans*. J Infect Dis 161:999–1005, 1990.

26. M Wang, H Friedman, JY Djeu. Enhancement of human monocyte function against *Candida albicans* by the colony-stimulating factors (CSF); IL-3, granulocyte-macrophage-CSF, and macrophage-CSF. J Immunol 143:671–677, 1989.

27. E Roilides, PA Pizzo. Modulation of host defenses by cytokines: evolving adjuncts in prevention and treatment of serious infections in immunocompromised hosts. Clin Infect Dis 15:508–524, 1992.

28. A Karbassi, JM Becker, JS Foster, RN Moore. Enhanced killing of *Candida albicans* by murine macrophages treated with macrophage colony-stimulating factor: evidence for augmented expression of mannose receptors. J Immunol 139:417–421, 1987.

29. E Cenci, A Bartocci, P Puccetti, S Mocci, ER Stanely and F Bistoni. Macrophage colony-stimulating factor in murine candidiasis: serum and tissue levels during infection and protective effect of exogenous administration. Infect Immun 59:868–872, 1991.

30. E Roilides, CA Lyman, T Sein, C Gonzalez, TJ Walsh. Antifungal activity of splenic, liver and pulmonary macrophages against *Candida albicans* and effects of macrophage colony-stimulating factor. Med Mycol 38:161–168, 2000.

31. E Roilides, CA Lyman, SD Mertins, DJ Cole, D Venzon, PA Pizzo, SJ Chanock, TJ Walsh. Ex vivo effects of macrophage colony-stimulating factor on human monocyte activity against fungal and bacterial pathogens. Cytokine 8:42–48, 1996.

32. A Khwaja, B Johnson, IE Addison, K Yong, K Ruthen, S Abramson, DC Linch. In vivo effects of macrophage colony-stimulating factor on human monocyte function. Br J Haematol 77:25–31, 1991.

33. E Roilides, PA Pizzo. Perspectives on the use of cytokines in the management of infectious complications of cancer. Clin Infect Dis 17:S385–S389, 1993.

34. JY Djeu, DK Blanchard, D Halkias, H Friedman. Growth inhibition of *Candida albicans* by human polymorphonuclear neutrophils: activation by interferon-gamma and tumor necrosis factor. J Immunol 137:2980–2984, 1986.

35. RD Diamond, CA Lyman and DR Wysong. Disparate effects of interferon-gamma and tumor necrosis factor-alpha on early neutrophil respiratory burst and fungicidal responses to *Candida albicans* hyphae in vitro. J Clin Invest 87:711–720, 1991.

36. CJ Morrison, E Brummer, RA Isenberg, DA Stevens. Activation of murine polymorphonuclear neutrophils for fungicidal activity by recombinant gamma interferon. J Leukoc Biol 41:434–440, 1987.

37. B Perussia, M Kobayashi, ME Rossi, I Anegon, G Trinchieri. Immune interferon enhances properties of human granulocytes: role of Fc receptors and effect of lymphotoxin, tumor necrosis factor, and granulocyte-macrophage colony-stimulating factor. J Immunol 138:765–774, 1987.

38. A Ferrante. Tumor necrosis factor alpha potentiates neutrophil antimicrobial activity: increased fungicidal activity against *Torulopsis glabrata* and *Candida albicans* and associated increases in oxygen radical production and lysosomal enzyme release. Infect Immun 57:2115–2122, 1989.

39. A Ferrante, M Nandoskar, A Walz, DHB Goh, IC Kowanko. Effects of tumour necrosis factor alpha and interleukin-1 alpha and beta on human neutrophil migration, respiratory burst and degranulation. Int Arch Appl Immunol 86:82–91, 1988.

40. YH Atkinson, WA Marasco, AF Lopez, AM Vadas. Recombinant human tumor necrosis factor alpha. Regulation of N-formyl methionyl leucyl phenylalanine receptor affinity and function on human neutrophils. J Clin Invest 81:759–765, 1988.

41. E Roilides, I Kadiltsoglou, A Dimitriadou, M Hatzistilianou, A Manitsa, J Karpouzas, PA Pizzo, TJ Walsh. Interleukin-4 suppresses antifungal activity of human mononuclear phagocytes against *Candida albicans* in association with decreased uptake of blastoconidia. FEMS Immunol Med Microbiol 19:169–180, 1997.

42. E Roilides, H Katsifa, S Tsaparidou, T Stergiopoulou, C Panteliadis, TJ Walsh. Interleukin-10 suppresses phagocytic and antihyphal activities of human neutrophils. Cytokine 12:379–387, 2000.

43. L Romani, A Mencacci, L Tonnetti, R Spaccapelo, E Cenci, P Puccetti, SF Wolf, F Bistoni. Interleukin-12 is both required and prognostic in vitro for T helper type 1 differentiation in murine candidiasis. J Immunol 152:5167–5175, 1994.

44. L Romani, A Mencacci, L Tonnetti, R Spaccapelo, E Cenci, S Wolf, P Puccetti, F Bistoni. Interleukin-12 but not interferon-gamma production correlates with induction of T helper type-1 phenotype in murine candidiasis. Eur J Immunol 24:909–915, 1994.

45. D Girard, ME Paquet, R Paquin, AD Beaulieu. Differential effects of interleukin-15 (IL-15) and IL-2 on human neutrophils: modulation of phagocytosis, cytoskeletal rearrangement, gene expression, and apoptosis by IL-15. Blood 88:3176–3184, 1996.

46. T Musso, L Calosso, M Zucca, M Millesimo, M Puliti, S Bulfone-Paus, C Merlino, D Savoia, R Cavallo, AN Ponzi, R Badolato. Interleukin-15 activates proinflammatory and antimicrobial functions in polymorphonuclear cells. Infect Immun 66:2640–2647, 1998.

47. N Vazquez, CA Lyman, SJ Chanock, D Friedman, TJ Walsh. Interleukin-15 augments superoxide production and microbicidal activity of human monocytes against *Candida albicans*. Infect Immun 66:145–150, 1998.

48. E Roilides, K Uhlig, D Venzon, PA Pizzo, TJ Walsh. Enhancement of oxidative response and damage caused by human neutrophils to *Aspergillus fumigatus* hyphae by granulocyte colony-stimulating factor and gamma interferon. Infect Immun 61:1185–1193, 1993.

49. E Roilides, K Uhlig, D Venzon, PA Pizzo, TJ Walsh. Prevention of corticosteroid-induced suppression of human polymorphonuclear leukocyte-induced damage of *Aspergillus fumigatus* hyphae by granulocyte colony-stimulating factor and interferon-γ. Infect Immun 61:4870–4877, 1993.

50. E Roilides, A Holmes, C Blake, PA Pizzo, TJ Walsh. Impairment of neutrophil fungicidal activity in HIV-infected children against *Aspergillus fumigatus* hyphae. J Infect Dis 167:905–911, 1993.

51. WC Liles, JE Huang, JA van Burik, RA Bowden, DC Dale. Granulocyte colony-stimulating factor administered in vivo augments neutrophil-mediated activity against opportunistic fungal pathogens. J Infect Dis 175:1012–1015, 1997.

52. S Vora, S Chauhan, E Brummer, DA Stevens. Activity of voriconazole combined with neutrophils or monocytes against *Aspergillus fumigatus*: effects of granulocyte colony-stimulating factor and granulocyte-macrophage colony-stimulating factor. Antimicrob Agents Chemother 42:2299–2303, 1998.

53. E Roilides, A Holmes, C Blake, D Venzon, PA Pizzo, TJ Walsh. Antifungal activity of elutriated human monocytes against *Aspergillus fumigatus* hyphae: enhancement by granulocyte-macrophage colony-stimulating factor and interferon-γ. J Infect Dis 170:894–899, 1994.

54. E Roilides, C Blake, A Holmes, PA Pizzo, TJ Walsh. Granulocyte-macrophage colony-stimulating factor and interferon-γ prevent dexamethasone-induced immunosuppression of antifungal monocyte activity against *Aspergillus fumigatus* hyphae. J Med Vet Mycol 34:63–69, 1996.

55. E Roilides, T Sein, A Holmes, C Blake, PA Pizzo, TJ Walsh. Effects of macrophage colony-stimulating factor on antifungal activity of mononuclear phagocytes against *Aspergillus fumigatus*. J Infect Dis 172:1028–1034, 1995.

56. A Schaffner. Therapeutic concentrations of glucocorticoids suppress the antimicrobial activity of human macrophages without impairing their responsiveness to gamma interferon. J Clin Invest 76:1755–1764, 1985.

57. A Schaffner, P Rellstab. Gamma-interferon restores listericidal activity and concurrently enhances release of reactive oxygen metabolites in dexamethasone-treated human monocytes. J Clin Invest 82:913–919, 1988.

58. JM Gaviria, JA van Burik, DC Dale, RK Root, WC Liles. Comparison of interferon-gamma, granulocyte colony-stimulating factor, and granulocyte-macrophage colony-stimulating factor for priming leukocyte-mediated hyphal damage of opportunistic fungal pathogens. J Infect Dis 179:1038–1041, 1999.

59. E Roilides, A Dimitriadou-Georgiadou, T Sein, I Kadiltzoglou, TJ Walsh. Tumor necrosis factor alpha enhances antifungal activities of polymorphonuclear and mononuclear phagocytes against *Aspergillus fumigatus*. Infect Immun 66:5999–6003, 1998.

60. E Roilides, A Dimitriadou, I Kadiltsoglou, P Pizzo, J Karpouzas, TJ Walsh. Effects of interleukin-4 on antifungal activity of mononuclear phagocytes against hyphae and conidia of *Aspergillus fumigatus* (Abstr. 999). Seventh European Congress of Clinical Microbiology and Infectious Diseases, Vienna, Austria, 1995, p 192.

61. E Roilides, A Dimitriadou, I Kadiltsoglou, T Sein, J Karpouzas, PA Pizzo, TJ Walsh. IL-10 exerts suppressive and enhancing effects on antifungal activity of mononuclear phagocytes against *Aspergillus fumigatus*. J Immunol 158:322–329, 1997.

62. CA Lyman, KF Garrett, PA Pizzo, TJ Walsh. Response of human polymorphonuclear leukocytes and monocytes to *Trichosporon beigelii*: host defense against an emerging opportunistic pathogen. J Infect Dis 170:1557–1565, 1994.

63. TJ Walsh, GP Melcher, JW Lee, PA Pizzo. Infections due to *Trichosporon* species: new concepts in mycology, pathogenesis, diagnosis and treatment. Curr Top Med Mycol 5:79–113, 1993.

64. P Martino, R Gastaldi, R Raccah, C Girmenia. Clinical patterns of *Fusarium* infections in immunocompromised patients. J Infect 28(suppl 1):7–15, 1994.

65. J Berenguer, JL Rodriguez-Tudela, C Richard, M Alvarez, MA Sanz, L Gaztelurrutia, J Ayats, JV Martenez-Surez. Deep infections caused by *Scedosporium prolificans*. A report of 16 cases in Spain and a review of the literature. Medicine 76:256–265, 1997.

66. RT Speilberger, BR Tegtmeier, MR O'Donnell, JI Ito. Fatal *Scedosporium prolificans* (*S. inflatum*) fungemia following allogeneic bone marrow transplantation: report of a case in the United States. Clin Infect Dis 21:1067, 1995.

67. M Rabodonirina, S Paulus, F Thevenet, R Loire, E Gueho, O Bastien, JF Mornex, M Celard, MA Piens. Disseminated *Scedosporium prolificans* (*S. inflatum*) infection after single-lung transplantation. Clin Infect Dis 19:138–142, 1994.

68. CE Gonzalez, K Kligys, D Shetty, R Torres, W Love, J Peter, CA Lyman, TJ Walsh. Characterization of host defenses against *Pseudallescheria boydii*: an emerging opportunistic pathogen (F-20). Abstracts of the Annual Meeting of the American Society for Microbiology, 1996, p 77.

69. C Gil Lamaignere, A Maloukou, JL Rodriguez-Tudela, E Roilides. Human phagocytic cell responses to *Scedosporium prolificans* hyphae. Med Mycol (2001, in press).

70. E Roilides, CA Lyman, T Sein, T Walsh. Human macrophage colony-stimulating factor (MCSF) enhances oxidative burst of elutriated human monocytes in response to *Penicillium marneffei*. Program and Abstracts of the 39th Interscience Conference on Antimicrobial Agents and Chemotherapy, San Francisco, 1999, p 555.

71. KS Antman, JD Griffin, A Elias, M Socinski, L Ryan, SA Cannista, D Oette, M Whittey, E Frei, L Schnipper. Effect of recombinant human granulocyte-macrophage colony-stimulating factor on chemotherapy-induced myelosuppression. N Engl J Med 319:593–598, 1988.

72. J Crawford, H Ozer, R Stoller, D Johnson, G Lyman, I Tabbara, M Kris, J Grous, V Picozzi, G Rausch, R Smith, W Gradishar, A Yahanda, M Vincent, M Stewart, J Glaspy. Reduction by granulocyte colony-stimulating factor of fever and neutropenia induced by chemotherapy in patients with small-cell lung cancer. N Engl J Med 325:164–170, 1991.

73. HH Gerhartz, M Engelhard, P Meusers, G Brittinger, W Wilmanns, G Schlimok, P Mueller, D Huhn, R Musch, W Siegert. Randomized, double-blind, placebo-controlled, Phase III study of recombinant human granulocyte-macrophage colony-stimulating factor as adjunct to induction treatment of high-grade malignant non-Hodgkin's lymphomas. Blood 82:2329–2339, 1993.

74. R Pettengell, H Gurney, JA Radford, DP Deakin, R James, PM Wilkinson, K Kane, J Bentley, D Crowther. Granulocyte colony-stimulating factor to prevent dose-limiting neutropenia in non-Hodgkin's lymphoma: a randomized controlled trial. Blood 80:1430–1436, 1992.

75. V Trillet-Lenoir, J Green, C Manegold, J Von Pawel, U Gatzemeier, B Lebeau, A Depierre, P Johnson, G Decoster, D Tomita, C Ewen. Recombinant granulocyte colony stimulating factor reduces the infectious complications of cytotoxic chemotherapy. Eur J Cancer 29A:319–324, 1993.

76. B Biesma, EGE de Vries, PHB Willemse, WJ Sluiter, PE Postmus, PC Limburg, AC Stern, E Vellenga. Efficacy and tolerability of recombinant human granulocyte-macrophage colony-stimulating factor in patients with chemotherapy-related leukopenia and fever. Eur J Cancer 26:932–936, 1990.

77. DW Maher, GJ Lieschke, M Green, J Bishop, R Stuart-Harris, M Wolf, WP Sheridan, RF Kefford, J Cebon, I Olver, J McKendrick, G Toner, K Bradstock, M Lieschke, S Bruickshank, DK Tomita, EW Hoffman, RM Fox, G Morstyn. Filgrastim in patients with chemotherapy-induced febrile neutropenia. A double-blind, placebo-controlled trial. Ann Intern Med 121:492–501, 1994.

78. Anonymous. American Society of Clinical Oncology recommendations for the use of hematopoietic colony-stimulating factors: evidence-based, clinical practice guidelines. J Clin Oncol 12:2471–2508, 1994.

79. A Polak-Wyss. Protective effect of human granulocyte colony-stimulating factor on *Cryptococcus* and *Aspergillus* infections in normal and immunosuppressed mice. Mycoses 34:205–215, 1991.

80. JR Graybill, R Bocanegra, LK Najvar, D Loebenberg, MF Luther. Granulocyte colony-stimulating factor and azole antifungal therapy in murine aspergillosis: role of immune suppression. Antimicrob Agents Chemother 42:2467–2473, 1998.

81. BJ Kullberg, JW van der Meer, JF Meis, M Keuter, JH Curfs, MG Netea. Recombinant murine granulocyte colony-stimulating factor protects against acute disseminated *Candida albicans* infection in nonneutropenic mice. J Infect Dis 177:175–181, 1998.

82. BJ Kullberg, MG Netea, AG Vonk, JW van der Meer. Modulation of neutrophil function in host defense against disseminated *Candida albicans* infection in mice. FEMS Immunol Med Microbiol 26:299–307, 1999.

83. H Muranaka, M Suga, K Nakagawa, K Sato, Y Gushima, M Ando. Effects of granulocyte and granulocyte-macrophage colony-stimulating factors in a neutropenic murine model of trichosporonosis. Infect Immun 65:3422–3429, 1997.

84. P Mayer, C Schutze, C Lam, F Kricek, E Liehl. Recombinant murine granulocyte-macrophage colony-stimulating factor augments neutrophil recovery and enhances resistance to infections in myelosuppressed mice. J Infect Dis 163:584–590, 1991.

85. T Walsh, C Gonzalez, C Lyman, S Lee, C Del Guercio, A Gehrt, T Sein, E Roilides, R Schafele, A Francesconi, P Pizzo. Human recombinant macrophage colony stimulating factor (M-CSF) augments pulmonary host defense against Aspergillus fumigatus [Abstract No. F-27]. Abstracts of the Annual Meeting of the American Society of Microbiology, Washington, 1994, p 593.

86. BJ Kullberg, JW Van't Wout, C Hoogstraten, R Van Furth. Recombinant interferon-γ enhances resistance to acute disseminated Candida albicans infection in mice. J Infect Dis 168:436–443, 1993.

87. RE Garner, U Kuruganti, CW Czarniecki, HH Chiu, JE Domer. In vivo immune responses to Candida albicans modified by treatment with recombinant murine gamma interferon. Infect Immun 57:1800–1808, 1989.

88. H Nagai, J Guo, H Choi, V Kurup. Interferon-γ and tumor necrosis factor-α protect mice from invasive aspergillosis. J Infect Dis 172:1554–1560, 1995.

89. A Louie, AL Baltch, RP Smith, MA Franke, WJ Ritz, JK Singh, MA Gordno. Tumor necrosis factor alpha has a protective role in a murine model of systemic candidiasis. Infect Immun 62: 2761–2772, 1994.

90. S Schelenz, DA Smith, GJ Bancroft. Cytokine and chemokine responses following pulmonary challenge with Aspergillus fumigatus: obligatory role of TNF-α and GM-CSF in neutrophil recruitment. Med Mycol 37:183–194, 1999.

91. B Mehrad, RM Strieter, TJ Standiford. Role of TNF-alpha in pulmonary host defense in murine invasive aspergillosis. J Immunol 162:1633–1640, 1999.

92. JW van't Wout, JWM van der Meer, M Barza, CA Dinarello. Protection of neutropenic mice from lethal Candida albicans infection by recombinant interleukin 1. Eur J Immunol 18:1143–1146, 1988.

93. BJ Kullberg, JW van't Wout, R van Furth. Role of granulocytes in enhanced host resistance to Candida albicans induced by recombinant interleukin-1. Infect Immun 58:3319–3323, 1990.

94. BJ Kullberg, JW van't Wout, RJ Poell, R van Furth. Combined effect of fluconazole and recombinant human interleukin-1 on systemic candidiasis in neutropenic mice. Antimicrob Agents Chemother 36:1225–1229, 1992.

95. JR Graybill, R Bocanerga, M Luther. Antifungal combination with G-CSF and fluconazole in experimental disseminated candidiasis. Eur J Clin Microbiol Infect Dis 14:700–703, 1995.

96. M Hamood, PF Bluche, C De Vroey, F Corazza, W Buzan, P Fondu. Effects of rhG-CSF on neutropenic mice infected with C. albicans: acceleration of recovery from neutropenia and potentiation of anti-Candida resistance. Mycoses 37:93–99, 1994.

97. K Uchida, Y Yamamoto, TW Klein, H Friedman, H Yamaguchi. Granulocyte-colony stimulating factor facilitates the restoration of resistance to opportunistic fungi in leukopenic mice. J Med Vet Mycol 30:293–300, 1992.

98. CR Vitt, JM Fidler, D Ando, RJ Zimmerman, SL Aukerman. Antifungal activity of rhM-CSF in models of acute and chronic candidiasis in the rat. J Infect Dis 169:369–374, 1994.

99. T Kuhara, K Uchida, H Yamaguchi. Therapeutic efficacy of human macrophage colony-stimulating factor, used alone and in combination with antifungal agents, in mice with systemic Candida albicans infection. Antimicrob Agents Chemother 44:19–23, 2000.

100. E Roilides, CA Lyman, T Sein, V Petraitis, T Walsh. Amphotericin B lipid complex synergizes with pulmonary alveolar macrophages to destroy conidia of A. fumigatus (Abstract No. 700). Program and Abstracts of the 39th Interscience Conference on Antimicrobial Agents and Chemotherapy, San Francisco, 1999, p 555.

101. E Cenci, A Mencacci, G Del Sero, A Bacci, C Montagnoli, C Fe d'Ostiani, P Mosci, M Bachmann, F Bistoni, M Kopf, L Romani. Interleukin-4 causes susceptibility to invasive aspergillosis through suppression of protective type 1 responses. J Infect Dis 180:1957–1968, 1999.

102. JA Winkelstein, MC Marino, RB Johnston, J Boyle, J Curnutte, JI Gallin, HL Malech, SM Holland, H Ochs, P Quie, RH Buckley, CB Foster, SJ Chanock, H Dickler. Chronic granulomatous disease. Report on a national registry of 368 patients. Medicine (Baltimore) 79:155–169, 2000.

103. RAB Ezekowitz, SH Orkin, PE Newburger. Recombinant interferon gamma augments phagocyte superoxide production and X-chronic granulomatous disease gene expression in X-linked variant chronic granulomatous disease. J Clin Invest 80:1009–1016, 1987.

104. RAB Ezekowitz, MC Dinauer, HS Jaffe, SH Orkin, PE Newburger. Partial correction of the phagocyte defect in patients with X-linked chronic granulomatous disease by subcutaneous interferon gamma. N Engl J Med 319:146–151, 1988.

105. JMG Sechler, HL Malech, CJ White, JI Gallin. Recombinant human interferon-gamma reconstitutes defective phagocyte function in patients with chronic granulomatous disease of childhood. Proc Natl Acad Sci USA 85:4874–4878, 1988.

106. International Chronic Granulomatous Disease Cooperative Study Group. A controlled trial of interferon gamma to prevent infection in chronic granulomatous disease. N Engl J Med 324:509–16, 1991.

107. JH Rex, JE Bennett, JI Gallin, HL Malech, ES Decarlo, DA Melnick. In vivo interferon-γ therapy augments the in vitro ability of chronic granulomatous disease neutrophils to damage *Aspergillus* hyphae. J Infect Dis 163:849–852, 1991.

108. A Ahlin, G Elinder, J Palmblad. Dose-dependent enhancements by interferon-gamma on functional responses of neutrophils from chronic granulomatous disease patients. Blood 89:3396–3401, 1997.

109. PR Williamson, KJ Kwon-Chung, JI Gallin. Successful treatment of *Paecilomyces varioti* infection in a patient with chronic granulomatous disease and a review of *Paecilomyces* species infections. Clin Infect Dis 14:1023–1026, 1992.

110. P Phillips, JC Forbes, DP Speert. Disseminated infection with *Pseudallescheria boydii* in a patient with chronic granulomatous disease: response to gamma-interferon plus antifungal chemotherapy. Pediatr Infect Dis J 10:536–539, 1991.

111. H Ozsahin, M von Planta, I Muller, HC Steinert, D Nadal, R Lauener, P Tuchschmid, UV Willi, M Ozsahin, NE Crompton, RA Seger. Successful treatment of invasive aspergillosis in chronic granulomatous disease by bone marrow transplantation, granulocyte colony-stimulating factor–mobilized granulocytes, and liposomal amphotericin-B. Blood 92:2719–2724, 1998.

112. E Roilides, L Sigler, E Bibashi, H Katsifa, N Flaris, C Panteliadis. Disseminated infection due to *Chrysosporium zonatum* in a patient with chronic granulomatous disease and review of non-*Aspergillus* infections in these patients. J Clin Microbiol 37:18–25, 1999.

113. A Cohen-Abbo, KM Edwards. Multifocal osteomyelitis caused by *Paecilomyces varioti* in a patient with chronic granulomatous disease. Infection 23:55–57, 1995.

114. S Pasic, M Abinun, B Pistignjat, B Vlajic, J Rakic, L Sarjanovic, N Ostojic. *Aspergillus* osteomyelitis in chronic granulomatous disease: treatment with recombinant gamma-interferon and itraconazole. Pediatr Infect Dis J 15:833–834, 1996.

115. F Touza Rey, C Martinez Vazquez, J Alonso, MJ Mendez Pineiro, M Rubianes Gonzalez, M Crespo Casal. The clinical response to interferon-gamma in a patient with chronic granulomatous disease and brain abcesses due to *Aspergillus fumigatus*. An Med Interna 17:86–87, 2000.

116. J Bernhisel-Broadbent, EE Camargo, HS Jaffe, HM Lederman. Recombinant human interferon-γ as adjunct therapy for *Aspergillus* infection in a patient with chronic granulomatous disease. J Infect Dis 163:908–911, 1991.

117. N Tsumura, Y Akasu, H Yamane, S Ikezawa, T Hirata, K Oda, Y Sakata, M Shirahama, A Inoue, H Kato. *Aspergillus* osteomyelitis in a child who has p67-phox-deficient chronic granulomatous disease. Kurume Med J 46:87–90, 1999.

118. O Lortholary, MC Meyohas, B Dupont, J Cadranel, D Salmon-Ceron, D Peyramond, D Simonin. Invasive aspergillosis in patients with acquired immunodeficiency syndrome: report of 33 cases. French Cooperative Study Group on Aspergillosis in AIDS. Am J Med 95:177–187, 1993.

119. SH Khoo, DW Denning. Invasive aspergillosis in patients with AIDS. Clin Infect Dis 19(suppl 1):S41–S48, 1994.

120. D Shetty, N Giri, CE Gonzalez, PA Pizzo, TJ Walsh. Invasive aspergillosis in human immunodeficiency virus–infected children. Pediatr Infect Dis J 16:216–221, 1997.

121. E Roilides, A Holmes, C Blake, PA Pizzo, TJ Walsh. Defective antifungal activity of monocyte-derived macrophages from HIV-infected children against *Aspergillus fumigatus*. J Infect Dis 168:1562–1565, 1993.

122. E Roilides, M Clerici, L DePalma, M Rubin, PA Pizzo, GM Shearer. T helper cell responses in children infected with human immunodeficiency virus type 1. J Pediatr 118:724–730, 1991.

123. DR Kuritzkes, D Parenti, DJ Ward, A Rachlis, RJ Wong, KP Mallon, WJ Rich, MA Jacobson. Filgrastim prevents severe neutropenia and reduces infective morbidity in patients with advanced HIV infection: results of a randomized, multicenter, controlled trial. AIDS 12:65–74, 1998.

124. ME Grauer, C Bokemeyer, W Bautsch, M Freund, H Link. Successful treatment of a *Trichosporon beigelii* septicemia in a granulocytopenic patient with amphotericin B and granulocyte colony-stimulating factor. Infection 22:283–286, 1994.

125. CH Poynton, RA Barnes, J Rees. Interferon gamma and granulocyte-macrophage colony-stimulating factor for the treatment of hepatosplenic candidosis in patients with acute leukemia. Clin Infect Dis 26:239–240, 1998.

126. RT Spielberger, MJ Falleroni, AJ Coene, RA Larson. Concomitant amphotericin B therapy, granulocyte transfusions, and GM-CSF administration for disseminated infection with *Fusarium* in a granulocytopenic patient. Clin Infect Dis 16:528–530, 1993.

127. CE Gonzalez, DR Couriel, TJ Walsh. Successful treatment of disseminated zygomycosis in a neutropenic patient with amphotericin B lipid complex and granulocyte colony-stimulating factor. Clin Infect Dis 24:192–196, 1997.

128. C Hennequin, M Benkerrou, JL Gaillard, S Blanche, S Fraitag. Role of granulocyte colony-stimulating factor in the management of infection with *Fusarium oxysporum* in a neutropenic child. Clin Infect Dis 18:490–491, 1994.

129. T Fukushima, R Sumazaki, M Shibasaki, H Saitoh, Y Fujigaki, M Kaneko, E Akaogi, K Mitsui, T Ogata, H Takita. Successful treatment of invasive thoracopulmonary mucormycosis in a patient with acute lymphoblastic leukemia. Cancer 76:895–899, 1995.

130. HJ Dornbusch, CE Urban, H Pinter, G Ginter, R Fotter, H Becker, T Miorini, C Berghold. Treatment of invasive pulmonary aspergillosis in severely neutropenic children with malignant disorders using liposomal amphotericin B, granulocyte colony-stimulating factor and surgery: report of 5 cases. Pediatr Hematol Oncol 12:577–586, 1995.

131. GP Bodey, E Anaissie, J Gutterman, S Vadhan Raj. Role of granulocyte-macrophage colony-stimulating factor as adjuvant therapy for fungal infection in patients with cancer. Clin Infect Dis 17:705–707, 1993.

132. J Maertens, H Demuynck, G Verhoef, P Vandenberhe, P Zachee, M Boogaerts. GM-CSF fails to improve outcome in invasive fungal infections in neutropenic cancer patients (Abstr. 560). 13th Congress of the International Society of Human and Animal Mycology, Parma, Italy, 1997.

133. BJ Kullberg, K van de Woude, M Aoun, F Jacobs, R Herbrecht, P Kujath, for the European Filgrastim Candidiasis Study Group. A double-blind, randomized, placebo-controlled Phase II study of filgrastim (recombinant granulocyte colony-stimulating factor) in combination with fluconazole for treatment of invasive candidiasis and candidemia in nonneutropenic patients (J-100). Program and Abstracts of the 38th Interscience Conference on Antimicrobial Agents and Chemotherapy, San Diego, 1998, p 479.

134. A Aviles, R Guzman, EL Garcia, A Talavera, JC Diaz-Maqueo. Results of a randomized trial of granulocyte colony-stimulating factor in patients with infection and severe granulocytopenia. Anti-Cancer Agents 7:392–397, 1996.

135. J Nemunaitis, CD Buckner, KS Dorsey, D Willis, W Meyer, F Appelbaum. Retrospective analysis of infectious disease in patients who received recombinant human granulocyte-macrophage colony-stimulating factor versus patients not receiving a cytokine who underwent autologous bone marrow transplantation for treatment of lymphoid cancer. Am J Clin Oncol 21:341–346, 1998.

136. JM Rowe, JW Anderson, JJ Mazza, JM Bennett, E Paietta, FA Hayes, R Oette, PA Cassileth, EA Stadtmauer, PH Wiernik. A randomized placebo-controlled Phase III study of granulocyte-macrophage colony-stimulating factor in adult patients (>55 to 70 years) with acute myelogenous leukemia: a study by the Eastern Cooperative Oncology Group (E1490). Blood 86:457–462, 1995.

137. JM Rowe, A Rubin, JJ Mazza. Incidence of infections in adult patients (>55) with acute myeloid leukemia treated with yeast-derived GM-CSF (sargramostim): results of a double-blind prospective study by the Eastern Cooperative Oncology Group. In: W Hiddemann, T Büchner, B Wörman, eds.

Acute Leukemias V: Prognostic Factors and Treatment Results. New York: Springer-Verlag, 1996, pp 178–184.

138.   JM Rowe. Treatment of acute myeloid leukemia with cytokines: effect on duration of neutropenia and response to infections. Clin Infect Dis 26:1290–1294, 1998.

139.   J Nemunaitis, K Shannon-Dorcy, FR Appelbaum, JD Meyers, A Owens, R Day, D Ando, C O'Neill, D Buckner, JW Singer. Long-term follow-up of patients with invasive fungal disease who received adjunctive therapy with recombinant human macrophage colony-stimulating factor. Blood 82: 1422–1427, 1993.

140.   R Ohno, S Miyawaki, K Hatake, K Kuriyama, K Saito, A Kanamaru, T Kobayashi, Y Kodera, K Nishikawa, S Matsuda, O Yamada, E Omoto, H Takeyama, K Tsukuda, N Asou, M Tanimoto, H Shiozaki, M Tomonaga, T Masaoka, Y Miura, F Takaku, Y Ohashi, K Motoyoshi. Human urinary macrophage colony-stimulating factor reduces the incidence and duration of febrile neutropenia and shortens the period required to finish three courses of intensive consolidation therapy in acute myeloid leukemia: a double-blind controlled study. J Clin Oncol 15:2954–2965, 1997.

141.   TN Flynn, SM Kelsey, DL Hazel, JF Guest. Cost effectiveness of amphotericin B plus G-CSF compared with amphotericin B monotherapy. Treatment of presumed deep-seated fungal infection in neutropenic patients in the UK. Pharmacoeconomics 16:543–550, 1999.

142.   E Roilides, MC Dignani, EJ Anaissie, JH Rex. The role of immunoreconstitution in the management of refractory opportunistic fungal infections. Med Mycol 36:12–25, 1998.

143.   MM Strumia. The effect of leukocyte cream injections in the treatments of the neutropenias. Am J Med Sci 187:527–544, 1934.

144.   CA Schiffer. Granulocyte transfusions: an overlooked therapeutic modality. Transfusion Med Rev IV:2–7, 1990.

145.   RB Epstein, RA Clift, DE Thomas. The effect of leukocyte transfusions on experimental bacteremia in the dog. Blood 34:782, 1969.

146.   RB Epstein, FJ Waxman, BT Bennett, BR Andersen. *Pseudomonas* septicemia in neutropenic dogs. Treatment with granulocyte transfusions. Transfusion 14:51, 1974.

147.   BC Dale, HY Reynolds, JE Pennington, RJ Elin, TW Pitts, RG Graw. Granulocyte transfusion therapy of experimental *Pseudomonas* pneumonia. J Clin Invest 54:664–671, 1974.

148.   JS Tobias, BL Brown, A Brivkalns, RA Yankee. Prophylactic granulocyte support in experimental septicemia. Blood 47:473–479, 1976.

149.   BC Dale, HY Reynolds, JE Pennington, RJ Elin, GP Herzig. Experimental *Pseudomonas* pneumonia in leukopenic dogs: comparison of therapy with antibiotics and granulocyte transfusions. Blood 47: 869–876, 1976.

150.   F Appelbaum, C Bowles, R Makuch, A Deisseroth. Granulocyte transfusion therapy of experimental *Pseudomonas* septicemia: study of cell dose and collection technique. Blood 52:323–331, 1978.

151.   SC Herbert, CS Suleyman, RB Epstein. Pathophysiology of *Candida albicans* meningitis in normal, neutropenic, and granulocyte transfused dogs. Blood 55:546–551, 1980.

152.   RG Graw, G Herzig, S Perry, E Henderson. Normal granulocyte transfusion therapy: treatment of septicemia due to gram-negative bacteria. N Engl J Med 287:367–371, 1972.

153.   JB Alavi, RK Root, I Djerassi, AE Evans, SJ Gluckman, RR MacGregor, D Guerry, AD Schreiber, JM Shaw, P Koch, RA Cooper. A randomized clinical trial of granulocyte transfusions for infection in acute leukemia. N Eng J Med 296:706–711, 1977.

154.   DJ Higby, JW Yates, ES Henderson, JF Holland. Filtration leukopheresis for granulocyte transfusion therapy: clinical and laboratory studies. N Engl J Med 292:761–766, 1975.

155.   IE Fortuny, CD Bloomfield, DC Hadlock, A Goldman, BJ Kennedy, JJ McCullough. Granulocyte transfusions: a controlled study in patients with acute nonlymphocytic leukemia. Transfusion 15: 548–558, 1975.

156.   RH Herzig, GP Herzig, RG Graw, MI Bull, KK Ray. Successful granulocyte transfusion therapy for gram-negative septicemia. N Engl J Med 13:701–705, 1977.

157.   WR Vogler, EF Winton. A controlled study of the efficacy of granulocyte transfusions in patients with neutropenia. Am J Med 63:548–555, 1977.

158. DJ Winston, GH Winston, RP Gale. Therapeutic granulocyte transfusions for documented infections: a controlled trial in ninety-five infectious granulocytopenic episodes. Ann Intern Med 97:509–515, 1982.

159. KB McCredie, EJ Freireich, JP Hester, C Vallejos. Leukocyte transfusion therapy for patients with host-defense failure. Transplant Proc 5:1285–1290, 1973.

160. DJ Higby, A Freeman, ES Henderson, L Sinks, E Cohen. Granulocyte transfusions in children using filter-collected cells. Cancer 38:1407–1413, 1976.

161. J Aisner, C Schiffer, P Wiernik. Granulocyte transfusions: evaluation of factors influencing results and a comparison of filtration and intermittent centrifugation leukapheresis. Br J Haematol 38: 121–129, 1978.

162. RG Strauss. Therapeutic granulocyte transfusions in 1993. Blood 81:1675–1678, 1993.

163. RG Strauss. Granulocyte transfusion therapy for hem/onc patients. Hem/Onc Ann 2:304–309, 1994.

164. JP Hester, KB McCredie, J Freireich. Granulocyte transfusions: analysis of donor, procedure, recipient variables. Blut 32:253–256, 1976.

165. JP Dutcher, C Riggs, J Fox, GS Johnston, D Norris, PH Wiernik, CA Schiffer. Effect of histocompatibility factors on pulmonary retention of indium-111-labeled granulocytes. Am J Hematol 33: 238–243, 1990.

166. JP Dutcher, CA Schiffer, GS Johnston, D Papenburg, PA Daly, J Aisner, PH Wiernik. Alloimmunization prevents the migration of transfused indium-111-labeled granulocytes to sites of infection. Blood 62:354–360, 1983.

167. EJ Freireich, RH Levin, J Whang, PP Carbone, W Bronson, EE Morse. The function and fate of transfused leukocytes from donors with chronic myelocytic leukemia in leukopenic recipients. Ann NY Acad Sci 113:1081–1089, 1964.

168. FT Sanel, J Aisner, CJ Tillman, CA Schiffer, PH Wiernik. Evaluation of granulocytes harvested by filtration leucapheresis: functional, histochemical and ultrastructural studies. In: JM Goldman, RM Lowenthal, eds. Leucocytes, Separation, Collection and Transfusion. London: Academic Press, 1975, pp 236–248.

169. GP Herzig, R Graw. Granulocyte transfusion for bacterial infections. Prog Hematol 9:207–228, 1975.

170. CA Schiffer. Granulocyte transfusion therapy. Can Treat Rep 67:113–119, 1983.

171. CG Craddock. Production, distribution and fate of granulocytes. In: WJ Williams, E Beutler, AJ Ersley, eds. Hematology. New York: McGraw Hill, 1972, pp 607–618.

172. DJ Higby, JM Mishler, W Rhomberg, RW Nicora, JF Holland. The effect of a single or double dose of dexamethasone on granulocyte collection with the continuous flow centrifugation. Vox Sang 28:243–248, 1975.

173. WI Bensinger, TH Price, DC Dale, FR Appelbaum, R Clift, K Lilleby, B Williams, R Storb, ED Thomas, CD Buckner. The effects of daily recombinant human granulocyte colony-stimulating factor administration on normal granulocyte donors undergoing leukapheresis. Blood 81:1883–1888, 1993.

174. LM Hollingshead, KL Goa. Recombinant granulocyte colony-stimulating factor: a review of its pharmacological properties and prospective role in neutropenic conditions. Drugs 42:300–330, 1991.

175. F Colotta, F Re, N Polentarutti, S Sozzani, A Mantovani. Modulation of granulocyte survival and programmed cell death by cytokines and bacterial products. Blood 80:2012–2020, 1992.

176. DM Cohen, JH Rex, CA Savary, JP Hester, EJ Anaissie, SC Bhalla. Effects of in vitro and in vivo cytokine treatment, leucapheresis and irradiation on the function of human neutrophils: implications for white blood cell transfusion therapy. Clin Lab Haematol 19:39–47, 1997.

177. JH Rex, SC Bhalla, DM Cohen, JP Hester, SE Vartivarian, EJ Anaissie. Protection of human polymorphonuclear leukocyte function from the deleterious effects of isolation, irradiation, and storage by interferon-γ and granulocyte colony-stimulating factor. Transfusion 35:605–611, 1995.

178. MC Dignani, EJ Freireich, BS Andersson, B Lichtiger, DB Jendiroba, H Kantarjian, JH Rex, SE Vartivarian, S O'Brien, JP Hester, EJ Anaissie. Treatment of neutropenia-related fungal infections with granulocyte colony-stimulating factor–elicited white blood cell transfusions: a pilot study. Leukemia 11:1621–1630, 1997.

179. C Peters, M Minkov, S Matthes-Martin, U Potschger, V Witt, G Mann, P Hocker, N Worel, J Stary, T Klingebiel, H Gadner. Leucocyte transfusions from rhG-CSF or prednisolone stimulated donors for treatment of severe infections in immunocompromised neutropenic patients. Br J Haematol 106: 689–696, 1999.

180. A Di Mario, S Sica, P Salutari, E Ortu La Barbera, R Marra, G Leone. Granulocyte colony-stimulating factor-primed leukocyte transfusions in *Candida tropicalis* fungemia in neutropenic patients. Haematologica 82:362–363, 1997.

181. L Catalano, R Fontana, N Scarpato, M Picardi, S Rocco, B Rotoli. Combined treatment with amphotericin-B and granulocyte transfusion from G-CSF-stimulated donors in an aplastic patient with invasive aspergillosis undergoing bone marrow transplantation. Haematologica 82:71–72, 1997.

182. S Bhatia, J McCullough, EH Perry, M Clay, NK Ramsay, JP Neglia. Granulocyte transfusions: efficacy in treating fungal infections in neutropenic patients following bone marrow transplantation. Transfusion 34:226–232, 1994.

183. C Pannuti, R Gingrich, MA Pfaller, C Kao, RP Wenzel. Nosocomial pneumonia in patients having bone marrow transplant. Attributable mortality and risk factors. Cancer 69:2653–2662, 1992.

184. A Wald, W Leisenring, JA van Burik, RA Bowden. Epidemiology of *Aspergillus* infections in a large cohort of patients undergoing bone marrow transplantation. J Infect Dis 175:1459–1466, 1997.

185. CA Savary, ML Grazziutti, B Melichar, D Przepiorka, R Freedman, RE Cowart, DM Cohen, EJ Anaissie, DG Woodside, BW McIntyre, DL Pierson, NR Pellis, JH Rex. Multidimensional flow cytometric analysis of dendritic cells in peripheral blood of normal donors and cancer patients. Cancer Immunol Immunother 45:234–240, 1998.

186. R Parkman, KI Weinberg. Immunological reconstitution following bone marrow transplantation. Immunol Rev 157:73–78, 1997.

187. ML Grazziutti, CA Savary, D Przepiorka, S Vadham-Raj, I Braunschwey, JH Rex. In vitro reconstitution of lymphocyte proliferative response against fungal antigens one month after peripheral blood stem cell transplantation (PBSCT) by addition of dendritic cells (Abstract 339). 36th Annual Meeting of Infectious Disease Society of America, 1998, p 139.

188. ML Grazziutti, CA Savary, A Ford, EJ Anaissie, RE Cowart, JH Rex. *Aspergillus fumigatus* conidia induce a Th1-type cytokine response. J Infect Dis 176:1579–1583, 1997.

189. E Cenci, A Mencacci, A Bacci, F Bistoni, VP Kurup, L Romani. T Cell Vaccination in mice with invasive pulmonary aspergillosis. J Immunol 165:381–388, 2000.

190. LJ Rodriguez-Adrian, ML Grazziutti, JH Rex, EJ Anaissie. The potential role of cytokine therapy for fungal infections in patients with cancer: is recovery from neutropenia all that is needed? Clin Infect Dis 26:1270–1278, 1998.

20
Mucosal Infection and Immunity in Candidiasis

Flavia De Bernardis and Maria Boccanera
Istituto Superiore di Sanità, Rome, Italy

I. MUCOSAL IMMUNE SYSTEM: GENERAL CONCEPTS

The mucosal immune system has long been recognized as the first line of defense against microbial invaders. The aerodigestive and the urogenital tracts as well as the eye conjunctiva and the ducts of all exocrine glands are all covered by mucous membranes with their highly specialized immune system. As described by Cwerkinsky et al. [1], this consists of an integrated and communicating network of lymphoid cells which works in concert with innate host factors to promote host defense. Major mucosal effector immune mechanisms include secretory antibodies, largely of immunoglobulin A (IgA) isotype, cytotoxic T-cells, and cytokines, chemokines, and their receptors. In a healthy human adult, this local immune system contributes almost 80% of all immunocytes [1]. These cells are accumulated in or transit between various mucosal organs and glands and together form the mucosa-associated lymphoid tissue (MALT), the largest mammalian lymphoid organ system. MALT comprises anatomically defined lymphoid microcompartments, such as the Peyer's patches in the small intestine, the appendix and solitary follicles in the large intestine and in the rectum, the nasal mucosa and the tonsils at the entrance of the aerodigestive tract, which serve as the principal mucosal inductive site where immune responses are being initiated. MALT contains diffuse accumulations of large numbers of lymphoid cells that do not associate into apparently organized structures. These cells are either distributed in the lamina propria or interspersed among epithelial cells in mucosal tissues and glands, giving rise to the mucosal effector sites where immune responses are being induced and/or expressed. The gut mucosa is particularly well endowed with such diffuse lymphoid tissues. More important for the elicitation of an immune response is the fact that immunization at certain inductive sites may give rise to a humoral immune response preferentially manifested at certain effector sites. Thus, a given inductive site may serve as a preferential but not exclusive source of precursor cells for certain mucosal tissues. MALT mainly functions to protect the mucous membranes against colonization by potentially dangerous microbes as well as to prevent the development of harmful responses to these antigens (Table 1).

MALT represents a well-known example of a compartmentalized immunological system. As opposed to the central and peripheral lymphoid organs, MALT contains inhomogenously distributed B-cells and T-cells whose phenotype, developmental origin, and secretion products (and probably also functions) are different. First, the prime immunoglobulin isotype produced

Table 1 Main Functions of Mucosa-Associated Lymphoid Tissue (MALT)

1. Protection of the mucous membranes against colonization by potentially dangerous microbes
2. Prevention of uptake of antigens including foreign protein derived from ingested food and commensal microorganisms
3. Prevention of development of harmful immune system responses to these antigens if they reach the internal milieu

and assembled in mucosal tissues is secretory immunoglobulin A (SIgA), which is only present in trace amounts and usually only early and transiently, in the intravascular compartment. Further, nonconventional lymphocytes rarely seen in the spleen and peripheral lymph nodes are encountered at appreciable frequency in different mucosal location such as, for instance, α/β TCR$^+$, CD4$^-$, CD8$^-$ cells, γ/δ TCR$^+$, CD4$^-$ CD8$^-$ cells, α^+ CD8$^-\beta^-$ and α/δ TCR$^+$, CD4,$^-$ CD8$^-$ α^-CD8 β^- T-cells in the gut epithelium. The antigen receptor reportoire is also different in each location. Furthermore, in contrast to the T- and B-lymphocytes found in central and peripheral lymphoid organ, certain γ/δ and α/β T-cells found in the mucosa, and presumably also the mucosal CD5$^+$ B-cells do not depend on the thymus or bone marrow for their development, respectively. Thus, by its cellular composition and compartmentalization, the MALT works with an essential independence from the systemic immune apparatus.

II. INDUCTIVE SITE IN THE GUT-ASSOCIATED LYMPHOID TISSUES (GALT)

The gut contains the most abundant lymphoid tissues and includes organized as well as diffuse lymphoid elements. Organized lymphoid tissues comprise two units: B-cell follicles and para- or interfollicular T-cell areas—assembled within a matrix of loose connective tissue. These follicles occur singly or in groups and harbor variable numbers of macrophages and T-cells [1].

Follicle-dome structures form the main lymphoid component of the Peyer's patches in the jejunum and ileum, and are also found in the large intestine, and especially in the appendix (Fig. 1). These structures appear to play an important role in the induction of immune responses to oral vaccines. Typically, the follicles contain a majority of B-cells, approximately half of which are activated. The T-cell zone comprises a majority of CD4$^+$ T-cells; CD8$^+\alpha\beta$ TCR T-cells are mainly located in the parafollicular area whereas CD8$^+$ $\gamma\delta$ TCR T-cells are rare. The dome is covered by a specialized epithelium or "follicle"-associated epithelium, containing antigen-transporting M cells.

Clusters of follicles are also found adjacent to the anorectal junction. The potential importance of the rectal lymphoid tissues as IgA-inductive sites and as a source of IgA plasma cell precursors is suggested by several studies [1]. First, the predominance of IgA$_2$ cells over IgA$_1$ cells in the lamina propria of the large intestine clearly diverges from the relative apportioning of the two in other mucosal tissues such as in the small intestine and in the upper large intestine. Further, rectal immunization of humans, nonhuman primates, and rodents has been shown to induce strong secretory antibody responses in the rectal mucosa. In some instances, rectal vaccination could induce specific antibodies in serum, and also in secretions from remote mucosal organs such as saliva and genital secretions.

PEYER'S PATCH

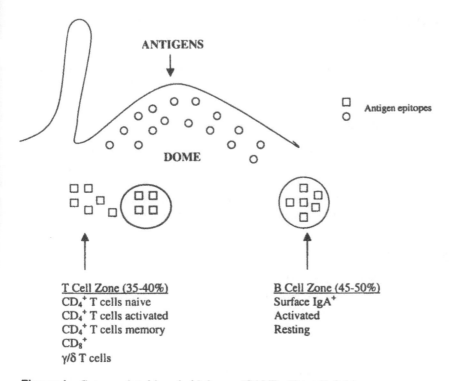

Figure 1 Gut-associated lymphoid tissues (GALT). (From Ref. 1.)

III. INDUCTIVE SITES IN THE GENITAL TRACT MUCOSA

Although long considered as immunologically incapable of supporting an active immune response against locally encountered antigens (perhaps as perception of the fact that it must tolerate semen antigens), the female reproductive tract mucosa has been shown to contain all cell populations required for initiating an immune response. HLA-DR+ Langerhans cells with elevated antigen presentation capability have been identified in the vaginal and cervical epithelia, being most abundant in the vulval epithelium. Intraepithelial T-cells have been found at all sites and comprise primarily CD8+ cells. A significant proportion of these cells express TIA-1 [1], suggestive of a cytolytic capacity. In contrast, CD4+ T-cells are rarely found in these epithelia but predominate in the submucosa of the vagina, cervix, and Fallopian tubes. Remarkably, the proportion of CD4+ and CD8+ T-cells may vary in the tissue districts. Although all components of the mucosal immune system are present in the female reproductive tract, the precise sites of induction of secretory immune responses in this organ are largely unknown. Moreover, even large differences in the cellular composition and distinct phenotypes have been reported in the same genital district of the different mammals. An example is the very different proportion of CD4+ and CD8+ T-lymphocytes in the subepithelial layers of mouse and rat vagina [2,3].

Interestingly, the CD4$^+$/CD8$^+$ T cell ratio in healthy women is much more similar to that of the rat than that of the mouse [4].

IV. PREDISPOSING FACTORS TO MUCOCUTANEOUS CANDIDIASIS

Physical or functional impairment of the mucosal barrier, deficient functioning of the SIgA, and T-cell deficiency syndromes make patients prone to develop mucosal or mucocutaneous candidiasis [5,6].

A. Breaches of the Integrity of the Mucosal Surface

Corticosteroid treatment and cytotoxic chemotherapy may result in impairment of mechanical integrity of the mucous membranes (mucosal erosion or disruption) and predispose to mucosal candidiasis. Outgrowth of *Candida* on mucosal surfaces can also occur because of the alteration of the bacterial flora. Elevation of the glucose concentration in saliva or vaginal fluid may increase the ability of *Candida* to adhere to the epithelium or to replicate at the mucosal surface.

B. Secretory IgA Deficiency

Patients with disorders of the SIgA immune system (deficiency of the secretory component or decreased production of SIgA) are prone to develop chronic diarrhea and recurrent intestinal candidiasis without an increased susceptibility to disseminated candidiasis.

C. Primary T-Cell Deficiency

Oropharyngeal thrush is a common complication of primary T-cell deficiency such as Di George syndrome, and persistent mucosal candidiasis is a hallmark in infants with combined immunodeficiency disorders. Patients with chronic mucocutaneous candidiasis with defective in T-lymphocytes have protracted candidal infections of the mucous membrantes, nails, and skin. These patients, however, are not susceptible to invasive candidiasis.

D. Acquired T-Cell Deficiency

Oral and gastrointestinal candidal infection are common manifestations of human immunodeficiency virus infection, and esophageal candidasis is one of the diagnostic criteria of the acquired immunodeficiency syndrome. Disseminated candidiasis is not frequent in this form of immunodeficiency and occurs only in the late stages of disease when other immune functions (for instance, the phagocytosis) are also severely impaired.

Mucocutaneous candidiasis is increasingly common in patients with defective cell-mediated immunity (CMI), including AIDS patients. The majority of human *Candida* infectious occur at mucosal surfaces. However, in contrast to systemic candidiasis, relatively little is known about the role of mucosal immunity in protection against *Candida*.

V. MUCOSAL IMMUNE RESPONSE TO *CANDIDA*

Candida species, especially *C. albicans*, are commensals of the oral cavity and the gastrointestinal and genitourinary tracts of healthy humans. Odds [5] has suggested that 40–50% of any given sample population temporarily or parmanently carries this fungus in their gastrointestinal tract. The shift of *C. albicans* from commensal to invasive pathogen is primarily the consequence of suppression of the host's cellular immune system as well as the virulence of the indigenous *C. albicans* strains [7–10]. Mucosal colonization with *C. albicans* induces both antibody-mediated immunity and CMI. However, *C. albicans* can persist on the mucosal surfaces of healthy individuals despite demonstrable adaptive anti-*Candida* immunity. In mere, alimentary tract colonization or oral immunization with *C. albicans* induced both A secretory and a cellular immune response in mice [11]. Oral immunization with *C. albicans* induced in mice a Th-1-type response in the spleen, but a Th-2 type-response in Peyer's patches. According to Xu-Amato et al. [12], this dichotomous T-helper cell response may explain why *Candida*-induced humoral and cellular immune response of the host does not result in clearance of *Candida* from the intestine and *C. albicans* continues to persist on the mucosal surfaces despite the persistence of systemic immunity.

Dissemination, which occurs under conditions that compromise the host immune system, such as quantitative or qualitative defects in phagocytic cell activity, results in a progressive disease associated with a high rate of mortality. In the majority of healthy human subjects, *Candida* commensalism is promptly revealed by the presence of appreciable levels of anti-*Candida* antibodies as well as by a state of T-cell-mediated delayed-type hypersensitivity (DTH) response to fungal antigens as an indicator of CMI responses in vivo. The DTH response to the fungus is presumed to prevent mucosal colonization from progression to symptomatic infection [13]. In fact, defective DTH reactivity and elevated levels of *Candida* specific IgE, IgA, and IgG [14] are often observed in immunocompromised individuals with recurrent or persistent *Candida* infection. As pointed out by Cassone et al. [15,16], *Candida albicans* possesses several molecules or molecular complexes in its cell wall which potentially have a great impact on the host. Since the cell wall and its polymers released into the external milieu are the main components which interact with the host, it is not surprising that *Candida* cell wall proteins and mannoproteins, particularly, are potent inducers of DTH in humans. Moreover, the ability of this fungus to develop, in vivo, into different growth forms is associated with its resistance to the host's clearance mechanisms. Thus, detectable mechanical obstacles imposed on phagocytes or more subtle mechanisms of immune evasion such as antigenic variation, binding of complement or other serum factors, secretion of immunoinhibitory constituents, low level of chemokine stimulation, and other factors may contribute to persistence of the organism [15,16] (Cassone et al., unpublished data).

Persistent but localized foci of colonization by *Candida* in the gastrointestinal (GI) tract of healthy individuals may be important reservoirs from which dissemination occurs under conditions that compromise the host immune defense system. For instance, this fungus is prone to invade the oropharyngeal and esophageal mucosa in AIDS patients [17] and internal organs in severly neutropenic subjects [8]. In such cases, *C. albicans* is likely to modify to a certain extent the nature of some of its cell wall constituents under the selective pressure of in vivo growth and the need for survival in new organ environments. Gastric candidiasis may also occur in these individuals especially if extensive esophageal *Candida* infection is involved. The expression of pH-regulated proteins in different environments, and its relationship to specific host tissue invasion, is a remarkable example of *C. albicans* adaptation [10].

Although the importance of T-cell-dependent immunity in chronic mucocutaneous candidiasis has been clearly established, the exact mechanisms of protection have not been clearly defined. Cellular immune responses at the mucosal level have been studied especially by La Casse et al. [18] in an experimental model of murine oral candidiasis. A massive migration of CD4+T cells and macrophages into the epithelium has been observed. The number of CD4+T-cells remained high in the basal layer for up to 30 days postinfection. After a second topical challenge, the infection subsided within 48 hr in contrast to the 5 days of local infection observed after the first challenge. Histologic studies showed a typical delayed-type hypersensitivity reaction. From these data it is inferred that CD4$^+$ T-lymphocytes may act locally against candidal infection by releasing cytokines and inducing delayed-type hypersensitivity reactions.

Clinical data and animal experiments generally support a role for CD4$^+$ T-cells in resistance to mucosal candidiasis [19]. AIDS patients with a reduced population of CD4$^+$ cells are susceptible to oropharyngeal candidiasis but they seldom develop disseminated candidiasis [17]. Overall, CD4$^+$ T cells-mediated resistance to candidiasis is thought to take place via the production of Th-1 cytokines and the subsequent activation of innate effector mechanisms (see below).

VI. SECRETORY IMMUNE RESPONSE

It is generally accepted that after oral immunization, antigens are taken up by Peyer's patches via a follicle-associated epithelium where *microfold* cells endocytose luminal antigen [20]. After antigen stimulation, locally induced specific secretory IgA$^+$ B-cells and T-cells migrate to mucosal effector tissues in the lamina propria regions of the gastrointestinal, upper respiratory, and genitourinary tract via a common mucosal immune pathway [21]. T-cells in intestinal lamina propria represent 30–40% of the total lymphocytes with a CD4$^+$/CD8$^+$ ratio of about 2:3. Both subgroups of CD3$^+$ CD4$^+$ Th-cells are present in lamina propria. Cytokines produced by Th-2 cells, especially interleukin (IL)-4, IL-5, and IL-6, may act directly on secretory IgA$^+$ B-cells and induce them to differentiate into IgA-producing cells [22]. Thus, an adequate number of functioning T-cells is mandatory for the development of a normal secretory immune response to *Candida*: in fact, T-cells seem to play a critical role in stimulating differentiation of IgA$^+$ B-lymphocytes to produce SIgA.

The mucosal tissues of human and animal reveal a remarkable preponderance of IgA-producing plasma cells which, as demonstrated in animal models, are delivered by precursors in the organized gut- and bronchus-associated lymphoid tissues (GALT, BALT) [23]. As reported by Maldoveanu et al. [24], it is plausible that GALT, BALT, and other sources (peritoneum, tonsil, and rectal lymphoid follicles) may preferentially supply IgA-committed antigen-sensitized cells to restricted (frequently adjacent) mucosal region. This possibility has considerable practical impact: inhalation or intranasal immunization might stimulate immune response preferentially in the upper respiratory and digestive tracts rather than in the genital or lower intestinal secretions, whereas ingestion of antigens or rectal immunization would stimulate responses preferentially in corresponding areas and less in nasal secretions or tears. However, our recent data show that immunization with *Candida* antigen intranasally or intravaginally in oophorectomized rats gives equivalent Ab-mediated protection against candidal vaginitis. Although the induction of antibodies in saliva, tears, and milk by mucosal immunization indicates that different inductive sites overlap in their ability to seed IgA-secreting cells to various effector sites, the extent to which the common mucosal immune system is compartmentalized or fully communicating is not clear.

As many pathogens display distinct tropisms for certain mucosal sites, exploitation of compartmentalization within the common mucosal immune system to direct an immune response to a particular site where the infection can be effectively counteracted would not only be efficient but also might avoid undesirable effects elsewhere, and this information may be valuable in the design of vaccines and appropriate immunization routes.

VII. ANIMAL MODELS OF GASTROINTESTINAL CANDIDIASIS

The mechanisms of mucosal colonization and survival of *C. albicans* in the alimentary tract and the nature of the mucosal immune and nonimmune cell response to the fungus are still poorly understood. Animal models have figured prominently in the examination of the nature of the host's innate and acquired immune defense mechanisms against gastrointestinal (GI) candidiasis.

Rodents, particularly mice, have been widely used as models of *Candida* infection of the GI tract. Pope et al. [25] were the first to describe an infant mouse model of GI candidiasis Guentzel and Herrera [26] suggested that *C. albicans* oral-intragastric inoculation of infant mice leads to persistent infection of the GI tract and provides an excellent model for studying the role of compromising agents and procedures in the exacerbation of GI and systemic candidiasis.

Other animal models have also focused on aspects of dissemination of *Candida* from the GI tract leading to systemic candidiasis [27–30] Problems with establishing experimental GI candidiasis were encountered when conventional adult animals were used without administration of immunocompromising agents, broad-spectrum antibiotics, and X-ray irradiation [31]. Cole and collaborators [32] have employed an infant mouse model of GI candidiasis which simulates the situations found both in normal humans with *C. albicans* as a component of GI microflora and in immunocompromised patients with resident *C. albicans* in the GI tract representing a reservoir of the pathogen for potential disseminated disease. CFW (SW) BR infant mice (6 days old) were oral-intragastric inoculated with 2×10^8 cells of *C. albicans* by a 24-gage animal feeding needle. In this model, oral-intragastric inoculation leads to colonization of the GI tract; in the absence of an immunocompromising treatment (by cyclophosphamide and cortisol), *Candida* was primarily localized in the stomach and intestines of mice. In immunocompromised animals proliferation of *C. albicans* in the stomach and intestine was evident with a high density of invasive hyphal growth and systemic spread. In most mice, abscesses containing both *C. albicans* hyphal and yeast cells were frequently observed in the liver and occasionally in the lungs and kidneys of immunocompromised mice 20 days postinoculation.

The authors suggest that the animal model described may be particularly useful for exploring methods which may prevent dissemination of *C. albicans* from localized foci of colonization in the GI tract after exposure of the host to immunocompromising drugs. In fact, the model permitted examination of the mechanisms of systemic spread of *C. albicans* from the gut. The model also provided for investigations of the histopathology of long-term *Candida* colonization of the GI tract and the nature of the mucosal immune and nonimmune cell responses to chronic yeast infection [33]. CFW infant mice inoculated orally with 2×10^8 yeast of *C. albicans* were examined for lesions of the oesophageal and gastric mucosa on 3 weeks postchallenge.

Histological sections of these regions revealed intraepithelial abscesses associated with infection by *C. albicans*. Thickening of the mucosal tissue (acanthosis) had occurred at the sites where fungal hyphae were visible. Indications of hyperkeratosis were also visible in these regions. Abundant PMN were found in cluster within the epithelial layer and adjacent to *C. albicans* cells, and comprised the majority of inflammatory cells present in the abscess.

The neutrophils had migrated to the site of *Candida* infection and were presumably entrapped between dividing epithelial cells and keratinocytes. Products released by the aggregation of the degranulated PMN within the abscess such as oxygen radicals associated with myeloperoxidase derived from the neutrophil granules and other antimicrobial compounds may contribute to the death of *Candida* cells trapped within the abscess.

Evidence has been presented that neutrophil lysis also causes the release of a cytoplasmic protein which can inhibit the growth of *C. albicans* without killing the organism [34]. Both candidacidal and candistatic activity may occur within and adjacent to the mucosal abscesses in the stomach of infected mice. These activities may contribute cooperatively to the gradual clearance of GI candidiasis from the immunocompetent mice.

To examine further the mobilization of the host cells to site of *Candida* infection in the gastric mucosa, Cole and collaborators [33] examined abscesses as well as various regions of the lamina propria of gastric mucosa of infected mice by immunofluorescence. Macrophages were also identified in the lamina propria of the gastric mucosa of infected mice, although in much lower number.

The numbers of T-lymphocytes and macrophages were significantly higher in infected than noninfected animals for all segments on regions of both lamina propria and submucosa. These observations led the authors to suggest that mobilization of both T-lymphocytes and macrophages occurred in the gastric mucosa in response to C. albicans infection, and that protection is associated with cell-mediated host response to persistent GI candidiasis along with the cooperative efforts of PMNs, macrophages, and monocytes.

Cole et al. [35] have also described a mouse model of GI candidiasis associated with an AIDS-related murine immunodeficiency syndrome. The model is based on infection of C57BL/6 mice with a viral complex known as murine AIDS (MAIDS virus) which demonstrates many characteristics of human AIDS. Both B-cell and T-cell functions have been found to be abnormal in these retroviral-infected mice. Impaired T-cell proliferative response has been suggested to be related to an intrinsic defect of the CD4$^+$ T-cells but not CD8$^+$ T-cells [19].

In the studies by Cole and collaborators [35], C57B2/6 infant mice were infected with C. albicans by oral-intragastric inoculation and 30 days later with the retrovirus via the *intraperitoneal* route. The infected mice developed murine AIDS and developed eruptive and suppurative lesions in the stomach containing C. albicans. In fact, the retrovirus induced immunosuppressive in mice which maintain C. albicans as an apparent commensal in their GI tract which promoted a significant increase in the number of C. albicans cells in the esophagus, stomach, and intestine and invasion of the gastric submucosal tissue. Large numbers of yeast cells and hyphae were visible between the outer keratin layers and epithelium. An intensive inflammatory cell response was evident in histological sections of the gastric mucosa of these animals in conjunction with the proliferation and invasiveness of the fungus.

Extensive mononuclear cell and neutrophil responses were evident in the lamina propria and submucosa of the infected mice adjacent to sites of C. albicans invasion. Immunofluorescence microscopy on sections of the gastric mucosa showed that a significant number of these mononuclear cells were macrophages and also revealed that abundant neutrophils were present.

Cole et al. [35] suggest that the inflammatory response in the immunodeficient murine model is due to several factors. *Candida* releases chemotactic factors as well as lytic enzymes capable of causing tissue damage. Lysis of keratinocytes occurs at the squamous epithelial surface of the GI tract, which potentiates the release of interleukin-1, proinflammatory cytokine for mucosal tissues, besides being a well-known chemotactic factor of PMN and mononuclear cells. In addition to these innate host defense mechanisms, C. albicans is known to release antigens which induce an immune response. This acquired host defense mechanism normally evokes local immunoprotection at the mucosal level, mediated by secretion of specific IgA and the presence of sensitized T-lymphocytes.

Several possible dysfunctions of the cell-mediated immunity exist in mice which have coexisting C. albicans and retrovirus (MAIDS) infections. These include T-cell activity in abscess formation as well as phagocytosis and killing of C. albicans by macrophages and neutrophils in the GI tract. The immunoregulatory disturbances associated with MAIDS exacerbate C. albicans infection of the gastric mucosa.

The murine models of GI candidiasis described by Cole et al. [32,33,35] permitted examination of the nature of C. albicans interaction with the gastric mucosa both in the immunocompetent host under conditions in which the yeast exists predominantly as a commensal organism and in the immunosuppressed host during progressive stages of AIDS induced by a retroviral infection.

It is well known that mucosal immunization in addition to protecting mucosal surfaces can also induce systemic immunity. In fact, Cantorna and Balish [36] demonstrated in a murine model of candidiasis that mucosal immunization with *C. albicans* confers protection against systemic challenge. They analyzed the protective immune response evoked by mucosal immunization.

The same observations were made by Fidel et al. [37] in the mouse model of vaginal candidiasis. Jones-Carson et al. [38] observed that B-cell-deficient $J_H D$ mice have normal T-cell responses to antigens inoculated systematically and manifest resistance to acute systemic candidiasis after oral colonization with *C. albicans*, and that the resistance correlated with CMI acquisition to *C. albicans*, thus inferring that the protective acquired immunity of mucosally immunized B-cell-deficient mice was due to B-cell-independent T-cell response evoked by colonizing the gastrointestinal tract with *C. albicans*. Their studies demonstrated that mucosal and not IV immunization with *C. albicans* leads to a B-cell-independent expansion and selection of memory T-cells which resulted in resistance to *C. albicans*. Thus, T-cells participate in mucosal immunity to *C. albicans* in the absence of B-cells, and CD4+ and thymic educated CD8+ TCR α/β memory subsets evoked by mucosal but not parenteral (IV) challenge, contribute to protective immunity to systemic candidiasis.

It is known that CD8+ T-cells are found in large numbers in most mucosal tissues. However, the role of CD8+ T-cells in resistance to mucosal candidiasis is scanty and somewhat controversial. For the most part, CD8+ intestinal intraepithelial lymphocytes are known to play important roles in suppression and cytolytic activities. The fact that the CD8+ T-cell population is not as severely depressed in AIDS patients as CD4+ T-cells indicates the former may be involved in resistance to systemic candidiasis of endogenous origin. Also, mice infected with retroviruses that reduce the population of their CD4+ T-cells, but not their CD8+ T-cells [38], also manifest increased gastric but not systemic candidiasis. In addition, studies by Fidel et al. [39] in mice treated with monoclonal antibodies (Mab) to deplete CD8+ T-cells suggest that they do not play a role in a resistance to vaginal candidiasis. There are also a few observations about a population of CD8+ T-cells capable of killing *C. albicans* hyphae upon stimulation in vitro with high doses of IL-2, but no in vivo evidence is available to substantiate this interesting finding [40].

Balish and coworkers [41] studied the effect of major histocompatibility complex class I (MHCI)-restricted responses to mucosal candidiasis in β2-microglobulin knockout mice. These mice express little MHCI protein on their cell surface and are deficient in CD8α/β T-cell receptor α/β (TcR α/β) T-cells. Their data demonstrated that immune defects that accompany the loss of β_2-microglobulin play an important role in murine resistance to gastric candidiasis of endogenous (intestinal tract) origin and that, in contrast, those mice had resistance to candidiasis in other intestinal, vaginal, and cutaneous tissues.

Mucosal tissues (such as the esophageal-gastric tract) in β2-microglobulin-deficient mice did show susceptibility to candidiasis and, in contrast other mucosal tissues (intestinal and vaginal), displayed resistance to candidiasis suggesting that CD4+, CD8α/β, TcRα/β, CD8α/α, TcRα/β T-cells, and phagocytic cells vary in their capacity to prevent candidiasis at different anatomic sites. The authors hypothesized that mucosal tissues may rely on different populations of thymic or extrathymic matured (β2m-dependent or -independent) T-cells for resistance to candidiasis.

Rahman and Challacombe [42] used an oral model of candidiasis in gnotobiotic mice to study the humoral and cellular immune responses after induction of mucosal infection by *C. albicans*. They observed significantly reduced salivary counts of *C. albicans* in orally immunized mice in comparison with controls. Anti-*Candida* SIgA antibodies were detected in intragastric (IG) immunized mice but not in intraperitoneal (IP) immunized mice while serum anti-*Candida* IgG antibodies were found in IP-immunized mice but not in IG-immunized mice.

Adoptive transfer of mesenteric lymph node cells after IG immunization led to significantly reduced numbers of *Candida* in the saliva of recipients compared with controls. Adoptive transfer of either mesenteric lymph node cells or CD8-enriched cells from orally immunized donors resulted in a significant reduction of oral *Candida* carriage, but transfer of CD4-enriched cells did not give the same protection. On the other hand, adoptive transfer of spleen, CD4, or CD8 cells from IP-immunized donors resulted in a reduction in oral *Candida* carriage. The authors suggest that both oral and systemic immunization can protect against *Candida* colonization at mucosal surfaces but that different mechanisms may operate.

Bistoni and collaborators [43,44] studied the course of mucosal *C. albicans* infections in various inbred strains of mice. In particular, these authors demonstrated that in genetically resistant BALB/c and C57BL/6 mice, an anticandidal protective immunization is achieved by mucosal infection with the fungus whereas the same primary mucosal infection fails to elicit a protective response in the susceptible DBA/2 strain. In fact, IG infection of adult DBA/2 mice resulted in significant yeast recovery from the stomach and small intestine for at least 4 weeks; however, the dissemination to visceral organs was not observed. The presence of yeast and filamentous forms of the fungus in the stomach instead was associated with localized self-limiting foci of mucosal involvement, which eventually cleared. On the basis of these findings, it appears that effective mechanisms of local anticandidal defense occur. These authors also observed that a second intragastric yeast inoculation 2 weeks after primary challenge did not result in more prolonged colonization. The analysis of patterns of Th responses by mucosal tissues in yeast-colonized mice showed impaired local Th-2-associated functions.

In fact, defective production of IgA, IL-4, and IL-5 by Peyer's patches CD4$^+$ lymphocytes and increased numbers of interferon-γ-producing T-cells in mesenteric lymph nodes and spleen along with high levels yeast-specific IgG-2$_A$ antibodies with lower IgG1, IgA, and IgE titers have been observed. Mice also exhibited strong footpad responses and increased resistance to systemic reinfection. The authors proposed that the absence of strong mucosal Th-2 reactivity associated with systemic expansion of Th-1 cells may prevent proliferation and invasiveness by the fungus, which remains, instead, confined to the outer layers of the stratified epithelium and is excluded from the mucosa. Thus, the authors point out that, unlike systemic challenge, gastrointestinal colonization by *C. albicans* of DBA/2 mice appears to be an effective stimulus for the systemic development of protective Th-1 responses.

The contribution of local Th-1 and Th-2-like responses to the course of primary and secondary gastrointestinal candidasis was also examined in Balb/c mice [45]. Gastrointestinal candidiasis in Balb/c mice appeared to be more severe than that occurring in the DBA/2 strain. High numbers of *Candida* cells were recovered from the stomach and, to a lesser extent, from the small intestine of Balb/c mice 1–8 weeks after challenge. Clearance of the fungus began ~8 weeks after infection and was complete within 10 weeks. *Candida*-colonized Balb/c mice had signs of mucosal invasion.

After 1 week, but more extensively 2 weeks after infection, multiple eruptive abscesses occurred throughout the keratinized epithelium. Large numbers of hyphal elements, yeast cells, and inflammatory cysts (mainly neutrophils) were visible between the outer keratin layers and the epithelium. Clearance of the fungus was accompanied by the disappearance of mucosal histopathologic changes.

Both Th-1 cytokines, such as interferon-γ, and Th-2 cytokines, IL-4 and IL-5, were produced by CD4$^+$ cells from Peyer's patches and mesenteric lymph nodes when the fungus was cleared from the stomach and intestine. Better induction of antigen-specific Th-2-like responses by treatment with the strong mucosal adjuvant cholera toxin did not modify the course of disease. In contrast, treatment with soluble IL-4 receptor, which tipped the cytokine balance toward Th-1 cells, was associated with enhanced yeast clearance. Moreover, the animals clearing the primary infection were resistant to the second challenge, as shown by the absence of yeast cells recovered from organs 2 weeks after challenge. IFN-γ but not IL-4 mRNA was detected in Peyers's patches and spleen CD4$^+$ cells in mice resistant to a second GI inoculation.

This study demonstrates that both Th-1-like and Th-2-like responses were present in mice recovering from GI candidiasis; however, protective immunity was associated with the induction of a local Th-1-like response that would ultimately be dominant over Th-2 reactivity. Activation of Th-1-like but not Th-2-like responses may be necessary locally for controlling GI candidiasis and generating protective immunity.

The same authors have shown that development of protective anticandidal Th-1 responses requires the concerted actions of several cytokines such as IFN-γ [46], TGF-β [47], IL-6 [43], TNF-α [48], and IL-12 in the relative absence of inhibitory Th-2 cytokines such as IL-4 and IL-10. Neutralization of Th-2 cytokines (IL-4 and IL-10) allows for the development of Th-1 rather than Th-2 cell responses. However, in highly susceptible mice, exogenous IL-12 did not exert beneficial effects on the course and outcome of mucosal infection [49]. Moreover, administration of IL-4 failed to convert an already established Th-1 response into a Th-2 response, and late IL-4 depletion exacerbated chronic infection, suggesting that in this model the Th differentiation may be locked in and persist.

To further delineate the role of cytokines in the development of murine candidiasis, these authors also analyzed the course of primary and secondary infections with distinct gene deletions, such as those involving cytokines, cytokine receptors, IL-1β-converting enzyme, or major histocompatibility complex class II antigen. Mutant and wild-type mice were injected intragastrically and after reinfected and the results showed that susceptibility to primary and secondary infections greatly varies in the different knockout mice.

Cytokine-gene deletion resulted either in specific defects of antifungal immunity or fully rescued immune mechanisms. For instance, TNF-α and IL-6 deficiencies rendered mice highly susceptible to primary gastric infection and impaired the successful development of innate and acquired immunity. In fact, both qualitative and quantitative changes in antifungal effector functions of phagocytic cells, particularly neutrophils and the activation of non protective Th-2 responses, were observed in these animals. In particular, more extensive fungal growth was detected in stomach and esophagus of cytokine-deficient mice compared to normal mice. All functions were improved upon administration of recombinant TNF-α. In gastric candidiasis protection relies on the induction of antifungal Th1 responses possibly occurring through stimulation of antifungal effector functions and the costimulatory activities of phagocytic cells [48]. In contrast, resistance to primary and secondary infections was not impaired in the absence of IL-1β or CD4$^+$ cells. In addition IL-12, IL-4, or IFN-γ deficiencies, while not affecting resistance to primary infections, rendered mice susceptible to reinfection. It was demonstrated that susceptibility to primary and secondary gastric infections in knockout mice correlates with the failure to develop anticandidal protective Th-1 responses and with the occurrence of nonprotective IL-4 and IL-10-producing Th-2 cells.

In conclusion, studies performed in genetically modified mice, including cytokine-deficient mice, have resulted in an understanding of cytokine-mediated regulation of Th cell development and effector functions in mucosal candidiasis and have revealed complex levels of immunoregulation that were previously unappreciated. Early in infection, production of some proinflammatory cytokines (TNF-α and IL-6) more than others (IL-1β) appears to be essential for the successful control of infection and the resulting protective Th-1-dependent immunity. Both IL-12 production and IL-12 responsiveness are required for the development of Th-1 cells, whose activation depends on the presence of physiological levels IL-4 and IL-10. Thus, a finely regulated balance of cytokines, such as IL-4, IL-10, and IL-12, appears to be required for optimal development and maintenance of Th-1 reactivity in mice with candidiasis. As a whole, these experiments go beyond the more usual paradigm of Th-1 and Th-2 cytokine-associated protection or susceptibility, respectively, to infection by *Candida*. Rather, they suggest that local and dose-dependent regulation of both Th-1 and Th-2 cytokines is needed for effective anti-*Candida* protection, a situation likely to be valid for many other models of infections.

VIII. MUCOSAL IMMUNITY IN *CANDIDA* VAGINITIS

Vulvovaginal candidiasis (VVC) is a widespread common mucosal infection caused by a few *Candida* species that affects a significant number of child-bearing-age women [40–52]. Acute VVC occurs under one or more predisposing factors, including antibiotic and oral contraceptive usage, pregnancy, uncontrolled diabetes mellitus, immunosuppressive therapy, and HIV infection [51,52]. Recurrent vulvovaginal candidiasis (RVVC) is usually defined as idiopathic with no known predisposing factors and seems to affect ~ 5% of all women [50]. In RVVC patients, antifungal therapy is highly effective for individual symptomatic attacks, but does not prevent subsequent recurrence. Also, there is little evidence that antifungal resistance plays a role in the pathogenesis of RVVC [53].

Susceptibility to RVVC is postulated to be immune based; some women experience repeated symptomatic episodes as a result of some immunological deficiencies. However, the pathogenesis of RVVC remains unknown. While, Fidel and collaborators have suggested that local rather than systemic cell-mediated immunity is an important defense against vaginal candidiasis [4,13,37,55], the specific host defense mechanism(s) functioning at the vaginal mucosa is (are) undefined. While it is clear that CD4$^+$ cells are an essential component of the homeostatic anti-*Candida* response, whether these or other T-cells (e.g., CD8$^+$) are ultimate anti-*Candida* effectors has never been proven. In contrast, animal models suggest that T-cell-dependent antibody responses against specific virulence determinants of the fungus may play a substantial effector role (see below).

IX. EXPERIMENTAL ANIMAL MODELS OF VAGINITIS

In order to investigate host defense mechanisms against *C. albicans* at the vaginal mucosa, animal models of vaginal candidiasis have been employed. The aim of using animal models is to study various aspects of the pathogenesis and immune response in *Candida* vaginitis under conditions possibly mimicking human infection. In these models *Candida* infections may be established under controlled and reproducible conditions. To date, many of the data have come from both a mouse and rat model; recently a primate model has also been explored.

The mouse model of candidal vaginitis has been used by Taschdjan et al. [56], Ryley and Mc Gregor [57], Valentin et al. [58], and Fidel et al. [37,54]. This model requires that animals be in a state of pseudoestrus at the time of intravaginal inoculation to obtain a persistent infection which can be studied for immunological or therapeutic studies.

A corresponding rat model was initially generated to study the efficacy of antifungal therapy [59–66] and has been adapted to study virulence factors of *Candida* and their role in the pathogenesis of vulvovaginal candidiasis [10,67–70] and also to investigate the immunological mechanisms of protection against *Candida* at the vaginal level [71,72].

The induction and the maintenance of "pseudoestrus" in rats require oophorectomy and estrogen administration [73]. Within 48 hr of estrogen administration, the vaginal mucosa is composed of stratified, squamous epithelium overlaid by keratin. *C. albicans* only colonizes the

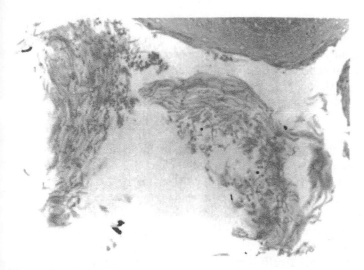

Figure 2 Section of rat vagina taken on day 7 from *C. albicans*–infected rat. The section is stained by the periodic acid Schiff method. Magnification ×400. (Courtesy of Dr. L. Morelli, Lab. Veterinary Medicine, Istituto Superiore di Sanità, Rome, Italy.)

vagina when the epithelium is fully keratinized (Fig. 2). Induction of germ tube (hyphal) formation, adherence, and induction of aspartyl proteinase activity are virulence factors which maybe essential for disease progression. In addition, during estrous phase there is a low-grade leukocyte infiltration, but the role of these cells in protection is not known. The hormonal dependence of candidal vaginitis in rats correlates well with the role of estrogen in human vaginitis. Although not exactly defined, this role can easily be inferred by the rarity of *Candida* vaginitis in premenarchal and postmenopausal ages, while its incidence greatly increases and is exacerbated in the week prior to menses [74].

Fidel and collaborators [13,39,75] using the mouse model, showed that systemic cell-mediated immunity (CMI) was not protective against vaginitis. In fact neither CD4+ Th-1 type CMI nor the presence of *Candida*-specific suppressor T-cells had any effect on vaginal fungal titers. The authors suggest that some form of locally acquired mucosal immunity (T-cell and/or antibody-mediated immunity) could be operative for protecting against vaginal *C. albicans* infection, and that some level of systemic immunological independence or immune compartmentalization exists. There data correlated well with clinical studies showing that RVVC patients had normal levels of *Candida*-specific Th-1-type CMI in the peripheral circulation during symptomatic episodes of vaginitis [52]. In addition, women with RVVC are generally not susceptible to oral, esophageal, or other forms of cutaneous candidiasis [50], and women with chronic mucocutaneous candidiasis are rarely susceptible to RVVC [5]. On the basis of the results obtained in murine model of vaginitis and from the observations made in clinical studies, Sobel and Fidel have postulated that the immune deficiency in RVVC patients is localized to the vaginal mucosa and does not involve systemic CMI [13,39,52,54,75].

Thus, Fidel and collaborators [76] focused on the presence of T-cell subpopulations at the vaginal mucosa and reported that vaginal lymphocytes are phenotypically distinct from those in periphery, confirming the observation of Ibraghimov et al. [77]. In fact, the authors, through flow cytometric analysis, found that although CD4+α/β TCR+ cells are predominant among T-cells in the vaginal mucosa, similar to lymph node cells, there is a higher percentage of γ/δ

TCR$^+$ cells (15%) and very few if any CD8$^+$ cells (normally 20% in blood). According to the authors [76,78–80] these observations further support the concept of immune compartmentaliza- tion of vaginal T-cells. Moreover, Fidel and collaborators have shown that with some anti-CD4 antibodies, vaginal CD4$^+$ T-cells express the CD4 protein in a different conformation compared to lymph node cells. In addition they showed that antibodies specific for T-cells given intrave- nously depleted T-cells in the peripheral blood but not in the vaginal mucosa, whereas the intravaginal administration of the same antibodies depleted T-cells in both the vagina and the peripheral blood [78].

These observations correlate with those of Jones-Carson et al. [81] who showed that mice depleted of γ/δ T-cell had an increased susceptibility to vaginitis. There data together suggest a potential role of γ/δ T-cells as a first-line defense mechanism of the vaginal mucosa. Moreover, in studying the genetic basis of anticandidal resistance in the vagina, Mulero-Marchese et al. [82] demonstrated that it is possible to adoptively transfer vaginal protection by splenic *Candida*-specific T-cells, mostly of the CD4$^+$ phenotype. However, when analyzing the vaginal cell populations by flow cytometry immunohistochemistry and RT-PCR, they found little or no change in the percentage or types of vaginal T-cells during either primary or secondary experi- mental vaginitis [2]. Furthermore, to evaluate the role of innate immunity in protection against vaginitis, these same authors examined the effects of polymorphonuclear leucocytes (PMNL) in nonestrogen and estrogen *C. albicans*–infected mice and showed that the depletion of PMNL had no effect on vaginal fungal burden under either hormonal condition [2].

Cantorna and coworkers [83] showed that animals immunodeficient in phagocytic cells (beige/beige) were not anymore susceptible to *C. albicans* vaginal infection under nonestrogen- ized conditions. These results indicate a lack of demonstrable effects by systemic CMI or innate immunity (mediated by PMN) against vaginitis. Fidel and collaborators have suggested that if local T-cells are important, they are functioning without showing significant increases in numbers within the vaginal mucosa during infection. However, the data obtained in the rat model of vaginitis do not support this conclusion (see below).

Fidel and collaborators also evaluated the role of "unconventional" immune cells such as epithelial and endothelial cells present in the vaginal mucosa as potential anti-*Candida* effector cells [86]. In fact, Hedges and collaborators [84] showed that epithelial cells produce cytokines and chemokines, and Fratti et al. [85] demonstrated that endothelial cells can phagocytize *Can- dida*. The results of Fidel and collaborators [86] demonstrated that vaginal epithelial cells of naive mice or from human and primate vaginal layers inhibited the growth of *C. albicans* in vitro, suggesting that epithelial cells may represent important innate resistance effector cells against *C. albicans* at the vaginal mucosa.

The local production of chemokines associated with the chemotaxis of T-cells, macro- phages (RANTES, MIP-1α, MCP-1), and polymorphonuclear neutrophils (MIP-2) during experi- mental vaginal candidiasis was analyzed [87]. The results showed significant increases in MCP-1 protein and mRNA in vaginal tissue of infected mice 2 and 4 days after *Candida* infection, respectively, which persisted through 21 days. In contrast RANTES, MIP-1α, and MIP-2 were not detected in the vaginal tissue of infected mice. Furthermore, intravaginal immunoneutraliza- tion of MCP-1 with anti MCP-1 antibodies resulted in a significant increase in vaginal *Candida* colonization early during the infection, suggesting that MCP-1 plays a role in reducing the fungal burden during vaginal infection [87].

In the rat model of vaginitis, we have shown that, after clearing the primary *C. albicans* infection, the animals were highly resistant to a second vaginal challenge with the fungus (Figs. 3, 4). The vaginal fluid of *Candida*-resistant rats contained antibodies directed against mannan constituents and aspartyl proteinase of *C. albicans* and was capable of transferring a degree of anti-*Candida* protection to naive, nonimmunized rats. Preabsorption of the antibody-containing

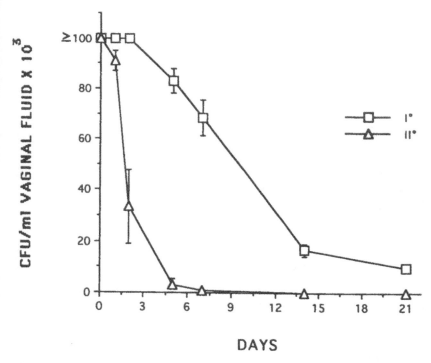

Figure 3 Outcome of vaginal infection in oophorectomized, estradiol-treated rats. The first and the second infections (denoted by □ and Δ, respectively) were performed with an intravaginal challenge of 10^7 cells on day 0. Each curve represents the mean (±SE) of the fungal CFU of five rats.

fluids with either or both mannan and proteinase sequentially reduced or abolished the level of protection [71]. A degree of protection against vaginitis was also conferred postinfection by intravaginal administration of antiproteinase and antimannan monoclonal antibodies and by intravaginal immunization with a mannan extract or a highly purified SAP2 preparation [71].

The protective role of antibodies against specific factors of the fungus that we observed is somewhat in contrast with previous suggestions [88]. In fact, there are few data to suggest that serum antibody and complement in vitro can independently kill *C. albicans* [88]. Additionally, there are no reports showing that patients with congenital or acquired B-cell deficiencies are susceptible to mucosal or systemic candidal infections [88]. However, this should not exclude the possibility that anti-*Candida* IgA or IgG antibodies have a protective role at the mucosal level, since seldom if ever have the specificity and isotype of the elicited Ab been established. It is quite clear that the above-mentioned factors need to be determined for assessing a protective Ab response.

Recently, this controversial role of humoral immunity has been reviewed by Casadevall [89] who postulated that protective, nonprotective, and immunologically indifferent antibodies exist in the milieu of antibody present during an infection. In addition, antibody-mediated anti-*Candida* protection was recently confirmed by Cutler and collaborators in an estrogen-dependent mouse vaginits model [90]. As a corollary, and possibly also as a link between CMI and Ab responses, we have recently demonstrated that CMI exerts a critical regulatory role in the anticandidal response, as witnessed by the lack of protection induced by active *Candida* antigen immunization in congenitally athymic nude rats [72].

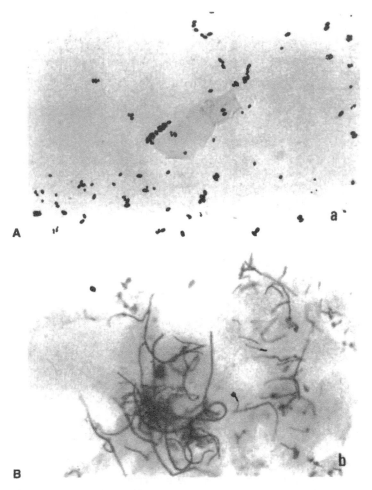

Figure 4 Vaginal scrapings taken from rats infected with *C. albicans* showing development from yeast (A, day 0) to mycelial (B, day 7) cells. The smears were stained by the periodic acid Schiff method. Magnification ×200. (Courtesy of Dr. L. Morelli, Lab. Veterinary Medicine, Istituto Superiore di Sanità, Rome, Italy.)

Thus, the study of T-cell activation at the vaginal level remains a key point for understanding local host immunity. In this context, we have recently attempted to identify T-cell populations in the vaginal mucosa of naive and *Candida*-infected rats, in line with the previously mentioned differences in the T-cell population between mouse and rats [3]. We found that in the rat vagina 60% of vaginal lymphocytes (VL) were CD3+ T-cells, while two-thirds of those expressed the α/β and one-third the γ/δ T-cell receptor (TCR), with some slight fluctuations occurring during *C. albicans* infection. Importantly, however, we noticed remarkable changes in the CD4+/CD8+ T-cell ratio, during subsequent fungal infections, particularly for the cell bearing the CD25 (IL-2 receptor α) activation marker. In fact, an increased number of both CD4+ α/β TcR and CD4+/CD25 + VL was observed after the second and the third *Candida* challenge, reversing the high initial CD8+ cell number of control, estrogen-treated but uninfected rats. During a third *Candida* challenge, VL showed proliferative activity, in a dose-dependent manner, with a mannoprotein fraction of *C. albicans*. These local T-cell modulations, which have no apparent counterparts in the local T-cell modulations in the mouse model, were not paralleled by similar changes at the systemic levels. However, these cells were fully responsive to polyclonal stimulants, demonstrating that in our model anti-*Candida* CMI at vaginal level was a true finding and was coupled with an alteration in T-cell vaginal repertoire (Fig. 5). This was also confirmed by the analysis of the cytokines secreted in the vaginal fluid of *Candida*-infected rats. In fact, high levels of IL-12 during the first, immunizing infection, followed by progressively increasing amounts of IL-2 and IFN-γ during the second and the third infections but not IL-4 or IL-5 were found. Altogether, these results show that a Th-1-restricted response can be induced at the vaginal level by repeated, protective *Candida* vaginal challenge. What remains to be established in future studies is the relationship between CMI, Th-1 elicitation, and protective Ab responses. In addition, our results open the way to addressing the question of adoptive transfer of vaginal T-cells.

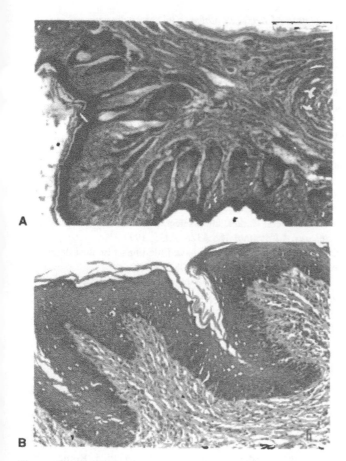

Figure 5 Section of rat vagina taken on day 7 after the first (A) and the third (B) infections by *C. albicans*. In the last infection, infiltration or local multiplication of mononucleate cells are visible. The sections are stained with hematoxylin-eosin. Magnification ×200. (Courtesy of Dr. L. Morelli, Lab. Veterinary Medicine, Istituto Superiore di Sanità, Rome, Italy.)

22. JR Mc Ghee, J Mestecky, CO Elson, H Kiyono. Regulation of IgA synthesis and immune response by T cells and interleukins. J Clin Immun 9:175–180, 1989.
23. AJ Husband, ed. Migration and Homing of Lymphoid Cells. Vol. II. Boca Raton, FL: CRC Press, 1988.
24. Z Modoveanu, MW Russell, HY Wu, WQ Huang, RW Compans, J Mestecky. Advances in Mucosal Immunology. New York: Plenum Press, 1995.
25. LM Pope, GT Cole, MN Guentzel, LJ Berry. Systemic and gastrointestinal candidiasis of infant mice after intragastric challenge. Infect Immun 25:702–707, 1979.
26. MN Guentzel, C Herrera. Effects of compromising agents on candidosis in mice with persistent infections initiated in infancy. Infect Immun 35:222–228, 1982.
27. A De Maria, H Buckley, F von Lichtensberg. Gastrointestinal candidiasis in rats treated with antibiotics, cortisone and azathioprine. Infect Immun 13:1761–1770, 1976.
28. RH Hector, JE Domer. Mammary gland contamination as a means of establishing long-term gastrointestinal colonization of infant mice with Candida albicans. Infect Immun 38:788–790, 1982.
29. MJ Kennedy, PA Volz. Ecology of Candida albicans colonization and dissemination from the gastrointestinal tract by bacterial antagonism. Infect Immun 49:654–663, 1985.
30. RL Myerowitz. Gastrointestinal and disseminated candidiasis. Arch Pathol Lab Med 105:138–143, 1981.
31. C Russel, JH Jones. Effects of oral inoculation of Candida albicans in tetracycline-treated rats. J Med Microbiol 6:275–279, 1973.
32. GT Cole, KT Lynn, KR Seshan, LM Pope. Gastrointestinal and systemic candidosis in immunocompromised mice. J Med Vet Mycol 27:363–80, 1989.
33. GT Cole, KR Seshan, KT Lynn, M Franco. Gastrointestinal candidasis: histopathology of Candida-host interactions in a murine model. Mycol Res 97:385–408, 1993.
34. PG Sohnle, C Collins-Lech. Comparison of candidacidal and candidastatic activities of human neutrophils. Infect Immun 58:2696–2698, 1990.
35. GT Cole, K Saha, KR Seshan, KT Lynn, M Franco, PKY Wong. Retrovirus-induced immunodeficiency in mice exacerbates gastrointestinal candidiasis. Infect Imm 60:4168–4178, 1992.
36. MT Cantorna, E Balish. Acquired immunity to systemic candidiasis in immunodeficient mice. J Infect Dis, 164:936, 1991.
37. PL Fidel, ME Lynch, JD Sobel. Candida-specific Th1-type responsiveness in mice with experimental vaginal candidiasis. Infect Immun 61:4202–4207, 1993.
38. J Jones-Carson, FA Vazquez-Torres, E Balish. B cell–independent selection of memory T cells after mucosal immunization with Candida albicans. J Immunol 158:4328–4335, 1997.
39. PL Fidel, ME Lynch, JD Sobel. Circulating CD4 and CD8 T-cells have little impact on host defence against experimental vaginal candidiasis. Infect Immun 63:2403–2408, 1995.
40. DW Beno, HL Mathews. Growth inhibition of Candida albicans by interleukin-2-induced lymph node cells. Cell Immunol 128:89–100, 1990.
41. E Balish, A Vazquez-Torres, J Jones-Carson, RD Wagner, T Warner. Importance of β_2-microglobulin in murine resistance to mucosal and systemic candidiasis. Infect Immun 64:5092–5097, 1996.
42. D Rahman, SJ Challacombe. Oral immunization against mucosal candidiasis in a mouse model. In: Advances in Mucosal Immunity. S Mesteeky, ed. New York: Plenum Press, 1995.
43. F Bistoni, E Cenci, A Mencacci, E Schiaffella, P Mosci, P Puccetti, L Romani. Mucosal and systemic T helper cell function after intragastric colonization of adult mice with Candida albicans. J Infect Dis 168:1449–1457, 1993.
44. A Mencacci, E Cenci, F Bistoni, A Bacci, G Del Sero, C Montagnoli, C Fé d'Ostiani, L Romani. Specific and non-specific immunity to Candida albicans: a lesson from a genetically modified animal. Res Immunol 149:352–360, 1998.
45. E Cenci, A Mencacci, R Spaccapelo, L Tonnetti, P Mosci, K-H Enssle, P Puccetti, L Romani; F Bistoni. T helper cell type 1 (Th1)- and Th2-like responses are present in mice with gastric candidiasis but protective immunity is associated with Th1 development. J Infect Dis 171:1279–88, 1995.
46. E Cenci, A Mencacci, G Del Sero, C Fe' d'Ostiani, P Mosci, M Kopf, L Romani. IFN-γ is required for IL-12 responsiveness in mice with Candida albicans infection. J Immunol 1161:3543–3550, 1998.

47. R Spaccapelo, L Romani, L Tonnetti, E Cenci, A Mencacci, R Tognellini, SG Reed, P Puccetti, F Bistoni. TGF-β is important in determining the in vivo susceptibility or resistance in mice infected with *Candida albicans*. J Immunol 155:1349–1360, 1995.

48. A Mencacci, E Cenci, G Del Sero, C Fé d'Ostiani, P Mosci, C Montagnoli, A Bacci, F Bistoni, VFJ Quesniaux, B Ryffel, L Romani. Defective co-stimulation and impaired Th 1 development in tumor necrosis factor/lymphotoxin-α double-deficient mice infected with *Candida albicans*. Int Immunol 10:37–48, 1998.

49. L Romani, F Bistoni, A Mencacci, E Cenci, R Spaccapelo, P Puccetti. IL 12 in *Candida albicans* infections. Res Immunol 146:532–538, 1996.

50. JD Sobel. Pathogenesis of *Candida* vulvovaginitis. In: MR Mc Ginnis Stuttgart: Springer-Verlag, M Borges, eds. Current Topics in Medical Mycology. 1989, pp 86–108.

51. JD Sobel. Pathogenesis and treatment of recurrent vulvovaginal candidiasis. Clin Infect Dis 14: S148–S153, 1992.

52. PL Fidel Jr, JD Sobel. Immunopathogenesis of recurrent vulvovaginal candidiasis. Clin Microbiol Rev 9:335–348, 1996.

53. ME Lynch, JD Sobel, PL Fidel. Role of antifungal drug resistance in the pathogenesis of recurrent vulvovaginal candidiasis. J Med Vet Mycol 34:337–339, 1996.

54. PL Fidel Jr, ME Lynch, JD Sobel. *Candida*-specific cell-mediated immunity is demonstrable in mice with experimental vaginal candidiasis. Infect Immun 61:1990–1995, 1993.

55. PL Fidel Jr, ME Lynch, DH Conaway, L Tait, JD Sobel. Mice immunized by primary vaginal *Candida albicans* infection develop acquired vaginal mucosal immunity. Infect Immun 63:547–553, 1995.

56. CL Taschdjian, F Reiss, PJ Kozinn. Experimental vaginal candidiasis in mice; its implications for superficial candidiasis in humans. J Invest Dermatol 34:89–94, 1960.

57. JF Ryley, S McGregor. Quantification of vaginal *Candida albicans* infections in rodents. J Med Vet Mycol 24:455–460, 1986.

58. A Valentin, C Bernard, M Mallie, M Huerre, JM Bastide. Control of *Candida albicans* vaginitis in mice by short-duration butoconazole treatment in situ. Mycoses 36:379–384, 1993.

59. HJ Scholer. Experimentalle vaginal candidiasis der ratte. Pathol Microbiol 23:62–68, 1960.

60. RJ McRipley, RA Schwind, PJ Erhard, RR Whitney. Evaluation of vaginal antifungous formulations in vivo. Postgrad Med J 55:648–652, 1979.

61. D Thienpont, J Van Cutsen, SM Borger. Ketoconazole in experimental candidosis. Rev Infect Dis 2:570–577, 1980.

62. JF Ryley, RG Wilson, MB Gravestock, JP Poyser. Experimental approaches to antifungal chemotherapy. Adv Pharm Chemother 18:49–176, 1981.

63. JD Sobel, G Muller. Comparison of ketoconazole, BAY N7133 and BAY L9139 in the treatment of experimental vaginal candidiasis. Antimicrob Agents Chemother 24:434–436, 1983.

64. OS Kinsman, AE Collard, TJ Savage, Ketoconazole in experimental vaginal candidosis in rats. Antimicrob Agents Chemother 30:771–773, 1986.

65. J Van Cutsen, F Van Gerven, AJ Janssen. Activity of orally, topically and parenterally administered itraconazole in the treatment of superficial and deep mycoses animal models. Rev Infect Dis 9(suppl 1):S15–S32, 1987.

66. TM Jansen, MA Van de Ven, MJ Borgers, FC Odds, JMP Van Cutsen. Fungal morphology after treatment with itraconazole as a single oral dose in experimental vaginal candidosis in rats. Am J Obstet Gynecol 165:1552–1557, 1991.

67. F De Bernardis, R Lorenzini, L Morelli, A Cassone. Experimental rat vaginal infection with *Candida parapsilosis* from outpatients with vaginitis. J Clin Microbiol 27:2598–2603, 1989.

68. L Agatensi, F Franchi, F Mondello, RL Bevilacqua, T Ceddia, F De Bernardis, A Cassone. Vaginopathic and proteolytic *Candida* species in outpatients attending a gynecology clinic. J Clin Pathol 44:826–830, 1991.

69. F De Bernardis, D Adriani, R Lorenzini, E Pontieri, G Carruba, A Cassone. Filamentous growth and elevated vaginopathic potential of a nongerminative variant of *Candida albicans* expressing low virulence in systemic infection. Infect Immun 61:1500–1508, 1993.

70. F De Bernardis, A Molinari, M Boccanera, AR Stringaro, R Robert, JM Senet, G Arancia, A Cassone. Modulation of cell surface-associated manoprotein antigen expression in experimental candidal vaginitis. Infect Immun 62:509–519, 1994.

71. A Cassone, M Boccanera, D Adriani, G Santoni, F De Bernardis. Rats clearing a vaginal infection by *Candida albicans* acquire specific, antibody-mediated resistance to vaginal reinfection. Infect Immun 63:2619–2624, 1995.

72. F De Bernardis, M Boccanera, D Adriani, E Spreghini, G Santoni, A Cassone. Protective role of antimannan and anti-aspartyl proteinase antibodies in an experimental model of *Candida albicans* vaginitis in rats. Infect Immun 65:3399–3405, 1997.

73. OS Kinsman, AE Collard. Hormonal factors in vaginal candidiasis in rats. Infect Immun 53:498–504, 1986.

74. JD Sobel, G Muller. Comparison of itraconazole and ketoconazole in the treatment of experimental candidal vaginitis. Antimicrob Agents Chemother 26:266–267, 1984.

75. PL Fidel, JL Cutright, JD Sobel. Effects of systemic cell-mediated immunity on vaginal candidiasis in mice resistant and susceptible to *Candida albicans* infections. Infect Immun 63:4191–4194, 1995.

76. PL Fidel, NA Wolf, MA Kukuruga. T lymphocytes in the murine vaginal mucosa are phenotypically distinct from those in the periphery. Infect Immun 64:3793–3799, 1996.

77. AR Ibraghimov, RE Sacco, M Sandor, LZ Iakanbov, RG Lynch. Resident CD4$^+$ $\alpha\beta$ T cells of the murine female genital tract: a phenotypically distinct T cell lineage that rapidly proliferates in response to systemic T cell activation stimuli. Int Immunol 7:1763–1769, 1995.

78. PL Fidel, W Luo, J Chabain, NA Wolf, E Vanburen. Use of cellular depletion analysis to examine circulation of immune effector function between the vagina and the periphery. Infect Immun 65:3939–3943, 1997.

79. PL Fidel, JD Sobel. Protective immunity in experimental *Candida* vaginitis. Res Immunol 146:361–373, 1998.

80. PL Fidel, KA Ginsburg, JL Cutright, NA Wolf, D Leaman, K Dunlops, JD Sobel. Vaginal-associated immunity in women with recurrent vulvovaginal candidiasis: evidence for vaginal Th1-type responses following intravaginal challenge with *Candida* antigen. J Infect Dis 176:728–739, 1997.

81. J Jones-Carson, A Vazquez-Torres, T van der Heyde, T Warner, RD Wagner, E Balish. $\gamma\delta$T cell-induced nitric oxide production enhances resistance to mucosal candidiasis. Nat Med 6:552–557, 1995.

82. RD Mulero-Marchese, KJ Blank, TG Siek. Genetic basis for protection against experimental vaginal candidiasis by peripheral immunization. J Infect Dis 178:227–234, 1998.

83. MT Cantorna, D Mook, E Balsh. Resistance of congenitally immuno-deficient gnotobiotic mice to vaginal candidiasis. Infect Immun 58:3813–3815, 1990.

84. SR Hedges, WW Agace, C Svanborg. Epithelial cytokine responses and mucosal cytokine networks. Trends Microbiol 3:266–270, 1995.

85. RA Fratti, MA Ghannoum, JE Edwards, SG Filler. Gamma interferon protects endothelial cells from damage by *Candida albicans* by inhibiting endothelial cell phagocytosis. Infect Immun 64:4714–4718, 1996.

86. H Ozenci, M Scott, W Luo, PL Fidel. Growth inhibition of *Candida albicans* by vaginal cells from naive mice. Presented at American Society for Microbiology Annual Meeting, Abstract F-107, 1997.

87. M Saavedra, B Taylor, N Lukacs, PL Fidel Jr. Local production of chemokines during experimental vaginal candidiasis. Infect Immun 67:5820–5826, 1999.

88. TJ Rogers, E Balish. Immunity to *Candida albicans*. Microbiol Rev 44:660–682, 1980.

89. A Casadevall. Antibody immunity and invasive fungal infections. Infect Immun 63:4211–4218, 1995.

90. Y Han, RP Morrison, JE Cutler. A vaccine and monoclonal antibodies that enhance mouse resistance to *Candida albicans* vaginal infection. Infect Immun 66:5771–5776, 1998.

91. TK Cain, RG Rank. Local Th1-like responses are induced by intravaginal infection of mice with the mouse pneumonitis biovar by *Chlamydia trachomatis*. Infect Immun 63:1784–1789, 1995.

92. KA Kelly, RG Rank. Identification of homing receptors that mediate the recruitment of CD4 T cells to the genital tract follozing intravaginal infection with *Chlamydia trachomatis*. Infect Immun 65:5198–5208, 1997.

93. L Perry, LK Feilwer, HD Caldwell. Immunity to *Chlamydia trachomatis* is mediated by T helper 1 cells through IFN-γ-dependent and independent pathways. J Immunol 158:3344–3351, 1997.

94. C Steele, M Ratteree, R Harrison, PL Fidel. Differential susceptibility to experimental vaginal candidiasis in macaque. Presented at ASM Annual Meeting (Abstract), 1998.

21
Systemic Immunity in Candidiasis

Luigina Romani and Francesco Bistoni
University of Perugia, Perugia, Italy

I. INTRODUCTION

Candida albicans is an opportunistic fungal pathogen commonly found in the human gastrointestinal tract and the female lower genital tract [1]. It is a unique parasite capable of colonizing, infecting, and persisting on mucosal surfaces, and also of stimulating mucosal responses. As a pathogen *C. albicans* is associated with a wide spectrum of diseases in humans, ranging from allergy, severe intractable mucocutaneous diseases, to life-threatening bloodstream infections [1]. Candidal infections are a significant clinical problem for a variety of immunocompetent and immunocompromised patients [1–5]. In the latter group, neutropenia and AIDS epitomize the two major conditions, i.e., the defect in antifungal effector function and the defect in T-cell directive immunity [6], that predispose to invasive and superficial infections, respectively. Although much progress has been made toward developing effective yet nontoxic antifungal therapies [7–10], the goal of finding an ideal agent for severe candidal infections is far from being achieved, and the morbidity and mortality rates associated with infections remain high.

 Candida spp. now rank fourth on the list of microbes most frequently isolated from blood cultures [11], and mortality rates from systemic candidiasis can be as high as 50%. Since early diagnosis of invasive infection is difficult, strategies for prevention would seem to be more attractive. Antifungal prophylaxis has been effective in reducing *Candida* infection in neutropenic patients with hematologic malignancies; however, there has been no proven benefit regarding the reduction in the overall mortality [12]. The beneficial effect of immunomodulators in clinical and preclinical models of the infection [13–15] suggests that maneuvers aimed at restoring the host defective antifungal immune responsiveness can be exploited in adjunctive antifungal therapies [10,15–18]. However, an essential point to be considered is the acknowledgment of the redundancy of the immune responses with respect to protection. This means that antigen targets, effector cells, and molecules must be precisely defined in terms of protection. Therefore, understanding the components of this host response at a basic level is likely to lead to a better understanding of the pathogenesis of *C. albicans* infections and diseases in humans, and result in optimization of preventive and therapeutic antifungal strategies.

 This chapter will review our knowledge of the nature of protective and nonprotective effector mechanisms in primary and acquired resistance to *C. albicans* infections, the role of the innate and adaptive immune systems, and their reciprocal regulation. The intention is to put forward a unitary view for cell-mediated immunity in anticandidal resistance without diminishing, but rather complementing, the roles that the different components of the innate and adaptive immunity play in specific forms of candidiasis.

II. PROTECTIVE AND NONPROTECTIVE IMMUNITY TO *C. ALBICANS*

A. Correlates of Protection in Humans

As a commensal, *C. albicans* is endowed with the ability to elude the host's immunological surveillance, thus allowing its persistence on mucosal surfaces. Antigens that encounter the intestinal immune system can initiate two types of immune response leading to the induction of immunity or tolerance [19]. In the case of *C. albicans*, underlying acquired immunity to the fungus, such as the expression of a positive delayed-type hypersensitivity (DTH), is demonstrable in adult immunocompetent individuals and is presumed to prevent mucosal colonization from progression to symptomatic infection [20–25]. Human lymphocytes show strong proliferative responses after stimulation with *Candida* antigens [26] and produce a number of different cytokines [27]. Due to their action on circulating leukocytes, the cytokines produced by *Candida*-specific T-cells may be instrumental in mobilizing and activating anticandidal effectors [6,28,29], thus providing prompt and effective control of infectivity once *Candida* has established itself (in mucosal tissue) or spread (to internal organs) via a combination of virulence factors and impairments to the host innate defense mechanisms [30–32]. The former include factors related to species and strains, adherence, dimorphism, molecular mimicry of mammalian adhesion molecules, toxin and enzyme production, and cell surface composition [30,31]. Among the latter, decreased neutrophil number and functions that work jointly with the breakdown of anatomic defenses are critical in systemic infections [32]. Therefore, host resistance to *Candida albicans* appears to be dependent on the induction of cellular immunity, mediated by T-lymphocytes and nonspecific effector cells, mainly granulocytes and tissue macrophages [6,33,34]. Not surprisingly, DTH to *Candida* antigens is a standard in vivo assay for T-cell-mediated immunity in humans. The expression of this cutaneous reactivity is often defective in immunocompromised individuals as well as in patients with recurrent or persistent infection with the yeast, or immunopathology associated with it [35]. In most cases of chronic mucocutaneous candidiasis [6,36] and recurrent vaginal candidiasis [37–39], antigen-specific T-cell anergy and altered production of cytokines are observed.

While representing a major factor predisposing to mucocutaneous candidiasis, downregulation of cell-mediated immunity may be a consequence of *C. albicans* infection, as demonstrated by much experimental and clinical evidence [40–43]. Polysaccharide antigens of *C. albicans* generate a complex series of interactions that result in the suppression of both T- and B-cell responses to the homologous antigen [44–46]. These effects are mediated by antigen-nonspecific inhibitory factors that are able to block the synthesis of cytokines by human monocytes [47] and T-cells [48]. In atopic subjects [49–55], in women with recurrent vaginal candidiasis [56,57], and in patients with chronic mucocutaneous candidiasis [36], the suppressed delayed skin responses to the fungus show an inverse relationship, with elevated levels of *Candida*-specific IgE, IgA, and IgG. These data argue for the reciprocal regulation of humoral and cell-mediated immunity to the fungus, and also point to a pathogenetic role of *Candida*-specific IgE in allergy [49–52], in asthma [53–55], and in vulvovaginal candidiasis [56,57].

If the ability of *C. albicans* to establish a disseminated infection involves a multiplicity of yeast and host-related factors, including neutropenia, its ability to persist in infected tissues may primarily involve downregulation of host cell-mediated (i.e., phagocytic-dependent) immunity [41]. Immediate hypersensitivity responses, including production of high antibody and IgE levels, will then become prominent characteristics of the response to *Candida*, with possible implications for immunopathology and autoimmunity [58]. Therefore, the yeast/host relationship in saprophytism and disease presents features compatible with T-helper (Th)1/Th2 paradigm of acquired immunity [59], as further discussed later. A Th1-type reactivity may characterize the saprophytic yeast carriage of healthy subjects, whereas Th2-type responses would be mostly associated with susceptibility to recurrent or persistent infection and allergy. Indeed, acquired immunity to the fungus correlates with the expression of local or peripheral anticandidal Th1 reactivity [27,60,61]. Th1 reactivity is downregulated in symptomatic infections where a biased Th2 response to the fungus is observed [62–66]. In addition, a pathogenic role of *Candida*-specific Th2-type responses may be operative in asthma [53–55], recurrent vaginitis [56,57], atopic dermatitis [49–52,67,68], and immunopathology associated with several diseases states [69], including some of the unusual skin disorders that are by contrast common in AIDS [70].

It is important to note that experimental [71,72] and clinical evidence [73,74] indicate that differences may exist in cell and cytokine requirement for antifungal resistance at mucosal or systemic levels, a notion exemplified by the long-recognized association between systemic candidiasis and qualitative [75] and quantitative [76] defects of neutrophils, and between chronic mucosal infections and T-cell abnormalities [6,20,35]. The concept of a reciprocal regulation between the phagocyte system and the T-cell compartment may provide a unifying thread between the systemic immune responses and events occurring on the mucosal surface. As observed previously [6,22], this notion emphasizes that the anticandidal responses characterized in systemic and mucosal infections are not unique to either condition. It is very likely that the relative contribution of each type of immune effector mechanism may vary depending on site [77] and duration of infection, eventually leading to the compartmentalization of host antifungal resistance.

B. Correlates of Protection in Mice

The clinical circumstances in which recurrent *Candida* infections occur definitely suggest an association with impaired cell-mediated immunity, yet only in the last decade has direct evidence of this relationship been obtained in experimental models of mucosal or systemic candidasis [6,20,21]. Studies in mice have clearly defined the role played by T-cells in infections caused by *C. albicans* [78–86] and have shed new light on the interplay between the different arms of the immune system in the control of infections [87,88].

In conditions where the animals are overwhelmed by a high dose of intravenous virulent yeast, mice of different inbred strains die within a few days. Therefore, any contribution of T-cell-mediated immunity could hardly be assessed in this setting. Thus, resistance of native mice to the acute systemic infection mostly reflects the activity of natural immune mechanisms (see below). However, upon intravenous challenge with sublethal fungal doses [89,90], or with low-virulence yeast cells [91], or after gastrointestinal infection [92,93], an enduring infection can be obtained that enables evaluation of the contribution of both nonspecific and specific effector immune mechanisms in the control of the infection and its pathology. Although with some controversial results [94–97], earlier studies indicated that mice rendered functionally T-cell deficient by various maneuvers [78,85,98–100], had increased susceptibility to *C. albicans* infections, particularly at the mucosal level. The clearance of mucosal candidiasis is impaired, and antigen-specific lymphoproliferation and footpad responses cannot be demonstrated in T-deficient mice. However, a combination of defects in both phagocytic cell function and T-cell-mediated immune responses predisposes mice to severe mucosal and systemic candidiasis of endogenous origin [101], thus emphasizing the interdependency between the innate and adaptive immune systems in candidiasis. Corroborative evidence was provided by the transfer of a *Candida*-specific T-cell line into sublethally irradiated mice [80]; the T-cell line confers resistance to primary systemic challenge. Further evidence indicated that both CD4+ and CD8+ cells contribute to the host response against the infection [79,81,82,85]. More recent evidence indicates that *C. albicans* induces selective expansion of T-lymphocytes bearing a particular T-cell receptor (TCR) Vβ specificity, raising the possibility that molecules with superantigen activity may be present on the fungus [102]. Differential expansion of T-lymphocytes of a particular Vβ specificity is also associated with a pattern of susceptibility and resistance to *C. albicans* infection (unpublished observations), and treatment with staphylococcal enterotoxin-B confers protection against systemic candidiasis [102]. These results point to a role for CD4+ TCR Vβ8+ cell anergy in the modulation of acquired immunity to *Candida*. Indeed, anergy reversal by treatment with IL-2 is associated with high susceptibility to the infection [103].

Beside T-lymphocytes bearing the αβ T-cell receptor (TCR), TCR γδ+ T-cells may play an important role in antifungal host defense by acting as a first line of defense at the mucosal level. Polyclonal γδ T-cells were found to be expanded in the gastric mucosa of mice orally infected with *C. albicans*, and TCR δ-chain-deficient mice were found to be susceptible to mucosal orogastric candidiasis, although to a lesser extent than αβ T-cell-deficient mice [104]. The general thrust of the above results indicates that the major function of the T-lymphocytes in *Candida* infections is the production of cytokines with activating and deactivating signals for fungicidal effector phagocytes.

The discovery that subsets of CD4+ Th cells could be distinguished according to their ability to produce discrete patterns of cytokines [105,106], offered the conceptual framework of adaptive immunity in *C. albicans* [20–25]. Th1 cells, by their production of interferon-gamma (IFN-γ) and lymphotoxin, are responsible for directing cell-mediated immune responses leading to the eradication of intracellular pathogens, but they may also cause immunopathology and organ-specific autoimmune diseases if dysregulated. Th2 cells, by producing interleukin (IL)-4 and and IL-5, known to activate mast cells and eosinophils, have been strongly implicated in atopy and humoral immunity. Helper T-cells and clones that produce both Th1- and Th2-type cytokines (termed Th0) have also been described [106]. It is possible that Th0 cells are involved in eliminating many pathogens, where a balance of both regulated cell-mediated immunity and an appropriate humoral response will eradicate an invading pathogen with minimum pathology [106]. However, chronic conditions may result in polarized Th1- and Th2-type responses which might not only be mutually exclusive [106] because of the counterregulatory effects of the cytokines brought about by the reciprocal subsets, but also pathogenic. Many studies now suggest that alternative regulatory populations exist which are somehow associated with, but distinct from, Th2 cells [107]. Regulatory T-cells, including Th3 [108] and Tr1 [109], can inhibit cell-mediated immune responses and/or inflammatory pathologies, and are involved in the maintenance of self-tolerance [110]. Much of this suppression has been attributed to transforming growth factor β (TGF-β) [107–110].

The Th1/Th2 paradigm, although somewhat simplistic, has allowed us a better understanding of the reciprocal regulation between the innate and adaptive immune systems in candidiasis. The relative importance of the different Th cells and of effector phagocytes to the outcome of the infection depends on numerous factors, including localization [92,93] and virulence [91,111,112] of the primary infection, early cytokine response of the host to the fungus [103,113–123], and predominant type of Th cells activated by challenge, namely host-protective Th1 or nonprotective Th2 cells [22–25,124]. The qualitative pattern of cytokines secreted by T-cells will ultimately determine the efficacy of the effector response mediated by phagocytes.

In evaluating the relative contributions of various Th-dependent effector mechanisms in murine candidiasis, it is important to distinguish between those responsible for recovery from the primary infection and those that mediate protection upon reinfection.

1. Primary Resistance

Studies done with two agerminative, avirulent variants of *C. albicans*, one of which induces protective immunity and the other nonprotective immunity, indicate that CD4$^+$ cells specific for the latter were found to inhibit the adoptive transfer of yeast-specific DTH when admixed with CD4$^+$ cells primed to the protective variant [112]. This confirmed the notion that DTH suppressor T-cells are induced by *C. albicans* constituents in mice [40]. Moreover, in studying the immune response to *C. albicans* in genetically distinct mice, Ashman [34] reported circumstantial evidence that active suppression of protective immunity might occur in inbred strains of mice that are classified as genetically susceptible to candidiasis and characterized by high antibody levels and poor DTH responses. More recent studies indicate that susceptibility and resistance to primary disseminated infection with low-virulence *C. albicans* vary among inbred strains of mice of different MHC haplotypes [84,88,122,125] (Table 1). C57BL/6 and BALB/c mice are resistant to primary infection and acquire resistance to reinfection; C3H/HeJ and CBA/J mice show moderate resistance to primary infection; and DBA/2, 129/Sv, SJL, and FVB mice are highly susceptible to primary infection.

Resistance or susceptibility to infection does not correlate with the antifungal effector functions of phagocytic cells, as no significant differences are observed among strains (Fig. 1). Likewise, resistance or susceptibility to infection does not correlate with the number of interferon (IFN)-γ-producing cells that are observed not only within strains, but also between resistant and susceptible strains, being actually higher in susceptible strains (Fig. 2). Thus, neither a defective antifungal effector function nor the extent of IFN-γ production correlates with resistance or susceptibility to the infection. However, the activation of protective Th1 and nonprotective Th2 CD4$^+$ cells is observed in these different strains, as revealed by the inverse correlation between IFN-γ and IL-4 production in the different strains (Fig. 3). One week after infection, CD4$^+$ cells from resistant strains produce more IFN-γ than susceptible strains, and the reverse is true for IL-4. Further studies reveal that induction of Th1 response requires IFN-γ [113], IL-12 [118,119], and CD4$^+$ cells [81] in the afferent induction, and involves IFN-γ-releasing CD4$^+$ cells expressing the β2 chain of the IL-12 receptor (IL-12R) [122], and CD8$^+$ [82] cells in the effector phase. At the same time, CD4$^+$ cells and IFN-γ are needed for elicitation of yeast-specific DTH [81],

Table 1 Strain Distribution of Resistance to Systemic Infection with Low-Virulence *Candida albicans*

| Strain | Haplotype | Resistance to infection | | Th status |
		MST	D/T	
C57BL/6	b	>60	0/22	Th1
BALB/c	d	>60	4/22	Th1
C3H/HeJ	k	28	12/20	Th1 + Th2
CBA/J	k	33	14/20	Th1 + Th2
DBA/2	d	7	20/20	Th2
129/Sv	b	14	12/12	Th2
SJL	s	14	8/8	Th2
FVB	q	8	8/8	Th2

Mice were intravenously infected from low-virulence *C. albicans* and assessed for parameters of infection (MST, median survival time in days, and D/T, number of dead animals over total animals infected) and Th immunity (antigen-specific cytokine and antibody production).

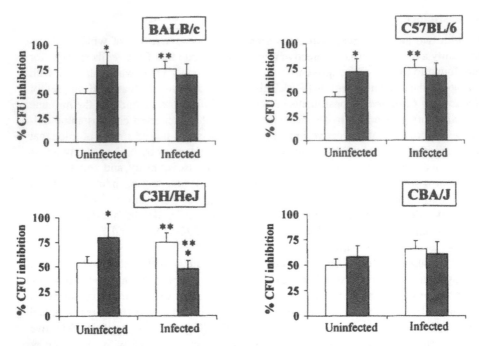

Figure 1 Candidacidal activity of effector phagocytes (splenic-adherent macrophages) from mice with different levels of resistance (BALB/c and C57BL/6 having greater resistance than C3H/HeJ and CBA/J) to primary infection with low-virulence *Candida albicans*. Splenic adherent cells were taken during the first week of the infection and assessed for candidacidal activity in vitro in the presence (■) or not (□) of IFN-γ. *P < .05 IFN-γ-exposed vs. non-IFN-γ-exposed cells. **P < .05 infected vs. uninfected mice.

and antigen-specific CD8[+] cells, with the capacity to lyse yeast-primed macrophages, can be detected in vitro [86]. Nonspecifically activated macrophages and granulocytes with microbicidal properties act as the final effector cells of this response [111,126–128].

The induction of CD4[+] Th1 cells requires the presence of an activated innate immune system, such as the ability of phagocytic cells to inhibit fungal growth and to release cytokines [88,122]. However, an activated innate immune system is not sufficient per se to induce the activation of CD4[+] Th1 cells [122]. Thus, proper integration between innate and antigen-specific immunity is required for efficient control of a subacute *C. albicans* infection. In contrast, high-level production of Th2 cytokines, such as IL-4 and IL-10, are observed early in infection and until the death of mice not surviving the primary infection. CD4[+] cells from these mice release Th2 cytokines in vitro in response to specific antigens [91].

In yeast-infected DBA/2 mice, neutralization of IL-10 early in infection restores "natural" antifungal effector functions, allows the activation of CD4[+] Th1 cells, and renders mice resistant to the primary and secondary infections [115]. These data indicate that IL-10 may have a pathogenic role in primary candidiasis. As similar results were obtained upon IL-4 neutralization [114], the general conclusion from these studies is that resistance to *C. albicans* infection is determined by nonspecific phagocytic mechanisms, the activity of which is augmented or reinforced by Th1 cytokines and impaired by Th2 cytokines. The results also suggest that susceptibility to candidal infection is linked to the occurrence of a dominant and suppressive Th2 response, rather than an intrinsically insufficient Th1 response. Such a conclusion may have important implications for immunotherapy of candidal infections (see below).

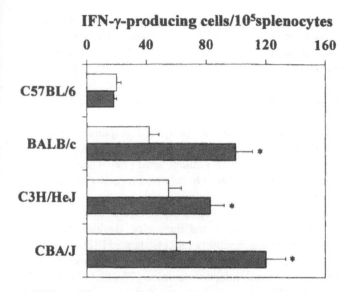

Figure 2 Number of IFN-γ-producing cells in mice with different levels of resistance (BALB/c and C57BL/6 having greater resistance than C3H/HeJ and CBA/J) to primary infection with low-virulence *Candida albicans*. The number of cytokine-producing cells in splenocytes taken during the first week of infection were determined by Elispot assay. (□) Uninfected and (■) infected mice. *$P < .05$, infected vs. noninfected mice.

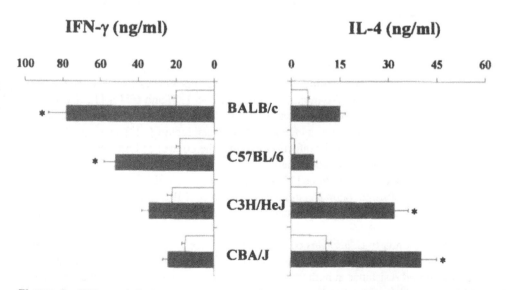

Figure 3 IFN-γ and IL-4 production by CD4+ T-cells from mice with different levels of resistance (BALB/c and C57BL/6 having greater resistance than C3H/HeJ and CBA/J) to primary infection with low-virulence *Candida albicans*. At 1 week of infection, CD4+ T-cells, purified from spleens, were cultured in vitro with irradiated splenocytes and heat-inactivated *Candida* cells for 48 hr before determination of the cytokine contents in culture supernatants by specific ELISA. *$P < .005$, infected (■) vs. noninfected (□) mice.

The systemic activation of host-protective CD4$^+$ Th1 cells, detected in mice with mucosal yeast colonization of limited grade and duration [92,93,129,130], is of interest. This condition, which simulates saprophytic yeast carriage by healthy humans with peripheral Th1-type reactivity [26], is associated with increased resistance to systemic reinfection [92,131]. Therefore, the expression of Th1-mediated resistance at both mucosal and nonmucosal effector sites may involve common mechanisms in humans who develop systemic immunity as a consequence of their mucosal colonization. This could explain why the absence of CD4$^+$ cells is consistently detrimental in mucosal colonization with the fungus (where a strong Th1 response typically develops). Paradoxically, ablation of T-cells or CD4$^+$ cells may be beneficial in conditions of vigorous Th2 cell activation, such as systemic infection with virulent yeast [79]. Therefore, T-cell or CD4$^+$ cell deficiency can be expected to affect resistance to infection largely conditioned by the Th1/Th2 ratio in the elicited response.

Recent studies on the genetic control of the host susceptibility and/or resistance to C. albicans infection highlight the role of antifungal cell-mediated immunity. Besides the fifth component of complement as a major influence on susceptibility and resistance [132], Ashman has recently provided evidence of the existence of at least two Mendelian-type, resistance genes that control the host response to systemic C. albicans infection in mice [77,133,134]. These genes appear to affect distinct parameters of infection, such as tissue destruction and kidney colonization. Allocation of presumptive "resistant" or "susceptible" alleles of these genes among various inbred strains correlates extremely well with the various measures of infection. One important implication of this finding is that effector mechanisms of antifungal resistance would be both genetically determined and site specific. This would be in line with the experimental and clinical evidence that induction of antifungal cell-mediated immunity is highly compartmentalized [20,73]. Indeed, the susceptibility of mouse strains to systemic candidiasis does not correlate with susceptibility to mucosal, gastrointestinal, or vaginal candidiasis [135]. Likewise, women who are susceptible to oral and esophageal candidiasis are generally not more susceptible to vaginal candidiasis [74].

2. Acquired Resistance

In models of sustained protective anticandidal responses, such as mucosal exposure to the fungus or vaccine-induced resistance, CD4$^+$ T-lymphocytes [85,129,136,137] and IFN-γ [138], but not B-lymphocytes [139], are required for the expression of acquired resistance to the fungus. Th1 reactivity is associated with acquired resistance to either disseminated [122] or vaginal [136] infection. Acquired resistance to the latter infection is under the control of host genetic factors, to such an extent that protection after subcutaneous immunization with C. albicans extracts is achieved in BALB/c, but not C57BL/6 mice [136]. Interestingly, C57BL/6 mice demonstrate a higher level of innate resistance to vaginal candidiasis than do BALB/c mice, a finding pointing to the complex levels of crossregulation between the innate and adaptive immune system in vaginal candidiasis.

That acquired immunity to C. albicans correlates with the expression of local or peripheral Th1 reactivity has also been confirmed by studies in healthy adult humans [27,60,61]. Therefore, the expression of acquired resistance at both mucosal and nonmucosal effector sites may involve common mechanisms in humans who develop systemic immunity as a consequence of their mucosal colonization. These mechanisms probably operate through the action of Th1 lymphocytes, the cytokines they release, and nonspecifically activated effector cells. The unitary role for cell-mediated immunity in anticandidal resistance would imply thatg, for example, eradication of an established infection, whether superficial or deep seated, could be accelerated by recruitment of antigen-specific cells which provide the high cytokine concentrations needed locally for optimal activation of the "natural" anticandidal effectors, granulocytes and monocytes/macrophages [140,141]. A mechanism of this type operates in DTH, where injection of antigen into the skin leads to activation of memory CD4$^+$ cells that act extravascularly to recruit circulating nonspecific effector leukocytes at the site of antigen challenge. A unitary role of cell-mediated immunity in candidiasis also implies that interference with the development or function of Th1 reactivity may greatly affect susceptibility to chronic or recurrent infection.

Recent findings have thrown light on the cytokine-mediated regulation of acquired resistance to *C. albicans*. Experiments in IL-4 knockout mice revealed that these mice, although capable of mounting anticandidal Th1 responses upon primary infection, are nevertheless unable to show resistance to reinfection [142]. Memory Th1 resistance is restored upon exogenous IL-4 administration in IL-4-deficient mice and impaired upon IL-4 neutralization in IL-4-sufficient mice, a finding suggesting the requirement for this cytokine in the expression of acquired immunity to the fungus. Experiments of cytokine neutralization performed in vaccine-induced resistant mice indicated that IL-10 and TGF-β are also required for the expression of acquired resistance (unpublished data). Thus, cytokines with inhibitory activity in developing Th1 cell responses are indeed required for sustained Th1 reactivity to the fungus. The possible regulatory cells and mechanisms underlying the expression of immunological memory to *C. albicans* are not understood for the time being. However, it is intriguing that different Th cells and cytokines are produced in response to different fungal antigens [143]. It is possible that the temporally distinct expressions of different antigens during fungal growth in vivo have a role in the expression of immunological memory responses to *C. albicans*, through the induction of regulatory T-cells.

III. REGULATION OF Th1- AND Th2-DEPENDENT IMMUNITY

A. The Role of Cytokines

Th1 and Th2 CD4$^+$ T-cells develop from a common naive CD4$^+$ T-cell precursor and several parameters have been reported to influence the pathway of differentiation of CD4$^+$ T-cell precursors [22,124]. Among these, cytokines appear to play a major role, acting not only as modulators of antifungal effector functions but also as key regulators in the development of the different Th subsets from precursor Th cells. Studies in mice have shown that development of protective anticandidal Th1 responses requires the concerted actions of several cytokines such as IFN-γ [113,138], TGF-β (117), IL-6 [120], tumor necrosis factor (TNF)-α [121], and IL-12 [118,119], in the relative absence of inhibitory Th2 cytokines, such as IL-4 and IL-10, which inhibit development of Th1 responses [123]. Early in infection, neutralization of Th1 cytokines (IFN-γ and IL-12) leads to the onset of Th2 rather than Th1 responses, while neutralization of Th2 cytokines (IL-4 and IL-10) allows for the development of Th1 rather than Th2 cell responses [113–115,119]. However, in highly susceptible mice, exogenous IL-12 does not exert beneficial effects on the course and outcome of disseminated and mucosal infections [119]. Moreover, administration of IL-4 fails to convert an already established Th1 response into a Th2 response [123,144], and late IL-4 depletion exacerbates chronic infection [144]. These findings indicate the existence of complex immunoregulatory circuits underlying cytokine activity in mice with candidiasis.

Studies performed in genetically modified mice, including cytokine-deficient mice, have furthered our understanding of cytokine-mediated regulation of Th cell development and effector

Table 2 Innate and Adaptive Immunity to *Candida albicans* Infection in Cytokine-Deficient Mice

Mice	Antifungal effector activity	Resistance to:		Th status	Reference
		Primary infection	Secondary infection		
TNF/LT-$\alpha^{-/-}$	↓	↓	↓	Th2	121
IL-6$^{-/-}$	↓	↓	↓	Th2	120
Caspase-1$^{-/-}$	↑	=	↓	Th2	unpublished
IFN-γR$^{-/-}$	↑	↑	↓	Th2	138
IL-4$^{-/-}$	↑	↑	↓	Th2	142
IL-12p40$^{-/-}$	↑	↑	↓	Th2	151
IL-10$^{-/-}$	↑	↑	↑	Th1	187

For details of the innate and adaptive Th immunity measurements, see the references. ↓, ↑, =, means decreased, increased, or unchanged compared to wild-type mice.

functions in candidiasis and have revealed complex levels of immunoregulation that were previously unappreciated (Table 2).

Caspase-1-deficient mice have revealed the important role of IL-18 in sustaining the expression of Th1 reactivity to the fungus. In the absence of IL-18, production of IFN-γ and IL-12 is not as impaired early in infection, as it is later on. Indeed, caspase-1 deficiency does not affect susceptibility to the acute systemic infection, but impairs the Th1-mediated resistance to reinfection, which is, however, restored by exogenous administration of IL-18 (unpublished observation). In contrast, TNF-α and IL-6 deficiencies render mice highly susceptible to primary *C. albicans* infections, by impairing the development of successful innate and acquired immunity [120,121]. Both qualitative and quantitative changes in antifungal effector functions of phagocytic cells, particularly neutrophils, and the activation of nonprotective Th2 responses are observed in these mice [120,121].

One likely mechanism of Th2 development in mice with candidiasis is the increased fungal burden in the organs, as the production of IL-4, and hence Th2 activation, is strictly dependent on the fungal dose [144]. Therefore, the defective antifungal effector functions of neutrophils, and in part of macrophages, and consequently, the unopposed fungal growth, may account for the failure of TNF/LT-α- and IL-6-deficient mice to mount a protective Th1 response. However, TNF-α was also found to be required for optimal expression of costimulatory molecules on phagocytic cells and for IL-12 responsiveness in CD4^{+} Th1 cells [121], whereas decreased production of IL-12 and increased production of IL-10 occurred in the absence of IL-6 [120]. A reduced production of IL-6 is observed in TNF/LT-α-deficient mice infected with the fungus [121], which may suggest that downstream production of IL-6 could also contribute to the absolute requirement of TNF-α in the development of protective anticandidal Th1 reactivity.

Studies in mice with IFN-γ deficiency have revealed that IFN-γ is not necessary for induction and expression of anticandidal Th2 responses [138], a finding in line with those obtained in mice with a disrupted IFN-γ gene and infected with *C. albicans* under conditions of Th2 cell activation [145,146]. Instead, the high susceptibility of IFN-γ-deficient mice to *C. albicans* infection correlates with the failure to mount protective anticandidal Th1 responses due to an impaired IL-12 responsiveness rather than IL-12 production [138]. The activities of IL-12 are mediated through a high-affinity receptor composed of two subunits, designated β1 and β2. The latter is more restricted in its distribution, and regulation of its expression is like a central mechanism by which IL-12 responsiveness is controlled [147]. Indeed, the IL-12Rβ2

subunit expression on activated CD4$^+$ Th1 cells is known to correspond to loss of IL-12 responsiveness and represents an early step in the commitment of T-cells to the Th2 pathway [148].
Because IFN-γ is necessary to override the IL-4-induced inhibition of IL-12β2 receptor expression in Th1 cells [148], it is likely that in murine candidiasis the Th1-promoting activity of IFN-
γ may rely on its ability to maintain IL-12 responsiveness on CD4$^+$ cells, by sustaining the IL-
12Rβ2 expression. As the production of IFN-γ is impaired in TNF/LT-α-deficient mice, this
may account for the impaired IL-12 responsiveness also observed in these mice [121].

The failure of IL-4-deficient mice to amount protective Th1 memory responses correlates
with a defective production of IL-12, a finding in line with the observation that exposure to IL-
12 restores IFN-γ production by CD4$^+$ T-cells from IL-4-deficient mice [149]. Although IL-4
induces anticandidal Th2 development [114] and exogenous IL-4 exacerbates candidiasis in
mice [123], the results obtained in IL-4-deficient mice reveal that endogenous IL-4 may also
participate in Th1 development by priming neutrophils for IL-12 synthesis [142]. Interestingly,
priming with IL-4 also resulted in the release of high levels of IL-6, which is known to regulate
IL-4 receptors on murine myeloid progenitor cells [150]. Therefore, it appears that a positive
amplification loop exists between IL-6 and IL-4 at the level of neutrophil response, which may
be one possible mechanism underlying the beneficial effect of IL-6 in mice with candidiasis
[120].

Defective production of IL-12 accounts for the inability of IL-12-deficient mice to develop
acquired protective Th1 immunity upon *C. albicans* infection. Importantly, the failure to develop
anticandidal Th1 responses correlated with a defective IL-10, rather than IFN-γ production, a
finding that points to the existence of a positive regulatory loop between IL-12 and IL-10 in
C. albicans infection. Thus, although IL-10-deficient mice develop an efficient Th1-mediated
resistance, IL-10 may also be required for optimal costimulation of IL-12-dependent CD4$^+$ Th1
development [151].

Altogether, these data suggest that (1) the cytokine requirement for the expression of
innate antifungal defense may be different from that required for the expression of Th1-mediated
protection; (2) a single cytokine may participate in both the innate and the adaptive phase of
the response to the fungus, sometimes exhibiting opposite effects; and (3) a hierarchic pattern
exists of cytokine-mediated regulation of antifungal Th cell development and effector function.
Early in infection, production of some proinflammatory cytokines (TNF-α and IL-6) appears
to be essential for the successful control of infection and the resulting protective Th1-dependent
immunity. Both IL-12 production and IL-12 responsiveness are required for the development
of Th1 cell responses, which are maintained in the presence of physiological levels of IL-4, IL-
10, and IL-18. Thus, a finely regulated balance of directive cytokines, such as IL-4, IL-10, IL-
12, and IL-18, rather than the relative absence of opposing cytokines, appears to be required
for optimal development and maintenance of Th1 reactivity in mice with candidiasis [152].

One further lesson from genetically disrupted mice concerns the complex level of crossregulation between the innate and adaptive immune systems to *C. albicans*. A defective innate
antifungal effector function predisposes mice to susceptibility to infection, eventually biasing
the adaptive specific response toward the Th2 pathway. In contrast, an activated innate immune
system correlates with resistance to the infection, which may or may not be associated with the
activation of antigen-specific Th1 cells in the adaptive compartment. Thus, a two-stage control
appears to be at work in *C. albicans* infection (Fig. 4). However, rather than behaving as two
separate entities, the two systems appear to be reciprocally regulated, and to work as an integrated
system, as will be discussed further in the next section.

B. The Role of the Innate Immune System

Innate and acquired cell-mediated immunity have been acknowledged as the primary mediators
of host resistance to *C. albicans* [6,23]. The observation that invasive candidiasis occurs in

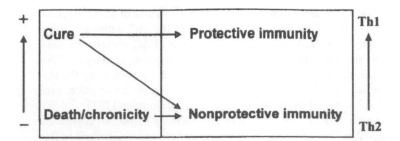

Figure 4 The two-stage control of *Candida albicans* infection. Studies done with genetically disrupted mice of different strains indicate that the level of activation of the innate immune system (indicated by + and − on the left) may efficiently oppose infectivity in the early stage of infection but also determines the quality of the subsequent adaptive Th immunity (on the right).

concomitance with defects in neutrophil number and functions [76], together with the detection of cells and mediators of the innate immune system with antifungal effector activities [6], has led to the central dogma of resistance to candidiasis, i.e., that resistance to invasive candidiasis is mediated by the innate immune system, which includes circulating neutrophils and monocytes, tissue macrophages, natural killer (NK) cells, and soluble molecules, including opsonins (specific antibodies, and components of classical and alternative complement pathways, see below), mannose-binding proteins, and defensins [6,76,87,153–158]. Data suggest, however, that platelets [159] endothelial [160] and epithelial [161] cells may serve as nonconventional antifungal effectors as well. While the effectiveness of this effector system is undoubtedly acknowledged, recent evidence would suggest a more complex and important role of the innate immune system. The data being gathered should not overshadow the traditional effector functions of neutrophils and macrophages in candidiasis, but rather magnify the importance of these cells in the overall antifungal immune resistance.

For many years, innate immunity has been considered as a separate entity from the adaptive immune response to pathogens and has been regarded as being of secondary importance in the hierarchy of immune functions. Evidence is accumulating on the essential role of the innate immunity in orchestrating the subsequent adaptive immunity to pathogens [162,163]. Through the involvement of different pattern recognition receptors, cells of the innate immune system not only discriminate between different forms of the fungus, but also contribute to discrimination between self and pathogens at the level of the adaptive Th immunity. In the vertebrate host, innate and adaptive immune responses are now considered to be integrated as a single immune system, with the innate response preceding, and being necessary for, the adaptive immune response. In the following section, we will first discuss the role of the innate immunity as an antifungal effector system, and then look at recent advances in revealing its function as both guide and instructor of the adaptive immunity.

1. The Innate Antifungal Effector Function

Fungal cells possess a variety of cell surface analogs of the integrin family and complement receptors [164–167] and adhesins [168,169], which may facilitate attachment (opsonization) and phagocytosis by phagocytic cells [170].

In accordance with human studies [171], polymorphonuclear neutrophils, more than monocyte/macrophages, represent the first line of defense in the control of experimental *C.*

albicans infections [6]. The evidence recently accumulating uniformly indicates that defects in neutrophil number or function increase susceptibility to disseminated [101,120,172–176], but also mucosal [177], *C. albicans* infection. Instead, the suppression of the antifungal activity of neutrophils by progesterone was taken to indicate a possible mechanism underlying the vulnerability of pregnant women to vaginal candidiasis [178]. In contrast, elimination of splenic macrophages, either increased [140] or decreased [179] susceptibility to experimental disseminated candidiasis. It has even been suggested that macrophages play a major role in the pathology of disseminated candidiasis [180]. Therefore, as binding of the fungus to macrophages occurs in vivo [181], macrophages clearly play a complex role in the host response to *Candida*, as suggested [182].

The killing of yeast and hyphal forms of *C. albicans* by activated macrophages and neutrophils is phagocytosis or contact dependent, and involves both oxidative and nonoxidative burst [170,183]. Production of nitric oxide (NO) also contributes to the antifungal activity of these cells [184,185]. Indeed, the ability of *C. albicans* to inhibit NO production by murine macrophages is one possible strategy to evade host defense mechanisms [186]. Moreover, production of IFN-γ-dependent NO mediates antifungal resistance at the mucosal [185,187] and peripheral levels [184,188]. However, due to the pleiotropic activities of NO on cells of the immune system, as well as on *C. albicans* itself [189], the effect of NO production in candidiasis appears to be quite complex.

Interactions of *C. albicans* yeast cells, germ tubes, and hyphae with neutrophils has been described in vitro [190,191]. Phagocytosis and killing of blastospores occur at a much higher rate than for small germ tubes. Longer germ tubes are not phagocytosed [190]. Receptors on neutrophils that are important for uptake of fungi include Fc receptors (FcR), complement receptors (CR1 and CR3), and receptors that interact with carbohydrate [192]. This last group, including scavenger and mannose receptors, is still poorly defined [193]. Effective phagocytosis and killing of *C. albicans* occurs through FcγRI (CD64) and FcαRI (CD89) receptors [194], a finding supporting the notion that opsonization is required for effective killing of the fungus after phagocytosis [195–198]. Phagocytosis of antibody-opsonized yeasts also induces TNF-α release [199], although beta-2 integrins have a major role in the regulation of cytokine gene expression in neutrophils [200]. However, a degree of cooperation between FcR- and CR-dependent processes appears to exist [201]. The beta-2 integrin CD11b/CD18 or CR3 or Mac-1 binds to iC3β-opsonized fungal particles, a finding indicating that candidal cells have receptors for cleavage fragments of C3 [202]. Stimulation of CR3-dependent phagocytosis or degranulation requires the simultaneous ligation of two distinct sites within CR3–one specific for iC3b, and a second specific for the fungal cell wall β-glucan [203].

Initiation of the oxidative metabolic burst and hydrogen peroxide production by neutrophils in response to unopsonized *Candida* blastospores is a property of the *C. albicans* mannan and occurs via a mannose-inhibitable mechanism [204]. Thus, the same mannose-specific mechanism may provide a means by which the fungus can suppress neutrophil activation as well. Opsonized or unopsonized hyphae also induce intracellular events and trigger a respiratory burst in neutrophils [205–207], although with some differences, such as the requirement of intracellular cytosolic calcium in the elicitation of the respiratory burst [208]. This means that if different events mediate neutrophil activation by opsonized and unopsonized hyphae, candidacidal activity in vivo may vary under divergent conditions with specific localized sites of infection. As IFN-γ and TNF-α have disparate effects on the fungicidal activity of neutrophils [209,210], the candidacidal activity in vivo also reflects the local cytokine microenvironment.

Macrophages are recruited in tissues, and yeast and filamentous forms of *Candida* can be detected both intracellularly and extracellularly [211]. Interestingly, the phagocytic and microbicidal activities of macrophages are enhanced by myeloperoxidase released by neutrophils [212], a finding suggesting possible cell cooperation in the activation of macrophages at sites of infection. Phagocytosis of *Candida* is mediated by a combination of lectin-like receptor-, FcR-, and CR3-dependent processes [213–215]. The most efficient uptake is dependent on CR-3-mediated phagocytosis of both serum opsonized and unopsonized yeasts, a process inhibitable by cell wall components of the fungus [213]. However, interaction with macrophage CR-3 leads to suppression of the immune response to the micro-organism [213]. Interestingly, signaling via CR-3 downregulates IL-12 production in response to the fungus *Histoplasma capsulatum* [216]. Through the different receptors, macrophages phagocytose opsonized and unopsonized *Candida* yeasts [156]. This has important consequences because the candidacidal activity and secretory function of macrophages largely depend on the cell source [217,218] and the mode of internalization: uptake of opsonized yeasts results in higher killing activity with increased activation of respiratory burst, compared with unopsonized yeasts [219–221]. Accordingly, the mannose receptor-mediated uptake of unopsonized *Candida* yeasts may lead to phagocyte abuse if not accompanied by the coordinate activation of the cell cytotoxic machinery [222]. Increased expression of the receptor with no induced cytotoxicity is induced upon exposure of macrophages to IL-4 [223]. In contrast, IFN-γ downregulates the expression of macrophage mannose receptors and nevertheless results in effective killing, presumably via increased coupling of the receptor to cytotoxic functions [221]. Moreover, production of some cytokines, such as IL-1β, IL-6, and granulocyte macrophage-colony stimulating factor (GM-CSF), are mediated by the mannose receptor, while some chemokine responses may be mediated by other receptors [224]. It is interesting that ingestion of *C. albicans* downregulates mannose receptor expression on macrophages [225], suggesting that subversion of phagocytic entry can be exploited as evasion strategy by the fungus.

Uptake of *Candida* yeast and filamentous forms occur differently in macrophages. Hyphae are not internalized by a macrophage cell line, although they are capable of inducing production of inflammatory cytokines to a greater extent than phagocytosed yeasts [226]. Instead, uptake of *Candida* yeasts and germ tubes by macrophages occurs through a phagocytic mechanism, requiring intact actin filament, protein kinase C activity, and participation of an nonmannose receptor [227]. Once inside the cells, *Candida* phagosomes rapidly fuse with late endosomes and lysosomes. However, the yeasts may develop germ tubes within phagolysosomes, escaping from them and destroying the macrophages [227]. Thus, the rapid recruitment of late endocytic/lysosomal compartments by *C. albicans* may favor its survival and virulence.

Although freshly isolated phagocytic cells clearly express intrinsic anti-*Candida* activity, numerous studies have demonstrated that expression of the full candidacidal function of both neutrophils and monocyte/macrophages requires activation by CSFs [6,14,228–230]. However, optimal expression of the antifungal effector activity of cells of the innate immune system is attained in association with antigen-driven immune responses [231,232]. This may occur through the release of cytokines with activating (IFN-γ/TNF-α) and deactivating (IL-4/IL-10) signals to effector phagocytes [28,29]. As already pointed out, increasing evidence indicates that effector mechanisms of antifungal resistance are site specific and genetically determined [77]. There is considerable heterogeneity in the effector function of macrophages derived from different anatomical sites [218], which probably reflects the response to local homeostatic mechanisms. For instance, alveolar macrophages produce arachidonic acid metabolism in response to alpha-mannan and beta-glucan constituents of *C. albicans* [217]. In addition, at infection sites, local control of the infection may be exerted by cells other than professional phagocytes, such as epithelial cells in the vagina [161] and, to some extent, endothelial cells [160].

In regard to the antifungal activity of NK cells, neither human [233] nor murine NK [234] cells have direct killing activity toward *C. albicans*, although the fungus may interact with NK cells and affect NK cell function [233]. This interaction, however, may initiate the release of cytokines affecting the activity of phagocytic effector cells [235–239].

Recent studies indicated that dendritic cells (DC) also behave as potent antifungal effector cells in *C. albicans* infection [152,240]. These cells may be of particular importance in candidiasis, considering that the fungus behaves as a commensal as well as a true pathogen of skin and mucosal surfaces [1] known to be highly enriched of DC. DC phagocytose both yeast and hyphal forms of the fungus. Apparently, DC phagocytose more yeasts than do neutrophils. Uptake of yeasts and hyphae occurs through different phagocytic mechanisms and are differently sensitive to inhibition by mannan. Engulfment of yeasts occurs via coiling overlapping phagocytosis, eventually leading to phagolysosome formation, where different stages of progressive yeast degradation are seen. In contrast, internalization of hyphae appears to occur through a more conventional, zipper-type phagocytosis. After phagocytosis of yeasts, DC produce elevated levels of NO and this correlates with the high fungicidal activity against yeasts. In contrast, the production of NO and the fungicidal activity are lower after phagocytosis of hyphae [240].

In vivo studies have provided convincing evidence that the complement system is an important component of innate resistance to fungi, through opsonization and induction of inflammatory response [158]. Mice with congenital deficiencies in C5 production show decreased resistance to disseminated candidiasis, probably due to a failure to mount an adequate inflammatory response [241,242]. *C. albicans* yeasts activate the complement cascade through both the classical and alternative pathways of initiation, eventually leading to deposition of C3 on the cell surface [158]. Indeed, C3 deposition is greatly accelerated through activation of the classical pathway by naturally occurring antimannan IgG antibody. This last observation also points to an important role of antibodies in the protective immune response to the fungus [157,243–253]. The expression of complement receptors is influenced by growth conditions and by morphology, being greater on hyphae than on blastoconidia. Although, the presence of complement receptors on most pathogenic *Candida* spp. suggests a role in pathogenesis [158], data indicating a role for complement receptors in virulence are largely correlative.

Besides the role played in initiation and regulation of complement activation by the fungus [158], inactivation of important virulence factors or augmentation of host immune defenses are possible mechanisms of antibody-mediated protection in candidiasis. For many years the role of antibodies in candidiasis has been neglected, mainly due to the observation that B-cell deficiency does not increase susceptibility to infection [71], and also because experiments with polyclonal sera have produced evidence for and against an important role of antibodies in host defense [157]. The use of monoclonal antibodies, however, has shown that protective and nonprotective antibodies can be detected in the course of *C. albicans* infection [157,245]. Antibodies recognizing a specific epitope within the phosphomannoprotein complex of *C. albicans* protect susceptible mice against disseminated candidiasis [247,248]. The protection afforded by a conjugate vaccine against disseminated and vaginal candidiasis is mediated by the induction of a protective antibody response [249]. Clinical observations suggest that naturally occurring fungicidal antibodies, mimicking the activity of microbial products, are present in humans and may participate in the host defense against mucosal candidiasis [243,250]. Moreover, the outcome of disseminated candidiasis correlates with the evolution over time of a specific antibody response directed against the 47-kDa antigen of HSP90 [251–253]. Sera from patients who had recovered from disseminated candidiasis confer protection upon passive transfer into infected mice [252]. All together, these results indicate that humoral immunity, either independently or through augmentation of complement activity or protective cell mediated immunity, may contribute to host resistance to *C. albicans*.

2. The Instructive Role of the Innate Immune System

Numerous studies of host defense against microbial pathogens have demonstrated that the functioning of effector lymphocytes bearing clonally rearranged receptors is absolutely dependent on signals provided by the innate recognition system. Through recognition of invariant molecular structures shared by large groups of pathogens (also known as PAMPs, pathogen-associated molecular patterns) by a set of germline-encoded receptors (referred to as pattern recognition receptors; PRRs) of host cells, the innate immune system fulfills the requirement of discrimination among different types of pathogens [162,163,254–257]. The products of these recognition events is the release of various antimicrobial peptides in mammals and insects [258,259] as well as the expression of cytokines, chemokines, and costimulatory molecules through which the innate immune system instructs and contributes to the discrimination of pathogens from self in adaptive immunity.

The instructive role of the innate immune system in the adaptive immune responses to *C. albicans* occurs at different levels [88]. The initial handling of the fungal pathogen by cells of the innate immune system plays a major role in determining CD4+ Th development. Indeed, qualitative or quantitative defects of antifungal effector and immunoregulatory functions of phagocytic cells results in the development of anticandidal Th2, rather than Th1, cell responses [124]. As CD4+ Th cell differentiation in vivo was found to be critically affected by the tissue fungal load [144], an important role of phagocytic cells relies on their ability to control the fungal growth through the antifungal effector mechanisms discussed above. In addition to this, the instructive role of the innate immune system in the adaptive immune response to the fungus is operative at the levels of chemokine [260,261] and cytokine production [124,174,175] and expression of costimulatory molecules [121,151,262].

An important immunoregulatory role has recently been attributed to neutrophils and dendritic cells. Neutrophils, more than macrophages, are endowed with the ability to produce directive cytokines, such as IL-10 and IL-12. Most importantly, IL-12 appears to be released in response to a *Candida* strain that initiates Th1 development in vivo, but IL-10 is released in response to a virulent strain [124,174,175]. By producing directive cytokines, such as IL-10 and IL-12, neutrophils influence antifungal Th cell development, as evidenced by the inability of neutropenic mice to mount protective anticandidal Th1 responses. However, Th1-mediated resistance is increased upon IL-12 administration in neutropenic mice or IL-10 neutralization in nonneutropenic mice. Thus, neutrophils, through the differential production of directive cytokines, may directly contribute to discriminative Th responses to virulent and nonvirulent forms of the fungus. Human neutrophils also produce bioactive IL-12 in response to a mannoprotein fraction of *C. albicans*, capable of inducing Th1 cytokine expression in peripheral blood mononuclear cells [263]. Production of IL-12 by neutrophils occurs independently of TNF-α (121) and IFN-γ [138] and is impaired upon iron overload [264], but increases upon in vitro priming with IL-4 [142], through upregulation of IL-4 receptor expression. Thus, the IL-12-promoting activity of IL-4 may account for its requirement in sustaining memory Th1 cell responses to the fungus [142]. Because of the large number of neutrophils present in the blood or inflammatory tissues in infection [124], it is likely that neutrophil production of cytokines may influence the development and/or maintenance of the Th repertoire to *C. albicans*. Interestingly, it has recently been reported that neutrophils quickly release *Candida* antigens upon phagocytosis [265]. More than that, neutrophil granulocyte-committed cells can be driven to acquire DC characteristics [266]. Thus, it is likely that the immunoregulatory role of neutrophils in candidiasis may go beyond their cytokine production, to include signaling through antigen presentation and costimulation. Ultimately, this would be a likely expectation shared with other cells of the innate immune system.

Human studies confirmed the multiple and complex role neutrophils may have in candidiasis. First, risk factors for invasive fungal infections are not the same in all neutropenic patients [267]. Second, chronic systemic candidiasis initiated by neutropenia may persist in spite of normal neutrophil counts and adequate antifungal therapy [268]. Third, some patients, particularly transplant recipients who have adequate or even normal neutrophil counts, may be at high risk for invasive mycoses [269].

Among the important virulence factors of *C. albicans* [30] is its ability to reversibly switch from a unicellular yeast form into various filamentous forms, all of which can be found in tissues [1,270,271]. Although recent studies have clearly shown that the ability to switch from yeast to filamentous form is required for virulence [167,272], whether it is the yeast or the hyphal form that is responsible for pathogenicity is still an open question. One possibility is that the filamentous growth form is required to evade cells of the immune system, whereas the yeast form may be the mode of proliferation in infected tissues. To make it likely, a cell had to exist that finely discriminated between the two forms of the fungus in terms of class of immune response elicited. Accumulating evidence points to a unique role of DC in infections, as they are regarded both as a sentinel for innate recognition and as initiator of Th cell differentiation and functional commitment [273,274].

In candidiasis, this performance required that DC be exquisitely sensitive to the different forms of the fungus, a finding in line with the increasingly recognized importance of pattern recognition receptors in host defense. This was indeed the case. DC are uniquely able to phagocytose both yeasts or hyphae of *C. albicans*. However, the downstream cellular events after internalization of either form of the fungus are clearly different. In vitro, ingestion of yeasts activates DC for IL-12 and NO production and priming of Th1 cells, while ingestion of hyphae inhibits IL-12, NO, and Th1 priming. In vivo, generation of antifungal protective immunity is observed upon injection of DC ex vivo pulsed with *C. albicans* yeasts but not hyphae. The immunization capacity of yeast-pulsed dendritic cells is lost in the absence of IL-12, whereas that of hypha-pulsed dendritic cells is gained in the absence of IL-4 [152,240]. These results indicate that dendritic cells fulfill the requirement of a cell uniquely capable of sensing the virulent and nonvirulent forms of *C. albicans* in terms of type of immune response elicited. Considering that the morphogenesis of *C. albicans* is activated in vivo by a wide range of signals [275] and that human DC also phagocytose *C. albicans* [276] and activate T-cell responses to the fungus [277], DC appear to meet the challenge of Th priming and education in *C. albicans* saprophytism and infections.

IV. THE Th PARADIGM IN CANDIDIASIS

A. The Shortcomings

As different Th cell subsets are endowed with the ability to release a distinct panel of cytokines, capable of activating and deactivating signals to effector phagocytes, the activation of an appropriate Th subset may be instrumental in the generation of a successful immune response to a fungal pathogen. In its basic conception, the paradigm calls for (1) an association between Th1 responses and the onset/maintenance of phagocyte-dependent immunity, critical for opposing infectivity of the commensals or clearing pathogenic fungi from infected tissue; (2) the occurrence of Th2 responses in fungal infections and fungal diseases; and (3) the reciprocal regulation of Th1 and Th2 cells, occurring either directly or through regulatory T cells, resulting in a dynamic balance between these two types of reactivity, which may operate from commensalism to infection and may contribute to the induction and maintenance of protective memory antifungal responses with minimum immunopathology. However, there is something beyond Th1 and Th2:

(1) individual cytokines can produce opposing effects depending on dose and timing of their participation in the immune response; (2) some cytokines exert their effects on both the innate and adaptive immunity; and (3) there are clearly redundant pathways in achieving a Th1 or a Th2 response.

B. The Implications for Therapy

As clinical resistance represents a significant component of the overall drug resistance of the antifungals [9], the first strategy to prevent antifungal drug resistance is to improve the immune functions of the immunocompromised host. Despite antifungals that appear to have excellent activity in vitro against the fungus, the host response is often weakened and failures occur. Although it is possible that problems with in vitro susceptibility testing methods contribute to this discrepancy, and some cases of relapse or failure may be due to compliance issues or pharmacokinetic factors, at least some of the cases simply represent the need for an adequate host response to succeed.

A variety of cytokines and growth factors proved to be beneficial in experimental fungal infections [10,13–18,278]. However, establishing the clinical utility of cytokines as therapy for fungal infections in patients has been difficult. The basic strategies pursued include the increase of number, function, and mobility of phagocytic effector cells, as it is now accepted that cytokines, effector cells [279], and antifungals work synergistically to oppose fungal growth. Recently, however, the Th1/Th2 balance itself was found to be the target of immunotherapy as well [280–283]. The therapeutic efficacy of antifungals in experimental candidiasis is significantly increased by the concomitant inhibition of Th2 cytokines, such as IL-4 and IL-10, more than the addition of Th1 cytokines, such as IL-12 [281,282]. Interestingly, the efficacy of combination therapy with the antifungals and selected cytokines appears to be largely dependent on the host immune reactivity. Thus, the efficacy of the combined treatment with amphotericin-B and IL-4 antagonists was superior in neutropenic, as compared to nonneutropenic, mice [282]. In contrast, the efficacy of combined treatment with fluconazole and IL-12 was superior in nonneutropenic, as compared to neutropenic, mice [282].

V. CONCLUSIONS

A large body of evidence in both preclinical settings and in humans supports the model of specific immunity to *C. albicans* infections as being highly susceptible to cytokine influences and reciprocally regulated by cells of the innate immune system. With the recognition of the reciprocal influences between the innate and the adaptive Th immunity, it appears that an integrated immune response determines the lifelong commensalism of the fungus at the mucosal level, as well as the transition from mucosal saprophyte to pathogen. The Th1/Th2 paradigm, although somewhat simplistic, has provided a better understanding of the reciprocal regulation between these components, which shed new light on previous uncertainties about the distinct roles of these components in resistance to *C. albicans* infection. Although much has still to be learned at the level of the host/pathogen interaction in candidiasis, the new advances may have important implications for pathogenesis and therapy.

First, an important corollary of the paradigm may be the possible combined effects on Th immunity of *Candida* carriage/infection and various disease states. While immune deficiency or dysregulation, resulting in an altered cytokine balance, as may occur in AIDS, can reasonably be expected to increase local infectivity of the fungus, it is even more intriguing that antifungal chemotherapy resolves some of the unusual skin disorders observed in patients with AIDS

[284,285]. Besides, an immunopathologic role for *Candida* has been suggested for atopic dermatitis [49–51], atopy [52], and other conditions, overtly associated [49–57] or not [69] with *Candida*. Thus the Th cell dichotomy to *Candida* may have implications not only for regulation of the balance between commensalism and infection, but may also contribute to the onset or dominance of Th2 response in other disease states.

Second, the Th1/Th2 paradigm appears to be useful in the identification of cytokines suitable for clinical intervention. Indeed, readdressing the Th balance by combination therapy with cytokines or cytokine antagonists represents a promising strategy for preventing and treating superficial and invasive candidiasis. It is our hope that such an integrated view of effector and regulatory mechanisms operating in *C. albicans* infection would expand the possibilities in the search for the cells, cytokines, and molecular pathways that are essential to control fungal infectivity or oppose fungus-associated immunopathology.

ACKNOWLEDGMENTS

Many thanks are due numerous collaborators over the years in different sections of the work described; in particular Drs. Antonella Mencacci and Elio Cenci for their dedicated work in all areas, and Jo-Anne Rowe for invaluable secretarial support. This work was supported by grants from the AIDS 50B.33 Project.

REFERENCES

1. FC Odds. *Candida* and Candidosis 2d ed. London: Baillière-Tindall, 1989, pp 68–92.
2. SP Fisher-Hoch, L Hutwagner. Opportunistic candidiasis. An epidemic of the 1980's. Clin Infect Dis 21:897–904, 1995.
3. MA Pfaller. Nosocomial candidiasis. Emerging species, reservoirs and models of transmission. Clin Infect Dis 22(suppl 2):S89–S94, 1996.
4. WL Wycliffe, RP Wenzel. Nosocomial *Candida*. Infect Dis Clin North Am 11:411–425, 1997.
5. TJ Walsh, JW Hiemenz, EJ Anaissie. Recent progress and current problems in treatment of invasive fungal infections in neutropenic patients. Infect Dis Clin North Am 10:365–400, 1996.
6. RB Ashman, JM Papadimitriou. Production and function of cytokines in natural and acquired immunity to *Candida albicans* infection. Microbiol Rev 59:646–672, 1995.
7. EJ Anaissie, RO Darouiche, D Abi-Said, O Uzun, J Mera, LO Gentry, T Williams, DP Kontoyiannis, CL Karl, GP Bodey. Management of invasive candidal infections: results of a prospective, randomized, multicenter study of fluconazole versus amphotericin B and review of the literature. Clin Infect Dis 23:964–972, 1996.
8. YF Berrouan, LA Herwaldt, MA Pfaller. Trends in antifungal use and epidemiology of nosocomial yeast infections in a university hospital. J Clin Microbiol 37:531–537, 1999.
9. BD Alexander, JR Perfect. Antifungal resistance trends towards the year 2000. Implications for therapy and new approaches. Drugs 54:657–678, 1997.
10. NH Georgopapadakou, TJ Wlash. Antifungal agents: chemotherapeutic targets and immunologic strategies. Antimicrob Agents Chemother 40:279–291, 1996.
11. WR Jarvis. Epidemiology of nosocomial fungal infections, with emphasis on *Candida* species. Clin Infect Dis 20:1526–1530, 1995.
12. A Bohme, M Karthaus, D Hoelzer. Antifungal prophylaxis in neutropenic patients with hematologic malignancies: is there a real benefit? Chemotherapy 45:224–232, 1999.
13. B Kullberg, MTE Vogels, and JWM van der Meer. Immunomodulators in bacterial and fungal infections. A review of their therapeutic potential. Clin Immunother 1:43–55, 1994.
14. BJ Kullberg, EJ Anaissie. Cytokines as therapy for opportunistic fungal infections. Res Immunol 149:478–488, 1998.

15. LJ Rodriguez-Adrian, ML Grazziutti, HJ Rex, EJ Anaissie. The potential role of cytokine for fungal infections in patients with cancer: is recovery from neutropenia all that is needed? Clin Infect Dis 26:1270–1278, 1998.

16. DA Stevens, TJ Walsh, F Bistoni, E Cenci, KV Clemons, G Del Sero, C Fè d'Ostiani, BJ Kullberg, A Mencacci, E Roilides, L Romani. Cytokines and mycoses. Med Mycol 36(S1):174–182, 1998.

17. DA Stevens. Combination immunotherapy and antifungal chemotherapy. Clin Infect Dis 26:1266–1269, 1998.

18. L Polonelli, A Cassone. Novel strategies for treating candidiasis. Curr Opin Infect Dis 12:61–69, 1999.

19. R Duchmann, MF Neurath, KH Meyer zum Büschenfelde. Responses to self and non-self intestinal microflora in health and inflammatory bowel disease. Res Immunol 148:589–494, 1998.

20. PL Fidel Jr, JD Sobel. The role of cell-mediated immunity in candidiasis. Trends Microbiol 6:202–206, 1994.

21. P Puccetti, L Romani, F Bistoni. A T_H1-T_H2-like switch in candidiasis: new perspectives for therapy. Trends Microbiol 3:237–240, 1995.

22. L Romani, P Puccetti, F Bistoni. Biological role of helper T-cell subsets in candidiasis. Chem Immunol 63:113–137, 1996.

23. L Romani, DH Howard. Mechanisms of resistance to fungal infections. Curr Opin Immunol 7:517–523, 1995.

24. L Romani. The T cell response to fungi. Curr Opin Immunol 9:484–490, 1997.

25. L Romani. Immunity to *Candida albicans*: Th1, Th2 cells and beyond. Curr Opin Microbiol 2:363–367, 1999.

26. CM Ausiello, GC Spagnoli, M Boccanera, I Casalinuovo, F Malavasi, CU Casciani, A Cassone. Proliferation of human peripheral blood mononuclear cells induced by *Candida albicans* and its cell wall fractions. J Med Microbiol 22:195–202, 1986.

27. CM Ausiello, F Urbani, S Gessani, GC Spagnoli, MJ Gomez, A Cassone. Cytokine gene expression in human peripheral blood mononuclear cells stimulated by mannoprotein constituents from *Candida albicans*. Infect Immun 61:4105–4111, 1986.

28. JY Djeu. Cytokines and anti-fungal immunity. In: H Friedman, TW Klein, H Yamaguchi, eds. Microbial Infections: Role of Biological Response Modifiers. New York: Plenum Press, 1992, pp 217–223.

29. JY Djeu. Modulators of immune response to fungi. In: JW Murphy, H Friedman, M Bendinelli, eds. Fungal Infections and Immune Response. New York: Plenum Press, 1993, pp 521–532.

30. JE Cutler. Putative virulence factors of *Candida albicans*. Annu Rev Microbiol 45:187–218, 1991.

31. MA Ghannoum, KH Abu-Elteen. Pathogenicity determinants of *Candida*. Mycoses 33:265–282, 1990.

32. EJ Anaissie. Opportunistic mycoses in the immunocompromized host: experience at a cancer center and review. Clin Infect Dis 14(S1):S43–S53, 1992.

33. RB Ashman, JM Papadimitriou. What's new in the mechanisms of host resistance to *Candida albicans* infection? Pathol Res Pract 186:527–534, 1990.

34. RB Ashman. Murine candidiasis: cell mediated immune responses correlate directly with susceptibility and resistance to infection. Immunol Cell Biol 68:15–20, 1990.

35. FC Odds. *Candida* and Candidosis, 2d ed. London: Baillière-Tindall, 1989, pp 124–152.

36. CH Kirkpatrick. Chronic mucocutaneous candidiasis. J Am Acad Dermatol 31:S14–S17, 1994.

37. SS Witkin, IR Yu, J Ledger. Inhibition of *Candida albicans*–induced lymphocyte proliferation by lymphocytes and sera from women with recurrent vaginitis. Am J Obstet Gynecol 147:809–811, 1983.

38. SS Witkin. Immunologic factors influencing susceptibility to recurrent candidal vaginitis. Clin Obstet Gynecol 34:662–668, 1991.

39. PL Fidel Jr, JD Sobel. Immunopathogenesis of recurrent vulvovaginal candidiasis. Clin Rev Microbiol 9:335–348, 1996.

40. RE Garner, AM Childress, LG Human, JE Domer. Characterization of *Candida albicans* mannan-induced, mannan-specific delayed hypersensitivity suppressor cells. Infect Immun 58:2613–2620, 1990.

41. RD Nelson, N Shibata, RP Podzorski, MJ Herron. *Candida* mannan: chemistry, suppression of cell mediated immunity, and possible mechanisms of action. Clin Microbiol Rev 4:1–19, 1991.

42. JE Domer, LJ Human, GB Andersen, JA Rudbach, GL Asherson. Abrogation of suppression of delayed hypersensitivity induced by *Candida albicans*–derived mannan by treatment with monophosphoryl lipid A. Infect Immun 61:2122–2130, 1993.

43. Y Wang, SP Li, SA Moser, KL Bost, JE Domer. Cytokine involvement in immunomodulatory activity affected by *Candida albicans* mannan. Infect Immun 66:1384–1391, 1998.

44. A Durandy, A Fischer, F Le Deist, E Drouhet, C Griscelli. Mannan-specific and mannan-induced T cell suppressive activity in patients with chronic mucocutaneous candidiasis. J Clin Immunol 7: 400–409, 1987.

45. A Fischer, JJ Ballet, C Griscelli. Specific inhibition of in vitro *Candida*-induced lymphocyte proliferation by polysaccharidic antigens present in the serum of patients with chronic mucocutaneous candidiasis. J Clin Invest 62:1005–1013, 1978.

46. E Piccolella, G Lombardi, R Morelli. Generation of suppressor cells in the response of human lymphocytes to a polysaccharide from *Candida albicans*. J Immunol 126:2151–2155, 1981.

47. G Lombardi, A Di Massimo, F Del Gallo, D Vismara, E Piccolella, O Pugliese, V Colizzi. Mechanism of action of an antigen nonspecific inhibitory factor produced by human T cells stimulated by MPPS and PPD. Cell Immunol 98:434–443, 1986.

48. G Lombardi, D Vismara, E Piccolella, V Colizzi, GL Asherson. A non-specific inhibitor produced by *Candida albicans*–activated T cells impairs cell proliferation by inhibiting interleukin-1 production. Clin Exp Immunol 60:303–313, 1985.

49. J Savolainen, A Koivikko, K Kalimo, E Nieminen, M Viander. IgE, IgA and IgG antibodies and delayed skin responses towards *Candida albicans* antigens in atopics with and without saprophytic growth. Clin Exp Allergy 20:549–554, 1990.

50. J Savolainen, K Lammintausta, K Kalimo, M Viander. *Candida albicans* and atopic dermatitis. Clin Exp Allergy 23:332–339, 1993.

51. E Morita, M Hide, Y Yoneya, M Kannbe, A Tanaka, S Yamamoto. An assessment of the role of *Candida albicans* antigen in atopic dermatitis. J Dermatol 26:282–287, 1999.

52. J Savolainen, O Kortekangas-Savolainen, M Nermes, M Viander, A Koivikko, K Kalimo, EO Terho. IgE, IgA, and IgG responses to common yeasts in atopic patients. Allergy 53:506–512, 1998.

53. Y Tanizaki, H Kitani, M Okazaki, T Mifune, F Mitsunobu. An increased level of specific IgG4 antibodies against *Candida albicans* in patients with bronchial asthma. J Asthma 29:343–348, 1992.

54. K Akiyama, T Shida, H Yasueda, H Mita, T Yamamoto, H Yamaguchi. Atopic asthma caused by *Candida albicans* acid protease: case reports. Allergy 49:778–781, 1994.

55. PS Moraes. Recurrent vaginal candidiasis and allergic rhinitis: a common association. Ann Allergy Asthma Immunol 81:165–169, 1998.

56. SS Witkin, J Jeremias, WJ Ledger. Vaginal eosinophilis and IgE antibodies to *Candida albicans* in women with recurrent vaginitis. J Med Vet Mycol 27:57–58, 1989.

57. P Regulez, JF Garcia Fernandez, MD Moragues, J Schneider, G Quintos, J Ponton. Detection of anti–*Candida albicans* IgE antibodies in vaginal washes from patients with acute vulvovaginal candidiasis. Gynecol Obstet Invest 37:110–114, 1994.

58. RC Mathews. *Candida albicans* HSP 90: link between protective and autoimmunity. J Med Microbiol 36:367–370, 1992.

59. S Romagnani. Lymphokine production by human T cells in disease states. Annu Rev Immunol 12: 367–370, 1992.

60. PL Fidel, KA Ginsburg, JL Cutright, NA Wolf, D Leaman, K Dunlap, JD Sobel. Vaginal-associated immunity in women with recurrent vulvovaginal candidiasis: evidence for vaginal Th1-type responses following intravaginal challenge with *Candida* antigen. J Infect Dis 176:728–739, 1997.

61. A La Sala, F Urbani, A Torosantucci, A Cassone, C Ausiello. Mannoproteins from *Candida albicans* elicit a Th1-type-1 cytokine profile in human *Candida* specific long-term T cell cultures. J Biol Regul Homeost Agents 10:8–12, 1996.

62. M Kobayashi, H Kobayashi, DN Herndon, RB Pollard, F Suzuki. Burn-associated *Candida albicans* infection caused by CD30$^+$ type 2 T cells. J Leukoc Biol 63:723–731, 1998.

63. JE Leigh, C Steele, FL Wormerly Jr, W Luo, W Gallaher, PL Fidel Jr. Th1/Th2 cytokine expression in saliva of HIV-positive and HIV-negative individuals: a pilot study in HIV-positive individuals with oropharyngeal candidiasis. J Acquir Immune Defic Syndr Hum Retrovirol 19:373–380, 1998.

64. D Lilic, AJ Cant, M Abinun, JE Calvert, GP Spickett. Chronic mucocutaneous candidiasis. I. Altered antigen-stimulated IL-2, IL-4, IL-6 and interferon-gamma (IFN-γ) production. Clin Exp Immunol 105:205–212, 1996.

65. LJ Kobrynski, L Tanimune, L Kilpatrick, DE Campbell, SD Douglas. Production of T-helper cell subsets and cytokines by lymphocytes from patients with chronic mucocutaneous candidiasis. Clin Diagn Lab Immunol 3:740–745, 1996.

66. E Roilides, T Sein, R Schaufele, SJ Chanock, TJ Walsh. Increased serum concentrations of interleukin-10 in patients with hepatosplenic candidiasis. J Infect Dis 178:589–592, 1998.

67. P Lintu, J Savolainen, K Kalimo, O Kortekangas-Savolainen, M Nermes, EO Terho. Cross-reacting IgE and IgG antibodies to *Pityrosporum ovale* mannan and other yeasts in atopic dermatitis. Allergy 54:1067–1073, 1999.

68. J Savolainen, J Kosonen, P Lintu, M Viander, J Pene, K Kalimo, EO Terho, J Bousquet. *Candida albicans* mannan- and protein-induced humoral, cellular and cytokine responses in atopic dermatitis patients. Clin Exp Allergy 29:824–831, 1999.

69. N Terada, A Konno, K Shirotory, T Fujisawa, J Atsuta, R Ichimi, Y Kikuchi, S Takaki, K Takatsu, K Togawa. Mechanism of eosinophil infiltration in the patient with subcutaneous angioblastic lymphoid hyperplasia with eosinophilia (Kimura's disease). Mechanism of eosinophil chemotaxis mediated by *Candida* antigen and IL-5. Int Arch Allergy Immunol 104:18–20, 1994.

70. D Nissen, H Nolte, H Permin, J Heining, PS Skov, S Norn. Evaluation of IgE-sensitization to fungi in HIV-positive patients with eczematous skin reactions. Ann Allergy Asthma Immunol 83:153–159, 1999.

71. RD Wagner, A Vasquez-Torres, J Jones-Carson, T Warner, E Balish. B-cell knockout mice are resistant to mucosal and systemic candidiasis of endogenous origin but susceptible to experimental systemic candidiasis. J Infect Dis 174:589–597, 1996.

72. E Balish, A Vasquez-Torres, J Jones-Carson, RD Wagner, T Warner. Importance of β2-microglobulin in murine resistance to mucosal and systemic candidiasis. Infect Immun 64:5092–5097, 1996.

73. P Schuman, JD Sobel, E Ohmit, KH Mayer, CJ Carpenter, A Rompalo, A Duerr, DK Smith, D Warren, RS Klein. Mucosal candidal colonization and candidiasis in women with or at risk for immunodeficiency virus infection. J Infect Dis 27:1161–1167, 1998.

74. PL Fidel Jr. Host defense against oropharyngeal and vaginal candidiasis: site-specific differences. Rev Iberoam Micol 16:8–15, 1999.

75. MH Kim, GE Rodey, RA Good, RA Chilgren, PG Quie. Defective candidacidal capacity of polymorphonuclear leukocytes in chronic granulomatous disease of childhood. J Pediatr 75:300–303, 1969.

76. GP Bodey, M Buckley, YS Sathe, EJ Freireich. Quantitative relationship between circulating leukocytes and infection in patients with acute leukemia. Ann Intern Med 64:328–340, 1986.

77. RB Ashman. *Candida albicans*: pathogenesis, immunity and host defense. Res Immunol 149:281–288, 1998.

78. M Cantorna, E Balish. Mucosal and systemic candidiasis in congenitally immunodeficient mice. Infect Immun 58:1093–1100, 1990.

79. LA Coker, CM Mercadal, BT Rouse, RN Moore. Differential effects of CD4+ and CD8+ cells in acute, systemic murine candidosis. J Leukoc Biol 51:305–306, 1992.

80. TG Sieck, MA Moors, HR Buckley, KJ Blank. Protection against disseminated candidiasis mediated by a *Candida albicans*–specific T-cell line. Infect Immun 61:3540–3543, 1993.

81. E Cenci, L Romani, A Vecchiarelli, P Puccetti, F Bistoni. Role of L3T4+ lymphocytes in protective immunity to systemic *Candida albicans* infection in mice. Infect Immun 57:3581–3587, 1989.

82. E Cenci, L Romani, A Vecchiarelli, P Puccetti, F Bistoni. T cell subsets and IFN-γ production in resistance to systemic candidosis in immunized mice. J Immunol 144:4333–4339, 1990.

83. RB Ashman. Mouse candidiasis. II. Host responses are T-cell dependent and regulated by genes in the major histocompatibility complex. Immunogenetics 25:200–203, 1987.

84. RB Ashman, EM Bolitho. Strain differences in the severity of lesions in murine systemic candidiasis correlate with the production of functional gamma interferon by *Candida*-activated lymphocytes in vitro. Lymphokine Cytokine Res 12:471–476, 1993.

85. L Romani, A Mencacci, E Cenci, P Mosci, G Vitellozzi, U Grohmann, P Puccetti, F Bistoni. Course of primary candidiasis in T-cell depleted mice infected with attenuated variant cells. J Infect Dis 166:1384–1392, 1992.

86. L Romani, S Mocci, E Cenci, R Rossi, P Puccetti, F Bistoni. *Candida albicans*-specific Lyt-2⁺ lymphocytes with cytolytic activity. Eur J Immunol 21:1567–1570, 1991.

87. L Romani, SHE Kaufmann. Immunity to fungi. Res Immunol 149:277–281, 1998.

88. A Mencacci, E Cenci, G Del Sero, C Fè d'Ostiani, C Montagnoli, A Bacci, F Bistoni, L Romani. Innate and adaptive immunity to *Candida albicans*: a new view of an old paradigm. Rev Iberoam Micol 16:4–7, 1999.

89. RF Hector, JE Domer, EW Carrow. Immune responses to *Candida albicans* in genetically distinct mice. Infect Immun 38:1020–1028, 1992.

90. G Marquis, S Montplaisir, M Pelletier, S Mousseau, P Auger. Strain-dependent differences in susceptibility of mice to experimental candidosis. J Infect Dis 154:906–908, 1986.

91. L Romani, A Mencacci, E Cenci, S Spaccapelo, P Mosci, P Puccetti, F Bistoni. CD4⁺ subset expression in murine candidiasis. Th responses correlate directly with genetically determined susceptibility or vaccine-induced resistance. J Immunol 150:925–931, 1993.

92. F Bistoni, E Cenci, A Mencacci, E Schiaffella, P Mosci, P Puccetti, L Romani. Mucosal and systemic T helper cell function after intragastric colonization of adult mice with *Candida albicans*. J Infect Dis 168:1449–1457, 1993.

93. E Cenci, A Mencacci, R Spaccapelo, L Tonnetti, K-H Enssle, P Puccetti, L Romani, F. Bistoni. T helper cell type 1 (Th1)- and Th2-like responses are present in mice with gastric candidiasis but protective immunity is associated with Th1 development. J Infect Dis 171:1279–1288, 1995.

94. JE Cutler. Acute systemic candidiasis in normal and congenitally thymic-deficient (nude) mice. J Reticuloendothel Soc 19:121–124, 1976.

95. TJ Rogers, E Balish, DD Manning. The role of thymus-dependent cell-mediated immunity in resistance to experimental disseminated candidiasis. J Reticuloendothel Soc 20:291–298, 1976.

96. E Balish, J Jensen, T Warner, J Brekke, B Leonard. Mucosal and disseminated candidiasis in gnotobiotic SCID mice. J Med Vet Mycol 31:143–154, 1993.

97. S Mahanty, RA Greenfield, WA Joyce, PW Kinkade. Inoculation candidiasis in a murine model of severe combined immunodeficiency syndrome. Infect Immun 56:3162–3166, 1988.

98. R Narayanan, WA Joyce, RA Greenfield. Gastrointestinal candidiasis in a murine model of severe combined immunodeficiency syndrome. Infect Immun 59:2116–2119, 1991.

99. SB Salvin, RDA Peterson, RA Good. The role of the thymus in resistance to infection and endotoxin toxicity. J Lab Clin Med 65:1004–1022, 1965.

100. M Cantorna, E Balish. Acquired immunity to systemic candidiasis in immunodeficient mice. J Infect Dis 164:936–943, 1991.

101. J Jensen, T Warner, E Balish. Resistance of SCID mice to *Candida albicans* administered intravenously or colonizing the gut: role of polymorphonuclear leukocytes and macrophages. J Infect Dis 167:912–919, 1993.

102. L Romani, P Puccetti, A Mencacci, R Spaccapelo, E Cenci, L Tonnetti, F Bistoni. Tolerance to staphylococcal enterotoxin B initiates Th1 cell differentiation in mice infected with *Candida albicans*. Infect Immun 62:4047–4053, 1994.

103. R Spaccapelo, G Del Sero, P Mosci, F Bistoni, L Romani. Early T cell unresponsiveness in mice with candidiasis and reversal by IL-2. Effect on T helper cell development. J Immunol 158: 2294–2302, 1997.

104. W Born, C Cady, J Jones-Carson, A Mukasa, M Lahn, R O'Brien. Immunoregulatory functions of gamma delta T cells. Adv Immunol 71:77–144, 1999.

105. TR Mosmann, RL Coffman. TH1 and TH2 cells: different patterns of lymphokine secretion lead to different functional properties. Annu Rev Immunol 7:145–173, 1989.

106. A O'Garra. Cytokines induce the development of functionally heterogeneous T helper cell subsets. Immunity 8:275–283, 1998.

107. D Mason, F Powrie. Control of immune pathology by regulatory T cells. Curr Opin Immunol 10: 649–655, 1998.

108. Y Chen, VK Kuchroo, J Inobe, DA Hafler, HL Weiner. Regulatory T cell clones induced by oral tolerance: suppression of autoimmune encephalomyelitis. Science 265:1237–1240, 1994.

109. H Groux, A O'Garra, M Bigler, M Rouleau, S Antonenko, JE de Vries, MG Roncarolo. A CD4$^+$ T-cell subset inhibits antigen-specific T-cell responses and prevents colitis. Nature 389:737–742, 1997.

110. HL Weiner. Oral tolerance for the treatment of autoimmune diseases. Annu Rev Med 48:341–351, 1997.

111. A Vecchiarelli, E Cenci, M Puliti, E Blasi, P Puccetti, P Marconi, A Cassone, F Bistoni. Protective immunity induced by low-virulence Candida albicans. Cytokine production in the development of the anti-infectious state. Cell Immunol 124:334–344, 1989.

112. L Romani, S Mocci, C Bietta, L Lanfaloni, P Puccetti, F Bistoni. Th1 and Th2 cytokine secretion patterns in murine candidiasis. Association of Th1 responses with acquired resistance. Infect Immun 59:4647–4654, 1991.

113. L Romani, E Cenci, A Mencacci, R Spaccapelo, U Grohmann, P Puccetti, F Bistoni. Gamma interferon modifies CD4$^+$ subset expression in murine candidiasis. Infect Immun 60:4950–4952, 1992.

114. L Romani, A Mencacci, U Grohmann, S Mocci, P Mosci, P Puccetti, F Bistoni. Neutralizing antibody to interleukin 4 induces systemic protection and T helper type 1–associated immunity in murine candidiasis. J Exp Med 176:19–25, 1992.

115. L Romani, P Puccetti, A Mencacci, E Cenci, R Spaccapelo, L Tonnetti, U Grohmann, F Bistoni. Neutralization of IL10 up-regulates nitric oxide production and protects susceptible mice from challenge with Candida albicans. J Immunol 152:3514–3521, 1993.

116. P Puccetti, A Mencacci, E Cenci, R Spaccapelo, P Mosci, K-H Enssle, L Romani, F Bistoni. Cure of murine candidiasis by recombinant soluble interleukin 4 receptor. J Infect Dis 169:13251331, 1994.

117. R Spaccapelo, L Romani, L Tonnetti, E Cenci, A Mencacci, R Tognellini, SG Reed, P Puccetti, F Bistoni. TGF-β is important in determining the in vivo susceptibility or resistance in mice infected with Candida albicans. J Immunol 155:1349–1360, 1995.

118. L Romani, A Mencacci, L Tonnetti, R Spaccapelo, E Cenci, S Wolf, P Puccetti, F Bistoni. Interleukin-12 but not interferon-γ production correlates with induction of T helper type-1 phenotype in murine candidiasis. Eur J Immunol 24:909–915, 1994.

119. L Romani, A Mencacci, L Tonnetti, R Spaccapelo, E Cenci, P Puccetti, SF Wolf, F Bistoni. Interleukin-12 is both required and prognostic in vivo for T helper type 1 differentiation in murine candidiasis. J Immunol 53:5157–5175, 1994.

120. L Romani, A Mencacci, E Cenci, R Spaccapelo, C Toniatti, P Puccetti, F Bistoni, V Poli. Impaired neutrophil response and CD4$^+$ T helper cell 1 development in interleukin-6-deficient mice infected with Candida albicans. J Exp Med 183:1–11, 1996.

121. A Mencacci, E Cenci, G Del Sero, C Fè d'Ostiani, P Mosci, F Bistoni, C Montagnoli, A Bacci, F Bistoni, VFJ Quesniaux, B Ryffel, L Romani. Defective co-stimulation and impaired Th1 development in tumor necrosis factor/lymphotoxin-α double-deficient mice infected with Candida albicans. Int Immunol 10:37–48, 1998.

122. A Mencacci, E Cenci, F Bistoni, G Del Sero, A Bacci, C Montagnoli, C Fe d'Ostiani, L Romani. Specific and non-specific immunity to Candida albicans: a lesson from genetically modified animals. Res Immunol 149:352–361, 1998.

123. L Tonnetti, R Spaccapelo, E Cenci, A Mencacci, P Puccetti, RL Coffman, F Bistoni, L Romani. Interleukin-4 and -10 exacerbate candidiasis in mice. Eur J Immunol 25:1559–1565, 1995.

124. L Romani, P Puccetti, F Bistoni. Initiation of T-helper cell immunity to Candida albicans by IL-12: the role of neutrophils. Chem Immunol 68:110–135, 1997.

125. RB Ashman, EM Bolitho, JM Papadimitriou. Pattern of resistance to Candida albicans in inbred mouse strains. Immunol Cell Biol 71:221–225, 1993.

126. F Bistoni, A Vecchiarelli, E Cenci, P Puccetti, P Marconi, A Cassone. Evidence for macrophage-mediated protection against lethal Candida albicans infection. Infect Immun 51:668–674, 1986.

127. F Bistoni, G Verducci, S Perito, A Vecchiarelli, P Puccetti, P Marconi, A Cassone. Immunomodula-
      tion by a low-virulence, agerminative variant of *Candida albicans*: further evidence for macrophage
      activation as one of the effector mechanisms of nonspecific anti-infectious protection. J Med Vet
      Mycol 26:285–299, 1988.

128. E Cenci, A Bartocci, P Puccetti, S Mocci, ER Stanley, F Bistoni. Macrophage colony-stimulating
      factor in murine candidiasis: serum and tissue levels during infection and protective effect of exoge-
      nous administration. Infect Immun 59:868–872, 1991.

129. MT Cantorna, E Balish. Role of CD4+ lymphocytes in resistance to mucosal candidiasis. Infect
      Immun 59:2447–2455, 1991.

130. PL Fidel Jr, ME Lynch, JD Sobel. *Candida*-specific cell-mediated immunity is demonstrable in
      mice with experimental vaginal candidiasis. Infect Immun 61:1990–1995, 1993.

131. JE Domer. Intragastric colonization of infant mice with *Candida albicans* induces systemic immunity
      demonstrable upon challenge as adults. J Infect Dis 157:950–958, 1988.

132. R Morelli, LT Rosenberg. Role of complement during experimental *Candida* infection in mice.
      Infect Immun 3:521–523, 1971.

133. RB Ashman. A gene (Carg1) that regulates tissue resistance to *Candida albicans* maps to chromo-
      some 14 of the mouse. Microb Pathog 25:333–335, 1998.

134. RB Ashman, A Fulurija, JM Papadimitriou. A second *Candida albicans* resistance gene (Carg2)
      regulates tissue damage, but not fungal clearance, in sub-lethal murine systemic infection. Microb
      Pathog 25:349–352, 1998.

135. L Romani. Animal models for candidiasis. Current Protocols in Immunology, Unit 19.6, Supplement
      30, 1999.

136. RD Mulero-Marchese, KJ Blank, TG Siek. Genetic basis for protection against experimental vaginal
      candidiasis by peripheral immunization. J Infect Dis 178:227–234, 1998.

137. PL Fidel Jr, ME Lynch, JD Sobel. *Candida*-specific Th1-type responsiveness in mice with experi-
      mental vaginal candidiasis. Infect Immun 61:4202–4207, 1993.

138. E Cenci, A Mencacci, G Del Sero, C Fè d'Ostiani, P Mosci, M Kopf, L Romani. IFN-γ is required
      for IL-12 responsiveness in mice with *Candida albicans* infection. J Immunol 161:3543–3550,
      1998.

139. J Jones-Carson, FA Vasquez-Torres, E Balish. B cell–independent selection of memory T cells
      after mucosal immunization with *Candida albicans*. J Immunol 158:4328–4335, 1997.

140. Q Qian, MA Jutila, N van Rooijen, JE Cutler. Elimination of mouse splenic macrophages correlates
      with increased susceptibility to experimental disseminated candidiasis. J Immunol 152:5000–5008,
      1994.

141. BJ Kullberg, JW Van't Wout, C Hoogstraten, R Van Furth. Recombinant interferon-γ enhances
      resistance to acute disseminated *Candida albicans* infection in mice. J Infect Dis 168:436–443,
      1993.

142. A Mencacci, G Del Sero, E Cenci, C Fè d'Ostiani, A Bacci, C Montagnoli, M Kopf, L Romani.
      IL-4 is required for development of protective CD4+ T helper type 1 cell responses to *Candida
      albicans*. J Exp Med 187:307–317, 1998.

143. A Cassone, F De Bernardis, CM Ausiello, MJ Gomez, M Boccanera, R La Valle, A Torosantucci.
      Immunogenic and protective *Candida albicans* constituents. Res Immunol 149:289–289, 1998.

144. A Mencacci, R Spaccapelo, G Del Sero, K-H Enssle, A Cassone, F Bistoni, L Romani. CD4+ T-
      helper-cell responses in mice with low-level *Candida albicans* infection. Infect Immun 64:
      4907–4914, 1996.

145. Q Qian, JE Cutler. Gamma interferon is not essential in host defense against disseminated candidiasis
      in mice. Infect Immun 65:1748–1753, 1997.

146. E Balish, RD Wagner, A Vasquez-Torres, C Pierson, T Warner. Candidiasis in interferon-γ knockout
      (IFN-γ) mice. J Infect Dis 178:478–487, 1998.

147. MK Gately, LM Renzetti, J Magram, AS Stern, L Adorini, U Gubler, DH Presky. The interleukin-
      12/interleukin-12-receptor system: role in normal and pathologic immune responses. Annu Rev
      Immunol 16:495–521, 1998.

148. SJ Szabo, AS Dighe, U Gubler, KM Murphy. Regulation of the interleukin (IL)-12Rβ2 subunit
      expression in developing T helper 1 (Th1) and Th2 cells. J Exp Med 185:817–824, 1997.

149. T Nakamura, Y Kamogawa, K Bottomly, RA Flavell. Polarization of IL-4- and IFN-γ-producing CD4⁺ T cells following activation of naive CD4⁺ T cells. J Immunol 158:1085–1094, 1997.

150. GM Feldman, S Ruhl, M Bickel, DS Finbloom, DH Pluznik. Regulation of interleukin-4 receptors on murine myeloid progenitor cells by interleukin-6. Blood 78:1678–1684, 1991.

151. A Mencacci, E Cenci, G Del Sero, C Fè d'Ostiani, P Mosci, F Bistoni, G Trinchieri, L Adorini, L Romani. IL-10 is required for development of protective CD4⁺ T helper type 1 cell responses to *Candida albicans*. J Immunol 161:6228–6237, 1998.

152. L Romani. Innate and adaptive immunity in *Candida albicans* infections and saprophytism. J Leuk Biol 68:175–179, 2000.

153. DJ Kitz, PD Stahl, JR Little. The effect of a mannose binding protein on macrophage interactions with *Candida albicans*. Cell Mol Biol 38:407–412, 1992.

154. JM Schroder, J Harder. Human beta-defensin-2. Int J Biochem Cell Biol 31:645–651, 1999.

155. S Rosseau, P Hammerl, U Maus, A Günther, W Seeger, F Grimminger, J Lohmeyer. Surfactant protein A downregulates proinflammatory cytokine production evoked by *Candida albicans* in human alveolar macrophages and monocytes. J Immunol 163:4495–4502, 1999.

156. L Marodi, HM Korchak, RB Johnston Jr. Mechanisms of host defense against *Candida* species. I. Phagocytosis by monocyte-derived macrophages. J Immunol 146:2783–2789, 1991.

157. A Casadevall. Antibody immunity and invasive fungal infections. Infect Immun 63:4211–4218, 1995.

158. TR Kozel. Complement activation by pathogenic fungi. Res Immunol 149:309–320, 1998.

159. R Robert, S Nail, A Marot-Leblond, J Cottin, M Miegeville, S Quenouillere, C Mahaza, JM Senet. Adherence of platelets to *Candida* species in vivo. Infect Immun 68:570–576, 2000.

160. SG Filler, AS Pfunder, BJ Spellberg, JP Spellberg, JE Edwards Jr. *Candida albicans* stimulates cytokine production and leukocyte adhesion molecule expression by endothelial cells. Infect Immun 64:2609–2617, 1996.

161. C Steele, H Ozenci, W Luo, M Scott, PL Fidel Jr. Growth inhibition of *Candida albicans* by vaginal cells from naive mice. Med Mycol 37:251–259, 1999.

162. JD Fearon, RM Locksley. The instructive role of innate immunity in the acquired immune response. Science 272:50–54, 1996.

163. RM Medzhitov, CA Janeway Jr. Innate immunity: the virtues of a nonclonal system of recognition. Cell 91:295–298, 1997.

164. RA Calderone. Recognition between *Candida albicans* and host cells. Trends Microbiol 1:55–58, 1993.

165. C Gale, D Finkel, N Tao, M Meinke, M McClellan, J Olson, Kendrick, M Hostetter. Cloning and expression of a gene encoding an integrin-like protein in *Candida albicans*. Proc Natl Acad Sci USA 93:357–361, 1996.

166. C Monteagudo, JL Lopez-Ribot, A Murgui, M Casanova, WL Chaffin, JP Martinez. Immunodetection of CD45 epitopes on the surface of *Candida albicans* cells in culture and infected human tissues. Am J Clin Pathol 113:59–63, 2000.

167. CA Gale, CM Bendel, M McClellan, M Hauser, JM Becker, J Berman, MK Hostetter. Linkage of adhesion, filamentous growth, and virulence in *Candida albicans* to a single gene, INT1. Science 279:1355–1358, 1998.

168. P Sundstrom. Adhesins in *Candida albicans*. Curr Opin Microbiol 2:353–357, 1999.

169. JF Staab, SD Bradway, PL Fidel, P Sundstrom. Adhesive mammalian transglutaminase substrate properties of *Candida albicans* Hwp1. Science 283:1535–1538, 1999.

170. RD Diamond. Fungal surfaces: effects of interactions with phagocytic cells. Rev Infect Dis 10:428–431, 1998.

171. A Schaffner, CE Davis, T Schaffner, M Markert. In vitro susceptibility of fungi to killing by neutrophil granulocytes discriminates between primary pathogenicity and opportunism. J Clin Invest 78:511–524, 1986.

172. FH Van Enckevort, MG Netea, AR Hermus, CG Sweep, JF Meis, JW Van der Meer, BJ Kullberg. Increased susceptibility to systemic candidiasis in interleukin-6 deficient mice. Med Mycol 37:419–426, 1999.

173. E Balish, RD Wagner, A Vazquez-Torres, J Jones-Carson, C Pierson, T Warner. Mucosal and systemic candidiasis in IL-8Rh-/- BALB/c mice. J Leukoc Biol 66:144–150, 1999.

174. L Romani, A Mencacci, E Cenci, G Del Sero, F Bistoni, P Puccetti. An immunoregulatory role for neutrophils in CD4[+] T helper subset selection in mice with candidiasis. J Immunol 158:2356–2362, 1997.

175. L Romani, A Mencacci, E Cenci, R Spaccapelo, G Del Sero, I Nicoletti, G Trinchieri, F Bistoni, P Puccetti. Neutrophils production of IL-12 and IL-10 in candidiasis and efficacy of IL-12 therapy in neutropenic mice. J Immunol 158:5349–5356, 1997.

176. SL Davis, EP Hawkins, EO Mason Jr, CW Smith, SL Kaplan. Host defenses against disseminated candidiasis are impaired in intercellular adhesion molecule 1–deficient mice. J Infect Dis 174: 435–439, 1996.

177. A Fulurija, RB Ashman, JM Papadimitriou. Neutrophil depletion increases susceptibility to systemic and vaginal candidiasis in mice, and reveals differences between brain and kidney in mechanisms of host resistance. Microbiology 142:3487–3496, 1996.

178. T Nohmi, S Abe, K Dobashi, S Tansho, H Yamaguchi. Suppression of anti-*Candida* activity of murine neutrophils by progesterone in vitro: a possible mechanism in pregnant women's vulnerability to vaginal candidiasis. Microbiol Immunol 39:405–409, 1995.

179. KW Lee, E Balish. Systemic candidosis in silica-treated athymic and euthymic mice. Infect Immun 41:902–907, 1983.

180. DA Hume, Y Denkins. The deleterious effect of macrophage colony-stimulating factor (CSF-1) on the pathology of experimental candidiasis in mice. Lymphokine Cytokine Res 11:95–98, 1992.

181. T Kanbe, MA Jutila, JE Cutler. Evidence that *Candida albicans* binds via a unique adhesion system on phagocytic cells in the marginal zone of the mouse spleen. Infect Immun 60:1972–1978, 1992.

182. A Vasquez-Torres, E Balish. Macrophages in resistance to candidiasis. Microbiol Mol Rev 61: 170–192, 1997.

183. RD Diamond, RA Clark, CC Haudenshild. Damage to *Candida albicans* hyphae and pseudohyphae by the myeloperoxidase system and oxidative products of neutrophil metabolism in vitro. J Clin Invest 66:908–917, 1980.

184. E Cenci, L Romani, A Mencacci, R Spaccapelo, E Schiaffella, P Puccetti, F Bistoni. Interleukin-4 and interleukin-10 inhibit nitric oxide–dependent macrophage killing of *Candida albicans*. Eur J Immunol 23:1034–1038, 1993.

185. J Jones-Carson, A Vazquez-Torres, HC van der Heyde, T Warner, RD Wagner, E Balish. Gamma delta T cell-induced nitric oxide production enhances resistance to mucosal candidiasis. Nat Med 1:552–557, 1995.

186. T Chinen, MH Qureshi, Y Koguchi, K Kawakami. *Candida albicans* suppresses nitric oxide (NO) production by interferon-gamma (IFN-gamma) and lipopolysaccharide (LPS)-stimulated murine macrophages. Clin Exp Immunol 115:491–497, 1999.

187. G Del Sero, A Mencacci, E Cenci, C Fé d'Ostiani, C Montagnoli, A Bacci, P Mosci, M Kopf, L Romani. Antifungal type 1 responses are upregulated in IL-10-deficient mice. Microbes Infect 1: 1169–1180, 1999.

188. A Vasquez-Torres, J Jones-Carson, RD Wagner, T Warner, E Balish. Early resistance of interleukin-10 knockout mice to acute systemic candidiasis. Infect Immun 67:670–674, 1999.

189. F Abaitua, A Rementeria, R San Millan, A Eguzkiza, JA Rodriguez, J Ponton, MJ Sevilla. In vitro survival and germination of *Candida albicans* in the presence of nitrogen compounds. Microbiology 145:1641–1647.

190. C Scherwitz, R Martin. The phagocytosis of *Candida albicans* blastospores and germ tubes by polymorphonuclear leukocytes. Dermatologica 159:12–23, 1979.

191. A Cockayne, FC Odds. Interactions of *Candida albicans* yeast cells, germ tubes and hyphae with human polymorphonuclear leucocytes in vitro. J Gen Microbiol 130:465–471, 1984.

192. S Greenberg, SC Silverstein. Phagocytosis. In: WE Paul, ed. Fundamental Immunology. 3d ed. New York: Raven Press, 1993, pp 941–964.

193. PD Stahl, RA Ezekowitz. The mannose receptor is a pattern recognition receptor involved in host defense. Curr Opin Immunol 10:50–55, 1992.

194. AB van Spriel, IE van den Herik-Oudijk, NM van Sorge, HA Vile, JA van Strijp, JG van de Winkel. Effective phagocytosis and killing of *Candida albicans* via targeting FcgammaRI (CD64) or FcalphaRI (CD89) on neutrophils. J Infect Dis 179:661–669, 1999.

195. K Kagaya, Y Fukazawa. Murine defense mechanism against *Candida albicans* infection. II. Opsonization, phagocytosis, and intracellular killing of *C. albicans*. Microbiol Immunol 25:807–818, 1981.

196. HA Pereira, CS Hosking. The role of complement and antibody in opsonization and intracellular killing of *Candida albicans*. Clin Exp Immunol 57:307:314, 1984.

197. M Yamamura, H Valdimarsson. Participation of C3 in intracellular killing of *Candida albicans*. Scand J Immunol 6:591–594, 1997.

198. JS Solomkin, EL Mills, GS Giebink, RD Nelson, RL Simmons, PG Quie. Phagocytosis of *Candida albicans* by human leukocytes: opsonic requirements. J Infect Dis 137:30–37, 1978.

199. F Bazzoni, MA Cassatella, C Laudanna, F Rossi. Phagocytosis of opsonized yeast induces tumor necrosis factor–alpha mRNA accumulation and protein release by human polymorphonuclear leukocytes. J Leukoc Biol 50:223–228, 1991.

200. B Walzog, P Weinmann, F Jeblonski, K Scharffetter-Kochanek, K Bommert, P Gaehtgens. A role for beta(2) integrins (CD11/CD18) in the regulation of cytokine gene expression of polymorphonuclear neutrophils during the inflammatory response. FASEB J 13:1855–1865, 1999.

201. T Tang, A Rosenkranz, KJM Assmann, MJ Goodman, JC Gutierrez-Ramos, MC Carroll, RS Cotran, TN Mayadas. A role for Mac-1 (CDIIb/CD18) in immune complex–stimulated neutrophil function in vivo: Mac-1 deficiency abrogates sustained Fcgamma receptor-dependent neutrophil adhesion and complement-dependent proteinuria in acute glomerulonephritis. J Exp Med 186:1853–1863, 1997.

202. JE Edwards Jr, TA Gaither, JJ O'Shea, D Rotrosen, TJ Lawley, SA Wright, MM Frank, I Green. Expression of specific binding sites on *Candida* with functional and antigenic characters of human complement receptors. J Immunol 137:3577–3583, 1986.

203. J Yan, V Vêtvicka, Y Xia, A Coxon, MC Carroll, TN Mayadas, G Ross. β-glucan, a "specific" biologic response modifier that uses antibodies to target tumors for cytotoxic recognition by leukocyte complement receptor type 3 (CD11b/CD18). J Immunol 163:3045–3052, 1999.

204. DL Danley, AE Hilger. Stimulation of oxidative metabolism in murine polymorphonuclear leukocytes by unopsonized fungal cells: evidence for a mannose-specific mechanism. J Immunol 127:551–556, 1981.

205. CA Lyman, ER Simons, DA Melnick, RD Diamond. Unopsonized *Candida albicans* hyphae stimulate a neutrophil respiratory burst and a cytosolic calcium flux without membrane depolarization. J Infect Dis 156:770–775, 1987.

206. SM Levitz, CA Lyman, T Murata, JA Sullivan, GL Mandell, RD Diamond. Cytosolic calcium changes in individual neutrophils stimulated by opsonized and unopsonized *Candida albicans* hyphae. Infect Immun 55:2783–2788, 1987.

207. MP Kolotila, RD Diamond. Stimulation of neutrophil actin polymerization and degranulation by opsonized and unopsonized *Candida albicans* hyphae and zymosan. Infect Immun 56:2016–2022, 1988.

208. DR Wysong, CA Lyman, RD Diamond. Independence of neutrophil respiratory burst oxidant generation from the early cytosolic calcium response after stimulation with unopsonized *Candida albicans* hyphae. Infect Immun 57:1499–1505, 1989.

209. RD Diamond, CA Lyman, DR Wysong. Disparate effects of interferon-gamma and tumor necrosis factor-alpha on early neutrophil respiratory burst and fungicidal responses to *Candida albicans* hyphae in vitro. J Clin Invest 87:711–720, 1991.

210. JY Djeu, DK Blanchard, D Halkias, H Friedman. Growth inhibition of *Candida albicans* by human polymorphonuclear neutrophils: activation by gamma-interferon and tumor necrosis factor. J Immunol 137:2980–2984, 1986.

211. R Kaposzta, P Tree, L Marodi, S Gordon. Characteristics of invasive candidiasis in gamma interferon– and interleukin-4-deficient mice: role of macrophages in host defense against *Candida albicans*. Infect Immun 66:1708–1717, 1998.

212. SS Lefkowitz, MP Gelderman, DL Lefkowitz, N Moguilevsky, A Bollen. Phagocytosis and intracellular killing of *Candida albicans* by macrophages exposed to myeloperoxidase. J Infect Dis 173: 1202–1207, 1996.

213. I Szabo, L Guan, TJ Rogers. Modulation of macrophage phagocytic activity by cell wall components of *Candida albicans*. Cell Immunol 164:182–188, 1995.

214. G Gaziri, LC Gaziri, R Kikuchi, J Scanavacca, I Felipe. Phagocytosis of *Candida albicans* by concanavalin-A activated peritoneal macrophages. Med Mycol 37:195–200, 1999.

215. CB Forsyth, EF Plow, L Zhang. Interaction of the fungal pathogen *Candida albicans* with integrin CD11b/CD18: recognition by the I domain is modulated by the lectin-like domain and the CD18 subunit. J Immunol 161:6198–6205, 1998.

216. T Marth, BL Kelsall. Regulation of interleukin-12 by complement receptor 3 signaling. J Exp Med 185:1987–1995, 1997.

217. M Castro, NV Ralston, TI Morgenthaler, MS Rohrbach, AH Limper. *Candida albicans* stimulates arachidonic acid liberation from alveolar macrophages through alpha-mannan and beta-glucan cell wall components. Infect Immun 62:3138–3145, 1994.

218. E Blasi, M Puliti, L Pitzurra, A Bartoli, F Bistoni. Heterogeneous secretory response of phagocytes from different anatomical districts to the dimorphic fungus *Candida albicans*. Cell Immunol 153: 239–247, 1994.

219. M Sasada, RB Johnston Jr. Macrophage microbicidal activity: correlation between phagocytosis-associated oxidative metabolism and the killing of *Candida* by macrophages. J Exp Med 152:85–98, 1980.

220. L Marodi, JR Forehand, RB Johnston Jr. Mechanisms of host defense against *Candida* species. II. Biochemical basis for the killing of *Candida* by mononuclear phagocytes. J Immunol 146: 2790–2794, 1991.

221. L Marodi, S Schreiber, DC Anderson, RP MacDermott, HM Korchak, RB Johnston Jr. Enhancement of macrophage candidacidal activity by interferon-γ increased phagocytosis, killing and calcium signal mediated by a decreased number of mannose receptors. J Clin Invest 91:2596–2601, 1993.

222. SH Kaufmann, MJ Reddehase. Infection of phagocytic cells. Curr Opin Immunol 2:43–49, 1989.

223. M Stein, S Keshav, N Harris, S Gordon. Interleukin 4 potently enhances murine macrophages mannose receptor activity: a marker of alternative immunologic macrophage activation. J Exp Med 176:287–292, 1992.

224. Y Yamamoto, TW Klein, H Friedman. Involvement of mannose receptor in cytokine interleukin-1β (IL-1β), IL-6, and granulocyte-macrophage colony-stimulating factor responses, but not in chemokine macrophage inflammatory protein 1β (MIP-1β), MIP-2, and KC responses, caused by attachment of *Candida albicans* to macrophages. Infect Immun 65:1077–1082, 1997.

225. VL Shepherd, KB Lane, R Abdolrasulnia. Ingestion of *Candida albicans* down-regulates mannose receptor expression on rat macrophages. Arch Biochem Biophys 344:350–356, 1997.

226. E Blasi, L Pitzurra, M Puliti, Lanfrancone, F Bistoni. Early differential molecular response of a macrophage cell line to yeast and hyphal forms of *Candida albicans*. Infect Immun 60:832–837, 1992.

227. R Káposzta, L Maródi, M Hollinshead, S Gordon, RP da Silva. Rapid recruitment of late endosomes and lysosomes in mouse macrophages ingesting *Candida albicans*. J Cell Sci 112:3237–3248, 1999.

228. R Calderone, J Sturtevant. Macrophage interactions with *Candida*. Immunol Ser 60:505–515, 1994.

229. WC Liles, JE Huang, JA van Burik, RA Bowden, DC Dale. Granulocyte colony-stimulating factor administered in vivo augments neutrophil-mediated activity against opportunistic fungal pathogens. J Infect Dis 175:1012–1015, 1997.

230. MD Richardson, CE Brownlie, GS Shankland. Enhanced phagocytosis and intracellular killing of *Candida albicans* by GM-CSF-activated human neutrophils. J Med Vet Mycol 30:433–441, 1992.

231. RB Ashman, JM Papadimitriou. What's new in the mechanisms of host resistance to *Candida albicans* infection? Pathol Res Pract 186:527534, 1990.

232. D Raveh, BA Kruskal, J Farland, RA Ezekowitz. Th1 and Th2 cytokines cooperate to stimulate mannose-receptor-mediated phagocytosis. J Leukoc Biol 64:108–113, 1998.

233. SJ Zunino, D Hudig. Interactions between human natural killer (NK) lymphocytes and yeast cells: human NK cells do not kill *Candida albicans*, although *C. albicans* blocks NK lysis of K562 cells. Infect Immun 56:564–569, 1988.
234. A Vecchiarelli, F Bistoni, E Cenci, S Perito, A Cassone. In vitro killing of *Candida* species by murine immunoeffectors and its relationship to the experimental pathogenicity. Sabouraudia 23: 377–387, 1985.
235. JY Djeu, DK Blanchard, AL Richards, H Friedman. Tumor necrosis factor induction by *Candida albicans* from human natural killer cells and monocytes. J Immunol 141:4047–4052, 1988.
236. DK Blanchard, MB Michelini-Norris, JY Djeu. Production of granulocyte macrophage colony-stimulating factor by large granular lymphocytes stimulated with *Candida albicans*: role in activation of human neutrophil function. Blood 77:2259–2265, 1991.
237. S Wei, DK Blanchard, S McMillen, JY Djeu. Lymphokine-activated killer cell regulation of T-cell-mediated immunity to *Candida albicans*. Infect Immun 60:3586–3595, 1992.
238. JY Djeu, JH Liu, S Wei, H Rui, CA Pearson, WJ Leonard, DK Blanchard. Function associated with IL-2 receptor-β of human neutrophils: mechanism of activation of antifungal activity against *Candida albicans* by IL-2. J Immunol 150:960–970, 1993.
239. E Rosati, L Scaringi, P Cornacchione, K Fettucciari, R Sabatini, L Mezzasoma, C Benedetti, S Cianetti, R Rossi, P Marconi. Activation of cytokine genes during primary and anamnestic response to inactivated *C. albicans*. Immunology 89:142–151, 1996.
240. C Fè d'Ostiani, G Del Sero, A Bacci, C Montagnoli, P Ricciardi-Castagnoli, A Spreca, L Romani. Dendritic cells discriminate between yeasts and hyphae of the fungus *Candida albicans*: implications for initiation of Th immunity in vitro and in vivo. J Exp Med 191:1661–1674, 2000.
241. RF Hector, JE Domer, EW Carrow. Immune responses to *Candida albicans* in genetically distinct mice. Infect Immun 38:1020–1028, 1982.
242. RF Hector, E Yee, MS Collins. Use of DBA/2N mice in models of systemic candidiasis and pulmonary and systemic aspergillosis. Infect Immun 38:1476–1478, 1990.
243. A Cassone, S Conti, F De Bernardis, L Polonelli. Antibodies, killer toxins and antifungal immunoprotection: a lesson from nature? Immunol Today 18:164–169, 1997.
244. A Casadevall, A Cassone, F Bistoni, JE Cutler, W Magliani, JW Murphy, L Polonelli, L Romani. Antibody and/or cell-mediated immunity, protective mechanisms in fungal diseases: an ongoing dilemma or an unnecessary dispute? Med Mycol 36(suppl. 1):95–244, 1998.
245. A Casadevall. Antibody-mediated protection against intracellular pathogens. Trends Microbiol 6: 5–10, 1998.
246. A Casadevall, MD Scharff. A return to the past: the case for antibody-based therapies in infectious diseases. Clin Infect Dis 21:150–161, 1995.
247. Y Han, JE Cutler. Antibody response that protects against disseminated candidiasis. Infect Immun 63:2714–2719, 1995.
248. Y Han, JE Cutler. Assessment of a mouse model of neutropenia and the effect of an anti-candidiasis monoclonal antibody in these animals. J Infect Dis 175:1169–1175, 1997.
249. Y Han, MA Ulrich, JE Cutler. *Candida albicans* mannan extract-protein conjugates induce a protective immune response against experimental candidiasis. J Infect Dis 179:1477–1484, 1999.
250. W Magliani, S Conti, F De Bernardis, M Gerloni, D Bertolotti, P Mozzoni, A Cassone, L Polonelli. Therapeutic potential of antiidiotypic chain antibodies with yeast killer toxin activity. Nat Biotechnol 15:155–158, 1997.
251. R Matthews, J Burnie. Acquired immunity to systemic candidiasis in immunodeficient mice: role of antibody to heat-shock protein. J Infect Dis 166:1193–1194, 1992.
252. RC Matthews, JP Burnie, D Smith, I Clark, J Midgley, M Conolly, B Gazzard. *Candida* and AIDS: evidence for protective antibody. Lancet 1:263–266, 1988.
253. Matthews R, Burnie J. The role of antibodies in protection against candidiasis. Res Immunol 149: 343–352, 1998.
254. R Medzhitov, CA Janeway Jr. Innate immunity: impact on the adaptive immune response. Curr Opin Immunol 9:4–9, 1997.
255. EB Kopp, R Medzhitov. The toll-receptor family and control of innate immunity. Curr Opin Immunol 11:13–18, 1999.

256. PD Stahl, RA Ezekowitz. The mannose receptor is a pattern recognition receptor involved in host defense. Curr Opin Immunol 10:50–55, 1998.
257. D Underhill, M Ozinsky, A Hajjar, AM Stevens, A Wilson, CB Bassetti, M Aderem. The Toll-like receptor 2 is recruited to macrophage phagosomes and discriminates between pathogens. Nature 401:811–815, 1999.
258. M Meister, B Lemaitre, JA Hoffmann. Antimicrobial peptide defense in Drosophilia. Bioessays 19:1019–1926, 1997.
259. RI Lehrer, T Ganz. Antimicrobial peptides in mammalian and insect host defense. Curr Opin Immunol 11:23–27, 1999.
260. M Saavedra, B Taylor, N Lukacs, PL Fidel Jr. Local production of chemokines during experimental vaginal candidiasis. Infect Immun 67:5820–5826, 1999.
261. HS Kim, DH Shin, SK Kim. Effects of interleukin-10 on chemokine KC gene expression by mouse peritoneal macrophages in response to Candida albicans. J Korean Med Sci 14:480–486, 1999.
262. AA Gaspari, B Burns, A Nasir, D Ramirez, RK Barth, CG Haidaris. CD86 (B7-2), but not CD80 (B7-1), expression in the epidermis of transgenic mice enhances the immunogenicity of primary cutaneous Candida albicans infections. Infect Immun 66:4440–4449, 1998.
263. A Cassone, P Chiani, I Quinti, A Torosantucci. A possible participation of polymorphonuclear cells stimulated by microbial immunomodulators in cytokine dysregulated patterns of AIDS patients. J Leukoc Biol 62:60–66, 1997.
264. A Mencacci, E Cenci, JR Boelaert, P Bucci, P Mosci, C Fè d'Ostiani, F Bistoni, L Romani. Iron overload alters innate and T helper cell responses to Candida albicans in mice. J Infect Dis 175: 1467–1476, 1997.
265. C Ashley, M Morhart, R Rennie, B Ziola. Release of Candida albicans yeast antigens upon interaction with human neutrophils in vitro. J Med Microbiol 46:747–755, 1997.
266. L Oehler, O Majdic, WF Pickl, J Stöckl, E Riedl, J Drach, K Rappersberger, K Geissler, W Knapp. Neutrophil granulocyte-committed cells can be driven to acquire dendritic cell characteristics. J Exp Med 187:1019–1028, 1998.
267. TJ Walsh, J Hiemenz, PA Pizzo. Evolving risk factors for invasive fungal infections: all neutropenic patients are not the same. Clin Infect Dis 18:793–798, 1994.
268. GP Bodey, EJ Anaissie. Chronic systemic candidiasis. Eur J Clin Microbiol Infect Dis 8:855–857, 1989.
269. EJ Bow, R Loewen, MS Cheang, B Schacter. Invasive fungal disease in adults undergoing remission-induction therapy for acute myeloid leukemia: the pathogenic role of the antileukemic regimen. Clin Infect Dis 21:361–369, 1995.
270. AP Mitchell. Dimorphism and virulence in Candida albicans. Curr Opin Microbiol 1:687–692, 1998.
271. SD Kobayashi, JE Cutler. Candida albicans hyphal formation and virulence: is there a clearly defined role? Trends Microbiol 6:2–4, 1998.
272. H Lo, JR Kohler, B DiDomenico, D Loebenberg, A Cacciapuoti, GR Fink. Nonfilamentous Candida albicans mutants are avirulent. Cell 90:939–949, 1997.
273. J Banchereau, RM Steinman. Dendritic cells and the control of immunity. Nature 392:245–252, 1998.
274. G Reis e Sousa, A Sher, P Kaye. The role of dendritic cells in the induction and regulation of immunity to microbial infection. Curr Opin Immunol 11:392–399, 1999.
275. AJP Brown, NA Gow. Regulatory networks controlling Candida albicans morphogenesis. Trends Microbiol 7:333–338, 1999.
276. B Chen, Y Shi, JD Smith, D Choi, JD Geiger, JJ Mulè. The role of tumor necrosis factor α in modulating the quantity of peripheral blood-derived, cytokine-driven human dendritic cells and its role in enhancing the quality of dendritic cell function in presenting soluble antigens to CD4+ T cells in vitro. Blood 91:4652–4661, 1998.
277. SL Newman, A Holly. Phagocytosis and killing of Candida albicans (Ca) by human dendritic cells. Society for Leukocyte Biology 15th International Congress, Cambridge, UK, 1999, Abs pp 18.
278. HC Poynton. Immune modulation by cytokines in the treatment of opportunistic infections. Curr Opin Infect Dis 10:275–280, 1997.

279. WC Liles, JE Huang, JA van Burik, RA Bowden, DC Dale. Granulocyte colony-stimulating factor administered in vivo augments neutrophil-mediated activity against opportunistic fungal pathogens. J Infect Dis 175:1012–1015, 1997.
280. A Mencacci, E Cenci, R Spaccapelo, L Tonnetti, L Romani, P Puccetti, F Bistoni. Rationale for cytokine and anticytokine therapy of *Candida albicans* infection. J Mycol Med 5:25–30, 1995.
281. E Cenci, A Mencacci, G Del Sero, F Bistoni, L Romani. Induction of protective Th1 responses to *Candida albicans* by antifungal therapy alone or in combination with an interleukin-4 antagonist. J Infect Dis 176:217–226, 1997.
282. A Mencacci, E Cenci, A Bacci, F Bistoni, L Romani. Host immune reactivity determines the efficacy of combination immunotherapy and antifungal chemotherapy in candidiasis. J Infect Dis 181:686–694, 2000.
283. CH Poynton, RA Barnes, J Rees. Interferon γ and granulocyte-macrophage colony-stimulating factor for the treatment of hepatosplenic candidosis in patients with acute leukemia. Clin Infect Dis 26:239–240, 1998.
284. M Hoashi, S Imayama, Y Hori, S Kashiwagi. An AIDS patient with atopic dermatitis-like eruption responsive to systemic anti-fungal treatment. J Dermatol 19:972–975, 1992.
285. CA Elmets. Management of common superficial fungal infections in patients with AIDS. J Am Acad Dermatol 31:S60–S63, 1994.

22

Cell Wall of *Aspergillus fumigatus*: Structure, Biosynthesis, and Role in Host–Fungus Interactions

Isabelle Mouyna and Jean-Paul Latgé
Pasteur Institute, Paris, France

I. INTRODUCTION

Aspergillus fumigatus is a saprophytic fungus that plays an essential role in recycling environmental carbon and nitrogen. It grows naturally on decaying organic material in the soil and sporulates abundantly. The conidia are present in all environments, indoors and outdoors, with a range of concentration between 0 to 100 conidia/m^3. They have a diameter small enough (2–3 μm) to reach all lung compartments. Inhalation of conidia by the immunocompetent host rarely has any adverse affect since they are eliminated relatively efficiently by innate immune mechanisms. Thus, until recently, *A. fumigatus* was viewed as a rather weak and infrequent pathogen responsible for aspergilloma, an overgrowth of the fungus in preexisting lung cavities, and allergic bronchopulmonary aspergillosis, a complication occurring in patients suffering from atopic asthma or cystic fibrosis [1,2]. Because of the increase in the number of immunocompromised patients and the degree of severity of modern immunosuppressive therapies, the situation has changed dramatically in recent years. Over the past 10 years, *A. fumigatus* has become the most prevalent airborne fungal pathogen, causing severe and usually fatal invasive infections mainly among hematology patients. This situation mainly results from (1) a difficult clinical and laboratory diagnosis; (2) a relatively ineffective antifungal therapy, mainly based on the use of amphotericin-B, which has severe secondary toxic effects for humans; and (3) a poor understanding of the physiopathology of invasive aspergillosis (IA).

The lack of knowledge of the host and fungal factors involved in the establishment of the disease can be seen at the level of the fungal cell wall (CW) which plays a central role in the pathogenic life of *A. fumigatus*. Several biological facts demonstrate the role of the CW during fungal infection: (1) the cell wall is continuously at the interface between the host and the fungal pathogen during the course of the disease; (2) it is rich in antigens and enzymes, the latter helping the fungus to invade the host tissues; and (3) it protects the fungus against an aggressive environment, in particular against phagocytic reactions and antifungal drugs.

This chapter will summarize our current knowledge of the cell wall of the conidia and mycelium of *A. fumigatus* with special emphasis on its role during infection. Comparative analysis of biochemical/genetic features found in the less pathogenic species of *Aspergillus* such

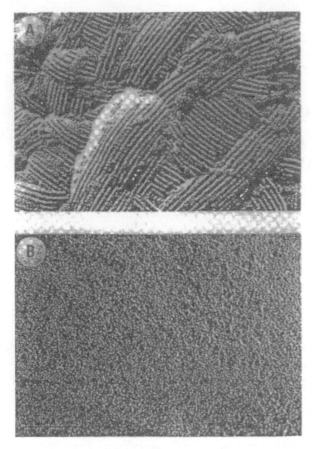

Figure 1 Surface carbon-platinum replicas of conidia of wild-type (A) and rodletless mutant (B) of *A. fumigatus*. Bar, 100 nm.

as *A. nidulans* and *A. niger* should help identify molecules or pathways specific for the pathogenic behavior of *A. fumigatus*.

II. CONIDIUM CELL WALL

A. Outer Layer and Adhesion

1. Hydrophobins

The outermost cell wall layer of *A. fumigatus* conidia is characterized by the presence of inter-woven rodlet fascicles. This layer, which is composed of hydrophobic proteins (hydrophobins), confers hydrophobic properties to *A. fumigatus* conidia. The hydrophobins are a family of homologous proteins, present on the surface of dry conidia of all aerial fungal species [3,4]. All hydrophobins have eight conserved cysteine residues, similar hydropathy patterns, low Mr in the range of 10–20 kDa, and they are extremely resistant to chemical degradation, which allows their extraction using concentrated acids. The gene encoding the rodlet protein RodAp of *A. fumigatus* has been cloned, and a rodletless mutant has been generated [3] (Fig. 1). In *A. fumiga-*

tus, the absence of this rodlet layer resulted in the apparition of conidia which become more hydrophilic than the conidia of the parental strain. Accordingly, conidia from this mutant bound less readily to proteins with hydrophobic pockets such as albumin or collagen. Binding to other proteins such as laminin and fibrinogen was not altered in the mutant, suggesting that the localization of specific host adhesins occurring on top of the rodlet layer is not modified in the Δ*rodA* mutant. In animal models of invasive pulmonary aspergillosis, mortality was comparable for the parent and rodletless strains, but the inflammatory response was lower with the Δ*rod* mutant [5]. Thus, RodAp appears to play a role in the hydrophobic interactions between the fungus and the host, but these interactions do not appear to be essential for the virulence of the fungus.

RODA genes from *A. nidulans* and *A. fumigatus* are highly homologous (75% identity). The *RODA* gene from *A. nidulans* can indeed complement the *A. fumigatus* Δ*rodA* mutant [6]. However, a comparative analysis using aqueous solvent partitioning assays, latex microsphere adhesion, and microelectrophoresis showed that the rodlet layer on surface of *A. nidulans* and *A. fumigatus* is associated with different physicochemical properties of the conidia of both species [7]. Microelectrophoresis experiments show that the conidia of *A. nidulans* are less negatively charged than those of *A. fumigatus*. Removal of the rodlet layer induced a greater loss of conidial hydrophobicity, as seen by a reduction in binding to latex microspheres for the *A. nidulans* Δ*rodA* mutant, whereas no difference was seen between *A. fumigatus* Δ*rodA* mutant and wild-type strain. In addition, in *A. fumigatus*, the absence of the rodlet layer led to the appearance of a weak basic and acid character, whereas it induces a strong basic character in *A. nidulans*. Although morphologically or structurally related, the rodlet layer organization is different and confers different surface properties to these two species of *Aspergillus*. Similarly, restoration of the rodlet structure and hydrophobicity of the conidia of *Magnaporthe grisea* can be obtained with hydrophobin genes from distantly unrelated species without fully restoring the biological activity due to the presence of the parental rodlet [8].

Associated with the different physicochemical properties of *A. fumigatus* and *A. nidulans* conidia, it was recently shown that the composition of hydrophobins differed in the two species. In *A. fumigatus*, a 14-kDa hydrophobin (named RodBp) which coextracts with the 16-kDa RodAp protein has been recently identified (Paris et al., unpublished). Sequence analysis has shown that *RODB*, which has not been found in *A. nidulans*, was also different from *DEWA*, a second hydrophobin present in *A. nidulans* [9], which has not been identified in *A. fumigatus*. Differences in the physicochemical properties of the conidia of the two species could also result from the occurrence of different molecules on the surface of the rodlet layer since immunocyto-chemical data indicated that adhesins and receptors to host proteins and cells occur on top of the rodlet layer [10]. Unfortunately, only *A. fumigatus* adhesins have been analyzed to date.

2. Adhesins

Conidia of *A. fumigatus* bind specifically to various circulating or basement membrane-associated host proteins (fibrinogen, laminin, complement, fibronectin, albumin, immunoglobulins, collagen, and surfactant proteins) and to lung epithelial and phagocytic cells [10–22]. Only a few of the existing adhesion systems have been characterized in *A. fumigatus* to a biochemical level (Table 1).

Carbohydrate and protein molecules on the conidial surface are involved in binding to host proteins. For example, an unknown carbohydrate molecule on conidia bound in a calcium-dependent manner to pulmonary surfactant proteins A and D. Fucose– and sialic acid–specific lectins also associated with adhesion have been identified on the conidial cell wall. Western blot analysis has shown that complement binds to a 54 to 58-kDa doublet protein found on the

Table 1 Adhesion of *A. fumigatus* to Host Proteins and Cells

Binding to host	Fungal cells	Fungal adhesins
Cells		
Alveolar (pneumocyte II) and bronchial epithelium	co[a]	?
Endothelium	co	?
Alveolar macrophage/monocyte	co/myc[b]	? (carbohydrate ?)
Polymorphonuclear neutrophil	co/myc	?
Proteins		
Fibrinogen, laminin	co	72 or 37 kDa[c]
Fibronectin	myc	23/30 kDa[d]
Immunoglobulins	co/myc	Proteins/polysaccharides
Complement	co/myc	54/58 kDa
Surfactant	co	?
Collagen, Albumin	co	14 kDa rodA

[a] co = conidium.
[b] myc = mycelium.
[c] Depending on the publication; not inhibited by RGD peptides but inhibited by sialic acid.
[d] Inhibited by RGD peptides.

surface of the conidia. The receptor for laminin is a 72-kDa glycoprotein also present on the surface of the conidium. Binding of fibrinogen, laminin, fibronectin, and complement is associated with the outer and inner wall layer of the conidia with a different localization for each protein, suggesting that different *Aspergillus* proteins bind specifically to unique host proteins.

Receptors involved in recognition and binding of conidia by alveolar macrophages have been poorly studied, although lectinlike interactions are thought to be primarily responsible for adherence and ingestion of conidia. This interaction would be expected since the alveolar environment of the resident macrophage is probably free of opsonic factors such as complement and immunoglobulins. The mannosyl-fucosyl receptor and two other receptors (inhibited by β-glucan and chito-oligosaccharides) have been suggested to mediate conidial binding, but specific receptors have not been identified.

Polymorphonuclear neutrophils (PMN) were thought to act exclusively on hyphae, as opposed to conidia, of *A. fumigatus*. Neutrophils also adhere to the surface of the hyphae, but the process of adhesion to the hyphae has been poorly studied. It is known that even though complement and antibodies bind avidly to hyphae, their presence is not required for the interaction between hyphae and neutrophils [20]. Conidial or mycelial fungal and PMN receptors involved in this interaction remain to be identified.

Although it has been repeatedly mentioned that adhesion of *A. fumigatus* to host cells and proteins is a prerequisite for infection, there is presently no evidence of such a role for *A. fumigatus* adhesins. None of the specific *A. fumigatus* adhesins have been purified to date, and their role in the establishment of disease will remain debatable until a mutant devoid of adhesive capacity is obtained. Another research avenue to assess the role of adhesins in the infectious process is a comparative analysis of adhesins of the pathogenic *A. fumigatus* with the least pathogenic species, *A. nidulans* or *A. niger*. A study by Henwick et al. [23] suggested that complement binding was increased in the nonpathogenic *Aspergillus* species. A careful reassessment of their data as well as recent studies in our laboratory indicated that nonpathogenic species such as *A. niger* are phagocytosed by alveolar macrophages as quickly and as efficiently as *A.*

fumigatus (B. Philippe, personal communication, 2000), suggesting that adhesion is not a species-specific event. Aldehyde-fixed conidia or 2-μm latex microspheres are engulfed by phagocytes, questioning the role of specific adhesin-receptors during phagocytosis. Moreover, the modification of the adhesin pattern and physicochemical surface properties of the conidial surface seen when the fungus is grown under different culture conditions also suggests that adhesion should not be a specific event required for host infection [11,24].

B. Inner Layer and Its Role in Resistance to Phagocytosis

1. Composition of the Inner Layer and Melanin Biosynthesis

Transmission electron microscopy observations have shown that the conidial cell wall is constituted of a dense pigmented outer layer and a translucent inner layer. The chemical composition of the conidial cell wall of *A. fumigatus* has not been characterized yet, with the exception of the melanin component whose biosynthesis has been the focus of several studies in the last few years [25–28]. The groups of Kwon-Chung and Brackhage have recently shown that *A. fumigatus* synthesizes its pigment through the dihydroxynaphthalene (DHN)-melanin pathway (Fig. 2). Molecular studies have identified to date six genes involved in the DHN-melanin biosynthetic pathway (Fig. 2). The six genes form a cluster spanning 19 kb and are all developmentally regulated and expressed during conidiation. *ALB1* (= *PKSP*) encodes a putative protein of 234 kDa which exhibits a high degree of similarity with a polyketide synthase and is homologous to *WA* of *A. nidulans*. The product of *ALB1* has not been identified. *ARP1* and *ARP2* encode a scytalone dehydratase and a 1, 3, 6, 8 THN reductase respectively, as shown by the analysis of accumulated intermediates of the DHN melanin using $\Delta arp1$ and $\Delta arp2$ mutants in media $\pm$ scytalone $\pm$ tricyclazole. *ABR1* and *ABR2* have signature sequences characteristic of oxidases: *ABR1* shares sequence homologies with multicopper oxidase while *ABR2* has a 65% similarity to a laccase encoded by *YA* of *A. nidulans*. The function of the sixth gene (*AYG1*) remains unknown. The inhibition of conidial pigmentation by tricyclazole and the identification of the function of *ALB1, ARP1,* and *ARP2* gene products suggest that the green pigment of *A. fumigatus* is synthesized via a DHN-melanin pathway. In contrast to *A. fumigatus, A. nidulans* pigmentation is not inhibited by tricyclazole, and no homolog of *ARP1* was identified in *A. nidulans,* suggesting the presence of dissimilar pathways in *A. nidulans* and *A. fumigatus* pigment synthesis [29].

Besides its interest in pigment biosynthesis, it is the first gene cluster identified in *A. fumigatus*. Clustering allows regulatory elements to be shared between genes but this suggestion has not been investigated. Clustering of genes involved in melanin and mycotoxin related pathways have already been identified in other fungi including non-*fumigatus Aspergillus* spp. [30].

In Figure 3 our current knowledge of the structure of the conidial cell wall with rodlets and melanin is shown. This figure illustrates how little we know about the interactions among the different components (adhesins, hydrophobins, constitutive polysaccharide and melanin, stored proteins) of the cell wall, and particularly their localization which is essential to host-pathogen interactions.

2. Cell Wall–Associated Resistance to Phagocytic Reactions

In humans, conidia of *A. fumigatus* are engulfed and killed by the alveolar macrophage. Even in the immunocompetent host, killing of conidia of *A. fumigatus* is extremely slow (3 days to reach 100%) (B. Philippe, personal communication, 2000). The slow conidial killing is at least partly associated with the protective role of the cell wall. The use of mutants with white conidia has shown that the melanin layer of the cell wall plays a major role in the protection against the phagocytic reactions. Both in vivo and in vitro studies have shown that white conidia were

Gene	Putative protein	Color of mutant conidia	Involvement in pathogenesis
ALB1 (=*PKSP*))	Polyketide synthase	White	+ complement binding increase
ARP2	HN reductase	dark reddish pink (= tricyclazole)	–
ARP1	Scytalone dehydratase	light reddish pink	+ complement binding increase
AYG1	?	yellow green	?
ABR1	oxidase	brown	?
ABR2	laccase	brown	?

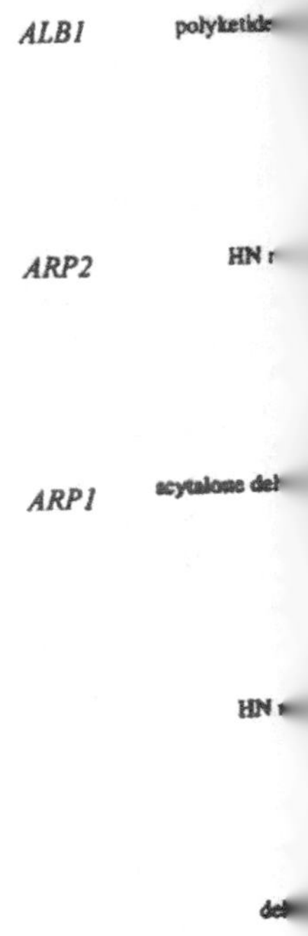

ALB1   polyketide

ARP2   HN r

ARP1   scytalone de

HN

de

Figure 2 Biosynthetic pathway of melanin in fungi. The solid arrows indicate the main melanin pathw branching pathways from the melanin pathway. The function of only three gene products of the melani

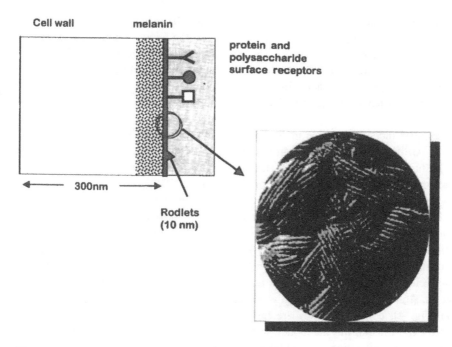

Figure 3 Hypothetical structure of the conidial cell wall.

more efficiently damaged by phagocytes than green conidia. When conidia were incubated in vitro with reactive oxygen species (ROS) like NaOCl or hydrogen peroxide, white conidia were about 10-fold more sensitive to ROS than green conidia [29]. This finding indicates that the pigment gives *A. fumigatus* conidia some protection against ROS, which could be responsible for the higher survival of green conidia in monocytes. Very similar results were obtained with the dematiaceous fungi or the melanin-producing yeast *C. neoformans*, where the ability to scavenge ROS has been already linked to the presence of the pigment. The conidia of *A. fumigatus* were also tested for their ability to cause ROS production in human monocytes and polymorphonuclear leukocytes (PMN) [29]. Conidia of the wild-type strain only led to a small release of ROS, which agreed well with previously reported results [13,31,32]. By contrast, conidia of the white mutant caused a 10-fold higher ROS release from both cell types. In fact, conidia of both strains induced an identical oxidative burst, but the resulting ROS was scavenged by the pigment present in wild-type conidia. Compared with *A. fumigatus*, similar results were obtained with *A. nidulans*; i.e., ΔWA conidia triggered a 10-fold increase in ROS release compared with wild-type conidia. These data indicate that ROS scavenging also occurs in pigmented *A. nidulans* conidia (B. Jahn, personal communication, 2000). Consequently, pigment confers a similar degree of protection against damage caused by ROS to both species and cannot account for the lower pathogenicity of *A. nidulans*.

Mutations in the melanin pathway resulted in *A. fumigatus* a modification or reorganization of the cell wall since white conidia have a smooth surface whereas wild-type conidia show a distinct ornamentation on their surface [26,33]. Although conidia of the white mutants and wild-type strains of *A. fumigatus* have very different surface morphologies, the global chemical composition of the conidial cell wall does not seem quantitavely altered in this species (J.P. Latgé, unpublished), and the chemical modifications associated with the change in surface ornamentation remain unknown. Hence, besides a lack of pigmentation, the mutants showed another morphological phenotype which might be important for the interplay between conidia and immune effector cells. The latter assumption was also supported by the findings that conidia of a white mutant exhibited significantly elevated complement component C3 binding capacity compared with wild-type conidia [26–28]. Also, white conidia seemed to be engulfed by alveolar macrophages much quicker than wild-type conidia, suggesting a modification of the conidial surface receptors recognized by the alveolar macrophage. A significant reorganization of the conidial cell wall has been indeed identified in *A. nidulans* as a consequence of a mutation in the melanin pathway. An increase in α1-3 glucan is noted in the cell wall of white conidia of the *WA* mutant [34]. Changes in conidial ornamentation are difficult to assess in the *A. nidulans* ΔWA since data are conflicting (B. Jahn, personal communication, 1999) [34].

Reduction in fungal pathogenicity due to the inhibition of the melanin pathway remains limited, however. In an intravenous murine infection model, the virulence of white conidia was only significantly reduced by approximately 20–50% of that observed with wild type. Nonpigmented conidia were still able to induce IA when inoculated intranasally into steroid-treated mice without any statistically significant difference with pigmented conidia (J.P. Latgé, unpublished). Nevertheless, in a study with mixed infections with wild-type white and green strains of *A. fumigatus*, strains with white conidia were recovered in lower amounts than strains with green conidia (Fig. 4), confirming they are less virulent than green strains even in an immunocompromised experimental murine model (unpublished). However, the role of melanin in virulence may have minor implications in our understanding and control of IA since >99.99% of airborne conidia are green while their invading mycelia are hyaline (a situation different from most plant fungal pathogens such as *Magnaporthe grisea* or the human pathogen *C. neoformans* where the invading organisms are pigmented or produce melanin).

Taken together, these results suggest that the conidial cell wall, and in particular pigment and/or associated structures, contributes to the resistance of *A. fumigatus* to killing by professional phagocytes. However, the role of melanin during *A. fumigatus* infection is not entirely clear. For example, Tsai et al. [28] indicated that an *ARP2* deletant does not have a significantly reduced virulence relative to the wild-type strain although it occurs upstream of *ARP1* in the

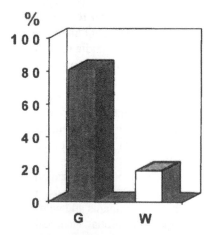

Figure 4   Recovery of strains from mice experimentally infected with a mixture of wild strains of *A. fumigatus* with white (W) and green (G) conidia. Strains were recovered after mouse death.

melanin cascade. The authors suggest that accumulated metabolites may be more important than the conidial pigment itself in influencing the virulence of *A. fumigatus*.

3. Phagocytic Mechanisms Involved in the Killing of *A. fumigatus* Conidia

The antimicrobial system(s) responsible for killing conidia have not been identified to date. Data obtained especially with white mutants suggest that reactive oxygen intermediates do not play an essential role in the killing of *A. fumigatus* wild-type conidia by macrophages [13,35–37]. The role of nitric oxide in the killing of *A. fumigatus* conidia has been insufficiently investigated. Fungicidal activity of alveolar macrophages was unaltered in the presence of the competitive inhibitor N-monomethyl L-arginine, suggesting that nitric oxide was not involved in the killing capacity of murine and human macrophages. Cationic peptides have been shown to have potent antifungal activity in vitro, but their role in vivo has not been assessed [38–40]. Lysosomal hydrolases could also play a main role in conidial alterations since the conidia are completely digested in the phagocyte (glycosyl hydrolases and in particular chitinases have been identified in phagocytes). These glycosylhydrolases, in conjunction with proteases which are known to be induced during macrophage phagocytosis, could play a main role in conidial killing [41]. The use of transgenic mice should be of major interest to identify the metabolic cascades involved in conidial killing.

III. THE MYCELIAL WALL

The mycelial wall has been long considered as an inert organelle. Recent studies, mainly based on the analysis of the yeast cell wall, suggest that the fungal cell wall is indeed a dynamic structure where constitutive polymers (polysaccharides and eventually proteins) are chemically modified and covalently linked together to form the mature cell wall [42]. In addition, in *A. fumigatus*, as in other fungi, the cell wall acts as a sieve and a reservoir for molecules such as antigens and enzymes playing an active role during infection [43]. Analysis of the biological properties of the *A. fumigatus* mycelial cell wall requires as a prerequisite a thorough understanding of the composition and arrangement of the structural components of the cell wall. Such studies have been undertaken only recently and are summarized below.

A. Structure

Cell wall polymers of fungi are classically divided into two groups depending on their solubility in hot alkali. In *A. fumigatus*, the alkali soluble fraction is composed of $\alpha(1\text{-}3)$ glucans and galactomannan [44]. The alkali-insoluble fraction of the *A. fumigatus* cell wall, which is the fraction believed to be responsible for fungal cell wall rigidity, has been extensively analyzed [45]. Using enzymatic digestion with recombinant endo-$\beta 1\text{-}3$ glucanase and chitinase, several fractions that contained specific interpolysaccharide covalent linkages were purified by liquid chromatography and characterized by GC-MS and NMR. Galactomannan, chitin, and $\beta 1\text{-}3$ glucan were the main components of the alkali-insoluble fraction. A galactosaminogalactan has been also found in the alkali-insoluble fraction of the cell wall but it does not seem to be linked to the structural cell wall polysaccharides [45,46]. This heteropolymer is reminiscent of a similar polysaccharide secreted by *A. parasiticus* [47,48]. A linear $\beta 1\text{-}3/1\text{-}4$ glucan, never previously described in fungi, was also found in *A. fumigatus*. The $\beta 1\text{-}3$ glucan is a branched polymer with 4% of $\beta 1\text{-}6$ branch points. Chitin, galactomannan, and the linear $\beta 1\text{-}3/1\text{-}4$ glucan were covalently linked to the nonreducing end of $\beta 1\text{-}3$ glucan side chains. As in *Saccharomyces cerevisiae* [49],

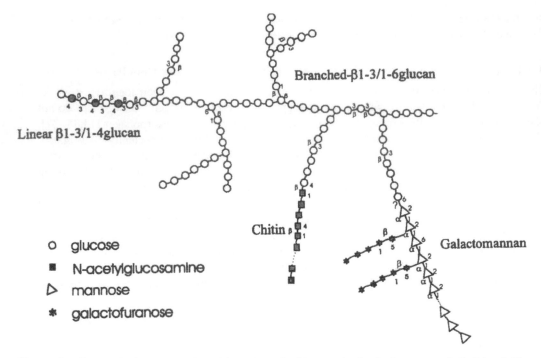

Figure 5 Hypothetical representation of the polysaccharide organization in the structural alkali-insoluble core of the *A. fumigatus* cell wall.

chitin was linked via a β1-4 linkage to β1-3 glucan. Branching of β1-3 glucan is a prerequisite and an early event in the construction of the cell wall resulting in an increase in potential acceptor sites; it precedes the formation of covalent linkages between the β1-3/1-6 glucan core and the other constitutive polysaccharides of the three-dimensional network of the *A. fumigatus* cell wall (Fig. 5). Such chronology in the biosynthetic events can be also applied to the yeast cell wall core. However, "decorating and finishing" are different in the cell wall of yeast and filamentous fungi. β1-6 glucan and proteins associated to β1-3 glucan or chitin in yeast [50] have not been found in *A. fumigatus*, suggesting that the entire model for the yeast cell wall architecture cannot be applied to filamentous fungi or at least to *A. fumigatus*. The mycelial cell wall of *A. nidulans* has been poorly investigated [51], but it appears that the polysaccharide composition of the cell wall is similar in both *Aspergillus* species.

B. Biosynthesis

Enzymes involved in the biosynthesis of the major polysaccharides of the alkali-insoluble fibrillar core of the cell wall, viz. β1-3 glucans and chitin, have been extensively investigated in *Aspergillus*.

Chitin synthases (CS) catalyze the polymerization of GlcNAc from UDP-GlcNAc. Because of the importance of chitin in the filamentous fungi, CS have been extensively studied in model organisms of moulds such as *N. crassa* or *A. nidulans* [52–54]. It has also been investigated in *A. fumigatus*, which is the fungal species with the greatest number of chitin synthases identified [55–57]. Chitin synthases are split into VI classes according to their amino acid sequences.

However, belonging to a chitin synthase class does not necessarily signify functional conservation, as was shown, for example, with chitin synthase C and G of *A. fumigatus* or for different mutant phenotypes seen after disruption of CS genes of the same family [57] (Table 2).

Each CS gene is composed of three regions, including a hydrophilic conserved region at the N-terminus, a neutral highly conserved region, and a hydrophobic region at the C-terminus, which is thought to anchor the enzyme at the membrane and contain the catalytic domain. The

Table 2 Chitin Synthases in *A. fumigatus* and in *A. nidulans*

Class	Chitin synthase genes		Expression/ localization	Phenotype of mutant[b]
	A. nidulans	*A. fumigatus*		
I (= ScCHS1)	C		Conidiophore/ hyphae	No
		A	?	No
II (= ScCHS2)	A		Conidiophore (hyphae)	No[c]
		B	?	No
III	B		Conidiophore (hyphae)	Severe defect in hyphal growth (hyperbranching with abnormal structure). Reduction (50%) of conidiation
		C	?	No
		G	?	Reduced growth rate. Increased hyphal branching
IV (= Sc CHS3)	E (= D. Takagi)		Conidiophore/ hyphae	No
		F	?	?
V	D (= CSmA Takagi)		Conidiophore	Normal growth rate but mycelium swollen; more sensitive to Calcofluour and Congo red and lysed in absence of osmotic protectant
		E	?	Abnormal hyphae (swollen structure) and conidiophore morphology (reduced conidiation). Slightly reduced growth rate
VI		D	?	No

[a] I, II, III zymogenic; IV, V, VI nonzymogenic.
[b] Phenotype not linked to the amount of chitin of cell wall mutant. CS activity difficult to assess due to the lack of measurement under the same conditions for all mutants.
[c] Phenotype (reduced conidiation) in the double mutants Δ*ChsAChsD* and Δ*ChsAChsC*.

N-terminal and central regions are thought to face the cytoplasmic side of the plasma membrane. No recombinant chitin synthase protein has been produced, and the organisation of this enzyme remains totally hypothetical. Biochemical data have suggested that classes I–III are zymogenic, i.e., stimulated in vitro by trypsin whereas classes IV–VI are nonzymogenic. However, in *Aspergillus* as in yeast, the protease(s) putatively involved in the activation of the zymogenic chitin synthase in vivo have not been identified. Class III and V have been only identified in filamentous fungi. Interestingly, mutants with the most altered phenotype belong to these two classes (Table 2) [54,55]. Mutations in *AfChSE*, *AfChSG*, and the *A. nidulans* homologs (*CHSD* and *CHSB*) result in the reduction of hyphal growth, periodic swellings along the length of hyphae and a block in conidiation that is partially restored by growth in presence of an osmotic stabilizer [52,54,57]. A direct correlation between the severity of the mutant phenotype and the amount of chitin left in the cell wall as a consequence of the chitin synthase gene disruption is not apparent (Table 2). Multiple (>2) gene disruptions have not been performed so as to really understand the additive or synergistic roles of the various chitin synthases in chitin deposition. Such a study should be done to identify the redundancy of genes in the chitin synthase gene family.

Studies in *S. cerevisiae* have shown that chitin incorporation in the cell wall follows an equilibrium between synthesis and lysis due to chitinases. In filamentous fungi, the dichotomy of synthases/hydrolases in the organization of chitin cell wall has been studied much less. Several chitinases have been detected in *A. fumigatus* [58,59]. Much of the cellular chitinase activity does not bind to Con A and is heat sensitive. In contrast, secreted chitinases are heat stable, bind to Con A, and are the most immunologically reactive components as judged by their binding to antibodies from sera of aspergillosis patients. Most chitinases detected have an acidic pI of 3.3–4.5. However, with the exception of a 45-kDa polypeptide, no chitinases have been biochemically purified or even identified by their Mr. Three chitinases have been cloned in *A. fumigatus*, but no mutant has been constructed yet (D. Adams, personal communication, 2000). Chitinases in *A. nidulans* have not been studied.

β(1-3) glucans of *D. fumigatus* are synthesized by a plasma membrane-bound glucan synthase complex, which uses UDP-glucose as a substrate and extrudes β(1-3) glucan chains through the membrane into the periplasmic space [42,60]. A gene homologous to the *FKS* genes of *S. cerevisiae* which encodes the putative catalytic subunit of β(1-3) glucan synthase has been identified in *A. fumigatus* (A. Beauvais, personal communication, 2000). The *FKS* ORF of *A. fumigatus* has 6.5 kb with 2 introns at the C and N termini and is almost identical to the *A. nidulans FKS* gene (90% amino acid identify and conservation of the intron positions). It is an integral membrane protein with a cytoplasmic N-terminus and 16 transmembrane domains. All attempts to express *FKS* have failed. No biochemical evidence has shown that *FKS* is the catalytic domain of the glucan synthase. In contrast, the regulatory *RHO1* subunit of the *A. fumigatus* glucan synthase has been identified. It is 949 bp long with 4 introns which encodes a 21.5-kDa protein highly homologous to *RHO1* of *S. pombe* and *S. cerevisiae* (85% and 79% identity, respectively).

β(1-3) glucan chains produced by the β(1-3) glucan synthase complex remain unorganized and alkali soluble until covalent linkages occur between β(1-3) glucans and other cell wall components. Enzymes present in the periplasmic space of the *A. fumigatus* cell wall which act on β1-3 glucans have been investigated. Two types of enzymatic activities were sought: (1) β1-3 glucanases, which should play a role in morphogenetic events such as germination and branching, both processes which require plasticization and cell wall expansion; and (2) transglycosidases responsible for the linkage between glucans and other polysaccharides which could contribute to the reticulation and stabilization of the cell wall skeleton.

A novel β(1-3) glucanosyltransferase (*GEL*) isolated from the cell wall of *A. fumigatus* was recently characterized [61]. This enzyme splits internally a β(1-3) glucan molecule and transfers the newly generated reducing end to the nonreducing end of another β(1-3) glucan molecule. The generation of a new β(1-3) linkage between the acceptor and donor molecules resulted in the elongation of β(1-3) glucan chains. The gene encoding Gel1p has been cloned and sequenced. The predicted amino acid sequence of Gel1p was homologous to several yeast protein families (Gasp in *Saccharomyces cerevisiae*, Phrp in *C. albicans*, Cggp in *C. glabrata*, and Epdp in *C. maltosa*) [62–69]. Although it was previously shown that the expression of these genes was required for correct morphogenesis in yeast, the biochemical function of the encoded proteins was unknown. The biochemical assays performed on purified recombinant Gas1p of *S. cerevisiae*, Phr1p and Phr2p of *C. albicans*, and complementation experiments have shown that these yeast proteins have a β(1-3) glucanosyltransferase activity similar to that of Gel1p [62,63]. Biochemical data and sequence analysis have shown that Gel1p is a glycosylated protein attached to the membrane through a glycosylphosphatidylinositol (GPI) anchor, similarly to the yeast homologous proteins. The activity has been also detected in membrane preparations, showing that this glucanosyltransferase is indeed active in vivo. Our results show for the first time that proteins anchored to the plasma membrane via a GPI have an enzymatic activity which plays an active role in cell wall synthesis and fungal morphogenesis (Table 3).

Mutations in the genes of these glucanosyltransferase families affect fungal virulence. For example, *PHR1* and *PHR2* are required for systemic and vaginal infection, respectively, by *C. albicans* [70]. The Δ*GEL2* mutant of *A. fumigatus*, which has a reduced mycelial growth, is also less pathogenic than wt strains in an experimental murine model of IA (I. Mouyna, unpublished).

Another glucanosyltransferase (Bgt1p) homologous to the *S. cerevisiae* and *C. albicans* Bgl2p [71,72] was found in *A. fumigatus* [73]. As in yeast, null mutants of *A. fumigatus* behave like the parental strains with respect to growth, osmotic stability, sporulation, cell wall composition, and sensitivity to cell wall inhibitors [74–76]. This transferase displayed the same physicochemical and enzymatic characteristics among different species. The protein is lightly glycosylated (10%), with a Mr around 30 kDa and is extremely resistant to heat. In *A. fumigatus*, after boiling for 10 min in a 2% SDS-containing buffer, transferase activity can still be detected (data not shown). Although present as a major cell wall associated protein in several species of yeast and filamentous fungi [77], the physiological role of Bgl2p/Bgt1p is unknown. Since the gene is present as a single copy in the genome, the absence of a phenotype for the null mutant suggests that this enzyme does not play a major role in cell wall morphogenesis or, in particular, cross-linking of cell wall polysaccharides as was suggested by Goldman et al. [71]. It also shows that not all cell wall–associated glucanosyltransferases will have a role in cell wall biosynthesis.

Surprisingly, although β(1-3) glucanases have been suggested for 30 years to play an essential role in fungal morphogenesis [78], very few studies have been directed toward this category of hydrolases. We have purified and biochemically characterized an endoβ(1-3) glucanase and 3 exoβ1-3 glucanases associated with the cell wall of *A. fumigatus* [79,80]. Molecular studies were exclusively centered on the endoglucanase for the following reasons:

1. In contrast to other fungal endoβ(1-3) glucanases reported in the litterature which are exocellular, this is the first cell wall–associated fungal β(1-3) endoglucanase identified.

2. Its cellular localization and mode of action (endosplitting activity efficient on a complex polysaccharide structure) suggested that this enzyme could play a role in hydrolyzing existing cell wall structures allowing for hyphal branching as well as for germ tube emergence or the formation of numerous free reducing and nonreducing ends necessary for the activity of β(1-3) glucanosyltransferase [80].

The gene encoding the endoglucanase has been sequenced and analyzed [81]. Expression studies have shown that *ENGL1* is constitutively expressed at all growth stages and that expres-

Table 3 GPI-Anchored β1-3 Glucanosyltransferases in *A. fumigatus* and Other Fungal Species

Gene	Expression	Regulation of expression	Mutant phenotype
A. fumigatus			
GEL1	+	Constitutive	No
GEL2	+	Constitutive	Slow growth, abnormal morphology (sporulation in liquid culture)
GEL3	−		?
S. cerevisiae			
GAS1	+	Cell cycle regulated	Slightly reduced growth with higher percentage of budded cells at stationary phase, increase in chitin and mannoprotein content of CW, modification of the βglucan organization
GAS2	+	Sporulation	No
GAS3	−		No
GAS4	+	Sporulation	No
GAS5	−		No
C. albicans			
PHR1	+	pH>5	No germ tube formation at neutral/ alkaline pH; increase in chitin and modification of βglucan of the cell wall
PHR2	+	pH < 5	Same phenotype as PHR1 but at acidic pH
PHR3	?	?	?
C. glabrata			
CGG1	+	constitutive	Reduced growth rate (no pH regulation)
CGG2	+	constitutive	Reduced growth rate (no pH regulation)
CGG3	?	?	?
C. maltosa			
EPD1	+	pH and nutritional composition of culture medium	No pseudohyphal growth at pH 7; reduced growth at pH 4 with morphological defects in the yeast cells
EPD2	+	pH and nutritional composition of culture medium; pattern of expression inverse of that of EPD1	Pseudohyphal growth reduced at pH 7 on hexadecane medium

sion of *ENGL1* was not stimulated by the addition of a β(1-3) glucan substrate such as laminarin or curdlan into the culture medium. The lack of differential expression of the endoβ(1-3) glucanase under different culture conditions would exclude a role for this enzyme in hydrolyzing exocellular soluble and insoluble β(1-3) glucans for fungal catabolism and would reinforce a putative role for this enzyme in cell wall morphogenesis. However, the absence of phenotype of the null *ENGL1* mutant (even in presence of compounds such as deoxycholate which increases branching in *A. fumigatus*, and the absence of *ENGL1* homolog) showed indeed that the endoβ(1-

3) glucanase activity of Engl1p does not play a morphogenetic role in *A. fumigatus*. This result is reminiscent of previous studies in yeast where disruption of exoβ(1-3) glucanase did not affect cellular growth [82]. Other endoβ(1-3) glucanases occur in *A. fumigatus* (T. Fontaine, personal communication); their role in cell wall morphogenesis is under study in our laboratory.

C. Role of the Mycelial Wall During Infection

1. Protection Against Host Defense Reaction

As the conidium cell wall protects the fungus in the alveolar macrophage, the cell wall protects the mycelium against phagocytosis by PMN [35]. Contact between neutrophils and hyphae triggers a respiratory burst, secretion of reactive-oxygen intermediates, and degranulation [35,83–85]. In contrast to the killing of conidia by macrophages, hyphal damage by PMN is rapid in that 50% of the hyphae are killed after a 2-hr incubation. Killing of hyphae required oxidants, but PMN oxidant release could not mediate hyphal killing without concomitant fungal damage by granule constituents. In addition, experiments with cells from patients with chronic granulomatous disease (CGD) or myeloperoxidase deficiencies have shown that at least two oxidative pathways are involved, but the target biomolecules of oxidant-mediated damage (lipid peroxidation, protein oxidation, or DNA degradation) are unknown. Functional defects in the oxidative mechanisms need not necessarily preclude killing since IFN-γ treatment of PMNs of patients with CGD increases hyphal killing without restoring oxidant production. Electron microscopic observations have shown that PMN-induced cell wall damage was detectable before killing occurred. Incubation of biotinylated hyphae with granules isolated from PMN resulted in a very rapid release of cell wall glycoproteins (<30 min). Hyphal surface glycoproteins that are released during attack by PMN have not been characterized. The enzymes responsible for cell wall injuries also have not been identified. However, polysaccharide hydrolases and proteases which have been isolated from phagocytes may play an essential role in this process. Aside from these mechanisms, cationic peptides, such as the defensins, may be active against hyphae and germinating conidia as well [38]. It is clear that the mechanisms responsible for killing have not been fully identified.

2. The Mycelium as a Source of Molecules That Play a Role During Infection

a. Enzymes. Several enzymes (Table 4) identified as playing a putative role in *A. fumigatus* infections are exocellular proteins with a typical signal peptide that are found in high concentration in the cell wall. For example, immunocytochemical detection indicated that the

Table 4 *A. fumigatus* Cell Wall–Associated Toxin and Enzymes with a Putative Role in Virulence

Category	Role *in vivo*	Molecule
Toxin	Host cell death	Ribonuclease (18 kDa)
Proteases	Promote lung matrix	Serine protease (33 kDa)
	Colonization and/or	Aspartic protease (38 kDa)
	Degrade humoral	Metalloprotease (40 kDa)
	Factors	Dipeptidylpeptidase (88 kDa)
Oxidases	Antioxidants during	Catalases (350 kDa, ?)
	phagocytosis	Superoxide dismutases (27, 67 kDa)

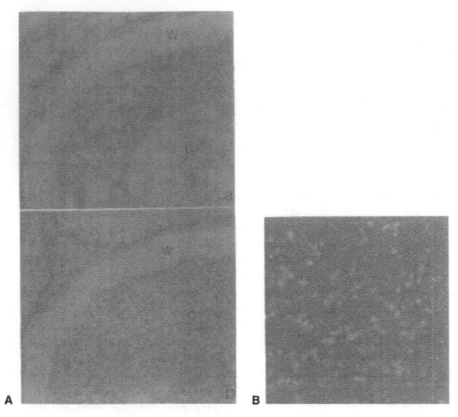

Figure 6 (A) Reactivity of the anti-ALP antiserum with the wall of a hyphal element in the lung of an infected mouse. (a) Anti-ALP and (b) preimmune sera were diluted 1/500 and antirabbit IgG immunoglobulin conjugated to colloidal gold (10 nm) was diluted 1/40 w, Mycelial wall; c, cytoplasm (×51 000). (B) Reactivity of mycelia in the lung of an immunocompromised mouse infected with A. *fumigatus*. Indirect immunofluorescence was achieved with anti-ALP rabbit antiserum diluted 1/100 and fluoresceinylated antirabbit IgG antiserum diluted 1/100.

18-kDa ribotoxin [86] or 33-kDa serine protease [87] were found in the cell wall of the mycelium during growth in vivo (Fig. 6). A similar situation was seen with the A. *fumigatus* catalases [88]. The cell wall localization of these enzymes can have two purposes: protection against destruction by the host, and intimate contact with the host cells and extracellular matrix proteins that these enzymes should attack to provide nutrients for fungal growth. However, disruption of genes encoding enzymes with a putative role in infection (proteases, catalases, ribotoxin) have led to the production of mutants which are as pathogenic as the parental wild-type strain [88–90]. These results suggest that either the target was wrongly selected or, more likely, virulence in A. *fumigatus* is polygenic and disruption of a single gene will never cause the loss of the fungal virulence. In agreement with the latest concept, a REMI-STM approach did not identify specific virulence genes in A. *fumigatus* [91]. Multiple gene involvement is presently studied by a transcriptome analysis of mycelium grown in vivo versus in vitro.

    b.   Antigens.   The most important antigens used in the diagnosis of aspergillosis in the immunocompetent or immunocompromised host are cell wall associated (Table 5). The most characterized antigens, which include a ribonuclease [92], a catalase [88], and a dipeptidylpepti-

Table 5 Cell Wall–Associated Molecules Reported in the Literature as Useful for the Diagnosis of Aspergillosis

Molecules	Antibody/antigen	Immune status of patient
(Glyco)proteins		
18-kDa ribonuclease[a]	Ab	Immunocompetent
28-kDa?	Ab	Immunocompetent
33-kDa serine protease	Ab	Immunocompetent
38-kDa aspartic protease	Ab	Immunocompetent
40-kDa metalloprotease	Ab	Immunocompetent
36 (70?)-kDa?	Ab	Immunocompetent
88-kDa dipeptidypeptidase V[a]	Ab	Immunocompetent
360-kDa catalase[a]	Ab	Immunocompetent
Polysaccharides		
Galactomannan[b]	Ag	Immunocompromised
β1-3 glucan	Ag	Immunocompromised

(1) Most useful for the detection of anti-*fumigatus* antibodies
(2) Most useful for antigen detection in IA

dase [93] (used for the detection of specific antibodies in immunocompetent patients), and the galactomannan [94,95] (used for the detection of antigens in immunocompromised patients) are described in the following section.

The mycelial catalase of *A. fumigatus* that has been characterized extensively is a tetrameric protein with monomeric subunits of 90 kDa. The oligomeric subunit contains an N-linked sugar moiety of 7 kDa which bears no antigen epitopes. The protein is remarkably stable, being relatively insensitive to high temperatures as well as to reducing and denaturing agents. The structural gene for the protein, *CAT1*, has been cloned and sequenced [88]. Analysis of the deduced amino acid sequence shows that *CAT1* has both a signal peptide of 15 amino acids and a propeptide of 12 amino acids, with a pair of basic amino acids Arg_{26}–Arg_{27} acting as a cleavage signal for a KEX2-like endopeptidase. Comparison of the *CAT1* sequence with other catalase genes suggests conservation of the tripeptide His_{102} Ser_{141} and Asn_{175} which is involved in the binding of proteins to its heme prosthetic group.

A dipeptidylpeptidase V which releases specific dipeptides from the N-terminus of polypeptides has been identified as the so called chymotrypsin antigen. It is a monomeric protein with a Mr of 88 kDa that contains a N-linked carbohydrate of 9 kDa. Like Cat1p, the peptide moiety bears antigenic epitopes [93]. Comparison of the *A. fumigatus DPPV* sequence with other DPPs shows the presence of a Gly_{558}-X-Ser_{562}-X-Gly_{562} consensus motif of serine hydrolases with a putative catalytic triad of the DPP arranged as $Ser_{560}Asp_{643}Thr_{675}$.

The ribonuclease of *A. fumigatus* is composed of 149 amino acids with a 27 amino acid leader sequence and a putative active site composed of the six amino acids H_{49}, E_{95}, F_{96}, P_{98}, R_{120}, H_{136}. This RNAase cleaves a single phosphodiester bond in a highly conserved region and releases a 300- to 400-base fragment from the 3′ end of the large ribosomal RNA [86].

Polysaccharide antigens are mostly used in the serodiagnosis of immunocompromised patients. In contrast to the immunocompetent host, growth of *A. fumigatus* in the tissues of an immunosuppressed host is not correlated with an increase in anti-*Aspergillus* antibody titers. In fact, the serological diagnosis of IA today is based on the detection of circulating polysaccharides in biological fluids, e.g., serum, urine, and bronchoalveolar lavages, from patients. Among these polysaccharides, galactomannan (GM) has been the only polysaccharide antigen characterized from *A. fumigatus*. Although data differ slightly, a consensus structure has been established: the mannan core has a linear configuration containing $\alpha(1-2)$- and (1-6)-linked residues in a ratio of 3:1, and the antigenic, acid-labile side chains, branched on two $\alpha(1-2)$-linked mannose residues, are composed of $\beta(1-5)$ galactofuranosyl residues with an average degree of polymerization of 4 [94]. Monoclonal antibodies directed against the galactofuran side chains of the galactomannan are the basis of the only ELISA antigen detection kit commercially available [95].

The other polysaccharide used for diagnosis is $\beta1-3$ glucan which can be detected at a very low concentration. $\beta1-3$ glucan, which is another component of the aspergilli fungal cell wall, can also be used diagnostically even though it is not an immunogenic molecule. In this case, the detection system is based on the activation of a proteolytic coagulation cascade, which is an essential defense mechanism of arthropods against the invasion of their hemocoele by pro- and eucaryotic microorganisms [96,97]. The components of the assay include factor G, which triggers specifically the $\beta1-3$ glucan–sensitive hemolymph clotting pathway, and a chromogenic Leu-Gly-Arg-pNA tripeptide, which is specifically cleaved by the last component of this proteolytic cascade. The assay can measure pg amounts of $\beta1-3$ glucans and has been used to show the presence of this polysaccharide during fungal systemic fungal infections [98]. The low amount of $\beta1-3$ glucans found in serum can be explained by the fact that $\beta1-3$ glucan is an integral component of the cell wall skeleton and, in contrast to GM, is not released normally from the fungal cell.

IV. CONCLUSIONS AND PERSPECTIVES

Obviously, the cell wall plays an essential role in host–*A. fumigatus* relationships. This role is directly associated with morphogenetic changes in dimorphic fungi where only one of the morphological forms is invasive for man. Interestingly, if *Candida albicans* is considered to be a very "smart" dimorphic fungus, nobody seems to recognize that all filamentous fungi are also dimorphic, the mycelial form being the invading structure. In *A. fumigatus*, the conidial cell wall seems to be mainly involved in the protection against phagocyte aggressions whereas the mycelial cell wall may function more as an active efflux pump, causing the release of enzymes into the external medium in the vicinity of the mycelial apex. The physicostructural organization of the cell wall has been poorly studied. Influx and efflux of molecules through the cell wall (including antifungals) have been totally ignored, although it should be a great area of research. When we begin to consider the cell wall as a living organelle, we could learn a lot from studies of its role in permeability.

The cell wall is a unique organelle of fungi. This review has shown how little we know of the biosynthetic events occurring at the cell wall level during growth. Comparative biochemistry and now genomics should allow us to identify species-specific or common pathways which should lead to a better understanding of cell wall morphogenesis and its adaptation to the surrounding environment. This area of research has been today scientifically neglected whereas signal reception and transduction should be different when *A. fumigatus* is submitted to 70°C in a compost or in the lysosome environment of a phagocyte.

REFERENCES

1. DW Denning. Therapeutic outcome in invasive aspergillosis. Clin Infect Dis 23:608–614, 1996.
2. JP Latgé. *Aspergillus fumigatus* and aspergillosis. Clin Microbiol Rev 12:310–350, 1999.

3. N Thau, M Monod, B Crestani, C Rolland, G Tronchin, JP Latgé, S Paris. *Rodletless* mutants of *Aspergillus fumigatus*. Infect Immun 62:4380–4388, 1994.

4. JGH Wessels. Wall growth, protein excretion, and morphogenesis in fungi. New Phytol 123:397–413, 1993.

5. K Shibuya, M Takaoka, K Uchida, M Wakayama, H Yamaguchi, K Takahashi, S Paris, JP Latgé, S Naoe. Histopathology of experimental invasive pulmonary aspergillosis in rats: pathological comparison of pulmonary lesions induced by specific virulent factor deficient mutants. Microb Pathog 27:123–131, 1999.

6. M Parta, Y Chang, S Rulong, P Pinto-Da Silva, KJ Kwon-Chung. *HYP1*, a hydrophobin gene from *Aspergillus fumigatus*, complements the *rodletless* phenotype in *Aspergillus nidulans*. Infect Immun 62:4389–4395, 1994.

7. H Girardin, S Paris, J Rault, MN Bellon-Fontaine, JP Latgé. The role of the rodlet structure on the physicochemical properties of *Aspergillus* conidia. Appl Environ Microbiol 29:364–369, 1999.

8. MJ Kershaw, G Wakley, NJ Talbot. Complementation of the Mpg1 mutant phenotype in *Magnaporthe grisea* reveals functional relationships between fungal hydrophobins. EMBO J 17:3849–3849, 1998.

9. MA Stringer, WT Timberlake. *dewA* encodes a fungal hydrophobin component of the *Aspergillus* spore wall. Mol Microbiol 16:33–44, 1995.

10. G Tronchin, JP Bouchara, JP Latgé, D Chabasse. Application of a Lowicryl K4M embedding technique for analysis of fungal adhesins. J Mycol Med 3:74–79, 1992.

11. JP Bouchara, M Sanchez, A Chevailler, A Marot-Leblond, JC Lissitzky, G Tronchin, D Chabasse. Sialic acid–dependent recognition of laminin and fibrinogen by *Aspergillus fumigatus* conidia. Infect Immun 65:2717–2724, 1997.

12. DJ DeHart, DE Agwu, JC Julian, RG Washburn. Binding and germination of *Aspergillus fumigatus* conidia on cultured A549 pneumocytes. J Infect Dis 175:146–150, 1996.

13. RD Diamond, R Krzesicki, B Epstein, W Jao. Damage to hyphal forms of fungi by human leukocytes in vitro. A possible host defense mechanism in aspergillosis and mucormycosis. Am J Pathol 91:313–323, 1978.

14. T Ishimaru, EM Bernard, S Tamada, D Armstrong. The fucose-specific lectin (FSL) produced by *Aspergillus fumigatus* may promote attachment of conidia to mammalian cells. Abstracts of the IDSA 34th Annual Meeting, 1996, p 206.

15. VL Kan, JE Bennett. Lectin-like attachment sites on murine pulmonary alveolar macrophages bind *Aspergillus fumigatus* conidia. J Infect Dis 158:407–414, 1988.

16. VL Kan, JE Bennett. β(1-4) oligoglucosides inhibit the binding of *Aspergillus fumigatus* conidia to human monocytes. J Infect Dis 163:1154–1156, 1991.

17. T Madan, P Eggleton, U Kishore, P Strong, S Aggrawal, PU Sarma, KBM Reid. Binding of pulmonary surfactant proteins A and D to *Aspergillus fumigatus* conidia enhances phagocytosis and killing by human neutrophils and alveolar macrophages. Infect Immun 65:3171–3179, 1997.

18. S Paris, E Boisvieux-Ulrich, B Crestani, O Houcine, D Taramelli, L Lombardi, JP Latgé. Internalization of *Aspergillus fumigatus* conidia by epithelial and endothelial cells. Infect Immun 65:1510–1514, 1997.

19. MC Penalver, JE O'Connor, JP Martinez, ML Gil. Binding of human fibronectin to *Aspergillus fumigatus* conidia. Infect Immun 64:1146–1153, 1996.

20. J Sturtevant, JP Latgé. Participation of complement in the phagocytosis of the conidia of *Aspergillus fumigatus* by human polymorphonuclear cells. J Infect Dis 166:580–586, 1992.

21. J Sturtevant, JP Latgé. Interactions between conidia of *Aspergillus fumigatus* and human complement component C3. Infect Immun 60:1913–1918, 1992.

22. G Tronchin, K Esnault, G Renier, R Filmon, D Chabasse, JP Bouchara. Expression and identification of laminin-binding protein in *Aspergillus fumigatus* conidia. Infect Immun 65:9–15, 1997.

23. S Henwick, SV Hetherington, CC Patrick. Complement binding to *Aspergillus* conidia correlates with pathogenicity. J Lab Clin Med 122:27–35, 1993.

24. J Bouchara, M Sanchez, K Esnault, G Tronchin. Interactions between *Aspergillus fumigatus* and Host matrix proteins. In: A Schmidt, AA Brakhage, B Jahn, A Schmidt, eds. *Aspergillus fumigatus*. Biology, Clinical Aspects and Molecular Approaches to Pathogenicity. Vol. 2. Wuppertal: Karger, 1999, pp 167–181.

25. K Langfelder, H Gehringer, A Schmidt, G Wanner, AA Brakhage. Identification of a polyketide synthase gene (*pksP*) of *Aspergillus fumigatus* involved in conidial pigment biosynthesis and virulence. Med Mol Immunol 187:79–89, 1998.

26. HF Tsai, RG Washburn, YC Chang, KJ Kwon-Chung. *Aspergillus fumigatus arp1* modulates conidial pigmentation and complement deposition. Mol Microbiol 26:175–183, 1997.

27. HF Tsai, YC Chang, RG Washburn, MH Wheeler, KJ Kwon-Chung. The developmentally regulated *alb1* gene of *Aspergillus fumigatus*: its role in modulation of conidial morphology and virulence. J Bacteriol 180:3031–3038, 1998.

28. HF Tsai, MH Wheeler, YC Chang, KJ Kwon-Chung. The developmentally regulated gene cluster involved in conidial pigment biosynthesis in *Aspergillus fumigatus*. J Bacteriol 181:6469–6477, 1999.

29. AA Brakhage, K Langfelder, G Wanner, A Schmidt, B Jahn. Pigment biosynthesis and virulence. In: A Schmidt, AA Brakhage, B Jahn, A Schmidt, eds. *Aspergillus fumigatus*. Biology, Clinical Aspects and Molecular Approaches to Pathogenicity. Vol. 2. Wuppertal: Karger, 1999, pp 205–215.

30. NP Keller, TM Hohn. Metabolic pathways gene clusters in filamentous fungi. Fungal Genet Biol 21:17–29, 1997.

31. RD Diamond, E Huber, CC Haudenschild. Mechanisms of destruction of *Aspergillus fumigatus* hyphae mediated by human monocytes. J Infect Dis 147:474–483, 1983.

32. SM Levitz, RD Diamond. Mechanisms of resistance of *Aspergillus fumigatus* conidia to killing by neutrophils in vitro. J Infect Dis 152:33–42, 1985.

33. B Jahn, A Koch, A Schmidt, G Wanner, H Gehringer, S Bhakdi, AA Brakhage. Isolation and characterization of a pigmentless-conidium mutant of *Aspergillus fumigatus* with altered conidial surface and reduced virulence. Infect Immun 65:5110–5117, 1997.

34. F Claverie-Martin, MR Diaz-Torres, MJ Geoghegan. Chemical composition and ultrastructure of wild-type and white mutant *Aspergillus nidulans* conidial walls. Curr Microbiol 16:281–287, 1988.

35. A Schaffner, H Douglas, AI Braude. Selective protection against conidia by mononuclear and against mycelia by polymorphonuclear phagocytes in resistance to *Aspergillus*. J Clin Invest 69:617–631, 1982.

36. A Schaffner, T Schaffner. Glucocorticoid-induced impairment of macrophage antimicrobial activity: mechanisms and dependence on the state of activation. Rev Infect Dis 9:S620–S629, 1987.

37. A Schaffner. Macrophage-*Aspergillus* interactions. In: BS Zwilling, TK Eisenstein, eds. Macrophage-Pathogen Interactions. Vol. 60. New York: Marcel Dekker, 1994, pp 545–552.

38. SM Levitz, ME Selsted, T Ganz, RI Lehrer, RD Diamond. In vitro killing of spores and hyphae of *Aspergillus fumigatus* and *Rhizopus oryzae* by rabbit neutrophil cationic peptides and bronchoalveolar macrophages. J Infect Dis 154:483–489, 1986.

39. E Michaliszyn, S Sénéchal, P Martel, L de Repentigny: Lack of involvement of nitric oxide in killing of *Aspergillus fumigatus* conidia by pulmonary alveolar macrophages Infect Immun 63:2075–2078, 1995.

40. D Taramelli, MG Malabarba, G Sala, N Basilico, G Cocuzza. Production of cytokines by alveolar and peritoneal macrophages stimulated by *Aspergillus fumigatus* conidia or hyphae. J Med Vet Mycol 34:49–56, 1996.

41. E Rodriguez, F Boudard, M Mallié, JM Bastide, M Bastide. Murine macrophage elastolytic activity induced by *Aspergillus fumigatus* strains in vitro: evidence of the expression of two macrophage-induced protease genes. Can J Microbiol 43:649–657, 1997.

42. FM Klis. Review: cell wall assembly in yeast. Yeast 10:851–869, 1994.

43. JP Debeaupuis, J Sarfati, H Kobayashi, D Boucias, P Gumowski, A Beauvais, S Paris, M Monod, JP Latgé. Antigens of *Aspergillus fumigatus* expressed during infection. Can J Bot 73:1087–1091, 1995.

44. T Fontaine, I Mouyna, RP Hartland, S Paris, JP Latgé. From the surface to the inner layer of the fungal cell wall. Biochem Soc Trans 25:194–199, 1997.

45. T Fontaine, C Simenel, G Dubreucq, O Adam, M Delepierre, J Lemoine, CE Vorgias, M Diaquin, JP Latgé. Molecular organization of the alkali-insoluble fraction of *Aspergillus fumigatus* cell wall. J Biol Chem 275:27594–27607, 2000.

46. JP Latgé, JP Debeaupuis, M Moutaouakil, M Diaquin, J Sarfati, MC Prévost, Y Leroy, JM Wieruszeski, B Fournet. Galactomannan and the circulating antigens of *Aspergillus fumigatus*. In: JP Latgé,

DG Boucias, eds. Fungal Cell Wall and Immune Response. Heidelberg: Springer-Verlag, 1991, pp 143–151.

47. Y Araki, H Takada, N Fujii, E Ito. A pathway of polygalactosamine formation in *Aspergillus parasiticus*: enzymatic deacetylation of N-acetylated polygalactosamine. Eur J Biochem 102:35–42, 1979.

48. H Takada, Y Ariki, E Ito. Structure of polygalactosamine produced by *Aspergillus parasiticus*. J Biochem 89:1265–1274, 1981.

49. R Kollar, E Petrakova, G Ashwell, PW Robbins, E Cabib. Architecture of the yeast cell wall. The linkage between chitin and β(1-3)-glucan. J Biol Chem 270:1170–1178, 1995.

50. T Fujii, H Shimoi. Structure of the glucan-binding sugar chain of Tip1p, a cell wall protein of *Saccharomyces cerevisiae*. Biochim Biophys Acta 1427:133–144, 1999.

51. BJM Zonneveld. Biochemical analysis of the cell wall of *Aspergillus nidulans*. Biochim Biophys Acta 249:506–514, 1971.

52. PT Borgia, N Iartchouk, PJ Riggle, KR Winter, Y Koltin, CE Bulawa. The *chsB* gene of *Aspergillus nidulans* is necessary for normal hyphal growth and development. Fungal Genet Biol 20:193–203, 1996.

53. H Horiuchi, M Takagi. Chitin synthase genes of *Aspergillus* species. In: A Schmidt, AA Brakhage, B Jahn, A Schmidt, eds. *Aspergillus fumigatus*. Biology, Clinical Aspects and Molecular Approaches to Pathogenicity. Vol. 2. Wuppertal: Karger, 1999, pp 193–204.

54. CA Specht, Y Liu, PW Robbins, CE Bulawa, N Iartchouk, KR Winter, PJ Riggle, JC Rhodes, CL Dodge, DW Culp, PT Borgias. The *chsD* and *chsE* genes of *Aspergillus nidulans* and their roles in chitin synthesis. Fungal Genet Biol 20:153–167, 1996.

55. A Aufauvre-Brown, E Mellado, NAR Gow, DW Holden. *Aspergillus fumigatus chsE*: a gene related to *CHS3* of *Saccharomyces cerevisiae* and important for hyphal growth and conidiophore development but not pathogenicity. Fungal Genet Biol 21:141–152, 1997.

56. E Mellado, A Aufauvre-Brown, CA Specht, PW Robbins, DW Holden. A multigene family related to chitin synthase genes of yeast in the opportunistic pathogen *Aspergillus fumgatus*. Mol Gen Genet 246:353–359, 1995.

57. E Mellado, A Aufauvre-Brown, NAR Gow, DW Holden. The *Aspergillus fumigatus chsC* and *chsG* genes encode class III chitin synthase with different functions. Mol Microbiol 20:667–679, 1996.

58. GM Escott, VM Hearn, DJ Adams. Inducible chitinolytic system of *Aspergillus fumigatus*. Microbiology 144:1575–1581, 1998.

59. VM Hearn, GM Escott, E Glyn, V Evans, DJ Adams. Complex chitinolytic system of *Aspergillus fumigatus*. Microbios 93:85–104, 1998.

60. A Beauvais, R Drake, K Ng, M Diaquin, JP Latgé. Characterization of the 1,3-beta-glucan synthase of *Aspergillus fumigatus*. J Gen Microbiol 39:3071–3078, 1993.

61. RP Hartland, T Fontaine, JP Debeaupuis, C Simenel, M Delepierre, JP Latgé. A novel β-(1-3)-glucanosyltransferase from the cell wall of *Aspergillus fumigatus*. J Biol Chem 271:26843–26849, 1996.

62. I Mouyna, T Fontaine, M Vai, M Monod, WA Fonzi, M Diaquin, L Popolo, RP Hartland, JP Latgé. GPI-anchored glucanosyltransferases play an active role in the biosynthesis of the fungal cell wall. J Biol Chem 275:14882–14889, 2000.

63. I Mouyna, M Monod, T Fontaine, B Henrissat, B Léchenne, JP Latgé. Identification of the catalytic residues of the first family of β(1-3)glucanosyltransferases identified in fungi. Biochem J 347:741–747, 2000.

64. FA Mühlschlegel, WA Fonzi. *PHR2* of *Candida albicans* encodes a functional homolog of the pH-regulated gene *PHR1* with an inverted pattern of pH-dependent expression. Mol Cell Biol 17:5960–5967, 1997.

65. T Nakazawa, H Horiuchi, A Ohta, M Takagi. Isolation and characterization of *EPD1*, an essential gene for pseudohyphal growth of a dimorphic yeast, *Candida maltosa*. J Bacteriol 180:2079–2086, 1998.

66. M Vai, E Gatti, E Lacana, L Popolo, L Alberghina. Isolation and deduced amino acid sequence of the gene encoding gp115, a yeast glycophophoslipid-anchored protein containing a serine-rich region. J Biol Chem 266:12252–12248, 1991.

67. SM Saporito-Irwin, CE Birse, PS Sypherd, WA Fonzi. *PHR1*, a pH-regulated gene of *Candida albicans*, is required for morphogenesis. Mol Cell Biol 15:601–613, 1995.
68. M Weig, M Frosch, K Haynes, F Muhlschlegel. The cell wall of *Candida glabrata*: *CGG1*, *CGG2* and *CGG3* constitute a family of cell surface proteins. Presented at Human Fungal Pathogens: Fungal Dimorphism and Disease, Grenada, Spain, 1999.
69. T Nakazawa, M Takahashi, H Horiuchi, A Ohta, M Takagi. Cloning and characterization of *EPD2*, a gene required for efficient pseudohyphal formation of a dimorphic yeast, *Candida maltosa*. Biosci Biotechnol Biochem 64:369–377, 2000.
70. F De Bernardis, A Cassone, W Fonzi. The pH of the host niche controls gene expression in and virulence of *Candida albicans*. Infect Immun 66:3317–3325, 1998.
71. RC Goldman, PA Sullivan, D Zakula, JO Capobianco. Kinetics of β-1,3 glucan interaction at the donor and acceptor sites of the fungal glucosyltransferase encoded by the *BGL2* gene. Eur J Biochem 227:372–378, 1995.
72. RP Hartland, GW Emerson, PA Sullivan. A secreted β-glucan-branching enzyme from *Candida albicans*. Proc R Soc Lond Ser B Biol Sci 246:155–160, 1991.
73. I Mouyna, RP Hartland, T Fontaine, M Diaquin, JP Latgé. A β(1-3) glucanosyltransferase isolated from the cell wall of *Aspergillus fumigatus* is an homolog of the yeast Bgl2p. Microbiology 144:3171–3180, 1998.
74. F Klebl, W Tanner. Molecular cloning of a cell wall exo-β-1,3-glucanase from *Saccharomyces cerevisiae*. J Bacteriol 171:6259–6264, 1989.
75. V Mrsa, F Klebl, W Tanner. Purification and characterization of the *Saccharomyces cerevisiae BGL2* gene product, a cell wall endo-β-1,3-glucanase. J Bacteriol 175:2102–2106, 1993.
76. AV Sarthy, T McGonigal, M Coen, DJ Frost, JA Meulbroek, RC Goldman. Phenotype in *Candida albicans* of a disruption of the *BGL2* gene encoding a 1,3-β-glucosyltransferase. Microbiology 143:367–376, 1997.
77. E Herrero, P Sanz, R Sentandreu. Cell wall proteins liberated by zymolyase from several ascomycetous and imperfect yeasts. J Gen Microbiol 133:2895–2903, 1987.
78. S Bartnicki-Garcia. Cell wall chemistry, morphogenesis, and taxonomy of fungi. Annu Rev Microbiol 22:87–108, 1968.
79. T Fontaine, RP Hartland, M Diaquin, C Simenel, JP Latgé. Differential patterns of activity displayed by two exo-β-1,3 glucanases associated with the *Aspergillus fumigatus* cell wall. J Bacteriol 179:3154–3163, 1997.
80. T Fontaine, RP Hartland, A Beauvais, M Diaquin, JP Latgé. Purification and characterization of an endo-1,3-β-glucanase from *Aspergillus fumigatus*. Eur J Biochem 243:315–321, 1997.
81. I Mouyna, T Fontaine, B Henrissat, J Sarfati, JP Latgé. Cloning and disruption of *ENGL1* of *Aspergillus fumigatus*, a gene encoding a β(1-3) endoglucanase belonging to a new family of glucanases. Submitted to Med Mycol, 2001.
82. G Larriba, E Andaluz, R Cueva, RD Basco. Molecular biology of yeast exoglucanases. FEMS Microbiol Lett 125:121–126, 1995.
83. RD Diamond, RA Clark. Damage to *Aspergillus fumigatus* and *Rhizopus oryzae* hyphae by oxidative and non-oxidative microbicidal products of human neutrophils in vitro. Infect Immun 38:487–495, 1982.
84. SM Levitz, TP Farrell. Human neutrophil degranulation stimulated by *Aspergillus fumigatus*. J Leukoc Biol 47:170–175, 1990.
85. A Schaffner, CE Davis, T Schaffner, M Markert, H Douglas, AI Braude. In vitro susceptibility of fungi to killing by neutrophil granulocytes discriminates between primary pathogenicity and opportunism. J Clin Invest 78:511–524, 1986.
86. B Lamy, M Moutaouakil, JP Latgé, J Davies. Secretion of a potential virulence factor, a fungal ribonucleotoxin, during human aspergillosis infections. Mol Microbiol 5:1811–1815, 1991.
87. M Moutaouakil, M Monod, M Prévost, J Bouchara, S Paris. Identification of the 33-kDa alkaline protease of *Aspergillus fumigatus* in vitro and in vivo. J Med Microbiol 39:393–399, 1993.
88. JA Calera, S Paris, M Monod, AJ Hamilton, JP Debeaupuis, M Diaquin, R Lopez-Medrano, F Leal, JP Latgé. Cloning and disruption of the antigenic catalase gene of *Aspergillus fumigatus*. Infect Immun 65:4718–4724, 1997.

89. M Monod, S Paris, J Sarfati, K Jaton-Ogay, P Ave, JP Latgé. Virulence of alkaline protease-deficient mutants of *Aspergillus fumigatus*. FEMS Microbiol Lett 106:39–46, 1993.
90. S Paris, M Monod, M Diaquin, B Lamy, L Arruda, P Punt, JP Latgé. A transformant of *Aspergillus fumigatus* deficient in the antigenic cytotoxin ASPFI. FEMS Microbiol Lett 111:31–36, 1993.
91. JS Brown, A Aufauvre-Brown, J Brown, JM Jennings, HJ Arst, DW Holden. Signature-tagged and directed mutagenesis identify PABA synthetase as essential for *Aspergillus fumigatus* pathogenicity. Mol Microbiol 36:1371–1380, 2000.
92. JP Latgé, M Moutaouakil, JP Debeaupuis, J Bouchara, K Haynes, MC Prévost. The 18-kilodalton antigen secreted by *Aspergillus fumigatus*. Infect Immun 59:2586–2594, 1991.
93. A Beauvais, M Monod, JP Debeaupuis, M Diaquin, H Kobayashi, JP Latgé. Biochemical and antigenic characterization of a new dipeptidyl-peptidase isolated from *Aspergillus fumigatus*. J Biol Chem 272: 6238–6244, 1997.
94. JP Latgé, H Kobayashi, JP Debeaupuis, M Diaquin, J Sarfati, JM Wieruszeski, E Parra, JM Bouchara, B Fournet. Chemical and immunological characterization of the extracellular galactomannan of *Aspergillus fumigatus*. Infect Immun 62:5424–5433, 1994.
95. D Stynen, A Goris, J Sarfati, JP Latgé. A new sensitive sandwich enzyme-linked immunosorbent assay to detect galactofuran in patients with invasive aspergillosis. Infect Immun 33:497–500, 1995.
96. T Miyazaki, S Kohno, K Mitsutake, S Maesaki, KI Tanaka, N Ishikawa, K Hara. Plasma (1-3)-beta-D-glucan and fungal antigenemia in patients with candidemia, aspergillosis, and cryptococcosis. J Clin Microbiol 33:3115–3118, 1995.
97. T Obayashi, M Yoshida, H Tamura, J Aketagawa, S Tanaka, T Kawai. Determination of plasma (1-3)-beta-D-glucan: a new diagnosis aid to deep mycosis. J Med Vet Mycol 30:275–280, 1992.
98. T Obayashi, M Yoshida, T Mori, H Goto, A Yasuoka, H Iwasaki, H Teshima, S Kohno, A Horiuchi, A Ito, H Yamaguchi, K Shimada, T Kawai. Plasma (1-3)-beta-D-glucan measurement in diagnosis of invasive deep mycosis and fungal febrile episodes. Lancet 345:17–20, 1995.

23

Cryptococcus neoformans and Macrophages

Thomas S. Harrison
St. George's Hospital Medical School, London, United Kingdom

Stuart M. Levitz
Boston University School of Medicine, Boston, Massachusetts

I. INTRODUCTION

Cryptococcosis has emerged as a major cause of morbidity and mortality in patients with impaired cell-mediated immunity, especially those with acquired immune deficiency syndrome (AIDS) [1]. Human immunodeficiency virus (HIV)-associated cryptococcosis is particularly common in Africa and Asia [2]. In the tropics, infection due to *C. neoformans* var. *gattii* in immunocompetent patients also results in significant mortality and blindness despite conventional therapy [3,4]. In developed countries, while antiretroviral therapy has contributed to a recent decrease in HIV-related infections, cases of cryptococcosis in immunocompromised cancer and transplant patients may continue to rise as the number of such patients increases. In addition to presenting a continuing clinical problem, *C. neoformans* is being used increasingly as a model organism to study the pathogenesis of and host defense against fungal infection.

Like many other fungal pathogens, exposure to *C. neoformans* is thought to result from the inhalation of airborne organisms. The infectious form may be yeasts or basidiospores. In nature, yeast cells tend to be small and poorly encapsulated compared with organisms growing in vivo [5]. Basidiospores are produced in the laboratory as a result of sexual reproduction between α and a mating types and by haploid fruiting of α strains [6,7]. Given the predominance of α strains in environmental and clinical isolates, the latter may be more important in nature. Although their production has yet to be observed outside the laboratory, basidiospores are easily aerosolized and a suitable size (1–3 μM) for alveolar deposition. Whatever the infecting cell type, it is likely that bronchoalveolar macrophages are the first host immune cells to interact with the organism.

Many components of innate and specific immunity have been shown to have activity against *C. neoformans* in vitro and assessing their relative importance in vivo is difficult. Exposure to the organism probably only rarely leads to disease, and there may be considerable redundancy of protective mechanisms. Nevertheless, the epidemiology of human infection and animal studies clearly demonstrates that specific T-cell-mediated immunity is critical in a protective immune response. Thus, cryptococcosis is seen with increased frequency in patients with AIDS, lymphomas, and sarcoidosis, and in those taking immunosuppressive medications such as corticosteroids, cyclophosphamide, or azathioprine that depress T-cell-mediated immunity.

Experiments comparing the course of infection in immunocompetent and T-cell-deficient nude mice have also unequivocally demonstrated the importance of specific T-cell-mediated immunity in protection against cryptococcosis [8].

As key cells in both innate and specific cell-mediated immunity, mononuclear phagocytes are central to a protective immune response to *C. neoformans*. Monocytes and resident macrophages, in particular bronchoalveolar macrophages, form a part of the first line of innate cell-mediated immunity. In addition, mononuclear phagocytes are critical in initiating a specific cell-mediated immune response through antigen presentation and costimulation of T-cells, cytokine and chemokine release, and once activated, also serving as effector cells contributing to the control of infection. On the other hand, it has long been known that under some circumstances *C. neoformans* can survive and replicate within macrophages in vitro [9]. In addition, recent molecular epidemiology studies suggest long-lasting dormant infection may be common in man [10], and recent studies of pulmonary infection in rats suggest such latency may result from long-term survival of organisms within macrophages and giant cells [11]. Together, this evidence suggests *C. neoformans* is an intracellular pathogen of macrophages as well as an extracellular pathogen.

Thus, the interaction of *C. neoformans* with host macrophages, the focus of this chapter, may be central to both the pathogenesis of, and host defense against, cryptococcosis. A more detailed understanding of this interaction may lead to new therapeutic approaches aimed at reconstituting or enhancing host defense, or depriving the organism of the intracellular sanctuary that macrophages may, in some circumstances, provide.

II. BINDING AND PHAGOCYTOSIS

Phagocytosis of any micro-organism involves two steps. First is a binding step where receptors on the phagocyte recognize ligands on the surface of the micro-organism. In many cases, binding is facilitated by the presence of host opsonins such as complement or immunoglobulins. Following binding, internalization of the micro-organism, or phagocytosis, occurs. Phagocytosis is an energy-dependent process that is accompanied by changes in the phagocyte's cytoskeleton. As discussed in more detail later, binding does not necessarily lead to phagocytosis.

C. neoformans is unique among medically important fungi in its possession of a polysaccharide capsule which is the organism's major virulence factor (for further details see chapter by Kwon-Chung and Wickes) [12]. In the environment, *C. neoformans* tends to have a small capsule which then thickens inside the host in response to physiological conditions such as bicarbonate and low iron [13,14]. As a consequence, following inhalation, resident bronchoalveolar macrophages may encounter thinly encapsulated yeasts (or unencapsulated basidiospores) whereas other macrophage populations may be challenged by fungi with large capsules [15]. In the presence of large capsules, binding and phagocytosis may be reduced [16,17].

Macrophages avidly bind and phagocytose mutant strains of *C. neoformans* that lack capsule, even in the absence of opsonins [18]. Recognition occurs via mannose and glucan receptors [19]. The physiological significance of this observation is uncertain as virtually all clinical isolates are encapsulated [15]. Capsule masks potential ligands on the cell wall of the fungus [20]. Thus, in the absence of opsonins, binding of encapsulated *C. neoformans* to macrophages is minimal although exceptions have been noted. For example, human and rat bronchoalveolar macrophages bind unopsonized *C. neoformans* [21,22]. Moreover, microglia (the resident macrophages of the brain) obtained from neonatal pigs phagocytose nonopsonized encapsulated *C. neoformans* by a mechanism that is inhibited by antibodies to CD14 [23].

Following incubation of *C. neoformans* with normal human serum, complement activation occurs resulting in deposition on the capsular surface of copious quantities of the third component of complement (C3), mainly in the form of iC3b [24]. Complement activation proceeds mostly through the alternative pathway, and can be reproduced using purified alternative pathway components in lieu of serum [25]. Roughly 10 times greater numbers of complement molecules get deposited on encapsulated *C. neoformans* than on acapsular *C. neoformans* or other fungi of comparative size. An in vivo consequence is that complement depletion can result [26]. In such situations, macrophage recognition of *C. neoformans* may not occur.

Macrophages possess three major complement receptors (CR) for C3 fragments, designated CD35 (CR1), CD11b/CD18 (CR3), and CD11c/CD18 (CR4). Monoclonal antibodies directed against any of the three receptors inhibit cryptococcal binding [18]. Moreover, Chinese hamster ovary (CHO) cells transfected with CR1, CR3, or CR4 bind serum-opsonized *C. neoformans* with the avidity of binding to CR3 being the greatest, followed in decreasing order by CR1 and CR4 [27]. Binding was not seen if *C. neoformans* was incubated in serum heat-inactivated to destroy complement or if CHO cells were transfected with an irrelevant receptor instead of a CR. Murine resident peritoneal cells also predominantly utilize CR3 to bind serum-opsonized, encapsulated *C. neoformans* [28]. The cytokines tumor necrosis factor-alpha and granulocyte-macrophage colony-stimulating factor (GM-CSF) enhance binding of serum-opsonized, encapsulated *C. neoformans* by directly upregulating CR3 [28]. Taken together, these data suggest that CR3, which primarily recognizes iC3b, is the predominant receptor responsible for binding of serum-opsonized *C. neoformans* but that CR1 and CR4 also participate.

With both human and murine macrophages, binding of encapsulated *C. neoformans* is inefficient compared with acapsular yeast cells, even in the presence of complement opsonization, [18,29]. Binding of encapsulated *C. neoformans* to monocyte-derived macrophages was profoundly inhibited by incubation in the cold or by inhibitors of receptor capping and actin microfilaments, suggesting an energy-dependent process [18]. In contrast, binding of acapsular yeasts proceeds in the cold and does not require actin.

Human serum from uninfected individuals as well as those with cryptococcosis commonly contains antibodies reactive with *C. neoformans* capsule [30–33]. However, anticapsular antibody found in uninfected individuals is not opsonic [18,33]. Thus, removal of antibody from normal human serum has no effect on C3 deposition on the fungus or on cryptococcal binding to macrophages. Capsule is poorly immunogenic, and a strong antibody response is rarely seen in experimental or clinical cryptococcosis. Moroever, specific antibodies that are generated are likely to form immune complexes with circulating capsular polysaccharide rather than binding to intact organisms.

The lack of an effective antibody response has prompted development of vaccination strategies. Conjugation of capsule to a carrier protein such as tetanus toxoid greatly increases its immunogenicity [34]. Numerous monoclonal antibodies have been generated following vaccination of mice, some of which are protective in murine models of cryptococcosis [35]. Interestingly, anticryptococcal antibodies of the IgG3 isotype are not protective. However, when switched to IgG1, IgG2a, and IgG2b isotypes of identical specificity, they become protective. Mouse macrophages have three different classes of receptors for IgG (FcγR). Two of these receptors, FcγRI and FcγRIII, share a common gamma chain. Opsonization of *C. neoformans* with anticapsular antibodies of the IgG1, IgG2a and IgG2b isotypes leads to macrophage phagocytosis and fungal growth inhibition. Anticapsular IgG1, IgG2a and IgG2b antibodies are not protective in mice that lack the common FcγR gamma chain (FcRγ −/−). Moreover, macrophages from FcRγ −/− mice do not phagocytose *C. neoformans* opsonized with such antibodies. In contrast, internalization of IgG3-treated organisms does not arrest fungal growth in macrophages from normal or FcRγ −/− mice [36]. In summary, the affinity of IgG subclasses for

FcγR receptors on macrophages is a key factor in determining whether anticapsular antibodies are protective. Protection appears to be mediated by opsonizing *C. neoformans* for phagocytosis and killing.

The major component of cryptococcal capsule, glucuronoxylomannan (GXM), is shed from *C. neoformans* and circulates in the blood and cerebrospinal fluid of patients with cryptococcosis. In addition to the virulence properties of capsule in situ on *C. neoformans*, free GXM has been postulated to contribute to virulence by diverse means including inhibiting leukocyte migration, modulating cytokine secretion, and causing loss of cellular receptors [37–39]. Interest has focused on defining cellular receptors for GXM that could be responsible for some of these effects. Neutrophils bind GXM at least in part via CD18, the beta chain of the beta-2 integrin family of adhesion molecules [40]. Macrophages also express CD18 [18], although it is not known whether macrophage CD18 binds GXM. GXM confers a strong negative charge on *C. neoformans* [41] and thus it is possible that rather than binding to specific receptors, some binding could be a result of charge interactions between the anionic GXM and cationic surface molecules on the cell. In addition to effects mediated by GXM, purified mannoprotein from *C. neoformans* stimulates human peripheral blood mononuclear cells to secrete TNF-α via a process that is dependent upon mannose binding protein and is inhibited by antibodies to CD14 [42].

Once binding of complement-opsonized *C. neoformans* takes place, phagocytosis (defined as internalization of the fungus into the macrophage) proceeds. The ability of a micro-organism to enter a macrophage can be a double-edged sword. Macrophage functions critical to the host, including antimicrobial activity and antigen processing and presentation, may require phagocytosis. Conversely, many organisms have adapted to intracellular parasitism and thrive within the macrophage. Compared with most other intracellular pathogens as well as acapsular *C. neoformans*, phagocytosis of encapsulated *C. neoformans* occurs very slowly. In this regard 30 min following binding, the majority of serum-opsonized encapsulated fungi still have not been phagocytosed [18]. However, by 2 hr, most fungi have been internalized [17]. *C. neoformans* strain 6, which has a particularly large capsule, binds to human monocyte-derived macrophages in the presence of serum but phagocytosis of bound organisms is <20% even after 2 hr [17]. Phagocytosis is considerably more efficient in the presence of specific anticapsular antibody. In vivo, capsule thickness can get so large that phagocytosis by individual macrophages is physically precluded [43]. In such cases, groups of macrophages may surrounded the organism. Cytokines derived from CD4+ T-cells may induce macrophages to fuse to form multinucleated giant cells containing internalized *C. neoformans* [44].

III. INTRACELLULAR SURVIVAL AND REPLICATION VERSUS INHIBITION AND KILLING

Once internalized within the macrophage, one of three outcomes of this host-fungal interaction is possible. First, the macrophage could fail to exert any antimicrobial activity, in which case intracellular growth of *C. neoformans* and eventual destruction of the macrophage would ensue. Second, the macrophage could inhibit cryptococcal growth but not kill the organism. If growth inhibition is transient, this can "buy time" for the host by allowing time for a specific immune response to develop before overwhelming infection occurs. However, prolonged microbial stasis may result in latency, with the potential for reactivation should host defenses falter. Finally, the macrophage may kill the fungal cell.

In vivo, all three possible outcomes may occur depending on the stage and site of infection, the expression of cryptococcal virulence factors, and the state of host immunity. Indeed, in vitro, the fate of *C. neoformans* inside the macrophage has been shown to vary depending upon macrophage species of origin, anatomical location, and state of activation and differentiation. Regarding the ability of *C. neoformans* to survive and replicate within macrophages, several investigators have been unable to show significant growth inhibition or killing of *C. neoformans* by mouse peritoneal macrophages in the absence of activating factors [29,45,46]. Within human monocyte-derived macrophages cultured on plastic surfaces, *C. neoformans* may replicate more rapidly than extracellular organisms cultured under the same conditions except for the absence of phagocytes [9]. Of particular relevance given the propensity of *C. neoformans* to infect the central nervous system, rabbit cerebrospinal fluid macrophages did not inhibit cryptococcal growth [47] and, in the absence of cytokine stimulation, *C. neoformans* was observed to replicate within human microglial cells forming large spacious vacuoles and leading to microglial cell lysis [48]. In addition, recent studies of pulmonary infection in mice and rats have provided evidence of replication and survival of *C. neoformans* within macrophages in vivo [11,49] (Fig. 1).

The mechanisms by which *C. neoformans* under some circumstances is able to survive and replicate within macrophages are incompletely understood. A key to the success of intracellular pathogens that parasitize macrophages is the ability to survive within the hostile acidic environment of the phagolysosome, to modify phagolysosome pH, or to evade the phagolysosome altogether. Phagosome-lysosome fusion appears to take place in macrophages that have ingested *C. neoformans* [50]. Furthermore, the cryptococcal phagolysosome is acidified normally with a resulting pH close to 5.0 [50]. However, at least extracellularly, *C. neoformans* replicates more rapidly at acidic than neutral pH, and this ability may consitute one factor favoring intracellular survival [50,51]. In addition, *C. neoformans* is a relatively weak stimulator of the respiratory burst, the process whereby phagocytes consume molecular oxygen to generate reactive oxygen species (e.g., hydrogen peroxide) with microbicidal activity [29]. Furthermore, like most fungi, *C. neoformans* is relatively resistant to the microbicidal oxidants generated as a consequence of the respiratory burst.

Unlike murine resident peritoneal macrophages, murine and rat resident bronchoalveolar macrophages will kill *C. neoformans*, an intrinsic antifungal activity that may play an important role in preventing the establishment of infection following inhalation of organisms [22,52].

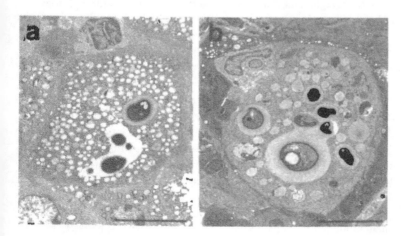

Figure 1 Electron micrographs taken (a) 28 days and (b) 7 days after intratracheal infection of C57BL6 mice with *C. neoformans*. Lung macrophages are shown with phagosomes containing multiple *C. neoformans* of differing size and including budding forms. Bars, 10 μm. (From Ref. 49.)

Paradoxically, encapsulated organisms are killed more efficiently by murine alveolar macrophages than acapsular mutants [52]. Preincubation of murine alveolar macrophages with M-CSF for 48–72 h increased anticryptococcal activity and in vivo administration of M-CSF by subcutaneous injection was also shown to increase ex vivo alveolar (and peritoneal) macrophage fungistasis compared with cells from control mice [53,54]. With rat alveolar macrophages, antifungal activity diminishes after overnight culture but is partially restored by addition of IFN-γ [55]. GM-CSF also increases the growth inhibitory activity of rat alveolar macrophages but requires preincubation for 5 days. However, the combination of IFN-γ and GM-CSF induces a rapid and sustained increase in anticryptococcal activity, unlike either cytokine alone [56].

A number of studies have defined conditions whereby murine peritoneal macrophages can be activated to inhibit *C. neoformans* growth. Peritoneal macrophages from mice infected with *L. monocytogenes* or *B. jellisoni* but not uninfected mice, were able to partially withstand a subsequent in vitro challenge with *C. neoformans* [45]. Granger and colleagues demonstrated complete fungistasis but only if macrophages were induced by peptone from BCG-infected mice, and endotoxin and a nonopsonic serum component were present in the culture medium [46,57]. In later experiments, fungistasis was obtained when IFN-γ was substituted for BCG [58]. Culture with IFN-γ also activated murine resident peritoneal cells to kill *C. neoformans* [29]. Granger and colleagues showed the effect of activated peritoneal macrophages to be dependent on nitric oxide, and perhaps other reactive nitrogen intermediaries, formed following oxidation of L-arginine to nitrite, nitrate, and L-citrulline [59,60]. Reagent nitric oxide was found to be fungistatic at low concentrations and fungicidal at high concentrations [61]. IFN-γ can also activate murine bone marrow macrophages to kill *C. neoformans*, although in this case, killing was mainly extracellular and apparently mediated by secreted proteins [62]. In mouse models of pulmonary and systemic cryptococcosis, IFN-γ-induced nitric oxide is necessary for optimal clearance of *C. neoformans* [63].

The situation with human macrophages is different. As in other species, freshly isolated human bronchoalveolar macrophages have some intrinsic anticryptococcal activity although this is fungistatic rather than fungicidal [64,65]. Freshly isolated human blood monocytes kill *C. neoformans* [66,67]. However, as described above, when monocytes are cultured in vitro on plastic surfaces so that they differentiate into macrophages, anticryptococcal activity is lost. Culture of monocytes on surfaces coated with fibronectin or poly-L-lysine, though not with laminin or collagen, resulted in macrophages that inhibited cryptococcal growth, although they still did not kill the fungus [68].

While it is presumed that cytokines produced during the course of a cell-mediated immune response are critical for activation of human macrophages to inhibit and kill *C. neoformans*, such cytokines have not been readily identified. Specifically, IFN-γ not only fails to increase anticryptococcal activity, but in some systems, treatment of human macrophages with IFN-γ has detrimental effects [68–70]. The likely reason for the disparity between the murine data, where IFN-γ is a powerful activator of fungicidal activity, and the human data relates to nitric oxide. Human macrophages have not been shown to produce significant quantities of nitric oxide when stimulated by *C. neoformans*, even if first activated by IFN-γ. Thus, fungistasis mediated by human macrophages, when it does occur, appears to proceed independently of nitric oxide production [69]. In contrast to results with IFN-γ, preincubation with GM-CSF for 7 days induces human PBMC, but not monocyte-derived macrophages, to effect complete fungistasis [68,71]. Preincubation for 3 days with GM-CSF and M-CSF has been shown to increase the anticryptococcal activity of human monocytes and monocyte-derived macrophages, respectively [72,73].

The mechanisms by which human macrophages can inhibit cryptococcal growth remain uncertain but may involve both reactive oxygen intermediates (ROI) and nonoxidative mechanisms. *C. neoformans* is sensitive to ROI in vitro and ROI are important in the activity of neutrophils against *C. neoformans*, in particular distal ROI, such as hydroxyl radical and hypochlorite, formed by the hydrogen peroxide-myeloperoxidase-halide system or by the metal ion-dependent Haber-Weiss reaction [66,74,75]. Since freshly isolated monocytes but not monocyte-derived macrophages possess myeloperoxidase activity, it is tempting to hypothesize that the reduction in anticryptococcal activity on in vitro culture of monocytes is related to this loss of myeloperoxidase. Another pathway present in macrophages, the polyamine oxidase-polyamine system, leads to production of hydrogen peroxide, aminaldehydes, and ammonia from spermine and spermidine and has been shown in an extracellular system to kill *C. neoformans* [76]. On the other hand, there is no evidence directly implicating such oxidative mechanisms in macrophage anticryptococcal activity—in human or murine systems [29,52,62].

Potential nonoxidative mechanisms include the physicochemical milieu of the phagosome and antimicrobial proteins and peptides. However, conditions within the phagosome could not account for the extracellular inhibition of cryptococcal growth by macrophages observed in a number of studies [46,62,64]. In addition, as discussed above, the acidity of the cryptococcal phagolysosome may actually promote cryptococcal growth. Iron restriction is an important host defense strategy against a range of pathogens. One means by which this may be achieved is by iron-free lactoferrin, released at sites of infection by neutrophils and other cells, that may be taken up into the phagolysosomes of macrophages thereby restricting the availability of iron to intracellular pathogens. The role of iron in macrophage activity against *C. neoformans* is unclear, but lactoferrin has been shown to kill *Candida albicans* in vitro [77], and bovine lactoferricin, a peptide derived from lactoferrin, has activity against *C. neoformans* [78]. Preliminary data suggest initial pulmonary clearance of *C. neoformans* and the outcome of cerebral infection in mice is adversely affected by prior iron overload [79]. Chloroquine reduces the availability of iron within macrophages and markedly enhances the anticryptococcal activity of macrophages [17,80]. Interestingly, chloroquine's effect in this case is independent of iron deprivation [17].

A number of other antimicrobial proteins or peptides, present in macrophages, that may play a role in activity against *C. neoformans* have been identified. Lysozyme has recently been shown to have anticryptococcal activity in vitro [81]. Defensins are amphipathic polypeptides of around 30 amino acids that affect pathogen membrane permeability and can kill *C. neoformans* [82]. They are found in the azurophilic granules of neutrophils from a range of species, but in macrophages defensins have thus far been identified only in rabbits. Murine macrophages contain at least three histonelike cationic proteins with activity against *C. neoformans* [83]. The molecular weights of these proteins (36,31, and 16 kDa) were not dissimilar from the apparent molecular weights (30 and 15 kDa) of proteins secreted by IFN-γ-activated bone marrow–derived murine macrophages and implicated in the extracellular killing of *C. neoformans* [62]. Human alveolar macrophages have been reported to contain cationic proteases with anticryptococcal activity that is upregulated by IFN-γ [84]. Calprotectin, a calcium-binding cytoplasmic protein present in a variety of human cells including monocytes and some macrophages has also been shown to kill *C. neoformans* in vitro, although the mechanism was unclear [85].

IV. ROLE OF MACROPHAGES IN THE INTEGRATED IMMUNE RESPONSE TO *C. NEOFORMANS*

Development of a specific cell-mediated immune response requires antigen-presenting cells (APC, of which mononuclear phagocytes are a major component), to process and present fungal antigen(s) to T-lymphocytes. As described above, exogenous antigens such as *C. neoformans* are taken up into acidic vesicles of the endosome-lysosome pathway where integral and secreted cryptococcal proteins may be processed into peptide fragments. These peptides bind MHC class II molecules within the vesicles and the MHC-peptide complexes are then expressed on the cell surface where they may be recognized by antigen-specific CD4 cells. For T cell activation and proliferation to occur, a second signal is also required that may be provided by costimulatory molecules on the APC, such as B7-1 and B7-2 that bind CD28/CTLA-4 on T cells, or by cytokines secreted by the APC. This antigen presentation and costimulation function of macrophages is reflected in vitro by their ability to support lymphocyte proliferation in response to cryptococcal antigens. Murine macrophages and human monocytes and bronchoalveolar macrophages challenged with *C. neoformans* stimulate T-cell proliferation [86–88]. Proliferation is greater in response to acapsular than encapsulated organisms but responses to encapsulated organisms can be enhanced by opsonizing antibody. In response to *C. neoformans*, human monocytes upregulate surface expression of B7-1 and, to a lesser extent, B7-2 [89]. Blockade of either, but particularly B7-2, reduces proliferation. Recent studies have shown that CD8 cells contribute to the proliferative response of human PBMC to *C. neoformans*, although this was dependent on the presence of CD4 cells for secondary signals [90]. In contrast to CD4 T-cells, CD8 T-cells recognize peptides, generated in the cytosol through the action of proteasomes, complexed with MHC class I molecules.

From animal studies, a protective T-cell-mediated response involves both CD4 and CD8 cells and increasing evidence supports the concept that protection is related to a TH1 pattern of cytokine release. Studies in knockout mice, and of depletion and administration of cytokines, have shown TNF-α, IFN-γ, IL-12, and more recently IL-18 to be involved in protection in the murine system [63,91–95]. As discussed above, IFN-γ is known to activate murine (but not human) macrophages to kill *C. neoformans* by a nitric oxide–dependent mechanism. In mice, both IL-12 and IL-18 contribute to protection by IFN-γ-dependent mechanisms [95]. IL-12 and IL-18 are important in the development of a TH1-type response and macrophages are major sources of both cytokines. Human monocytes or PBMC release only low levels of IL-12 in response to whole *C. neoformans*, although higher concentrations have been measured in response to certain mannoprotein fractions [96,97]. Blockade of endogenous IFN-γ reduces, and priming with IFN-γ prior to stimulation greatly enhances, IL-12 release [98,99]. Release is also increased by coculture of monocytes with activated T-cells or anti-CD40 and anti-MHC class II antibodies, especially together [99].

Studies by Huffnagle and colleagues of pulmonary cryptococcosis in mice have shown that early production of TNF-α, together with the chemokines MCP-1 and MIP-1α, is important for the recruitment of monocytes and lymphocytes to the lung [100–102]. TNF-α, with GM-CSF, also enhances complement-mediated phagocytosis [103]. Whether TNF-α enhances effector cell function is less clear, although it does not increase the anticryptococcal activity of monocyte-derived macrophages [68]. TNF-α is produced by human monocytes and alveolar macrophages stimulated with whole *C. neoformans* [16] or components of the capsule and cell wall, especially a mannoprotein fraction [104]. In response to *C. neoformans*, MCP-1 is also produced by human monocytes in vitro, although constitutive production of MCP-1 by alveolar macrophages was not increased [105]. In mice, CCR2, expressed on monocytes and activated T-cells and the major receptor for MCP-1 [106], is required for normal leukocyte recruitment, development of a TH1-type cytokine response, and the restriction of cryptococcal growth [107]. Interestingly, mice lacking CCR5 (also expressed on monocytes and activated T-cells and the receptor for MIP-1α, MIP-1β, and RANTES) have recently been shown to have increased susceptibility to *C. neoformans* due to a defect of leucocyte recruitment to the brain whereas recruitment to the lungs was unaffected [108]. Since CCR5 is a known coreceptor for HIV gp 120, it is possible that HIV may interfere with such brain recruitment.

Other cytokines produced by macrophages in response to *C. neoformans* in vitro include IL-1 and IL-6 [109]. Their role in vivo is less clear, although they may contribute to resistance to cerebral infection. Intracerebral administration of IL-1 and IL-6, but not TNF-α, prior to intracerebral infection with *C. neoformans*, has been shown to reduce CFU and prolong survival of mice [110]. *C. neoformans* also stimulates release of IL-15 from monocytes, which may then, with IL-2, induce lymphocyte proliferation and lymphocyte anticryptococcal activity [111]. IL-10 produced by PBMC and monocytes in reponse to whole organisms [112] and purified glucuronoxylomannan [113], the major capsule constituent, may be important in immune regulation and the immunosuppressive effects of cryptococcal infection and capsular polysaccharide in particular. In vitro, IL-10 reduces release of TNF-α, IL-1, and IL-12, expression of MHC class II, and lymphoproliferation in response to *C. neoformans* [112,114].

Thus, cytokines produced by macrophages in response *C. neoformans*, are involved in leukocyte recruitment and determination of the pattern of immune response as well as in immune regulation. The organism and host factors that affect the production of these cytokines will clearly impact the outcome of infection and are the subject of ongoing research. Compared with acapsular strains, encapsulated *C. neoformans* induce less TNF-α, IL-1β, and IL-12, but more IL-10 production by human monocytes or PBMC [99,115,116]. Addition of a monoclonal antibody against capsular polysaccharide increased TNF-α and IL-1β and reduced IL-10 production to levels seen with acapsular organisms [117]. Antibody also caused a more modest increase in IL-12 production [99]. In addition, high melanin production, catalyzed by the cryptococcal laccase, has been associated with reduced production of TNF-α by murine alveolar macrophages in vitro and a reduced and ineffective immune response in vivo [118]. Thus, there is evidence that two of the established virulence factors of *C. neoformans*, capsule and laccase, may subvert a protective immune response by altering the interaction of the organism with macrophages. Host factors that influence the balance between Th1 and Th2-type cytokines and the outcome of infection have also been examined. Relative sensitivity of BALB/c compared with C.B-17 mice to cryptococcal infection is associated with increased Th2 cytokine and specific antibody levels that appear to be determined by genes linked to the Ig heavy-chain complex [119]. Responses of murine macrophages to *C. neoformans* may differ depending on mouse strain [120]. In this regard, studies of the influence of the *bcg/Nramp* locus on the interaction of macrophages with *C. neoformans* are under way.

A variety of effector cells in addition to macrophages, including neutrophils, NK cells, and T lymphocytes, have been shown to inhibit or kill *C. neoformans* under defined conditions [75,121,122]. The relative importance of these cell types in vivo is not clear, but the importance of macrophages in the effector arm of the human immune response to *C. neoformans* is supported by the pathology of human cryptococcosis. This is highly variable, reflecting both the immune status of the host and organism factors, and is often notable for the paucity of inflammation with large numbers of extracellular organisms forming cysticlike spaces in the tissues. Nevertheless, when present, inflammation is granulomatous with many fungi in close apposition or actually inside macrophages and giant cells [123,124]. Similarly, resolution of infection in animal models is invariably associated with granulomatous inflammation. For example, following an intratracheal challenge of mice with *C. neoformans*, multinucleated giant cells surround the fungus. Mice depleted of CD4 T-cells did not form multinucleated giant cells and were unable to clear the *C. neoformans* from the lungs, suggesting that CD4 T-cells produce cytokines at the site of infection which promote granuloma formation and activate macrophages [44].

V. EFFECTS OF HIV ON MACROPHAGE–*C. NEOFORMANS* INTERACTIONS

There is evidence that HIV infection may adversely affect the effector function of monocytes and macrophages against *C. neoformans*. Following a 2-hr incubation, PBMC from persons with

AIDS bound and killed *C. neoformans* as well as control cells [125]. However, when the incubation was extended overnight, PBMC and monocytes obtained from persons with HIV infection had impaired anticryptococcal activity compared with cells from HIV-seronegative donors [126,127]. Again, phagocytosis appeared to be unaffected, but there was a marked reduction in oxidative burst as assessed by production of hydrogen peroxide in response to the organism. In vivo, such functional defects could result from a combination of the direct effects of HIV infection of blood monocytes (although the percentage of peripheral blood monocytes that are infected with HIV appears to be low), effects of HIV products such as gp 120, and reduced T-cell stimulation secondary to the loss of CD4 cells. There is some in vitro evidence to support each of these mechanisms. Normal human monocytes and peritoneal macrophages (but not alveolar macrophages) had temporarily reduced anticryptococcal activity after infection with a monocytotropic strain of HIV [128]. Alveolar macrophages from HIV-infected donors were not found to have reduced anticryptococcal activity compared with cells from uninfected donors [129]. However, addition of gp 120 to human bronchoalveolar macrophages significantly reduced the capacity of human bronchoalveolar macrophages to phagocytose and inhibit the growth of *C. neoformans* [65]. Lastly, in a study of rhesus monkeys infected with SIV, alveolar macrophages had reduced anticryptoccocal activity associated with reduced superoxide anion production late, but not early, in the course of the disease. These defects could be restored by supernatants of mitogen-stimulated PBMC from uninfected animals. Restoration was dependent in part on IFN-γ, suggesting that in this case the defect in macrophage function was due largely to defective activation [130]. In a more recent study, GM-CSF was shown to increase the anticryptococcal activity of monocytes from AIDS patients toward that of cells from uninfected donors [72]. GM-CSF increased complement receptor expression, phagocytosis, and superoxide anion generation.

Antigen presentation and cytokine production by macrophages in response to *C. neoformans* may also be affected in HIV disease but there is some evidence that the major effects of HIV infection on lymphoproliferation and the pattern of cytokine release result from the marked reduction in IFN-γ-producing CD4 cells. Proinflammatory cytokine and chemokine release by macrophages does not appear to be impaired in HIV disease [105]. In contrast, there is evidence for a deficit in IL-12 production [131]. As discussed above, IL-12 release from monocytes in response to whole *C. neoformans* is at the lower limits of detection in cells from HIV-infected and uninfected donors. However, when primed with INF-γ prior to stimulation, cells from both HIV-infected and uninfected donors release substantial and similar levels of IL-12 [98]. IL-12 was reported to restore lymphoproliferative responses to influenza and HIV envelope antigens in PBMC from HIV-infected donors [132]. As with other recall antigens, lymphoproliferation in response to *C. neoformans* progressively wanes during the course of HIV infection [96,133] but this defect was not corrected by the addition of exogenous IL-12 [96]. The affect of HIV gp120 on cryptococcal antigen presentation by monocytes has recently been studied by Vecchiarelli and colleagues [134]. HIV gp120 inhibits lymphoproliferation in response to *C. neoformans* if present throughout the assay but enhances lymphoproliferation if it is removed from the monocyte–*C. neoformans* culture prior to addition of autologous T-cells. Interestingly, under both conditions, gp120 decreased IFN-γ and increased IL-4 production, although the effects were small.

Some studies have started to look at the levels of macrophage-derived (and other) cytokines in AIDS patients with cryptococcosis. In one report, CSF TNF-α levels were found to be markedly raised in comparison with HIV-infected patients with other neurological disorders [135]. However, a second study found relatively low levels of TNF-α with raised levels of IL-1, IL-6, IL-8, and IL-10 [136]. Plasma TNF-α and IL-10 in HIV patients with cryptococcosis were found to be higher in patients with fungemia and disseminated infection than in those with isolated meningitis [137].

Conversely, macrophage–*C. neoformans* interactions may have effects on HIV. *C. neoformans* induced viral replication in T-cells freshly infected with HIV by a mechanism that required monocytic cells but was independent of TNF-α [138]. Furthermore, *C. neoformans* enhanced HIV expression in a latently HIV-infected myelomonocytic cell line through a TNF-α- and NF-κB-dependent mechanism [139].

VI. RELEVANCE OF MACROPHAGE–*C. NEOFORMANS* INTERACTIONS FOR PREVENTION AND THERAPY

Given that most patients with cryptococcosis are immunosuppressed, a clear rationale exists for immunomodulatory therapies. An increased understanding of the details of the interaction of *C. neoformans* with macrophages may lead to new approaches aimed at reconstituting or enhancing host defense. The beneficial effects of selected monoclonal antibodies on phagocytosis, growth inhibition, cytokine secretion, granuloma formation, and clearance of cryptococcal antigen by macrophages has been described and one such antibody is going forward to Phase I studies in AIDS patients recovering from cryptococcal meningitis [140]. The macrophage activating cytokines IFN-γ and GM-CSF are commercially available. Enthusiasm for IFN-γ, based on studies in murine systems, is perhaps tempered by the lack of clear beneficial effects of IFN-γ using human cells in vitro, although this may relate to the artificiality of the conditions. GM-CSF has been shown to have some beneficial effects on human cells and it has been used in a small number of patients with cryptococcosis with no apparent adverse effects [141,142]. In one small comparative study of amphotericin-B with or without GM-CSF in AIDS patients with cryptococcal meningitis, adjunctive GM-CSF was associated with a trend toward more rapid sterilization of the CSF [143]. Both agents deserve further study. Of the cytokines produced by macrophages, IL-12, although at a much earlier stage of development, is of therapeutic interest, given its role in the development of Th1-type responses and its deficiency in HIV disease.

Study of the effects of antifungal agents on macrophage function has given additional insights into their possible modes of action in vivo. Amphotericin-B has been shown to induce TNF-α production [144] and enhance the oxidative burst [145] and IFN-γ-induced nitric oxide production by murine macrophages [146,147]. The latter effect appeared to be mediated in part by increased TNF-α and IL-1β. Amphotericin-B also increases superoxide production by human macrophages [148] and accumulates in human monocytes [149]. Furthermore, subinhibitory concentrations of amphotericin-B appear to reduce cryptococcal capsule size, perhaps by inducing shedding of polysaccharide, and thereby facilitate phagocytosis [150]. Itraconazole and, to a lesser extent, ketoconazole accumulate within macrophages. In contrast, fluconazole does not appear to activate or accumulate within macrophages [58].

Given increasing evidence for the role of intracellular infection and latency in cryptococcosis, drugs that target intracellular *C. neoformans* may find a place in future treatment combinations or prophylaxis. In this regard, the antimalarial drugs chloroquine and quinacrine, being diprotic weak bases, have been shown to accumulate within the acidic cryptococcal phagolysosome and within acidic vacuoles within the fungal cell [151] (Fig. 2). They have direct antifungal activity [151] and, at pharmacologically relevant concentrations, markedly enhance the anticryptococcal activity of macrophages [17,80]. The latter effect probably results from the accumulation of high drug concentrations within the cryptococcal phagosome and neutralization of phagosome pH that may slow cryptococcal growth. Chloroquine given intracerebrally has been shown to prolong the survival of mice subsequently infected intracerebrally with *C. neoformans* [80]

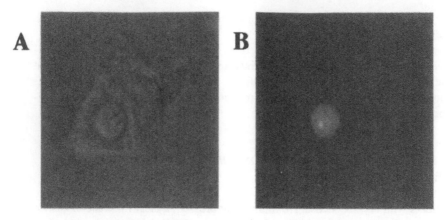

Figure 2 Accumulation of quinacrine within phagocytosed *C. neoformans*. Macrophages were incubated with *C. neoformans* in PBS containing 10% PHS and 1 μM quinacrine. After 1 hr, slides were washed and immediately examined on a laser confocal scanning microscope. (A) Bright field view of a macrophage containing a single internalised *C. neoformans*. (B) Fluorescence image of the same cell. (From Ref. 150.)

and, given by intraperitoneal injection, to prolong the survival of mice subsequently infected intravenously [17].

Regarding immunomodulatory agents, it may be that combinations of two or more will be needed with complementary or synergistic effects to result in significant benefit for patients. In addition, some immunomodulating agents, such as cytokines, may have qualitatively different effects on the immune response depending on the schedule and doses used. These issues pose a considerable challenge to the ingenuity and perseverance of researchers. Much further work will need to be done if increased knowledge of the interaction of *C. neoformans* with macrophages and other host cells is to be translated into more effective therapies.

REFERENCES

1. A Casadevall, JR Perfect. *Cryptococcus neoformans*. Washington: ASM Press, 1998.
2. KP Kayembe, M Na, RL Colebunders. Opportunistic infections and diseases. In: M Essex, S Mboup, PJ Kanki, eds. AIDS in Africa. New York: Raven Press, 1994, pp 373–391.
3. RA Seaton, S Naraqi, JP Wembri, DA Warrell. Predictors of outcome in *Cryptococcus neoformans* var. *gattii* meningitis. Q J Med 89:423–428, 1996.
4. RA Seaton, N Verma, S Naraqi, JP Wembri, DA Warrell. Visual loss in immunocompetent patients with *Cryptococcus neoformans* var. *gattii* meningitis. Trans R Soc Trop Med Hyg 91:44–49, 1997.
5. JB Neilson, RA Fromtling, GS Bulmer. Cryptococcus neoformans: size range of infectious particles from aerosolized soil. Infect Immun 17:634–638, 1977.
6. KJ Kwon-Chung. A new genus, filobasidiella, the perfect state of *Cryptococcus neoformans*. Mycologia 67:1197–1200, 1975.
7. BL Wickes, ME Mayorga, U Edman, JC Edman. Dimorphism and haploid fruiting in *Cryptococcus neoformans*: association with the alpha-mating type. Proc Natl Acad Sci USA 93:7327–7331, 1996.
8. JR Graybill, L Mitchell, DJ Drutz. Host defense in cryptococcosis. III. Protection of nude mice by thymus transplantation. J Infect Dis 140:546–552, 1979.
9. RD Diamond, JE Bennett. Growth of *Cryptococcus neoformans* within human macrophages in vitro. Infect Immun 7:231–236, 1973.

10. D Garcia-Hermoso, G Janbon, F Dromer. Epidemiological evidence for dormant *Cryptococcus neoformans* infection. J Clin Microbiol 37:3204–3209, 1999.

11. DL Goldman, SC Lee, AJ Mednick, L Montella, A Casadevall. Persistent *Cryptococcus neoformans* pulmonary infection in the rat is associated with intracellular parasitism, decreased inducible nitric oxide synthase expression, and altered antibody responsiveness to cryptococcal polysaccharide. Infect Immun 68:832–838, 2000.

12. R Cherniak, JB Sundstrom. Polysaccharide antigens of the capsule of *Cryptococcus neoformans*. Infect Immun 62:1507–1512, 1994.

13. DL Granger, JR Perfect, DT Durack. Virulence of *Cryptococcus neoformans*. Regulation of capsule synthesis by carbon dioxide. J Clin Invest 76:508–516, 1985.

14. SE Vartivarian, EJ Anaissie, RE Cowart, HA Sprigg, MJ Tingler, ES Jacobson. Regulation of cryptococcal capsular polysaccharide by iron. J Infect Dis 167:186–190, 1993.

15. SM Levitz. The ecology of Cryptococcus neoformans and the epidemiology of cryptococcosis. Rev Infect Dis 13:1163–1169, 1991.

16. SM Levitz, A Tabuni, H Kornfeld, CC Reardon, DT Golenbock. Production of tumor necrosis factor alpha in human leukocytes stimulated by *Cryptococcus neoformans*. Infect Immun 62:1975–1981, 1994.

17. SM Levitz, TS Harrison, A Tabuni, X Liu. Chloroquine induces human mononuclear phagocytes to inhibit and kill *Cryptococcus neoformans* by a mechanism independent of iron deprivation. J Clin Invest 100:1640–1646, 1997.

18. SM Levitz, A Tabuni. Binding of *Cryptococcus neoformans* by human cultured macrophages. Requirements for multiple complement receptors and actin. J Clin Invest 87:528–535, 1991.

19. CE Cross, GJ Bancroft. Ingestion of acapsular *Cryptococcus neoformans* occurs via mannose and beta-glucan receptors, resulting in cytokine production and increased phagocytosis of the encapsulated form. Infect Immun 63:2604–2611, 1995.

20. TR Kozel, EC Gotschlich. The capsule of cryptococcus neoformans passively inhibits phagocytosis of the yeast by macrophages. J Immunol 129:1675–1680, 1982.

21. SM Levitz, A Tabuni, R Wagner, H Kornfeld, EH Smail. Binding of unopsonized *Cryptococcus neoformans* by human bronchoalveolar macrophages: inhibition by a large-molecular-size serum component. J Infect Dis 166:866–873, 1992.

22. B Bolaños, TG Mitchell. Phagocytosis and killing of *Cryptococcus neoformans* by rat alveolar macrophages in the absence of serum. J Leukoc Biol 46:521–528, 1989.

23. MM Lipovsky, G Gekker, WR Anderson, TW Molitor, PK Peterson, AI Hoepelman. Phagocytosis of nonopsonized *Cryptococcus neoformans* by swine microglia involves CD14 receptors. Clin Immunol Immunopathol 84:208–211, 1997.

24. TR Kozel, GS Pfrommer. Activation of the complement system by *Cryptococcus neoformans* leads to binding of iC3b to the yeast. Infect Immun 52:1–5, 1986.

25. TR Kozel, MA Wilson, GS Pfrommer, AM Schlageter. Activation and binding of opsonic fragments of C3 on encapsulated *Cryptococcus neoformans* by using an alternative complement pathway reconstituted from six isolated proteins. Infect Immun 57:1922–1927, 1989.

26. AM Macher, JE Bennett, JE Gadek, MM Frank. Complement depletion in cryptococcal sepsis. J Immunol 120:1686–1690, 1978.

27. SM Levitz, A Tabuni, TR Kozel, RS MacGill, RR Ingalls, DT Golenbock. Binding of *Cryptococcus neoformans* to heterologously expressed human complement receptors. Infect Immun 65:931–935, 1997.

28. CE Cross, HL Collins, GJ Bancroft. CR3-dependent phagocytosis by murine macrophages: different cytokines regulate ingestion of a defined CR3 ligand and complement-opsonized *Cryptococcus neoformans*. Immunology 91:289–296, 1997.

29. SM Levitz, DJ DiBenedetto. Differential stimulation of murine resident peritoneal cells by selectively opsonized encapsulated and acapsular *Cryptococcus neoformans*. Infect Immun 56:2544–2551, 1988.

30. RD Diamond, JE Bennett. Prognostic factors in cryptococcal meningitis. A study in 111 cases. Ann Intern Med 80:176–181, 1974.

31. F Dromer, P Aucouturier, JP Clauvel, G Saimot, P Yeni. *Cryptococcus neoformans* antibody levels in patients with AIDS. Scand J Infect Dis 20:283–285, 1988.

32. R Fleurodor, RH Lyles, L Pirofski. Quantitative and qualitative differences in the serum antibody profiles of human immunodeficiency virus-infected persons with and without *Cryptococcus neoformans* meningitis. J Infect Dis 180:1526–1535, 1999.

33. DC Houpt, GS Pfrommer, BJ Young, TA Larson, TR Kozel. Occurrences, immunoglobulin classes, and biological activities of antibodies in normal human serum that are reactive with *Cryptococcus neoformans* glucuronoxylomannan. Infect Immun 62:2857–2864, 1994.

34. SJ Devi, R Schneerson, W Egan, TJ Ulrich, D Bryla, JB Robbins, JE Bennett. *Cryptococcus neoformans* serotype A glucuronoxylomannan-protein conjugate vaccines: synthesis, characterization, and immunogenicity. Infect Immun 59:3700–3707, 1991.

35. A Casadevall. Antibody immunity and invasive fungal infections. Infect Immun 63:4211–4218, 1995.

36. R Yuan, R Clynes, J Oh, JV Ravetch, MD Scharff. Antibody-mediated modulation of *Cryptococcus neoformans* infection is dependent on distinct Fc receptor functions and IgG subclasses. J Exp Med 187:641–648, 1998.

37. ZM Dong, JW Murphy. Intravascular cryptococcal culture filtrate (CneF) and its major component, glucuronoxylomannan, are potent inhibitors of leukocyte accumulation. Infect Immun 63:770–778, 1995.

38. ZM Dong, JW Murphy. Cryptococcal polysaccharides induce L-selectin shedding and tumor necrosis factor receptor loss from the surface of human neutrophils. J Clin Invest 97:689–698, 1996.

39. W Chaka, AF Verheul, VV Vaishnav, R Cherniak, J Scharringa, J Verhoef, H Snippe, AI Hoepelman. Cryptococcus neoformans and cryptococcal glucuronoxylomannan, galactoxylomannan, and mannoprotein induce different levels of tumor necrosis factor alpha in human peripheral blood mononuclear cells. Infect Immun 65:272–278, 1997.

40. ZM Dong, JW Murphy. Cryptococcal polysaccharides bind to CD18 on human neutrophils. Infect Immun 65:557–563, 1997.

41. JD Nosanchuk, A Casadevall. Cellular charge of *Cryptococcus neoformans*: contributions from the capsular polysaccharide, melanin, and monoclonal antibody binding. Infect Immun 65:1836–1841, 1997.

42. W Chaka, AF Verheul, VV Vaishnav, R Cherniak, J Scharringa, J Verhoef, H Snippe, AI Hoepelman. Induction of TNF-alpha in human peripheral blood mononuclear cells by the mannoprotein of *Cryptococcus neoformans* involves human mannose binding protein. J Immunol 159:2979–2985, 1997.

43. M Kalina, Y Kletter, M Aronson. The interaction of phagocytes and the large-sized parasite *Cryptococcus neoformans*: cytochemical and ultrastructural study. Cell Tissue Res 152:165–174, 1974.

44. JO Hill. CD4 + T cells cause multinucleated giant cells to form around *Cryptococcus neoformans* and confine the yeast within the primary site of infection in the respiratory tract. J Exp Med 175:1685–1695, 1992.

45. LO Gentry, JS Remington. Resistance against *Cryptococcus* conferred by intracellular bacteria and protozoa. J Infect Dis 123:22–31, 1971.

46. DL Granger, JR Perfect, DT Durack. Macrophage-mediated fungistasis in vitro: requirements for intracellular and extracellular cytotoxicity. J Immunol 136:672–680, 1986.

47. JR Perfect, MM Hobbs, DL Granger, DT Durack. Cerebrospinal fluid macrophage response to experimental cryptococcal meningitis: relationship between in vivo and in vitro measurements of cytotoxicity. Infect Immun 56:849–854, 1988.

48. SC Lee, Y Kress, ML Zhao, DW Dickson, A Casadevall. *Cryptococcus neoformans* survive and replicate in human microglia. Lab Invest 73:871–879, 1995.

49. M Feldmesser, Y Kress, P Novikoff, A Casadevall. *Cryptococcus neoformans* is a facultative intracellular pathogen in murine pulmonary infection. Infect Immun July 68:4225–4237, 2000.

50. SM Levitz, SH Nong, KF Seetoo, TS Harrison, RA Speizer, ER Simons. *Cryptococcus neoformans* resides in an acidic phagolysosome of human macrophages. Infect Immun 67:885–890, 1999.

51. DH Howard. Some factors that affect the initiation of growth of *Cryptococcus neoformans*. J Bacteriol 82:430–435, 1961.

52. SM Levitz, DJ DiBenedetto. Paradoxical role of capsule in murine bronchoalveolar macrophage-mediated killing of *Cryptococcus neoformans*. J Immunol 142:659–665, 1989.

53. E Brummer, F Nassar, DA Stevens. Effect of macrophage colony-stimulating factor on anticryptococcal activity of bronchoalveolar macrophages: synergy with fluconazole for killing. Antimicrob Agents Chemother 38:2158–2161, 1994.

54. F Nassar, E Brummer, DA Stevens. Effect of in vivo macrophage colony-stimulating factor on fungistasis of bronchoalveolar and peritoneal macrophages against *Cryptococcus neoformans*. Antimicrob Agents Chemother 38:2162–2164, 1994.

55. CH Mody, CL Tyler, RG Sitrin, C Jacksoon, GB Toews. Interferon-gamma activates rat alveolar macrophages for anticryptococcal activity. Am J Respir Cell Mol Biol 5:19–26, 1991.

56. GH Chen, JL Curtis, CH Mody, PJ Christensen, LR Armstrong, GB Toews. Effect of granulocyte-macrophage colony-stimulating factor on rat alveolar macrophage anticryptococcal activity in vitro. J Immunol 152:724–734, 1994.

57. DL Granger, JR Perfect, DT Durack. Macrophage-mediated fungistasis: requirement for a macromolecular component in serum. J Immunol 137:693–701, 1986.

58. JR Perfect, DL Granger, DT Durack. Effects of antifungal agents and gamma interferon on macrophage cytotoxicity for fungi and tumor cells. J Infect Dis 156:316–323, 1987.

59. DL Granger, JBJ Hibbs, JR Perfect, DT Durack. Specific amino acid (L-arginine) requirement for the microbiostatic activity of murine macrophages. J Clin Invest 81:1129–1136, 1988.

60. DL Granger, JBJ Hibbs, JR Perfect, DT Durack. Metabolic fate of L-arginine in relation to microbiostatic capability of murine macrophages. J Clin Invest 85:264–273, 1990.

61. JA Alspaugh, DL Granger. Inhibition of *Cryptococcus neoformans* replication by nitrogen oxides supports the role of these molecules as effectors of macrophage-mediated cytostasis. Infect Immun 59:2291–2296, 1991.

62. IE Flesch, G Schwamberger, SH Kaufmann. Fungicidal activity of IFN-gamma-activated macrophages. Extracellular killing of *Cryptococcus neoformans*. J Immunol 142:3219–3224, 1989.

63. JA Lovchik, CR Lyons, MF Lipscomb. A role for gamma interferon–induced nitric oxide in pulmonary clearance of *Cryptococcus neoformans*. Am J Respir Cell Mol Biol 13:116–124, 1995.

64. PB Weinberg, S Becker, DL Granger, HS Koren. Growth inhibition of *Cryptococcus neoformans* by human alveolar macrophages. Am Rev Respir Dis 136:1242–1247, 1987.

65. RP Wagner, SM Levitz, A Tabuni, H Kornfeld. HIV-1 envelope protein (gp120) inhibits the activity of human bronchoalveolar macrophages against *Cryptococcus neoformans*. Am Rev Respir Dis 146:1434–1438, 1992.

66. RD Diamond, RK Root, JE Bennett. Factors influencing killing of *Cryptococcus neoformans* by human leukocytes in vitro. J Infect Dis 125:367–376, 1972.

67. MF Miller, TG Mitchell. Killing of *Cryptococcus neoformans* strains by human neutrophils and monocytes. Infect Immun 59:24–28, 1991.

68. SM Levitz, TP Farrell. Growth inhibition of *Cryptococcus neoformans* by cultured human monocytes: role of the capsule, opsonins, the culture surface, and cytokines. Infect Immun 58:1201–1209, 1990.

69. ML Cameron, DL Granger, JB Weinberg, WJ Kozumbo, HS Koren. Human alveolar and peritoneal macrophages mediate fungistasis independently of L-arginine oxidation to nitrite or nitrate. Am Rev Respir Dis 142:1313–1319, 1990.

70. CC Reardon, SJ Kim, RP Wagner, H Kornfeld. Interferon-gamma reduces the capacity of human alveolar macrophages to inhibit growth of *Cryptococcus neoformans* in vitro. Am J Respir Cell Mol Biol 15:711–715, 1996.

71. SM Levitz. Activation of human peripheral blood mononuclear cells by interleukin-2 and granulocyte-macrophage colony-stimulating factor to inhibit *Cryptococcus neoformans*. Infect Immun 59:3393–3397, 1991.

72. C Tascini, A Vecchiarelli, R Preziosi, D Francisci, F Bistoni, F Baldelli. Granulocyte-macrophage colony-stimulating factor and fluconazole enhance anti-cryptococcal activity of monocytes from AIDS patients. AIDS 13:49–55, 1999.

73. F Nassar, E Brummer, DA Stevens. Macrophage colony-stimulating factor (M-CSF) induction of enhanced anticryptococcal activity in human monocyte-derived macrophages: synergy with fluconazole for killing. Cell Immunol 164:113–118, 1995.

74. ES Jacobson, SB Tinnell. Antioxidant function of fungal melanin. J Bacteriol 175:7102–7104, 1993.

75. V Chaturvedi, B Wong, SL Newman. Oxidative killing of *Cryptococcus neoformans* by human neutrophils. Evidence that fungal mannitol protects by scavenging reactive oxygen intermediates. J Immunol 156:3836–3840, 1996.

76. SM Levitz, DJ DiBenedetto, RD Diamond. Inhibition and killing of fungi by the polyamine oxidase-polyamine system. Antifungal activity of the PAO-polyamine system. Antonie Van Leeuwenhoek 58:107–114, 1990.

77. T Soukka, J Tenovuo, M Lenander-Lumikari. Fungicidal effect of human lactoferrin against *Candida albicans*. FEMS Microbiol Lett 69:223–228, 1992.

78. LH Vorland, H Ulvatne, J Andersen, H Haukland, O Rekdal, JS Svendsen, TJ Gutteberg. Lactoferricin of bovine origin is more active than lactoferricins of human, murine and caprine origin. Scand J Infect Dis 30:513–517, 1998.

79. JR Boelaert, E Blasi. Cryptococcosis and smoking: the potential role for iron. J Infect Dis 180:1412, 1999.

80. R Mazzolla, R Barluzzi, A Brozzetti, JR Boelaert, T Luna, S Saleppico, F Bistoni, E Blasi. Enhanced resistance to *Cryptococcus neoformans* infection induced by chloroquine in a murine model of meningoencephalitis. Antimicrob Agents Chemother 41:802–807, 1997.

81. Y Nakamura, R Kano, S Watanabe, H Takahashi, A Hasegawa. Enhanced activity of antifungal drugs by lysozyme against *Cryptococcus neoformans*. Mycoses 41:199–202, 1998.

82. T Ganz, ME Selsted, D Szklarek, et al. Defensins. Natural peptide antibiotics of human neutrophils. J Clin Invest 76:1427–1435, 1985.

83. PS Hiemstra, PB Eisenhauer, SS Harwig, MT Barselaar, R van Furth, RI Lehrer. Antimicrobial proteins of murine macrophages. Infect Immun 61:3038–3046, 1993.

84. A Vecchiarelli, D Pietrella, M Dottorini, C Monari, C Retini, T Todisco, F Bistoni. Encapsulation of *Cryptococcus neoformans* regulates fungicidal activity and the antigen presentation process in human alveolar macrophages. Clin Exp Immunol 98:217–223, 1994.

85. M Steinbakk, CF Naess-Andresen, E Lingaas, I Dale, P Brandtzaeg, MK Fagerhol. Antimicrobial actions of calcium binding leucocyte L1 protein, calprotectin. Lancet 336:763–765, 1990.

86. HL Collins, GJ Bancroft. Encapsulation of *Cryptococcus neoformans* impairs antigen-specific T-cell responses. Infect Immun 59:3883–3888, 1991.

87. A Vecchiarelli, M Dottorini, D Pietrella, C Monari, C Retini, T Todisco, F Bistoni. Role of human alveolar macrophages as antigen-presenting cells in *Cryptococcus neoformans* infection. Am J Respir Cell Mol Biol 11:130–137, 1994.

88. RM Syme, TF Bruno, TR Kozel, CH Mody. The capsule of *Cryptococcus neoformans* reduces T-lymphocyte proliferation by reducing phagocytosis, which can be restored with anticapsular antibody. Infect Immun 67:4620–4627, 1999.

89. A Vecchiarelli, C Monari, C Retini, D Pietrella, B Palazzetti, L Pitzurra, A Casadevall. *Cryptococcus neoformans* differently regulates B7-1 (CD80) and B7-2 (CD86) expression on human monocytes. Eur J Immunol 28:114–121, 1998.

90. RM Syme, CJ Wood, H Wong, CH Mody. Both CD4 + and CD8 + human lymphocytes are activated and proliferate in response to *Cryptococcus neoformans*. Immunology 92:194–200, 1997.

91. K Aguirre, EA Havell, GW Gibson, LL Johnson. Role of tumor necrosis factor and gamma interferon in acquired resistance to *Cryptococcus neoformans* in the central nervous system of mice. Infect Immun 63:1725–1731, 1995.

92. KV Clemons, E Brummer, DA Stevens. Cytokine treatment of central nervous system infection: efficacy of interleukin-12 alone and synergy with conventional antifungal therapy in experimental cryptococcosis. Antimicrob Agents Chemother 38:460–464, 1994.

93. K Kawakami, M Tohyama, Q Xie, A Saito. IL-12 protects mice against pulmonary and disseminated infection caused by *Cryptococcus neoformans*. Clin Exp Immunol 104:208–214, 1996.

94. K Kawakami, MH Qureshi, T Zhang, H Okamura, M Kurimoto, A Saito. IL-18 protects mice against pulmonary and disseminated infection with *Cryptococcus neoformans* by inducing IFN-gamma production. J Immunol 159:5528–5534, 1997.

95. MH Qureshi, T Zhang, Y Koguchi, K Nakashima, H Okamura, M Kurimoto, K Kawakami. Combined effects of IL-12 and IL-18 on the clinical course and local cytokine production in murine pulmonary infection with *Cryptococcus neoformans*. Eur J Immunol 29:643–649, 1999.

96. TS Harrison, SM Levitz. Role of IL-12 in peripheral blood mononuclear cell responses to fungi in persons with and without HIV infection. J Immunol 156:4492–4497, 1996.

97. L Pitzurra, R Cherniak, M Giammarioli, S Perito, F Bistoni, A Vecchiarelli. Early induction of interleukin-12 by monocytes exposed to *Cryptococcus neoformans* mannoproteins. Infect Immun 68:558–563, 2000.

98. TS Harrison, SM Levitz. Priming with IFN-gamma restores deficient IL-12 production by peripheral blood mononuclear cells from HIV-seropositive donors. J Immunol 158:459–463, 1997.

99. C Retini, A Casadevall, D Pietrella, C Monari, B Palazzetti, A Vecchiarelli. Specific activated T cells regulate IL-12 production by human monocytes stimulated with *Cryptococcus neoformans*. J Immunol 162:1618–1623, 1999.

100. GB Huffnagle, RM Strieter, TJ Standiford, RA McDonald, MD Burdick, SL Kunkel, GB Toews. The role of monocyte chemotactic protein-1 (MCP-1) in the recruitment of monocytes and CD4 + T cells during a pulmonary *Cryptococcus neoformans* infection. J Immunol 155:4790–4797, 1995.

101. GB Huffnagle, GB Toews, MD Burdick, MB Boyd, KS McAllister, RA McDonald, SL Kunkel, RM Strieter. Afferent phase production of TNF-alpha is required for the development of protective T cell immunity to *Cryptococcus neoformans*. J Immunol 157:4529–4536, 1996.

102. GB Huffnagle, RM Strieter, LK McNeil, RA McDonald, MD Burdick, SL Kunkel, GB Toews. Macrophage inflammatory protein-1alpha (MIP-1alpha) is required for the efferent phase of pulmonary cell-mediated immunity to a *Cryptococcus neoformans* infection. J Immunol 159:318–327, 1997.

103. HL Collins, GJ Bancroft. Cytokine enhancement of complement-dependent phagocytosis by macrophages: synergy of tumor necrosis factor-alpha and granulocyte-macrophage colony-stimulating factor for phagocytosis of *Cryptococcus neoformans*. Eur J Immunol 22:1447–1454, 1992.

104. D Delfino, L Cianci, M Migliardo, G Mancuso, V Cusumano, C Corradini, G Teti. Tumor necrosis factor–inducing activities of *Cryptococcus neoformans* components. Infect Immun 64:5199–5204, 1996.

105. SM Levitz, EA North, Y Jiang, S Nong, H Kornfeld, TS Harrison. Variables affecting production of monocyte chemotactic factor 1 from human leukocytes stimulated with *Cryptococcus neoformans*. Infect Immun 65:903–908, 1997.

106. AD Luster. Chemokines—chemotactic cytokines that mediate inflammation. N Engl J Med 338: 436–445, 1998.

107. TR Traynor, WA Kuziel, GB Toews, GB Huffnagle. CCR2 expression determines T1 versus T2 polarization during pulmonary *Cryptococcus neoformans* infection. J Immunol 164:2021–2027, 2000.

108. GB Huffnagle, LK McNeil, RA McDonald, JW Murphy, GB Toews, N Maeda, WA Kuziel. Role of C-C chemokine receptor 5 in organ-specific and innate immunity to *Cryptococcus neoformans*. J Immunol 163:4642–4646, 1999.

109. D Delfino, L Cianci, E Lupis, A Celeste, ML Petrelli, F Curro, V Cusumano, G Teti. Interleukin-6 production by human monocytes stimulated with *Cryptococcus neoformans* components. Infect Immun 65:2454–2456, 1997.

110. E Blasi, R Barluzzi, R Mazzolla, L Pitzurra, M Puliti, S Salappico, F Bistoni. Biomolecular events involved in anticryptococcal resistance in the brain. Infect Immun 63:1218–1222, 1995.

111. CH Mody, JC Spurrell, CJ Wood. Interleukin-15 induces antimicrobial activity after release by *Cryptococcus neoformans*–stimulated monocytes. J Infect Dis 178:803–814, 1998.

112. SM Levitz, A Tabuni, SH Nong, DT Golenbock. Effects of interleukin-10 on human peripheral blood mononuclear cell responses to *Cryptococcus neoformans*, *Candida albicans*, and lipopolysaccharide. Infect Immun 64:945–951, 1996.

113. A Vecchiarelli, C Retini, C Monari, C Tascini, F Bistoni, TR Kozel. Purified capsular polysaccharide of *Cryptococcus neoformans* induces interleukin-10 secretion by human monocytes. Infect Immun 64:2846–2849, 1996.

114. C Monari, C Retini, B Palazzetti, F Bistoni, A Vecchiarelli. Regulatory role of exogenous IL-10 in the development of immune response versus *Cryptococcus neoformans*. Clin Exp Immunol 109: 242–247, 1997.

115. A Vecchiarelli, C Retini, D Pietrella, C Monari, C Tascini, T Beccari, TR Kozel. Downregulation by cryptococcal polysaccharide of tumor necrosis factor alpha and interleukin-1 beta secretion from human monocytes. Infect Immun 63:2919–2923, 1995.

116. C Retini, A Vecchiarelli, C Monari, F Bistoni, TR Kozel. Encapsulation of *Cryptococcus neoformans* with glucuronoxylomannan inhibits the antigen-presenting capacity of monocytes. Infect Immun 66:664–669, 1998.

117. A Vecchiarelli, C Retini, C Monari, A Casadevall. Specific antibody to *Cryptococcus neoformans* alters human leukocyte cytokine synthesis and promotes T-cell proliferation. Infect Immun 66: 1244–1247, 1998.

118. GB Huffnagle, GH Chen, JL Curtis, RA McDonald, RM Strieter, GB Toews. Down-regulation of the afferent phase of T cell–mediated pulmonary inflammation and immunity by a high melanin-producing strain of *Cryptococcus neoformans*. J Immunol 155:3507–3516, 1995.

119. JA Lovchik, JA Wilder, GB Huffnagle, R Riblet, CR Lyons, MF Lipscomb. Ig heavy chain complex-linked genes influence the immune response in a murine cryptococcal infection. J Immunol 163: 3907–3913, 1999.

120. E Brummer, DA Stevens. Anticryptococcal activity of macrophages: role of mouse strain, C5, contact, phagocytosis, and L-arginine. Cell Immunol 157:1–10, 1994.

121. MR Hidore, N Nabavi, F Sonleitner, JW Murphy. Murine natural killer cells are fungicidal to *Cryptococcus neoformans*. Infect Immun 59:1747–1754, 1991.

122. SM Levitz, MP Dupont, EH Smail. Direct activity of human T lymphocytes and natural killer cells against *Cryptococcus neoformans*. Infect Immun 62:194–202, 1994.

123. JM McDonnell, GM Hutchins. Pulmonary cryptococcosis. Hum Pathol 16:121–128, 1985.

124. SC Lee, DW Dickson, A Casadevall. Pathology of cryptococcal meningoencephalitis: analysis of 27 patients with pathogenetic implications. Hum Pathol 27:839–847, 1996.

125. RG Washburn, CU Tuazon, JE Bennett. Phagocytic and fungicidal activity of monocytes from patients with acquired immunodeficiency syndrome. J Infect Dis 151:565, 1985.

126. TS Harrison, H Kornfeld, SM Levitz. The effect of infection with human immunodeficiency virus on the anticryptococcal activity of lymphocytes and monocytes. J Infect Dis 172:665–671, 1995.

127. TS Harrison, SM Levitz. Mechanisms of impaired anticryptococcal activity of monocytes from ffrom donors infected with human immunodeficiency virus. J Infect Dis 176:537–540, 1997.

128. ML Cameron, DL Granger, TJ Matthews, JB Weinberg. Human immunodeficiency virus (HIV)-infected human blood monocytes and peritoneal macrophages have reduced anticryptococcal activity whereas HIV-infected alveolar macrophages retain normal activity. J Infect Dis 170:60–70, 1994.

129. CC Reardon, SJ Kim, RP Wagner, H Koziel, H Kornfeld. Phagocytosis and growth inhibition of *Cryptococcus neoformans* by human alveolar macrophages: effects of HIV-1 infection. AIDS 10: 613–618, 1996.

130. SJ Brodie, VG Sasseville, KA Reimann, MA Simon, PK Sehgal, DJ Ringler. Macrophage function in simian AIDS. Killing defects in vivo are independent of macrophage infection, associated with alterations in Th phenotype, and reversible with IFN-gamma. J Immunol 153:5790–5801, 1994.

131. J Chehimi, SE Starr, I Frank, A D'Andrea, X Ma, RR MacGregor, J Sennelier, G Trinchieri. Impaired interleukin-12 production in human immunodeficiency virus–infected patients. J Exp Med 179: 1361–1365, 1994.

132. M Clerici, DR Lucey, JA Berzofsky, LA Pinto, TA Wynn, SP Blatt, MJ Dolan, CW Hendrix, SF Wolf, GM Shearer. Restoration of HIV-specific cell-mediated immune responses by interleukin-12 in vitro. Science 262:1721–1726, 1993.

133. JF Hoy, DE Lewis, GG Miller. Functional versus phenotypic analysis of T cells in subjects seropositive for the human immunodeficiency virus: a prospective study of in vitro responses to *Cryptococcus neoformans*. J Infect Dis 158:1071–1078, 1988.

134. D Pietrella, C Monari, C Retini, B Palazzetti, TR Kozel, A Vecchiarelli. HIV type 1 envelope glycoprotein gp 120 induces development of a T helper type 2 response to *Cryptococcus neoformans*. AIDS 13:2197–2207, 1999.

135. CM Mastroianni, M Lichtner, F Mengoni, P Santopadre, V Vullo, S Delia. Marked activation of the tumour necrosis factor system in AIDS-associated cryptococcosis. AIDS 10:1436–1438, 1996.

136. W Chaka, R Heyderman, I Gangaidzo, V Robertson, P Mason, J Verhoef, A Verhuel, AIM Hoepelman. Cytokine profiles in cerebrospinal fluid of human immunodeficiency virus–infected patients with cryptococcal meningitis: no leukocytosis despite high interleukin-8 levels. University of Zimbabwe Meningitis Group. J Infect Dis 176:1633–1636, 1997.

137. O Lortholary, L Improvisi, N Rayhane, F Gray, C Fitting, JM Cavaillon, F Dromer. Cytokine profiles of AIDS patients are similar to those of mice with disseminated *Cryptococcus neoformans* infection. Infect Immun 67:6314–6320, 1999.

138. JM Orendi, HS Nottet, MR Visser, AF Verheul, H Snippe, J Verhoef. Enhancement of HIV-1 replication in peripheral blood mononuclear cells by *Cryptococcus neoformans* is monocyte-dependent but tumour necrosis factor–independent. AIDS 8:423–429, 1994.

139. TS Harrison, S Nong, SM Levitz. Induction of human immunodeficiency virus type 1 expression in monocytic cells by *Cryptococcus neoformans* and *Candida albicans*. J Infect Dis 176:485–491, 1997.

140. A Casadevall, W Cleare, M Feldmesser, A Glatman-Freedman, DL Goldman, TR Kozel, N Lendvai, J Mukerjee, LA Pirofski, J Rivera, AL Rosas, MD Scharff, P Valadon, K Westin, Z Zhong. Characterization of a murine monoclonal antibody to *Cryptococcus neoformans* polysaccharide that is a candidate for human therapeutic studies. Antimicrob Agents Chemother 42:1437–1446, 1998.

141. DA Price, JL Klein, M Fisher, J Main, JS Bingham, RJ Coker. Potential role for granulyte-macrophage colony-stimulating factor in the treatment of HIV-associated cryptococcal meningitis. AIDS 11:693–694, 1997.

142. R Manfredi, OV Coronado, A Mastroianni, F Chiodo. Liposomal amphotericin B and recombinant human granulocyte-macrophage colony-stimulating factor (rHuGM-CSF) in the treatment of paediatric AIDS-related cryptococcosis. Int J STD AIDS 8:406–408, 1997.

143. IR Torres, UC Villareal, RM Robles, P Aparicio, DC Cano. Comparative study between two treatment schedules in AIDS patients with meningitis caused by *Cryptococcus neoformans*: GM-CSF plus amphotericin B versus amphotericin B alone. In: Leucomax, Current Use and Future Applications. Lucerne: Adelphi Communications, 1993, p 64.

144. A Louie, AL Baltch, MA Franke, RP Smith, MA Gordon. Comparative capacity of four antifungal agents to stimulate murine macrophages to produce tumour necrosis factor alpha: an effect that is attenuated by pentoxifylline, liposomal vesicles, and dexamethasone. J Antimicrob Chemother 34:975–987, 1994.

145. JE Wolf, SE Massof. In vivo activation of macrophage oxidative burst activity by cytokines and amphotericin B. Infect Immun 58:1296–1300, 1990.

146. M Tohyama, K Kawakami, A Saito. Anticryptococcal effect of amphotericin B is mediated through macrophage production of nitric oxide. Antimicrob Agents Chemother 40:1919–1923, 1996.

147. N Mozaffarian, JW Berman, A Casadevall. Enhancement of nitric oxide synthesis by macrophages represents an additional mechanism of action for amphotericin B. Antimicrob Agents Chemother 41:1825–1829, 1997.

148. E Wilson, L Thorson, DP Speert. Enhancement of macrophage superoxide anion production by amphotericin B. Antimicrob Agents Chemother 35:796–800, 1991.

149. E Martin, A Stüben, A Görz, U Weller, S Bhakdi. Novel aspect of amphotericin B action: accumulation in human monocytes potentiates killing of phagocytosed *Candida albicans*. Antimicrob Agents Chemother 38:13–22, 1994.
150. JD Nosanchuk, W Cleare, SP Franzot, A Casadevall. Amphotericin B and fluconazole affect cellular charge, macrophage phagocytosis, and cellular morphology of *Cryptococcus neoformans* at subinhibitory concentrations. Antimicrob Agents Chemother 43:233–239, 1999.
151. TS Harrison, GE Griffin, SM Levitz. Conditional lethality of the diprotic weak bases chloroquine and quinacrine against *Cryptococcus neoformans*. J Infect Dis 182:283–289, 2000.

24

Antifungals Currently Used in the Treatment of Invasive Fungal Diseases

Xiao-jiong Zhao and Richard A. Calderone
Georgetown University Medical Center, Washington, D.C.

I. INTRODUCTION

The systemic mycoses and especially those fungi that cause opportunistic infections, such as *Candida albicans* and *Aspergillus fumigatus*, pose important problems for the clincan who must choose from a short-list of antifungals to achieve cure. Why are there only a few choices of drugs to use and what other problems surface in treating these infections? First, as both the fungus and host are eukaryotic, the number of compounds specifically toxic for the fungus is small. Second, amphotericin-B, the "gold standard" of antifungals for the treatment of the systemic mycoses, invariably causes some degree of toxicity because it also binds to similar targets of host cells. Third, successful cure is, in part, often compromised by the low sensitivity of existing detection methods, especially in the case of invasive aspergillosis where laboratory diagnosis by blood culture most often fails [1]. Fourth, resistance to fluconazole, the azole which is commonly used to treat several mycoses, is encountered with increasing frequency. For example, pathogens such as *A. fumigatus* are resistant to fluconazole, so that again, there are few choices other than amphotericin B in treating invasive infections caused by this organism [1]. In the case of *C. albicans*, resistance to fluconazole has been reported in greater frequency, and the emergence of non-*albicans Candida* spp. has become a major problem in a number of clinical settings because of the inherent resistance of these species to fluconazole. Thus, the answers to the questions raised above are fairly straightforward, but solutions are not at hand. Nevertheless, in spite of these problems in treatment, drugs which target ergosterol (amphotericin-B/azoles) of the human pathogenic fungi remain the logical choices for treatment.

The intent of this chapter is to review the literature in regard to the antifungal drugs which are used to treat systemic fungal infections; many of these compounds are also active against the superficial mycoses and are important in the treatment of dermatophytosis, for example. For almost all fungal infections, the choice of therapeutics is restricted to either inhibitors of ergosterol synthesis or compounds which act by binding to ergosterol causing perturbations and loss of functional activity of the plasma membranes of pathogenic fungi. The former group includes the imidazoles and triazoles; the latter group, the polyene antifungals. Following a general description of each of these drugs, we will present some of the current concepts on dealing with the problems mentioned above. In Chapter 25, new antifungals and the components of fungi that are targeted by these new drugs will be discussed in detail. Excellent reviews on polyenes, azoles, and delivery systems have also been published recently [2–5].

II. GENERAL CONSIDERATIONS

A. Classes of Antifungal Drugs

Systemic antifungals currently in use belong to one of four different classes of compounds (Fig. 1). As stated above, polyenes (inhibitors of plasma membrane function through binding to

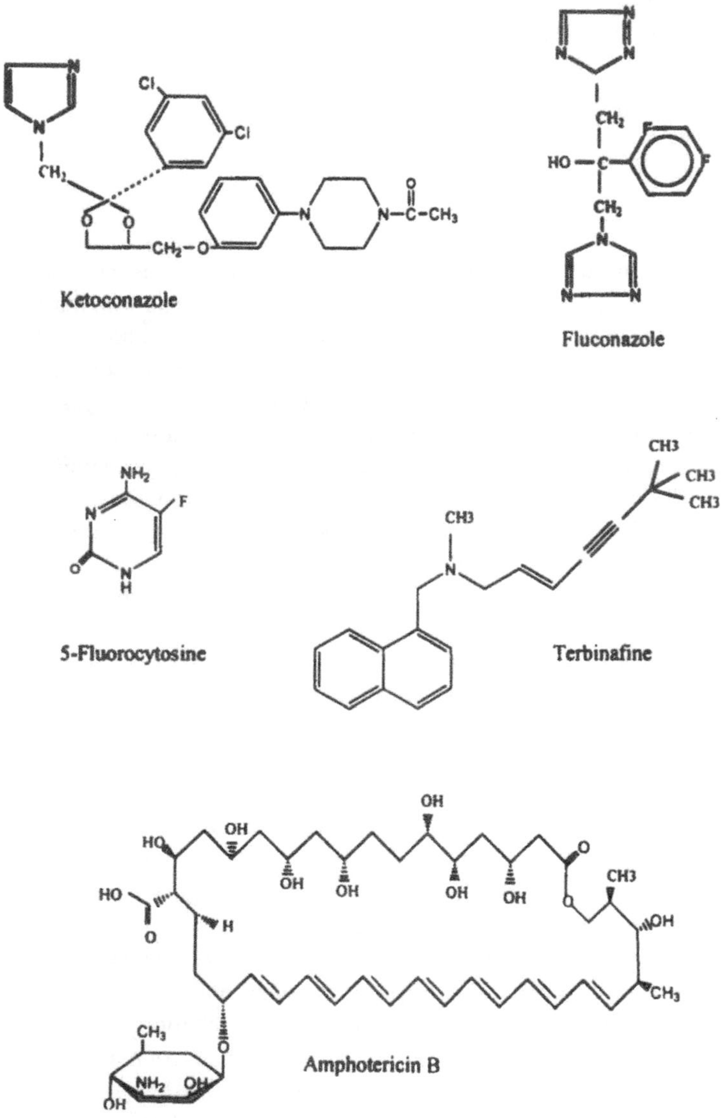

Figure 1 Structures of the four principal groups of antifungals: amphotericin-B; selected azoles (ketoconazole, an imidazole and fluconazole, a triazole); 5-fluorocytosine; and terbinafine.

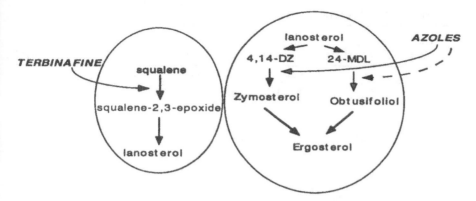

Figure 2 Pathway for ergosterol synthesis and targets for the azoles and terbinafine antifungals are indicated. The dashed arrow indicates a second target of the azoles as suggested by others [5]. 4,14-DZ = 4,14-dimethylzymosterol; 24-MDL = 24-methylenedihydrolanosterol. The pathway does not include the obtusifoliol conversion to ergosterol via a 15-methylfecosterol intermediate.

ergosterol) and azoles (inhibit the conversion of lanosterol to its demethylated form during ergosterol synthesis) are the two most commonly used types of antifungals (Fig. 2). The third group of inhibitors, although used much less frequently in a clinical setting than the first two, is the fluoropyrimidines (Fig. 1). The only example of a fluoropyrimidine in clinical usage is 5-fluorocytosine (5-FC), and although fungicidal, this compound has a rather limited spectrum of activity so it is used almost exclusively in combined therapy regimens with amphotericin-B. A fourth group of compounds, the allylamines/thiocarbamates, are also inhibitors of ergosterol synthesis. However, their mode of action in inhibiting ergosterol synthesis is different than the azoles in that the allylamines/thiocarbamates inhibit the enzyme squalene epoxidase, which together with the enzyme (2,3)-oxidosqualene cyclase, catalyzes a cyclization of squalene to lanosterol (Fig. 2). In general, however, the allylamines, such as terbinafine, have almost exclusively been used to treat superficial fungal infections, especially dermatophytosis. Their use in the treatment of systemic disease is still experimental. Specific modes of action of each of these four groups of inhibitors (azoles, polyenes, fluoropyrimidine, and allylamines) will be discussed in detail below. Some of their general features are described and summarized in Table 1. The compounds listed in Table 1 under azoles represent examples of either imidazoles or triazoles but is not intended to include all known azoles. A detailed discussion of azoles has been published by Sheehan et al. [4] and of antifungals by Ghannoum and Rice [5] and Georgopapadakou [2].

B. General Features of Antifungals

Table 1 depicts several common characteristics of antifungal agents. First, most of the compounds listed are fungistatic rather than fungicidal. The exceptions, amphotericin-B and 5-FC, are fungicidal. Second, resistance to all the major groups of compounds has been described, but the extent of resistance to any particular drug is both organism and drug dependent. In the case of terbinafine, a drug-resistant strain of *Candida glabrata* has been described, and clinical failures have been reported [5]. Thus, resistance among fungi to drugs that are more commonly used (fluconazole) is greater than resistance to drugs less commonly used (ketoconazole). On the other hand, clinical resistance to amphotericin-B is uncommon even though this drug is widely used for treating

Table 1 General Characteristics of the Major Antifungals

Classes of compounds examples	Trade name	Clinical use[a]	Cidal or static	Target	Other activities[b]	Resistance[b]
Azoles			Static Broad-spectrum	$P450_{DM}$	Yes	Common
Imidazoles						
Ketoconazole	Nizoral	system/super				
Miconazole	Monistat	super				
Clotrimazole	Lotrimin	super				
Triazoles						
Itraconazole	Sporonox	system/super				
Fluconazole	Diflucon	system/super				
Terconazole	Terazol	vulvovag				
Ticonazole	Vagistat	vulvovag				
Allylamines		super	Static	Squalene epoxidase	No	Uncommon
Terbinafine						
Polyenes	Fungizone	System	Broad-spectrum Cidal	Ergosterol	Yes	Uncommon
Amphotericin B[c]						
Fluoropyrimidines		System	Cidal	Protein synthesis	No	Common
5-fluorocytosine[d]						

[a] System = systemic use; super = superficial; vulvovag = vulvovaginal.
[b] Except for fluconazole and the polyenes, few data are available.
[c] Primarily binds to ergosterol, causing membrane perturbations; also causes oxidative damage to susceptible cells.
[d] Limited usefulness, primarily in the treatment of cryptococcal meningitis and some types of candidiasis.

systemic fungal infections. One possible explanation for the scarcity of drug resistance to amphotericin-B in comparison to an azole such as fluconazole is that the former drug is fungicidal while fluconazole is fungistatic and more likely to result in selection of resistant strains during treatment. Some species of *Candida*—e.g., *Candida lusitaniae*—are inherently resistant to amphotericin-B. Interestingly, of those mechanisms of resistance which have been described for pathogenic fungi, modification of a drug resulting in its inactivation has not been described in fungi. This is in stark contrast to the case in bacteria where such mechanisms are common. A more thorough discussion of resistance mechanisms by fungi, especially *Candida* spp., is presented in Chapters 26 and 27. The reader is also directed to reviews on resistance [2,5–7].

Third, the compounds listed in Table 1 either have a very narrow activity (narrow spectrum) or have a rather broad spectrum of activity. For example, amphotericin B and itraconzaole exhibit a broad specificity while 5-FC is rather limited in its range of activity against pathogenic fungi. Fourth, the inhibitory activity of a drug might reflect several activities in addition to affecting a target common to a specific group of compounds. For example, the imidazoles (miconazole and ketoconazole) affect not only the activity of the P450-dependent 14α-demethylase ($P450_{DM}$), but additionally, other membrane-bound enzymes and membrane lipid synthesis are inhibited [5]. The azoles also inhibit two distinct reactions in ergosterol synthesis (Fig. 2), although the $P450_{DM}$ target is the most studied. Amphotericin-B not only disrupts membrane activity in susceptible fungi but also causes oxidative damage to cells which may at least partly contribute to its fungicidal activity [6]. Fifth, because of the rather narrow biochemical differences between fungal and mammalian cells, toxicity accompanying antifungal therapy is likely, but the extent of toxicity (the therapeutic index) varies from drug to drug. Thus, amphotericin-B has a narrow therapeutic index, since binding of the compound to mammalian cell cholesterol results in toxicity to the patient. Likewise, mammalian cholesterol biosynthesis is at least partially blocked by azoles, and this activity is also associated with binding of the azoles to $P450_{DM}$. However, the dose required to inhibit mammalian cell activity is much higher than that required for the inhibition of fungal enzyme activity [8,9]. For example, voriconazole at 7.4 µM inhibits 50% of the activity of rat liver $P450_{DM}$, but its activity against fungi is about 250-fold higher (0.03 µM). It can be expected, therefore, that the higher degree of toxicity of azoles such as ketoconazole compared to voriconazole is related in part to the greater affinity of the former drug for the mammalian cell enzyme than the latter one.

Sixth, as to be expected, many factors must be taken into account when decisions are made in regard to the use of an antifungal in treating, severe *Candida* infections. This point is clearly illustrated in a recent publication which summarizes the general recommendations of a panel of 22 international experts in the management of candidemias and other *Candida* infections [4,10] (see Fig. 3). Among the recommendations for the management of severe candidal infections, the panel considered fluconazole, amphotericin-B (alone or in combination) and intraconazole among the agents to be used in treating candidemia, candiduria, peritonitis, chronic disseminated disease (formerly hepatosplenic), and *Candida* endophthalmitis. Management strategies emphasized not only the site of candidiasis but also the presence/absence of neutropenia, drug susceptibilities of the *Candida* isolates, the general condition of the patient, and whether the patient had undergone a solid organ or bone marrow transplantation. The consensus opinions among these experts included the following: (1) Any patient with candidemia should be treated and IV catheters should be removed or changed. (2) The choice of a primary therapy in treating

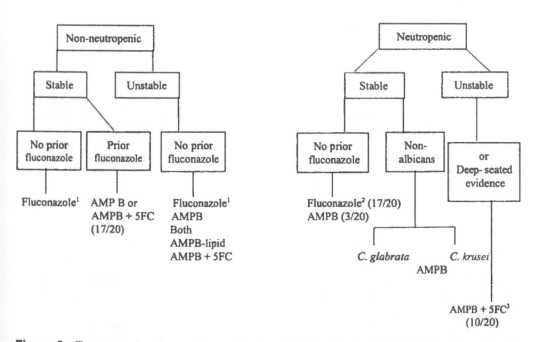

Figure 3 Treatment of patients with candidemias. Selection of an antifungal is dependent upon the stability of the patient, presence/absence of neutropenia, prior exposure to fluconazole, and the offending species of *Candida*. The fractions (i.e., 10/20, etc.) reflect participant's opinions. (From Refs. 4, 10.)

candidemia depended upon the stability of the patient, whether neutropenic or nonneutropenic, the presence or lack of *C. krusei* (or other known resistant species) from a patient culture, and whether or not the patient had prior use of fluconazole. Thus, fluconazole was recommended by a majority of the clinicans in the stable neutropenic, or nonneutropenic, patient unlikely to have *C. krusei*, who also had not received prior treatment with fluconazole. Fluconazole was chosen over amphotericin-B due to its reduced toxicity. (3) If the candidemia is due to a fluconazole-resistant organism such as *C. krusei* or *C. glabrata*, or occurs in patients receiving fluconazole, then amphotericin-B is the preferred choice among drugs. (4) For the unstable neutropenic or nonneutropenic patient, several drugs are suggested including amphotericin-B with or without 5-FC, although some preference was expressed for fluconazole if in this same patient setting, fluconazole had not been previously used or *C. krusei* was not isolated. (5) For patients with a solid-organ transplant, the choice of amphotericin-B or fluconazole depended upon some of the same parameters discussed above, i.e., stability of the patient or prior use of fluconazole. (6) In the clinical setting of *Candida* peritonitis, chronic disseminated candidiasis (formerly hepatosplenic candidiasis, in the nonneutropenic), candiduria, or endophthalmitis, fluconazole was recommended as a first-line drug by a majority of the experts. This was especially true for the treatment of candiduria where fluconazole was overwhelmingly chosen as the drug of choice, provided that a non-*albicans Candida* sp. is not the pathogen. While the protocol shown in Figure 3 will likely be modified as new drugs become available, the insights from this study certainly speak to the complexity of treatment. In addition, the data also point out the importance of the clinical mycology laboratory in the determination of both the species causing the infection and the drug susceptibility of that species to various antifungals. These data are then used in the selection of the most efficacious drug for treating this important group of infections, which is reported to be third in frequency among all nosocomial diseases [11].

III. SUSCEPTIBILITY TESTING

The determination of a minimal inhibitory concentration (MIC) of a drug for a clinical isolate would be a useful laboratory test if it were predictive of a therapeutic response. In fact, this parameter is useful when standardized drug sensitivity testing is limited to several of the azoles, amphotericin-B, and 5-fluorocytosine with isolates of *Candida albicans*. It is also recommended that patient isolates should be limited to mucosal or bloodborne infections [12]. Additional data must be gathered before correlating MIC with clinical outcome in regard to non-*albicans* isolates as well as isolates of *Cryptococcus neoformans*. As pointed out by Rex et al. [12], antimicrobial susceptibility testing is a measurement that needs to be interpreted in light of other factors, most of which often are more important than susceptibility testing. These factors can be of host origin (underlying disease, phagocytic function, site of the infection as, for example, an abscess, necrosis, or presence of a foreign body), drug pharmacokinetics (inadequate drug dosing, inadequate penetration, chemical instability of the drug, drug interactions, inactivity of the drug due to protein binding), or the pathogen itself (virulence factors, protected isolation of the pathogen). While susceptibility testing methods (see below) have been developed for *Candida albicans* and *Cryptococcus neoformans*, standardization has not been achieved with filamentous fungi and other yeast pathogens. Therefore, the discussion which follows will focus upon methods that have been developed for these two organisms.

The National Committee for Clinical Laboratory Standards (NCCLS)-approved version (M27-A) for susceptibility testing of clinical isolates employs either a broth macrodilution (1 mL final volume) or microdilution (0.3 mL final volume) method and must include two quality control isolates and QC ranges established from a previous procedure (M23) for amphotericin-

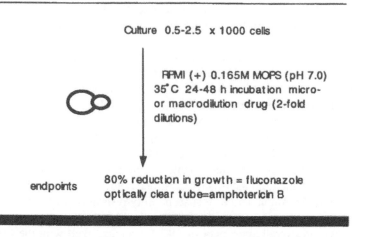

Candida Susceptibility Testing for
Fluconazole, itraconazole and amphotericin B

Culture 0.5-2.5 x 1000 cells

RPMI (+) 0.165M MOPS (pH 7.0)
35°C 24-48 h incubation micro-
or macrodilution drug (2-fold
dilutions)

endpoints   80% reduction in growth = fluconazole
optically clear tube=amphotericin B

for improved correlation of MIC and clinical outcome, a 24-hr
and 50% endpoint inhibition is used with fluconazole

Figure 4 The protocol for in vitro susceptibility testing of *Candida* spp. isolates as recommended by the NCCLS2M committee (version M27-A).

B, flucytosine, ketoconazole, itraconazole, and fluconazole. Figure 4 illustrates the parameters for a typical susceptibility test, including inoculum size, medium, temperature of incubation, incubation time and end point determinations. Modifications most likely will occur which will optimize the test even more. For example, with fluconazole it would appear that for isolates of *Candida* species, a better correlation between the MIC and an in vivo response is obtained when a 50% reduction in growth after a 24-hr incubation is used rather than the 80% reduction in growth after a 48-hr incubation, as indicated in Figure 4 [13].

Once a standardized assay was established, tentative interpretive breakpoints (μg/mL for *Candida* isolates against fluconazole and itraconazole) for clinical isolates were developed based upon an analysis of data packages, which included MIC determinations as well as outcome data (primarily for oropharyngeal candidiasis [OPC] in AIDS patients), pharmacology, correlation between MICs and results of animal studies, and clinical data correlating MIC with outcome. Over 500 *Candida* isolates (of which 77% were isolates of *C. albicans*) were tested by the NCCLSM27-A method. Interpretive breakpoints included susceptible (S), susceptible-dose dependent (S-DD), intermediate (I), and resistant (R) reactions of isolates for both fluconazole and itraconazole [12]. Conclusions from the study with fluconazole included the following: (1) Response to fluconazole does vary with MIC [a similar observation was made with studies of itraconazole]; (2) higher doses of fluconazole can be used to treat patients infected with isolates for which MICs are higher; and (3) failure of fluconazole therapy is likely when the MIC determined by the NCCLS M27-A method substantially exceeds the predicted serum level of fluconazole for a given dosage regimen. As stated above, interpretative breakpoints are not yet established for *C. neoformans*.

In addition to the micro/macrodilution standardized assay described above, MIC determinations can be assessed using antimicrobial gradient strips (Etest) which are commercially available [14–16]. In this regard, the NCCLS M27-A assay was compared to the Etest for the identification of amphotericin-B-resistant strains of *Cryptococcus neoformans* [14]. In this study, both assays were performed using three different media, including RPMI, yeast nitrogen base-glucose, and antibiotic 3-glucose media. With the NCCLS M-27 assay, reliable discrimination between susceptible and resistant isolates occurred using the antibiotic 3 medium only. On the other hand, the identification of amphotericin-B-resistant strains was possible with the Etest using both the antibiotic 3 medium and RPMI [14]. Of importance, the data from this study does indicate that the Etest strip assay may be a simple and reliable method to determine the MICs of *C. neoformans*. Similarly, in a multicenter evaluation of 18 isolates of *Candida* spp. and two of *Cryptococcus neoformans*, the Etest method was found to be suitable for routine use with *Candida* spp. and amphotericin-B and flucytosine [15]. On the other hand, the Etest assay misclassified both itraconazole and fluconazole-susceptible isolates of *Candida* spp. as resistant in 5–62% (itraconazole) and 1.5–15% (fluconazole) of the tests. Although only two isolates of *C. neoformans* were tested, the Etest incorrectly identified both as resistant to flucytosine, fluconazole, and itraconazole [15]. Amphotericin-B susceptibility/resistance has been further evaluated in *Candida* spp. and *C. neoformans* using the Etest method [16]. When compared to the NCCLS microdilution assay, an overall agreement of 98.3% was discerned with the Etest [16].

The data described above are encouraging in regard to the development of MIC assays which can be used to predict success/failure in treating patients with oropharyngeal and other forms of candidiasis. However, the determination of breakpoint numbers for other pathogens, especially the aspergilli, needs to be established. For example, recent data indicate that in vitro susceptibility testing of *Aspergillus* spp. was predictive of therapeutic success or failure [17]. Thus, isolates from six patients with an MIC of <2 mg/L survived whereas 22/23 patients with isolates of MIC ≥2 mg/L died of invasive aspergillosis.

Likewise, a good correlation of in vitro resistance/susceptibility of *A. fumigatus* isolates was observed with the outcome of invasive disease in a mouse model of aspergillosis [18]. Thus, itraconazole prolonged the survival of mice infected with a strain that had a low MIC, while the drug did not protect animals against a strain with a high MIC (16 mg/L). On the other hand, other investigators were unable to correlate in vivo (clinical) outcomes with susceptibility testing of three isolates of *A. fumigatus*, although a correlation with treatment failure with a high MIC (2 mg/L) was observed for *Aspergillus terreus* [19].

IV. MODE OF ACTION OF ANTIFUNGALS

A. Polyenes

The two polyenes which have been most utilized in the treatment of the mycoses are nystatin and amphotericin B (Fig. 1). Of the two, historically, nystatin has been limited to the treatment of mucosal forms of candidiasis, but more recently has been replaced for the most part by azoles. Amphotericin-B remains a broad spectrum fungicidal compound which is used in the treatment of several of the systemic mycoses. Therefore, most of the discussion below will focus upon amphotericin-B, whose structure is indicated in Figure 1.

A number of observations support the hypothesis that polyenes bind to ergosterol in the plasma membrane of susceptible fungi and cause perturbations in membrane function [20–22]. These proofs include the following observations: First, all susceptible organisms contain sterols in the plasma membranes, and resistant organisms lack sterols in their membranes. This observation applies to fungi as well as algae and protozoa and also at least partially explains the toxicity of amphotericin-B for human cells, which also binds to the principal sterol of mammalian cells, which is cholesterol. Second, susceptible species can be protected against the effect of amphotericin-B by providing sterol in the growth medium. Here, it is presumed that the exogenous sterol binds amphotericin-B, or at least enough of it to prevent growth inhibition or cell death. Third, and perhaps most important, is the observation that binding of polyenes to sterols has been detected by spectrophotometric analysis. A model for the interaction of polyenes and membrane sterols, which was proposed several years ago but which is still widely accepted, is that polyene molecules form an aqueous pore in the membranes of susceptible fungi through hydrophobic interactions of the polyene (the hydroxy residues are on the inner side of the pore which is formed within the membrane) with membrane sterols [5]. The consequence of pore formation is a change in permeability of cells, resulting in leakage of cations initially, and subsequent cell death.

While amphotericin-B has widespread use in treating many of the systemic fungal infections, its toxicity has resulted in the search for other drugs which are as effective but less toxic. However, in spite of its toxicity, one of the virtues of amphotericin-B is the relatively low numbers of patients from which resistant isolates of fungi have been isolated. On the other hand, there are several studies on amphotericin-B-resistant fungi, especially among the aspergilli and *Candida* spp. [23–29]. Importantly, in cases of invasive aspergillosis where treatment failures with amphotericin-B are documented, one could speculate that, in part, failures may be a consequence of infections caused by resistant organisms. However, this observation may be difficult to prove since, as stated above, in some studies the correlation of MIC of several aspergilli with clinical failure has yielded disparate results [19]. For example, Johnson et al. [19] tested three strains of *A. fumigatus*, two of which (AF210 and AF294) were considered susceptible in vivo to amphotericin-B while a third strain (AF65) was resistant in vivo to the drug. An isolate of *A. terreus* also was tested and shown to be resistant in vivo (animal model) to amphotericin-B. Of the four strains tested for in vitro sensitivities, only the *A. terreus* isolate had a consistently elevated MIC (2 mg/L), while the MIC determinations for the *A. fumigatus* isolates were variable and not correlated with in vivo results. Changes in cell wall composition have been associated with the development of resistance to amphotericin B in *Aspergillus flavus* [28]. Serial transfer of cells of the organism on agar plates containing increasing concentrations of the drug resulted in the development of resistant strains. When analyzed biochemically, such strains had higher levels of 1,3-alpha glucan than the susceptible strains.

Studies on the mechanisms which result in resistance to amphotericin-B have been carried out with *C. albicans* or *C. lusitaniae* [23,26,29–32]. In this chapter, emphasis will be placed upon recent studies of resistance which develop as a consequence of (1) exposure of the organism to other antifungals; (2) resistance as a function of the formation of biofilms; and (3) changes in the cell wall of an organism which renders the cell resistant. Other specific mechanisms of resistance (changes in uptake of the drug, mutations in target enzymes, over expression of transport pumps) are the subject of Chapters 26 and 27.

1.   Resistance to amphotericin-B was associated with preexposure of cells of *Candida albicans* to subinhibitory levels of azoles [26]. In this study, strains susceptible to amphotericin-B as measured by in vitro MIC determinations became resistant to the drug when incubated overnight with fluconazole or itraconazole. Depending upon growth conditions, resistance persisted for several days following removal of the azole drug. This conclusion provides a cautionary suggestion that resistance to amphotericin B may develop clinically in therapies which include combinations of drugs, i.e., azole and amphotericin-B. In fact, others have suggested that this combination of drugs may be antagonistic [33]. The second important point of this study is that patients who fail to respond to amphotericin-B may be infected with isolates which are thought

to be sensitive to amphotericin-B by in vitro standards, but transient exposure to other antifungals could induce a resistant clinical phenotype.

2.   Resistance to amphotericin-B has also been associated with the formation of biofilms of *C. albicans*, as for example, on indwelling catheters [30–32]. Biofilms of *C. albicans* were made by incubating yeast cells of the organism with small disks of polyvinyl chloride central venous catheters [30]. Disks with adhering organisms were placed in microtiter tissue culture plates and cultured for 48 hr in a growth medium. Subsequently, the medium was replaced with one (or a combination) of antifungals in media, including, amphotericin-B, flucytosine, or the azoles fluconazole, itraconazole, or ketoconazole. After 5 hr of incubation, the effect of the antifungal on the biofilm organisms was measured by (^{3}H) leucine incorporation or tetrazolium reduction. The effect of each antifungal was measured as the percentage inhibition of leucine incorporation or MTT-formazan formation by the biofilms. Inhibition of cells of the biofilm by each drug was compared to planktonic cells (cells in suspension) which was used as a control. These investigators found that all of the antifungal agents tested were much less active against 48-hr biofilms than against planktonic cells [30]. For example, IL_{50} determinations were five to eight times higher for biofilms than planktonic cells and 30–2000 times higher than the relevant MICs for each drug. The same general trend was observed with other species of *Candida* tested, including *C. tropicalis*, *C. parapsilosis*, and *C. kefyr*. These observations imply that resistance can develop clinically under conditions where biofilm formation occurs. As IV catherization is a common predisposing factor for invasive candidiasis, it would seem imperative to consider clinical failures as a consequence of a nitus of organism growing in this manner. The mechanism(s) which might explain the development of resistance in this situation are uncertain. Insufficient penetration of the drug into the biofilm is certainly one possibility. It has been observed that biofilms generated in vitro are embedded in a matrix which is formed along with the aggregated (adhering) yeast cells of *C. albicans*. The production of this matrix is increased substantially when the biofilm is subjected to the flow of a growth medium, as in a chemostat [31]. On the other hand, statically incubated biofilms produced a minimal amount of matrix. Thus, a flow of nutrients is required to enhance biofilm formation, perhaps like the conditions which exist in vivo where it is likely that biofilms are exposed to a constant flow of blood. The identity of this matrix is unknown, but conceptually its role in protecting cells might be to serve as a barrier which prevents the penetration of drugs such as amphotericin-B into the biofilm. In another study, the effect of various antifungals on biofilms of *C. albicans* was investigated using cellulose acetate filters coated with adhering yeast cells [32]. The washed filters coated with adhering organisms were transferred to a perfused biofilm fermentor and subjected to various flow rates of growth media with or without drugs. During incubation of the biofilms, naturally eluted cells were collected and plated to determine the effect of the drug on the biofilm. The authors observed that amphotericin-B caused a greater reduction in the number of eluted cells than the other drugs tested, including flucytosine, fluconazole, and ketoconazole. These initial experiments were done with drug concentrations that were 20 times the MIC for each drug. Interestingly, however, at lower drug concentrations of amphotericin-B, biofilms were much more resistant to the drug at all growth rates than planktonic cells grown under similar rates. In addition, cells resuspended from biofilms were less resistant than intact biofilms but more resistant than daughter cells which were obtained as cells eluted from the biofilm. In fact, planktonic cells displayed some resistance but only at low growth rates. Thus, resistance to amphotericin-B in this experimental system is dependent upon the growth rate of the organism but also is enhanced because of feature(s) of the biofilm. The factor(s) associated with increased resistance of cells as part of a biofilm are not known, but again, the secreted matrix may offer one possible reason for enhanced resistance to amphotericin-B and possibly other antifungals.

B. Azoles

Azoles are classified as imidazoles if they have two nitrogens in the azole ring (ketoconazole, miconazole, or clotrimazole) or triazoles, which contain three nitrogens in the azole ring (fluconazole, itraconazole, terconazole) (Fig. 1). In addition to the difference in structure, these two groups of antifungals differ in regard to their use in clinical applications. For instance, except for ketoconazole, the other imidizoles have found use only in the treatment of the superficial mycoses, while at least two of the triazoles are used to treat both the superficial and systemic mycoses (fluconazole and itraconazole) [4]. In the case of itraconazole, efficacy studies in the treatment of several mycoses are under way; this drug is potentially important in the treatment of invasive aspergillosis. The triazoles appear to have less toxicity than the imidazoles since they are more specific for the fungal P450-cytochrome-heme enzymes than those of mammalian cells [4]. This observation was stated earlier in the chapter in reference to the activity of voriconazole (see Sec. II, General Considerations). As stated in Table 1 (and see Fig. 2), the azoles are inhibitors of ergosterol biosynthesis. The inhibition is accomplished by the binding of these drugs to the heme protein of the cytochrome P450-dependent 14α-demethylase which converts lanosterol to demethylated zymosterol in the ergosterol pathway. It should be pointed out that while the latter statement is true for all azoles, additional activities are noted among the azoles and among susceptible fungi. For example, the imidazoles inhibit other enzyme activities in addition to the one described above for ergosterol biosynthesis [4] (Fig. 2). As another example of differences in the mode of action among the azoles, both $C.$ $albicans$ and $C.$ $krusei$ are inhibited in vitro by voriconazole (MIC$_{80}$ of 0.003 and 0.5 μg/mL, respectively) [34,35]. However, the sterols which accumulated in treated cells were different in that 24-methylenedihydrolanosterol (MDL) accumulation was two times greater in $C.$ $albicans$ than in $C.$ $krusei$ while lanosterol accumulation was higher in cells of $C.$ $krusei$ [34]. Further, the level of zymosterol was lower in $C.$ $krusei$ than in $C.$ $albicans$. These latter data indicate that this drug may have subtle differences on enzymes of ergosterol biosynthesis from different $Candida$ spp. Equally important, voriconazole appears to be an alternative to fluconazole in treating infections caused by $C.$ $krusei$, which is resistant to fluconazole (MIC$_{80}$ of 32 μg/mL) [34]. Also, in this same study, resistant strains of $C.$ $albicans$ (MIC$_{80}$ of >64 μg/mL) were relatively susceptible to voriconazole [34]. In $Cryptococcus$ $neoformans$, both itraconazole and fluconazole, in addition to inhibiting P450$_{DM}$ activity, also effect the conversion of obtusifolione to obtusifoliol in ergosterol synthesis, resulting in the accumulation of methylated sterol precursors [35] (Fig. 2). Thus, it is fairly clear that differences in activity among the azoles are observed which appear to be species specific. Nevertheless, the common activity of all azoles is the inhibition of P450$_{DM}$.

As stated above, the azoles are thought to target the heme protein (cytochrome) coenzyme of the P450-dependent 14α-demethylation (demethylase) of lanosterol [4,5]. The sterol 14-demethylase P450 (P450$_{DM}$) is also an essential enzyme for cholesterol synthesis of mammalian cells. The mono-oxygenase P450 proteins comprise a large gene superfamily consisting of at least several hundred or more species found both in eukaryotes and several prokaryotes including $Bacillus$ spp. and $Mycobacterium$ $tuberculosis$ [36–39]. Amino acid sequence analysis of various P450$_{DM}$ indicates that these proteins are strongly conserved in distinct kingdoms, such as the Fungi and Mammalia [36]. Further, the identity between mammalian and fungal P450$_{DM}$ proteins is comparable to that observed for other corresponding mammalian and fungal housekeeping enzymes. Conservation between the mammalian and yeast proteins seems to occur in six clusters of amino acids, distributed equally along the proteins [36]. Of these clusters of amino acids (designated as CR-1 to CR-6), four (CR-1, 2, 3, and 4) overlap with four of the seven putative substrate recognition sites, indicating their importance in the recognition of their common substrates and conservation of function. Nevertheless, while both have similar substrate recognition

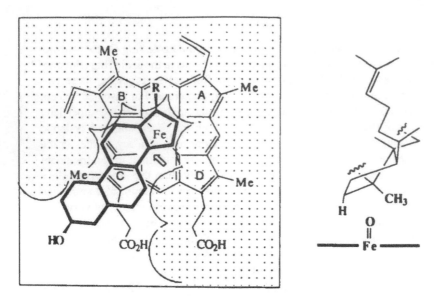

Figure 5 (Left) The proposed binding orientation of lanosterol, indicated by a darker line which outlines its molecular structure and the active site of lanosterol 14α-demethylase. The dotted surface indicates the portions of the active site that are inaccessible from the iron atom. The four pyrrole rings of the P450 heme are indicated as A–D. Ring C is believed to be involved in binding of the sterol to the active site. (Right) A structural representation of the sterol-heme interaction showing the 14α-methyl and 15α-hydrogens near the ferryl oxygen. Data from several analyses indicate that the azoles bind at the active site of the P450$_{DM}$. (From Ref. 37.)

sites and can catalyze 14α-demethylation, each has slightly higher activities for their respective endogenous substrates [38], which may account for differences in regard to the inhibitory activity of azoles against fungal and mammalian cells.

Molecular models of substrate and azole interactions with P450$_{DM}$ have been proposed [40–44]. These models indicate that the substrate of the demethylase lanosterol is oriented such that the sterol is bound over the pyrrole ring C of the P450 heme (Fig. 5), which constitutes the active site of the enzyme [44]. In cells treated with azoles, the predicted model suggests that the N3 of imidazole and N4 of triazole rings have sufficient bond forming distances with the heme iron of P450$_{DM}$, and that the azoles most likely interact with the amino acid residues Phe87, Tyr96, Val295, Val396, and Ile395 at a hydrophobic site of the P450 protein [40]. Additionally, amino acid residue Tyr96 may play an important role in docking of azoles to the substrate protein [40]. On the other hand, other data indicate that the N1 ring of azoles may also play an important role in the activity of this group of antifungals [42]. Other molecular modeling experiments also indicate that azoles such as ketoconazole fit into the active site of the P450$_{DM}$ for its substrate, lanosterol [43]. Interactions of the azole with the P450$_{DM}$ include heme ligation, hydrogen bonding, pi-pi stacking, and hydrophobic interactions within the enzyme's heme environment [43].

The choice of azoles in the treatment of fungal infections is, in general, dependent upon the fungus which causes the infection (Table 2) [4,10,45]. Among the major observations on the use of azoles:

Table 2 Azole Antifungal Agents and Their Use in Treating the Systemic Mycoses

Clinical use	Ketoconazole	Itraconazole	Fluconazole
Aspergillosis		X	
Candidemia		X	X
Cryptococcosis		X	X
Blastomycosis	X	X	X
Histoplasmosis	X	X	X
Coccidioidomycosis	X	X	X
Paracoccidioidomycosis	X	X	X
Sporotrichosis		X	X
Pseudoallescheriasis	X	X	

Qualifications:

Ketoconazole and itraconazole: contraindicated in meningeal infections of aspergillosis, blastomycosis, histoplasmosis, coccidioidomycosis.

Itraconazole: clinical studies needed for afficacy determinations of candidemia.

Fluconazole: Same for histoplasmosis, paracoccidioidomycosis, and sporotrichosis.

Itraconazole: no IV formulation limits usefulness in seriously ill patients with candidemia.

Ketoconazole, itraconazole, fluconazole: second-line azole in treatment of blastomycosis, histoplasmosis, paracoccidioidomycosis, cryptococcosis.

Ketoconazole: nonimmunocompromised blastomycosis and histopplasmosis patients only.

Source: Ref. 4.

1. Fluconazole is inactive against the aspergilli. As an alternative to amphotericin-B in the treatment of invasive aspergillosis, however, itraconazole appears to be as effective as amphotericin B [4].

2. At present only itraconazole should be considered for primary prophylaxis against aspergillosis.

3. The use of fluconazole in treating candidemias and mucosal infections is dependent upon a number of factors, as stated above, including patient stability, prior fluconazole usage, and whether or not the offending pathogen is an *albicans* or non-*albicans Candida* species. On the other hand, for the treatment of candidiasis in the neonate, it is recommended that amphotericin B (88% of respondents) be used if a single blood culture is positive for *Candida* spp. [46]. Thus, the choice of the preferred drug may also be dependent upon the age of the patient population.

4. Combinations of antifungals offer a useful approach to therapy, but current data are in conflict in regard to the therapeutic benefits of combined amphotericin-B/azole therapy. The suggestion is that antagonism between these two drugs may interfere with any therapeutic benefit [4].

5. In combination, amphotericin-B and 5-fluorocytosine remain as an important approach in the treatment of cryptococcosis in both the AIDS and non-AIDS patient, although long-term prophylaxis with fluconazole is recommended in the AIDS patient following primary treatment [45].

C. Fluoropyrimidines (5-Fluorocytosine)

5-Fluorocytosine (5-FC) (Fig. 1) is the only fluoropyrimidine which has been used in the treatment of the systemic mycoses. The drug is taken up by sensitive strains and is thought to be initially converted to 5-fluorouracil (5-FU) by a cytosine deaminase [47]. Strains lacking this enzyme render the drug ineffective. 5-FU is then acted upon by a UMP-pyrophosphorylase forming 5-fluorouridylic acid (FUMP). Following an additional phosphorylation, FUMP is incorporated into RNA. As a consequence of this event, protein synthesis by susceptible strains is inhibited. Other investigations have revealed that thymidylate synthetase activity in crude extracts of sensitive strains could be inhibited by coincubation of the enzyme with 5-fluoro-2′-deoxyuridylic acid [47]. Additionally, coincubation of the enzyme with 5-FC reduced the activity of the enzyme by 80% [47]. These latter observations indicate that 5-FC inhibits DNA synthesis in sensitive cells.

5-FC appears to have its primary use therapeutically when combined with amphotericin-B, as in the treatment of cryptococcal meningoencephalitis. The drug often results in treatment failure when used alone; the explanation for this is related (most likely) to the high degree of resistance to the drug which develops in isolates. For example, in *C. albicans*, partial or high resistance, quantitated by the number of resistant colonies which grew in the presence of the drug on MFC50 medium compared to sensitive isolates, was observed in 43% of 137 independent clinical isolates [48]. While most of the primary screening indicated that strains were partially resistant, all such strains gave rise to variants which were highly resistant to 5-FC [48].

D. Allylamines (Terbinafine/Naftifine)

The allylamines (Fig. 1) like the azoles act by inhibiting the synthesis of ergosterol. However, in this case, the enzyme inhibited in the ergosterol pathway coverts squalene to squalene-2,3-epoxide (Fig. 2) [5]. Terbinafine is effective in the treatment of dermatophytosis especially tinea capitis caused by *Trichophyton tonsurans* [49]. Nevertheless, further evaluation to confirm these observations is necessary in randomized double-blind controlled studies. As yet, the compound has not been approved for clinical usage for systemic therapy by the Food and Drug Administration [49]. The reader is directed to other observations on the use of terbinafine and other inhibitors in the treatment of dermatophytosis [50]. Terbinafine is highly inhibitory *in vitro* to several dermatophytes as well as *C. albicans*, including strains of *C. albicans* which are resistant to some azoles, and *C. neoformans* [50,51]. However, the use of terbinafine treatment in systemic infections caused by these fungi awaits further study.

V. SOLVING PROBLEMS IN ANTIFUNGAL CHEMOTHERAPY

In this chapter, a discussion of several of the most commonly used antifungals has pointed out several of the major problems which the clinican has to overcome to obtain a successful cure. Often, a cure does not occur, as in the allogeneic bone marrow transplant patient with invasive aspergillosis. Resistance to fluconazole both in *C. albicans* and non-*albicans Candida* spp. is on the rise. Toxicity accompanies therapy with amphotericin-B and with azoles such as ketoconazole. Laboratory efforts to identify pathogens from invasive infections is often too late or fails. It is obvious that solving many of these problems awaits the development of antifungals which are both safe and broad-spectrum as well as the establishment of improved diagnostic tests; however, these goals remain long-term objectives. On the other hand, are there immediate solutions to some of the current problems? In this section, some of these issues are addressed.

A. What to Do About the Toxicity of Antifungals

At least two approaches have been used to address the problem of toxicity, and both have been implemented with some success. First, the use of amphotericin-B in combination with another antifungal may obviate the need for higher amounts of amphotericin-B when used singly. In this regard, 5-FC has been used most often with amphotericin-B. Other combinations with amphotericin-B have been tried experimentally. The suggestion is that the combination of amphotericin-B and an azole induces antagonism, based upon the idea that both drugs target ergosterol. The companion compound used with amphotericin-B need not be an antifungal, although in this instance additional studies are required to prove efficacy. Nevertheless, there is agreement that immune reconstitution along with amphotericin-B might provide cures. Recent studies by Lyman et al. [52] indicate that a hematoregulatory peptide (SK&F 107647) potentiates cure when used in combination with amphotericin-B. This peptide had no effect on the dissemination of candidiasis in a neutropenic rabbit model when used alone. However, in combination with low dosages of amphotericin-B, a significant reduction in organism burden in the lungs, spleen, and kidneys occurred when compared to untreated controls as well as to animals treated with amphotericin-B alone. An immunoregulatory antifungal approach, at least conceptually, is feasible but very much in need of further study. It would seem that any antifungal (as long as the organism is sensitive to the drug) would be more likely to cure if the immunity of the host were returned to at least near normal levels. This issue is addressed further in the chapters on immunotherapy and passive protection of patients with immunoprotective antibodies (Chaps. 15 and 19). The reader is also directed to a review on the subject of combination immunotherapy and antifungal chemotherapy [53].

A second approach to the problem of amphotericin-B toxicity is to improve drug delivery. This has been accomplished by various forms of lipid encapsulation of amphotericin-B. The theory is that solubility (and hence, absorption) of the drug is improved, and thus lower dosages of the drug can be used. Again, the problem of treating invasive aspergillosis has resulted in several studies which evaluate the efficacy of encapsulated versus nonencapsulated drug. In a recent study of this nature, treatment of invasive aspergillosis was monitored using two dosages of amphotericin-B in the AmBisome lipid formulation, 1 mg/(kg/d) and 4 mg/(kg/d) [54]. Clinical responses (radiologic response rates, 6-month survival rates) were observed in 64% and 48% of patients receiving the 1 mg and 4 mg dosage of amphotericin B, respectively. These data support the observations made by other investigative teams [55], although the former study [54] did not include a cohort treated only with amphotericin-B.

Reducing the toxicity of antifungal drugs is an important consideration in achieving a therapeutic cure. Toxicity occurs not only because of the inherent toxicity of the drugs themselves, but also as a consequence of drug-drug interactions, especially for the azoles, since they are potent inhibitors of some of the isoforms of human P450. For example, ketoconazole inhibits two P450 proteins, 3A and 2C19, while fluconazole inhibits the 2C9 isoform, and itraconazole binds to the 3A human P450 (http://www.Drug-interactions.com). Each of these P450 isoforms is required for the metabolism of other drugs. For example, P4503A is known to metabolize cisapride (a prokinetic drug which is used for reducing gastric reflux due to nocturnal heartburn) and simvastatin (an oral antilipemic agent, HMG-CoA reductase inhibitor which is used in the treatment of primary hypercholesterolemia). Thus, patients using cisapride that are treated with azoles (itraconazole or ketoconazole) have elevated levels of cisapride because its metabolism by P4503A is inhibited by these azoles [56]. As a consequence, patients are at higher risk for developing cardiac arrhythmias. Similarly, itraconazole increases the risk of skeletal muscle toxicity by simvastatin [57]. Thus, knowledge of drug-drug interactions by the clinican may avoid toxicity problems.

B. Emerging Fungal Pathogens Pose Significant Problems in Their Eradication

In Chapter 32, new and emerging fungal pathogens are discussed. The most common of these pathogens is *Fusarium* spp. [58]. Further, infections (fusariosis) caused by this fungus can mimic aspergillosis, and because of this, amphotericin-B is usually the drug of choice. Unfortunately, this compound has poor activity against fusariosis. A recent study of 40 immunocompromised patients (bone marrow transplants, leukemias, cytotoxic therapy) with an antemortem diagnosis of fusariosis revealed that 13 patients responded to therapy, although relapses were observed in two of the 13 patients [58]. A response was associated with granulocyte transfusions and lipid-amphotericin-B formulation, and one patient received itraconazole; however, resolution of the infection was only observed in patients who recovered from myelosuppression. This study again emphasizes the potential of a combined use of immunotherapy and an antifungal.

VI. SUMMARY

The four classes of antifungals that are primarily used to treat the systemic mycoses are either inhibitors of ergosterol synthesis/function or target protein synthesis. These drugs can be broad or narrow spectrum, are cidal or static in their activity, and probably have more than a single activity. Problems with the antifungal therapy include toxicity to the patient and drug resistance, both of which are drug and organism specific. Antifungal sensitivity testing has progressed and, in the case of candidiasis caused by *C. albicans*, breakthrough numbers can be correlated with predicting whether or not a specific antifungal used in treatment is in fact effecting a cure. On the other hand, similar protocols are not yet standardized for one of the most important groups of human pathogens, the aspergilli.

REFERENCES

1. J-P Latge. *Aspergillus fumigatus* and aspergillosis. Clin Microbiol Rev 12:310–350, 1999.
2. NF Georgopapadakou. Antifungals: mechanisms of action and resistance, established and novel drugs. Current Opin Microbiol 1:547–557, 1998.
3. JR Graybill. The future of antifungal therapy. Clin Infect Dis 22(suppl 2):S166–S178, 1996.
4. DJ Sheehan, CA Hitchock, CM Sibley. Current and emerging azole antifungals. Clin Microbiol Rev 12:40–79, 1999.
5. MA Ghannoum, LB Rice. Antifungal agents: mode of action, mechanisms of resistance and correlation of these mechanisms with bacterial resistance. Clin Microbiol Rev 12:501–517, 1999.
6. H Vanden Bossche, P Marichal, F Odds. Molecular mechanisms of drug resistance in fungi. Trends Microbiol 2:393–400, 1994.
7. T White. Clinical, cellular and molecular factors that contribute to antifungal drug resistance. Clin Microbiol Rev 11:382–402, 1998.
8. C Hitchcock, K Dickinson, SB Brown, EGV Evans, DJ Adams. Interaction of azole antifungal antibiotics with cytochrome P-450 dependent 14α-sterol methylase purified from *Candida albicans*. J Bacteriol 266:475–480, 1990.
9. H Vanden Bossche, G Willemsens. Effect of the antimycotics, miconazole and ketoconazole on cytochrome P450 in yeast microsomes and rat liver microsomes. Arch Physiol Biochem 90: B218–B219, 1982.
10. J Edwards et al. International conference for the development of a consensus on the management and prevention of severe candidal infections. Clin Infect Dis 25:43–59, 1997.
11. EJ Anaissie. Opportunistic mycoses in the immunocompromised host: experience at a cancer center and review. Clin Infect Dis 14:43–53, 1992.
12. JH Rex et al. Development of interpretative breakpoints for antifungal susceptibility testing: conceptual framework and analysis of in vitro–in vivo correlation data for fluconazole, itraconazole and *Candida* infections. Clin Infect Dis 24:235–247, 1997.

13. JH Rex et al. Optimizing the correlation between results of testing in vitro and therapeutic outcome in vivo for fluconazole by testing critical isolates in a murine model of invasive candidiasis. Antimicrob Agents Chemother 42:129–134, 1998.

14. M Lozano-Chiu, VL Paetznick, MA Ghannoum, JH Rex. Detection of resistance to amphotericin B among *Cryptococcus neoformans* clinical isolates: performance of three different media assessed by using E-test and National Committee for Clinical Laboratory standards M27-A methodologies. J Clin Microbiol 36:2817–2822, 1998.

15. DW Warnock, EM Johnson, TR Rogers. Multi-center evaluation of the E-Test method for antifungal drug susceptibility testing of *Candida* spp. and *Cryptococcus neoformans*. BSAC Working Party on Antifungal Chemotherapy. J Anitmicrob Chemother 42:321–331, 1998.

16. MA Pfaller, SA Messer, A Bolstrom. Evaluation of Etest for determining in vitro susceptibility of yeast isolates to amphotericin B. Diag Microbiol Infect Dis 32:223–227, 1998.

17. C Lass-Florl, G Kofler, G Kropshofer, J Hermans, A Kreczy, MP Dierich, D Niederwieser. In-vitro testing of susceptibility to amphotericin B is a reliable predictor of clinical outcome in invasive aspergillosis. J Antimicrob Chemother 42:497–502, 1998.

18. E Dannaoui, E Borel, F Persat, MF Monier, MA Piens. In-vivo itraconazole resistance of *Aspergillus fumigatus* in systemic murine aspergillosis. EBGA Network. European research group on biotypes and genotypes of *Aspergillus fumigatus*. J Med Microbiol 48:1087–1093, 1999.

19. EM Johnson, KL Oakley, SA Radford, CB Moore, P Warn, DW Warnock, DW Denning. Lack of correlation of in vitro amphotericin B susceptibility testing with outcome in a murine model of *Aspergillus* infection. J Antimicrob Chemother 45:85–93, 2000.

20. AW Norman, RA Demel, B de Kruijff, WSM Guerts–van Kessek, LLM Van Deenen. Studies on the biological properties of polyene antibiotics; comparison of other polyenes with filipin in their ability to interact specifically with sterol. Biochim Biophys Acta 290:1–14, 1972.

21. JO Lampen, PM Arnow, RS Safferman. Mechanism of protection by sterol against polyene antibiotics. J Bacteriol 80:200–206, 1960.

22. RW Holz. The effects of the polyene antibiotics nystatin and amphotericin B on thin lipid membranes. Ann NY Acad Sci 235:469–479, 1973.

23. CA Hitchcock, NJ Russell, KJ Barrett-Bee. Sterols in *Candida albicans* mutants resistant to polyene or azole antifungals, and of a double mutant *C. albicans* 6.4. Crit Rev Microbiol 15:111–115, 1987.

24. PE Verweij, KL Oakley, J Morrissey, DW Denning. Efficacy of LY303366 against amphotericin B–susceptible and–resistant *Aspergillus fumigatus* in a murine model of invasive aspergillosis. Antimicrob Agents Chemother 42:873–878, 1998.

25. EK Manavathu, GJ Alangaden, PH Chandrasekar. In-vitro isolation and resistance of amphotericin B–resistant mutants of *Aspergillus fumigatus*. J Antimicrob Chemother 41:615–619, 1998.

26. JA Vazquez, MT Arganoza, D Boikov, S Yoon, JD Sobel, RA Akins. Stable phenotypic resistance of *Candida* species to amphotericin B conferred by preexposure to subinhibitory levels of azoles. J Clin Microbiol 36:2690–2695, 1998.

27. DA Sutton, SE Sanche, SG Revankar, AW Fothergill, MG Rinaldi. In vitro amphotericin B resistance in clinical isolates of *Aspergillus terreus*, with a head-to-head comparison to voriconazole. J Clin Microbiol 37:2343–2345, 1999.

28. K Seo, H Okiyoshi, Y Ohnishi. Alteration of cell wall composition leads to amphotericin B resistance in *Aspergillus flavus*. Microbiol Immunol 43:1017–1025, 1999.

29. SA Yoon, JA Vazquez, PE Steffan, JD Sobel, RA Akins. High-frequency, in vitro reversible switching of *Candida lusitaniae* clinical isolates from amphotericin B susceptibility to resistance. Antimicrob Agents Chemother 43:836–945, 1999.

30. SP Hawser, LJ Douglas. Resistance of *Candida albicans* to antifungal agents in vitro. Antimicrob Agents Chemother 39:2128–2131, 1995.

31. SP Hawser, GS Baillie, LJ Douglas. Production of extracellular matrix by *Candida albicans* biofilms. J Med Microbiol 47:253–256, 1998.

32. GS Baillie, LJ Douglas. Effect of growth rate on resistance of *Candida albicans* biofilms to antifungal agents. Antimicrob Agents Chemother 42:1900–1905, 1998.

33. A Louie, P Banerjee, GL Drusano, M Shayegani, MH Miller. Interaction between fluconazole and amphotericin B in mice with systemic infection due to fluconazole-susceptible or -resistant strains of *Candida albicans*. Antimicrob Agents Chemother. 43:2841–2847, 1999.

34. H Sanati, P Belanger, R Fratti, M Ghannoum. A new triazole voriconazole (UK-109,946) blocks sterol biosynthesis in *Candida albicans* and *Candida krusei*. Antimicrobiol Agents Chemother 41: 2492–2496.

35. CA Hitchcock, GW Pye, GP Oliver, P Troke. UK-109,496; a novel, wide-spectrum triazole derivative for the treatment of fungal infections: antifungal activity and selectivity in vitro, abstr. 2739. *In*: Program and Abstracts of the 35h Interscience Conference on Antimicrobial Agents and Chemotherapy. Washington: American Society for Microbiology.

36. Y Aoyama et al. Sterol 14-demethylase P450 (P45014DM) is one of the most ancient and conserved P450 species. J Biochem 119:926–922, 1996.

37. A Bellamine, AT Mangla, WD Nes, MR Waterman. Characterization and catalytic properties of the sterol 14α-demethylase from *Mycobacterium tuberculosis*. Proc Natl Acad Sci USA 96:8937–8942, 1999.

38. DR Nelson. Cytochrome P450 and the individuality of species. Arch Biochem Biophys 369:1–10, 1999.

39. DC Lamb, DE Kelly, SL Kelly. Molecular diversity of sterol 14alpha-demethylase substrates in plants, fungi and humans. FEBS Lett 425:263–265, 1998.

40. TT Talele, V Hariprasad, VM Kulkarni. Docking analysis of a series of cytochrome P-450 (14) alpha DM inhibiting azole antifungals. Drug Des Discov 15:181–190, 1997.

41. HD Holtje, C Fattorusso. Construction of a model of the *Candida albicans* lanosterol 14-alpha-demethylase active site using homology modeling technique. Pharm Acta Helv 72:271–277, 1998.

42. TT Talele, VM Kulkarni. Three-dimensional quantitative structure-activity relationship (QSAR) and receptor mapping of cytochrome P-450 (14 alpha DM) inhibiting azole antifungal agents. J Chem Inf Comput Sci 39:204–210, 1999.

43. DF Lewis, A Wiseman, MH Tarbit. Molecular modeling of lanosterol 14 alpha-demethylase (CYP51) from *Saccharomyces cerevisiae* via homology with CYP102, a unique bacterial cytochrome P450 isoform: quantitative structure-activity relationships (QSARs) within two related series of antifungal azole derivatives. J Enzyme Inhib 14:175–192, 1999.

44. SF Tuck, Y Aoyama, Y Yoshida, PR Ortiz de Montellano. Active site topology of *Saccharomyces cerevisiae* lanosterol 14αdemethylase (CYP51) and its G310D mutant (cytochrome P-450 *SGI*). J Biol Chem 267:13175–13179, 1992.

45. TG Mitchell, JR Perfect. Cryptococcosis in the era of AIDS—100 years after the discovery of *Cryptococcus neoformans*. Clin Microbiol Rev 8:515–548, 1995.

46. JL Rowen, JM Tate. Management of neonatal candidiasis. Neonatal Candidiasis Study Group. Pediatr Infect Dis J 17:1007–1011, 1998.

47. RA Diasio, JE Bennett, CE Myers. Mode of action of 5-fluorocytosine. Biochem Pharm 27:703–707, 1978.

48. KS Defever, WL Whelan, AL Rogers, ES Beneke, JM Veselenak, DR Soll. *Candida albicans* resistance to 5-fluorocytosine: frequency of partially resistant strains among clinical isolates. Antimicrob Agents Chemother 22:810–815, 1982.

49. AK Gupta, TR Einarson, RC Summerbell, NH Shear. An overview of topical therapy in dermatomycoses. A North American perspective. Drugs 55:645–674, 1998.

50. SF Friedliner. The evolving role of itraconazole, fluconazole and terbinafine in the treatment of tinea capitis. Pediatr Infect Dis J 18:205–210, 1999.

51. N Ryder, B Favre. Antifungal activity and mechanism of action of terbinafine. Rev Contemp Pharmacother 8:275–287, 1997.

52. CA Lyman, C Gonzalez, M Schneider, J Lee, TJ Walsh. Effect of the hematoregulatory peptide SK&F 107647 alone and in combination with amphotericin B against disseminated candidiasis in persistently neutropenic rabbits. Antimicrob Agents Chemother 43:2165–2169, 1999.

53. DA Stevens. Combination immunotherapy and antifungal chemotherapy. Clin Infect Dis 26: 1266–1269, 1998.

54. M Ellis et al. An EORTC International Multicenter Randomized Trial (EORTC No. 19923) comparing two dosages of liposomal amphotericin B for treatment of invasive aspergillosis. Clin Infect Dis 27: 1406–1412, 1998.

55. TTC Ng, DW Denning. Liposomal amphotericin B (AmBisome) therapy in invasive fungal infections. Evaluation of United Kingdom compassionate use data. Arch Intern Med 155:1093–1098, 1995.

56. L Carlsson, GJ Amos, B Andersson, L Drews, G Duker, G Wadstedt. Electrophysiological characterization of the prokinetic agents cisapride and mosapride in vivo and in vitro: implications for proarrhythmic potential. J Pharmacol Exp Ther 282:220–227, 1997.

57. PJ Neuvonen, T Kantola, KT Kivisto. Simvastatin but not pravastatin is very susceptible to interaction with the CYP3A4 inhibitor, itraconazole. Clin Pharmacol Ther 63:332–341, 1998.

58. EI Boutati, EJ Anaissie. *Fusarium*, a significant emerging pathogen in patients with hematologic malignancy: ten years' experience at a cancer center and implications for management. Blood 90: 999–1008, 1997.

25

Antifungal Drug Targets: Discovery and Selection

Ronald L. Cihlar and Christina Kellogg
Georgetown University Medical Center, Washington, D.C.

Sheldon Broedel, Jr.
Dorlin Pharmaceuticals, Inc., Baltimore, Maryland

I. INTRODUCTION

The occurrence of fungal infections has escalated significantly in recent years, and this increase is particularly profound in those immunocompromised by disease or therapies. Unfortunately, the number of antifungal drugs available for combating fungal infections is limited. The paucity of effective agents is due in large part to the high degree of relatedness between the biochemical machinery of fungi and the mammalian host. In this regard, only a few suitable targets with the necessary specificity have been successfully utilized in antifungal development, and these are primarily associated with fungal cell wall and membrane biosynthesis. The most effective antifungal agents against these targets are azole derivatives and amphotericin B. While these agents are commonly employed, their usage presents a number of shortcomings including toxicity, fungistatic versus fungicidal activity, and the appearance of drug-resistant organisms. Thus, there is an urgent need to augment the number of suitable targets in order to detect and develop novel and effective antifungal agents. This need is being addressed by major research efforts to uncover potential targets and companion agents. Such investigations are supported in part by new approaches, as well as more traditional methodologies, designed to more rapidly identify and verify novel targets. The aim of the following discussion is to present an overview of such approaches and promising antifungal targets.

II. STRATEGIES FOR TARGET DISCOVERY

There are at least four criteria that should be applied when judging whether a protein, say, may be a good target for antifungal chemotherapeutic intervention. First, the objective of any antibiotic treatment is to eradicate the disease-causing organism without harming the host. As such, an antibiotic drug should be highly selective, acting only on the pathogen and not interfering with host functions. Thus, targets that are unique to the pathogen or sufficiently different from the host permit the development of selective inhibitors. Second, the target should be essential for

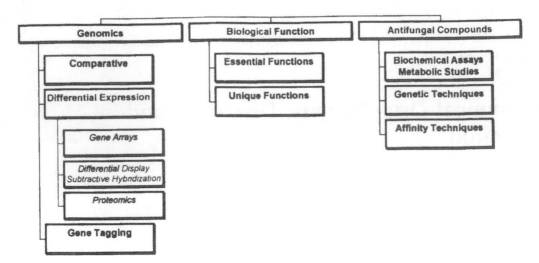

Figure 1 Strategies for identifying new antifungal targets.

growth or viability. This must be placed within the context of the pathogen growing in the host, i.e., a function essential during in vitro growth does not necessarily mean it will be essential in vivo. Conversely, some gene products nonessential to growth in culture may be essential for growth in a host organism. Third, the target should be reasonably well characterized. In order to develop a mechanism-based drug screen, the target's role in fungal physiology and pathogenesis should be understood and its biochemical nature, enzymatic activity, etc., defined. Further, the measurement of its function should be amenable to high-throughput assay platforms, though this latter property is not always essential. Finally, where possible, the target should be shared by a broad spectrum of pathogenic fungi, as commercial viability may be dependent on this factor.

Only three cellular processes, sterol biosynthesis and function, cell wall biosynthesis, and nucleic acid metabolism, have been successfully exploited commercially as targets for chemotherapeutic intervention (reviewed in [1]). The azole family of drugs, which inhibit sterol biosynthesis, and polyenes, which disrupt membrane integrity, are the most widely used classes of antifungals. Fluorocytosine, due to its toxic side effects, is typically used in combination therapies. More recently, antifungal drugs which inhibit enzymes required for cell wall biosynthesis, ceramide synthase, and translation have been evaluated for clinical efficacy [2], though it remains to be seen if these new drugs will have a clinical impact. Despite such advances, there remain a limited number of targets described and developed. With the advent of high-throughput screens, antimicrobial drug discovery and development relies more and more on mechanism-based bioassays to identify lead candidates from very large numbers of compounds. There are three basic strategies for identification of potential targets (Fig. 1): the use of known antifungal compounds; the use of knowledge of cellular physiology and metabolism; and the use of molecular genetics and bioinformatics.

A. Target Definition by Use of Antifungal Agents

Certainly the oldest and perhaps the most direct approach used to define a target is through the use of a known antifungal agent. Before the development of high-throughput biochemical assays, natural-product, semisynthetic, and synthetic compound libraries were screened using whole-cell growth inhibition assays. This led to the discovery of many of the antifungal drugs currently in use. Only after these compounds were discovered was the basis of the antifungal activity elucidated. Once the cellular targets were identified, more specific bioassays were developed to improve the drug development process. Thus, new classes of antifungal compounds should be useful to identify new targets.

The strategy relies on identifying and characterizing antifungal compounds. Any number of source libraries can be used. Natural products are arguably the most diverse, presenting perhaps the best opportunity for finding new agents. However, with the advent of combinatorial chemistry techniques, semisynthetic and synthetic libraries have begun to rival natural-product libraries in their degree of diversity. The advantage of synthetic libraries is that they are highly defined and thus do not require extensive fractionation or chemical characterization to identify the structure of an inhibitor. In either case, the library is screened to identify compounds with antifungal activity using whole-cell growth inhibition assays. A recent review of phytochemical antimicrobial agents provides an example of the potential diversity such a screen can uncover [3].

1. Biochemical Analysis

Once a compound has been selected, it can be used as a tool to uncover the target. There are several approaches that can be used to elucidate the target, including biochemical analyses to determine the cellular process that is being inhibited, and/or the use of radioisotopes in labeling experiments. By tracking label incorporation, a general idea of which cellular process is being inhibited by the compound can be obtained. Depending on the result, different biochemical assays can then be designed to narrow the field to more specific targets. This approach has been used for many years and formed the basis of early studies on the mechanism of inhibition. It remains a useful tool, but has fallen into disfavor in recent years with the development of methods based in molecular biology and molecular genetics.

2. Molecular Biology

The use of molecular and genetic techniques are often coupled with biochemical analysis in defining a putative target. For example, the characterization of mutant strains resistant to the inhibitor, or to knockout mutants of suspected targets, can help to confirm the target. Another approach relies on recent advances in gene array and protein analysis technology. These techniques have all but eclipsed the use of biochemical labeling studies in situations where they can be applied. DNA-based arrays [4–6] can be used to examine global changes in gene expression during exposure to sublethal doses of the compound [7]. For example, the signature pattern of the new inhibitor can be compared to the expression pattern caused by known antifungals. A similar pattern would indicate a mode of action related to the reference inhibitor, whereas a different pattern would suggest a different mechanism. Proteins that are overexpressed under appropriately designed conditions are also candidate targets. Identification of the overexpressed gene is based on the array composition. The subsequent identification of the protein can then lead to the design of a cell-free biochemical assay that is then used to verify inhibitory activity.

Potential problems in using an expression array include that it provides an assay of mRNA, which does not necessarily reflect protein activity, and that a highly defined DNA array is required. The latter requirement will not easily be met for most pathogenic fungi until genomic sequencing efforts progress further, although the use of random oligonucleotides in constructing the arrays is a possibility. An alternative approach to DNA-expression arrays is the use of proteomic analysis [8]. This technique employs two-dimensional gel electrophoresis to separate proteins with high resolution [9,10]. Protein expression patterns between reference and test cells can be compared and pattern similarities and differences determined. Protein sequencing tools, such as Edman sequencing [11], or mass spectroscopy detection systems [12] are then used to identify proteins (see Sec. II.D.1).

3. Affinity Separation

The most technically challenging, though most direct, approach of identifying a target using an antifungal agent is affinity separation. This technique involves the attachment of the inhibitor to a solid support. The target protein, which binds the ligand, is separated from a crude or partially purified cell extract. After recovery of the bound protein from the solid support, its identity is determined by protein microsequencing. A classic example of this approach was the use of novobiocin-Sepharose affinity chromatography to identify *E. coli* DNA gyrase as the major target of the antibacterial activity of novobiocin [13].

While potentially very powerful, affinity separation using tethered ligands has serious drawbacks. First, the ligand in question must be amenable to chemical coupling to solid support resins. Not all candidate compounds have the functional groups necessary to perform the requisite chemistry. Further, steric hindrance may preclude receptor binding and the altered spatial relationships may affect the ligand binding affinity and kinetics. Finally, the cellular fraction to be employed must be carefully prepared to ensure that the receptor retains function. This is not always possible, particularly for membrane-bound proteins. Nevertheless, in planning a strategy this approach should be explored as it could lead directly to target identification.

An alternative approach is to affinity-label cellular fractions with photoaffinity analogs of a model inhibitor [14]. The radiolabeled inhibitor is chemically modified to contain a photoactivated crosslinking moiety. When incubated with a cell extract or fraction, the modified inhibitor binds to the target protein. Exposure to light then activates the crosslinking reaction, and the inhibitor is covalently attached to the target protein. For this technique to be successful, however, two key criteria must be met: First, the compound to be modified must be amenable to the derivatization needed; and second, it must have specific, high binding affinity, as low affinity or specificity would result in a complex labeling pattern. Despite these potential drawbacks, the technique was successfully applied to the identification of echinocandin targets in *Candida albicans* [14].

B. Identity by Biological Function

A basic understanding of the biology of pathogenic fungi suggests two general gene classes useful in target discovery: genes essential for survival, and genes found in fungi but not in humans. Some of the specific targets being exploited within these categories are discussed in detail in Section III below.

1. Obvious Essential Functions

Based on our knowledge of general and fungal physiology, there are several cellular functions that are obviously essential for survival. These include replication, transcription, translation, cell membrane synthesis, some aspects of intermediary metabolism, signal transduction, and cell division. Mutations resulting in loss of function of these genes are generally lethal. For opportunistic pathogens, including many fungi, the concept of lethal mutation must be placed within the context of infection. For example, functions essential to the organism growing in a host environment can be very different from those required for growth outside of a host, and this factor must be considered in target selection.

2. Functions Unique to Fungi

Another set of obvious target candidates are genes that encode proteins involved in functions not shared in mammalian cells. Cell wall biosynthesis is one example, and investigations in this area have led to discovery of the echinocandin class of glucan synthase inhibitors and the nikkomycin class of chitin synthase inhibitors [2]. However, not all genes that specify fungal-specific proteins are essential. For example, in the case of proteins involved in cell wall metabolism, construction of certain null mutants have shown that phenotypic changes affecting growth, morphology, infectivity, etc., can occur without being lethal [15,16]. Further complicating this issue, some genes are expressed only under specific niche conditions. For instance in the *PHR* (encoding beta-glucanosyltransferase) gene family of *C. albicans*, a *PHR1* null mutant was avirulent in the mouse model of systemic candidiasis, but still virulent in a rat model of vaginitis [17]. The pattern was reversed when the effect of *PHR2* mutation on virulence was assessed [18]. These phenotypes reflected response of the organism to environmental cues, with expression of *PHR1* under neutral pH conditions [17,19] and *PHR2* in the more acidic environment of the vaginal cavity [20].

C. Identity by Molecular Genetics

The recent explosion in the science of genomics has led to its preeminent position as a strategy for discovering drug targets. Genomes of several key bacterial pathogens have been completely sequenced [21], as has the genome of the model yeast, *S. cerevisiae* (e.g., speedy.mips.biochem.mpg.de/mips/yeast/yeast-genome.htmlx or genome-www.stanford.edu/ Saccharomyces). Pathogenic fungi whose genomes are nearly complete include *C. albicans*, *A. fumigatus*, and *C. neoformans*. Table 1 lists some of the Web site addresses where sequence information is available for these fungi.

1. Comparative Genomics

Directly stemming from the availability of sequence information is the potential for comparative genomics. Completely sequenced genomes, and to some extent, partial sequences, allow for direct comparison of genes between two or more organisms. Among bacteria, genomewide comparisons have led to lists of candidate targets specific to bacteria [22–25]. Similarly, by comparing fungi to fungi, and fungi to human, one should be able to develop a list of candidate

Table 1 Internet Addresses for Fungal Genome Sequence Projects and Data

Species	Web address
Candida albicans	http://alces.med.umn.edu/Candida.html
Saccharomyces cerevisiae	speedy.mips.biochem.mpg.de/mips/yeast/yeast_genome.htmlx *or* genome-www.stanford.edu/Saccharomyces
Neurospora crassa	//biology.umn.edu/~npg/home.html, *or* www.mips.biiochem.mpg.de/proj/neurospora/
Pneumocystis carinii	www.uky.edu/Projects/Pneumocystis
Aspergillus nidulans	www.genome.ou.edu/fungal.html
Aspergillus parasiticus	www.aspergillus.man.ac.uk
Aspergillus fumigatus	www.aspergillus.man.uk/
Saccharomyces pombe	www.sanger.ac.uk/Projects/S_pombe/

Table 2 Parameters Used to Quantify Ranking of Potential Antifungal Targets Based on Genome Data

Parameter	Description
Quality	Importance of the gene to the cells. Essential genes, those where the null mutant is nonviable, get the maximum score, while nonessential genes get a zero score.
Occurrence	Scored based on the type and number of organisms where the gene is found.
Specificity	Quantifies the degree of similarity between homologs in other organisms.
Assay development	Assesses the effort required to develop and operate a high-throughput screen.
Homology	Measures the degree of homology to other genes.

See Ref. 30 for detailed description of the computer-aided target selection scheme.

gene products that are unique to fungi, as well as genes that are common to a wide range of fungi. This latter group would ideally permit development of broad spectrum antifungals. Fungi and humans share many metabolic pathways and corresponding enzymatic activities. However, it is not clear that shared catalytic activities preclude the development of selective inhibitors with potential for a good therapeutic index (e.g., fatty acid synthase inhibitors [26]), and care should be taken not to exclude potential candidate targets solely on the basis of shared biochemical function. Moreover, the search parameters can affect the quality of data obtained. In other words, how much homology is significant? Several examples of highly selective antimicrobial agents (trimethoprim, quinolones, and fluconazole) target proteins where there are variable degrees of homology between the pathogen and human proteins—28% [27], 20% [28], and 37% [29], respectively. In these instances, a low-degree homology is indicative of a possible target.

A potentially more useful technique for selecting antifungal targets from genome information is to use criteria critical to the drug development process. Spaltmann et al. [30] developed a computer-aided target selection scheme using five parameters as criteria for selection of target candidates: quality, occurrence, specificity, assay development, and homology. Table 2 lists general descriptions of the scoring parameters used in the algorithm. By assigning a numerical score for each parameter relative to the importance of the gene and adding up the scores, a total score is obtained. This value is then used to rank the genes. When the analysis was performed on *S. cerevisiae*, only two known antifungal targets, elongation factor 3 and H^+-transporting P-type ATPase, were in the top 25 candidate genes. A more comprehensive analysis including the pathogenic fungi must await completion of the genome sequencing efforts.

D. Differential Expression

In order for fungi to colonize and infect a host, it has been presumed that a unique and essential set of genes is required. Further, it is likely that some of these functions are not expressed, or are expressed at lower levels, during growth outside the host. Likewise, expression of certain genes in vivo may be detrimental to the establishment of infection. Therefore, examination of the set of genes differentially expressed (up- or downregulated) by the pathogen while growing in the host represents a potential source of antifungal targets. There are several approaches to identify such differentially expressed genes. Most require the use of an animal model where infectious material can be obtained and, to some extent, separated or differentiated from host tissue, or the use of in vitro models.

1. Whole Genome Expression Analysis

DNA microarrays composed of oligonucleotide, cDNA or expressed sequence tags (EST), and genomic DNA can be used to screen expression of genes under a variety of conditions. Once a complete genome sequence is available, oligonucleotide arrays present the best approach to DNA-based expression analysis. Sets of oligonucleotides representing each ORF sequence are used to build the array. Single–base pair (bp) mismatched oligos are placed adjacent to the wild-type oligo as a means of calibrating the hybridization signal. This approach reduces overlaps and redundancies that would be inherent in genomic or EST DNA arrays. To examine expression, mRNA is used to prepare linearly amplified cDNA. The labeled cDNA is then hybridized to the array and the signal quantified.

Wodicka et al. [4] used this technique to perform expression analysis in yeast grown in rich and minimal medium. Differentially and uniquely expressed genes were readily identified. Pertinent to this review, others have developed approaches utilizing microarrays for drug target validation and identification of secondary drug target effects [31], and for examining transcription patterns related to multidrug resistance in yeast [32]. In the former investigation, the method allowed for direct confirmation of drug targets as well as for detection of drug-dependent changes in gene expression occurring in pathways that are not directly related to the target of the drug. It was suggested that this approach will allow for greater efficiency in development of antifungals. Another way of applying this technique to define potential antifungal targets is through comparison of gene expression in cells obtained from the site of infection to the expression pattern of cells grown in culture. In theory, those genes differentially expressed or expressed solely in vivo may represent target candidates. A key aspect of employing this approach would be to eliminate, or render insignificant, signals due to contamination of fungal samples with host tissue or other resident microbial flora. This problem might be partially addressed by comparing the signal pattern of infected to noninfected host tissue. Alternatively, the array can be designed to limit the degree of homology in choosing the array composition. Similarly, methods to remove contaminants may reduce this as an obstacle to obtaining meaningful results.

Another set of techniques for whole-genome expression analysis is known as proteomics [33,34]. This involves the large-scale analysis of the proteins expressed under a given set of conditions by the test organism using two-dimensional gel electrophoresis and protein identification techniques. Proteomics provides information on the identity, quantity, state of modification, and association of proteins. This approach depends on the ability to prepare two-dimensional gel maps of the pathogen protein profile coupled with a number of different protein and peptide detection techniques. While this has been accomplished for *S. cerevisiae* [35–37] and coupled with the yeast protein database [38], it has only recently been applied to *C. albicans* [39].

The basic technology platform is the 2-D gel. Typically, a large gel set up capable of resolving 10,000 individual proteins is employed. Covalently immobilized pH gradients have greatly improved the run to run consistency of the gels [40]. Whole cell extracts or partially fractionated materials are separated first by isoelectric focusing and then by molecular mass, usually under denaturing conditions. The protein spots are detected by silver staining, or when very low detection limits are required, by labeling the proteins with fluorescent dyes. Image recording devices are used to record the profiles. To manage the large amount of information generated by these gels, database management systems are used to aid in the collection, storage, organization, and interpretation of the data.

To identify a specific protein on a gel, several techniques can be used. A protein spot can be excised from the gel or recovered after blotting, and subjected to protein sequencing. This approach is laborious and not amenable to whole-genome analysis. An alternative approach for whole-genome analysis has been to use peptide fingerprinting. The recovered protein spot is digested with a peptidase and the resulting fragment size is determined by MALDI or ESI mass spectrometry [39,41]. The peptide data are compared to a database derived from known protein sequences and ambiguities resolved by protein sequencing. When configured with automated systems, entire gels can be evaluated within a relatively short time. An emerging tool for peptide and protein analysis is capillary electrophoresis which, when coupled with tandem MS, permits femptomole detection levels. This technology should allow for the detection and analysis of proteins with expression levels below the limit of current instrumentation.

When applied to pathogenic fungi for the identification of new targets, this approach must be used with caution. There is not a perfect in vitro model of infection that can be used in place of animal model systems. However, material from infected tissue is difficult to obtain, and will always be contaminated with host proteins. In vitro models for selected functions require rigorous verification. Technically, the resolving power of the gel system limits the number of proteins that can be analyzed. Overlapping physicochemical properties of the proteins, microheterogeneity, sample load limits, and disparate expression levels all contribute to the level of resolution. Some of these limitations can be overcome by using narrow pH ranges on multiple gels or by examining subcellular fractions. It should also always be kept in mind that some proteins are inherently difficult to extract, particularly membrane and cell wall proteins.

2. Differential Display

One technique for identifying differentially expressed genes is differential display [42,43]. The technique, as applied to in vivo gene expression, involves the isolation of RNA from infected tissue, followed by preparation of single-stranded cDNA. PCR amplification with radiolabeled nucleotides is then performed using sets of primers designed such that the 3′ primers are anchored while the 5′ primers are random sequences. The products are resolved on denaturing acrylamide gels and visualized by autoradiography. Putative differentially expressed sequences are identified by comparison of results from in vivo and in vitro samples, and determining reproducibility in duplicate RT-PCR samples. Positive products are excised from the gel and subcloned into an appropriate plasmid vector for sequencing or are sequenced directly. It is important that each fragment isolated be verified as being fungal in origin and differentially expressed. In application, using animal host/pathogen models, the RNA samples are likely to be contaminated with host RNA as well as with resident microbial flora. Fragment sequence determination or Southern blots can be used to verify origin. Additionally, Northern blot analysis and/or RT-PCR techniques are used to verify differential expression. When using dimorphic fungi, such as C. albicans, it is important to examine expression in hyphal and yeast forms of the organism. For fungi, the technique has been applied in studies including C. albicans gene expression in a rat oral model of candidiasis [44], and macrophage-induced gene expression by Histoplama capsulatum [45].

3. Subtractive Hybridization

Another technique designed to detect differential gene expression is subtractive hybridization [46,47]. In this procedure, cDNA is prepared from RNA purified from cells grown under the target growth conditions, e.g., from fungal cells grown in a host animal. In parallel, cDNA is made using appropriate control cultures. The primer for the cDNA synthesis has a restriction site for an enzyme with a 4-bp recognition sequence, which after cleavage leaves a 4-bp overhang. The cDNAs are digested with the restriction enzyme to produce small fragments. The target cDNA is split into two pools and each pool ligated with a different oligonucleotide adapter. The adapters are made without 5′ phosphates so that only the 5′ end of the cDNA is modified. In addition, the adapter sequences incorporate convenient but different restriction site(s), and a 3′ end that is complementary to the restriction site in the cDNA synthesis. Reference cDNA is not modified with adapters. The two pools for target cDNA are hybridized with excess reference cDNA. Any target cDNAs common to the reference cDNA are hybridized with reference cDNA. Unique cDNAs are left single stranded. Without denaturing, a second hybridization is performed by pooling the two prehybridized target cDNA pools. In this step, the single-stranded unique sequences hybridize with their respective complement from the opposite pool. After filling in the DNA ends, the mixture is subjected to PCR amplification. The primers for PCR are designed to amplify only those cDNA hybrids that contain one strand from one target cDNA pool and the other strand from the second target cDNA pool. All other potential hybrids are not amplified. The resulting PCR products are subcloned into an appropriate vector. As for the differential display technique, each isolate must be verified as being of fungal origin (e.g., sequence or Southern blot) and differentially expressed (e.g., Northern blot or RT-PCR).

4. Gene Disruption/Tagging

Gene tagging has proved to be a useful tool to identify potential targets in bacterial pathogens. Referred to as signature-tagged mutagenesis, transposon-based mutant strains have been used to identify bacterial genes essential to infection [48,49]. This approach biases the genes identified, because those that are not required during infection are eliminated by the selection techniques employed. Because of the diploid nature of most fungal genomes, the direct application of the technique to fungal systems has been limited to only a few organisms [50]; however, a related technique relying on gene disruption has been widely used to assess the function of target genes [51,52]. The technique has been most often exploited in *C. albicans*, where selectable cassettes have been constructed for use in creating the desired heterozygotic or null mutant strain. More recently PCR-based protocols have been employed that offer some advantages over the original methodology, including faster mutant isolation and requirements for smaller disruption cassettes [53,54]. Use of these techniques has been most widespread in determining whether a gene is essential for growth or infectivity, where positive results have been interpreted as providing validation that a particular gene may be a suitable target for antifungal development. It is expected that gene disruption techniques will continue to be utilized for such purposes with *Candida* and other fungi [52].

E. Targets

As discussed earlier, a variety of methods are being utilized to identify novel targets suitable for detection and characterization of new antifungals. Examples of promising targets under investigation are presented below, but the topics discussed are not all-inclusive. The reader is referred to recent reviews for additional information [55–60].

1. Cell Wall Targets

Obvious potential antifungal targets include components required for synthesis and/or maintenance of the structural integrity of the cell wall, for which there is no counterpart in mammalian cells [61]. The fungal cell wall is usually composed primarily of glucan and chitin, which confer organization and rigidity to the wall, as well as of mannoproteins/mannan, which are found primarily in the outer cell wall as a more amorphous, fluid component. Importantly, it has been well documented that enzymes required in the biosynthesis of the former two moieties are necessary for normal cell growth and viability. As a result, the search for compounds with antifungal activity has historically been directed at the enzymes involved in their synthesis. For example, chitin synthase and β-(1-3)-glucan synthase served as targets leading to the identification of promising drug candidates; the lipopeptides of the echiocandin and papulacandin structural families in the latter case [2,61], and the peptide-nucleosides nikkomycins and polyoxins in the former instance [2,61]. To date, however, the compounds have not lived up to early expectations.

Mannoproteins have been shown to play a role in the antifungal properties of the pradimicin drug group [62]. Rather than serving as the direct target, mannoproteins interact with the agent in an association that is Ca^+ dependent. Antifungal activity of the resulting complex is through effects to the structural integrity of the plasma membrane [62,63]. Importantly, the class of drugs has been reported to have activity against a wide array of fungi [62,63].

More recently, attention has turned to the possibility that the enzymes required for mannosylation can be exploited as targets. The fact that mechanisms of O-linked glycosylation differ significantly between fungal and mammalian cells reinforces this notion. Since comprehensive analyses of many counterpart pathways and companion genes have been performed in S. cerevisiae, investigators have begun analysis of analogous genes by exploiting this knowledge base. In particular, C. albicans gene counterparts to S. cerevisiae PMT1 [64], MNT1 [65], and MNN9 [66] have been described. In each investigation strains disrupted in both alleles were constructed with the eventual aim of determining any resultant phenotypic effects. Disruption of CaPMT1 resulted in reduced growth rates, tendency for aggregate formation, and loss of CaPmt1 activity. In addition, Pmt1 mutants were incapable of undergoing hyphal morphogenesis, showed reduced adherence in in vitro assays, and were hypersensitive to several compounds with antifungal activity (e.g., hygromycin-B, calcofluor white). Importantly, the latter phenotypic changes are not observed in strains of S. cerevisae in which PMT1 has been disrupted. This reinforces the concept that gene function(s) determined in S. cerevisiae cannot be ascribed a priori to the "identical" gene in pathogenic fungi. Failure to recognize such functional differences may, therefore, hinder the detection of potential antifungal targets. Finally, using a mouse model of systemic candidiasis it was demonstrated that disruption of both CaPMT1 alleles resulted in loss of strain virulence, and inability to colonize the kidney.

Similar analysis were performed during study of CaMNT1 [65]. CaMnt1 is an α-1,2-mannosyl transferase that is responsible for addition of the second mannose residue in a trimannose structure finally found as O-linked mannan in C. albicans. It was demonstrated that the disruption of both CaMNT1 alleles resulted in a dramatic alteration in adherence capabilities both in vitro and in vivo. Virulence studies in both guinea pig and mouse models of systemic candidiasis showed a CaMNT1 null mutant strain to be only relatively avirulent when compared to the wild type. Likewise, tissue census demonstrated that despite relative avirulence, internal organs could still be colonized, albeit at lower frequency and in lower numbers [65].

In the case of CaMNN9, a gene required for the synthesis of N-linked outer chain mannan, abolishment of gene function also resulted in major phenotypic changes including aberrant hyphal formation and osmotic sensitivity. Although virulence studies were not reported, the phenotypic changes observed were interpreted as likely being incompatible with normal pathogenicity [66]. In summary, studies concerning all of these genes offer encouragement toward their eventual use as antifungal targets.

2. Fatty Acid Synthase

Housekeeping enzymes are often overlooked as potential drug targets. This is with good reason as there is usually a similar, structurally related mammalian enzyme, and it would thus be expected that agents directed at such targets would be toxic. On the other hand, some fungal housekeeping enzymes differ significantly structurally from their human counterpart and may offer an underexamined source of untapped targets. One such enzyme that meets this criterion is fatty acid synthase (Fas). Fas has been isolated and characterized from a diverse group of organisms, and comparison of the fungal and mammalian enzymes shows that major structural differences exist. The fungal enzyme is comprised of two nonidentical polypeptides each of about 200 kDa, designated α and β [67]. The two polypeptides contain the seven component activities necessary for fatty acid synthesis, and the active enzyme is an $\alpha_6\beta_6$ hexamer. In contrast, mammalian Fas is composed of a single polypeptide and is active as a homodimer configured in a head-to-tail arrangement [68]. All activities are present on each polypeptide. It has also been reported that Fas activity is required for successful *C. albicans* infection in both the mouse systemic and rat oroesophageal models of candidiasis, suggesting the enzyme can be exploited as a valid drug target [69].

As a result of the above findings, searches for compounds with differential effects on the fungal and human enzymes have been performed [26]. Several derivatives of the antilipogenic agent cerulenin were reported to differentially inhibit fungal Fas, but not significantly affect human Fas. These agents were subsequently shown to be inactivated in human serum and it is therefore unlikely that they will be of use as antifungal agents. Other compounds active primarily against the fungal enzyme have been isolated from natural product extracts and remain under study concerning their potential efficacy [71].

3. Two-Component Signal Transduction Pathways

Fungal cells adapt to environmental changes through altered gene expression, and in general, responses to external stimuli are mediated via signal transduction pathways. Of particular interest for exploitation as possible drug targets are certain two-component signal transduction regulatory systems, as such systems have not been observed in higher eukaryotes. First described in prokaryotes, two-component regulatory systems play a key role in allowing bacteria to detect and adjust to changes in their environment [72]. More recently, two-component systems have been described in fungi, where they mediate responses to heat shock, oxidative stress, and changes in osmolarity, among others. Two-component regulator systems consist of a histidine protein kinase, which serves as sensor protein, and a protein that serves as a regulator of the internal response [72]. Upon activation, the sensor autophosphorylates a histidine residue that is found within a conserved amino acid sequence. The phosphorylated sensor in turn serves as a phosphodonor to a conserved aspartate in the response regulator, which modulates regulator activity. This is discussed further by Calera and Calderone in this volume.

Most importantly, two-component regulator systems have not been observed in mammalian cells, suggesting such systems may be attractive target candidates. A number of laboratories have addressed this issue. Calera et al. [73] have studied the role of both *CaHK1* and *CaSSK1*, sensor and response regulator genes, respectively, in the pathogenesis and developmental programs of *C. albicans*. In both cases it was demonstrated that null strains in either gene were avirulent in a mouse model of systemic candidiasis. Analysis of the developmental patterns of the null strains also showed defects in hyphal development. Other groups have isolated a different two-component histidine kinase, designated *COS1* [74] or *CaNIK1* [75]. Studies with deletion mutants by these groups have demonstrated a role for the gene products in hyphal development. Since *C. albicans* germination is important to pathogenesis, the results strengthen the case that these systems may be useful in drug discovery. In this regard, the efficacy of three inhibitors of bacterial histidine kinases has been evaluated against *S. cerevisiae* and *C. albicans* [76]. All three drugs were found to inhibit growth of both organisms in growth assays; however, it was shown that one of the compounds failed to inhibit *S. cerevisiae* Sln1p (histidine kinase), and furthermore the compounds still showed antifungal activity against a *SLN1* deletion strain. Thus, it was concluded that antifungal activity was not directly related to histidine kinase inhibition. Nonetheless, such systems may prove to be useful antifungal targets.

4. Plasma Membrane H$^+$-ATPases

Plasma membrane H$^+$-ATPases serve as proton pumps and through their action an electrochemical gradient is generated that is essential for pH regulation as well as nutrient uptake [77]. The enzymes belong to the P class of ATPases, and possess several attributes considered attractive for an antifungal target including location at the cell surface and their performance of an essential function for cell viability. Both the enzyme and corresponding genes (*PMA*) have been analyzed from a number of fungi including *S. cerevisae* [78,79], *C. albicans* [80], and *Aspergillus nidulans* [81], thereby providing a large database for comparative studies. Importantly, despite the fact that mammalian cells also specify P-type ATPases, structural analysis has shown significant differences between the fungal and mammalian enzymes. Concerning drug development, because of the large amount of structural data available it is possible that highly specific antifungal drugs can be designed. More traditional compound screening from natural products may also prove productive. In this regard, a complex carbohydrate from the cell wall of *Mucor rouxii* with apparent activity against plasma membrane H$^+$-ATPases from several *Candida* spp. has recently been investigated [82]. It was shown that the compound was fungicidal and inhibited proton pumping of all strains tested, while other antifungal drugs that affect membrane integrity (i.e., amphotericin-B, itraconazole) had no effect. Further work, however, demonstrated the affects on H$^+$-ATPase activity was only indirect. Nonetheless, the result lends further evidence toward the validity of the protein as a potential drug target.

5. Myristoyl-CoA:Protein N-Myristoyltransferase

Myristoyl-CoA:protein N-myristoyltransferase (Nmt) is responsible for covalent addition of myristate to the N-terminal glycine of various proteins in eukaryotic cells. Such modification is often necessary for protein function, and thus inhibition of Nmt can prove to be a lethal event. The enzyme has also been investigated in *C. albicans* where it has been shown to modify a small number of proteins [83]. CaNmt shares ~47% sequence identity with the human counterpart [83]. Analysis has shown that the acylCoA binding site of the *C. albicans* and human Nmt are highly conserved; however, significant differences have been reported in peptide binding sites. This has led to the suggestion that Nmt may prove a useful antifungal drug target as compounds might be designed with differential effects against the respective enzymes. In this regard, a *C. albicans NMT* null mutant strain has been constructed and it was shown that the gene is essential for cell viability [84]. Furthermore, experiments demonstrated the null strain was avirulent in the mouse model of systemic candidiasis, thereby providing additional evidence for the validity of CaNmt as a suitable drug target. Studies that will be aimed at identification of lead compounds with activity against Nmt have begun. In particular, an in vivo assay to measure CaNmt activity has been reported [85]. The method detects N-myristoylization of *C. albicans* ADP-ribosylation factor (CaArf) by a shift in position of the protein in a gel mobility shift assay. The investigators have been successful in using the assay to demonstrate the efficacy of a peptidomimetic inhibitor of CaNmt, and more recently nonpeptide CaNmt inhibitors have also been reported [86]. Finally, the structure of CaNmt has been obtained [87,88], and these data should facilitate the design of more potent compounds effective against the enzyme.

6. Prenylation

Protein posttranslational modification by prenylation has been proposed as a potential antifungal target [89], since prenylation can be vital for protein function and for correct membrane localization of certain proteins. Enzymes that mediate prenylation include farnesyltrasferase (FTase) or geranylgeranyltrasferase 1 (GGTase 1), and in fungi, most studies concerning the enzymes have been performed with *S. cerevisiae* [89]. Geranylgeranylated proteins are often responsible for modification of GTP-binding proteins required in cell wall/membrane-mediated processes (e.g., morphogenesis, signal transduction, etc.). In addition, it has been demonstrated that one substrate of GGTase1, Rhop1 has been implicated in several roles in morphogenesis including serving as the GTP-binding regulatory subunit of 1,3-β-D-glucan synthase [90], and at least this function is conserved in *C. albicans* [90,91]. For these reasons, attention has turned to examining *C. albicans* GGTase. Studies to date have been limited to cloning of the genes encoding the GGTase-1 subunit genes *CaRAM* and *CaCDC4*. Key results have shown that the encoded polypeptides share 42% and 34% identity to the *S. cerevisiae* counterparts and 30% identity to the human homologues [92]. The respective genes were expressed in *S. cerevisiae*, and coexpression resulted in a functional protein; however, expression of functional heterodimers was not detected. Finally, the investigators suggest that analysis of sequence data will lead to discovery of inhibitors with appropriate specificity for the fungal versus human enzyme.

7. Sphingolipid Biosynthesis

Sphingolipids have been extensively studied in *S. cerevisiae*, and 13 known genes have been identified as being required in their biosynthesis, with others likely to be discovered [93,94]. Biochemical pathways are at least partially shared among various fungi including the pathogens *C. albicans*, *Cryptococcus neoformans*, *Aspergillus* spp., and *Histoplama capsulatum*. Significantly, certain fungal enzymes responsible for synthesis of phosphoinositol-containing sphingolipids are present in fungi but not in humans. In addition, it has been demonstrated that sphingolipids are essential for fungal growth, required in an array of functions including resistance to various environmental stresses, and may play a role as second messengers regulating extracellular signal transduction pathways [94]. For these reasons, attention has been given to identifying key enzymes that may serve as antifungal targets. In this regard, inositol phosphorylceramide synthase (IPCase) has garnered the most attention [95]. The enzyme catalyzes the transfer of phosphoinositol from phosphatidylinositol to ceramide, yielding IPC. IPCase has been detected in all fungi and plants in which appropriate investigations have been performed, but IPC is not found in mamalian cells. The gene (*AURI*) has been characterized from several fungi, and furthermore, it has been demonstrated that the target of the antifungal agent aureobasidin A is IPCase [95,96]. In addition to aureobasidin-A, other IPCase inhibitors have been reported recently. The agent khafrefungin [97], as well as the macrolides rustimicin and galbonolide B [98] inhibit sphinolipid biosynthesis, and showed antifungal activity against *Candida* spp. and *C. neoformans*, but not *A. fumigatus*. It was also reported that khafrefungin did not inhibit sphingolipid synthesis in mammalian cells [97]. These results provide further validation of the exploitation of the enzyme as an antifungal target.

8. Elongation Factors

Translation in fungi requires the participation of three soluble elongation factors—EF-1α, EF-2, and EF-3. EF-3 is unique to, and essential for, the growth of fungi, and as such has generated interest as a potential antifungal target [99,100]. The protein has a MW of approximately 116 kDa and facilitates ribosomal binding of the ternary complex, perhaps by direct interaction with EF-1α, and also furnishes the ATPase activity required [101]. While data suggest its attractiveness as a target, it has yet to be fully exploited, as no suitable inhibitors have been reported. In contrast, EF-2 is conserved in eukaryotes and because of similarities between the fungal and mammalian proteins, might not usually be thought of as a possible drug target; however, studies have shown it to be the target of a class of compounds designated sordarins, which have specificity for fungal EF-2 [102]. Sodarins have been known as inhibitors of protein synthesis since

1970 [103], but EF-2 has only recently been identified through both biochemical and genetic analysis as the target [102,103]. In particular, EF-2 facilitates ribosomal translocation during protein synthesis, and it is precisely this process that was inhibited in the former studies. The latter studies correlated mutations conferring sordarin resistance in *S. cerevisiae* to genes encoding EF-2. More recent studies have demonstrated that mutations to the ribosomal protein L10e also confer sordarin resistance, suggesting a functional relationship between EF-2 and L10e [104]. These results have led to experiments using sordarin as template for producing more active analogs [105]. Likewise, additional sordarin-related compounds have been identified in high-throughput screens of natural products [106].

9. Secreted Aspartic Proteinases

Aspartic proteinases are widely distributed and have been implicated to play a role in virulence for a variety of organisms [107]. They have been particularly well studied in *C. albicans* where a large gene family encoding secreted aspartyl proteinases (SAP) has been described [108]. This work has led to an accumulation of evidence suggesting that the enzymes play an important role in pathogenesis of the organism including (1) *SAP* null mutants have reduced virulence and adherence capabilities [109–111], and (2) differential protease expression correlates with virulence [112]. Differential expression is particularly intriguing in the implication that different Sap family members are required during phenotypic switching, the yeast to hyphal morphologic transition and in vivo at different sites of infection (e.g., systemic vs. mucosal) [113]. At the same time this diversity suggests that use of Saps as targets for the development of antifungal drugs could prove difficult as such agents may not prove equally effective against all family members, and similarly, activity of less affected species may compensate for reduced activity of others. Nonetheless, progress on at least two fronts suggests that antifungals directed against Saps may be of therapeutic value.

First, the three-dimensional structure of Sap2, complexed with an inhibitor, has been determined [114], and this information has been applied to understanding the relationship between the enzyme family. The investigations identified specific differences between certain *C. albicans* Saps that allowed their grouping into classes, while at the same time revealing more generally conserved features. This may allow for more effective drug design that might minimize concerns expressed above. In addition, clear distinctions between the fungal and some mammalian aspartic proteinases suggest that an agent with an appropriate therapeutic index may be designed.

Second, it has been noted that a reduction in oropharyngeal candidiasis in AIDS patients has accompanied patient usage of HIV aspartic protease inhibitors [115,116]. In order to examine the basis of this observation, investigations were performed that demonstrated four inhibitors of the HIV enzyme also inhibited *C. albicans* Saps 1–3, and that ritonavir and saquinavir, but not indinavir, reduced adherence capabilities [115]. Other experiments yielded similar results and in addition demonstrated that saquinavir could be candicidal at high concentrations; however, none of the agents effected *C. albicans* viability at concentrations at or below 0.1 mg/mL [116]. The in vivo efficacy of drug usage remains under study.

10. Topoisomerases

Topoisomerases are ubiquitous enzymes found in both prokaryotes and eukaryotes. They are responsible for resolution of topological constraints that arise during such processes as DNA replication, transcription, recombination, etc. [117]. There are two enzyme types: type I catalyzes transient nicks in one DNA strand resulting in relaxation, while type II mediates topological crossing of two double-strand DNA segments as a result of two double-stranded breaks [117].

Topoisomerases have already served as useful targets for several anticancer [118,119] and anti-bacterial drugs [120]. Action of such drugs can result in stabilization of the enzyme-DNA cleavage complex, and eventually lead to cell death [117,119]. Studies of topoisomerase I and II as possible antifungal targets have been performed with fungal pathogens, primarily *C. albicans* [121,122], and more recently with *C. neoformans* [123]. In the former case, it was found that *CaTOP1* was not essential for cell growth; however, effects on cell morphology and rate of cell growth were noted for a *CaTOP1* null mutant [124]. In addition, it was shown that the strain was relatively avirulent in the mouse model of systemic candidiasis, but tissue census from the kidney revealed no significant decrease in colonization when compared to the wild type. In contrast, *TOP1* was deemed essential for survival of *C. neoformans*, as reported by Del Poeta et al. [125]. It has also been observed that both the *C. albicans* and *C. neoformans* topoisomerases contain a linker amino acid insertion that differs significantly from that of the human homolog. Such structural differences suggest that drugs with differential effects on the human and fungal enzyme may be found. In this regard, activity of the drug camptothecin and derivatives are active and synergistic with known antifungals against the *C. neoformans* enzymes [125]. Other studies have identified a number of potential lead compounds effective against the *C. albicans* topoisomerase I [126]. Similar studies with *C. albicans* topoisomerase II have been reported that also suggest the enzyme may serve as an antifungal target [127].

11. Group I Intron Splicing

Group I introns are ribozymes capable of mediating their own excision from precursor RNA [128]. Group I introns have been found in a variety of organisms including some fungi, but have not been reported in the human genome. In this regard, they have been reported in rRNA genes of *Pneumocytis carinii* [129] as well as rRNA genes of ~40% of strains of *C. albicans* and all *C. dubliniensis* strains examined [130,131]. Since the inability to remove introns from rRNA might be expected to be deleterious, it has been suggested that the ribozyme might serve as an antifungal target. Liu et al. [132] have recently shown that pentamidine and pentamidine analogues inhibit splicing of the group 1 intron found in *P. carinii* 26S RNA in in vitro experiments and speculated that such inhibition could at least in part explain the mechanism of action of the drug in vivo. Similar experiments have been performed using *C. albicans* rRNA with the same results [133]. In this study it was also shown that pentamidine inhibited growth in culture of intron-containing strains, greater than the percentage inhibition observed for strains that did not contain the intron. The results lend support to the idea that group 1 introns might prove useful in drug discovery; however, since (1) multiple copies of rDNA are found in *C. albicans* as well as in other fungi, (2) not every gene copy may harbor the intron, and (3) only a percentage of *C. albicans* strains harbor the intron, it is not clear that any such drug would prove practical as a therapeutic agent. Of course, some of these concerns may prove different for other fungi, and/or the discovery of uniform intron distribution in another essential gene may enhance the appeal of the target.

12. Virulence Genes

The array of fungal genes that are required for virulence have come under increasing scrutiny as potential antifungal targets. The availability of methods discussed above for detection of genes or their products required for virulence has led to many investigations to define fungal virulence factors or genes that more indirectly contribute to virulence. Defining such targets and then discovering agents that attenuate virulence but that are not fungicidal, may have the added advantage of reducing problems related to development of drug resistance. An in-depth discussion of this topic is beyond the scope of this chapter, and the reader is referred to a recent review for further information [55].

III. CONCLUDING REMARKS

Clearly, the need exists for additional safe and effective antifungal drugs. Many attractive targets, some of which have been discussed, are being exploited in the search for novel agents. Identification of novel antifungal targets has suffered due in part to the similarity between the fungal and mammalian cell. However, the application of new methodologies coupled with the skill and ingenuity of investigators in the field points to significant advances in the near future.

REFERENCES

1. MA Ghannoum, LB Rice. Antifungal agents: mode of action, mechanisms of resistance, and correlation of these mechanism with bacterial resistance. Clin Microbiol Rev 12:501–517, 1999.
2. B DiDomenico. Novel antifungal drugs. Curr Opin Microbiol 2:509–515, 1999.
3. MM Cowan. Plant products as antimicrobial agents. Clin Microbiol Rev 12:564–582, 1999.
4. L Wodicka, H Dong, M Mittmann, M-H Ho, DJ Lockhart. Genome-wide expression monitoring in *Saccharomyces cerevisiae*. Nat Biotechnol 15:1359–1367, 1997.
5. G Ramsay. DNA chips: state-of-the-art. Nat Biotechnol 16:40–44, 1998.
6. C Debouck, PN Goodfellow. DNA microarrays in drug discovery and development. Nat Genet 21: 48–50, 1999.
7. MJ Martin, JL DeRisi, HA Bennett, VR Iyer, MR Meyer, CJ Roberts, R Stoughton, J Burchard, D Slade, H Dai, DE Bassett Jr, LH Hartwell, PO Brown, SH Friend. Drug target validation and identification of secondary drug target effects using DNA microarrays. Nat Med 4:1293–1301, 1998.
8. MR Wilkins, C Pasquali, RD Appel, K Ou, O Golaz, J-C Sanchez, JX Yan, AA Gooley, G Hughes, I Humphery-Smith, KL Williams, DF Hochstrasser. From proteins to proteomes: large scale protein identification by two-dimensional electrophoresis and amino acid analysis. Biotechnology 14:61–65, 1996.
9. PH O'Farrell. High resolution two-dimensional electrophoresis of proteins. J Biol Chem 250: 4007–4021, 1975.
10. T Barrett, HJ Gould. Tissue and species specificity of non-histone chromatin proteins. Biochim Biophys Acta 294:165–170, 1973.
11. VC Wasinger, SJ Cordwell, A Poljak, JX Yan, AA Gooley, MR Wilkins, M Duncan, R Harris, KL Williams, I Humphery-Smith. Progress with gene-product mapping of the Mollicutes: *Mycoplasma genitalium*. Electrophoresis 16:1090–1094, 1995.
12. A Shevchenko, ON Jensen, AV Podtelejnikov, F Sagiocco, M Wilm, O Vorm, P Mortensen, A Shevchenko, N Boucherie, M Mann. Linking genome and proteome by mass spectrometry: large-scale identification of yeast proteins from two-dimensional gels. Proc Natl Acad Sci USA 93: 14440–14445, 1996.
13. WL Stuadenbauer, E Orr. DNA gyrase: affinity chromatography on novobiocin-Sepharose and catalytic properties. Nucleic Acids Res 9:3589–3603, 1981.
14. JA Radding, SA Heidler, WW Turner. Photoaffinity analog of the semisynthetic echinocadin LY303366: identification of echinocandin targets in *Candida albicans*. Antimicrob Agents Chemother 42:1187–1194, 1998.
15. CE Bulawa, DW Miller, LK Henry, JM Becker. Attenuated virulence of chitin-deficient mutants of *Candida albicans*. Proc Natl Acad Sci USA 92:10570–10574, 1995.
16. M Mio, T Yabe, M Sudoh, Y Satoh, T Nakajima, M Arisawa, H Yamada-Okabe. Role of three chitin synthase genes in the growth of *Candida albicans*. J Bacteriol 178:2416–2419, 1996.

17. MA Ghannoum, B Spellberg, SM Saporito-Irwin, WA Fonzi. Reduced virulence of *Candida albicans* *PHR1* mutants. Infect Immun 63:4528–4530, 1995.

18. F De Bernardis, FA Muhlschlegal, A Cassone, WA Fonzi. The pH of the host niche controls gene expression in and virulence of *Candida albicans*. Infect Immun 66:3317–3325, 1998.

19. M Saporito-Irwin, CE Birse, PS Sypherd, WA Fonzi. PHR1, a pH-regulated gene of *Candida albicans*, is required for morphogenesis. Mol Cell Biol 15:601–613, 1995.

20. FA Muhlschlegel, WA Fonzi. PHR2 of *Candida albicans* encodes a functional homolog of the pH-regulated gene PHR1 with an inverted pattern of pH-dependent expression. Mol Cell Biol 17: 5960–5970, 1997.

21. DT Moir, KJ Shaw, RS Hare, GF Vovis. Genomics and antimicrobial drug discovery. Antimicrob Agents Chemother 43:439–446, 1999.

22. A Goffeau, BG Barrell, H Bussey, RW Davis, B Dujon, H Feldman, et al. Life with 6000 genes. Science 274:546–567, 1996.

23. F Arigoni, F Talabot, M Peitsch, MD Edgerton, E Meldrum, E Allet, R Fish, T Jamotte, M-L Curchod, H Loferer. A genome-based approach for the identification of essential bacterial genes. Nat Biotechnol 16:851–856, 1998.

24. AR Mushegian, EV Koonin. A minimal gene set for cellular life derived by comparison of complete bacterial genomes. Proc Natl Acad Sci USA 93:10268–10273, 1996.

25. L Tatusov, EV Koonin, DJ Lipman. A genomic perspective on protein families. Science 278: 631–637, 1997.

26. SE Broedel Jr, X-J Zhao, RL Cihlar. Fatty acid synthase as a target in the development of new antifungals. Recent Res Dev Antimicrob Agents Chemother 1:25–33, 1996.

27. I Schweitzer, AP Dicker, JR Bertino. Dihydrofolate reductase as a therapeutic target. FASEB J 4: 2441–2352, 1990.

28. K Hoshino, K Sato, T Une, Y Osada. Inhibitory effects of quinolones on DNA gyrase of *Escherichia coli* and topoisomerase II of fetal calf thymus. Antimicrob Agents Chemother 33:1816–1818, 1989.

29. Y Aoki, F Yoshihara, M Kondoh, Y Nakamura, N Nakayama, M Arisawa. RO 09-1470 is a selective inhibitor of P-450 lanosterol C-14 demethylase of fungi. Antimicrob Agents Chemother 37: 2662–2667, 1993.

30. F Spaltmann, M Blunck, K Ziegelbauer. Computer-aided target selection—prioritizing targets for antifungal drug discovery. Drug Disc Today 4:17–26, 1999.

31. MJ Marton, JL DeRisi, HA Bennett, VR Iyer, MR Meyer, CJ Roberts, R Stoughton, J Burchard, D Slade, H Dai, DE Bassett Jr, LH Hartwell, PO Brown, SH Friend. Drug target validation and identification of secondary drug target effects using DNA microarrays. Nat Med 4:1293–301, 1998.

32. J Derisi, B van den Hazel, P Marc, E Balzi, P Brown, C Jacq, A Goffeau. Genome microarray analysis of transcriptional activation in multidrug resistance. FEBS Lett 470:156–160, 2000.

33. M Mann. Quantitative proteomics? Nat Biotechnol 17:954–955, 1999.

34. A Persidis. Proteomics: an ambitious drug development platform attempts to link gene sequence to expressed phenotype under various physiological states. Nat Biotechnol 16:393–394, 1998.

35. H Boucherie, G Dujardin, M Kermorgant, C Monribot, P Sonimski, M Perrot. Two-dimensional protein map of *Saccharomyces cerevisiae*: construction of a gene-protein index. Yeast 11:601–603, 1995.

36. I Mailet, G Lagniel, M Perrot, H Boucherie, J Babarre. Rapid identification of yeast protein in two-dimensional gels. J Biol Chem 271:10263–10270, 1996.

37. F Sagilocco, J-C Guillemot, C Monribot, J Capdevielle, M Perrot, E Ferran, P Ferrara, H Boucherie. Identification of proteins of the yeast protein map using genetically manipulated strains and peptide mass fingerprinting. Yeast 12:1519–1533, 1996.

38. JI Garrels, B Futcher, R Kobayashi, GI Latter, B Schwender, T Volpe, JR Warner, CS McLaughlin. Protein identifications for a *Saccharomyces cerevisiae* protein database. Electrophoresis 15: 1466–1486, 1994.

39. M Niimi, RD Cannon, BC Monk. *Candida albicans* pathogenicity: a proteomic perspective. Electrophoresis 20:2299–2308, 1999.

40. PA Haynes, SP Gygi, D Figeys, R Aebersold. Proteome analysis: biological assay or data archive. Electrophoresis 19:1862–1871, 1997.

41. P James. Of genomes and proteomes. Biochem Biophys Res Commun 231:1–6, 1997.

42. P Liang, AB Pardee. Differential display of eukaryotic RNA by means of the polymerase chain reaction. Science 257:967–971, 1992.

43. P Liang, AB Pardee. Recent advances in differential display. Curr Opin Immunol 7:274–280, 1995.

44. X-J Zhao, JT Newsome, RL Cihlar. Up-regulation of two *Candida albicans* genes in the rat model of oral candidiasis detected by differential display. Microb Pathogen 25:121–129, 1998.

45. S Colonna-Romano, A Porta, A Franco, GS Kobayashi, B Maresca. Identification and isolation of genes differentially expressed by *Histoplasma capsulatum* during macrophage infection. Microb Pathogen 25:55–66, 1998.

46. L Diatchenko, Y-FC Lau, AP Campbell, A Chenchik, F Moqadam, B Huang, S Lukyanov, K Lukyanov, N Gurskaya, ED Sverdlov, PD Siebert. Suppression subtractive hybridization: a method for generating differentially regulated or tissue-specific cDNA probes and libraries. Proc Natl Acad Sci USA 93:6025–6030, 1996.

47. NG Gurskaya, L Diatchenko, A Chenchik, PD Siebert, GL Khaspekov, K Lukyanov, LL Vagner, OD Ermolaeva, S Lukyanov, ED Sverdlov. Equalizing cDNA subtraction based on selective suppression of polymerase chain reaction: cloning of Jurkat cell transcripts induced by phytohemaglutinin and phorbol 12-myristate 13-acetate. Anal Biochem 240:90–97, 1996.

48. M Hensel, JE Shae, C Gleeson, MD Jones, E Dalton, DW Holden. Simultaneous identification of bacterial virulence genes by negative selection. Science 269:400–403, 1995.

49. JM Mei, F Nourbakhsh, CW Ford, DW Holden. Identification of *Staphylococcus aureus* virulence genes in a murine model of bacteriaemia using signature-tagged mutagenesis. Mol Microbiol 26:399–407, 1997.

50. DP Kontoyiannis. Genetic analysis of azole resistance by transposon mutagenesis in *Saccharomyces cerevisisae*. Antimicrob Agents Chemother 43:2731–2735, 1999.

51. WA Fonzi, MY Irwin. Isogenic strain construction and gene mapping in *Candida albicans*. Genetics 134:717–728, 1993.

52. K Kwon-Chung. Gene disruption to evaluate the role of fungal candidate virulence genes. Curr Opin Microbiol 1:381–389, 1998.

53. RB Wilson, D Davis, BM Enloe, AP Mitchell. A recyclable *Candida albicans* URA3 cassette for PCR product directed gene disruptions. Yeast 15:65–70, 2000.

54. RB Wilson, D Davis, AP Mitchell. Rapid hypothesis testing with *Candida albicans* through gene disruption with short homology regions. J Bacteriol 181:1868–1874, 1999.

55. JR Perfect. Fungal virulence genes as targets for antifungal chemotherapy. Antimicrob Agents Chemother 40:1577–1583, 1996.

56. NH Georgopapadakou, TJ Walsh. Antifungal agents: chemotherapeutic targets and immunologic strategies. Antimicrob Agents Chemother 40:297–291, 1996.

57. AH Groll, SC Piscitelli, TJ Walsh. Clinical pharmacology of systemic antifungal agents: a comprehensive review of agents in clinical use, current investigational compounds, and putative targets for antifungal drug development. Adv Pharmacol 44:343–500, 1998.

58. VT Andriole. Current and future antifungal therapy: new targets for antifungal agents. J Antimicrob Chemother 44:151–162, 1999.

59. JM Fostel, PA Larty. Emerging novel antifungal agents. Drug Disc Today 5:25–32, 2000.

60. CC Chiou, AH Groll, TJ Walsh. New drugs and novel targets for treatment of invasive fungal infections in patients with cancer. Oncologist 5:120–135, 2000.

61. NH Georgopapadakou, JS Tkacz. The fungal cell wall as drug target. Trends Microbiol 3:98–104, 1995.

62. TJ Walsh, N Giri. Pradimicins: a novel class of broad-spectrum antifungal compounds. Eur J Clin Microbiol Infect Dis 16:93–97, 1997.

63. KL Oakley, CB Moore, DW Denning. Activity of pradimicin BMS-181184 against *Aspergillus* spp. Int J Antimicrob Agents 12:267–269, 1999.

64. C Timpel, S Strahl-Bolsinger, K Ziegelbauer, JF Ernst. Multiple functions of Pmt1p-mediated protein O-mannosylation in the fungal pathogen *Candida albicans*. J Biol Chem 273:20837–46, 1998.

65. ET Buurman, C Westwater, B Hube, AJP Brown, FC Odds, NAR Gow. Molecular analysis of CaMnt1p, a mannosyl transferase important for adhesion and virulence of *Candida albicans*. Proc Natl Acad Sci USA 95:7670–7675, 1998.

66. SB Southard, CA Specht, C Mishra, J Chen-Weiner, PW Robbins. Molecular analysis of the *Candida albicans* homolog of *Saccharomyces cerevisiae* MNN9, required for glycosylation of cell wall mannoproteins. J Bacteriol 181:7439–7438, 1999.

67. E Schweizer, G Muller, LM Roberts, M Schweizer, J Rosch, P Weisner, J Beck, D Stratman, I Zauner. Genetic control of fatty acid biosynthesis and structure in lower fungi. Fat Sci Technol 89: 570–577, 1997.

68. S Smith. The animal fatty acid synthase: one gene, one polypeptide, seven enzymes. FASEB J 8: 1248–1259, 1994.

69. X-J Zhao, GE McElhaney-Feser, MJ Sheridan, SE Broedel Jr, RL Cihlar. Avirulence of *Candida albicans* FAS2 mutants in a mouse model of systemic candidiasis. Infect Immun 65:829–832, 1997.

70. X-J Zhao, GE McElhaney-Feser, WH Bowen, MF Cole, SE Broedel Jr, RL Cihlar. Requirement for the *Candida albicans* FAS2 gene for infection in a rat model of oropharyngeal candidiasis. Microbiology 142:2509–3514, 1996.

71. SE Broedel, RL Cihlar. Manuscript in preparation.

72. TW Grebe, JB Stock. The histidine protein kinase superfamily. Adv Microb Physiol 41:139–227, 1999.

73. JA Calera, R Calderone. Histidine kinase two-component signal transduction proteins of *Candida albicans* and the pathogenesis of candidosis. Mycoses 42(suppl 2):49–53, 1999.

74. LA Alex, C Korch, DP Selitrennikoff, MI Simon. COS1 a two-component histidine kinase that is involved in hyphal development in the opportunistic pathogen *Candida albicans*. Proc Natl Acad Sci USA 95:7069–7073, 1998.

75. T Yamada-Okabe, T Mio, N Ono, Y Kashima, M Matsui, M Arisawa, H Yamada-Okabe. Roles of three histidine kinase genes in hyphal development and virulence of the pathogenic fungus *Candida albicans*. J Bacteriol 181:7243–7247, 1999.

76. RJ Deschenes, H Lin, AD Ault. Antifungal properties and target evaluation of three putative bacterial histidine kinase inhibitors. Antimicrob Agents Chemother 43:1700–1703, 1999.

77. B Monk, DS Perlin. Fungal plasma membrane proton pumps as promising new antifungal targets. Crit Rev Microbiol 20:209–223, 1994.

78. BC Monk, AB Mason, G Abramochkin, JE Haber, D Seto-Young, DS Perlin. The yeast plasma membrane proton pumping ATPase is a viable antifungal target. I. Effects of the cysteine-modifying reagent omeprazole. Biochim Biophys Acta 1239:81–90, 1995.

79. D Seto-Young, B Monk, AB Mason, DS Perlin. Exploring an antifungal target in the plasma membrane H$^+$-ATPase of fungi. Biochim Biophys Acta 1326:249–256, 1997.

80. BC Monk, AB Mason, TB Kardos, DS Perlin. Targeting the fungal plasma membrane proton pump. Acta Biochim Pol 42:481–96, 1995.

81. E Reoyo, EA Espeso, MA Penalva, T Suarez. The essential *Aspergillus nidulans* gene pmaA encodes an homologue of fungal plasma membrane H(+)-ATPases. Fungal Genet Biol 23:288–99, 1998.

82. AM Ben-Josef, EK Manavathu, D Platt, JD Sobel. Proton translocating ATPase mediated fungicidal activity of a novel complex carbohydrate: CAN-296. Int J Antimicrob Agents 13:287–295, 2000.

83. JA Sikorski, B Devadas, ME Zupec, SK Freeman, DL Brown, HF Lu, S Nagarajan, PP Mehta, AC Wade, NS Kishore, ML Bryant, DP Getman, CA McWherter, JI Gordon. Selective peptidic and peptidomimetic inhibitors of *Candida albicans* myristoylCoA:protein N-myristoyltransferase: a new approach to antifungal therapy. Biopolymers 43:43–71, 1997.

84. RA Weinberg, CA McWherter, SK Freeman, DC Wood, JI Gordon, SC Lee. Genetic studies reveal that myristoylCoA:protein N-myristoyltransferase is an essential enzyme in *Candida albicans*. Mol Microbiol 16:241–250, 1995.

85. JK Lodge, E Jackson-Machelski, B Devadas, ME Zuppec, DP Getman, N Kishore, SK Freeman, CA McWherter, JA Sikorski, JI Gordon. N-myristoylation of Arf proteins in *Candida albicans*: an in vivo assay for evaluating antifungal inhibitors of myristoyltransferase. Microbiology 143: 357–366, 1997.

86. DB Devadas, SK Freeman, CA McWherter, NS Kishore, JK Lodge, E Jackson, JI Gordon, JA Sikorski. Novel biologically active nonpeptide inhibitors of myristoylCoA:protein N-myristoyltransferase. J Med Chem 6:996–1000, 1998.

87. RS Bhatnagar, K Futterer, G Waksman, JI Gordon. The structure of myristoylCoA:protein N-myristoyltransferase. Biochim Biophys Acta 23:162–17, 1999.

88. SA Weston, R Camble, J Colls, G Rosenbrock, I Taylor, M Egerton, AD Tucker, A Tunnicliffe, A Mistry, F Mancia, E de la Fortelle, J Irwin, G Bricogne, RA Pauptit. Crystal structure of the anti-fungal target N-myristoyl transferase. Nat Struct Biol 5:213–221, 1998.

89. WR Schafer, J Fihe. Protein prenylation: genes, enzymes, targets, and functions. Annu Rev Genet 30:209–237, 1992.

90. J Drgonova, T Drgon, K Tanaka, R Kollar, G-C Chen, RA Ford CSM Chan, Y Takai, E Cabib. Rho 1p, a yeast protein at the interface between cell polarization and morphogenesis. Science 272:277–279, 1996.

91. O Kondoh, Y Tachibana, Y Ohya, N Arisawa, T Watanabae. Cloning of the RHO1 gene from *Candida albicans* and its regulation of β-1-3-glucan syntheses. J Bacteriol 179:7734–7741, 1997.

92. P Mazur, E Register, CA Bonfiglio, X Yuan, MB Kurtz, JM Williamson, R Kelly. Purification of geranylgeranyltransferase I from *Candida albicans* and cloning of the CaRAM2 and CaCDC43 genes encoding its subunits. Microbiology 145:1123–1135, 1999.

93. RC Dickson, RL Lester. Yeast sphingolipids. Biochim Biophys Acta 1426:347–357, 1999.

94. RC Dickson. Sphingolipid functions in Saccharomyces cerevisiae: comparison to mammals. Annu Rev Biochem 67:24–48, 1998.

95. NM Nagiec, EE Nagiec, JA Baltisberger, GB Wells, RL Lester, RC Dickson. Sphingolipid synthesis as a target for antifungal drugs: complementation of the inositol phosphorylceramide synthase defect in a mutant strain of *Saccharomyces cerevisiae* by the AUR1 gene. J Biol Chem 272:9809–9817, 1997.

96. M Kuroda, T Hashida-Okado, R Yasumoto, K Gomi, I Kato, K Takesako. An aureobasidin A resistance gene isolated from *Aspergillus* is a homolog of yeast AUR1, a gene responsible for inositol phosphorylceramide (IPC) synthase activity. Mol Gen Genet 261:290–296, 1999.

97. SM Mandala, RA Thornton, M Rosenbach, J Milligan, M Garcia-Calvo, HG Bull, MB Kurtz. Khafrefungin, a novel inhibitor of sphingolipid synthesis. J Biol Chem 272:32709–32714, 1997.

98. GH Harris, A Shafiee, MA Cabello, JE Curotto, O Genilloud, KE Goklen, MB Kurtz, M Rosenbach, PM Salmon, RA Thornton, DL Zink, SM Mandala. Inhibition of fungal sphingolipid biosynthesis by rustmicin, galbonolide B and their new 21-hydroxy analogs. J Antibiot 51:837–844, 1998.

99. G Belfield, MF Tuite. Translation elongation factor-3: a fungus-specific translation factor? Mol Microb 9:411–418, 1993.

100. GP Belfield, NJ Ross, MF Tuite. Translation elongation factor-3 (EF-3): an evolving eukaryotic ribosomal protein? J Mol Evol 41:376–387, 1995.

101. K Chakraburtty. Functional interaction of yeast elongation factor-3 with yeast ribosomes Int J Biochem Cell Biol 31:163–173, 1999.

102. JM Dominguez, JJ Martin. Identification of elongation factor 2 as the essential protein targeted by sordarins in *Candida albicans*. Antimicrob Agents Chemother 42:2279–2283, 1998.

103. D Hauser HP Sigg. Isolierung und Abbau von Sordarin. Helv Chim Acta 54:1178–1190, 1970.

104. MC Justice, T Ku, MJ Hsu, K Carniol, D Schmatz, J Nielsen. Mutations in ribosomal protein L10e confer resistance to the fungal-specific eukaryotic elongation factor 2 inhibitor sordarin. J Biol Chem 274:4869–4875, 1999.

105. JM Dominguez, VA Kelly, OS Kinsman, MS Marriott, F Gomez de las Heras, JJ Martin. Sordarins: a new class of antifungals with selective inhibition of the protein synthesis elongation cycle in yeasts. Antimicrob Agents Chemother 42:2274–2278, 1998.

106. TC Kennedy, G Webb, RJP Cannell, OS Kinsman, RF Middleton, PJ Sidebottom, NL Taylor, MJ Dawson, AD Buss. Novel inhibitors of fungal protein synthesis produced by a strain of *Graphium putredinis*. Isolation, characterization and biological properties. J Antibiotics 51:1012–1018, 1998.

107. DM Ogrydziak. Yeast extracellular proteases. Crit Rev Biotechnol 13:1–55, 1993.

108. B Hube. *Candida albicans* secreted aspartyl proteinases. Curr Top Med Mycol 7:55–69, 1996.

109. HJ Watts, FSH Cheah, B Hube, D Sanglard, NAR Gow. Altered adherence in strains of *Candida albicans* harboring null mutations in secreted asparatic proteinase genes. FEMS Microb Lett 159: 129–135, 1998.

110. B Hube, D Sanglard, FC Odds, D Hess, M Monod, W Schafer, AJP Brown, NAR Gow. Gene disruption of each of the secreted aspartyl proteinase genes SAP1, SAP2, and SAP3 in *Candida albicans* attenuates virulence. Infect Immun 65:3529–3538, 1997.

111. K Fallon, K Bausch, J Noonan, E Huguenel, P Tamburini. Role of aspartic proteases in disseminated *Candida albicans* infection in mice. Infect Immun 65:551–556, 1997.

112. M Schaller, W Schafer, HC Korting, B Hube. Differential expression of secreted aspartyl proteinases in a model of human oral candidosis and in patient samples from the oral cavity. Mol Microbiol 29:605–616, 1998.

113. P Staib, M Kretschmar, T Nichterlein, H Hof, J Morschhauser. Differential activation of a *Candida albicans* virulence gene family during infection. Proc Natl Acad Sci USA 97:6102–6107, 2000.

114. SM Cutfield, EJ Dodson, BF Anderson, PC Moody, CJ Marshall, PA Sullivan, JF Cutfield. The crystal structure of a major secreted aspartic proteinase from *Candida albicans* in complexes with two inhibitors. Structure 3:1261–1271, 1995.

115. M Borg-von Zepelin, I Meyer, R Thomssen, R Wurzner, D Sanglard, A Telenti, M Monod. HIV-protease inhibitors reduce cell adherence of *Candida albicans* strains by inhibition of yeast secreted aspartic proteases. J Invest Dermatol 113:747–751, 1999.

116. A Gruber, J Berlit, C Speth, C Lass-Flori, G Kofler, M Nagl, M Borg-von Zepelin, MP Dierich, R Wurzner. Dissimilar attenuation of *Candida albicans* virulence properties by human immunodeficiency virus type 1 protease inhibitors. Immunobiology 201:133–144, 1999.

117. JC Wang. DNA topoisomerases. Annu Rev Biochem 65:635–692, 1996.

118. SM Guichard, MK Danks. Topoisomerase enzymes as drug targets. Curr Opin Oncol 6:482–489, 1999.

119. B Gatto, C Capranico, M Palumbo. Drugs acting on DNA topoisomerases: recent advances and future perspectives. Curr Pharm Des 5:195–215, 1999.

120. DC Hooper. Clinical application of quinolones. Biochim Biophys Acta 1400:45–61, 1998.

121. LL Shen, J Baranowski, J Fostel, DA Montgomery, PA Lartey. DNA topoisomerases from pathogenic fungi: targets for the discovery of antifungal drugs. Antimicrob Agents Chemother 36: 2778–2784, 1992.

122. JM Fostel, DA Montgomery, LL Shen. Characterization of DNA topoisomerase I from *Candida albicans* as a target for drug discovery. Antimicrob Agents Chemother 36:2131–8, 1992.

123. M Del Poeta, DL Toffaletti, TH Rude, CC Dykstra, J Heitman, JR Perfect. Topoisomerase I is essential in *Cryptococcus neoformans*: role in pathobiology and as an antifungal target. Genetics 152:167–178, 1999.

124. W Jiang, D Gerhold, EB Kmiec, M Hauser, JM Becker, Y Koltin. The topoisomerase I gene from *Candida albicans*. Microbiology 143:377–386, 1997.

125. M Del Poeta, SF Chen, D Von Hoff, CC Dykstra, MC Wani, G Manikumar, J Heitman, ME Wall, JR Perfect. Comparison of in vitro activities of camptothecin and nitidine derivatives against fungal and cancer cells. Antimicrob Agents Chemother 43:2862–2868, 1999.

126. JM Fostel, D Montgomery. Identification of the ammino-catechol A-3253 as an in vitro poison of DNA topoisomerase I from *Candida albicans*. Antimicrob Agents Chemother 39:586–592, 1995.

127. BA Keller, S Patel, LM Fisher. Molecular cloning and expression of the *Candida albicans* TOP2 gene allows study of fungal DNA topoisomerase II inhibitors in yeast. Biochem J 324:329–339, 1997.

128. TR Cech. Self-splicing of group I introns. Annu Rev Biochem 59:543–568, 1990.

129. Y Liu, M Roucourt, S Pan, C Liu, MJ Liebowitz. Sequence and variability of the 5.8S and 26S rRNA genes of *Pneumocystis carinii*. Nucleic Acids Res 20:3763–3772, 1992.

130. S Mercure, S Montplaisir, G Leay. Correlation between the presence of a self-splicing intron in the 25S rDNA of *C. albicans* and strains susceptibility to 5-fluoruracil. Nucleic Acids Res 21: 6020–6027, 1993.

131. H Boucher, S Mercure, S Montplaisir, G LeMay. A novel group I intron in *Candida dubliniensis* is homologous to a *Candida albicans* intron. Gene 180:189–196, 1996.
132. Y Liu, RR Tidwell, MJ Liebowitz. Inhibition of in vitro splicing of a group I intron of *Pneumocystis carinii*. J Euk Microbiol 41:31–38, 1994.
133. KM Miletti, MJ Leibowitz. Pentamide inhibition of group I intron splicing in *Candida albicans* correlates with growth inhibition. Antimicrob Agents Chemother 44:958–966, 2000.

26
Drug Resistance Mechanisms of Human Pathogenic Fungi

Rajendra Prasad, Sneh Lata Panwar, and Shankarling Krishnamurthy
Jawaharlal Nehru University, New Delhi, India

I. INTRODUCTION

Over 100 fungal species have been recorded as human pathogens that can cause superficial to life-threatening infections. While infections due to *Candida* and *Aspergillus* spp. are the most common, recently a number of otherwise rarely encountered opportunistic fungi have emerged as significant pathogens. As examples, *Trichosporon beigelii, Fusarium* spp. *Pseudallescheria boydii*, and molds of the class Zygomycetes can cause invasive infections. In addition to the opportunistic fungi, a limited number of fungi, viz., *Histoplasma capsulatum, Paracoccidioides brasiliensis, Penicillium marneffei*, and *Coccidioides immitis*, occur in various endemic regions with pathogenic potential for healthy hosts and may cause life-threatening infections [1–4].

Because of the increasing number of immunocompromised patients, fungal infections caused by common, new, and emerging human pathogenic fungi have recently become more rampant. Among the various human fungal pathogens, *Candida albicans* accounts for the majority of systemic infections with mortality rates ranging from 50% to 100%. *C. albicans* is the most common form of septicaemia in western hospitals, and >80% of the HIV-infected population develops oropharyngeal and clinical thrush. *Candida* spp. rank fifth among causes of nosocomial bloodstream infections. Furthermore, although *C. albicans* is the predominant pathogenic species, infections caused by non-*albicans* species, such as *C. glabrata, C. parapsilosis, C. tropicalis*, and *C. krusei*, have been increasing, especially in neutropenic patients and neonates [1–4].

The increasing threat of fungal infections has stimulated the search for better antifungals with a distinct mode of action [1–6]. However, the effective use of these antifungals has often been minimized by their toxicity and their narrow spectrum. The azoles and the polyenes are the safest antifungals in use to fight systemic fungal infections. In spite of the problem of nephrotoxicity associated with amphotericin B, it has been the most extensively used polyene [2,3]. The introduction of orally active antifungal azole drugs, particularly fluconazole, was a significant development, allowing treatment of systemic fungal infections without the problem of nephrotoxicity. Nonetheless, the recent rise in the emergence of fungi that develop resistance to a number of these antifungals has, however, compounded the problem [1–8]. There have been several recent articles dealing with the clinical and cellular aspects of multidrug resistance (MDR) in human pathogenic fungi [1–9]. Therefore, in this chapter we have focused on the molecular mechanisms of drug resistance and physiological relevance of MDR proteins known in human pathogenic fungi in general and *C. albicans* in particular.

II. MDR IN *CANDIDA*

Multidrug resistance is defined as resistance of an organism against a spectrum of drugs that share neither a common target nor a common structure [10–12]. MDR is not only limited to humans but is spread throughout the evolutionary scale [13–15]. A general theme that has emerged over the years from earlier investigations is that the mechanisms of multidrug resistance are the result of modification of normal pathways of cellular homeostasis [10,16]. Some of the major mechanisms resulting in MDR are: (1) decreased accumulation of drugs, which is the dominant feature of MDR; (2) changes in expression of some cellular proteins, e.g., P-glycoprotein, MRP, GST, catalase, and topoisomerase; and (3) changes in cellular physiology affecting the structure of plasma membrane, cytosolic pH, and lysosomal structure and function [6,17]. Various resistance mechanisms found in many microbes, including fungi and bacteria, parasitic pathogens, and cancerous cells, often arise out of different synergistic combinations of the above-mentioned mechanisms [18].

Dimorphic, opportunistic, and the most predominant human pathogenic yeast, *C. albicans* is naturally more resistant to several drugs than *Saccharomyces cerevisiae* [19–22]. In addition, the incidence of *C. albicans* cells acquiring resistance to antifungals like azoles has increased considerably in recent years, which has posed serious problems toward successful chemotherapy [5,23–26]. The incidence of antifungal resistance has also increased in the non-*albicans* species, such as *C. glabrata, C. parapsilosis, C. tropicalis,* and *C. krusei* [1,27].

Candida infections are treated with antifungal agents, particularly with the triazole derivatives fluconazole, itraconazole, and ketoconazole. To combat the attack of antifungals, *Candida* has evolved a variety of mechanisms to acquire resistance to these drugs. The resistance to azoles in *C. albicans* was earlier thought to occur primarily through an alteration or an overexpression of the target enzyme 14α-lanosterol demethylase (P45014DM) involved in sterol biosynthesis [28–30]. However, the characterization of the ATP-Binding Cassette (ABC) proteins Cdr1p and Cdr2p [31,32] and CaMdr1p, a transporter of the major facilitator superfamily (MFS) [19,20], led to the suggestion that efflux mechanisms represent an important determinant of antifungal susceptibilities in *C. albicans.*

In spite of the use of polyene antibiotics for several years, there are limited instances of *Candida* cells becoming resistant to amphotericin-B (AmB) and nystatin; however, intrinsic resistance to AmB in some *Candida* spp., viz, *C. lusitaniae* and *T. beigelii,* is common [1–3]. The strains resistant to AmB, which have been isolated from patients with candidiasis, belong mostly to non-*albicans* species such as *C. lusitaniae* and *C. tropicalis.* The resistance to AmB has been mainly associated with changes in sterol content of the cell [33]. The clinical isolates of *C. albicans* resistant to AmB were shown to lack ergosterol and accumulated 3-β-ergosta-7, 22-dienol and 3-β-ergosta-8-enol due to a defect in $\Delta^{5,6}$-desaturase enzyme. The decreased ergosterol content resulted in reduced binding of polyenes as compared to susceptible cells. In certain instances of AmB resistance, an increased catalase activity has also been shown as a part of the mechanism of resistance to control oxidative damage caused by the drug [1]. In another pathogenic isolate of *Cryptocococcus neoformans* which was isolated from AIDS patients who failed antifungal therapy, a similar correlation between polyene resistance and sterol contents was reported [34]. There are also reports to suggest that polyene action could be affected by alteration in components other than sterols—viz., cell wall component and membrane phospholipids [35–40].

III. MDR IN OTHER FUNGI

Aspergillus fumigatus is the most common species of *Aspergillus* causing pulmonary disease. AmB and itraconazole are the two commonly used drugs to which this fungus is sensitive. Notwithstanding the variations in MIC value determination, higher MICs have been recorded for some clinical isolates. The role of P45014DM and efflux pumps has been implicated in itraconazole resistance in *A. fumigalus* [41]. Two ABC transporter genes *atrA* and *atrB* have recently been cloned from the filamentous fungus *A. nidulans*. The proteins encoded by these genes share the same topology as the ABC transporters *CDR1* from *C. albicans* and *PDR5* and *SNQ2* from *S. cerevisiae*. Similar to transcriptional activation of *CDR1* by a variety of stresses, the transcription of *atrA* and *atrB* is also upregulated by several drugs, azoles, and fungicides. In order to dissect the functional role of *atrA* and *atrB* in drug resistance, a *pdr5* mutant (hypersensitive to various drugs) was functionally complemented with them. While *atrB* could complement the *pdr5* defect and elicited resistance to several drugs, *atrA* failed to complement the hypersensitive phenotype of *pdr5* mutant strain. It is assumed that both *atrA* and *atrB* play a role in protecting the fungus from natural toxic compounds by effluxing the drugs and thereby reducing their intracellular concentrations [42,43]. Another ABC transporter gene, *atrC*, has also been cloned from *A. nidulans*. Northern analysis revealed that *atrC* mRNA levels increased in response to cycloheximide. There is evidence suggesting the presence of at least eight additional ABC transporters in *A. nidulans*. The contribution of these new genes in azole resistance needs to be elucidated [44].

Another ABC pump, *adr1*, which is highly similar to *PDR5* and *CDR1*, has recently been characterized in *A. fumigatus*. The transcript of *adr1* was found to be enhanced in an itraconazole-resistant isolate, which also showed less accumulation of the drug [45]. ABC transporters *AfuMDR1* and *AfuMDR2* from *A. fumigatus* and *AflMDR1* from *A. flavus* have also been identified. *AfuMDR1* and *AflMDR1* bear a high degree of similarity to the *S. pombe* leptomycin B resistance protein and to human *MDR1* [42]. On the other hand, *AfuMDR2* encodes a protein containing four putative transmembrane domains and shows a high degree of similarity to the *S. cerevisiae* genes *MDL1* and *MDL2* [46]. Expression of *AfuMDR1* in *S. cerevisiae* conferred increased resistance to the antifungal agent cilofungin [42].

Cryptococcus neoformans is another pathogenic fungus that infects immunocompromised patients. Current treatment of this fungal infection involves the use of azole antifungal drugs. Recently fluvastatin, a cholesterol-lowering agent, has also been shown to exhibit an inhibitory effect against *C. neoformans*. Fluvastatin combined with fluconazole and itraconazole has a synergistic as well as an additive effect. Although the mechanism for fluvastatin action is not clear, it is assumed that it could be inhibiting one of the enzymes of the ergosterol biosynthetic pathway [47]. The development of resistance to these drugs in *C. neoformans* led to a search for mechanisms involved in this phenomenon. Subsequently, *CneMDR1*, a gene encoding a protein related to the other MDR proteins, has been cloned and characterized from *C. neoformans*. Evidence also shows the presence of a second MDR-like gene (*CneMDR2*) in this fungus [48].

In one report, an increase in MIC values in strains of *H. capsulatum* recovered from AIDS patients has been documented [49]. Interestingly, the enhanced resistance to fluconazole in posttreatment isolates was accompanied by higher sensitivity to itraconazole. Another example of negative cross-resistance has also been reported for *Penicillium italicum* strains. These strains are highly resistant to 14α-demethylase inhibitors but are sensitive to the morpholine fenpropimorph [50,51]. Therefore, although the number of reports of antifungal resistance in fungal species other than *Candida* are still limited, the mechanisms of resistance reported so far appear to be common.

Table 1 Targets and Mechanisms of Resistance of Some Antifungals

Antifungal	Target	Mechanism of resistance
Pyrimidine 5-Flucytosine	Thymidylate synthetase	Failure to metabolize 5-FC to 5 FUTP and 5 FdUMP Loss of feedback control of pyrimidine biosynthesis Defect in cytosine permease
Polyenes Nystatin, amphotericin-B	Membrane ergosterol	Alteration in membrane lipids, mainly ergosterol (resistant clinical isolates lack ergosterol and accumulate 3-β-ergosta-7,22-dienol and 3-β-ergosta-8-enol, due to defect in $\Delta^{5,6}$-desaturase gene (ERG3) Enhanced catalase activity
Azoles Fluconazole, ketoconazole, itraconazole, voriconazole, clotrimazole	14α-demethylase (ERG11, also designated ERG16 earlier)	Mutations in the target enzyme cytochrome P450 14α-demethylase which alters the affinity of this enzyme to the azoles Overexpression of 14α-demethylase Failure to accumulate azoles due to rapid efflux mediated by ABC and MFS family of MDR transporters Alteration of sterol $\Delta^{5,6}$-desaturase (ERG3)
Allylamines Naftifine, terbinafine, tolnaftate	Squalene epoxidase (ERG1)	Overexpression of CDR1, CDR2, and CaMDR1
Morpholines Amorolfine	Δ^{14}-reductase (ERG24), $\Delta^{8,7}$-isomerase (ERG2)	Overexpression of Δ^{14}-reductase (ERG24) or sterol C-24 (28) reductase (ERG4) genes Overexpression of CDR1 and CDR2
Lipopeptides Echinocandins, pneumocandins, aculcacins, cyclopeptamine	β-1,3-glucan synthetase (encoded by FKS1 and RHO1)	Mutations in FKS1 gene alters affinity of the enzyme

In order to understand the mechanism of antifungal resistance, it is essential that the site and mechanisms of action of these drugs are elucidated. Table 1 lists most of the known antifungals with their proven targets and mechanisms of resistance. The reader is referred to some very informative reviews for further details on these aspects [1–8].

IV. MECHANISMS OF RESISTANCE TO ANTIFUNGALS

Although the molecular basis of drug resistance in *Candida* is not very clear, evidence accumulated so far suggests that MDR is a multifactorial phenomenon where a combination of mecha-

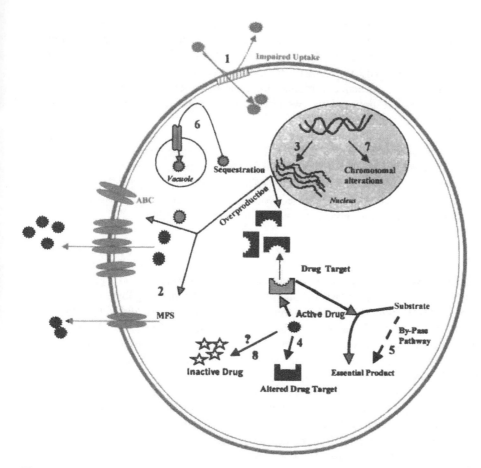

Figure 1 Mechanisms of azole resistance in fungi. (1) Changes in the cell wall/plasma membrane leading to impaired azole uptake. (2) Efflux of drugs mediated by the ABC or the MFS class of efflux pumps. (3) Overexpression of the drug target or of the efflux pumps. (4) Mutation in the drug target (P45014DM) does not allow the drug to bind or allows drug to bind with low affinity. (5) Activation of alternate pathways such as $\Delta^{5,6}$-desaturase. (6) Sequestration of the drug into an organelle-like vacuole by organellar pump. (7) Chromosome alterations or changes in chromosome number as a means to maintain more copies of the required gene. (8) Modification of azoles (Δ^{22}-desaturase and CYP52) to an inactive form.

nisms could contribute to drug resistance (Fig. 1) [29,52,53]. Some of the well-known mechanisms of drug resistance in pathogenic fungi are discussed below.

A. P45014DM

Azole resistance in *C. albicans* was earlier thought to occur primarily through an alteration or an overexpression of the *ERG11* (earlier designated as *ERG16*)-encoded gene product 14α-lanosterol demethylase (CYP51: also known as P45014DM) involved in sterol biosynthesis. Azoles inhibit a specific step in the ergosterol biosynthesis in fungi by binding to and inhibiting P45014DM [54–56]. This leads to high levels of 14-methylated sterols, causing disruption of membrane structures. Azole derivatives have been shown to interact with the heme molecule in P45014DM where an unhindered nitrogen atom of the azole ring (N3 in imidazole or N4 in derivatives) binds to the heme iron at its sixth coordinate position. The blocking of this position, which is normally occupied by activated oxygen, prevents initiation of the hydroxylation reaction. The affinity and the selectivity of the azoles for their target(s) are also determined by their structure, lipophilicity, and stereochemical orientation of the N-1 side chain. Azole derivatives also fit in the P45014DM substrate pocket, which normally accepts lanosterol as a natural substrate [30,57,58].

The import of azoles inside the cell still remains unresolved although it has been suggested that the hydrophobicity of this drug could facilitate its entry. Once the drug enters the cell its interaction with P45014DM can be modified in two ways—target alteration and overexpression. Certain point mutations in the gene for P45014DM (*ERG11*) make the enzyme less sensitive to drug. The overexpression of enzyme, which necessitates the need for a higher dose of azoles, as compared to a susceptible strain, represents another mechanism [2,8,59].

1. Alterations in P45014DM

Several publications have reported point mutations in the P45014DM (*ERG11*) gene, which change the affinity of the azoles to its target protein leading to resistance [28,30,60–63]. White analyzed a series of *C. albicans* strains which were earlier isolated by Redding et al. from a single HIV patient [64] and identified a single amino acid substitution—R467K in Erg11p. This mutation lies between two residues presumably involved in interactions with the heme moiety. Using site-directed mutagenesis, Lamb et al. introduced the mutation T315A and expressed the mutated protein in *S. cerevisiae*. The mutated protein showed higher MIC values for fluconazole and ketoconazole. The purified mutated protein exhibited reduced enzyme activity and affinity for azoles, thus providing the first example of a single base change in the target enzyme leading to azole resistance through reduced affinity [28]. Sanglard et al. [30], using a similar strategy, expressed *ERG11* genes from sequential clinical isolates of *C. albicans* in *S. cerevisiae*. This strategy enabled them to isolate five *ERG11* genes with mutations—G129A, Y132H, S405F, G464S, and R467K. The presence of these mutations in the *C. albicans* isolates led to an increase in the MIC values for azoles [30]. Loffler et al. have identified additional mutations—E266D, F105L, K287R, G448E, G450E, G464S, and V488I in fluconazole-resistant isolates [63]. The mutations G464S and R467K in the heme-binding domain of P45014DM in clinical strains have also been shown to cause resistance by reducing the affinity for fluconazole [62,65]. The mutation Y132H does not permit normal binding of fluconazole to protein, as was revealed from the spectral studies. Under normal circumstances binding of fluconazole to the native protein results in type II spectra due to its coordination with heme as a sixth ligand while a type I spectrum was observed for Y132H mutation. However, there is no effect of this mutation on ergosterol biosynthesis. This suggested that Y132H substitution occurred without significant perturbation of the heme environment and thus represents a novel change in protein leading to fluconazole resistance [66]. Recently, Marichal et al. have identified new mutations linked to azole resistance in *ERG11*—A149V, D153E, E165Y, S279F, V452A, and G465S [60] and have compiled all the known 29 substitutions of *C. albicans ERG11* in a graphical representation to show the frequency and positions of these amino acid substitutions (Fig. 2) [60]. Four mutations—D116E, K128T, E266D, and G464S—occurred with highest frequency; G464S was the only substitution seen exclusively in azole-resistant isolates. I-helix stretch of Erg11p, which is highly conserved in cytochromes P450 family, did not show any spontaneous mutations. The exact placement of these mutations in a 3D model of the protein show that these mutations are not randomly distributed but rather clustered in three hot spot regions between 105–165, 266–287, and 405–488 amino acid residues [60].

Molecular modeling of P45014DM has provided an additional tool to understand its interactions with its natural substrate (viz., lanosterol) and the inhibitors (viz., azoles). Two homology-based models of P45014DM have been built to illustrate these interactions. In one of the model, which has been built on the basis of the known crystal structure of P450cam [67], the interaction of lanosterol with the amino acid residues Gly-310, Leu-307, Met-313, and Ser-508 has been predicted [57]. In another homology-based model which was built by using the available coordinates of P450BM-3 (CYP102) as a template, Lewis et al. have predicted interaction of lanosterol with some more additional amino acid residues, viz., Val-138, Ile-139, Leu-257, Met-313, Met-509, Val-510, Tyr-140, Ser-382, Phe-153, Thr-318, His-381, Tyr-126, and Leu-228. The putative interactions of an inhibitor like ketoconazole with CYP51 have been shown to be quite similar to that predicted for lanosterol [68]. However, none of the substitutions identified in azole-resistant isolates so far include a change in these interactive amino acid residues. The two 3D models of P45014DM built are too distant from Erg11p of *C. albicans*. Indeed, a closer model to Erg11p is necessary to illustrate the interaction among inhibitors, substrate, and protein.

The increasingly important pathogen *C. krusei* is less susceptible to fluconazole and ketoconazole than itraconazole. The interaction of P45014DM with itraconazole, fluconazole, and ketoconazole has been confirmed in *C. krusei* by in vitro ergosterol biosynthesis and by using difference spectroscopy [69]. In *C. krusei* the demethylase activity was found to be severalfold resistant to inhibition by fluconazole as compared to the activity of this enzyme from fluconazole-susceptible isolates of *C. albicans* [70]. It appears that fluconazole resistance in *C. krusei* is predominantly mediated by reduced susceptibility of 14α-demethylase to the drug [70]. The interactions between azoles and Erg11p have also been demonstrated in filamentous fungi like *A. fumigatus* and *Penicillium italicum* [71,72].

2. Overexpression of P45014DM

Resistance to fluconazole in many clinical isolates has often been associated with the overexpression of P45014DM. However, it has been difficult to correlate the overexpression with the resistance observed mainly because of the simultaneous existence of mutations in P45014DM or overexpression of the efflux pumps in the same isolates. Gene amplification is one of the common mechanisms of resistance in eukaryotic cells [73,74]. However, overexpression of P45014DM in *C. albicans* has not been linked to gene amplification [29,75,76]. In a clinical isolate of *C. glabrata*, increased levels of P45014DM were shown to be associated with the amplification of the *ERG11* gene [77–79]. Amplification of the *ERG11* gene in this isolate has also been linked to chromosomal duplication, which in turn results in high levels of P45014DM [78]. That gene conversion or mitotic recombination could also play a role in fluconazole resistance in *C. albicans*, was apparent from a study done by White where additional genetic variations in a clinical isolate of *C. albicans* with R467K substitution have been reported. It was shown that the allelic differences present in sensitive isolates of *C. albicans* were eliminated in the resistant isolates from the region of the *ERG11* promoter, ORF and terminator region and into the immediate downstream gene *THR1*, by gene conversion or mitotic recombination [80]. The resulting strain has R467K mutation in both copies of *ERG11* and was more resistant to azoles than a strain with single allelic substitution.

B. Δ^{5,6}-Desaturase

A defect in Δ^{5,6}-desaturase (*ERG3*), another enzyme of the ergosterol biosynthesis pathway, has also been shown to contribute to azole resistance. A defect in this enzyme leads to the accumulation of 14α-methylfecosterol instead of 14α-methylergosta-8,24(28)-dien-3β,6-α diol.

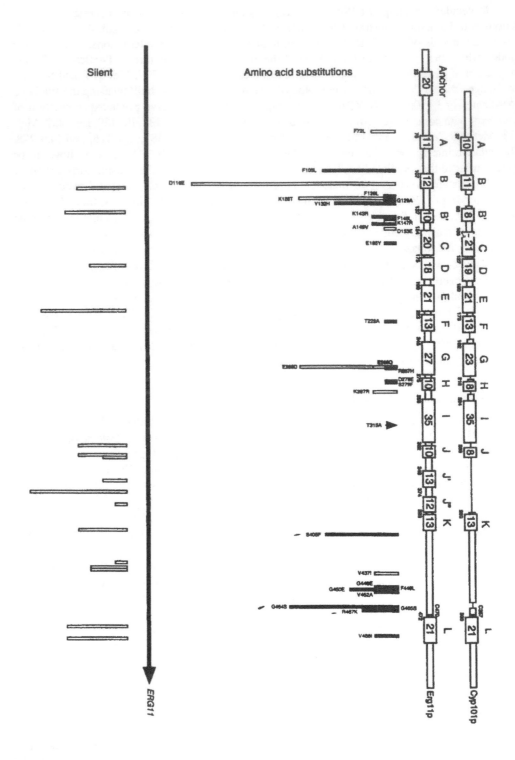

Accumulation of sufficient amounts of 14α-methylfecosterol compensates for ergosterol in the membranes and thus contributes to azole resistance in *C. albicans* [1–3]. It has also been shown that the lethality of *S. cerevisiae* disruptant of *CYP51* (*ERG11*) can be suppressed by $\Delta^{5,6}$-desaturase [81]. The decrease in ergosterol contents due to a defect in $\Delta^{5,6}$-desaturase in fluconazole resistant clinical isolates of *C. albicans* resulted in cross-resistance to AmB [82].

Recently another cytochrome P450, Δ^{22}-desaturase (CYP61 and also *ERG5*), has been purified from an *ERG11* (P45014DM) disruptant strain of *C. glabrata* [83]. The purified enzyme showed desaturase activity in a reconstituted system. Δ^{22}-destaurase, and its homologs have also been identified in *C. albicans* and *Schizosaccharomyces pombe*. The type II spectra obtained with the azole antifungal compounds ketoconazole, fluconazole, and itraconazole in reconstituted Δ^{22}-desaturase suggested that these drugs directly interact with the cytochrome heme. These results thereby suggest the potential for Δ^{22}-desaturase to be an antifungal target [83].

C. Efflux Pumps

In addition to an alteration or an overexpression of 14α-lanosterol demethylase involved in sterol biosynthesis, azole resistance in *C. albicans* is also elicited by other mechanisms. The characterization of ABC proteins, e.g., *CDR1* [31], *CDR2* [32], and *CaMDR1* [19,20], a MFS transporter, as efflux pumps of *C. albicans* and their overexpression in certain instances of azole-resistant isolates, has confirmed that these transporters represent another mechanism involved in the MDR scenario of *C. albicans* [32,53,76,84,85].

1. ABC Transporters

CDR1 of *C. albicans* was the first ABC efflux pump implicated in conferring resistance to cycloheximide in a *PDR5* disruptant hypersensitive strain of *S. cerevisiae* [31]. *CDR1* encodes a protein of 1501 amino acid residues (169.9 kDa) whose predicted structural organization is characterized by two homologous halves, each comprising a hydrophobic region with a set of six transmembrane stretches, preceded by a hydrophilic nucleotide binding fold. The structure is identical to that of the *S. cerevisiae* ABC proteins Pdr5p and Snq2p. It mirrors the architecture of the yeast **a**-mating pheromone transporter *STE6*, as well as mammalian drug resistance P-glycoprotein (P-gp or *MDR1*) and cystic fibrosis factor CFTR [26]. The significance of inversion of domains in some of the ABC drug transporters is not understood and may be related to their

Figure 2 Visualization of the localization and frequency of mutations in the Erg11p (Cyp51p) sequence. At the top of the figure, the aligned sequences of C. *albicans* Erg11p and P. *putida* Cyp101p and their secondary structure are visualized. Gaps are represented as thin lines and predicted α-helices are represented by boxes, for which the starting position and length are indicated. The letter code for the helices is indicated above the boxes. The amino acid mutations found are represented by bars, for which the length is proportional to the frequency of occurrence. Different filling patterns are used to categorize the mutation: mutations found both in azole-sensitive and -resistant strains are indicated by open bars; filled bars depict the mutations for which it has been experimentally demonstrated that they were important for the affinity of an azole for the cytochrome P450; hatched bars indicate noncharacterized mutations. The PCR-induced mutation is shown with an arrow. The fifth ligand of the haem, C470, is also indicated. At the bottom of the figure, the *ERG11* ORF is represented by a horizontal arrow. Silent mutations found in this study are represented by open bars. Again the length of the bars is proportional to the frequency of occurrence. (From Ref. 60.)

physiological functions. Cdr1p is remarkably similar to Pdr5p of *S. cerevisiae*. The similarity is not limited to ATP binding motif but is conserved along the entire length of the protein. Despite the high homology between *CDR1* and *PDR5*, and their encoded products, some functional features tend to distinguish them. For example, while single or low copies of *CDR1* are sufficient to increase drug resistance in *S. cerevisiae*, multiple copies of *PDR5* are required to yield a similar level of drug resistance. In addition to differences in the efficiencies in conferring drug resistance to the same drugs, the spectrum of drugs to which *CDR1* and *PDR5* confer resistance is also partly distinct. For example, both genes share overlapping specificities for cycloheximide and chloramphenicol, but *CDR1* affects sensitivity to oligomycin while neither amplification nor disruption of *PDR5* alter susceptibilities to it. It is worth mentioning here that some of the close homologs of *CDR1* in *C. albicans* are also functionally distinct (discussed below).

That efflux pumps other than *CDR1* could be contributing to drug resistance became apparent after isolation of its close homolog *CDR2*. Cdr2p exhibits 84% identity with Cdr1p and confers resistance to fluconazole and several other drugs [32]. In spite of the close identity between *CDR1* and *CDR2*, their flanking promoter regions exhibit considerable divergence. While the *CDR1* promoter is modular and highly regulated with putative elements like HSE, DRE, SRE, AP1, and YAP1, the *CDR2* promoter is devoid of such elements, which suggests that they may be differentially regulated [86,87]. Interestingly, the promoter of both genes, though a close homolog of *PDR5*, lacks the characteristic *PDR1/PDR3* transcription factor binding sites, which further suggests divergence in the regulation between close homologs of two yeasts [31].

Since azole resistance appears to be a multifactorial phenomenon (discussed later), this led to the search for more homologs of *CDRs*. Using PCR-based cloning, other homologs of *CDR1* and *CDR2*, namely *CDR3*, *CDR4*, and *CDR5*, were identified [88–90]. Cdr3p and Cdr4p show the highest homology to Cdr1p and Cdr2p; however, compared to Cdr1p and Cdr2p, which are >90% similar, Cdr3p and Cdr4p are only 75% similar to Cdr1p and Cdr2p. Interestingly, overexpression of *CDR3* and deletion of *CDR3* and *CDR4* could not affect drug susceptibilities [88,89]. Why some of the known Cdrps are unable to elicit a multidrug resistance phenomenon is not clear. The ball model of Cdr1p and Cdr3p drawn from a hydropathy plot (Fig. 3a, b) shows that the two proteins have similar topological arrangements where a hydrophilic domain containing nucleotide binding motif precedes hydrophobic transmembrane stretches. The only apparent difference between the two proteins appears to be in the C-terminal where Cdr3p has an extended loop connecting TM11 and TM12. In addition, the last 21 amino acids in the C-terminal of Cdr3p are totally different from Cdr1p.

Keeping in view the importance of these regions in drug binding in human Mdr1p and Cdr1p, these subtle differences in structure of these proteins could affect substrate specificity, thereby rendering them unable to bind and transport drugs [91–94]. A high-resolution 3D protein structure of Cdrps would finally be able to resolve the molecular basis of drug transport by these transporters; however, mutational analysis of homologous and nonhomologous regions between different Cdrps can provide valuable information. This situation is reminiscent of the observation that two types of P-glycoproteins are found in mammals: one which transports hydrophobic drugs, and one that cannot transport drugs. Thus, mouse mdr1 and mdr3 and human MDR1 can transport and confer drug resistance while mouse mdr2 and human MDR2 (also known as MDR3) cannot [95–101]. Further studies employing MDR1-MDR2 chimeric proteins have led to the identification of some amino acid residues in the TM6 of MDR1 which are sufficient to allow an MDR2 backbone in the N-terminal half of P-gp to transport several MDR1 substrates [102]. These studies indicate a close relationship between MDR1, a multidrug transporter, and MDR2, a phosphotidylcholine flippase. Since Cdr1p, Cdr2p, and Cdr3p have similar domain structure, their substrate preferences are most likely to be determined by some nonidentical amino acid residues. However, this remains to be investigated.

A search for more homologs of MDR genes in *C. albicans* revealed at least 14 new partial gene sequences, which displayed homology with human *MDR1* ATP binding region. These sequences showed no significant homology to known Cdrps but nonetheless could have an effect on drug susceptibilities of *C. albicans* [103]. There is a great possibility that *C. albicans* will have several drug efflux pumps.

Clinical resistance to fluconazole as a result of reduced intracellular accumulation has also been reported for other pathogenic *Candida* species including *C. tropicalis*, *C. glabrata*, *C. Krusei*, and *C. dubliniensis*. Table 2 lists the known MDR transporters in *Candida* spp. Among non-*albicans* species, *C. glabrata* has emerged as an important nosocomial pathogen. Homologs of *CDR1*, *CgCDR1*, and *CgCDR2* have been isolated by functionally complementing a hypersensitive mutant of *S. cerevisiae* [104,105]. *CgCDR1* was shown to be upregulated in the azole-resistant isolates. Another putative ABC transporter encoded by *PDH1* has also been implicated in azole resistance of *C. glabrata* [106]. *C. dubliniensis* is phylogenetically closely related to *C. albicans* and is associated with oral candidiasis. Fluconazole-resistant isolates of *C. dubliniensis* showed reduced accumulation of the drug as compared to susceptible strains. This led to the identification of two ABC transporters, viz., *CdCDR1* and *CdCDR2*, which mediate fluconazole resistance in clinical isolates of *C. dubliniensis* [107]. A potential role of two putative ABC transporters *ABC1* and *ABC2* in drug resistance has also been suggested for *C. krusei* [108].

2. MFS Transporters

Another class of efflux pumps/transporters which are structurally quite similar to ABC pumps but do not contain ATP binding domains are known as major facilitators (MFS). The MFS has originally been defined as a superfamily of permeases that are characterized by two structural units of six transmembrane spanning α-helical segments, linked by a cytoplasmic loop. The structure-function relationships of MFS have not been generalized in details due to the diversity in their nucleotide and amino acid sequences. But it has been postulated that the N-terminal halves of different major facilitator families share greater similarities than their C-terminal halves, which suggests that C-terminal regions are involved in substrate recognition and N-terminal regions are involved in proton translocation [109,110].

The MFS proteins, which are involved in symport, antiport, or uniport of various substrates, have been found to be ubiquitously present from bacteria to higher eukaryotes [111]. These transporters have been classified into five distinct clusters or families of membrane transport proteins within the MFS involved in (1) drug resistance, (2) sugar uptake, (3) uptake of Krebs cycle intermediates, (4) phosphate ester/phosphate antiport, and (5) oligosaccharide uptake [109–111]. The drug resistance proteins are proton motive force (PMF)-dependent antiporters which efflux out drugs in exchange of one or more H^+ with a substrate molecule [110]. On the basis of hydropathy and phylogenetic analyses, MFS drug efflux proteins, which have >100 members, can be divided into two distinct groups containing either 12 or 14 TMS (transmembrane segments) [110].

In *S. cerevisiae*, 28 MFS proteins have been identified of which a few have been implicated to play a role in drug resistance [112]. *FLR1* of *S. cerevisiae*, which encodes for an MFS protein, has been shown to cause resistance to cycloheximide, fluconazole, cadmium, and H_2O_2 [113]. *CaMDR1* (previously known as BEN^R) and *FLU1* are the two MFS genes identified in *C. albicans*. *CaMDR1* was initially identified as a gene which conferred resistance to the tubulin-binding agent benomyl, and the tetrahydrofolate reductase inhibitor methotrexate [19,20]. *FLU1*,

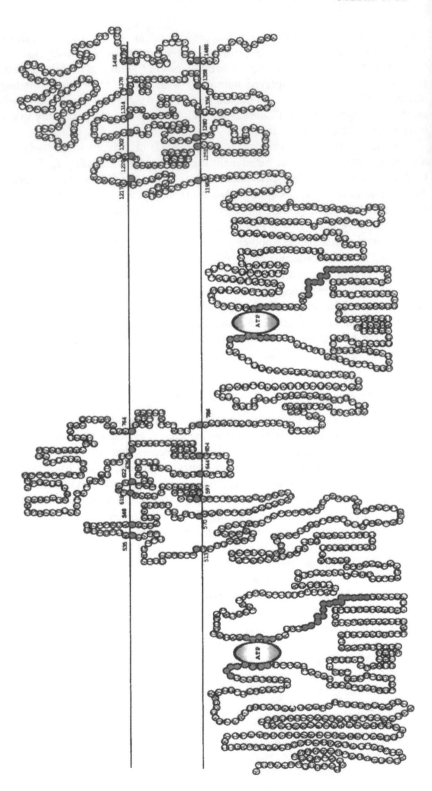

A

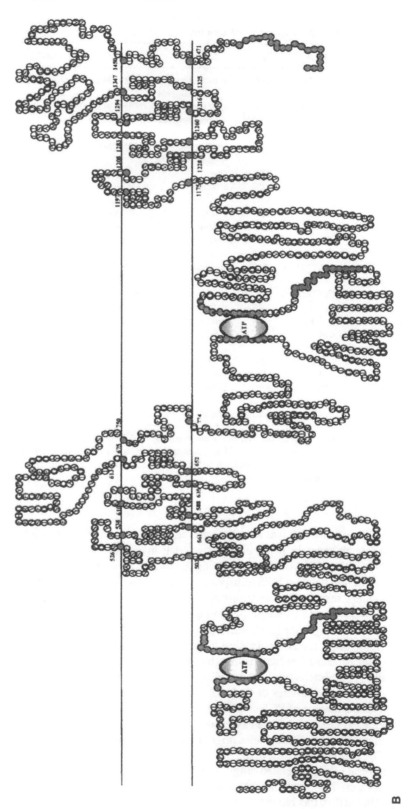

Figure 3 A hypothetical two-dimensional model of Cdr1p (A) and Cdr3p (B). The model is based on the hydrophobicity profiles of amino acid sequences and functional domains. Small circles represent amino acid residues, which are filled with single-letter codes of amino acid. The numbers indicate the beginning and the end of the transmembrane (TM) domains. The putative ATP-binding sites and the signature sequences are shown in *gray*. The first and the last amino acid

Table 2   Multidrug Transporters of *Candida*

Species	Gene	Type	Substrate	Topology	Localization	Chr	Knockout	Ref.
Candida albicans	*CaCDR1*	ABC	antifungals, phospholipids	$(NBF\text{-}TMS_6)_2$	PM	3	viable	31,141
	CaCDR2	ABC	antifungals, phospholipids	$(NBF\text{-}TMS_6)_2$	?	3	viable	32
	CaCDR3	ABC	phospholipids	$(NBF\text{-}TMS_6)_2$	?	4	viable	88
	CaCDR4	ABC	antifungals, phospholipids	$(NBF\text{-}TMS_6)_2$	?	1	viable	89,169
	CaCDR5	ABC	drugs?	?	?	5	?	90
	HST6	ABC	a-factor, drugs?	$(TMS_6\text{-}NBF)_2$	?	3	ND	139
	CaMDR1	MFS	antifungals	$(TMS_6)_2$	?	6	viable	19,20
	FLU1	MFS	antifungals	$(TMS_6)_2$	?	7	viable	114
	CaYOR1	MRP	drugs?	?	?	ND	ND	170
	CaYCF1	MRP	drugs?	?	?	7	viable	171
Candida glabrata	*CgCDR1*	ABC	antifungals	$(NBF\text{-}TMS_6)_2$	?	ND	viable	104
	CgCDR2	ABC	antifungals	$(NBF\text{-}TMS_6)_2$	?	ND	viable	105
	PDH1	ABC	drugs?	$(NBF\text{-}TMS_6)_2$	?	ND	ND	106
	CgMDR1	MFS	antifungals	$(TMS_6)_2$	?	ND	ND	104
Candida dubliniensis	*CdCDR1*	ABC	drugs?	$(NBF\text{-}TMS_6)_2$	?	ND	ND	107
	CdCDR2	ABC	drugs?	$(NBF\text{-}TMS_6)_2$	?	ND	ND	107
	CdMDR1	MFS	antifungals	$(TMS_6)_2$	?	ND	ND	107
Candida krusei	*ABC1*	ABC	drugs?	?	?	ND	ND	108
	ABC2	ABC	drugs?	?	?	ND	ND	108

Abbreviations: Chr, chromosome; ND, not determined; ABC, ATP-binding cassette; MFS, major facilitator superfamily; MRP, multidrug resistance-associated protein; *CDR*, *Candida* drug resistance; *PDH*, pleomorphic drug resistance homolog; *YOR*, yeast oligomycin resistance; *YCF*, yeast cadmium factor, NBF, nucleotide binding fold; TMS, transmembrane segment; PM, plasma membrane; *MDR*, multidrug resistance.

on the other hand, was initially detected as a clone which could confer resistance to fluconazole, although recently it has been shown that mycophenolic acid is a specific substrate for Flu1p [114]. *CaMDR1* is highly homologous to *FLR1*, while *FLU1* revealed high similarity to an *S. cerevisiae* putative MFS transporter, YLL028wp [113,114]. *CaMDR1* expression in *S. cerevisiae* confers resistance to several unrelated drugs, and its overexpression has been linked to azole resistance in *C. albicans*. Seven polymorphic mutant alleles of *CaMDR1* (*CaMDR1-1* to *1–7*) have recently been identified. The complete sequencing of *CaMDR1* alleles revealed several in frame point mutations leading to changes in amino acid residues where insertion/replacement of an aspartate residue in a stretch of serine-asparagine-aspartate-rich domain was most noteworthy [115]. Interestingly, these alleles showed distinct drug resistance profiles (Fig. 4). The relevance of such alleles of *CaMDR1* in azole resistance in *C. albicans* remains to be ascertained. The sequencing of PCR amplified serine-asparagine-aspartate-rich domains of fluconazole-resistant and laboratory isolates revealed that these isolates harbor different alleles. This would mean that the polymorphic alleles might exist in nature in response to different environmental stresses. The expression of *CaMDR1* in *C. albicans* cells was enhanced by benomyl, methotrexate, and several other, unrelated drugs, and was more pronounced in some of the azole-resistant clinical isolates. This confirms that while *CaMDR1* overexpression is linked to fluconazole resistance, other efflux mechanisms are equally important [115].

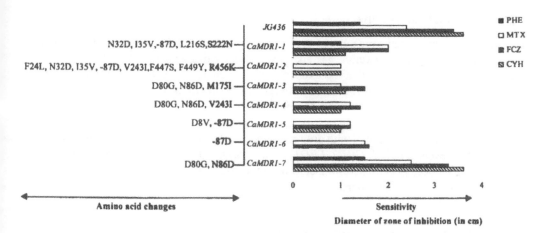

Figure 4   Drug resistance profile and the amino acid changes in the mutant alleles of *CaMDR1*. The bars in histogram represent the sensitivity of mutant alleles (*CaMDR1-7*) to indicated drugs. The increase in height of the bars represents increase in the diameter of inhibition zone on a filter disk assay implying enhanced sensitivity. The left panel of the figure indicates the mutations found in different alleles. The substitution and insertion of amino acids are shown as single-letter codes. (From Ref. 115.)

In order to study the efflux mechanism mediated by CaMdr1p, we have (unpublished observation) overexpressed it in a baculovirus expression system. Overexpression of several membrane proteins including human Mdr1p and Cdr1p has also been achieved by using baculovirus expression systems [94,116]. Since insect cells are able to accomplish many of the higher eukaryotic posttranslation modifications, a recombinant protein produced by this expression system is expected to be functionally active. The *Sf9* cells overexpressing CaMdr1p (~13% of total viral protein) were used to study the accumulation of methotrexate and fluconazole. Reduced accumulation of radiolabeled drugs in infected cells suggested that these drugs are effluxed out by CaMdr1p. The extent of reduction in the accumulation of both the drugs in *Sf9* cells expressing *CaMDR1* was more pronounced (rapid efflux) than when it was expressed in yeast. Interestingly, while both methotrexate and fluconazole could be effluxed out by *CaMDR1* expressing *Sf9* cells, methotrexate appeared to be a better substrate since it was least accumulated. The competition experiments revealed some interesting features of CaMdr1p. For example, a 100-fold excess of benomyl could prevent the efflux of fluconazole and methotrexate, which suggests overlapping binding site(s). However, excess of fluconazole had no effect on the accumulation of methotrexate and vice versa, which indicated that their binding sites on CaMdr1p are probably different. Cycloheximide was also unable to inhibit the transport of both the drugs, again indicating independent binding sites.

Recently, homologs of *CaMDR1* have been identified from *C. dubliniensis* and *C. glabrata* which are termed as *CdMDR1* and *CgMDR1*, respectively [104,107]. It appears that increased expression of *CdMDR1* is the main mechanism of fluconazole resistance involved in *C. dubliniensis* clinical isolates [107]. Since *CgMDR1* confers specific resistance to fluconazole, its constitutive expression in *C. glabrata* may be responsible for the intrinsically low susceptibility of this yeast species to fluconazole [104]. Interestingly, the ser-asn-asp-rich domain present in CaMdr1p and its alleles was not found in CdMdr1p, Cyhrp (cycloheximide resistance protein from *C. maltosa*) and Flr1p from *S. cerevisiae*, which are 93%, 57% and 46% identical to CaMdr1p, respectively, but do not confer resistance to benomyl [107,115].

D. Chromosomal Alterations

An alteration in chromosomal copy number in response to selective pressure, a regulatory princi-ple of gene expression in lower fungi, has also been recently discovered in *C. albicans* [115]. Perepnikhatka et al. have shown that the exposure of *C. albicans* cells to fluconazole resulted in the nondisjunction of two specific chromosomes in drug-resistant mutants. Drug exposure for different time periods led to the gain of one copy of chromosome 3 and in the loss of a homolog of chromosome 4. While at least two genes, *CDR1* and *CDR2*, are localized on chromo-some 3, none of the genes associated with drug resistance are situated on chromosome 4. Interest-ingly, the mRNA levels of *CDR1, CDR2, ERG11*, and *CaMDR1* in these mutants either remained the same or were reduced [117]. Therefore, chromosomal nondisjunction could represent another possible mechanism of drug resistance.

V. PHYSIOLOGICAL RELEVANCE OF DRUG TRANSPORTERS

The rapidly growing family of transporters particularly belonging to ABC superfamily (traffic ATPases) comprises an extremely diverse class of membrane transport proteins that couple ATP hydrolysis to the translocation of solutes across the membrane. The diversity of these proteins is also reflected in their ever-emerging additional roles in absorption, excretion, signal transduction, bacterial pathogenesis, and most importantly, drug and antibiotic resistance [118,119]. Under-standably, considerable attention is being given to the physiological relevance of ABC transport-ers. The most intriguing aspect of these transporters relates to their wider specificity, where a single transporter can recognize a variety of unrelated xenobiotics. As discussed above, there are already several energy-dependent transporters, which have been identified in *Candida* and other pathogenic fungi and some of them have been shown to be involved in drug transport [31,32,88,89]. The *S. cerevisiae* genome completion revealed that it has 29 ABC transporters and an almost equal number of MFS transporters [112,120]. Therefore, it is very likely that *Candida* could also harbor a lot more transporters than what have been discovered so far. Considering the diversity and large size of such proteins [121,122] it is believed that such a large family of transporters may not exclusively export drugs in fungi. Understanding the molecular mechanism of transmembrane transport in general and of antifungal resistance in particular is important for the improved management of *C. albicans* infection. Recent evidence suggests that ABC transporters of fungi, *C. albicans* in particular, are multifunctional proteins with important physiological functions. The following sections describe the so far identified physiological roles for these proteins.

A. Cdr1p Is a Multifunctional Protein

1. Cdr1p as Drug Transporter

The mechanism of entry of drugs into a fungal cell is not very well understood. Several studies have used intracellular accumulation of the drug as an index of efflux, where less accumulation implies more efflux. Many laboratories using this parameter have shown that azole-resistant clinical isolates of *C. albicans* accumulated lower amount of drug than sensitive strains. Notwith-standing the mechanism of efflux, these studies do demonstrate that the effluxing ability of ABC pumps is an energy-dependent process. Cdr1p and Cdr2p are able to efflux out fluconazole, which can be competed out with several unrelated drugs. A causal relationship between efflux mediated by these two proteins was established when Δ*cdr1* and Δ*cdr1*Δ*cdr2* showed much less accumulation of fluconazole than wild-type strain of *C. albicans* [32].

We have shown that the deletion of 79 amino acids from the carboxy terminal of Cdr1p ($\Delta CDR1$; which encompasses the TM12 of this transporter) did not result in total loss of its ability to efflux cytotoxic agents. While the expression of $\Delta CDR1$ in *S. cerevisiae* resulted in impaired sensitivity to drugs like cycloheximide, anisomycin, sulfomethuron methyl, and nystatin, its ability to confer resistance to drugs like o-phenanthroline, 4-nitrosoquinoline-N-oxide, cerulenin, azoles, oligomycin, erythromycin, chloramphenicol, and benomyl remained unaltered. It appears that similar to human Mdr1p, Cdr1p also has localized drug-binding sites in the TM12 stretch. TM12 deletion did not lead to any significant impairment in NTPase activities or in its ability to efflux Rh123 and steroid hormones like β-estradiol. To dissect the functionality of Cdr1p, its truncated version was overexpressed in baculovirus-insect cell expression system. The deletion of TM12 did not affect the targeting of the protein, and ΔCdr1p was exclusively localized in plasma membrane of *Sf9* cells as detected by imunofluorescence. Interestingly, the expression of ΔCdr1p in the baculovirus-insect expression system generated a high drug-stimulated plasma membrane-bound ATPase activity which was not demonstrable when Cdr1p was expressed in *S. cerevisiae* [94]. It has been observed for human Mdr1p that different stimulatory effect of drugs on ATPase activity is closely related to the drug transport (resistance) activities [10,123]. However, we were unable to find such a distinct correlation with Cdr1p. Nevertheless, a cross-talk between the nucleotide-binding domain (NBD) and drugs is demonstrable in the case of Cdr1p.

Using a combination of random mutagenesis and phenotypic screening, Egner et al. have isolated several mutants of the *Pdr5* transporter and have identified amino acids important for substrate recognition and drug transport. Some of the point mutations could also affect the folding of Pdr5p, suggesting that the folded structure is the major substrate specificity determinant [124]. How and where azoles would bind to Cdr1p and Cdr2p before being effluxed out can be predicted by inducing point mutations in these proteins. Why Cdr3p or Cdr4p, which are very similar proteins to Cdr1p and Cdr2p, do not efflux drugs could also become clear from these mutational studies.

2. Cdr1p as Human Steroid Transporter

Cdr1p can specifically transport human steroid hormones—namely, β-estradiol and corticosterone. An *S. cerevisiae* transformant harboring *CDR1* accumulated about threefold less β-estradiol and about twofold less corticosterone than the nontransformed strain. Efflux of β-estradiol and corticosterone was inhibited by 100-fold concentration of β-estradiol, corticosterone, ergosterol, and dexamethasone, but progesterone, which is not transported by Cdr1p, did not affect the efflux or thus the accumulation. Interestingly, some of the drugs—viz., cycloheximide, chloramphenicol, fluconazole, and o-phenanthroline—to which *CDR1* confers resistance could also prevent to some extent the efflux of β-estradiol and corticosterone, thus suggesting commonality among binding site(s) [125]. The homolog of Cdr1p, Pdr5p, and Snq2p from *S. cerevisiae* can also mediate transmembrane transport of steroids and glucosteroids [126,127].

Rhodamine-123 and rhodamine-6G are taken up by *S. cerevisiae* cells, and quenching of its fluorescence has been linked to Pdr5p activity [126]. It has been demonstrated that the dye was accumulated much more abundantly (increase in relative fluorescence) in a *PDR5* disrupted hypersensitive *S. cerevisiae* strain JG436. Strains expressing Cdr1p showed 50% less accumulation of the dye than the control cells. It has recently been observed that, compared to Cdr2p and Cdr3p, Cdr1p is a better efflux pump for β-estradiol, rhodamine, and fluconazole (Prasad et al., unpublished observations).

Since Cdr1p, a multidrug transporter, can selectively mediate energy-dependent transport of human steroid hormones with high affinity and specificity, it is possible that these hormones could be physiological substrates of this protein. In this regard, it is pertinent to mention that corticosteroid- and estrogen-binding proteins in *C. albicans* and other species of *Candida* have already been identified [128–130]. The interaction of some of the azoles such as ketoconazole with corticosteroid-binding protein has been observed [131]. Moreover, the presence of a steroid-responsive element in the promoter of *CDR1* and the upregulation of *CDR1* transcription by β-estradiol strongly suggests the possibility of a steroid receptor cascade, linked to multidrug resistance in *C. albicans* [76]. A steroid efflux system mediated by Cdr1p, which could be a part of the total sterol homeostasis of *Candida* cells, also merits further attention.

3. Cdr1p as Membrane Phospholipid Translocator

Except for human MDR1, which is a general phospholipid translocator and steroid transporter [132,133], and MDR2, which is a specific phosphatidylcholine translocator between the lipid monolayers [134], none of the other characterized MDR genes are so distinctly identified with its functions. Yeast MDRs are no exception and from a host of genes identified as putative efflux pumps, none have been related with their physiological roles, the only exception being *STE6*, which codes for an ABC protein involved in the export of the a-mating factor in *S. cerevisiae* [135–138]. A homolog of *STE6*, *HST6*, has also been isolated from *C. albicans* which is capable of complementing a *S. cerevisiae* ste6 null mutation by restoring the ability of the cells to export the a-factor [139]. There is a report suggesting the involvement of *PDR5* and *YOR1* in lipid translocation in *S. cerevisiae* [140].

   CDR1, which is a homolog of *PDR5*, has energy-dependent floppase activity, which translocates phosphatidylethanolamine (PtdEtn) to the outer leaflet of the membrane [141]. The involvement of Cdr1p in phospholipid translocation was further evident from the fact that increased amount of PtdEtn in the outer leaflet of plasma membrane (PM) of the mycelial form correlated well with *CDR1* expression, which was also more pronounced in mycelial than in bud form of *C. albicans*. The decrease in the availability of PtdEtn in the outer half of PM of a homozygous *CDR1* disruptant confirmed its involvement in PtdEtn translocation. Interestingly, *S. cerevisiae* transformant expressing *CaMDR1* of *C. albicans* was found to have no effect on the PtdEtn distribution pattern between the two leaflets, thus suggesting that phospholipid translocation activity is probably a feature of ABC transporters [141].

   Using NBD fluorescent tagged phospholipid analogs, it has been shown that Cdr1p and its homologs Cdr2p and Cdr3p are general phospholipid translocators (Prasad et al., unpublished observations). Interestingly, Cdr1p and Cdr2p, whose overexpression leads to MDR, elicit outwardly directed phospholipid transbilayer exchange (floppases), while Cdr3p, which does not confer MDR, is shown to be involved in inwardly directed translocation of phospholipids (flippase). In addition to the difference in the directionality of phospholipid translocation, the floppase activities of Cdr1p and Cdr2p and flippase activity of Cdr3p are further distinguishable. The flippase and floppase activities respond differently to inhibitors like NEM and cytochalasinE (Fig. 5). Most importantly, drugs like fluconazole, cycloheximide, and miconazole could affect transbilayer movement of phospholipids mediated by Cdr1p and Cdr2p but had no effect on Cdr3p-mediated transbilayer exchange. These results point out that Cdr1p and Cdr2p presumably have common binding sites for drugs and phospholipids while flippase activity of Cdr3p could be independent of drug binding. The difference in the directionality of phospholipid transfer between Cdrps could be linked to their ability to efflux cytotoxic drugs. If the activity of Cdr3p pump is inwardly directed (flippase), then its inability to participate in drug transport can be explained. However, comprehension of the molecular basis of functional differences between these transporters will have to wait for more experimentation.

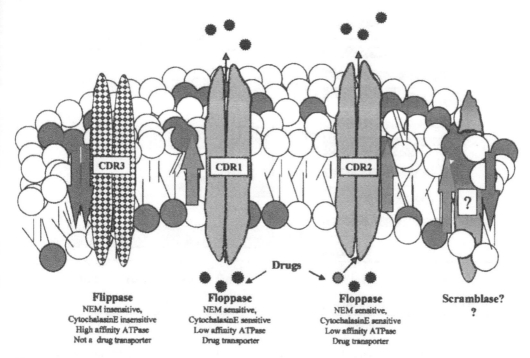

Figure 5 Hypothetical model depicting physiological functioning of Cdr1p, Cdr2p, and Cdr3p. Cdr1p and Cdr2p could translocate membrane phospholipids from inner to outer leaflet (floppase) while evidence suggest that Cdr3p translocate phospholipids from outer to inner leaflet (flippase). Both flippase and floppase activities of these ABC transporters have distinct sensitivities to specific inhibitors (as indicated). Another protein depicted as scramblase is also shown which can translocate phospholipids in either direction; however, such protein which has been identified in mammalian system, has not been detected in any pathogenic fungi.

B. Membrane Lipid Composition and Drug Resistance

Prior studies from our laboratory and others have shown that alterations in the physical state of plasma membrane lipids can influence a number of transport processes. The physical state of a membrane also affects gene expression and signal transduction in yeast [142,143]. A possible role of membrane lipids in the modulation of drug-binding activity of P-glycoprotein (P-gp) has been suggested, where functions such as drug transport and ATPase activities were shown to be affected by the lipid environment and exogenous lipids [144–146]. The ATP binding folds of P-gp could also interact with membrane phospholipids, particularly of the inner leaflet [145]. In vitro experiments involving mammalian P-gp indicated that it required fluid phospholipids in its immediate surroundings for optimal ATPase activity since the membrane-binding domains interact with lipid bilayer, and probably their integrity depends on the presence of certain phospholipids [140,146]. As a result, considerable importance is attached to membrane lipids vis-a-vis functioning of multidrug transporters [141,146,147].

Smriti et al. have analyzed Cdr1p functions by expressing it in different isogenic *S. cerevisiae erg* mutants, which accumulated various intermediates of ergosterol biosynthesis and thus had altered membrane fluidity [148]. Interestingly, the effect of lipid phase alterations due to the accumulation of various intermediates of sterol biosynthesis in *erg* mutants had different

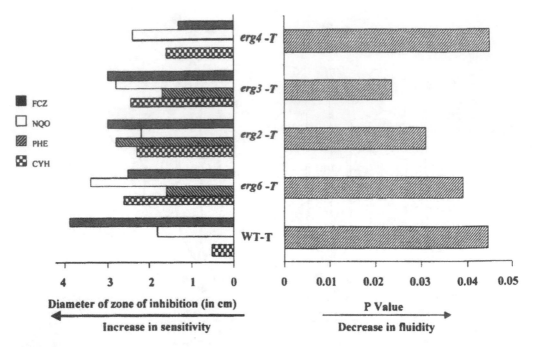

Figure 6 Membrane fluidity affects drug resistance conferred by Cdr1p. The right panel shows bars representing P values (fluorescence anisotropy). The increase in P value indicates higher membrane order (rigidity). *Erg* mutants expressing *CDR1* which have altered membrane fluidity show different susceptibilities to indicated drugs (bars shown on left panel) which is shown as diameter of inhibition zones. (From Ref. 149.)

effects on various functional aspects of Cdr1p. This study provided an understanding of how a yeast multidrug transporter protein, Cdr1p, could change its behavior in response to an altered membrane environment [149]. Earlier, Kaur et al., using the above-mentioned *erg* mutants, showed the effect of an altered lipid environment on the functioning of *PDR5*, a homolog of *CDR1* in *S. cerevisiae* [148]. Given the intimate relationship among P-gp, its hydrophobic substrates, and the surrounding membrane environment, it is expected that membrane lipids will have considerable influence over the functioning of Cdr1p. For instance, the floppase activity associated with Cdr1p could sustain the fluctuation in membrane fluidity of various *erg* mutants. The ability of the transformants to efflux drugs was, however, severely hampered since all became more sensitive to most of the tested drugs to which their wild type was resistant (Fig. 6). CaMdr1p, the MFS drug pump, also elicited reduced resistance and effluxing ability of methotrexate and fluconazole in cells with altered membrane fluidity. Our preliminary results suggest that azole-resistant isolates of *C. albicans* have altered membrane fluidity (Prasad et al., unpublished observation). Therefore, it is important to consider the physical state of membranes while assessing the functioning of membrane-bound drug transporters.

VI. EMERGING FUTURE TARGETS

The understanding of molecular mechanisms of drug resistance is still evolving. There are definitely undefined mechanisms yet to be discovered, especially when multidrug resistance in human pathogenic fungi appears to be contributed by many attributes. In this regard, we discuss here some recent findings worth consideration, which could not only lead to an understanding of newer mechanism of drug susceptibilities but could also lead to new antifungal targets.

A. Protein Glycosylation

It has been known that cell surface components are involved in drug permeability as well as in adhesion, hypha formation, and protein secretion of *C. albicans*. Recent results obtained in *S. cerevisiae* and in *C. albicans* indicate that glycosylation of secreted proteins is a key factor in basal antifungal resistance. In *S. cerevisiae*, defects in *N*-glycosylation enhance sensitivities, especially for aminoglycosides [150,151]. Timpel et al. have identified the *CaPMT1* gene (dolichylphosphate-D mannose; protein O-D-mannosytransferase) and analyzed its function in *C. albicans*. Any defect in *O*-glycosylation of *C. albicans*, in addition to reduced growth and tendency to form cellular aggregates, also has dramatic effects on antifungal sensitivity. *Capmt1* mutants lacking the protein mannosyltransferase that initiates *O*-glycosylation are supersensitive to aminoglyocosides, ketoconazole, and drugs affecting cell wall integrity. No alterations in susceptibilities to drugs like nystatin, AmB, fluconazole, sulfanilamide, and caffeine were observed in *Capmt1* mutants of *C. albicans*. *Capmt1* homozygous mutant strain is avirulent in a mouse model of systemic infection. While deletion of both the alleles of *PMT1* was necessary to see increased sensitivity to most of the drugs, supersensitivity to hygromycin-B was observed even with the heterozygous *CaPMT1/Capmt1* strain. Interestingly, deletion of several *PMT* genes of *S. cerevisiae* is necessary to see altered drug susceptibilities, while the deletion of *CaPMT1* alone was sufficient to see enhanced sensitivity to antifungals [152]. To date, the mechanisms by which glycosylation affects antifungal sensitivities are unknown. In this regard it is worth mentioning some preliminary results where fluconazole was shown to be much less accumulated in *pmt* mutants of *C. albicans*, thus demonstrating a direct effect of defective glycosylation on efflux pump (Prasad et al., unpublished results). The *Candida* Dbase (http://alces.med.umn.edu/candida.html) lists at least five *PMT* genes; therefore, it would be worthwhile to identify and define their role in glycosylation and determine drug susceptibilities in *C. albicans*.

B. *CaALK8*—an Alkane-Inducible Cytochrome P450

The modification of drugs to their nontoxic forms mediated by cytochrome P450 represents an important mechanism by which a cell could confer resistance to different drugs. The role of cytochrome P450 as the detoxifying enzymes in prokaryotes as well as in eukaryotes is well established [153,154]. Although in yeasts, the existence of two different classes of cytochromes P450—viz., P45014DM and P450alk (alkane-inducible)—have been shown, neither has been linked to xenobiotic metabolism [57,58]. P450alk genes represent a large family of genes in *Candida* which make them unique in utilizing straight-chain hydrocarbons. *S. cerevisiae*, on the other hand, lacks this class of cytochromes P450 and thus is unable to assimilate hydrocarbons. Recently, Panwar et al. (manuscript submitted) characterized an alkane-inducible cytochrome P450 gene of *C. albicans* and showed that it is involved in MDR. This gene, designated *CaALK8*, shows sequence homology to a family of alkane-inducible cytochrome P450 genes involved in hydrocarbon assimilation. Interestingly, when *CaALK8* is expressed in *S. cerevisiae* or *C. albicans*, it confers resistance to fluconazole, cycloheximide, o-phenanthroline, nitrosoquinoline oxide, miconazole, and itraconazole. Eight members of P450alk genes have already been identified in *C. maltosa* and *C. tropicalis*, and availability of partial sequences in the *Candida* dbase (http://alces.med.umn.edu/candida.html) also suggests the existence of a multigene family of *ALK* genes in *C. albicans* [155,156]. The involvement of all *CaALK* genes in conferring MDR and the mechanism by which they render the drug nontoxic are two important aspects which need to be investigated. An interesting possibility could be that the incoming drug is modified by *CaALK8* like the alkanes. Interestingly, drugs like o-phenanthroline have been shown to compete for the hydroxylation of lauric acid in *C. tropicalis* [157]. So far, however, the metabolic conversion of drugs has not been shown as part of the drug resistance mechanism in *Candida*. It is pertinent to mention a report from Kelly et al. that demonstrates CYP61 (Δ^{22}-desaturase), which is involved in 22-desaturation in ergosterol biosynthesis in *S. cerevisiae*, can also metabolize xenobiotics. Thus, metabolism of benzo(a)pyrene to 3-hydoxybenzo(a)pyrene mediated by CYP61 was demonstrated. CYP61 appears to be a low-activity hydroxylase as compared to its V_{max} for sterol 22-desaturation but nevertheless, a CYP superfamily member could show metabolism of aflatoxins, dimethylnitrosaminc, and various cyclopenta(a)phenanthrenes [158]. In the light of this, our finding that *CaALK8* (a member of the *CYP52* gene family) could confer MDR in *C. albicans* acquires considerable significance. In view of the large family of such genes present specifically in *Candida* species, these probably may represent new loci contributing to antifungal resistance and thus could be of considerable therapeutic value.

C. Signal Transduction and Drug Resistance

In spite of the widespread occurrence of the ABC and MFS transporter genes and their involvement in conferring drug resistance in *C. albicans* and other pathogenic fungi, the molecular mechanism controlling the expression of MDR genes is poorly understood. Recently, a family of transcription regulators have been identified in *C. albicans* [113,159–161]. It is also well established that at least two of the *C. albicans* MDR genes, *CDR1* and *CDR2*, are transcriptionally activated by different environmental stresses [76]. The precise molecular mechanism by which the signals generated by a divergent family of toxicants activate this small group of transcription factors which, in turn, results in the upregulation of a battery of drug resistance genes, is not known. In order to understand the regulatory mechanisms, which could be controlling the expression of MDR genes in *C. albicans*, it is essential to unravel the regulatory circuits controlling such genes. It is anticipated that uncovering such mechanism(s) may lead to the development of novel therapeutic measures to cope with the induction of drug resistance in the pathogenic variants. Identification of genes, which are involved in drug resistance or identification of transcription factors, which regulate the expression of these drug-resistant genes would be an important step toward such a dissection.

Recently, multiple-antibiotic resistance in *pneumococci* has been shown to be linked to signal transduction system involving specific kinases [162]. It is likely that some signal cascade may also regulate the expression of MDR genes in *C. albicans* and thus could play a role in determining drug susceptibilities of these pathogens. The up/downregulation of the MDR genes in response to various external stresses point out such a possibility. Considering the existence of regulators and their signaling pathways which control phenotypic divergence and virulence of *Candida* [163–168], existence of similar pathways regulating the expression of the MDR genes is a possibility.

VII. CONCLUDING REMARKS

It is evident from the wealth of data recently accumulated that ABC and MFS types of transporters in pathogenic fungi have a very definite role in antifungal resistance. However, an investigation of the roles played by such large family of transporters in fungal physiology is still in its infancy.

Given the vast number of such proteins that are expected to be present in these pathogens, their relevance to overall cellular physiology must be understood. In view of this we have particularly concentrated on the structural and functional analyses as well as on the physiological aspects of drug transporters. Needless to mention, a great deal of investigation is required before mechanistic aspects of substrate specificity, transport, and normal physiological roles of these proteins are resolved. The permeability constrains in antifungal resistance is indeed one of the several mechanisms that contribute to resistance. It is apparent from studies so far that MDR in fungi is a multifactorial phenomenon where still unknown mechanism(s) may be involved. The intensive ongoing research in understanding the molecular mechanism of resistance is expected to improve diagnostics, treatment, and prevention strategies to combat drug resistance in pathogenic fungi.

ACKNOWLEDGMENTS

The authors are thankful to Shyamal Goswami, Neerja, Avmeet, and Neeti for their suggestions during the preparation of the manuscript. We also thank Nandkumar, Sudhakar, Naseem, and Suneet in finalizing the figures. The work reported from the authors' laboratory was supported in part by grants to one of us (R.P.) from the Department of Biotechnology (DBT-BT/PRO798/HRD20/8/98), Department of Science and Technology (SP/SO/D57/97), Council of Scientific and Industrial Research (60(0028)/98-EMR-II), India.

REFERENCES

1. H Vanden Bossche, F Dromer, L Improvisi, M Lozano-Chiu, JH Rex, D Sanglard. Antifungal drug resistance in pathogenic fungi. Med Mycol 36:119–128, 1998.
2. M Ghannoum, LB Rice. Antifungal agents: mode of action, mechanism of resistance, and correlation of these mechanisms with bacterial resistance. Clin Microbiol Rev 12:501–512, 1999.
3. TC White, KA Marr, RA Bowden. Clinical, cellular, and molecular factors that contribute to antifungal drug resistance. Clin Microbiol Rev 11:382–402, 1998.
4. H Vanden Bossche, P Marichal, FC Odds. Molecular mechanisms of drug resistance in fungi. Trends Microbiol 2:393–400, 1994.
5. H Vanden Bossche. Chemotherapy of human fungal infections. In: H Lyr, ed. Modern Selective Fungicides. Properties, Applications, Mechanisms of Action. Jena: Gustav Fisher Verlag, 1995, pp 431–484.
6. MB Frosco, JF Barrett. Importance of antifungal drug-resistance: clinical significance and need for novel therapy. Exp Opin Invest Drugs 7:175–198, 1998.
7. JH Rex, MG Rinaldi, MA Pfaller. Resistance of Candida species to fluconazole. Antimicrob Agents Chemother 39:1–8, 1995.
8. P Marichal. Mechanisms of resistance to azole antifungal compounds. Curr Opin Anti-infect Invest Drugs 1:318–333, 1999.
9. H Yamaguchi. Molecular and biochemical mechanisms of drug resistance in fungi. Nippon Ishinkin Gakkai Zasshi 40:199–208, 1999.
10. MM Gottesman, I Pastan, SV Ambudkar. P-glycoprotein and multidrug resistance. Curr Opin Genet Dev 6:610–617, 1996.
11. S Ruetz, P Gros. A mechanism of P-glycoprotein action in multidrug resistance: are we there yet? Trends Pharmacol Sci 15:260–263, 1994.
12. T Skovsgaard, D Nielsen, C Maare, K Wasserman. Cellular resistance to cancer chemotherapy. Int Rev Cytol 156:77–157, 1994.
13. JS Blanchard. Molecular mechanisms of drug resistance in Mycobacterium tuberculosis. Annu Rev Biochem 65:215–239, 1996.

14. P Borst, M Ouelette. New mechanisms of drug resistance in parasitic protozoa. Annu Rev Microbiol 49:427–460, 1995.

15. BF Dupont, F Dromer, L Improvisi. The problem of azole resistance in *Candida*. J Mycol Med 6: 12–18, 1996.

16. MM Gottesman, CA Hrycyna, PV Schoenlein, UA Germann, I Pastan. Genetic analysis of the multidrug transporter. Annu Rev Genet 29:607–649, 1995.

17. SM Simon, M Schindler. Cell biological mechanisms of multidrug resistance in tumors. Proc Natl Acad Sci USA 91:3497–3504, 1994.

18. CF Higgins. ABC transporters: from microorganisms to man. Annu Rev Cell Biol 8:67–113, 1992.

19. ME Fling, J Kopf, A Tamarkin, JA Gorman, HA Smith, Y Koltin. Analysis of a *Candida albicans* gene that encodes a novel mechanism for resistance to benomyl and methotrexate. Mol Gen Genet 227:318–329, 1991.

20. R Ben-Yaacov, S Knoller, GA Caldwell, JM Becker, Y Koltin. *Candida albicans* gene encoding resistance to benomyl and methotrexate is a multidrug resistance gene. Antimicrob Agents Chemother 38:648–652, 1994.

21. R Prasad. *Candida albicans*: Cellular and Molecular Biology. Berlin: Springer-Verlag, 1991.

22. FC Odds. *Candida* and Candidosis: A Review and Bibliography. 2nd ed. London: Ballière Tindall, 1988.

23. HJ Scholer, A Polak. Resistance to systemic antifungal agents. In: L Bryan, ed. Antimicrobial drug resistance. Orlando: Academic Press, 1984, pp 393–460.

24. CA Hitchcock. Resistance of *Candida albicans* to azole antifungal agents. Biochem Soc Trans 21: 1039–1047, 1993.

25. S Sternberg. The emerging fungal threat. Science 266:1632–1634, 1994.

26. R Prasad, S Krishnamurthy, R Prasad, V Gupta, S Lata. Multidrug resistance: an emerging threat. Curr Sci 71:205–213, 1996.

27. DC Coleman, MG Rinaldi, KA Haynes, JH Rex, RC Summerbell, E Anaissic, A Li, DJ Sullivan. Importance of *Candida* species other than *Candida albicans* as opportunistic pathogens. Med Mycol 36:156–165, 1998.

28. DC Lamb, DE Kelly, W-H Schunck, AZ Shyadehi, M Akhtar, DJ Lowe, BC Baldwin, SL Kelly. The mutation T315A in *Candida albicans* sterol 14α-demethylase causes reduced enzyme activity and fluconazole resistance through reduced affinity. J Biol Chem 272:5682–5688, 1997.

29. TC White. Increased mRNA levels of *ERG16*, *CDR*, and *MDR1* correlate with increased azole resistance in *Candida albicans* isolates from a patient infected with human immunodeficiency virus. Antimicrob Agents Chemother 41:1482–1487, 1997.

30. D Sanglard, F Ischer, L Koymans. Amino acid substitutions in the cytochrome P-450 lanosterol 14-α-demethylase (CYP51A1) from azole-resistant *Candida albicans* clinical isolates contribute to resistance to azole antifungals. Antimicrob Agents Chemother 42:241–255, 1998.

31. R Prasad, PD Worgifosse, A Goffeau, E Balzi. Molecular cloning and characterisation of a novel gene of *C. albicans, CDR1*, conferring multiple resistance to drugs and antifungals. Curr Genet 27: 320–329, 1995.

32. D Sanglard, F Ischer, M Monod, J Bille. Cloning of *Candida albicans* genes conferring resistance to azole antifungal agents: characterisation of *CDR2*, a new multidrug ABC transporter gene. Microbiology 143:405–416, 1997.

33. JMT Hamilton-Miller. Physiological properties of mutagen induced variants of *Candida albicans* resistant to polyene antibiotics. Bacteriol Rev 37:166–196, 1973.

34. SL Kelly, DC Lamb, M Taylor, AJ Corran, BC Baldwin, WG Powderly. Resistance to amphotericin B associated with defective sterol Δ^{8-7} isomerase in a *Cryptococcus neoformans* strain from an AIDS patient. FEMS Microbiol Lett 122:39–42, 1994.

35. TVG Rao, A Trivedi, R Prasad. Phospholipid enrichment of *Saccharomyces cerevisiae* and its effect on polyene sensitivity. Can J Microbiol 31:322–326, 1985.

36. TVG Rao, S Das, R Prasad. Effect of phospholipid enrichment on nystatin action: differences in antibiotic sensitivity between in vivo and in vitro conditions. Microbios 42:145–153, 1985.

37. R Prasad. Lipids in the structure and function of yeast membrane. Adv Lipid Res 21:187–242, 1985.

38. P Mishra, R Prasad. Lipids of *Candida albicans*. In: R Prasad, ed. *Candida albicans*: Cellular and Molecular Biology. Berlin: Springer-Verlag, 1991, pp 128–143.

39. R Prasad, A Koul, PK Mukherjee, MA Ghannoum. Lipids of *Candida albicans*. In: R Prasad, MA Ghannoum, eds. Lipids of Pathogenic Fungi. Boca Raton: CRC Press, 1996, pp 105–137.

40. AS Ibrahim, R Prasad, MA Ghannoum. Antifungals. In: R Prasad, MA Ghannoum, eds. Lipids of Pathogenic Fungi. Boca Rotan: CRC Press, 1996, pp 235–252.

41. DW Denning, K Venkateswarlu, KL Oakley, MJ Anderson, NJ Manning, DA Stevens, DW Warnock, SL Kelly. Itraconazole resistance in *Aspergillus fumigatus*. Antimicrob Agents Chemother 41: 1364–1368, 1997.

42. MB Tobin, RB Peery, PL Skatrud. Genes encoding multiple drug resistance-like proteins in *Aspergillus fumigatus* and *Aspergillus flavus*. Gene 200:11–23, 1997.

43. G Del Sorbo, AC Andrade, JG Van Nisterlooy, JA Van Kan, E Balzi, MA De Waard. Multidrug resistance in *Aspergillus nidulans* involves novel ATP-binding cassette transporters. Mol Gen Genet 254:417–426, 1997.

44. K Angermayr, W Parson, G Stoffler, H Haas. Expression of *atrC*—encoding a novel member of the ATP binding cassette transporter family in *Aspergillus nidulans*—is sensitive to cycloheximide. Biochim Biophys Acta 1453:304–310, 1999.

45. JW Slaven, MJ Anderson, D Sanglard, GK Dixon, J Bille, IS Roberts, DW Denning. Induced expression of a novel ABC transporter gene *adr1* from *Aspergillus fumigatus* in response to itraconazole (abstr). ICAAC 36.J447:555, 1999.

46. M Dean, R Allikmets, B Gerrard, C Stewart, A Kistler, B Shafer, S Michaelis, J Strathern. Mapping and sequencing of two yeast genes belonging to the ATP-binding cassette superfamily. Yeast 10: 377–383, 1994.

47. N-X Chin, I Weitzman, P Della-Latta. In vitro activity of fluvastatin, a cholesterol-lowering agent, and synergy with fluconazole and itraconazole against *Candida* species and *Cryptococcus neoformans*. Antimicrob Agents Chemother 41:850–852, 1997.

48. SJ Thornewell, RB Peery, PL Skatrud. Cloning and chracterisation of *CneMDR1*: a *Cryptococcus neoformans* gene encoding a protein related to multidrug resistance proteins. Gene 201:21–29, 1997.

49. J Wheat, P Marichal, H Vanden Bossche, A Le Monte, P Connolly. Hypothesis on the mechanism of resistance to fluconazole in *Histoplasma capsulatum*. Antimicrob Agents Chemother 41:410–414, 1997.

50. MA De Waard, H Groeneweg, JGM Van Nistelrooy. Laboratory resistance to fungicides which inhibit ergosterol synthesis in *Pencillium italicum*. Neth J Plant Pathol 88:99–112, 1982.

51. MA De Waard, JGM Van Nistelrooy. Toxicity of fenpropimorph to fenarimol-resistant isolates of *Pencillium italicum*. Neth J Plant Pathol 88:231–236, 1982.

52. R Franz, SL Kelly, DC Lamb, DE Kelly, M Ruhnke, J Morschhauser. Multiple molecular mechanisms contribute to a stepwise development of fluconazole resistance in clinical *Candida albicans* strains. Antimicrob Agents Chemother 42:3065–3072, 1998.

53. JL Lopez-Ribol, RK McAtee, LN Lee, WR Kirkpatrick, TC White, D Sanglard, TF Patterson. Distinct patterns of gene expression associated with development of fluconazole resistance in serial *Candida albicans* isolates from human immunodeficiency virus–infected patients with oropharyngeal candidiasis. Antimicrob Agents Chemother 42:2932–2937, 1998.

54. H Vanden Bossche, P Marichal, J Gorrens, M-C Coene, G Willemsens, D Bellens, I Roels, H Moereels, PAJ Janssen. Biochemical approaches to selective antifungal activity, focus on azole antifungals. Mycoses 32:35–52, 1989.

55. CF Wilkinson, K Hetnarski, LJ Hicks. Substituted imidazoles as inhibitors of microsomal oxidation and insecticide synergists. Pest Biochem Physiol 4:299–312, 1974.

56. CF Wilkinson, K Hetnarski, TO Yellin. Imidazole derivatives-a new class of microsomal enzyme inhibitors. Biochem Phamacol 21:3187–3192, 1972.

57. H Vanden Bossche, L Koymans. Cytochromes P450 in fungi. Mycoses 41:32–38, 1998.

58. O Kappeli. Cytochromes P-450 of yeasts. Microbiol Rev 50:244–258, 1996.

59. TC White. Antifungal drug resistance in *Candida albicans*. ASM News 63:427–433, 1997.

60. P Marichal, L Koymans, S Willemsens, D Bellens, P Verhasselt, W Luyten, M Borgers, FCS Ramackers, FC Odds, H Vanden Bossche. Contribution of mutations in the cytochrome P450 14α-

demethylase (Erg11p,Cyp51p) to azole resistance in *Candida albicans*. Microbiology 145: 2701–2713, 1999.

61. B Favre, M Didmon, NS Ryder. Multiple amino acid substitutions in lanosterol 14α-demethylase contribute to azole resistance in *Candida albicans*. Microbiology 145:2715–2725, 1999.

62. SL Kelly, DC Lamb, L Juergen, H Einsele, DE Kelly. The G464S amino acid substitution in *Candida albicans* sterol 14α-demethylase causes fluconazole resistance in the clinic through reduced affinity. Biochem Biophys Res Commun 262:174–179, 1999.

63. J Loffler, SL Kelly, H Hebart, U Schumacher, C Lass-Florl, H Einsele. Molecular analysis of *cyp51* from fluconazole-resistant *Candida albicans* strains. FEMS Microbiol Lett 151:263–268, 1997.

64. SJ Redding, J Smith, G Farinacci, M Rinaldi, A Fothergill, J Rhine-Chalberg, M Pfaller. Resistance of *Candida albicans* to fluconazole during treatment of oropharyngeal candidiasis in a patient with AIDS: documentation by in vitro susceptibility testing and DNA subtype analysis. Clin Infect Dis 18:240–242, 1994.

65. DC Lamb, DE Kelly, TC White, SL Kelly. The R467K amino acid substitution in *Candida albicans* sterol 14α-demethylase causes drug resistance through reduced affinity. Antimicrob Agents Chemother 44:63–67, 2000.

66. SL Kelly, DC Lamb, DE Kelly. Y132H substitution in *Candida albicans* sterol 14α-demethylase confers fluconazole resistance by preventing binding to haem. FEMS Microbiol Lett 180:171–175, 1999.

67. PE Boscott, GH Grant. Modeling cytochrome P450 14 alpha-demethylase (*Candida albicans*) from P450cam. J Mol Graph 12:185–192, 1994.

68. DFV Lewis, A Wiseman, MH Tarbit. Molecular modelling of lanosterol 14α-demethylase (CYP51) from *Saccharomyces cerevisiae* via homology with CYP102, a unique bacterial cytochrome P450 isoform: quantitative structure-activity relationships (QSARs) within two related series of antifungal azole derivatives. J Enzyme Inhib 14:175–192, 1999.

69. K Venkateswarlu, DW Denning, SL Kelly. Inhibition and interaction of cytochrome P450 of *Candida krusei* with azole antifungal drugs. J Med Vet Mycol 35:19–25, 1997.

70. AS Orozco, LM Higginbotham, CA Hitchcock, T Parkinson, DJ Falconer, AS Ibrahim, MA Ghannoum, SG Filler. Mechanism of fluconazole resistance in *Candida krusei*. Antimicrob Agents Chemother 42:2645–2649, 1998.

71. SA Ballard, SL Kelly, SW Ellis, PF Troke. Interaction of microsomal cytochrome P-450 isolated from *Aspergillus fumigatus* with fluconazole and itraconazole. J Med Vet Mycol 28:327–334, 1990.

72. J Guan, HMJ Braks, A Kerkenaar, MA De Waard. Interaction of microsomal cytochrome P450 isozymes isolated from *Pencillium italicum* with DMI fungicides. Pest Biochem Physiol 42:24–34, 1992.

73. AM Van der Bleik, F Baas, T Van der Velde-Koerts, JL Biedler, MB Meyers, RF Ozols, TC Hamilton, H Joenje, P Borst. Genes amplified and overexpressed in human multidrug-resistant cell lines. Cancer Res 48:5927–5932, 1988.

74. GR Stark, GM Wahl. Gene amplification. Annu Rev Biochem 53:447–491, 1984.

75. D Sanglard, K Kuchler, F Ischer, J-L Pagani, M Monod, J Bille. Mechanisms of resistance to azole antifungal agents in *Candida albicans* isolates from AIDS patients involve specific multidrug transporters. Antimicrob Agents Chemother 39:2378–2386, 1995.

76. S Krishnamurthy, V Gupta, R Prasad, SL Panwar, R Prasad. Expression of *CDR1*, a multidrug resistance gene of *Candida albicans*: In vitro transcriptional activation by heat shock, drugs and human steroid hormones. FEMS Microbiol Lett 160:191–197, 1998.

77. H Vanden Bossche, P Marichal, FC Odds, L Le Jeune, MC Coene. Characterization of an azole resistant *Candida glabrata* isolate. Antimicrob Agents Chemother 36:2602–2610, 1992.

78. P Marichal, H Vanden Bossche, FC Odds, G Nobels, DW Warnock, V Timmerman, C Van Broeckhoven, S Fay, P Mose-Larsen. Molecular biological characterization of an azole-resistant *Candida glabrata* isolate. Antimicrob Agents Chemother 41:2229–2237, 1997.

79. H Vanden Bossche, DG Warnock, B Dupont, D Kerridge, S Sengupta, L Improvisi, P Marichal, FC Odds, F Provost, O Ronin. Mechanisms and clinical impact of antifungal drug resistance. J Med Vet Mycol 32:189–202, 1994.

80. TC White. The presence of an R467K aminoacid substitution and loss of allelic variation correlate with an azole-resistant lanosterol 14α demethylase in *Candida albicans*. Antimicrob Agents Chemother 41:1488–1494, 1997.

81. Y Aoyama, Y Yoshida, S Hata, T Nishino, H Katsuki, US Maitra, VP Mohan, DB Sprinson. Altered cytochrome P-450 in a yeast mutant blocked in demethylating C-32 of lanosterol. J Biol Chem 258: 9040–9042, 1983.

82. SL Kelly, DC Lamb, DE Kelly, NJ Manning, J Loeffler, H Hebart, U Schumacher, H Einsele. Resistance to fluconazole and cross-resistance to amphotericin B in *Candida albicans* from AIDS patients caused by defective sterol $\Delta^{5,6}$-desaturation. FEBS Lett 400:80–82, 1997.

83. DC Lamb, S Maspahy, DE Kelly, NJ Manning, A Geber, JE Bennett, SL Kelly. Purification, reconstitution, and inhibition of cytochrome P-450 sterol Δ^{22}-desaturase from the pathogenic fungus *Candida glabrata*. Antimicrob Agents Chemother 43:1725–1728, 1999.

84. GD Albertson, M Niimi, RD Cannon, HF Jenkinson. Multiple efflux mechanisms are involved in *Candida albicans* fluconazole resistance. Antimicrob Agents Chemother 40:2835–2841, 1996.

85. KA Marr, CN Lyons, RA Bowden, TC White. Rapid, transient fluconazole resistance in *Candida albicans* is associated with increased mRNA levels of CDR. Antimicrob Agents Chemother 42: 2584–2589, 1998.

86. N Puri, S Krishnamurthy, S Habib, SE Hasnain, SK Goswami, R Prasad. *CDR1*, a multidrug resistance gene from *Candida albicans*, contains multiple regulatory domains in its promoter and the distal AP-1 element mediates its induction by miconazole. FEMS Microbiol Lett 180:213–219, 1999.

87. M de Micheli, J Bille, D Sanglard. Regulation of ATP-binding cassette (ABC) transporter genes in *Candida albicans* (abstr). Presented at ASM Conference on Candida and Candidiasis C32:57, 1999.

88. I Balan, AM Alarco, M Raymond. The *Candida albicans CDR3* gene codes for an opaque-phase ABC transporter. J Bacteriol 179:7210–7218, 1997.

89. R Franz, S Michel, J Morschhauser. A fourth gene from the *Candida albicans CDR* family of ABC transporters. Gene 220:91–98, 1998.

90. D Sanglard, F Ischer, D Calabrese, M de Micheli, J Bille. Multiple resistance mechanism to azole antifungals in yeast clinical isolates. Drug Resist Updates 1:255–265, 1998.

91. TW Loo, DM Clarke. Identification of residues in the drug binding domain of human P-glycoprotein. J Biol Chem 274:35388–35392, 1999.

92. TW Loo, DM Clarke. Mutations to amino acids located in predicted transmembrane segment 6 (TM6) modulate the activity and substrate specificity of human P-glycoprotein. Biochemistry 33: 14049–14057, 1994.

93. X Zhang, KI Collins, LM Greenberger. Functional evidence that transmembrane 12 and the loop between transmembrane 11 and 12 form part of the drug-binding domain in P-glycoprotein encoded by *MDR1*. J Biol Chem 270:5441–5448, 1995.

94. S Krishnamurthy, U Chatterjee, V Gupta, R Prasad, P Das, P Snehlata, SE Hasnain, R Prasad. Deletion of transmembrane domain 12 of *CDR1*, a multidrug transporter from *Candida albicans*, leads to altered drug specificity: expression of a yeast multidrug transporter in baculovirus expression system. Yeast 14:535–550, 1998.

95. AH Schinkel, JJ Mit, OV Tellingen, JH Beijnen, E Wagenaar, L Van Deemter, CA Mol, VM Van Der, EC Robanus-Maandag, HP te Riele, AJ Berns, P Borst. Disruption of the mouse *mdr1a* P-glycoprotein gene leads to a deficiency in the blood-brain barrier and to increased sensitivity to drugs. Cell 77:491–502, 1994.

96. K Ueda, DP Clark, C-J Chen, IB Roninson, MM Gottesman, I Pastan. The human multidrug resistance (mdr1) gene. cDNA cloning and transcription initiation. J Biol Chem 262:505–508, 1987.

97. CR Leveille-Webster, IM Arias. The biology of the P-glycoproteins. J Membr Biol 143:89–102, 1995.

98. F Buschman, RJ Arceci, JM Croop, M Che, IM Arias, DE Housman, P Gros. *mdr2* encodes P-glycoprotein expressed in the bile canalicular membrane as determined by isoform-specific antibodies. J Biol Chem 267:18093–18099, 1992.

99. E Buschman, P Gros. The inability of the mouse *mdr2* gene to confer multidrug resistance is linked to reduced drug binding to the protein. Cancer Res 54:4892–4898, 1994.
100. S Ruetz, M Raymond, P Gros. Functional expression of P-glycoprotein encoded by the mouse *mdr3* gene in yeast. Proc Natl Acad Sci USA 90:11588–11592, 1993.
101. AJ Smith, JLPM Timmermans-Hereijgers, B Roelofsen, KWA Wirtz, WJ Van Blitterswijk, JJM Smit, AH Schinkel, P Borst. The human MDR3 P-glycoprotein promotes translocation of phosphatidylcholine through the plasma membrane of fibroblasts from transgenic mice. FEBS Lett 354: 263–266, 1994.
102. Y Zhou, MM Gottesman, I Pastan. Studies of human MDR1-MDR2 chimeras demonstrate the functional exchangeability of a major transmembrane segment of the multidrug transporter and phosphatidylcholine flippase. Mol Cell Biol 19:1450–1459, 1999.
103. TJ Walsh, M Kasai, A Francesconi, D Landsman, SJ Chanock. New evidence that *Candida albicans* possesses additional ATP-binding cassette MDR-like genes: implications for antifungal azole resistance. J Med Vet Mycol 35:133–137, 1997.
104. D Sanglard, F Ischer, D Calabrese, PA Majcherczyk, J Bille. The ATP binding cassette transporter gene *CgCDR1* from *Candida glabrata* is involved in the resistance of clinical isolates to azole antifungal agents. Antimicrob Agents Chemother 43:2753–2765, 1999.
105. D Sanglard, F Ischer, J Bille. The role of the ATP binding cassette (ABC)-transporter genes *CgCDR1* and *CgCDR2* in the resistance of *Candida glabrata* to azole antifungals (abstr). Presented at ASM Conference on Candida and Candidiasis C25:55, 1999.
106. H Miyazaki, Y Miyazaki, A Geber, T Parkinson, CA Hitchcock, DJ Falconer, DJ Ward, K Marsden, JE Bennett. Fluconazole resistance associated with drug efflux and increased transcription of a drug transporter gene, *PDH1*, in *Candida glabrata*. Antimicrob Agents Chemother 42:1695–1701, 1998.
107. GP Moran, D Sanglard, SM Donnelly, DB Shanley, DJ Sullivan, DC Coleman. Identification and expression of multidrug transporters responsible for fluconazole resistance in *Candida dubliniensis*. Antimicrob Agents Chemother 42:1819–1830, 1998.
108. SK Katiyar, TD Edlind. Identification of *Candida krusei* multidrug resistance genes: potential role in azole resistance (abstr). ICAAC C153:38, 1998.
109. MHJ Saier, J Reizer. Families and super familes of transport proteins common to prokaryotes and eukaryotes. Curr Opin Struct Biol 1:362–368, 1991.
110. IT Paulsen, MH Brown, RA Skurray. Proton-dependent multidrug efflux systems. Microbiol Rev 60:575–608, 1996.
111. MD Marger, MHJ Saier. A major superfamily of transmembrane facilitators that catalyse uniport, symport and antiport. Trends Biochem Sci 18:13–20, 1993.
112. A Goffeau, J Park, IT Paulsen, J-L Jonniaux, T Dinh, P Mordant, MHJ Saier. Multidrug-resistant transport proteins in yeast: complete inventory and phylogenetic characterization of yeast open reading frames within the major facilitator superfamily. Yeast 13:43–54, 1997.
113. AM Alarco, I Balan, D Talibi, N Mainville, M Raymond. AP1-mediated multidrug resistance in *Saccharomyces cerevisiae* requires *FLR1* encoding a transporter of the major facilitator superfamily. J Biol Chem 272:19304–19313, 1997.
114. DC Calabrese, J Bille, D Sanglard. Characterization of *FLU1*, a novel multidrug eflux transporter gene from *C. albicans* (abstr). Presented at ASM conference on Candida and Candidiasis C56:63, 1999.
115. V Gupta, AK Kohli, S Krishnamurthy, N Puri, SA Aalamgeer, SL Panwar, R Prasad. Identification of polymorphic mutant alleles of *CaMDR1*, a major facilitator of *Candida albicans* which confers multidrug resistance and its in vitro transcriptional activation. Curr Genet 34:192–199, 1998.
116. UA Germann, MC Willingham, I Pastan, MM Gottesman. Expression of the human multidrug transporter in insect cells by a recombinant baculovirus. Biochemistry 29:2295–2303, 1990.
117. V Perepnikhatka, FJ Fischer, M Niimi, RA Baker, RD Cannon, Y-K Wang, F Sherman, E Rustchenko. Specific chromosome alterations in fluconazole-resistant mutants of *Candida albicans*. J Bacteriol 181:4041–4049, 1999.
118. BE Bauer, H Wolfger, K Kuchler. Inventory and function of yeats ABC proteins: about sex, stress, and pleiotropic drug and heavy metal resistance. Biochim Biophys Acta 1461:217–236, 1999.

119. GFL Ames, CS Mimura, SR Holbrook, V Shyamala. Traffic ATPases: a superfamily of transport proteins operating from *Escherichia coli* to humans. Adv Enzymol 65:1–47, 1992.
120. A Decottignies, A Goffeau. Complete inventory of the yeast ABC proteins. Nat Genet 15:137–145, 1997.
121. R Prasad. Nutrient transport in *Candida albicans*, a pathogenic yeast. Yeast 3:209–221, 1987.
122. A Varma, BB Singh, N Karnani, H Lichtenberg-Fraté, M Hofer, BB Magee, R Prasad. Molecular cloning and functional characterisation of a glucose transporter, *CaHGT1*, of *Candida albicans*. FEMS Microbiol Lett 182:15–21, 2000.
123. SV Ambudkar, IH Lelong, J Zhang, CO Cardarelli, MM Gottesman, I Pastan. Partial purification and reconstitution of the human multidrug-resistance pump: characterization of the drug-stimulatable ATP hydrolysis. Proc Natl Acad Sci USA 89:8472–8476, 1992.
124. R Egner, FE Rosenthal, A Kralli, D Sanglard, K Kuchler. Genetic separation of FK506 susceptibility and drug transport in the yeast Pdr5 ATP-binding cassette multidrug resistance transporter. Mol Biol Cell 9:523–543, 1998.
125. S Krishnamurthy, V Gupta, P Snehlata, R Prasad. Characterisation of human steroid hormone transport mediated by Cdr1p, multidrug transporter of *Candida albicans*, belonging to the ATP binding cassette superfamily. FEMS Microbiol Lett 158:69–74, 1998.
126. M Kolaczkowski, ME van der Rest, A Cybularz-Kolaczkowska, J-P Soumillion, WN Konings, A Goffeau. Anticancer drugs, ionophoric peptides, and steroids as substrates of the yeast multidrug transporter Pdr5p. J Biol Chem 271:31543–31548, 1996.
127. Y Mahe, Y Lemoine, K Kuchler. The ATP binding cassette transporters Pdr5 and Snq2 of *Saccharomyces cerevisiae* can mediate transport of steroids in vivo. J Biol Chem 271:26167–26172, 1996.
128. DS Loose, DJ Schurman, D Feldman. A corticosteroid binding protein and endogenous ligand in *C. albicans* indicating a possible steroid-receptor system. Nature 293:477–479, 1981.
129. PJ Malloy, X Zhao, ND Madani, D Feldman. Cloning and expression of the gene from *Candida albicans* that encodes a high-affinity corticosteroid-binding protein. Proc Natl Acad Sci USA 90: 1902–1906, 1993.
130. R Skowronski, D Feldman. Characterisation of an estrogen-binding protein in the yeast *Candida albicans*. Endocrinology 124:1965–1972, 1989.
131. EP Stover, DS Loose, DA Stevens, D Feldman. Ketoconazole binds to the intracellular corticosteroid-binding protein in *Candida albicans*. Biochem Biophys Res Commun 117:43–50, 1983.
132. A Van Helvoort, AJ Smith, H Sprong, I Fritzche, AH Schinkel, P Borst, G Van Meer. MDR1 P-glycoprotein is a lipid translocase of broad specifity, while MDR3 P-glycoprotein specifically translocates phosphatidylcholine. Cell 87:507–517, 1996.
133. K Ueda, N Okamura, M Hirai, Y Tanigawara, T Saeki, N Kioka, T Komano, R Hori. Human P-glycoprotein transports cortisol, aldosterone and dexamethasone, but not progesterone. J Biol Chem 267:24248–24252, 1992.
134. S Ruetz, P Gros. Phosphatidylcholine translocase: a physiological role for the *mdr2* gene. Cell 77: 1071–1081, 1994.
135. E Balzi, A Goffeau. Genetics and biochemistry of yeast multidrug resistance. Biochim Biophys Acta 1187:152–162, 1994.
136. E Balzi, A Goffeau. Yeast multidrug resistance: the PDR network. J Bioenerget Biomembr 27: 71–76, 1995.
137. K Kuchler, RE Sterne, J Thorner. *Saccharomyces cerevisiae STE6* gene product: a novel pathway for protein export in eukaryotic cells. EMBO J 8:3973–3984, 1989.
138. S Michaelis. STE6, the yeast a-factor transporter. Semin Cell Biol 4:17–27, 1993.
139. M Raymond, D Dignard, AM Alarco, N Mainville, BB Magee, DY Thomas. A Ste6p/P-glycoprotein homologue from the asexual yeast *Candida albicans* transports the a-factor mating pheromone in *Saccharomyces cerevisiae*. Mol Microbiol 27:587–598, 1998.
140. A Decottignies, AM Grant, JW Nichols, H De Wet, DB McIntosh, A Goffeau. ATPase and multidrug transport activities of the overexpressed yeast ABC protein Yor1p. J Biol Chem 273:12612–12622, 1998.

141. S Dogra, S Krishnamurthy, V Gupta, BL Dixit, CM Gupta, D Sanglard, R Prasad. Asymmetric distribution of phosphatidylethanolamine in *C. albicans*: possible mediation by *CDR1*, a multidrug transporter belonging to ATP binding cassette (ABC) superfamily. Yeast 15:111–121, 1999.

142. E Moskvina, E-M Imre, H Ruis. Stress factors acting at the level of plasma membrane induce transcription via the stress response element (STRE) of the yeast *Saccharomyces cerevisiae*. Mol Microbiol 32:1263–1272, 1999.

143. H Ruis, C Schuller. Stress signaling in yeast. BioEssays 17:959–965, 1995.

144. T Saeki, AM Shimabuku, Y Azuma, Y Shibano, T Komano, T Ueda. Expression of human P-glycoprotein in yeast cells: effects of membrane component sterols on the activity of P-glycoprotein. Agric Biol Chem 55:1859–1865, 1991.

145. IL Urbatsch, AE Senior. Effects of lipids on ATPase activity of purified Chinese hamster P-glycoprotein. Arch Biochem Biophys 316:135–140, 1995.

146. FJ Sharom. The P-glycoprotein multidrug transporter: interactions with membrane lipids, and their modulation of activity. Biochem Soc Trans 25:1088–1096, 1996.

147. S Ruetz, M Brault, WS Dalton, P Gros. Functional interactions between synthetic alkyl phospholipids and the ABC transporters P-glycoprotein, ste6, MRP, and pgh1. Biochemistry 36:8180–8188, 1997.

148. R Kaur, AK Bachhawat. The yeast multidrug resistance pump, Pdr5p, confers reduced drug resistance in erg mutants of *Saccharomyces cerevisiae*. Microbiology 145:809–818, 1999.

149. Smriti, S Krishnamurthy, R Prasad. Membrane fluidity affects functions of Cdr1p, a multidrug ABC transporter of *Candida albicans*. FEMS Microbiol Lett 173:475–481, 1999.

150. M Gentzsch, W Tanner. The *PMT* gene family: protein *O*-glycosylation in *Saccharomyces cerevisiae* is vital. EMBO J 15:5752–5759, 1996.

151. M Gentzsch, W Tanner. Protein-O-glycosylation in yeast: protein-specific mannosyltransferases. Glycobiology 7:481–486, 1997.

152. C Timpel, S Strahl-Bolsinger, K Ziegelbauer, JF Ernst. Multiple functions of Pmt1p-mediated protein *O*-mannosylation in the fungal pathogen *Candida albicans*. J Biol Chem 273:20837–20846, 1998.

153. S Graham-Lorence, JA Peterson. P450s: structural similarities and functional differences. FASEB J 10:206–214, 1996.

154. DW Nebert, FJ Gonzalez. P450 genes: structure, evolution, and regulation. Annu Rev Biochem 56: 945–993, 1987.

155. M Ohkuma, S-I Muraoka, T Tanimoto, M Fujii, A Ohta, M Takagi. *CYP52* (cytochrome P450alk) multigene family in *Candida maltosa*: identification and characterisation of eight members. DNA Cell Biol 14:163–173, 1995.

156. W Seghezzi, C Meili, R Ruffiner, R Kuenzi, D Sanglard, A Fiechter. Identification and characterisation of additional members of the cytochrome P450 multigene family *CYP52* of *Candida tropicalis*. DNA Cell Biol 11:767–780, 1992.

157. W Duppel, J-M Lebeult, MJ Coon. Properties of a yeast cytochrome P-450-containing enzyme system which catalyses the hydroxylation of fatty acids, alkanes and drugs. Eur J Biochem 36: 583–592, 1973.

158. SL Kelly, DC Lamb, DE Kelly. Sterol 22-desaturase, cytochrome P45061, possesses activity in xenobiotic metabolism. FEBS Lett 412:233–235, 1997.

159. AM Alarco, M Raymond. The bZip transcription factor Cap1p is involved in multidrug resistance and oxidative stress response in *Candida albicans*. J Bacteriol 181:700–708, 1999.

160. D Talibi, M Raymond. Isolation of a putative *Candida albicans* transcriptional regulator involved in pleiotropic drug resistance by functional complementation of a *pdr1 pdr3* mutation in *Saccharomyces cerevisiae*. J Bacteriol 181:231–240, 1999.

161. S Wirsching, S Michel, G Kohler, J Morschhauser. Activation of the multiple drug resistance gene MDR1 in fluconazole-resistant clinical *Candida albicans* strains is caused by mutations in a trans regulatory factor. J Bacteriol 182:400–404, 2000.

162. MS Gilmore, JA Hoch. A vancomycin surprise. Nature 399:524–527, 1999.

163. BR Braun, AD Johnson. Control of filament formation in *Candida albicans* by the transcriptional repressor TUP1. Science 277:105–109, 1997.

164. VR Stoldt, A Sonneborn, CE Leuker, JF Ernst. Efg1p, an essential regulator of morphogenesis of the human pathogen *Candida albicans*, is a member of a conserved class of bHLH proteins regulating morphogenetic processes in fungi. EMBO J 16:1982–1991, 1997.

165. JA Calera, GH Choi, RA Calderone. Identification of a putative histidine kinase two-component phosphorelay gene (*CaHK1*) in *Candida albicans*. Yeast 14:665–674, 1998.

166. H Liu, J Kohler, GR Fink. Suppression of hyphal formation in *Candida albicans* by mutation of a *STE12* homolog. Science 266:1723–1726, 1994.

167. A Porta, AM Ramon, WA Fonzi. *PRR1*, a homolog of *Aspergillus nidulans palF*, controls pH-dependent gene expression and filamentation in *Candida albicans*. J Bacteriol 181:7516–7523, 1999.

168. AM Ramon, A Porta, WA Fonzi. Effect of environmental pH on morphological development of *Candida albicans* is mediated via the PacC-related transcription factor encoded by *PRR2*. J Bacteriol 181:7524–7530, 1999.

169. D Sanglard, F Ischer, M Monod, S Dogra, R Prasad, J Bille. Analysis of the ATP-binding cassette (ABC)-transporter gene of *CDR4* from *Candida albicans* (abstr) Presented at ASM Conference on Candida and Candidiasis C27:56, 1999.

170. A Ogawa, T Hashida-Okado, M Endo, T Tsuruo, K Takesako, I Kato. Role of ABC transporters in aureobasidin A resistance. Antimicrob Agents Chemother 42:755–761, 1998.

171. S Theiss, N Agabian, G Kohler. Molecular characterization of a new ATP-dependent transporter in *Candida albicans* (abstr). Presented at ASM Conference on Candida and Candidiasis C29:56–57, 1999.

27

Antifungal Susceptibility Testing: Clinical Interpretations and Utility

Sofia Perea and Thomas F. Patterson
University of Texas Health Science Center at San Antonio, San Antonio, Texas

I. INTRODUCTION

During the past two decades the prevalence and severity of fungal infections in humans have dramatically increased along with the incidence of drug-resistant disease in immunocompromised patient populations [1]. Infections due to genera *Candida* spp. and *Aspergillus* spp. are important causes of morbidity and mortality in these patients, representing 90% of all fungal infections [2]. In case of *Candida*, the rates of candidemia are increasing worldwide, representing the fourth most common nosocomial bloodstream infection in the United States [3,4]. Systemic *Candida* infections are associated with a high mortality rate (38%) and a prolongation of hospital stay. Currently, *C. albicans* accounts for ~50–60% of all nosocomial infections, although a noticeable shift in the species toward yeasts other than *C. albicans* (*C. tropicalis, C. krusei, C. parapsilosis, C. glabrata*) has occurred. Even though non-*albicans* species are considered less invasive and virulent than *C. albicans*, some species arc inherently less susceptible to common antifungals, which makes them less amenable to treatment [5–8].

In HIV-infected patients, other fungal infections such as mucosal *Candida* infection, cryptococcal meningitis, and systemic endemic mycoses are also present, having an increased incidence with the progression of the HIV infection and reduction of the CD4 lymphocyte count [9,10]. *C. neoformans* is the cause of the most common life-threatening fungal infection in AIDS patients. Given the high incidence of relapse after initial antifungal therapy with amphotericin and flucytosine, the current management includes lifelong suppressive fluconazole therapy [11]. Chronic use of azoles for long-term suppressive therapy may become a factor for the selection of cryptococcal isolates that are more resistant to azoles, although the development of the fluconazole resistance has been very uncommon [12]. Aspergillosis, although less frequent than candidiasis, is associated with the highest mortality rate (27–77%) in severely immunocompromised populations such as patients undergoing allogenic bone marrow transplantation [13–16]. Several other, less common fungi are becoming increasingly recognized as the source of deep fungal infections such as zygomycetes (*Rhizopus arrhizus, Absidia corymbifera,* and *Rhizomucor pusillus*), *Fusarium* spp., *Trichosporon beigelii, Blastoschizomyces capitatus, Scedosporium* spp., *Acremonium* spp., *Malassezia furfur,* and dematiaceous fungi [17–23].

This increase in the rate of opportunistic fungal infections has been accompanied by the development of new, less toxic, and systemically active antifungal agents that represent therapeutic alternatives to amphotericin-B such as fluconazole, itraconazole, the various amphotericin lipid formulations (amphotericin-B lipid complex [Abelcet], amphotericin-B colloidal dispersion [Amphocil], liposomal amphotericin-B [Ambisome]); the new triazoles (voriconazole, posaconazole [SCH 56592], ravuconazole [BMS-207147]); antifungals with new targets such as echinocandins and pneumocandins (LY303366 and MK-0991); pradimicins (BMS 181184); and nikkomycin [24,25]. With the proliferation of antifungal agents, therapeutic options are more numerous, and the clinician must select the agent that represents the best treatment strategy for a given patient. However, the growing number of reports on the development of drug resistance to one or more antifungal agents makes this decision more difficult [26–29].

All these factors have increased the interest in developing standardized tests to determine antifungal drug susceptibility as well as in optimizing these tests to accurately predict clinical outcome. A decade ago, antifungal susceptibility testing was only occasionally performed and had not been carefully developed and standardized. This fact was translated into a very poor reproducibility, and agreement of results obtained in intralaboratory and interlaboratory testing was not acceptable. In 1983, the National Committee for Clinical Laboratory Standards (NCCLS) responded to these problems by establishing a subcommittee to develop standardized antifungal susceptibility testing procedures that focused on broth-based methodologies with defined media. A number of investigators collaborated both independently and in cooperation with the subcommittee to determine the role of different variables such as inoculum size, inoculum preparation, medium composition (liquid vs. solid media), incubation time, temperature, volume, and endpoint definition in the standardization of the technique [30–35].

As a result of all these experiments, in 1997, the National Committee for Clinical Laboratory Standards approved the methodology for the standardization of broth-based macrodilution and microdilution methods for determination of the susceptibility of *Candida* spp. and *Cryptococcus neoformans* against amphotericin-B, flucytosine, ketoconazole, itraconazole, and fluconazole [36]. This document, called M27-A and titled "Reference Method for Broth Dilution Antifungal Susceptibility Testing of Yeast," addressed the selection and preparation of antifungal agents, the implementation and interpretation of test procedures, and the quality control requirements for susceptibility testing of yeasts that cause invasive fungal infections (Table 1). Subsequent studies showed a good agreement between both macrodilution and microdilution methods, established that the incubation at 35°C for 48 and 72 hr provided the most consistent results for *Candida* spp. and *C. neoformans*, respectively, and demonstrated that the 48-hr microdilution method was in close agreement with the macrodilution MIC [37,38].

The next step was to correlate the clinical outcome with the in vitro results. Ideally, the results of in vitro antifungal susceptibility tests should provide a reliable prediction of in vivo response to therapy in human infections. However, the limitation of these highly artificial testing methods is such that only modest correlation exists between in vitro susceptibility testing and the outcome of the complex biological process that clinical infection represents. Predicting clinical outcome is an extraordinarily difficult issue, where the MIC is just a piece of the puzzle in which other factors, such as the drug used, the host, and the isolate itself, participate (Table 2) [39–49]. The problem is, then, to determine the approximate relationship between the MIC and the likelihood of successful outcome despite the influence of these factors.

Prior to the establishment of interpretative breakpoints for antifungal susceptibility testing using the results obtained in clinical trials, attempts have been made to correlate the clinical outcome with in vitro results using animal models [50–55]. This method offers the advantage of fully integrating the effects of both the antimicrobial and the host factors. By performance of parallel studies of organisms that differ only in their susceptibility in vitro, the effect of MIC can be studied in detail. This technique is powerful, and studies with antifungal agents have often demonstrated general correlation with MIC and outcome. Unfortunately, animal models

Table 1 Summary of the M27-A Methodology Developed by the NCCLS for Standardization of Antifungal Susceptibility Testing for Yeasts

Factor	M27-A methodology
Methodology	Broth macrodilution, 1 mL final volume; or broth microdilution, 0.2 mL final volume
Medium	RPMI-1640 containing 0.165 M MOPS (morpholinepropanesulfonic acid) pH, 7.0 [a]Antibiotic medium 3 for amphotericin-B [b]Yeast nitrogen base for *C. neoformans* [c]Supplementation of test medium with glucose to a final concentration of 20 g/L
Fungal inoculum	0.5–2.5 × 10³ organisms
Incubation temperature	35°C
Incubation time	[d]48 hr (*Candida* spp.) or 72 hr (*Cryptococcus neoformans*)
Endpoint	Amphotericin-B; optically clear tube; [e]azoles and 5-flucytosine: 80% reduction in turbidity by comparison with growth control
Drugs and quality control (QC) isolates	Two QC isolates and corresponding QC ranges established via the M23 procedure are specified for amphotericin-B, flucytosine, ketoconazole, itraconazole, and fluconazole

Some modifications were included for special circumstances:

[a] The use of Antibiotic Medium 3 may enhance detection of resistance, but this medium is not standardized and substantial lot-to-lot variability is possible [102,103].

[b] The use of yeast nitrogen base may enhance the growth of *C. neoformans* and improve the clinical relevance of antifungal MICs [66].

[c] Supplementation of the test medium so that it contains glucose at a final concentration of 20 g/L may simplify endpoint determination [104].

[d] Improved interlaboratory reproducibility was observed when reading at 48 hr vs. 24 hr in case of amphotericin-B, flucytosine, ketoconazole, and fluconazole [105].

[e] This endpoint was used to handle the trailing growth phenomenon seen with the azole antifungal agents [35,57]. Trailing growth cause the MICs for fluconazole for some *Candida* isolates to be low (<1 µg/mL) after 24 hr of growth but much higher (>64 µg/mL) after 48 hr.

Source: Ref. 57.

Table 2 Factors Other Than Susceptibility In Vitro Influencing the Clinical Outcome of a Fungal Infection

Factors	Variables
Drug pharmacokinetics	Dosing regimen, biodisponibility, drug stability, metabolism, drug interactions, protein binding, metabolites, tissue penetration, postantifungal effect.
Host factors	Patient compliance, immune system, type of infection, underlying disease.
Site of infection	Source of infection, drug penetration, presence of foreign body (prosthetic devices, intravascular catheters), abscess formation.
Pathogen	Virulence factors; evasion of host immflamatory response, biofilm formation, decrease virulence by the acquisition of resistance.

Source: Ref. 57.

of infection may not necessarily mimic human infection. In addition, drug kinetics often differ substantially between man and other animals. However, these types of results can provide a very useful starting point for the establishment of the correlation between in vitro results and in vivo clinical outcome.

II. YEASTS

A. *Candida* spp.

In case of *Candida* spp., an early attempt to correlate in vitro antifungal susceptibility data with clinical outcome was published by Ghannoum et al. in 1996 [56]. In their review, they examined all the studies published in the literature that contained data about MIC and clinical outcomes in humans, and concluded that in vitro susceptibility testing could predict outcome only in selected clinical situations, such as fluconazole-treated AIDS patients with oropharyngeal candidiasis. In the case of more complex clinical situations, such as heterogeneous patients with invasive candidiasis, no such clear-cut correlation was observed. One year later, tentative breakpoints for susceptibility testing of fluconazole and itraconazole MICs against *Candida* spp. were established largely using data from oropharyngeal candidiasis [57]. Three principles of interpretation of antimicrobial susceptibility testing were employed in the method: (1) a MIC is not a physical or chemical measurement; (2) host factors are often more important than susceptibility test results in determining clinical outcome; and (3) in vitro susceptibility may not always predict success of a particular therapy, but could indicate the possibility of failure for a particular drug or dosage when an infection is caused by a resistant isolate.

The data packages developed by the manufacturers of the antifungals fluconazole (Pfizer, New York) and itraconazole (Janssen Pharmaceutica, Titusville, NJ) that contained MICs of *Candida* isolates and outcome data from trials of therapy with either fluconazole and itraconazole for oropharyngeal candidiasis in patients with AIDS, and also, in the case of fluconazole, from patients with invasive *Candida* infections were analyzed. In case of fluconazole, 636 *Candida* isolates from patients enrolled in six trials of fluconazole as therapy for oropharyngeal candidiasis in patients with AIDS (528 isolates: 77% *C. albicans*, 13% *C. glabrata*, 5% *C. tropicalis*, 3% *C. krusei*, and 2% other *Candida* spp.) and from three trials of fluconazole as therapy for nonneutropenic patients with bloodstream and visceral candidiasis infection (108 isolates) were evaluated [58–64]. Based on the data analyzed, tentative breakpoints of ≤8 μg/mL as susceptible and ≥64 μg/mL as resistant were established. Isolates inhibited by fluconazole at concentrations of 16–32 μg/mL that respond to increased doses of fluconazole, were placed in the new category called susceptible dose dependent (S-DD). These studies suggested that a dose response to fluconazole varied with the MIC with higher doses of fluconazole used to successfully treat patients infected with isolates for which MICs are higher, and that failure of fluconazole therapy becomes likely when the MIC determined by NCCLS methodology exceeds the predicted peak serum levels of fluconazole expected for a given dosing regimen.

Recent additional studies in experimental models [64] have shown that AUC/MIC ratio is important in predicting clinical response to fluconazole. The correlation between fluconazole MIC and response to therapy is strongest for patients with oropharyngeal candidiasis and *C. albicans* infection. More limited data are available that correlate MIC or dose with outcome for non-*albicans Candida* infections or for invasive *Candida* infections. For *C. krusei*, the definition of susceptible and resistant does not apply since this organism is considered to be intrinsically resistant to fluconazole.

In case of itraconazole, 355 *Candida* spp. isolates (87% *C. albicans*, 9% *C. glabrata*, 2% *C. krusei*, and 2% other *Candida* spp.) from HIV patients enrolled in four trials of itraconazole

Table 3 Tentative Breakpoints for Fluconazole and Itraconazole When the MIC is Determined by M27-A[a,b]

Antifungal agent	Range of MICs (μg/mL) per category		
	Susceptible (S)	Susceptible-dose dependent (S-DD)[c]	Resistant (R)
Fluconazole	≤8	16–32	≥64
Itraconazole	≤0.125	0.25–0.5	≥1

[a] Isolates for *Candida krusei* should be considered resistant to fluconazole regardless of the reported MIC to fluconazole.
[b] Breakpoint values are applicable for MICs determined according to NCCLS-approved methods only.
[c] Isolates having an MIC in the susceptible-dose-dependent range should be treated with fluconazole 400–800 mg or an appropriate dose of itraconazole with results in serum concentration of ≥0.5 μg/mL.

solution as therapy for oropharyngeal candidiasis were studied [65]. As with fluconazole, consideration of the overall clinical data and their correlation with the pharmacokinetics of itraconazole indicated that the response of oropharyngeal candidiasis to itraconazole varies with MIC. Based on the data analyzed, tentative itraconazole breakpoints of ≤0.125 μg/mL as susceptible and ≥1.0 μg/mL as resistant were established. Because infections due to isolates for which the itraconazole MIC are 0.25–0.5 μg/mL were observed to respond more often if higher itraconazole plasma levels were ensured, these isolates were placed in the susceptible dose-dependent category, which means that susceptibility is dependent on achieving the maximal possible blood level. These data were developed only in patients with mucosal infection, so the extrapolation of these data to patients with invasive candidal infection is not established.

The guidelines for interpreting the MIC of fluconazole and itraconazole proposed represented a substantial advance in the process of making antifungal susceptibility a clinically useful tool (Table 3). However, it is important to comment on the limitations of the approach:

1. The breakpoints proposed are only valid for two drugs, fluconazole and itraconazole, and only for *Candida*.

2. The in vivo–in vitro correlation for isolates at the higher MIC values obtained was not as strong in the case of yeasts other than *C. albicans* or in the case of systemic mycoses.

B. *Cryptococcus neoformans*

In case of *C. neoformans* it was clear throughout the development of the M27 methodology that this approach was suboptimal, because of the slow growth rate obtained in the RPMI 1640 broth, requiring 72 hr of incubation, and because some strains did not even grow. On the basis of these findings, a modification of the NCCLS method, using yeast nitrogen-based medium buffered to a pH of 7.0, an inoculum of 10^4 cells/mL, and incubation at 35°C for 48 hr in a microdilution format was developed. The MIC endpoint was read spectrophotometrically and for fluconazole was defined as 50% inhibition at 420 nm [66]. A multicenter evaluation of this method demonstrated an excellent level of interlaboratory agreement (96%) and an overall agreement of 90% with the M27 microdilution method [67]. The number of studies that establish the value of MICs as predictors of the clinical response to therapy in patients with *C. neoformans* infections are scarce. Because of that, the establishment of interpretative breakpoints for this pathogen are not available.

Casadevall et al. found an increase in the fluconazole MICs for serial *C. neoformans* isolates that were recovered from five patients with recurrent cryptococcal meningitis [68]. Paugam et al. and Birley et al. reported clinical and in vitro fluconazole resistance in three AIDS patients with recurrent crypococcal meningitis (increases from 4 to 64, 16 to 128, and 0.25 to 16 μg/mL, respectively) [69,70]. Armengou et al. described another possible case of fluconazole resistance development during suppressive therapy in patients with AIDS-associated cryptococcal meningitis (an increase in the MIC to 64 μg/mL) [71]. Witt et al. studied the isolates from 76 patients with acute AIDS-associated cryptococcal meningitis treated with fluconazole ± flucytosine. It was observed that those cases where the MIC of fluconazole was 0.25 μg/mL had a 25% treatment failure. This probability of failure increased to >80% for those whose MIC of fluconazole was 16 μg/mL [72]. Aller et al. studied the isolates from 25 patients with AIDS-associated cryptococcal disease treated with fluconazole. CSF and serum cryptococcal antigen levels were higher for the five patients with treatment failure than for the other 20 patients that responded to treatment. The fluconazole MICs for the infecting isolates from the five treatment failure patients were 16 μg/mL, while the fluconazole MICs for the other patients isolates were lower (<16 μg/mL) [73].

III. FILAMENTOUS FUNGI

The importance of susceptibility testing with pathogenic filamentous fungi has been less thoroughly studied than in yeasts. A proposed method (M38-P) to standardize the in vitro antifungal susceptibility testing for molds has recently been established (Table 4) [74]. Prior to this, studies to examine the role of different variables (inoculum size, type of inocula, incubation temperature, time of reading) as well as collaborative interlaboratory evaluations of the M27 reference method adapted to the testing of molds were developed. The initial collaborative six-center study evaluated the use of macrodilution and microdilution broth methods to determine the in vitro susceptibility of 25 isolates of filamentous fungi (*Aspergillus fumigatus*, *Aspergillus flavus*, *Pseudallescheria boydii*, *Rhizopus arrhizus*, and *Sporothrix schenckii*) to amphotericin-B, fluconazole, itraconazole, miconazole, and ketoconazole [75]. The results of this study were very encouraging

Table 4 Summary of the Methodology Developed for Standardization of Antifungal Susceptibility Testing for Filamentous Fungi (M38-P)

Factor	M38-P methodology
Methodology	Broth macrodilution, 1 mL final volume; or broth microdilution, 0.2 mL final volume
Medium	RPMI-1640 containing 0.165 M MPOS (morpholinepropanesulfonic acid), pH 7.0
Fungal inoculum	0.4–5 × 10⁴ CFU/mL
Incubation temperature	35°C
Incubation time	21–26 hr (*Rhizopus spp.*)
	46–50 hr (most other opportunistic, filamentous fungi: *Fusarium* spp., *Aspergillus* spp., and *Sporothrix schenkii*)
	70–74 hr (*P. boydii*)
Endpoint	Amphotericin-B; optically clear tube; azoles, flucytosine: ≥50% reduction in turbidity by comparison with growth control
Quality control isolates	*C. parapsilosis* ATCC 22019, *C. krusei* ATCC 6258, *A. flavus*[a] and *A. fumigatus*[a]

[a] ATCC numerical designation still pending.

Table 5 Evaluation of Correlation Between Antifungal Susceptibilities of Filamentous Fungi In Vitro and Antifungal Treatment Outcomes in Animal Infection Models[a]

	Results for amphotericin-B			Results for itraconazole		
Fungus isolate (n)	MIC (μg/mL)	Response	Fungus isolate (n)	MIC (μg/mL)	Response
Rhizopus arrhizus (2)	0.25	Active	Aspergillus fumigatus (1)	0.25	Active
Aspergillus flavus (1)	1	Not active	Aspergillus flavus (1)	0.5	Active
Aspergillus fumigatus (1)	1	Active	Pseudallescheria boydii (2)	1	Not active
Fusarium solani (1)	1	Not active	Rhizopus arrhizus (2)	2	Not active
Fusarium oxysporium (2)	2	Not active/?[b]	Fusarium oxysporium (2)	>16	Not active/?[b]
Pseudallescheria boydii (2)	4	Not active	Fusarium solani (1)	>16	Not active

[a]The isolates are those used in Espinel-Ingroff et al. [75] for the study of standardization of antifungal susceptibility testing.
[b]Results were too inconclusive for interpretation.
Source: Ref. 78.

and demonstrated excellent intralaboratory and interlaboratory agreement (90–100%) for macrodilution and microdilution methods in the testing of amphotericin-B, fluconazole, miconazole, and ketoconazole. A lower level of agreement (70–90%) was observed in the testing of itraconazole. A subsequent large-scale study involving 11 laboratories and 30 isolates representing six species of opportunistic mold pathogens showed a high level of interlaboratory agreement among the MICs determined by a broth microdilution adaptation of the M27 method [76]. Similar results were obtained by other independent investigators using microdilution techniques following the M27 reference method with minor modifications (incubation temperature 25°C) [77].

The clinical importance of mold infections in immunocompromised hosts cannot be overstated; however, the incidence of infections with most opportunistic mold pathogens is too low to permit a large-scale prospective comparison of antifungal MICs for molds with the clinical results of antifungal treatment. For this reason, to date, there are minimal clinical data to support the relevance of filamentous fungi susceptibility testing in vitro. However, several studies in animal models have correlated efficacy with susceptibility results for some genera of molds (Table 5) [78]. Animal experiments were carried out by Odds et al. in which the activities of amphotericin-B and itraconazole were determined in relation to previously calculated MICs of the infecting isolates. For both drugs the treatment responses judged as showing some activity were associated with lower MICs than the responses considered as showing no activity in vivo. For the fungi for which the amphotericin-B or the itraconazole MIC was <1 μg/mL, a response of some kind was seen in the experimental infections. For the fungi for which the amphotericin-B MICs were ≥2 μg/mL, or the itraconazole MICs were ≥1 μg/mL, no response was seen. However, the overlap (amphotericin-B) and a 1-dilution difference (itraconazole) in MICs associated with response judged as active and inactive suggest that such MICs could not be interpreted as predicting treatment outcome in these animal models. The conclusion of the study was that a limited association between MIC and treatment outcome was seen, but such an association could be determined with confidence for less than half of the isolates studied because of the limitations of the animal model used.

A. Endemic Mycoses

Although the endemic mycoses are restricted to certain geographical areas, they are evolving into opportunistic infections that may be encountered elsewhere [1]. Immunocompromised patients are especially susceptible to all but paraccodiomycosis, and those with depleted CD4+ counts are at high risk for disseminated infection and CNS involvement. Until now very few studies have been conceived to determine the role of different variables, such as inoculum size, type of inoculum, incubation temperature and time of reading, in order to develop a reproducible method for antifungal susceptibility testing for endemic fungi (*Coccidioidesimmitis*, *Histoplasma capsulatum*, *Blastomyces dermatitidis*, *Paracoccidioides brasiliensis*) [79–83].

B. Dermatophytes

In the last two decades the incidence of infection caused by dermatophytes has increased considerably [84]. With an increasing variety of drugs available for the treatment of dermatophytoses, the need for a reference method for the testing of the antifungal susceptibility testing of dermatophytes has become apparent. In developing this method many variables need to be considered, such as inoculum size, temperature, duration of incubation, medium, and endpoint determination. It has been established that the use of RPMI 1640, an incubation temperature of 35°C for 4 days, inoculum of 10^3 conidia/mL, and the use of oatmeal cereal agar to promote conidial formation are optimal for determination of the antifungal susceptibilities of dermatophytes [85]. A total of 251 isolates using this methodology were evaluated to determine their susceptibility to fluconazole, griseofulvin, itraconazole, and terbinafine. A larger number of dermatophytes need to be tested to determine the inter- and intralaboratory agreements of such a method. Additionally, MICs need to be correlated with the clinical outcome to develop interpretative breakpoints for dermatophyte susceptibility testing [86].

IV. CONCLUSIONS AND GUIDELINES FOR USE

The field of antifungal susceptibility testing has progressed considerably since 1982. The development of standardized susceptibility testing procedures, quality control strains, and the optimization of the M27 method have placed antifungal susceptibility testing well within the reach of many clinical microbiology laboratories. Currently, the recommended guidelines for studying clinical fungal isolates and antifungal susceptibility testing are as follows:

    1.   Periodic batch antifungal susceptibility testing of clinical isolates to establish the susceptibility for fluconazole and 5-FC against *Candida* spp.

    2.   Oropharyngeal candidiasis in patients with AIDS unresponsive to azole therapy (fluconazole and itraconazole).

    3.   Testing of isolates from deep sites, especially non-*albicans* isolates (fluconazole, itraconazole, and flucytosine).

    In case of cryptococcal isolates, even though the results obtained with the broth microdilution method appear to be superior to those obtained with the NCCLS reference method, additional studies will be necessary to standardize this method, and to allow its use in testing *C. neoformans* against other antifungal agents. In the case of mold infection, the routine testing for any class of isolates is still not recommended.

    Evidence to support the clinical relevance of antifungal susceptibility testing will continue to grow as standardized methodology for yeasts and filamentous fungi evolves and large-scale surveys of clinical isolates are completed. Future efforts must be directed toward different issues such as:

    1.   Alternative approaches that are more convenient and easy to perform, in order to reduce the amount of work and the subjectivity and improve the results of current procedures, such as the E-test, colorimetric tests, and broth microdilution panels. The E-test is a novel

strip diffusion method utilizing a stable gradient of an antimicrobial agent, and has been well documented as an accurate and simple method for bacterial susceptibility testing [87]. The E-test has been recently been shown in several studies to produce results comparable to those obtained with the NCCLS method when testing the susceptibility of *Candida* spp. isolates to ketoconazole, itraconazole, and fluconazole [88]. In the case of amphotericin-B, the E-test readily identified the resistant isolates on glucose-supplemented RPMI 1640 agar as well as on undefined antibiotic medium 3 [89,90]. Colorimetric methods, which are based on the measurement of metabolic activity, may facilitate the determination of a MIC, since they have have the potential to generate clear-cut endpoints based on visually detectable color change. Such a method overcomes the difficulty in making an MIC determination in the case of azoles because of the trailing phenomenon, caused by the partial inhibition of fungal growth, and the subjectivity of visual reading with the aid of a magnifying mirror. Several colorimetric methods have been developed using different dyes such as Alamar blue, XTT (tetrazolium salt 2,3-bis(2-methoxy-4-nitro-5-sulfophenyl)-5-[(phenylamino)carbonil]-2H-tetrazolium hydroxide), and MTT (3-(4,5-dimethyl-2-thiazyl)-2,5-diphenyl-2H-tetrazolium bromide). All these methods have been shown to have comparable levels of agreement and reproducibility when compared to the broth-macrodilution and microdilution NCCLS methods [91–93]. The broth microdilution panels are new antifungal susceptibility testing systems which are faster and easier to use than the reference broth microdilution method test from the NCCLS. The advantages of these methods appear to be the inclusion of quality-controlled, premade antifungal agents containing six (or 10), twofold serial dilutions of the drug and a one-step inoculation system whereby all the wells are simultaneously inoculated in a single step, showing a good agreement in comparison to the NCCLS methodology [94,95].

2.   Improvement of the proposed methodology for filamentous fungi. A number of future modifications of the M27-A are under investigation. To provide for isolates exhibiting trailing growth when tested against azole antifungal agents, the method established 48 hr as the appropriate time for reading MICs for *Candida* spp. and the endpoint criterion as an 80% reduction in growth. However, the correlation between MICs and outcome in vivo may be improved by shortening the incubation time to 24 hr and by relaxing the endpoint criterion, at least for fluconazole, to the lowest drug concentration producing a 50% reduction in growth [96]. Another strategy used with the broth microdilution method to eliminate the trailing phenomenon is the mechanical agitation of the microdilution tray before the MIC is read [97]. More recently, it has also been demonstrated that adjustment of the medium pH eliminates trailing in azole drug susceptibility testing [98].

3.   Establishment of interpretative breakpoints for the new antifungal agents under development and for the currently available amphotericin-B lipid formulations. Each of these new agents will pose additional challenges to the existing methodology, which may require additional adjustments to accurately reflect their clinically relevant antifungal activity [99].

4.   Improvement of the correlation of in vitro results with in vivo clinical outcome in cases of invasive *albicans* and non-*albicans Candida* infections, as well as in filamentous fungal invasive infections.

V. SUMMARY

The current interest in fungal infections has been stimulated by a raise in their incidence in immunocompromised patients and has led to increased interest to antifungal susceptibility testing. Additional studies are needed to generate more data to support the correlation between clinical outcome and MIC. The microbiology laboratory and clinicians must work to clarify the relative value of the antifungal susceptibility testing in the management of fungal infections. A decision not to use a particular antifungal agent might be determined solely on the finding of *in vitro* resistance—and might have significant clinical consequences for the patient. In the case of invasive fungal infections, the determination of in vitro susceptibility may become very useful in determining optimal use of antifungal agents [100,101].

REFERENCES

1. O Lortholary, B Dupont. Antifungal prophylaxis during neutropenia and immunodeficiency. Clin Microbiol Rev 13:477–504, 1997.
2. DW Warnock. Fungal infections in neutropenia: current problems and chemotherapeutic control. J Antimicrob Chemother 41:95–105, 1998.
3. MA Pfaller, RN Jones, GV Doern, HS Sader, SA Messer, A Houston, S Coffman, RJ Hollis. International surveillance of blood stream infections due to *Candida* species from the SENTRY antimicrobial surveillance program in North America and Latin America, 1997 and 1998 [abstract no. F115, p. 319]. In program and abstract of the 99th General Meeting of the American Society for Microbiology, Chicago, 1999.
4. C Viscoli, C Girmenia, A Marinus, P Martino, F Meunier. A surveillance study of fungemia in cancer patients in Europe. Invasive Fungal Infections Cooperative Group (IFIG of EORTC) [abstract 2]; Trends in Invasive Fungal Infections 3 (Brussels), 1995.
5. RE Lewis, ME Klepser, MA Pfaller. Update on clinical antifungal susceptibility testing for *Candida* species. Pharmacotherapy 18:509–515, 1998.
6. JR Wingard. Importance of *Candida* species other than *C. albicans* as pathogens in oncology patients. Clin Infect Dis 20:115–25, 1995.
7. MA Pfaller. Nosocomial candidiasis: emerging species, reservoirs, and modes of transmision. Clin Infect Dis 22(suppl 2):S89–S94, 1996.
8. DC Coleman, MG Rinaldi, KA Haynes, JH Rex, RC Summerbell, EJ Anaissie, A Li, DJ Sullivan. Importance of *Candida* species other than *Candida albicans* as opportunistic pathogens. Med Mycol 1:156–165, 1998.
9. NM Ampel. Emerging disease issues and fungal pathogens associated with HIV infection. Emerg Infect Dis 2:109–116, 1996.
10. LJ Wheat. Diagnosis and management of fungal infections in AIDS. Curr Opin Infect Dis 6:617–627, 1993.
11. CM Van der Host, MS Saag, GA Cloud, et al. Treatment of cryptococcal meningitis associated with the acquired immunodeficiency syndrome. N Engl J Med 337:15–21, 1997.
12. BP Currie, M Ghannoum, L Bessen, A Casadevall. Decreased fluconazole susceptibility of a relapse *Cryptococcus neoformans* isolate after fluconazole treatment. Infect Dis Clin Pract 4:318–319, 1995.
13. JP Latgé. *Aspergillus fumigatus* and aspergillosis. Clin Microbiol Rev 12:310–350, 1999.
14. R Patel, CV Paya. Infections on solid organ transplant recipients. Clin Microbiol Rev 10:86–124, 1997.
15. I Varthalitis, F Meunier. Prophylaxis of fungal infections. In: F Meunier, ed. Bailliere's Clinical Infectious Diseases. Vol. 2. London: Bailliere Tindall, 1995, pp 157–177.
16. DW Warnock. Fungal complications of transplantation: diagnosis, treatment and prevention. J Antimicrob Chemother 36:73–90, 1995.
17. JK Van den Saffele, JR Boelaert. Zygomycosis in HIV-positive patients: review of the literature. Mycoses 39:77–84, 1996.
18. E Annaissie, GP Bodey, H Kantarjian, J Ro, SE Vartivarian, R Hopfer, J Hoy, K Rolston. New spectrum of fungal infections in patients with cancer. Rev Infect Dis 2:369–378, 1989.
19. JR Perfect, WA Schell. The new fungal opportunists are coming. Clin Infect Dis 22(suppl 2): S112–S118, 1996.
20. AS Gamis, T Gudnason, GS Gieink. Disseminated infection with *Fusarium* in recipients of bone marrow transplants. Rev Infect Dis 13:1077–1088, 1991.

21. E Anaissie, H Kantarajan, J Ro, R Hopfer, K Rolston, V Fainstein, G Bodey. The emerging role of *Fusarium* infections in patients with cancer. Medicine 67:77–83, 1988.

22. TJ Walsh, KR Newman, M Moody, RC Wharon, JC Wade. Trichosporonosis in patients with neoplastic disease. Medicine (Baltimore) 65:268–279, 1986.

23. E Annaissie. Opportunistic mycoses in the immunocompromised host: experience at a cancer center and review. Clin Infect Dis 14(suppl 1):S43–S53, 1992.

24. NH Georgepapadakou, TJ Walsh. Antifungal agents: chemotherapeutic targets and immunologic strategies. Antimicrob Agents Chemother 40:279–291, 1996.

25. DJ Sheehan, CA Hitchcock, CM Sibley. Current and emerging azole antifungal agents. Clin Microbiol Rev 12:40–79, 1999.

26. JH Rex, MG Rinaldi, MA Pfaller. Resistance of *Candida* species to fluconazole. Antimicrob Agents Chemother 39:1–8, 1995.

27. J Fichtenbaum Carl, WG Powderly. Refractory mucosal candidiasis in patients with human immunodeficiency virus infection. Clin Infect Dis 26:556–565, 1998.

28. R Franz, SL Kelly, DC Lamb, DE Kelly, M Ruhnke, J Morschhauser. Multiple molecular mechanisms contribute to stepwise development of fluconazole resistance in clinical *Candida albicans* strains. Antimicrob Agents Chemother 42:3065–3072, 1998.

29. TC White, KA Marr, RA Bowden. Clinical, cellular, and molecular factors that contribute to antifungal drug resistance. Clin Microbiol Rev 11:382–402, 1998.

30. A Espinel-Ingroff, TM Kerkering, PR Goldson, S Shadomy. Comparison study of broth macrodilution and microdilution antifungal susceptibility tests. J Clin Microbiol 29:1089–1094, 1991.

31. R Guinet, D Nerson, F DeClosets, et al. Collaborative evaluation in seven laboratories of a standardized micromethod for yeast susceptibility testing. J Clin Microbiol 26:2307–2312, 1988.

32. R Shawar, V Paetnick, Z Witte, LG Ensign, E Anaissie, M LaRocco. Collaborative investigation of broth microdilution and semisolid agar dilution for in vitro susceptibility testing of *Candida albicans*. J Clin Microbiol 30:1976–1981, 1992.

33. JN Galgiani. Susceptibility testing of fungi: current status of the standardization progress. Antimicrob Agents Chemother 37:2517–2521, 1993.

34. MA Pffaler, MG Rinaldi, JN Galgiani, et al. Collaborative investigation of variables in susceptibility testing of yeasts. Antimicrob Agents Chemother 34:1648–1654, 1990.

35. JH Rex, MA Pfaller, MG Rinaldi, N Polak, JN Galgiani. Antifungal susceptibility testing. Clin Microbiol Rev 6:367–381, 1993.

36. National Committee for Clinical and Laboratory Standards. Reference method for broth dilution antifungal susceptibility of yeasts. Approved standard M27-A. Wayne, PA: National Committee for Clinical and Laboratory Standards, 1997.

37. MA Pfaller, M Bale, B Buschelman, M Lacaster, A Espinel-Ingroff, JH Rex, MG Rinaldi, CR Cooper, MR McGinnis. Quality control guidelines for National Committee for Clinical Laboratory Standards recommended broth macrodilution testing of amphotericin B, fluconazole, and flucytosine. J Clin Microbiol 33:1104–1107, 1995.

38. JH Rex, MA Pfaller, M Lancaster, FC Odds, Bolmstrom, MG Rinaldi. Quality control guidelines for National Committee for Clinical Laboratory Standards recommended broth macrodilution testing of ketoconazole and itraconazole. J Clin Microbiol 34:816–817, 1996.

39. MA Pfaller, JH Rex, MG Rinaldi. Antifungal susceptibility testing: technical advances and potential clinical applications. Clin Infect Dis 24:776–784, 1997.

40. GL Drusano. Role of pharmacokinetics in the outcome of infections. Antimicrob Agents Chemother 32:289–297, 1988.

41. D Debruyne, JP Ryckelnyck. Clinical pharmacokinetics of fluconazole. Clin Pharm 24:10–27, 1993.

42. SM Grant, SP Clissold. Itraconazole. Drugs 37:310–344, 1989.

43. M Barza, G Cuchural. General principles of antibiotic tissue penetration. J Antimicrob Chemother 15:59–75, 1985.

44. R Wise. The relevance of pharmacokinetics to in vitro models: protein binding-does it matter? J Antimicrob Chemother 15:77–83, 1985.

45. SP Hawser, LJ Douglas. Resistance of *Candida albicans* biofilms to antifungal agents. Antimicrob Agents Chemother 39:2128–2131, 1995.

46. JH Rex. Catheters and candidemia. Clin Infect Dis 22:467–470, 1996.

47. F Miguez, ML Chiu, JE Lima, R Ñique, J Prieto. Activity of fluconazole: postantifungal effect, effects of low concentrations and of pretreatment on the susceptibility of Candida albicans to leucocytes. J Antimicrob Chemother 34:93–100, 1994.

48. JR Graybill, E Montalbo, WR Kirkpatrick, MF Luther, SG Revankar, TF Patterson. Fluconazole versus Candida albicans: a complex relationship. Antimicrob Agents Chemother 42:2938–2942, 1998.

49. CC Sanders. ART versus ASTs: where are we going? J Antimicrob Chemother 28:621–625, 1991.

50. EJ Anaissie, NC Karyotakis, R Hachem, MC Dignani, JH Rex, V Paetznick. Correlation between in vitro and in vivo activity of antifungal agents against Candida species. J Infect Dis 170:384–389, 1994.

51. JW Van't Wout, H Mattie, R Van Futh. Comparison of the efficacies of amphotericin B, fluconazole, and itraconazole against a systemic Candida albicans infection in normal and neutropenic mice. Antimicrob Agents Chemother 33:147–151, 1989.

52. TE Rogers, JN Galgiani. Activity of fluconazole (UK49,858) and ketoconazole against Candida albicans in vitro and in vivo. Antimicrob Agents Chemother 30:418–422, 1986.

53. RL Stiller, JE Bennett, HJ Scholer, M Wall, A Polak, DA Stevens. Correlation of in vitro susceptibility test results with in vivo response: flucytosine therapy in a systemic candidiasis model. J Infect Dis 147:1070–1077, 1983.

54. MA Fisher, SH Shen, J Haddad, WF Tarry. Comparison of in vivo activity of fluconazole with that of amphotericin B against Candida tropicalis, Candida glabrata, and Candida krusei. Antimicrob Agents Chemother 33:1443–1446, 1989.

55. F Barchiesi, LK Najvar, MF Luther, G Scalise, MG Rinaldi, JR Graybill. Variation in fluconazole efficacy for Candida albicans strains sequentially isolated from oral cavities of patients with AIDS in an experimental murine candidiasis model. Antimicrob Agents Chemother 40:1317–1320, 1996.

56. MA Ghannoum, JH Rex, JN Galgiani. Susceptibility testing of fungi: current status of correlation of in vitro data with clinical outcome. J Clin Microb 34:489–495, 1996.

57. JH Rex, MA Pfaller, JN Galgiani, et al. Development of interpretative breakpoints for antifungal susceptibility testing: conceptual framework and analysis of in vitro–in vivo correlation data for fluconazole, itraconazole, and Candida infections. Clin Infect Dis 24:235–247, 1997.

58. JA Sangeorzan, SF Bradley, X He, et al. Epidemiology of oral candidiasis in HIV-infected patients: colonization, infection, treatment, and emergence of fluconazole resistance. Am J Med 97:339–346, 1994.

59. TF Patterson, SG Revankar, WR Kirkpatrick, et al. Simple method for detecting fluconazole-resistant yeast with chromogenic agar. J Clin Microbiol 34:1794–1797, 1996.

60. ML Webb, JL Gerberding, WK Hadley, BL Lee, JD Stansell, MA Sande. The prevalence of oral thrush and fluconazole resistant thrush in HIV + patients [abstract No. 483] Clin Infect Dis 21: 800, 1995.

61. S Redding, J Smith, G Farinacci, et al. Resistance of Candida albicans to fluconazole during treatment of oropharingeal candidiasis in a patient with AIDS: documentation of in vitro susceptibility testing and DNA subtype analysis. Clin Infect Dis 18:240–242, 1994.

62. JH Rex, MA Pfaller, AL Barry, PW Nelson, CD Webb. Antifungal susceptibility testing of isolates from a randomized, multicenter trial of fluconazole versus amphotericin B as treatment of nonneutropenic patients with candidemia. Antimicrob Agents Chemother 39:40–44, 1995.

63. JR Graybill, J Vazquez, RO Darouiche, R Morhart, BL Moskovitz, I Malego. Itraconazole oral solution versus fluconazole treatment of oropharyngeal candidiasis [abstract No. I200]. In program and abstract of the 35th Interscience Conference on Antimicrobial Agents and Chemotherapy, San Francisco, 1995, American Society for Microbiology.

64. D Andes, M van Ogtrop. Characterization and quantitation of the pharmacodynamics of fluconazole in a neutropenic murine disseminated candidiasis infection model. Antimicrob Agents Chemother 43:2116–2120, 1999.

65. MA Ghannoum, AS Ibrahim, Y Fu, MC Shafiq, JE Edwards, RS Criddle. Susceptibility testing of Cryptococcus neoformans: a microdilution technique. J Clin Microbiol 30:2881–2886, 1992.

66. H Sanati, SA Messer, M Pfaller, W Mallory, R Larsen, A Espinel-Ingroff, M Ghannoum. Multicenter evaluation of broth microdilution method for susceptibility testing of *Cryptococcus neoformans* against fluconazole. J Clin Microbiol 34:1280–1282, 1996.

67. A Casadevall, ED Spitzer, D Webb, MG Rinaldi. Susceptibilities of serial *Cryptococcus neoformans* isolates from patients with recurrent cryptococcal meningitis. Antimicrob Agents Chemother 37: 1383–1386, 1993.

68. A Paugam, P Dupovy-Camet, JP Gaugneux, C Tourte-Schafer, D Sicard. Increased fluconazole resistance of *Cryptococcus neoformans* isolated from a patient with AIDS and recurrent meningitis. Clin Infect Dis 19:975–976, 1994.

69. HDL Birley, EM Johnson, P McDonald, C Parry, PB Carey, DW Warnock. Azole drug resistance as a cause of clinical relapse in AIDS patients with cryptococcal meningitis. Int J STD AIDS 6: 353–355, 1995.

70. A Armengou, C Porcar, J Mascaro, F Garcia Bragado. Possible development of resistance to fluconazole during suppressive therapy for AIDS-associated cryptococcal meningitis. Clin Infect Dis 23: 1337–1338, 1996.

71. MD Witt, RJ Lewis, RA Larsen, EN Milefchik, MAE Leal, RH Haubrich, JA Richie, JE Edwards, MA Ghannoum. Identification of patients with acute AIDS-associated cryptococcal meningitis who can be effectively treated with fluconazole: the role of antifungal susceptibility testing. Clin Infect Dis 22:322–328, 1996.

72. AI Aller, E Martin-Mazuelos, F Lozano, J Gomez-Mateos, L Steele-Moore, WS Holloway, MS Gutierrez, FJ Recio, A Espinel-Ingroff. Correlation of fluconazole MICs with clinical outcome in cryptococcal meningitis. Antimicrob Agents Chemother 44:1544–1548, 2000.

73. National Committee for Clinical and Laboratory Standards. Reference method for broth dilution antifungal susceptibility of conidium-forming filamentous fungi; proposed standard Wayne, PA: NCCLS, document M38-P, 1998.

74. A Espinel-Ingroff, K Dawson, M Pfaller, E Anaissie, B Breslin, D Dixon, A Fothergill, V Paetznick, J Peter, MG Rinaldi. Comparative and collaborative evaluation of standarization of antifungal susceptibility testing for filamentous fungi. Antimicrob Agents Chemother 39:314–319, 1995.

75. A Espinel-Ingroff, M Barlett, R Bowden, NX Chin, C Cooper, A Fothergill, MR McGinnis, P Menezes, SA Messer, PW Nelson, FC Odds, L Pasarell, MA Pfaller, JH Rex, MG Rinaldi, GS Shankland, TJ Walsh, I Weitzman. Multicenter evaluation of proposed standarized procedure for antifungal susceptibility testing of filamentous fungi. J Clin Microbiol 35:139–143, 1997.

76. I Pujol, J Guarro, C Llop, L Soler, J Fernandez-Ballart. Comparison study of broth macrodilution and microdilution antifungal susceptibility tests for the filamentous fungi. Antimicrob Agents Chemother 40:2106–2110, 1996.

77. FC Odds, F Van Gerven, A Espinel-Ingroff, MS Barlett, MA Ghannoum, MV Lancaster, MA Pfaller. Evaluation of possible correlations between antifungal susceptibilities of filamentous fungi in vitro and antifungal treatment outcomes in animal infection models. Antimicrob Agents Chemother 42: 282–288, 1998.

78. RF Hector, BL Zimmer, D Pappagianis. Microtiter method for MIC testing with spherule-endospore-phase of *Coccidiodes immitis*. J Clin Microbiol 26:2667–2668, 1988.

79. GS Kobayashi, SJ Travis, MG Rinaldi, G Medoff. In vitro and in vivo activities of Sch 390304, fluconazole, and amphotericin B against *Histoplasma capsulatum*. Antimicrob Agents Chemother 34:524–528, 1990.

80. VP Kurup, A Resnick, HD Rose. Medium for susceptibility testing and yeast phase conversion of *Blastomyces dermatitidis*. Mycopathologia 95:25–28, 1986.

81. RC Hahn, JS Hamdan. Effects of amphotericin B and three azole derivatives on the lipids of yeast cells of *Paracoccidioides brasiliensis*. Antimicrob Agents Chemother 44:1997–2000, 2000.

82. RK Li, MA Cibblak, N Wordoff, L Pasarell, DW Warnock, MR McGinnis. In vitro activities of voriconazole, itraconazole, and amphotericin B against *Blastomyces dermatitidis*, *Coccidioides immitis* and *Histoplasma capsulatum*. Antimicrob Agents Chemother 44:1734–1736, 2000.

83. I Weitzman, RC Summerbell. The dermatophytes. Clin Microbiol Rev 8:240–259, 1995.

84. HA Norris, BE Elewski, MA Ghannoum. Optimal growth conditions for the determination of the antifungal susceptibility of three species of dermatophytes with the use of a microdilution method. J Am Acad Dermatol 40:59–63, 1999.

85. CJ Jessup, J Warner, N Isham, I Hasan, MA Ghannoum. Antifungal susceptibility testing of dermatophytes: establishing a medium for inducing conidial growth and evaluation of susceptibility of clinical isolates. J Clin Microbiol 38:341–344, 2000.

86. ML Sanchez, RN Jones. Etest, an antimicrobial susceptibility testing method with broad clinical and epidemiological application. Antimicrob Newsl 8:1–8, 1992.

87. AL Colombo, DA Barchiesi, DA McGough, MG Rinaldi. Comparison of Etest and National Committee for Clinical and Laboratory Standards broth macrodilution method for azole antifungal susceptibility testing. J Clin Microbiol 33:535–540, 1995.

88. A Wanger, K Mills, PW Nelson, JH Rex. Comparison of Etest and National Committee for Clinical Standards Broth macrodilution method for antifungal susceptibility testing: enhanced ability to detect amphotericin B–resistant *Candida* isolates. Antimicrob Agents Chemother 39:2520–2522, 1995.

89. CJ Clancy, MH Nguyen. Correlation between in vitro susceptibility determined by E test and response to therapy with amphotericin B: results from a multicenter prospective study of candidemia. Antimicrob Agents Chemother 43:1289–1290, 1999.

90. A Espinel-Ingroff, M Pfaller, SA Messer, CC Knapp, S Killian, HA Norris, MA Ghannoum. Multicenter comparison of the Sensititre YeastOne colorimetric antifungal panel with the National Committee for Clinical Laboratory Standards M27-A reference method for testing clinical isolates of common and emerging *Candida* spp., *Cryptococcus* spp., and other yeasts and yeast-like organisms. J Clin Microbiol 37:591–595, 1999.

91. SP Hawser, H Norris, CJ Jessup, MA Ghannoum. Comparison of a 2,3-bis(2-methoxy-4-nitro-5-sulfophenyl)-5-[(phenylamino)~~~ carbonil]-2*H*-tetrazolium hydroxide (XTT) colorimetric method with the standardized National Committee for Clinical Laboratory Standards method of testing clinical yeast isolates for susceptibility to antifungal agents. Antimicrob Agents Chemother 38:1450–1452, 1998.

92. J Meletiadis, JFGM Meis, JW Mouton, P Donnelly, PE Verweij. Comparison of NCCLS and 3-(4,5-dimethyl-2-thiazyl)-2,5-diphenyl-2H-tetrazolium bromide (MTT) methods of in vitro susceptibility testing of filamentous fungi and development of a new simplified method. Antimicrob Agents Chemother 38:2949–2954, 2000.

93. BA Arthington-Skags, M Motley, DW Warnock, CJ Morrison. Comparative evaluation of PASCO and National Committee for Clinical Laboratory Standards M27-A broth microdilution methods for antifungal drug susceptibility testing of yeasts. J Clin Microbiol 38:2254–2260, 2000.

94. KG Davey, AD Holmes, EM Johnson, A Szekely, DW Warnock. Comparative evaluation of FUNGITEST and broth microdilution methods for antifungal drug susceptibility testing of *Candida* species and *Cryptococcus neoformans*. J Clin Microbiol 36:926–930, 1998.

95. JH Rex, PW Nelson, VL Paetznick, M Lozano-Chiu, A Espinel-Ingroff, EJ Anaissie. Optimizing the correlation between results of testing in vitro and therapeutic outcome in vivo for fluconazole by testing critical isolates in a murine model of invasive candidiasis. Antimicrob Agents Chemother 42:129–134, 1998.

96. EJ Anaissie, VL Paetznick, LG ensign, A Espinel-Ingroff, JN Galgiani, CA Hitchcock, M LaRocco, TF Patterson, MA Pfaller, JH Rex. Microdilution antifungal susceptibility testing of *C. albicans* and *C. neoformans* with and without agitation: an eight-center collaborative study. Antimicrob Agents Chemother 40:2387–2391, 1996.

97. KA Marr, TR Rustad, JH Rex, TC White. The trailing end point phenotype in antifungal susceptibility testing is pH dependent. Antimicrob Agents Chemother 43:1383–1386, 1999.

98. KL Oakley, CB More, DW Denning. Comparison of in vitro activity of liposomal nystatin against *Aspergillus* species with those of nystatin, amphotericin B desoxycholate, AB colloidal dispersion, liposomal AB lipid complex, and itraconazole. Antimicrob Agents and Chemother 43:1264–1266, 1999.

99. L Verbist. Relevance of antibiotic susceptibility testing for clinical practice. Eur J Clin Microbiol Infect Dis 12:2–5, 1993.

100. FC Odds. Personal opinion: can antifungal sensitivity tests predict clinical treatment outcomes? Rev Iberoam Micol 14:83–84, 1997.
101. JH Rex, CR Cooper, WG Mertz, JN Galgiani, EJ Anaisse. Detection of amphotericin B-resistant *Candida* isolates in a broth-based system. Antimicrob Agents Chemother 39:906–909, 1995.
102. M Lozano-Chiu, PW Nelson, M Lancaster, MA Pfaller, JH Rex. Lot-to-lot variability of antibiotic medium 3 used for testing susceptibility of *Candida* isolates to amphotericin B. Antimicrob Agents Chemother 35:270–272, 1997.
103. JL Rodriguez-Tudela, JV Martinez-Suarez. Improved medium for fluconazole susceptibility testing of *Candida albicans*. Antimicrob Agents Chemother 38:45–48, 1994.
104. SG Revankar, WR Kirkpatrick, RK McAtee, AW Fothergill, SW Redding, MG Rinaldi, TF Patterson. Interpretation of trailing endpoints in antifungal susceptibility testing by the National Committee for Clinical Laboratory Standards Method. J Clin Microbiol 36:153–156, 1998.

28

Molecular Diagnosis of Fungal Infections

Roy L. Hopfer and Darius Amjadi
University of North Carolina, Chapel Hill, North Carolina

I. INTRODUCTION

Fungal infections have become of increasing medical importance over the last several decades. This has mainly been due to the increasing emergence of two groups of immunocompromised patients, HIV-infected individuals, and cancer patients, although surgical patients and newborn infants are also at risk. Additional predisposing factors identified have been the widespread use of broad spectrum antibiotics (surveillance cultures of GI and/or GYN tracts often become positive for yeast after 1 or more weeks of antibiotic therapy), use of long-term indwelling venous and urethral catheters, malnutrition, diabetes, obstructive uropathy, and renal failure. The HIV pandemic has created millions of individuals susceptible to opportunistic pathogens, including many fungi that are not usually associated with human pathogenesis. The primary example in the setting of AIDS is *Pneumocystis carinii*, which causes pulmonary illness and can cause systemic infection. With the advent of antiretroviral therapy, the life span of these patients has been increased, thus creating a larger window of opportunity for infection [1]. Cancer patients are the other group of immunocompromised patients that are neutropenic due to chemotherapy or to bone marrow transplantation. These patients are highly susceptible to fungal infection; however, the incidence of infection caused by any one species is quite low. For example, infections caused by specific species of *Candida* may vary between <1–10%, *Aspergillus* 5–15%, and *Fusarium* <2% of serious fungal infections in these patients [2].

The stakes for affected patients are quite high. Most disseminated fungal infections carry an extremely high level of morbidity and mortality. Even with appropriate therapy, systemic or disseminated candidal infections have a crude mortality rate of 50–80% [3]. Clinicians are loath to begin treatment with amphotericin B without a firm diagnosis because of its severe side effects on already critically ill patients. In addition, some of these fungi are resistant to clinically achievable concentrations of amphotericin B, the "gold standard" for treating life-threatening fungal infections. The clinical presentation in many of these patients is soft, often being suspected following prolonged fever refractory to broad-spectrum antibiotic therapy [4]. Effective treatment depends on rapid detection and definitive identification of the causative agent.

Traditionally, blood, wound culture, and tissue biopsy have served as the gold standards of laboratory diagnosis of systemic or deep-seated fungal infections. However, fungal cultures from blood and other fluids are very insensitive, failing to become culture positive in up to 50–75% of cases of candidiasis and 65% of cases of endopthalmitis. Tissue biopsy is often precluded due to the patient's cytopenias. When blood or fluid cultures are positive, it is often indicative of advanced and disseminated disease, often too far advanced for treatment to be effective. Autopsy tissue diagnosis has shown that blood cultures can remain negative even in the face of widespread dissemination to the solid organs. Depending on the fungus involved, delays in appropriate treatment can be due to the need for several days incubation before growth is detected. Additionally, delays may be increased while awaiting biochemical testing or development of specific morphological structures for identification. During such delays the patient may have suffered significant morbidity or even death. One study showed that 57% of patients with nosocomial candidiasis died, and of those, 46% died within 1 week of onset [5]. Mortality for patients with invasive aspergillosis may approach 100%. Finally, even when empiric treatment is initiated, innate or drug-induced resistance to the antifungal chosen can lead to treatment failures, again increasing the likelihood of a worse outcome in an already poor clinical setting.

Ideal molecular methods for detection of fungal infection would provide early detection of serious infection allowing timely treatment of the patient and would: (1) distinguish colonization and easily treatable superficial infections from invasive or widespread life threatening disseminated infection; (2) detect the majority of known pathogenic fungi; and (3) distinguish to the genus and species level in order to better address appropriate therapy. In addition, lack of cross-reactivity with human, bacterial, viral, and protozoan DNA sequences would be a major factor required in an effective molecular assay. Finally, quantitative methods that allowed the disease course and effectiveness of treatment to be followed would increase the ability of clinicians to modify treatment appropriate to disease response and allow better predictions of clinical outcome.

A number of methods for detection of systemic mycoses have been studied including detection of antibodies, antigens, breakdown products produced by the fungus, and unique metabolites produced by the fungus and not the host. Each method has intrinsic limitations. Antibodies may crossreact causing false positives, may indicate local rather than systemic infection, or may indicate past rather than present infection. Additionally, many opportunistic infections occur because of the host's inability to mount an adequate immune reaction to the organism, and thus antibodies may only be present in undetectably low concentrations. With the exception of assays for cryptococcal antigen, and perhaps histoplasmal antigen, antigen detection is hampered by low serum concentrations even during fulminant infection, may indicate phagocytosis rather than systemic infection and often cannot be used to identify the precise fungal agent. Additionally, the presence of antigen cannot be used to distinguish between dead and living fungi.

II. POLYMERASE CHAIN REACTION

Early attempts to make infectious diagnoses using DNA methods were hampered by technical limitations related to the small amount of target DNA present in the clinical specimen. In the 1980s, this obstacle was overcome by the development of polymerase chain reaction (PCR) chemistry. The PCR technique allows for amplification of small amounts of specific sequences to detectable levels. To date, the use of PCR for diagnostic purposes of pathogenic fungi has been most studied for *Pneumocystis carinii*, *Candida albicans*, and *Aspergillus fumigatus*. There are a limited number of reports of using PCR for detecting DNA from other fungal pathogens.

As fungal DNA sequences have been studied over the last few decades, probes for both highly conserved regions and species-specific variable regions have been discovered and described. This offers the potential for sensitive panfungal markers for detection of fungal infection followed by the use of species-specific markers for identification. One could even argue that in addition to panfungal markers, it would be ideal to have mutually exclusive panbacterial, panparasitic, and panviral markers. Clearly a panfungal PCR method would be an important laboratory tool since >150 different fungi have been recognized as human pathogens. Although

Candida albicans and the other candidal species are responsible for a majority of the fungal infections, other fungi and fungal species can have similar clinical presentations, but may require different treatments. This has led to a focus on molecular tests that not only detect the fungal DNA but can use it to speciate the infectious agent. Most proposed panfungal markers interact with either ribosomal DNA (rDNA) (portions of which have been highly conserved in most known human pathogens), mitochondrial DNA, or intrinsic molecules like actin. Up to 100 copies of rDNA are present in each fungal cell, which increases the target material available from each organism [6–8]. The sensitivity of most molecular assays is great enough to detect 10 fg of purified fungal DNA, but require from 1–10 fungal cells per mL blood for reliable PCR results [6,9,10].

Three major techniques have been investigated for differentiating or identifying specific fungi that are detected by universal primers: restriction fragment length polymorphism (RFLP), hybridization of the amplicon with a species-specific probe, and single-strand conformational polymorphism (SSCP) analysis [11]. Depending on the target or the restriction enzyme used, all of these techniques can provide fungal DNA fingerprinting similar to that used in fingerprinting of human DNA [12]. RFLP techniques cleave amplified DNA products producing fragments of varying lengths based on slight differences in location of restriction sites found in the different species or groups. The basic strategy of SSCP is detection of base pair differences in the amplicon. The amplified product is subjected to denaturation and a gel electrophoresis method that differentiates the amplicons based on gel migration changes resulting from subtle conformational changes caused by the single base pair change. SSCP methods do not require further enzymatic digestion, results can be obtained sooner with less technical involvement, and the risk of potential contamination may be reduced [11,12]. Use of restriction enzymes in addition to PCR has been shown to be useful in identifying particular species of fungi [1,6,13,14]. Hybridization of species-specific probes with PCR-derived amplicons is probably the most specific method of discriminating fungal species [10,15]. However, the requirement of probes for each potential pathogen limits the usefulness of this technology to the most commonly seen species [11,16].

Regardless of the specific technology employed, the use of PCR or other amplification methods is being investigated by numerous laboratories and most reports are encouraging in that the sensitivity of PCR is no worse, and often better, than other currently used diagnostic technologies. Almost all of the reports on amplification of fungal diagnosis have used PCR; therefore, the term PCR will be used throughout the chapter with the understanding that the same advantages and limitations associated with PCR are also applicable to the other amplification technologies.

III. *CANDIDA*

Candida is the most common cause of nosocomial fungal infections, accounting for 8% of hospital acquired septicemias. More than 12% of bone marrow transplant patients develop candidiasis with a mortality rate of 50–80%, even after appropriate treatment. Untreated patients have mortality rates approaching 100%. In patients with prolonged granulocytopenia, endogenous *Candida* spp. often colonize mucosal membranes, or pass through intact GI mucosa by persorption or by penetrative growth, as a prelude to vascular space invasion and hematologic dissemination. Indwelling catheters can serve as either a source of candidal entry or be seeded from remote sites.

The traditional method of detection of candidiasis, by blood culture, has been shown to be too insensitive (missing more than one-half of cases in some autopsy studies). This is especially true in patients with chronic disseminated candidiasis and patients with hepatosplenic lesions. The turnaround time for cultures to become positive is usually 2–5 days. This contributes to the grim nature of these infections since almost half of candidial fatalities occur within the first week of diagnosis of disseminated infection. Serologic assays have focused on antibody and antigen detection. Antibody detection is generally not used because the patient population most likely to have invasive or disseminated infection are the highly immunosuppressed. A variety of antigens including enolase, cell wall mannan, and crude cell lysates as well as metabolic markers such as D-arabinitol have been investigated for diagnostic purposes [3,17]. Most of these assays have sensitivities of <75%, although cell wall mannan testing approaches 90% sensitivity for *C. albicans*, but is lower for other *Candida* spp. [3]. A few commercial kits for detecting antigen have been available since the mid-1980s but are not widely accepted for diagnosis of serious candidal infection.

The first molecular approach for detecting and identifying fungal DNA in clinical specimens was reported in 1990 [18]. Those investigators chose a fragment from the cytochrome P450 lanosterol-14 alpha-demethylase (current nomenclature ERG11) which was present in species of Candida but presumably not in bacteria or humans. This gene is present in single copy within the *C. albicans* (actually two copies since the organism is diploid), and the investigators were able to demonstrate the presence of similar fragments in several, but not all, clinically relevant species of Candida. This paper was also very encouraging in that their PCR methods led to positive results from a variety of clinical specimens including urine, sputa, and blood. The primary limitation of the test related to the inability of the test to detect DNA from other potentially invasive fungi such a aspergilli and cryptococci. The use of ethidium bromide staining of electrophoresed product perhaps limited the sensitivity of the test; however, the method was able to detect DNA from as few as 10 organisms in seeded samples. This investigation was instrumental in demonstrating the feasibility of using PCR for diagnosis of fungal infection; at the same time, it demonstrated or directed attention to many of the shortcomings that remained to be resolved.

Shortly thereafter, other investigators [19] attempted to improve PCR sensitivity using the multicopy gene of a mitochondrial DNA fragment (E03). In addition, rather than use of ethidium-stained gels for analysis, a ^{32}P-labeled probe for amplicon detection, following Southern blotting. The sensitivity was not greatly improved; however, the use of labeled probe demonstrated the highly specific nature of the PCR amplicon in that it recognized *C. albicans*, but not other, closely related *Candida* spp.

Our laboratory approached the problem from a different perspective that would enable detection of fungal DNA in clinical specimens regardless of the identity of the infecting organism [6]. We used a multicopy rDNA fragment that had been described by other investigators interested in taxonomy of fungi. We selected universal primers for all fungi in that this particular fragment of rDNA had been conserved throughout evolution. In addition, we were unable to demonstrate presence of the fragment in bacterial and human genomic preparations. The sensitivity (10–100 cfu) was similar to that reported earlier by Buchman et al. [18]. All fungi tested produced a similar-size product (310 bp) except the Zygomycetes, which had a somewhat larger amplicon (~341 bp). Based on known sequence differences present in the 310-bp fragment, we also showed that medically important groups of fungi could be differentiated from one another by RFLP analysis of the PCR amplicon. Five distinct RFLP patterns could differentiate or separate *Candida* and related yeasts, *Cryptococcus* and *Trichosporon* species, septate molds, aseptate molds, and the dimorphic fungi into distinct groups. Further, additional enymes could be used to differentiate species within individual groups. The major limitations were twofold: (1) the sensitivity of the test was greatly reduced following enzyme digestion of the amplified product, and (2) in the presence of high magnesium concentrations, the primers were able to initiate the PCR reaction using human DNA as template.

Many of the obstacles encountered by earlier investigators [6] were overcome using a similar approach with an rDNA fragment present in all medically important fungi, except *Mucor* spp. [20]. Based on sequence data collected from numerous clinical isolates of fungi, they selected a highly conserved segment (687 bp) from 18S-ribosomal DNA and prepared probes that could detect as little as 100 fg fungal DNA in Southern analysis with a chemiluminescence detection system.

About this same time, another interesting approach was reported [21] using multiple primer sets (which recognized 5S rDNA and adjacent nontranscribed spacer region) in the same PCR reaction tube. Using this technology, all *Candida* spp. tested produced a small PCR fragment (105 bp). *C. albicans* produced this small fragment as well a larger (1015 bp) fragment, thereby allowing detection and differentiation of *C. albicans* from all other *Candida* spp. Such an approach suggested that perhaps multiple levels of information could be gleaned from single PCR reaction mixes.

The next major advance in diagnosis of fungal infection using molecular approaches utilized conserved regions of the 28S rRNA gene [15]. Sandhu et al. [15] first sequenced large portions of the 28S rDNA from 50 medically important fungi. Using these data, universal primers were prepared that recognized a highly conserved 260-bp fragment. There was considerable sequence variability within the universal primers, which enabled synthesis of 21 species-specific probes. Although the hybridization technology utilized in this report is cumbersome and not suited to clinical laboratories in its current form, the concept of detection and species identification of fungal amplicons present in clinical specimens was validated.

Another approach, single-strand conformational polymorphism (SSCP) analysis, was introduced as a means of identifying species following amplification of a 197-bp fragment of 18S rRNA gene which was common to all medically relevant fungi tested [11]. The investigators were able to demonstrate that following PCR, SSCP could be used to identify a number of clinical isolates. This method could reduce the number of steps needed to determine which organism was responsible for infection, but the denaturing gel technology does require some expertise and experience. Nonetheless, the method may prove to be a valuable alternative to the use of restriction enzyme or hybridization probe technology.

One of the more promising reports regarding sensitivity of detection of fungal DNA in blood was reported using a combination of seven different dioxigenin labeled probes following amplification of a 482- to 503-bp fragment of 18S rRNA genes [10]. The primers could detect DNA from seven species of Candida and the six species of Aspergillus, another variety of fungi, but not bacteria or viruses. The seven probes could be used to identify five of the seven species of Candida and could group the *Aspergillus* spp. into two different groups. DNA detection sensitivity ranged between 10 to 50 fg DNA. Clinically, the PCR was positive in the first blood sample obtained in 88% of 21 untreated patients with invasive fungal infection and 100% of patients with two or more blood samples.

Another approach using universal fungal primers (termed panfungal by these investigators) that recognized a 580-bp fragment of the small-subunit rRNA gene was used in conjunction with a digoxigenin-labeled probe to improve sensitivity [2]. The time course of PCR and cultural diagnosis of fungal infection in several bone marrow transplant patients was also presented. The PCR time to positivity compared to culture positivity was variable and did not show clear advantage; i.e., PCR was not always positive before culture results became available. PCR results varied from positive to negative and back to positive in several of the patients studied which may support the transient nature of organisms of their DNA in clinical specimens.

Species-specific probes containing reporter dyes to identify amplicons produced by fungus-specific primers have recently been reported [22]. These investigators used universal primers that recognized a portion of the ITS rRNA genes for target DNA. Expanding upon earlier work using enzyme immunoassay to detect PCR products, three different fluorescent dyes were tagged to the species-specific probes. This allows detection of PCR product in a luminescence spectrometer. Although this report was limited to testing liquid from positive blood culture bottles, the concept of using differentially tagged probes lends itself to automation and could also be used for direct detection from clinical specimens.

In a recent report [14], a fragment of the ERG11 (lanosterol demethylase) gene was used as target followed by restriction enzyme analysis (REA) with three restriction enzymes. The REA allowed identification of 1 of 7 commonly identified *Candida* spp. Based on a limited number (13 PCR-positive) of 14 proven cases of disseminated candidiasis, the investigators felt that this PCR method was more sensitive than blood culture methods; however, samples from several patients showed what could be false-positive PCR results. These patients were in a less well defined group with possible candidal infection. Interestingly, the negative predictive value of the PCR-REA result was quite high (97%) compared to blood cultures (84%).

Because of chapter length considerations, this review of the state of the art regarding PCR as a tool to detect candidal DNA in clinical specimens precludes extensive discussion of many other interesting studies. For example during the past decade PCR assays have been reported for a variety of genes, including cytochome P450 lanosterol-alpha-demethylase [3,4,14,23], actin [24], beta-tubulin [25], chitin synthetase [26], heat shock protein 90 [27], secreted aspartic proteinase [28], mitochondrial DNA [19], and number of rRNA gene fragments derived from internal transcribed regions [17,29], 5S rRNA [2,6,21], 18S rRNA [10,11,20], and 28S rRNA [15]. Undoubtedly, other gene fragments are being investigated for their potential as diagnostic tools. Assay methods to improve sensitivity of detecting amplified product is also being addressed in many research laboratories. One major aspect of laboratory concern not addressed in this chapter is the general area of how specimens should be processed prior to performing the PCR assay. The literature is replete with different methods for removing contaminating nonfungal DNA, breaking fungal cells, and extracting target DNA while at the same time preventing destruction of target DNA prior to PCR.

IV. *ASPERGILLUS*

Aspergillus spp. are ubiquitous saprophytic fungi that can colonize the respiratory tract in humans, cause allergic bronchopulmonary aspergillosis (not an infection), and cause a fungus ball in certain tissues, all of which can be benign. However, fungus balls may become invasive, and invasive Aspergillus (IA) infection is a particular problem for patients with prolonged granulocytopenia secondary to cancer treatment or steroid therapy. IA is a life-threatening infection in these patients, and is responsible for a significant percentage (>40%) of deaths in acute leukemia patients. Successful therapy is directly related to early diagnosis and treatment. Unfortunately, the clinical features are often vague, and because of the severity of the underlying illness, it may be difficult to obtain the invasive samples (open lung or transbronchial biopsy) necessary for histologic diagnosis. Culture results take days to weeks, have low sensitivity, and may represent contamination due to the ubiquitous nature of this fungus. Various methods of detection have been tried in the past, but most require high fungal burdens, associated with advanced disease, by which time clinical intervention is ineffective [30]. Antibody screens have not been of clinical use, often because of the poor immune status of the patient. Antigenuria and antigenemia tends to be transient in nature, and initial testing methods for antigen such as radioimmunoassay, immunoblotting, enzyme immunoassay, and latex agglutination did not show the necessary sensitivity [31,32]. The most successful antigen-based attempts to diagnose IA have focused on the detection of a major cell wall constituent and exoantigen, galactomannan, released during invasive disease [30,33–39]. Use of sandwich ELISA increased the sensitivity of the test, with autopsy-verified cases showing sensitivity and specificity of serial GM monitoring to both exceed 90% [30,34,38].

The use of PCR to detect *Aspergillus* DNA in respiratory specimens has an inherent problem of potential false-positive results because of possible contamination or colonization. PCR was first attempted on respiratory specimens using a 401-bp fragment of the 26S/intergenic spacer region of rDNA [40]. The primers were specific for *A. fumigatus* but not other related *Aspergillus* spp. or several other medically important fungi. In a limited patient population, PCR was positive in all six patients with proven IA or colonization and positive in specimens from several culture-negative high- and low-risk patients.

Soon thereafter others applied PCR for detecting Aspergillus DNA in bronchoalveolar lavage (BAL) specimens using a fragment of the alkaline protease (ALP) genes [41]. Their PCR could detect as little as 500 fg and differentiate DNA from *A. fumigatus* and *A. flavus* based on size of the amplicon. Again, a low number of cases were studied and PCR was found to be positive in proven cases of IA as well as from many patients presumably colonized with the organism.

Genus-specific gene fragments from 18S rRNA genes were then used for fungal detection in BAL specimens [42]. The primers were specific for *Aspergillus* spp. and *Penicillin marneffei*, and the products could be distinguished by enzyme digestion. Between 1 and 10 fg *Aspergillus* DNA could be detected. Similar to the earlier studies, the PCR was positive in BAL collected from all four proven cases of IA, in a few patients with suspected IA, and in three of eight neutropenic patients that developed pulmonary infiltrates. In spite of the possibility of false positives from colonization, the authors felt the PCR would be helpful when testing specimens from high-risk patients. The following year, these same investigators compared their PCR to an ELISA for galactomannan, a cell wall polysaccharide antigen of *Aspergillus* [33]. The antigen test had the same problems with potential false positives as the PCR. The serum ELISA tended to be positive before either ELISA or PCR of BAL fluid. The investigators felt that both *Aspergillus* genus-specific PCR and sandwich ELISA might be beneficial for use in early diagnosis of IA in patients with hematological malignancies.

Subsequent studies using these latter primer systems with and without nested PCR, and studies using mitochondrial DNA from *Aspergillus* as target have compared the efficacy of PCR with antigen detection and have generally come to the same conclusion [32,34,35,38]. Both PCR and antigen are of value in establishing IA when performed in the appropriate clinical setting. All of these studies using PCR for detection of DNA from Aspergillus (or Candida) point out the promising nature of a new diagnostic approach while at the same time remind us of the challenges associated with interpreting PCR results from patients at high risk of fungal infection. This is especially true for these two groups of fungi compared to other fungi, such as *C. neoformans*, which is more likely to be present as a pathogen rather than a colonizer or contaminant.

V. PNEUMOCYSTIS CARINII

Pneumocystis carinii was first recognized as a respiratory pathogen in the early 1900s. It first emerged as a cause of pneumonia among malnourished populations after World War II. More recently, *P. carinii* has become recognized as the most frequent cause of opportunistic pulmonary infection in AIDS patients, and is a major cause of pneumonia in patients receiving immunosuppressive therapy. Prior to the advent of highly active antiretroviral therapy, or HAART, >50% of AIDS patients developed *Pneumocystis carinii* pneumonia (PCP) during the course of their disease. Clinical presentations are frequently subtle or vague, with often insidious onset of nonproductive cough, progressive dyspnea, malaise, and low-grade fever. The overall mortality from PCP is 20% for AIDS patients and 40% for non-HIV patients [43]. Early diagnosis and treatment of PCP has been shown to significantly reduce morbidity and mortality [44]. Traditionally, because *P. carinii* has yet to be successfully cultured from human specimens, diagnosis was made on the basis of examination of stained BAL samples for the presence of cysts or trophozoites. Inherent problems in this method are variation in observer skill, staining technique, and inability to detect organisms in low number, causing sensitivity of the traditional Toluidine Blue O staining (at least on induced sputa) to range from 21% to 95% [45]. Use of direct immunofluorescence staining lead to an increase in sensitivity of detection of PCP of up to 90–95% using BAL specimens [46]. The downside to BAL analysis is that it is an invasive as well as an expensive procedure. Less invasive methods for obtaining respiratory samples, such as induced sputum (IS), are more desirable as a test sample but have proven less sensitive (70–90%) than BAL using fluorescent stain. Although expectorated sputum samples are even less sensitive using fluorescent staining than IS, sputa may prove more reliable for PCR analysis [47].

With the advent of anti-PCP antibiotic prophylaxis and HAART, there have been changes in the clinical presentation and a reduction in the number of organisms were reported present in respiratory samples [48]. Because of these diagnostic problems, *P. carinii* was one of the earliest fungal infections in which the potential of PCR was studied as a diagnostic tool [49]. The original PCR methodology required specialized equipment, radioactive probes, and lengthy extraction and hybridization procedures which made it difficult for most clinical laboratories to perform routinely [50]. More recently, nested or "touchdown" PCR methods have made the procedure rapid and relatively simple [51]. Six different PCR targets have been studied, including sequences encoding mitochondrial large-subunit rRNA (mtLSU rRNA), both 5S rRNA, 16S rRNA, dihydrofolate reductase (DHFR), internal transcribed spacers (ITS) of the rRNA operon, and major surface glycoprotein (MSG) [52]. Most studies have looked at mitochondria rDNA and DHFR in BAL and IS. Latouche used ITS-1 and ITS-2 sequences in combination to type strains of *P. carinii*, and found 11 types among 36 samples taken from 16 patients [53]. Of six patients with several episodes of PCP, one had identical sequence results indicating reactivation while the other five had different sequence results indicating de novo infection by a different *P. carinii* strain. These results show the potential value of PCR for epidemiologic study of PCP.

Because many of the fungi, bacteria, and viruses that cause pneumonia are found in the upper respiratory tract, attempts have been made to use other, less invasive material such as nasal, pharyngeal, and blood or serum samples. PCR studies using a rat model demonstrated a sensitivity of 93% for nasal aspirate and 75% for pharyngeal aspirate samples. These same aspirates were negative as determined by both silver and Giemsa stains for pneumocystis [47]. Human studies comparing oral samples obtained by rinse and gargle have shown a sensitivity of ~90% when compared to BAL-positive specimens [51,54]. The touchdown PCR method used by these investigators appears advantageous, because it is a single, round PCR providing earlier results while at the same time having reduced risk of contamination with no loss of sensitivity compared to other nested methods.

The incidence of extrapulmonary *P. carinii* infection is rare, accounting for only 1–2.5% of all *P. carinii* infections. The majority of these cases occur in patients receiving aerosolized pentamidine prophylaxis [43]. Therefore, it is not clear why there might be circulating *P. carinii* antigens or DNA in patients with pulmonary disease. The studies using PCR to detect *P. carinii* DNA in blood have had mixed results, with sensitivities ranging from 0% to 100% [51]. Schluger, amplifying the DHFR gene, found DNA in serum from 12 of 14 AIDS patients with PCP for a sensitivity of 86% [55]. Atzori et al., amplifying rRNA internal transcribed spacers (Pc-ITS-PCR) found at least one positive serum in 27 patients tested (100% sensitivity) by nested PCR.

These were patients with PCP, not extrapulmonary disease. In contrast, only two of 20 (10% sensitivity) patient serum samples were positive using DHFR as the PCR target [56]. In a later study, Atzori et al. found that the same technique applied to serum samples was 71% sensitive, but when combined with gargled oropharyngeal washes (79% sensitive), the combined sensitivity was 100% [57]. These investigators noted that serum-positive PCR results were more often found during the acute clinically overt phase of PCP and rapidly disappeared from serum following treatment. Wagner et al. [58], using an mtrRNA PCR, reported the sensitivity of serum PCR to be 8%, and Tumburrini et al. [59], using similar methods, found 0% sensitivity for serum diagnosis. It remains to be determined whether these large discrepancies in sensitivity of serum PCR for detecting *Pneumocystis* DNA in patients with PCP are reflections of the different DNA targets used, are differences in laboratory procedures in processing specimens and subsequent PCR methods, or reflect differences in the patient populations studied such as stage of disease. The concept that detecting fungal DNA in serum from a patient with pulmonary disease (or some other organ involvement such as liver) need not require fully disseminated disease is both intriguing and bothersome. In patients with diseases such as PCP or histoplasmosis, finding fungal DNA in serum might provide meaningful clinical data regardless of the extent of infection. However, in other fungal infections, such as *Aspergillus* and *Candida* infection, it is imperative to know whether the circulating DNA is from mere colonization, superficial cutaneous infection, invasive infection, or widespread dissemination. These and related issues will require much more investigation before being resolved.

Another major problem with use of PCR for diagnosis of PCP stems from the reports of positive PCR results with lack of histologic or clinical correlates, despite long-term follow-up. This has been generally interpreted, primarily because a number of these patients have positive follow-up tests, as representing subclinical *P. carinii* infection or as colonization. This is problematic, because it reduces the positive predictive value of PCR in the evaluation of PCP [48,60].

VI. MISCELLANEOUS FUNGAL PCR

A. Dimorphic Fungi

There are very few reported investigations of using PCR to detect fungal DNA in clinical specimens within the dimorphic fungi. Primary pulmonary infection may not be clinically apparent and may only be diagnosed if severe pulmonary symptoms develop. Some infections may be detected by skin test reactivity in groups of patients exposed in a particular environmental setting. In individuals with progressive pulmonary disease, the yeast, or yeastlike, organisms can often be found in respiratory specimens. The observed morphology of *Blastomyces dermatitidis*, *Coccidioides immitis*, and *Paracoccidioides brasiliensis* are indicative of the etiologic agent causing the disease. Since the yeast form of *Histoplasma capsulatum* is similar to other yeasts, the diagnosis is usually made from culture or by antigen testing. Disseminated disease caused by any of these organisms is often suspected and diagnosed by microscopic examination and culture of a variety of clinical specimens. PCR reports using universal or panfungal PCR [6,15,20] mention that PCR can detect DNA from these organisms, but data from few, if any, actual clinical specimens are presented. A nested PCR designed to amplify a 185-bp fragment of the ITS1 region of rDNA of *H. capsulatum* has been used to help establish diagnosis in a retrospective study of liver tissue containing granulomas from 22 patients [61]. PCR was useful in demonstrating the presence of *H. capsulatum* in 15 cases compared to only five patients by conventional testing. A recent report uses nested primer PCR to detect a fragment of the ITS region of the 5.8S rRNA gene of *H. capsulatum* from soil specimens. There have been occasional preliminary reports of detection of DNA from the dimorphic fungi, but few publications have appeared in the literature. Finally, PCR has been used to detect a DNA fragment specific for *Paracoccidioides brasiliensis* in an animal model [62].

B. *Cryptococcus* and *Trichosporon*

Perhaps the existence of a reliable, inexpensive, and rapid latex agglutination test for cryptococcal antigen has lessened the interest in use of PCR for detecting cryptococcal DNA in clinical specimens. Most, it not all, of the universal primers systems that detect conserved rDNA sequences from the medically relevant fungi demonstrate the detection of cryptococci in the "me-too" category, and data specific for cryptococcal infection have not been presented. One study using primers that produced a 343-bp amplicon from the 18S rDNA of *C. neoformans* was able to detect as little as 100 fg of DNA after Southern blotting [63]. DNA from *Trichosporon* and *Klebsiella pneumoniae* produced PCR product but did not react with the specific probe used for confirmation. Results of PCR applied to CSF specimens from patients with cryptococcal meningitis were 100% sensitive compared to culture but only 89% and 84% sensitive compared to antigen and India ink, respectively. Other investigators [64], using nested PCR for a 415-bp fragment of ITS regions of rDNA specific for *C. neoformans*, compared the result of 21 known positive CSF specimens. PCR was positive in all 21 cases and when compared to culture, antigen, and India ink results, PCR was felt most likely to be positive in patients with living fungal cells present. Therefore, PCR might be a useful tool to follow treatment since cryptococcal antigen tends to remain positive for extended periods of time in CSF of patients with treated cryptococcal meningitis.

A nested PCR for detection of a 259-bp fragment of 26S rRNA genes specific for *Trichosporon* spp. was used to assay serum from patients with disseminated trichosporonosis [65]. Although capable of detecting as little as 5 fg of DNA, PCR was positive in blood samples of seven of 11 patients with proven disseminated infection while cultures and antigen testing were positive in eight and six patients, respectively. As with many of the other fungal PCR findings, the cryptococcal and trichosporon results support the concept that PCR assays can be clinically useful in some patients but, to date, lack sufficient sensitivity to stand alone as a diagnostic tool.

C. *Fusarium*

Since the histopathology of tissue infected with *Fusarium* spp. is indistinguishable from many other septate organisms, a diagnostic PCR for these organisms would be of value, especially when cultures are not positive or when the entire specimen was placed in formalin. In a case report, PCR was used to detect a 189-bp fragment from the cutinase gene of *F. solani* for diagnosis of a Fusarium infection from postmortem ocular tissue [66]. These same investigators established the PCR test using an animal model of keratitis [67]. Others have used PCR to detect a 329-bp fragment containing ITS2 and a portion of 5.8S and 28S rDNA in spiked human blood samples and tissue specimens obtained from an infected mouse model [68].

VII. MOLECULAR METHODS: PAST, PRESENT, AND FUTURE

One of us (R.L.H.) was introduced to the difficulty of laboratory diagnosis of life-threatening fungal infection, especially disseminated candidiasis and invasive aspergillosis, over 25 years ago. Others, both clinicians and laboratorians, had been aware of these problems for several decades before me. Like many others, I tackled these problems with high enthusiasm and great expectation that the problem would soon be resolved. It soon becomes apparent to all investigators who enter the diagnostic arena that if diagnosis were easy it would have been satisfactorily resolved years ago. The challenges that face all investigators in the past, present, and future remain numerous. There are three general areas of concern that contribute to the challenges facing investigators interested in resolving these diagnostic issues.

A. Host Parasite Factors

1. Organisms can be colonizers, part of normal flora, or environmental contaminants, so merely culturing Candida or Aspergillus from many clinical specimen sites is not diagnostic for life-threatening infection.

2. Blood cultures are insensitive even in cases of widespread dissemination. This implies intermittent shedding or a very low level fungemia in the blood. The pseudohyphal and hyphal forms of Candida and the hyphal forms of Aspergillus present in the infected host are probably infrequently shed into the host circulation. Even though Candida produces blastospores, they are either not readily shed into surround milieu, or if so, the host clearance system is effective in preventing most from entering the bloodstream.

3. Many fungal antigens are poorly immunogenic, eliciting low-level antibody in patients (patients are also often immunosuppressed); however, this low immunogenicity also increases difficulty in preparing high titer antisera needed for fungal antigen detection.

4. Many patients with life-threatening fungal infection have overriding medical problems which preclude collection of deep tissue samples for establishing the diagnosis.

5. Circulation of fungal antigen/DNA, much like blood cultures, occurs either intermittently or in extremely low concentrations or both. Low sensitivity of antigen/DNA detection may also be due to rapid destruction or removal of antigen/DNA from circulation by host factors such as enzymatic degradation, phagocytic removal and destruction, and/or absorption to host tissue.

B. Diagnostic Test and Study Design Problems

1. Since Candida and Aspergillus are easy to culture, insensitivity issues are most likely due to timing and/or volume of blood collected. Typically, 20 mL blood is collected for most automated blood culture systems. If seeding of a viable fungal unit occurs with a frequency that provides on average, one unit per 50 or 100 mL, then a positive culture result is based on luck of the draw.

2. There is no gold standard diagnostic test to which newly developed PCR methods can be compared. To date, most PCR results are compared to blood culture, although some reports also include antigen analysis. Either way, the insensitivity of the standard leaves much to be desired.

3. Most molecular studies that have been published, or reported at scientific meetings, involve a minimum number of patients, often including 10 or fewer proven cases of severe fungal infection. Tissue proof is often missing, which is one of the main reasons for needing a better diagnostic test. Unfortunately, short of histologic demonstration of tissue invasion as proof of deep seated infection, investigators accept different criteria for inclusion or exclusion of patients into different categories of infection. Besides making comparative analysis between different laboratories difficult, these biases can influence perceived sensitivity of the test. This further frustrates the investigator because the PCR findings end up being correlated to ill-defined "high-risk" patient populations or groups of patients "highly suspected" of having invasive disease.

4.   Sensitivity results of new PCR methods are based only on proven case results. The PCR test is usually relatively sensitive, 80–90% or higher, on patients that typically have over-whelming infection. Most reports have somewhat inflated estimates of sensitivity since the sensitivity is based on requiring that multiple specimens be tested from each patient. Additionally, many of the positive PCR results, just like many of the antigen and blood culture results, turn positive too late in the course of the infection to improve patient management.

5.   There is no gold standard or consensus on the most efficient, practical, or cost-effective method for processing the clinical specimen in order to release and purify target DNA prior to PCR amplification. The basic methods include lysis of the fungal cell using cell wall lytic enzymes, glass beads with vortexing, or heating. The variations and specific methods used tend to be based on the favorite method developed within each investigating laboratory. Similar differences exist in methods for DNA preparation following breakage of fungal cells such as use of proteinases, phenol with or without chloroform extraction, use of commercial DNA capture devices, ethanol precipitation, etc. Therefore, it is hard to determine whether apparent increases in sensitivity are the result of primer selection, specimen preparation methods, PCR conditions, amplicon detection method, or some combination of all the variables.

6.   Most reports support the premise that there is justification for continued development of new PCR tests. Unfortunately, subsequent reports from the same group of investigators are rarely forthcoming (see funding problems), or the newly reported PCR method is never pursued by other laboratories. Alternatively, the new method may have been examined in a secondary laboratory, but the results were much less promising and the data were never published. Failure to reproduce PCR results could be due to minor changes in procedures, thought to be insignifi-cant, but that were in fact less effective than the original methods.

7.   Given the tissue load of fungal cells and debris observed in cases of disseminated disease, it is expected that antigen/DNA should be shed directly from the fungal cell or shed following release by host clearance mechanisms, making the detection of antigen/DNA relatively easy. The failure of any antigen or PCR test to become positive early in the disease process could be used as evidence that the newly described diagnostic test detects the wrong antigen or wrong target DNA.

8.   If fungal DNA is present in clinical specimens only within intact fungal cells (alive or dead), then PCR might be more sensitive than culture because dead cells would provide target DNA. On the other hand, if, fungal cells release DNA into the host circulation as they die or are attacked by host cells, one might hope to improve the sensitivity of the PCR test over culture. Upon release from the fungal cell, small (<200 bp) target pieces of DNA would seem more likely to escape host nucleases than larger (>500 bp) targets, thereby making their detection by PCR more feasible. The downside of using smaller DNA fragments as target is that smaller fragments provide less opportunity for species identification. A small cell-free panfungal target with a long serum half-life would seem ideal for diagnostic purposes.

9.   Finally, in this day of cost containment, it is unlikely that many hospital laboratories will be able to institute a relatively expensive molecular test for a disease that is of low incidence in their institution. This increases the difficulty in collecting data needed for clinical evaluation and acceptance of promising diagnostic methods.

C.   Funding Mechanisms Are Scarce to Nonexistent

1.   One of the primary reasons for insufficient funds being available for the development of diagnostic tests for invasive fungal infection is because the incidence of life-threatening fungal infection is below the necessary threshold needed to generate long-term commitment

from funding agencies. Furthermore, invasive fungal infections pose no serious public health threat as do HIV infection or tuberculosis, and thus do not generate significant public concern.

2. Potential sources of government-based and industry-supported monies for applied research tend to be directed toward clinical trials for prophylaxis and treatment of fungal infection. Ironically, these prophylaxis and empiric therapy trials are justified because diagnosis of these infections is so insensitive. One could argue that, in the long run, money spent on development of diagnostics would ultimately decrease the number of patients placed unnecessarily on empiric therapy or antifungal prophylaxis, reducing overall expenses. Reducing the number of patients receiving unnecessary antifungals would also decrease the emerging threat of antifungal resistance.

3. Potential research monies from major commercial diagnostic corporations is virtually nonexistent because of the perceived low volume testing. For example it is easier to justify huge research expenditures for developing diagnostic service products for HIV or HCV than for similar expenditures for developing much lower volume fungal diagnostic tests.

4. PCR projects designed to detect fungal DNA in clinical specimens usually arise as a secondary research project assigned to a postdoctoral fellow or graduate student. The principal investigators have funding to perform basic research on one or more of their favorite fungal genes, and the student/postdoc is asked to see if PCR for their favorite gene could be applied as a diagnostic tool. Alternatively, someone with an interest in the fungal diagnosis problem, usually from a clinical laboratory or an infectious disease department, obtains limited startup funding for 12–18 months to demonstrate feasibility of a PCR approach. In either case the study is performed, perhaps with the promising results, then the funds dry up and the lead personnel graduate or move on to other research programs that will serve to improve their academic careers. As mentioned earlier, investigators acquire a certain level of frustration associated with obtaining PCR results from patients that are at "high risk" or are "highly suspected" of having life threatening fungal infections but have no proof of level of fungal involvement. This, often coupled with receipt of small numbers of samples for PCR testing, further discourages investigators from continuing such studies in their future endeavors. Finally, a publication may be generated from the data collected, the study is deemed successful, personnel are reassigned to other tasks, and the project is "laid to rest."

VIII. SOLUTIONS

Investigators with interest in the diagnosis of invasive fungal infection have struggled with all the previously mentioned problems and will likely continue making progress despite these continuing limitations. Since adequate solutions of the problems have not occurred in the past 5 decades, it is unlikely there will be a satisfactory solution or quick fix in the near future. Most likely, resolution of all the problems will require many years. The only hope for longterm funding would be some level of government funding to be set aside for individual research grants or for a government-sponsored contract. A laboratory or consortium of interested laboratories could then systematically address all of the issues related to improved diagnostic testing. This would require a long-term commitment for funding as well as commitment from the investigators because some of the initial work may seem rather mundane and not very glamorous. Multiple laboratories should be involved to ensure that the methods are simple and straightforward so that results are reproducible from one laboratory to another. Involvement of large numbers of interested laboratory researchers and clinicians, perhaps through organizations such as the Mycoses Study Group, that are seeing these types of patients will also be necessary for evaluation of results and specimen acquisition.

REFERENCES

1. A Velegraki, ME Kambouris, G Skiniotis, M Savala, A Mitroussia-Ziouva, NJ Legakis. Identification of medically significant fungal genera by polymerase chain reaction followed by restriction enzyme analysis. FEMS Immunol Med Microbiol 23:303–312, 1999.
2. JA van Burik, D Myerson, RW Schreckhise, RA Bowden. Panfungal PCR assay for detection of fungal infection in human blood specimens. J Clin Microbiol 36:1169–1175, 1998.
3. E Reiss, CJ Morrison. Nonculture methods for diagnosis of disseminated candidiasis. Clin Microbiol Rev 6:311–323, 1993.
4. E Chryssanthou, B Andersson, B Petrini, S Lofdahl, J Tollemar. Detection of *Candida albicans* DNA in serum by polymerase chain reaction. Scand J Infect Dis 26:479–485, 1994.
5. SB Wey, M Mori, MA Pfaller, RF Woolson, RP Wenzel. Hospital-acquired candidemia: the attributable mortality and excess length of stay. Arch Intern Med 148:2642–2645, 1988.
6. RL Hopfer, P Walden, S Satterquist, WE Highsmith. Detection and differentiation of fungi in clinical specimens using polymerase chain reaction (PCR) amplification and restriction enzyme analysis. J Med Vet Mycol 31:65–75, 1993.
7. TJ White, TD Bruns, SB Lee, JW Taylor. Amplification and direct sequencing of fungal ribosomal RNA genes for phylogenetics. In: MA Innis, DH Gelfand, JJ Sninsky, TJ White, eds. PCR Protocols. A Guide to Methods and Applications. San Deigo: Academic Press, 1990, pp 315–322.
8. R Maleszka, GD Clark-Walker. Yeasts have a four-fold variation in ribosomal DNA copy number. Yeasts 9:53–58, 1993.
9. FX Hue, M Huerre, MA Rouffault, C deBievre. Specific detection of *Fusarium* species in blood and tissues by PCR technique. J Clin Microbiol 37:2434–2438, 1999.
10. H Einsele, H Hebart, G Roller, J Loffler, I Rothenhofer, CA Muller, RA Bowden, JA van Burik, D Engelhard, L Kanz, U Schumacher. Detection and identification of fungal pathogens in blood by using molecular probes. J Clin Microbiol 35:1353–1360, 1997.
11. TJ Walsh, A Francesconi, M Kasai SJ Chanock. PCR and single-strand conformational polymorphism for recognition of medically important opportunistic fungi. J Clin Microbiol 33:3216–3220, 1995.
12. A van Belkum. DNA Fingerprinting of medically important microorganisms by use of PCR. Clin Microbiol Rev 7:174–184, 1994.
13. AO Ahmed, MM Mukhtar, M Kools-Sijmons, AH Fahal, S de Hoog, BG van den Ende, EE Zijlstra, H Verbrugh, ESAM Abugroun AM Elhassan A van Belkum. Development of a species-specific PCR-restriction fragment polymorphism analysis procedure for identification of *Madurella mycetomatis*. J Clin Microbiol 37:3175–3178, 1999.
14. G Morace, L Pagano, M Sanguinetti, B Posteraro, L Mele, F Equitani, G D' Amore, G Leone, G Fadda. PCR-restriction enzyme analysis for detection of *Candida* DNA in blood from febrile patients with hematological malignancies. J Clin Microbiol 37:1871–1875, 1999.
15. GS Sandhu, BC Kline, L Stockman, GD Roberts. Molecular probes for diagnosis of fungal infections. J Clin Microbiol 33:2913–2919, 1995.
16. CY Turenne, SE Sanche, DJ Hoban, JA Karlowsky, AM Kabani. Rapid identificatin of fungi by using the ITS2 genetic region and an automated fluorescent capillary electrophoresis system. J Clin Microbiol 37:1846–1853, 1999.
17. JP Burnie, N Golbang, RC Matthews. Semiquantitative polymerase chain reaction enzyme immunoassay for diagnosis of disseminated candidiasis. Eur J Microbiol Infect Dis 16:346–350, 1997.
18. TG Buchman, M Rossier, WG Merz, P Charache. Detection of surgical pathogens by in vitro DNA amplification. Part I. Rapid identification of *Candida albicans* by in vitro amplification of a fungus-specific gene. Surgery 108:338–347, 1990.
19. Y Miyakawa, T Mabuchi, K Kagaya, Y Fukazawa. Isolation and characterization of a species-specific DNA fragment for detection of *Candida albicans* by polymerase chain reaction. J Clin Microbiol 30:894–900, 1992.
20. K Makimura, SY Murayama, H Yamaguchi. Detection of a wide range of medically important fungi by the polymerase chain reaction. J Med Microbiol 40:358–364, 1994.

21. AR Holmes, RD Cannon, MG Shepherd, HF Jenkinson. Detection of *Candida albicans* and other yeasts in blood by PCR. J Clin Microbiol 32:228–231, 1994.
22. JH Shin, FS Nolte, BP Holloway, CJ Morrison. Rapid identification of up to three *Candida* species in a single reaction tube by a 5′ exonuclease assay using fluorescent DNA probes. J Clin Microbiol 37:165–170, 1999.
23. G Deng, Y Zhang, G Xiao. Experimental study and clinical application of rapid diagnosis of systemic *Candida albicans* infection in burns by polymerase chain reaction. Chin J Plastc Surg Burns 11: 323–326, 1995.
24. VL Kan. Polymerase chain reaction for the diagnosis of candidemia. J Infect Dis 168:779–783, 1993.
25. H Muramatsu. Detection of *Candida albicans* by nested PCR. J Jpn Assoc Infect Dis 68:1465–1471, 1994.
26. JA Jordan. PCR identification of four medically important *Candida* species by using a single primer pair. J Clin Microbiol 32:2962–2967, 1994.
27. AC Crampin, RC Matthews. Application of the polymerase chain reaction to the diagnosis of candidosis by amplification of an HSP 90 gene fragment. J Med Microbiol 39:233–238, 1993.
28. M Flahaut, D Sanglard, M Monod, J Bille, M Rossier. Rapid detection of *Candida albicans* in clinical samples by DNA amplification of common regions from *C. albicans*–secreted aspartic proteinase genes. J Clin Microbiol 36:395–401, 1998.
29. S Fugita, BA Lasker, TJ Lott, E Reiss, CJ Morrison. Microtitration plate enzyme immunoassay to detect PCR-amplified DNA from *Candida* species in blood. J Clin Microbiol 33:962–967, 1995.
30. J Maertens, J Verhaegen, H Demuynck, P Brock, G Verhoef, P Vandenberghe, J van Eldere, L Verbist, M Boogaerts. Autopsy-controlled prospective evaluation of serial screening for circulating galactomannan by a sandwich enzyme-linked immunosorbent assay for hematological patients at risk for invasive aspergillosis. J Clin Microbiol 37:3223–3228, 1999.
31. VM Hearn, C Pinel, S Blachier, P Ambroise-Thomas, R Grillot. Specific antibody detection in invasive aspergillosis by analytical isoelectrofucosing and immunoblotting methods. J Clin Microbiol 33:982–986, 1995.
32. Y Yamakami, A Hashimoto, I Tokimatsu, M Nasu. PCR detection of DNA specific for *Aspergillus* species in serum of patients with invasive aspergillosis. J Clin Microbiol 34:2464–2468, 1996.
33. PE Verweij, JP Latge, AJMM Rijs, WJG Melchers, BE dePauw, JAA Hoogkamp-Korstanje, JFGM Meis. Comparison of antigen detection and PCR Assay using bronchoalveolar lavage fluid for diagnosing invasive pulmonary aspergillosis in patients receiving teatment for hematological malignancies. J Clin Microbiol 33:3150–3153, 1995.
34. PE Verweij, EC Dompeling, JP Donnelly, AVMB Schattenberg, JFGM Meis. Serial monitoring of *Aspergillus* antigen in the early diagnosis of invasive aspergillosis. Preliminary investigations with two examples. Infection 25:86–89, 1997.
35. S Bretagne, JM Costa, E Bart-Delabesse, N Dhedin, C Rieux, C Cordonnier. Comparison of serum galactomannan antigen detection and competitive polymerase chain reaction for diagnosisn invasive aspergillosis. Clin Infect Dis 26:1407–1412, 1998.
36. A Paugam, J Sarfati, R Romieu, M Viguier, J Dupouy-Camet, JP Latge. Detection of *Aspergillus* galactomannan: comparison of an enzyme-linked immunoassay and a europium-linked time-resolved fluoroimmunoassay. J Clin Microbiol 36:3079–3080, 1998.
37. Y Yamakami, A Hashimoto, E Yamagata, P Kamberi, R Karashima, H Nagai, M Nasu. Evaluation of PCR for detection of DNA specific for *Aspergillus* species in sera of patients with various forms of pulmonary aspergillosis. J Clin Microbiol 36:3619–3623, 1998.
38. S Kawamura, S Maesaki, T Noda, Y Hirakata, K Tomono, J Jashiro, S Kohno. Comparison between PCR and detection of antigen in sera for diagnosis of pulmonary aspergillosis. J Clin Microbiol 37: 218–220, 1999.
39. PE Verweij, K Brinkman, HPH Kremer, BJ Kullberhg, JFGM Meis. *Aspergillus* meningitis: diagnosis by non-culture-based microbiological methods and management. J Clin Microbiol 37:1186–1189, 1999.
40. C Spreadbury, D Holden, A Aufauvre-Brown, B Bainbridge, J Cohen. Detection of *Aspergillus fumigatus* by polymerase chain reaction. J Clin Microbiol 31:615–621, 1993.

41. CM Tang, DW Holden, A Aufauvre-Brown, J Cohen. The detection of *Aspergillus* spp. by the polymerase chain reaction and its evaluation in bronchoalveolar lavage fluid. Am Rev Respir Dis 148:1313–1317, 1993.

42. WJG Melchers, PE Verweij, P van den Hurk, A van Belkum, PE dePauw, JAA Hoogkamp-Korstanje, JFGM Meis. General primer-mediated PCR for detection of *Aspergillus* species. J Clin Microbiol 32:1710–1717, 1994.

43. CF Thomas, AH Limper. *Pneumocystis* pneumonia: Clinical presentation and diagnosis in patients with and without acquired immune deficiency syndrome. Semin Respir Infect 13:289–295, 1998.

44. A Mathis, R Weber, H Kuster, R Speich. Simplified sample processing combined with a sensitive one-tube nested PCR assay for detection of *Pneumocystis carinii* in respiratory specimens. J Clin Microbiol 35:1691–1695, 1997.

45. D Eisen, BC Ross, J Fairbairn, RJ Warren, RW Baird, B Dwyer. Comparison of *Pneumocystis carinii* detection by Toluidine Blue O staining, direct immunofluorescence and DNA amplification in sputum specimens from HIV positive patients. Pathology 26:198–200, 1994.

46. C Chouaid, P Roux, I Lavard, JL Poirot, B Housset. Use of the polymerase chain reaction technique on induced-sputum samples for the diagnosis of *Pneumocystis carinii* pneumonia in HIV-infected patients. A clinical and cost-analysis study. Clin Microbiol Infect Dis 104:72–75, 1995.

47. HS Oz, WT Hughes. DNA amplification of nasopharyngeal aspirates in rats: a procedure to detect *Pneumocystis carinii*. Microb Pathog 27:119–121, 1999.

48. JA Ribes, AH Limper, MJ Espy, TF Smith. PCR detection of *Pneumocystis carinii* in bronchoalveolar lavage specimens: analysis of sensitivity and specificity. J Clin Microbiol 35:830–835, 1997.

49. K Kitada, S Oka, S Kimura, K Shiada, T Serikawa, J Yamada, H Tsunoo, K Egawa, Y Nakamura. Detection of *Pneumocystis carinii* sequences by polymerase chain reaction: animal models and clinical application to non invasive specimens. J Clin Microbiol 29:1895–1990, 1991.

50. M Rabodonirina, D Raffenot, L Cotte, A Boibieux, M Mayencon, G Bayle, F Persat, F Rabatel, C Trepo, D Peyramond, MA Piens. Rapid Detection of *Pneumocystis carinii* in bronchoalveolar lavage specimens from human immunodeficiency virus–infected patients: use of a simple DNA extraction procedure and nested PCR. J Clin Microbiol 35:2748–2751, 1997.

51. J Helweg-Larsen, JS Jensen, T Benfield, UG Svendsen, JD Lundgren, B Lundgren. Diagnostic use of PCR for detection of Pneumocystis carinii in oral wash samples. J Clin Microbiol 36:2068–2072, 1998.

52. M Rabodontrina, L Cotte, A Boibieux, K Kaiser, M Mayencon, D Raffenot, C Trepo, D Peyramond, S Picot. Detection of *Pneumocystis carinii* DNA in blood specimens from human immunodeficiency virus–infected patients by nested PCR. J Clin Microbiol 37:127–131, 1999.

53. S Latouche, JL Poirot, C Bernard, P Roux. Study of internal transcribed spacer and mitochondrial large-subunit genes of *Pneumocystis carinii hominis* isolated by repeated bronchoalveolar lavage from human immunodeficiency virus–infected patients during one or several episodes of pneumonia. J Clin Microbiol 35:1687–1690, 1997.

54. C Atzori, F Agostoni, G Gubertini, A Cargnel. Diagnosis of PCP by ITSs nested PCR on noninvasive oropharyngeal samples. J Eukaryot Microbiol 43:41S, 1996.

55. N Schluger. Application of DNA amplification of pneumocystosis: presence of serum *Pneumocystis carinii* DNA during human and experimental induced *Pneumocystis carinii* pneumonia. J Exp Med 176:1327–1333, 1992.

56. C Atzori, JJ Lu, B Jiang, MS Bartlett, G Orlando, SF Queener, JW Smith, A Cargnel, CH Lee. Diagnosis of *Pneumocystis carinii* pneumonia in AIDS patients by using polymerase chain reactions on serum specimens. J Infect Dis 172:1623–1626, 1995.

57. C Atzori, F Agostoni, E Angeli, A Mainini, G Orlando, A Cargnel. Combined use of blood and oropharyngeal samples for noninvasive diagnosis of *Pneumocystis carinii* pneumonia using the polymerase chain reaction. Eur J Microbiol Infect Dis 17:241–246, 1998.

58. D Wagner, J Koniger, WV Kern, P Kern. Serum PCR of *Pneumocystis carinii* DNA in immunocompromised patients. Scand J Infect Dis 29:159–164, 1997.

59. E Tamburrini, P Mencarini, E Visconti, M Zolfo, A deLuca, A Siracusano, E Ortona, AE Wakefield. Detection of *Pneumocystis carinii* DNA in blood by PCR is not of value for diagnosis of *P. carinii* pneumonia. J Clin Microbiol 34:1586–1588, 1996.

60. M Weig, H Klinker, BH Bogner, A Meier, U Gross. Usefulness of PCR for diagnosis of *Pneumocystis carinii* pneumonia in different patient groups. J Clin Microbiol 35:1445–1449, 1997.

61. MH Collins, B Jiang, JM Croffie, SKF Chong, CH Lee. Hepatic granulomas in children. A clinicopathologic analysis of 23 cases including polymerase chain reaction for histoplasma. Am J Surg Pathol 20:332–338, 1996.

62. LZ Goldani, AM Sugar. Short report: Use of the polymerase chain reaction to detect *Paracoccidioides brasiliensis* in muring paracoccidioidomycosis. Am J Trop Hyg 58:152–153, 1998.

63. C Prariyachatigul, A Chaiprasert, V Meevootisom, S Pattanakitsakul. Assessment of a PCR technique for the detection and identification of *Cryptococcus neoformans*. J Med Vet Mycol 34:251–258, 1996.

64. P Rappelli, R Are, G Casu, PL Fiori, P Cappuccinelli, A Aceti. Development of nested PCR for detection of *Cryptococcus neoformans* in cerebrospinal fluid. J Clin Microbiol 36:3438–3440, 1998.

65. H Nagai, Y Yamakami, A Hashimoto, I Tokimatsu, M Nasu. PCR detection of DNA specific for *Trichosporon* species in serum of patients with disseminated trichosporonosis. J Clin Microbiol 37: 694–699, 1999.

66. G Alexandrakis, M Sears, P Gloor. Postmortem diagnosis of *Fusarium* panophthalmitis by the polymerase chain reaction. A J Ophthalmol 121:221–223, 1996.

67. G Alexandrakis, S Jalali, P Gloor. Diagnosis of *Fusarium* keratitis in an animal model using the polymerase chain reaction. Br J Ophthalmol 82:306–311, 1998.

68. FX Hue, M Huerre, MA Rouffault, C de Bievre. Specific detection of fusarium species in blood and tissues by a PCR technique. Journal of Clinical Microbiology 37:2434–2438, 1999.

29

Serological Approaches to the Diagnosis of Invasive Fungal Infections

Christine J. Morrison and Mark D. Lindsley
Centers for Disease Control and Prevention, Atlanta, Georgia

I. INTRODUCTION

The incidence of invasive opportunistic fungal infections continues to increase as sophisticated technologies to prolong the lives of severely ill patients are implemented [1–5]. Opportunistic fungi are often found as either normal commensal organisms of the human gastrointestinal tract, skin, and/or mucosa or as ubiquitous organisms in the soil and environment. Extended granulocytopenia, the use of broad-spectrum antibacterial antibiotics, indwelling central venous catheters, immunosuppression, and disruption of mucosal barriers by chemotherapy and radio-therapy have all been implicated as risk factors allowing these normally harmless fungi to invade host tissues [1,6–8]. Once invasion occurs, these infections cause significant morbidity and mortality in patients with underlying diseases such as cancer, diabetes, and AIDS [1,6,7,9].

The prognosis in these patient populations is poor but may be improved if infections are diagnosed and treated promptly after onset [10–12]. Unfortunately, a clinical diagnosis is difficult because signs and symptoms are often nonspecific. Fever unresponsive to antibacterial antibiotics may be the only clinical sign. Whereas recent advances in diagnostic radiology and computed tomography, as well as ultrasound examination, have improved clinical detection, these methods are only effective when invasive organisms cause typically recognizable alterations in the host [13,14]. Therefore, a specific diagnosis relies on a combination of clinical, microbiological, histological, serological, and, more recently, molecular biological methods.

The incidence of endemic mycoses, like that of the opportunistic mycoses, has also risen dramatically in recent years as a result of the AIDS epidemic [15–18]. Disseminated penicilliosis (*Penicillium marneffei*) and histoplasmosis in HIV-positive patients are now AIDS-defining illnesses in their endemic areas [9,15,17,19]. The endemic fungi are not known to colonize humans, and therefore infections with these organisms are derived from exogenous environmental sources. Disease is usually acquired as a result of inhalation of fungal elements. Therefore, outbreaks continue to occur as natural niches for these organisms are disturbed by (1) construction in endemic areas (histoplasmosis) [20]; (2) hiking near water sources (blastomycosis) [21]; (3) outdoor activities or cave exploration in areas where exposure to contaminated bird or bat guano occurs (histoplasmosis) [22–24]; and (4) natural disasters, such as earthquakes, where dust clouds are raised and infectious particles are disseminated (coccidioidomycosis) [25]. Also, movement of people into and out of the endemic areas has resulted in an increased occurrence of these infections outside of their endemic areas [26–28].

After inhalation of conidia, *Histoplasma capsulatum*, *P. marneffei*, and *Paracoccidiodes brasiliensis* generally cause localized granulomatous responses in their hosts in contrast to *Coccidioides immitis* and *Blastomyces dermatitidis*, where a combination of acute and chronic inflammation is often observed. In immunocompetent hosts, calcifications are often seen on chest radiograms of patients after resolution of granulomatous lesions. However, a positive DTH skin test, when available, and production of specific antibodies may often be the only evidence of infection in an immunocompetent patient. In the immunocompromised host, such as AIDS patients or those receiving corticosteroids, such an immune response may not occur, and the disease often disseminates [10]. Therefore, in a manner similar to that for the diagnosis of opportunistic fungal infections, a specific diagnosis of the endemic mycoses relies on laboratory confirmation of clinical signs and symptoms.

Laboratory confirmation of infection includes microbiological, histological, serological, and molecular biological methods. Microbiological methods include direct visualization of the organism in body fluids (such as CSF), in scrapings of lesions, or in biopsies of patient tissues, and culture of blood, other body fluids, or tissues, which may require up to 4 weeks for the detection of some slower-growing fungi. However, absence of the organism in culture does not necessarily rule out infection. For example, blood cultures can be negative in up to 50% of patients with invasive candidiasis [10,29–31]. Similarly, blood cultures for *Aspergillus* spp. are routinely negative, and sputum and bronchoalveolar lavage fluid from patients with proven invasive pulmonary aspergillosis may yield recoverable fungi in only 30% and 50% of cases, respectively [32–37]. Because opportunistic fungi often colonize normal human skin and mucosal surfaces, isolation of the organism does not necessarily indicate infection. For example, blood cultures positive for *Candida* spp. may represent transient, catheter-related candidemia rather than invasive disease [29,30]. In addition, recovery of an opportunistic mold from a nonsterile site often represents contamination or colonization rather than true infection. On the other hand, in one study, a strong correlation was found between recovery of *Aspergillus* spp. from respiratory sites in immunocompromised patients and invasive disease [37]. Nonetheless, to circumvent the potential misdiagnosis of invasive fungal disease as a result of contamination or simple colonization, tissue biopsies for culture and histology are used to confirm that an organism is actively invading tissue rather than simply colonizing it [38–40]. Such invasive procedures, however, are potentially hazardous in patients who are severely ill, granulocytopenic, and thrombocytopenic. Therefore, such procedures are used only when absolutely necessary. Serological tests, on the other hand, provide a rapid, less invasive means to diagnose fungal infections. Serological tests have their limitations, however, and, depending upon the disease in question, these tests can lack sensitivity and/or specificity and must be interpreted in the context of other clinical and laboratory findings [28,41].

Serological tests can be roughly divided into two major categories: those that detect antibodies and those that detect antigens. Antibodies are useful in the confirmation of endemic mycoses such as histoplasmosis and coccidioidomycosis in immunocompetent hosts [41–43]. These tests are especially valuable when appropriate timing of sample collection is performed so that acute and convalescent serum can be obtained to determine whether antibody titers are rising or falling, a valuable and often necessary diagnostic tool in many instances [28]. The time of first sample collection is also important in that if body fluids are collected to detect antibodies too early in the course of the infection, before a humoral response has developed, the antibody test would give false-negative results [41]. In addition, serological diagnosis of opportunistic fungal infections must rely almost exclusively on antigen detection tests rather than antibody detection tests because the immunosuppressed patient population most at risk for these infections produce variable or no antibody responses [41,43,44].

Therefore, in immunosuppressed as well as in immunocompetent patients, antigen detection offers a potential alternative. Limitations of antigen detection assays include: a lack of sensitivity resulting from rapid clearance of the agent or poor sensitivity of the detector antibodies; few or no standardized reagents or commercial tests available, resulting in laboratory-to-laboratory variability; and formation of antigen-antibody complexes [41,44]. Antigen-antibody complexes can be dissociated through the use of enzymes or heat if antigens are carbohydrate in nature [28,30], but such procedures cannot generally be used with protein antigens, which may be denatured or antigenically altered by such treatment. Although strides have been made to obtain purified antigens through chemical and molecular biological means [45–52], very often unpurified mixtures of antigenic components are still being used as the gold standard in many antibody detection tests and as immunogens to produce antibodies for antigen detection tests. Monoclonal detector antibodies are generally more specific than polyclonal antibodies, which often need to be adsorbed with cross-reacting micro-organisms before they are usefully specific [28,29]. However, the increase in specificity gained by tests using monoclonal antibodies may be counterbalanced by a concomitant decrease in test sensitivity [53,54].

Despite their limitations, serological tests remain a useful adjunct to clinical and other laboratory findings for the diagnosis of invasive fungal infections. Therefore, in this chapter, an overview of currently used tests for the serodiagnosis of the most prevalent fungal infections will be presented for which immunodiagnostic tests exist. In addition, tests under development which hold promise for the rapid and specific diagnosis of fungal infections will be addressed. A discussion of the clinical presentation and risk factors associated with newly emerging fungal infections for which serological tests are rarely available will also be presented.

Serological tests currently used to diagnose diseases caused by the most common etiologic agents include complement fixation (CF), counterimmunoelectrophoresis (CIE), enzyme immunoassay (EIA), immunodiffusion (ID), latex agglutination (LA), reverse passive latex agglutination (RPLA), radioimmunoassay (RIA), tube agglutination (TA), and, less frequently, immunoblot (or Western blot) analysis.

Exoantigen tests will not be addressed in this chapter because these have been almost exclusively replaced by commercial molecular probe assays [54] (AccuProbe, GenProbe, San Diego, Calif.), nor will biochemical tests such as the detection of D-arabinitol, β-1,3-glucan, or D-mannitol be discussed as they are not immunological in nature. Immunohistochemical detection of fungi in tissue sections by immunofluorescent or immunoperoxidase staining will also not be addressed to any extent in this chapter. Readers are referred to previous reviews which discuss the above topics in detail [10,28,30,38–40,55,56]. In addition, the current status of molecular biological approaches to the diagnosis of fungal infections will not be discussed here as this topic is addressed in another chapter of this book. Whereas significant progress has occurred in the past few years in the field of molecular diagnosis, serological approaches, despite their limitations, remain the most common in clinical laboratory practice today.

II. DIAGNOSIS OF OPPORTUNISTIC MYCOSES

Opportunistic fungi are often normal commensal organisms of the human skin, mucosal surfaces, and gastrointestinal tract or are ubiquitously found in the environment in soil, on plants, or in bird or bat guano. Normally these fungi are harmless saprophytes. However, if the host's immune system should become debilitated, these organisms may cause severe or life-threatening illnesses. Therefore, opportunistic fungi are named as such because of their propensity to take the "opportunity'' to cause invasive disease when the debilitation of the host allows invasion to occur. Debilitation can result from infections such as AIDS, by underlying diseases such as diabetes or leukemia, or by chemotherapy that decreases immune function as part of drug regimens administered to treat cancer or to facilitate solid organ or bone marrow transplantation [10,40].

Opportunistic fungi include both filamentous molds such as *Aspergillus* spp., *Fusarium* spp., and Zygomycetes, to name only a few, and yeasts such as *Candida* spp. *Trichosporon asahii* (*beigelii*), and *Malassezia* spp., [40]. Although *Cryptococcus neoformans* can cause invasive disease in normal persons, the incidence of cryptococcosis is much higher in immunosuppressed hosts, and this organism has been listed as an opportunist for this reason. Although *Pneumocystis carinii* has been recently reclassified as a fungus based on DNA homology studies [57–60], no routine serological tests for the diagnosis of this opportunist are currently available. Only monoclonal antibodies which identify the cyst and trophozooite forms have been developed for direct and indirect immunofluorescent antibody assays [28]. Therefore, no further discussion of this organism will be presented in this chapter, and readers are referred to other reviews for descriptions of clinical and microbiological identification of this organism [28,58,61–65] as well as Chapter 11 in this text.

The number and variety of organisms causing opportunistic infections are rapidly increasing [66]. Because the host's capacity to mount an immune response is debilitated in patients susceptible to opportunistic infections and, as a result, the disease course is often fulminant and rapidly fatal, serological tests which rely on a functioning immune system (i.e., antibody detection tests) are of limited value. Therefore, antigen detection tests, which can be used in the absence of a functioning immune system, have gained favor [41]. Nonetheless, invasive opportunistic infections can occasionally occur in less severely immunocompromised patients, resulting in a more prolonged, localized disease progression. In these cases, or in cases where patients are recovering from immunosuppression but are still generally debilitated, antibody detection can sometimes be useful. Therefore, both antigen and antibody detection tests for the diagnosis of opportunistic fungal infections will be reviewed in the following section.

A. Aspergillosis

1. Background

Aspergillosis has traditionally been caused by two species of *Aspergillus*—*A. fumigatus* and *A. flavus*. *A. terreus* is a newly emerging cause of invasive disease which is resistant to amphotericin-B [10,33,67]. Other species, including *A. niger, A. ustus, A. versicolor,* and very rarely *A. nidulans,* have also been reported to cause invasive disease [33,41,67]. *Aspergillus* spp. are commonly isolated from the soil worldwide and are variously pigmented. Species are identified by a combination of color and colony type and by characteristic microscopic features such as vesicle size and shape, conidiophore morphology, metulae, phialides, and conidia. In tissue, typical septate, hyaline hyphae with dichotomous 45° branching may be seen although *Aspergillus* spp. can often be mistaken for *Fusarium* spp., *Pseudallescheria boydii,* or even agents of zygomycosis, and *vice versa,* in histological tissue sections [40]. Differentiation of agents of aspergillosis from *P. boydii* infections can be critical because pseudallescheriasis is often resistant to amphotericin-B therapy [10].

Aspergillosis is second only to candidiasis as an agent of invasive fungal infection in severely granulocytopenic patients [6,68–70], and its incidence may soon surpass that of invasive candidiasis [67,71,72]. It has been speculated that whereas the introduction of fluconazole for the prophylaxis of *Candida* infections in severely immunocompromised patients has reduced the incidence of systemic candidiasis, the relative incidence of invasive aspergillosis has increased concomitantly because (1) *Aspergillus* spp. are relatively unaffected by this azole, and (2) fluconazole prophylaxis has effectively eliminated competition from infecting *Candida* spp. [34,73]. This increased trend in aspergillosis cases is particularly disturbing given that the attributable mortality from invasive aspergillosis can exceed 80% [74–77].

2. Acquisition/Risk Factors

Aspergillosis is acquired through the inhalation of airborne spores derived from soil, decaying plant materials, and compost. Cases of invasive disease in immunocompromised hosts have also been attributed to the use of contaminated teas and spices. Aspergillosis can manifest itself in a broad spectrum of disease states depending upon the immune status of the host including cutaneous, mucosal, cerebral, ocular, and invasive. *Aspergillus* spp. cause allergic bronchopulmonary aspergillosis, aspergilloma, chronic necrotizing aspergillosis of the lung, sinusitis, tracheobronchitis, endocarditis/myocarditis, osteomyelitis, and acute invasive aspergillosis of the lung [40,78]. The most common of the manifestations are allergic bronchopulmonary and aspergilloma, which occur among immunocompetent hosts, and acute invasive aspergillosis of the lung, with or without dissemination to other organs, which occurs primarily in the immunocompromised host. The focus of this section will be primarily on invasive aspergillosis of the lung and other organs as it is the most life-threatening and most difficult of the syndromes to diagnose.

The incidence of invasive aspergillosis (IA) has increased dramatically in recent years as the number of patients receiving bone marrow and organ transplants has risen [7,70,79]. The aggressive immunosuppressive regimens used in these patients often result in severe granulocytopenia. IA has been reported to occur in 70% of patients who remain granulocytopenic for 34 or more days [7,41,77]. Elaborate HEPA filtration and laminar air flow systems have been implemented in the hospital setting in the hope of reducing the granulocytopenic patient's contact with airborne conidia from this ubiquitous filamentous fungal micro-organism [80]. However, at least one report suggests that many cases of aspergillosis may also be community acquired [81].

A disease that has been predominant in bone marrow transplant patients and significantly higher in patients receiving allogeneic as compared to autologous bone marrow transplants [6,76,77,82], invasive pulmonary aspergillosis had been relatively rare until recently. The incidence of invasive pulmonary aspergillosis in patients with HIV infection has been suggested to be increasing [83–85] as a result of several factors including: a greater number of AIDS patients surviving into late-stage HIV-disease [86–88], increased use of corticosteroids and neutropenia-inducing drug regimens [86,87,89], and increased therapeutic or recreational exposure to inhaled substances such as marijuana, which are often contaminated with aspergilli [86,90]. However, in at least one report describing IA in AIDS patients, no classical risk factors for aspergillosis (i.e., granulocytopenia or corticosteroid treatment) were present in 50% of 33 cases. The only risk factor found in almost all of the HIV-positive aspergillosis patients was a CD4 cell count of <50/mm^3 [84].

3. Clinical Presentation/Diagnosis

Symptoms of invasive pulmonary disease are often nonspecific and the most common clinical manifestations are fever and pulmonary infiltrates unresponsive to broad-spectrum antibiotics. Patients treated with corticosteroids may present without fever but may have chest pain [91]. Other symptoms can include pleural friction rub with or without chest pain, and cough [92,93]. Focal pulmonary disease may appear on radiographs as small nodular lesions that cavitate with larger peripheral lesions of lower attenuation around these nodular lesions (also known as "halo signs") which are characteristic for aspergillosis. If cavitation of lesions occurs, a crescent of air ("crescent sign") will appear in lesions. If radiographs show only diffuse infiltrates without these characteristic radiologic signs, other means will be needed for a correct diagnosis. Whereas bronchoscopy is seldom useful for the diagnosis of focal infection, microscopic exam and culture of bronchoalveolar lavage fluid in patients with diffuse radiographic findings are often helpful [40].

Blood cultures are notoriously negative, despite the use of lysis centrifugation tubes and despite the common dissemination of *Aspergillus* spp. to deep tissues via the hematogenous route [67,92]. Even when rigorous attempts are made to isolate *Aspergillus* spp. from other clinical specimens, positive cultures are recovered in only 12–34% of cases [41]. Disseminated disease will occur in approximately one-third of neutropenic patients with *Aspergillus* pneumonia [94] and may affect the brain, heart, kidneys, liver, thyroid, spleen, bones, or skin [67]. The primary manifestation of hematogenous spread is dissemination to the central nervous system [94,95], where signs of encephalitis or stroke may indicate cerebral hemorrhage or necrosis due to obliteration of blood vessels. In less severely immunocompromised hosts, cerebral aspergillosis may present as an abscess, but meningitis is rarely seen. Hematogenous spread to the heart may result in endocarditis, focal myocarditis, pericarditis, or pancarditis. Dissemination from the gastrointestinal tract has also been reported with involvement of the esophagus and the bowel, and direct extension to the brain from the paranasal sinuses has been demonstrated [67,92,95].

Recovery of *Aspergillus* species from respiratory secretions is considered to be a relatively insensitive and nonspecific diagnostic indicator because of the ubiquitous nature of the organisms [92,96,97]. However, the repeated recovery of an *Aspergillus* spp. from the respiratory tract of a febrile immunocompromised patient with pulmonary infiltrates warrants further examination, as continued colonization has been associated with invasive disease [37,67,78,98]. In contrast, positive cultures from nongranulocytopenic patients demonstrated a low predictive value for invasive disease [37]. Similarly, the significance of nasal colonization is less clear although during an outbreak of nosocomial aspergillosis, Aisner and colleagues found that positive nasal surveillance cultures of *A. flavus* in granulocytopenic patients correlated with invasive pulmonary aspergillosis [11]. On the other hand, the absence of positive nasal cultures in a persistently febrile neutropenic patient with a pulmonary infiltrate should not be used to exclude a diagnosis of invasive aspergillosis [10].

Therefore, culture of tissue from a sterile site or histopathological evidence of tissue invasion showing nonpigmented, septate fungal hyphae with dichotomous 45° branching is used to confirm invasive aspergillosis [93]. However, procedures to obtain adequate biopsy material are very invasive and are not recommended in severely granulocytopenic patients, who may also be thrombocytopenic. Other drawbacks to the histological identification of *Aspergillus* spp. include cross-reactivity of immunofluorescent or immunochemical reagents with other fungi and morphologies in tissue similar to those of *Pseudallescheria boydii* and *Fusarium* spp. [40]. Therefore, much effort has been placed on the diagnosis of invasive aspergillosis using serological assays.

4. Immunoserology and Interpretation

a. Antibody Detection. Immunodiffusion (ID) and counterimmunoelectrophoresis (CIE) have proved valuable tests for the diagnosis of aspergilloma and allergic bronchopulmonary aspergillosis in immunocompetent individuals. The contribution of specific antibody detection tests for the diagnosis of invasive aspergillosis in immunosuppressed patients, however, who either lack a sufficient antibody response or who mount variable antibody responses, remains controversial [28,99–102]. Differences in antigen preparations and timing of sample collection by different research laboratories may be responsible, in part, for reported variations in the success of these tests [41,103–105]. For example, precipitins were found in the sera of 14 of 16 patients in one study [105] whereas other authors did not find antibodies in invasive aspergillosis patients even though the same immunodiffusion test format was used [106]. There appears, however, to be agreement among several studies that seroconversion can be successfully used to monitor disease when serial serum samples are tested. In addition, an increase in antibody titer at the end of immunosuppression indicates a good prognosis whereas absent or declining antibodies suggest a poor outcome [99,102,107]. For example, in a study of post–lung transplant patients with *A. fumigatus* infections, increasing specific IgG antibody levels corresponded with impairment of lung function and with cytological and microbiological recovery of the organism; recovery of lung function correlated with a decrease in antibody titer [108,109]. Therefore, antibody detection in immunocompromised hosts may be used for determining a prognosis, if not a diagnosis.

Tests to detect *Aspergillus* spp. antibodies include complement fixation (CF), counterimmunoelectrophoresis (CIE), immunodiffusion (ID), indirect hemagglutination (IHA), and enzyme immunoassay (EIA) [28,100]. The micro-ID test, recommended for detection of *Aspergillus* antibodies, uses buffered phenolized agar and antigens produced from 5-week-old stationary Sabouraud glucose broth cultures of *A. fumigatus*, *A. flavus*, *A. terreus*, and *A. niger*. Only serum that produces a line or lines of identity with reference serum from a proven human case of aspergillosis is considered positive. One or more precipitin bands indicate aspergilloma, allergic bronchopulmonary aspergillosis, or infection whereas three or more precipitin bands indicate aspergilloma or invasive disease. Nonspecific bands may occur as a result of C-reactive protein in patient's serum and these may be removed by treatment with sodium citrate. Precipitin bands are reported to occur in 90% of patients with aspergilloma, 70% of patients with allergic bronchopulmonary aspergillosis, and less frequently in patients with invasive aspergillosis [28].

The sensitivity of CIE equals or exceeds that of ID, but because its specificity is equal to or less than that of ID, CIE is recommended for screening purposes only. An EIA to detect *Aspergillus* antibody has also been developed [110] and is reported to give results which correlate with those of ID. It was hypothesized that the sensitivity of the EIA might be improved if a battery of antigens for multiple *Aspergillus* spp. were included in the test. Others have also developed EIA [104,111–114] or RIA [115] formats to detect specific IgG antibodies and found these formats to be more sensitive than ID or CIE although serial serum samples were required to optimize detection.

Several commercial tests including CIE, CF, IHA, and EIA formats have been evaluated; sensitivities were reported to range from 14% to 36% and specificities to range from 72% to 99% [100,116]. Reagents or kits for these tests and/or for the ID test may be obtained from multiple European sources (Roche Serologie, Munich, Germany; Labor Diagnostika, Heiden, Germany; Fumouze Diagnostics, Asnierrs France; DDV Diagnostika, Marburg, Germany; Institut Virion, Wurzberg, Germany) and from sources within the United States (Beckman Instruments, Brea, CA; Gibson Laboratories, Lexington, KY; Greer Laboratories, Lenoir, NC; Hollistier-Stier Laboratories, Spokane, WA; Immuno-Mycologics, Norman, OK; and Meridian Diagnostics, Cincinnati, OH).

Immunoblot analyses have also been used to identify antibodies directed against protein antigens of *Aspergillus* spp. [103,117–123]. For example, 14 patients with invasive aspergillosis demonstrated antibody reactivity to proteins ranging in molecular weight from 33 to 88 kDa whereas control patient's sera reacted with only two antigens, a 63-kDa and an 88-kDa component [124]. In another study, using a rabbit model of invasive aspergillosis, seroreactivity developed to a 41-kDa antigen in 50% of animals that survived for >10 days; reactivity to this antigen was absent prior to immunosuppression or infection, indicating that seroconversion was occurring [118]. It is unfortunate that aspergillosis is so rapidly fatal and that antibody production is so severely limited in the immunocompromised patient population that even if antigen-specific antibodies could be identified, a test would need to be optimized to detect very low levels of antibody at a very early stage of disease to be most valuable.

   b. Antigen Detection.

   Carbohydrate Antigens.   Because the predictive value of antibody detection for the diagnosis of invasive aspergillosis in immunosuppressed patients has been so poor, much research has been devoted to the detection of *Aspergillus* spp. antigens in serum and urine [53,125–129]. Of the carbohydrate antigens, galactomannan (GM), a major cell wall component of *Aspergillus* spp., has received the most attention. Initial studies used counterimmunoelectrophoresis to detect GM in the serum and urine of patients and in rabbits experimentally infected with *A. fumigatus* [127,130]. This was followed by the development of increasingly more sensitive and more rapid RIA [125,129,131,132], LA [128,133], and EIA [53,126,134–137] test formats. It was reported by Talbot et al. [138] that the RIA had a sensitivity of 70–80%, a specificity of 90%, a positive predictive value of 82%, and a negative predictive value of 85% when patient serum was used as the source of antigen. Antigen was detected before invasive disease was suspected in 30% of patients and before laboratory confirmation of invasive aspergillosis in 46% of patients [138]. Unfortunately, antigen was not detectable in approximately 25% of proven cases. The use of bronchoalveolar lavage fluid, instead of serum, may provide increased sensitivity for detection in cases where sera are negative but pulmonary involvement is present [139].

   It was reported that detection of GM in urine using either an RIA or an inhibition EIA format was more rapid and more sensitive than detection in serum. Whereas serum was positive in only 33% of rabbits (and 17% of patients), urine was positive in 100% of these rabbits and in 54% of patients. In addition, serum did not become detectably positive until 48 hr after infection whereas urine was positive as early as 1 day postinfection. Also, urine antigen levels increased proportionately with disease progression [125]. It has been speculated that detection of GM in urine may be more useful than detection in serum because of the prolonged excretion of GM in urine, unlike the transient presence of GM in serum [44,140]. In addition, because detection often requires multiple serial sampling for optimal GM recovery, obtaining multiple urine samples would be less invasive than obtaining multiple serum samples from granulocytopenic and often thrombocytopenic patients [126].

   LA and sandwich EIA test formats to detect circulating GM are commercially available in Europe (Sanofi Diagnostics Pasteur, Marnes-la-Coquette, France). Both assays utilize the same anti-GM monoclonal antibody, EB-A2 [137,141,142]. The LA test was the first to be developed [128,143,144], and despite its ease of use, it lacked sufficient sensitivity. For example, Verweij and colleagues demonstrated that the LA test was positive in eight patients with proven aspergillosis but was negative in the initial serum sample from five of these patients [128]. In addition, no positive reaction was found in six samples from two patients with histological evidence of *Aspergillus* infection [128]. GM was not detected in a rabbit model of aspergillosis until a mean of 3 days after infection, and multiple serum samples were required for optimal detection [145]. Another drawback to the use of the EB-A2 monoclonal antibody is that it has been shown to cross-react with the airborne contaminants *Penicillium chrysogenum, Acremonium* spp., *Alternaria alternata*, and *Cladosporium herbarum* as well as with the fungal opportunists *Fusarium oxysporum, Penicillium marneffei, Wangiella dermatitidis*, and *Rhodotorula rubra* [100]. The LA test could not be used with urine samples due to nonspecific reactivity [126,146]. For example, Haynes and Rogers reported that whereas the sensitivity and specificity of the LA test were 95% and 90%, respectively, for serum, the same test gave false-positive results in 42% of urine specimens from control patients [146].

More recently, a sandwich EIA format to detect circulating GM using the same monoclonal antibody was developed [137,147]. This assay appears to be more sensitive than the LA test in that the EIA can detect 1 ng of GM per mL in contrast to the LA test which could detect only 15 ng/mL or greater. The EIA could also detect GM in serum at an earlier stage of infection than the LA test [137,148] and before clinical signs and symptoms had become apparent [147,149]. Although the use of urine specimens in the EIA still lacked sufficient specificity [126], detection of GM in bronchoalveolar lavage fluid (BAL) from patients with invasive aspergillosis was positive and correlated with serum positivity [150]. In a retrospective study of hematologic patients, of whom 36 were granulocytopenic, Caillot and colleagues found BAL to be positive by the sandwich EIA in 83% of 23 histologically proven and 14 highly probable cases of invasive aspergillosis [12]. False-positive results in serum were still problematic, however; 8–11% of samples gave false-positive results. This was especially evident when samples were collected within 30 days of bone marrow transplant or within 10 days of cytotoxic chemotherapy [126,148,151]. The reasons for false positivity have been speculated to be the result of the use of cyclophosphamide immunosuppression [126,152] or from a carbohydrate breakdown product, occurring in serum or urine, which shares cross-reactive epitopes with the antigenic galactofuranose molecule [126].

Most tests require the dissociation of antigen-antibody complexes for optimal detection of *Aspergillus* species GM. This may be accomplished by diluting serum and heating it [125,136], by precipitation of proteins with trichloroacetic acid followed by dialysis [132,134], by treatment with citric acid, heat, pepsin, and dilution of serum [129], or by simple boiling in the presence of EDTA [126,135].

Protein Antigens. A series of immunoreactive proteins have been associated with somatic and cell wall components of *Aspergillus* spp. [103,111,117,119,121,122]. These include 58-kDa [119], 88-kDa [117], and 18-kDa [103,146] proteins, an alkaline protease [121], a serine protease [122], a superoxide dismutase [153,154], and a catalase [123]. Whereas patient antiserum has been shown to react with these compounds in immunoblots, the number of tests developed to detect these protein antigens, rather than reactive antibodies, in the urine or serum of patients with invasive aspergillosis has been more limited.

The protein moiety receiving the most attention has been an 18-kDa protein [103,146] which has sequence homology with a ribonucleotoxin or "restrictocin" of *Aspergillus fumigatus* [155–158]. The 18-kDa protein is a specific RNA nuclease which cleaves a phosphodiester bond in a conserved region of the large ribosomal subunit of RNA releasing a 400 bp fragment from the 3′ end [157]. Urine from 10 bone marrow transplant patients contained several protein antigens of low molecular mass (11–44 kDa) detected using rabbit polyclonal antiserum directed against a cell wall extract from *A. fumigatus*. These antigens were distinct from GM in that an anti-GM monoclonal antibody (EB-A1) reacted diffusely and only to those >45 kDa in mass. Urine specimens from 23 patients with no evidence of invasive aspergillosis did not react with either the polyclonal or monoclonal antibodies [143].

Circulating proteins have also been reported to occur in the serum and urine of rats experimentally infected with *A. fumigatus*, where an 80-kDa antigen was detected [159]. Unfortunately, the 80-kDa antigen was not found in the serum of three human cases of aspergillosis. Others [160] used a rat model of experimental aspergillosis and examined urine from infected animals for the presence of *A. fumigatus* antigens by immunoblotting. Antigenic bands were detected at 88, 40, 27, and 20 kDa. Burnie and Matthews determined that the 88-kDa band could be detected using a monoclonal antibody directed against the heat shock 90 protein (HSP90) of *C. albicans* [161]. It was hypothesized that there may be an *Aspergillus* antigen, analogous to the HSP90 of *C. albicans*, circulating in invasive aspergillosis.

B. Candidiasis

1. Background

Candidiasis is one of the most prevalent invasive fungal diseases world-wide. It is caused by yeast-like fungi of the genus *Candida*. Whereas *Candida albicans* remains the dominant causative agent of invasive candidiasis, infections caused by other *Candida* species, most notably *C. glabrata, C. tropicalis, C. parapsilosis*, and *C. krusei*, have increased in incidence over the past few years [162–164]. It is not known why such a shift in *Candida* spp. infections has occurred, but it has been speculated that the introduction of fluconazole prophylaxis and treatment of patients has increased infections caused by *Candida* spp. which are either innately less susceptible to azoles, such as *C. krusei* [163–165], or which rapidly convert to a more resistant phenotype upon exposure to azoles, such as *C. glabrata* [162,166,167]. The differing susceptibilities of the various *Candida* spp. to azole antifungal drugs makes identification of the organism to the species level more important so that appropriately targeted antifungal therapy can be initiated.

Microscopically, *Candida* spp. generally exhibit a combination of blastoconidia, pseudohyphae, and/or true hyphae. *C. glabrata* produces only blastoconidia, and *C. albicans* and *C. dublinensis* produce germ tubes in serum. Sugar assimilation and fermentation tests are often used to differentiate most of the medically important *Candida* spp. when specimens are culture positive. However, 50% or less of blood cultures from patients with disseminated candidiasis are positive even when lysis centrifugation tubes, which are reported to increase yeast recovery from blood, are used [30,31]. When blood cultures are positive, this may only indicate transient candidemia, although even a single positive blood culture in patients at high risk of invasive disease should not be ignored. Transient candidemia generally resolves upon removal of central venous catheters. Therefore, multiple blood cultures on successive days, especially after removal of central lines, are an indication of truly invasive disease. Positive urine cultures, in the absence of indwelling urinary catheters, which yield $>1 \times 10^4$ cfu/mL should raise suspicion of infection. On the other hand, *Candida* spp. may be absent from the urine even in disseminated infection [40].

2. Acquisition/Risk Factors

Candida spp. are found as colonizers and as opportunistic pathogens worldwide. The mouth and gastrointestinal tract of 30–50% of normal individuals are colonized with *Candida* spp. and colonization is even greater in HIV-positive patients and in women with chronic recurrent vaginitis. *Candida* spp. infections are generally acquired via hematogenous spread or dissemination from the gastrointestinal tract. *Candida* spp. have emerged as a major nosocomial bloodstream pathogen causing 10% of all bloodstream infections and 25% of all urinary tract infections [8,40].

Predisposing factors for the dissemination of this normally commensal yeast include low birth weight in infants, vascular catheterization, total parenteral nutrition, tracheal intubation, use of broad-spectrum antibacterial antibiotics, intravenous drug abuse, neutropenia in bone marrow transplant and cancer patients, organ transplantation, and housing in a surgical intensive care unit [7,8,40]. Outbreaks of candidiasis caused by *C. albicans* have occurred in neonatal and intensive care units [168], and nosocomial acquisition of *C. parapsilosis* infections has been associated with isolation of the organism from the hands of health care workers in the hospital setting [169].

3. Clinical Presentation/Diagnosis

Disseminated candidiasis occurs in two distinct forms: acute and chronic (or hepatosplenic). Acute disseminated candidiasis presents as a fulminant, life-threatening disease in neutropenic patients or in patients recovering from neutropenia, and often the only clinically apparent sign is persistent fever unresponsive to antibacterial antibiotics. Macronodular lesions of the skin can occur in up to 10% of neutropenic patients and in up to 30% of patients recovering from neutropenia [40]. Complications include meningitis, brain abscess, renal abscess, myositis, myocarditis, and endocarditis.

Chronic candidiasis, in contrast to acute disseminated candidiasis, presents as an indolent infection. Although it is often referred to as hepatosplenic candidiasis, other major organs can be involved as well. Chronic candidiasis most often occurs in patients after recovery of neutrophil counts following remission induction therapy. As in the acute form, the patient presents with a fever unresponsive to antibacterial antibiotics and has no discernible organ lesions in the early stages of the infection. As the disease progresses, signs of infection may include weight loss, abdominal pain, enlargement of the liver and spleen, and numerous small, radiolucent lesions in the liver or spleen which can be observed upon CT scan [10,34].

4. Immunoserology and Interpretation

a. Antibody Detection. The clinical usefulness of antibody detection for the diagnosis of systemic candidiasis has been limited by false-negative results in immunosuppressed patients who produce low or undetectable levels of antibody, and by false-positive results in patients colonized at superficial sites. For example, positive agglutination or CF tests are of little value in the serodiagnosis of candidiasis because positive responses have been detected in healthy persons and in persons colonized or superficially infected without systemic involvement. On the other hand, ID and CIE tests to detect antibodies to *Candida* can be a valuable diagnostic tool in the immunocompetent host [28] and have been reported to have a sensitivity of 80% in proven cases [40]. Patient serum containing antibodies that react with homogenate antigens of *C. albicans* in the ID test may produce one or several lines of identity. Disseminated candidiasis should be suspected if serial serum specimens show an increase in the number of reactive bands over time or if a seronegative patient becomes seropositive. Positive results may also indicate colonization or infection caused by *C. glabrata* [28]. ID antigens, control sera, and kits are commercially available from Meridian Diagnostics, Immuno-Mycologics, and Gibson Laboratories.

Latex agglutination tests can be helpful as well. A fourfold increase in LA antibody titer, conversion from seronegativity to seropositivity (LA titer of 1:4 or higher), or a single serum titer of 1:8 or higher can be considered presumptive evidence for infection in immunocompetent hosts; conversely, a fourfold decrease may denote successful antifungal therapy. However, a single positive LA titer of only 1:4 is not diagnostic because the somatic antigens used to coat the latex particles cannot differentiate between antibodies formed as a result of mucosal colonization versus those of true infection. Instead, a patient who shows an LA titer of only 1:4 may have early disease, may be colonized by *Candida* spp. including *C. glabrata*, or may be falsely positive [28,40].

In an attempt to reduce false-positive results, several researchers have developed tests to detect antibodies directed against cytoplasmic antigens, based on the assumption that the host would not be exposed to intracellular antigens except during invasive disease. Unfortunately, in a study of patients undergoing induction chemotherapy for acute leukemia, antibody to a major 54-kDa cytoplasmic antigen described by Jones and colleagues was infrequently (25%) detected in cases of disseminated candidiasis [170–173]; others found increases in antibody titers in 10% of patients without candidiasis [174]. Therefore, it would seem that immunosuppressed patients often fail to produce antibodies or their antibody production can be variable making diagnostic tests to detect antibodies less useful for the diagnosis of systemic candidiasis in this patient population [28,41,44]. However, such patients may be in antigen excess, making the detection of antigens a potentially more successful strategy for the diagnosis of candidiasis in this patient group.

  b. *Antigen Detection.*

  Aspartyl Proteinase.   An inducible, secreted aspartyl proteinase (Sap) was first described by Staib in 1965 and has since been studied extensively as a virulence factor in the invasion and dissemination of *C. albicans* in animal models of infection [175–178]. The theoretical usefulness of Sap as a diagnostic antigen stems from the hypothesis that because Sap is an inducible enzyme, produced during active tissue invasion [179,180], its production should correlate with invasive disease and not simple colonization. In a rabbit model of disseminated candidiasis, Sap antigenuria was followed during disease progression using a competitive binding inhibition EIA [30]. After 24 hr, urine from eight rabbits demonstrated significant inhibition in the EIA (15 ± 7%), and inhibition increased daily in direct proportion to disease severity to a peak of 46% by day 3 postinfection. The EIA was negative when urine was tested from rabbits with gastrointestinal colonization with *C. albicans*, or from rabbits infected with *A. fumigatus, C. neoformans*, other *Candida* spp., or bacteria [30]. In addition, women with invasive vulvovaginal candidiasis (VVC) demonstrated the presence of Sap in vaginal wash fluids by Western blot analysis whereas those from culture-negative individuals had no demonstrable Sap in wash fluids [181]. Finally, Ruchel et al. [182] examined serum samples from patients for the utility of Sap detection as an aid to the diagnosis of invasive disease. Using anti-Sap antibodies in an EIA format, the sensitivity for detection was low (proteinase antigen was detected in only 50% of suspected plus confirmed cases) [182] and may potentially be compromised by the formation of complexes between Sap and alpha-2-macroglobulin in the circulation [183]. Therefore, detection of Sap in serum does not appear to be as promising as detection in urine or vaginal wash fluids. Nonetheless, further studies using serum are warranted, especially regarding methods to optimize dissociation of serum proteins from Sap prior to assay.

  Cell Wall Mannoprotein.   Dissociation of antigen-antibody complexes is necessary for the optimal detection of circulating cell wall mannoprotein, or mannan, from *C. albicans*. This antigen is heat stable and resists boiling, proteinase treatment, and acidic pH [30]. Therefore, antigen-antibody complexes are routinely dissociated by boiling in the presence of EDTA [28]. Bailey et al. [184] detected mannan in the serum of 17 of 21 patients with disseminated candidiasis when specimens were treated with pronase and heat, whereas only three of 21 patients were positive if no dissociation step was included. Mannan is rapidly cleared from the circulation resulting in serum concentrations of 100 ng/mL or less. Therefore, multiple serial serum samples are required for optimal detection [10,30]. Mannanemia occurs in ~31–90% of patients with disseminated candidiasis depending upon (1) the frequency of sampling; (2) the spectrum of the underlying disease; (3) the degree of immunosuppression; (4) the serotype of *C. albicans* and the *Candida* spp. involved; (5) the definition of disseminated candidiasis; (6) the specificity and titer of the capture antibodies; and (7) the immunoassay method employed. Multiple laboratories have attempted to use RIA, EIA, LA or reverse passive latex agglutination (RPLA) to detect circulating mannan [30,44]. Methods developed in research laboratories using sandwich EIA [185–188] and RPLA [184,189] formats have been shown to have moderate sensitivity but good specificity for disseminated disease. In a retrospective study of patients with cancer, the sandwich EIA showed a sensitivity of 65% and a specificity of 100% [187]. Fujita and Hashimoto compared the sensitivity of the sandwich EIA format to that of the RPLA format using the same capture antibodies. They found that whereas the sensitivity of the RPLA was 38%, that of the sandwich EIA was 74% [188]. In other studies, the RPLA test detected serum mannan in 78% of leukemia patients with disseminated candidiasis [189] and, in another study, in 13 of 18 (72%) patients for whom disseminated candidiasis was confirmed by biopsy, autopsy, or persistent candidemia during granulocytopenia [184]. Commercial LA kits are available (LA-*Candida* Antigen Detection System, Immuno-Mycologics, Norman, OK; Pastorex *Candida*, Diagnostics Pasteur,

Marnes-la-Coquette, France), but the sandwich EIA is only available in research laboratories [28,30]. Hybridtech Inc. commercialized a membrane immunoassay that has since been discontinued and a sandwich EIA is under development commercially in Japan.

Cytoplasmic Antigens. Cytoplasmic proteins of *C. albicans* have been detected by a number of researchers using a variety of test formats [124,190–193]. The two predominant cytoplasmic proteins described to date include a 47-kDa protein which is a breakdown product of a 90-kDa heat-shock protein (HSP-90) and a 48-kDa protein later found to be a *Candida* enolase [52,190,193,194]. Western blot analysis does not resolve the antigens in the 47- to 52-kDa range unless monoclonal antibodies that recognize the enolase antigen are applied [195]. The 47-kDa antigen can be detected in the serum of 77% of neutropenic patients with disseminated candidiasis using an enzyme-linked dot immunobinding assay [190] which proved to be more sensitive than a reverse passive latex agglutination test for the same antigen [196].

Preliminary studies in mice and rabbits using an EIA format revealed that the presence of the 48-kDa antigen in serum correlated with disseminated disease, was positive in the absence of candidemia, and declined with antifungal therapy [10]. The assay was then commercialized as a double-sandwich liposomal assay using murine IgA monoclonal antibody adsorbed to a nitrocellulose membrane (Directigen$_{1-2-3}$ Disseminated Candidiasis Test, Becton Dickinson, Philadelphia, PA). Patient serum is added to the membrane for testing and then polyclonal rabbit anti–*C. albicans* enolase is applied. Bound rabbit antibody is then detected with a liposome-containing rhodamine dye and coated with goat anti-rabbit IgG. Results of a multicenter study conducted at cancer centers over a 2-year period revealed a sensitivity per sample of 54% whereas when multiple samples were tested, detection of antigenemia was improved to 85% [197]. No doubt multiple sampling will be necessary to optimize antigenemia detection. Unfortunately, this test is no longer available commercially.

Heat-Labile Antigens. Unlike the previously described antigens which have been identified and chemically purified, Gentry et al. [198] described detection of a structurally uncharacterized, 56°C-labile antigen, by reverse passive latex agglutination (RPLA). Latex particles were sensitized with serum from rabbits immunized with whole, heat-killed *C. albicans* blastoconidia. The test was commercialized as the Cand-Tec test (Ramco Laboratories, Houston, TX) and has been the subject of several investigations [185,186,189,199–202]. The circulating antigen was not only heat sensitive but also susceptible to pronase, 2-mercaptoethanol, and sodium periodate treatment, suggesting that the molecule may be a glycoprotein [44]. The sensitized latex particles could not agglutinate mannan [198], and it has been suggested that the assay may detect a neoantigen derived from *C. albicans* after host processing or a host antigen which crossreacts with those of *C. albicans* [44]. Although relatively easy to perform, the test appears to lack sensitivity when an antigen titer of ≥1:8, which excludes most false-positive results, is used as the cutoff value for positivity [185,186,189,199–202]. In a study of 10 patients with disseminated candidiasis, only 28 out of 108 serum samples (26%) were positive by this test at a titer of >1:8 [199]. Unfortunately, test specificity is also low. For example, in a retrospective study, the test gave false-positive results (≥1:8) in four of six patients with transient candidemia, in one of 20 healthy individuals with rheumatoid factor, and in one patient who had a positive cryptococcal latex agglutination test [185]. Therefore, studies to date suggest that the Cand-Tec test does not provide sufficient predictive value for a reliable diagnosis of disseminated candidiasis.

C. Cryptococcosis

1. Background

Cryptococcosis is a common life-threatening mycosis in AIDS patients with between 3% and 13% of these patients becoming infected [40,203]. It is caused by the fungus *Cryptococcus neoformans*, which characteristically demonstrates spherical, budding yeast cells with thin, dark cell walls. The cells are often encapsulated by a mixture of at least three heteroglycans which allow *C. neoformans* isolates to be divided into four serological groups—A, B, C, and D—with the serodominant polysaccharide being glucuronoxylomannan [101]. Polysaccharide capsules can be visualized during microscopic examination of CSF or other body fluids after India ink staining, but small or acapsular forms of the organism can appear to be other yeasts, leading to misidentification [204]. The organism has been cultured from CSF but multiple specimens may be necessary, and large volumes (4–8 mL) may need to be plated out and incubated for up to 2 weeks for detection. Cultures may also be positive from blood, urine, sputum, bronchoalveolar lavage fluid, and prostatic fluid, and lysis centrifugation tubes have been shown to optimize recovery from blood [40,101].

2. Acquisition/Risk Factors

Cryptococcosis is acquired through the inhalation of small dessicated cells found in air, pigeon droppings, or contaminated soil. Two varieties of *Cryptococcus neoformans* exist, *C. neoformans* var. *neoformans*, which has worldwide distribution, and *C. neoformans* var. *gatti*, which is found in tropical and subtropical regions only and which has as its natural habitat the red gum tree *Eucalyptus camaldulensis* [40,205,206].

C. neoformans* is a major cause of illness and death in AIDS patients and in those with severe T-cell defects such as cancer and organ transplantation patients and those undergoing corticosteroid therapy [40,203,207]. Immunocompetent hosts may also be infected but the majority of cases occur among the immunocompromised. Interestingly, *C. neoformans* var. *gatti* rarely causes cryptococcosis in AIDS patients [40,208].

3. Clinical Presentation/Diagnosis

The lungs are the initial site of infection and it may take weeks or months for its dissemination to other body sites. Meningitis is the most common clinical presentation, but pneumonia and disseminated infections also occur [207,209]. Pulmonary cryptococcosis in immunocompetent hosts presents as chest pain, sputum production, weight loss, and fever. Radiographs demonstrate well-defined, noncalcified single or multiple nodular lesions. In immunocompromised hosts, however, pneumonia can be more fulminant and the organism can disseminate rapidly [210]. Fever, malaise, cough, and night sweats can occur and chest radiographs demonstrate diffuse, interstitial, or alveolar infiltrates with occasional nodular lesions. Meningitis occurs when organisms disseminate from the lungs. Headache, drowsiness, and confusion have been associated with meningeal disease, but fever can be minimal or absent until late in infection. CT scans can show single or multiple enhancing or nonenhancing lesions [10,40,211].

4. Immunoserology and Interpretation

a. Antibody Detection. Antibodies can be detected in patients with early or localized infection, but antibody detection is not useful for diagnostic screening for disseminated cryptococcosis because high titers of freely circulating antibodies rarely occur in active disease. It is not known whether this is because of neutralization by antigen excess or whether their generation is suppressed. Antibodies have been demonstrated in patients after recovery from infection, and antibody production is associated with clinical recovery and a good prognosis [40,101].

b. Antigen Detection. Cryptococcal polysaccharide antigen detection in CSF, serum, and other body fluids has been a very useful and reliable method for the diagnosis of cryptococcosis. Unlike other polysaccharide antigens such as mannan from *Candida* spp. and galactomannan from *Aspergillus* spp. which rapidly clear from the circulation, the glucuronoxylomannan (GXM) of *C. neoformans* contains gluronic acid and xylose side chains which prevent rapid receptor-mediated clearance [101]. Extremely high concentrations of GXM may therefore be detected in patient CSF, serum, and urine. In AIDS patients, levels of antigen in CSF are substantially higher than in other immunocompromised patients, and LA titers often exceed 1:1024 [212]. Titers remain high in AIDS patients even after therapy, and there is a potential for relapse upon withdrawal of therapy. Both LA and EIA formats for antigen detection are available commercially (LA—CALAS latex agglutination test; EIA—PREMIER Cryptococcal Antigen test, Meridian Diagnostics). The LA test is by far the most commonly employed because of its ease of performance and reliability.

Greater than 90% of patients with untreated meningitis will be positive by the LA test [10,40]. Titers of 1:8 or greater in CSF, serum, or urine is strong evidence of active infection; titers of 1:4 arc suggestive [28]. Serial CSF specimens from non-HIV-infected patients may be useful for therapeutic monitoring because titers decrease as the disease resolves; serial serum specimens are also useful if evaluated in tandem with the clinical presentation of the patient [10,28]. False-negative results can occur from either the formation of antigen-antibody complexes or by a prozone effect [44]. The former can be corrected by pronase treatment of the samples [213] and the latter by dilution of the specimen [214]. Pronase treatment is recommended as its use has been reported to increase the sensitivity of the LA test in some CSF and most serum samples [213,215,216]. False-positive results have been reported occasionally for CSF from patients with septicemia caused by DF-2 [217], patients with cancer [218], patients with rheumatoid arthritis, and patients with disseminated infections with the newly emerging pathogen *Trichosporon asahii* (formerly *T. beigelii*) [28,219,220]. False positives can be removed by boiling specimens in Na_2EDTA or by pronase treatment [28].

The EIA can detect antigen earlier and at lower concentrations (6 ng/mL) than the LA (35 ng/mL) and, unlike the LA, no pronase treatment of samples is required [28,221,222]. In addition, the increased sensitivity of the EIA suggests a possible role for this test in the detection of infections by capsule-deficient mutants of *C. neoformans* [44]. The EIA is not applicable to urine specimens, however [28]. Therefore, despite some advantages provided by the EIA, the LA assay will probably remain the most widely used test format because of its rapidity and ease of performance, specificity, and good diagnostic and prognostic value.

III. DIAGNOSIS OF ENDEMIC AND DIMORPHIC MYCOSES

The following section discusses fungi which differ from the true opportunistic organisms in that they are pathogenic for both immunocompetent as well as immunosuppressed patients. Included in this group is *B. dermatitidis*, *C. immitis*, *H. capsulatum*, *P. brasiliensis*, *P. marneffei*, and *Sporothrix schenckii*. All except *S. schenckii* are limited to discrete regions of the world where they are considered endemic pathogens. Persons living in and/or traveling through these endemic regions have a greater risk of exposure to and infection by these fungi.

All of these fungi are thermally dimorphic. When found in the environment, or when cultured at 25–30°C, these organisms grow in a filamentous, mycelial form. Mycelia may then evolve to produce infectious conidia or arthroconidia which can become aerosolized. Typically, the infectious spores enter the body through the respiratory route. When found in the invasive tissue form, or when cultured at 37°C, the fungi grow in the yeast form. Yeast phase organisms are typically not infectious and cannot be transmitted from person to person. These fungi cause a wide range of disease states including asymptomatic, chronic localized, and acute. Infections may become disseminated, especially in immunocompromised patients. Further details regarding these organisms and methods for the immunodiagnosis of the diseases they cause are discussed below.

A. Blastomycosis

1. Background

B. dermatitidis is primarily endemic to the eastern portion of the United States and particularly to the Mississippi and Ohio River valleys. Infections due to *B. dermatitidis* have also been described in other parts of the world, such as in Africa, where endemic regions exist [223]. However, because an adequate skin test antigen is lacking [224], the range of endemicity has not been fully elucidated. In the environment, *B. dermatitidis* is associated with soil and, particularly, with soil near water sources such as rivers and lakes [21]. Unfortunately, *B. dermatitidis* has been very difficult to isolate from the environment and therefore its exact ecological niche has not been determined. Blastomycosis also occurs in animals other than humans [225,226]. Dogs are the most frequently infected nonhuman animal and are the most susceptible.

2. Acquisition/Risk Factors

Primary infection occurs as a result of inhalation of infectious conidia released into the environment. Conidia enter the lungs where they convert to the yeast phase. As the yeasts replicate, they form characteristic broad-based buds that are diagnostic for this organism in a primary specimen or in histopathologic sections of biopsied tissue. Other modes of infection include cutaneous inoculation which may occur in a laboratory setting. A needlestick may introduce the organism subcutaneously and result in a localized infection [227,228]. Rare cases of transmission through dog bites have been reported [229]. However, dissemination from this mode of infection has not been documented.

Rates of infection by *B. dermatitidis* have been difficult to determine. Unlike infections with the other endemic mycoses, no good skin test antigens or highly sensitive and specific serologic assays are available for the diagnosis of blastomycosis. Of patients who are symptomatic, disease parameters can range from severe pulmonary infections to disseminated, systemic disease involving multiple organs [230]. Chronic cutaneous infection, and those that involve bone, may represent dissemination from an unapparent pulmonary infection.

As opposed to other fungal infections such as cryptococcosis, histoplasmosis, and coccidioidomycosis, blastomycosis has not been found to be a prevalent cause of infection in immunocompromised patients. However, Pappas reported that the number of patients seen at the University of Alabama with blastomycosis that were immunosuppressed increased almost 10-fold between the pre- and post-AIDS eras [231]. These patients usually had more severe disease, had disease which more often involved the CNS, and had a much greater mortality rate than immunocompetent patients.

3. Clinical Presentation/Diagnosis

A specific diagnosis of blastomycosis is complicated by the fact that the signs and symptoms of disease are vague and nondescriptive. Most infections develop gradually, often over many years, into a chronic condition. The clinical manifestations are such that they may be confused with a variety of other diseases. Those at risk for infection are individuals that, through either employment or pleasure, participate in outdoor activities in areas endemic for blastomycosis. Microbiologists, pathologists, and veterinarians are at high risk for professional acquisition and, particularly, for cutaneous inoculation [227,228].

As with other fungal infections, definitive diagnosis of blastomycosis is through either microscopic detection of the typical 8- to 15-μm broad-based budding yeast and/or culture of *B. dermatitidis* from a clinical specimen. However, the number of yeast forms in a specimen may be too small to detect in a direct smear and the time to detection and identification in culture may be too long for a timely diagnosis. Therefore, serological techniques have been employed to detect antibody responses in an attempt to shorten the time to diagnosis. Histopathologic diagnosis from tissue biopsies is definitive when broad-based budding yeasts of typical size are present [38].

4. Immunoserology and Interpretation

a. Antibody Detection.   Standard serologic assays for the diagnosis of blastomycosis are the complement fixation (CF) and immunodiffusion (ID) assays. However, these assays suffer from low sensitivity and specificity due to the use of a crude, unpurified antigen [28]. A radioimmunoassay that uses a yeast antigen has been developed and results in an increased sensitivity over CF (95% vs. 50%). However, this test still has very low specificity [232]. Use of a partially purified antigen resulted in improved sensitivity and specificity; however, significant crossreactivity was observed when sera from patients with histoplasmosis were tested [233]. Incorporation of a DEAE-column-purified antigen (A antigen) into the CF and ID assays improved specificity.

The A antigen was also incorporated into an EIA format resulting in a sensitivity ranging between 86% and 100% and a specificity of 87–92% [234–236]. Two studies reported using a commercial version of the EIA. Sekhon et al. [237] reported this test to have a sensitivity of 100% but a specificity of 85.6%. Bradshear et al. [238], however, reported a sensitivity of only 85% and specificity of only 47%. It was concluded from these studies that the EIA increased the sensitivity and specificity of the diagnosis, but it was recommended that a positive result should be confirmed using the standard immunodiffusion assay [235,237].

b. Characterization of Immunoreactive Antigens.   Klein and Jones [48] performed further analyses to characterize the immunodominant antigen of *B. dermatitidis*. Western blot revealed a 120-kDa protein which they named WI-1. It was determined that this was a unique surface antigen found only in *B. dermatitidis* and not in *H. capsulatum* or *C. albicans*. An RIA using WI-1 antigen was 85% sensitive and >97% specific. Biochemically, WI-1 differed from A antigen in that WI-1 was a 120-kDa protein containing no carbohydrate whereas the A antigen was characterized as a 135-kDa protein containing 37% carbohydrate. Also, antibody reactivity against WI-1 was reduced after pretreatment of WI-1 with protease whereas antibody reactivity to the A antigen was not affected by this treatment [49].

Immunologically, WI-1 and A antigen are very similar in that antibodies made to WI-1 recognize A antigen and vice versa, indicating antigenic relatedness. Klein et al. produced cDNA clones of *B. dermatitidis* and screened them using rabbit antisera specific for WI-1 [239]. A 942-bp sequence was identified as that portion of *B. dermatitidis* DNA that encoded the 120-kDa protein WI-1. A 25 amino acid repeat near the carboxyl-terminus was shown to be the major antigenic epitope of WI-1. A 17-kDa recombinant peptide, consisting of 4.5 copies of the 25 amino acid tandem repeat, was produced for use in an RIA. Monoclonal antibodies specific for A antigen reacted strongly to the 25 amino acid repeat molecule in the RIA. It was also determined that an A antigen EIA could be inhibited by preincubating the anti-A antisera with the tandem repeat antigen [49]. These results suggest that the A antigen and WI-1 are closely related, if not identical.

Scalarone and colleagues developed and refined a serologic assay for blastomycosis using an experimental dog model of infection. The best antigen source was derived from a yeast extract rather than a mycelial extract of *B. dermatitidis* [240]. This antigen was obtained from 10 different strains of *B. dermititidis* [241], and all antigens, regardless of the originating strain, had equal reactivity. An EIA test using antigens purified by isoelectric focusing was shown to be superior in sensitivity and specificity compared to an EIA using yeast lysate or yeast filtrate antigens [242]. Antigens found in fraction No. 5 (pH 4.3) produced the best EIA results giving a sensitivity of 100% and a specificity of 93% [243]. The use of isoelectric focusing was able to segregate specific antigens from those with crossreactive epitopes. Crossreactive epitopes were localized in fractions 9 and 10 (pH 5.54–5.97) as determined by their reactivity with sera from histoplasmosis patients.

B. Coccidioidomycosis

1. Background

C. immitis is primarily found in the arid and semiarid regions of the southwestern United States, Central America, and South America. In the environment, *C. immitis* is located in the soil and grows in a mycelial, filamentous state which can then produce infectious propagules, termed arthroconidia. After inhalation, arthroconidia enter the lungs where they produce spherical structures appropriately termed "spherules." As the spherule matures, multiple endospores are produced which, upon rupture of the spherule, are released into the surrounding tissue. Each endospore is then capable of producing a new spherule.

2. Acquisition/Risk Factors

Unlike other deeply invasive molds, *C. immitis* hyphae directly fragment into small, light, and highly infectious arthroconidia that can easily become airborne. After the rainy season in the desert, an increased growth of *C. immitis* occurs and arthroconidia are produced. The arthroconidia can then become airborne after a disturbance of the soil either naturally, by winds or earthquakes [25], or by man, through construction.

Those most at risk include people that live and work in endemic areas, and people who travel through or to endemic areas. The desert Southwest has become a popular retirement area resulting in an influx of seronegative senior citizens that are more likely to develop symptomatic and potentially more severe infections [26,244]. Immunocompromised patients are also more likely to develop disseminated disease as are those of the Asian and Black races, who are genetically predisposed [245].

3. Clinical Presentation/Diagnosis

Sixty percent of those infected are asymptomatic or present with a mild upper respiratory tract infection [40]. Forty percent develop a more classical infection which results in initial flu-like symptoms: fever, chills, arthralgia, and cough. This may progress to the formation of a pulmonary cavity that may resemble an infection with *Mycobacterium tuberculosis*. If the acute process does not resolve, the disease may persist in a chronic state which may last for years. Fewer than 1% of infections in immunocompetent patients disseminate to other areas of the body including bone, skin, and brain. Dissemination may occur from a primary infection or from reactivation of a previous infection [40].

A definitive diagnosis of *C. immitis* infection is determined by culture. However, depending on the concentration of organisms in the specimen, culture and confirmation of infection may be delayed for as long as 2–3 weeks. Serologic methods are therefore used to aid in the diagnosis of infection when culture has not been performed or when growth has been delayed [28].

4. Immunoserology and Interpretation

a. Antibody Detection. Standard specimens for serologic diagnosis are sera and, if confirming meningeal involvement, CSF. Currently, the gold standard for serologic diagnosis is a combination of complement fixation (CF) and immunodiffusion (IDCF) using a crude antigen preparation, coccidioidin. The tests detect predominantly IgG antibodies and become positive 2–6 weeks after the onset of illness. These assays are positive in 98% of disseminated infections. Localized infections result in lower titers and reduced sensitivity of the assay [246]. Antibodies (IgG) typically disappear after 6 months but will persist in patients with disseminated infection.

To detect IgM antibodies, it is necessary to first heat the coccidioidin for 30 min at 60°C prior to the assay to alter the test antigen. Initially, this antigen preparation was used in a tube precipitation assay (TP), a liquid-based assay which, in the presence of specific antibodies, results in clumping of the antigen. The use of this antigen in an immunodiffusion format (IDTP) provided similar results [28]. Antibodies (IgM) using this antigen are detected within 1–3 weeks after the onset of symptoms [246]. Together, the CF and the IDTP assays for the diagnosis of coccidioidomycosis have a combined sensitivity of 90% [28]. A latex agglutination assay is also available for the detection of IgM antibodies. This assay is faster to perform and more sensitive than the IDTP test [246]. However, the assay has a >5% false-positive rate [247]. This false-positive rate is even greater if sera have been diluted before testing [248]. Therefore, results which are positive by this method must be confirmed using the IDTP assay.

The CF test is very labor-intensive and technically challenging, requiring highly trained and experienced personnel to adequately perform this assay. Conversion of this assay to a more user-friendly format has been the goal of many researchers. In a commercially available assay (Meridian Diagnostics, Cincinnati, OH), microwell plates are coated with purified CF or TP antigens for detection of IgG or IgM antibodies, respectively. Kaufman et al. [249] reported that the greatest sensitivity (100%) could be achieved only if both tests were used. Sera from patients with blastomycosis, however, showed crossreactivity in this assay resulting in a reduced specificity (96%). It was therefore recommended that immunodiffusion assays should be used for confirmation of positive results. Martins et al. [250] reported similar findings where their results gave a somewhat lower sensitivity (94.8%) but a higher specificity (98.5%). More recently, Zartarian et al. also reported similar findings: 100% sensitivity and 98% specificity [251]. It must be noted that ~15% of the sera from patients with cystic fibrosis nonspecifically react with IgM but not IgG in the absence of a positive culture for *C. immitis* [252].

b. Characterization of Immunoreactive Antigens. Intense efforts have been undertaken to define the major coccidioidal antigens that are specifically recognized during infection so as to increase the specificity of serologic assays. The antigen with which CF antibodies react is a 110-kDa protein that reduces to a 48-kDa protein in the presence of SDS [253]. The CF antigen activity is destroyed by proteolytic enzyme treatment and has been identified as a chitinase [254,255]. The IDTP antigens are 110-kDa and 120-kDa proteins [256] which are heavily glycosolated with 3-O-methyl mannose [257]. Pronase has no effect on the activity of the IDTP antigens whereas periodate treatment eliminates the reactivity. These data suggest that glycosylation is important for antigenicity [258].

Cloning and expression of the chitinase gene resulted in the production of a 427 amino acid, 47-kDa protein which was shown to contain CF epitopes and chitinase activity [259,260]. The recombinant CF/chitinase antigen produced by Johnson et al. [261] reacted similarly to the standard culture filtrate antigen in both the ID and CF assays. When used in an EIA format, the assay was as sensitive as and more specific than the commercial assay [261]. However, Yang et al. [262], using their recombinant protein, also demonstrated a high sensitivity but a lower specificity in that the test demonstrated crossreactivity with serum from patients with histoplasmosis and blastomycosis. The recombinant clone was refined to a 190 amino acid peptide [263]. Use of this antigen in an EIA format resulted in the detection of antibody in the serum of 21 of 22 patients with coccidioidomycosis and did not react with serum from histoplasmosis or blastomycosis patients or with serum from healthy subjects.

Galgiani et al. [264,265] isolated a 33-kDa immunoreactive protein found in the walls of mature spherules which appears to be different than the antigen detected by the CF assay (i.e., chitinase). The antigen was used in an EIA format to test CSF from patients with suspected coccidioidal meningitis. Whereas one of 73 (1.4%) patients without meningitis reacted positively, 74 of 103 (71.8%) patients with meningitis yielded a positive result. Of the 103 CSF specimens from meningitis patients, 58 were positive and 45 were negative by CF. The EIA was positive in 53 of 58 CF positive and 21 of 45 CF negative specimens. This resulted in a 91.4% correlation with CF and a 46.7% rate of false-positive reactions when the CF test was used as the gold standard. However, these results may also indicate an increased sensitivity of the EIA compared to CF when CSF is tested.

C. Histoplasmosis

1. Background

Histoplasmosis is caused by two varieties of *H. capsulatum*: var. *capsulatum* and var. *duboisii*. Both are filamentous molds producing microconidia and tuberculate macroconidia in culture at 25°C and are yeasts at 37°C in culture or in tissue. However, *H. capsulatum* var. *capsulatum* appears as small oval yeasts (2–4 μm) in tissue whereas *H. capsulatum* var. *duboisii* appears as larger yeasts (8–15 μm) with thicker cell walls [40,42]. *H. capsulatum* var. *duboisii* are also characteristically uninucleate and have narrow-based buds which attain the same size as the parent cell prior to cell separation [42].

H. capsulatum var. *capsulatum* is global in distribution but is most commonly found in the central Mississippi and Ohio River valleys of North America and in Central and South America. In contrast, *H. capsulatum* var. *duboisii* has been restricted in distribution to central Africa. Hence, the name "African histoplasmosis" for infections caused by *H. capsulatum* var. *duboisii*. *H. capsulatum* var. *capsulatum* has also been the cause of histoplasmosis in central Africa, however, and has been known to cause infections in Australia, India, Malaysia, and the Caribbean as well [15,40]. Differentiation of *H. capsulatum* var. *capsulatum* from *H. capsulatum* var. *duboisii* has been attempted using subspecies-specific monoclonal antibodies [266,267] and this information may be epidemiologically useful in regions where both varieties occur; however, patient therapy would not differ for either infection [42].

H. capsulatum has been recovered from sputum, pus, bone marrow, tissue, and blood, and recovery from the last is best accomplished using lysis centrifugation tubes [40,268–270]. Because species of *Sepedonium, Chrysosporium, Arthroderma*, and *Renispora* also form tuberculate conidia, conversion of mycelial colonies to the yeast form is required for specific culture confirmation [40,42]. Isolation from clinical specimens may require 2–3 weeks [271] and conversion may require an additional 3–6 weeks [40]. Therefore, exoantigen testing (requiring 48–72 h for specific identification) and, more recently, molecular probe identification [272] have become preferred means for confirmation [28,40,42].

2. Acquisition/Risk Factors

Histoplasmosis is estimated to occur in one-half million persons per year, making it the most common of the endemic mycoses [40]. It is estimated that 2–5% of AIDS patients living in endemic areas are infected [273], and up to 95% of these patients present with disseminated disease [40]. The natural habitat of *H. capsulatum* is in soil enriched with bird or bat guano, and outbreaks have been associated with demolition of chicken coops [24], with construction in endemic areas [20], and with exploration of bat caves [15,22,23,40]. Several outbreaks of histoplasmosis have been reported in Indianapolis alone, and in one recent outbreak AIDS patients accounted for >50% of culture-proven cases of histoplasmosis suggesting that many of these cases were caused by recent exposure rather than reactivation of latent infection [15,274]. Cases of histoplasmosis in HIV-infected patients living outside the endemic areas, however, have also been reported and, at least in some instances, represent reactivation of infection acquired during residence in or travel through endemic areas [27,275,276]. Approximately 80% of normal individuals are histoplasmin skin test positive in the endemic area signifying exposure to *H. capsulatum*. There is generally a 1- to 3-week incubation period following inhalation of conidia before symptoms of infection occur.

3. Clinical Presentation/Diagnosis

Whereas inhalation of conidia initiates either an asymptomatic or a mild, self-limited pulmonary infection in most normal hosts, it may advance to cause chronic infection of the lungs or disseminate. Dissemination in immunocompromised patients often leads to a progressive, fatal illness [18,40]. Acute disease in nonimmunocompromised hosts presents with nonspecific flulike symptoms. Fever, chills, headache, myalgia, loss of appetite, cough, and chest pain are the most common symptoms. Ten percent of those infected acquire aseptic arthritis associated with erythema multiforme or nodosum. Normal persons recover in ~1–3 weeks although general weakness may persist for several months. Chest radiographs are usually normal although small, scattered nodular infiltrates may be present. Infiltrates may leave scattered calcifications throughout the lungs which may result in the development of a round nodule, termed a "histoplasmoma," that may enlarge as fibrous material becomes deposited around the lesion [40]. "Cryptic dissemination" to multiple organs in asymptomatic infections can lead to reactivation at pulmonary and extrapulmonary sites at a later time in a manner similar to reactivation of tuberculosis, particularly if individuals become immunocompromised [10,27].

Chronic pulmonary histoplasmosis usually occurs in middle-aged men with a history of chronic obstructive lung disease and is manifested by either transient pneumonia or progression to fibrosis, cavitation, and tissue damage which may lead to death. Symptoms are similar to those for acute histoplasmosis although chest radiographs often show interstitial infiltrates in the apical segments of the lung with fibrosis and cavitation [18]. Hemoptysis occurs in >30% of patients with chronic pulmonary histoplasmosis [40].

Disseminated histoplasmosis occurs primarily in immunocompromised hosts, and is associated with those with T-cell immunodeficiencies, but may also occur in normal individuals. In immunocompromised patients and infants, the disease is often progressive and may be fatal [10,40,277]. Symptoms include high fever, chills, malaise, prostration, and loss of weight and appetite. Liver function tests may be abnormal and the liver and spleen may be enlarged. Chest radiographs are commonly normal but may display diffuse interstitial infiltrates. In normal hosts, disseminated disease is more chronic and indolent. Hepatic infection and adrenal gland destruction often occur but enlargement of the liver and spleen is less pronounced than in acute forms of the disease. Meningitis may be a complication of chronic disseminated disease as well as endocarditis and mucosal ulcerations of the gastrointestinal tract, and bone or joint involvement may occur in infants and children [18,277–279].

A clinical diagnosis of acute pulmonary histoplasmosis is complicated by the nonspecific signs and symptoms associated with this disease. Chronic histoplasmosis often presents clinically and radiographically similar to tuberculosis or coccidioidomycosis, and cutaneous lesions are similar to those of tuberculosis, syphilis, or paracoccidioidomycosis. Organisms are often more abundant in peripheral blood smears and in bronchial washings from AIDS patients than from nonimmunocompromised patients. Microscopic examination of Wright-stained peripheral blood smears or stained tissue sections may reveal small, oval budding yeasts associated with macrophages, particularly in specimens from persons residing in, or who have visited, endemic areas [10,40]. Because other organisms such as atypical small cells of *B. dermatitidis*, nonencapsulated *C. neoformans*, or *P. marneffei* and *S. schenckii* may appear very similar to yeast cells of *H. capsulatum*, means besides microscopic examination must be used to confirm the diagnosis [28,40].

4. Immunoserology and Interpretation

a. Antibody Detection. Detection of antibody responses in patients with histoplasmosis by CF, ID, CIE, and LA tests has been used as strong evidence of infection [17,28]. Antigens used in these tests include (1) histoplasmin, the soluable filtrate antigen of 25°C mycelial cultures of *H. capsulatum* grown in synthetic broth medium for 4 to 5 weeks [28] or longer [17], and (2) a suspension of merthiolate-killed, whole yeast form cells [28]. These antigens may be obtained commercially from Meridian Diagnostics or Immuno-Mycologics.

Histoplasmin contains two antigens of particular interest—the H antigen, against which antibodies are formed during active disease, and the M antigen, against which antibodies are formed during active as well as chronic histoplasmosis [280,281]. In contrast to the common carbohydrate "C" antigen, which crossreacts with sera from patients with coccidioidomycosis and blastomycosis [17,281], the H and M antigens were thought to be specific proteins for the detection of anti–*H. capsulatum* antibodies. The M antigen, however, was later found to be a catalase that was not specific in Western blot analysis for the diagnosis of histoplasmosis unless the molecule was deglycosylated [46,282]. Nonetheless, histoplasmin has been used as a valuable skin test antigen and as an immunoreactive component in the ID and CF assays [283,284].

The CF test can provide both diagnostic and prognostic information. However, although the CF test is highly sensitive (>90% of culture-proven cases of histoplasmosis were positive by the CF test if serum samples were collected at 2- to 3-week intervals), it is not entirely specific. Crossreactions have been reported with sera from patients with coccidioidomycosis, blastomycosis, and other mycoses as well as from patients with leishmaniasis when the yeast form antigen was used. Antibodies to the yeast form antigen usually appear within 4 weeks after exposure whereas antibodies to the histoplasmin antigen occur later and have titers which are considerably lower [28].

Antibody titers of 1:8 and greater against either the yeast or histoplasmin antigens are considered to be presumptive evidence of infection, and titers >1:32 to be strong presumptive evidence. However, false-positive reactions can occur at titers >1:32, so titer changes observed in serial samples are often more helpful diagnostically. Fourfold changes of titer in either direction can be of assistance in determining the progression or remission of disease, although other clinical and laboratory data should also be considered. This is especially the case when titers of 1:8 to 1:32 are obtained because patients with proven histoplasmosis have also been found to have titers in this range [28].

The ID test is more specific than the CF test but has lower sensitivity. Reactivity with the H antigen is specific for active disease but only occurs in 10–20% of proven cases. On the other hand, reactivity with the M antigen occurs in 85% of active cases but may also be positive in patients with past infections or recent skin test administration [28,40]. Therefore, positive reactivity to the M antigen alone, which may occur earlier in infection, is presumptive evidence for infection whereas reactivity to both the H and M antigens, which may occur later in infection, is more convincing evidence [40].

The LA test, where latex particles are sensitized with histoplasmin, is primarily used for the early diagnosis of acute histoplasmosis and is less helpful for the diagnosis of chronic infection [17,28]. A positive LA test can be obtained as early as 2–3 weeks after exposure, and a titer >1:16 is considered to be significant; a titer of 1:32 is considered strong evidence for active or very recent infection. False-positive results can occur, however, and it is recommended that results be confirmed with the ID or other laboratory tests [28]. Commercially prepared sensitized latex particles are available from Immuno-Mycologics.

A number of attempts to improve antibody detection assays have been proposed including RIA [232,285], immunoblot [46,286], and EIA [287–291] formats. The RIA was shown to be twice as sensitive as the CF test, with detection of IgG responses providing greater sensitivity than detection of IgM responses using the RIA format [285], but crossreactivity remained problematic [232]. An EIA format proved to be as sensitive as the RIA but had no better specificity [292] until modified by incorporation of ferrous beads [293].

Most attempts to replace the CF test with an EIA format have been frustrated even when antigens have been column purified because of the presence of crossreactive carbohydrate moieties associated with the H and M antigens [28,46]. Periodate treatment of the M antigen, however, was demonstrated to remove crossreactivity in an immunoblot format and resulted in a specificity of 91% [17,46]. Nonglycosylated, recombinant H and M molecules have been recently produced which may overcome problems of crossreactivity [47,288]. However, problems may still exist in that false positives may continue to occur due to prior exposure to the organism or to skin testing [17,28] and false negatives may occur in immunosuppressed patients with histoplasmosis [291]. Therefore, antigen detection tests may ultimately have wider utility.

b. Antigen Detection.   Several test formats to detect *H. capsulatum* antigens in body fluids of patients with immunodeficiencies have been developed [294–297]. Initial studies by Wheat et al. used an RIA format where a rabbit polyclonal antibody raised against formalinized yeast cells was used as an immunogen. This antibody was used as both the capture antibody and the radiolabeled detector antibody in a solid-phase RIA format [298]. The test was particularly successful for detecting antigen in urine from patients with disseminated histoplasmosis and especially from AIDS patients. It was an effective test for monitoring responses to therapy and for the detection of relapses in AIDS patients [18,294]. Serum detection was somewhat less successful than urine detection with ~90–95% of urine samples and 50–85% of serum samples positive from patients with disseminated disease [28,40]. Antigen detection was less effective for the diagnosis of patients with self-limited disease or for those with cavitary histoplasmosis than for those with disseminated disease [28,40,298]. For example, antigen was detected in 92%, 21%, and 39% of patients with disseminated, chronic pulmonary, and self-limited forms of the disease, respectively [299]. The RIA has been successful in detecting antigen in bronchoalveolar lavage fluid from most patients with disseminated histoplasmosis and in the cerebrospinal fluid of patients with *H.* meningitis [41,300].

The antigen detected, although poorly defined, was shown to be heat stable (as it could tolerate heating at 100°C for 30 min) and was a polysaccharide in nature (as it bound to concanavalin A) [17,28,40]. *Histoplasma* polysaccharide antigen, HPA for short, was not affected by repeated freeze-thawing (five times) or by storage at room temperature, 4°C, or at −40°C for an indefinite time [41]. Polyclonal antibodies used in the RIA were particularly difficult to generate [28] and up to 40 rabbits were immunized before one immunoreactive animal could be successfully used for antibody production.

The HPA assay can quantitate antigen if standardized procedures are used. Results are recorded as counts per minute (cpm) and converted to RIA units by division with a negative control value (usually normal human urine). RIA results of ≥ 1.0 unit are considered to be positive. An increase of 2 RIA units appears to be clinically significant for monitoring relapse to amphotericin-B therapy in AIDS patients with disseminated histoplasmosis [41].

Unfortunately, there are some drawbacks to the RIA procedure. Day-to-day variability has been attributed to differences in how well the assay plates were coated on a given day, the use of different lots of radiolabelled detector antibodies, and decay of the radiolabel on the antibodies [41]. There is, of course, the inherent danger in the use of radiolabeled compounds [17] as well as issues regarding their disposal. Perhaps the greatest drawback to the use of this assay stems from the large degree of crossreactivity of the test. Initial studies used control specimens from patients with *C. albicans* and *C. neoformans* infections and the specificity of the test appeared to be excellent [294,300]. However, later testing, using specimens from patients with more closely related infections, showed that crossreacting antigen was detected in 63% of patients with blastomycosis, 89% of patients with paracoccidioidomycosis, and 94% of patients with penicilliosis marneffei; however, no crossreactions were observed for specimens from six patients with coccidioidomycosis [10]. Also, there is limited accessibility of the test in that the only laboratory currently performing the HPA test is the Histoplasmosis Reference Laboratory in Indianapolis.

Therefore, because of these limitations, other test formats to detect HPA were developed [293,296,297]. Initial attempts [293] to develop a sensitive and specific EIA to replace the RIA were not successful until an EIA using a biotin-conjugated detector antibody, recognized by a streptavidin–horseradish peroxidase probe, was incorporated into the test format. Assay results for the two formats were comparable (correlation coefficient = 0.974) and demonstrated measurable antigen levels in 50 of 56 patients (89%) with disseminated histoplasmosis and 11 of 30 patients (37%) with self-limited disease [296]. Most recently, Garringer et al. developed an inhibition EIA format in an attempt to conserve the amount of polyclonal antibody used for each assay [297]. This assay used the same polyclonal rabbit antibody as did the sandwich EIA described above. The sensitivity and specificity were comparable in the two test formats when serum obtained from culture proven cases of histoplasmosis was tested (sensitivity for sandwich EIA = 82.5% vs. 80% for the inhibition EIA; specificity for sandwich EIA = 95% vs. 92.5% for the inhibition EIA). However, whereas the sensitivity to detect antigen in urine was comparable in the two formats (sandwich EIA = 92.5% and inhibition EIA = 87.5%), the specificity was significantly lower for the inhibition EIA than the sandwich EIA (80% vs. 97.5%). The reproducibility of the sandwich EIA was also greater for control samples relative to the inhibition EIA, and no reduction in crossreactivity to antigens from patients with blastomycosis, paracoccidioidomycosis, or penicilliosis was observed by the inhibition EIA [297]. Therefore, the inhibition EIA was not recommended as a replacement for the sandwich EIA.

Specific detection of *H. capsulatum* antigens using monoclonal, instead of polyclonal, antibodies was also attempted [281,301]. Gomez et al. [302] developed an inhibition EIA to detect a 69- to 70-kDa antigen distinct from HPA [296,300]. The overall sensitivity of the EIA for various forms of histoplasmosis was 71%, and for acute and chronic histoplasmosis 89% and 57%, respectively [302]. The specificity was 98% when normal human serum was tested and 85% when specimens from chronic fungal or bacterial infections were tested. In contrast to the sandwich EIA, the inhibition EIA detected antigen more frequently in serum than in urine [10,296].

D. Paracoccidioidomycosis

1. Background

P. brasiliensis is found almost exclusively in Latin America and the greatest number of cases are found in Brazil. It is the most common of the endemic fungal infections in this region. The specific ecological niche of this fungus has not been fully elucidated; however, it is believed to be a soil-associated micro-organism. Infection is through inhalation of conidia which converts to a typical yeast form in tissues. The morphology of the yeast cell is diagnostic when seen in its characteristic form: a 5- to 30-μm, oval to round yeast cell that reproduces by budding at multiple sites (morphologically often referred to as a "mariner's wheel").

2. Acquisition/Risk Factors

Infection is acquired after the inhalation of spores and may remain subclinical, only detected by a positive skin test, or may remain latent and reactivate at a later date, especially following immunosuppression. Whereas *P. brasiliensis* appears to infect both sexes equally before puberty, there is an increased incidence of infection in adult men compared with adult women. Restrepo et al. have speculated that transformation from the mycelial to yeast forms in vivo is inhibited by the presence of estrogen in postpubertal women [303].

Infection in persons with AIDS has been described; however, the rate of infection is surprisingly low. Goldani and Sugar reported in 1995 that of the ~500,000 HIV-infected persons in South America, only 27 cases of paracoccidioidomycosis in this patient group had been reported in the literature [304]. Several reasons were offered to explain this paucity of cases. First, many HIV-infected patients receive trimethoprim-sulfamethoxazole as prophylaxis against *P. carinii* infection, and *P. brasiliensis* is also susceptible to this drug combination. Second, the majority of HIV-infected persons are located in urban areas, whereas paracoccidioidomycosis occurs primarily in rural areas. Third, because of the difficulty in diagnosing paracoccidioidomycosis premortem, and because autopsies are not commonly performed on AIDS patients, the incidence may be underreported.

3. Clinical Presentation/Diagnosis

The lung is the initial site of infection. A subacute to acute disease may occur particularly in children and young adults. However, a slow, indolent progression to a chronic unifocal disease, affecting only a single organ, or further progression to a multifocal infection, may occur after many years. Typically, the sites of infection include the lung, lymph nodes, adrenal glands, and mucous membranes of the mouth. Other organs are less likely to be involved but may include liver, spleen, kidney, skin, brain, and intestines. Infection of the skin and mucous membranes can be quite disfiguring and is usually found on the face and mouth.

4. Immunoserology and Interpretation

a. Antibody Detection. As with the other dimorphic fungi, standard serologic diagnosis is accomplished through the use of the CF and ID assays [305–309]. Antigen preparations include crude yeast cell extract or culture filtrate which provide adequate sensitivity and good specificity in the ID assay. Crossreactivity of serum from patients infected with *H. capsulatum* reduces the specificity of the assays when these same antigens are used in CF tests. Casotto [310] analyzed the cellular proteins of *P. brasiliensis* from 3-day-old cultures and concluded that of the 13 proteins detected by immunoblot, five were considered specific for the diagnosis of paracoccidioidomycosis. It was also determined that the immunodominant protein varied depending upon the strain of *P. brasiliensis* used [311]. Researchers were therefore unable to correlate reactivity against the different proteins with the clinical forms of disease in each patient [310]. Almeida et al. [312] attempted to purify the antigens and compared them in a quantitative chemiluminescent EIA. They found that use of a purified glycoprotein, termed gp43, provided greater test specificity than use of the crude antigens. Bueno et al. [313] used gp43 in an EIA format to detect IgG, IgM, and IgA titers to *P. brasiliensis* and found that the IgG and IgA EIA had a 100% positive and negative predictive value. They also determined that the assay could be used to follow treatment as titers declined after therapy. A 27-kDa recombinant protein has also been used in an EIA format giving a sensitivity of 73.4% and a specificity of 87.5% [314,315].

  b. Antigen Detection. Antibodies are not always produced or detected in immunosuppressed patients. Therefore, assays to detect antigen have been developed to aid in the diagnosis of *P. brasiliensis* infections in these patients. Ferreira-da-Cruz et al. developed an immunoradiometric assay which employed polyclonal rabbit IgG for the detection of *P. brasiliensis* antigens in serum [316]. Two different antigen preparations (cellular and metabolic) were used to compare the radiometric assay to the immunodiffusion assay. The radiometric assay was considered more sensitive since it was able to detect 3.6 ng of cellular antigen and 360 ng of metabolic antigen, whereas the immunodiffusion assay could detect only 12 μg of either antigen preparation. Gomez et al. [317] developed a monoclonal antibody that was specific for an 87-kDa antigen and used it in an inhibition EIA. This assay was able to detect 5.8 ng of antigen per mL serum. The sensitivity of the assay was 80.4% with a specificity of 81.4%. Further studies indicated that the effect of treatment could be followed using this assay as titers decreased with effective therapy [318]. Salina et al. [319] used two methods—immunoblotting, which detected gp43 and gp70, and competitive enzyme immunoassay, to antigen in urine. The sensitivities were 92% and 75%, respectively. The specificity of both tests was 100%.

E. Penicilliosis Marneffei

1. Background

P. marneffei is the only known dimorphic mold in the genus *Penicillium*. Mycelial colonies produce a red pigment which diffuses into the growth agar which is characteristic, but not diagnostic, because some nonpathogenic species of *Penicillium* also produce a similar pigment [40]. Yeastlike structures of *P. marneffei* found in vivo can easily be mistaken for *H. capsulatum* in histological sections. However, replication of these organisms is not through budding but by fission which, when seen in tissue, is diagnostic. It is endemic to the regions of Southeast Asia and China.

2. Acquisition/Risk Factors

The ecological niche of this mold has not been totally elucidated. It is believed to be associated with soil and with bamboo rats and/or their burrows. Infection begins in the lungs after inhalation of conidia. There is some evidence for a seasonal variation in the incidence of disease in Thailand with the greatest incidence occurring during the rainy season, from May to October [40]. While infection can occur in normal, healthy individuals, disease is much more prominent in immunosuppressed persons. This is particularly true as cases of AIDS increase in the region. *P. marneffei* infection is the third most prominent opportunistic infection in Thailand [19].

3. Clincial Presentation/Diagnosis

Infection is characterized by a chronic, productive cough, pulmonary infiltrates, and generalized symptoms of lymphadenopathy, hepatosplenomegaly, and intermittent fever. Disseminated infection may often result in the development of umbilicate skin lesions on the face, upper trunk, and limbs. The skin is affected in ~68% of patients with disseminated disease and may include generalized rash, papules, and progression to subcutaneous abscesses and ulcers. These symptoms may be confused with other infections such as tuberculosis, mulloscum contagiosum, and fungal infections such as histoplasmosis and cryptococosis. Disseminated infection with *P. marneffei* is almost exclusively observed in immunosuppressed patients.

4. Immunoserology and Interpretation

a. Antibody Detection. Diagnosis of *P. marneffei* infections has relied mainly on culture and/or histopathologic identification. All assays have been developed using very crude antigen extracts. Research to determine the specific antigen(s) to which the immune response reacts has elucidated a 90-kDa cell wall mannoprotein and a 38-kDa protein.

Serologic diagnosis, while predominately experimental, can be made through the detection of antibody in the serum. Immunodiffusion (ID) and indirect fluorescent antibody (IFA) assays have been used for the detection of antibody in suspected cases [320–322]. The ID assay as described by Sekhon et al. [320] used rabbit anti–*P. marneffei* antibodies tested against antigens made from 10 species of *Penicillium*. The serum reacted with only *P. marneffei*–derived antigen. Conversely, rabbit antisera made against five *Aspergillus* spp., four systemic dimorphic fungi, and three thermophilic actinomycetes showed no crossreactivity with the *P. marneffei* antigen preparation. Viviani et al. used the ID assay [321] to follow the serological response of an AIDS patient infected with *P. marneffei*. They followed this patient over a 5-month period and collected 13 sera. They then compared undiluted serum to two-fold concentrated serum and found that the concentrated serum was more sensitive (10/13) than the unconcentrated serum (3/13). They were unable to determine the true sensitivity of this assay as this patient was being treated and was subsequently cured of his infection. There is no known information on the length of time antibody is detectable after treatment or cure. Yuen et al. [322] later developed an IFA assay using whole-cell antigens. All eight confirmed *P. marneffei* cases were serologically positive with titers of 1:160. None of the sera from patients with other diseases (0/95) or from healthy controls (0/78) had titers >1:40. This report, however, did not determine the assay's specificity against other dimorphic fungi.

b. Antigen Detection. Assays used in the detection of antigen in suspected cases include the ID and LA tests for assaying serum [323], and EIA for assaying urine [324]. The ID and LA tests were used by Kaufman et al. [323] to detect *P. marneffei* antigen in the serum of 17 culture-confirmed, HIV-infected patients from Thailand. Of the 17 culture-confirmed cases, the LA test was more sensitive (13/17 positive; 76.5%) than the ID assay (10/17 positive; 58.8%). Treatment of serum to dissociate antibody-antigen complexes prior to assaying was not performed which may have decreased assay sensitivity. Desakorn et al. [324] developed an EIA for the detection of antigen in the urine of HIV + patients. The urine was boiled, centrifuged, and the supernatant was removed and tested. Using a cutoff titer of 1:40 for positivity, the assay was 97% sensitive and 98% specific.

F. Sporotrichosis

1. Background

S. schenckii is a dimorphic mold that has worldwide distribution. *S. schenckii* infects and causes a chronic disease of the cutaneous and sub-cutaneous layers of the skin, usually of the extremities, and may cause extracutaneous infections in susceptible hosts [325]. This mold is unique in relation to the other thermally dimorphic molds in that the primary route of infection is not through inhalation but through dermal inoculation. The natural habitat of *S. schenckii* is in soil, plants, and plant material of temperate, subtropical, and tropical regions of the world. Sphagnum moss has been a classic source of infection. A variety of *S. schenckii, S. schenckii* var. *luriei*, has been described to cause subcutaneous sporotrichosis [326] as well as a case of fatal pulmonary sporotrichosis [327]. Outbreaks occur when people are exposed to a common source. One of the largest outbreaks occurred in the early 1940s where 3000 miners in South Africa were infected with *S. schenckii* after exposure to the mine timber on which the organism was growing [325].

2. Acquisition/Risk Factors

Persons most commonly infected are those with outdoor occupations or hobbies that would bring them into contact with soil and possible dermal inoculation such as farming and gardening. In fact, a common name for the disease is "rose gardener's disease." There is no particular demographic predilection for the acquisition of cutaneous sporotrichosis except that it is more common in adults than in children [40]. Extracutaneous disease, however, is more often found in men >30 years old. Extracutaneous disease can be found as pulmonary, osteoarticular, or disseminated visceral infections [325]. While not common, disseminated infection is more likely to occur in the immunosuppressed patient.

3. Clinical Presentation/Diagnosis

Most *S. schenckii* infections are confined to cutaneous or subcutaneous lesions [325]. The incubation period after inoculation is ~1–12 weeks after which an initial lesion forms at the site of inoculation. This lesion is characteristically a papule which enlarges to become more nodular with periodic draining of a serous to purulent exudate. Ulceration of the lesion may occur. The infection may remain as a single lesion or form one or more lesions up the extremity along the lymphatic channels. Lesions may come and go but infection is never resolved without treatment. In children, the lesion is often on the face.

Extracutaneous disease may occur and inhalation of conidia may result in pulmonary sporotrichosis [325]. Typically, pulmonary sporotrichosis occurs in alcoholic, middle-aged men with chronic obstructive pulmonary disease. The disease frequently mimics that of pulmonary tuberculosis which is included in the differential diagnosis along with the other systemic endemic mycoses. Disseminated infection may spread to other parts of the body including lungs, joints, bones, and meninges. Osteoarticular sporotrichosis is an infection of the joint, typically the knee; however, the ankle, wrist, and elbow may also be infected. Often no other sites of infection are detected. Diagnosis of *S. schenckii* infection is routinely by culture or histopathology where the round to oval to sometimes cigar-shaped yeast forms are seen in tissue.

4. Immunoserology and Interpretation

Serologic testing has had little impact on the diagnosis of sporotrichosis. The development of newer, more sensitive and specific assays has not been given as high a priority as for other fungal diseases. A CF test using culture supernatant antigen was initially examined. The specificity of the assay was good (no crossreaction with sera from patients infected with other fungi); however, the assay lacked sensitivity [328]. The assay was more effective for the diagnosis of extracutaneous disease (positive in eight of nine patients) than for the diagnosis of cutaneous disease (positive in only three of nine patients). Karlin and Nielsen [329] compared four different methods for the diagnosis of sporotrichosis: CF, ID, TA, and LA. Whole yeast cells antigens were used for CF and TA, whereas culture supernatants from *S. schenckii*, grown in either yeast or hyphal form, were used in the LA and ID assays. They concluded that the LA assay using the yeast culture supernatant provided the best results. Blumer et al. [330] compared five different methods and showed that the LA and TA assays had sensitivities of 94% (75/80) and 96% (77/80), respectively. The specificity of the LA assay was 100% (0/86) whereas that of the TA assay was 98% (2/85). False positives occurred in 2 of 10 cases of leishmaniasis. An enzyme immunoassay was described and used in a study of *S. schenckii* meningitis [331]. A comparison between the LA and EIA was performed using both serum and CSF from meningitis patients. Antibody titers were typically higher using the EIA than the LA and, for a few cases, the EIA was more sensitive.

IV. EMERGING PATHOGENS

A. Emerging Yeast Pathogens

Less common yeast infections have been described in AIDS patients including osteomyelitis resulting from *C. glabrata* infection, meningitis caused by *Cryptococcus clavatus*, and even fungemias due to *Saccharomyces cerevisiae* or *Rhodotorula rubra* [9,332]. Other yeasts, which have traditionally been considered to be part of the normal cutaneous or mucosal flora of humans, such as *Trichosporon asahii* (formerly *T. beigelii*) and *Malassezia furfur*, have been reported to cause life-threatening infections in immunocompromised hosts or low-birth-weight infants [34,40,66,333–336]. Unfortunately, immunoserology has little to offer for the diagnosis of these infections.

1. Trichosporonosis

Disseminated trichosporonosis, although uncommon, is often fatal and occurs in patients with granulocytopenia, cancer, and AIDS and in those undergoing bone marrow or solid-organ transplantation [34,336,337]. It is also a common cause of sepsis in low-birth-weight infants [333]. Signs and symptoms are similar to those for disseminated candidiasis and, also similar to candidiasis, the disease occurs in either an acute or chronic form. Fever unresponsive to broad-spectrum antibacterial antibiotics is the most common symptom. *T. asahii* (*beigelii*) can be cultured from blood, urine, or cutaneous lesions but, despite this, diagnosis is often made postmortem [40]. Although monoclonal antibodies have been produced for the detection of *T. asahii* in tissues [338,339], no reliable serological tests are available for the diagnosis of this infection. *T. asahii* does, however, share a heat-stable antigenic determinant with the capsular antigen of *C. neoformans*; serum from trichosporonosis patients may therefore crossreact in the *Cryptococcus* LA test [28,40,219,340–342]. However, negative *Cryptococcus* LA test results have also been obtained using serum from patients with disseminated trichosporonosis [335,342].

2. Malassezia Species Infections

Malassezia spp. (*M. furfur* and *M. pachydermatis*) infections occur most often in low-birth-weight infants, or in debilitated adults or children, receiving total parenteral nutrition via central venous catheters. Nutritional solutions containing lipids, or catheters supplying these solutions, become contaminated and provide a nidus for replication of the organism [40,343]. *M. furfur* is difficult to recover from peripheral blood although contaminated catheter tips may yield viable organisms. Subculture requires inclusion of lipid in the growth medium or overlayering of agar with lipid and incubation for 4–6 days at 32°C. In infants, symptoms may include fever, apnea, tachycardia, interstitial pneumonia, and thrombocytopenia [334,343]. In adults, symptoms may include fever, but also may include mycotic thrombi around catheter insertion sites, endocarditis with vegetations, or inflammatory lesions of the lungs [40]. No serological tests exist to diagnose *Malassezia* spp. infections.

B. Emerging Mold Infections

Unusual infections caused by molds other than *Aspergillus* spp. have been reported with increased frequency [9,40,66,344]. These include endocarditis and pneumonia caused by *Fusarium* spp., rhinocerebral zygomycosis caused by *Saksenaea vasiformis* or *Rhizopus* spp. [9,345,346], cases of zygomycosis caused by *Cunninghamella bertholletiae* and *Absidia corymbifera*, and gastrointestinal infections with *Basidiobolus ranarum* [347–349]. In addition, other infections have been reported including endocarditis and bone and central nervous system infections caused by *Pseudallescheria boydii* [9,350–352], cutaneous as well as pulmonary infections caused by *Alternaria alternata* [353], and infections of the esophagus and other anatomical sites by *Acremonium* spp. [354–357]. *Scedosporum prolificans* (*inflatum*) has been associated with infections of bone and the respiratory tract [9,358], and disseminated cases have been reported in immunocompromised hosts [9,359–361]. A nosocomial outbreak of fatal *S. prolificans* infections has also been reported [362]. Perhaps the most unusual case of emerging mold infections involved an immunosuppressed patient with sinusitis caused by the "mushroom" fungus *Schizophyllum commune* [9]. Few or no serological tests are available for many of these infections or, when available, are in research laboratories only and have not been fully evaluated in clinical trials. Therefore, the following is primarily an overview of the infections themselves, with relevant references to diagnostic tests when available.

1. Hyalohyphomycosis

Hyalohyphomycotic agents are hyaline (colorless or lightly pigmented) molds with septate, branched, and often mycelial forms in tissue. Generally found in soil, in water, or as plant pathogens, infection can be initiated following inhalation, traumatic implantation, or aspiration of contaminated water. Worldwide in distribution, major agents of hyalohyphomycoses include species of *Acremonium, Beauveria, Chrysosporium, Fusarium, Paecilomyces, Pseudallescheria, Scedosporium*, and *Trichoderma*. Organisms can invade bone, lung, spleen, liver, brain, and the lymphatic system. Diseases caused by the hyalophyphomycetes include pneumonia, endocarditis, endopthalmitis, sinusitis, pleural effusion, keratitis, osteomyelitis, and cutaneous lesions [6,40,363].

   Fusariosis can be caused by at least 12 different species of *Fusarium* although the most common are *F. solani, F. moniliforme*, and *F. oxysporum*. Disseminated infections with *Fusarium* spp. are most common among granulocytopenic cancer and bone marrow transplant patients and are accompanied by a high mortality rate [34,40,363]. These infections are difficult to treat with currently available antifungal agents as many *Fusarium* spp. isolates are refractory to therapy with amphotericin-B, fluconazole, and itraconazole [364–367]. Clinical symptoms resemble those for aspergillosis and, in tissue, *Fusarium* spp. are often mistaken for *Aspergillus* spp. Both organisms have a predilection for vascular invasion and cause thrombosis and tissue necrosis. Unlike *Aspergillus* spp., however, *Fusarium* can be more easily cultured from blood and can be recovered from 60% of cases [40]. There are currently no serological tests for the diagnosis of *Fusarium* spp. infections.

   Pseudallescheriasis is becoming a more frequent cause of systemic disease [351,368]. It is caused by the hyalohyphomycete *P. boydii* (imperfect form, named *Scedosporium apiospermum*; also has a *Graphium* synanamorph). Infections with this organism, as with those caused by *Fusarium* spp., are difficult to differentiate from *Aspergillus* spp. infections based on tissue morphology alone. In culture, the dichotomously branching, septate hyaline hyphae produce dark, round cleistothecia containing asci and ascospores [10,40].

Although *P. boydii* can cause infection in immunocompetent hosts, these infections are generally confined to localized, chronic, mycetoma-like infections of the feet and hands [10,40]. In immunocompromised hosts, however, infections may disseminate to the lungs, bone, brain, and multiple systemic sites [350,352]. Clinical symptoms of disseminated infection often mimic aspergillosis where endocarditis or dissemination from the lungs to the central nervous system can often be fatal [351,360–362,365,369]. Similar to fusariosis, pseudallescheriasis is also refractory to antifungal drug therapy [370] and no serological tests are available for diagnosis. However, fluorescent antibodies to mycelial and conidial antigens of *P. boydii* have been reported which crossreact with various *Aspergillus* spp., *Fusarium* spp., and *Scopulariopsis* spp. but which react specifically with *P. boydii* when absorbed out with these organisms [28].

S. prolificans (not to be confused with *S. apiospermum*, the imperfect form of *P. boydii*) has become particularly problematic among neutropenic patients because infections are frequently fatal [360–362] and many isolates have proven resistant to multiple antifungal agents [371,372].

2. Phaeohyphomycosis

Agents of phaeohyphomycosis have a worldwide distribution and have become newly emerging pathogens in immunocompromised hosts with as many as 100 different molds belonging to 60 different genera being implicated in infection [40,344,358,359]. The most significant disease-causing agents include species of *Alternaria, Bipolaris, Cladophialophora, Curvularia, Dactylaria, Exophiala, Exserohilum, Phialemonium, Phialophora, Phoma*, and *Wangiella*. In contrast to the hyalohyphomycotic agents, these molds are darkly pigmented (dematiaceous) and often appear as moniliform, beadlike, toruloid, or swollen septate hyphae which may or may not be branched. In culture, most genera display mycelial forms but at least two genera (*Exophiala, Wangiella*) display mucoid, yeast-like forms [373].

Similar to the hyalohyphomycetes, these organisms are often found associated with plants and soil, and infections can be acquired via inhalation or traumatic implantation [40,374]. Clinical manifestations include subcutaneous abscesses, sinus infections, mycetoma, or, in the immunocompromised patient, dissemination to the bone, lung, brain, and other organs [353,360–362,371]. No serological or fluorescent antibody tests are available for identification of the phaeohyphomycotic agents.

3. Zygomycosis

The agents of zygomycotic infections are classified in two orders—the Mucorales and the Entomophthorales [375,376]. Agents belonging to the Mucorales are more likely to cause invasive or disseminated infections, whereas those of the Entomophthorales cause primarily subcutaneous zygomycoses that rarely disseminate [10,377]. Primary agents of zygomycosis of the order Mucorales include species of *Absidia, Apophysomyces, Cunninghamella, Mucor, Rhizopus, Rhizomucor*, and *Saksenaea*, and those of the order Entomophthorales include *Conidiobolus* and *Basidiobolus* [10,40]. In tissue, these organisms appear as sparsely septate, irregularly branched, broad (10–20 μm), ribbonlike hyphae [10,38]. Rhinocerebral disease is often caused by zygomycetes and may involve the palate, the nasal mucosa and sinuses, and the orbit, and may invade into the brain resulting in a fatal infection [10,40]. Pulmonary, gastrointestinal, and cutaneous forms of zygomycosis also occur, and any of these forms, as well as the rhinocerebral form, may disseminate in susceptible hosts [40,378,379]. Dissemination may occur via vascular invasion to the lungs, liver, spleen, kidneys, and gastrointestinal tract. In patients with hematological malignancy, zygomycosis is the third most frequent fungal infection, and patients with diabetic ketoacidosis are also at high risk for infection. In these patients, zygomycosis presents as a fulminant disease requiring aggressive antifungal therapy and often surgical debridement [10,40].

Although early diagnosis of this rapidly progressing disease is desirable, >90% of disseminated cases of zygomycosis are diagnosed at autopsy [28]. A fatal case of rhinocerebral infection caused by *Saksenaea vasiformis* was successfully diagnosed by a combination of tissue morphology and detection of zygomycotic antibodies by EIA [380]. In this study, antigens were derived from *Rhizopus arrhizus* and *S. vasiformis*. Antibodies could be detected early during infection, 1 month before isolation, and 3 months before identification of the organism [29]. CSF titers against *S. vasiformis* were diagnostic for infection and increased during clinical deterioration [380].

Antibodies to *R. arrhizus* have been detected in the sera of patients with brain infections by this organism [346,381]. In addition, antibodies to *Absidia corymbifera* were found in the CSF and serum of a patient with culture-proven *Absidia* meningitis [382] and in the serum of a patient with histologically proven *A. corymbifera* brain abscess [346].

There appears to be significant crossreactivity among the major antigens of the genera *Rhizomucor*, *Absidia*, and *Rhizopus* as a result of the common peptido-L-fuco-D-mannan on their cell surfaces [28,383]. Preliminary work using homogenized antigens from *A. corymbifera*, *Rhizomucor pusillus*, and *R. arrhizus* demonstrated a sensitivity of 70% and a specificity of 90% for these zygomycetes in an ID test format [28,384]. Apparently whereas patients generate antibodies to certain zygomycetes during infection, normal healthy volunteers and patients with other infections had very low amounts of antibodies to these agents [101]. An EIA format was also tested and, using a 1:400 antigen titer as positive, the EIA was 81% sensitive and 94% specific for zygomycosis [28,380].

Experimental ID tests to detect antibodies to *Conidiobolus coronatus* and *Basidiobolus ranarum* in patients with subcutaneous infections or tumorous masses of nasal or adjacent tissues have been developed using culture filtrate antigens from these organisms. The tests appear to be sensitive and specific and were useful for monitoring disease resolution [384]. Further evaluations of the ID and EIA tests for the diagnosis of zygomycosis are needed to determine their true clinical usefulness.

V. CONCLUDING REMARKS

Despite a tremendous amount of work over the past 30 years, much research is still needed to optimize existing serological tests and to develop new tests. Standardization and commercialization of existing tests would help to ensure greater availability and to improve interlaboratory reproducibility. Purification of antigens through chemical and recombinant molecular biological means should help to better define immunoreactive components and allow tests to have greater specificity. Similarly, the development of monoclonal antibodies to replace polyclonal antibodies should the development of tests with greater specificity and for a larger supply of identically reactive reagents. What may be lost in sensitivity by the use of monoclonal antibodies might be recouped via the use of more sensitive detection systems using chemiluminescent or other detector reagents which may amplify chemical signals. Purification and characterization of antigens for the specific detection of antibodies against the endemic fungi are under way and should yield improved diagnostic tests in the near future. Finding new antigens for the identification of opportunistic fungal infections that are not transient in nature, and developing tests to detect the newly emerging fungal pathogens, are greatly needed. Perhaps a combination of molecular biological and immunological methods will finally be needed to improve the status of diagnosing fungal infections in a timely and accurate manner.

REFERENCES

1. F Anaissie. Opportunistic mycoses in the immunocompromised host: experience at a cancer center and review. Clin Infect Dis 14(suppl 1):S43–53, 1992.
2. AE Brown. Overview of fungal infections in cancer patients. Semin Oncol 17:2–5, 1990.
3. R Horn, B Wong, TE Kiehn, D Armstrong. Fungemia in a cancer hospital: changing frequency, earlier onset, and results of therapy. Rev Infect Dis 7:646–655, 1985.
4. JD Meyers. Fungal infections in bone marrow transplant patients. Semin Oncol 17:10–13, 1990.
5. Anonymous. National Nosocomial Infections Surveillance (NNIS) System report, data summary from January 1990 to May 1999, issued June 1999. Am J Infect Control 27:520–532, 1999.
6. VA Morrison, RJ Haake, DJ Weisdorf. The spectrum of non-*Candida* fungal infections following bone marrow transplantation. Medicine (Baltimore) 72:78–89, 1993.
7. JR Rees, RW Pinner, RA Hajjeh, ME Brandt, AL Reingold. The epidemiological features of invasive mycotic infections in the San Francisco Bay area, 1992–1993: results of population-based laboratory active surveillance. Clin Infect Dis 27:1138–1147, 1998.
8. SK Fridkin, WR Jarvis. Epidemiology of nosocomial fungal infections. Clin Microbiol Rev 9: 499–511, 1996.
9. B Dupont, DW Denning, D Marriott, A Sugar, MA Viviani, T Sirisanthana. Mycoses in AIDS patients. J Med Vet Mycol 32(suppl 1):65–77, 1994.
10. TJ Walsh, SJ Chanock. Diagnosis of invasive fungal infections: advances in nonculture systems. Curr Clin Top Infect Dis 18:101–153, 1998.
11. J Aisner, PH Wiernik, SC Schimpff. Treatment of invasive aspergillosis: relation of early diagnosis and treatment to response. Ann Intern Med 86:539–543, 1977.
12. D Caillot, O Casasnovas, A Bernard, JF Couaillier, C Durand, B Cuisenier, E Solary, F Piard, T Petrella, A Bonnin, G Couillault, M Dumas, H Guy. Improved management of invasive pulmonary aspergillosis in neutropenic patients using early thoracic computed tomographic scan and surgery. J Clin Oncol 15:139–147, 1997.
13. JE Kuhlman, EK Fishman, PA Burch, JE Karp, EA Zerhouni, SS Siegelman. CT of invasive pulmonary aspergillosis. Am J Roent 150:1015–1020, 1988.
14. M Mori, JR Galvin, TJ Barloon, RD Gingrich, W Stanford. Fungal pulmonary infections after bone marrow transplantation: evaluation with radiography and CT. Radiology 178:721–726, 1991.
15. NM Ampel. Emerging disease issues and fungal pathogens associated with HIV infection. Emerg Infect Dis 2:109–116, 1996.
16. TA Duong. Infection due to *Penicillium marneffei*, an emerging pathogen: review of 155 reported cases. Clin Infect Dis 23:125–130, 1996.
17. AJ Hamilton. Serodiagnosis of histoplasmosis, paracoccidioidomycosis and penicilliosis marneffei; current status and future trends. Med Mycol 36:351–364, 1998.
18. LJ Wheat, PA Connolly-Stringfield, RL Baker, MF Curfman, ME Eads, KS Israel, SA Norris, DH Webb, ML Zeckel. Disseminated histoplasmosis in the acquired immune deficiency syndrome: clinical findings, diagnosis and treatment, and review of the literature. Medicine (Baltimore) 69: 361–374, 1990.
19. W Nittayananta. Penicilliosis marneffei: another AIDS defining illness in Southeast Asia. Oral Dis 5:286–293, 1999.
20. TF Jones, GL Swinger, AS Craig, MM McNeil, L Kaufman, W Schaffner. Acute pulmonary histoplasmosis in bridge workers: a persistent problem. Am J Med 106:480–482, 1999.
21. BS Klein, JM Vergeront, RJ Weeks, UN Kumar, G Mathai, B Varkey, L Kaufman, RW Bradsher, JF Stoebig, JP Davis. Isolation of *Blastomyces dermatitidis* in soil associated with a large outbreak of blastomycosis in Wisconsin. N Engl J Med 314:529–534, 1986.

22. H Valdez, RA Salata. Bat-associated histoplasmosis in returning travelers: case presentation and description of a cluster. J Travel Med 6:258–260, 1999.
23. DA Ashford, RA Hajjeh, MF Kelley, L Kaufman, L Hutwagner, MM McNeil. Outbreak of histoplasmosis among cavers attending the National Speleological Society Annual Convention, Texas, 1994. Am J Trop Med Hyg 60:899–903, 1999.
24. DS McKinsey, RA Spiegel, L Hutwagner, J Stanford, MR Driks, J Brewer, MR Gupta, DL Smith, MC O'Connor, L Dall. Prospective study of histoplasmosis in patients infected with human immunodeficiency virus: incidence, risk factors, and pathophysiology. Clin Infect Dis 24:1195–1203, 1997.
25. E Schneider, RA Hajjeh, RA Spiegel, RW Jibson, EL Harp, GA Marshall, RA Gunn, MM McNeil, RW Pinner, RC Baron, RC Burger, LC Hutwagner, C Crump, L Kaufman, SE Reef, GM Feldman, D Pappagianis, SB Werner. A coccidioidomycosis outbreak following the Northridge, Calif, earthquake. JAMA 277:904–908, 1997.
26. Anonymous. Coccidioidomycosis—Arizona, 1990–1995. MMWR 45:1069–1073, 1996.
27. DN Fredricks, N Rojanasthien, MA Jacobson. AIDS-related disseminated histoplasmosis in San Francisco, California. West J Med 167:315–321, 1997.
28. L Kaufman, JA Kovacs, E Reiss. Clinical immunomycology. In: NR Rose, EC de Macario, JD Folds, HC Lane, RM Nakamura, eds. Manual of Clinical Laboratory Immunology. Washington: ASM Press, pp 585–604, 1997.
29. L De Repentigny, L Kaufman, GT Cole, D Kruse, JP Latge, RC Matthews. Immunodiagnosis of invasive fungal infections. J Med Vet Mycol 32(suppl 1):239–252, 1994.
30. E Reiss, CJ Morrison. Nonculture methods for diagnosis of disseminated candidiasis. Clin Microbiol Rev 6:311–323, 1993.
31. JM Jones. Laboratory diagnosis of invasive candidiasis. Clin Microbiol Rev 3:32–45, 1990.
32. R Duthie, DW Denning. *Aspergillus* fungemia: report of two cases and review. Clin Infect Dis 20:598–605, 1995.
33. R Kappe, D Rimek. Laboratory diagnosis of *Aspergillus fumigatus*–associated diseases. Contrib Microbiol 2:88–104, 1999.
34. JH Rex, TJ Walsh, EJ Anaissie. Fungal infections in iatrogenically compromised hosts. Adv Intern Med 43:321–371, 1998.
35. M von Eiff, N Roos, R Schulten, M Hesse, M Zuhlsdorf, J van de Loo. Pulmonary aspergillosis: early diagnosis improves survival. Respiration 62:341–347, 1995.
36. H Saito, EJ Anaissie, RC Morice, R Dekmezian, GP Bodey. Bronchoalveolar lavage in the diagnosis of pulmonary infiltrates in patients with acute leukemia. Chest 94:745–749, 1988.
37. VL Yu, RR Muder, A Poorsattar. Significance of isolation of *Aspergillus* from the respiratory tract in diagnosis of invasive pulmonary aspergillosis. Results from a three-year prospective study. Am J Med 81:249–254, 1986.
38. FW Chandler, W Kaplan, L Ajello. Color atlas and text of the histopathology of mycotic diseases. Chicago: Year Book, 1980.
39. CN Powers. Diagnosis of infectious diseases: a cytopathologist's perspective. Clin Microbiol Rev 11:341–365, 1998.
40. RD Richarson, DW Warnock. Fungal Infection. Diagnosis and Management. 2nd ed. Malden, MA: Blackwell Science, 1997.
41. HR Buckley, MD Richardson, EG Evans, LJ Wheat. Immunodiagnosis of invasive fungal infection. J Med Vet Mycol 30(suppl 1):249–260, 1992.
42. L Kaufman. Laboratory methods for the diagnosis and confirmation of systemic mycoses. Clin Infect Dis 14(suppl 1):S23–S29, 1992.
43. P Martino, C Girmenia. Diagnosis and treatment of invasive fungal infections in cancer patients. Support Care Cancer 1:240–244, 1993.
44. L de Repentigny. Serodiagnosis of candidiasis, aspergillosis, and cryptococcosis. Clin Infect Dis 14(suppl 1):S11–22, 1992.
45. CJ Morrison, SF Hurst, SL Bragg, RJ Kuykendall, H Diaz, DW McLaughlin, E Reiss. Purification and characterization of the extracellular aspartyl proteinase of *Candida albicans*: removal of extraneous proteins and cell wall mannoprotein and evidence for lack of glycosylation. J Gen Microbiol 139:1177–1186, 1993.

46. RM Zancope-Oliveira, SL Bragg, E Reiss, JM Peralta. Immunochemical analysis of the H and M glycoproteins from *Histoplasma capsulatum*. Clin Diagn Lab Immunol 1:563–568, 1994.

47. RM Zancope-Oliveira, E Reiss, TJ Lott, LW Mayer, GS Deepe Jr. Molecular cloning, characterization, and expression of the M antigen of *Histoplasma capsulatum*. Infect Immun 67:1947–1953, 1999.

48. BS Klein, JM Jones. Isolation, purification, and radiolabeling of a novel 120-kD surface protein on *Blastomyces dermatitidis* yeasts to detect antibody in infected patients. J Clin Invest 85:152–161, 1990.

49. BS Klein, JM Jones. Purification and characterization of the major antigen WI-1 from *Blastomyces dermatitidis* yeasts and immunological comparison with A antigen. Infect Immun 62:3890–3900, 1994.

50. JG McEwen, BL Ortiz, AM Garcia, AM Florez, S Botero, A Restrepo. Molecular cloning, nucleotide sequencing, and characterization of a 27-kDa antigenic protein from *Paracoccidioides brasiliensis*. Fungal Genet Biol 20:125–131, 1996.

51. PS Cisalpino, R Puccia, LM Yamauchi, Ml Cano, JF da Silveira, LR Travassos. Cloning, characterization, and epitope expression of the major diagnostic antigen of *Paracoccidioides brasiliensis*. J Biol Chem 271:4553–4560, 1996.

52. P Sundstrom, GR Aliaga. Molecular cloning of cDNA and analysis of protein secondary structure of *Candida albicans* enolase, an abundant, immunodominant glycolytic enzyme. J Bacteriol 174: 6789–6799, 1992.

53. TF Patterson, P Miniter, JE Patterson, JM Rappeport, VT Andriole. *Aspergillus* antigen detection in the diagnosis of invasive aspergillosis. J Infect Dis 171:1553–1558, 1995.

54. TR Rogers, KA Haynes, RA Barnes. Value of antigen detection in predicting invasive pulmonary aspergillosis. Lancet 336:1210–1213, 1990.

55. L Kaufman, PG Standard, M Jalbert, DE Kraft. Immunohistologic identification of *Aspergillus* spp. and other hyaline fungi by using polyclonal fluorescent antibodies. J Clin Microbiol 35:2206–2209, 1997.

56. L Kaufman, PG Standard. Specific and rapid identification of medically important fungi by exoantigen detection. Annu Rev Microbiol 41:209–225, 1987.

57. AE Wakefield, SE Peters, S Banerji, PD Bridge, GS Hall, DL Hawksworth, LA Guiver, AG Allen, JM Hopkin. *Pneumocystis carinii* shows DNA homology with the ustomycetous red yeast fungi. Mol Microbiol 6:1903–1911, 1992.

58. AE Wakefield, JR Stringer, E Tamburrini, E Dei-Cas. Genetics, metabolism and host specificity of *Pneumocystis carinii*. Med Mycol 36(suppl 1):183–193, 1998.

59. FJ Pixley, AE Wakefield, S Banerji, JM Hopkin. Mitochondrial gene sequences show fungal homology for *Pneumocystis carinii*. Mol Microbiol 5:1347–1351, 1991.

60. JC Edman, JA Kovacs, H Masur, DV Santi, HJ Elwood, ML Sogin. Ribosomal RNA sequence shows *Pneumocystis carinii* to be a member of the fungi. Nature 334:519–522, 1988.

61. A Wilkin, J Feinberg. *Pneumocystis carinii* pneumonia: a clinical review. Am Fam Physician 60: 1699–1708, 1713.

62. CF Thomas Jr, AH Limper. *Pneumocystis* pneumonia: clinical presentation and diagnosis in patients with and without acquired immune deficiency syndrome. Semin Respir Infect 13:289–295, 1998.

63. WT Hughes. Current issues in the epidemiology, transmission, and reactivation of *Pneumocystis carinii*. Semin Respir Infect 13:283–288, 1998.

64. DM Kroe, CM Kirsch, WA Jensen. Diagnostic strategies for *Pneumocystis carinii* pneumonia. Semin Respir Infect 12:70–78, 1997.

65. JA Fishman. Treatment of infection due to *Pneumocystis carinii*. Antimicrob Agents Chemother 42:1309–1314, 1998.

66. JR Perfect, WA Schell. The new fungal opportunists are coming. Clin Infect Dis 22(suppl 2): S112–S118, 1996.

67. R Ruchel, U Reichard. Pathogenesis and clinical presentation of aspergillosis. Contrib Microbiol 2:21–43, 1999.

68. G Bodey, B Bueltmann, W Duguid, D Gibbs, H Hanak, M Hotchi, G Mall, P Martino, F Meunier, S Milliken. Fungal infections in cancer patients: an international autopsy survey. Eur J Clin Microbiol Infect Dis 11:99–109, 1992.

69. JP Burnie. Developments in the serological diagnosis of opportunistic fungal infections. J Antimicrob Chemother 28(suppl A):23–33, 1991.

70. A Wald, W Leisenring, JA van Burik, RA Bowden. Epidemiology of *Aspergillus* infections in a large cohort of patients undergoing bone marrow transplantation. J Infect Dis 175:1459–1466, 1997.

71. CH Debusk, R Daoud, MC Thirumoorthi, FM Wilson, R Khatib. Candidemia: current epidemiologic characteristics and a long-term follow-up of the survivors. Scand J Infect Dis 26:697–703, 1994.

72. AH Groll, PM Shah, C Mentzel, M Schneider, G Just-Nuebling, K Huebner. Trends in the postmortem epidemiology of invasive fungal infections at a university hospital. J Infect 33:23–32, 1996.

73. JL Goodman, DJ Winston, RA Greenfield, PH Chandrasekar, B Fox, H Kaizer, RK Shadduck, TC Shea, P Stiff, DJ Friedman. A controlled trial of fluconazole to prevent fungal infections in patients undergoing bone marrow transplantation. N Engl J Med 326:845–851, 1992.

74. L Kaiser, T Huguenin, PD Lew, B Chapuis, D Pittet. Invasive aspergillosis. Clinical features of 35 proven cases at a single institution. Medicine (Baltimore) 77:188–194, 1998.

75. C Pannuti, R Gingrich, MA Pfaller, C Kao, RP Wenzel. Nosocomial pneumonia in patients having bone marrow transplant. Attributable mortality and risk factors. Cancer 69:2653–2662, 1992.

76. P Saugier-Veber, A Devergie, A Sulahian, P Ribaud, F Traore, H Bourdeau-Esperou, E Gluckman, F Derouin. Epidemiology and diagnosis of invasive pulmonary aspergillosis in bone marrow transplant patients: results of a 5 year retrospective study. Bone Marrow Transplant 12:121–124, 1993.

77. DW Denning, DA Stevens. Antifungal and surgical treatment of invasive aspergillosis: review of 2,121 published cases [published erratum appears in Rev Infect Dis 1991;13(2):345]. Rev Infect Dis 12:1147–1201, 1990.

78. OP Sharma, R Chwogule. Many faces of pulmonary aspergillosis. Eur Respir J 12:705–715, 1998.

79. J Salonen, J Nikoskelainen. Lethal infections in patients with hematological malignancies. Eur J Haematol 51:102–108, 1993.

80. RJ Sherertz, A Belani, BS Kramer, GJ Elfenbein, RS Weiner, ML Sullivan, RG Thomas, GP Samsa. Impact of air filtration on nosocomial *Aspergillus* infections. Unique risk of bone marrow transplant recipients. Am J Med 83:709–718, 1987.

81. JE Patterson, A Zidouh, P Miniter, VT Andriole, TF Patterson. Hospital epidemiologic surveillance for invasive aspergillosis: patient demographics and the utility of antigen detection. Infect Control Hosp Epidemiol 18:104–108, 1997.

82. JR Wingard, SU Beals, GW Santos, WG Merz, R Saral. *Aspergillus* infections in bone marrow transplant recipients. Bone Marrow Transplant 2:175–181, 1987.

83. DJ Addrizzo-Harris, TJ Harkin, G McGuinness, DP Naidich, WN Rom. Pulmonary aspergilloma and AIDS. A comparison of HIV-infected and HIV-negative individuals. Chest 111:612–618, 1997.

84. O Lortholary, MC Meyohas, B Dupont, J Cadranel, D Salmon-Ceron, D Peyramond, D Simonin. Invasive aspergillosis in patients with acquired immunodeficiency syndrome: report of 33 cases. French Cooperative Study Group on Aspergillosis in AIDS. Am J Med 95:177–187, 1993.

85. JM Wallace, R Lim, BL Browdy, PC Hopewell, J Glassroth, MJ Rosen, LB Reichman, PA Kvale. Risk factors and outcomes associated with identification of *Aspergillus* in respiratory specimens from persons with HIV disease. Pulmonary Complications of HIV Infection Study Group. Chest 114:131–137, 1998.

86. DW Denning, SE Follansbee, M Scolaro, S Norris, H Edelstein, DA Stevens. Pulmonary aspergillosis in the acquired immunodeficiency syndrome. N Engl J Med 324:654–662, 1991.

87. GY Minamoto, TF Barlam, NJ Vander Els. Invasive aspergillosis in patients with AIDS. Clin Infect Dis 14:66–74, 1992.

88. N Singh, VL Yu, JD Rihs. Invasive aspergillosis in AIDS. South Med J 84:822–827, 1991.

89. EJ Bow, R Loewen, MS Cheang, B Schacter. Invasive fungal disease in adults undergoing remission-induction therapy for acute myeloid leukemia: the pathogenetic role of the antileukemic regimen. Clin Infect Dis 21:361–369, 1995.

90. R Hamadeh, A Ardehali, RM Locksley, MK York. Fatal aspergillosis associated with smoking contaminated marijuana, in a marrow transplant recipient. Chest 94:432–433, 1988.

91. DW Denning. Invasive aspergillosis. Clin Infect Dis 26:781–803, 1998.

92. G Maartens, MJ Wood. The clinical presentation and diagnosis of invasive fungal infections. J Antimicrob Chemother 28(suppl A):13–22, 1991.

93. S Schwartz, E Thiel. Clinical presentation of invasive aspergillosis. Mycoses 40(suppl 2):21–24, 1997.

94. GP Bodey, S Vartivarian. Aspergillosis. Eur J Clin Microbiol Infect Dis 8:413–437, 1989.

95. JS Salaki, DB Louria, H Chmel. Fungal and yeast infections of the central nervous system. A clinical review. Medicine (Baltimore) 63:108–132, 1984.

96. BD Fisher, D Armstrong, B Yu, JW Gold. Invasive aspergillosis. Progress in early diagnosis and treatment. Am J Med 71:571–577, 1981.

97. PA Herbert, AS Bayer. Fungal pneumonia (Part 4): invasive pulmonary aspergillosis. Chest 80: 220–225, 1981.

98. TR Treger, DW Visscher, MS Bartlett, JW Smith. Diagnosis of pulmonary infection caused by *Aspergillus*: usefulness of respiratory cultures. J Infect Dis 152:572–576, 1985.

99. E Manso, M Montillo, G De Sio, S D'Amico, G Discepoli, P Leoni. Value of antigen and antibody detection in the serological diagnosis of invasive aspergillosis in patients with hematological malignancies. Eur J Clin Microbiol Infect Dis 13:756–760, 1994.

100. R Kappe, A Schulze-Berge, HG Sonntag. Evaluation of eight antibody tests and one antigen test for the diagnosis of invasive aspergillosis. Mycoses 39:13–23, 1996.

101. R Kappe, HP Seeliger. Serodiagnosis of deep-seated fungal infections. Curr Top Med Mycol 5: 247–280, 1993.

102. JP Latge. Tools and trends in the detection of *Aspergillus fumigatus*. Curr Top Med Mycol 6: 245–281, 1995.

103. JP Latge, M Moutaouakil, JP Debeaupuis, JP Bouchara, K Haynes, MC Prevost. The 18-kilodalton antigen secreted by *Aspergillus fumigatus*. Infect Immun 59:2586–2594, 1991.

104. J Brouwer. Detection of antibodies against *Aspergillus fumigatus*: comparison between double immunodiffusion, ELISA and immunoblot analysis. Int Arch Allergy Appl Immunol 85:244–249, 1988.

105. RM Coleman, L Kaufman. Use of the immunodiffusion test in the serodiagnosis of aspergillosis. Appl Microbiol 23:301–308, 1972.

106. RC Young, JE Bennett. Invasive aspergillosis. Absence of detectable antibody response. Am Rev Respir Dis 104:710–716, 1971.

107. JC Schaefer, B Yu, D Armstrong. An aspergillus immunodiffusion test in the early diagnosis of aspergillosis in adult leukemia patients. Am Rev Respir Dis 113:325–329, 1976.

108. JF Tomee, GP Mannes, W van der Bij, TS van der Werf, WJ de Boer, GH Koeter, HF Kauffman. Serodiagnosis and monitoring of *Aspergillus* infections after lung transplantation. Ann Intern Med 125:197–201, 1996.

109. GE Westney, S Kesten, A De Hoyos, C Chapparro, T Winton, JR Maurer. *Aspergillus* infection in single and double lung transplant recipients. Transplantation 61:915–919, 1996.

110. JH Froudist, GB Harnett, R McAleer. Comparison of immunodiffusion and enzyme linked immunosorbent assay for antibodies to four *Aspergillus* species. J Clin Pathol 42:1215–1221, 1989.

111. VM Hearn, GC Donaldson, MJ Healy. A method to determine significant levels of immunoglobulin G to *Aspergillus fumigatus* antigens in an ELISA system and a comparison with counterimmunoelectrophoresis and double diffusion techniques. J Immunoassay 6:137–158, 1985.

112. K Holmberg, M Berdischewsky, LS Young. Serologic immunodiagnosis of of invasive aspergillosis. J Infect Dis 141:656–664, 1980.

113. SK Mishra, S Falkenberg, KN Masihi. Efficacy of enzyme-linked immunosorbent assay in serodiagnosis of aspergillosis. J Clin Microbiol 17:708–710, 1983.

114. AK Trull, J Parker, RE Warren. IgG enzyme linked immunosorbent assay for diagnosis of invasive aspergillosis: retrospective study over 15 years of transplant recipients. J Clin Pathol 38:1045–1051, 1985.

115. R Marier, W Smith, MJansen, VT Andriole. A solid-phase radioimmunoassay for the measurement of antibody to *Aspergillus* in invasive aspergillosis. J Infect Dis 140:771–779, 1979.

116. F Persat, M Gari-Toussaint, B Lebeau, M Cambon, H Raberin, A Addo, S Picot, MA Piens, A Blancard, M Mallie, JM Bastide, R Grillot. Specific antibody detection in human aspergillosis: a GEMO multicentre evaluation of a rapid immunoelectrophoresis method (Paragon). Group d'Etude des Mycoses Opportunistes. Mycoses 39:427–432, 1996.

117. JP Burnie, RC Matthews. Heat shock protein 88 and *Aspergillus* infection. J Clin Microbiol 29: 2099–2106, 1991.

118. L de Repentigny, E Kilanowski, L Pedneault, M Boushira. Immunoblot analyses of the serologic response to *Aspergillus fumigatus* antigens in experimental invasive aspergillosis. J Infect Dis 163: 1305–1311, 1991.

119. PM Fratamico, HR Buckley. Identification and characterization of an immunodominant 58-kilodalton antigen of *Aspergillus fumigatus* recognized by sera of patients with invasive aspergillosis. Infect Immun 59:309–315, 1991.

120. R Matthews, JP Burnie, A Fox, S Tabaqchali. Immunoblot analysis of serological responses in invasive aspergillosis. J Clin Pathol 38:1300–1303, 1985.

121. M Monod, G Togni, L Rahalison, E Frenk. Isolation and characterisation of an extracellular alkaline protease of *Aspergillus fumigatus*. J Med Microbiol 35:23–28, 1991.

122. U Reichard, S Buttner, H Eiffert, F Staib, R Ruchel. Purification and characterisation of an extracellular serine proteinase from *Aspergillus fumigatus* and its detection in tissue. J Med Microbiol 33: 243–251, 1990.

123. H Schonheyder, P Andersen, JC Petersen. Rapid immunoelectrophoretic assay for detection of serum antibodies to *Aspergillus fumigatus* catalase in patients with pulmonary aspergillosis. Eur J Clin Microbiol 4:299–303, 1985.

124. RC Matthews, JP Burnie, S Tabaqchali. Isolation of immunodominant antigens from sera of patients with systemic candidiasis and characterization of serological response to *Candida albicans*. J Clin Microbiol 25:230–237, 1987.

125. B Dupont, M Huber, SJ Kim, JE Bennett. Galactomannan antigenemia and antigenuria in aspergillosis: studies in patients and experimentally infected rabbits. J Infect Dis 155:1–11, 1987.

126. SF Hurst, GH Reyes, DW McLaughlin, E Reiss, CJ Morrison. Comparison of commercial latex agglutination and sandwich enzyme immunoassays with a competitive binding inhibition enzyme immunoassay for detection of antigenemia and antigenuria in a rabbit model of invasive aspergillosis. Clin Diagn Lab Immunol 7:477–485, 2000.

127. PF Lehmann, E Reiss. Invasive aspergillosis: antiserum for circulating antigen produced after immunization with serum from infected rabbits. Infect Immun 20:570–572, 1978.

128. PE Verweij, AJ Rijs, BE De Pauw, AM Horrevorts, JA Hoogkamp-Korstanje, JF Meis. Clinical evaluation and reproducibility of the Pastorex *Aspergillus* antigen latex agglutination test for diagnosing invasive aspergillosis. J Clin Pathol 48:474–476, 1995.

129. MH Weiner, M Coats-Stephen. Immunodiagnosis of systemic aspergillosis. I. Antigenemia detected by radioimmunoassay in experimental infection. J Lab Clin Med 93:111–119, 1979.

130. E Reiss, PF Lehmann. Galactomannan antigenemia in invasive aspergillosis. Infect Immun 25: 357–365, 1979.

131. PJ Shaffer, G Medoff, GS Kobayashi. Demonstration of antigenemia by radioimmunoassay in rabbits experimentally infected with *Aspergillus*. J Infect Dis 139:313–319, 1979.

132. PJ Shaffer, GS Kobayashi, G Medoff. Demonstration of antigenemia in patients with invasive aspergillosis by solid phase (protein A-rich *Staphylococcus aureus*) radioimmunoassay. Am J Med 67:627–630, 1979.

133. J Van Cutsem, L Meulemans, F Van Gerven, D Stynen. Detection of circulating galactomannan by Pastorex *Aspergillus* in experimental invasive aspergillosis. Mycoses 33:61–69, 1990.

134. JR Sabetta, P Miniter, VT Andriole. The diagnosis of invasive aspergillosis by an enzyme-linked immunosorbent assay for circulating antigen. J Infect Dis 152:946–953, 1985.

135. L de Repentigny, M Boushira, L Ste-Marie, G Bosisio. Detection of galactomannan antigenemia by enzyme immunoassay in experimental invasive aspergillosis. J Clin Microbiol 25:863–867, 1987.

136. EV Wilson, VM Hearn, DW Mackenzie. Evaluation of a test to detect circulating *Aspergillus fumigatus* antigen in a survey of immunocompromised patients with proven or suspected invasive disease. J Med Vet Mycol 25:365–375, 1987.

137. PE Verweij, D Stynen, AJ Rijs, BE de Pauw, JA Hoogkamp-Korstanje, JF Meis. Sandwich enzyme-linked immunosorbent assay compared with Pastorex latex agglutination test for diagnosing invasive aspergillosis in immunocompromised patients. J Clin Microbiol 33:1912–1914, 1995.

138. GH Talbot, MH Weiner, SL Gerson, M Provencher, S Hurwitz. Serodiagnosis of invasive aspergillosis in patients with hematologic malignancy: validation of the *Aspergillus fumigatus* antigen radioimmunoassay. J Infect Dis 155:12–27, 1987.

139. CP Andrews, MH Weiner. *Aspergillus* antigen detection in bronchoalveolar lavage fluid from patients with invasive aspergillosis and aspergillomas. Am J Med 73:372–380, 1982.

140. JE Bennett, MM Friedman, B Dupont. Receptor-mediated clearance of *Aspergillus* galactomannan. J Infect Dis 155:1005–1010, 1987.

141. R Kappe, A Schulze-Berge. New cause for false-positive results with the Pastorex *Aspergillus* antigen latex agglutination test. J Clin Microbiol 31:2489–2490, 1993.

142. D Stynen, J Sarfati, A Goris, MC Prevost, M Lesourd, H Kamphuis, V Darras, JP Latge. Rat monoclonal antibodies against *Aspergillus* galactomannan. Infect Immun 60:2237–2245, 1992.

143. K Haynes, TR Rogers. Retrospective evaluation of a latex agglutination test for diagnosis of invasive aspergillosis in immunocompromised patients. Eur J Clin Microbiol Infect Dis 13:670–674, 1994.

144. DW Warnock, AB Foot, EM Johnson, SB Mitchell, JM Cornish, A Oakhill. *Aspergillus* antigen latex test for diagnosis of invasive aspergillosis [letter]. Lancet 338:1023–1024, 1991.

145. P Francis, JW Lee, A Hoffman, J Peter, A Francesconi, J Bacher, J Shelhamer, PA Pizzo, TJ Walsh. Efficacy of unilamellar liposomal amphotericin B in treatment of pulmonary aspergillosis in persistently granulocytopenic rabbits: the potential role of bronchoalveolar D-mannitol and serum galactomannan as markers of infection. J Infect Dis 169:356–368, 1994.

146. KA Haynes, JP Latge, TR Rogers. Detection of *Aspergillus* antigens associated with invasive infection. J Clin Microbiol 28:2040–2044, 1990.

147. D Stynen, A Goris, J Sarfati, JP Latge. A new sensitive sandwich enzyme-linked immunosorbent assay to detect galactofuran in patients with invasive aspergillosis. J Clin Microbiol 33:497–500, 1995.

148. A Sulahian, M Tabouret, P Ribaud, J Sarfati, E Gluckman, JP Latge, F Derouin. Comparison of an enzyme immunoassay and latex agglutination test for detection of galactomannan in the diagnosis of invasive aspergillosis. Eur J Clin Microbiol Infect Dis 15:139–145, 1996.

149. P Rohrlich, J Sarfati, P Mariani, M Duval, A Carol, C Saint-Martin, E Bingen, JP Latge, E Vilmer. Prospective sandwich enzyme-linked immunosorbent assay for serum galactomannan: early predictive value and clinical use in invasive aspergillosis. Pediatr Infect Dis J 15:232–237, 1996.

150. PE Verweij, JP Latge, AJ Rijs, WJ Melchers, BE De Pauw, JA Hoogkamp-Korstanje, JF Meis. Comparison of antigen detection and PCR assay using bronchoalveolar lavage fluid for diagnosing invasive pulmonary aspergillosis in patients receiving treatment for hematological malignancies. J Clin Microbiol 33:3150–3153, 1995.

151. CM Swanink, JF Meis, AJ Rijs, JP Donnelly, PE Verweij. Specificity of a sandwich enzyme-linked immunosorbent assay for detecting *Aspergillus* galactomannan. J Clin Microbiol 35:257–260, 1997.

152. K Hashiguchi, Y Niki, R Soejima. Cyclophosphamide induces false-positive results in detection of aspergillus antigen in urine [letter]. Chest 105:975–976, 1994.

153. AJ Hamilton, MD Holdom, RJ Hay. Specific recognition of purified Cu,Zn superoxide dismutase from *Aspergillus fumigatus* by immune human sera. J Clin Microbiol 33:495–496, 1995.

154. MD Holdom, RJ Hay, AJ Hamilton. Purification, N-terminal amino acid sequence and partial characterization of a Cu,Zn superoxide dismutase from the pathogenic fungus *Aspergillus fumigatus*. Free Radic Res 22:519–531, 1995.

155. LK Arruda, TA Platts-Mills, JW Fox, MD Chapman. *Aspergillus fumigatus* allergen I, a major IgE-binding protein, is a member of the mitogillin family of cytotoxins. J Exp Med 172:1529–1532, 1990.

156. TT Brandhorst, WR Kenealy. Production and localization of restrictocin in *Aspergillus restrictus*. J Gen Microbiol 138:1429–1435, 1992.

157. B Lamy, J Davies. Isolation and nucleotide sequence of the *Aspergillus restrictus* gene coding for the ribonucleolytic toxin restrictocin and its expression in *Aspergillus nidulans*: the leader sequence protects producing strains from suicide. Nucleic Acids Res 19:1001–1006, 1991.

158. B Lamy, M Moutaouakil, JP Latge, J Davies. Secretion of a potential virulence factor, a fungal ribonucleotoxin, during human aspergillosis infections. Mol Microbiol 5:1811–1815, 1991.

159. P Phillips, G Radigan. Antigenemia in a rabbit model of invasive aspergillosis. J Infect Dis 159: 1147–1150, 1989.

160. B Yu, Y Niki, D Armstrong. Use of immunoblotting to detect *Aspergillus fumigatus* antigen in sera and urines of rats with experimental invasive aspergillosis. J Clin Microbiol 28:1575–1579, 1990.

161. JP Burnie, RC Matthews, I Clark, LJ Milne. Immunoblot fingerprinting *Aspergillus fumigatus*. J Immunol Methods 118:179–186, 1989.

162. MH Nguyen, JE Peacock Jr, AJ Morris, DC Tanner, ML Nguyen, DR Snydman, MM Wagener, MG Rinaldi, VL Yu. The changing face of candidemia: emergence of non–*Candida albicans* species and antifungal resistance. Am J Med 100:617–623, 1996.

163. MA Pfaller, RN Jones, GV Doem, HS Sader, SA Messer, A Houston, S Coffman, RJ Hollis. Bloodstream infections due to *Candida* species: SENTRY antimicrobial surveillance program in North America and Latin America, 1997–1998. Antimicrob Agents Chemother 44:747–751, 2000.

164. JR Wingard, WG Merz, MG Rinaldi, TR Johnson, JE Karp, R Saral. Increase in *Candida krusei* infection among patients with bone marrow transplantation and neutropenia treated prophylactically with fluconazole. N Engl J Med 325:1274–1277, 1991.

165. JR Wingard. Infections due to resistant *Candida* species in patients with cancer who are receiving chemotherapy. Clin Infect Dis 19(suppl 1)S49–S53, 1994.

166. CA Hitchcock, GW Pye, PF Troke, EM Johnson, DW Warnock. Fluconazole resistance in *Candida glabrata*. Antimicrob Agents Chemother 37:1962–1965, 1993.

167. JR Wingard, WG Merz, MG Rinaldi, CB Miller, JE Karp, R Saral. Association of *Torulopsis glabrata* infections with fluconazole prophylaxis in neutropenic bone marrow transplant patients. Antimicrob Agents Chemother 37:1847–1849, 1993.

168. JP Burnie, FC Odds, W Lee, C Webster, JD Williams. Outbreak of systemic *Candida albicans* in intensive care unit caused by cross infection. Br Med J (Clin Res Ed) 290:746–748, 1985.

169. V Sanchez, JA Vazquez, D Barth-Jones, L Dembry, JD Sobel, MJ Zervos. Nosocomial acquisition of *Candida parapsilosis*: an epidemiologic study. Am J Med 94:577–582, 1993.

170. RA Greenfield, JM Jones. Purification and characterization of a major cytoplasmic antigen of *Candida albicans*. Infect Immun 34:469–477, 1981.

171. RA Greenfield, MJ Bussey, JL Stephens, JM Jones. Serial enzyme-linked immunosorbent assays for antibody to *Candida* antigens during induction chemotherapy for acute leukemia. J Infect Dis 148:275–283, 1983.

172. JM Jones. Kinetics of antibody responses to cell wall mannan and a major cytoplasmic antigen of *Candida albicans* in rabbits and humans. J Lab Clin Med 96:845–860, 1980.

173. JM Jones. Quantitation of antibody against cell wall mannan and a major cytoplasmic antigen of *Candida* in rabbits, mice, and humans. Infect Immun 30:78–89, 1980.

174. S Fujita, F Matsubara, T Matsuda. Enzyme-linked immunosorbent assay measurement of fluctuations in antibody titer and antigenemia in cancer patients with and without candidiasis. J Clin Microbiol 23:568–575, 1986.

175. F Staib. Serum proteins as nitrogen source for yeastlike fungi. Sabouraudia 4:187–193, 1965.

176. KJ Kwon-Chung, D Lehman, C Good, PT Magee. Genetic evidence for role of extracellular proteinase in virulence of *Candida albicans*. Infect Immun 49:571–575, 1985.

177. F Macdonald, FC Odds. Virulence for mice of a proteinase-secreting strain of *Candida albicans* and a proteinase-deficient mutant. J Gen Microbiol 129:431–438, 1983.

178. B Hube. *Candida albicans* secreted aspartyl proteinases. Curr Top Med Mycol 7:55–69, 1996.

179. F Macdonald, FC Odds. Inducible proteinase of *Candida albicans* in diagnostic serology and in the pathogenesis of systemic candidosis. J Med Microbiol 13:423–435, 1980.

180. TL Ray, CD Payne. Scanning electron microscopy of epidermal adherence and cavitation in murine candidiasis: a role for *Candida* acid proteinase. Infect Immun 56:1942–1949, 1988.

181. F de Bernardis, L Agatensi, IK Ross, GW Emerson, R Lorenzini, PA Sullivan, A Cassone. Evidence for a role for secreted aspartate proteinase of *Candida albicans* in vulvovaginal candidiasis. J Infect Dis 161:1276–1283, 1990.

182. R Ruchel, B Boning-Stutzer, A Mari. A synoptical approach to the diagnosis of candidosis, relying on serological antigen and antibody tests, on culture, and on evaluation of clinical data. Mycoses 31:87–106, 1988.

183. R Ruchel, B Boning. Detection of *Candida* proteinase by enzyme immunoassay and interaction of the enzyme with alpha-2-macroglobulin. J Immunol Methods 61:107–116, 1983.

184. JW Bailey, E Sada, C Brass, JE Bennett. Diagnosis of systemic candidiasis by latex agglutination for serum antigen. J Clin Microbiol 21:749–752, 1985.

185. C Lemieux, G St-Germain, J Vincelette, L Kaufman, L de Repentigny. Collaborative evaluation of antigen detection by a commercial latex agglutination test and enzyme immunoassay in the diagnosis of invasive candidiasis. J Clin Microbiol 28:249–253, 1990.

186. ME Bougnoux, C Hill, D Moissenet, M Feuilhade de Chauvin, M Bonnay, I Vicens-Sprauel, F Pietri, M McNeil, L Kaufman, J Dupouy-Camet. Comparison of antibody, antigen, and metabolite assays for hospitalized patients with disseminated or peripheral candidiasis. J Clin Microbiol 28:905–909, 1990.

187. L de Repentigny, RJ Kuykendall, FW Chandler, JR Broderson, E Reiss. Comparison of serum mannan, arabinitol, and mannose in experimental disseminated candidiasis. J Clin Microbiol 19:804–812, 1984.

188. S Fujita, T Hashimoto. Detection of serum *Candida* antigens by enzyme-linked immunosorbent assay and a latex agglutination test with anti–*Candida albicans* and anti–*Candida krusei* antibodies. J Clin Microbiol 30:3132–3137, 1992.

189. FW Kahn, JM Jones. Latex agglutination tests for detection of *Candida* antigens in sera of patients with invasive candidiasis. J Infect Dis 153:579–585, 1986.

190. R Matthews, J Burnie. Diagnosis of systemic candidiasis by an enzyme-linked dot immunobinding assay for a circulating immunodominant 47-kilodalton antigen. J Clin Microbiol 26:459–463, 1988.

191. GF Araj, RL Hopfer, S Chesnut, V Fainstein, GP Bodey Sr. Diagnostic value of the enzyme-linked immunosorbent assay for detection of *Candida albicans* cytoplasmic antigen in sera of cancer patients. J Clin Microbiol 16:46–52, 1982.

192. P Stevens, S Huang, LS Young, M Berdischewsky. Detection of candida antigenemia in human invasive candidiasis by a new solid phase radioimmunoassay. Infection 8(suppl 3):S334–S338, 1980.

193. NA Strockbine, MT Largen, SM Zweibel, HR Buckley. Identification and molecular weight characterization of antigens from *Candida albicans* that are recognized by human sera. Infect Immun 43:715–721, 1984.

194. AB Mason, ME Brandt, HR Buckley. Enolase activity associated with a *C. albicans* cytoplasmic antigen. Yeast 5(Spec No):S231–S239, 1989.

195. NA Strockbine, MT Largen, HR Buckley. Production and characterization of three monoclonal antibodies to *Candida albicans* proteins. Infect Immun 43:1012–1018, 1984.

196. J Burnie. A reverse passive latex agglutination test for the diagnosis of systemic candidosis. J Immunol Methods 82:267–280, 1985.

197. TJ Walsh, JW Hathorn, JD Sobel, WG Merz, V Sanchez, SM Maret, HR Buckley, MA Pfaller, R Schaufele, C Sliva. Detection of circulating *Candida* enolase by immunoassay in patients with cancer and invasive candidiasis. N Engl J Med 324:1026–1031, 1991.

198. LO Gentry, ID Wilkinson, AS Lea, MF Price. Latex agglutination test for detection of *Candida* antigen in patients with disseminated disease. Eur J Clin Microbiol 2:122–128, 1983.

199. JP Burnie, JD Williams. Evaluation of the Ramco latex agglutination test in the early diagnosis of systemic candidiasis. Eur J Clin Microbiol 4:98–101, 1985.

200. JC Fung, ST Donta, RC Tilton. *Candida* detection system (CAND-TEC) to differentiate between *Candida albicans* colonization and disease. J Clin Microbiol 24:542–547, 1986.

201. P Phillips, A Dowd, P Jewesson, G Radigan, MG Tweeddale, A Clarke, I Geere, M Kelly. Nonvalue of antigen detection immunoassays for diagnosis of candidemia. J Clin Microbiol 28:2320–2326, 1990.

202. MF Price, LO Gentry. Incidence and significance of *Candida* antigen in low-risk and high-risk patient populations. Eur J Clin Microbiol 5:416–419, 1986.

203. RH Eng, E Bishburg, SM Smith, R Kapila. Cryptococcal infections in patients with acquired immune deficiency syndrome. Am J Med 81:19–23, 1986.

204. EJ Bottone, GP Wormser. Capsule-deficient cryptococci in AIDS [letter]. Lancet 2:553, 1985.

205. D Ellis, T Pfeiffer. The ecology of *Cryptococcus neoformans*. Eur J Epidemiol 8:321–325, 1992.

206. T Pfeiffer, D Ellis. Environmental isolation of *Cryptococcus neoformans gattii* from California [letter]. J Infect Dis 163:929–930, 1991.

207. JR Sabetta, VT Andriole. Cryptococcal infection of the central nervous system. Med Clin North Am 69:333–344, 1985.

208. YC Chen, SC Chang, CC Shih, CC Hung, KT Luhbd, YS Pan, WC Hsieh. Clinical features and in vitro susceptibilities of two varieties of *Cryptococcus neoformans* in Taiwan. Diagn Microbiol Infect Dis 36:175–183, 2000.

209. JW van't Wout. Clinical manifestations of systemic fungal infections. Mycoses 31(suppl 2):9–14, 1988.

210. TM Kerkering, RJ Duma, S Shadomy. The evolution of pulmonary cryptococcosis: clinical implications from a study of 41 patients with and without compromising host factors. Ann Intern Med 94:611–616, 1981.

211. RA Thompson. Clinical features of central nervous system fungus infection. Adv Neurol 6:93–100, 1974.

212. WG Powderly, GA Cloud, WE Dismukes, MS Saag. Measurement of cryptococcal antigen in serum and cerebrospinal fluid: value in the management of AIDS-associated cryptococcal meningitis. Clin Infect Dis 18:789–792, 1994.

213. JR Hamilton, A Noble, DW Denning, DA Stevens. Performance of *Cryptococcus* antigen latex agglutination kits on serum and cerebrospinal fluid specimens of AIDS patients before and after pronase treatment. J Clin Microbiol 29:333–339, 1991.

214. AM Stamm, SS Polt. False-negative cryptococcal antigen test. JAMA 244:1359, 1980.

215. L Stockman, GD Roberts. Specificity of the latex test for cryptococcal antigen: a rapid, simple method for eliminating interference factors. J Clin Microbiol 16:965–967, 1982.

216. LD Gray, GD Roberts. Experience with the use of pronase to eliminate interference factors in the latex agglutination test for cryptococcal antigen. J Clin Microbiol 26:2450–2451, 1988.

217. MA Westerink, D Amsterdam, RJ Petell, MN Stram, MA Apicella. Septicemia due to DF-2. Cause of a false-positive cryptococcal latex agglutination result. Am J Med 83:155–158, 1987.

218. RL Hopper, EV Perry, V Fainstein. Diagnostic value of cryptococcal antigen in the cerebrospinal fluid of patients with malignant disease. J Infect Dis 145:915, 1982.

219. EJ McManus, MJ Bozdech, JM Jones. Role of the latex agglutination test for cryptococcal antigen in diagnosing disseminated infections with *Trichosporon beigelii*. J Infect Dis 151:1167–1169, 1985.

220. GP Melcher, KD Reed, MG Rinaldi, JW Lee, PA Pizzo, TJ Walsh. Demonstration of a cell wall antigen cross-reacting with cryptococcal polysaccharide in experimental disseminated trichosporonosis. J Clin Microbiol 29:192–196, 1991.

221. TF Eckert, TR Kozel. Production and characterization of monoclonal antibodies specific for *Cryptococcus neoformans* capsular polysaccharide. Infect Immun 55:1895–1899, 1987.

222. W Gade, SW Hinnefeld, LS Babcock, P Gilligan, W Kelly, K Wait, D Greer, M Pinilla, RL Kaplan. Comparison of the PREMIER cryptococcal antigen enzyme immunoassay and the latex agglutination assay for detection of cryptococcal antigens. J Clin Microbiol 29:1616–1619, 1991.

223. AF DiSalvo. The ecology of *Blastomyces dermatitidis*. In: Y Al-Doory, AP DiSalvo, eds. Blastomycosis. New York: Plenum, 1992, pp 1–7.

224. SF Davies, GA Sarosi. Epidemiological and clinical features of pulmonary blastomycosis. Semin Respir Infect 12:206–218, 1997.

225. Y Al-Doory. Introduction. In: Y Al-Doory, AF DiSalvo, eds. Blastomycosis. New York: Plenum, pp 1–7, 1992.

226. AF DiSalvo. The epidemiology of blastomycosis. In: Y Al-Doory, AF DiSalvo, eds. Blastomycosis. New York: Plenum, 1992, pp 1–7.

227. DL Sewell. Laboratory-associated infections and biosafety. Clin Microbiol Rev 8:389–405, 1995.

228. CK Campbell. Hazards to laboratory staff posed by fungal pathogens. J Hosp Infect 30(suppl): 358–363, 1995.

229. JW Gnann Jr, GS Bressler, CA3d Bodet, CK Avent. Human blastomycosis after a dog bite. Ann Intern Med 98:48–49, 1983.

230. RW Bradsher. Clinical features of blastomycosis. Semin Respir Infect 12:229–234, 1997.

231. PG Pappas. Blastomycosis in the immunocompromised patient. Semin Respir Infect 12:243–251, 1997.

232. RB George, RS Lambert, MJ Bruce, JW Pickering, RM Wolcott. Radioimmunoassay: a sensitive screening test for histoplasmosis and blastomycosis. Am Rev Respir Dis 124:407–410, 1981.

233. N Khansari, D Segre, M Segre. Diagnosis of histoplasmosis and blastomycosis by an antiglobulin hemagglutination test. Am J Vet Res 43:2279–2283, 1982.

234. S Turner, L Kaufman, M Jalbert. Diagnostic assessment of an enzyme-linked immunosorbent assay for human and canine blastomycosis. J Clin Microbiol 23:294–297, 1986.

235. BS Klein, JN Kuritsky, WA Chappell, L Kaufman, J Green, SF Davies, JE Williams, GA Sarosi. Comparison of the enzyme immunoassay, immunodiffusion, and complement fixation tests in detecting antibody in human serum to the A antigen of *Blastomyces dermatitidis*. Am Rev Respir Dis 133:144–148, 1986.

236. BS Klein, JM Vergeront, L Kaufman, RW Bradsher, UN Kumar, G Mathai, B Varkey, JP Davis. Serological tests for blastomycosis: assessments during a large point-source outbreak in Wisconsin. J Infect Dis 155:262–268, 1987.

237. AS Sekhon, L Kaufman, GS Kobayashi, NH Moledina, M Jalbert. The value of the Premier enzyme immunoassay for diagnosing *Blastomyces dermatitidis* infections. J Med Vet Mycol 33:123–125, 1995.

238. RW Bradsher, PG Pappas. Detection of specific antibodies in human blastomycosis by enzyme immunoassay. South Med J 88:1256–1259, 1995.

239. BS Klein, LH Hogan, JM Jones. Immunologic recognition of a 25-amino acid repeat arrayed in tandem on a major antigen of *Blastomyces dermatitidis*. J Clin Invest 92:330–337, 1993.

240. ML Orr, JL Bono, RO Abuodeh, SJ Williams, AM Legendre, GM Scalarone. Comparative stability, sensitivity and specificity studies with different lots of *Blastomyces dermatitidis* yeast and mycelial lysate antigens. Mycoses 37:155–160, 1994.

241. A Wakamoto, BM Fryer, MA Fisher, TJ Johnson, DK Lundgren, JD Knickerbocker, SL Rounds, GM Scalarone. Detection of antibodies and delayed dermal hypersensitivity with different lots of *Blastomyces dermatitidis* yeast lysate antigen: stability and specificity evaluations. Mycoses 40: 303–308, 1997.

242. MA Fisher, AM Legendre, GM Scalarone. Immunological and chemical characterization of glycoproteins in IEF fractions of *Blastomyces dermatitidis* yeast lysate antigen. Mycoses 40:83–90, 1997.

243. JL Bono, AM Legendre, GM Scalarone. Detection of antibodies and delayed hypersensitivity with Rotofor preparative IEF fractions of *Blastomyces dermatitidis* yeast phase lysate antigen. J Med Vet Mycol 33:209–214, 1995.

244. JA Leake, DG Mosley, B England, JV Graham, BD Plikaytis, NM Ampel, BA Perkins, RA Hajjeh. Risk factors for acute symptomatic coccidioidomycosis among elderly persons in Arizona, 1996–1997. J Infect Dis 181:1435–1440, 2000.

245. L Louie, S Ng, R Hajjeh, R Johnson, D Vugia, SB Werner, R Talbot, W Klitz. Influence of host genetics on the severity of coccidioidomycosis. Emerg Infect Dis 5:672–680, 1999.

246. D Pappagianis, BL Zimmer. Serology of coccidioidomycosis. Clin Microbiol Rev 3:247–268, 1990.

247. M Huppert, ET Peterson, SH Sun, PA Chitjian, WJ Derrevere. Evaluation of a latex particle agglutination test for coccidioidomycosis. Am J Clin Pathol 49:96–102, 1968.

248. D Pappagianis, RI Krasnow, S Beall. False-positive reactions of cerebrospinal fluid and diluted sera with the coccidioidal latex-agglutination test. Am J Clin Pathol 66:916–921, 1976.

249. L Kaufman, AS Sekhon, N Moledina, M Jalbert, D Pappagianis. Comparative evaluation of commercial Premier EIA and microimmunodiffusion and complement fixation tests for *Coccidioides immitis* antibodies. J Clin Microbiol 33:618–619, 1995.

250. TB Martins, TD Jaskowski, CL Mouritsen, HR Hill. Comparison of commercially available enzyme immunoassay with traditional serological tests for detection of antibodies to *Coccidioides immitis*. J Clin Microbiol 33:940–943, 1995.

251. M Zartarian, EM Peterson, LM de la Maza. Detection of antibodies to *Coccidioides immitis* by enzyme immunoassay. Am J Clin Pathol 107:148–153, 1997.

252. A Dosanjh, J Theodore, D Pappagianis. Probable false positive coccidioidal serologic results in patients with cystic fibrosis. Pediatr Transplant 2:313–317, 1998.

253. BL Zimmer, D Pappagianis. Characterization of a soluble protein of *Coccidioides immitis* with activity as an immunodiffusion-complement fixation antigen. J Clin Microbiol 26:2250–2256, 1988.

254. SM Johnson, D Pappagianis. The coccidioidal complement fixation and immunodiffusion-complement fixation antigen is a chitinase. Infect Immun 60:2588–2592, 1992.

255. SM Johnson, CR Zimmermann, D Pappagianis. Amino-terminal sequence analysis of the *Coccidioides immitis* chitinase/immunodiffusion-complement fixation protein. Infect Immun 61:3090–3092, 1993.

256. D Kruse, GT Cole. Isolation of tube precipitin antibody-reactive fractions of *Coccidioides immitis*. Infect Immun 58:169–178, 1990.

257. BL Zimmer, D Pappagianis. Immunoaffinity isolation and partial characterization of the *Coccidioides immitis* antigen detected by the tube precipitin and immunodiffusion-tube precipitin tests. J Clin Microbiol 27:1759–1766, 1989.

258. GT Cole, D Kruse, SW Zhu, KR Seshan, RW Wheat. Composition, serologic reactivity, and immunolocalization of a 120-kilodalton tube precipitin antigen of *Coccidioides immitis*. Infect Immun 58: 179–188, 1990.

259. C Yang, Y Zhu, DM Magee, RA Cox. Molecular cloning and characterization of the *Coccidioides immitis* complement fixation/chitinase antigen. Infect Immun 64:1992–1997, 1996.

260. CR Zimmermann, SM Johnson, GW Martens, AG White, D Pappagianis. Cloning and expression of the complement fixation antigen-chitinase of *Coccidioides immitis*. Infect Immun 64:4967–4975, 1996.

261. SM Johnson, CR Zimmermann, D Pappagianis. Use of a recombinant *Coccidioides immitis* complement fixation antigen-chitinase in conventional serological assays. J Clin Microbiol 34:3160–3164, 1996.

262. MC Yang, DM Magee, RA Cox. Mapping of a *Coccidioides immitis*–specific epitope that reacts with complement-fixing antibody. Infect Immun 65:4068–4074, 1997.

263. MC Yang, DM Magee, L Kaufman, Y Zhu, RA Cox. Recombinant *Coccidioides immitis* complement-fixing antigen: detection of an epitope shared by *C. immitis, Histoplasma capsulatum*, and *Blastomyces dermatitidis*. Clin Diagn Lab Immunol 4:19–22, 1997.

264. JN Galgiani, SH Sun, KO Dugger, NM Ampel, GG Grace, J Harrison, MA Wieden. An arthroconidial-spherule antigen of *Coccidioides immitis*: differential expression during in vitro fungal development and evidence for humoral response in humans after infection or vaccination. Infect Immun 60:2627–2635, 1992.

265. JN Galgiani, T Peng, ML Lewis, GA Cloud, D Pappagianis. Cerebrospinal fluid antibodies detected by ELISA against a 33-kDa antigen from spherules of *Coccidioides immitis* in patients with coccidioidal meningitis. National Institute of Allergy and Infectious Diseases Mycoses Study Group. J Infect Dis 173:499–502, 1996.

266. AJ Hamilton, MA Bartholomew, LE Fenelon, J Figueroa, RJ Hay. A murine monoclonal antibody exhibiting high species specificity for *Histoplasma capsulatum* var. *capsulatum*. J Gen Microbiol 136:331–335, 1990.

267. AJ Hamilton, MA Bartholomew, L Fenelon, J Figueroa, RJ Hay. Preparation of monoclonal antibodies that differentiate between *Histoplasma capsulatum* variant *capsulatum* and *H. capsulatum* variant *duboisii*. Trans R Soc Trop Med Hyg 84:425–428, 1990.

268. J Bille, L Stockman, GD Roberts, CD Horstmeier, DM Ilstrup. Evaluation of a lysis-centrifugation system for recovery of yeasts and filamentous fungi from blood. J Clin Microbiol 18:469–471, 1983.

269. J Bille, RS Edson, GD Roberts. Clinical evaluation of the lysis-centrifugation blood culture system for the detection of fungemia and comparison with a conventional biphasic broth blood culture system. J Clin Microbiol 19:126–128, 1984.

270. CV Paya, GD Roberts, FRD Cockerill. Laboratory methods for the diagnosis of disseminated histo-plasmosis: clinical importance of the lysis-centrifugation blood culture technique. Mayo Clin Proc 62:480–485, 1987.

271. MG Hove, GL Woods. Duration of fungal culture incubation in an area endemic for *Histoplasma capsulatum*. Diagn Microbiol Infect Dis 28:41–43, 1997.

272. AA Padhye, G Smith, PG Standard, D McLaughlin, L Kaufman. Comparative evaluation of chemilu-minescent DNA probe assays and exoantigen tests for rapid identification of *Blastomyces dermatit-idis* and *Coccidioides immitis*. J Clin Microbiol 32:867–870, 1994.

273. J Wheat. Histoplasmosis in the acquired immunodeficiency syndrome. Curr Top Med Mycol 7: 7–18, 1996.

274. LJ Wheat. Histoplasmosis in Indianapolis. Clin Infect Dis 14(suppl 1):S91–S99, 1992.

275. CT Huang, T McGarry, S Cooper, R Saunders, R Andavolu. Disseminated histoplasmosis in the acquired immunodeficiency syndrome. Report of five cases from a nonendemic area. Arch Intern Med 147:1181–1184, 1987.

276. SH Salzman, RL Smith, CP Aranda. Histoplasmosis in patients at risk for the acquired immunodefi-ciency syndrome in a nonendemic setting. Chest 93:916–921, 1988.

277. CM Odio, M Navarrete, JM Carrillo, L Mora, A Carranza. Disseminated histoplasmosis in infants. Pediatr Infect Dis J 18:1065–1068, 1999.

278. CC Hung, JM Wong, PR Hsueh, SM Hsieh, MY Chen. Intestinal obstruction and peritonitis resulting from gastrointestinal histoplasmosis in an AIDS patient. J Formos Med Assoc 97:577–580, 1998.

279. PG Scapellato, J Desse, R Negroni. Acute disseminated histoplasmosis and endocarditis. Rev Inst Med Trop Sao Paulo 40:19–22, 1998.

280. WE Bullock. *Histoplasma capsulatum*. In: GL Mandell, JE Bennett, R Dolin, eds. Principles and Practices of Infectious Diseases. New York: Churchill Livingstone, 1994, pp 2340–2353.

281. E Reiss, JB Knowles, SL Bragg, L Kaufman. Monoclonal antibodies against the M-protein and carbohydrate antigens of histoplasmin characterized by the enzyme-linked immunoelectrotransfer blot method. Infect Immun 53:540–546, 1986.

282. AJ Hamilton, MA Bartholomew, J Figueroa, LE Fenelon, RJ Hay. Evidence that the M antigen of *Histoplasma capsulatum* var. *capsulatum* is a catalase which exhibits cross-reactivity with other dimorphic fungi. J Med Vet Mycol 28:479–485, 1990.

283. RB George, RS Lambert. Significance of serum antibodies to *Histoplasma capsulatum* in endemic areas. South Med J 77:161–163, 1984.

284. J Wheat, ML French, RB Kohler, SE Zimmerman, WR Smith, JA Norton, HE Eitzen, CD Smith, TG Slama. The diagnostic laboratory tests for histoplasmosis: analysis of experience in a large urban outbreak. Ann Intern Med 97:680–685, 1982.

285. LJ Wheat, RB Kohler, ML French, M Garten, M Kleiman, SE Zimmerman, W Schlech, J Ho, A White, Z Brahmi. Immunoglobulin M and G histoplasmal antibody response in histoplasmosis. Am Rev Respir Dis 128:65–70, 1983.

286. M Torres, H Diaz, T Herrera, E Sada. Evaluation of enzyme linked immunosorbent-assay and Western blot for diagnosis of histoplasmosis. Rev Invest Clin 45:155–160, 1993.

287. R Chandrashekar, KC Curtis, BW Rawot, GS Kobayashi, GJ Weil. Molecular cloning and characteri-zation of a recombinant *Histoplasma capsulatum* antigen for antibody-based diagnosis of human histoplasmosis. J Clin Microbiol 35:1071–1076, 1997.

288. GS Deepe Jr, GG Durose. Immunobiological activity of recombinant H antigen from *Histoplasma capsulatum*. Infect Immun 63:3151–3157, 1995.

289. FJ Gomez, AM Gomez, GS Deepe Jr. Protective efficacy of a 62-kilodalton antigen, HIS-62, from the cell wall and cell membrane of *Histoplasma capsulatum* yeast cells. Infect Immun 59:4459–4464, 1991.

290. FJ Gomez, AM Gomez, GS Deepe Jr. An 80-kilodalton antigen from *Histoplasma capsulatum* that has homology to heat shock protein 70 induces cell-mediated immune responses and protection in mice. Infect Immun 60:2565–2571, 1992.

291. LJ Wheat. Diagnosis and management of histoplasmosis. Eur J Clin Microbiol Infect Dis 8:480–490, 1989.
292. RS Lambert, RB George. Evaluation of enzyme immunoassay as a rapid screening test for histoplasmosis and blastomycosis. Am Rev Respir Dis 136:316–319, 1987.
293. SE Zimmerman, PC Stringfield, LJ Wheat, ML French, RB Kohler. Comparison of sandwich solid-phase radioimmunoassay and two enzyme-linked immunosorbent assays for detection of *Histoplasma capsulatum* polysaccharide antigen. J Infect Dis 160:678–685, 1989.
294. LJ Wheat, P Connolly-Stringfield, RB Kohler, PT Frame, MR Gupta. *Histoplasma capsulatum* polysaccharide antigen detection in diagnosis and management of disseminated histoplasmosis in patients with acquired immunodeficiency syndrome. Am J Med 87:396–400, 1989.
295. LJ Wheat, P Connolly-Stringfield, B Williams, K Connolly, R Blair, M Bartlett, M Durkin. Diagnosis of histoplasmosis in patients with the acquired immunodeficiency syndrome by detection of *Histoplasma capsulatum* polysaccharide antigen in bronchoalveolar lavage fluid. Am Rev Respir Dis 145:1421–1424, 1992.
296. MM Durkin, PA Connolly, LJ Wheat. Comparison of radioimmunoassay and enzyme-linked immunoassay methods for detection of *Histoplasma capsulatum* var. *capsulatum* antigen. J Clin Microbiol 35:2252–2255, 1997.
297. TO Garringer, LJ Wheat, EJ Brizendine. Comparison of an established antibody sandwich method with an inhibition method of *Histoplasma capsulatum* antigen detection. J Clin Microbiol 38:2909–2913, 2000.
298. LJ Wheat, RB Kohler, RP Tewari. Diagnosis of disseminated histoplasmosis by detection of *Histoplasma capsulatum* antigen in serum and urine specimens. N Engl J Med 314:83–88, 1986.
299. B Williams, M Fojtasek, P Connolly-Stringfield, J Wheat. Diagnosis of histoplasmosis by antigen detection during an outbreak in Indianapolis, Ind. Arch Pathol Lab Med 118:1205–1208, 1994.
300. J Wheat. Histoplasmosis. Experience during outbreaks in Indianapolis and review of the literature. Medicine (Baltimore) 76:339–354, 1997.
301. SM Kamel, LJ Wheat, ML Garten, MS Bartlett, MR Tansey, RP Tewari. Production and characterization of murine monoclonal antibodies to *Histoplasma capsulatum* yeast cell antigens. Infect Immun 57:896–901, 1989.
302. BL Gomez, JI Figueroa, AJ Hamilton, BL Ortiz, MA Robledo, A Restrepo, RJ Hay. Development of a novel antigen detection test for histoplasmosis. J Clin Microbiol 35:2618–2622, 1997.
303. A Restrepo, ME Salazar, LE Cano, EP Stover, D Feldman, DA Stevens. Estrogens inhibit mycelium-to-yeast transformation in the fungus *Paracoccidioides brasiliensis*: implications for resistance of females to paracoccidioidomycosis. Infect Immun 46:346–353, 1984.
304. LZ Goldani, AM Sugar. Paracoccidioidomycosis and AIDS: an overview. Clin Infect Dis 21:1275–1281, 1995.
305. LA Yarzabal, MB De Albornoz, NA De Cabral, AR Santiago. Specific double diffusion microtechnique for the diagnosis of aspergillosis and paracoccidioidomycosis using monospecific antisera. Sabouraudia 16:55–62, 1978.
306. T Mistreta, MJ Souza, LG Chamma, SZ Pinho, M Franco. Serology of paracoccidioidomycosis. I. Evaluation of the indirect immunofluorescent test. Mycopathologia 89:13–17, 1985.
307. A Restrepo, LE Cano, MT Ochoa. A yeast-derived antigen from *Paracoccidioides brasiliensis* useful for serologic testing. Sabouraudia 23:23–29, 1985.
308. GM Del Negro, NM Garcia, EG Rodrigues, MI Cano, MS de Aguiar, VdS Lirio, CdS Lacaz. The sensitivity, specificity and efficiency values of some serological tests used in the diagnosis of paracoccidioidomycosis. Rev Inst Med Trop Sao Paulo 33:277–280, 1991.
309. AM Siqueira, CS Lacaz. Serologic characterization of *Paracoccidioides brasiliensis* E2 antigen. Braz J Med Biol Res 24:807–813, 1991.
310. M Casotto. Characterization of the cellular antigens of *Paracoccidioides brasiliensis* yeast form. J Clin Microbiol 28:1188–1193, 1990.
311. M Casotto, S, Paris, ZP Camargo. Antigens of diagnostic value in three isolates of *Paracoccidioides brasiliensis*. J Med Vet Mycol 29:243–253, 1991.

312. IC Almeida, EG Rodrigues, LR Travassos. Chemiluminescent immunoassays: discrimination between the reactivities of natural and human patient antibodies with antigens from eukaryotic pathogens, *Trypanosoma cruzi* and *Paracoccidioides brasiliensis*. J Clin Lab Anal 8:424–431, 1994.

313. JP Bueno, MJ Mendes-Giannini, GM Del Negro, CM Assis, CK Takiguti, MA Shikanai-Yasuda. IgG, IgM and IgA antibody response for the diagnosis and follow-up of paracoccidioidomycosis: comparison of counterimmunoelectrophoresis and complement fixation. J Med Vet Mycol 35: 213–217, 1997.

314. BL Ortiz, AM Garcia, A Restrepo, JG McEwen. Immunological characterization of a recombinant 27-kilodalton antigenic protein from *Paracoccidioides brasiliensis*. Clin Diagn Lab Immunol 3: 239–241, 1996.

315. BL Ortiz, S Diez, ME Uran, JM Rivas, M Romero, V Caicedo, A Restrepo, JG McEwen. Use of the 27-kilodalton recombinant protein from *Paracoccidioides brasiliensis* in serodiagnosis of paracoccidioidomycosis. Clin Diagn Lab Immunol 5:826–830, 1998.

316. MF Ferreira-da-Cruz, B Galvao-Castro, CT Daniel-Ribeiro. Sensitive immunoradiometric assay for the detection of *Paracoccidioides brasiliensis* antigens in human sera. J Clin Microbiol 29: 1202–1205, 1991.

317. BL Gomez, JI Figueroa, AJ Hamilton, B Ortiz, MA Robledo, RJ Hay, A Restrepo. Use of monoclonal antibodies in diagnosis of paracoccidioidomycosis: new strategies for detection of circulating antigens. J Clin Microbiol 35:3278–3283, 1997.

318. BL Gomez, JI Figueroa, AJ Hamilton, S Diez, M Rojas, AM Tobon, RJ Hay, A Restrepo. Antigenemia in patients with paracoccidioidomycosis: detection of the 87-kilodalton determinant during and after antifungal therapy. J Clin Microbiol 36:3309–3316, 1998.

319. MA Salina, MA Shikanai-Yasuda, RP Mendes, B Barraviera, MJ Mendes Giannini. Detection of circulating *Paracoccidioides brasiliensis* antigen in urine of paracoccidioidomycosis patients before and during treatment. J Clin Microbiol 36:1723–1728, 1998.

320. AS Sekhon, JS Li, AK Garg. Penicillosis marneffei: serological and exoantigen studies. Mycopathologia 77:51–57, 1982.

321. MA Viviani, AM Tortorano, G Rizzardini, T Quirino, L Kaufman, AA Padhye, L Ajello. Treatment and serological studies of an Italian case of penicilliosis marneffei contracted in Thailand by a drug addict infected with the human immunodeficiency virus. Eur J Epidemiol 9:79–85, 1993.

322. KY Yuen, SS Wong, DN Tsang, PY Chau. Serodiagnosis of *Penicillium marneffei* infection. Lancet 344:444–445, 1994.

323. L Kaufman, PG Standard, M Jalbert, P Kantipong, K Limpakarnjanarat, TD Mastro. Diagnostic antigenemia tests for penicilliosis marneffei. J Clin Microbiol 34:2503–2505, 1996.

324. V Desakorn, MD Smith, AL Walsh, AJ Simpson, D Sahassananda, A Rajanuwong, V Wuthiekanun, P Howe, BJ Angus, P Suntharasamai, NJ White. Diagnosis of *Penicillium marneffei* infection by quantitation of urinary antigen by using an enzyme immunoassay. J Clin Microbiol 37:117–121, 1999.

325. CA Kauffman. Sporotrichosis. Clin Infect Dis 29:231–236; quiz 237, 1999.

326. F Alberici, CT Paties, G Lombardi, L Ajello, L Kaufman, F Chandler. *Sporothrix schenckii* var *luriei* as the cause of sporotrichosis in Italy [published erratum appears in Eur J Epidemiol 1989; 5(4):537]. Eur J Epidemiol 5:173–177, 1989.

327. AA Padhye, L Kaufman, E Durry, CK Banerjee, SK Jindal, P Talwar, A Chakrabarti. Fatal pulmonary sporotrichosis caused by *Sporothrix schenckii* var. *luriei* in India. J Clin Microbiol 30:2492–2494, 1992.

328. RD Jones, GA Sarosi, JD Parker, RJ Weeks, FE Tosh. The complement-fixation test in extracutaneous sporotrichosis. Ann Intern Med 71:913–918, 1969.

329. JV Karlin, HS Nielsen Jr. Serologic aspects of sporotrichosis. J Infect Dis 121:316–327, 1970.

330. SO Blumer, L Kaufman, W Kaplan, DW McLaughlin, DE Kraft. Comparative evaluation of five serological methods for the diagnosis of sporotrichosis. Appl Microbiol 26:4–8, 1973.

331. EN Scott, L Kaufman, AC Brown, HG Muchmore. Serologic studies in the diagnosis and management of meningitis due to *Sporothrix schenckii*. N Engl J Med 317:935–940, 1987.

332. H Papadogeorgakis, E Frangoulis, C Papaefstathiou, A Katsambas. *Rhodotorula rubra* fungaemia in an immunosuppressed patient. J Eur Acad Dermatol Venereol 12:169–170, 1999.

333. BS Yoss, RL Sautter, HJ Brenker. *Trichosporon beigelii*, a new neonatal pathogen. Am J Perinatol 14:113–117, 1997.

334. HM Richet, MM McNeil, MC Edwards, WR Jarvis. Cluster of *Malassezia furfur* pulmonary infections in infants in a neonatal intensive-care unit. J Clin Microbiol 27:1197–1200, 1989.

335. TJ Walsh, KR Newman, M Moody, RC Wharton, JC Wade. Trichosporonosis in patients with neoplastic disease. Medicine (Baltimore) 65:268–279, 1986.

336. TJ Walsh, GP Melcher, JW Lee, PA Pizzo. Infections due to *Trichosporon* species: new concepts in mycology, pathogenesis, diagnosis and treatment. Curr Top Med Mycol 5:79–113, 1993.

337. J Hoy, KC Hsu, K Rolston, RL Hopfer, M Luna, GP Bodey. *Trichosporon beigelii* infection: a review. Rev Infect Dis 8:959–967, 1986.

338. M Kobayashi, S Kotani, M Fujishita, H Taguchi, T Moriki, H Enzan, I Miyoshi. Immunohistochemical identification of *Trichosporon beigelii* in histologic section by immunoperoxidase method. Am J Clin Pathol 89:100–105, 1988.

339. T Takeuchi, M Kobayashi, T Moriki, I Miyoshi. Application of a monoclonal antibody for the detection of *Trichosporon beigelii* in paraffin-embedded tissue sections. J Pathol 156:23–27, 1988.

340. CA Lyman, SJ Devi, J Nathanson, CE Frasch, PA Pizzo, TJ Walsh. Detection and quantitation of the glucuronoxylomannan-like polysaccharide antigen from clinical and nonclinical isolates of *Trichosporon beigelii* and implications for pathogenicity. J Clin Microbiol 33:126–130, 1995.

341. EJ McManus, JM Jones. Detection of a *Trichosporon beigelii* antigen cross-reactive with *Cryptococcus neoformans* capsular polysaccharide in serum from a patient with disseminated *Trichosporon* infection. J Clin Microbiol 21:681–685, 1985.

342. TJ Walsh, JW Lee, GP Melcher, E Navarro, J Bacher, D Callender, KD Reed, T Wu, G Lopez-Berestein, PA Pizzo. Experimental *Trichosporon* infection in persistently granulocytopenic rabbits: implications for pathogenesis, diagnosis, and treatment of an emerging opportunistic mycosis. J Infect Dis 166:121–133, 1992.

343. SF Welbel, MM McNeil, A Pramanik, R Silberman, AD Oberle, G Midgley, S Crow, WR Jarvis. Nosocomial *Malassezia pachydermatis* bloodstream infections in a neonatal intensive care unit. Pediatr Infect Dis J 13:104–108, 1994.

344. IF Salkin, MR McGinnis, MJ Dykstra, MG Rinaldi. *Scedosporium inflatum*, an emerging pathogen. J Clin Microbiol 26:498–503, 1988.

345. L Kaufman, AA Padhye, S Parker. Rhinocerebral zygomycosis caused by *Saksenaea vasiformis*. J Med Vet Mycol 26:237–241, 1988.

346. PF Pierce Jr, SL Solomon, L Kaufman, VF Garagusi, RH Parker, L Ajello. Zygomycetes brain abscesses in narcotic addicts with serological diagnosis. JAMA 248:2881–2882, 1982.

347. ZU Khan, B Prakash, MM Kapoor, JP Madda, R Chandy. Basidiobolomycosis of the rectum masquerading as Crohn's disease: case report and review. Clin Infect Dis 26:521–523, 1998.

348. TM Pasha, JA Leighton, JD Smilack, J Heppell, TV Colby, L Kaufman. Basidiobolomycosis: an unusual fungal infection mimicking inflammatory bowel disease. Gastroenterology 112:250–254, 1997.

349. GM Lyon, JD Smilack, KK Komatsu, TM Pasha, JA Leighton, J Guarner, TV Colby, MD Lindsley, M Phelan, DW Warnock, RA Hajjeh. Gastrointestinal basidiobolomycosis in Arizona: clinical and epidemiologic characteristics and review of the literature. Clin Infect Dis 32(in press), 2001.

350. NY Busaba, M Poulin. Invasive *Pseudallescheria boydii* fungal infection of the temporal bone. Otolaryngol Head Neck Surg 117:S91–S94, 1997.

351. J Berenguer, J Diaz-Mediavilla, D Urra, P Munoz. Central nervous system infection caused by *Pseudallescheria boydii*: case report and review. Rev Infect Dis 11:890–896, 1989.

352. FK Welty, GX McLeod, C Ezratty, RW Healy, AW Karchmer. *Pseudallescheria boydii* endocarditis of the pulmonic valve in a liver transplant recipient. Clin Infect Dis 15:858–860, 1992.

353. LH Lerner, EA Lerner, YM Bello. Co-existence of cutaneous and presumptive pulmonary alternariosis. Int J Dermatol 36:285–288, 1997.

354. MR Mascarenhas, KL McGowan, E Ruchelli, B Athreya, SM Altschuler. *Acremonium* infection of the esophagus. J Pediatr Gastroenterol Nutr 24:356–358, 1997.

355. NM Brown, EL Blundell, SR Chown, DW Warnock, JA Hill, RR Slade. *Acremonium* infection in a neutropenic patient. J Infect 25:73–76, 1992.

356. WR Jeffrey, JE Hernandez, AL Zarraga, GE Oley, LW Kitchen. Disseminated infection due to *Acremonium* species in a patient with Addison's disease [letter]. Clin Infect Dis 16:170, 1993.

357. J Trupl, M Majek, J Mardiak, Z Jesenska, V Krcmery, Jr. *Acremonium* infection in two compromised patients [letter]. J Hosp Infect 25:299–301, 1993.

358. CM Wilson, EJ O'Rourke, MR McGinnis, IF Salkin. *Scedosporium inflatum*: clinical spectrum of a newly recognized pathogen. J Infect Dis 161:102–107, 1990.

359. J Marin, MA Sanz, GF Sanz, J Guarro, ML Martinez, M Prieto, E Gueho, JL Menezo. Disseminated *Scedosporium inflatum* infection in a patient with acute myeloblastic leukemia. Eur J Clin Microbiol Infect Dis 10:759–761, 1991.

360. M Rabodonirina, S Paulus, F Thevenet, R Loire, E Gueho, O Bastien, JF Mornex, M Celard, MA Piens. Disseminated *Scedosporium prolificans* (*S. inflatum*) infection after single-lung transplantation. Clin Infect Dis 19:138–142, 1994.

361. P Nenoff, U Gutz, K Tintelnot, A Bosse-Henck, M Mierzwa, J Hofmann, LC Horn, UF Haustein. Disseminated mycosis due to *Scedosporium prolificans* in an AIDS patient with Burkitt lymphoma. Mycoses 39:461–465, 1996.

362. M Alvarez, B Lopez Ponga, C Rayon, J Garcia Gala, MC Roson Porto, M Gonzalez, JV Martinez-Suarez, JL Rodriguez-Tudela. Nosocomial outbreak caused by *Scedosporium prolificans* (*inflatum*): four fatal cases in leukemic patients. J Clin Microbiol 33:3290–3295, 1995.

363. E Anaissie, H Kantarjian, J Ro, R Hopfer, K Rolston, V Fainstein, G Bodey. The emerging role of *Fusarium* infections in patients with cancer. Medicine (Baltimore) 67:77–83, 1988.

364. E Cofrancesco, C Boschetti, MA Viviani, C Bargiggia, AM Tortorano, M Cortellaro, C Zanussi. Efficacy of liposomal amphotericin B (AmBisome) in the eradication of *Fusarium* infection in a leukaemic patient. Haematologica 77:280–283, 1992.

365. E Anaissie, GP Bodey, H Kantarjian, J Ro, SE Vartivarian, R Hopfer, J Hoy, K Rolston. New spectrum of fungal infections in patients with cancer. Rev Infect Dis 11:369–378, 1989.

366. E Anaissie, H Kantarjian, P Jones, B Barlogie, M Luna, G Lopez-Berestein, GP Bodey. *Fusarium*. A newly recognized fungal pathogen in immunosuppressed patients. Cancer 57:2141–2145, 1986.

367. P Martino, R Gastaldi, R Raccah, C Girmenia. Clinical patterns of *Fusarium* infections in immunocompromised patients. J Infect 28(suppl 1):7–15, 1994.

368. JF Fisher, S Shadomy, JR Teabeaut, J Woodward, GE Michaels, MA Newman, E White, P Cook, A Seagraves, F Yaghmai, JP Rissing. Near-drowning complicated by brain abscess due to *Petriellidium boydii*. Arch Neurol 39:511–513, 1982.

369. SG Alsip, CG Cobbs. *Pseudallescheria boydii* infection of the central nervous system in a cardiac transplant recipient. South Med J 79:383–384, 1986.

370. TJ Walsh, J Peter, DA McGough, AW Fothergill, MG Rinaldi, PA Pizzo. Activities of amphotericin B and antifungal azoles alone and in combination against *Pseudallescheria boydii*. Antimicrob Agents Chemother 39:1361–1364, 1995.

371. GM Wood, JG McCormack, DB Muir, DH Ellis, MF Ridley, R Pritchard, M Harrison. Clinical features of human infection with *Scedosporium inflatum*. Clin Infect Dis 14:1027–1033, 1992.

372. M Wedde, D Muller, K Tintelnot, GS De Hoog, U Stahl. PCR-based identification of clinically relevant *Pseudallescheria/Scedosporium* strains. Med Mycol 36:61–67, 1998.

373. DH Larone. Medically Important Fungi. A Guide to Identification. 3rd ed. Washington: ASM Press, 1995.

374. GS de Hoog, FD Marvin-Sikkema, GA Lahpoor, JC Gottschall, RA Prins, E Gueho. Ecology and physiology of the emerging opportunistic fungi *Pseudallescheria boydii* and *Scedosporium prolificans* [published erratum appears in Mycoses 1994;37(5–6):223–224]. Mycoses 37:71–78, 1994.

375. J Kwon-Chung, JE Bennett. Medical Mycology. Melvern, PA: Lea and Febiger, 1994.

376. JW Rippon. Medical Mycology: The Pathogenic Fungi and the Pathogenic Actinomycetes. 3rd ed. Philadelphia: W.B. Saunders, 1988.

377. TJ Walsh, G Renshaw, J Andrews, J Kwon-Chung, RC Cunnion, HI Pass, J Taubenberger, W Wilson, PA Pizzo. Invasive zygomycosis due to *Conidiobolus incongruus*. Clin Infect Dis 19:423–430, 1994.

378. J Rozich, D Oxendine, J Heffner, W Brzezinski. Pulmonary zygomycosis. A cause of positive lung scan diagnosed by bronchoalveolar lavage. Chest 95:238–240, 1989.

379. DM Yousem, SL Galetta, DA Gusnard, HI Goldberg. MR findings in rhinocerebral mucormycosis. J Comput Assist Tomogr 13:878–882, 1989.

380. L Kaufman, LF Turner, DW McLaughlin. Indirect enzyme-linked immunosorbent assay for zygomycosis. J Clin Microbiol 27:1979–1982, 1989.

381. KW Jones, L Kaufman. Development and evaluation of an immunodiffusion test for diagnosis of systemic zygomycosis (mucormycosis): preliminary report. J Clin Microbiol 7:97–103, 1978.

382. DW Mackenzie, JF Soothill, JH Millar. Meningitis caused by *Absidia corymbifera*. J Infect 17: 241–248, 1988.

383. DR Wysong, AR Waldorf. Electrophoretic and immunoblot analyses of *Rhizopus arrhizus* antigens. J Clin Microbiol 25:358–363, 1987.

384. L Kaufman, L Mendoza, PG Standard. Immunodiffusion test for serodiagnosing subcutaneous zygomycosis. J Clin Microbiol 28:1887–1890, 1990.

30

Molecular Typing and Epidemiology of *Candida* spp. and Other Important Human Fungal Pathogens

Derek J. Sullivan and David C. Coleman
University of Dublin, Trinity College, Dublin, Ireland

I. INTRODUCTION

During the 1980s and 1990s the vast majority of articles on the epidemiology of fungal infections introduced the topic with a statement highlighting the increased incidence of fungal infections during the last two decades of the 20th century. This was not just pointless exaggeration to attract the attention of readers and grant reviewers; it is simply very difficult to ignore the fact that there are now far more identified cases of fungal infections caused by a greater number of fungal species than ever before. What we hope to achieve in this chapter is to explain why the study of fungal epidemiology is important and necessary and to outline our current understanding of the epidemiology of infections caused by *Candida* and other fungal species. We will describe the molecular techniques currently in use and will discuss the factors that have influenced the changes that have occurred in the epidemiology of selected pathogenic fungal species.

II. EPIDEMIOLOGY OF FUNGAL DISEASES

Fungi are ubiquitous. Some of them are commensal organisms living on our skin, on mucosal surfaces, and in our intestines and urinary tracts; they are in the air we breathe and the food we eat, and they inhabit practically every surface we come into contact with. Given the availability of nutrients in the human body, it is not surprising that for some fungal species this represents a prime location for colonization and infection. Fortunately, for the most part the human immune system is particularly effective at keeping these interlopers at bay—so much so in fact that, until recently, fungi were assumed to be largely innocuous and were relatively rarely identified as human pathogens. However, during the 1980s there was a significant increase in the incidence of fungal infections. In one study based on hospital discharge data from the United States, the incidence of oropharyngeal candidosis was seen to increase by 4.7-fold, while the incidence of disseminated candidosis increased by a factor of 11 [1]. Fungal species are now responsible for ~10% of all nosocomial bloodstream infections in the United States [2]. More than 75% of these infections are caused by *Candida* spp. which now represent the fourth most common cause

717

Table 1 Risk Factors for Fungal Infection

Acquired immunodeficiency syndrome (AIDS)
Hematologic malignancy
Immunosuppressive therapy due to organ transplantation
Anticancer chemotherapy
Prolonged hospitalization
Extensive surgery
Burns
Head and neck irradiation
Old age
Premature birth
Prior colonization with fungus
Catheterization
Parenteral feeding
Broad-spectrum antibiotic therapy
Corticosteroid therapy
Diabetes mellitus

of nosocomial disease [3,4]. Other fungal species frequently associated with disease include *Aspergillus fumigatus*, *Cryptococcus neoformans*, Zygomycetes, and dermatophytes. In addition, endemic mycotic infections, such as histoplasmosis, penicilliosis, and coccidioidomycosis, occur in specific geographic locations throughout the world.

There are a number of reasons that may help explain the increased significance of fungi and the emergence of some novel fungal species as human pathogens. The most significant underlying cause of this phenomenon is the growing number of individuals with impaired or dysfunctional immune systems. Patients can be immunocompromised for a wide variety of reasons. Clearly individuals with hematologic malignancies and neutropenia and those undergoing bone marrow and solid-organ transplantion are at increased risk of acquiring disseminated fungal infections such as candidemia and aspergillosis. Similarly, patients receiving chemotherapeutic agents for cancer treatment and patients in intensive care units, especially those with catheters or intravenous lines, are also particularly susceptible to systemic disease. Individuals at risk of acquiring superficial fungal diseases, such as oropharyngeal candidosis, include the very young and very old (particularly those receiving broad-spectrum antibiotics or steroid treatment), individuals receiving head and neck irradiation, and individuals infected with HIV. Risk factors that predispose individuals to fungal infection are listed in Table 1.

III. EPIDEMIOLOGICAL TECHNIQUES

To understand an infectious disease fully and to unravel its epidemiology, there is an absolute requirement to be able to identify individual strains of the species responsible for the infection. Thus, given the increased importance of fungal diseases during the past two decades, and with a view to improving the treatment and prognosis of these infections, a variety of strain typing techniques have been developed to improve our understanding of how and why fungi cause disease. Information obtained using such techniques can (1) aid in the identification of the source of an infection, (2) identify if more than one strain is present in a clinical specimen and determine the relative proportions of the individual strains present, (3) identify if relapses in disease are caused by reinfection with a novel organism or by persistence of the original infecting strain, (4) determine if specific strains (or clones) are more likely to cause disease than others, (5) monitor the epidemiology of drug-resistant strains, and (6) yield insights into the population dynamics of specific species. The latter has particular significance for the development of vaccines and in furthering our understanding of the emergence of antifungal drug resistance.

To be effective, strain typing techniques must first and foremost have excellent discriminatory power (i.e., have the ability to distinguish between unrelated isolates), must yield reproducible results, and should be applicable to all strains within a species and, if possible, to a wide range of species. Furthermore, to be suitable for use in routine clinical laboratories and in large-scale epidemiological studies, effective typing techniques must be rapid and easy to perform, have a large sample throughput volume capability, and, very importantly, be as inexpensive as possible. Very few of the techniques available currently fulfill all of these criteria. However, a concerted effort by a large number of laboratories using a wide range of techniques has revealed much about the epidemiology of fungal infections, particularly those caused by *Candida albicans* and other *Candida* spp. In the following section the advantages and disadvantages of specific techniques will be discribed.

A. Phenotypic Strain Typing Methods

The original methods used to discriminate between individual strains of micro-organisms were based on the comparison of specific phenotypic characteristics of individual isolates. The tests developed to compare strains frequently relied upon colonial morphology or color on specific culture media, biochemical tests, substrate assimilation reactions, resistotyping, serology, and killer toxin susceptibility. Although many of these methods are easy to perform and amenable to a high sample volume throughput, they all have limited discriminatory power and are frequently poorly reproducible. The poor discriminatory ability is largely due to the fact that these methods usually rely on the examination of a limited number of specific phenotypic traits that are often shared by large numbers of unrelated isolates within individual species and even between isolates belonging to different species. The lack of reproducibility can be ascribed to phenomena such as phenotypic switching, which can result in variation of certain phenotypic traits in a number of fungal species, particularly *C. albicans*. Because of the inherent limitations of these techniques much effort has been spent during the past 20 years in developing more efficient methods to discriminate between fungal strains using molecular techniques.

B. Molecular Strain Typing Methods

Although phenotypic characteristics provide the basis for species and strain discrimination of macrobiota, they are of limited use when applied to the comparison of micro-organisms. In recent years, rather than relying on phenotypic differences, microbial strain typing has focused on the comparison of differences in macromolecules such as proteins and, most importantly, DNA. In particular, polymorphisms in genomic DNA have proved to be excellent markers or signatures of specific strains. The most commonly used of the techniques that exploit these genotypic differences are described briefly below. For a more in-depth description and assessment of these methodologies, the reader is referred to a recent comprehensive review by Soll [5].

1. Restriction Enzyme Analysis

One of the simplest and earliest molecular techniques developed to compare interstrain relationships between microbial strains, including a wide variety of fungal species, was restriction

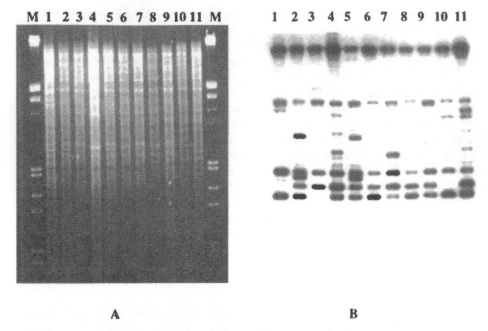

Figure 1   (A) Agarose gel showing restriction enzyme analysis (REA) profiles of genomic DNA from 11 clinical isolates of *C. albicans* generated with the restriction endonuclease *Eco*RI. The profiles shown in the lanes marked M correspond to molecular size reference markers. (B) Autoradiogram showing DNA fingerprint profiles of the digested DNA patterns shown in panel A following hybridization with the *C. albicans*–specific fingerprinting probe 27A.

enzyme analysis (REA). This method relies on the analysis of strain-specific band patterns generated by the electrophoresis of DNA fragments resulting from the digestion of genomic DNA with specific restriction endonucleases, such as *Eco*RI (see Fig. 1, panel A). These strain-specific patterns result from polymorphisms in the recognition sites of individual restriction enzymes. Although applicable to a wide variety of species and relatively easy to perform, this technique usually results in very complex fingerprint patterns which are difficult to compare objectively and are not readily amenable to computer-assisted fingerprint comparison analysis. However, the variable mobility of strongly stained repeat sequence-containing fragments and the use of enzymes that cleave DNA infrequently in conjunction with pulsed-field gel electropho-resis (see below) improve the discriminatory power of this technique. REA has been used extensively in the analysis of *C. albicans* [6–17] as well as a wide variety of other pathogenic fungal species [18].

2.   Restriction Fragment Length Polymorphisms and Species-Specific DNA Fingerprinting Probes

To improve the sensitivity of REA the DNA fingerprint patterns generated using enzymes such as *Eco*RI can be probed in Southern hybridization experiments with labeled species-specific DNA probes that contain repetitive elements (see Fig. 1, panel B). This results in less complex hybridization banding patterns (analogous to bar codes) which are readily amenable to further analysis. The best known of these probes for fungal species are the related *C. albicans*–specific probes 27A [19] and Ca3 [20]. These probes contain the repetitive element RPS which is dispersed throughout the *C. albicans* genome. More recently, species-specific fingerprinting probes have also been developed for the analysis of other *Candida* spp. such as *C. glabrata* [21], *C. tropicalis* [22], *C. krusei* [23], and *C. dubliniensis* [24] as well as for other fungal species such as *Aspergillus fumigatus* [25,26] and *Cryptococcus neoformans* [27].

The most commonly used strategies to identify sequences for use as fingerprinting probes include probing genomic DNA libraries with restriction endonuclease-generated genomic DNA fragments to detect strongly hybridizing clones containing repeat sequences [20] and the differential screening of genomic libraries with labeled DNA from related species [19]. DNA fingerprinting using species-specific probes is time-consuming and expensive and involves complex labeling and hybridization procedures that are tedious to apply to large numbers of strains. However, the data obtained are unambiguous, are reproducible, and, most importantly, in a majority of cases allow the discrimination of unrelated strains. In addition, the fingerprint patterns, which usually comprise 10–20 hybridization bands, have a high degree of contrast and separation and are therefore amenable to computer-assisted analyses. Studies using these techniques have proven invaluable in the analysis of the epidemiology and population dynamics of a wide variety of fungal species. Details of some these studies are discussed below.

3. Oligonucleotide Fingerprinting

A variation of the previous methods is to use radiolabeled oligonucleotides, rather than species-specific sequences, as DNA fingerprinting probes (see Fig. 2) [28,29]. Although there is a wide range of target sequences based on which oligonucleotide primers can be designed, the most commonly used are homologous to microsatellite sequences. Microsatellites, also referred to as simple sequences, are 1- to 6-bp repeats scattered randomly throughout the genome of eukaryotic organisms, including fungi, and generally consist of mono-, di-, tri-, and tetranucleotide motifs in multiple tandem repeats—i.e., $(A)_n$, $(GT)_n$, $(GTG)_n$, $(GATA)_n$, $(GACA)_n$, and $(GGAT)_n$. The functions of microsatellite sequences have yet to be determined. One advantage of using these probes in fingerprinting studies with fungi is that the same labeled probes can be used on a wide range of species and the results can be verified by hybridizing the same target DNA sample consecutively with a range of different oligonucleotides. However, although they have been applied to a variety of eukaryotic species, including fungi, they have only been used in a limited number of studies on medically important fungal species [30–34]. As well as being used in DNA hybridization experiments, these or similar oligonucleotides can also be used as primers in PCR experiments, resulting in strain specific amplimer patterns [35,36].

4. Pulsed-Field Gel Electrophoresis

One of the techniques used most widely in the determination of relationships between fungal strains relies on the comparison of the molecular karyotype profiles of individual strains. The separation of chromosomes and other high-molecular-weight chromosomal-size DNA fragments is performed using pulsed-field gel electrophoresis (PFGE) in agarose gels [37]. The basis of this technique is the alternation of the electric field between pairs of electrodes during electrophoresis. This allows very large DNA molecules and fragments to reorient and migrate through the pores of the agarose gel at varying rates, and following staining of the DNA fragments, results in a specific karyotypic banding profile (see Fig. 3). A number of variations of PFGE have been developed and applied to the analysis of fungal species with varying success, including clamped homogeneous electric field (CHEF) electrophoresis, orthogonal-field-alternation gel electrophoresis (OFAGE), and field-inversion gel electrophoresis (FIGE). As mentioned previously, PFGE has also been used to separate very large restriction endonuclease-generated fragments obtained

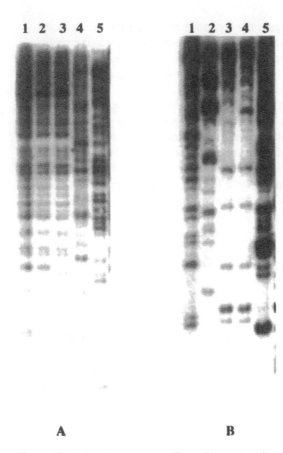

A   B

Figure 2   DNA fingerprint profiles of *C. parapsilosis* isolates (A) and *C. neoformans* isolates (B) generated by hybridization of *Eco*RI-digested genomic DNA with the oligonucleotide probe (GT)₈.

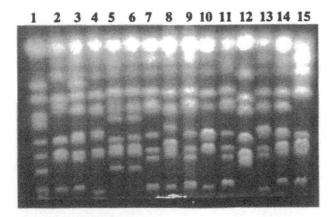

Figure 3   Agarose gel showing karyotype profiles of clinical *C. dubliniensis* isolates generated by pulsed-field gel electrophoresis.

using restriction endonucleases, such as *Sfi*I, *Sma*I, and *Not*I, which cleave DNA infrequently [38–40]. Although PFGE has been demonstrated to be very effective at discriminating between separate strains of a wide range of medically important fungi [13,41–50], the apparatus required is expensive and the sample preparation and electrophoresis procedures are time-consuming.

5. Randomly Amplified Polymorphic DNA (RAPD) Analysis

RAPD analysis relies on the use of oligonucleotide primers of arbitrarily chosen sequence (either singly or in pairs) in low-stringency PCR amplification reactions [51]. Fingerprints or profiles of amplimers are generated by electrophoresis of the amplified products on agarose or acrylamide gels. Minor sequence differences between strains can result in increased or decreased annealing of primers, resulting in the presence or absence of specific amplimers and therefore differences in fingerprint pattern. Owing to the simplicity and speed of this technique, it has been used in a wide variety of studies. However, because of the low-stringency conditions used in RAPD, the results obtained are very sensitive to minor changes in experimental conditions and separate experiments can yield amplimer bands of varying intensity [52]. This lack of reproducibility makes it difficult to compare data generated using the same primers and conditions in different laboratories. Despite this, however, the ease of use of RAPD has allowed it to be used, with varying degrees of success, in a large number of studies of fungal species [33,53–60].

6. Multilocus Enzyme Electrophoresis (MLEE)

This method analyses allelic polymorphisms that lead to the altered electrophoretic mobility of proteins in polyacrylamide gels. MLEE has been used successfully and widely applied in population studies of fungi, particularly in the analysis of reproductive strategies [61–64]. One drawback of the technique is that it is unable to detect silent mutations that do not lead to changes in the electrophoretic mobility of proteins. In addition, MLEE is quite time-consuming and the protein markers used must be chosen empirically.

IV. EPIDEMIOLOGY OF CANDIDOSIS

Candida spp. are the most common and versatile human fungal pathogens. They can cause superficial infections such as oropharyngeal and vaginal candidiasis as well as nosocomially acquired disseminated candidiasis or candidemia, during which multiple organs can be infected. Candidemia frequently results in prolonged hospitalization and has an attributable mortality estimated to be as high as 35% [65]. The species of *Candida* most frequently associated with disease, and the most pathogenic *Candida* sp., is *C. albicans*. During the 1980s *C. albicans* accounted for ~75% of cases of oral and disseminated candidiasis [66,4]. However, during the last decade there has been a significant shift in the *Candida* spp. identified as the etiological agents of candidiasis [67]. Throughout the 1990s there were a number of large-scale investigations into the epidemiology of nosocomial bloodstream infections in the United States and further afield. One such surveillance study, the SENTRY Antimicrobial Surveillance Program, determined the distribution of *Candida* spp. in nosocomial infections in 72 medical centers in the United States, Canada, Latin America, and, more recently, Europe [68–70]. The findings of this and similar studies indicated that although *C. albicans* remains the most commonly identified cause of systemic candidiasis, it was only identified in ~53% of 476 isolates examined [68].

Although regional variation was observed within the United States and Europe, the second most commonly identified species in the United States was *C. glabrata*; in the other locations it was *C. parapsilosis* [68,69]. Epidemiological studies of oral candidiasis have also cataloged a similar decrease in the incidence of *C. albicans* in relation to other *Candida* spp. [66]. However, it has been shown that in the elderly there is a marked increase in the levels of colonization with *C. albicans* and other species, especially *C. glabrata* [71]. The reasons for the epidemiological shifts observed in superficial and systemic forms of candidiasis have yet to be adequately explained. However, species such as *C. glabrata* and *C. krusei*, which have inherent decreased susceptibility to antifungal drugs relative to *C. albicans*, appear to have emerged as significant causes of infection since the introduction of the azole antifungal drug fluconazole in the early 1990s. This has led to the suggestion that the emergence of these species may be associated with the subsequent widespread use of this agent [67,72,73]. However, since *C. parapsilosis* isolates are usually susceptible to fluconazole, the emergence of this species as a significant cause of infection may be linked to the increased number of critically ill patients with indwelling venous catheters and poor implementation of cross-infection control procedures [68]. Other species identified, albeit rarely, in these studies include *C. guilliermondii*, *C. famata*, and *C. inconspicua*. Recent studies using molecular epidemiological methods have also documented the emergence of the newly discovered species *C. dubliniensis* as a significant cause of superficial and systemic disease, especially in HIV-infected individuals and in other immunocompromised patients. The significance of *C. dubliniensis* and its incidence as an agent of disease in humans have yet to be fully assessed [50,74–77].

In order to facilitate infection prevention and control, the source of the infection must be determined. The application of molecular typing systems in conjunction with computer-assisted data analysis systems to allow accurate strain comparisons during the last decade has provided a wealth of information concerning how *Candida* infections are acquired and transmitted. In most studies, different combinations of techniques have been used for confirmatory purposes and have revealed that strains within each *Candida* species are genetically very diverse and distinguishable using adequate techniques. In a majority of studies of nosocomial candidemia, it has been demonstrated that infections are most likely acquired endogenously (i.e., from commensal organisms) and that each patient is infected with a single strain that is specific for the patient [13,18,78]. Indeed, it is likely that prior colonization of oral mucosal surfaces and/or the gastrointestinal tract with *Candida* (along with previous treatment with fluconazole) is a significant risk factor for the development of candidemia in susceptible individuals [65,79,80]. In addition, in cases where multiple samples have been taken during recurrent episodes of disease, it has been shown that patients are, for the most part, infected by the same strain, although on rare occasions multiple strains and strain replacement have been observed [13].

Although most candidemia cases result from infection with endogenous commensal strains, there have been reports of multiple patients being infected by *Candida* strains with identical genotypes, suggesting that nosocomial transmission of strains does occur [13,81]. In particular, studies investigating the origin of outbreaks of candidemia in neonatal intensive care units have sometimes strongly implicated transmission of strains from health care workers or catheters to neonates [82–84]. Indeed, the reported increased incidence of exogenously acquired infections caused by *C. parapsilosis* represents a particularly worrying trend and suggests that infection control procedures and/or their implementation in many institutions require improvement. In a recent study it was demonstrated that health care workers are frequently contaminated by infected or colonized patients [85]. Clearly these workers are a major potential source of cross-infection in the absence of adequate handwashing and without adequate implementation of general cross-infection control precautions.

The origin of superficial cases of candidiasis is almost always endogenous; however, there have been reports of transmission between individuals [86]. In one particular study, it was shown that an azole-resistant strain of *C. albicans* was transmitted from the oral cavity of an HIV-positive mother to her children, who subsequently developed oropharyngeal candidiasis [87].

Vaginal and oral candidiasis are often recalcitrant to therapy, and relapses or recurrences of infection occur frequently. In the case of vaginal infections it has been shown that recurrent episodes are most often due to maintenance of the originally infecting strain (or substrain) [88,89]. However, in some cases of oral candidiasis, particularly in patients with AIDS, the situation is often far more complex and dynamic with the populations of individual strains and species altering significantly from one episode of infection to another [66,90].

In addition to the analysis of the source of infections and outbreaks, the genotyping techniques described above have also been applied to the analysis of the population biology of *C. albicans*. The genetic diversity present within *Candida* spp. has already been referred to. However, the application of sophisticated DNA fingerprint profile comparison computer software programs to large numbers of isolates recovered from cases of bloodstream infection caused by *C. albicans* has led to the suggestion that, at least in some instances, particular clones of *C. albicans* are found more frequently in certain geographic areas of the United States [91]. The additional finding that some strains appear to be endemic within some institutions is further evidence for lapses in, or inadequate, cross-infection control practices [91]. More recently, a DNA fingerprinting study on 266 *C. albicans* isolates, obtained from a variety of anatomical sites in different patient types from around the world, suggested the existence of a cluster of related isolates, termed *general-purpose genotype* isolates, which predominate as the main cause of candidiasis worldwide.

The significance of this cluster (which accounted for 37% of the strains examined) has yet to be determined [92]. However, if strains with the general-purpose genotype can be shown to be more pathogenic than other *C. albicans* strains, comparisons between these groups of strains could prove helpful in improving our understanding of how *C. albicans* colonizes and infects susceptible individuals. Given the fact that the population structure of *C. albicans* is primarily clonal, and that to date no sexual reproductive cycle has been identified in this species, it is probably not surprising that some clones are more prevalent than others. However, there is some genotypic evidence for limited recombination in *C. albicans* suggesting that this species has, or has had in the past, the ability to undergo sexual reproduction [93]. In addition, the recent identification of *C. albicans* sequences homologous to the *Saccharomyces cerevisiae* mating loci (*MATa* and *MATα*) suggests that the reproductive strategy of this species may be more complex than has previously been considered [94].

Analysis of *Candida* populations is complicated further by the phenomenon of microevolution, in which clonally related populations exhibit subtle differences in genotype, manifested as minor band differences in DNA fingerprint and karyotype profiles. To date, microevolution has been demonstrated to occur in *C. albicans* [19,88], *C. glabrata* [21], and *C. dubliniensis* [24,32,95], although its significance in vivo has yet to be assessed. However, the association of microevolutaonary events with the development of azole resistance in isolates of *C. albicans* [96] and *C. dubliniensis* [97] exposed to fluconazole suggests that microevolution may be associated with the ability of these species to adapt to harsh environmental conditions, including the presence of toxic compounds.

V. EPIDEMIOLOGY OF ASPERGILLOSIS

Aspergillus spp. are filamentous fungi found widely in the environment and are significant opportunistic pathogens of immunocompromised individuals. Patients particularly at risk of contracting invasive aspergillosis are those who are neutropenic as a result of acute myeloid leukemia and those who have received bone marrow transplants [98,99]. As a consequence of the increased numbers of these patients in our hospitals, it is not surprising that during the last two decades there has been a parallel increase in the incidence of *Aspergillus* infections [100]. The *Aspergillus* sp. most commonly associated with human disease is *A. fumigatus*. It is believed that *Aspergillus* is mainly acquired by inhalation of airborne fungal spores that are ubiquitous in the environment. Decaying vegetation has been suggested as the main source of these fungi, which produce copious quantities of spores. Owing to their small size and aerodynamic properties, spores can remain airborne for prolonged time periods and are thus easily dispersed [101]. These spores can then be inhaled and subsequently colonize the upper or lower respiratory tract and act as a reservoir for infection, or directly infect the lower respiratory tract, depending on the level of immunocompetence of the individual. The ubiquitous nature of *Aspergillus* spores in the air has led to the suggestion that activities such as building construction, which can cause agitation of air currents, may be an important contributory factor to outbreaks of nosocomial aspergillosis [2,101]. Because of the high mortality rate associated with aspergillosis and the expense of treating invasive infection it is essential that the epidemiology of these infections be thoroughly investigated and that adequate prophylactic and preventive measures be determined and implemented. The techniques used to investigate the epidemiology of *Aspergillus* infections are essentially the same as those used for to the molecular analysis of *Candida* infections, as described above.

Since isolates of *A. fumigatus* are virtually impossible to distinguish on the basis of phenotypic characteristics, the majority of studies on the epidemiology of *Aspergillus* infections have relied on DNA-based fingerprinting analysis. To date the most frequently used methods have included RAPD and REA analysis [18,102–109]. In one of the most extensive studies performed to date, more than 700 isolates obtained from clinical and environmental samples from four hospitals over a 12-to-24-month period were fingerprinted using the highly discriminatory and effective DNA fingerprinting probe *Afut*1 [110]. This probe contains a semirepetitive retrotransposon-like sequence. Computer-aided analysis of the fingerprint patterns obtained using this probe demonstrated that the vast majority of strains identified were unique. These findings confirmed the very high level of genetic diversity present among *A. fumigatus* isolates identified previously and the absence of clones with increased virulence. However, a small proportion (15%) of strains were isolated in particular institutions on more than one occasion, suggesting an ability of the organisms to persist in the environment for prolonged time periods. Interestingly, most patients examined were colonized or infected with more than one strain. In addition, in a number of cases, the same strain was recovered from separate patients or from a patient and an environmental sample, strongly suggesting that aspergillosis can be acquired nosocomially.

Similar findings to those described above have also been obtained with *A. fumigatus* isolates using microsatellite markers [111]. However, whether some cases of aspergillosis may result from outgrowth of spores acquired in the community prior to hospitalization has yet to be investigated. Given the high degree of genetic variation present within *A. fumigatus* and the absence of an identified teleomorph, the population biology of *A. fumigatus* may be more complicated than has previously been thought. Evidence from MLEE and DNA fingerprinting analyses suggests that *A. fumigatus* is haploid and clonal in nature. However, as with *C. albicans*, there is some evidence suggesting that limited recombination may take place within this species [112].

VI. EPIDEMIOLOGY OF CRYPTOCOCCOSIS

Cryptococcus neoformans is a basidiomycetous, heterothallic, encapsulated yeast species which is a significant cause of morbidity in immunocompromised individuals [113]. The most common manifestation of cryptococcal disease is meningoencephalitis in HIV-infected and AIDS patients.

In a recent epidemiological study of the incidence of cryptococcosis in four areas of the United States during the period 1992–94, it was found that 86% of cases were in HIV-infected individuals, with the remainder occurring in patients with cancer or diabetes and in patients receiving corticosteroid therapy [114]. The effect of the introduction of highly active antiretroviral therapy (HAART) in recent years on the incidence of cryptococcosis has yet to be addressed adequately. However, despite the clear improvement in the immune status of individuals receiving HAART, cases of cryptococcosis are still occurring [115]. The species *C. neoformans* has been divided into four groups A-D (serotype A/D has also been described), originally on the basis of serotyping studies [113].

The majority of isolates associated with human disease belong to serotype A; the remainder (mainly recovered in Europe) belong predominantly to serotype D. Strains belonging to serotypes B and C are mainly found in environmental samples in tropical regions (they are particularly associated with eucalyptus trees) and have only been associated with human disease in a small number of cases. On the basis of phenotypic and genotypic analyses, isolates belonging to serotype B and C have been assigned variety status and now belong to variety *C. neoformans* var. *gatii*, while serotype A and D belong to variety *C. neoformans* var. *neoformans* [113]. It has been suggested recently that these groupings may actually represent separate species [53]. In addition, on the basis of clinical and epidemiological data, DNA fingerprinting analysis, and nucleotide sequence analysis of the *URA5* gene, it has been proposed that serotype A strains should be assigned to a new variety called *C. neoformans* var. *grubii* [116].

A wide variety of molecular epidemiological techniques have been applied to the analysis of cryptococcal epidemiology and population biology. The most frequently used methods include DNA fingerprinting with species-specific probes (CNRE-1, UT-4p, and CND1.7) [117–122]. In addition, DNA fingerprinting with oligonucleotide probes homologous to microsatellite sequences [30,33], RAPD [123–125,33,126], PFGE [127–130], MLEE [131–133], and *URA5* nucleotide sequence analysis [134,135] have also been used to varying effect. As with other fungal pathogens, these methods have revealed a high degree of genotypic heterogeneity in clinical and environmental isolates. However, in a number of cases the close similarity of DNA fingerprints of strains (obtained using CNRE-1 and UT-4p) recovered from pigeon guano and human infections supports the suggestion that the environmental source of serotype A and D strains is likely to be the pigeon [136,137]. The demonstration of zoonotic transmission of a *C. neoformans* strain from a pet cockatoo to its owner suggests that it is likely that other avian species may also act as reservoirs of infection [138]. In another recent study using DNA fingerprinting with the CNRE-1 probe and RAPD it was found that strains causing cryptococcal meningitis in African emigrants in France were more similar to African than French environmental isolates, despite the fact that the infection was diagnosed following prolonged periods of residence in France. These data strongly suggest that, in these cases at least, the strains responsible for infection were dormant in the host (perhaps in alveolar macrophages) and only began to cause symptoms and disease once the patients immune system became compromised [124].

One of the main problems in treating cryptococcal meningitis is that despite apparent resolution of symptoms following treatment with antifungal agents, many patients suffer from recurrent episodes of disease. DNA fingerprinting studies suggest that in the majority of cases recurrent infections are caused by the persistence of the originally infecting strain [30,33,120,121,128]. However, in two separate studies RAPD and DNA fingerprint analysis using microsatellite probes suggested that reinfection with a novel strain can occur [30,33]. In addition, these studies demonstrated that patients can be infected with more than one strain simultaneously. In the same study, analysis of DNA fingerprints of multiple colonies from primary isolation plates suggested that cryptococcal populations can show a significant degree of heterogeneity, indicating that *C. neoformans* isolates may exhibit microevolution in vivo

[30,33]. Minor karyotype changes have also been shown in a strain passaged in vitro [139]. Interestingly, the parental strain and daughter derivatives differed in a number of phenotypic characteristics, including virulence. This suggests that genetic instability/plasticity may be related to the recently described phenomenon of *C. neoformans* phenotypic switching [140,141].

MLEE and genotypic analysis suggests that *C. neoformans* populations are clonal in origin with the possible existence of widely distributed pathogenic clones. However, data also suggest that a limited degree of recombination may occur in natural populations [112].

VII. OTHER FUNGAL SPECIES

The previous discussion has been limited to the most clinically significant fungal pathogens. However, there exist a wide variety of other fungi that have the capacity to cause human infections but which have not been subjected to extensive molecular epidemiological analysis. There are a number of reasons for this. For instance, dermatophytes are mainly responsible for superficial infections, Zygomycetes, although responsible for very serious systemic infections, are still relatively rare, and species such as *Coccidioides immitis*, *Histoplasma capsulatum*, and *Penicillium marneffei* are largely endemic in relatively limited geographical areas. However, molecular tools are being deployed to aid in the investigation of the epidemiology and population biology of a number of these species [142–146]. Recent analysis of *Trichophyton rubrum* strains from around the world using PCR-based techniques has revealed that this species shows only limited genetic variation. This is indicative of a strictly clonal mode of reproduction, which might be dictated by the adaptive evolution of this species to survive in the specialized environment of the human skin [147].

Molecular analysis of *C. immitis*, a pathogenic species which is endemic in the U.S. Southwest, has revealed recombination in natural populations and the existence of two cryptic species which are geographically isolated [148,149]. In a recent study using analysis of single-nucleotide polymorphisms to compare strains isolated from infected patients in 1993–94, strains were found to exhibit a high degree of genetic variability [150]. Based on these data the authors of this report concluded that outbreaks of coccidioidomycosis in endemic areas are not the result of the emergence of pathogenic clones. Instead, statistical data indicate that outbreaks are very likely due to a combination of environmental conditions (e.g., lengths of droughts and the amounts of rainfall). These data suggest that, at least in some cases, climatic and other environmental variables should be considered in the epidemiological analysis of epidemics of fungal infections and that these conditions could be monitored and used to predict possible outbreaks of disease.

VIII. CONCLUSIONS

The significant increased incidence of fungal infections observed during the last decade is likely to continue for the foreseeable future. The main reason for this is that the patient population with risk factors for the acquisition of opportunistic infections, namely the immunocompromised and the hospitalized, continues to grow. Also despite the introduction of highly active antiretroviral therapy (HAART) for the treatment of HIV infection and AIDS and the effect that this has had on the reduction of mycoses in HIV-infected individuals, there have been reports of non-compliance and therapeutic failure [151,152]. In addition, on a world-wide scale the proportion of HIV-infected patients receiving HAART therapy is very small. As well as the commonly identified fungal species described above, new and emerging fungal species (e.g.,

C. dubliniensis and *P. marneffeii*), the epidemiology of which require extensive further analysis, continue to emerge as human pathogens. Finally, there are increasing numbers of reports of cases of fungal infections that are recalcitrant to treatment with the currently available range of antifungal agents, largely due to developed or inherent resistance to these drugs. Therefore it is clear that the need for the development of inexpensive and informative epidemiological tools for the analysis of fungal diseases will continue. In addition, in cases of fungal species where there may be scope for the development of vaccines, it is essential that the population dynamics of these species are investigated rigorously to determine if there are significant levels of recombination within the population.

The continuing development and refinement of currently used and novel molecular epidemiological techniques coupled with the application of automated and computer-assisted technologies will certainly increase our understanding of how fungal diseases are acquired and spread within human populations. However, these techniques in isolation will not address all of the questions that require answering. No effective epidemiological survey can take place without adequate attention to the collection of the isolates to be typed and the choice of epidemiologically unrelated control or reference strains for the purposes of comparison. Despite the increasing volume of data showing the presence of multiple strains (and sometimes species) in particular cases of infection and the phenomenon of microevolution, many epidemiologists continue to assume that single colonies from primary isolation plates are representative of the entire microbial population present in particular clinical samples. As discussed previously, this is not always the case. Although the analysis of multiple colonies from individual samples greatly increases the fingerprinting workload, the data generated will be more accurate and reliable. Therefore, with the combined application of improved typing methods and sample collection, it is to be hoped that the acquisition of epidemiological information will ultimately lead to the improved care and prognosis of patients suffering from or at risk of acquiring mycotic infections.

REFERENCES

1. SP Fischer-Hoch, L Hutwagner. Opportunistic candidiasis: an epidemic of the 1980s. Clin Infect Dis 21:897–904, 1995.
2. SK Fridkin, WR Jarvis. Epidemiology of nosocomial fungal infections. Clin Microbiol Rev 9: 499–511, 1996.
3. C Beck-Sague, WR Jarvis. Secular trends in the epidemiology of nosocomial fungal infections in the United States, 1980–1990. J Infect Dis 167:1247–1251, 1993.
4. WR Jarvis. Epidemiology of nosocomial fungal infections, with emphasis on *Candida* species. Clin Infect Dis 20:1526–1530, 1995.
5. DR Soll. The ins and outs of DNA fingerprinting the infectious fungi. Clin Microbiol Rev 13: 332–370, 2000.
6. E Bart-Delabesse, P Boiron, A Carlotti, B Dupont. *Candida albicans* genotyping in studies with patients with AIDS developing resistance to fluconazole. J Clin Microbiol 31:2933–2937, 1993.
7. A Carlotti, G Zambardi, A Couble, N Lefrancois, X Martin, J Villard. Nosocomial infection with *Candida albicans* in a pancreatic transplant recipient investigated by means of restriction enzyme analysis. J Infect 29:157–164, 1994.
8. PR Hunter. A critical review of typing methods for *Candida albicans* and their applications. Crit Rev Microbiol 17:4176–4434, 1991.
9. R Khatib, MC Thirumoothi, KM Riederer, L Sturm, LA Oney, J Baran Jr. Clustering of *Candida* infections in the neonatal intensive care unit: concurrent emergence of multiple strains simulating intermittent outbreaks. Pediatr Infect Dis 17:130–134, 1998.

10. BB Magee, TM D'Souza, PT Magee. Strain and species identification by restriction fragment length polymorphisms in the ribosomal DNA repeat of *Candida albicans*. J Bacteriol 169:1639–1643, 1987.

11. R Matthews, J Burnie. Assessment of DNA fingerprinting for rapid identification of outbreaks of systemic candidiasis. BMJ 298:354–357, 1989.

12. MA Pfaller, I Cabezudo, R Hollis, B Huston, RP Wenzel. The use of biotyping and DNA fingerprinting in typing *Candida albicans* from hospitalised patients. Diagn Microbiol Infect Dis 13:481–489, 1990.

13. MA Pfaller, SA Messer, A Houston, MS Rangel-Frausto, T Wiblin, HM Blumberg, JE Edwards, W Jarvis, MA Martin, HC Neu, L Saiman, JE Patterson, JC Dibb, CM Roldsn, MG Rinaldi, RP Wenzel. National epidemiology of mycoses survey: a multicenter study of strain variation and antifungal susceptibility among isolates of *Candida* species. Diagn Microbiol Infect Dis 31:289–296, 1998.

14. F Romano, G Ribera, M Giuliano. A study of a hospital cluster of systemic candidosis using DNA typing methods. Epidemiol Infect 112:393–398, 1994.

15. S Scherer, DA Stevens. Application of DNA typing methods to epidemiology and taxonomy of *Candida* species. J Clin Microbiol 675–679, 1987.

16. DA Stevens, FC Odds, S Scherer. Application of DNA typing methods to *Candida albicans* epidemiology and correlations with phenotype. Rev Infect Dis 12:258–266, 1990.

17. JA Vazquez, A Beckley, JD Sobel, MJ Zervos. Comparison of restriction enzyme analysis and pulsed-field gradient gel electrophoresis as typing systems for *Candida albicans*. J Clin Microbiol 29:962–967, 1991.

18. E Reiss, K Tanaka, G Bruker, V Chazalet, D Coleman, JP Debeaupuis, R Hanazawa, JP Latge, J Lortholary, K Makimura, CJ Morrison, SY Murayama, S Naoe, S Paris, J Sarfati, K Shibuya, D Sullivan, K Uchida, H Yamaguchi. Molecular diagnosis and epidemiology of fungal infections. Med Mycol 36(suppl1):249–257, 1998.

19. S Scherer, DA Stevens. A *Candida albicans* dispersed, repeated gene family and its epidemiologic applications. Proc Natl Acad Sci USA 85:1542–1465, 1988.

20. C Sadhu, MJ McEachern, EP Rustchenko-Bulgac, J Schmid, DR Soll, JB Hicks. Telomeric and dispersed repeat sequences in *Candida* yeasts and their use in strain identification. J Bacteriol 173:842–850, 1991.

21. SR Lockhart, S Joly, C Pujol, JD Sobel, MA Pfaller, DR Soll. Development and verification of fingerprinting probes for *Candida glabrata*. Microbiology 143:3733–3746, 1997.

22. S Joly, C Pujol, K Schroppel, DR Soll. Development and verification of two species fingerprinting probes for *Candida tropicalis* amenable for computer analysis. J Clin Microbiol 34:3063–3071, 1996.

23. A Carlotti, R Grillot, A Couble, J Villard. Typing of *Candida krusei* clinical isolates by restriction endonuclease analysis and hybridization with the CkF1,2 DNA probe. J Clin Microbiol 32:1691–1699, 1994.

24. S Joly, C Pujol, M Rysz, K Vargas, DR Soll. Development and characterization of complex DNA fingerprinting probes for the infectious yeast *Candida dubliniensis*. J Clin Microbiol 37:1035–1044, 1999.

25. H Girardin, J-P Latge, T Srikantha, B Morrow, DR Soll. Development of DNA probes for fingerprinting *Aspergillus fumigatus*. J Clin Microbiol 31:1547–1554, 1993.

26. C Neuveglise, J Safarti, JP Latge, S Paris. *Afut*1, a retrotransposon-like element from *Aspergillus fumigatus*. Nucleic Acids Res 24:1428–1434, 1996.

27. ED Spitzer, SG Spitzer. Use of a dispersed repetitive DNA element to distinguish clinical isolates of *Cryptococcus neoformans*. J Clin Microbiol 30:1094–1097, 1992.

28. C Epplen, G Melmer, I Siedlaczk, F-W Schwaiger, W Mauler, JT Epplen. On the essence of "meaningless" simple repetitive DNA in eukaryotic genomes. In: SDJ Pena, R Chakraborty, JT Epplen, AJ Jeffreys, eds. DNA Fingerprinting: State of the Science. Basel: Birkauser Verlag, 1993, pp 29–45.

29. D Field, C Wills. Long, polymorphic microsatellites in simple organisms. Proc R Soc Lond B 263:209–215, 1996.

30. KA Haynes, DJ Sullivan, DC Coleman, JCK Clarke, R Emilianus, C Atkinson, KJ Cann. Involvement of multiple *Cryptococcus neoformans* strains in a single episode of cryptococcosis and reinfection with novel strains in recurrent infection demonstrated by random amplication of polymorphic DNA and DNA fingerprinting. J Clin Microbiol 33:99–102, 1995.

31. W Meyer, A Koch, C Niemann, B Beyermann, JT Epplen, T Borner. Differentiation of species and strains among filamentous fungi by DNA fingerprinting. Curr Genet 19:239–242, 1991.

32. D Sullivan, D Bennett, M Henman, P Harwood, S Flint, F Mulcahy, D Shanley, D Coleman. Oligonucleotide fingerprinting of isolates of *Candida* species other than *C. albicans* and of atypical *Candida* species from human immunodeficiency virus-positive and AIDS patients. J Clin Microbiol 31:2124–2133, 1993.

33. D Sullivan, K Haynes, G Moran, D Shanley, D Coleman. Persistence, replacement, and microevolution of *Cryptococcus neoformans* strains in recurrent meningitis in AIDS patients. J Clin Microbiol 34:1739–1744, 1996.

34. BM Wilkinson, L Morris, DJ Adams, EG Evans, CJ Lacey, RM Walmsley. A new, sensitive polynucleotide probe for distinguishing *Candida albicans* strains and its use with a computer assisted archiving and pattern comparison system. J Med Vet Mycol 30:123–131, 1992.

35. W Meyer, TG Mitchell. Polymerase chain reaction fingerprinting in fungi using single primers specific to minisatellites and simple repetitive DNA sequences: strain variation in *Cryptococcus neoformans*. Electrophoresis 16:1648–1656, 1995.

36. K Weissing, R Atkinson, R Gardner. Genomic fingerprinting by microsatellite-primed PCR: a critical evaluation. PCR Methods App 4:249–255, 1995.

37. DC Schwartz, CR Cantor. Separation of yeast chromosome-sized DNAs by pulsed-field gradient gel electrophoresis. Cell 37:67–75, 1984.

38. M Doi, M Homma, S Iwaguchi, K Horibe, K Tanaka. Strain relatedness of *Candida albicans* strains isolated from children with leukemia and their bedside parents. J Clin Microbiol 32:2253–2259, 1994.

39. E Pontieri, L Gregori, M Gennarelli, T Ceddia, G Novelli, B Dallapiccola, F De Bernardis, G Carruba. Correlation of *Sfi*I macrorestriction endonuclease fingerprint analysis of *Candida parapsilosis* isolates with source of isolation. J Med Microbiol 45:173–178, 1996.

40. K Riederer, P Fozo, R Khatib. Typing of *Candida albicans* and *Candida parapsilosis*: species-related limitations of electrophoretic karyotyping and restriction endonuclease analysis of genomic DNA. Mycoses 41:397–402, 1998.

41. K Asakura, S-I Iwaguchi, M Homma, T Sukai, K Higashide, K Tanaka. Electrophoretic karyotypes of clinically isolated yeasts of *Candida albicans* and *C. glabrata*. J Gen Microbiol 137:2531–2538, 1991.

42. F Barchiesi, RJ Hollis, SA Messer, G Scalise, MG Rinaldi, MA Pfaller. Electrophoretic karyotype and in vitro antifungal susceptibility of *Cryptococcus neoformans* isolates from AIDS patients. Diagn Microbiol Infect Dis 23:99–103, 1995.

43. G Carruba, E Pontieri, F De Bernardis, P Martino, A Cassone. DNA fingerprinting and electrophoretic karyotypes of environmental and clinical isolates of *Candida parapsilosis*. J Clin Microbiol 29:916–922, 1991.

44. A Espinel-Ingroff, A Quart, L Steele-Moore, I Metcheva, GA Buck, VL Bruzzese, D Reich. Molecular karyotyping of multiple yeast species isolated from nine patients with AIDS during prolonged fluconazole therapy. J Med Vet Mycol 34:111–116, 1996.

45. BB Magee, PT Magee. Electrophoretic karyotypes and chromosome numbers in *Candida* species. J Gen Microbiol 133:425–430, 1987.

46. WG Merz, U Khazan, MA Jabra-Rizk, L-C Wu, GJ Osterhout, PF Lehmann. Strain delineation and epidemiology of *Candida (Clavispora) lusitaniae*. J Clin Microbiol 30:449–454, 1992.

47. M Monod, S Porchet, F Baudraz-Rosselet, E Frenk. The identification of pathogenic yeast strains by electrophoretic analysis of their chromosomes. J Med Microbiol 32:123–129, 1990.

48. S Pan, GT Cole. Electrophoretic karyotypes of clinical isolates of *Coccidioides immitis*. Infect Immun 60:4872–4880, 1992.

49. V Sanchez, JA Vasquez, D Barth-Jones, L Dembry, JD Sobel, MJ Zervos. Epidemiology of nosocomial acquisition of *Candida lusitaniae*. J Clin Microbiol 30:3005–3008, 1992.

50. DJ Sullivan, TJ Westerneng, KA Haynes, DE Bennett, DC Coleman. *Candida dubliniensis* sp. nov.: phenotypic and molecular characterisation of a novel species associated with oral candidosis in HIV-infected individuals. Microbiology 141:1507–1521, 1995.

51. J Welsh, M McClelland. Fingereprinting genomes using PCR with arbitrary primers. Nucleic Acids Res 18:7213–7224, 1990.

52. DL Ellsworth, KD Rittenhouse, RL Honeycutt. Artifactual variation in randomly amplified polymorphic DNA banding patterns. BioTehniques 14:214–217, 1993.

53. T Boekhout, A Van Belkum, AC Leenders, HA Verbrugh, P Mukamurangkwa, D Swinne, WA Scheffers. Molecular typing of *Cryptococcus neoformans*: taxonomic and epidemiologic aspects. Int J Syst Bacteriol 47:432–442, 1997.

54. A Bostok, MN Khattak, R Matthews, J Burnie. Comparison of DNA fingerprinting, by random amplification of polymorphic DNA, with other molecular typing methods for *Candida albicans*. J Gen Microbiol 139:2179–2184, 1993.

55. K Holmberg, F Feroze. Evaluation of an optimised system for random amplified polumorphic DNA (RAPD)-analysis for genotypic mapping of *Candida albicans* strains. J Clin Lab Anal 10:59–69, 1996.

56. A Leenders, A Van Belkum, S Janssen, S De Marie, J Kluytmans, J Wielenga, B Lowenberg, H Verbrugh. Molecular epidemiology of apparent outbreak of invasive aspergillosis. J Clin Microbiol 34:345–351, 1996.

57. PF Lehman, D Lin, BA Lasker. Genotypic identification and characterization of species and strains within the genus *Candida* by using random amplified polymorphic DNA. J Clin Microbiol 30: 3249–3254, 1992.

58. D Lin, PF Lehmann. Random amplified polymorphic DNA for strain delineation within *Candida tropicalis*. J Med Vet Mycol 33:241–246, 1995.

59. C Pujol, S Joly, SR Lockhart, S Noel, M Tibayrenc, DR Soll. Parity among the randomly amplified polymorphic DNA method, multilocus enzyme electrophoresis, and Southern blot hybridization with the moderately repetitive probe Ca3 for fingerprinting *C. albicans*. J Clin Microbiol 35: 3248–3258, 1997.

60. F Robert, F Lebreton, ME Bougnoux, A Paugam, D Wassermann, M Schlotterer, C Tourte-Schaefer, J Dupouy-Camet. Use of random amplified polymorphic DNA as a typing method for *Candida albicans* in epidemiological surveillance of a burns unit. J Clin Microbiol 33:2366–2371, 1995.

61. P Boerlin, F Boerlin-Petzold, J Goudet, C Durussel, JL Pagani, JP Chave, J Bille. Typing *Candida albicans* oral isolates from human immunodeficiency virus–infected patients by multilocus electrophoresis and DNA fingerprinting. J Clin Microbiol 34:1235–1248, 1996.

62. ME Brandt, LC Hutwagner, RJ Kuykendall, RW Pinner. Comparison of multilocus enzyme electrophoresis and random amplified polymorphic DNA analysis for molecular subtyping of *Cryptococcus neoformans*. J Clin Microbiol 33:1890–1895, 1995.

63. C Pujol, F Renaud, M Mallie, T de Meeus, J-M Bastide. Atypical strains of *Candida albicans* recovered from AIDS patients. J Med Vet Mycol 35:115–121, 1997.

64. E Rodriguez, T De Meeus, M Mallie, F Renaud, F Symoens, P Mondon, MA Piens, B Lebeau, MA Viviani, R Grillot, N Nolard, F Chapuis, AM Tortorano, JM Bastide. Multicentric epidemiological study of *Aspergillus fumigatus* isolates by multilocus enzyme electrophoresis. J Clin Microbiol 34:2559–2568, 1996.

65. RP Wenzel. Nosocomial candidemia: risk factors and attributable mortality. Clin Infect Dis 20: 1531–1534, 1995.

66. DC Coleman, DE Bennett, PJ Gallagher, SR Flint, A Nolan, FM Mulcahy, DJ Sullivan, MC Henman, RJ Russell, DB Shanley. Oral candidiasis and HIV infection: antifungal drug resistance and changes in *Candida* population dynamics. In: JS Greenspan, D Greenspan, eds. Oral Manifestations of HIV Infection. Chicago: Quintessence, 1995, pp 112–118.

67. MH Nguyen, JE Peacock, AJ Morris, DC Tanner, ML Nguyen, DR Snydman, MM Wagener, MG Rinaldi, LY Yu. The changing face of candidemia: emergence of non–*Candida albicans* species and antifungal resistance. Am J Med 100:617–623, 1996.

68. MA Pfaller, RN Jones, GV Doern, AC Fluit, J Verhoef, HS Sader, SA Messer, A Houston, S Coffman, RJ Hollis. International surveillance of blood stream infections due to *Candida* species

in the European SENTRY program: species distribution and antifungal susceptibility including the investigational triazole and echinocandin agents. Diagn Microbiol Infect Dis 35:19–25, 1999.

69. MA Pfaller, RN Jones, GV Doern, HS Sader, RJ Hollis, SA Messer, Sentry Participant Group. International surveillance of bloodstream infections due to *Candida* species: frequency of occurrence and antifungal susceptibilities of isolates collected in 1997 in the United States, Canada, and South America for the SENTRY program. J Clin Microbiol 36:1886–1889, 1998.

70. MA Pfaller, RN Jones, GV Doern, HS Sader, SA Messer, A Houston, S Coffman, RJ Hollis. Bloodstream infections due to *Candida* species: SENTRY antimicrobial surveillance program in North America and Latin America, 1997–1998. Antimicrob Agents Chemother 44:747–751, 2000.

71. S Lockhart, S Joly, K Vargas, J Swails-Wenger, L Enger, DR Soll. Natural defenses against *Candida* colonization break down in the oral cavity of the elderly. J Dent Res 78:857–868, 1998.

72. D Abi-Said, E Anaissie, O Uzun, I Raad, H Pinzcowski, S Vartivarian. The epidemiology of hematogenous candidiasis caused by different *Candida* species. Clin Infect Dis 24:1122–1128, 1997.

73. MF Price, MT La Rocco, LO Gentry. Fluconazole susceptibilities of *Candida* species and distribution of species recovered from blood cultures over a 5-year period. Antimicrob Agents Chemother 38: 1422–1424, 1994.

74. ME Brandt, LH Harrison, M Pass, AN Sofair, S Huie, R-K Li, CJ Morrison, DW Warnock, RH Hajjeh. *Candida dubliniensis* fungemia: the first four cases in North America. Emerg Infect Dis 6: 46–49, 2000.

75. JF Meis, M Ruhnke, BE De Pauw, FC Odds, W Siegert, PE Verweij. *Candida dubliniensis* candidemia in patients with chemotherapy-induced neutropenia and bone marrow transplantation. Emerg Infect Dis 5:150–153, 1999.

76. I Polacheck, J Strahilevitz, D Sullivan, S Donnelly, IF Salkin, DC Coleman. Recovery of *Candida dubliniensis* from non–human immunodeficiency virus–infected patients in Israel. J Clin Microbiol 38:170–174, 2000.

77. D Sullivan, K Haynes, J Bille, P Boerlin, L Rodero, S Lloyd, M Henman, D Coleman. Widespread geographic distribution of oral *Candida dubliniensis* strains in human immunodeficiency virus–infected individuals. J Clin Microbiol 35:960–964, 1997.

78. A Voss, RJ Hollis, MA Pfaller, RP wenzel, BN Doebbeling. Investigation of the sequence of colonization and candidemia in non-neutropenic patients. J Clin Microbiol 32:975–980, 1994.

79. CA Pertowski, RC Baron, BA Lasker, SB Werner, WR Jarvis. Nosocomial outbreak of *Candida albicans* sternal wound infections following cardiac surgery traced to a scrub nurse. J Infect Dis 172:817–822, 1995.

80. D Pittet, M Monod, PM Suter, E Frenk, R Auckenthaler. *Candida* colonization and subsequent infections in critically ill surgical patients. Ann Surg 220:751–758, 1994.

81. A Voss, MA Pfaller, RJ Hollis, J Rhine-Chalberg, BN Doebbeling. Investigation of *Candida albicans* transmission in a surgical intensive care unit cluster by using genomic DNA typing methods. J Clin Microbiol 33:576–580, 1995.

82. P Betremieux, S Chevrier, G Quindos, D Sullivan, L Polonelli, C Guiguen. Use of DNA fingerprinting and biotyping methods to study a *Candida albicans* outbreak in a neonatal intensive care unit. Pediatr Infect Dis J 13:899–905, 1994.

83. DR Reagan, MA Pfaller, RJ Hollis, RP Wenzel. Evidence of nosocomial spread of *Candida albicans* causing bloodstream infection in a neonatal intensive care unit. Diagn Microbiol Infect Dis 21: 191–194, 1995.

84. LA Waggoner-Fountain, MW Walker, RJ Hollis, MA Pfaller, JE Ferguson 2d, RP Wenzel, LG Donowitz. Vertical and horizontal transmission of unique *Candida* species to premature newborns. Clin Infect Dis 22:803–808, 1996.

85. F Marco, SR Lockhart, MA Pfaller, C Pujol, MS Rangel-Frausto, T Wiblin, HM Blumberg, JE Edwards, W Jarvis, L Saiman, JE Patterson, MG Rinaldi, RP Wenzel, DR Soll. Elucidating the origins of nosocomial infections with *Candida albicans* by DNA fingerprinting with the complex probe Ca3. J Clin Microbiol 37:2817–2828, 1999.

86. F Dromer, L Improvisi, B Dupont, M Eliaszewicz, G Pialoux, S Fournier, V Feuillie. Oral transmission of *Candida albicans* between partners in HIV-infected couples could contribute to dissemination of fluconazole-resistant isolates. AIDS 11:1095–1101, 1997.

87. FM Muller, M Kasai, A Francesconi, B Brillante, M Roden, J Peter, SJ Chanock, TJ Walsh. Transmission of an azole-resistant isogenic strain of *Candida albicans* among human immunodeficiency virus–infected family members with oropharyngeal candidiasis. J Clin Microbiol 37:3405–3408, 1999.

88. SR Lockhart, BD Reed, CL Pierson, DR Soll. Most frequent scenario for recurrent *Candida* vaginitis is strain maintenance with "substrain shuffling": demonstration by sequential DNA fingerprinting with probes Ca3, C1, and CARE2. J Clin Microbiol 34:767–777, 1996.

89. K Schroppel, M Rotman, R Galask, K Mac, DR Soll. Evolution and replacement of *Candida albicans* strains during recurrent vaginitis demonstrated by DNA fingerprinting. J Clin Microbiol 32:2646–2654, 1994.

90. MA Pfaller. Epidemiology and control of fungal infections. Clin Infect Dis 19(suppl 1):S8–S13, 1994.

91. MA Pfaller, SR Lockhart, C Pujol, JA Swails-Wenger, SA Messer, MB Edmond, RN Jones, RP Wenzel, DR Soll. Hospital specificity, region specificity, and fluconazole resistance of *Candida albicans* bloodstream isolates. J Clin Microbiol 36:1518–1529, 1998.

92. J Schmid, S Herd, PR Hunter, RD Cannon, MS Yasin, S Samad, M Carr, D Parr, W McKinney, M Schousboe, B Harris, R Ikram, M Harris, A Restrepo, G Hoyos, KP Singh. Evidence for a general-purpose genotype in *Candida albicans*, highly prevalent in multiple geographical regions, patient types and types of infection. Microbiology 145:2405–2413, 1999.

93. Y Graser, M Volovsek, J Arrington, G Schonian, W Presber, TG Mitchell, R Vilgalys. Molecular markers reveal that population structure of the human pathogen *Candida albicans* exhibits both clonality and recombination. Proc Natl Acad Sci USA 93:12473–12477, 1996.

94. CM Hull, AD Johnson. Identification of a mating type–like locus in the asexual pathogenic yeast *Candida albicans*. Science 285:1271–1275, 1999.

95. GP Moran, DJ Sullivan, MC Henman, CE McCreary, BJ Harrington, DB Shanley, DC Coleman. Antifungal drug susceptibilities of oral *Candida dubliniensis* isolates from human immunodeficiency virus (HIV)-infected and non-HIV-infected subjects and generation of stable fluconazole-resistant derivatives in vitro. Antimicrob Agents Chemother 41:617–623, 1997.

96. TC White, MA Pfaller, MG Rinaldi, J Smith, SW Redding. Stable azole drug resistance associated with a substrain of *Candida albicans* from an HIV-infected patient. Oral Dis 3(suppl 1):S102–S109, 1997.

97. GP Moran, D Sanglard, SM Donnelly, DB Shanley, DJ Sullivan, DC Coleman. Identification and expression of multidrug transporters responsible for fluconazole resistance in *Candida dubliniensis*. Antimicrob Agents Chemother 42:1819–1830, 1998.

98. DW Denning. Invasive aspergillosis in immunocompromised patients. Curr Opin Infect Dis 7: 456–462, 1994.

99. DW Denning. Invasive aspergillosis. Clin Infect Dis 26:781–803, 1998.

100. AH Groll, PM Shah, C Mentzel, M Schneider, G Just-Nuebling, K Huebner. Trends in the postmortem epidemiology of invasive fungal infections at a university hospital. J Infect 33:23–32, 1996.

101. MF Vandenbergh, PE Verweij, A Voss. Epidemiology of nosocomial fungal infections: invasive aspergillosis and the environment. Diagn Microbiol Infect Dis 34:221–227, 1999.

102. MJ Anderson, K Gull, DW Denning. Molecular typing by random amplification of polymorphic DNA and M13 Southern hybridization of related paired isolates of *Aspergillus fumigatus*. J Clin Microbiol 34:87–93, 1996.

103. A Aufauvre-Brown, J Cohen, DW Holden. Use of randomly amplified polymorphic DNA markers to distinguish isolates of *Aspergillus fumigatus*. J Clin Microbiol 30:2991–2993, 1992.

104. M Birch, MJ Anderson, DW Denning. Molecular typing of *Aspergillus* species. J Hosp Infect 30(suppl):339–351, 1995.

105. A Leenders, A van Belkum, S Janssen, S de Marie, J Kluytmans, J Wielenga, B Lowenberg, H Verbrugh. Molecular epidemiology of apparent outbreak of invasive aspergillosis in a hematology ward. J Clin Microbiol 34:345–351, 1996.

106. AC Leenders, A van Belkum, M Behrendt, A Luijendijk, HA Verbrugh. Density and molecular epidemiology of *Aspergillus* in air and relationship to outbreaks of *Aspergillus* infection. J Clin Microbiol 37:1752–1757, 1999.

107. D Lin, PF Lehmann, BH Hamory, AA Padhye, E Durry, RW Pinner, BA Lasker. Comparison of three typing methods for clinical and environmental isolates of *Aspergillus fumigatus*. J Clin Microbiol 33: 1596–1601, 1995.

108. P Mondon, J Thelu, B Lebeau, P Ambroise-Thomas, R Grillot. Virulence of *Aspergillus fumigatus* strains investigated by random amplified polymorphic DNA analysis. J Med Microbiol 42:299–303, 1995.

109. E Rodriguez, F Symoens, P Mondon, M Mallie, MA Piens, B Lebeau, AM Tortorano, F Chaib, A Carlotti, J Villard, MA Viviani, F Chapuis, N Nolard, R Grillot, JM Bastide. Combination of three typing methods for the molecular epidemiology of *Aspergillus fumigatus* infections. European Research Group on Biotype and Genotype of *Aspergillus*. J Med Microbiol 48:181–194, 1999.

110. V Chazalet, JP Debeaupuis, J Sarfati, J Lortholary, P Ribaud, P Shah, M Cornet, H Vu Thien, E Gluckman, G Brucker, JP Latge. Molecular typing of environmental and patient isolates of *Aspergillus fumigatus* from various hospital settings. J Clin Microbiol 36:1494–1500, 1998.

111. E Bart-Delabesse, C Cordonnier, S Bretagne. Usefulness of genotyping with microsatellite markers to investigate hospital-acquired invasive aspergillosis. J Hosp Infect 42:321–327, 1999.

112. JW Taylor, DM Geiser, A Burt, V Koufopanou. The evolutionary biology and population genetics underlying fungal strain typing. Clin Microbiol Rev 12:126–146, 1999.

113. TG Mitchell, JR Perfect. Cryptococcosis in the era of AIDS—100 years after the discovery of *Cryptococcus neoformans*. Clin Microbiol Rev 8:515–548, 1995.

114. RA Hajjeh, LA Conn, DS Stephens, W Baughman, R Hamill, E Graviss, PG Pappas, C Thomas, A Reingold, G Rothrock, LC Hutwagner, A Schuchat, ME Brandt, RW Pinner. Cryptococcosis: population-based multistate active surveillance and risk factors in human immunodeficiency virus–infected persons. Cryptococcal Active Surveillance Group. J Infect Dis 179:449–454, 1999.

115. ML Woods 2d, R MacGinley, DP Eisen, AM Allworth. HIV combination therapy: partial immune restitution unmasking latent cryptococcal infection. AIDS 12:1491–1494, 1998.

116. SP Franzot, IF Salkin, A Casadevall. *Cryptococcus neoformans* var. *grubii*: separate varietal status for *Cryptococcus neoformans* serotype A isolates. J Clin Microbiol 37:838–840, 1999.

117. F Chen, BP Currie, LC Chen, SG Spitzer, ED Spitzer, A Casadevall. Genetic relatedness of *Cryptococcus neoformans* clinical isolates grouped with the repetitive DNA probe CNRE-1. J Clin Microbiol 33:2818–2822, 1995.

118. I Polacheck, G Lebens, JB Hicks. Development of DNA probes for early diagnosis and epidemiological study of cryptococcosis in AIDS patients. J Clin Microbiol 30:925–930, 1992.

119. ED Spitzer, SG Spitzer. Use of a dispersed repetitive DNA element to distinguish clinical isolates of *Cryptococcus neoformans*. J Clin Microbiol 30:1094–1097, 1992.

120. ED Spitzer, SG Spitzer, LF Freundlich, A Casadevall. Persistence of initial infection in recurrent *Cryptococcus neoformans* meningitis. Lancet 341:595–596, 1993.

121. A Varma, KJ Kwon-Chung. DNA probe for strain typing of *Cryptococcus neoformans*. J Clin Microbiol 30:2960–2967, 1992.

122. A Varma, D Swinne, F Staib, JE Bennett, KJ Kwon-Chung. Diversity of DNA fingerprints in *Cryptococcus neoformans*. J Clin Microbiol 33:1807–1814, 1995.

123. ME Brandt, LC Hutwagner, RJ Kuykendall, RW Pinner. Comparison of multilocus enzyme electrophoresis and random amplified polymorphic DNA analysis for molecular subtyping of *Cryptococcus neoformans*. Cryplococcal Disease Active Surveillance Group. J Clin Microbiol 33:1890–1895, 1995.

124. D Garcia-Hermoso, G Janbon, F Dromer. Epidemiological evidence for dormant *Cryptococcus neoformans* infection. J Clin Microbiol 37:3204–3209, 1999.

125. W Meyer, K Marszewska, M Amirmostofian, RP Igreja, C Hardtke, K Methling, MA Viviani, A Chindamporn, S Sukroongreung, MA John, DH Ellis, TC Sorrell. Molecular typing of global isolates of *Cryptococcus neoformans* var. *neoformans* by polymerase chain reaction fingerprinting and randomly amplified polymorphic DNA—a pilot study to standardize techniques on which to base a detailed epidemiological survey. Electrophoresis 20:1790–1799, 1999.

126. Y Yamamoto, S Kohno, H Koga, H Kakeya, K Tomono, M Kaku, T Yamazaki, M Arisawa, K Hara. Random amplified polymorphic DNA analysis of clinically and environmentally isolated *Cryptococcus neoformans* in Nagasaki. J Clin Microbiol 33:3328–3332, 1995.

127. BC Fries, F Chen, BP Currie, A Casadevall. Karyotype instability in *Cryptococcus neoformans* infection. J Clin Microbiol 34:1531–1534, 1996.

128. JR Perfect, N Ketabchi, GM Cox, CW Ingram, CL Beiser. Karyotyping of *Cryptococcus neoformans* as an epidemiological tool. J Clin Microbiol 31:3305–3309, 1993.

129. JR Perfect, BB Magee, PT Magee. Separation of chromosomes of *Cryptococcus neoformans* by pulsed-field gel electrophoresis. Infect Immun 57:2624–2627, 1989.

130. M Pfaller, J Zhang, S Messer, M Tumberland, E Mbidde, C Jessup, M Ghannoum. Molecular epidemiology and antifungal susceptibility of *Cryptococcus neoformans* isolates from Ugandan AIDS patients. Diagn Microbiol Infect Dis 32:191–199, 1998.

131. S Bertout, F Renaud, D Swinne, M Mallie, JM Bastide. Genetic multilocus studies of different strains of *Cryptococcus neoformans*: taxonomy and genetic structure. J Clin Microbiol 37:715–720, 1999.

132. ME Brandt, SL Bragg, RW Pinner. Multilocus enzyme typing of *Cryptococcus neoformans*. J Clin Microbiol 31:2819–2823, 1993.

133. ME Brandt, LC Hutwagner, LA Klug, WS Baughman, D Rimland, EA Graviss, RJ Hamill, C Thomas, PG Pappas, AL Reingold, RW Pinner. Molecular subtype distribution of *Cryptococcus neoformans* in four areas of the United States. J Clin Microbiol 34:912–917, 1996.

134. A Casadevall, LF Freundlich, L Marsh, MD Scharff. Extensive allelic variation in *Cryptococcus neoformans*. J Clin Microbiol 30:1080–1084, 1992.

135. SP Franzot, JS Hamdan, BP Currie, A Casadevall. Molecular epidemiology of *Cryptococcus neoformans* in Brazil and the United States: evidence for both local genetic differences and a global clonal population structure. J Clin Microbiol 35:2243–2251, 1997.

136. BP Currie, LF Freundlich, A Casadevall. Restriction fragment length polymorphism analysis of *Cryptococcus neoformans* isolates from environmental (pigeon excreta) and clinical sources in New York City. J Clin Microbiol 32:1188–1192, 1994.

137. D Garcia-Hermoso, S Mathoulin-Pelissier, B Couprie, O Ronin, B Dupont, F Dromer. DNA typing suggests pigeon droppings as a source of pathogenic *Cryptococcus neoformans* serotype D. J Clin Microbiol 35:2683–2685, 1997.

138. JD Nosanchuk, S Shoham, BC Fries, DS Shapiro, SM Levitz, A Casadevall. Evidence of zoonotic transmission of *Cryptococcus neoformans* from a pet cockatoo to an immunocompromised patient. Ann Intern Med 132:205–208, 2000.

139. SP Franzot, J Mukherjee, R Cherniak, LC Chen, JS Hamdan, A Casadevall. Microevolution of a standard strain of *Cryptococcus neoformans* resulting in differences in virulence and other phenotypes. Infect Immun 66:89–97, 1998.

140. BC Fries, DL Goldman, R Cherniak, R Ju, A Casadevall. Phenotypic switching in *Cryptococcus neoformans* results in changes in cellular morphology and glucuronoxylomannan structure. Infect Immun 67:6076–6083, 1999.

141. DL Goldman, BC Fries, SP Franzot, L Montella, A Casadevall. Phenotypic switching in the human pathogenic fungus *Cryptococcus neoformans* is associated with changes in virulence and pulmonary inflammatory response in rodents. Proc Natl Acad Sci USA 95:14967–14972, 1998.

142. SA Howell, RJ Barnard, F Humphreys. Application of molecular typing methods to dermatophyte species that cause skin and nail infections. J Med Microbiol 48:33–40, 1999.

143. G Kac, ME Bougnoux, M Feuilhade De Chauvin, S Sene, F Derouin. Genetic diversity among *Trichophyton mentagrophytes* isolates using random amplified polymorphic DNA method. Br J Dermatol 140:839–844, 1999.

144. MR Reyes-Montes, M Bobadilla-Del Valle, MA Martinez-Rivera, G Rodriguez-Arellanes, E Maravilla, J Sifuentes-Osornio, ML Taylor. Relatedness analyses of *Histoplasma capsulatum* isolates from Mexican patients with AIDS-associated histoplasmosis by using histoplasmin electrophoretic profiles and randomly amplified polymorphic DNA patterns. J Clin Microbiol 37:1404–1408, 1999.

145. K Voigt, E Cigelnik, K O'Donnell. Phylogeny and PCR identification of clinically important Zygomycetes based on nuclear ribosomal-DNA sequence data. J Clin Microbiol 37:3957–3964, 1999.

146. KE Yates-Siilata, DM Sander, EJ Keath. Genetic diversity in clinical isolates of the dimorphic fungus *Blastomyces dermatitidis* detected by a PCR-based random amplified polymorphic DNA assay. J Clin Microbiol 33:2171–2175, 1995.

147. Y Graser, J Kuhnisch, W Presber. Molecular markers reveal exclusively clonal reproduction in *Trichophyton rubrum*. J Clin Microbiol 37:3713–3717, 1999.

148. A Burt, DA Carter, GL Koenig, TJ White, JW Taylor. Molecular markers reveal cryptic sex in the human pathogen *Coccidioides immitis*. Proc Natl Acad Sci USA 93:770–773, 1996.

149. V Koufopanou, A Burt, JW Taylor. Concordance of gene genealogies reveals reproductive isolation in the pathogenic fungus *Coccidioides immitis*. Proc Natl Acad Sci USA 94:5478–5482, 1997.

150. MC Fisher, GL Koenig, TJ White, JW Taylor. Pathogenic clones versus environmentally driven population increase: analysis of an epidemic of the human fungal pathogen *Coccidioides immitis*. J Clin Microbiol 38:807–813, 2000.

151. L Hoegl, E Thoma-Greber, M Rocken, HC Korting. Shift from persistent oral pseudomembranous to erythematous candidosis in a human immunodeficiency virus (HIV)-infected patient upon combination treatment with an HIV protease inhibitor. Mycoses 41:213–217, 1998.

152. C Michelet, C Arvieux, C Francois, JM Besnier, JP Rogez, F Souala, C Allavena, F Raffi, M Garre, F Cartier. Opportunistic infections occurring during highly active antiretroviral treatment. AIDS 12:1815–1822, 1998.

31

Strain Variation and Clonality in *Candida* spp. and *Cryptococcus neoformans*

Jianping Xu
McMaster University, Hamilton, Ontario, Canada

Thomas G. Mitchell
Duke University Medical Center, Durham, North Carolina

I. INTRODUCTION

In the past two decades, there has been an explosion of information on the epidemiology, strain typing, and population genetics of fungi pathogenic to humans and other animals. This interest can be attributed to several factors, including the increase in mycotic infections, the expanding variety and availability of strain typing techniques, significant collaborative studies involving diverse geographic areas, and the conceptual advancements in methodology and development of improved software programs for analysis of the data. As described in previous chapters and recent reviews, a variety of molecular typing methods have been developed to analyze the similarity among individual strains [1,2]. Studies of the population dynamics of a pathogenic fungus will clarify epidemiological trends and assist researchers to select appropriate strains in the search for virulence factors and target molecules for novel antifungal drugs, vaccines, or diagnostic tests [3,4].

This chapter reviews approaches based on population genetic analyses to understand the patterns and mechanisms of genetic variation among pathogenic fungi. Specifically, we will show evidence for clonality in *Candida* spp. and *Cryptococcus neoformans*. This chapter does not include a comprehensive review of the population genetic studies of all medical fungi. Rather, the chapter introduces basic concepts, issues, and rationales for studying clonality and presents examples of the application of this approach to a few common pathogenic yeast species.

II. MAKING SENSE OF THE GENETIC VARIATION WITHIN A COLLECTION OF STRAINS

Populations of species of pathogenic fungi and other micro-organisms vary greatly. A significant question is the extent to which the mode of reproduction—sexual or asexual—contributes to this variation [3,5]. Asexual reproduction results in progeny that are genetically identical to each other and to their parent (except for the occurrrence of spontaneous mutation). This process

739

produces a clonal population structure. This section will discuss the importance and detection of clonality.

The evolution and practical impact of clonality can best be understood in parallel with recombination. The process of recombination results in new combinations of genes or chromosomes that did not exist in the progenitors. Since all microbes are able to reproduce asexually (through mitosis in eukaryotic micro-organisms), it is expected that most microbial species would exhibit some evidence of clonality in nature. Hence, one goal of the population genetic studies of fungi has been to determine to what extent, if any, recombination contributes to the genetic variation of natural populations.

The amount of clonality and recombination that occurs in natural populations of microorganisms affects their long-term evolution and the management strategies of human infectious diseases. Sexual reproduction and recombination pose an evolutionary paradox. While an organism that reproduces asexually passes on all its genes to each of its progeny, one that reproduces sexually contributes only half its genes to each progeny. Therefore, under uniform conditions, natural selection favors the organism that reproduces asexually because with an equal number of progeny, the asexual individual has double the fitness of the sexually reproducing one.

Sexual reproduction affords micro-organisms two possible advantages: panmixia and DNA repair [6,7]. The panmictic argument suggests that without the mixing of genes associated with sexual recombination, adaptive evolution is limited to the accumulation of favorable mutations that occur successively in each independently evolving lineage. Sexual recombination allows favorable mutations that arise in separate lineages to become combined in the same individual, providing an advantage in the adaptation to different environments. The repair argument points out that the two haplotypes associated with diploidy during sex provide an error-correction mechanism for repairing DNA damage. Genetic damage can occur spontaneously and continuously during replication and possibly transcription. The intact DNA of one haplotype can serve as a template for correcting any damaged DNA in the other haplotype. Moreover, deleterious mutations in one haplotype can be overcome by compensatory dominant mutations in diploids. Whether one or both of these purported advantages of sexual recombination account for the origin and maintenance of sexual reproduction is subject to debate and investigation.

Another reason to compare the levels of clonality and recombination in natural populations pertains to issues of epidemiology, sampling, research strategies, and treatment. If recombination occurs among natural populations, there may be significant ramifications for the evolution and dissemination of genes related to antibiotic resistance, pathogenicity, and host specificity. A recombining population structure implies that the selected properties reflect distinct genes or non-recombining genetic elements. (Discrete genes are usually assumed not to undergo recombination, even though intragenic recombination has been demonstrated in many organisms that were critically examined.) Conversely, a clonal population structure implies that the selected units are clones or clonal lineages, not individual genes. To study medically relevant traits in a clonal population, representatives of every clonal lineage should be evaluated. However, in a recombining population, it would be more profitable to focus primarily on individual genes.

To determine whether a microbial population structure is predominantly clonal or recombining, population microbiologists employ genetic tests. The next section will discuss the tests used to dissect clonality and recombination within a population of a pathogenic fungus. These tests differ somewhat for haploid and diploid species.

Many of the methods used in analysis of population genetics structure require rigorous statistical measurements [4,5,8–19]. These statistical tests, such as those used to test null hypotheses of random mating and recombination within a population, are selected to address specific questions and evaluated according to their appropriateness and accuracy.

Statistical tests should be examined for their propensity to error. Generally, statistical tests are prone to two distinct and opposing types of error, designated type I or type II error. A type I error occurs if a test is too rigorous for the data, perhaps by not accounting for errors due to random sampling, which results in rejection of the hypothesis, even when it is true. Conversely, with a type II error, the statistical test permits excessive leeway in the data, and the hypothesis will seldom be rejected, even when it is false. The balance between type I and type II errors is such that reducing the probability of one type will increase the probability of the other. For null hypotheses, the 5% level of statistical significance is applied, indicating there is a 95% probability that the hypothesis is true and a 5% chance that a true hypothesis has been rejected (type I error). The chance of a type II error, failure to reject a false hypothesis, will vary [9].

III. ANALYSIS OF CLONALITY AND RECOMBINATION

A. Molecular Markers

Clonality and recombination are best evaluated when allelic information from genetically well-defined individual loci are available. For purposes of population genetics, a genetic locus may be defined as (1) the site of a change or substitution in a single base, i.e., a single nucleotide polymorphism (SNP); (2) a polymorphic restriction endonuclease recognition site, which usually involves a region of 4–6 base pairs; (3) a single nucleotide insertion or deletion; (4) any nucleotide sequence difference within a region of continuous DNA of variable length; (5) variation in the staining profile of an enzyme, detected by multilocus enzyme electrophoresis (MLEE); (6) an arbitrary DNA fragment generated by a PCR method, including amplified fragment length polymorphism (AFLP), random amplified polymorphic DNA (RAPD), and PCR fingerprinting; and (7) DNA fingerprinting generated through Southern hybridization of repetitive elements [2]. Table 1 summarizes the most common molecular typing methods used and their advantages and disadvantages. Because the theoretical bases and practical techniques for generating these molecular markers have been recently reviewed [1,2,4], these methods will not be detailed here.

Since the size of the genetic element that may be recognized as a locus varies, analytical methods may be different and should be carefully selected. All analytical methods rely on two basic assumptions: (1) once a locus is defined, recombination within the locus is assumed to occur only negligibly, if at all, and (2) each distinct allele is the result of a unique mutational event that occurred only once in the population. With these assumptions, indistinguishable alleles are considered to be "identical by descent" [3].

The usual criteria for clonality and recombination are presented in Table 2. Essentially, clonality in microbial populations is synonymous with nonrandom association of genetic markers within a sample of strains. Statistical tests for the nonrandom association vary depending on the ploidy of the study organism.

B. Haploid Species

Because each cell of haploid species contains only one set of chromosomes, each isolate has only one allele for each locus. Consequently, tests of recombination in natural haploid populations involve examining the allelic associations among different loci.

In tests for the occurrence of recombination in natural populations, there are two distinct questions. First, is the population panmictic? A panmictic population structure implies that the alleles at all loci are randomly associated with each other. If the hypothesis of panmixia is statistically rejected and a clonal population structure is assumed, the second question arises: Can all or part of the genetic variation in natural populations be attributed to recombination?

Table 1 Comparison of Molecular Methods Currently Used to Study Strain Variation in Fungal Pathogens

Techniques considerations	MLEE	EK (PFGE/ CHEF)	Whole DNA-DNA hybrid- ization	RFLP- Southern hybrid- ization	RAPD/AP- PCR PCR fingerprint	AFLP	SSCP	PCR-RFLP	DNA sequencing
Typical applications									
Population structure	yes	no	yes/no	yes	yes/no	yes	yes	yes	yes
Identify taxa	no	yes/no	yes	yes	yes/no	yes/no	yes/no	yes/no	yes
Phylogenetic analysis	yes/no	no	no	yes/no	yes/no	yes/no	yes/no	yes/no	yes
Practical factors									
Pure cultures required	yes	yes	yes/no	yes/no	yes	yes	no	yes/no	not with PCR
Sample preparation	minimal	medium to high	medium to high	medium	minimal	minimal	minimal	medium	maximal
Reproducibility	good	good	good	Very good	good to poor	good	good	very good	best
Cost	least expensive	moderate	moderate	moderate	low to moderate	moderate	moderate	low to moderate	most expensive
Turnaround time	moderate	slow	slow	slow	rapid	moderate	slow	moderate	slowest
Analytical factors									
Sensitivity to detect polymorphism	low (but often adequate)	low to high	high	moderate	high	very high	extremely high	moderate	highest
Utility of genetic markers (dominant or codominant)	codominant	chromosome markers codominant/ but often unknown	NA	usually codominant	dominant	dominant	Usually codominant	codominant	codominant

Abbreviations: MLEE, multilocus enzyme electrophoresis; RFLP, restriction fragment length polymorphism; EK, electrophoretic karyotype; PFGE, pulsed-field gel electrophoresis; CHEF, contour-clamped homogeneous electric field; RAPD, random amplified polymorphic DNA; AP-PCR arbitrarily primed-PCR; SSCP, single-strand conformational polymorphism.

Table 2 Common Criteria for Distinguishing Clonality and Recombination in Microbial Populations with Large Population Size and Examined with Neutral, Genetically Unlinked Markers

Criteria	Ploidy	Clonality	Recombination
Allelic association within a locus	≥Diploidy		
Hardy-Weinberg equilibrium		No	Yes
Excess homozygosity		Yes	No
Excess heterozygosity		Yes	No
Allelic association between loci	All ploidy		
Random		No	Yes
Nonrandom		Yes	No
Gene genealogies	All ploidy (but mostly for haploids)		
Congruence		Yes	No
Incongruence		No	Yes
Over- or underrepresentation of certain multilocus genotype	All ploidy		
Overrepresentation		Yes	No
Underrepresentation		Yes	No
Balanced		No	Yes

Since all medical fungi are capable of reproducing asexually through mitosis, identical genotypes (i.e., clonality) are expected among the samples. To assess the contribution or extent of recombination in the population, representatives of each different multilocus genotype are analyzed to distinguish between the null hypothesis of recombination and the alternative hypothesis of clonality.

A small number of genotypes are often dominant or overrepresented within a population [12,20]. It is common in these analyses to include the genotypes of all isolates in the population as well as only analyzing the unique genotypes within the population; the subgroup of unique genotypes is called the "clone-corrected" sample. Rejection of panmixia for the total sample but acceptance of panmixia for the clone-corrected sample is usually regarded as evidence for clonal expansion with evidence of random mating in the genetic structure. However, there are several caveats related to the truncation of a sample by clone correction before testing. Even though generally assumed, it is usually not confirmed that isolates with identical multilocus genotypes are actually clonal in origin. The justification for clone correction of the sample should be based on biological and ecological considerations. Furthermore, decreasing the sample size may also decrease the power of the statistical test for rejection of the null hypothesis of recombination, thus increasing type II error [9].

Tests for panmixia in a haploid population involve comparing observed allelic associations with those derived under the null hypothesis of random mating [15]. The two most widely used population genetic tests for haploid genomes are tests for allelic association (linkage equilibrium) between pairs of loci [3,7,17], and the overall index of association (I_A) involving alleles at all loci [15]. A third test compares the observed genotypic diversity with that expected under the null hypothesis of random mating [11]. A fourth test uses phylogenetic analysis of DNA sequence-based characters [3,4,21]. In this test, incongruences between different gene genealogies for a set of strains are considered to be consistent with the hypothesis of recombination.

C. Diploid Species

Since each diploid strain has two alleles at every locus, tests for recombination in diploid organisms differ somewhat from haploid species. In diploid species, the association of alleles within a locus is quite often used as a measure of recombination. A chi-square goodness-of-fit test can be performed to compare the observed and expected genotypic counts at each locus and summed across loci [11,12,17,20,22]. This computation is the test for Hardy-Weinberg equilibrium (HWE). To illustrate, a single locus, A, with two alleles, A1 and A2, presents three possible genotypes: A1A1, A1A2, and A2A2. The expected Hardy-Weinberg frequencies of genotypes A1A1, A1A2, and A2A2 are p^2, $2pq$, and q^2, respectively, where p is the frequency of allele A1 and q is the frequency of allele A2, and $A1 + A2 = 1$. If the observed counts of genotypes are not significantly different from the expected counts, then the population is in HWE and assumed to have a recombining structure. It must be stressed that Hardy-Weinberg frequencies are valid only when populations comply with several common but crucial assumptions: that the population size is large, that the markers being analyzed are free of selective pressure, that rates of migration and mutation are negligible, and that the population undergoes random mating [17].

It is injudicious to infer the population structure from HWE tests alone. First, violation of any of the first three assumptions above may cause significant deviation from the expected frequencies, even if the population is randomly mating. Second, failure to reject the null hypothesis of HWE does not guarantee that the population is indeed randomly mating (type II error). Third, different populations of the same species may reveal different population structures.

The "exact test" for allelic association among loci was developed by Zaykin et al. [19]. In this test, based on allelic counts, the probability of the set of multilocus genotypes in a sample is calculated from the multinomial theory under the hypothesis of no association. Alleles are then permuted and the conditional probability is calculated for the permuted genotypic array. The level of significance of the exact test is judged by the proportion of arrays that are no more probable than those provided by the original sample. The exact test is not restricted by the number of loci. It also allows one to calculate the probability of multilocus genotypes conditional on the genotypic counts of individual loci. Separating the allelic association test *within* a locus from genotypic association *among* loci can be very useful for diploid species, such as *C. albicans*, because deleterious recessive mutations may be common and the genotypic counts of many loci may deviate significantly from Hardy-Weinberg expectations.

Tests identical or similar to those described for diploid species can be applied to polyploid species as well. However, as ploidy increases, the number of expected genotypic classes increases, which will require larger sample sizes than those used in diploids to obtain statistically and biologically meaningful results in the tests of allelic associations.

IV. EVIDENCE OF CLONALITY IN *C. NEOFORMANS* AND SPECIES OF *CANDIDA*

Surprisingly, despite their medical significance, the issues of clonality and recombination have not been studied for many pathogenic fungal species. From the limited evidence so far, there is evidence for clonality in almost all species of genus *Candida* and in *Cryptococcus neoformans*. Conversely, the evidence for recombination is very limited for these species.

According to the common tests listed above and presented in Table 2, current knowledge of clonality and recombination is confined to *C. neoformans* and some of the common *Candida* species. These findings are presented in Table 3 and briefly summarized below.

A. *Cryptococcus neoformans*

This significant pathogen is currently represented by three varieties—*C. neoformans* var. *neoformans*, *C. neoformans* var. *grubii*, and *C. neoformans* var. *gattii* [23,25]. These varieties include four major serotypes. Despite the ability to cross-hybridize in the laboratory [13,26–28], various strain typing studies have revealed clear differences among these varieties and serotypes [13,29–33]. When populations of *C. neoformans* were analyzed through multilocus enzyme electrophoresis, abundant evidence supported a predominantly clonal genetic structure (Table 3) [13,34,35]. Significant clonal components were still observed when only representatives of unique multilocus genotypes were analyzed separately for individual serotypes [4]. However, recent evidence indicated limited, but statistically significant incongruences among four gene genealogies for the three varieties, providing evidence for recent natural hybridization and recombination in this biological species [13].

B. *Candida albicans*

Many studies have examined the population biology and epidemiology of *C. albicans* infections. However, only a few investigations used codominant genetic markers [10–12,18,36]. Codominant markers are important for exploring clonality and recombination in this apparently asexual, diploid fungus.

Table 3 Summary of Population Genetic Evidence for Clonality and Recombination in *Cryptococcus neoformans* and *Candida* Species

Criteria	Species									
	Cn	Ca	Cd	Cgl	Cgu	Ck	Cl	Cp	Cr	Ct
Allelic association within a locus (≥diploids)										
Hardy-Weinberg equilibrium	NA	Y	NT	NA	NT	NT	NT	NT	NT	NT
Excess homozygosity	NA	Y	NT	NA	NT	NT	NT	NT	NT	NT
Excess heterozygosity	NA	Y	NT	NA	NT	NT	NT	NT	NT	NT
Allelic association between loci										
Random	Y	Y	NT	Y	NT	NT	NT	NT	NT	NT
Nonrandom	Y	Y	NT	Y	NT	NT	NT	NT	NT	NT
Gene genealogy										
Congruence	Y	NT	NT	NT	NT	NT	NT	NT	NT	NT
Incongruence	N	NT	NT	NT	NT	NT	NT	NT	NT	NT
Over- or underrepresentation of certain multilocus genotype										
Overrepresentation	Y	Y	Y	Y	Y	Y	Y	Y	Y	Y
Underrepresentation	Y	Y	Y	Y	Y	Y	Y	Y	Y	Y
Balanced	UK	UK	UK	UK	UK	UK	UK	UK	UK	UK

Abbreviations: Cn, *Cryptococcus neoformans* [13,30,32,34,35]; Ca, *Candida albicans* [10–12,18,36]; Cd, *Candida dubliniensis* [47]; Cgl, *Candida glabrata* [37,38]; Cgu, *Candida guilliermondii* [54]; Ck, *Candida krusei* [39]; Cl, *Candida lusitaniae* [49,51,52]; Cp, *Candida parapsilosis* [40–42,48,53,55,58]; Cr, *Candida rugosa* [43]; Ct, *Candida tropicalis* [45]; NA, not applicable; Y, yes; NT, not tested; Y/N, yes/no; UK, unknown.

Pujol et al. [18] used 21 isozyme loci to characterize the genotypes of 55 isolates of *C. albicans* from patients infected with the human immunodeficiency virus (HIV). Among these strains, 13 of 21 enzymatic loci were polymorphic. Six of the 13 polymorphic loci deviated significantly from HWE. The expected counts for some of the multilocus genotypes were also much lower than the observed counts. They concluded that *C. albicans* has a clonal population structure [18].

Gräser et al. [10] used 12 SNPs to characterize a mixed sample of *C. albicans* isolates from Durham, NC. Among the 52 strains analyzed, 27 unique multilocus genotypes were detected. Similar to the findings of Pujol et al. [18], about half of the markers showed significant deviation from HWE. However, when they calculated the associations of alleles at different loci, >70% of the comparisons of pairwise loci were not significantly different from random association. They concluded that the population structure of *C. albicans* included both clonal and recombinational components. Similar patterns have been confirmed among populations of *C. albicans* from different groups of hosts, including HIV-infected patients, non-HIV patients, and healthy persons [12,20].

More recently, through the use of control samples, a group of 78 strains of *C. albicans* from a single geographic area were examined for the relationship between genetic relatedness and resistance to fluconazole [14]. The strains comprised two samples from Durham, NC: one from patients infected with HIV, and the other from healthy volunteers. For each strain, the minimum inhibitory concentration (MIC) to fluconazole was determined. Genotypes were obtained by PCR fingerprinting with five separate primers. The analysis revealed little evidence for genotypic clustering according to the HIV status or body site. However, a small group of fluconazole-resistant strains isolated from patients infected with HIV formed a genetically distinct cluster. Statistical analysis suggested that this cluster is most likely to represent a single origin of fluconazole resistance, followed by the subsequent spread of the resistant genotype among different hosts. In addition, two fluconazole-resistant strains were isolated from individuals who never took fluconazole or any other antifungal drug, one from a patient infected with HIV and the other from a healthy person. These results suggest both clonal and spontaneous origins of fluconazole resistance in *C. albicans* and that fluconazole resistance could be a significant problem in *C. albicans* in the future.

C. Other *Candida* Species

Preliminary evidence for clonality has been found in *Candida* species other than *C. albicans* (Table 3). For example, based on isozyme surveys, *C. (Torulopsis) glabrata* may have both a clonal and a recombining population genetic component [37,38] (J.-M. Bastide, personal commication). Various epidemiological surveys have also found genetically identical or similar strains of *C. krusei* from different hosts [1,39]. As shown in Table 3, overrepresented genotypes were found in *C. parapsilosis*, *C. stellatoidea*, *C. tropicalis*, *C. dubliniensis*, *C. lusitaniae*, *C. rugosa*, and *C. guilliermondii* [40–59]. While these results are consistent with the clonal expansion and spread of certain genotypes, the lack of information on a genetically defined set of polymorphic loci precludes the more stringent statistical analysis for evidence of clonality and recombination. More critical tests are required to demonstrate clearly the contributions of clonality and recombination in these species.

V. EPIDEMIC CLONES AND FUTURE DIRECTIONS

Analyzing the genetic structure of populations to determine the extent of clonality versus recombination is only the first, albeit critical, step toward understanding population history and the evolution of the pathogenicity of medical fungi. Subsequent questions concern the reason(s) that a particular genetic structure exists for certain populations and species. Approaches to address this issue are emerging, but definitive answers are lacking. Many factors can contribute to the genetic structure of a population, including the founding history of the population (founder effects), the mutation rate, the mutation spectrum (i.e., the distribution of mutations that are lethal, deleterious, neutral, and advantageous), genetic drift, population size, the mating system (or its absence), the specific environmental conditions under which infection occurs, and the selection pressure from antifungal drug treatment [e.g., 14]. It is a challenge to examine at the same time the relative contributions of all of these factors to the genetic structure of natural populations. Therefore, we recommend the appropriate sampling strategies to control confounding factors in attempts to dissect the contributions of specific factors to the population genetic structure with regard to the extent of clonality and the dispersal of clonal populations among hosts in clinical and natural environments.

One common approach has been to examine whether and why certain genotypes are more prevalent than expected. This information can be obtained by inspecting the relative frequencies of different multilocus genotypes or by calculating the expected frequencies of individual genotypes based on allelic frequencies and under the hypothesis of random mating (i.e., random associations of alleles within and among loci). When predominant genotypes are identified, the medically important traits of those isolates can be compared with less common genotypes in the population to explore the potential associations between the genotype and clinically relevant phenotypes. This information will be highly valuable for making predictions about the clinical outcome of a mycosis, identifying potential antifungal drug targets, designing clinically effective treatment strategies, and investigating the molecular mechanisms of fungal pathogenicity.

REFERENCES

1. DR Soll. The ins and outs of DNA fingerprinting the infectious fungi. Clin Microbiol Rev 13: 332–370, 2000.
2. TG Mitchell, J Xu. Molecular methods to identify pathogenic fungi. In: D Howard, ed. Pathogenic Fungi of Humans and Animals. New York: Marcel Dekker, in press.
3. J Xu, TG Mitchell. Population genetic analyses of medical fungi. In: D Howard, ed. Pathogenic Fungi of Humans and Animals. New York: Marcel Dekker, in press.
4. JW Taylor, DM Geiser, A Burt, V Koufopanou. The evolutionary biology and population genetics underlying fungal strain typing. Clin Microbiol Rev 12:126–146, 1999.
5. JM Maynard Smith, NH Smith, M O'Rourke, BG Spratt. How clonal are bacteria? Proc Natl Acad Sci USA 90:4384–4388, 1993.
6. RE Michod, BR Levin. The Evolution of Sex: An Examination of Current Ideas. Sunderland, MA Sinauer Associates, 1987.
7. JM Maynard Smith. The Evolution of Sex. Cambridge: Cambridge University Press, 1978.
8. BS Weir. Genetic Data Analysis. 2d ed. Sunderland, MA: Sinauer Associates, 1996.
9. RR Sokal, FJ Rohlf. Biometry. 2d ed. New York: W.H. Freeman, 1988.
10. Y Gräser, M Volovsek, J Arrington, G Schönian, W Presber, TG Mitchell, RJ Vilgalys. Molecular markers reveal that population structure of the human pathogen *Candida albicans* exhibits both clonality and recombination. Proc Natl Acad Sci USA 93:12473–12477, 1996.
11. J Xu, CM Boyd, E Livingston, W Meyer, JF Madden, TG Mitchell. Species and genotypic diversities and similarities of pathogenic yeasts colonizing women. J Clin Microbiol 37:3835–3843, 1999.
12. J Xu, TG Mitchell, RJ Vilgalys. PCR-restriction fragment length polymorphism (RFLP) analyses reveal both extensive clonality and local genetic differences in *Candida albicans*. Mol Ecol 8:59–73, 1999.
13. J Xu, R Vilgalys, TG Mitchell. Multiple gene genealogies reveal recent dispersion and hybridization in the human pathogenic fungus *Cryptococcus neoformans*. Mol Ecol 9:1471–1481, 2000.
14. J Xu, AR Ramos, RJ Vilgalys, TG Mitchell. Clonal and spontaneous origins of fluconazole resistance in *Candida albicans*. J Clin Microbiol 38:1214–1220, 2000.
15. AHD Brown, MW Feldman, E Nevo. Multilocus structure of natural populations of *Hordeum spontaneum*. Genetics 96:523–536, 1980.
16. DE Dykhuizen, L Green. Recombination in *Escherichia coli* and the definition of biological species. J Bacteriol 173:7257–7268, 1991.
17. DL Hartl, AG Clark. Principles of Population Genetics. 2d ed. Sunderland, MA: Sinauer Associates, 1989.
18. C Pujol, J Reynes, F Renaud, M Raymond, M Tibayrenc, FJ Ayala, F Janbon, M Mallié, J-M Bastide. The yeast *Candida albicans* has a clonal mode of reproduction in a population of infected human immunodeficiency virus-positive patients. Proc Natl Acad Sci USA 90:9456–9459, 1993.
19. D Zaykin, L Zhivotovsky, BS Weir. Exact tests for association between alleles at arbitrary numbers of loci. Genetica 96:169–178, 1995.
20. J Xu, RJ Vilgalys, TG Mitchell. Lack of genetic differentiation between two geographically diverse samples of *Candida albicans* isolated from patients infected with human immunodeficiency virus. J Bacteriol 181:1369–1373, 1999.
21. JP Huelsenbeck. Combining data in phylogenetic analysis. TREE 11:152–158, 1996.
22. J Xu, RW Kerrigan, P Callac, PA Horgen, JB Anderson. Genetic structure of natural populations of *Agaricus bisporus*, the commercial button mushroom. J Hered 88:482–488, 1997.
23. A Casadevall, JR Perfect. *Cryptococcus neoformans*. Washington: ASM Press, 1999.
24. TG Mitchell, JR Perfect. Cryptococcosis in the era of AIDS—100 years after the discovery of *Cryptococcus neoformans*. Clin Microbiol Rev 8:515–548, 1995.

25. SP Franzot, BC Fries, W Cleare, A Casadevall. Genetic relationship between *Cryptococcus neoformans* var. *neoformans* strains of serotypes A and D. J Clin Microbiol 36:2200–2204, 1998.

26. KJ Kwon-Chung. A new genus, *Filobasidiella*, the perfect state of *Cryptococcus neoformans*. Mycologia 67:1197–1200, 1975.

27. KJ Kwon-Chung. Morphogenesis of *Filobasidiella neoformans*, the sexual state of *Cryptococcus neoformans*. Mycologia 68:821–833, 1976.

28. KA Schmeding, S-C Jong, R Hugh. Sexual compatibility between serotypes of *Filobasidiella neoformans (Cryptococcus neoformans)*. Curr Microbiol 5:133–138, 1981.

29. ME Brandt, SL Bragg, RW Pinner. Multilocus enzyme typing of *Cryptococcus neoformans*. J Clin Microbiol 31:2819–2823, 1993.

30. T Boekhout, A van Belkum, ACAP Leenders, HA Verbrugh, P Mukamurangwa, D Swinne, WA Scheffers. Molecular typing of *Cryptococcus neoformans*: taxonomic and epidemiological aspects. Int J Syst Bacteriol 47:432–442, 1997.

31. BP Currie, LF Freundlich, A Casadevall. Restriction fragment length polymorphism analysis of *Cryptococcus neoformans* isolates from environmental (pigeon excreta) and clinical sources in New York City. J Clin Microbiol 32:1188–1192, 1994.

32. SP Franzot, JS Hamdan, BP Currie, A Casadevall. Molecular epidemiology of *Cryptococcus neoformans* in Brazil and the United States: evidence for both local genetic differences and a global clonal population structure. J Clin Microbiol 35:2243–2251, 1997.

33. W Meyer, TG Mitchell. PCR fingerprinting in fungi using single primers specific to minisatellites and simple repetitive DNA sequences: strain variation in *Cryptococcus neoformans*. Electrophoresis 16:1648–1656, 1995.

34. ME Brandt, LC Hutwagner, RJ Kuykendall, RW Pinner, Cryptococcal Disease Active Surveillance Group. Comparison of multilocus enzyme electrophoresis and random amplified polymorphic DNA analysis for molecular subtyping of *Cryptococcus neoformans*. J Clin Microbiol 33:1890–1895, 1995.

35. ME Brandt, LC Hutwagner, LA Klug, WS Baughman, D Rimland, EA Graviss, RJ Hamill, C Thomas, PG Pappas, AL Reingold, RW Pinner, Cryptococcal Disease Active Surveillance Group. Molecular subtype distribution of *Cryptococcus neoformans* in four areas of the United States. J Clin Microbiol 34:912–917, 1996.

36. A Forche, G Schönian, Y Gräser, R Vilgalys, TG Mitchell. Genetic structure of typical and atypical populations of *Candida albicans* from Africa. Fungal Genet Biol 28:107–125, 1999.

37. U Schwab, F Chernomas, L Larcom, JJ Weems Jr. Molecular typing and fluconazole susceptibility of urinary *Candida glabrata* isolates from hospitalized patients. Diagn Microbiol Infect Dis 28:11–17, 1997.

38. PL Fidel Jr, JA Vazquez, JD Sobel. *Candida glabrata*: review of epidemiology, pathogenesis, and clinical disease with comparison to *C. albicans*. Clin Microbiol Rev 12:80–96, 1999.

39. SM Essayag, GG Baily, DW Denning, JP Burnie. Karyotyping of fluconazole-resistant yeasts with phenotype reported as *Candida krusei* or *Candida inconspicua*. Int J Syst Bacteriol 46:35–40, 1996.

40. ML Branchini, MA Pfaller, J Rhine-Chalberg, T Frempong, HD Isenberg. Genotypic variation and slime production among blood and catheter isolates of *Candida parapsilosis*. J Clin Microbiol 32:452–456, 1994.

41. G Carruba, E Pontieri, F De Bernardis, P Martino, A Cassone. DNA fingerprinting and electrophoretic karyotype of environmental and clinical isolates of *Candida parapsilosis*. J Clin Microbiol 29:916–922, 1991.

42. RS Dassanayake, LP Samaranayake. Characterization of the genetic diversity in superficial and systemic human isolates of *Candida parapsilosis* by randomly amplified polymorphic DNA (RAPD). Acta Path Microbiol Immunol Scand 108:153–160, 2000.

43. JC Dib, M Dube, C Kelly, MG Rinaldi, JE Patterson. Evaluation of pulsed-field gel electrophoresis as a typing system for *Candida rugosa*: comparison of karyotype and restriction fragment length polymorphisms. J Clin Microbiol 34:1494–1496, 1996.

44. BN Doebbeling, RJ Hollis, HD Isenberg, RP Wenzel, MA Pfaller. Restriction fragment analysis of a *Candida tropicalis* outbreak of sternal wound infections. J Clin Microbiol 29:1268–1270, 1991.

45. BN Doebbeling, PF Lehmann, RJ Hollis, L-C Wu, AF Widmer, A Voss, MA Pfaller. Comparison of pulsed-field gel electrophoresis with isoenzyme profiles as a typing system for *Candida tropicalis*. Clin Infect Dis 16:377–383, 1993.

46. A Espinel-Ingroff, A Quart, L Steele-Moore, I Metcheva, GA Buck, VL Bruzzese, D Reich. Molecular karyotyping of multiple yeast species isolated from nine patients with AIDS during prolonged fluconazole therapy. J Med Vet Mycol 34:111–116, 1996.

47. S Joly, C Pujol, M Rysz, K Vargas, DR Soll. Development and characterization of complex DNA fingerprinting probes for the infectious yeast *Candida dubliniensis*. J Clin Microbiol 37:1035–1044, 1999.

48. YC Huang, TY Lin, HS Leu, HL Peng, JH Wu, HY Chang. Outbreak of *Candida parapsilosis* fungemia in neonatal intensive care units: clinical implications and genotyping analysis. Infection 27:97–102, 1999.

49. D King, J Rhine-Chalberg, MA Pfaller, SA Moser, WG Merz. Comparison of four DNA-based methods for strain delineation of *Candida lusitaniae*. J Clin Microbiol 33:1467–1470, 1995.

50. PF Lehmann, BJ Kemker, C-B Hsiao, S Dev. Isoenzyme biotypes of *Candida* species. J Clin Microbiol 27:2514–2521, 1989.

51. WG Merz, U Khazan, MA Jabra-Rizk, L-C Wu, GJ Osterhout, PF Lehmann. Strain delineation and epidemiology of *Candida (Clavispora) lusitaniae*. J Clin Microbiol 30:449–454, 1992.

52. MA Pfaller, SA Messer, RJ Hollis. Strain delineation and antifungal susceptibilities of epidemiologically related and unrelated isolates of *Candida lusitaniae*. Diagn Microbiol Infect Dis 20:127–133, 1994.

53. MA Pfaller, SA Messer, RJ Hollis. Variations in DNA subtype, antifungal susceptibility and slime production among clinical isolates of *Candida parapsilosis*. Diagn Microbiol Infect Dis 21:9–14, 1995.

54. RM San Millan, L-C Wu, IF Salkin, PF Lehmann. Clinical isolates of *Candida guilliermondii* include *Candida fermentati*. Int J Syst Bacteriol 47:385–393, 1997.

55. SL Solomon, RF Khabbaz, RH Parker, RL Anderson, MA Geraghty, RM Furman, WJ Martone. An outbreak of *Candida parapsilosis* bloodstream infections in patients receiving parenteral nutrition. J Infect Dis 149:98–102, 1984.

56. JA Vazquez, A Beckley, S Donabedian, JD Sobel, MJ Zervos. Comparison of restriction enzyme analysis versus pulsed-field gradient gel electrophoresis as a typing system for *Torulopsis glabrata* and *Candida* species other than *C. albicans*. J Clin Microbiol 31:2021–2030, 1993.

57. LA Waggoner-Fountain, MW Walker, RJ Hollis, MA Pfaller, JE Ferguson, II, RP Wenzel, LG Donowitz. Vertical and horizontal transmission of unique *Candida* species to premature newborns. Clin Infect Dis 22:803–808, 1996.

58. SF Welbel, MM McNeil, RJ Kuykendall, TJ Lott, A Pramanik, R Silberman, AD Oberle, LA Bland, S Aguero, M Arduino, S Crow, WR Jarvis. *Candida parapsilosis* bloodstream infections in neonatal intensive care unit patients: epidemiologic and laboratory confirmation of a common source outbreak. Pediatr Infect Dis J 15:998–1002, 2000.

59. S Zeng, L-C Wu, PF Lehmann. Random amplified polymorphic DNA analysis of culture collection strains of *Candida* species. J Med Vet Mycol 34:293–297, 1996.

32

New and Emerging Pathogens: What's a Lab to Do?

Chester R. Cooper, Jr.
Youngstown State University, Youngstown, Ohio

> In science there are no authorities; at most, there are experts.
> —Carl Sagan (1996): *The Demon-Haunted World. Science as a Candle in the Dark*

I. INTRODUCTION

To paraphrase a quote favored by a renowned medical mycologist, the human body is a living, breathing culture medium waiting to be exploited by both true and opportunistic pathogens. Fungi categorized as true, or primary, pathogens have historically encompassed those species capable of routinely causing disease in apparently healthy persons. Conversely, the term opportunistic pathogen has been circumscribed to include those fungi that afflict individuals having a compromised immune status. Both of these concepts seemed to have held true for numerous decades as indicated by the medical literature. For example, prior to the antibiotic era (ca. 1945), the reported incidence of fungal infections as well as the limited spectrum of etiologic agents suggests that only the true pathogens were capable of growth in the "human culture dish."

However, the distinction between pathogens that are naturally virulent and those that are characteristically exploitative has become blurred. Multitudes of fungal species now find sustenance in the human culture dish. Many of these species previously caused few, if any, infections. The rise of these fungi as pathogens directly correlates with the dramatic alterations in the dynamics of human life on earth during the past 50 years. Similar increases have been noted for other types of microbial pathogens, particularly those who have been considered nontraditional etiologic agents. Hence, the phrase "emerging and reemerging pathogens" was born.

When viewed in retrospect, the past five decades have fostered developments that collectively and unwittingly promoted significant increases in infectious diseases, including those caused by fungi [1]. For instance, the antibiotic era not only introduced the possibility of the antimicrobial "silver bullet," but it also marked the beginning for the evolution of microbial drug resistance. Similarly, the overuse of antifungal agents has led to appearance of drug-resistant fungi. In addition, the decades following World War II were accompanied by new economic and technological developments that included advances in health care. This has been a two-edge sword in that human life can be prolonged and diseases can be stifled, but at the cost of compromising an individual's short- or long-term innate ability to ward off infection.

Again, fungi that were once considered nominally opportunistic pathogens have become notorious public health threats. This is especially true for hospitalized populations in which patients are being treated for unrelated ailments. During this same period, the world's population exploded resulting in demographic changes. Such changes have had economic and political effects upon the delivery of health care. Combined with the present relative ease of international travel, legal and illegal immigration has been facilitated making our world a true global village. As a consequence, the worldwide spread of particular infectious diseases is now being realized as more people are being exposed to new and exotic microbes.

These microbes include fungal species previously viewed more as mycological curiosities than as pathogens. A perfect example would be the rise of penicilliosis due to *Penicillium marneffei* [2]. Moreover, the importation of mycotic diseases in nonendemic regions seems to be increasing dramatically [3–7]. For example, the incidence of histoplasmosis and blastomycosis is rising in Europe where such diseases have rarely been noted previously. Even coccidioidomycosis and paracoccidioidomycosis, mycoses limited to specific regions of the Western hemisphere, have now been diagnosed in Japan. The advent of the human immunodeficiency virus (HIV) pandemic has further compounded this situation. Again, those infected with the HIV are susceptible to infections by fungi once considered relatively benign, e.g., *Candida* spp. (other than *C. albicans*), *Aspergillus* spp., and *P. marneffei* [1,2,8]. Finally, for a variety of reasons, the laboratory infrastructure supporting public health measures has been and continues to be severely challenged. This has been especially true of mycology reference laboratories, many of which have disappeared or restricted their activities during the past two decades [1,9,10]. Consequently, initiatives and programs designed to protect public health teeter at the abyss of failure.

This solemn perspective should provoke despair among physicians and medical mycologists alike. With concurrent trends in managed health care restricting the amount and types of diagnostic procedures that may be performed by clinicians, not to mention the extent of therapeutic courses available to patients, the future does indeed look bleak for those afflicted with common fungal infections. When the apparent emergence and reemergence of other pathogenic fungi are added into the clinical equation, the outlook seems even more grim. What, then, is the laboratory to do?

It is quite clear that medical mycology has entered a new era. The old traditional methods seem archaic and impotent at facing the new challenges in the delivery of quality health care. Perhaps, then, it is time to reevaluate the role of the clinical mycology laboratory and the direction it needs in the next century.

II. REEVALUATION OF CLINICAL MYCOLOGY

The world of infectious diseases has dramatically changed in the past three decades [11]. These changes have spurred new developments and approaches in the way infectious diseases are diagnosed and treated. Unfortunately, for a number of reasons, medical mycology and the diagnostic means used to detect fungal infections has lagged behind the progress made in other fields. A reevaluation of the present status of clinical mycology is desperately needed in order to keep pace with current trends in mycotic diseases. Yet, the following discussion is not meant to be an all-encompasssing discourse on the problems faced by the clinical mycology laboratory. Rather, selected topics are addressed. For a comprehensive picture that goes beyond the thoughts shared below, readers are directed to other sources [1,9,10].

A. Emerging and Reemerging Fungal Pathogens

The term "emerging and reemerging pathogens" has been the mantra of infectious disease research for the past decade. Moreover, this aphorism and what it represents promises to influence the driving forces in both applied and basic biomedical investigations for at least part of the new century. However, these terms tend to be used quite frequently. This poses the inherent danger that they may become trite and virtually meaningless. Hence, the true concept of emerging and reemerging pathogens might be, and in some instances already has been, erroneously reduced to refer to the etiologic agents causing minor episodic incidents or epidemics of limited scope and significance.

Given current trends, medical mycologists could easily be tempted to apply these terms improperly. For example, there is no doubt that the spectrum of fungi causing infections in humans has greatly expanded [1,12,13]. However, despite the fact that many of the case reports detail only a few incidents of infection involving specific fungal species, some authors often feel compelled to use the term "emerging" or "reemerging" to describe what they assess is a new disease or recurrence of a rare pathogen. Usage of these terms in this manner can be misleading. Hence, the medical mycology community would be best served if the definition of an "emerging pathogen" was limited to describing an etiological agent that causes a significant and sustained increase in infections. By comparison, a "reemerging pathogen" would describe those pathogenic fungi that once caused significant, sustained numbers of infections, but have until recently remained quiescent with regard to the number of recognized cases. These defini- tions would then aptly distinguish true emerging pathogens from those fungi responsible for spurious, discontinuous epidemics, that is, clusters of infections that occur over a very short, defined period of time. Examples of "emerging and reemerging pathogens" would include *Coccidioides immitis*, *P. marneffei*, and *Fusarium* spp. among others [2,8,14] whereas fungi incorrectly labelled as such might include *Candida krusei*, *Malassezia furfur*, and members of the Zygomycota [1,12,13].

That is not to say, however, that these latter fungi are not significant pathogens. For example, *C. krusei* and *M. furfur* remain important pathogens to consider among immunosup- pressed individuals receiving fluconazole treatment or lipid hyperalimenation, respectively. However, by the above definitions, these fungi would not be "emerging" or "reemerging" pathogens since the medical literature reports only sporadic and finite numbers of infections caused by these etiologic agents. Conversely, once a disease has become firmly established over a sustained period of time, it cannot any longer be considered as caused by an emerging pathogen. Rather, it would be regarded as an endemic disease.

B. Current Status of the Reference Laboratory

The relative absence of reference laboratories is particularly critical in the field of medical mycology. Sadly enough, many of the traditionally strong clinical mycology reference laborato- ries that graciously, and often freely, served the public good have disappeared. An example of the latter was the Mycology Laboratories at the New York State Department of Health in Albany [9]. Once a vibrant and dynamic resource boasting several respected doctoral-level members of the clinical mycology community, it now stands as a shell of its former self. For the reasons discussed below, the mission of this institution is now focused mainly on enforcing proficiency testing regulations and grant-funded research rather than classical medical mycology.

The situation in Albany symbolizes the status of the mycology reference laboratory today. Few institutions in North America exist that accept outside requests for help with the processing of fungal pathogens and the primary clinical specimens suspected of harboring them [9,10]. In fact, in this author's experience, routine processing of clinical specimens by either private or government-sponsored laboratories is decidedly not cost-effective for a laboratory of modest resources. Often, the most notable resource in short supply is qualified support staff. Outside the Western hemisphere, even fewer distinguished laboratories exist in Europe, Asia, and the subcontinents that have the personnel and resources to sustain a fully operational reference laboratory. Collectively, the programs that exist today either are overwhelmed with requests, have restricted their services, or have understandably redirected their efforts to for-profit activities.

Obviously, the increased pressure to manage costs and to be accountable for health-related expenditures has significantly affected the reference laboratory in developed countries. To contain costs, hospital laboratories are moving toward performing only the simplest and most financially beneficial of diagnostic tests. More complicated and expensive tests are being subcontracted to commercial laboratories. Rural laboratories having small volumes are doing the same since the cost-to-profit ratio is small. Commercial laboratories, in theory, could do larger volumes of tests for less cost. In general, this would seem to be a good idea in that any savings would be reflected in lower insurance premiums and a lighter financial burden on the delivery of public health care. Many might debate this point. What is not debatable, however, is that most commercial laboratories typically do not have individuals on staff specifically trained to identify uncommon fungi.

Government policies and regulations also have affected how business is conducted in the clinical mycology laboratory in developed nations. For instance, certain diagnostic tests employing colored differential media or reactants are recognized in Europe as valuable tools in the identification of pathogenic fungi. Yet, these same methods are not approved for use in the United States as diagnostic tests, but rather as experimental procedures. In addition, quality control issues have promulgated a diverse number of regulations. Meeting the standards posed by this legislation is not only costly, but it is often superfluous. Where common sense should rule, regulatory bodies interfere with grandiose schemes meant to increase the accuracy and efficiency of diagnoses. Not only do these schemes fail to achieve their goal, but also their cumbersome and dogmatic requirements tend to discourage individuals to strive for excellence. In the final analysis, many laboratories have opted not to perform some, if any, procedures related to the investigation of mycotic diseases.

As bad as the situation is in developed nations, the lack of adequate reference laboratories is profoundly worse in low-income countries. Those that do exist offer minimal services due to the costs involved. Routine tests performed in clinical mycology laboratories for nominal fees can be exorbitant, even prohibitive, in underdeveloped nations. For their particular circumstances, however, most reference laboratories in these countries provide a valuable service to their citizens.

C. Nomenclature of Fungal Pathogens

To the nonmedical mycologists, one of the most irritating aspects of this field is the seemingly constant change in the names used to identify fungi. Even to medical mycologists, these changes can go beyond being irritating and escalate into a verbal conflagration between individuals having opposing taxonomic opinions. Some of the very best gossip and political intrigue in medical mycology has come from such arguments. However, fungal taxonomists ascribing to either the so-called "splitter" and "lumper" philosophy are well within their purview to change the name of a fungus so long as they follow rules contained in the International Code of Botanical Nomenclature, the accepted standard for this activity [10]. This confusion in names had often led to an outcry for establishing a different system based upon the most basic of characters—the genotype [15]. This approach has merit, but is not practical—yet. Advances in molecular phylogenetics are occurring daily, and the moment may eventually arrive that will permit a more quantitative means of describing the fungi. This would be a sad day for most traditional medical mycologists, with the reasons rightfully understood. The innate beauty of fungal form and structure would be reduced to mere numbers or symbols. A part of history might be lost and not correlated with the new methodology for establishing taxonomy and nomenclature. That time, though, is probably still quite a long way off. For now, despite the intellectual torture it evokes, clinical laboratory personnel are encouraged to keep apprised of current medically relevant mycological nomenclature. Such information is usually summarized annually in a leading medical-oriented journal [e.g., 16].

D. Diagnosis and Management of Fungal Diseases

Given that immune compromised patients comprise a significant population in the clinical realm, fungal infections must be a component of the differential diagnosis contemplated by physicians. Again, however, economic concerns have been a driving force in changing the means by which diagnoses are derived [9,10]. Today, presumptive diagnoses as well as inferred pathogen identifications are in vogue. To some degree, this has resulted in more rapid and accurate treatment of disease. Yet, it has also come with the cost in that the quality of care for some patients has been poor, particularly for those whom presumptive methods have effected incorrect diagnoses and subsequent ineffectual treatment.

III. MEDICAL MYCOLOGY IN THE NEXT MILLENNIUM

The twenty-first century is upon us. With it comes the challenge to implement changes that would vastly improve the clinical mycology laboratory as well as advance the field in areas of diagnosis and treatment of mycotic diseases. The needs are great and can be divided into different targeted areas. However, it has become quite obvious to this author that the most significant challenges are reflected in the grave paucity of reference laboratories and well-trained individuals capable of meeting the clinical needs of medical mycology, let alone monitoring current trends in mycotic diseases.

Undoubtedly, there needs to be a greater emphasis placed upon maintaining the few reference laboratories that exist in North America and elsewhere. Creating new centers of excellence would be ideal, but fiscal issues are the main constraints prohibiting these efforts. Hence, resources need to be invested in what institutions we have at present. The current activities of these laboratories must be supported as well as expanded. Expansion would be best served by supporting the establishment of an infrastructure devoted to contemporary research endeavors. Both basic and applied research should be embraced in terms of more instrumentation, an emphasis on molecular and cellular research, computerized databases, and preserved culture collections. Some of these goals can be met by cooperative agreements between institutions as well as commercial concerns.

Equally important is the need to recruit, educate, and train more individuals in medical mycology. This needs to be initiated at the most fundamental level of our educational system—the microbiology classroom. Too often, fungi are dismissed by microbiology instructors as eukaryotic trespassers in a prokaryotic world. And it is not only the instructor with a prokaryotic bias that commits this act of ignorance. Many "card-carrying" fungal biologists do so as well by not deviating from the standard topic outline used by most microbiology textbooks. Often, a single chapter or less is devoted to fungal biology and the role of fungi in human affairs. This is not acceptable.

For medical mycology to prosper, the traditional teaching of microbiology with a major focus on bacteria needs to be more appropriately balanced against the importance of eukaryotic microbes. Also, just as important, more courses that focus on fungal biology and medical mycology must be integrated into both the baccalaureate and graduate educational system. This would include laboratory-based courses in clinical and experimental mycology. Experiences in general or medical mycology research for undergraduates need also to be offered. Moreover, graduate and medical students also need to be encouraged to seek careers involving medical mycology. Such training has been recognized by some pharmaceutical companies as necessary for both the public and themselves. Fellowships and special educational gifts have been generously provided in recent years by these businesses. For its part, the federal government is also contributing to this goal. The National Institutes of Health currently supports research and clinical training in medical mycology. More support is needed, however, from both private and public sources to adequately meet the challenge to train more medical mycologists. It is also incumbent upon the medical mycology community to foster the public's perception that fungi play important roles in the quality of human life and health.

IV. WHAT IS THE LAB TO DO?

Despite the earlier discussions and recommendations, the main problem of this chapter has yet to addressed. Just what is the lab to do with the problem of emerging and reemerging pathogens until the promises of the new century are fulfilled? The issue of whether or not the pathogen is an emerging or reemerging pathogen is relatively insignificant except at the epidemiological level. Identification and treatment of the infection need to be the main concern. In light of the needs and the possibilities of help for the present-day clinical mycology laboratory, a firm and positive answer appears elusive. To answer this question, this author surveyed several prominent medical mycologists for their responses. Uniformly, having a clear vision of the challenges faced by medical mycology, each had the same basic recommendations.

Until implementation of the changes needed in the clinical mycology, the following advice is presented for dealing with suspected fungal pathogens. The two key issues that a lab must face are pathogen identification and antifungal susceptibility. The former is important for a physician to recognize the type of disease being faced; the latter is needed for treatment strategies.

With respect to the first issue, the typical clinical laboratory that has isolated a suspected mold or yeast pathogen should identify the fungus to the genus level, and if possible, to the species level. The report should be submitted to the responsible physician as soon as possible. It is the physician's responsibility to act on this information, not that of the clinical mycologist. If requested, factual information regarding the disease potential/incidence of the fungus should be given. An opinion may also be solicited. However, this opinion should be based solely on firmly established fact. Again, the choice of the course of treatment rests with the physician. Recommendations by nonmedically trained personnel should be avoided.

Perhaps most important, however, is that a laboratory realize its limitations in fungal identification. Should there be a lack of expertise, it would best serve the patient to send the specimen immediately to a reference laboratory capable of making the identification in a reasonable period of time. Cost versus patient benefit may be a concern, but adherence to the old adage that "the patient comes first" should ideally govern all decisions.

Regarding susceptibility testing, enough data exist for some fungal species that the physician may find it more efficient to treat the infection empirically. Should this strategy fail, then susceptibility testing is indeed warranted. What is difficult to justify is the routine testing of clinical isolates for antifungal susceptibility. This is an expensive proposition that requires enormous resources and results in very little gain for the typical laboratory. In most cases, it would be more expeditious and fiscally prudent to submit isolates for testing by established laboratories where methodological and quality control issues have already been addressed.

In summation, "emerging and reemerging" fungal pathogens should be handled like any other specimen that might be considered a common etiologic agent. It is irrelevant, particularly to the patient, whether the fungus represents the cause of a new or reappearing pathology. What is relevant is the prompt and efficacious identification and treatment of the disease. Certainly, "emerging and reemerging" pathogens pose particular challenges, but none greater than those faced today by the individuals responsible for providing the diagnoses. For these individuals, the challenge is knowing one's limitations and the appropriate resources to seek out when help is needed.

ACKNOWLEDGMENTS

The author wishes to thank Drs. Michael McGinnis, Michael Rinaldi, and Ira Salkin for graciously providing their opinions and insights into the topic of this chapter. The author appreciates their sharing some of their outstanding career experiences as true medical mycologists. However, taking comfort in Dr. Sagan's quote at the beginning of this chapter, the author is solely responsible for any opinions expressed in this chapter which may be found disagreeable by others.

REFERENCES

1. DM Dixon, MM McNeil, ML Cohen, BG Gellin, JR La Montagne. Fungal infections: a growing threat. Public Health Rep 111:226–235, 1996.
2. CR Cooper Jr, NG Haycocks. *Penicillium marneffei*: an insurgent species among the Penicillia. J Eukaryot Microbiol 47:24–28, 2000.
3. E Drouhet, B Dupont. Histoplasmoses et autres mycoses d'importation en 1989. Rev Prat 39: 1675–1682, 1989.
4. B Dupont. Exotic pulmonary mycoses. Rev Pneumol Clin 54:301–308, 1998.
5. J Fujio, K Nishimura, M Miyaji. Epidemiological survey of the imported mycoses in Japan. Nippon Ishinkin Gakkai Zasshi 40:103–109, 1999.
6. R Manfredi, A Mazzoni, A Nanetti, F Chiodo. *Histoplasmosis capsulati* and *duboisii* in Europe: the impact of the HIV pandemic, travel and immigration. Eur J Epidemiol 10:675–681, 1994.
7. WB Warnock, B Dupont, CA Kauffman, T Sirisanthana. Imported mycoses in Europe. Med Mycol 36(suppl 1):87–94, 1998.
8. NM Ampel. Emerging disease issues and fungal pathogens associated with HIV infection. Emerg Infect Dis 2:109–116, 1996.
9. A Espinel-Ingroff. History of medical mycology in the United States. Clin Microbiol Rev 9:235–272, 1996.
10. MR McGinnis. Medical mycology in the next century. Japa J Med Mycol 39:1–9, 1998.
11. ML Cohen. Resurgent and emergent disease in a changing world. Br Med Bull 545:523–532, 1998.
12. JA Alspaugh, JR Perfect. Infections due to zygomycetes and other rare fungal opportunits. Semin Resp Crit Care Med 18:265–279, 1997.
13. JR Perfect, WA Schell. The new fungal opportunists are coming. Clin Infect Dis 22(suppl 2): S112–S118, 1996.
14. TJ Walsh. Emerging fungal pathogens: evolving challenges to immunocompromised patients. In: WM Scheld, D Armstrong, JM Hughes, eds. Emerging Infections 1. Washington: ASM Press, 1998, pp 221–232.
15. DR Reynolds, JW Taylor, eds. The Fungal Holomorph: Mitotic, Meiotic, and Pleomorphic Speciation in Fungal Systematics. Wallingford, England: CAB International, 1993, p 375.
16. MR McGinnis, L Sigler, MG Rinaldi. Some medically important fungi and their common synomyms and names of uncertain application. Clin Infect Dis 29:728–730, 1999.

Index

Printed and bound by CPI Group (UK) Ltd, Croydon, CR0 4YY

23/10/2024

01778245-0017